6 JR   # 22.15

**UNIT OPERATIONS
OF CHEMICAL
ENGINEERING**

**McGRAW-HILL
BOOK COMPANY**
New York
St. Louis
San Francisco
Auckland
Düsseldorf
Johannesburg
Kuala Lumpur
London
Mexico
Montreal
New Delhi
Panama
Paris
São Paulo
Singapore
Sydney
Tokyo
Toronto

WARREN L. McCABE

Dean Emeritus
Polytechnic Institute of New York

R. J. Reynolds Professor, Emeritus, in Chemical Engineering
North Carolina State University

JULIAN C. SMITH

Professor of Chemical Engineering
Director, School of Chemical Engineering
Cornell University

Unit Operations of Chemical Engineering

THIRD EDITION

This book was set in Times Roman.
The editors were B. J. Clark and Madelaine Eichberg;
the production supervisor was Dennis J. Conroy.
New drawings were done by J & R Services, Inc.
Kingsport Press, Inc., was printer and binder.

Library of Congress Cataloging in Publication Data

McCabe, Warren Lee, date
    Unit operations of chemical engineering.

    (McGraw-Hill chemical engineering series)
    Includes bibliographies and index.
    1. Chemical processes. I. Smith, Julian Cleveland,
date   joint author. II. Title.
TP155.7.M3   1976   660.2′842   75-12941
ISBN 0-07-044825-6

**UNIT OPERATIONS
OF CHEMICAL
ENGINEERING**

1234567890KPKP798765

CONTENTS

PREFACE

This book is a beginning text on the unit operations. It is written for undergraduate students in the junior or senior years who have the usual training in mathematics, physics, chemistry, and mechanics. An elementary knowledge of material-balance calculations and thermodynamics is helpful but not essential.

We have continued the traditions of earlier books on this subject. During the past generation the unit-operation approach has demonstrated its usefulness, both in education and in engineering practice. Although certain operations—notably gas absorption, distillation, and extraction—have tended to fuse together, we believe it wise to protect the integrity of the individual operation by giving it separate treatment, as each one has a definite task to perform.

The limits of the text are fixed by space and by level of treatment. Many of the less conventional operations, such as adsorption, dialysis, expression, colloid milling, ion exchange, freezing, sublimation, and specialized methods of mechanical separation, are omitted because of space. Other important topics have been left out because of space and because any adequate treatment of them would raise the level of the book to a point inaccessible to the undergraduate. For these reasons, multicomponent separations, transient phenomena, and two- and three-dimensional flow patterns have been omitted or touched on but lightly. Every effort has been made to keep a uniform level of treatment throughout the book.

An effort has also been made to carry the treatment to a point where the student will find an easy transition to more advanced and specialized chemical engineering handbooks, texts, and monographs.

Two ideas have influenced the choice of equipment for discussion in the text. Most of the standard "bread-and-butter" types are shown, and a selection of newer and more specialized devices has also been included. The choice of the latter is quite arbitrary, but those illustrating important engineering principles have been favored.

The basic philosophy, level of treatment, and approach to the unit operations used in earlier editions of the text remain unchanged. Many small alterations in the arrangement of topics and in specific subjects have been made, and some new problems and examples have been added. The major additions of subject matter are the following:

1. A thorough discussion is given of the three unit systems, cgs, SI, and fps, used in present practice. The approach is from the basic proportionalities of macroscopic mechanics, thermodynamics, and physical chemistry. The objective is to prepare the chemical engineer to use with skill and confidence any unit system that may become dominant during his professional future.

2. Since chemical engineers are receiving more fundamental instruction in thermodynamics and transfer processes, a basic treatment of fugacity and activity coefficients, based on the chemical potential, has been added.

3. A completely new chapter on the fundamentals of multicomponent distillation has been written. This chapter gives a thorough discussion of the Kwauk method of counting the independent variables in complicated separation plants.

4. The recent spectacular breakthrough in the understanding of contact nucleation is discussed in the chapter on crystallization, and a model for incorporating this knowledge in crystallization practice is included.

5. It is the opinion of the authors that the strong position of chemical engineering in environmental problems should be recognized. Many problems in this field deal with particulate systems, and the chapters on these operations have been revised and strengthened. Mixing also is given a more complete treatment.

6. Finally, the extensive use of the text in many foreign countries continues to influence the authors. The needs of these students, many of whom must study basic chemical engineering with little or no access to instructors, is always in our minds.

WARREN L. McCABE
JULIAN C. SMITH

**UNIT OPERATIONS
OF CHEMICAL
ENGINEERING**

1

INTRODUCTION

Chemical engineering has to do with industrial processes in which raw materials are changed or separated into useful products. The chemical engineer must develop, design, and engineer both the complete process and the equipment used in it. He must choose the proper raw materials; he must operate his plants efficiently, safely, and economically; and he must see to it that his products meet the requirements set by his customers. Consistent with engineering generally, chemical engineering is both an art and a science. Whenever science helps the engineer to solve his problems, he should use science. When, as is usually the case, science does not give him a complete answer, he must use experience and judgment. His professional stature depends on his skill in combining all sources of information to reach practical solutions to processing problems.

The range and variety of processes and industries that call for the services of chemical engineers are both great. The field is not one that is easy to define. The processes described in standard treatises on chemical technology and the process industries give the best idea of the field of chemical engineering.[6]†

† Superior numerals in the text correspond to the numbered references at the end of each chapter.

Because of the variety and complexity of modern processes, it is not practicable to cover the entire subject matter of chemical engineering under a single head. The field is divided into convenient, but arbitrary, sectors. This text covers that portion of chemical engineering known as the unit operations.

UNIT OPERATIONS

An economical method of organizing much of the subject matter of chemical engineering is based on two facts: (1) although the number of individual processes is great, each one can be broken down into a series of steps, called operations, each of which in turn appears in process after process; (2) the individual operations have common techniques and are based on the same scientific principles. For example, in most processes solids and fluids must be moved, heat or other forms of energy must be transferred from one substance to another, and tasks like drying, size reduction, distillation, and evaporation must be performed. The unit-operation concept is this: by studying systematically these operations themselves—operations which clearly cross industry and process lines—the treatment of all processes is unified and simplified.

The strictly chemical aspects of processing are studied in a companion area of chemical engineering called reaction kinetics. The unit operations are largely used to conduct the primarily physical steps of preparing the reactants, separating and purifying the products, recycling unconverted reactants, and controlling the energy transfer into or out of the chemical reactor.

The unit operations are as applicable to many physical processes as to chemical ones. For example, the process used to manufacture common salt consists of the following sequence of the unit operations: transportation of solids and liquids, transfer of heat, evaporation, crystallization, drying, and screening. No chemical reaction appears in these steps. On the other hand, the cracking of petroleum, with or without the aid of a catalyst, is a typical chemical reaction conducted on an enormous scale. Here the unit operations—transportation of fluids and solids, distillation, and various mechanical separations—are vital, and the cracking reaction could not be utilized without them. The chemical steps themselves are conducted by controlling the flow of material and energy to and from the reaction zone.

Because the unit operations are a branch of engineering, they are based on both science and experience. Theory and practice must combine to yield designs for equipment that can be fabricated, assembled, operated, and maintained. A balanced discussion of each operation requires that theory and equipment be considered together. An objective of this book is to present such a balanced treatment.

Scientific foundations of unit operations A number of scientific principles and techniques are basic to the treatment of the unit operations. Some are elementary physical and chemical laws such as the conservation of mass and energy, physical equilibria, kinetics, and certain properties of matter. Their general use is described in the remainder of this chapter. Other special techniques important in chemical engineering are considered at the proper places in the text.

UNIT SYSTEMS

The official international system of units is the SI system (Système International d'Unités). Strong efforts are underway for its universal adoption as the exclusive system for all engineering and science, but older systems, particularly the cgs and fps engineering gravitational systems, are still in use and probably will be around for some time. The chemical engineer finds many of his physiochemical data given in cgs units, and he makes most of his calculations in fps units. But more and more he will encounter the SI units in both science and engineering, and so he should be expert in the use of all three systems.

In the following treatment, the SI system is discussed first, and the other systems are then derived from it. The procedure reverses the historical order, as the SI units evolved from the cgs system. Because of the growing importance of the SI system and its probable survival value, logically it should be given preference. If, in time, the other systems are phased out, they can be ignored and the SI system then used exclusively.

Physical Quantities

Any physical quantity consists of two parts: a unit, which tells what the quantity is and gives the standard by which it is measured, and a number, which tells how many units are needed to make up the quantity. For example, the statement that the distance between two points is 8 ft means all this: a definite length has been measured; to measure it, a standard length, called the foot, has been chosen as a unit; and eight 1-ft units, laid end to end, are needed to cover the distance. If an integral number of units is either too few or too many to cover a given distance, submultiples, which are fractions of the unit, are defined by dividing the unit into fractions, so that a measurement can be made to any degree of precision in terms of the fractional units. No physical quantity is defined until both the number and the unit are given.

SI Units

The SI system covers the entire field of science and engineering, including electromagnetics and illumination. For the purposes of this book, a subset of the SI units covering chemistry, gravity, mechanics, and thermodynamics is sufficient. The units are derivable from (1) four proportionalities of chemistry and physics, (2) arbitrary standards for mass, length, time, temperature, and the mole, and (3) arbitrary choices for the numerical values of two proportionality constants.

Basic Equations   The basic proportionalities, each written as an equation with its own proportionality factor, are

$$F = k_1 \frac{d}{dt}(mu) \tag{1-1}$$

$$F = k_2 \frac{m_a m_b}{r^2} \tag{1-2}$$

$$Q_c = k_3 W_c \tag{1-3}$$

$$T = k_4 \lim_{p \to 0} \frac{pV}{m} \tag{1-4}$$

where†
$$
\begin{aligned}
F &= \text{force} \\
t &= \text{time} \\
m &= \text{mass} \\
r &= \text{distance} \\
W &= \text{work} \\
Q &= \text{heat} \\
p &= \text{pressure} \\
V &= \text{volume} \\
T &= \text{thermodynamic absolute temperature} \\
k_1, k_2, k_3, k_4 &= \text{proportionality factors}
\end{aligned}
$$

Equation (1-1) is Newton's second law of motion, showing the proportionality between the resultant of all the forces acting on a particle of mass m and the time rate of increase in momentum of the particle in the direction of the resultant force.

Equation (1-2) is Newton's law of gravitation, giving the force of attraction between two particles of masses m_a and m_b a distance r apart.

Equation (1-3) is one statement of the first law of thermodynamics. It affirms the proportionality between the work performed by a closed system during a cycle and the heat absorbed by that system during the same cycle.

Equation (1-4) shows the proportionality between the thermodynamic absolute temperature and the zero-pressure limit of the pressure-volume product of a definite mass of any gas.

Each equation states that if means are available for measuring the values of all variables in that equation and if the numerical value of k is calculated, then the value of k is constant and depends only on the units used for measuring the variables in the equation.

Standards   By international agreement, standards are fixed arbitrarily for the quantities mass, length, time, temperature, and the mole. These are five of the *base units* of the SI system. Currently, the standards are as follows.

The standard of mass is the kilogram (kg), defined as the mass of the international kilogram, a platinum cylinder preserved at Sèvres, France.

† A list of symbols is given at the end of each chapter.

The standard of length is the meter (m), defined† as 1,650,763.73∗ wavelengths of a certain spectral line emitted by ^{86}Kr.

The standard of time is the second (s), defined as 9,192,631.770∗ frequency cycles of a certain quantum transition in an atom of ^{133}Ce.

The standard of temperature is the kelvin (K), defined by assigning the value 273.16∗ K to the temperature of pure water at its triple point, the unique temperature at which liquid water, ice, and steam can exist at equilibrium.

The mole (abbreviated mol) is defined[4] as the amount of a substance comprising as many elementary units as there are atoms in 12∗ g of ^{12}C. The definition of the mole is equivalent to the statement that the mass of one mole of a pure substance in grams is numerically equal to its molecular weight calculated from the standard table of atomic weights, in which the atomic weight of carbon is given as 12.01115. This number differs from 12∗ because it applies to the natural isotopic mixture of carbon rather than to pure ^{12}C.

Evaluation of constants   From the basic standards, values of m, m_a, and m_b in Eqs. (1-1) and (1-2) are measured in kilograms, r in meters, and u in meters per second. Constants k_1 and k_2 are not independent but are related by eliminating F from Eqs. (1-1) and (1-2). This gives

$$\frac{k_1}{k_2} = \frac{d(mu)/dt}{m_a m_b / r^2}$$

Either k_1 or k_2 may be fixed arbitrarily. Then the other constant must be found by experiments in which inertial forces calculated by Eq. (1-1) are compared with gravitational forces calculated by Eq. (1-2). In the SI system, k_1 is fixed at unity and k_2 found experimentally. Equation (1-1) then becomes

$$F = \frac{d}{dt}(mu) \tag{1-5}$$

The force defined by Eq. (1-5) and also used in Eq. (1-2) is called the *newton* (N). From Eq. (1-5),

$$1 \text{ N} \equiv 1 \text{ kg-m/s}^2 \tag{1-6}$$

Constant k_2 is denoted by G and called the *gravitational constant*. Its experimental value, as accepted by the United States Bureau of Standards,[4] is

$$G = 6.6732 \times 10^{-11} \text{ N-m}^2/\text{kg}^2 \tag{1-7}$$

Work, energy, and power   In the SI system both work and energy are measured in newton-meters, a unit called the *joule* (J), and so

$$1 \text{ J} \equiv 1 \text{ N-m} = 1 \text{ kg-m}^2/\text{s}^2 \tag{1-8}$$

Power is measured in joules per second, a unit called the *watt* (W).

---

† The asterisk at the end of a number signifies that the number is exact, by definition.

Heat   The constant k_3 in Eq. (1-3) may be fixed arbitrarily. In the SI system it, like k_1, is set at unity. Equation (1-3) becomes

$$Q_c = W_c \tag{1-9}$$

Heat, like work, is measured in joules.

Temperature   The quantity pV/m in Eq. (1-4) may be measured in $(N/m^2)(m^3/kg)$, or J/kg. With an arbitrarily chosen gas, this quantity can be determined by measuring p and V of m kg of gas while it is immersed in a thermostat. In this experiment, only constancy of temperature, not magnitude, is needed. Values of pV/m at various pressures and at constant temperature can then be extrapolated to zero pressure to obtain the limiting value required in Eq. (1-4) at the temperature of the thermostat. For the special situation when the thermostat contains water at its triple point, the limiting value is designated by $(pV/m)_0$. For this experiment Eq. (1-4) gives

$$273.16 = k_4 \lim_{p \to 0} \left(\frac{pV}{m} \right)_0 \tag{1-10}$$

For an experiment at temperature T K Eq. (1-4) can be used to eliminate k_4 from Eq. (1-10), giving

$$T \equiv 273.16 \frac{\lim_{p \to 0} (pV/m)_T}{\lim_{p \to 0} (pV/m)_0} \tag{1-11}$$

Equation (1-11) is the definition of the Kelvin temperature scale from the experimental pressure-volume properties of a real gas.

Celsius temperature   In practice, temperatures are expressed on the Celsius scale, in which the zero point is set at the ice point, defined as the equilibrium temperature of ice and air-saturated water at a pressure of one atmosphere. Experimentally, the ice point is found to be 0.01 K below the triple point of water, and so it is at 273.15 K. The Celsius temperature (°C) is defined by

$$T°C \equiv T\,K - 273.15 \tag{1-12}$$

On the Celsius scale, the experimentally measured temperature of the steam point, which is the boiling point of water at a pressure of one atmosphere, is 100.00°C.

Decimal units   In the SI system a single unit is defined for each quantity, but named decimal multiples and submultiples also are recognized. They are listed in Appendix 1. Time may be expressed in the nondecimal units: minutes (min), hours (h), or days (d).

Standard gravity   For certain purposes, the acceleration of free fall in the earth's gravitational field is used. From deductions based on Eq. (1-2), this quantity, denoted by g, is nearly constant. It varies slightly with latitude and height above sea level. For precise calculations, an arbitrary standard g_n has been set, defined by

$$g_n \equiv 9.80665* \text{ m/s}^2 \tag{1-13}$$

Pressure units The natural unit of pressure in the SI system is the newton per square meter. This unit, called the *pascal* (Pa), is inconveniently small, and a multiple, called the *bar*, also is used. It is defined by

$$1 \text{ bar} \equiv 1 \times 10^5 \text{ Pa} = 1 \times 10^5 \text{ N/m}^2 \tag{1-14}$$

A more common empirical unit for pressure, used with all systems of units, is the *standard atmosphere* (atm), defined by

$$1 \text{ atm} \equiv 1.01325* \times 10^5 \text{ Pa} = 1.01325 \times 10^5 \text{ N/m}^2 \tag{1-15}$$

CGS Units

The older centimeter-gram-second (cgs) system can be derived from the SI system by making certain arbitrary decisions.

The standard for mass is the gram (g), defined by†

$$1 \text{ g} \equiv 1 \times 10^{-3} \text{ kg} \tag{1-16}$$

The standard for length is the centimeter (cm), defined by‡

$$1 \text{ cm} \equiv 1 \times 10^{-2} \text{ m} \tag{1-17}$$

Standards for time, temperature, and the mole are unchanged.

As in the SI system, constant k_1 in Eq. (1-1) is fixed at unity. The unit of force is called the *dyne* (dyn), defined by

$$1 \text{ dyn} \equiv 1 \text{ g-cm/s}^2 \tag{1-18}$$

The unit for energy and work is the *erg*, defined by

$$1 \text{ erg} \equiv 1 \text{ dyn-cm} = 1 \times 10^{-7} \text{ J} \tag{1-19}$$

Constant k_3 in Eq. (1-3) is not unity. A unit for heat, called the *calorie* (cal), is used to convert the unit for heat to ergs. Constant $1/k_3$ is replaced by J, which denotes the quantity called the mechanical equivalent of heat and which is measured in joules per calorie. Equation (1-3) becomes

$$W_c = JQ_c \tag{1-20}$$

Two calories are defined.[4] The *thermochemical calorie* (cal), used in chemistry, is defined by

$$1 \text{ cal} \equiv 4.1840* \times 10^7 \text{ ergs} = 4.1840* \text{ J} \tag{1-21}$$

The *international steam table calorie* (cal$_{IT}$), used in engineering, is defined by

$$1 \text{ cal}_{IT} \equiv 4.1868* \times 10^7 \text{ ergs} = 4.1868* \text{ J} \tag{1-22}$$

The calorie is so defined that the specific heat of water is approximately 1 cal/g-°C.

The standard acceleration of free fall in cgs units is

$$g_n \equiv 980.665 \text{ cm/s}^2 \tag{1-23}$$

† See Appendix 1.
‡ See Appendix 1.

Gas Constant

If mass is measured in kilograms or grams, constant k_4 in Eq. (1-4) differs from gas to gas. But when the concept of the mole as a mass unit is used, k_4 can be replaced by the universal gas constant R, which, by Avogadro's law, is the same for all gases. The numerical value of R depends only on the units chosen for energy, temperature, and mass. Then Eq. (1-4) is written

$$\lim_{p \to 0} \frac{pV}{nT} = R \qquad (1\text{-}24)$$

where n is the number of moles. This equation applies also to mixtures of gases if n is the total number of moles of all the molecular species that make up the volume V.

The accepted experimental value of R is[4]

$$R = 8.31434 \text{ J-K/mol} = 8.31434 \times 10^7 \text{ erg-K/mol} \qquad (1\text{-}25)$$

Values of R in other units for energy, temperature, and mass are given in Appendix 2.

Although the mole is defined as a mass in grams, the concept of the mole is easily extended to other mass units. Thus, the kilogram mole (kg mol) is the usual molecular or atomic weight in kilograms, and the pound mole (lb mol) is that in avoirdupois pounds. When the mass unit is not specified, the gram mole (g mol) is intended. Molecular weight M is a pure number.

FPS Engineering Units

In English-speaking countries a nondecimal gravitational unit system has long been used in commerce and engineering. The system can be derived from the SI system by making the following decisions.

The standard for mass is the avoirdupois pound (lb), defined by

$$1 \text{ lb} \equiv 0.45359237* \text{ kg} \qquad (1\text{-}26)$$

The standard for length is the inch (in.), defined as 2.54* cm. This is equivalent to defining the foot (ft) as

$$1 \text{ ft} \equiv 2.54 \times 12 \times 10^{-2} \text{ m} = 0.3048* \text{ m} \qquad (1\text{-}27)$$

The standard for time remains the second (s).

The thermodynamic temperature scale is called the Rankine scale, in which temperatures are denoted by °R and defined by

$$1°R \equiv \frac{1}{1.8} \text{ K} \qquad (1\text{-}28)$$

The ice point on the Rankine scale is $273.15 \times 1.8 = 491.67$ K.

The analog of the Celsius scale is the Fahrenheit scale, in which readings are denoted by °F. It is derived from the Rankine scale by setting its zero point exactly 32°F below the ice point on the Rankine scale, so that

$$T°F \equiv T°R - (491.67 - 32) = T°R - 459.69 \qquad (1\text{-}29)$$

The relation between the Celsius and the Fahrenheit scales is given by the exact equation

$$T°F = 32 + 1.8°C \qquad (1\text{-}30)$$

From this equation, temperature differences are related by

$$\Delta T°C = 1.8\Delta T°F = \Delta T\,K \qquad (1\text{-}31)$$

The steam point is 212.00°F.

Pound force   The fps system is characterized by a gravitational unit of force, called the *pound force* (lb_f). The unit is so defined that a standard gravitational field exerts a force of one pound on a mass of one avoirdupois pound. The standard acceleration of free fall in fps units is, to five significant figures,

$$g_n = \frac{9.80665 \text{ m/s}^2}{0.3048 \text{ m/ft}} = 32.174 \text{ ft/s}^2 \qquad (1\text{-}32)$$

The pound force is defined by

$$1 \text{ lb}_f \equiv 32.174 \text{ lb-ft/s}^2 \qquad (1\text{-}33)$$

Then Eq. (1-1) gives

$$F \text{ lb}_f \equiv \frac{d(mu)/dt}{32.174} \quad \text{lb-ft/s}^2 \qquad (1\text{-}34)$$

Equation (1-1) can also be written with $1/g_c$ in place of k_1:

$$F = \frac{d(mu)/dt}{g_c} \qquad (1\text{-}35)$$

Comparison of Eqs. (1-34) and (1-35) shows that to preserve both numerical equality and consistency of units in these equations it is necessary to define g_c, called the *Newton's-law proportionality factor for the gravitational force unit*, by

$$g_c \equiv 32.174 \text{ lb-ft/s}^2\text{-lb}_f \qquad (1\text{-}36)$$

The unit for work and mechanical energy in the fps system is the *foot-pound force* ($ft\text{-}lb_f$). Power is measured by an empirical unit, the *horsepower* (hp), defined by

$$1 \text{ hp} \equiv 550 \text{ ft-lb}_f\text{/s} \qquad (1\text{-}37)$$

The unit for heat is the *British thermal unit* (Btu), defined by the implicit relation

$$1 \text{ Btu/lb-°F} \equiv 1 \text{ cal}_{IT}\text{/g-°C} \qquad (1\text{-}38)$$

As in the cgs system, constant k_3 in Eq. (1-3) is replaced by $1/J$, where J is the mechanical equivalent of heat, equal to 778.17 $ft\text{-}lb_f$/Btu.

The definition of the Btu requires that the numerical value of specific heat be the same in both systems, and in each case the specific heat of water is approximately 1.0.

Conversion of Units

Since three unit systems are in common use, it is often necessary to convert the magnitudes of quantities from one system to another. This is accomplished by using conversion factors. Only the defined conversion factors for the base units are required since conversion factors for all other units can be calculated from them. Interconversions between the SI and cgs systems are simple. Both use the same standards for time, temperature, and the mole, and only the decimal conversions defined by Eqs. (1-16) and (1-17) are needed. The SI and fps systems also use the second as the standard for time; the three conversion factors defined for mass, length, and temperature by Eqs. (1-26), (1-27), and (1-28), respectively, are sufficient for all conversions of units between these two systems.

Example 1-1 demonstrates how conversion factors are calculated from the exact numbers used to set up the definitions of units in the SI and fps systems. In conversions involving g_c in fps units, the use of the exact numerical ratio 9.80665/0.3048 in place of the fps number 32.1740 is recommended to give maximum precision in the final calculation and to take advantage of possible cancellations of numbers during the calculation.

EXAMPLE 1-1 Using only exact definitions and standards, calculate factors for converting (a) newtons to pounds force, (b) British thermal units to IT calories, (c) atmospheres to pounds force per square inch, and (d) British thermal units to foot-pounds force.

SOLUTION (a) From Eqs. (1-6), (1-26), and (1-27),

$$1 \text{ N} = 1 \text{ kg-m/s}^2 = \frac{1 \text{ lb-ft/s}^2}{0.45359237 \times 0.3048}$$

From Eq. (1-32)

$$1 \text{ lb-ft/s}^2 = \frac{0.3048}{9.80665} \text{ lb}_f$$

and so

$$1 \text{ N} = \frac{0.3048}{9.80665 \times 0.45359237 \times 0.3048} \text{ lb}_f$$

$$= \frac{1}{9.80665 \times 0.45359237} \text{ lb}_f = 0.224809 \text{ lb}_f$$

In Appendix 3 it is shown that to convert newtons to pound force one should multiply by 0.224809. Clearly, to convert from pounds force to newtons, multiply by $9.80665 \times 0.45359237 = 4.448222$.

(b) From Eq. (1-38)

$$1 \text{ Btu} = 1 \text{ cal}_{IT} \frac{1 \text{ lb}}{1 \text{ g}} \frac{1°\text{F}}{1°\text{C}}$$

$$= 1 \text{ cal}_{IT} \frac{1 \text{ lb}}{1 \text{ kg}} \frac{1 \text{ kg}}{1 \text{ g}} \frac{1°\text{F}}{1°\text{C}}$$

From Eqs. (1-16), (1-26), and (1-31)

$$1 \text{ Btu} = 1 \text{ cal}_{IT} \frac{0.45359237 \times 1,000}{1.8} = 251.996 \text{ cal}_{IT}$$

(c) From Eqs. (1-6) and (1-15)

$$1 \text{ atm} = 1.01325 \times 10^5 \text{ kg-m/s}^2\text{-m}^2$$

From Eqs. (1-26), (1-27), and (1-32), since 1 ft = 12 in.,

$$1 \text{ atm} = 1.01325 \times 10^5 \frac{1 \text{ lb-ft/s}^2}{0.45359237} \frac{0.3048^2}{\text{ft}^2}$$

$$= \frac{1.01325 \times 10^5 \times 0.3048^2}{9.80665 \times 0.45359237 \times 12^2} \text{ lb}_f/\text{in.}^2$$

$$= 14.6959 \text{ lb}_f/\text{in.}^2$$

(d)   From Eqs. (1-16), (1-22), and (1-38)

$$1 \text{ Btu/lb-}°\text{F} = 4.1868 \text{ J/g-}°\text{C} \times 10^3 \text{ g/kg}$$

Using Eq. (1-8) gives

$$1 \text{ Btu} = 4.1868 \times 10^3 \text{ m}^2/\text{s}^2\text{-lb-}(\Delta T\,°\text{F}/\Delta T\,°\text{C})$$

Using Eqs. (1-27) and (1-31) leads to

$$1 \text{ Btu} = \frac{4.1868 \times 10^3}{1.8} \left(\frac{1 \text{ ft}}{0.3048}\right)^2 \frac{1 \text{ lb}}{s^2}$$

From Eq. (1-33)

$$1 \text{ Btu} = \frac{4.1868 \times 10^3}{1.8} \left(\frac{1 \text{ ft}}{0.3048}\right)^2 \frac{0.3048}{9.80665} \frac{1 \text{ lb}_f}{\text{ft-lb}}$$

$$= \frac{4.1868 \times 10^3}{0.3048 \times 9.80665 \times 1.8} \text{ ft-lb}_f$$

$$= 778.17 \text{ ft-lb}_f \qquad\qquad\qquad ////$$

Although conversion factors may be calculated as needed, it is more efficient to use tables of the common factors. A table for the factors used in this book is given in Appendix 3.

Units and Equations

Although Eqs. (1-1) to (1-4) are sufficient for the description of unit systems, they are but a small fraction of the equations needed in this book. Many such equations contain terms that represent properties of substances, and these are introduced as needed. All new quantities are measured in combinations of units already defined, and all are expressible as functions of the five base units for mass, length, time, temperature, and mole.

Precision of calculations   In the above discussion, the values of experimental constants are given with the maximum number of significant digits consistent with present estimates of the precision with which they are known, and all digits in the values of defined constants are retained. In practice, such extreme precision is seldom necessary, and defined and experimental constants can be truncated to the number of digits appropriate to the problem at hand, although the advent of the digital computer makes it possible to retain maximum precision at small cost. The engineer should use his own judgment in setting the level of precision suitable for his purpose. In the problems and examples in this book a precision of one significant figure beyond slide-rule precision is generally used.

General equations   Except for the appearance of the proportionality factors g_c and J, the equations for all three unit systems are alike. In the SI system, neither constant appears; in the cgs system, g_c is omitted and J retained; in the fps system both constants appear. In this text, to obtain equations in a general form for all systems, g_c and J are included for use with fps units; then either g_c or both g_c and J may be equated to unity when the equations are used in the cgs or SI systems.

Dimensionless Equations and Consistent Units

Equations derived directly from the basic laws of the physical sciences consist of terms which either have the same units or which can be written in the same units by using the definitions of derived quantities to express complex units in terms of the five base ones. Equations meeting this requirement are called *dimensionally homogeneous equations*. When such an equation is divided by any one of its terms, all units in each term cancel and only numerical magnitudes remain. These equations are called *dimensionless equations*.

A dimensionally homogeneous equation can be used as it stands with any set of units provided that the same units for the five base units are used throughout. Units meeting this requirement are called *consistent units*. No conversion factors are needed when consistent units are used.

For example, consider the usual equation for the vertical distance Z traversed by a freely falling body during time t when the initial velocity is u_0:

$$Z = u_0 t + \frac{gt^2}{2} \tag{1-39}$$

Examination of Eq. (1-39) shows that the units in each term reduce to that for length. Dividing the equation by Z gives

$$1 = \frac{u_0 t}{Z} + \frac{gt^2}{2Z} \tag{1-40}$$

A check of each term in Eq. (1-40) shows that the units in each term cancel and each term is dimensionless. A combination of variables for which all dimensions cancel in this manner is called a *dimensionless group*. The numerical value of a dimensionless group for given values of the quantities contained in it is independent of the units used, provided they are consistent. Both terms on the right-hand side of Eq. (1-40) are dimensionless groups.

Dimensional equations   Equations derived by empirical methods, in which experimental results are correlated by empirical equations without regard to dimensional consistency, usually are not dimensionally homogeneous and contain terms in several different units. Equations of this type are *dimensional equations*, or dimensionally nonhomogeneous equations. In these equations there is no advantage in using consistent units, and two or more length units, e.g., inches and feet, or two or more time units, e.g., seconds and minutes, may appear in the same equation. For example, a formula for the rate of heat loss from a horizontal pipe to the atmosphere by conduction and convection is

$$\frac{q}{A} = 0.50 \frac{\Delta T^{1.25}}{(D'_o)^{0.25}} \tag{1-41}$$

where q = rate of heat loss, Btu/h

A = area of pipe surface, ft^2

ΔT = excess of temperature of pipe wall over that of ambient (surrounding atmosphere), °F

D'_o = outside diameter of pipe, in.

Obviously, the units of q/A are not those of the right-hand side of Eq. (1-41), and the equation is dimensional. Quantities substituted in Eq. (1-41) must be expressed in the units as given, or the equation will give the wrong answer. If other units are to be used, the coefficient must be changed. To express ΔT in degrees Celsius, for example, the numerical coefficient must be changed to $0.50 \times 1.8^{1.25} = 1.045$ since there are 1.8 Fahrenheit degrees in 1 Celsius degree of temperature difference.

In this book all equations are dimensionally homogeneous *unless otherwise noted*.

DIMENSIONAL ANALYSIS

Many important engineering problems cannot be solved completely by theoretical or mathematical methods. Problems of this type are especially common in fluid-flow, heat-flow, and diffusional operations. One method of attacking a problem for which no mathematical equation can be derived is that of empirical experimentation. For example, the pressure loss from friction in a long, round, straight, smooth pipe depends on all these variables: the length and diameter of the pipe, the flow rate of the liquid, and the density and viscosity of the liquid. If any one of these variables is changed, the pressure drop also changes. The empirical method of obtaining an equation relating these factors to pressure drop requires that the effect of each separate variable be determined in turn by systematically varying that variable while keeping all others constant. The procedure is laborious, and it is difficult to organize or correlate the results so obtained into a useful relationship for calculations.

There exists a method intermediate between formal mathematical development and a completely empirical study.[1-3] It is based on the fact that if a theoretical equation does exist among the variables affecting a physical process, that equation must be dimensionally homogeneous. Because of this requirement it is possible to group many factors into a smaller number of dimensionless groups of variables. The groups themselves rather than the separate factors appear in the final equation.

Dimensional analysis does not yield a numerical equation, and experiment is required to complete the solution of the problem. The result of a dimensional analysis is valuable in guiding experiments and is useful in pointing a way to correlations of experimental data suitable for engineering use. A discussion of the technique of making a dimensional analysis, with an example of the method, is given in Appendix 4.

Named dimensionless groups   Some dimensionless groups occur with such frequency that they have been given names and special symbols. A list of the most important ones is given in Appendix 5.

Equations of State of Gases

A pure gas consisting of n mol and held at a temperature T and pressure p will fill a volume V. If any of the three quantities are fixed, the fourth also is determined and only three of these quantities are independent. This fact can be expressed by the functional equation

$$f(p, T, V, n) = 0 \tag{1-42}$$

Specific forms of this relation are called *equations of state*. Many such equations have been proposed, and several are in common use. The most satisfactory equations of state can be written in the form

$$\frac{pV}{nRT} = 1 + \frac{B}{V/n} + \frac{C}{(V/n)^2} + \frac{D}{(V/n)^3} + \cdots \tag{1-43}$$

This equation, known as the *virial equation*, is well substantiated by molecular gas theory. Coefficients B, C, and D are called the second, third, and fourth virial coefficients, respectively. Each is a function of temperature and is independent of pressure. Additional coefficients may be added, but numerical values for coefficients beyond D are so little known that more than three are seldom used. The virial equation also applies to mixtures of gases. Then the virial coefficients depend on temperature and composition of the mixture. Rules are available for estimating the values of B, C, and D for mixtures from those of the individual pure gases.[5]

Virial coefficients have a definite physical meaning. Each molecule of a gas is surrounded by a force field which depends on its atomic structure and on how the molecule is constructed from its atoms. The forces associated with a molecule act on all other molecules nearby, and mutual attractions and repulsions are generated. The reactions are called *collisions*. They affect the behavior of the molecules and so determine the values of the virials. The second virial, B, accounts for collisions of pairs of molecules, C for collisions involving three molecules, and so on.

Compressibility factor and molar density   For engineering purposes, Eq. (1-43) often is written

$$z = \frac{p}{\rho RT} = 1 + \rho B + \rho^2 C + \rho^3 D \tag{1-44}$$

where z is the compressibility factor and ρ the molar density, defined by

$$\rho = \frac{n}{V} \tag{1-45}$$

Ideal-gas law Real gases under high pressure require the use of all three virials for accuracy in calculating z from ρ and T. As the density is reduced by lowering the pressure, the numerical values of the virial terms fade out although the values of the coefficients remain unchanged. As the effect of D becomes negligible, the series is truncated by dropping its term, then the term in C, and so on, until at low pressures (about 1 or 2 atm for ordinary gases), all three virials can be neglected. The result is the simple gas law

$$z = \frac{pV}{nRT} = \frac{p}{\rho RT} = 1 \tag{1-46}$$

This equation clearly is consistent with Eq. (1-24), which contains the definition of the absolute temperature. The limiting process indicated in Eq. (1-24) rigorously eliminates the virial coefficients to provide a precise definition; Eq. (1-46) covers a useful range of densities for practical calculations and is called the *ideal-gas law*.

Equation (1-46) is seldom used in the above form. It states three facts, however: (1) the volume of a gas is directly proportional to the number of moles; (2) the volume is directly proportional to the absolute temperature; and (3) the volume is inversely proportional to the pressure. These facts are valuable in converting a gas volume at one temperature and pressure to that at another temperature and pressure. For example, V_a ft^3 of an ideal gas at an absolute temperature of T_a and a pressure of p_a will occupy V_b ft^3 at an absolute temperature of T_b and a pressure of p_b, where V_b is given by

$$V_b = V_a \frac{p_a}{p_b} \frac{T_b}{T_a} \tag{1-47}$$

MOLAR VOLUME Equation (1-2) shows that a mole of gas under definite conditions of temperature and pressure always occupies a definite volume, regardless of the nature of the gas. The gas law applies to a mixture of gases as well as to a pure gas. Thus a gram mole of ideal gas, or a mixture of ideal gases, occupies 22.41 l (0.02241 m^3) at 0°C and 760 mm Hg. This volume is called the gram-molar volume. A pound mole of gas, either pure or a mixture, occupies 359 ft^3 at 32°F and 1 atm. The pound-molar volume at any other combination of temperature and pressure can be calculated by Eq. (1-47).

The mass of a gas mixture that occupies the molar volume is the *average molecular weight* of the gas mixture.

PARTIAL PRESSURES A useful quantity for dealing with the individual components in a gas mixture is partial pressure. The partial pressure of a component in a mixture, e.g., component A, is defined by the equation

$$\bar{p}_A \equiv p y_A \tag{1-48}$$

where $\bar{p}_A$ = partial pressure of component A in mixture
y_A = mole fraction of component A in mixture
p = total pressure on mixture

This definition does not involve the concept of ideal gas.

Each component in the mixture has its own partial pressure. For example, for components B and C

$$\bar{p}_B = y_B p \qquad \bar{p}_C = y_C p$$

If all the partial pressures for a given mixture are added, the result is

$$\bar{p}_A + \bar{p}_B + \bar{p}_C + \cdots = p(y_A + y_B + y_C + \cdots)$$

Since the sum of the mole fractions is unity,

$$\bar{p}_A + \bar{p}_B + \bar{p}_C + \cdots = p \qquad \text{(1-49)}$$

All partial pressures in a given mixture add to the total pressure. This applies to mixtures of both ideal and nonideal gases.

DALTON'S LAW   A relationship concerning partial pressures that is restricted to ideal gases is Dalton's law. This states that in a mixture of ideal gases each component gas fills the entire volume of the mixture at the temperature of the mixture and at the partial pressure of that component. Then, for component A, for example,

$$\bar{p}_A = \frac{y_A R T n}{V} \qquad \text{(1-50)}$$

Equivalent equations may be written for all the other components.
    A generally useful consequence of the ideal-gas law is

$$\text{Volume percent} = \text{pressure percent} = \text{mole percent} \qquad \text{(1-51)}$$

For example, since air contains 79 percent nitrogen and 21 percent oxygen by volume, 1 ft^3 of air under a pressure of p atm may be considered to be a mixture of 0.21 ft^3 of oxygen and 0.79 ft^3 of nitrogen, both measured at p atm and the temperature of the mixture. It may also be considered to be 1 ft^3 of oxygen at 0.21 atm and 1 ft^3 of nitrogen at 0.79 atm, so that 21 percent of the pressure is from oxygen and 79 percent from nitrogen. Finally, 1 mol of air contains 0.21 mol of oxygen and 0.79 mol of nitrogen at all temperatures and pressures.

EXAMPLE 1-2   A mixture of 25 percent ammonia gas and 75 percent air (dry basis) is passed upward through a vertical scrubbing tower, to the top of which water is pumped. Scrubbed gas containing 0.5 percent ammonia leaves the top of the tower, and an aqueous solution containing 10 percent ammonia by weight leaves the bottom. Both entering and leaving gas streams are saturated with water vapor. The gas enters the tower at 100°F (37.8°C) and leaves at 70°F (21.1°C). The pressure of both streams and throughout the tower is 15 lb$_f$/in.^{2} (1.02 atm) gauge. The air-ammonia mixture enters the tower at a rate of 1,000 ft^3/min (28.32 m^3/min), measured as dry gas at 60°F (15.6°C) and 1 atm. What percentage of the ammonia entering the tower is not absorbed by the water? How many gallons of water per minute are pumped to the top of the tower?
    SOLUTION   Because the problem concerns mixtures of gases, the use of Eq. (1-51) is indicated. Also, the temperatures and pressures vary, and the calculations are simplified if pound moles are used in the material balances. Again, since the air in the entering gas

passes through the absorber unchanged in mass, the dry, ammonia-free air provides a convenient basis for calculating the moles of ammonia absorbed.

The molar volume at 60°F and 1 atm is 359(460 + 60)/492 = 379 ft³/lb mol and the total ammonia and air entering the tower is 1,000/379 = 2.64 lb mol/min. Gas analyses are always by volume, and by Eq. (1-51) the volume analyses can be used directly as molal analyses. The ammonia entering the absorber is, therefore, 0.25 × 2.64 = 0.660 mol/min, and the dry air is 0.75 × 2.64 = 1.98 mol/min. The molal ratio of ammonia to air in the effluent gas is 0.005/0.995, and the unabsorbed ammonia is 1.98 × 0.005/0.995 = 0.0099 mol/min. The fraction of the entering ammonia unabsorbed is

$$\frac{0.0099}{0.66} 100 = 1.5 \text{ percent}$$

The vapor pressure of water at 100°F and at 70°F is 0.949 and 0.363 $lb_f/in.^2$, respectively (see Appendix 8). The total pressure of the gas is 14.7 + 15.0 = 29.7 $lb_f/in.^2$ The partial pressure of water in the entering gas is 0.949 $lb_f/in.^2$, and that of the dry gas is 29.7 − 0.949 = 28.75 $lb_f/in.^2$ Because a partial-pressure ratio equals a molal ratio, the water in the inlet gas is (0.949/28.75)2.64 = 0.0871 mol/min. Likewise, since the air and ammonia in the effluent gas are 1.98 + 0.0099 = 1.99 mol/min, the water in this stream is 1.99 × 0.363/(29.70 − 0.363) = 0.0246 mol/min. The water condensed from the gas stream is (0.0871 − 0.0246)18 × 60 = 67.5 lb/h.

The ammonia liquor leaving the tower is 10 percent by weight, and carries 0.660 − 0.0099 = 0.650 mol of ammonia per minute. The total water in the liquor is therefore 0.650 × 17 × 0.90/0.10 = 99.4 lb/min. Of this, 67 lb/h, or 1.1 lb/min, is absorbed from the entering air, and 99.4 − 1.1 = 98.3 lb/min is pumped to the tower. The mass of 1 gal water is 8.33 lb, and the volume of water to the tower is 98.3/8.33 = 11.8 gal/min (2.68 m³/h).

Note that the water condensed from the gas is approximately 1 percent of the total and could be neglected in the calculation of the water consumption of the absorber.   ////

Material balances   The law of conservation of matter states that matter cannot be created or destroyed. This leads to the concept of mass, and the law may be stated in the form that the mass of the materials taking part in any process is constant. It is known now that the law is too restricted for matter moving at velocities near that of light or for substances undergoing nuclear reactions. Under these circumstances energy and mass are interconvertible, and the sum of the two is constant, rather than only one. In most engineering, however, this transformation is too small to be detected, and in this book it is assumed that mass and energy are independent.

Conservation of mass requires that the materials entering any process must either accumulate or leave the process. There can be neither loss nor gain during the process. Most of the processes considered in this book involve neither accumulation nor depletion, and the law of conservation of matter takes the simple form that input equals output. The law is often applied in the form of material balances. The process is debited with everything that enters it and is credited with everything that leaves it. The sum of the credits must equal the sum of the debits. Material balances must hold over the entire process or equipment and over any part of it. They must apply to all the material that enters and leaves the process and to any single material that passes through the process unchanged.

EXAMPLE 1-3 An evaporator is fed continuously with 25 tons/hr (22,680 kg/h) of a solution consisting of 10 percent NaOH, 10 percent NaCl, and 80 percent H_2O. During evaporation, water is boiled off, and salt precipitates as crystals, which are settled and removed from the remaining liquor. The concentrated liquor leaving the evaporator contains 50 percent NaOH, 2 percent NaCl, and 48 percent H_2O.

Calculate (a) the pounds of water evaporated per hour, (b) the pounds of salt precipitated per hour, and (c) the pounds of concentrated liquor produced per hour.

SOLUTION (a) Because the NaOH is the only constituent passing through the evaporator unchanged, it is convenient to base the concentrations of the H_2O and the NaCl on the NaOH. The total input of NaOH is

$$25 \times 2{,}000 \times 0.10 = 5{,}000 \text{ lb/h}$$

The ratio of H_2O to NaOH in the feed is $\frac{80}{10}$, and that in the concentrated liquor is $\frac{48}{50}$. Since these ratios are on a common basis, they can be subtracted to give the mass of water evaporated per pound of NaOH. The total evaporation is then

$$5{,}000(\tfrac{80}{10} - \tfrac{48}{50}) = 35{,}200 \text{ lb/h (15,966 kg/h)}$$

(b) By the same method, the ratios of NaCl to NaOH in feed and concentrated liquors are $\frac{10}{10}$ and $\frac{2}{50}$, respectively. The NaCl precipitated is then

$$5{,}000(\tfrac{10}{10} - \tfrac{2}{50}) = 4{,}800 \text{ lb/h (2,177 kg/h)}$$

(c) The concentrated liquor can be calculated in two ways. The first method uses a NaOH balance. The ratio of concentrated liquor per pound of NaOH is $\frac{100}{50}$, and the total liquor is then

$$5{,}000 \times \tfrac{100}{50} = 10{,}000 \text{ lb/h (4,536 kg/h)}$$

The second method uses an overall balance, which states that the mass of the concentrated liquor equals that of the feed liquor less the sum of the masses of the salt precipitated and the water evaporated. The feed liquor is $25 \times 2{,}000 = 50{,}000$ lb/h. The concentrated liquor is then

$$50{,}000 - (4{,}800 + 35{,}200) = 10{,}000 \text{ lb/h}$$

The two methods are not independent, and must give identical results. ////

EXAMPLE 1-4 In the distillation technique called *rectification* (see Chap. 19), condensing vapor and boiling liquid streams are passed through a rectifying column in intimate countercurrent contact. The more volatile components tend to transfer from the liquid to the vapor and the less volatile components from the vapor to the liquid. Often the energy relations are such that the number of moles of both liquid and vapor is constant during the process, and for every mole of liquid vaporized there is a mole of vapor condensed. In a specific rectification process, operating on ethanol-and-water mixtures, the analyses of the vapor and liquid streams entering and leaving the column are those given in Table 1-1. Calculate

Table 1-1 DATA FOR EXAMPLE

	Ethanol, mol %	Water, mol %
Liquid entering top	75.1	24.9
Liquid leaving bottom	25.1	74.9
Vapor entering bottom	40.2	59.8
Vapor leaving top	77.3	22.7

the number of moles of vapor flowing up the column per mole of liquid flowing down the column.

SOLUTION In this problem, since the moles of both vapor and liquid are constant, concentrations are best expressed in mole fractions of the total stream. The mole fractions are found by dividing the mole percentages by 100. When a basis of 1 mol of liquid is chosen and V is used to represent the number of moles of vapor, the ethanol lost by the liquid is $0.751 - 0.251$ mol and the ethanol gained by the vapor is $V(0.773 - 0.402)$ mol. Equating these quantities gives

$$V(0.773 - 0.402) = 0.751 - 0.251$$

This gives $V = 1.35$ mol of vapor per mole of liquid.

The transfer of $0.751 - 0.251 = 0.500$ mol of ethanol from liquid to vapor is balanced by a transfer of 0.500 mol of water from vapor to liquid.   ////

TOTAL-ENERGY EQUATION OF STEADY FLOW

The most important processes in unit operations are flow processes, in which materials flow into, through, and out of, pieces of equipment. Useful general equations, based on the laws of conservation of mass and energy, apply to flow systems. The following equations are restricted to the class of processes characterized by steady flow.

Steady-flow process In a steady-flow process, the flow rates and the properties of the flowing materials, such as temperature, pressure, composition, density, and velocity, at each point in the apparatus, including all entrance and exit ports, are constant with time. These quantities can, and usually do, vary from point to point in the system, but at any one location they do not change. Because of this constancy of local conditions, there is no accumulation or depletion of either mass or energy within the apparatus, and all material and energy balances are of the simple type

$$\text{Input} = \text{output}$$

Energy balance for single-stream process As an example of steady-flow process in which a single stream of material is treated, consider the process shown in Fig. 1-1. The equipment is any device through which the material is passing. Assume the material is flowing through the system at a constant mass rate. Consider the flow of m lb of material. The entering stream has a velocity of u_a ft/s and is Z_a ft above the horizontal datum above which heights are measured. Its enthalpy (a quantity discussed later) is H_a Btu/lb. The corresponding quantities for the leaving stream are u_b, Z_b, and H_b. Heat in the amount of Q Btu is being transferred through the boundaries of the equipment to the material flowing through it during the time m lb of fluid enters the equipment. If the equipment includes a turbine or engine, it may do work, usually by means of a turning shaft, on the outside. If the unit includes a pump, work from the outside must be done on the material, again through the agency of a turning shaft. Work effects of this kind are called *shaft work*. Assume that shaft work equal to W_s ft-lb$_f$ is being done on the outside by the equipment. For this process, the

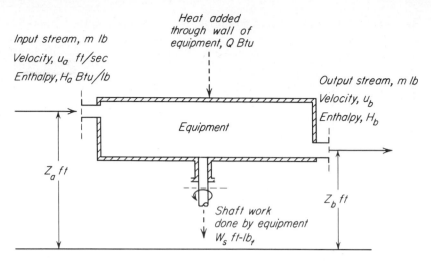

FIGURE 1-1
Diagram for steady-flow process.

following equation, which is derived in standard texts on technical thermodynamics, applies:[7]

$$m\left[\frac{u_b^2 - u_a^2}{2g_cJ} + \frac{g(Z_b - Z_a)}{g_cJ} + H_b - H_a\right] = Q - \frac{W_s}{J} \qquad (1\text{-}52)$$

where J, g, and g_c have their usual meanings.

Discussion of Eq. (1-52)   Several limitations and interpretations of Eq. (1-52) must be understood.

   1   Changes in electric, magnetic, surface, and mechanical-stress energies are not taken into account. Except in rare situations, these are absent or unimportant.
   2   To apply Eq. (1-52) to a specific situation, a precise choice of the boundaries of the equipment must be made. The inlet and outlet streams must be identified, the inlet and outlet ports located, and rotating shafts noted. All heat-transfer areas between the equipment and its surroundings must be located. The boundaries of the equipment and the cross sections of all shafts and inlet and outlet ports form the *control surface*. This must be a closed envelope, without gaps. Equation (1-52) applies to the equipment and material inside the control surface. For example, the control surface of the process of Fig. 1-1 is bounded by the walls of the equipment and the cross sections of the shaft and inlet and outlet ports, as shown by the dotted lines. The space enclosed by the control surface is called the *control volume*.

3   The heat effect Q is, by convention, positive when heat flows from the outside of the control surface into the equipment and negative when heat flows in the opposite direction. The shaft work W_s is taken as positive when the work is done on the outside of the control surface by the equipment and is negative when the work is supplied to the equipment from outside the control surface. Thus work required by a pump located within the control surface is negative. Both Q and W_s are net effects. If there is more than one heat flow or shaft work, the individual values are added algebraically and the net values of Q and W_s used in Eq. (1-52).

4   No term appears in Eq. (1-52) for friction. Friction is an internal transformation of mechanical energy into heat and occurs inside the control surface. Its effects are included in the other terms of the equation.

Enthalpy   The quantities H_a and H_b in Eq. (1-52), the enthalpies of the inlet and outlet streams, respectively, are physical properties of the material. The enthalpy of a unit mass of a pure substance is a function of pressure and temperature. Tables and diagrams provide numerical values of this property at various temperatures and pressures. These are given in texts and handbooks. Enthalpies of liquid water and saturated steam are given in the steam table, a shortened form of which is given in Appendix 8.

Absolute enthalpies are not obtainable, and numerical values of this property for a given substance are based on an arbitrarily defined datum, or standard state, for that substance. The method is analogous to that of specifying heights above sea level. The datum chosen in the steam table, for example, is liquid water at 32°F (0°C) and in equilibrium with its own vapor at that temperature. Steam-table enthalpies for liquid water, saturated steam, and superheated steam at other temperatures and pressures are excess values over that of water at the datum condition. An independent choice of datum must be made for each substance for which numerical enthalpies are desired. Because all enthalpies are relative to an arbitrary datum, only enthalpy differences have physical significance.

The enthalpy change accompanying the vaporization or condensation of a pure substance at constant pressure (and therefore at constant temperature) is the ordinary latent heat of vaporization λ. Then

$$H_y - H_x = \lambda \tag{1-53}$$

where H_y and H_x are the enthalpies of vapor and liquid, respectively. Latent heats are also given in tables of properties of substances.

Although enthalpy is, in general, a function of both temperature and pressure, two special cases are met where the pressure effect may be neglected: (1) the enthalpy of an ideal gas is independent of pressure, and the pressure effect can be ignored when the ideal-gas law is used; (2) the enthalpies of liquids and solids are not greatly affected by pressure, and under ordinary conditions, unless pressures of several atmospheres are involved, the effect of pressure on the enthalpy of a liquid or solid may be neglected.

At constant pressure, or under conditions where the effect of pressure on enthalpy may be neglected, the enthalpy difference over the temperature range from T_a to T_b is given by the equation

$$H_b - H_a = \int_{T_a}^{T_b} c_p \, dT = \bar{c}_p (T_b - T_a) \tag{1-54}$$

where c_p = specific heat (at constant pressure)
$\bar{c}_p$ = mean specific heat over temperature range T_a to T_b

Equation (1-54) cannot be used if a phase change occurs in the temperature range covered by the equation. For temperature ranges of less than 50 to 100°C it is satisfactory to use a constant value of c_p in place of $\bar{c}_p$ and to choose the value of c_p at a temperature midway between T_a and T_b.

Flow equation for several streams Equation (1-52) can be generalized for use in treating steady-flow processes where there are several streams entering and leaving the equipment. For each stream calculate the quantity

$$E = m \left(\frac{u^2}{2g_cJ} + \frac{gZ}{g_cJ} + H \right) \tag{1-55}$$

Put the quantity E for each leaving stream on the left-hand side of Eq. (1-52), using a plus sign, and put the quantity E for each entering stream on the same side of the equation with a minus sign; then equate this algebraic sum of the E's to $Q - W_s/J$. All this can be expressed mathematically by the equation

$$\Sigma E = \Sigma \left[m \left(\frac{u^2}{2g_cJ} + \frac{gZ}{g_cJ} + H \right) \right] = Q - \frac{W_s}{J} \tag{1-56}$$

where the operator Σ means to add algebraically all the E values, using positive values for leaving streams and negative values for entering streams. The overall mass balance written this way is

$$\Sigma m = 0 \tag{1-57}$$

Since the process is one of steady flow, Eq. (1-57) must hold as well as Eq. (1-56).

It is seldom that all the terms for all streams appear in a specific problem. Also, Q is zero if the process is adiabatic, and W_s is zero if no shaft work is done. The most common special case in unit operations is that where the kinetic energies $u^2/2g_cJ$ and potential energies gZ/g_cJ are negligible in comparison with Q and W_s is zero. Then Eq. (1-56) becomes simply

$$\Sigma mH = Q \tag{1-58}$$

EXAMPLE 1-5 Air is flowing steadily through a horizontal heated tube. The air enters at 40°F (4.4°C) and at a velocity of 50 ft/s (15.2 m/s). It leaves the tube at 140°F (60°C) and 75 ft/s (22.9 m/s). The average specific heat of air, from Appendix 15, is 0.24 Btu/lb-°F (1.00 J/g-°C). How many Btu per pound of air are transferred through the wall of the tube?

SOLUTION The quantities for use in Eq. (1-52) are

$$Z_a = Z_b \qquad W_s = 0 \qquad u_a = 50 \text{ ft/s} \qquad u_b = 75 \text{ ft/s}$$

$$H_b - H_a = 0.24(140 - 40) = 24 \text{ Btu/lb}$$

With a basis of 1 lb of air ($m = 1.0$), Eq. (1-52) gives

$$\frac{75^2 - 50^2}{2 \times 32.2 \times 778} + 24 = Q$$

$$Q = 24 + 0.06 = 24.1 \text{ Btu/lb } (56.06 \text{ J/kg})$$

It is clear that the effect of the kinetic-energy terms is negligible at these velocities when gas is heated. In high-speed flow, where gas velocities approach the speed of sound, the kinetic-energy change is important.

SYMBOLS

In general, quantities are given in SI, cgs, and fps units; quantities given only in either cgs or fps systems are limited to that system; quantities used in only one equation are identified by the number of the equation.

A Area of heating surface, ft² [Eq. (1-41)]

B Second virial coefficient, equation of state, m³/kg mol, cm³/g mol, or ft³/lb mol

C Third virial coefficient, equation of state, m⁶/(kg mol)², cm⁶/(g mol)², or ft⁶(lb mol)²

c_p Specific heat at constant pressure, J/kg-°C, cal/g-°C, or Btu/lb-°F; $\bar{c}_p$, average value of c_p

D Fourth virial coefficient, equation of state, m⁹/(kg mol)³, cm⁹/(g mol)³, or ft⁹/(lb mol)³ D'_0, outside diameter of pipe, in. [Eq. (1-41)]

E Total energy, J, cal, or Btu, defined by Eq. (1-55)

F Force, N, dyn, or lb$_f$

f Function

G Gravitational constant, N-m²/kg², dyn-cm²/g², or ft-lb$_f$-ft²/lb²

g Acceleration of free fall, m/s², cm/s², or ft/s²; g_n, standard value, 9.80665* m/s², 980,665 cm/s², 32.1740 ft/s²

g_c Proportionality factor, $1/k_1$ in Eq. (1-35), 32.1740 ft-lb/lb$_f$-s²

H Enthalpy, J/kg, cal/g, or Btu/lb; H_a, enthalpy of entering stream; H_b, enthalpy of leaving stream; H_x, enthalpy of saturated liquid; H_y, enthalpy of saturated vapor

J Mechanical equivalent of heat, 4.1868 J/cal$_{IT}$, 778.17 ft-lb$_f$/Btu

k Proportionality factor, k_1, in Eq. (1-1); k_2 in Eq. (1-2); k_3, in Eq. (1-3); k_4, in Eq. (1-4)

M Molecular weight

m Mass, kg, g, or lb; m_a, m_b, masses of particles [Eq. (1-2)]

n Number of moles

p Pressure, Pa, dyn/cm², or lb$_f$/ft²; $\bar{p}_A$, partial pressure of component A; $\bar{p}_B$, partial pressure of component B; $\bar{p}_C$, partial pressure of component C; p_a, initial pressure or pressure of entering stream; p_b, final pressure or pressure of leaving stream

Q Quantity of heat, J, cal, or Btu; Q_c, heat absorbed by system during cycle

q Rate of heat transfer, Btu/h [Eq. (1-41)]

R Gas-law constant, 8.31434 × 10³ J-K/kg mol, 8.31434 × 10⁷ erg-°C/g mol, or 1.98585 Btu-°R/lb mol

r Distance between two mass points m_a and m_b [Eq. (1-2)]

T Temperature, K, °C, °R, or °F

t Time, s

u Linear velocity, m/s, cm/s, or ft/s; u_a, velocity of entering stream; u_b, velocity of leaving stream; u_0, initial velocity of falling body

V Volume, m³, cm³, or ft³

v Molar volume, 1/g mol [Eq. (1-59)]

W Work, J, ergs, or ft-lb$_f$: W_c, work delivered by system during cycle; W_s, shaft work

y Mole fraction in gas mixture; y_A, mole fraction of component A; y_B, mole fraction of component B; y_C, mole fraction of component C

Z Height above datum plane, m, cm, or ft; Z_a, for entering stream; Z_b, for leaving stream

z Compressibility factor, dimensionless

Greek letters

ΔT Temperature difference, °F [Eq. (1-41)]

λ Latent heat, J/kg, cal/g, or Btu/lb

ρ Molar density, kg mol/m³, g mol/cm³, lb mol/ft³

Σ Operator, meaning "algebraic sum of"

PROBLEMS

1-1 Using defined constants and conversion factors for mass, length, time, and temperature, calculate conversion factors for (*a*) horsepower to kilowatts, (*b*) foot-pounds force to kilowatthours, (*c*) gallons (1 gal = 231 in.³) to liters (10³ cm³), (*d*) Btu per pound mole to joules per kilogram mole.

1-2 The Beattie-Bridgman equation, a famous equation of state for real gases, may be written

$$p = \frac{RT\,[1 - (c/vT^3)]}{v^2}\left[v + B_0\left(1 - \frac{b}{v}\right)\right] - \frac{A_0}{v^2}\left(1 - \frac{a}{v}\right) \tag{1-59}$$

where a, A_0, b, B_0, and c are experimental constants. (*a*) Show that this equation can be put into the form of Eq. (1-44) and derive equations for the virial coefficients B, C, and D in terms of the constants in Eq. (1-59). (*b*) For air the constants are $a = 0.01931$, $A_0 = 1.3012$, $b = -0.01101$, $B_0 = 0.04611$, $c \times 10^{-4} = 66.00$, all in cgs units (atmospheres, liters, gram moles, kelvins, with $R = 0.08206$). Calculate values of the virial coefficients for air in SI units. (*c*) Calculate z for air at a temperature of 300K and a molar volume of 0.200 m³/kg mol.

1-3 Dry gas containing 75 percent air and 25 percent ammonia vapor enters the bottom of an absorbing column, which is constructed as follows. A welded steel cylinder, 20 ft high and 2 ft in diameter, is filled with lumps of coke from 1½ to 2 in. in diameter. This coke tower filling is supported by a grid, positioned 2 ft from the bottom of the tower, and the inlet gas enters below the grid. The depth of the filling is 16 ft. In the top of the tower nozzles distribute fresh water over the coke. A solution of ammonia and water is drawn from the bottom of the column, and scrubbed gas leaves the top. The gas enters at 80°F and 760 mm Hg pressure. It leaves at 60°F and 730 mm. The leaving gas contains, on the dry basis, 1.0 percent ammonia. (*a*) If the entering gas flows through the empty bottom of the column at an average velocity (upward) of 1.5 ft/s, how many cubic feet of entering gas are treated per hour? (*b*) How many pounds of ammonia are absorbed per hour?

1-4 Prepare a freehand sketch of the top of the tower showing the connections for water and gas and the distribution device for the water.

1-5 Air flows steadily and adiabatically through a horizontal straight pipe. The air enters the pipe at an absolute pressure of 100 lb$_f$/in.², a temperature of 100°F, and a linear velocity of 10 ft/s. The air leaves at 2 $lb_f/in.^2$ abs. What are the temperature and velocity of the leaving air?

1-6 Water enters the bottom of a vertical evaporator tube 20 ft long at a temperature of 130°F and a velocity of 1.5 ft/s. The tube is 2.00 in. OD, and has a wall 0.065 in. thick. Heat amounting to 284,000 Btu/h flows through the wall of the tube and is absorbed by the water. The pressure at the exit of the tube is 4.0 $lb_f/in.^2$ abs. Assuming the liquid-and-vapor mixture leaving the tube is in equilibrium, what fraction of the liquid water entering the bottom of the tube is vaporized?

REFERENCES

1 Bridgman, P. W.: "Dimensional Analysis," rev. ed., Yale University Press, New Haven, Conn., 1931.
2 Ellis, Brian: "Basic Concepts of Measurement," Cambridge University Press, Cambridge, 1966.
3 Focken, C. M.: "Dimensional Methods and Their Application," Arnold, London, 1952.
4 *Natl. Bur. Stand. Tech. News Bull.,* **55**:3 (March 1971).
5 Prausnitz, J. M.: "Molecular Thermodynamics of Fluid-Phase Equilibria," Prentice-Hall, Englewood Cliffs, N.J., 1969.
6 Shreve, R. N.: "The Chemical Process Industries," 3d ed., McGraw-Hill, New York, 1967.
7 Smith, J. M., and H. C. Van Ness: "Introduction to Chemical Engineering Thermodynamics," 2d ed., pp. 34–38, McGraw-Hill, New York, 1959.

Fluid mechanics

The behavior of fluids is important to process engineering generally and constitutes one of the foundations for the study of the unit operations. An understanding of fluids is essential, not only for accurately treating problems on the movement of fluids through pipes, pumps, and all kinds of process equipment but also for the study of heat flow and the many separation operations that depend on diffusion and mass transfer.

The branch of engineering science that has to do with the behavior of fluids— and fluids are understood to include liquids, gases, and vapors—is called *fluid mechanics*. Fluid mechanics in turn is a part of a larger discipline called *continuum mechanics*, which also includes the study of stressed solids.

Fluid mechanics has two branches important to the study of unit operations: *fluid statics*, which treats fluids in the equilibrium state of no shear stress, and *fluid dynamics*, which treats fluids when portions of the fluid are in motion relative to other parts.

The chapters of this section deal with those areas of fluid mechanics which are important to unit operations. The choice of subject matter is but a sampling of the huge field of fluid mechanics generally. Chapter 2 treats fluid statics and some of its important applications. Chapter 3 discusses the important phenomena appearing in flowing fluids. Chapter 4 deals with the basic quantitative laws and equations of fluid flow. Chapter 5 treats flow of incompressible fluids through pipes and in thin layers, Chap. 6 is on compressible fluids in flow, and Chap. 7 describes flow past solids immersed in the flowing fluid. Chapter 8 deals with the important engineering tasks of moving fluids through process equipment and of measuring and controlling fluids in flow. Finally, Chap. 9 covers mixing and agitation, a process which essentially is applied fluid mechanics.

FLUID STATICS AND ITS APPLICATIONS

Nature of fluids   A fluid is a substance that does not permanently resist distortion. An attempt to change the shape of a mass of fluid results in layers of fluid sliding over one another until a new shape is attained. During the change in shape, shear stresses exist, the magnitudes of which depend upon the viscosity of the fluid and the rate of sliding, but when a final shape has been reached, all shear stresses will have disappeared. A fluid in equilibrium is free from shear stresses.

At a given temperature and pressure, a fluid possesses a definite density, which in engineering practice is usually measured in pounds per cubic foot or kilograms per cubic meter. Although the density of a fluid depends on temperature and pressure, the variation of density with changes in these variables may be large or small. If the density is but little affected by moderate changes in temperature and pressure, the fluid is said to be *incompressible*, and if the density is sensitive to changes in these variables, the fluid is said to be *compressible*. Liquids are considered to be incompressible and gases compressible. The terms are relative, however, and the density of a liquid can change appreciably if pressure and temperature are changed over wide limits. Also, gases subjected to small percentage changes in pressure and temperature act as incompressible fluids, and density changes under such conditions may be neglected without serious error.

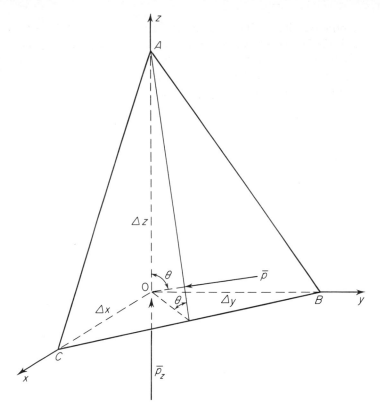

FIGURE 2-1
Forces on static element of fluid.

Pressure concept   The basic property of a static fluid is pressure. Pressure is familiar as a surface force exerted by a fluid against the walls of its container. Pressure also exists at every point within a volume of fluid. A fundamental question is: What kind of quantity is pressure? Is pressure independent of direction, or does it vary with direction? For a static fluid, as shown by the following analysis, pressure turns out to be independent of the orientation of any internal surface on which the pressure is assumed to act.

Choose any point O in a mass of static fluid and, as shown in Fig. 2-1, construct a cartesian system of coordinate axes with O as the origin. The x and y axes are in the horizontal plane, and the z axis points vertically upward. Construct a plane ABC cutting the x, y, and z axes at distances from the origin of Δx, Δy, and Δz, respectively. Planes ABC, AOC, COB, and AOB form a tetrahedron. Let θ be the angle between planes ABC and COB. This angle is less than 90° but otherwise is chosen at random. Imagine that the tetrahedron is isolated as a free body and consider all forces acting on it in the direction of the z axis, either from outside the fluid or from the surrounding fluid. Three forces are involved: (1) the force of gravity acting downward, (2) the pressure force on plane COB acting upward, and (3) the vertical component of the pressure force on plane ABC acting downward. Since the fluid is in equilibrium, the resultant of these forces is zero. Also, since a fluid in equilibrium cannot support shear stresses, all pressure forces are normal to the surface on which they act. Otherwise there would be shear-force components parallel to the faces.

The area of face COB is $\Delta x\,\Delta y/2$. Let the average pressure on this face be $\bar{p}_z$. Then the upward force on the face is $\bar{p}_z\,\Delta x\,\Delta y/2$. Let $\bar{p}$ be the average pressure on face ABC. The area of this face is $\Delta x\,\Delta y/(2\cos\theta)$, and the total force on the face is $\bar{p}\,\Delta x\,\Delta y/(2\cos\theta)$. The angle between the force vector of pressure $\bar{p}$ with the z axis is also θ, so the vertical component of this force acting downward is

$$\frac{p\,\Delta x\,\Delta y\cos\theta}{2\cos\theta} = \frac{p\,\Delta x\,\Delta y}{2}$$

The volume of the tetrahedron is $\Delta x\,\Delta y\,\Delta z/6$. If the fluid density is ρ, the force of gravity acting on the fluid in the tetrahedron is $\rho\,\Delta x\,\Delta y\,\Delta z\,g/6g_c$. This component acts downward. The force balance in the z direction becomes

$$\frac{\bar{p}_z\,\Delta x\,\Delta y}{2} - \frac{\bar{p}\,\Delta x\,\Delta y}{2} - \frac{\rho\,\Delta x\,\Delta y\,\Delta z\,g}{6g_c} = 0$$

Dividing by $\Delta x\,\Delta y$ gives

$$\frac{\bar{p}_z}{2} - \frac{\bar{p}}{2} - \frac{\rho\,\Delta z\,g}{6g_c} = 0 \tag{2-1}$$

Now, keeping angle θ constant, let plane ABC move toward the origin O. As the distance between ABC and O approaches zero as a limit, Δz approaches zero also and the gravity term vanishes. Also, the average pressures $\bar{p}$ and $\bar{p}_z$ approach p and p_z, the local pressures at point O, and Eq. (2-1) shows that p_z becomes equal to p.

Force balances parallel to the x and y axes can also be written. Each gives an equation like Eq. (2-1) but containing no gravity term, and in the limit

$$p_x = p_y = p_z = p$$

Since both point O and angle θ were chosen at random, the pressure at any point is independent of direction.

Hydrostatic equilibrium   In a stationary mass of a single static fluid, the pressure is constant in any cross section parallel to the earth's surface but varies from height to height. Consider the vertical column of fluid shown in Fig. 2-2. Assume the cross-sectional area of the column is S. At a height Z above the base of the column let the pressure be p and the density be ρ. As in the preceding analysis, the resultant of all forces on the small volume of fluid of height dZ and cross-sectional area S must be zero. Three vertical forces are acting on this volume: (1) the force from pressure p acting in an upward direction, which is pS; (2) the force from pressure $p + dp$ acting

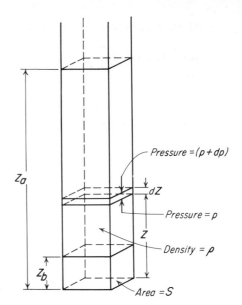

FIGURE 2-2
Hydrostatic equilibrium.

in a downward direction, which is $(p + dp)S$; (3) the force of gravity acting downward, which is $(g/g_c)\rho S\ dZ$. Then

$$+pS - (p + dp)S - \frac{g}{g_c}\rho S\ dZ = 0 \qquad (2\text{-}2)$$

In this equation, forces acting upward are taken as positive and those acting downward as negative. After simplification and division by S Eq. (2-2) becomes

$$dp + \frac{g}{g_c}\rho\ dZ = 0 \qquad (2\text{-}3)$$

Equation (2-3) cannot be integrated for compressible fluids unless the variation of density with pressure is known throughout the column of fluid. However, it is often satisfactory for engineering calculations to consider ρ to be essentially constant. The density is constant for incompressible fluids and, except for large changes in height, is nearly so for compressible fluids. Integration of Eq. (2-3) on the assumption that ρ is constant gives

$$\frac{p}{\rho} + \frac{g}{g_c}Z = \text{const} \qquad (2\text{-}4)$$

or, between the two definite heights Z_a and Z_b shown in Fig. 2-2,

$$\frac{p_b}{\rho} - \frac{p_a}{\rho} = \frac{g}{g_c}(Z_a - Z_b) \qquad (2\text{-}5)$$

Equation (2-4) expresses mathematically the condition of hydrostatic equilibrium.

APPLICATIONS OF FLUID STATES

Barometric equation   For an ideal gas the density and pressure are related by the equation

$$\rho = \frac{pM}{RT} \tag{2-6}$$

where M = molecular weight
$\quad\quad T$ = absolute temperature

Substitution from Eq. (2-6) into Eq. (2-3) gives

$$\frac{dp}{d} + \frac{gM}{g_cRT}dZ = 0 \tag{2-7}$$

Integration of Eq. (2-7) between levels a and b, on the assumption that T is constant, gives

$$\ln\frac{p_b}{p_a} = -\frac{gM}{g_cRT}(Z_b - Z_a)$$

or

$$\frac{p_b}{p_a} = \exp\left[-\frac{gM(Z_b - Z_a)}{g_cRT}\right] \tag{2-8}$$

Equation (2-8) is known as the *barometric equation*.

Methods are available in the literature[3] for estimating the pressure distribution in situations, for example, a deep gas well, in which the gas is not ideal and the temperature is not constant.

As stated on page 14, these equations are used in SI units by setting $g_c = 1$.

Hydrostatic equilibrium in a centrifugal field   In a rotating centrifuge a layer of liquid is thrown outward from the axis of rotation and is held against the wall of the bowl by centrifugal force. The free surface of the liquid takes the shape of a paraboloid of revolution,[2] but in industrial centrifuges the rotational speed is so high and the centrifugal force so much greater than the force of gravity that the liquid surface is virtually cylindrical and coaxial with the axis of rotation. This situation is illustrated in Fig. 2-3, in which r_1 is the radial distance from the axis of rotation to the free liquid surface and r_2 is the radius of the centrifuge bowl. The entire mass of liquid indicated in Fig. 2-3 is rotating like a rigid body, with no sliding of one layer of liquid over another. Under these conditions the pressure distribution in the liquid may be found from the principles of fluid statics.

The pressure drop over any ring of rotating liquid is calculated as follows. Consider the ring of liquid shown in Fig. 2-3 and the volume element of thickness dr at a radius r.

$$dF = \frac{\omega^2 r\, dm}{g_c}$$

where dF = centrifugal force
$\quad\quad dm$ = mass of liquid in element
$\quad\quad \omega$ = angular velocity, rad/s

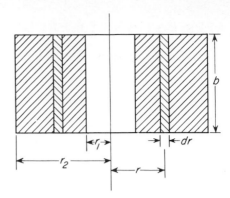

FIGURE 2-3
Single liquid in centrifuge bowl.

If ρ is the density of the liquid, and b the breadth of the ring,

$$dm = 2\pi\rho r b \, dr$$

Eliminating dm gives

$$dF = \frac{2\pi\rho b\omega^2 r^2 \, dr}{g_c}$$

The change in pressure over the element is the force exerted by the element of liquid divided by the area of the ring

$$dp = \frac{dF}{2\pi r b} = \frac{\omega^2 \rho r \, dr}{g_c}$$

The pressure drop over the entire ring is

$$p_2 - p_1 = \int_{r_1}^{r_2} \frac{\omega^2 \rho r \, dr}{g_c}$$

Assuming the density is constant and integrating gives

$$p_2 - p_1 = \frac{\omega^2 \rho (r_2^2 - r_1^2)}{2g_c} \tag{2-9}$$

Manometers   The manometer is an important device for measuring pressure differences. Figure 2-4 shows the simplest form of manometer. Assume that the shaded portion of the U tube is filled with liquid A, having a density ρ_A, and that the arms of the U tube above the liquid are filled with fluid B, having a density ρ_B. Fluid B is immiscible with liquid A and less dense than A.

A pressure p_a is exerted in one arm of the U tube and a pressure p_b in the other. As a result of the difference in pressure $p_a - p_b$, the meniscus in one branch of the U tube is higher than in the other, and the vertical distance between the two meniscuses R_m may be used to measure the difference in pressure. To derive a relationship between $p_a - p_b$ and R_m, start at the point 1, where the pressure is p_a; then, as shown by Eq. (2-4), the pressure at point 2 is $p_a + (g/g_c)(Z_m + R_m)\rho_B$. By the principles of

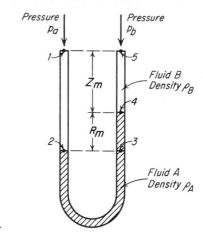

FIGURE 2-4
Simple manometer.

hydrostatics, this is also the pressure at point 3. The pressure at point 4 is less than that at point 3 by the amount $(g/g_c)R_m\rho_A$, and the pressure at point 5, which is p_b, is still less by the amount $(g/g_c)Z_m\rho_B$. These statements can be summarized by the equation

$$p_a + \frac{g}{g_c}\left[(Z_m + R_m)\rho_B - R_m\rho_A - Z_m\rho_B\right] = p_b$$

Simplification of this equation gives

$$p_a - p_b = \frac{g}{g_c}R_m(\rho_A - \rho_B) \tag{2-10}$$

Note that this relationship is independent of the distance Z_m and of the dimensions of the tube provided that pressures p_a and p_b are measured in the same horizontal plane.

EXAMPLE 2-1   A manometer of the type shown in Fig. 2-4 is used to measure the pressure drop across an orifice (see Fig. 8-21). Liquid A is mercury (specific gravity 60°F/60°F = 13.6), and fluid B, flowing through the orifice and filling the manometer leads, is brine (specific gravity 60°F/60°F = 1.26). When the pressures at the taps are equal, the level of the mercury in the manometer is 3 ft (0.914 m) below the orifice taps. Under operating conditions, the pressure at the upstream tap is 2.0 lb$_f$/in.² (13,790 N/m²) gauge,† and that at the downstream tap is 10 in. (254 mm) Hg below atmospheric. What is the reading of the manometer in millimeters?

SOLUTION   Call atmospheric pressure zero and consider g/g_c equal to unity; then the numerical data for substitution in Eq. (2-10) are

$$p_a = 2 \times 144 = 288 \ \text{lb}_f/\text{ft}^2$$

† Gauge pressure is pressure measured above the prevailing atmospheric pressure.

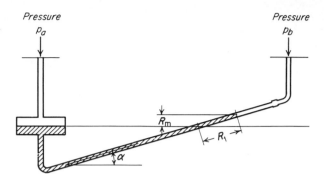

FIGURE 2-5
Inclined manometer.

From Appendix 14, the density of water at 60°F is 62.37 lb/ft³;

$$p_b = -\tfrac{10}{12} \times 13.6 \times 62.37 = -707 \text{ lb}_f/\text{ft}^2$$
$$\rho_A = 13.6 \times 62.37 = 848.2 \text{ lb/ft}^3$$
$$\rho_B = 1.26 \times 62.37 = 78.6 \text{ lb/ft}^3$$

Substituting in Eq. (2-10) gives

$$288 + 707 = R_m(848.2 - 78.6)$$
$$R_m = 1.29 \text{ ft}$$

Since 1 ft = 304.8 mm, the reading of the manometer is 1.29 × 304.8 = 0.393 m.   ////

For measuring small differences in pressure, the *inclined manometer* shown in Fig. 2-5 may be used. In this type, one leg of the manometer is inclined in such a manner that, for a small magnitude of R_m, the meniscus in the inclined tube must move a considerable distance along the tube. This distance is R_m divided by the sine of α, the angle of inclination. By making α small, the magnitude of R_m is multiplied into a long distance R_1, and a large reading becomes equivalent to a small pressure difference; so

$$p_a - p_b = \frac{g}{g_c} R_1(\rho_A - \rho_B) \sin \alpha \qquad (2\text{-}11)$$

In this type of pressure gauge, it is necessary to provide an enlargement in the vertical leg so that the movement of the meniscus in the enlargement is negligible within the operating range of the instrument.

Continuous gravity decanter   A gravity decanter of the type shown in Fig. 2-6 is used for the continuous separation of two immiscible liquids of differing densities. The feed mixture enters at one end of the separator; the two liquids flow slowly through the vessel, separate into two layers, and discharge through overflow lines at the other end of the separator.

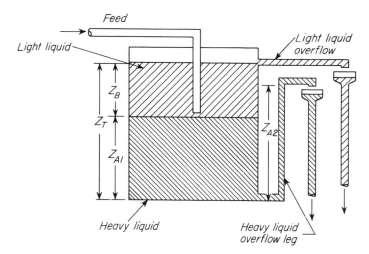

FIGURE 2-6
Continuous gravity decanter for immiscible liquids.

Provided the overflow lines are so large that frictional resistance to the flow of the liquids is negligible, and provided they discharge at the same pressure as that in the gas space above the liquid in the vessel, the performance of the decanter can be analyzed by the principles of fluid statics.

For example, in the decanter shown in Fig. 2-6 let the density of the heavy liquid be ρ_A and that of the light liquid be ρ_B. The depth of the layer of heavy liquid is Z_{A1} and that of the light liquid is Z_B. The total depth of liquid in the vessel Z_T is fixed by the position of the overflow line for the light liquid. Heavy liquid discharges through an overflow leg connected to the bottom of the vessel and rising to a height Z_{A2} above the vessel floor. The overflow lines and the top of the vessel are all vented to the atmosphere.

Since there is negligible frictional resistance to flow in the discharge lines, the column of heavy liquid in the heavy-liquid overflow leg must balance the somewhat greater depth of the two liquids in the vessel. A hydrostatic balance leads to the equation

$$Z_B\rho_B + Z_{A1}\rho_A = Z_{A2}\rho_A \qquad (2\text{-}12)$$

Solving Eq. (2-12) for Z_{A1} gives

$$Z_{A1} = Z_{A2} - Z_B\frac{\rho_B}{\rho_A} = Z_{A2} - (Z_T - Z_{A1})\frac{\rho_B}{\rho_A} \qquad (2\text{-}13)$$

where the total depth of liquid in the vessel is $Z_T = Z_B + Z_{A1}$. From this

$$Z_{A1} = \frac{Z_{A2} - Z_T(\rho_B/\rho_A)}{1 - \rho_B/\rho_A} \qquad (2\text{-}14)$$

Equation (2-14) shows that the position of the liquid-liquid interface in the separator depends on the ratio of the densities of the two liquids and on the elevations of the overflow lines. It is independent of the rates of flow of the liquids. Equation (2-14) shows that as ρ_A approaches ρ_B, the position of the interface becomes very sensitive to changes in Z_{A2}, the height of the heavy-liquid leg. With liquids that differ widely in density this height is not critical, but with liquids of nearly the same density it must be set with care. Often the top of the leg is made movable so that in service it can be adjusted to give the best separation.

EXAMPLE 2-2 A vertical cylindrical continuous decanter is to separate 1,500 bbl/d (9.93 m³/h) of a liquid petroleum fraction from an equal volume of wash acid. The density of the petroleum fraction is 54 lb/ft³ (865 kg/m³); that of the acid is 72 lb/ft³ (1,153 kg/m³). The required settling time is 20 min. Compute (a) the size of the vessel, (b) the height of the acid overflow above the floor of the vessel.

SOLUTION (a) The size of the vessel is found as follows. Since 1 bbl = 42 gal, the rate of flow of each stream is

$$\frac{1,500 \times 42}{24 \times 60} = 43.8 \text{ gal/min}$$

The total liquid holdup is

$$2 \times 43.8 \times 20 = 1,752 \text{ gal}$$

The vessel size should be about 1.1 × 1,752 or 2,000 gal (7.57 m³).

The height of the tank should approximately equal its diameter. A tank 7 ft (2.13 m) in diameter and 7 ft (2.13 m) deep would be satisfactory.

(b) The volume of the liquid in the separator is 1,752 gal. In a 7-ft vessel the depth of the liquid Z_T is

$$Z_T = \frac{1,752 \times 4}{7.48\pi 7^2} = 6.08 \text{ ft}$$

Set the interface halfway between the vessel floor and the liquid surface. Then $Z_{A1} = 3.04$ ft. Solving Eq. (2-14) for Z_{A2}, the height of the heavy-liquid overflow, gives

$$Z_{A2} = Z_{A1} + (Z_T - Z_{A1})\frac{\rho_B}{\rho_A}$$

$$= 3.04 + (6.08 - 3.04)\tfrac{54}{72} = 5.32 \text{ ft (1.62 m)} \qquad ////$$

Centrifugal decanter When the difference between the densities of the two liquids is small, the force of gravity may be too weak to separate the liquids in a reasonable time. The separation may then be accomplished in a liquid-liquid centrifuge, shown diagrammatically in Fig. 2-7. It consists of a cylindrical metal bowl, usually mounted vertically, which rotates about its axis at high speed. In Fig. 2-7a the bowl is at rest and contains a quantity of two immiscible liquids of differing densities. The heavy liquid forms a layer on the floor of the bowl beneath a layer of light liquid. If the bowl is now rotated, as in Fig. 2-7b, the heavy liquid forms a layer, denoted as zone A in the figure, next to the inside wall of the bowl. A layer of light liquid, denoted as zone

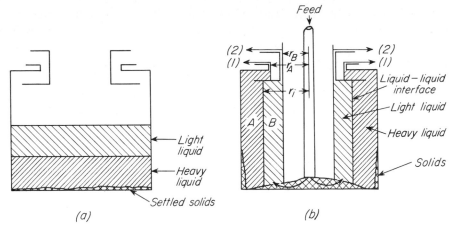

FIGURE 2-7
Centrifugal separation of immiscible liquids: (*a*) bowl at rest; (*b*) bowl rotating.
Zone *A*, separation of light liquid from heavy; zone *B*, separation of heavy liquid
from light. (1) Heavy-liquid drawoff. (2) Light-liquid drawoff.

B, forms inside the layer of heavy liquid. A cylindrical interface of radius r_i separates
the two layers. Since the force of gravity can be neglected in comparison with the
much greater centrifugal force, this interface is vertical. It is called the *neutral zone*.

In operation of the machine the feed is admitted continuously near the bottom
of the bowl. Light liquid discharges at (2) through ports near the axis of the bowl;
heavy liquid passes under a ring, inward toward the axis of rotation, and discharges
over a dam at (1). If there is negligible frictional resistance to the flow of the liquids
as they leave the bowl, the position of the liquid-liquid interface is established by a
hydrostatic balance and the relative "heights" (radial distances from the axis) of the
overflow ports at (1) and (2).

Assume that the heavy liquid, of density ρ_A, overflows the dam at radius r_A,
and the light liquid, of density ρ_B, leaves through ports at radius r_B. Then if both
liquids rotate with the bowl and friction is negligible, the pressure difference in the
light liquid between r_B and r_i must equal that in the heavy liquid between r_A and r_i.
The principle is exactly the same as in a continuous gravity decanter.

Thus

$$p_i - p_B = p_i - p_A \tag{2-15}$$

where p_i = pressure at the liquid-liquid interface
$\quad p_B$ = pressure at free surface of light liquid at r_B
$\quad p_A$ = pressure at free surface of heavy liquid at r_A

From Eq. (2-9)

$$p_i - p_B = \frac{\omega^2 \rho_B (r_i^2 - r_B^2)}{2g_c} \quad \text{and} \quad p_i - p_A = \frac{\omega^2 \rho_A (r_i^2 - r_A^2)}{2g_c}$$

Equating these pressure drops and simplifying lead to

$$\rho_B(r_i^2 - r_B^2) = \rho_A(r_i^2 - r_A^2)$$

Solving for r_i gives

$$r_i = \sqrt{\frac{r_A^2 - (\rho_B/\rho_A)r_B^2}{1 - \rho_B/\rho_A}} \tag{2-16}$$

Equation (2-16) is analogous to Eq. (2-14) for a gravity settling tank. It shows that r_i, the radius of the neutral zone, is sensitive to the density ratio, especially when the ratio is nearly unity.[1] If the densities of the fluids are too nearly alike, the neutral zone may be unstable even if the speed of rotation is sufficient to separate the liquids quickly. The difference between ρ_A and ρ_B should not be less than approximately 3 percent for stable operation.

Equation (2-16) also shows that if r_B is held constant and r_A, the radius of the discharge lip for the heavier liquid, increased, the neutral zone is shifted toward the wall of the bowl. If r_A is decreased, the zone is shifted toward the axis. An increase in r_B, at constant r_A, also shifts the neutral zone toward the axis, and a decrease in r_B causes a shift toward the wall. The position of the neutral zone is important practically. In zone A, the lighter liquid is being removed from a mass of heavier liquid, and in zone B, heavy liquid is being stripped from a mass of light liquid. If one of the processes is more difficult than the other, more time should be provided for the more difficult step. For example, if the separation in zone B is more difficult than that in zone A, zone B should be large and zone A small. This is accomplished by moving the neutral zone toward the wall, by increasing r_A or decreasing r_B. To obtain a larger time factor in zone A, the opposite adjustments would be made. Many centrifugal separators are so constructed that either r_A or r_B can be varied to control the position of the neutral zone.

Flow through continuous decanters Equations (2-14) and (2-16) for the interfacial position in continuous decanters are based entirely on hydrostatic balances. Nothing has been said about the rates of flow of the liquids or about the relative quantities of the two liquids in the feed. In practice, however, the rate of separation of one liquid phase from the other is an important variable, for it fixes the size of the separator and determines whether or not a high centrifugal force is needed in place of the force of gravity. The rate of separation depends on the relative densities of the liquids, the viscosity of the continuous liquid phase, the size of the drops of dispersed liquid, and, in a centrifuge, on the speed of rotation. The sedimentation rates of a dispersed phase in a continuous phase are discussed in Chap. 7.

SYMBOLS

b	Breadth, ft or m
F	Force, lb_f or N
g	Gravitational acceleration, ft/s^2 or m/s^2

g_c Newton's-law proportionality factor, 32.174 ft-lb/lb$_f$-s^2
M Molecular weight
p Pressure, lb$_f$/ft^2 or N/m^2; p_a, at location a; p_b, at location b; p_i, at liquid-liquid interface; p_z, in z direction; p_1, at free liquid surface; p_2, at wall of centrifuge bowl; $\bar{p}$, average pressure
R Gas-law constant, 1,545 ft-lb$_f$/lb mol-°R or 8,314.34 J/kg mol-K
R_m Reading of manometer, ft or m
r Radial distance from axis, ft or m; r_A, to heavy-liquid overflow; r_B, to light-liquid overflow; r_i, to liquid-liquid interface; r_1, to free liquid surface in centrifuge; r_2, to wall of centrifuge bowl
S Cross-sectional area, ft^2 or m^2
T Absolute temperature, °R or K
Z Height, ft or m; Z_{A1}, of layer of heavy liquid in decanter; Z_{A2}, of heavy liquid in overflow leg; Z_B, of layer of light liquid in decanter; Z_T, total depth of liquid; Z_m, of pressure connections in manometer above measuring liquid
Greek letters
α Angle with horizontal
ρ Density, lb/ft^3 or kg/m^3; ρ_A, of fluid A; ρ_B, of fluid B
ω Angular velocity, rad/s

PROBLEMS

2-1 A simple U-tube manometer is installed across an orifice meter. The manometer is filled with mercury (specific gravity = 13.6), and the liquid above the mercury is carbon tetrachloride (specific gravity = 1.6). The manometer reads 200 mm. What is the pressure difference over the manometer in newtons per cubic meter?

2-2 A differential manometer of the type shown in Fig. 2-8 is sometimes used to measure small pressure differences. When the reading is zero, the levels in the two reservoirs are equal. Assume that fluid A is methane at atmospheric pressure and 60°F, that liquid B in the reservoirs is kerosene (specific gravity = 0.815), and that liquid C in the U tube is water. The inside diameters of the reservoirs and U tube are 2.0 and $\frac{1}{4}$ in., respectively. If the reading of the manometer is 5.72 in., what is the pressure difference over the instrument in inches of water (*a*) when the change in levels in the reservoirs is neglected, (*b*) when the change in levels in the reservoirs is taken into account? What is the percent error in the answer to part (*a*)?

2-3 The temperature of the earth's atmosphere drops about 5°C for every 1,000 m of elevation above the earth's surface. If the air temperature at ground level is 15°C and the pressure is 760 mm Hg, at what elevation is the pressure 380 mm Hg? Assume that the air behaves as an ideal gas.

2-4 A continuous gravity decanter is to separate nitrobenzene, with a density of 75 lb/ft^3, from an aqueous wash liquid having a density of 64 lb/ft^3. If the total depth in the separator is 3 ft and the interface is to be $1\frac{1}{2}$ ft from the vessel floor, (*a*) what should the height of the heavy-liquid overflow leg be; (*b*) how much would an error of 2 in. in this height affect the position of the interface?

2-5 A centrifuge bowl 10 in. ID is turning at 4,000 r/min. It contains a layer of chlorobenzene 2 in. thick. If the density of the chlorobenzene is 1,100 kg/m^3 and the pressure at the liquid surface is atmospheric, what gauge pressure is exerted on the wall of the centrifuge bowl?

2-6 The liquids described in Prob. 2-4 are to be separated in a tubular centrifuge bowl with an inside diameter of 4 in., rotating at 15,000 r/min. The free-liquid surface inside the bowl is 1 in. from the axis of rotation. If the centrifuge bowl is to contain equal volumes of the two liquids, what should be the radial distance from the rotational axis to the top of the overflow dam for heavy liquid?

REFERENCES

1   Ambler, C. M.: *Chem. Eng. Prog.,* **48:**150 (1952).
2   Bird, B., W. E. Stewart, and E. N. Lightfoot: "Transport Phenomena," pp. 96–98, Wiley, New York, 1959.
3   Knudsen, J. G., and D. L. Katz: "Fluid Dynamics and Heat Transfer," pp. 69-71, McGraw-Hill, New York, 1958.

3

FLUID-FLOW PHENOMENA

The behavior of a flowing fluid depends strongly on whether or not the fluid is under the influence of solid boundaries. A moving fluid uninfluenced by stationary solid walls is not subject to shear, and shear stresses do not exist within it. The flow of incompressible fluids with no shear is called *potential flow* and is completely described by the principles of Newtonian mechanics and conservation of mass. The mathematical theory of potential flow is highly developed[5] but is outside the scope of this book. Potential flow has two important characteristics: (1) neither circulations nor eddies can form within the stream, so that potential flow is also called irrotational flow, and (2) friction cannot develop, so that there is no dissipation of mechanical energy into heat.

Potential flow can exist at distances not far from a solid boundary. A fundamental principle of fluid mechanics, originally stated by Prandtl in 1904, is that, except for fluids moving at low velocities or possessing high viscosities, the effect of the solid boundary on the flow is confined to a layer of the fluid immediately adjacent to the solid wall. This layer is called the *boundary layer*, and shear and shear forces are confined to this part of the fluid. Outside the boundary layer, potential flow survives. Most technical flow processes are best studied by considering the fluid stream as two parts, the boundary layer and the remaining fluid. In some situations the boundary layer may be neglected, and in others, such as flow through pipes, the boundary layer fills the entire channel, and there is no potential flow.

Within the current of an incompressible fluid under the influence of solid boundaries, four important effects appear: (1) the coupling of velocity-gradient and shear-stress fields, (2) the onset of turbulence, (3) the formation and growth of boundary layers, and (4) the separation of boundary layers from contact with the solid boundary.

In the flow of compressible fluids past solid boundaries additional effects appear, arising from the significant density changes characteristic of compressible fluids. These are considered in Chap. 6, on flow of compressible fluids.

The velocity field   When a stream of fluid is flowing in bulk past a solid wall, the fluid adheres to the solid at the actual interface between solid and fluid. The adhesion is a result of the force fields at the boundary, which are also responsible for the interfacial tension between solid and fluid. If, therefore, the wall is at rest in the reference frame chosen for the solid-fluid system, *the velocity of the fluid at the interface is zero.* Since at distances away from the solid the velocity is finite, there must be variations in velocity from point to point in the flowing stream. Therefore, the velocity at any point is a function of the space coordinates of that point, and a velocity field exists in the space occupied by the fluid. The velocity at a given location may also vary with time. When the velocity at each location is constant, the field is invariant with time and the flow is said to be *steady*.

ONE-DIMENSIONAL FLOW   Velocity is a vector, and in general the velocity at a point has three components, one for each space coordinate. In many simple situations all velocity vectors in the field are parallel or practically so, and only one velocity component, which may be taken as a scalar, is required. This situation, which obviously is much simpler than the general vector field, is called *one-dimensional flow*; examples are flows through closed conduits and past plates parallel to the flow. The following discussion is based on the assumptions of steady one-dimensional flow.

Laminar flow   At low velocities fluids tend to flow without lateral mixing, and adjacent layers slide past one another like playing cards. There are neither cross-currents nor eddies. This regime is called *laminar flow*. At higher velocities turbulence appears, and eddies form, which, as discussed later, lead to lateral mixing.

The velocity-gradient and rate-of-shear fields   Consider the steady one-dimensional laminar flow of an incompressible fluid along a solid plane surface. Figure 3-1a shows the velocity profile for such a stream. The abscissa u is the velocity, and the ordinate y is the distance measured perpendicularly from the wall and therefore at right angles to the direction of the velocity. At $y = 0$, $u = 0$, and u increases with distance from the wall but at a decreasing rate. Focus attention on the velocities on two nearby planes, plane A and plane B, distance Δy apart. Let the velocities along

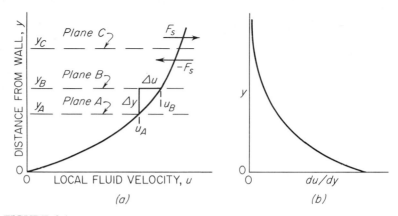

FIGURE 3-1
Profiles of velocity and velocity gradient in layer flow: (*a*) velocity; (*b*) velocity gradient or rate of shear.

the planes be u_A and u_B, respectively, and assume that $u_B > u_A$. Call $\Delta u = u_B - u_A$. Define the velocity gradient at y_A, du/dy, by

$$\frac{du}{dy} = \lim_{\Delta y \to 0} \frac{\Delta u}{\Delta y} \tag{3-1}$$

The velocity gradient is clearly the reciprocal of the slope of the velocity profile of Fig. 3-1*a*. Since the gradient is a function of position in the stream, as shown in Fig. 3-1*b*, the gradient also defines a field. Assume that x is the distance measured parallel with the flow. Then, by definition of velocity,

$$\frac{du}{dy} = \frac{d(dx/dt)}{dy} = \frac{d(dx/dy)}{dt} \tag{3-2}$$

The quantity dx/dy is the shear at plane B. Equation (3-2) shows that the gradient may be considered to be the time rate of shear, or *shear rate*, and it is often so designated. It is clear from Eq. (3-2) that when the shear vanishes ($dx/dy = 0$), the gradient vanishes also.

The shear-stress field   Since an actual fluid resists shear, a shear force must exist wherever there is a time rate of shear. In one-dimensional flow the shear force acts parallel to the plane of the shear. For example, at plane C, at distance y_C from the wall, the shear force F_s, shown in Fig. 3-1*a*, acts in the direction shown in the figure. This force is exerted by the fluid outside of plane C on the fluid between plane C and the wall. By Newton's third law, an equal and opposite force, $-F_s$, acts on the fluid outside of plane C, from the fluid inside of plane C. It is convenient to use, not total force F_s, but the force per unit area of the shearing plane, called the *shear stress* and denoted by τ, or

$$\tau = \frac{F_s}{A_s} \tag{3-3}$$

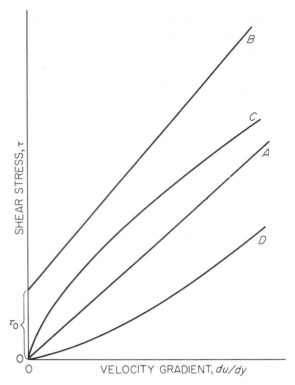

FIGURE 3-2
Shear stress versus velocity gradient for
Newtonian and non-Newtonian fluids.

where A_s is the area of the plane. Since τ depends on y, the shear stress also constitutes a field. Shear forces are generated in both laminar and turbulent flow. The shear stress arising from laminar flow is denoted by τ_v. The effect of turbulence is described later.

Newtonian and non-Newtonian fluids   The fact that at each point in a flowing fluid both a rate of shear and a shear stress exist suggests that these quantities may be related. In fact they are; the science of rheology deals with their coupling and the complex relations between them.

Figure 3-2 shows several examples of the rheological behavior of fluids. The curves are plots of shear stress vs. rate of shear and apply at constant temperature and pressure. The simplest behavior is that shown by curve A, which is a straight line passing through the origin. Fluids following this simple linearity are called Newtonian fluids. Gases, true solutions, and noncolloidal liquids are Newtonian. The other curves shown in Fig. 3-2 represent the rheological behavior of liquids called non-Newtonian. Some liquids, e.g., sewage sludge, do not flow at all until a threshold shear stress, denoted by τ_0, is attained and then flow linearly at shear stresses greater than τ_0. Curve B is an example of this relation. Liquids acting this way are called Bingham plastics. Line C represents a pseudoplastic fluid. The curve passes through the origin, is concave downward at low shears, and becomes linear at high shears. Rubber latex is an example of such a fluid. Curve D represents a dilatant fluid. The curve is concave upward at low shears and becomes linear at high shears. Quicksand and some sand-filled emulsions show this behavior. Pseudo plastics are said to be *shear-rate-thinning* and dilatant fluids *shear-rate-thickening*.

TIME-DEPENDENT FLOW   None of the curves in Fig. 3-2 depend on the history of the fluid, and a given sample of material shows the same behavior no matter how long the shearing stress has been applied. Such is not the case for some non-Newtonian liquids, whose stress-vs.-rate-of-shear curves depend on how long the shear has been active. *Thixotropic* liquids break down under continued shear and on mixing give lower shear stresses for a given shear rate. *Rheopectic* substances behave in the reverse manner, and the shear stress at constant shear rate increases with time. The original structures are usually recovered on standing.

Viscosity   In a Newtonian fluid the shear rate obviously is proportional to the shear stress. The proportionality constant μ is called the viscosity and is defined by the equation

$$\tau_v = \frac{\mu}{g_c}\frac{du}{dy} \tag{3-4}$$

Viscosity and momentum flux   It is useful to interpret Eq. (3-4) as follows. The moving fluid above plane C shown in Fig. 3-1 possesses momentum in the x-direction. Through the shear stress τ, the fluid above the plane acts on the fluid below the plane, and, by the momentum principle given by Eq. (1-1), the fluid immediately above the plane transfers some of its momentum to the fluid immediately below. This fluid in turn transfers momentum to the next lower layer, and so on. Hence x momentum can be imagined to flow in the $-y$ direction clear to the wall bounding the fluid, where $u = 0$, and is delivered to the wall as wall shear, denoted by τ_w. In SI units, τ is measured in N/m^2, which is identical to the momentum flux,† measured in $kg/m\text{-}s^2$. The existence of Newtonian viscosity, as shown by Eq. (3-4), can be interpreted as a statement that the shear stress in laminar flow represents a flux of momentum, in the direction perpendicular to the direction of flow of the fluid, proportional to the velocity gradient $-du/dy$. Then the viscosity μ is a proportionality factor, as shown in Eq. (3-4).

Momentum transfer is analogous to conductive heat transfer resulting from a temperature gradient, where the proportionality factor between the heat flux and temperature gradient is called the *thermal conductivity*. In viscous flow, momentum is transferred as a result of the velocity gradient, and the viscosity may be regarded as the conductivity of momentum transferred by this mechanism.

† Flux is defined generally as any quantity passing through a unit area in unit time.

In c.g.s units, viscosity is expressed in grams per centimeter-second. The unit called the *poise* (P).

In English units absolute viscosity is expressed either in pounds per foot-second or pounds per foot-hour. From standard conversion factors the ratios between the poise (P) and the unnamed English units are

$$1 \text{ P} = 0.0672 \text{ lb/ft-s} = 242 \text{ lb/ft-h}$$

$$1 \text{ cP} = 6.72 \times 10^{-4} \text{ lb/ft-s} = 2.42 \text{ lb/ft-h}$$

Viscosities of gases and liquids The viscosity of a Newtonian fluid depends primarily on temperature and to a lesser degree on pressure. The viscosity of a gas increases with temperature approximately in accordance with an equation of the type

$$\frac{\mu}{\mu_0} = \left(\frac{T}{273}\right)^n \tag{3-5}$$

where μ = viscosity at absolute temperature T, K
μ_0 = viscosity at 0°C (273 K)
n = constant

Viscosities of gases have been intensively studied in kinetic theory, and accurate and elaborate tables of temperature coefficients are available. Constant n is a complicated parameter and ranges in magnitude from 0.65 to 1.00.

At high pressures the viscosities of gases increase with pressure, especially in the neighborhood of the critical point.

The viscosities of liquids are sensitive to temperature and decrease with increasing temperature. For example, the viscosity of water at 0°C is more than 6 times that at 100°C. The viscosities of liquids increase moderately with pressure, although water is an exception in that its viscosity first decreases and then increases as the pressure increases.

Viscosity data are given in all standard tables of physical data. Appendixes 9 and 10 give data for common gases and liquids over a range of temperature at 1 atm.

The absolute viscosities of fluids vary over an enormous range of magnitudes, from virtually zero for liquid helium II to more than 10^{18} P for glass at room temperature. Most extremely viscous materials are non-Newtonian and possess no single viscosity independent of shear rate.

KINEMATIC VISCOSITY   The ratio of the absolute viscosity to the density of a fluid, μ/ρ, is often useful. This property is called the *kinematic viscosity* and designated by v. To distinguish μ from v, the former may be called the *dynamic viscosity*. In the SI system, the unit for v is square meters per second. In the cgs system, the kinematic viscosity is called the stoke (St), defined as 1 cm²/s. The fps unit is square feet per second. Conversion factors are

$$1 \text{ m}^2/\text{s} = 10^4 \text{ St} = 10.7639 \text{ ft}^2/\text{s}$$

Kinematic viscosities vary with temperature over a narrower range than absolute viscosities.

Rate of shear vs. shear stress for non-Newtonian fluids Bingham plastics like that represented by curve B in Fig. 3-2 follow a rheological equation of the type

$$\tau_v g_c = \tau_0 g_c + K \frac{du}{dy} \tag{3-6}$$

where K is a constant. Dilatant and pseudoplastic fluids often follow a power law, also called the *Ostwald-de Waele* equation,

$$\tau_v g_c = K' \left(\frac{du}{dy}\right)^{n'} \tag{3-7}$$

where K' and n' are constants called the *flow consistency index* and the *flow behavior index* respectively. For pseudo plastics (curve C) $n' < 1$, and for dilatant fluids (curve D) $n' > 1$. Clearly $n' = 1$ for Newtonian fluids.

TURBULENCE

The distinction between laminar and turbulent flow was first demonstrated in a classic experiment by Osborne Reynolds, reported in 1883. The equipment used by Reynolds is shown in Fig. 3-3. A horizontal glass tube was immersed in a glass-walled tank filled with water. A controlled flow of water could be drawn through the tube by opening a valve. The entrance to the tube was flared, and provision was made to introduce a fine filament of colored water from the overhead flask into the stream at the tube entrance. Reynolds found that, at low flow rates, the jet of colored water flowed intact along with the mainstream and no cross mixing occurred. The behavior of the color band showed clearly that the water was flowing in parallel straight lines and that the flow was laminar. When the flow rate was increased, a velocity, called the *critical velocity*, was reached at which the thread of color disappeared and the color diffused uniformly throughout the entire cross section of the stream of water. This behavior of the colored water showed that the water no longer flowed in laminar motion but moved erratically in the form of crosscurrents and eddies. This type of motion is *turbulent flow*.

Reynolds number and transition from laminar to turbulent flow Reynolds studied the conditions under which one type of flow changes into the other and found that the critical velocity, at which laminar flow changes into turbulent flow, depends on four quantities: the diameter of the tube, and the viscosity, density, and average linear velocity of the liquid. Furthermore, he found that these four factors can be combined into one group and that the change in kind of flow occurs at a definite value of the group. The grouping of variables so found was

$$N_{Re} = \frac{D\overline{V}}{\mu} = \frac{D\overline{V}}{\nu} \tag{3-8}$$

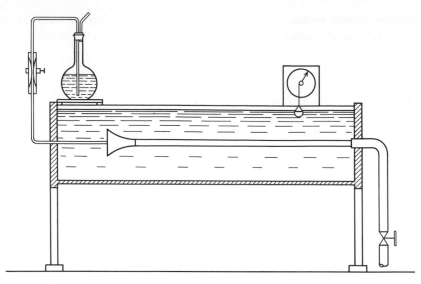

FIGURE 3-3
Reynolds' experiment.

where D = diameter of tube
$\bar{V}$ = average velocity of liquid [Eq. (4-4)]
μ = viscosity of liquid
ρ = density of liquid
v = kinematic viscosity of liquid

The dimensionless group of variables defined by Eq. (3-8) is called the *Reynolds number* N_{Re}. It is one of the named dimensionless groups listed in Appendix 5. Its magnitude is independent of the units used, provided the units are consistent.

Additional observations have shown that the transition from laminar to turbulent flow actually may occur over a wide range of Reynolds numbers. Laminar flow is always encountered at Reynolds numbers below 2,100, but it can persist up to Reynolds numbers of several thousand under special conditions of well-rounded tube entrance and very quiet liquid in the tank. Under ordinary conditions of flow, the flow is turbulent at Reynolds numbers above about 4,000. Between 2,100 and 4,000 a *transition*, or *dip*, *region* is found, where the type of flow may be either laminar or turbulent, depending upon conditions at the entrance of the tube and on the distance from the entrance.

Reynolds' experiments were conducted with water, but they have long since been shown to apply to other liquids and also to gases.

Reynolds number for non-Newtonian fluids   Since non-Newtonian fluids do not have a single-valued viscosity independent of shear rate, Eq. (3-8) for the Reynolds number cannot be used. The definition of a Reynolds number for such fluids is some-what arbitrary; a widely used definition for power-law fluids is

$$N_{Re,n} = 2^{3-n'} \left(\frac{n'}{3n' + 1} \right)^{n'} \frac{D^{n'} \rho \bar{V}^{2-n'}}{K'} \tag{3-9}$$

The basis of this somewhat complicated definition is discussed on page 49. The following equation for critical Reynolds number at which transition to turbulent flow begins has been proposed:[3]

$$N_{Re,nc} = 2,100 \frac{(4n' + 2)(5n' + 3)}{3(3n' + 1)^2} \tag{3-10}$$

Equation (3-10) is in accord with observations that the onset of turbulence occurs at Reynolds numbers above 2,100 with pseudoplastic fluids, for which $n' < 1$.

Nature of turbulence   Because of its importance in many branches of engineering, turbulent flow has been extensively investigated in recent years, and a large literature has accumulated on this subject.[1] Refined methods of measurement have been used to follow in detail the actual velocity fluctuations of the eddies during turbulent flow, and the results of such measurements have shed much qualitative and quantitative light on the nature of turbulence.

Turbulence may be generated in other ways than by flow through a pipe such as that shown in Fig. 3-3. In general, it can result either from contact of the flowing stream with solid boundaries or from contact between two layers of fluid moving at different velocities. The first kind of turbulence is called *wall turbulence* and the second kind *free turbulence*. Wall turbulence appears when the fluid flows through closed or open channels or past solid shapes immersed in the stream. Free turbulence appears in the flow of a jet into a mass of stagnant fluid or when a boundary layer separates from a solid wall and flows through the bulk of the fluid. Free turbulence is especially important in mixing, which is the subject of Chap. 9.

Turbulent flow consists of a mass of eddies of various sizes coexisting in the flowing stream. Large eddies are continually formed. They break down into smaller eddies, which in turn evolve still smaller ones. Finally, the smallest eddies disappear. At a given time, and in a given volume, a wide spectrum of eddy sizes exists. The size of the largest eddy is comparable with the smallest dimension of the turbulent stream; the diameter of the smallest eddies is about 1 mm. Smaller eddies than this are rapidly destroyed by viscous shear. Flow within an eddy is laminar. Since even the smallest eddies contain about 10^{16} molecules, all eddies are of macroscopic size, and turbulent flow is not a molecular phenomenon.

Any given eddy possesses a definite amount of mechanical energy, much like that of a small spinning top. The energy of the largest eddies is supplied by the potential energy of the bulk flow of the fluid. From an energy standpoint turbulence is a transfer process in which large eddies, formed from the bulk flow, pass energy of rotation along a continuous series of smaller eddies. Mechanical energy is not appreciably dissipated into heat during the breakup of large eddies into smaller and smaller

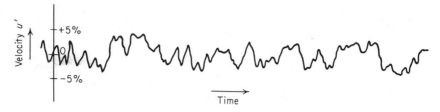

FIGURE 3-4
Velocity fluctuations in turbulent flow. The percentages are based on the
constant net velocity. [*From F. L. Wattendorf and A. M. Kuethe, Physics,* **5:**153
(1934).]

ones, but such energy is not available for maintaining pressure or overcoming resistance to flow and is worthless for practical purposes. This mechanical energy is
finally converted to heat when the smallest eddies are obliterated by viscous action.

Deviating velocities in turbulent flow   A typical picture of the variations in the
instantaneous velocity at a given point in a turbulent-flow field is shown in Fig. 3-4.
This velocity is really a single component of the actual velocity vector, all three components of which vary rapidly in magnitude and direction. Also, the instantaneous
pressure at the same point fluctuates rapidly and simultaneously with the fluctuations
of velocity. Oscillographs showing these fluctuations provide the basic experimental
data on which modern theories of turbulence are based.

   Although at first sight turbulence seems to be chaotic, structureless, and ran-
domized, further analysis of oscillographs like that in Fig. 3-4[6] shows that this is not
quite so. There *is* a distinct and recognizable repetitive pattern of velocities with time
at a given location and also a repetitive pattern of velocities at various points within a
field at a specific time. It is only because such patterns exist in a statistical sense that
turbulence is amenable to scientific study.

   The instantaneous local velocities at a given point and at variable times can be
analyzed by splitting each component of the total instantaneous velocity into two
parts, one a constant part that is the time average, or mean value, of the component
in the direction of flow of the stream, and the other, called the *deviating velocity,* which
is the instantaneous fluctuation of the component around the mean. The net velocity
is that measured by ordinary flowmeters, such as a pitot tube, which are too
sluggish to follow the rapid variations of the fluctuating velocity. The split of a
velocity component can be formalized by the following method. Let the three components (in cartesian coordinates) of the instantaneous velocity in directions x, y,
and z be u_i, v_i, and w_i, respectively. Assume also that the x axis is oriented in the
direction of flow of the stream and that components v_i and w_i are the y and z components, respectively, both perpendicular to the direction of bulk flow. Then the
equations defining the deviating velocities are

$$u_i = u + u' \qquad v_i = v' \qquad w_i = w' \qquad (3\text{-}11)$$

where u_i, v_i, w_i = instantaneous total velocity components in x, y, and z directions, respectively

u = constant net velocity of stream in x direction

u', v', w' = deviating velocities in x, y, and z directions, respectively

Terms v and w are omitted in Eqs. (3-11) because there is no net flow in the directions of the y and z axes in one-dimensional flow, and so v and w are zero.

The deviating velocities u', v', and w' all fluctuate about zero as an average. Figure 3-4 is actually a plot of the deviating velocity u'; a plot of the instantaneous velocity u_i, however, would be identical in appearance, since the ordinate would everywhere be increased by the constant quantity u.

For pressure,

$$p_i = p + p' \tag{3-12}$$

where p_i = variable local pressure

p = constant average pressure as measured by ordinary manometers or pressure gauges

p' = fluctuating part of pressure due to eddies

Because of the repetitive pattern of turbulent velocities and pressure, the time averages of the fluctuating components of velocity and pressure vanish when averaged over a time period t_0 of the order of a few seconds. Therefore

$$\frac{1}{t_0} \int_0^{t_0} u' \, dt = 0 \qquad \frac{1}{t_0} \int_0^{t_0} w' \, dt = 0$$

$$\frac{1}{t_0} \int_0^{t_0} v' \, dt = 0 \qquad \frac{1}{t_0} \int_0^{t_0} p' \, dt = 0$$

The reason these averages vanish is that for every positive value of a fluctuation there is an equal negative value, and the algebraic sum is zero.

Although the time averages of the fluctuating components themselves are zero, this is not necessarily true of other functions or combinations of these components. For example, the time average of the mean square of any one of these velocity components is not zero. This quantity for component u' is defined by

$$\frac{1}{t_0} \int_0^{t_0} (u')^2 \, dt = \overline{(u')^2} \tag{3-13}$$

Thus the mean square is not zero, since u' takes on a rapid series of positive and negative values, which, when squared, always give a positive product. Therefore $\overline{(u')^2}$ is inherently positive and vanishes only when turbulence does not exist.

In laminar flow there are no eddies; the deviating velocities and pressure fluctuations do not exist; the total velocity in the direction of flow u_i is constant and equal to u; and v_i and w_i are both zero.

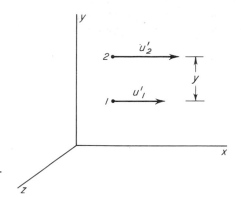

FIGURE 3-5
Fluctuating velocity components in measurement of scale of turbulence.

Statistical nature of turbulence The distribution of deviating velocities at a single point reveals that the value of the velocity is related to the frequency of occurrence of that value and that the relationship between frequency and value is Gaussian and therefore follows the error curve characteristic of completely random statistical quantities. This result establishes turbulence as a statistical phenomenon, and the most successful treatments of turbulence have been based upon its statistical nature.[1]

By measuring u', v', and w' at different places and over varying time periods two kinds of data are obtained: (1) the three deviating velocity components at a single point can be measured, each as a function of time, and (2) the values of a single deviating velocity (for example, u') can be measured at different positions over the same time period. Figure 3-5 shows values of u' measured simultaneously at two points separated by vertical distance y. Data taken at different values of y show that the correspondence between the velocities at the two stations varies from a very close relationship at very small values of y to complete independence when y is large. This is to be expected, because when the distance between the measurements is small with respect to the size of an eddy, it is a single eddy that is being measured, and the deviating velocities found at the two stations are strongly correlated. This means that when the velocity at one station changes either in direction or in magnitude, the velocity at the other station acts in practically the same way (or exactly the opposite way). At larger separation distances the measurements are being made on separate eddies, and the correlation disappears.

When the three components of the deviating velocities are measured at the same point, in general any two of them are also found to be correlated and a change in one is accompanied by a change in the other two.

These observations are quantified by defining correlation coefficients.[2] One such coefficient, which corresponds to the situation shown in Fig. 3-5, is defined as follows:

$$R_{u'} = \frac{\overline{u'_1 u'_2}}{\sqrt{\overline{(u'_1)^2}\ \overline{(u'_2)^2}}}$$

(3-14)

where u_1' and u_2' are the values of u' at stations 1 and 2, respectively. Another correlation coefficient which applies at a single point is defined by the equation

$$R_{u'v'} = \frac{\overline{u'v'}}{\sqrt{\overline{(u')^2}\ \overline{(v')^2}}} \tag{3-15}$$

where u' and v' are measured at the same point at the same time. In general, the average of the product of any two components is a negative number, even though both components average to zero individually, so that correlation coefficients of this type are inherently negative.

Intensity and scale of turbulence Turbulent fields are characterized by two average parameters. The first measures the intensity of the field and refers to the speed of rotation of the eddies and the energy contained in an eddy of a specific size. The second measures the size of the eddies. Intensity is measured by the root mean square of a velocity component. It is usually expressed as a percentage of the mean velocity or as $100\sqrt{\overline{(u')^2}}/u$. Very turbulent fields, such as those immediately below turbulence-producing grids, may reach an intensity of 5 to 10 percent. In unobstructed flow intensities are less and of the order of 0.5 to 2 percent. A different intensity usually is found for each component of velocity.

The scale of turbulence is based on correlation coefficients such as $R_{u'}$ measured as a function of the distance between stations. By determining the values of $R_{u'}$ as a function of y, the scale L_y of the eddy in the y direction is calculated by the integral

$$L_y = \int_0^\infty R_{u'}\,dy \tag{3-16}$$

Each direction usually gives a different value of L_y, depending upon the choice of velocity components used in the definition. For air flowing in pipes at 40 ft/s, the scale is about 0.4 in., and this is a measure of the average size of the eddies in the pipe.

Isotropic turbulence Although correlation coefficients generally depend upon the choice of component, in some situations this is not true, and the root-mean-square components are equal for all directions at a given point. In this situation the turbulence is said to be *isotropic*, and

$$\overline{(u')^2} = \overline{(v')^2} = \overline{(w')^2}$$

Other situations exist where this is not true, and the phenomenon is then called *anisotropic turbulence*, where

$$\overline{(u')^2} \neq \overline{(v')^2} \neq \overline{(w')^2}$$

Also, in isotropic turbulence there are no correlations of the type defined in Eq. (3-15), and

$$\overline{u'v'} = \overline{u'w'} = \overline{v'w'} = 0$$

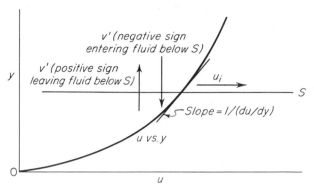

FIGURE 3-6
Reynolds stress.

Isotropic turbulence occurs only where there is no velocity gradient. Thus, for one-dimensional flow if $du/dy = 0$, the turbulence is isotropic. This situation is found at the centerline of a pipe, at the outer edge of a boundary layer, or immediately behind a grid. Otherwise, turbulent motions are anisotropic.

   The anisotropy of actual flows is largely restricted to the larger eddies. Small eddies, especially those near obliteration from viscous action, are practically isotropic.

Reynolds stresses   It has long been known that shear forces much larger than those occurring in laminar flow exist in turbulent flow wherever there is a velocity gradient across a shear plane. The mechanism of turbulent shear depends upon the deviating velocities in anisotropic turbulence. Turbulent shear stresses are called *Reynolds stresses*. They are measured by the correlation coefficients of the type $R_{u'v'}$ defined in Eq. (3-15).

   To relate Reynolds stresses to correlations of deviating velocities, the momentum principle may be used. Consider a fluid in turbulent flow moving in a positive x direction, as shown in Fig. 3-6. Plane S is parallel to the flow. The instantaneous velocity in the plane is u_i, and the mean velocity is u. Assume that u increases with y, the positive direction measured perpendicularly to the layer S, so that the velocity gradient du/dy is positive. Eddies having deviating velocities v' cross plane S from both directions. Consider arbitrarily the fluid below plane S and calculate the rate of momentum increase in this fluid from eddies per unit area of plane S. An eddy moving out of the fluid across unit area of plane S from below the plane represents a mass flow rate of $\rho v'$ *out* of the fluid, or $-\rho v'$ *into* the fluid. The velocity of the eddy in the x direction is u_i, and the rate of momentum increase of the fluid from eddy outflow is $-\rho v' u_i$. An eddy moving into the fluid across plane S from above the plane carries momentum in the fluid at a rate of $\rho v' u_i$. However, since v' is inherently negative for flow in this direction, the momentum increase for inflow of eddies is also negative. Thus, no matter which direction an eddy flows across the plane S, the eddy increases the momentum of the fluid below plane S at the rate $-\rho v' u_i$ when v' is given its proper vectorial sign. The total rate of momentum flow into the fluid, after time averaging, is then $-\rho\overline{v'u_i}$ for all eddies, and, by the momentum principle, a shear stress τ_t is generated in surface S, given by the equation

$$\tau_t g_c = -\rho\overline{v'u_i} \tag{3-17}$$

Eliminating u_i by substitution from Eq. (3-11) gives

$$\frac{\tau_t g_c}{\rho} = -\overline{uv'} - \overline{u'v'}$$

Since u is constant and $\overline{v'} = 0$, the first term on the right-hand side vanishes, and

$$\frac{\tau_t g_c}{\rho} = -\overline{u'v'} \tag{3-18}$$

The relation between turbulent shear stress and the correlation coefficient $R_{u'v'}$ is found by eliminating $-\overline{u'v'}$ from Eqs. (3-15) and (3-18)

$$\frac{\tau_t g_c}{\rho} = -R_{u'v'}\sqrt{\overline{(u')^2}\overline{(v')^2}} \tag{3-19}$$

Eddy viscosity As stated above, in isotropic turbulence $-\overline{u'v'}$, $-R_{u'v'}$, and du/dy all vanish, and Eqs. (3-18) and (3-19) show that τ_t also is zero. In anisotropic turbulence all four of these quantities exist, and there is a one-to-one correspondence between τ_t and du/dy. It is customary to linearize the relation between τ_t and du/dy in the form

$$\tau_t g_c = E_v \frac{du}{dy} \tag{3-20}$$

The close analogy between Eq. (3-20) for turbulent flow and Eq. (3-4) for laminar flow is obvious. For this reason the quantity E_v defined in Eq. (3-20) is called the *eddy viscosity* and is analogous to μ, the absolute viscosity. Also, in analogy with the kinematic viscosity v the quantity ε_M, called the *eddy diffusivity of momentum*, is defined as $\varepsilon_M = E_v/\rho$.

The total shear stress in a turbulent fluid is the sum of the viscous stress and the turbulent stress, or

$$\tau g_c = (u + E_v) \frac{du}{dy} \tag{3-21}$$

$$\tau g_c = (v + \varepsilon_M) \frac{d(\rho u)}{dy} \tag{3-22}$$

Although E_v and ε_M are analogous to μ and v, respectively, in that all these quantities are coefficients relating shear stress and velocity gradient, there is a basic difference between the two kinds of quantity. The viscosities μ and v are true properties of the fluid and are the macroscopic result of averaging motions and momenta of myriads of molecules. The eddy viscosity E and the eddy diffusivity ε_M are not

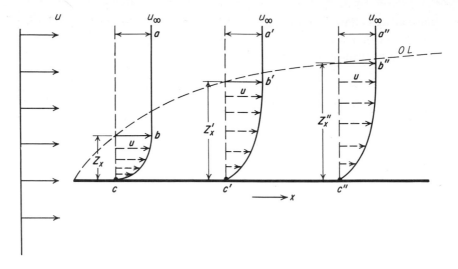

FIGURE 3-7
Prandtl boundary layer: x, distance from leading edge; u_∞, velocity of undisturbed stream; Z_x, thickness of boundary layer at distance x; u, local velocity;
abc, $a'b'c'$, $a''b''c''$, velocity-versus-distance-from-wall curves at points c, c', c'';
OL, outer limit of boundary layer.

properties of the fluid but depend on the fluid velocity and the geometry of the system.
They are functions of all factors that influence the detailed patterns of turbulence and
the deviating velocities, and they are especially sensitive to location in the turbulent
field and the local values of the scale and intensity of the turbulence. Viscosities can
be measured on isolated samples of fluid and presented in tables or charts of physical
properties, as in Appendixes 9 and 10. Eddy diffusivities are determined (with
difficulty, and only by means of special instruments) by experiments on the flow itself.

Flow in boundary layers A boundary layer is defined as that part of a moving
fluid in which the fluid motion is influenced by the presence of a solid boundary. As a
specific example of boundary-layer formation consider the flow of fluid parallel with
a thin plate, as shown in Fig. 3-7. The velocity of the fluid upstream from the leading
edge of the plate is uniform across the entire fluid stream. The velocity of the fluid
at the interface between the solid and fluid is zero. The velocity increases with distance from the plate, as shown in Fig. 3-7. Each of these curves corresponds to a
definite value of x, the distance from the leading edge of the plate. The curves change
slope rapidly near the plate; they also show that the local velocity approaches asymptotically the velocity of the bulk of the fluid stream.

In Fig. 3-7 the dotted line OL is so drawn that the velocity changes are confined
between this line and the trace of the wall. Because the velocity lines are asymptotic
with respect to distance from the plate, it is assumed, in order to locate the dotted line
definitely, that the line passes through all points where the velocity is 99 percent of the

bulk fluid velocity u_∞. Line OL represents an imaginary surface which separates the fluid stream into two parts: one in which the fluid velocity is constant and the other in which the velocity varies from zero at the wall to a velocity substantially equal to that of the undisturbed fluid. This imaginary surface separates the fluid that is directly affected by the plate from that in which the local velocity is constant and equal to the initial velocity of the approach fluid. The zone, or layer, between the dotted line and the plate constitutes the boundary layer.

The formation and behavior of the boundary layer are important, not only in the flow of fluids but also in the transfer of heat, discussed in Chap. 12, and mass, discussed in Chap. 22.

Laminar and turbulent flow in boundary layers   The fluid velocity at the solid-fluid interface is zero, and the velocities close to the solid surface are, of necessity, small. Flow in this part of the boundary layer very near the surface therefore is laminar. Farther away from the surface the fluid velocities, though less than the velocity of the undisturbed fluid, may be fairly large, and flow in this part of the boundary layer may become turbulent. Between the zone of fully developed turbulence and the region of laminar flow is a transition, or buffer, layer of intermediate character. Thus a turbulent boundary layer is considered to consist of three zones: the viscous sublayer, the buffer layer, and the turbulent zone. In some cases the boundary layer may be entirely laminar, but in most cases of engineering interest it is part laminar, part turbulent.

Near the leading edge of a flat plate immersed in a fluid of uniform velocity, the boundary layer is thin, and nearly all the fluid in the boundary layer is moving at a low velocity. Under these conditions the flow in the boundary layer is entirely laminar. As the layer thickens, however, at distances farther from the leading edge, a point is reached where turbulence appears. The onset of turbulence is characterized by a sudden rapid increase in the thickness of the boundary layer, as shown in Fig. 3-8.

When flow in the boundary layer is completely laminar, the thickness Z_x of the layer increases with $x^{0.5}$, where x is the distance from the leading edge of the plate.[4] For a short time after turbulence appears, Z_x increases with $x^{1.5}$ and then, after turbulence is fully developed, with $x^{0.8}$.

The initial, fully laminar part of the boundary layer may grow to a moderate thickness of perhaps 2 mm with air or water moving at moderate velocities. Once turbulence begins, however, the thickness of the laminar part of the boundary layer diminishes considerably, typically to about 0.2 mm.

TRANSITION FROM LAMINAR TO TURBULENT FLOW; REYNOLDS NUMBER   The factors that determine the point at which turbulence appears in a laminar boundary layer are coordinated by the dimensionless Reynolds number defined by the equation

$$N_{\text{Re},x} = \frac{xu_\infty\rho}{\mu} \tag{3-23}$$

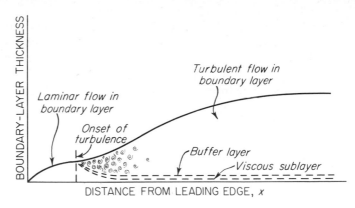

FIGURE 3-8
Development of turbulent boundary layer on a flat plate.

where x = distance from leading edge of plate

u_∞ = bulk fluid velocity

ρ_∞ = density of fluid

μ = viscosity of fluid

With parallel flow along a plate, turbulent flow first appears at a critical Reynolds number between about 10^5 and 3×10^6. The transition occurs at the lower Reynolds numbers when the plate is rough and the intensity of turbulence in the approaching stream is high and at the higher values when the plate is smooth and the intensity of turbulence in the approaching stream is low.

Boundary-layer formation in straight tubes Consider a straight tube having a bell-shaped entrance, like that used in the Reynolds experiment shown in Fig. 3-3. As shown in Fig. 3-9, a boundary layer begins to form at the entrance to the tube, and as the fluid moves through the first part of the channel, the layer thickens. During this stage the boundary layer occupies only part of the cross section of the tube, and the total stream consists of a core of fluid flowing in a rodlike manner at constant velocity and an annular boundary layer between the wall and the core. In the boundary layer the velocity increases from zero at the wall to the constant velocity existing in the core. As the stream moves farther down the tube, the boundary layer occupies an increasing portion of the cross section. Finally, at a point well downstream from the entrance, the boundary layer reaches the center of the tube, the rodlike core disappears, and the boundary layer occupies the entire cross section of the stream. At this point the velocity distribution in the tube reaches its final form, as shown by the last curve at the right of Fig. 3-9, and remains unchanged during the remaining length of the tube. Such flow with an unchanging velocity distribution is called *fully developed flow*.

TRANSITION LENGTH FOR LAMINAR AND TURBULENT FLOW The length of the entrance region of the tube necessary for the boundary layer to reach the center of the tube and for fully developed flow to be established is called the *transition length*. Since the

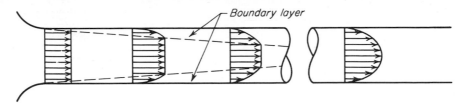

FIGURE 3-9
Development of boundary-layer flow in pipe.

velocity varies not only with length of tube but with radial distance from the center of the tube, flow in the entrance region is two-dimensional.

The approximate length of straight pipe necessary for completion of the final velocity distribution is, for laminar flow,[3]

$$\frac{x_t}{D} = 0.05 N_{Re} \tag{3-24}$$

where x_t = transition length
D = diameter of pipe

Thus for a 2-in.-ID pipe and a Reynolds number of 1,500 the transition length is 12.5 ft. If the fluid entering the pipe is turbulent and the velocity in the tube is above the critical, the transition length is nearly independent of the Reynolds number and is about 40 to 50 pipe diameters, with little difference between the distribution at 25 diameters and that at greater distances from the entrance. For a 2-in.-ID pipe, 6 to 8 ft of straight pipe is sufficient. If the fluid entering the tube is in laminar flow and becomes turbulent on entering the tube, a longer transition length, as large as 100 pipe diameters, is needed. The transition length for turbulent flow is less than that for laminar flow unless the Reynolds number is small.

Boundary-layer separation and wake formation   In the preceding paragraphs the growth of boundary layers has been discussed. Now consider what happens at the far side of a submerged object, where the fluid leaves the solid surface.

At the trailing edge of a flat plate which is parallel to the direction of flow, the boundary layers on the two sides of the plate have grown to a maximum thickness. For a time after the fluid leaves the plate, the layers and velocity gradients persist. Soon, however, the gradients fade out; the boundary layers intermingle and disappear, and the fluid once more moves with a uniform velocity. This is shown in Fig. 3-10a.

Suppose, now, the plate is turned at right angles to the direction of flow, as in Fig. 3-10b. A boundary layer forms as before in the fluid flowing over the upstream face. When the fluid reaches the edge of the plate, however, its momentum prevents it from making the sharp turn around the edge and it separates from the plate and proceeds outward into the bulk of the fluid. Behind the plate is a backwater zone of strongly decelerated fluid, in which large eddies, called *vortices*, are formed. This zone is known as the *wake*. The eddies in the wake are kept in motion by the shear

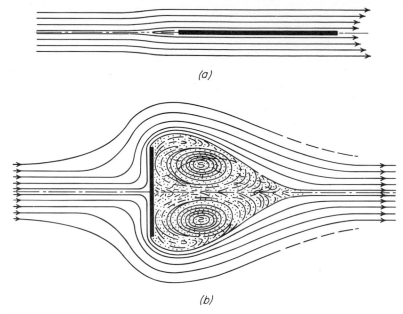

FIGURE 3-10
Flow past flat plate: (*a*) flow parallel with plate; (*b*) flow perpendicular to plate.

stresses between the wake and the separated current. They consume considerable mechanical energy and may lead to a large pressure loss in the fluid.

Boundary-layer separation occurs whenever the change in velocity of the fluid, either in magnitude or direction, is too large for the fluid to adhere to the solid surface. It is most frequently encountered when there is an abrupt change in the flow channel, like a sudden expansion or contraction, a sharp bend, or an obstruction around which the fluid must flow. As discussed in Chap. 5, page 108, however, separation may also occur from velocity decrease in a smoothly diverging channel. Because of the large energy losses resulting from the formation of a wake, it is often desirable to minimize or prevent boundary-layer separation. In some cases this can be done by suction, i.e., by drawing part of the fluid into the solid surface at the area of potential separation. Most often, however, separation is minimized by avoiding sharp changes in the cross-sectional area of the flow channel and by streamlining any objects over which the fluid must flow. For some purposes, such as the promotion of heat transfer or mixing in a fluid, boundary-layer separation may be desirable.

SYMBOLS

A	Area, ft^2 or m^2; A_s, of plane on which shear force acts
D	Diameter, ft or m
E_v	Eddy viscosity, lb/ft-s or p

F_s	Shear force, lb_f or N
g_c	Newton's-law proportionality factor, 32.174 ft-lb/lb_f-s^2
K	Constant in Eq. (3-6)
K'	Flow consistency index, lb/ft-s$^{2-n'}$ or g/m-s$^{2-n'}$ [Eq. (3-7)]
L_y	Scale of turbulence, ft or m
N_{Re}	Reynolds number, $L\rho u/\mu$; $N_{Re,n}$, for non-Newtonian fluids; $N_{Re,nc}$, critical value
n	Exponent in Eq. (3-5)
n'	Flow behavior index, dimensionless [Eq. (3-8)]
p	Pressure, lb_f/ft^2 or N/m^2; p_i, variable local pressure; p', fluctuating component
$R_{u'}$, $R_{u'v'}$	Correlation coefficients defined by Eqs. (3-14) and (3-15)
t	Time, s; t_0, time interval for averaging
u	Velocity, ft/s or m/s; velocity component in x direction; u_i, instantaneous value; u_0, constant velocity of plate; u_∞, bulk velocity of undisturbed fluid; u', deviating velocity
$\bar{V}$	Average velocity, ft/s or m/s
v, w	Velocity components in y and z directions, respectively; v_i, w_i, instantaneous values; v', w', deviating velocities
x	Distance measured parallel with flow direction, ft or m; x_t, transition length
y	Distance perpendicular to wall, ft or m
Z_x	Thickness of boundary layer, ft or m

Greek letters

ε_M	Eddy diffusivity of momentum, ft^2/s or m^2/s
μ	Viscosity, absolute, lb/ft-s or P; μ_0, at $T = 273$ K
ν	Kinematic viscosity μ/ρ, ft^2/s or m^2/s
ρ	Density, lb/ft^3 or kg/m^3
τ	Shear stress, lb_f/ft^2 or N/m^2; τ_t, turbulent shear stress; τ_v, laminar shear stress; τ_w, stress at wall; τ_0, threshold stress for Bingham plastic

REFERENCES

1 Hinze, J. O.: "Turbulence," McGraw-Hill, New York, 1959.
2 Knudsen, J. G., and D. L. Katz: "Fluid Dynamics and Heat Transfer," pp. 115–120, McGraw-Hill New York, 1958.
3 Rothfus, R. R., and R. S. Prengle: *Ind. Eng. Chem.,* **44:**1683 (1952).
4 Schlichting, H.: "Boundary Layer Theory," 6th ed., p. 38, McGraw-Hill, New York, 1968.
5 Streeter, V. L.: "Fluid Mechanics," 5th ed., McGraw-Hill, New York, 1971.
6 Wattendorf, F. L., and A. M. Kuethe: *Physics,* **5:**153 (1934).

4

BASIC EQUATIONS OF FLUID FLOW

The principles of physics most useful in the applications of fluid mechanics are the mass-balance, or continuity, equations; the linear- and angular-momentum-balance equations; and the mechanical-energy balance. They may be written in differential form, showing conditions at a point within a volume element of fluid, or in integrated form applicable to a finite volume or mass of fluid. Only the integral equations are discussed in this chapter.

Certain simple forms of the differential equations of fluid flow are given in later chapters. A complete treatment requires vector and tensor equations outside the scope of the text. They underlie more advanced study of the phenomena, and they are discussed in many texts dealing with applied mechanics and transport processes.[1-3]

Mass balance In steady flow the mass balance is particularly simple. The rate of mass entering the flow system equals that leaving, as mass can be neither accumulated nor depleted within a flow system under steady conditions.

STREAMLINES AND STREAM TUBES Discussions of fluid-flow phenomena are facilitated by visualizing, in the stream of fluid, fluid paths called *streamlines*. A streamline is an imaginary curve in a mass of flowing fluid so drawn that at every point on the curve

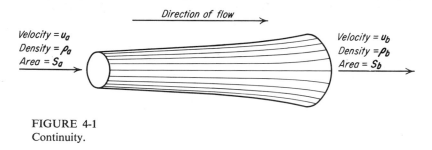

FIGURE 4-1
Continuity.

the net-velocity vector u is tangent to the streamline. No net flow takes place across such a line. In turbulent flow eddies do cross and recross the streamline, but, as shown in Chap. 3, the net flow from such eddies in any direction other than that of flow is zero.

A *stream tube*, or *stream filament*, is a tube of small or large cross section and of any convenient cross-sectional shape that is entirely bounded by streamlines. A stream tube can be visualized as an imaginary pipe in the mass of flowing fluid through the walls of which no net flow is occurring.

The mass balance gives an important relation concerning flow through a stream tube. Since flow cannot take place across the walls of the tube, the rate of mass flow into the tube during a definite period of time must equal the rate of mass flow out of the tube. Consider the stream tube shown in Fig. 4-1. Let the fluid enter at a point where the area of the cross section of the tube is S_a and leave where the area of the cross section is S_b. Let the velocity and density at the entrance be u_a and ρ_a, respectively, and let the corresponding quantities at the exit be u_b and ρ_b. Assume that the density in a single cross section is constant. Assume also that the flow through the tube is nonviscous, or potential, flow. Then the velocity u_a is constant across the area S_a, and the velocity u_b is constant across the area S_b. The mass of fluid entering and leaving the tube in unit time is

$$\dot{m} = \rho_a u_a S_a = \rho_b u_b S_b \tag{4-1}$$

where $\dot{m}$ is the rate of flow in mass per unit time. From this equation it follows for a stream tube

$$\dot{m} = \rho u S = \text{const} \tag{4-2}$$

Equation (4-2) is called the *equation of continuity*. It applies either to compressible or incompressible fluids. In the latter case, $\rho_a = \rho_b = \rho$.

AVERAGE VELOCITY   If the flow through the stream tube is not potential flow but lies wholly or in part within a boundary layer in which shear stresses exist, the velocity u will vary from point to point across the area S_a and u_b will vary from point to point across area S_b. Then it is necessary to distinguish between the local and the average velocity.

If the fluid is being heated or cooled, the density of the fluid also varies from point to point in a single cross section. In this text density variations in a single cross section of a stream tube are neglected, and both ρ_a and ρ_b are independent of location within the cross section.

The mass flow rate through a differential area in the cross section of a stream tube is

$$dm = \rho u \, dS$$

and the total mass flow rate through the entire cross-sectional area is

$$\dot{m} = \rho \int_S u \, dS \tag{4-3}$$

The integral $\int_S$ signifies that the integration covers the area S.

The average velocity $\overline{V}$ of the entire stream flowing through cross-sectional area S is defined by

$$\overline{V} \equiv \frac{\dot{m}}{\rho S} = \frac{1}{S} \int_S u \, dS \tag{4-4}$$

$\overline{V}$ also equals the total volumetric flow rate of the fluid divided by the cross-sectional area of the conduit; in fact, it is usually calculated this way. Thus

$$\overline{V} = \frac{\dot{V}}{S} \tag{4-5}$$

where $\dot{V}$ is the volumetric flow rate.

Comparison of Eqs. (4-2) and (4-4) shows that $\overline{V}$ and u are equal only when the local velocity is the same at all points in area S.

The continuity equation for flow through a finite stream tube in which the velocity varies within the cross section is

$$\dot{m} = \rho_a \overline{V}_a S_a = \rho_b \overline{V}_b S_b = \rho \overline{V} S \tag{4-6}$$

For the important special case where the flow is through channels of circular cross section

$$\dot{m} = \tfrac{1}{4}\pi D_a^2 \rho_a \overline{V}_a = \tfrac{1}{4}\pi D_b^2 \rho_b \overline{V}_b$$

from which

$$\frac{\rho_a \overline{V}_a}{\rho_b \overline{V}_b} = \left(\frac{D_b}{D_a}\right)^2 \tag{4-7}$$

where D_a and D_b are the diameters of the channel at the upstream and downstream stations, respectively.

Mass velocity   Equation (4-4) can be written

$$\overline{V}\rho = \frac{\dot{m}}{S} \equiv G \tag{4-8}$$

This equation defines the mass velocity G, calculated by dividing the mass flow rate by the cross-sectional area of the channel. In practice the mass velocity is expressed in pounds per second per square foot, pounds per hour per square foot, or kilograms per second per square meter. The advantage of using G is that it is independent of

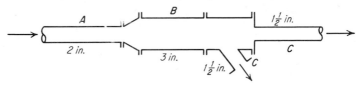

FIGURE 4-2
Piping system for Example 4-1.

temperature and pressure when the flow is steady (constant $\dot{m}$) and the cross section is unchanged (constant S). This fact is especially useful when compressible fluids are considered, for both $\overline{V}$ and ρ vary with temperature and pressure. Also, certain relationships appear later in this book in which $\overline{V}$ and ρ are associated as their product, so that the mass velocity represents the net effect of both variables. The mass velocity G can also be described as the mass current density or mass flux, where flux is defined generally as any quantity passing through a unit area in unit time. The average velocity $\overline{V}$, as shown by Eq. (4-5), can be described as the volume flux of the fluid.

EXAMPLE 4-1   Crude oil, specific gravity 60°F/60°F = 0.887, flows through the piping shown in Fig. 4-2. Pipe A is 2-in. (50-mm) Schedule 40, pipe B is 3-in. (75-mm) Schedule 40, and each of pipes C is $1\frac{1}{2}$-in. (38-mm) Schedule 40. An equal quantity of liquid flows through each of the pipes C. The flow through pipe A is 30 gal/min (6.65 m³/h). Calculate (a) the mass flow rate in each pipe, (b) the average linear velocity in each pipe, and (c) the mass velocity in each pipe.

SOLUTION   Dimensions and cross-sectional areas of standard pipe are given in Appendix 6. Cross-sectional areas needed are for 2-in. pipe, 0.0233 ft²; for 3-in. pipe, 0.0513 ft²; for $1\frac{1}{2}$-in. pipe, 0.01414 ft².
   (a)   The density of the fluid is

$$\rho = 0.887 \times 62.37 = 55.3 \text{ lb/ft}^3$$

Since there are 7.48 gal in 1 ft³ (Appendix 3), the total volumetric flow rate is

$$q = \frac{30 \times 60}{7.48} = 240.7 \text{ ft}^3/\text{h}$$

The mass flow rate is the same for pipes A and B and is the product of the density and the volumetric flow rate, or

$$\dot{m} = 240.7 \times 55.3 = 13,300 \text{ lb/h}$$

The mass flow rate through each of pipes C is one-half the total or 13,300/2 = 6,650 lb/h (0.8379 kg/s).
   (b)   Use Eq. (4-4). The velocity through pipe A is

$$\overline{V}_A = \frac{240.7}{3,600 \times 0.0233} = 2.87 \text{ ft/s}$$

through pipe B is

$$\overline{V}_B = \frac{240.7}{3,600 \times 0.0513} = 1.30 \text{ ft/s}$$

and through each of pipes C is

$$\overline{V}_C = \frac{240.7}{2 \times 3,600 \times 0.01414} = 2.36 \text{ ft/s}$$

(c) Use Eq. (4-8). The mass velocity through pipe A is

$$G_A = \frac{13,300}{0.0233} = 571,000 \text{ lb/ft}^2\text{-h (744 kg/m}^2\text{-s)}$$

through pipe B is

$$G_B = \frac{13,300}{0.0513} = 259,000 \text{ lb/ft}^2\text{-h (351 kg/m}^2\text{-s)}$$

and through each of pipes C is

$$G_C = \frac{13,300}{2 \times 0.01414} = 470,000 \text{ lb/ft}^2\text{-h (637 kg/m}^2\text{-s)} \qquad \text{////}$$

Macroscopic momentum balance A momentum balance, similar to the overall mass balance, can be written for the control volume shown in Fig. 4-1, assuming that flow is steady and unidirectional in the x direction. The sum of all forces acting on the fluid in the x direction, by the momentum principle, equals the increase in the time rate of momentum of the flowing fluid. That is to say, the sum of forces acting in the x direction equals the difference between the momentum leaving with the fluid per unit time and that brought in per unit time by the fluid, or

$$\sum F = \frac{1}{g_c} (\dot{M}_b - \dot{M}_a) \tag{4-9}$$

Momentum of total stream; momentum correction factor The momentum flow rate $\dot{M}$ of a fluid stream having a mass flow rate $\dot{m}$ and all moving at a velocity u equals $\dot{m}u$. If u varies from point to point in the cross section of the stream, however, the total momentum flow does not equal the product of the mass flow rate and the average velocity, or $\dot{m}\overline{V}$; in general it is somewhat greater than this.

The necessary correction factor is best found from the convective momentum flux, i.e., the momentum carried by the moving fluid through a unit cross-sectional area of the channel in a unit time. This is the product of the linear velocity normal to the cross section and the mass velocity (or mass flux). For a differential cross-sectional area dS, then, the momentum flux is

$$\frac{d\dot{M}}{dS} = (\rho u)u = \rho u^2 \tag{4-10}$$

The momentum flux of the whole stream, for a constant-density fluid, is

$$\frac{\dot{M}}{S} = \frac{\rho \int_S u^2 \, dS}{S} \tag{4-11}$$

The momentum correction factor β is defined by the relation

$$\beta \equiv \frac{\dot{M}/S}{\rho \overline{V}^2} \tag{4-12}$$

Substituting from Eq. (4-11) gives

$$\beta = \frac{1}{S} \int_S \left(\frac{u}{\overline{V}}\right)^2 dS \tag{4-13}$$

To find β for any given flow situation the variation of u with position in the cross section must be known. Once β is known, the momentum flow rate associated with a given stream is found from the relation

$$\dot{M} = \beta \overline{V} \dot{m} \tag{4-14}$$

Thus Eq. (4-9) may be written

$$\sum F = \frac{\dot{m}}{g_c} (\beta_b \overline{V}_b - \beta_a \overline{V}_a) \tag{4-15}$$

In using this relation, care must be taken to identify and include in $\sum F$ all force components acting on the fluid in the direction of the velocity component in the equation. Several such forces may appear: (1) pressure change in the direction of flow; (2) shear stress at the boundary between the fluid stream and the conduit or (if the conduit itself is considered to be part of the system) external forces acting on the solid wall; (3) if the stream is inclined, the appropriate component of the force of gravity. Assuming one-dimensional flow in the x direction, a typical situation is represented by the equation

$$\sum F = p_a S_a - p_b S_b + F_w - F_g \tag{4-16}$$

where p_a, p_b = inlet and outlet pressures, respectively
S_a, S_b = inlet and outlet cross sections, respectively
F_w = net force of wall of channel on fluid
F_g = component of force of gravity (written for flow in upward direction)

When flow is not unidirectional, the appropriate vector components of momentum and force are used in applying Eqs. (4-15) and (4-16).

Momentum balance in potential flow: the Bernoulli equation without friction   An important relation, called the Bernoulli equation without friction, can be derived by applying the momentum balance to the steady flow of a fluid in potential flow.

Consider a volume element of a stream tube within a larger stream of fluid in steady potential flow, as shown in Fig. 4-3. Assume that the cross section of the tube increases continuously in the direction of flow. Also, assume that the axis of the tube is straight and inclined upward at an angle ϕ from the vertical. Denote the cross section, pressure, linear velocity, and elevation at the tube entrance by S, p, u, and Z,

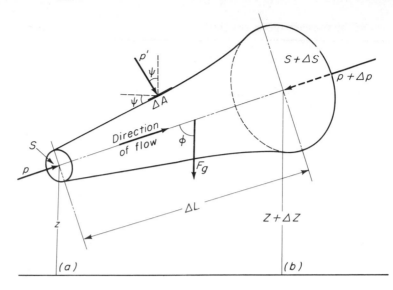

FIGURE 4-3
Forces acting on volume element of stream tube, potential flow.

respectively; and let the corresponding quantities at the exit be $S + \Delta S$, $p + \Delta p$, $u + \Delta u$, and $Z + \Delta Z$. The axial length is ΔL, and the constant fluid density is ρ. The constant mass flow rate through the tube is $\dot{m}$.

The rate of momentum flow at the tube entrance $\dot{M}_a$ is $\dot{m}u/g_c$, and that at the exit, $\dot{M}_b$, is $\dot{m}(u + \Delta u)/g_c$. Equation (4-9) can therefore be written

$$\sum F = \frac{\dot{m}\,\Delta u}{g_c} \tag{4-17}$$

The pressure forces normal to the cross section of the tube at the inlet and outlet of the tube, with the terminology of Eq. (4-16), are

$$p_a S_a = pS \qquad p_b S_b = (p + \Delta p)(S + \Delta S) \tag{4-18}$$

Since the side of the tube is not parallel to the axis, the pressure at the side possesses a component in the axial direction acting to increase momentum. Let dA be an element of side area. Since the flow is potential, there is no shear force and the local pressure p' is normal to the surface element. The pressure force is $p'\,dA$. Its component in the direction of flow is $p'\,dA \sin \psi$, where ψ is the angle between the axis and the pressure vector at element dA. But $dA \sin \psi$ is also the projection of area dA on the cross section at the discharge, so that the pressure acting in the direction of flow is $p'\,dS$. Since the total projected area from the side of the tube is exactly ΔS, the total side force is

$$F_w = \int_0^{\Delta S} p'\,dS = \bar{p}'\,\Delta S \tag{4-19}$$

where $\bar{p}'$, the average value of the pressure surrounding the tube, has a value between p and $p + \Delta p$.

The only other force acting on the flowing fluid is the component of gravity acting along the axis. The volume of the tube may be written as $\bar{S}\,\Delta L$, where $\bar{S}$ is the average cross section, which has a value between S and $S + \Delta S$. Then the mass of fluid in the tube is $\bar{S}\rho\,\Delta L$. The component of the gravitational force in the direction of flow is

$$F_g = \frac{g}{g_c}\,\bar{S}\rho\,\Delta L\cos\phi$$

Since $\cos\phi = \Delta Z/\Delta L$,

$$F_g = \frac{g}{g_c}\,\bar{S}\rho\,\Delta L\,\frac{\Delta Z}{\Delta L} = \frac{g}{g_c}\,\bar{S}\rho\,\Delta Z \tag{4-20}$$

Substitution from Eqs. (4-17) to (4-20) in Eq. (4-16) gives

$$\frac{\dot{m}}{g_c}\,\Delta u = \Delta S(\bar{p}' - p) - S\,\Delta p - \Delta p\,\Delta S - \frac{g}{g_c}\,\bar{S}\rho\,\Delta Z \tag{4-21}$$

Dividing Eq. (4-21) by $\rho S\,\Delta L$ yields

$$\frac{\dot{m}}{g_c\rho S}\,\frac{\Delta u}{\Delta L} = \frac{\bar{p}' - p}{\rho S}\,\frac{\Delta S}{\Delta L} - \frac{1}{\rho}\,\frac{\Delta p}{\Delta L} + \frac{\Delta S}{\rho S}\,\frac{\Delta p}{\Delta L} - \frac{g}{g_c}\,\frac{\bar{S}}{S}\,\frac{\Delta Z}{\Delta L} \tag{4-22}$$

Now, find the limits of all terms in Eq. (4-22) as $L \to 0$. Then, $\Delta S \to 0$, $\bar{S} \to S$, $\bar{p}' - p \to 0$, and the ratios of increments all become the corresponding differential coefficients, so that, in the limit,

$$\frac{\dot{m}}{g_c\rho S}\,\frac{du}{dL} = -\frac{1}{\rho}\,\frac{dp}{dL} - \frac{g}{g_c}\,\frac{dZ}{dL} \tag{4-23}$$

The mass flow rate is

$$\dot{m} = u\rho S$$

Substituting in Eq. (4-23) gives

$$\frac{u\rho S}{G_c\rho S}\,\frac{du}{dL} = -\frac{1}{\rho}\,\frac{dp}{dL} - \frac{g}{g_c}\,\frac{dZ}{dL} = 0$$

and

$$\frac{1}{\rho}\,\frac{dp}{dL} + \frac{g}{g_c}\,\frac{dZ}{dL} + \frac{d(u^2/2)}{g_c\,dL} = 0 \tag{4-24}$$

Equation (4-24) is the point form of the Bernoulli equation without friction. Although derived for the special situation of an expanding cross section and an upward flow, the equation is applicable to the general case of constant or contracting cross section and horizontal or downward flow (the sign of the differential dZ corrects for change in direction).

When the cross section is constant, u does not change with position, the term $d(u^2/2)/dL$ is zero, and Eq. (4-24) becomes identical with Eq. (2-3) for a stationary fluid. In unidirectional potential flow at a constant velocity, then, the magnitude of the velocity does not affect the pressure drop in the tube; the pressure drop depends only on the rate of change of elevation. In a straight horizontal tube, in consequence, there is *no* pressure drop in steady constant-velocity potential flow.

The differential form of Eq. (4-24) is

$$\frac{dp}{\rho} + \frac{g}{g_c}\,dZ + \frac{1}{g_c}\,d\left(\frac{u^2}{2}\right) = 0 \tag{4-25}$$

Between two definite points in the tube, say stations a and b, Eq. (4-25) can be integrated, since ρ is constant, to give

$$\frac{p_a}{\rho} + \frac{gZ_a}{g_c} + \frac{u_a^2}{2g_c} = \frac{p_b}{\rho} + \frac{gZ_b}{g_c} + \frac{u_b^2}{2g_c} \tag{4-26}$$

Mechanical-energy equation   The Bernoulli equation is a special form of a mechanical-energy balance, as shown by the fact that all the terms in Eq. (4-26) are scalar and have the dimensions of energy per unit mass. Each term represents a mechanical-energy effect based on a unit mass of flowing fluid. Terms $(g/g_c)Z$ and $u^2/2g_c$ are the mechanical potential and mechanical kinetic energy, respectively, of a unit mass of fluid, and p/ρ represents mechanical work done by forces, external to the stream, on the fluid in pushing it into the tube or the work recovered from the fluid leaving the tube. For these reasons Eq. (4-26) represents a special application of the principle of conservation of energy.

DISCUSSION OF THE BERNOULLI EQUATION   Equation (4-26) also shows that in the absence of friction, when the velocity u is reduced, either the height above datum Z or the pressure p, or both, must increase. When the velocity increases, it does so only at the expense of Z or p. If the height is changed, compensation must be found in a change of either pressure or velocity.

The Bernoulli equation has a greater range of validity than its derivation implies. As derived, it applies to a streamline, but since in a stream tube in potential flow the velocity in any cross section is constant, the equation may also be used for a stream tube, as implied in the early steps in its derivation. Also, although in the derivation the assumption was made that the stream tube is straight and of constant cross section, the principle of conservation of energy permits the extension of the equation to potential flow taking place in curved stream tubes of variable cross section. If the tube is curved, the direction of the velocity changes and in the Bernoulli equation the scalar speed, rather than the vector velocity, is used. Finally, by the use of correction factors the equation can be modified for use in boundary-layer flow, where velocity variations within a cross section occur and friction effects are active. The corrections are discussed in the next sections.

To apply the Bernoulli equation to a specific problem it is essential to identify the streamline or stream tube and to choose definite upstream and downstream stations. Stations a and b are chosen on the basis of convenience and are usually taken at locations where the most information about pressures, velocities, and heights is available.

EXAMPLE 4-2   Brine, specific gravity 60°F/60°F = 1.15, is draining from the bottom of a large open tank through a standard 2-in. (50-mm) Schedule 40 pipe. The drainpipe ends at a point 15 ft (4.57 m) below the surface of the brine in the tank. Considering a streamline starting at the surface of the brine in the tank and passing through the center of the drain line to the point of discharge, and assuming that friction along the streamline is negligible, calculate the velocity of flow along the streamline at the point of discharge from the pipe.

SOLUTION   To apply Eq. (4-26), choose station a at the brine surface and station b at the end of the streamline at the point of discharge. Since the pressure at both stations is atmospheric, p_a and p_b are equal, and $p_a/\rho = p_b/\rho$. At the surface of the brine, u_a is negligible, and the term $u_a^2/2g_c$ is dropped. The datum for measurement of heights can be taken through station b, so $Z_b = 0$ and $Z_a = 15$ ft. Substitution in Eq. (4-26) gives

$$\frac{15g}{g_c} = \frac{u_b^2}{2g_c}$$

and the velocity on the streamline at the discharge is

$$u_b = \sqrt{15 \times 2 \times 32.17} = 31.1 \text{ ft/s (9.48 m/s)}$$

Note that this velocity is independent of density and of pipe size.   ////

Bernoulli equation: correction for effects of solid boundaries   Most fluid-flow problems encountered in engineering involve streams that are influenced by solid boundaries and therefore contain boundary layers. This is especially true in the flow of fluids through pipes and other equipment, where the entire stream may be in boundary-layer flow.

To extend the Bernoulli equation to cover these practical situations two modifications are needed. The first, usually of minor importance, is a correction of the kinetic-energy term for the variation of local velocity u with position in the boundary layer, and the second, of major importance, is the correction of the equation for the existence of fluid friction, which appears whenever a boundary layer forms.

Also, the usefulness of the corrected Bernoulli equation in solving problems of flow of incompressible fluids is enhanced if provision is made in the equation for the work done on the fluid by a pump.

Kinetic energy of stream   The term $u^2/2g_c$ in Eq. (4-26) is the kinetic energy of a unit mass of fluid all of which is flowing at the same velocity u. When the velocity varies across the stream cross section, the kinetic energy is found in the following manner. Consider the element of cross-sectional area dS. The mass flow rate through this is $\rho u\, dS$. Each unit mass of fluid flowing through area dS carries kinetic energy in amount $u^2/2g_c$, and the energy flow rate through area dS is therefore

$$d\dot{E}_k = (\rho u\, dS)\frac{u^2}{2g_c} = \frac{\rho u^3\, dS}{2g_c}$$

where $\dot{E}_k$ represents the time rate of flow of kinetic energy. The total rate of flow of kinetic energy through the entire cross section S is, assuming constant density within the area S,

$$\dot{E}_k = \frac{\rho}{2g_c} \int_S u^3 \, dS \qquad (4\text{-}27)$$

The total rate of mass flow is given by Eqs. (4-3) and (4-8), and the kinetic energy per pound of flowing fluid, which replaces $u^2/2g_c$ in the Bernoulli equation, is

$$\frac{\dot{E}_k}{\dot{m}} = \frac{(1/2g_c) \int_S u^3 \, dS}{\int_S u \, dS} = \frac{(1/2g_c) \int_S u^3 \, dS}{\overline{V}S} \qquad (4\text{-}28)$$

KINETIC-ENERGY CORRECTION FACTOR   It is convenient to eliminate the integral of Eq. (4-28) by a factor operating on $\overline{V}^2/2g_c$ to give the correct value of the kinetic energy as calculated from Eq. (4-28). This factor, called the kinetic-energy correction factor, is denoted by α and is defined by

$$\frac{\alpha \overline{V}^2}{2g_c} \equiv \frac{\dot{E}_k}{\dot{m}} = \frac{\int_S u^3 \, dS}{2g_c \overline{V}S}$$

$$\alpha = \frac{\int_S u^3 \, dS}{\overline{V}^3 S} \qquad (4\text{-}29)$$

If α is known, the average velocity can be used to calculate the kinetic energy from the average velocity by using $\alpha \overline{V}^2/2g_c$ in place of $u^2/2g_c$. To calculate the value of α from Eq. (4-29), the local velocity must be known as a function of location in the cross section, so that the integrals in the equation can be evaluated. The same knowledge of velocity distribution is needed to calculate the value of $\overline{V}$ by Eq. (4-4).

Correction of Bernoulli equation for fluid friction   Friction manifests itself by the disappearance of mechanical energy. In frictional flow the quantity

$$\frac{p}{\rho} + \frac{u^2}{2g_c} + \frac{g}{g_c} Z$$

is not constant along a streamline, as called for by Eq. (4-26), but always decreases in the direction of flow, and, in accordance with the principle of conservation of energy, an amount of heat equivalent to the loss in mechanical energy is generated. Fluid friction can be defined as any conversion of mechanical energy into heat in a flowing stream.

For incompressible fluids, the Bernoulli equation is corrected for friction by adding a term to the right-hand side of Eq. (4-26). Thus, after introducing the kinetic-energy correction factors α_a and α_b, Eq. (4-26) becomes

$$\frac{p_a}{\rho} + \frac{gZ_a}{g_c} + \frac{\alpha_a \overline{V}_a^2}{2g_c} = \frac{p_b}{\rho} + \frac{gZ_b}{g_c} + \frac{\alpha_b \overline{V}_b^2}{2g_c} + h_f \qquad (4\text{-}30)$$

The units of h_f and those of all other terms in Eq. (4-30) are energy per unit mass. The term h_f represents all the friction generated per unit mass of fluid (and therefore all the conversion of mechanical energy into heat) that occurs in the fluid between stations a and b. It differs from all other terms in Eq. (4-30) in two ways: (1) The

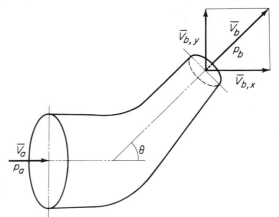

FIGURE 4-4
Flow through reducing fitting, viewed from the top. Example 4-4.

mechanical terms represent conditions *at* specific locations, namely, the inlet and outlet stations *a* and *b*, whereas h_f represents the loss of mechanical energy at all points *between* stations *a* and *b*. (2) Friction is not interconvertible with the mechanical-energy quantities. The sign of h_f, as defined by Eq. (4-30), is always positive. It is zero, of course, in potential flow.

Friction appears in boundary layers because the work done by shear forces in maintaining the velocity gradients in both laminar and turbulent flow is eventually converted into heat by viscous action. Friction generated in unseparated boundary layers is called *skin friction*. When boundary layers separate and form wakes, additional energy dissipation appears within the wake and friction of this type is called *form friction* since it is a function of the position and shape of the solid. Form friction can be quite large.

In a given situation both skin friction and form friction may be active in varying degrees. In the case of Fig. 3-10a, the friction is entirely skin; in that of Fig. 3-10b, the friction is largely form friction, because of the large wake, and skin friction is relatively unimportant. The total friction h_f in Eq. (4-30) includes both types of frictional loss.

EXAMPLE 4-3   Water with a density of 998 kg/m³ (62.3 lb/ft³) enters a 50-mm (1.969-in.) pipe fitting horizontally, as shown in Fig. 4-4, at a steady velocity of 1.0 m/s (3.28 ft/s) and a gauge pressure of 100 kN/m² (2,088.5 lb/ft²). It leaves the fitting horizontally, at the same elevation, at an angle of 45° with the entrance direction. The diameter at the outlet is 20 mm (0.787 in.). Assuming the fluid density is constant, that the kinetic-energy and momentum correction factors at both entrance and exit are unity, and that the friction loss in the fitting is negligible, calculate (a) the gauge pressure at the exit of the fitting and (b) the forces in the x and y directions exerted by the fitting on the fluid.

SOLUTION   (a) $V_a = 1.0$ m/s. From Eq. (4-7),

$$V_b = V_a \frac{D_a^2}{D_b} = 1.0 \left(\frac{50}{20}\right)^2 = 6.25 \text{ m/s}$$

$$p_a = 100 \text{ kN/m}^2$$

The outlet pressure p_b is found from Eq. (4-30). Since $Z_a = Z_b$ and h_f may be neglected, Eq. (4-30) becomes

$$\frac{p_a - p_b}{\rho} = \frac{\overline{V}_b^2 - \overline{V}_a^2}{2}$$

from which

$$p_b = p_a - \frac{\overline{V}_b^2 - \overline{V}_a^2}{2} = 100 - \frac{998(6.25^2 - 1.0^2)}{1,000 \times 2}$$

$$= 100 - 18.99 = 81.01 \text{ kN/m}^2 \ (11.75 \text{ lb}_f/\text{in.}^2)$$

Note that g_c is omitted when working in SI units.

(b) The forces acting on the fluid are found by combining Eqs. (4-15) and (4-16). For the x direction, since $F_g = 0$ for horizontal flow, this gives

$$\dot{m}(\beta_b \overline{V}_{b,x} - \beta_a \overline{V}_{a,x}) = p_a S_{a,x} - p_b S_{b,x} F_{w,x} \tag{4-31}$$

where $S_{a,x}$ and $S_{b,x}$ are the projected areas of S_a and S_b on planes normal to the flow direction. (Recall that pressure p is a scalar quantity.) Since the flow enters in the x direction, $\overline{V}_{a,x} = \overline{V}_a$ and

$$S_{a,x} = S_a = \frac{\pi}{4} 0.050^2 = 0.001964 \text{ m}^2$$

From Fig. 4-4

$$\overline{V}_{b,x} = \overline{V}_b \cos \theta = 6.25 \cos 45° = 4.42 \text{ m/s}$$

Also

$$S_{b,x} = S_b \cos \theta = \frac{\pi}{4} 0.020^2 \cos 45° = 0.000222 \text{ m}^2$$

From Eq. (4-6)

$$\dot{m} = \overline{V}_a \rho S_a = 1.0 \times 998 \times 0.001964 = 1.960 \text{ kg/s}$$

Substituting in Eq. (4-31) and solving for $F_{w,x}$ gives

$$F_{w,x} = 1.96(4.42 - 1.0) - 100,000 \times 0.001964 + 81,010 \times 0.000222$$

$$= 6.7 - 196.4 + 18.0 = -171.7 \text{ N} \ (-38.6 \text{ lb}_f)$$

Similarly, for the y direction, $\overline{V}_{a,y} = 0$ and $S_{a,y} = 0$, and

$$\overline{V}_{b,y} = \overline{V}_b \sin \theta = 4.42 \text{ m/s} \qquad S_{b,y} = S_b \sin \theta = 0.000222 \text{ m}^2$$

Hence

$$F_{w,y} = \dot{m}(\beta_b \overline{V}_{b,y} - \beta_a \overline{V}_{a,y}) - p_a S_{a,y} + p_b S_{b,y}$$

$$= 1.96(4.42 - 0) - 0 + 81.01 \times 0.000222 \times 1000$$

$$= 8.66 + 17.98 = 26.64 \text{ N} \ (5.99 \text{ lb}_f) \qquad\qquad ////$$

Pump work in Bernoulli equation   A pump is used in a flow system to increase the mechanical energy of the flowing fluid, the increase being used to maintain flow. Assume that a pump is installed between the stations a and b linked by Eq. (4-30). The work supplied to the pump is an example of shaft work W_s, discussed in Chap. 1 and used in Eq. (1-52). Because of the sign convention adopted for shaft work, however, pump work is inherently negative, and it is convenient to denote it by W_p, where

$$W_p = -\frac{W_s}{m} \tag{4-32}$$

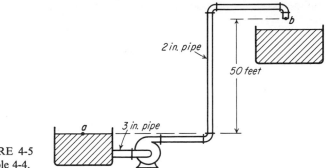

FIGURE 4-5
Example 4-4.

Thus W_p is the negative of the shaft work per unit mass of fluid. Since the Bernoulli equation is a balance of mechanical energy only, account must be taken of friction occurring within the pump. In an actual pump not only are all the sources of fluid friction active, but mechanical friction occurs as well, in bearings, pistons, and stuffing boxes. The mechanical energy supplied to the pump as negative shaft work must be discounted by these frictional losses to give the net mechanical energy actually available to the flowing fluid. Let h_{fp} be the total friction in the pump per unit mass of fluid. Then the net work to the fluid is $W_p - h_{fp}$. In practice, in place of h_{fp} a pump efficiency denoted by η is used, defined by the equation

$$W_p - h_{fp} \equiv \eta W_p$$

or
$$\eta = \frac{W_p - h_{fp}}{W_p} \tag{4-33}$$

The mechanical energy delivered to the fluid is, then, ηW_p, where $\eta < 1$. Equation (4-30) corrected for pump work is

$$\frac{p_a}{\rho} + \frac{gZ_a}{g_c} + \frac{\alpha_a \bar{V}_a^2}{2g_c} + \eta W_p = \frac{p_b}{\rho} + \frac{gZ_b}{g_c} + \frac{\alpha_b \dot{V}_b^2}{2g_c} + h_f \tag{4-34}$$

Equation (4-34) is a final working equation for problems on the flow of incompressible fluids.

EXAMPLE 4-4   In the equipment shown in Fig. 4-5, a pump draws a solution, specific gravity = 1.84, from a storage tank through a 3-in. (75-mm) Schedule 40 steel pipe. The efficiency of the pump is 60 percent. The velocity in the suction line is 3 ft/s (0.914 m/s). The pump discharges through a 2-in. (50-mm) Schedule 40 pipe to an overhead tank. The end of the discharge pipe is 50 ft (15.2 m) above the level of the solution in the feed tank. Friction losses in the entire piping system are 10 ft-lb$_f$/lb (29.9 J/kg). What pressure must the pump develop? What is the power of the pump?

SOLUTION   Use Eq. (4-34). Take station a at the surface of the liquid in the tank and station b at the discharge end of the 2-in. pipe. Take the datum plane for elevations through the station a. Since the pressure at both stations is atmospheric, $p_a = p_b$. The velocity at station a is negligible because of the large diameter of the tank in comparison with that of the pipe. The kinetic-energy factor α can be taken as 1.0 with negligible error. Equation (4-34) becomes

$$W_p \eta = \frac{g}{g_c} Z_b + \frac{\overline{V}_b^2}{2g_c} + h_f$$

By Appendix 6, the cross-sectional areas of the 3- and 2-in. pipes are 0.0513 and 0.0233 ft², respectively. The velocity in the 2-in. pipe is

$$\overline{V}_b = \frac{3 \times 0.0513}{0.0233} = 6.61 \text{ ft/s}$$

Then

$$0.60 W_p = 50 \frac{g}{g_c} + \frac{6.61^2}{64.34} + 10 = 60.68$$

and

$$W_p = \frac{60.68}{0.60} = 101.1 \text{ ft-lb}_f/\text{lb}$$

The pressure developed by the pump can be found by writing Eq. (4-34) over the pump itself. Station a is in the suction connection and station b in the pump discharge. The difference in level between suction and discharge can be neglected, so $Z_a = Z_b$, and Eq. (4-34) becomes

$$\frac{p_b - p_a}{\rho} = \frac{\overline{V}_a^2 - \overline{V}_b^2}{2g_c} + W_p \eta$$

The pressure developed by the pump is

$$p_b - p_a = 1.84 \times 62.37 \left(\frac{3^2 - 6.61^2}{2 \times 32.17} + 60.68 \right)$$

$$= 6,902 \text{ lb}_f/\text{ft}^2 \text{ or } \frac{6,902}{144} = 47.9 \text{ lb}_f/\text{in.}^2 \text{ (330 kN/m}^2\text{)}$$

The power used by the pump is the product of W_p and the mass flow rate divided by the conversion factor, 1 hp $= 550$ ft-lb$_f$/s. The mass flow rate is

$$\dot{m} = 0.0513 \times 3 \times 1.84 \times 62.37 = 17.66 \text{ lb/s}$$

and the power is

$$P = \frac{\dot{m} W_p}{550} = \frac{17.66 \times 101.1}{550} = 3.25 \text{ hp } (2.42 \text{ kW}) \qquad\qquad ////$$

Angular-momentum equation Analysis of the performance of rotating fluid-handling machinery such as pumps, turbines, and agitators is facilitated by the use of force moments and angular momentum. The moment of a force $\mathbf{F}$ about point O is the vector product of $\mathbf{F}$ and the position vector $\mathbf{r}$ of a point on the line of action of the vector from O. When a force, say F_θ, acts at right angles to the position vector, at a radial distance r from point O, the moment of the force equals the torque T, or

$$F_\theta r = T \qquad\qquad (4\text{-}35)$$

The *angular momentum* (also called the *moment of momentum*) of an object moving about a center of rotation is the vector product of the position vector and the

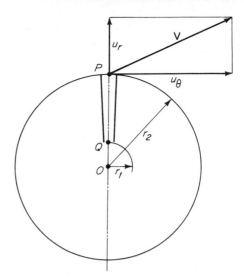

FIGURE 4-6
Angular momentum of flowing fluid.

tangential momentum vector of the object (its mass times its tangential component of velocity). Figure 4-6 shows the rotation for a situation involving two-dimensional flow: fluid at point P is moving about point O at a velocity V, which has radial and tangential components u_r and u_θ, respectively. The angular momentum of a mass m of fluid at point P is therefore rmu_θ.

Suppose that Fig. 4-6 represents part of the impeller of a centrifugal pump or turbine through which fluid is flowing at a constant mass rate $\dot{m}$. It enters at point Q near the center of rotation, at a radial distance r_1 from point O, and leaves at radial distance r_2. Its tangential velocities at these points are $u_{\theta 1}$ and $u_{\theta 2}$, respectively. The tangential force F_θ on the fluid at point P is proportional to the rate of change of angular momentum of the fluid; hence the torque is given, from Eq. (4-35), by the relation

$$T = F_\theta r_2 = \frac{\dot{m}}{g_c}(r_2 u_{\theta 2} - r_1 u_{\theta 1}) \tag{4-36}$$

Equation (4-36) is the angular-momentum equation for steady two-dimensional flow. It is analogous to Eq. (4-15), the momentum equation. It is assumed in deriving Eq. (4-36) that at any given radial distance r all the fluid is moving with the same velocity, so $\beta_1 = \beta_2 = 1$. Applications of Eq. (4-36) are given in Chaps. 8 and 9.

SYMBOLS

A Area, ft^2 or m^2
D Diameter of circular channel, ft or m; D_a, at station a; D_b, at station b
E_k Kinetic energy of fluid, ft-lb$_f$ or J; E_k, time rate of flow of kinetic energy, ft-lb$_f$/s or J/s

F Force, lb_f or N; F_g, component of gravity force; F_w, net force of channel wall on fluid; $F_{w,x}$, component of F_w in x direction; $F_{w,y}$, component in y direction; F_θ, tangential force or force component

G Mass velocity, lb/s-ft^2 or kg/s-m^2

g Acceleration of gravity, ft/s^2 or m/s^2

g_c Newton's-law proportionality factor, 32.174 ft-lb/lb$_f$-s^2

h Friction loss, ft-lb$_f$/lb or J/kg; h_f, friction loss in conduit between stations a and b; h_{fp}, total friction loss in pump

L Length, ft or m

M Momentum, ft-lb/s or kg-m/s; $\dot{M}$, time rate of flow of momentum, ft-lb/s^2 or kg-m/s^2; $\dot{M}_a$, at station a; $\dot{M}_b$, at station b

m Mass, lb or kg; $\dot{m}$, mass flow rate, lb/s or kg/s

P Power, hp or kW

p Pressure, lb_f/ft^2 or N/m^2; p_a, at station a; p_b at station b; p', normal to surface; $\bar{p}'$, average value of p'

q Volumetric flow rate, ft^3/s or m^3/s

r Radial distance. ft or m; r_1, at station 1; r_2, at station 2

S Cross-sectional area, ft^2 or m^2; S_a, at station a; S_b, at station b; $S_{a,x}$, $S_{b,x}$, projections of S_a and S_b on planes normal to x axis; $S_{a,y}$, $S_{b,y}$, projections on planes normal to y axis; $\bar{S}$, average value

T Torque, ft-lb$_f$ or N-m

u Velocity or velocity component in x direction, ft/s or m/s; u_a, at station a; u_b, at station b; u_r, in radial direction; u_θ, in tangential direction

V Total velocity vector, ft/s or m/s; $\bar{V}$, average velocity; $\bar{V}_a$, at station a; $\bar{V}_b$, at station b; $\bar{V}_x$, component of average velocity in x direction; $\bar{V}_y$, component in y direction

W_p Pump work per unit mass of fluid, ft-lb$_f$/lb or J/g

W_s Shaft work, ft-lb$_f$ or J

Z Height above datum plane, ft or m; Z_a, at station a; Z_b, at station b

Greek letters

α Kinetic-energy correction factor defined by Eq. (4-29)

β Momentum correction factor defined by Eq. (4-12); β_a, at station a; β_b, at station b; β_1, at station 1; β_2, at station 2

η Overall efficiency of pump, dimensionless

θ Angle of discharge pipe, Fig. 4-4

ρ Density, lb/ft^3 or kg/m^3; ρ_a, at station a; ρ_b, at station b

Σ Operator, meaning "algebraic sum of"

ϕ Angle with vertical

ψ Angle between axis and pressure vector, in Fig. 4-3

PROBLEMS

4-1   A liquid is flowing in steady flow through a 75-mm pipe. The local velocity varies with distance from the pipe axis as shown in Table 4-1. Calculate (a) average velocity $\bar{V}$, (b) kinetic-energy correction factor α, (c) momentum correction factor β.

4-2   Water at 68°F is pumped at a constant rate of 5 ft^3/min from a large reservoir resting on the floor to the open top of an experimental absorption tower. The point of discharge is 15 ft above the floor, and frictional losses in the 2-in. pipe from the reservoir to the tower amount to 0.8 ft-lb$_f$/lb. At what height in the reservoir must the water level be kept if the pump can develop only $\frac{1}{8}$ hp?

4-3 Water enters a 100-mm-ID 90° elbow, positioned in a horizontal plane, at a velocity of 6 m/s and a pressure of 70 kN/m² gauge. Neglecting friction, what are the magnitude and the direction of the force that must be applied to the elbow to keep it in position without moving?

Table 4-1 DATA FOR PROB. 4-1

Local velocity u, m/s	Distance from pipe axis, mm	Local velocity u, m/s	Distance from pipe axis, mm
1.042	0	0.919	22.50
1.033	3.75	0.864	26.25
1.019	7.50	0.809	30.00
0.996	11.25	0.699	33.75
0.978	15.00	0.507	35.625
0.955	18.75	0	37.50

REFERENCES

1 Bennett, C. O., and J. E. Myers: "Momentum, Heat, and Mass Transfer," 2d ed., McGraw-Hill, New York, 1964.
2 Bird, R. B., W. E. Stewart, and E. N. Lightfoot: "Transport Phenomena," Wiley, New York, 1960.
3 Streeter, V. L., "Fluid Mechanics," 5th ed., McGraw-Hill, New York, 1971.

5

FLOW OF INCOMPRESSIBLE FLUIDS IN CONDUITS AND THIN LAYERS

Industrial processes necessarily require the flow of fluids through pipes, conduits, and processing equipment. The chemical engineer most often is concerned with flow through closed pipes filled with the moving fluid. He also encounters problems requiring the flow of fluids in partially filled pipes, in layers down vertically inclined surfaces, through beds of solids, and in agitated vessels.

This chapter is concerned with flow through closed pipes and in layers on surfaces. Other types of flow are discussed in later chapters.

FLOW OF INCOMPRESSIBLE FLUIDS IN PIPES

Flow in circular conduits is important not only in its own right as an engineering operation but as an example of the quantitative relationships involving fluid flow in general. It is therefore discussed in some detail in this chapter. This discussion is restricted to steady flow.

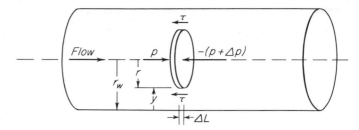

FIGURE 5-1
Fluid element in steady flow through pipe.

Shear-stress distribution in a cylindrical tube   Consider the steady flow of a viscous fluid at constant density in fully developed flow through a horizontal tube. Visualize a disk-shaped element of fluid, concentric with the axis of the tube, of radius r and length ΔL, as shown in Fig. 5-1. Assume the element is isolated as a free body. Let the fluid pressure on the upstream and downstream faces of the disk be p and $p + \Delta p$, respectively. Since the fluid possesses a viscosity, a shear force opposing flow will exist on the rim of the element. Apply the momentum equation (4-15) between the two faces of the disk. Since the flow is fully developed, $\beta_b = \beta_a$, and $\overline{V}_b = \overline{V}_a$, so that $\sum F = 0$. The quantities for substitution in Eq. (4-16) are

$$S_a = S_b = \pi r^2 \qquad p_a = p \qquad p_a S_a = \pi r^2 p \qquad p_b S_b = (\pi r^2)(p + \Delta p)$$

The shear force F_s acting on the rim of the element is the product of the shear stress and the cylindrical area, or $(2\pi r \, \Delta L)\tau$. Since the channel is horizontal, F_g is zero. Substituting these quantities into Eq. (4-16) gives

$$\sum F = \pi r^2 p - \pi r^2 (p + \Delta p) - (2\pi r \, \Delta L)\tau = 0$$

Simplifying this equation and dividing by $\pi r^2 \, \Delta L$ gives

$$\frac{\Delta p}{\Delta L} + \frac{2\tau}{r} = 0 \tag{5-1}$$

In steady flow, either laminar or turbulent, the pressure at any given cross section of a stream tube is constant, so that $\Delta p / \Delta L$ is independent of r. Equation (5-1) can be written for the entire cross section of the tube by taking $\tau = \tau_w$ and $r = r_w$, where τ_w is the shear stress at the wall of the conduit and r_w is the radius of the tube. Equation (5-1) then becomes

$$\frac{\Delta p}{\Delta L} + \frac{2\tau_w}{r_w} = 0 \tag{5-2}$$

Subtracting Eq. (5-1) from Eq. (5-2) gives

$$\frac{\tau_w}{r_w} = \frac{\tau}{r} \tag{5-3}$$

Also, when $r = 0$, $\tau = 0$. The simple linear relation between τ and r in Eq. (5-3) is shown graphically in Fig. 5-2.

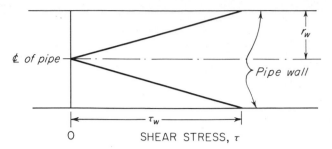

FIGURE 5-2
Variation of shear stress in pipe.

Relation between skin friction and wall shear   Equation (4-30) can be written over the length ΔL of the complete stream. Here $p_a = p, p_b = p + \Delta p, Z_b - Z_a = 0$, and the two kinetic-energy terms cancel. Also, the only kind of friction is skin friction between the wall and the fluid stream. Denote this by h_{fs}. Then Eq. (4-30) becomes

$$\frac{p}{\rho} = \frac{p + \Delta p}{\rho} + h_{fs}$$

or

$$-\frac{\Delta p}{\rho} = h_{fs} \tag{5-4}$$

Eliminating Δp from Eqs. (5-2) and (5-4) gives the following relation between h_{fs} and τ_w:

$$h_{fs} = \frac{2}{\rho} \frac{\tau_w}{r_w} \Delta L = \frac{4}{\rho} \frac{\tau_w}{D} \Delta L \tag{5-5}$$

where D is the diameter of the pipe.

The friction factor   Another common parameter, especially useful in the study of turbulent flow, is the friction factor, denoted by f and defined as the ratio of the wall shear stress to the product of the velocity head $\overline{V}^2/2g_c$ and the density

$$f \equiv \frac{\tau_w}{\rho \overline{V}^2/2g_c} = \frac{2g_c\tau_w}{\rho \overline{V}^2} \tag{5-6}†$$

Relations between skin-friction parameters   The four common quantities used to measure skin friction in pipes, $h_{fs}, \Delta p_s, \tau_w$, and f, are related by the equations

$$h_{fs} = \frac{2}{\rho} \frac{\tau_w}{r_w} \Delta L = -\frac{\Delta p_s}{\rho} = 4f\frac{\Delta L}{D} \frac{\overline{V}^2}{2g_c} \tag{5-7}$$

from which

$$f = \frac{-\Delta p_s \, g_c D}{2\Delta L \, \rho \overline{V}^2} \tag{5-7a}$$

† The friction factor f, defined by Eq. (5-6), is called the *Fanning friction factor*. Another friction factor common in fluid-mechanics literature and called the *Blasius* or *Darcy friction factor* is $4f$.

The subscript s is used in Δp_s and h_{fs} to call attention to the fact that in Eqs. (5-7) and (5-7a) these quantities, when they are associated with the Fanning friction factor, relate *only to skin friction*. If other terms in the Bernoulli equation are present, or if form friction is also active, $p_b - p_a$ differs from Δp_s. If boundary-layer separation occurs, h_f is greater than h_{fs}. The last term in Eq. (5-7), which includes the friction factor, is written in a manner to show the relation of h_{fs} to the velocity head $\overline{V}^2/2g_c$.

The various forms in Eq. (5-7) are convenient for specific purposes, and all are frequently encountered in the literature and in practice.

Laminar Flow in Pipes

Equations (5-1) to (5-7a) apply both to laminar and turbulent flow provided the fluid is incompressible and the flow is steady and fully developed. Use of the equations for more detailed calculations depends upon the mechanism of shear and therefore also upon whether the flow is laminar or turbulent. Because the shear-stress law for laminar flow is simple, the equations can be applied most readily to laminar flow. The treatment is especially straightforward for a Newtonian fluid.

Laminar flow of Newtonian fluids   In the discussion of general fluid-flow relations in Chap. 4 it was shown that the key step in their derivations is that of relating local velocity u to position in the stream tube. In circular channels, because of symmetry about the axis of the tube, the local velocity u depends only on the radius r. Also, the element of area dS is that of a thin ring, of radius r and width dr. The area of this elementary ring is

$$dS = 2\pi r \, dr \tag{5-8}$$

The desired velocity-distribution relation is u as a function of r.

A direct method of obtaining the velocity distribution for Newtonian fluids is to use the definition of viscosity [Eq. (3-4)] written as

$$\mu = -\frac{\tau g_c}{du/dr} \tag{5-9}$$

The negative sign in the equation accounts for the fact that in a pipe u decreases as r increases. Eliminating τ from Eqs. (5-3) and (5-9) provides the following ordinary differential equation relating u and r:

$$\frac{du}{dr} = -\frac{\tau g_c}{\mu} = -\frac{\tau_w g_c}{r_w \mu} r \tag{5-10}$$

Integration of Eq. (5-10) with the boundary condition $u = 0$, $r = r_w$ gives

$$\int_0^u du = -\frac{\tau_w g_c}{r_w} \int_{r_w}^r r \, dr$$

$$u = \frac{\tau_w g_c}{2r_w \mu} (r_w^2 - r^2) \tag{5-11}$$

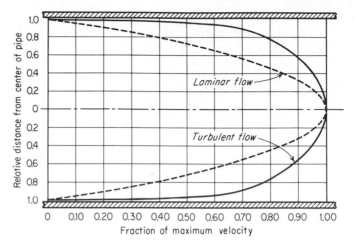

FIGURE 5-3
Distribution of velocity across pipe, fully developed flow of Newtonian fluid.

The maximum value of the local velocity is denoted by u_{max} and is located at the center of the pipe. The value of u_{max} is found from Eq. (5-11) by substituting 0 for r, giving

$$u_{max} = \frac{\tau_w g_c r_w}{2\mu} \tag{5-12}$$

Dividing Eq. (5-11) by Eq. (5-12) gives the following relationship for the ratio of the local velocity to the maximum velocity:

$$\frac{u}{u_{max}} = 1 - \left(\frac{r}{r_w}\right)^2 \tag{5-13}$$

The form of Eq. (5-13) shows that in laminar flow the velocity distribution with respect to radius is a parabola with the apex at the centerline of the pipe. This distribution is shown as the dashed line in Fig. 5-3.

Average velocity, kinetic-energy factor, and momentum correction factor for laminar flow of Newtonian fluids   Exact formulas for the average velocity $\bar{V}$, the kinetic-energy correction factor α, and the momentum correction factor β are readily calculated from the defining equations given in Chap. 4 and the velocity distribution shown in Eq. (5-11).

AVERAGE VELOCITY   Substitution of dS from Eq. (5-8), u from Eq. (5-11), and πr_w^2 for S into Eq. (4-4) gives

$$\bar{V} = \frac{\tau_w g_c}{r_w^3 \mu} \int_0^{r_w} (r_w^2 - r^2) r \, dr = \frac{\tau_w g_c r_w}{4\mu} \tag{5-14}$$

Comparison of Eqs. (5-12) and (5-14) shows that

$$\frac{\overline{V}}{u_{max}} = 0.5 \tag{5-15}$$

The average velocity is precisely one-half the maximum velocity.

KINETIC-ENERGY CORRECTION FACTOR The kinetic-energy factor α is calculated from Eq. (4-29), using Eqs. (5-8) for dS, (5-11) for u, and (5-14) for $\overline{V}$. The mathematical process is straightforward. The final result is $\alpha = 2.0$. The proper term for kinetic energy in the Bernoulli equation for laminar flow is therefore $\overline{V}^2/g_c$.

MOMENTUM CORRECTION FACTOR Again to obtain the value of β for laminar flow, the defining equation (4-13) is used. The result is $\beta = \frac{4}{3}$.

Hagen-Poiseuille equation For practical calculations, Eq. (5-14) is transformed by eliminating τ_w in favor of Δp_s by use of Eq. (5-7) and using pipe diameter in place of pipe radius. The result is

$$\overline{V} = -\frac{\Delta p_s\, g_c\, r_w}{\Delta L}\frac{r_w}{2}\frac{r_w}{4\mu} = -\frac{\Delta p_s\, g_c D^2}{32 \Delta L\, \mu}$$

Solving for $-\Delta p_s$ gives

$$-\Delta p_s = \frac{32\Delta L\, \overline{V}\mu}{g_c D^2} \tag{5-16}$$

and since $-\Delta p_s = 4\tau_w/D\,\Delta L$,

$$\tau_w = \frac{8\overline{V}\mu}{g_c D} \tag{5-17}$$

Substitution from Eq. (5-17) into Eq. (5-6) gives

$$f = \frac{16\mu}{D\overline{V}\rho} = \frac{16}{N_{Re}} \tag{5-18}$$

Equation (5-16) is the Hagen-Poiseuille equation. One of its uses is in the experimental measurement of viscosity, by measuring the pressure drop and volumetric flow rate through a tube of known length and diameter. From the flow rate $\overline{V}$ is calculated by Eq. (4-4) and μ calculated by Eq. (5-16). In practice, corrections for kinetic-energy and entrance effects are necessary.

Laminar flow of non-Newtonian liquids Because of the difference in the relation between shear stress and velocity gradient, the shape of the velocity profile for non-Newtonian liquids differs from that of a Newtonian liquid. In the more complicated situations of non-Newtonian flow the shape of the profile is determined experimentally. For the simpler cases such as the Ostwald–de Waele model [Eq. (3-7)] or the Bingham model [Eq. (3-6)] the same methods used for determining the flow parameters of a Newtonian fluid can be used for non-Newtonian fluids in these categories.

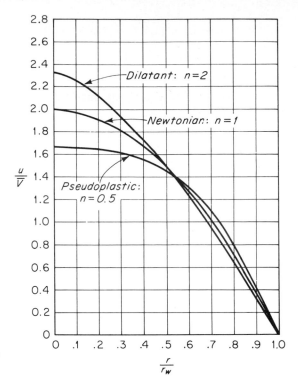

FIGURE 5-4
Velocity profiles in the laminar flow of
Newtonian and non-Newtonian liquids.

For fluids following the Ostwald–de Waele model the velocity variation with
radius follows the formula

$$u = \left(\frac{\tau_w g_c}{r_w K'}\right)^{1/n'} \frac{r_w^{1+1/n'} - r^{1+1/n'}}{1 + 1/n'} \tag{5-19}$$

Velocity profiles defined by Eq. (5-19) when $n' = 0.5$ (a pseudoplastic fluid),
1.0 (a Newtonian fluid), and 2.0 (a dilatant fluid) are shown in Fig. 5-4. In all cases K'
is assumed to be the same. The curve for the dilatant fluid is narrower and more
pointed than a true parabola; that for pseudoplastic fluid is blunter and flatter.

The pressure difference for the flow of an Ostwald–de Waele fluid is found by the
methods used in deriving Eq. (5-16) for a Newtonian fluid. The result is

$$-\Delta p_s = \frac{2K'}{g_c} \left(\frac{3n' + 1}{n'}\right)^{n'} \frac{\bar{V}^{n'}}{r_w^{n'+1}} \Delta L \tag{5-20}$$

Equation (5-20) corresponds to Eq. (5-16) for a Newtonian fluid.

The behavior of fluids following the Bingham-plastic flow model is somewhat
more complicated. The general shape of the curve of u vs. r is shown in Fig. 5-5a. In
the central portion of the tube there is no velocity variation with the radius, and the
velocity gradient is confined to an annular space between the central portion and tube
wall. The center portion is moving in plug flow. In this region the shear stress that

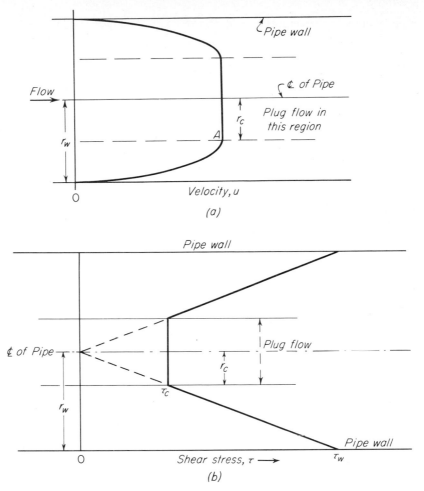

FIGURE 5-5
(a) Velocity profile and (b) shear diagram for Bingham-plastic flow.

would be generated in other types of flow is too small to overcome the threshold shear τ_0. The shear diagram is shown in Fig. 5-5b. For the velocity variation in the annular space between the tube wall and the plug, the following equation applies:

$$u = \frac{g_c}{K}(r_w - r)\left[\frac{\tau_w}{2}\left(1 + \frac{r}{r_w}\right) - \tau_0\right] \qquad (5\text{-}21)$$

The boundary between the plug and the remaining fluid is found by differentiating Eq. (5-21) and setting the velocity gradient equal to zero, or more simply by reading the value from Fig. 5-5b. The result is

$$r_c = \frac{\tau_0}{\tau_w} r_w \qquad (5\text{-}22)$$

The velocity in the central core u_c, the speed at which the plug is moving, is found by substituting the value of r_c from Eq. (5-22) for r in Eq. (5-21) and rearranging. This gives

$$u_c = \frac{g_c \tau_0}{2Kr_c} (r_w - r_c)^2 \tag{5-23}$$

Turbulent Flow in Pipes and Closed Channels

In flow of fluid through a closed channel turbulence cannot exist permanently at the boundary between the solid and the flowing fluid. The velocity at the interface is zero because of the adherence of the fluid to the solid, and velocity components normal to the wall cannot exist. It is probable that within a thin volume immediately adjacent to the wall eddies seldom occur. This layer is called the *viscous sublayer*. Formerly it was assumed that the sublayer possessed a definite thickness and was permanently free from eddies. It has been established, however, that statistically the layer is invaded by eddies that at times may reach the wall of the channel. Thus the thickness of the viscous sublayer at any point must be wildly fluctuating as a result of the impact of eddies of various sizes from adjacent fluid. Within the viscous sublayer only viscous shear is important, and eddy diffusion, if present at all, is minor. The viscous sublayer occupies a very small fraction of the total cross section of the stream, and the remaining stream is in turbulent flow. A transition layer exists immediately adjacent to the viscous sublayer in which both viscous shear and shear due to eddy diffusion exist. The transition layer, which is sometimes called a *buffer layer*, also is relatively thin. The bulk of the cross section of the flowing stream is occupied by flow which is entirely turbulent and which is called the *turbulent core*. In a turbulent core viscous shear is negligible in comparison with that from eddy viscosity.

Velocity distribution for turbulent flow   Because of the dependence of important flow parameters on velocity distribution, considerable study, both theoretical and experimental, has been devoted to determining the velocity distribution in turbulent flow. Although the problem has not been completely solved, useful relationships are available that may be used to calculate the important characteristics of turbulence; and the results of theoretical calculations check with experimental data reasonably well.

A typical velocity distribution for a Newtonian fluid moving in turbulent flow in a smooth pipe is shown in Fig. 5-3. The figure also shows the velocity distribution of laminar flow at the same maximum velocity at the center of the pipe. The distribution curve for turbulent flow is clearly much flatter than that for laminar flow and the difference between the average velocity and the maximum velocity is considerably less. The velocity gradient, like that in laminar flow, is zero at the center line. It is known that the eddies in the turbulent core are large but of low intensity, and those in the transition zone are small but intense. Most of the kinetic-energy content of the eddies lies in the buffer zone and the outer portion of the turbulent core. At the centerline the turbulence is isotropic. In all other areas of the turbulent-flow regime, turbulence is anisotropic; otherwise, there would be no shear.

It is customary to express the velocity distribution in turbulent flow not as velocity vs. distance but in terms of dimensionless parameters defined by the following equations:

$$u^* \equiv \bar{V} \sqrt{\frac{f}{2}} = \sqrt{\frac{\tau_w g_c}{\rho}} \qquad (5\text{-}24)$$

$$u^+ \equiv \frac{u}{u^*} \qquad (5\text{-}25)$$

$$y^+ \equiv \frac{yu^*\rho}{\mu} = \frac{y}{\mu} \sqrt{\tau_w g_c \rho} \qquad (5\text{-}26)$$

where u^* = friction velocity
   u^+ = velocity quotient, dimensionless
   y^+ = distance, dimensionless
   y = distance from wall of tube

The relationship between y, r, and r_w, the radius of the tube, is

$$r_w = r + y \qquad (5\text{-}27)$$

Equations relating u^+ to y^+ are called universal velocity-distribution laws.

Universal velocity-distribution equations Since the viscous sublayer is very thin, $r \approx r_w$, and Eq. (5-10) can be written, with the substitution of $-dy$ for dr, as

$$\frac{du}{dy} = \frac{\tau_w g_c}{\mu} \qquad (5\text{-}28)$$

Substituting u^* from Eq. (5-24), u^+ from Eq. (5-25), and y^+ from Eq. (5-26) into Eq. (5-28) gives

$$\frac{du^+}{dy^+} = 1$$

Integrating, with the lower limits $u^+ = y^+ = 0$, gives, for the velocity distribution in the laminar sublayer,

$$u^+ = y^+ \qquad (5\text{-}29)$$

An empirical equation for the so-called buffer layer is

$$u^+ = 5.00 \ln y^+ - 3.05 \qquad (5\text{-}30)$$

THE LOGARITHMIC VELOCITY LAW FOR THE TURBULENT CORE   For practical purposes the generalized relationship for the turbulent core is the most important of the three velocity equations. The buffer layer and the viscous sublayer are so thin that the amount of fluid flowing through these layers is small in comparison with the flow through the turbulent core, and it is the turbulent core that is numerically significant. Because of its importance, a simplified treatment of the relationship of this law to the eddy diffusivity ε_M, defined in Eq. (3-22), is of interest. The treatment is essentially that of Prandtl,[5] who originally proposed the logarithmic law in 1933.

Since the law applies to the turbulent core, where viscous effects are negligible, the kinematic viscosity v in Eq. (3-22) can be neglected and the total shear stress at position y in the tube is

$$\frac{\tau g_c}{\rho} = \varepsilon_M \frac{du}{dy} \qquad (5\text{-}31)$$

The first assumption is that ε_M is a function only of position in the pipe y and velocity gradient du/dy. By simple dimensional analysis this assumption leads to the formation of the dimensionless ratio $(\varepsilon_M/y^2)(du/dy)$. The second assumption is that this function is a constant, denoted by κ^2, so that

$$\varepsilon_M = \kappa^2 y^2 \frac{du}{dy} \qquad (5\text{-}32)$$

Elimination of ε_M from Eqs. (5-31) and (5-32) gives

$$\sqrt{\frac{\tau g_c}{\rho}} = \kappa y \frac{du}{dy} \qquad (5\text{-}33)$$

Strictly, τ varies with r (and therefore with y) in accordance with Eq. (5-3). The third assumption is that τ in Eq. (5-33) may be replaced by τ_w, the shear stress at the wall, and any approximation inherent in this assumption is absorbed in κ. With this simplification, and use of Eq. (5-24), Eq. (5-33) may be written

$$du = \frac{u^*}{\kappa} \frac{dy}{y} \qquad (5\text{-}34)$$

Integrating this equation gives

$$u = \frac{u^*}{\kappa} \ln y + \text{const} \qquad (5\text{-}35)$$

This may be written

$$\frac{u}{u^*} = \frac{1}{\kappa} \ln \left(\frac{yu^*}{v} \frac{v}{u^*} \right) + \text{const}$$

Using Eq. (5-25) for u^+ and Eq. (5-26) for y^+ leads to

$$u^+ = \frac{1}{\kappa} \ln y^+ + \frac{1}{\kappa} \ln \frac{v}{u^*} + \text{const}$$

which finally becomes, using c_1 for the sum of the last two constants,

$$u^+ = \frac{1}{\kappa} \ln y^+ + c_1 \qquad (5\text{-}36)$$

Actual velocity-distribution data confirm the form of Eq. (5-36). Because of the many assumptions made in its derivation it is best to consider the constants κ and c_1 in the logarithmic law as empirical and to measure their values by experiment.

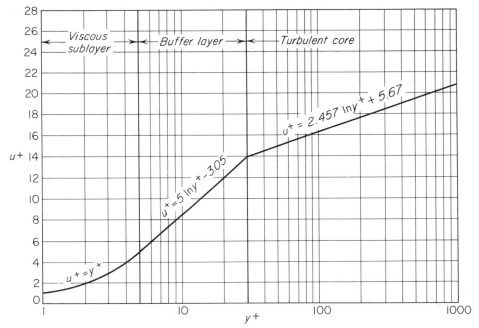

FIGURE 5-6
Universal velocity distribution, turbulent flow of Newtonian fluid in smooth pipe.

With κ taken as 0.407 and c_1 as 5.67, Eqs. (5-29), (5-30), and (5-36) are plotted on semilogarithmic coordinates in Fig. 5-6. From the two intersections of the three lines representing the equations the ranges covered by the equations are

Equation (5-29), for the viscous sublayer: $y^+ < 5$
Equation (5-30), for the buffer zone: $5 < y^+ < 30$
Equation (5-36), for the turbulent core: $30 < y^+$

Limitations of universal velocity-distribution laws   The universal velocity equations have a number of limitations. It is certain that the buffer zone has no independent existence and that there is no discontinuity between the buffer zone and the turbulent core. Also, there is doubt as to the reality of the existence of the viscous sublayer. The equations do not apply well for Reynolds numbers from the critical to approximately 10,000, and it is known that a simple y^+-u^+ relationship is not adequate for the turbulent core near the buffer zone or in the buffer zone itself. Finally, Eq. (5-36) calls for a finite velocity gradient at the centerline of the pipe although it is known that the gradient at this point must be zero. In spite of the imperfections of the equations, however, they represent the results of experiment reasonably well when integrated to give practical flow equations and provide a reasonably sound basis for actual engineering design.

Much research has been devoted to improving the velocity-distribution equations and eliminating or reducing some of their deficiencies. This work is reported in advanced texts[3,6] but is beyond the scope of this book.

Flow quantities for turbulent flow in smooth round pipes   With a velocity-distribution relation at hand, the important flow quantities can be calculated by the usual methods. The quantities of interest are the average velocity in terms of the maximum velocity at the center of the pipe; relations linking the flow-resistance parameters τ_w and f with the average velocity, the maximum velocity, and the Reynolds number; the kinetic-energy correction factor α; and the momentum correction factor β.

Calculation of the flow quantities requires an integration with pipe radius from the centerline of the pipe to the wall. Strictly, the integration should be conducted in three parts, the first over the range $y^+ = 0$ to $y^+ = 5$, using Eq. (5-29); the second from $y^+ = 5$ to $y^+ = 30$, using Eq. (5-30); and the third from $y^+ = 30$ to its value at the center of the pipe, using Eq. (5-36). Because of the thinness of the layers covered by the first two integrations, they may be neglected; and a single integration based on Eq. (5-36) over the entire range $r = 0$ to $r = r_w$ is justified.

AVERAGE VELOCITY   Equation (5-36) may be written for the centerline of the pipe as

$$u_c^+ = \frac{1}{\kappa} \ln y_c^+ + c_1 \tag{5-37}$$

where u_c^+ and y_c^+ are the values of u^+ and y^+ at the centerline, respectively. Also, by Eqs. (5-25) and (5-26)

$$u_c^+ = \frac{u_{max}}{u^*} \tag{5-38}$$

$$y_c^+ = \frac{r_w u^*}{\nu} \tag{5-39}$$

where u_{max} is the maximum velocity at the centerline.
Subtracting Eq. (5-37) from Eq. (5-36) yields

$$u^+ = u_c^+ + \frac{1}{\kappa} \ln \frac{y^+}{y_c^+} \tag{5-40}$$

The average velocity $\overline{V}$ is, from Eq. (4-4) after substituting πr_w^2 for S and $2\pi r \, dr$ for dS,

$$\overline{V} = \frac{2\pi}{\pi r_w^2} \int_0^{r_w} ur \, dr \tag{5-41}$$

From Eq. (5-27) $r = r_w - y$ and $dr = -dy$. Also, when $r = 0$, $y = r_w$, and when $r = r_w$, $y = 0$. Equation (5-41) becomes, after elimination of r,

$$\overline{V} = \frac{2}{r_w^2} \int_0^{r_w} u(r_w - y) \, dy \tag{5-42}$$

Equation (5-42) can be written in dimensionless parameters by substituting u from Eq. (5-25), y from Eq. (5-26), and u^+ from Eq. (5-40). This gives

$$\bar{V} = \frac{2v^2}{\kappa r_w^2 u^*} \int_0^{y_c{}^+} \left(\kappa u_c^+ + \ln \frac{y^+}{y_c^+} \right) (y_c^+ - y^+) \, dy^+ \tag{5-43}$$

Formal integration of Eq. (5-43)† gives

$$\frac{\bar{V}}{u^*} = u_c^+ - \frac{3}{2\kappa} = \frac{1}{\sqrt{f/2}} \tag{5-44}$$

Substituting u_c^+ from Eq. (5-38) and u^* from Eq. (5-24) into Eq. (5-44) gives

$$\frac{\bar{V}}{u_{max}} = \frac{1}{1 + (3\sqrt{f/2})/2\kappa} \tag{5-45}$$

THE REYNOLDS NUMBER; FRICTION-FACTOR LAW FOR SMOOTH TUBES   The equations at hand can be used to derive an important relation between N_{Re} and f for turbulent flow in smooth round pipes. This equation is derived by appropriate substitutions into Eq. (5-37). From the defining equation of y_c^+ [Eq. (5-39)] and from the definition of u^* [Eq. (5-24)]

$$y_c^+ = \frac{r_w \bar{V}}{v} \sqrt{\frac{f}{2}} = \frac{D\bar{V}\sqrt{f/2}}{2v} = \frac{N_{Re}}{2} \sqrt{\frac{f}{2}} = N_{Re} \sqrt{\frac{f}{8}} \tag{5-46}$$

From Eq. (5-44)

$$u_c^+ = \frac{1}{\sqrt{f/2}} + \frac{3}{2\kappa} \tag{5-47}$$

Substituting u_c^+ from Eq. (5-47) and y_c^+ from Eq. (5-46) into Eq. (5-37) gives, after rearrangement,

$$\frac{1}{\sqrt{f/2}} = \frac{1}{\kappa} \ln \left(N_{Re} \sqrt{\frac{f}{8}} \right) - \frac{3}{2\kappa} + c_1 \tag{5-48}$$

THE KINETIC-ENERGY AND MOMENTUM CORRECTION FACTORS   Values of α and β for turbulent flow are closer to unity than for laminar flow, and for most purposes both can be considered to be unity. Equations for these correction factors are readily obtained, however, by integrating equations (4-13) and (4-29) defining them and using the logarithmic velocity law. The equations so found are

$$\alpha = 1 + \frac{f}{8\kappa^2} \left(15 - \frac{9}{\kappa} \sqrt{f} \right) \tag{5-49}$$

$$\beta = 1 + \frac{5}{8\kappa^2} f \tag{5-50}$$

† Substitution of $x = y^+/y_c^+$ and use of standard integral tables suffices for this integration. Also, for the lower limit, $x \ln x = 0$, and $x^2 \ln x = 0$, when $x = 0$.

The kinetic-energy correction factor may be important in applying Bernoulli's theorem between stations when one is in laminar flow and the other in turbulent. Also, factors α and β are of some importance in certain types of compact heat-exchange equipment, where there are many changes in size of the fluid channel, and where the tubes or heat-transfer surfaces themselves are short.[2] In most practical situations both are taken as unity in turbulent flow.

EMPIRICAL VALUES OF CONSTANTS κ AND c_1 To use the equations in the last section, numbers must of course be assigned to the empirical constants. The original magnitudes specified by Prandtl were

$$\kappa = 0.40 \quad \text{and} \quad c_1 = 5.5$$

From these Eq. (5-48) becomes the von Kármán equation

$$\frac{1}{\sqrt{f}} = 4.06 \log (N_{Re} \sqrt{f}) - 0.60 \tag{5-51}$$

A more common equation proposed by Nikuradse on the basis of experimental data is

$$\frac{1}{\sqrt{f}} = 4.0 \log (N_{Re} \sqrt{f}) - 0.40 \tag{5-52}$$

Equation (5-52) corresponds to Eq. (5-48) with κ taken as 0.407 and c_1 as 5.67. Other empirical equations relating f and N_{Re} are given in Chap. 12 [Eqs (12-63) and (12-68)].

Relations between maximum velocity and average velocity   The quantities N_{Re} and ratio $\overline{V}/u_{max}$ are useful in relating the average velocity to the maximum velocity in the center of the tube as a function of flow conditions; for example, an important method of measuring fluid flow is the pitot tube (page 213), which can be used to measure u_{max}, and this relationship is then used to determine the average velocity from this single observation.

Experimentally measured values of $\overline{V}/u_{max}$ as a function of the usual Reynolds number and of $Du_{max}\, \rho/\mu$ are shown in Fig. 5-7, which covers the range from laminar flow to turbulent flow. For laminar flow the ratio is exactly 0.5, in accordance with Eq. (5-15). The ratio changes rapidly from 0.5 to about 0.8, when laminar flow changes to turbulent.

Equation (5-44) can be used to relate $\overline{V}/u_{max}$ to f, and Eq. (5-48) to relate f to N_{Re}. For the usual values of κ and c_1 this calculation yields values of $\overline{V}/u_{max}$ that lie above those shown in the experimental curve of Fig. 5-7. The discrepancy results from the fact that the logarithmic velocity law does not apply accurately at the intermediate ranges of Reynolds numbers from about 4,000 to 10,000 and from the other approximations inherent in the simplified theory.

Effect of roughness   The discussion thus far has been restricted to smooth tubes, without defining smoothness. It has long been known that in turbulent flow a rough pipe leads to a larger friction factor for a given Reynolds number than a smooth pipe

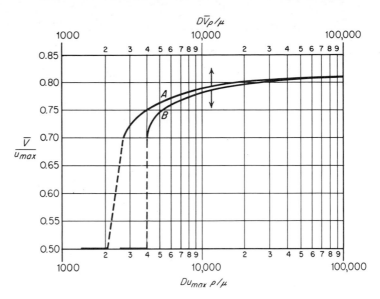

FIGURE 5-7
$\bar{V}/u_{max}$ vs. N_{Re}.

does. If a rough pipe is smoothed, the friction factor is reduced. When further smooth-ing brings about no further reduction in friction factor for a given Reynolds number, the tube is said to be *hydraulically smooth.* Equations (5-51) and (5-52) refer to a hydraulically smooth tube.

Figure 5-8 shows several idealized kinds of roughness. The height of a single unit of roughness is denoted by k and is called the *roughness parameter.* From dimen-sional analysis, f is a function of both N_{Re} and the relative roughness k/D, where D is the diameter of the pipe. For a given kind of roughness, e.g., that shown in Fig. 5-8a and b, it can be expected that a different curve of f vs. N_{Re} would be found for each magnitude of the relative roughness and also that for other types of roughness, such as those shown in Fig. 5-8c and d, a different family of curves of N_{Re} vs. f would be found for each type of roughness. Experiments on artificially roughened pipe have confirmed these expectations. It has also been found that all clean, new commercial pipes seem to have the same type of roughness and that each material of construction has its own characteristic roughness parameter.

Old, foul, and corroded pipe can be very rough, and the character of the rough-ness differs from that of clean pipe.

Roughness has no appreciable effect on the friction factor for laminar flow unless k is so large that the measurement of the diameter becomes uncertain.

The friction-factor chart   For design purposes, the frictional characteristics of round pipe, both smooth and rough, are summarized by the friction-factor chart, Fig. 5-9, which is a plot of $\log f$ vs. $\log N_{Re}$. For laminar flow Eq. (5-18) relates the friction

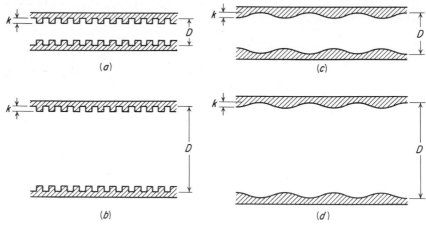

FIGURE 5-8
Types of roughness.

factor to the Reynolds number. A log-log plot of Eq. (5-18) is a straight line with a slope of -1. This plot line is shown on Fig. 5-9 for Reynolds numbers less than 2,100.

For turbulent flow the lowest line represents the friction factor for smooth tube and is consistent with Eq. (5-52). The other curved lines in the turbulent-flow range represent the friction factors for various types of commercial pipe, each of which is characterized by a different value of k. The parameters for several common metals are given in the figure. Clean wrought-iron or steel pipe, for example, has a k value of 1.5×10^{-4}, regardless of the diameter of the pipe. Drawn copper and brass pipe may be considered hydraulically smooth.

The use of Fig. 5-9 is simple. If the rate of flow, the density and viscosity of the fluid, and the diameter of the pipe are known, the Reynolds number is calculated. The Reynolds number is used as the abscissa; the friction factor is read as the ordinate and is used in one of the forms of Eq. (5-7) to calculate the desired friction loss or wall shear. The friction loss so calculated, combined with other forms of friction within the flow system, may then be substituted for h_f in the Bernoulli equation.

CALCULATION OF FLOW RATE FROM KNOWN SKIN FRICTION   Figure 5-9 is useful for calculating h_{fs} from a known pipe size and conditions of flow. It is not convenient for the calculation of the flow rate from a known h_{fs} because $\overline{V}$, the average linear velocity, appears in both N_{Re} and f. A trial-and-error solution can be used for this calculation, but a more convenient solution can be obtained by plotting Eq. (5-52) directly, using $N_{Re} \sqrt{f}$ as the abscissa and f as the ordinate. A plot of this type is shown in Fig. 5-10. To use this figure the abscissa is calculated from the known friction loss, the ordinate f is read, and the velocity $\overline{V}$ calculated from f by means of Eq. (5-7a).

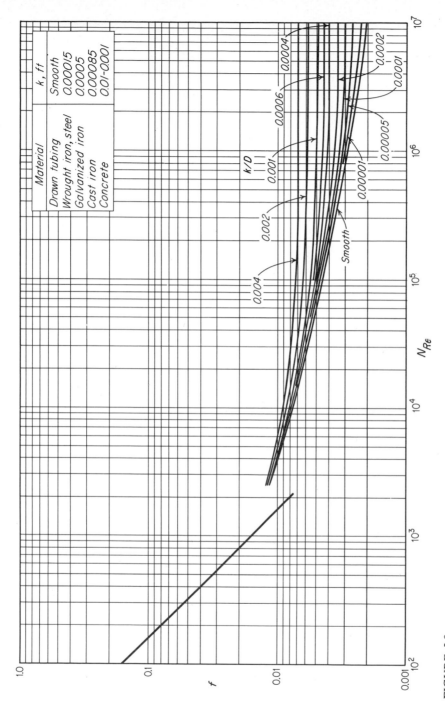

FIGURE 5-9
Friction-factor chart.

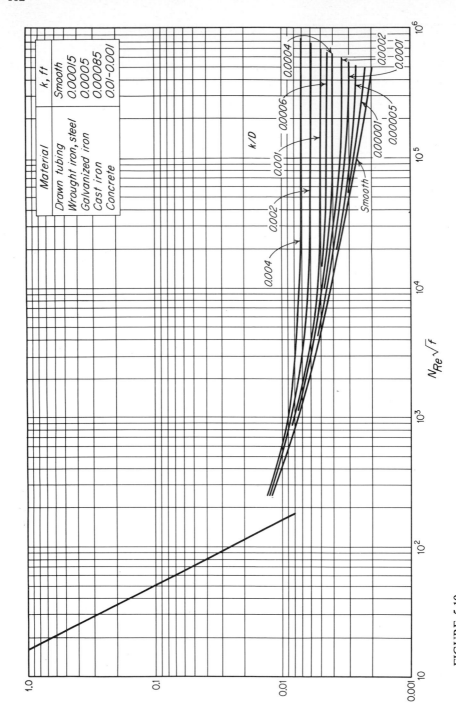

FIGURE 5-10

Friction-factor chart, $N_{Re}\sqrt{f}$ vs. f.

Reynolds numbers and friction factor for non-Newtonian fluids   Equation (5-7a), the relation of friction factor to pressure drop, and Eq. (5-20), the equation for the friction pressure loss of an Ostwald–de Waele fluid, may be used to calculate the friction factor for pseudoplastic fluids. If $-\Delta p_s$ is eliminated from these two equations the equation for f can be written

$$f = \frac{2^{n'+1}K'}{D^{n'}\rho\overline{V}^{2-n'}}\left(3 + \frac{1}{n'}\right)^{n'} \tag{5-53}$$

From this equation a Reynolds number $N_{Re,n}$ for non-Newtonian fluids can be defined, on the assumption that for laminar flow

$$f = \frac{16}{N_{Re,n}} \tag{5-54}$$

Combining Eqs. (5-53) and (5-54) yields

$$N_{Re,n} = 2^{3-n'}\left(\frac{n'}{3n'+1}\right)^{n'}\frac{D^{n'}\rho\overline{V}^{2-n'}}{K'} \tag{5-55}$$

This is the definition of the Reynolds number $N_{Re,n}$ given in Eq. (3-9). This Reynolds number is useful for a friction-factor plot, as it forces the log-log plot of f vs. N_{Re} to be a straight line with a slope of -1, to correspond to the laminar flow of Newtonian fluids.

Figure 5-11 is a friction-factor chart in which f is plotted against $N'_{Re,n}$ for the flow of Ostwald–de Waele fluids in smooth pipes.[1] A series of lines depending on the magnitude of n' is needed for turbulent flow. For these lines the following equation, analogous to Eq. (5-52), for Newtonian fluids, has been suggested[1]

$$\frac{1}{\sqrt{f}} = \frac{4.0}{(n')^{0.75}}(\log N'_{Re,n}f^{1-0.5n'}) - \frac{0.4}{(n')^{1.2}} \tag{5-56}$$

Figure 5-11 shows that for pseudoplastic fluids ($n' < 1$), laminar flow persists to higher Reynolds numbers than with Newtonian fluids.

Effect of heat transfer on friction factor   The methods for calculating friction described thus far apply only where there is no transfer of heat between the wall of the channel and the fluid. When the fluid is either heated or cooled by a conduit wall hotter or colder than the fluid, the velocity field is modified by the temperature gradients thus created within the fluid. The effect on the velocity gradients is especially pronounced with liquids where viscosity is a strong function of temperature. Quite elaborate theories have been developed for the effect of heat transfer on the velocity distribution of gases. For ordinary engineering practice the following simple method is empirically justified for both gases and liquids. (1) The Reynolds number is calculated on the assumption that the fluid temperature equals the *mean bulk temperature*, which is defined as the arithmetic average of the inlet and outlet temperatures.

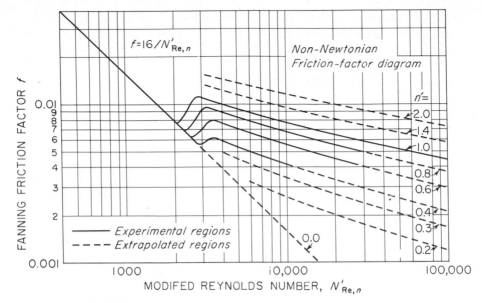

FIGURE 5-11
Friction-factor chart, Ostwald-de Waele liquid. [Data from Dodge, D. W., and A. B. Metzner: *A.I.Ch.E.J.*, **5**:189 (1959).]

(2) The friction factor corresponding to the mean bulk temperature is divided by a factor ψ, which in turn is calculated from the following equations:[4]

For $N_{Re} > 2,100$:
$$\psi = \begin{cases} \left(\dfrac{\mu}{\mu_w}\right)^{0.17} & \text{for heating} & (5\text{-}57a) \\[2ex] \left(\dfrac{\mu}{\mu_w}\right)^{0.11} & \text{for cooling} & (5\text{-}57b) \end{cases}$$

For $N_{Re} < 2,100$:
$$\psi = \begin{cases} \left(\dfrac{\mu}{\mu_w}\right)^{0.38} & \text{for heating} & (5\text{-}58a) \\[2ex] \left(\dfrac{\mu}{\mu_w}\right)^{0.23} & \text{for cooling} & (5\text{-}58b) \end{cases}$$

where μ = viscosity of fluid at mean bulk temperature
μ_w = viscosity at temperature of wall of conduit

Friction factor in flow through channels of noncircular cross section   The friction in long straight channels of constant noncircular cross section can be estimated by using the equations for circular pipes if the diameter in the Reynolds number and in the definition of the friction factor is taken as an *equivalent diameter*, defined as

4 times the hydraulic radius. The hydraulic radius is denoted by r_H and in turn is defined as the ratio of the cross-sectional area of the channel to the wetted perimeter of the channel:

$$r_H \equiv \frac{S}{L_p} \tag{5-59}$$

where S = cross-sectional area of channel
L_p = perimeter of channel in contact with fluid

Thus, for the special case of a circular tube, the hydraulic radius is

$$r_H = \frac{\pi D^2/4}{\pi D} = \frac{D}{4}$$

The equivalent diameter is $4r_H$ or, simply, D.

An important special case is the annulus between two concentric pipes. Here the hydraulic radius is

$$r_H = \frac{\pi D_o^2/4 - \pi D_i^2/4}{\pi D_i + \pi D_o} = \frac{D_o - D_i}{4} \tag{5-60}$$

where D_i and D_o are the inside and outside diameters of the annulus, respectively. The equivalent diameter of an annulus is therefore the difference of the diameters. Also, the equivalent diameter of a square duct with a width of side b is $4(b^2/4b) = b$.

The hydraulic radius is a useful parameter for generalizing fluid-flow phenomena in turbulent flow. Equation (5-7) can be so generalized by substituting $4r_H$ for D or $2r_H$ for r_w:

$$h_{fs} = \frac{\tau_w}{\rho r_H} \Delta L = -\frac{\Delta p_s}{\rho} = f \frac{\Delta L}{r_H} \frac{\bar{V}^2}{2g_c} \tag{5-61}$$

Also

$$N_{Re} = \frac{4 r_H \bar{V} \rho}{\mu} \tag{5-62}$$

$$N_{Re} \sqrt{f} = \frac{4 r_H \rho}{\mu} \sqrt{\frac{2 h_{fs} r_H g_c}{\Delta L}} = \frac{4 r_H \rho}{\mu} \sqrt{\frac{2(-\Delta p_s) r_H g_c}{\rho \Delta L}} \tag{5-63}$$

The simple hydraulic-radius rule does not apply to laminar flow through non-circular sections. For laminar flow through an annulus, for example, f and N_{Re} are related by the equation[3]

$$f = \frac{16}{N_{Re}} \phi_a \tag{5-64}$$

where ϕ_a is the function of D_i/D_o shown in Fig. 5-12. The value of ϕ_a is unity for circular cross section and 1.5 for parallel planes. Equations for laminar flow through other cross sections are given in various texts.[3]

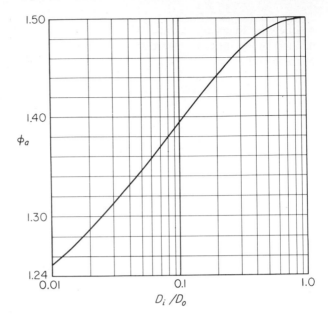

FIGURE 5-12
Value of ϕ_a in Eq. (5-64).

Friction from changes in velocity or direction   Whenever the velocity of a fluid is changed, either in direction or magnitude, by a change in the direction or size of the conduit, friction additional to the skin friction from flow through the straight pipe is generated. Such friction includes form friction resulting from vortices which develop when the normal streamlines are disturbed and when boundary-layer separation occurs. In most situations, these effects cannot be calculated precisely, and it is necessary to rely on empirical data. Often it is possible to estimate friction of this kind in specific cases from a knowledge of the losses in known arrangements.

FRICTION LOSS FROM SUDDEN EXPANSION OF CROSS SECTION   If the cross section of the conduit is suddenly enlarged, the fluid stream separates from the wall and issues as a jet into the enlarged section. The jet then expands to fill the entire cross section of the larger conduit. The space between the expanding jet and the conduit wall is filled with fluid in vortex motion characteristic of boundary-layer separation, and considerable friction is generated within this space. This effect is shown in Fig. 5-13.

The friction loss h_{fe} from a sudden expansion of cross section is proportional to the velocity head of the fluid in the small conduit and can be written

$$h_{fe} = K_e \frac{\overline{V}_a^2}{2g_c} \tag{5-65}$$

where K_e is a proportionality factor called the *expansion-loss coefficient* and $\overline{V}_a$ is the average velocity in the smaller, or upstream, conduit. In this case the calculation of

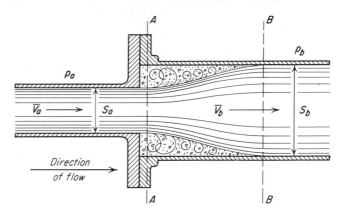

FIGURE 5-13
Flow at sudden enlargement of cross section.

K_e can be made theoretically and a satisfactory result obtained. The calculation utilizes the continuity equation (4-1), the steady-flow momentum-balance equation (4-15), and the Bernoulli equation (4-30). Consider the control volume defined by sections AA and BB and inner surface of the larger downstream conduit between these sections, as shown in Fig. 5-13. Gravity forces do not appear because the pipe is horizontal, and wall friction is negligible because the wall is relatively short and there is almost no velocity gradient at the wall between the sections. The only forces, therefore, are pressure forces on sections AA and BB. The momentum equation gives

$$(p_a S_b - p_b S_b)g_c = \dot{m}(\beta_b \overline{V}_b - \beta_a \overline{V}_a) \tag{5-66}$$

Since $Z_a = Z_b$, Eq. (4-30) may be written for this situation as

$$\frac{p_a - p_b}{\rho} = \frac{\alpha_b \overline{V}_b^2 - \alpha_a \overline{V}_a^2}{2g_c} + h_{fe} \tag{5-67}$$

For usual flow conditions, $\alpha_a = \alpha_b = 1$, and $\beta_a = \beta_b = 1$, and these correction factors are disregarded. Also, elimination of $p_a - p_b$ between Eqs. (5-66) and (5-67) yields

$$h_{fe} = \frac{(\overline{V}_a - \overline{V}_b)^2}{2g_c} \tag{5-68}$$

From Eq. (4-5), $\overline{V}_b = \overline{V}_a(S_a/S_b)$, and Eq. (5-68) can be written

$$h_{fe} = \frac{\overline{V}_a^2}{2g_c}\left(1 - \frac{S_a}{S_b}\right)^2 \tag{5-69}$$

Comparison of Eqs. (5-65) and (5-69) shows that

$$K_e = \left(1 - \frac{S_a}{S_b}\right)^2 \tag{5-70}$$

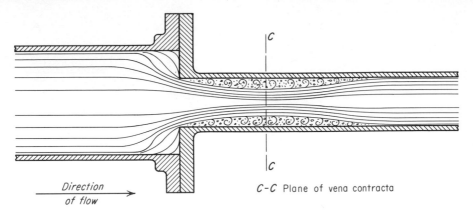

FIGURE 5-14
Flow at sudden contraction of cross section.

A correction for α and β should be made if the type of flow between the two sections differs. For example, if the flow in the larger pipe is laminar and that in the smaller pipe turbulent, α_b should be taken as 2 and β_b as $\frac{4}{3}$ in Eqs. (5-66) and (5-67).

FRICTION LOSS FROM SUDDEN CONTRACTION OF CROSS SECTION   When the cross section of the conduit is suddenly reduced, the fluid stream cannot follow around the sharp corner and the stream breaks contact with the wall of the conduit. A jet is formed, which flows into the stagnant fluid in the smaller section. The jet first contracts and then expands to fill the smaller cross section, and downstream from the point of contraction the normal velocity distribution eventually is reestablished. The cross section of minimum area at which the jet changes from a contraction to an expansion is called the *vena contracta*. The flow pattern of a sudden contraction is shown in Fig. 5-14. Section CC is drawn at the vena contracta. Vortices appear as shown in the figure.

The friction loss from sudden contraction is proportional to the velocity head in the smaller conduit and can be calculated by the equation

$$h_{fc} = K_c \frac{\overline{V}_b^2}{2g_c} \tag{5-71}$$

where the proportionality factor K_c is called the *contraction-loss coefficient* and $\overline{V}_b$ is the average velocity in the smaller, or downstream, section. Experimentally, for laminar flow, $K_c < 0.1$, and the contraction loss h_{fc} is negligible. For turbulent flow, K_c is given by the empirical equation

$$K_c = 0.4\left(1 - \frac{S_b}{S_a}\right) \tag{5-72}$$

where S_a and S_b are the cross-sectional areas of the upstream and downstream conduits, respectively.

Effect of fittings and valves Fittings and valves disturb the normal flow lines and cause friction. In short lines with many fittings, the friction loss from the fittings may be greater than that from the straight pipe. The friction loss h_{ff} from fittings is found from an equation similar to Eqs. (5-65) and (5-71)

$$h_{ff} = K_f \frac{\bar{V}_a^2}{2g_c}$$ (5-73)

where K_f = loss factor for fitting
$\bar{V}_a$ = average velocity in pipe leading to fitting

Factor K_f is found by experiment and differs for each type of connection. A short list of factors is given in Table 5-1.

FORM-FRICTION LOSSES IN THE BERNOULLI EQUATION Form-friction losses are incorporated in the h_f term of Eq. (4-34). They are combined with the skin-friction losses of the straight pipe to give the total friction loss. Consider, for example, the flow of incompressible fluid through the two enlarged headers, the connecting tube, and the open globe valve shown in Fig. 5-15. Let $\bar{V}$ be the average velocity in the tube, D the diameter of the tube, and L the length of the tube. The skin-friction loss in the straight tube is, by Eq. (5-7), $4f(L/D)(\bar{V}^2/2g_c)$; the contraction loss at the entrance to the tube is, by Eq. (5-71), $K_c(\bar{V}^2/2g_c)$; the expansion loss at the exit of the tube is, by Eq. (5-65), $K_e(\bar{V}^2/2g_c)$; and the friction loss in the globe valve is, by Eq. (5-73), $K_f(\bar{V}^2/2g_c)$. When skin friction in the entrance and exit headers is neglected, the total friction is

$$h_f = \left(4f\frac{L}{D} + K_c + K_e + K_f\right)\frac{\bar{V}^2}{2g_c}$$ (5-74)

Table 5-1 LOSS COEFFICIENTS FOR STANDARD THREADED PIPE FITTINGS†

Fitting	K_f
Globe valve, wide open	10.0
Angle valve, wide open	5.0
Gate valve, wide open	0.2
Half open	5.6
Return bend	2.2
Tee	1.8
Elbow, 90°	0.9
45°	0.4

†From J. K. Vennard, in V. L. Streeter (ed.), "Handbook of Fluid Dynamics," p. 3-23, McGraw-Hill Book Company, New York, 1961.

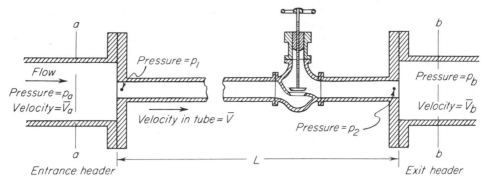

FIGURE 5-15
Flow of incompressible fluid through typical assembly.

To write the Bernoulli equation for this assembly, take station a in the inlet header and station b in the outlet header. Because there is no pump between station a and b, $W_p = 0$; also α_a and α_b can be taken as 1.0, the kinetic-energy term can be canceled, and Eq. (4-34) becomes

$$\frac{p_a - p_b}{\rho} + \frac{g}{g_c}(Z_a - Z_b) = \left(4f\frac{L}{D} + K_c + K_e + K_f\right)\frac{\bar{V}^2}{2g_c} \tag{5-75}$$

Practical use of velocity heads in design As shown by Eq. (5-75), the friction loss in a complicated flow system is expressible as a number of velocity heads for losses in pipes and fittings and for expansion and contraction losses. This fact is the basis for a rapid practical method for the estimation of friction. By setting $4f(L/D) = 1.0$, it follows that a length equal to a definite number of pipe diameters generates a friction loss equal to one velocity head. Since, in turbulent flow, f varies from about 0.01 to about 0.002, the number of pipe diameters equivalent to a velocity head is from $1/(4 \times 0.01) = 25$ to $1/(4 \times 0.002) = 125$, depending upon the Reynolds number. For ordinary practice, 50 pipe diameters is assumed for this factor. Thus, if the pipe in the system of Fig. 5-15 is standard 2-in. steel (actual ID = 2.07 in.) and is 100 ft long, the skin friction is equivalent to $(100 \times 12)/(2 \times 50) = 12$ velocity heads. In this case, the friction from the single fitting and the expansion and contraction are negligible in comparison with that in the pipe. In other cases, where the pipes are short and the fittings and expansion and contraction losses numerous, the friction loss in the pipes only may be negligible.

Separation from velocity decrease Boundary-layer separation can occur even where there is no sudden change in cross section if the cross section is continuously enlarged. For example, consider the flow of a fluid stream through the conical expander shown in Fig. 5-16. Because of the increase of cross section in the direction of flow, the velocity of the fluid decreases, and, by the Bernoulli equation, the pressure must increase. Consider two stream filaments, one, aa, very near the wall, and the

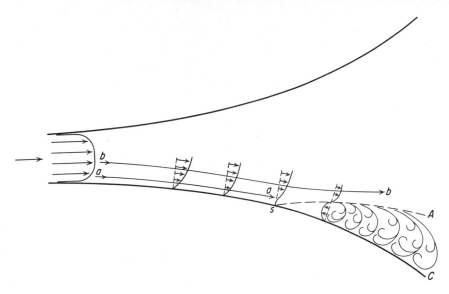

FIGURE 5-16
Separation of boundary layer in diverging channel.

other, *bb*, a short distance from the wall. The pressure increase over a definite length of conduit is the same for both filaments, because the pressure throughout any single cross section is uniform. The loss in velocity head is, then, the same for both filaments. The initial velocity head of filament *aa* is less than that of filament *bb*, however, because filament *aa* is the nearer to the wall. A point is reached, at a definite distance along the conduit, where the velocity of filament *aa* becomes zero but where the velocities of filament *bb* and of all other filaments farther from the wall than *aa* are still positive. This point is point *s* in Fig. 5-16. Beyond point *s* the velocity at the wall changes sign, a backflow of fluid between the wall and filament *aa* occurs, and the boundary layer separates from the wall. In Fig. 5-16, several curves are drawn of velocity *u* vs. distance from the wall *y*, and it can be seen how the velocity near the wall becomes zero at point *s* and then reverses in sign. The point *s* is called a *separation point*. Line *sA* is called the *line of zero tangential velocity*.

The vortices formed between the wall and the separated fluid stream, beyond the separation point, cause excessive form-friction losses. Separation occurs in both laminar and turbulent flow. In turbulent flow, the separation point is farther along the conduit than in laminar flow. Separation can be prevented if the angle between the wall of the conduit and the axis is made small. The maximum angle that can be tolerated in a conical expander without separation is 7°.

Minimizing expansion and contraction losses   A contraction loss can be nearly eliminated by reducing the cross section gradually rather than suddenly. For example, if the reduction in cross section shown in Fig. 5-14 is obtained by a conical reducer or by a trumpet-shaped entrance to the smaller pipe, the contraction coefficient K_c can be reduced to approximately 0.05 for all values of S_b/S_a. Separation and vena-contracta formation do not occur unless the decrease in cross section is sudden.

An expansion loss can also be minimized by substituting a conical expander for the flanges shown in Fig. 5-13. The angle between wall and axis of the cone must be less than 7°, however, or separation may occur. For angles of 30° or more, the loss through a conical expander can become greater than that through a sudden expansion for the same area ratio S_a/S_b because of the excessive form friction from the vortices caused by the separation.

FLOW OF LIQUIDS IN THIN LAYERS

Couette flow   In one form of layer flow, illustrated in Fig. 5-17, the fluid is bounded between two very large, flat, parallel plates separated by distance B. The lower plate is stationary, and the upper plate is moving to the right at a constant velocity u_0. For a Newtonian fluid the velocity profile is linear, and the velocity u is zero at $y = 0$ and u_0 at $y = B$, where y is the vertical distance measured from the lower plate. The velocity gradient is constant and equals u_0/B. Considering an area A of both plates, the shear force needed to maintain the motion of the top plate is, from Eqs. (3-3) and (3-4),

$$F_s = \frac{\mu A u_0}{g_c B} \tag{5-76}$$

Flow under these conditions is called *Couette flow*.

Layer flow with free surface   In another form of layer flow the liquid layer has a free surface and flows under the force of gravity over an inclined or vertical surface. Such flow is in steady state, with fully developed velocity gradients, and the thickness of the layer is assumed to be constant. Flow usually is laminar, and often there is so little drag at the free liquid surface that the shear stress there can be ignored. Under these assumptions and the further assumption that the liquid surface is flat and free from ripples, the fluid motion can be analyzed mathematically.

Consider a layer of a Newtonian liquid flowing in steady flow at constant rate and thickness over a flat plate as shown in Fig. 5-18. The plate is inclined at an angle β with the vertical. The breadth of the layer in the direction perpendicular to the plane of the figure is b, and the thickness of the layer in the direction perpendicular to the plate is δ. Isolate a control volume as shown in Fig. 5-18. The upper surface of the control volume is in contact with the atmosphere, the two ends are planes perpendicular to the plate at a distance L apart, and the lower surface is the plane parallel with the wall at a distance r from the upper surface of the layer.

Since the layer is in steady flow with no acceleration, by the momentum principle the sum of all forces on the control volume is zero. The possible forces acting on the control volume in a direction parallel to the flow are the pressure forces on the ends,

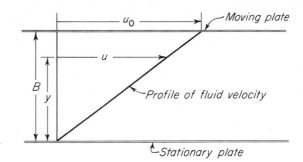

FIGURE 5-17
Velocities and velocity gradient in a
fluid between flat plates.

the shear forces on the upper and lower faces, and the component of the force of
gravity in the direction of flow. Since the pressure on the outer surface is atmospheric,
the pressures on the control volume at the ends of the volume are equal and oppositely
directed. They therefore vanish. Also, by assumption, the shear on the upper surface
of the element is neglected. The two forces remaining are therefore the shear force
on the lower surface of the control volume and the component of gravity in the direction of flow. Then

$$F_g \cos \beta - \tau A = 0 \tag{5-77}$$

where F_g = gravity force
τ = shear stress on lower surface of control volume

From this equation, noting that $A = bL$ and $F_g = \rho r L b g / g_c$,

$$\rho r L B \frac{g}{g_c} \cos \beta = \tau L b$$

or

$$\tau g_c = \rho r g \cos \beta \tag{5-78}$$

Since the flow is laminar, $\tau g_c = -\mu \, du/dr$ and

$$-\mu \frac{du}{dr} = g \rho r \cos \beta \tag{5-79}$$

Rearranging and integrating between limits gives

$$\int_0^u du = -\frac{g\rho \cos \beta}{\mu} \int_\delta^r r \, dr$$

$$u = \frac{\rho g \cos \beta}{2\mu} (\delta^2 - r^2) \tag{5-80}$$

since δ is the total thickness of the liquid layer. Equation (5-80) shows that in laminar
flow on a plate the velocity distribution is parabolic, as it is in a pipe.

Consider now a differential element of cross-sectional area dS, where $dS = b \, dr$.
The differential mass flow rate $d\dot{m}$ through this element equals $\rho u b \, dr$. The total mass

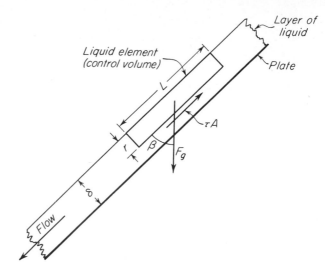

FIGURE 5-18
Forces on liquid element in layer flow.

flow rate of the fluid then is

$$\dot{m} = \int_0^\delta \rho u b \, dr \tag{5-81}$$

Substituting from Eq. (5-80) into Eq. (5-81) and integrating gives

$$\frac{\dot{m}}{b} = \frac{\delta^3 \rho^2 g \cos \beta}{3\mu} = \Gamma \tag{5-82}$$

where $\Gamma = \dot{m}/b$ and is called the liquid loading. The dimensions of Γ are pounds per second per foot of width or kilograms per second per meter of width.

Rearrangement of Eq. (5-82) gives, for the thickness of the layer,

$$\delta = \left(\frac{3\mu\Gamma}{\rho^2 g \cos \beta} \right)^{1/3} \tag{5-83}$$

Layer flow in pipes   Equations (5-80) and (5-83) can be applied to the laminar flow of a liquid down either the inside or the outside of a vertical tube, provided δ is small in comparison with the diameter of the tube. For a vertical surface $\cos \beta = 1$. The Reynolds number for layer flow is found by using for D 4 times the hydraulic radius defined by Eq. (5-59), or $4(S/L_p)$, where S is the cross-sectional area of the layer and L_p is the wetted perimeter. For a circular tube, the width b becomes the wetted perimeter and $\Gamma = \dot{m}/L_p$. Since $\overline{V}_p = \dot{m}/S$, the Reynolds number is, by Eq. (4-8),

$$N_{\text{Re}} = \frac{4r_H \overline{V} \rho}{\mu} = \frac{4(S/L_p)(\dot{m}/S)}{\mu} = \frac{4(\dot{m}/L_p)}{\mu} = \frac{4\Gamma}{\mu} \tag{5-84}$$

It has been found experimentally that flow in a layer becomes turbulent when $4\Gamma/\mu >$ 2,100. Equations (5-80) to (5-83) do not apply for turbulent flow.

Equation (5-84) also applies when the liquid fills the pipe; i.e., for a given mass flow rate the Reynolds number is the same whether the pipe is full or the fluid is flowing in a layer form. The last equality in Eq. (5-84) provides a convenient method of calculating N_{Re} from the mass rate of flow for either layer or full-pipe flow.

SYMBOLS

A	Area, ft^2 or m^2
b	Width of square duct, ft or m; also breadth of liquid layer, ft or m
c_1	Constant in Eq. (5-36)
D	Diameter, ft or m; D_i, inside diameter of annulus; D_o, outside diameter of annulus
F	Force, lb$_f$ or N; F_g, gravity force; F_s, shear force
f	Fanning friction factor, dimensionless
G	Mass velocity, lb/s or kg/s-m^2
g	Gravitational acceleration, ft/s or m/s^2
g_c	Newton's-law proportionality factor, 32.174 ft-lb/lb$_f$-s^2
h_f	Friction loss, ft-lb$_f$/lb or J/kg; h_{fc}, from sudden contraction; h_{fe}, from sudden expansion; h_{ff}, in flow through fitting or valve; h_{fs}, skin friction
K_c	Contraction-loss coefficient, dimensionless
K_e	Expansion-loss coefficient, dimensionless
K_f	Loss factor for fitting or valve, dimensionless
K'	Flow consistency index, lb/ft-s^{2-n} or kg/m-s^{2-n}
k	Roughness parameter, ft or m
L	Length, ft or m
L_p	Wetted perimeter, ft or m
$\dot{m}$	Mass flow rate, lb/s or kg/s
N_{Re}	Reynolds number, dimensionless; $N_{Re,n}$, modified Reynolds number for non-Newtonian fluids, defined by Eq. (5-55)
n'	Flow behavior index, dimensionless
p	Pressure, lb$_f$/ft^2 or N/m^2; p_a, at station a; p_b, at station b; $-\Delta p$, pressure loss, $p_a - p_b$
r	Radial distance from pipe axis, ft or m; also distance from surface of liquid layer; r_c, radius of cylinder of plastic fluid in plug flow; r_w, radius of pipe
r_H	Hydraulic radius of conduit, ft or m
S	Cross-sectional area, ft^2 or m^2; S_a, at station a; S_b, at station b
u	Net or time-average local fluid velocity in x direction, ft/s or m/s; u_c, velocity of cylinder of plastic fluid in plug flow; u_{max}, maximum local velocity
u^*	Friction velocity, dimensionless, $\bar{V}\sqrt{f/2}$
u^+	Dimensionless velocity quotient, u/u^*; u_c^+, at pipe axis
v	Time-average local fluid velocity in y direction, ft/s or m/s
W_p	Pump work, ft-lb$_f$/lb or J/kg
x_t	Transition length, ft or m
y	Radial distance from pipe wall, ft or m
y^+	Dimensionless distance, yu^*/v; y_c^+, at pipe axis
Z	Height above datum plane, ft or m; Z_a, at station a, Z_b, at station b

Greek letters

α	Kinetic-energy correction factor, dimensionless; α_a, at station a; α_b, at station b
β	Angle with vertical; also momentum correction factor, dimensionless; β_a, at station a; β_b, at station b

Γ Liquid loading in layer flow, lb/s-ft or kg/s-m

$-\Delta p_s$ Pressure loss from skin friction

δ Thickness of liquid layer, ft or m

ε_M Eddy diffusivity of momentum, ft^2/s or m^2/s

κ Constant in Eq. (5-32)

μ Absolute viscosity, lb/ft-s or p; μ_w, at temperature of pipe wall

v Kinematic viscosity, μ/ρ, ft^2/s or m^2/s

ρ Density, lb/ft^3 or kg/m^2

τ Shear stress, lb$_f$/ft^2 or N/m^2; τ_w, at pipe wall; τ_0, threshold stress in plastic fluid

ϕ_a Factor in Eq. (5-64) for laminar flow in annulus, dimensionless

ψ Temperature correction factor for skin friction, dimensionless

PROBLEMS

5-1 Prove that the flow of a liquid in laminar flow between infinite parallel flat plates is given by

$$p_a - p_b = \frac{12\mu \overline{V} L}{a^2 g_c}$$

where L = length of plate in direction of flow

    a = distance between plates

Neglect end effects.

5-2 Calculate the power required per meter of width of stream to force lubricating oil through the gap between two horizontal flat plates under the following conditions:

Distance between plates, 6 mm
Flow rate of oil per meter of width, 100 m^3/h
Viscosity of oil, 25 cP
Density of oil, 0.88 g/cm^3
Length of plates, 3 m

Assume that the plates are very wide in comparison with the distance between them and that end effects are neglected.

5-3 A liquid with a specific gravity of 4.7 and a viscosity of 1.3 cP flows through a smooth pipe of unknown diameter, resulting in a pressure drop of 0.183 lb$_f$/in.^{2} for 1.73 mi. What is the pipe diameter in inches if the mass rate of flow is 5,900 lb/h?

5-4 Water flows through a 100-mm steel pipe at an average velocity of 2 m/s. Downstream the pipe divides into a 100-mm main and a 25-mm bypass. The equivalent length of the bypass is 10 m; the length of the 100-mm pipe in the bypassed section is 8 m. Neglecting entrance and exit losses, what fraction of the total water flow passes through the bypass?

5-5 Water at 60°F is pumped from a reservoir to the top of a mountain through a Schedule 120 6-in. pipe at an average velocity of 10 ft/s. The pipe discharges into the atmosphere at a level 4,000 ft above the level in the reservoir. The pipeline itself is 5,000 ft long. If the overall efficiency of the pump and the motor driving it is 70 percent, and if the cost of electric energy to the motor is 1½ cents per kilowatthour, what is the hourly energy cost for pumping this water?

5-6 Crude oil having a specific gravity of 0.80 and a viscosity of 4 cP is draining by gravity from the bottom of a tank. The depth of liquid above the drawoff connection in the tank is 6 m. The line from the drawoff is Schedule 40 4-in. pipe. Its length is 45 m, and it contains one ell and two gate valves. The oil discharges into the atmosphere 9 m below the drawoff connection of the tank. What flow rate, in cubic meters per hour, can be expected through the line?

5-7 Water is being heated in a horizontal shell-and-tube condenser like that shown in Fig. 11-1. The condenser has 500 tubes having an inside diameter of 0.620 in. and a length of 12 ft. The total flow of water into the condenser is 2,400 gal/min. The inlet water temperature is 70°F, and the outlet temperature is 150°F. The temperature of the walls of the tubes is 230°F. The inlet and outlet connections are both 6 in. ID, and the area of each header is 2.5 times that of the total inside cross section of all the tubes. Estimate the pressure drop in this condenser in inches of water.

5-8 The condenser of Prob. 5-7 is used to heat air. The air enters at 20 in. H_2O above atmospheric pressure and 70°F. It leaves at 200°F. The air flow through the unit is 3,500 ft³/min, measured at entrance conditions. The tube-wall temperature may be taken as 205°F. What is the pressure drop in the condenser in inches of water?

5-9 Water at 68°F is to be pumped at a constant rate of 5 ft³/min to the top of an experimental absorber from a supply tank resting on the floor. The point of discharge is 15 ft above the floor, and the frictional losses in the Schedule 40 2-in. pipe are estimated to be 0.8 ft-lb$_f$/lb. At what height in the supply tank must the water level be kept if the pump can develop a net power of only $\frac{1}{8}$ hp?

5-10 A centrifugal pump takes brine from the bottom of a supply tank and delivers it into the bottom of another tank. The brine level in the discharge tank is 200 ft above that in the supply tank. The line between the tanks is 700 ft of Schedule 40 6-in. pipe. The flow rate is 810 gal/min. In the line are two gate valves, four standard tees, and four ells. What is the energy cost for running this pump for one 24-h day? The specific gravity of brine is 1.18; the viscosity of brine is 1.2 cP; and the energy cost is $150 per horsepower-year on a basis of 300 d/year. The overall efficiency of pump and motor is 60 percent.

5-11 A fan draws air at rest and sends it through a 200- by 300-mm rectangular duct 55 m long. The air enters at 15°C and 750 mm Hg abs pressure at a rate of 0.5 m³/s. What is the theoretical power required?

REFERENCES

1 Dodge, D. W., and A. B. Metzner: *AIChE J.,* **5:**189 (1959).
2 Kays, W. M., and A. L. London: "Compact Heat Exchangers," 2d ed., McGraw-Hill, New York, 1964.
3 Knudsen, J. G., and D. L. Katz: "Fluid Dynamics and Heat Transfer," pp. 97, 101–105, 158–171, McGraw-Hill, New York, 1958.
4 Perry, J. H. (ed.): "Chemical Engineers' Handbook," 5th ed., p. **5**-23, McGraw-Hill, New York, 1973.
5 Prandtl, L.: *VDI Z.,* **77:**105 (1933).
6 Schlichting, H.: "Boundary Layer Theory," 6th ed., McGraw-Hill, New York, 1968.

6

FLOW OF COMPRESSIBLE FLUIDS

Many important applications of fluid dynamics require that density variations be taken into account. The complete field of compressible fluid flow has become very large, and it covers wide ranges of pressure, temperature, and velocity. Chemical engineering practice involves a relatively small area from this field. For incompressible flow the basic parameter is the Reynolds number, a parameter also important in some applications of compressible flow. In compressible flow at ordinary densities and high velocities a more basic parameter is the Mach number. At very low densities, where the mean free path of the molecules is appreciable in comparison with the size of the equipment or solid bodies in contact with the gas, other factors must be considered. This type of flow is not treated in this text.

The Mach number, denoted by N_{Ma}, is defined as the ratio of u, the speed of the fluid, to a, the speed of sound in the fluid under conditions of flow,

$$N_{\text{Ma}} \equiv \frac{u}{a} \tag{6-1}$$

By speed of the fluid is meant the magnitude of the relative velocity between the fluid and a solid bounding the fluid or immersed in it, whether the solid is considered to be stationary and the fluid flowing past it, or whether the fluid is assumed to be stationary and the solid moving through it. The former situation is the more common in chemical engineering, and the latter is of greatest importance in aeronautics, for the motion of missiles, rockets, and other solid bodies through the atmosphere. By definition the Mach number is unity when the speed of the fluid equals that of sound in the same fluid at the pressure and temperature of the fluid. Flow is called *subsonic*, *sonic*, or *supersonic*, according to whether the Mach number is less than unity, at or near unity, or greater than unity, respectively. The most interesting problems in compressible flow lie in the high-velocity range, where Mach numbers are comparable with unity or where flow is supersonic.

Other important technical areas in fluid dynamics include chemical reactions, electromagnetic phenomena, ionization, and phase change[1] which are excluded from this discussion.

In this chapter the following simplifying assumptions are made. Although they may appear restrictive, many actual engineering situations may be adequately represented by the mathematical models obtained within the limitations of the assumptions.

1 The flow is steady.
2 The flow is one-dimensional.
3 Velocity gradients within a cross section are neglected, so that $\alpha = \beta = 1$ and $\overline{V} = u$.
4 Friction is restricted to wall shear.
5 Shaft work is zero.
6 Gravitational effects are negligible, and mechanical-potential energy is neglected.
7 The fluid is an ideal gas of constant specific heat.

The following basic relations are used:

1 The continuity equation
2 The steady-flow total-energy balance
3 The mechanical-energy balance with wall friction
4 The equation for the velocity of sound
5 The equation of state of the ideal gas

Each of these equations must be put into a suitable form.

Continuity equation For differentiation, Eq. (4-2) may be written in logarithmic form

$$\ln \rho + \ln S + \ln u = \text{const}$$

Differentiating this equation gives

$$\frac{d\rho}{\rho} + \frac{dS}{S} + \frac{du}{u} = 0 \qquad (6\text{-}2)$$

Total-energy balance   The steady-flow total-energy equation (1-52) may be written, in light of the simplifying assumptions, by omitting terms for potential energy and shaft work. Then

$$\frac{Q}{m} = H_b - H_a + \frac{u_b^2}{2g_cJ} - \frac{u_a^2}{2g_cJ} \tag{6-3}$$

This equation written differentially is

$$\frac{dQ}{m} = dH + d\left(\frac{u^2}{2g_cJ}\right) \tag{6-4}$$

Mechanical-energy balance   Equation (4-30) may be written over a short length of conduit in the following differential form:

$$\frac{dp}{\rho} + d\left(\frac{\alpha \overline{V}^2}{2g_c}\right) + \frac{g}{g_c}dZ + dh_f = 0 \tag{6-5}$$

In the light of the assumptions, this equation is simplified by omitting the potential-energy terms, noting that $\alpha_a = \alpha_b = 1.0$, $u = \overline{V}$, and restricting the friction to wall shear. Equation (6-5) then becomes

$$\frac{dp}{\rho} + d\left(\frac{u^2}{2g_c}\right) + dh_{fs} = 0 \tag{6-6}$$

From Eq. (5-61)

$$dh_{fs} = \frac{u^2}{2g_c}\frac{f\,dL}{r_H} \tag{6-7}$$

Eliminating dh_{fs} from Eqs. (6-6) and (6-7) gives the form of the mechanical-energy equation suitable for treatment of compressible flow:

$$\frac{dp}{\rho} + d\left(\frac{u^2}{2g_c}\right) + \frac{u^2}{2g_c}\frac{f\,dL}{r_H} = 0 \tag{6-8}$$

Velocity of sound   The velocity of sound through a continuous material medium, also called the *acoustical velocity*, is the velocity of a very small compression-rarefaction wave moving adiabatically and frictionlessly through the medium. Thermodynamically, the motion of a sound wave is a constant-entropy, or isentropic, process. The magnitude of the acoustical velocity in any medium is shown in physics texts to be

$$a = \sqrt{g_c\left(\frac{dp}{d\rho}\right)_S} \tag{6-9}$$

where the subscript S calls attention to the isentropic restraint on the process.

Ideal-gas equations   Subject to assumptions 1 to 6, Eqs. (6-2) to (6-9) apply to any fluid. In fact, they may be used for incompressible flow simply by assuming that the density ρ is constant. To apply them to compressible flow, it is necessary that the density be related to temperature and pressure. The simplest relation, and one of considerable engineering utility, is the ideal-gas law [Eq. (1-24)], which for the present purpose may be written in the form

$$p = \frac{R}{M} \rho T \tag{6-10}$$

where R = molar gas-law constant, in units of mechanical energy per mole per degree absolute

M = molecular weight

The gas may either be pure or a mixture, but if it is not pure, the composition should not change. Equation (6-10) may be written logarithmically and then differentiated to give

$$\frac{dp}{p} = \frac{d\rho}{\rho} + \frac{dT}{T} \tag{6-11}$$

Since the specific heat c_p is assumed to be independent of temperature, the enthalpy of the gas at temperature T is, by Eq. (1-54),

$$H = H_0 + c_p(T - T_0) \tag{6-12}$$

where H = enthalpy per unit mass at temperature T

H_0 = enthalpy at arbitrary temperature T_0

The differential form of Eq. (6-12) is

$$dH = c_p \, dT \tag{6-13}$$

ACOUSTICAL VELOCITY AND MACH NUMBER OF IDEAL GAS   For an ideal gas, an isentropic path follows the equations

$$p\rho^{-\gamma} = \text{const} \tag{6-14}$$

$$Tp^{-(1-1/\gamma)} = \text{const} \tag{6-15}$$

where γ is the ratio of c_p, the specific heat at constant pressure, to c_v, the specific heat at constant volume. For an ideal gas,

$$\gamma \equiv \frac{c_p}{c_v} = \frac{c_p}{c_p - R/MJ} \tag{6-16}$$

Since, by assumption, c_p is independent of temperature, so are c_v and γ.

The quantity $(dp/d\rho)_s$ can be calculated by differentiating the logarithmic form of Eq. (6-14), giving

$$\frac{dp}{p} - \gamma \frac{d\rho}{\rho} = 0 \quad \text{and} \quad \left(\frac{dp}{d\rho}\right)_s = \gamma \frac{p}{\rho}$$

Substituting into Eq. (6-9) yields

$$a = \sqrt{\frac{g_c \gamma p}{\rho}} = \sqrt{\frac{g_c \gamma T R}{M}} \tag{6-17}$$

Equation (6-10) is used to establish the second equality in Eq. (6-17). The second equality shows that the acoustical velocity of an ideal gas is a function of temperature only. From Eqs. (6-1) and (6-17) the square of the Mach number of a gas is

$$N_{Ma}^2 = \frac{\rho u^2}{g_c \gamma p} = \frac{u^2}{g_c \gamma TR/M} \tag{6-18}$$

The asterisk condition   The state of the fluid moving at its acoustic velocity is important in some processes of compressible-fluid flow. The condition where $u = a$ and $N_{Ma} = 1$ is called the *asterisk condition*, and the pressure, temperature, density, and enthalpy are denoted by p^*, T^*, ρ^*, and H^* at this state.

Stagnation temperature   The stagnation temperature of a high-speed fluid is defined as the temperature the fluid would attain were it brought to rest adiabatically without the development of shaft work. The relation between the actual fluid temperature, the actual fluid velocity, and the stagnation temperature is found by using the total-energy equation (6-3) and the enthalpy equation (6-12). The terminal a in Eq. (6-3) and reference state 0 in Eq. (6-12) are identified with the stagnation condition, and stagnation is denoted by subscript s. Also, terminal b in Eq. (6-3) is chosen as the state of the flowing gas, and this subscript is dropped. Then, since the process is adiabatic and $Q = 0$, Eq. (6-3) becomes

$$H - H_s = -\frac{u^2}{2g_c J} = H - H_0 \tag{6-19}$$

Eliminating $H - H_0$ from Eq. (6-19) by substitution from Eq. (6-12) gives, for the stagnation temperature T_s,

$$T_s = T + \frac{u^2}{2g_c J c_p} \tag{6-20}$$

The *stagnation enthalpy* H_s is defined by the equation

$$H_s = H + \frac{u^2}{2g_c J} \tag{6-21}$$

Equation (6-3) can be written

$$\frac{Q}{m} = H_{sb} - H_{sa} = (T_{sb} - T_{sa})c_p \tag{6-22}$$

where H_{sa} and H_{sb} are the stagnation enthalpies at states a and b, respectively. For an adiabatic process, $Q = 0$, $T_{sa} = T_{sb}$, and the stagnation temperature is constant.

PROCESSES OF COMPRESSIBLE FLOW

The individual processes to be considered in this chapter are shown diagrammatically in Fig. 6-1. It is assumed that a very large supply of gas at specified temperature and pressure and at zero velocity and Mach number is available. The origin of the gas is

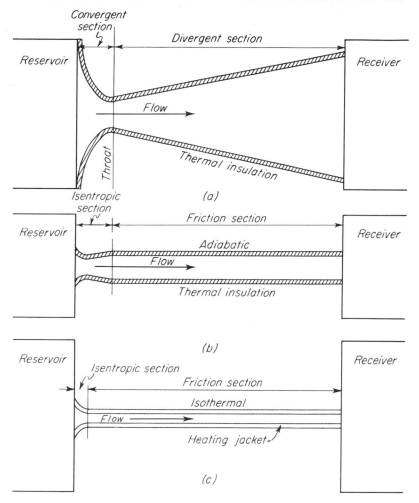

FIGURE 6-1
(*a*) Isentropic expansion in convergent-divergent nozzle. (*b*) Adiabatic frictional flow. (*c*) Isothermal frictional flow.

called the *reservoir*, and the temperature and pressure of the gas in the reservoir are called *reservoir conditions*. The reservoir temperature is a stagnation value, which does not necessarily apply at other points in the flow system.

From the reservoir the gas is assumed to flow, without friction loss at the entrance, into and through a conduit. The gas leaves the conduit at definite temperature, velocity, and pressure and goes into an exhaust receiver, at which the pressure may be independently controlled at a constant value less than the reservoir pressure.

Within the conduit any one of the following processes may occur:

1   An isentropic expansion. In this process the cross-sectional area of the conduit must change, and the process is described as one of variable area. Because the process is adiabatic, the stagnation temperature does not change in the conduit. Such a process is shown diagrammatically in Fig. 6-1*a*.

2   Adiabatic frictional flow through a conduit of constant cross section. This process is irreversible, and the entropy of the gas increases, but, as shown by Eq. (6-22), since $Q = 0$, the stagnation temperature is constant throughout the conduit. This process is shown in Fig. 6-1*b*.

3   Isothermal frictional flow through a conduit of constant cross-sectional area, accompanied by a flow of heat through the conduit wall sufficient to keep the temperature constant. This process is nonadiabatic and nonisentropic; the stagnation temperature changes during the process, since T is constant and, by Eq. (6-20), T_s changes with u. The process is shown in Fig. 6-1*c*.

The changes in gas temperature, density, pressure, velocity, and stagnation temperature are predictable from the basic equations. The purpose of this section is to demonstrate how these three processes can be treated analytically on the basis of such equations.†

FLOW THROUGH VARIABLE-AREA CONDUITS

A conduit suitable for isentropic flow is called a *nozzle*. As shown in Fig. 6-1*a*, a complete nozzle consists of a convergent section and a divergent section joined by a throat, which is a short length where the wall of the conduit is parallel with the axis of the nozzle. For some applications, a nozzle may consist of a divergent section only, and the throat connects directly with the receiver. The configuration of an actual nozzle is controlled by the designer, who fixes the relation between S, the cross-sectional area, and L, the length of the nozzle measured from the entrance. Nozzles are designed to minimize wall friction and to suppress boundary-layer separation. The convergent section is rounded and can be short, since separation does not occur in a converging channel. To suppress separation the angle in the divergent section is small, and this section is therefore relatively long. The nozzle entrance is sufficiently large relative to the throat to permit the velocity at the entrance to be taken as zero and the temperature and pressure at the entrance to be assumed equal to those in the reservoir.

The purpose of the convergent section is to increase the velocity and decrease the pressure of the gas. At low Mach numbers the process conforms essentially to the usual Bernoulli relation for incompressible flow [Eq. (4-26)]. In the convergent section flow is always subsonic, but it may become sonic at the throat. Mach numbers

† A generalized treatment, including general heat transfer to and from the gas, injection of gas into the conduit, variations in specific heat and molecular weight, chemical reactions, the drag of internal bodies, and change of phase, is given in Ref. 3.

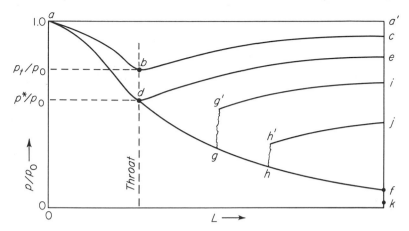

FIGURE 6-2
Variation of pressure ratio with distance from nozzle inlet.

greater than unity cannot be generated in a convergent nozzle. In the divergent section, the flow may be subsonic or supersonic. The purpose of the divergent section differs sharply in the two situations. In subsonic flow the purpose of the section is to reduce the velocity and regain pressure, in accordance with the Bernoulli equation. An important application of these nozzles is the measurement of fluid flow, which is discussed in Chap. 8. In supersonic flow, the usual purpose of the divergent section is to obtain Mach numbers greater than unity for use in experimental devices such as wind tunnels.

Flow through a given nozzle is controlled by fixing the reservoir and receiver pressures. For a given flow through a specific nozzle, a unique pressure exists at each point along the axis of the nozzle. The relation is conveniently shown as a plot of p/p_0 vs. L, where p_0 is the reservoir pressure and p the pressure at point L. Figure 6-2 shows how the pressure ratio varies with distance and how changes in receiver pressure at constant reservoir pressure affect the pressure distribution. The pressures at the throat and in the receiver are denoted by p_t and p_r, respectively.

If p_r and p_0 are equal, no flow occurs and the pressure distribution is represented by the line aa'. If the receiver pressure is slightly below the reservoir pressure, flow occurs and a pressure distribution such as that shown by line abc is established. Pressure recovery in the convergent section is shown by line bc. The maximum velocity occurs at the throat. If the receiver pressure is further reduced, the flow rate and the velocity throughout the nozzle increase. A limit is attained when the velocity at the throat becomes sonic. This case is shown by line ade, where $p_t = p^*$, $u_t = a$, and $N_{Ma} = 1$. The ratio p^*/p_0 is called the *critical pressure ratio* and is denoted by r_c. Flow is subsonic at all other points on line ade.

As the receiver pressure decreases from that of point a' to that of point e, the mass flow rate through the nozzle increases. The flow rate is not affected by reduction of pressure below that corresponding to critical flow. Figure 6-3 shows how mass flow

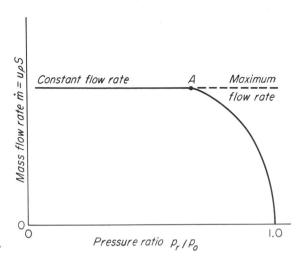

FIGURE 6-3
Mass flow rate through nozzle.

rate varies with the pressure ratio p_r/p_0. The flow rate attains its maximum at point A, which is reached when the pressure ratio in the throat reaches its critical value. Further reduction in pressure p_r does not change the flow rate.

The reason for the phenomenon of the maximum flow rate is this: when the velocity in the throat is sonic and the cross-sectional area of the conduit is constant, sound waves cannot move upstream into the throat and the gas in the throat has no way of receiving a message from downstream. Further reduction of the receiver pressure cannot be transmitted into the throat.

If the receiver pressure is reduced to the level shown by point f in Fig. 6-2, the pressure distribution is represented by the continuous line $adghf$. This line is unique for a given gas and nozzle. Only along the path $dghf$ is supersonic flow possible. If the receiver pressure is reduced below that of point f, for example, to point k, the pressure at the end of the nozzle remains at that of point f and flow through the nozzle remains unchanged. On issuing from the nozzle into the receiver the gas suffers a sudden pressure drop from that of point f to that of point k. The pressure change is accompanied by wave phenomena in the receiver. If the receiver pressure is held at a level between points e and f, pressure-distribution curves of the type $dgg'i$ and $dhh'j$ are found. Sections dg and dh represent isentropic supersonic flow. The sudden pressure jumps gg' and hh' represent wave motions and shocks, where the flow changes suddenly from supersonic to subsonic. Shocks are thermodynamically irreversible and are accompanied by an increase in entropy in accordance with the second law of thermodynamics. Curves $g'i$ and $h'j$ represent subsonic flows in which ordinary pressure recovery is taking place.

Isentropic flow is confined to the subsonic area $adea'a$ and the single line $dghf$. The area below line $adghf$ is not accessible to any kind of adiabatic flow.

The qualitative discussions of Figs. 6-2 and 6-3 apply to the flow of any compressible fluid. Quantitative relations are most easily found for ideal-gas flow.

Equations for isentropic flow   The phenomena occurring in the flow of ideal gas through nozzles are described by equations derivable from the basic equations given earlier in this chapter.

CHANGE IN GAS PROPERTIES DURING FLOW   The density and temperature paths of the gas through any isentropic flow are given by Eqs. (6-14) and (6-15). The constants are evaluated from the reservoir condition. This gives

$$\frac{p}{\rho^{\gamma}} = \frac{p_0}{\rho_0^{\gamma}} \tag{6-23}$$

$$\frac{T}{p^{1-1/\gamma}} = \frac{T_0}{p_0^{1-1/\gamma}} \tag{6-24}$$

These equations apply both to frictionless subsonic and supersonic flow, but they must not be used across a shock front.

VELOCITY IN NOZZLE   In the absence of friction, the mechanical-energy balance [Eq. (6-6)] becomes simply

$$\frac{dp}{\rho} = -d\left(\frac{u^2}{2g_c}\right) \tag{6-25}$$

Eliminating ρ from Eq. (6-25) by substitution from Eq. (6-23) and integrating from a lower limit based on the reservoir, where $p = p_0$, $\rho = \rho_0$, and $u = 0$, gives

$$\int_0^u d\left(\frac{u^2}{2g_c}\right) = -\frac{p_0^{1/\gamma}}{\rho_0} \int_{p_0}^p \frac{dp}{p^{1/\gamma}}$$

Integrating and substituting the limits yields, after rearrangement,

$$u^2 = \frac{2\gamma g_c p_0}{(\gamma - 1)\rho_0}\left[1 - \left(\frac{p}{p_0}\right)^{1-1/\gamma}\right] \tag{6-26}$$

A Mach-number form of Eq. (6-26) is convenient. It is derived by substituting u^2 from Eq. (6-26) into the first Eq. (6-18) and eliminating ρ/ρ_0 by substitution from Eq. (6-23). This gives

$$N_{\text{Ma}}^2 = \frac{2}{\gamma - 1}\frac{p_0}{p}\frac{\rho}{\rho_0}\left[1 - \left(\frac{p}{p_0}\right)^{1-1/\gamma}\right] = \frac{2}{\gamma - 1}\left[\left(\frac{p_0}{p}\right)^{1-1/\gamma} - 1\right] \tag{6-27}$$

Solved explicitly for the pressure ratio, Eq. (6-27) becomes

$$\frac{p}{p_0} = \frac{1}{\{1 + [(\gamma - 1)/2]N_{\text{Ma}}^2\}^{1/(1-1/\gamma)}} \tag{6-28}$$

The critical pressure ratio, denoted by r_c, is found from Eq. (6-28) by substituting p^* for p and 1.0 for N_{Ma}

$$r_c = \frac{p^*}{p_0} = \left(\frac{2}{\gamma + 1}\right)^{1/(1-1/\gamma)} \tag{6-29}$$

The mass velocity is found by calculating the product of u and ρ, using Eqs. (6-23) and (6-26)

$$G = u\rho = \sqrt{\frac{2\gamma g_c \rho_0\, p_0}{\gamma - 1}} \left(\frac{p}{p_0}\right)^{1/\gamma} \sqrt{1 - \left(\frac{p}{p_0}\right)^{1-1/\gamma}} \tag{6-30}$$

Effect of cross-sectional area   The relation between the change in cross-sectional area, velocity, and Mach number is useful in correlating the various cases of nozzle flow. Substitution of ρ from Eq. (6-25) into the continuity equation (6-2) gives

$$\frac{du}{u} + \frac{dS}{S} - \left(\frac{d\rho}{dp}\right)_s \frac{u\, du}{g_c} = 0 \tag{6-31}$$

Subscript S is used to call attention to the fact that the flow is isentropic. From Eq. (6-9)

$$\left(\frac{dp}{d\rho}\right)_s = \frac{a^2}{g_c} \tag{6-32}$$

Eliminating $(dp/d\rho)_S$ from Eqs. (6-31) and (6-32) gives

$$\frac{du}{u}\left(1 - \frac{u^2}{a^2}\right) + \frac{dS}{S} = 0$$

and, by substituting N_{Ma} from Eq. (6-1),

$$\frac{du}{u}(N_{Ma}^2 - 1) = \frac{dS}{S} \tag{6-33}$$

Equation (6-33) shows that for subsonic flow, where $N_{Ma} < 1$, the velocity increases with decreasing cross section (converging conduit) and decreases with increasing cross section (diverging conduit). This corresponds to the usual situation of incompressible flow. Lines abc, ade, $g'i$, and $h'j$ of Fig. 6-2 represent examples. For supersonic flow, where $N_{Ma} > 1$, the velocity increases with increasing cross section, as in the diverging section of the nozzle. This conforms to line $dghf$ of Fig. 6-2. The apparent anomaly of supersonic flow is a result of the variation in density and velocity along an isentropic path. Since the mass flow rate is the same at all points in the nozzle, by continuity the cross-sectional area of the nozzle must vary inversely with the mass velocity $u\rho$. The velocity steadily increases with Mach number, and the density decreases. However, at $N_{Ma} = 1$, the value of G goes through a maximum: in the subsonic regime the velocity increases faster than the density decreases, the mass velocity increases, and S decreases. In the supersonic regime, the increase in velocity is overcome by a sharper decrease in density, the mass velocity decreases, and S increases to accommodate the total mass flow. This behavior of G is demonstrated by studying the first and second derivatives of Eq. (6-30) in the usual manner for investigating maxima and minima.

EXAMPLE 6-1   Air enters a convergent-divergent nozzle at a temperature of 1000°R (555.6 °K) and a pressure of 20 atm. The throat area is one-half that of the discharge of the divergent section. (a) Assuming the Mach number in the throat is 0.8, what are the values of the following quantities at the throat: pressure, temperature, linear velocity, density, and mass velocity? (b) What are the values of p^*, T^*, u^*, G^* corresponding to reservoir conditions? (c) Assuming the nozzle is to be used supersonically, what is the maximum Mach number at the discharge of the divergent section? For air $\gamma = 1.4$ and $M = 29$.

SOLUTION   (a)   The pressure at the throat is calculated by use of Eq. (6-28):

$$\frac{p_t}{20} = \frac{1}{\{1 + [(1.4 - 1)/2]0.8^2\}^{1/(1-1/1.4)}} = 0.656$$

$$p_t = 13.12 \text{ atm}$$

From Eq. (6-10) since $R = 1,545$ ft-lb$_f$/lb mol-°R, the reservoir density ρ_0 is

$$\rho_0 = \frac{20 \times 144 \times 14.7 \times 29}{1,545 \times 1000} = 0.795 \text{ lb/ft}^3 \text{ (12.73 kg/m}^3)$$

Substituting p_0/ρ_0 from Eq. (6-10) into Eq. (6-26) gives, for the velocity in the throat,

$$u_t = \sqrt{\frac{2\gamma g_c R T_0}{M(\gamma - 1)} \left[1 - \left(\frac{p_t}{p_0}\right)^{1-1/\gamma}\right]}$$

$$= \sqrt{\frac{2 \times 1.4 \times 32.17 \times 1,545 \times 1,000}{29(1.4 - 1)}} \sqrt{1 - 0.656^{1-1/1.4}}$$

$$= 1,167 \text{ ft/s (355.7 m/s)}$$

The density at the throat is, from Eq. (6-23),

$$\rho_t = \rho_0 \left(\frac{p_t}{p_0}\right)^{1/\gamma} = 0.795 \times 0.656^{1/1.4} = 0.588 \text{ lb/ft}^3 \text{ (9.42 kg/m}^3)$$

The mass velocity at the throat is

$$G_t = u_t \rho_t = 1,167 \times 0.588 = 686 \text{ lb/ft}^2\text{-s (3,349 kg/m}^2\text{-s)}$$

[The mass velocity can also be calculated directly by use of Eq. (6-30).] The temperature at the throat is, from Eq. (6-24),

$$T_t = T_0 \left(\frac{p_t}{p_0}\right)^{1-1/\gamma} = 1,000 \times 0.656^{1-1/1.4} = 886.5°R \text{ (492.5 K)}$$

(b)   From Eq. (6-29)

$$r_c = \frac{p^*}{p_0} = \left(\frac{2}{1.4 + 1}\right)^{1/(1-1/1.4)} = 0.528 \qquad p^* = 20 \times 0.528 = 10.56 \text{ atm}$$

From Eqs. (6-24) and (6-29)

$$T^* = 1,000 \left(\frac{2}{1.4 + 1}\right)^{1-1/1.4} = 833°R \text{ (462.8 K)}$$

From Eq. (6-23)

$$\left(\frac{\rho_0}{\rho^*}\right)^\gamma = \frac{p_0}{p^*} \qquad \frac{\rho_0}{\rho^*} = \left(\frac{p^*}{p_0}\right)^{1/\gamma}$$

$$\rho^* = 0.795 \times 0.528^{1/1.4} = 0.504 \text{ lb/ft}^3 \text{ (8.07 kg/m}^3)$$

From Eq. (6-30)

$$G^* = \sqrt{\frac{2 \times 1.4 \times 32.17 \times 0.795 \times 20 \times 144 \times 14.7}{1.4 - 1}} \, 0.528^{1/1.4}\sqrt{1 - 0.528^{1-1/1.4}}$$

$$= 713 \text{ lb/ft}^2\text{-s} \ (348.1 \text{ kg/m}^2\text{-s})$$

$$u^* = \frac{G^*}{\rho^*} = \frac{713}{0.504} = 1,415 \text{ ft/s} \ (431.3 \text{ m/s})$$

(c) Since, by continuity, $G \propto 1/S$, the mass velocity at the discharge is

$$G_r = \tfrac{713}{2} = 356.5 \text{ lb/ft}^2\text{-s} \ (1,741 \text{ kg/m}^2\text{-s})$$

From Eq. (6-30)

$$356.5 = \sqrt{\frac{2 \times 1.4 \times 32.17 \times 0.795 \times 20 \times 144 \times 14.7}{0.4} \left[1 - \left(\frac{p_r}{p_0}\right)^{1-1/1.4}\right] \left(\frac{p_r}{p_0}\right)^{1/1.4}}$$

$$\left(\frac{p_r}{p_0}\right)^{1/1.4} \sqrt{\left[1 - \left(\frac{p_r}{p_0}\right)^{1-1/1.4}\right]^{0.4}} = 0.1295$$

This equation is solved for p_r/p_0 to give

$$\frac{p_r}{p_0} = 0.0938$$

From Eq. (6-27) the Mach number at the discharge is

$$N_{\text{Ma},r} = \sqrt{\frac{2}{1.4 - 1} \left(\frac{1}{0.0938^{1-1/1.4}} - 1\right)} = 2.20 \qquad \qquad ////$$

ADIABATIC FRICTIONAL FLOW

Flow through straight conduits of constant cross section is adiabatic when heat transfer through the pipe wall is negligible. The process is shown diagrammatically in Fig. 6-1b. The typical situation is a long pipe into which gas enters at a given pressure and temperature and flows at a rate determined by the length and diameter of the pipe and the pressure maintained at the outlet. In long lines and with a low exit pressure, the speed of the gas may reach the sonic velocity. It is not possible, however, for a gas to pass through the sonic barrier from the direction of either subsonic or supersonic flow; if the gas enters the pipe at a Mach number greater than 1, the Mach number will decrease but will not become less than 1. If an attempt is made, by maintaining a constant discharge pressure and lengthening the pipe, to force the gas to change from subsonic to supersonic flow or from supersonic to subsonic, the mass flow rate will decrease to prevent such a change. This effect is called choking.

THE FRICTION PARAMETER   The basic quantity that measures the effect of friction is the friction parameter fL/r_H. This arises from the integration of Eq. (6-8). In adiabatic frictional flow, the temperature of the gas changes. The viscosity also varies, and the

Reynolds number and friction factor are not actually constant. In gas flow, however, the effect of temperature on viscosity is small, and the effect of Reynolds number on the friction factor f is still less. Also, unless the Mach number is nearly unity, the temperature change is small. It is satisfactory to use an average value for f as a constant in calculations. If necessary, f can be evaluated at the two ends of the conduit and an arithmetic average used as a constant.

Friction factors in supersonic flow are not well established. Apparently, they are approximately one-half those in subsonic flow for the same Reynolds number.[2]

In all the integrated equations in the next section it is assumed that the entrance to the conduit is rounded to form an isentropic convergent nozzle. If supersonic flow in the conduit is required, the entrance nozzle must include a divergent section to generate a Mach number greater than 1.

Equations for adiabatic frictional flow   Equation (6-8) is multiplied by ρ/p, giving

$$\frac{dp}{p} + \frac{\rho}{p}\frac{u\,du}{g_c} + \frac{\rho u^2}{2pg_c}\frac{f\,dL}{r_H} = 0 \tag{6-34}$$

It is desired to obtain an integrated form of this equation. The most useful integrated form is one containing the Mach number as the dependent variable and the friction parameter as an independent variable. To accomplish this, the density factor is eliminated from Eq. (6-34) using Eq. (6-18), and relationships between N_{Ma}, dp/p, and du/u are found from Eqs. (6-2) and (6-11). Quantity dT/T, when it appears, is eliminated by using Eqs. (6-4), (6-13), and (6-18). The results are

$$\frac{dp}{p} = -\frac{1 + (\gamma - 1)N_{\text{Ma}}^2}{1 + [(\gamma - 1)/2]N_{\text{Ma}}^2}\frac{dN_{\text{Ma}}}{N_{\text{Ma}}} \tag{6-35}$$

Also

$$\frac{du}{u} = \frac{dp}{p} + 2\frac{dN_{\text{Ma}}}{N_{\text{Ma}}} \tag{6-36}$$

Substitution from Eqs. (6-35) and (6-36) into Eq. (6-34) and rearrangement gives the final differential equation

$$f\frac{dL}{r_H} = \frac{2(1 - N_{\text{Ma}}^2)\,dN_{\text{Ma}}}{\gamma N_{\text{Ma}}^3\{1 + [(\gamma - 1)/2]N_{\text{Ma}}^2\}} \tag{6-37}$$

Formal integration of Eq. (6-37) between an entrance station a and exit station b gives

$$\int_{L_a}^{L_b} \frac{f}{r_H}\,dL = \int_{N_{\text{Ma},a}}^{N_{\text{Ma},b}} \frac{2(1 - N_{\text{Ma}}^2)\,dN_{\text{Ma}}}{\gamma N_{\text{Ma}}^3\{1 + [(\gamma - 1)/2]N_{\text{Ma}}^2\}}$$

$$\frac{\bar{f}}{r_H}(L_b - L_a) = \frac{\bar{f}\,\Delta L}{r_H} = \frac{1}{\gamma}\left(\frac{1}{N_{\text{Ma},a}^2} - \frac{1}{N_{\text{Ma},b}^2}\right.$$

$$\left. -\frac{\gamma + 1}{2}\ln\frac{N_{\text{Ma},b}^2\{1 + [(\gamma - 1)/2]N_{\text{Ma},a}^2\}}{N_{\text{Ma},a}^2\{1 + [(\gamma - 1)/2]N_{\text{Ma},b}^2\}}\right) \tag{6-38}$$

where $\bar{f}$ is the arithmetic average value of the terminal friction factors, $(f_a + f_b)/2$, and $\Delta L = L_b - L_a$.

PROPERTY EQUATIONS   For calculating the changes in pressure, temperature, and density, the following equations are useful.

The pressure ratio over an adiabatic frictional flow is found by direct integration of Eq. (6-35) between the limits p_a, p_b, and $N_{\text{Ma},a}$, $N_{\text{Ma},b}$ to give

$$\frac{p_a}{p_b} = \frac{N_{\text{Ma},b}}{N_{\text{Ma},a}} \sqrt{\frac{1 + [(\gamma - 1)/2]N_{\text{Ma},b}^2}{1 + [(\gamma - 1)/2]N_{\text{Ma},a}^2}} \tag{6-39}$$

The temperature ratio is calculated from Eq. (6-20), noting that $T_{0a} = T_{0b}$; so

$$T_a + \frac{u_a^2}{2g_cJc_p} = T_b + \frac{u_b^2}{2g_cJc_p} \tag{6-40}$$

From the temperature form of the equation (6-18) for the Mach number of an ideal gas,

$$T_a + \frac{\gamma RT_aN_{\text{Ma},a}^2}{2MJc_p} = T_b + \frac{\gamma RT_bN_{\text{Ma},b}^2}{2MJc_p} \tag{6-41}$$

From Eq. (6-16)

$$\frac{Jc_pM}{R} = \frac{\gamma}{\gamma - 1} \tag{6-42}$$

Substituting Jc_pM/R from Eq. (6-42) into Eq. (6-41) and solving for the temperature ratio gives

$$\frac{T_a}{T_b} = \frac{1 + [(\gamma - 1)/2]N_{\text{Ma},b}^2}{1 + [(\gamma - 1)/2]N_{\text{Ma},a}^2} \tag{6-43}$$

The density ratio is calculated from the gas equation of state (6-10) and the pressure and temperature ratios given by Eqs. (6-39) and (6-43), respectively,

$$\frac{\rho_a}{\rho_b} = \frac{p_a}{p_b}\frac{T_b}{T_a} = \frac{N_{\text{Ma},b}}{N_{\text{Ma},a}}\sqrt{\frac{1 + [(\gamma - 1)/2]N_{\text{Ma},a}^2}{1 + [(\gamma - 1)/2]N_{\text{Ma},b}^2}} \tag{6-44}$$

MAXIMUM CONDUIT LENGTH   To ensure that the conditions of a problem do not call for the impossible phenomenon of a crossing of the sonic barrier, an equation is needed giving the maximum value of $\bar{f}L/r_H$ consistent with a given entrance Mach number. Such an equation is found from Eq. (6-38) by choosing the entrance to the conduit as station a and identifying station b as the asterisk condition, where $N_{\text{Ma}} = 1.0$. Then the length $L_b - L_a$ represents the maximum length of conduit that can be used for a fixed value of $N_{\text{Ma},a}$. This length is denoted by L_{max}. Equation (6-38) then gives

$$\frac{\bar{f}L_{\text{max}}}{r_H} = \frac{1}{\gamma}\left(\frac{1}{N_{\text{Ma},a}^2} - 1\right) - \frac{\gamma + 1}{2}\ln\frac{2\{1 + [(\gamma - 1)/2]N_{\text{Ma},a}^2\}}{N_{\text{Ma},a}^2(\gamma + 1)} \tag{6-45}$$

Corresponding equations for p/p^*, T/T^*, and ρ/ρ^* are found from Eqs. (6-39), (6-43), and (6-44).

Mass velocity   To calculate the Reynolds number for evaluating the friction factor, the mass velocity is needed. From Eq. (6-18) and the definition of G

$$N_{Ma}^2 = \frac{(\rho u)^2}{p^2 g_c \gamma TR/M} = \frac{G^2}{p^2 g_c \gamma TR/M} = \frac{(\rho u)^2}{\rho g_c \gamma p} = \frac{G^2}{\rho g_c \gamma p}$$

and

$$G = \rho N_{Ma} \sqrt{\frac{g_c \gamma TR}{M}} = N_{Ma} \sqrt{\rho g_c \gamma p} \tag{6-46}$$

Since, for constant-area flow, G is independent of length, the mass velocity can be evaluated at any point where the gas properties are known. Normally the conditions at the entrance to the conduit are used.

EXAMPLE 6-2   Air flows from a reservoir through an isentropic nozzle into a long straight pipe. The pressure and temperature in the reservoir are 20 atm and 1000°R (555.6 K), respectively, and the Mach number at the entrance of the pipe is 0.05. (a) What is the value of $\bar{f}L_{max}/r_H$? (b) What are the pressure, temperature, density, linear velocity, and mass velocity when $L_b = L_{max}$? (c) What is the mass velocity when $\bar{f}L_{max}/r_H = 400$?

SOLUTION   (a)   Values for substitution in Eq. (6-45) are $\gamma = 1.4$ and $N_{Ma,a} = 0.05$. Then

$$\frac{\bar{f}L_{max}}{r_H} = \frac{1}{1.4}\left(\frac{1}{0.05^2} - 1\right) - \frac{1.4 + 1}{2} 2.3026 \log \frac{2\{1 + [(1.4 \quad 1)/2]0.05^2\}}{(1.4 + 1)0.05^2}\right) = 280$$

(b) The pressure at the end of the isentropic nozzle p_a is given by Eq. (6-28)

$$p_a = \frac{20}{\{1 + [(1.4 - 1)/2]0.05^2\}^{1.4/(1.4-1)}} = \frac{20}{1.0016} \approx 20 \text{ atm}$$

The pressure, temperature, and density change in the nozzle are negligible, and except for linear velocity, the reservoir conditions also pertain to the pipe entrance. The density of air at 20 atm and 1000°R is 0.795 lb/ft³. The acoustical velocity is, from Eq. (6-17),

$$a_a = \sqrt{1.4 \times 32.174 \times 1,000 \frac{1,545}{29}} = 1,550 \text{ ft/s (472.4 m/s)}$$

The velocity at the entrance of the pipe is

$$u_a = 0.05 \times 1,550 = 77.5 \text{ ft/s (23.6 m/s)}$$

When $L_b = L_{max}$, the gas leaves the pipe at the asterisk condition, where $N_{Ma,b} = 1.0$. From Eq. (6-43)

$$\frac{1,000}{T^*} = \frac{2.4}{2\{1 + [(1.4 - 1)/2]0.05^2\}} = 1.2$$

$$T^* = 834°R (463.3 K)$$

From Eq. (6-44)

$$\frac{0.795}{\rho^*} = \frac{1}{0.05}\sqrt{\frac{2\{1 + [(1.4 - 1)/2]0.05^2\}}{2.4}}$$

$$\rho^* = 0.0435 \text{ lb/ft}^3 (0.697 \text{ kg/m}^3)$$

From Eq. (6-39)

$$\frac{20}{p^*} = \frac{1}{0.05} \sqrt{1.2}$$

$$p^* = 0.913 \text{ atm}$$

The mass velocity through the entire pipe is

$$G = 0.795 \times 77.5 = 0.0435u^* = 61.61 \text{ lb/ft}^2\text{-s (300.8 kg/m}^2\text{-s)}$$

$$a = u^* = 1,416 \text{ ft/s (431.6 m/s)}$$

Since the exit velocity is sonic, u^* can also be calculated from Eq. (6-17) using $T = T^* = 834°R$ (463.3 K)

$$a = u^* = 1,550 \sqrt{\frac{834}{1,000}} = 1,416 \text{ ft/s (431.6 m/s)}$$

(c) Using Eq. (6-45) with $\bar{f}L_{max}/r_H = 400$ gives

$$400 = \frac{1}{1.4} \left(\frac{1}{N_{Ma,a}^2} - 1 - \frac{1.4 + 1}{2} 2.3026 \log \frac{1.2}{1/N_{Ma,a}^2 + 0.7} \right)$$

This equation must be solved for $N_{Ma,a}$ by trial and error. The log term is much smaller than the quantity $1/N_{Ma,a}^2 - 1$, and as a first approximation it may be neglected. The final corrected result is $N_{Ma,a} = 0.0425$. Then

$$u_a = \frac{0.0425}{0.05} 77.5 = 65.8 \text{ ft/s (20.1 m/s)}$$

$$G = 65.8 \times 0.795 = 52.4 \text{ lb/ft}^2\text{-s (255.8 kg/m}^2\text{-s)} \qquad ////$$

ISOTHERMAL FRICTIONAL FLOW

The temperature of the fluid in compressible flow through a conduit of constant cross section may be kept constant by a transfer of heat through the conduit wall. Long, small, uninsulated pipes in contact with air transmit sufficient heat to keep the flow nearly isothermal. Also, for small Mach numbers, the pressure pattern for isothermal flow is nearly the same as that for adiabatic flow for the same entrance conditions, and the simpler equations for isothermal flow may be used.

The maximum velocity attainable in isothermal flow is

$$a' = \sqrt{\frac{g_c p}{\rho}} = \sqrt{\frac{g_c T R}{M}} \tag{6-47}$$

The dimensions of a' are those of velocity. Comparing Eq. (6-47) with Eq. (6-17) for the acoustical velocity shows that

$$a = a' \sqrt{\gamma} \tag{6-48}$$

Thus, for air where $\sqrt{\gamma} = \sqrt{1.4} \approx 1.2$ the acoustical velocity is approximately 20 percent larger than a' Corresponding to the true Mach number, an isothermal Mach number may be defined:

$$N_i \equiv \frac{u}{a'} \qquad (6\text{-}49)$$

The parameter N_i plays the same part in isothermal flow as the Mach number does in adiabatic flow. An isothermal process cannot pass through the limiting condition where $N_i = 1.0$. If the flow is initially subsonic, it must remain so and a maximum conduit length for a given inlet velocity is reached at the limiting condition. The velocity obtainable in isothermal flow is less than that in adiabatic flow because a' is less than a.

The basic equation for isothermal flow is simple. It is obtained by introducing the mass velocity into the mechanical-energy balance [Eq. (6-8)] and integrating directly.

Multiplying Eq. (6-8) by ρ^2 gives

$$\rho \, dp + \frac{\rho^2 u \, du}{g_c} + \frac{\rho^2 u^2 f \, dL}{2g_c r_H} = 0 \qquad (6\text{-}50)$$

Since $\rho u = G$, $u \, du = -(G^2 \rho^3) \, d\rho$, and $\rho = Mp/RT$, Eq. (6-50) can be written

$$\frac{M}{RT} p \, dp - \frac{G^2}{g_c} \frac{d\rho}{\rho} + \frac{G^2 f \, dL}{2g_c r_H} = 0 \qquad (6\text{-}51)$$

Integrating between stations a and b gives

$$\frac{M}{2RT} (p_a^2 - p_b^2) - \frac{G^2}{g_c} \ln \frac{\rho_a}{\rho_b} = \frac{G^2 f \, \Delta L}{2g_c r_H} \qquad (6\text{-}52)$$

The pressure ratio p_a/p_b may be used in place of ρ_a/ρ_b in Eq. (6-52).

Equation (6-52) can also be used when the temperature change over the conduit is small. Then, in place of T, an arithmetic average temperature may be used. For example, adiabatic flow at low Mach numbers (below about 0.3) follows the equation closely.

Heat transfer in isothermal flow The steady-flow energy equation (6-22) and Eq. (6-20) for the stagnation temperature combine to give, after noting that $T_a = T_b$,

$$\frac{Q}{m} = \frac{u_b^2 - u_a^2}{2g_c J} \qquad (6\text{-}53)$$

Substituting G/ρ for u gives the mass-velocity form of Eq. (6-53),

$$\frac{Q}{m} = \frac{G^2}{2g_c J} \left(\frac{1}{\rho_b^2} - \frac{1}{\rho_a^2} \right) \qquad (6\text{-}54)$$

In these equations Q/m is the heat flow into the gas in Btu per pound or kilojoules per kilogram.

Equations (6-52) to (6-54) are used for subsonic flow only.

SYMBOLS

a Acoustical velocity in fluid, ft/s or m/s; a_a, at pipe entrance

a' Maximum attainable velocity in isothermal flow in conduit, ft/s or m/s

c Specific heat, Btu/lb · °F or J/g − °C; c_p, at constant pressure; c_v, at constant volume

f Fanning friction factor, dimensionless; f_a, at station a; f_b, at station b; $\bar{f}$, average value

G Mass velocity, lb/s-ft² or kg/s-m²; G_r, in receiver; G_t, at throat; G^*, value when $N_{\text{Ma}} = 1.0$

g Gravitational acceleration, ft/s² or m/s²

g_c Newton's-law proportionality factor, 32.174 ft-lb/lb$_f$-s²

H Enthalpy, Btu/lb or J/g; H_a, at station a; H_b, at station b; H_0, at reference temperature, also at stagnation; H^*, value when $N_{\text{Ma}} = 1.0$

h_f Friction loss, ft-lb$_f$/lb or N-m/g; h_{fs}, loss from skin friction

J Mechanical equivalent of heat, 778.17 ft-lb$_f$/Btu

L Length, ft or m; L_a, from entrance to station a; L_b, to station b; $L_{\max}$, length of conduit when $N_{\text{Ma}} = 1.0$ at outlet

M Molecular weight of fluid

m Mass, lb or kg

N_i Mach number for isothermal flow, u/a'

N_{Ma} Mach number, u/a; $N_{\text{Ma},a}$, at station a; $N_{\text{Ma},b}$, at station b; $N_{\text{Ma},r}$, at nozzle discharge

p Pressure, lb$_f$/ft² or N/m²; p_a, at station a; p_b, at station b; p_r, in receiver; p_t, at throat of convergent-divergent nozzle; p_0, in reservoir; p^*, value when $N_{\text{Ma}} = 1.0$

Q Quantity of heat, Btu or J

R Gas-law constant, 1,545 ft-lb$_f$/lb mol °R or 8.314 N-m/g mol K

r_c Critical pressure ratio, p^*/p_0

r_H Hydraulic radius of conduit, ft or m

S Cross-sectional area of conduit, ft² or m²

T Temperature, °R or K; T_a, at station a; T_b, at station b; T_s, stagnation value; T_t, at nozzle throat; T_0, reference value; T^*, value when $N_{\text{Ma}} = 1.0$

u Fluid velocity, ft/s or m/s; u_a, at station a; u_b, at station b; u_t, at throat of convergent-divergent nozzle; u^*, value when $N_{\text{Ma}} = 1.0$

$\bar{V}$ Average velocity of fluid stream, ft/s or m/s

Z Height above datum plane, ft or m

Greek letters

α Kinetic-energy correction factor; α_a, at station a; α_b, at station b

β Momentum correction factor

γ Ratio of specific heats, c_p/c_v

ρ Density of fluid, lb/ft³ or kg/m³; ρ_0, in reservoir; ρ^*, value when $N_{\text{Ma}} = 1.0$

Subscripts

S Isentropic flow

s Stagnation value

0 Reference value, reservoir conditions

PROBLEMS

6-1 Air at 1.7 atm gauge and 15°C enters a horizontal 75-mm pipe 70 m long. The velocity at the entrance of the pipe is 60 m/s. (*a*) Assuming isothermal flow, what is the pressure at the discharge end of the line? (*b*) What is the maximum length of pipe for these inlet conditions?

6-2 Natural gas consisting essentially of methane is to be transported through a 20-in.-ID pipeline over flat terrain. Each pumping station increases the pressure to 100 lb$_f$/in.² abs, and the pressure drops to 25 lb$_f$/in.² abs at the inlet to the next pumping station 50 mi away. What is the gas flow rate in cubic feet per hour measured at 60°F and 30 in. Hg pressure?

6-3 A standard 1-in. Schedule 40 horizontal steel pipe is used to conduct argon gas. The gas enters the pipe through a rounded entrance at a pressure of 7 atm abs, a temperature of 100°C, and a velocity of 30 m/s. (*a*) What is the maximum possible length of the pipe? (*b*) What is the pressure and stagnation temperature of the gas at the end of the pipe at maximum length? Assume adiabatic flow. For argon, $\gamma = 1.67$, and $M = 40.0$.

6-4 Assuming that the flow process in Prob. 6-3 occurs isothermally, (*a*) what is the maximum pipe length? (*b*) How much heat must be transferred to the gas at maximum length?

6-5 A divergent-convergent nozzle has the proportions shown in Table 6-1. Air ($\gamma = 1.40$, $M = 29.0$) enters the nozzle from a reservoir in which the pressure is 20 atm abs and the temperature is 550 K. Plot the pressure, the temperature, and the Mach number vs. the length of the nozzle when (*a*) the flow rate is a maximum without shock and with subsonic discharge, (*b*) the flow is isentropic and the discharge is supersonic.

Table 6-1 DATA IN PROB. 6-1

Length		Diameter	Length	Diameter
Reservoir	0	∞	Throat 0.30	0.25
	0.025	0.875	0.40	0.28
	0.050	0.700	0.50	0.35
	0.075	0.575	0.60	0.45
	0.100	0.500	0.70	0.56
	0.150	0.375	0.80	0.68
	0.200	0.300	0.90	0.84
			Receiver 1.00	1.00

REFERENCES

1 Cambel, A. B.: pp. **8**-5 to **8**-12 in V. L. Streeter, "Handbook of Fluid Dynamics," McGraw-Hill, New York, 1961.
2 Cambel, A. B., and B. H. Jennings: "Gas Dynamics," pp. 85–86, McGraw-Hill, New York, 1958.
3 Shapiro, A. H., and W. R. Hawthorne: *J. Appl. Mech.,* **14:** *A*-317 (1947).

7

FLOW PAST IMMERSED BODIES

The discussion in Chaps. 4 and 5 centered about the laws of fluid flow and factors that control changes of pressure and velocity of fluids flowing past solid boundaries and was especially concerned with flow through closed conduits. Emphasis during the discussion was placed primarily on the fluid. In many problems, however, the effect of the fluid on the solid is of interest. The fluid may either be at rest and the solid moving through it, or the solid may be at rest and the fluid flowing past it. The situation where the solid is immersed in, and surrounded by, fluid is the subject of this chapter.

It is generally immaterial which phase, solid or fluid, is assumed to be at rest, and it is the relative velocity between the two that is important. An exception to this is met in some situations when the fluid stream has been previously influenced by solid walls and is in turbulent flow. The scale and intensity of turbulence then may be important parameters in the process. In wind tunnels, for example, where the solid shape is at rest and the stream of air is in motion, turbulence may give different forces on the solid than if the solid were moving at the same relative velocity through a quiescent and turbulence-free mass of air. Objects in free fall through a continuous medium may move in spiral patterns or rotate about their axis, or both; again the

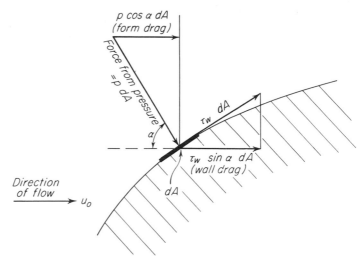

FIGURE 7-1
Wall drag and form drag on immersed body.

forces acting on them are not the same as when they are held stationary and the fluid is passed over them.

Drag The force in the direction of flow exerted by the fluid on the solid is called *drag*. By Newton's third law of motion, an equal and opposite net force is exerted by the body on the fluid. When the wall of the body is parallel with the direction of flow, as in the case of the thin flat plate shown in Fig. 3-10a, the only drag force is the wall shear τ_w. More generally, however, the wall of an immersed body makes an angle with the direction of flow. Then the component of the wall shear in the direction of flow contributes to drag. An extreme example is the drag of the flat plate perpendicular to the flow, shown in Fig. 3-10b. Also, the fluid pressure, which acts in a direction normal to the wall, possesses a component in the direction of flow, and this component also contributes to the drag. The total drag on an element of area is the sum of the two components. Figure 7-1 shows the pressure and shear forces acting on an element of area dA inclined at an angle of $90° - \alpha$ to the direction of flow. The drag from wall shear is $\tau_w \sin \alpha \, dA$, and that from pressure is $p \cos \alpha \, dA$. The total drag on the body is the sum of the integrals of these quantities each evaluated over the entire surface of the body in contact with the fluid.

The total integrated drag from wall shear is called *wall drag*, and the total integrated drag from pressure is called *form drag*.

In potential flow, $\tau_w = 0$, and there is no wall drag. Also, the pressure drag in the direction of flow is balanced by an equal force in the opposite direction, and the integral of the form drag is zero. There is no net drag in potential flow.

The phenomena causing both wall and form drag in actual fluids are complicated and cannot in general be calculated. For spheres and other regular shapes at low

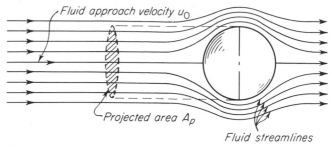

FIGURE 7-2
Flow past immersed sphere.

fluid velocities, the flow patterns and drag forces can be estimated by numerical methods;[12] for irregular shapes and high velocities they must be determined by experiment.

Drag coefficients In treating fluid flow through conduits, a friction factor, defined as the ratio of the shear stress to the product of the velocity head and density, was shown to be useful. An analogous factor, called a *drag coefficient*, is used for immersed solids. Consider a smooth sphere immersed in a flowing fluid and at a distance from the solid boundary of the stream sufficient for the approaching stream to be in potential flow. Define the projected area of the solid body as the area obtained by projecting the body on a plane perpendicular to the direction of flow, as shown in Fig. 7-2. Denote the projected area by A_p. For a sphere, the projected area is that of a great circle, or $(\pi/4)D_p^2$, where D_p is the diameter. If F_D is the total drag, the average drag per unit projected area is F_D/A_p. Just as the friction factor f is defined as the ratio of τ_w to the product of the density of the fluid and the velocity head, so the drag coefficient C_D is defined as the ratio of F_D/A_p to this same product, or

$$C_D \equiv \frac{F_D/A_p}{\rho u_0^2/2g_c} \tag{7-1}$$

where u_0 is the velocity of the approaching stream (by assumption $u_0 = \overline{V}_0$).

For particles having shapes other than spherical, it is necessary to specify the size and geometric form of the body and its orientation with respect to the direction of flow of the fluid. One major dimension is chosen as the characteristic length, and other important dimensions are given as ratios to the chosen one. Such ratios are called *shape factors*. Thus for short cylinders, the diameter is usually chosen as the defining dimension, and the ratio of length to diameter is a shape factor. The orientation between the particle and the stream also is specified. For a cylinder, the angle formed by the axis of the cylinder and the direction of flow is sufficient. Then the projected area is determinate and can be calculated. For a cylinder so oriented that its axis is perpendicular to the flow, A_p is LD_p, where L is the length of the cylinder. For a cylinder with its axis parallel to the direction of flow, A_p is $(\pi/4)D_p^2$, the same as for a sphere of the same diameter.

From dimensional analysis, the drag coefficient of a smooth solid in an incompressible fluid depends upon a Reynolds number and the necessary shape factors. For a given shape

$$C_D = \phi(N_{Re,p})$$

The Reynolds number for a particle in a fluid is defined as

$$N_{Re,p} \equiv \frac{G_0 D_p}{\mu} \qquad (7\text{-}2)$$

where D_p = characteristic length
$G_0 = u_0\rho$.

A different C_D-vs.-$N_{Re,p}$ relation exists for each shape and orientation. The relation must in general be determined experimentally, although a well-substantiated theoretical equation exists for smooth spheres at low Reynolds numbers.

Drag coefficients for compressible fluids increase with increase in the Mach number when the latter becomes more than about 0.6. Coefficients in supersonic flow are generally greater than in subsonic flow.

Drag coefficients of typical shapes   In Fig. 7-3, curves of C_D vs. $N_{Re,p}$ are shown for spheres, long cylinders, and disks. The axis of the cylinder and the face of the disk are perpendicular to the direction of flow; these curves apply only when this orientation is maintained. If, for example, a disk or cylinder is moving by gravity or centrifugal force through a quiescent fluid, it will twist and turn as it moves freely through the fluid.

From the complex nature of drag, it is not surprising that the variation of C_D with $N_{Re,p}$ is more complicated than that of f with N_{Re}. The variations in slope of the curves of C_D vs. $N_{Re,p}$ at different Reynolds numbers are the result of the interplay of the various factors that control form drag and wall drag. Their effects can be followed by discussing the case of the sphere.

For low Reynolds numbers the drag force for a sphere conforms to a theoretical equation called *Stokes' law*, which may be written

$$F_D = 3\pi \frac{\mu u D_p}{g_c} \qquad (7\text{-}3)$$

Of the total drag one-third is form drag and two-thirds wall drag. From Eq. (7-3), the drag coefficient predicted by Stokes' law, using Eq. (7-1), is

$$C_D = \frac{24}{N_{Re,p}} \qquad (7\text{-}4)$$

In theory, Stokes' law is valid only when $N_{Re,p}$ is considerably less than 1. Practically, as shown by the left-hand portion of the graph of Fig. 7-3, Eqs. (7-3) and (7-4) may be used with small error for Reynolds numbers less than about 2. At the low velocities at which the law is valid, the sphere moves through the fluid by deforming it. The wall shear is the result of viscous forces only, and inertial forces are negligible. The motion of the sphere affects the fluid at considerable

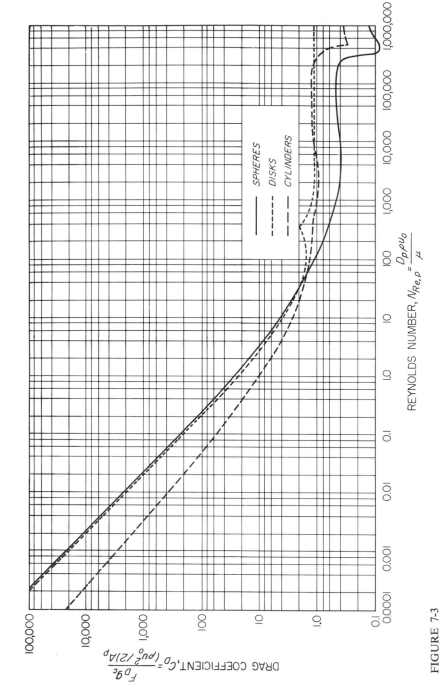

FIGURE 7-3

Drag coefficients for spheres, disks, and cylinders. [*By permission from J. H. Perry (ed.), "Chemical Engineers' Handbook," 5th ed., pp. 5-62. Copyright, © 1973, McGraw-Hill Book Company.*]

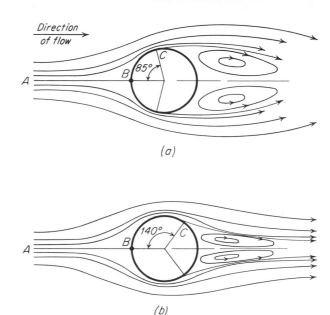

FIGURE 7-4
Flow past single sphere, showing separation and wake formation: (*a*) laminar flow in boundary layer; (*b*) turbulent flow in boundary layer; *B*, stagnation point; *C*, separation point. [*By permission from "Engineering Application of Fluid Mechanics," pp. 202–203. Copyright, 1947, McGraw-Hill Book Company.*]

distances from the body, and if there is a solid wall within 20 or 30 diameters of the sphere, Stokes' law must be corrected for the wall effect. The type of flow treated in this law is called *creeping flow*. The law is especially valuable for calculating the resistance of small particles, such as dusts or fogs, moving through gases or liquids of low viscosity, or for the motion of larger particles through highly viscous liquids.

As the Reynolds number is increased beyond the range of Stokes' law, separation occurs at a point just forward of the equatorial plane, as shown in Fig. 7-4*a*, and a wake, covering the entire rear hemisphere, is formed.[5] It has been shown in Chap. 3 that a wake is characterized by a large friction loss. It also develops a large form drag, and, in fact, most form drag occurs as a result of wakes. In a wake, the angular velocity of the vortices, and therefore their kinetic energy of rotation, is large. The pressure in the wake is, by the Bernoulli principle, less than that in the separated boundary layer; a suction develops in the wake, and the component of the pressure vector acts in the direction of flow. The pressure drag, and hence the total drag, is large. This is shown by the upward trend of the drag coefficient in Fig. 7-3.

At larger Reynolds numbers, a true boundary layer forms, originating at the apex, point *B* in Fig. 7-4. The boundary layer grows and separates, flowing freely around the wake after separation. At first the boundary layer is in laminar flow, both before and after separation. As the Reynolds number is further increased, transition to turbulence takes place, first in the free boundary layer and then in the boundary layer still attached to the front hemisphere of the sphere. When turbulence occurs in the latter, the separation point moves toward the rear of the body and the wake shrinks, as shown in Fig. 7-4b. Both friction and drag decrease, and the remarkable drop in drag coefficient from 0.45 to 0.10 at a Reynolds number of about 250,000 is a result of the shift in separation point when the boundary layer attached to the sphere becomes turbulent. At Reynolds numbers above 300,000, the drag coefficient is nearly constant.

The Reynolds number at which the attached boundary layer becomes turbulent is called the critical Reynolds number for drag. The curve for spheres shown in Fig. 7-3 applies only when the fluid approaching the sphere is nonturbulent or when the sphere is moving through a stationary static fluid. If the approaching fluid is turbulent, the critical Reynolds number is sensitive to the scale of turbulence and becomes smaller as the scale increases. For example, if the scale of turbulence, defined as $100\sqrt{(u')^2}/u$, is 2 percent, the critical Reynolds number is about 140,000.[6] One method of measuring the scale of turbulence is to determine the critical Reynolds number and use a known correlation between the two quantities.

The curve of C_D vs. $N_{Re,p}$ for a cylinder is much like that for a sphere, as the same phenomena occur in both cases. Disks do not show the drop in drag coefficient at a critical Reynolds number, because once the separation occurs at the edge of the disk, the separated stream does not return to the back of the disk and the wake does not shrink when the boundary layer becomes turbulent. Bodies which show this type of behavior are called *bluff bodies*. For a disk the drag coefficient C_D is approximately unity at Reynolds numbers above 2,000.

Form drag and streamlining   Form drag can be minimized by forcing separation toward the rear of the body. This is accomplished by streamlining. The usual method of streamlining is to so proportion the rear of the body that the increase in pressure in the boundary layer, which is the basic cause of separation, is sufficiently gradual to delay separation. Streamlining usually calls for a pointed rear, like that of an airfoil. A typical streamlined shape is shown in Fig. 7-5. A perfectly streamlined body would have no wake and little or no form drag.

STAGNATION POINT   The streamlines in the fluid flowing past the body in Fig. 7-5 show that the fluid stream in the plane of the section is split by the body into two parts, one passing over the top of the body and the other under the bottom. Streamline AB divides the two parts and terminates at a definite point B at the nose of the body. This point is called a stagnation point. The velocity at a stagnation point is zero. Equation (4-26) may be written on the assumption that the flow is horizontal and friction along the streamline is negligible. The undisturbed fluid is identified with station a in Eq. (4-26) and the stagnation point with station b. Then

$$\frac{p_s - p_0}{\rho} = \frac{u_0^2}{2g_c} \tag{7-5}$$

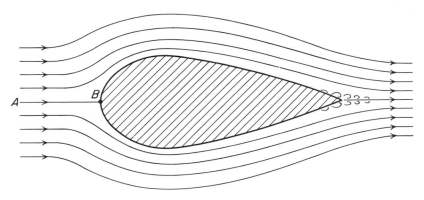

FIGURE 7-5
Streamlined body. *AB*, streamline to stagnation point *B*.

where p_s = pressure on body at stagnation point
$\quad\quad p_0$ = pressure in undisturbed fluid
$\quad\quad u_0$ = velocity of undisturbed fluid

The pressure increase $p_s - p_0$ for the streamline passing through a stagnation point is larger than that for any other streamline, because at that point the entire velocity head of the approaching stream is converted into pressure head.

Stagnation pressure   Equation (7-5) may be used for compressible fluids at low Mach numbers but becomes increasingly inaccurate when N_{Ma} is larger than about 0.4. The proper pressure to use for p_s then is the isentropic stagnation pressure, defined as the pressure of the gas when the stream is brought to rest isentropically, and calculated from the equations of Chap. 6 as follows. From Eq. (6-24)

$$\left(\frac{p_s}{p_0}\right)^{1-1/\gamma} = \frac{T_s}{T_0} \tag{7-6}$$

where p_s = stagnation pressure
$\quad\quad T_s$ = stagnation temperature
$\quad\quad T_0$ = temperature of approaching stream.

From Eq. (6-43), using $T_a = T_s$, $T_b = T_0$, $N_{\text{Ma},a} = 0$, and $N_{\text{Ma},b} = N_{\text{Ma},0}$,

$$\frac{T_s}{T_0} = 1 + \frac{\gamma - 1}{2} N_{\text{Ma},0}^2 \tag{7-7}$$

and
$$\frac{p_s}{p_0} = \left(1 + \frac{\gamma - 1}{2} N_{\text{Ma},0}^2\right)^{1/(1-1/\gamma)} \tag{7-8}$$

Equation (7-8) can be brought into the same form as Eq. (7-5) by subtracting unity from both sides and expanding the right-hand side of Eq. (7-8) by the binomial theorem. After substitution and simplification, using the relationship for N_{Ma} given in Eq. (6-18), and multiplication by p_0/ρ_0, the result is

$$\frac{p_s - p_0}{\rho_0} = \frac{u_0^2}{2g_c}\left(1 + \frac{N_{Ma,0}^2}{4} + \frac{2 - \gamma}{24}N_{Ma,0}^4 + \cdots\right) \tag{7-9}$$

Comparison of Eq. (7-9) with Eq. (7-5) shows that the quantity in the parentheses is the correction factor for converting Eq. (7-5) into a form suitable for compressible fluids, in the range $0 \le N_{Ma} < 1.0$.

FRICTION IN FLOW THROUGH BEDS OF SOLIDS

In several technical processes, liquids or gases flow through beds of solid particles. In the unit operations, important examples are filtration and the two-phase counter-current flow of liquid and gas through packed towers. In filtration, the bed of solids consists of small particles that are removed from the liquid by a filter cloth or fine screen. In other processes, such as ion-exchange or catalytic reactors, a single fluid (liquid or gas) flows through a bed of granular solids. Filtration is discussed in Chap. 30 and packed towers in Chap. 23. The present treatment is restricted to the flow of a single fluid phase through a column of stationary solid particles.

The resistance to the flow of a fluid through the voids in a bed of solids is the resultant of the total drag of all the particles in the bed. Depending on the Reynolds number, $D_p G_0/\mu$, laminar flow, turbulent flow, form drag, separation, and wake formation occur. As in the drag of a single solid particle, there is no sharp transition between laminar and turbulent flow like that occurring in flow through conduits of constant cross section.

Although progress has been reported in relating the total pressure drop through a bed of solids to the drag of the individual particles,[15] the most common methods of correlation are based on estimates of the total drag of the fluid on the solid boundaries of the tortuous channels through the bed particles. It is assumed that the actual channels may be effectively replaced by a set of identical parallel conduits, each of variable cross section. It also is assumed that the mean hydraulic radius of the channels is adequate to account for the variations in channel cross-sectional size and shape and that the total drag per unit area of channel wall is the sum of two kinds of force: (1) viscous drag forces and (2) inertial forces. In view of the complexity of the phenomena underlying drag, this is probably an oversimplification, which is justified by the success of the model in predicting the form of the final equation that is verified by experiment. It is also assumed that the particles are packed at random, with no preferred orientation of individual particles; that roughness effects are unimportant; that all particles are of the same size and shape; and that end and wall effects are negligible. By this last assumption is meant that the number of particles adjacent to the entrance of the fluid and to the wall of the vessel or tower containing the bed of solids are few relative to the total number of particles in the bed. This assumption is valid when the diameter and depth of the bed are large in comparison with the diameter of the individual particles.

It was shown in Eq. (5-17) that viscous shear force per unit area of the channel walls can be written in the form

$$\tau_w = \frac{F_v}{A_s} = \frac{k_1 \mu \overline{V}}{g_c r_H}$$

where k_1 = constant
$\quad r_H$ = hydraulic radius
$\quad \mu$ = viscosity
$\quad \overline{V}$ = velocity in channel
$\quad A_s$ = area of channel boundaries

In this equation, the hydraulic radius r_H is used in place of D, and the constant 8 is absorbed in the empirical constant k_1. At high velocities, inertial forces become large, viscous forces become negligible, and viscosity is no longer a parameter. This is equivalent to a constant friction factor; hence from Eq. (5-7) the inertial force per unit area can be written in the form

$$\tau_w = \frac{F_i}{A_s} = \frac{k_2 \rho \overline{V}^2}{g_c}$$

where k_2 is an empirical proportionality constant which absorbs the friction factor and other constants.

The assumption of additivity of viscous and inertial forces gives for F_D, the total force exerted in the fluid on the bed of solids,

$$\frac{F_D}{A_s} = \frac{F_v}{A_s} + \frac{F_i}{A_s} = \frac{1}{g_c} \left(\frac{k_1 \mu \overline{V}}{r_H} + k_2 \rho \, \overline{V}^2 \right) \tag{7-10}$$

Manipulation of Eq. (7-10) is facilitated by the introduction of a parameter, denoted by ε and defined as the ratio of the volume of voids in the bed to the total volume (voids plus solid) of the bed. Then the total volume of the channels is the product of ε and the total bed volume. The numerical value of ε is quite sensitive to the method used in building the bed of particles. It is usually measured *in situ* after the bed has been formed, by filling the bed with water, draining, and comparing the volume of water to the total volume of the bed as calculated from its depth and cross section. The usual range of values for ε is from 0.35 to 0.70. Quantity ε is called the *porosity*.

The velocity $\overline{V}$ in Eq. (7-10) is the average velocity through the channels in the bed. It is more convenient to use $\overline{V}_0$, the velocity of the stream just before it encounters the first layer of solids. The usual configuration of a packed bed is a large cylindrical shell, with a supporting grid some distance above the bottom of the tower with the bed or solid dumped on the grid. The velocity $\overline{V}_0$ is often called the *superficial* or *empty-tower velocity*, as it is the upward (or downward) velocity in the open section below the grid (or in the empty tower above the bed). Since the average free cross section of all the channels has the same ratio to the total cross section of the empty tower as the volume of the channels has to the total volume of the bed of solids, $\overline{V}_0$ and $\overline{V}$ are related by the equation

$$\overline{V} = \frac{\overline{V}_0}{\varepsilon} \tag{7-11}$$

The total area A_s is given as

$$A_s = N_p s_p \tag{7-12}$$

where N_p = total number of particles in bed
s_p = area of one particle

Also the number of particles is given by the total volume of the solids in the tower divided by v_p, the volume of one particle. If L is the total bed depth and S_0 the cross section of the empty tower, the total volume of the solids is $S_0 L(1 - \varepsilon)$, and

$$N_p = \frac{S_0 L(1 - \varepsilon)}{v_p} \tag{7-13}$$

Eliminating N_p from Eq. (7-12) by Eq. (7-13) gives

$$A_s = \frac{S_0 L(1 - \varepsilon)s_p}{v_p} \tag{7-14}$$

The hydraulic radius r_H was defined in Chap. 5 as the ratio of the cross section of the conduit to the perimeter of the conduit. If the numerator and denominator of this ratio are both multiplied by L, r_H becomes the ratio of the volume of the voids in the tower to the total area of the solids. The volume of voids is $S_0 L \varepsilon$. Therefore, using Eq. (7-14) yields

$$r_H = \frac{S_0 L \varepsilon}{A_s} = \frac{S_0 L \varepsilon}{S_0 L(1 - \varepsilon)s_p/v_p} = \frac{\varepsilon v_p}{(1 - \varepsilon)s_p} \tag{7-15}$$

Now $\bar{V}$ from Eq. (7-11), A_s from Eq. (7-14), and r_H from Eq. (7-15) can be substituted into Eq. (7-10), to give

$$F_D g_c = \frac{S_0 \rho L(1 - \varepsilon)s_p}{\varepsilon_c^2 v_p}\left[\frac{k_1 \mu \bar{V}_0(1 - \varepsilon)s_p}{v_p \rho} + k_2 \bar{V}_0^2\right] \tag{7-16}$$

The pressure drop over the bed $-\Delta p$ may be used in place of the total force F_D by noting that this force must equal the product of the pressure drop and the cross-sectional area of the channels $S_0 \varepsilon$. Therefore

$$-\Delta p\, \varepsilon S_0 = F_D \tag{7-17}$$

Eliminating F_D from Eqs. (7-16) and (7-17) yields

$$-\frac{\Delta p\, g_c}{\rho L} = \frac{1 - \varepsilon}{\varepsilon^3}\frac{s_p}{v_p}\left[\frac{k_1 \mu \bar{V}_0(1 - \varepsilon)s_p}{\rho v_p} + k_2 \bar{V}_0^2\right] \tag{7-18}$$

Equivalent particle diameter; sphericity   The equivalent diameter of a non-spherical particle is defined as the diameter of a sphere having the same volume as the particle. The *sphericity* Φ_s is the ratio of the surface area of this sphere to the actual surface area of the particle. Since for a sphere $s_p = \pi D_p^2$ and $v_p = \frac{1}{6}\pi D_p^3$, it follows that for any particle

$$\frac{s_p}{v_p} = \frac{6}{\Phi_s D_p} \tag{7-19}$$

Values of the sphericity for several materials are given in Table 26-1.

Ergun[2] has correlated experimental data to show that, in Eq. (7-18), satisfactory values of k_1 and k_2 are 150/36 and 1.75/6, respectively. Substitution of these values and elimination of s_p/v_p from Eq. (7-18) by Eq. (7-19) gives

$$\frac{-\Delta p\, g_c}{\rho \overline{V}_0^2} \frac{\Phi_s D_p}{L} \frac{\varepsilon^3}{1-\varepsilon} = \frac{150(1-\varepsilon)}{\Phi_s D_p G_0/\mu} + 1.75 \tag{7-20}$$

where G_0 equal to $\rho \overline{V}_0$ has been substituted. Equation (7-20) is called the *Ergun equation*. It does not apply well to such extreme shapes as long thin needles or the rings and saddle-shaped solids used as tower packings.

A *friction factor* for a packed bed is defined by the left-hand side of Eq. (7-20):

$$f_p \equiv \frac{-\Delta p\, g_c \Phi_s D_p \varepsilon^3}{\rho \overline{V}_0^2 L (1-\varepsilon)} \tag{7-21}$$

Except for a constant 2, the sphericity, and the porosity function $\varepsilon^3/(1-\varepsilon)$, f_p has the same form as the friction factor f for a conduit, as defined by Eq. (5-7a). By using Eq. (7-2) for particle Reynolds number, Eq. (7-20) can therefore be written

$$f_p = \frac{150(1-\varepsilon)}{N_{\mathrm{Re},p}} + 1.75 \tag{7-22}$$

Equation (7-22) is subject to a simple interpretation. At low Reynolds numbers, the quantity 1.75 is negligible in comparison with the Reynolds-number term. This implies that viscous forces control and that inertial forces are unimportant. Equation (7-22) can be written for this case as

$$- \frac{\Delta p\, g_c \Phi_s^2 D_p^2 \varepsilon^3}{L \overline{V}_0 \mu (1-\varepsilon)^2} = 150 \tag{7-23}$$

This is called the *Kozeny-Carman equation* and for obvious reasons is a laminar-flow equation. For a given system it indicates that the flow rate is directly proportional to the pressure drop. This statement is also known as *Darcy's law*.

For large Reynolds numbers, the first term on the right-hand side of Eq. (7-22) fades out as viscous forces become negligible and inertial forces control. Equation (7-22) then becomes

$$\frac{-\Delta p}{\rho L} \frac{g_c}{\overline{V}_0^2} \frac{\Phi_s D_p \varepsilon^3}{1-\varepsilon} - 1.75 \tag{7-24}$$

This is called the *Blake-Plummer equation*.

MIXTURES OF PARTICLES   Equation (7-20) can be used for beds consisting of a mixture of different particle sizes by using, in place of s_p/v_p, a mean value, defined as the volumetric weighted mean of the specific surface. Thus, for mixture of particles of three sizes,

$$\left(\frac{s_p}{v_p}\right)_m = x_1 \left(\frac{s_p}{v_p}\right)_1 + x_2 \left(\frac{s_p}{v_p}\right)_2 + x_3 \left(\frac{s_p}{v_p}\right)_3 = \sum_1^3 x_i \left(\frac{s_p}{v_p}\right)_i \qquad (7\text{-}25)$$

where x_i is the ratio of the volume of particles i to the total volume of all particles, and so on. By definition $x_1 + x_2 + x_3 = 1.0$.

The mean effective diameter is, then, for a mixture of n different sizes,

$$D_{pm} = \frac{6}{\left(\dfrac{s_p}{v_p}\right)_m} = \frac{6}{\displaystyle\sum_{i=1}^n x_i \dfrac{6}{\Phi_s D_{pi}}} = \frac{1}{\displaystyle\sum_{i=1}^n \dfrac{x_i}{\Phi_s D_{pi}}} \qquad (7\text{-}26)$$

COMPRESSIBLE FLUIDS   When the density change of the fluid is small—and seldom is the pressure drop large enough to change the density greatly—Eq. (7-20) may be used by calculating the inlet and outlet values of $\overline{V}_0$ and using the arithmetic mean for $\overline{V}_0$ in the equation.

MOTION OF PARTICLES THROUGH FLUIDS

Many processing steps, especially mechanical separations, involve the movement of solid particles or liquid drops through a fluid. The fluid may be gas or liquid, and it may be flowing or at rest. Examples are the elimination of dust and fumes from air or flue gas, the removal of solids from liquid wastes to allow discharge into public drainage systems, and the recovery of acid mists from the waste gas of an acid plant.

Mechanics of particle motion   The movement of a particle through a fluid requires that a density difference exist between the particles and the fluid. Also, an external force is needed to impart motion to the particle relative to the fluid. The external force is usually gravity, but when gravity is not sufficiently strong, centrifugal force, which can be many times that of gravity, is used. If the densities of particle and fluid are equal, the buoyant force from the immersion of the particle in the fluid will counterbalance an external force, however large, and the particle will not move through the fluid. The greater the density difference, the more effective the process.

Three forces act on a particle moving through a fluid: (1) the external force, gravitational or centrifugal; (2) the buoyant force, which acts parallel with the external force but in the opposite direction; and (3) the drag force, which appears whenever there is relative motion between the particle and the fluid. The drag force acts to oppose the motion and acts parallel with the direction of movement but in the opposite direction.

In the general case, the direction of movement of the particle relative to the fluid may not be parallel with the direction of the external and buoyant forces, and the drag force then makes an angle with the other two. In this situation, which is called *two-dimensional motion*, the drag must be resolved into components, which complicates the treatment of particle mechanics. Equations are available for two-dimensional motion,[8] but only the one-dimensional case, where the lines of action of all forces acting on the particle are collinear, will be considered in this book.

Equations for one-dimensional motion of particle through fluid Consider a particle of mass m moving through a fluid under the action of an external force F_e. Let the velocity of the particle relative to the fluid be u. Let the buoyant force on the particle be F_b, and let the drag be F_D. Then the resultant force on the particle is $F_e - F_b - F_D$, the acceleration of the particle is du/dt, and, by Eq. (1-35), since m is constant,

$$\frac{m}{g_c}\frac{du}{dt} = F_e - F_b - F_D \qquad (7\text{-}27)$$

The external force can be expressed as a product of the mass and the acceleration a_e of the particle from this force, and

$$F_e = \frac{ma_e}{g_c} \qquad (7\text{-}28)$$

The buoyant force is, by Archimedes' principle, the product of the mass of the fluid displaced by the particle and the acceleration from the external force. The volume of the particle is m/ρ_p, where ρ_p is the density of the particle, and the particle displaces this same volume of fluid. The mass of fluid displaced is $(m/\rho_p)\rho$, where ρ is the density of the fluid. The buoyant force is, then,

$$F_b = \frac{m\rho a_e}{\rho_p g_c} \qquad (7\text{-}29)$$

The drag force is, from Eq. (7-1),

$$F_D = \frac{C_D u_0^2 \rho A_p}{2g_c} \qquad (7\text{-}30)$$

where C_D = dimensionless drag coefficient
 A_p = projected area of particle measured in plane perpendicular to direction of motion of particle
 $u_0 = u$

Substituting the forces from Eqs. (7-28) to (7-30) into Eq. (7-27) gives

$$\frac{du}{dt} = a_e - \frac{\rho a_e}{\rho_p} - \frac{C_D u^2 \rho A_p}{2m} = a_e \frac{\rho_p - \rho}{\rho_p} - \frac{C_D u^2 \rho A_p}{2m} \qquad (7\text{-}31)$$

MOTION FROM GRAVITATIONAL FORCE If the external force is gravity, a_e is g, the acceleration of gravity, and Eq. (7-31) becomes

$$\frac{du}{dt} = g\frac{\rho_p - \rho}{\rho_p} - \frac{C_D u^2 \rho A_p}{2m} \qquad (7\text{-}32)$$

MOTION IN A CENTRIFUGAL FIELD   A centrifugal force appears whenever the direction of movement of a particle is changed. The acceleration from a centrifugal force from circular motion is

$$a_e = r\omega^2 \tag{7-33}$$

where r = radius of path of particle
ω = angular velocity, rad/s

Substituting into Eq. (7-31) gives

$$\frac{du}{dt} = r\omega^2 \frac{\rho_p - \rho}{\rho_p} - \frac{C_D u^2 \rho A_p}{2m} \tag{7-34}$$

In this equation, u is the velocity of the particle relative to the fluid, and is directed outwardly along a radius.

Terminal velocity   In gravitational settling, g is constant. Also, the drag always increases with velocity. Equation (7-32) shows that the acceleration decreases with time and approaches zero. The particle quickly reaches a constant velocity, which is the maximum attainable under the circumstances, and which is called the *terminal velocity*. The equation for the terminal velocity u_t is found, for gravitational settling, by taking $du/dt = 0$. Then, from Eq. (7-32),

$$u_t = \sqrt{\frac{2g(\rho_p - \rho)m}{A_p \rho_p C_D \rho}} \tag{7-35}$$

In motion from a centrifugal force, the velocity depends on the radius, and the acceleration is not constant if the particle is in motion with respect to the fluid. In many practical uses of centrifugal force, however, du/dt is small in comparison with the other two terms in Eq. (7-34), and if du/dt is neglected, a terminal velocity at any given radius can be defined by the equation

$$u_t = \omega \sqrt{\frac{2r(\rho_p - \rho)m}{A_p \rho_p C_D \rho}} \tag{7-36}$$

Drag coefficients   The quantitative use of Eqs. (7-31) to (7-36) requires that numerical values be available for the drag coefficient C_D. Figure 7-3, which shows the drag coefficient as a function of Reynolds number, indicates such a relationship. A portion of the curve of C_D vs. $N_{Re,p}$ for spheres is reproduced in Fig. 7-6. The drag curve shown in Fig. 7-6 applies, however, only under restricted conditions. The particle must be a solid sphere, it must be free to move without being influenced by other particles or by the wall or bottom of the vessel, it must be moving at a constant velocity, it must not be too small, and the fluid through which it is moving must be still.

Variations in particle shape can be accounted for by obtaining separate curves of C_D vs. $N_{Re,p}$ for each shape, as shown in Fig. 7-3 for cylinders and disks. As pointed out earlier, however, the curves for cylinders and disks in Fig. 7-3 apply only to a specified orientation of the particle. In the free motion of nonspherical particles

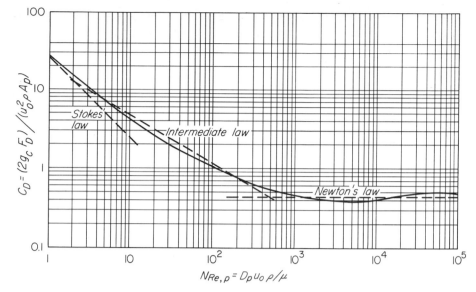

FIGURE 7-6
Drag coefficients for spheres.

through a fluid the orientation is constantly changing. This change consumes energy, increasing the effective drag on the particle, and C_D is greater than for the motion of the fluid past a stationary particle. As a result the terminal velocity, especially with disks and other platelike particles, is less than would be predicted from curves for a fixed orientation.

In the following treatment the particles will be assumed to be spherical, for once the drag coefficients for free particle motion are known, the same principles apply to any shape.[3,14]

When the particle is at sufficient distance from the boundaries of the container and from other particles, so that its fall is not affected by them, the process is called *free settling*. If the motion of the particle is impeded by other particles, which will happen when the particles are near each other even though they may not actually be colliding, the process is called *hindered settling*. The drag coefficient in hindered settling is greater than in free settling.

If the particle is accelerating, the drag is influenced by the changes in velocity gradients near the surface of the particle.[4] This causes a drag greater than that shown in Fig. 7-6 for the same Reynolds number. Also, if the particles are liquid drops, circulatory currents are generated in them, and rapid oscillations or changes in drop shape may occur. Additional drag is necessary to supply the energy required to maintain these motions in the drop itself.

If the particles are very small, Brownian movement appears. This is a random motion imparted to the particle by collisions between the particle and the molecules of the surrounding fluid. This effect becomes appreciable at a particle size of about

2 to 3 μm and predominates over the force of gravity with a particle size of 0.1 μm or less. The random movement of the particle tends to suppress the effect of the force of gravity, so settling does not occur. Application of centrifugal force reduces the relative effect of Brownian movement.

MOTION OF SPHERICAL PARTICLES   If the particles are spheres of diameter D_p,

$$m = \frac{\pi D_p^3 \rho_p}{6} \tag{7-37}$$

and

$$A_p = \frac{\pi D_p^2}{4} \tag{7-38}$$

Substitution of m and A_p from Eqs. (7-37) and (7-38) into Eq. (7-31) gives

$$\frac{du}{dt} = a_e \frac{\rho_p - \rho}{\rho_p} - \frac{3C_D u^2 \rho}{4\rho_p D_p} \tag{7-39}$$

At the terminal velocity, $du/dt = 0$, and

$$a_e(\rho_p - \rho) = \frac{3C_D u_t^2 \rho}{4D_p} \tag{7-40}$$

APPROXIMATE EQUATIONS FOR DRAG COEFFICIENTS OF SPHERES   Although the relationship of C_D vs. $N_{Re,p}$ in Fig. 7-3 is a continuous curve, for use in calculations it can be replaced by three straight lines without serious loss in accuracy. These lines, each of which covers a definite range of Reynolds numbers, are shown as dotted lines in Fig. 7-6. The equations for the lines and the ranges of the Reynolds number over which each applies are, for $N_{Re,p} < 2$,

$$C_D = \frac{24}{N_{Re,p}} \tag{7-41}$$

and

$$F_D = \frac{3\pi \mu u_t D_p}{g_c} \tag{7-42}$$

This is the Stokes'-law range. For $2 < N_{Re,p} < 500$, the equations are

$$C_D = \frac{18.5}{N_{Re,p}^{0.6}} \tag{7-43}$$

and

$$F_D = \frac{2.31\pi (u_t D_p)^{1.4} \mu^{0.6} \rho^{0.4}}{g_c} \tag{7-44}$$

This is the intermediate range. For $500 < N_{Re,p} < 200,000$,

$$C_D = 0.44 \tag{7-45}$$

and

$$F_D = \frac{0.055\pi (u_t D_p)^2 \rho}{g_c} \tag{7-46}$$

This is Newton's-law range.

Equations (7-41), (7-43), and (7-45) can be written in the single general form

$$C_D = \frac{b_1}{N_{\text{Re},p}^n}$$ (7-47)

Then Eqs. (7-42), (7-44), and (7-46) all can be written as

$$F_D = \frac{\mu^n b_1 \pi (D_p u_t)^{2-n} \rho^{1-n}}{8 g_c}$$ (7-48)

where b_1 and n are constants, summarized in Table 7-1.

A general equation for the terminal velocity of spheres is obtained by substituting C_D from Eq. (7-47) into Eq. (7-40) and solving for u_t.

$$a_e(\rho_p - \rho) = \frac{3 u_t^2 b_1 \rho \mu^n}{4 D_p (D_p u_t \rho)^n}$$

and

$$u_t = \left[\frac{4 a_e D_p^{1+n}(\rho_p - \rho)}{3 b_1 \mu^n \rho^{1-n}} \right]^{1/(2-n)}$$ (7-49)

For gravity settling, $a_e = g$, and for centrifugal settling, $a_e = \omega^2 r$. The special cases of Eq. (7-49) are for Stokes'-law range

$$u_t = \frac{a_e D_p^2(\rho_p - \rho)}{18 \mu}$$ (7-50)

for intermediate range

$$u_t = \frac{0.153 a_e^{0.71} D_p^{1.14}(\rho_p - \rho)^{0.71}}{\rho^{0.29} \mu^{0.43}}$$ (7-51)

and for Newton's-law range

$$u_t = 1.75 \sqrt{\frac{a_e D_p(\rho_p - \rho)}{\rho}}$$ (7-52)

If the terminal velocity of a particle of known diameter is desired, the Reynolds number is unknown and a choice of equation cannot be made. To identify the range

Table 7-1 CONSTANTS IN EQUATIONS FOR DRAG COEFFICIENTS

Range	b_1	n
Stokes'-law	24	1
Intermediate	18.5	0.6
Newton's-law	0.44	0

in which the motion of the particle lies, the velocity term is eliminated from the Reynolds number by substituting u_t from Eq. (7-50) to give, for the Stokes'-law range,

$$N_{Re,p} = \frac{D_p u_t \rho}{\mu} = \frac{D_p^3 a_e \rho(\rho_p - \rho)}{18\mu^2} \tag{7-53}$$

If Stokes' law is to apply, $N_{Re,p}$ must be less than 2.0. To provide a convenient criterion K, let

$$K = D_p \left[\frac{a_e \rho(\rho_p - \rho)}{\mu^2}\right]^{1/3} \tag{7-54}$$

Then, from Eq. (7-53), $N_{Re,p} = K^3/18$. Setting $N_{Re,p}$ equal to 2.0 and solving gives $K = 36^{1/3} = 3.3$. If the size of the particle is known, K can be calculated from Eq. (7-54). If K so calculated is less than 3.3, Stokes' law applies.

Substitution for u_t from Eq. (7-52) shows that for the Newton's-law range $N_{Re,p} = 1.74K^{1.5}$. Setting this equal to 500 and solving gives $K = 43.6$. Thus if K is greater than 3.3 but less than 43.6, the intermediate law applies; if K is between 43.6 and 2,360, Newton's Law must be chosen. When K is greater than 2,360, the drag coefficient may change abruptly with small changes in fluid velocity. Under these conditions the terminal velocity is calculated from Eq. (7-40), using a value of C_D found by trial from Fig. 7-3.

EXAMPLE 7-1   Drops of oil 15 μm in diameter are to be settled from their mixture with air. The specific gravity of the oil is 0.90, and the air is at 70°F (21.1°C) and 1 atm. A settling time of 1 min is available. How high should the chamber be to allow settling of these particles?

SOLUTION   The effects of flow within the drops and of the initial period of acceleration are neglected. Also, the density of the air is so small in comparison with that of the drops that ρ_p can be used in place of $\rho_p - \rho$. The quantities for use in Eq. (7-54) are

$$D_p = \frac{15 \times 10^{-4}}{30.48} = 4.92 \times 10^{-5} \text{ ft}$$

The density of air at 70°F and 1 atm is 0.075 lb/ft^3. From Appendix 9, the viscosity of the air is 0.018 cP, and

$$\mu = 0.018 \times 6.72 \times 10^{-4} = 1.21 \times 10^{-5} \text{ lb/ft-s}$$

Also $\rho_p = 0.90 \times 62.37 = 56.1 \text{ lb/ft}^3$

From Eq. (7-54)

$$K = 4.92 \times 10^{-5} \left(\frac{32.17 \times 56.1 \times 0.075}{1.21^2 \times 10^{-10}}\right)^{1/3} = 0.479$$

The motion of the drops is well within the Stokes'-law range, and Eq. (7-50) can be used to calculate the terminal velocity.

$$u_t = \frac{32.17 \times 4.92^2 \times 10^{-10} \times 56.1}{18 \times 1.21 \times 10^{-5}} = 0.020 \text{ ft/s (0.0061 m/s)}$$

In 1 min the particles will settle $0.02 \times 60 = 1.2$ ft (0.37 m), so the height of the chamber should not be greater than this.   ////

Hindered settling In hindered settling the particles are sufficiently close together to cause the velocity gradients surrounding each particle to be affected by the presence of the neighboring units. Also, the particles in settling displace liquid and generate an appreciable upward velocity. The velocity of the liquid is, then, greater with respect to the particle than with respect to the apparatus. The effective density of the fluid can be taken as that of the slurry itself and can be calculated from the composition of the slurry and the densities of the particles and of the fluid.

In hindered settling of spherical particles the drag force given by Eq. (7-48) is too small. The true drag force corresponds to that which would exist in free settling through a fluid of higher viscosity than that actually used. To find this higher effective viscosity the actual viscosity μ is divided by an empirical correction factor ψ_p, which depends on the fractional volume of the slurry occupied by the liquid. This is equivalent to the porosity of the aggregation of particles, and is denoted by ε. Then, mathematically,

$$F_D = \frac{\pi b_1 (D_p u_{tr})^{2-n} \rho_m^{1-n} (\mu/\psi_p)^n}{8 g_c} \tag{7-55}$$

where u_{tr} = terminal velocity of particle *relative to fluid*
ψ_p = function of the porosity ε
ρ_m = density of slurry

The terminal velocity with respect to the apparatus u_t is less than that with respect to the fluid u_{tr}, and the relationship between the velocities is

$$u_t = \varepsilon u_{tr} \tag{7-56}$$

The drag is equal to the difference between the external and buoyant forces, so, from Eqs. (7-28), (7-29), (7-37), (7-55), and (7-56),

$$\frac{\pi a_e D_p^3 (\rho_p - \rho_m)}{6 g_c} = \frac{\pi b_1 (D_p u_t)^{2-n} \mu^n \rho_m^{1-n}}{8 \varepsilon^{2-n} g_c (\psi_p)^n}$$

Solution of this equation for u_t gives

$$u_t = \left[\frac{4 a_e D_p^{1+n} \varepsilon^{2-n} (\psi_p)^n (\rho_p - \rho_m)}{3 b_1 \rho_m^{1-n} \mu^n} \right]^{1/(2-n)} \tag{7-57}$$

A general relation between ε and ψ_p over the entire range of Reynolds numbers has not been determined. A relationship for the settling of spheres in the Stokes'-law range has been found[16] and is given by the empirical equation

$$\psi_p = e^{-4.19(1-\varepsilon)} \tag{7-58}$$

This relationship does not apply to sharp-edged particles.

Assuming that the Stokes'-law range terminates at a Reynolds number of 2.0, based on velocity u_{tr}, that of the particle relative to the fluid, the criterion for hindered settling in the Stokes'-law range is

$$K = D_p \left[\frac{a_e \rho_m (\rho_p - \rho_m)(\psi_p)^2}{\mu^2} \right]^{1/3} \tag{7-59}$$

If $K < 3.3$, the settling is in the Stokes'-law range.

EXAMPLE 7-2   Particles of sphalerite (specific gravity = 4.0) are settling under the force of gravity through a slurry consisting of 25 percent by volume of quartz particles (specific gravity = 2.65) and water. The diameter of the sphalerite particles is 0.006 in. (0.15 mm). The volumetric ratio of sphalerite to slurry is 0.25. The temperature is 50°F (10°C). What is the terminal velocity of the sphalerite?

   SOLUTION   The density of the settling medium is that of the slurry. This can be calculated from the composition. The specific gravity of the slurry is $0.25 \times 2.65 + 0.75 = 1.41$, and the specific-gravity difference between particles and medium is $4.00 - 1.41 = 2.59$. The density difference $\rho_p - \rho_m$ is $62.37 \times 2.59 = 161.6$ lb/ft^3. The density of the medium ρ_m is $62.37 \times 1.41 = 87.9$ lb/ft^3. The viscosity of water at 50°F is, from Appendix 10, 1.310 cP. The porosity is $\varepsilon = 1 - 0.25 = 0.75$. Correction factor ψ_p, from Eq. (7-58), is

$$\psi_p = e^{-4.19(1-0.75)} = 0.35$$

The criterion K is, from Eq. (7-59),

$$K = \frac{0.006}{12} \left[\frac{32.17 \times 161.6 \times 87.9 \times 0.35^2}{(6.72 \times 1.31 \times 10^{-4})^2} \right]^{1/3} = 2.1$$

The settling occurs in the Stokes'-law range.

   The terminal velocity can be calculated from Eq. (7-57), taking $n = 1$ and $b_1 = 24$:

$$u_t = \frac{4 \times 32.17(0.006/12)^2 \times 0.75 \times 0.35 \times 161.6}{3 \times 24 \times 1.31 \times 6.72 \times 10^{-4}} = 0.022 \text{ ft/s (67 mm/s)} \qquad ////$$

Settling and rise of bubbles and drops   Unlike solid particles, dispersed drops of liquid or bubbles of gas may change shape as they move through a continuous phase. With liquid drops the distortion is relatively small: small drops tend to be spherical, while larger ones, especially when moving through a gas at high velocities, are somewhat flattened in the direction of flow. (The familiar teardrop shape of the cartoonist is entirely imaginary.) When the drops are large, the terminal velocity of drops falling or rising through a gas or another liquid can normally be estimated with sufficient accuracy from Eqs. (7-50) to (7-52) for spheres.

   Very small gas bubbles, smaller than 0.2 mm in diameter, are spherical and move in accordance with Stokes' law [Eq. (7-50)]. Larger bubbles, up to about 2 mm in diameter, are also spherical but tend to rise in a spiral path. Still larger bubbles become flattened ellipsoids; very large ones, more than 10 mm in diameter, are lens-shaped, helmet-shaped, or hemispherical. The shape of ellipsoidal or lens-shaped bubbles changes as they rise through the liquid.

   Large bubbles in thin liquids all rise at more or less the same velocity, regardless of their size. The buoyant force on a large bubble is greater than on a small one, but since their spiral motion and changes of shape consume additional energy, they rise much more slowly than rigid spheres under otherwise similar conditions. Figure 7-7 shows how single air bubbles in water, even bubbles as small as 1 mm in diameter, rise more slowly than predicted from Stokes' law and how their rise velocity peaks at a diameter of about 1.8 mm then falls and levels off at about 0.24 m/s (0.8 ft/s) for all bubble diameters greater than about 3 mm. As expected, swarms or clouds of bubbles 4 mm in average diameter rise more slowly than single bubbles, but swarms

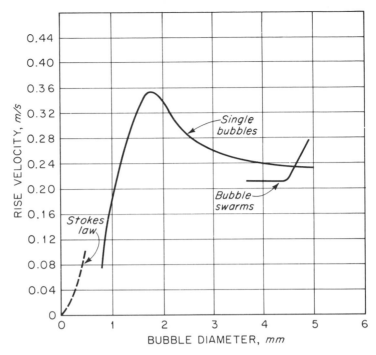

FIGURE 7-7
Rise velocity of air bubbles in water. [*By permission, from J. H. Perry (ed.), "Chemical Engineers' Handbook," 5th ed., pp.* **18**–71, *Copyright,* © 1973, *McGraw-Hill Book Company.*]

of larger bubbles rise more rapidly than expected. Bubble behavior is complex, depending on the nature of the continuous liquid phase and on the amount and nature of materials dissolved in it. Results similar to those shown in Fig. 7-7, however, have been reported for a variety of gases and liquids.[13]

FLUIDIZATION

A liquid or a gas flowing at low velocities through a porous bed of solid particles, as in a packed tower, does not cause the particles to move. The fluid passes through the small, tortuous channels, losing pressure energy. The pressure drop in a stationary bed of solids is given by the Ergun equation (7-20). If the fluid velocity is steadily increased, however, a point is eventually reached at which the particles no longer remain stationary but "fluidize" under the action of the liquid or gas.

Mechanism of fluidization   Visualize a short vertical tube partly filled with a granular material such as fine sand. Air is being admitted at a very low rate to the

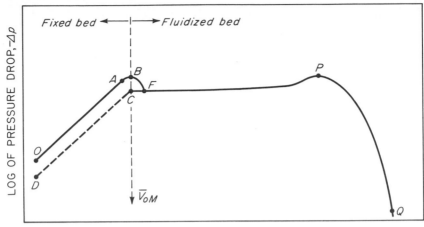

FIGURE 7-8
Pressure drop versus superficial gas velocity in a bed of solids. [*By permission, from D. Kunii and O. Levenspiel, "Fluidization Engineering," p. 74. Copyright, © 1969, John Wiley & Sons.*]

bottom of the tube and flows upward through the sand without causing the grains to move. The air rate is now slowly and progressively increased. As the velocity rises, the pressure drop in the air passing through the bed increases, as is shown by the linear segment OA in Fig. 7-8. Eventually the pressure drop equals the force of gravity on the particles, and the grains begin to move. This is at point A on the graph. First the bed expands slightly with the grains still in contact. The porosity increases, and the pressure drop rises more slowly than before. When point B is reached, the bed is in the loosest possible condition with the grains still in contact. As the velocity is still further increased, the grains separate and true fluidization begins. The pressure drop sometimes diminishes a little from point B to point F. From point F onward the particles move more and more vigorously, swirling about and traveling in random directions. The contents of the tube strongly resemble a boiling liquid, and the name *boiling bed* has been given to solids fluidized in this way.

If the gas velocity is now reduced to and below $\bar{V}_{OM}$, the minimum velocity required for fluidization, the bed collapses once more; the pressure drop again follows the relationship for a fixed bed. The porosity may be higher than in the original bed, however, and the pressure drop for a given gas velocity lower than it was before. In the expanded bed the pressure drop is given by the line DC.

In a fluidized bed the linear velocity of the fluid between the particles is much higher than the velocity in the space above the bed. Consequently nearly all the particles drop out of the fluid above the bed. Even with vigorous fluidization only the smallest grains are entrained in the fluid and carried away. Suppose now, however, that the fluid velocity is still further increased. The porosity of the bed rises, the bed of solids expands, and its density falls.

Beyond point P entrainment becomes appreciable, then severe, then complete. At point Q the porosity approaches unity, the bed ceases to exist, and the phenomenon becomes the simultaneous flow of two phases. From point F to point P the pressure drop is nearly constant, as explained later.

Fluidization without solids entrainment is called *dense-phase fluidization*; when entrainment is complete, the process is called *disperse-phase* or *lean-phase fluidization with pneumatic transport*.

Dense-phase fluidization and boiling beds   The behavior of a mass of fluidized solids depends on the particle size of the solids and the nature of the fluid. With large particles, fluidized by either a gas or a liquid, fluidization begins as a gentle rocking or oscillation of the solid particles. In the fully fluidized bed the particles move in random directions through all parts of the liquid. There are strong transient currents in the bed, with many particles traveling temporarily in the same direction, but in general the particles move as individuals. This mode of action is called *particulate fluidization*.

When the fluid is a gas, the action in the bed is somewhat different and is strongly influenced by the particle size. Under conditions for good fluidization (with particles of the proper size and density) some of the gas travels through the bed between individual particles, but much of it travels through in bubbles, or pockets, containing almost no solids. At the bed surface the bubbles break, splashing individual particles or streamers into the space above. In the bed itself the particles move in distinct aggregates, which are lifted by the bubbles or which move aside to let the bubbles past. A fluidizing action of this kind is known as *aggregative fluidization*.

Fluidized beds are most used with fine particles, smaller than 20-mesh (0.8 mm) and containing appreciable amounts of fines smaller than 200-mesh (0.07 mm). In large units the bed may be 10 to 15 m deep. Coarser particles, larger than 0.25 mm and containing few fines, may be fluidized in a shallow bed, perhaps 1 m deep, in what is called a *teeter bed*, used most in roasters and some kinds of gas generators. Still larger particles, such as grains and peas, may be dried in a *spouted bed*, in which a jetlike central column of gas raises the particles to the top of the vessel; from there they percolate slowly downward near the vessel wall and are reentrained by the rising column of gas.

When particles are fluidized in a tall narrow vessel, *slugging* may occur. Bubbles of gas tend to coalesce and grow as they rise through the fluidized bed. The rate of growth depends on the size and density of the particles: it is rapid when the particles are large and heavy, slow when they are small and light. If a vessel that is small in diameter contains a deep bed of solids, the bubbles may grow until they fill the entire cross section of the vessel. Successive bubbles then travel up the vessel, separated by slugs of solid particles. Operation is erratic and unstable. Slugging can be avoided by the proper choice of particle size and by using shallow beds of solids.

Minimum porosity   In the discussion of Fig. 7-8 it was pointed out that at the onset of fluidization the bed expands somewhat from its static condition before true fluidization begins. The porosity increases over that for the static bed. The porosity

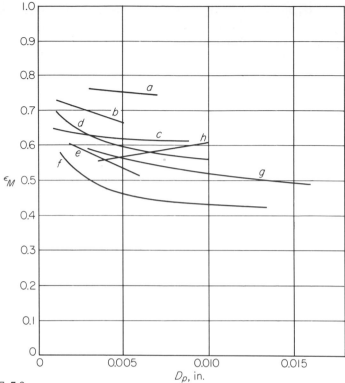

FIGURE 7-9

Minimum porosity for fluidization versus particle size: (*a*) soft brick; (*b*) adsorption carbon; (*c*) broken Raschig rings; (*d*) coal and glass powder; (*e*) carborundum; (*f*) round sand; (*g*) sharp sand; (*h*) coke. [*By permission, from M. Leva, "Fluidization," p. 21. Copyright, © 1959, McGraw-Hill Book Company.*]

of the bed when true fluidization begins is called the *minimum porosity for fluidization* and is designated by ε_M. Figure 7-9 shows the minimum porosity for various materials. It depends on the shape and size of the particles, usually (though not always) decreasing as the particle diameter increases.

Bed height   When the fluid velocity is increased above the minimum required for fluidization, the bed expands and the porosity increases. If the cross-sectional area of the vessel does not change with height, the porosity is a direct function of the height of the bed. Let L_0 be the height the bed would have if the porosity were zero, i.e., if the solids existed as a single lump containing no voids. If L is the height of the fluidized bed, the porosity is given by

$$\varepsilon = \frac{L - L_0}{L} = 1 - \frac{L_0}{L} \tag{7-60}$$

Often the porosity at one condition, such as the minimum porosity for fluidization or the porosity of the static bed, is known. If the corresponding bed height is also known, the bed height for any new value of the porosity can be found from

$$L_2 = L_1 \frac{1 - \varepsilon_1}{1 - \varepsilon_2} \tag{7-61}$$

where ε_1 and ε_2 are the porosities at heights L_1 and L_2, respectively.

Pressure drop in fluidized bed   When fluidization just begins, the pressure drop through the bed counterbalances the force of gravity on the solids. The actual pressure drop may be a little greater than this would indicate because of friction between particles, energy loss by collisions, or electrostatic effects[7a,11] but as a first approximation the pressure drop at incipient fluidization can be found by equating the force it exerts on the solids to the force of gravity minus the buoyant force of the displaced fluid. If L_M is the bed height at incipient fluidization, S its cross-sectional area, and ε_M the minimum porosity for fluidization,

$$(p_a - p_b)S = \frac{g}{g_c} [\rho_p(1 - \varepsilon_M)L_M S - \rho(1 - \varepsilon_M)L_M S] \tag{7-62}$$

Solving Eq. (7-62) for the pressure drop per foot of height gives

$$-\frac{\Delta p}{L_M} = \frac{p_a - p_b}{L_M} = \frac{g}{g_c} (1 - \varepsilon_M) (\rho_p - \rho) \tag{7-63}$$

As the fluid velocity increases above that required to initiate fluidization, the pressure drop increases slightly, but the change is small, and it is usually satisfactory to consider the pressure drop to be constant and equal to that given by Eq. (7-63). Since the bed expands, however, the pressure drop per foot of expanded bed height falls. If L and ε are the height and porosity, respectively, of the expanded bed, the pressure drop per foot of bed height is found by combining Eqs. (7-61) and (7-63), to give

$$-\frac{\Delta p}{L} = \frac{g}{g_c} (1 - \varepsilon) (\rho_p - \rho) \tag{7-64}$$

Rearranging Eq. (7-64) shows that

$$-\frac{\Delta p}{L(1 - \varepsilon)} = \frac{g}{g_c} (\rho_p - \rho) = \text{const} \tag{7-65}$$

Minimum fluidization velocity   The gas velocity $\bar{V}_{OM}$ at which fluidization begins can be estimated by a small extrapolation of the Ergun equation for packed beds, combined with the pressure-drop equation (7-63), to give

$$\frac{1.75 D_p^2 \rho^2}{\Phi_s \varepsilon^3 \mu^2} \bar{V}_0^2 + \frac{150 D_p \rho(1 - \varepsilon)}{\Phi_s^2 \varepsilon^3 \mu} \bar{V}_0 - \frac{D_p^3 g \rho(\rho_p - \rho)}{\mu^2} = 0 \tag{7-66}$$

If Φ_s and ε_M are known, $\overline{V}_{OM}$ (at which $\varepsilon = \varepsilon_M$) is found by solving this quadratic expression. If Φ_s and ε_M are not known, the following equations may be used.[17] For conditions at the onset of fluidization, Eq. (7-66) may be written

$$\frac{1.75}{\Phi_s \varepsilon_M^3} N_{\text{Re},pM}^2 + \frac{150(1 - \varepsilon_M)}{\Phi_s^2 \varepsilon_M^3} N_{\text{Re},pM} - K^3 = 0 \tag{7-67}$$

where K is the settling criterion defined by Eq. (7-54). For many types and sizes of solid particles, quantity $\Phi_s \varepsilon_M^3$ is approximately $\frac{1}{14}$ and $(1 - \varepsilon_M)/\Phi_s^2 \varepsilon_M^3$ is approximately 11. Substituting these values in Eq. (7-67) gives

$$N_{\text{Re},pM} = (33.67^2 + 0.0408K^3)^{1/2} - 33.67 \tag{7-68}$$

For small particles, for which $N_{\text{Re},pM} < 20$, the first term in Eq. (7-66) is negligible, and

$$\overline{V}_{OM} = \frac{D_p^2 g (\rho_p - \rho)}{1,650\mu} \tag{7-69}$$

With large particles, for which $N_{\text{Re},pM} > 1,000$, the second term in Eq. (7-66) may be neglected, and

$$\overline{V}_{OM} = \left[\frac{D_p g (\rho_p - \rho)}{24.5\rho} \right]^{1/2} \tag{7-70}$$

Expansion of fluidized beds   The variation of porosity (and hence of bed height) with the superficial fluid velocity through a fluidized bed can be estimated as follows, with certain restrictions. Consider the fluidization of a bed of small particles, for which $N_{\text{Re},pM} < 20$. Assume that Eq. (7-66), with the first term neglected, applies over the entire range of fluidizing velocities. Then, since all terms except ε are constant,

$$\overline{V}_0 = \frac{D_p^2 g (\rho_p - \rho)\Phi_s^2}{150\mu} \frac{\varepsilon^3}{1 - \varepsilon} = k_3 \frac{\varepsilon^3}{1 - \varepsilon} \tag{7-71}$$

where k_3 is a constant for the system.

Equation (7-71) has been found by experiment to hold very well for particulate fluidization of solids with liquids, provided $\varepsilon < 0.80$. It applies much less well to aggregative fluidization of fine solids with a gas, for two reasons: (1) Small particles may clump together to form aggregates which are larger in effective diameter than the individual particles, and the amount of this aggregation depends on the fluid velocity. (2) In aggregative fluidization much of the fluidizing gas passes through the bed in bubbles containing virtually no solids. To this type of flow system the Ergun equation cannot be expected to apply.

Empirically it has been found that the porosity varies with some power of the fluid velocity, giving a linear logarithmic plot of ε vs. $\overline{V}_0$ of the type shown in Fig. 7-10. Similar plots apply to both particulate and aggregative fluidization.[9a] Because the depth of the fluidized bed is not constant, however, but fluctuates considerably, the measured porosity is a somewhat indefinite time-averaged value and the various correlations that have been proposed do not agree well with one another.[7c] With

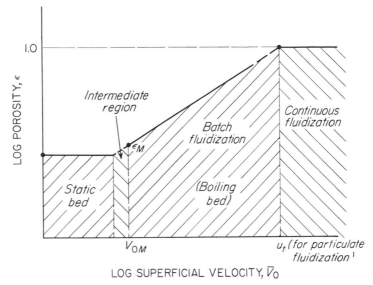

FIGURE 7-10
Porosity of fluidized beds.

coarse particles fluidized by a liquid the value of $\overline{V}_0$ at which ε becomes unity, as indicated by extrapolation of the graph, agrees well with the terminal settling velocity u_t of a single particle through the liquid.[10] For fluidization by a gas the relationships between minimum fluidization velocity and terminal settling velocity are discussed in the following section.

Range of fluidization velocities The terminal settling velocity of fluidized particles is difficult to estimate with precision, depending as it does on the sphericity of the particles. For spheres, however, the limiting ratios of terminal velocity u_t to the minimum fluidization velocity V_{OM} are readily calculated. For small particles, to which Stokes' law applies, combining Eqs. (7-50) and (7-69) gives $u_t/V_{OM} = 1,650/18 = 91.7$. For larger particles, Newton's law [Eq. (7-52)] is combined with Eq. (7-70) to give $u_t/V_{OM} = 1.74\sqrt{24.5} = 8.6$. The operating range of fluidization velocities is thus much greater with small particles than with large ones. This is illustrated in Fig. 7-11, in which u_t/V_{OM} is plotted against $C_D N_{Re,p}^2$ or $4D_p^3 g\rho(\rho_p - \rho)/3\mu^2$.

In aggregative fluidization it is sometimes possible to operate at superficial velocities several times the settling velocity of the particles without serious particle entrainment. This is because most of the gas flows through the bed in large bubbles which contain almost no solids while the particles in the dense phase are suspended by relatively slow-moving gas. Operating under these conditions sacrifices much of the benefit from fluidization, and the bubbles act largely as a waste of reactor volume.

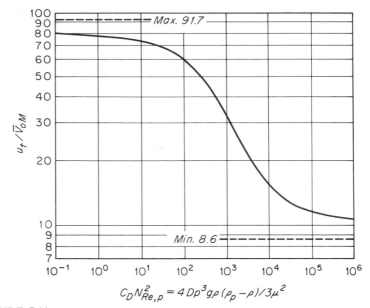

$$C_D N_{Re,p}^2 = 4 D p^3 g \rho (\rho_p - \rho)/3\mu^2$$

FIGURE 7-11
Ratio of terminal settling velocity to minimum fluidization velocity. [*By permission, from D. Kunii and O. Levenspiel, "Fluidization Engineering," p. 78. Copyright,* © *1969, John Wiley & Sons.*]

EXAMPLE 7-3   A bed of sized pulverized coal 35-mesh in particle size is to be fluidized with a liquid petroleum fraction having a viscosity of 15 cP. The height of the static bed is 6 ft (1.83 m); the porosity of the static bed is 0.38. The density of the coal particles is 84 lb/ft³ (1,346 kg/m³); that of the liquid is 55 lb/ft³ (881 kg/m³). Calculate the pressure drop required for fluidization.

SOLUTION   Since the particles are relatively coarse and are fluidized by a liquid, the minimum porosity for fluidization ε_M may be taken as that for the static bed, 0.38. The pressure drop is found from Eq. (7-63). Since $L_M = 6$ ft and g/g_c is unity,

$$-\Delta p = 6(1 - 0.38)(84 - 55) = 108 \; lb_f/ft^2 = 0.75 \; lb_f/in.^2 \qquad ////$$

EXAMPLE 7-4   A bed containing 36 tons (32,700 kg) of 100-mesh sharp sand is to be fluidized with air at 400°C and a pressure of 17 atm abs in a cylindrical vessel 10 ft (3.05 m) in diameter. The ultimate density of the sand particles is 168 lb/ft³ (2,690 kg/m³). The viscosity of air at the operating conditions is 0.032 cP. Call g/g_c unity. Calculate (a) the minimum porosity for fluidization, (b) the minimum height of the fluidized bed, (c) the pressure drop in the bed, (d) the critical superficial air velocity.

SOLUTION   The diameter of 100-mesh particles (Appendix 20) is 0.0058 in., or 4.83×10^{-4} ft. The density of air is

$$\rho = \frac{29}{359} \frac{273}{273 + 400} \frac{17}{1} = 0.557 \; lb/ft^3$$

The viscosity of air is

$$\mu = 0.032 \times 0.000672 = 2.15 \times 10^{-5} \text{ lb/ft-s}$$

(a) For a particle diameter of 0.0058 in. the minimum porosity ε_M, from curve g of Fig. 7-9, is 0.55.

(b) The minimum height of the fluidized bed is found as follows.

$$\text{Volume of solids} = \frac{36 \times 2,000}{168} = 428.6 \text{ ft}^3$$

The height this volume of solids would occupy in the tower, with a porosity of zero, is

$$L_0 = \frac{4 \times 428.6}{\pi 10^2} = 5.45 \text{ ft}$$

The height of the fluidized bed, from Eq. (7-61) with $\varepsilon_0 = 0$, is

$$L_M = 5.45 \frac{1}{1 - 0.55} = 12.1 \text{ ft (3.69 m)}$$

(c) The pressure drop, from Eq. (7-63), is

$$-\Delta p = 12.1(1 - 0.55)(168 - 0.557) = 912 \text{ lb}_f/\text{ft}^2 = 0.431 \text{ atm}$$

(d) The critical velocity for the fluidization of small particles is given by Eq. (7-71):

$$\bar{V}_{OM} = \frac{32.17(168 - 0.557)(4.83 \times 10^{-4})^2 \times 0.55^3}{150 \times 2.15 \times 10^{-5}(1 - 0.55)} = 0.144 \text{ ft/s (0.0439 m/s)} \quad ////$$

Applications of batch fluidization Extensive use of fluidization began in the catalytic-cracking reactors in the petroleum industry. Fluidization now finds application in many catalytic processes and in other operations, such as the drying of solids, as well.[1,9,18] The chief advantages of fluidization are that it ensures contact of the fluid with all parts of the solid particles, prevents segregation of the solids by thoroughly agitating the bed, and minimizes temperature variations even in a large reactor, again by virtue of the vigorous agitation.

When a fluid passes through a static bed of solids, some of the surface of the particles is screened from direct contact with the fluid. If the solid is a catalyst, some of its effectiveness is lost. In addition, if heat is generated in a static bed by an exothermic reaction, strong temperature gradients and often a localized zone of very high temperature may be formed. Such a zone is known as a *hot spot*. Since the hot-spot temperature must not exceed a set value, which is usually close to the optimum temperature for the reaction, this means that much of the catalyst is operating at a temperature well below the optimum. In a fluidized bed, by contrast, the temperature throughout the mass rarely varies by more than 3 to 5°C. Because of contact of the fluid with all the catalyst surface and the elimination of hot spots, a pound of fluidized catalyst may convert from 2 to 10 times as much as a pound of the same catalyst in a static condition. In drying crystalline solids, for the same reasons, the drying rate may be several times that in a static bed.

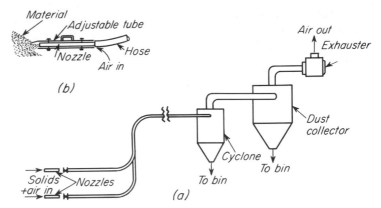

FIGURE 7-12
Pneumatic conveying system: (*a*) typical multiple-inlet system; (*b*) nozzle detail.

Offsetting these advantages are the increased power needed because of the higher pressure drop in a fluidized bed, the increased size of the vessel or reactor, and increased breakage of the solid particles. Facilities must usually be provided for recovering the fine solids carried out of the fluidized bed.

Continuous fluidization and pneumatic transport   When the fluid velocity through a bed of solids becomes large enough, all the particles are entrained in the fluid and are carried along with it, to give *continuous fluidization*. Its principal application is in transporting solids from point to point in a factory or plant. Sometimes the fluid is a liquid in which solids are suspended to form a pumpable slurry. More commonly, however, the suspending fluid is a gas, usually air, flowing at velocities between 50 and 100 ft/s in pipes ranging from 2 to 16 in. in diameter. A typical conveyor system is illustrated in Fig. 7-12.

Pressure drop in pneumatic conveyors   To pass air alone through a pneumatic conveying system at the velocities normally used requires a relatively small amount of energy. To lift and move the solid requires considerable additional energy. For a mass ratio of solids to air, r, this additional energy requirement, by a mechanical-energy balance based on Eq. (4-34), is

$$E_s = r\left[\frac{p_b - p_a}{\rho_s} + \frac{V_{sb}^2 - V_{sa}^2}{2g_c} + \frac{g}{g_c}(Z_b - Z_a)\right] \tag{7-72}$$

where V_{sa} = velocity of solids at inlet
  V_{sb} = velocity of solids at outlet
  ρ_s = density of solid

The energy E_s is supplied by the air. It is transmitted to the solid particles through the action of drag forces between the air and the solid. The energy E_s is a work term, and it must appear in the mechanical-energy balance for the air.

Assuming the pressure drop is a small fraction of the absolute pressure, the air can be considered to be an incompressible fluid of constant density $\bar{\rho}$, the average density of the air between the inlet and outlet. When the change in velocity head is neglected, when the kinetic-energy factor is assumed to be unity, and when E_s is allowed for, the Bernoulli equation (4-34) becomes, for a unit mass of air,

$$\frac{p_b - p_a}{\bar{\rho}} + \frac{g}{g_c}(Z_b - Z_a) = -E_s - h_f \tag{7-73}$$

where h_f is the total friction in the stream. Eliminating E_s from Eqs. (7-72) and (7-73) and solving for $p_a - p_b$ gives

$$p_a - p_b = \frac{(g/g_c)(1 + r)(Z_b - Z_a) + r(V_{sb}^2 - V_{sa}^2)/2g_c + h_f}{1/\bar{\rho} + r/\rho_s} \tag{7-74}$$

Methods of calculating the friction loss h_f are discussed in the literature.[9b] The problem of simultaneous flow of two phases is complex, and the friction loss can rarely be calculated with high accuracy. In many conveying systems, however, the friction loss is small compared with the losses resulting from elevation and acceleration of the solids, and the total pressure drop $p_a - p_b$ as given by Eq. (7-74) is usually fairly accurate despite the uncertainty in h_f.

SYMBOLS

A	Area, ft^2 or m^2; A_p, projected area of particle; A_s, surface area of channel boundaries, also total surface area of particles in bed
a_e	Acceleration of particle from external force, ft/s^2 or m/s^2
b_1	Constant in Eq. (7-48)
C_D	Drag coefficient, $2F_D g_c/u_0^2 \rho A_p$, dimensionless
D	Diameter, ft or m; D_p, diameter of spherical particle or of a sphere having the same volume as the particle; D_{pm}, mean effective diameter for mixture of particles
E_s	Energy supplied to solids by air in a pneumatic conveyor, ft-lb$_f$ or J
F	Force, lb$_f$ or N; F_D, total drag force; F_b, buoyant force; F_e, external force; F_i, inertial force; F_v, drag force from viscous shear
f	Fanning friction factor, dimensionless; f_p, friction factor for packed bed
G_0	Mass velocity of fluid approaching particle, lb/ft^2-s or kg/m^2-s; also, superficial mass velocity in packed bed
g	Gravitational acceleration, ft/s^2 or m/s^2
g_c	Newton's-law proportionality factor, 32.174 ft-lb/lb$_f$-s^2
h_f	Total friction loss in fluid, ft-lb$_f$/lb or J/kg
K	Criterion for settling, defined by Eq. (7-54), dimensionless
k_1, k_2	Constants in Eq. (7-10)
k_3	Constant in Eq. (7-71)
L	Length of cylindrical particle, ft or m; also total height of packed or fluidized bed; L_M, bed height at incipient fluidization; L_0, height of bed of zero porosity
m	Mass, lb or kg
N_{Ma}	Mach number, dimensionless; $N_{Ma,a}$, $N_{Ma,b}$, at stations a and b; $N_{Ma,0}$, of approaching fluid
N_p	Number of particles in bed
N_{Re}	Reynolds number, dimensionless

$N_{Re,p}$ Particle Reynolds number, $D_p G_0/\mu$, dimensionless; $N_{Re,pM}$, at the onset of fluidization
n Constant in Eq. (7-48)
p Pressure, lb_f/ft^2 or N/m^2; p_a, p_b, at stations a and b; p_s, at stagnation point; p_0, in undisturbed fluid
r Radius of particle path, ft or m; also, mass ratio of solids to air in pneumatic conveyor
r_H Hydraulic radius of channel, ft or m
S Cross-sectional area of bed, ft^2 or m^2; S_0, of empty tower
s_p Surface area of single particle, ft^2 or m^2
T Temperature, °F, °R, °C, or K; T_a, T_b, at stations a and b; T_s, at stagnation point; T_0, of approaching stream
t Time, s
u Velocity of fluid or particle, ft/s or m/s; u_t, terminal velocity of particle; u_{tr}, terminal velocity relative to fluid in hindered settling; u_0, of approaching stream; u', fluctuating component
V_s Velocity of solids in pneumatic conveyor, ft/s or m/s; V_{sa}, V_{sb}, at stations a and b
$\bar{V}$ Volumetric average velocity of fluid, ft/s or m/s; $\bar{V}_0$, superficial or empty-tower velocity; $\bar{V}_{OM}$, minimum superficial velocity for fluidization
v_p Volume of single particle, ft^3 or m^3
x_l Volume fraction of particles of size i in bed of mixed particles
Z Height above datum plane, ft or m; Z_a, Z_b, at stations a and b

Greek letters

α Angle with perpendicular to flow direction
γ Ratio of specific heats, c_p/c_v
$-\Delta p$ Pressure drop in packed or fluidized bed
ε Porosity or volume fraction of voids in bed of solids; ε_M, minimum porosity for fluidization
μ Absolute viscosity, lb/ft-s or cP
ρ Density, lb/ft^3 or kg/m^3; ρ_m, of slurry; ρ_p, of particle; ρ_s, of conveyed solid; ρ_0, of approaching stream; $\bar{\rho}$, average density of air in pneumatic conveyor
τ_w Shear stress at channel boundary, lb_f/ft^2 or N/m^2
Φ_s Sphericity, defined by Eq. (7-19)
ϕ Function
ψ_p Viscosity correction factor for hindered settling [Eq. (7-55)]
ω Angular velocity, rad/s

PROBLEMS

7-1   A catalyst tower 50 ft high and 20 ft in diameter is packed with 1-in.-diameter spheres. Gas enters the top of the bed at a temperature of 500°F and leaves at the same temperature. The pressure at the bottom of the catalyst bed is 30 $lb_f/in.^2$ abs. The bed porosity is 0.40. If the gas has average properties similar to propane and the time of contact between the gas and the catalyst is 10 s, what is the inlet pressure?

7-2   How long will it take for the spherical particles in Table 7-2 to settle, at their terminal velocities under free-settling conditions, through 2 m of water at 20°C?

7-3   Galena spheres 0.001 in. in diameter are rotated in a water suspension by a centrifuge at 70°F. The speed of the centrifuge is 600 r/min. The inside diameter of the rotating liquid is 6 in., and the outside diameter is 12 in. Assuming the particles are at all times at the terminal velocities corresponding to their locations, what time is required to separate the particles completely from the liquid? The initial distribution of the particles in the suspension is uniform, and the settling is under free-settling conditions.

7-4   Urea pellets are made by spraying drops of molten urea into cold gas at the top of a tall tower and allowing the material to solidify as it falls. Pellets 6 mm in diameter are to be made in a tower 25 m high containing air at 20°C. The density of urea is 1,330 kg/m^3.

(a) What would be the terminal velocity of the pellets, assuming free-settling conditions? (b) Would the pellets attain 99 percent of this velocity before they reached the bottom of the tower?

7-5 Catalyst pellets 0.2 in. in diameter are to be fluidized with 100,000 lb/h of air at 1 atm and 170°F in a vertical cylindrical vessel. The density of the catalyst particles is 60 lb/ft^3; their sphericity is 0.86. (a) If the given quantity of air is just sufficient to fluidize the solids, what is the vessel diameter? (b) If the air velocity is now increased to twice the minimum required for fluidization and the height of the expanded bed is 8 ft, what will be the total mass of solid particles in the fluidized bed?

Table 7-2 DATA FOR PROB. 7-2

Substance	Sp. gr.	Diameter, mm
Galena	7.5	0.25
		0.025
Quartz	2.65	0.25
		0.025
Coal	1.3	6
Steel	7.7	25

REFERENCES

1 Clark, W. E.: *Chem. Eng., 74* (3):177 (1967).
2 Ergun, S.: *Chem. Eng. Prog., 48*:89 (1952).
3 Heiss, J. F., and J. Coull: *Chem. Eng. Prog., 48*:133 (1952).
4 Hughes, R. R., and E. R. Gilliland: *Chem. Eng. Prog., 48*:497 (1952).
5 Hunsaker, J. C., and B. G. Rightmire: "Engineering Applications of Fluid Mechanics," pp. 202–203, McGraw-Hill, New York, 1947.
6 Knudsen, J. G., and D. L. Katz: "Fluid Mechanics and Heat Transfer," p. 317, McGraw-Hill, New York, 1958.
7 Kunii, D., and O. Levenspiel: "Fluidization Engineering," Wiley, New York, 1969; (a) p. 74, (b) p. 78, (c) pp. 90–91.
8 Lapple, C. E., and C. B. Shepherd: *Ind. Eng. Chem., 32*:605 (1940).
9 Leva, M.: "Fluidization," McGraw-Hill, New York, 1959; (a) pp. 90–106, (b) pp. 132–166.
10 Lewis, E. W., and E. W. Bowerman: *Chem. Eng. Prog., 48*:603 (1952).
11 Lewis, W. K., E. R. Gilliland, and W. C. Bauer: *Ind. Eng. Chem., 41*:1104 (1949).
12 Masliyah, J. H., and N. Epstein: *J. Fluid Mech., 44*:493 (1970).
13 Perry, J. H. (ed.): "Chemical Engineers' Handbook," 5th ed., p. 18-71, McGraw-Hill, New York, 1973.
14 Pettyjohn, E. S., and E. B. Christiansen: *Chem. Eng. Prog., 44*:157 (1948).
15 Ranz, W. E.: *Chem. Eng. Prog., 48*:247 (1952).
16 Steinour, H. H.: *Ind. Eng. Chem., 36*:618 (1944).
17 Wen, C. Y., and Y. H. Yu: *AIChE J., 12*:610 (1966).
18 Zenz, F. A., and D. F. Othmer: "Fluidization and Fluid-Particle Systems," Reinhold, New York, 1960.

8

TRANSPORTATION AND METERING OF FLUIDS

Preceding chapters have dealt with theoretical aspects of fluid motion. The engineer is concerned, also, with practical problems in transporting fluids from one place to another and in measuring their rates of flow. Such problems are the subject of this chapter.

The first part of the chapter deals with the transportation of fluids, both liquids and gases. Solids are sometimes handled by similar methods by suspending them in a liquid to form a pumpable slurry or by conveying them in a high-velocity gas stream. It is cheaper to move fluids than solids, and materials are transported as fluids whenever possible. In the process industries, fluids are nearly always carried in closed channels, sometimes square or rectangular in cross section but much more often circular. The second part of the chapter discusses common methods of measuring flow rate.

PIPE, FITTINGS, AND VALVES

Pipe and tubing Fluids are usually transported in pipe or tubing, which is circular in cross section and available in widely varying sizes, wall thicknesses, and materials of construction. There is no clear-cut distinction between the terms *pipe* and *tubing*. Generally speaking, pipe is heavy-walled, relatively large in diameter, and comes in moderate lengths of 20 to 40 ft; tubing is thin-walled and often comes in coils several hundred feet long. Metallic pipe can be threaded; tubing usually cannot. Pipe walls are usually slightly rough; tubing has very smooth walls. Lengths of pipe are joined by screwed, flanged, or welded fittings; pieces of tubing are connected by compression fittings, flare fittings, or soldered fittings. Finally, tubing is usually extruded or cold-drawn, while pipe is made by welding, casting, or piercing a billet in a piercing mill.

Pipe and tubing are made from many materials, including metals and alloys, plastics, wood, ceramics, and glass. Most common is low-carbon steel, fabricated into what is sometimes called black-iron pipe. Wrought-iron and cast-iron pipes are also used for a number of special purposes.

SIZES Pipe and tubing are specified in terms of their diameter and their wall thickness. With steel pipe the standard nominal diameters, in American practice, range from $\frac{1}{8}$ to 30 in. For large pipe, more than 12 in. in diameter, the nominal diameters are the actual outside diameters; for small pipe the nominal diameter does not correspond to any actual dimension. The nominal value is close to the actual inside diameter for 3- to 12-in. pipe, but for very small pipe this is not true. Regardless of wall thickness, the outside diameter of all pipe of a given nominal size is the same, to ensure interchangeability of fittings. Standard dimensions of steel pipe are given in Appendix 6. Pipe of other materials is also made with the same outside diameters as steel pipe, to permit interchanging parts of a piping system. These standard sizes for steel pipe, therefore, are known as IPS (iron pipe size) or NPS (normal pipe size). Thus the designation "2-in. nickel IPS pipe" means nickel pipe having the same outside diameter as standard 2-in. steel pipe.

The wall thickness of pipe is indicated by the *schedule number*, given by the approximation equation

$$\text{Schedule number} = \frac{1{,}000p'}{S} \tag{8-1}$$

where p' = internal working pressure, $\text{lb}_f/\text{in.}^2$
S = allowable stress, $\text{lb}_f/\text{in.}^2$, for alloy used

Ten schedule numbers, 10, 20, 30, 40, 60, 80, 100, 120, 140, and 160, are in use, but with pipe less than 8 in. in diameter only numbers 40, 80, 120, and 160 are common. For steel pipe the actual wall thicknesses corresponding to the various schedule numbers are given in Appendix 6; with other alloys the wall thickness may be greater or less than that of steel pipe, depending on the strength of the alloy. With steel at ordinary temperatures the allowable stress S is one-fourth the ultimate strength of the metal.

The size of tubing is indicated by the outside diameter. The nominal value is the actual outer diameter, to within very close tolerances. Wall thickness is ordinarily given by the BWG (Birmingham wire gauge) number, which ranges from 24 (very light) to 7 (very heavy). Sizes and wall thicknesses of heat-exchanger tubing are given in Appendix 7.

SELECTION OF PIPE SIZES   The optimum size of pipe for a specific situation depends upon the relative costs of investment, power, maintenance, and stocking pipe and fittings. In small installations rules of thumb are sufficient. For example, representative ranges of velocity in pipes are shown in Table 8-1. The values in the table are representative of ordinary practice, and special conditions may dictate velocities outside the ranges. Low velocities should ordinarily be favored, especially in gravity flow from overhead tanks. The relation between pipe size, volumetric flow rate, and velocity is shown in Appendix 6.

For large complex piping systems the cost of piping may be a substantial fraction of the total investment, and elaborate methods of optimizing pipe sizes are justified. Computers are used in these cases.

Joints and fittings   The methods used to join pieces of pipe or tubing depend in part on the properties of the material but primarily on the thickness of the wall. Thick-walled tubular products are usually connected by screwed fittings, by flanges, or by welding. Pieces of thin-walled tubing are joined by soldering or by compression or flare fittings. Pipe made of brittle materials like glass or carbon or cast iron is joined by flanges or bell-and-spigot joints.

When screwed fittings are used, the ends of the pipe are threaded externally with a threading tool. The thread is tapered, and the few threads farthest from the end of the pipe are imperfect, so that a tight joint is formed when the pipe is screwed into a fitting. Threading weakens the pipe wall, and the fittings, in general, are not so strong as the pipe itself; when screwed fittings are used, therefore, the schedule number computed from Eq. (8-1) should be doubled. Screwed fittings are standardized for pipe sizes up to 12 in., but because of the difficulty of threading and handling large pipe, they are rarely used in the field with pipe larger than 3 in.

Lengths of pipe larger than about 2 in. are usually connected by flanges or by welding. Flanges are matching disks or rings of metal bolted together and compressing

Table 8-1   FLUID VELOCITIES IN PIPE

Fluid	Type of flow	Velocity ft/s	Velocity m/s
Thin liquid	Gravity flow	0.5–1	0.15–0.30
	Pump inlet	1–3	0.3–0.9
	Pump discharge	4–10	1.2–3
	Process line	4–8	1.2–2.4
Viscous liquid	Pump inlet	0.2–0.5	0.06–0.15
	Pump discharge	0.5–2	0.15–0.6
Steam		30–50	9–15
Air or gas		30–100	9–30

a gasket between their faces. The flanges themselves are attached to the pipe by screwing them on or by welding or brazing. A flange with no opening, used to close a pipe, is called a *blind flange* or a *blank flange*. For joining pieces of large steel pipe in process piping, especially for high-pressure service, welding has become the standard method. Welding makes stronger joints than screwed fittings do, and since it does not weaken the pipe wall, lighter pipe can be used for a given pressure. Properly made welded joints are leakproof. Almost the only disadvantage is that a welded joint cannot be opened without destroying it.

Allowances for expansion   Almost all pipe is subjected to varying temperatures, and in some high-temperature lines the temperature change is very large. Such changes cause the pipe to expand and contract. If the pipe is rigidly fixed to its supports, it may tear loose, bend, or even break. In large lines, therefore, fixed supports are not used; instead the pipe rests loosely on rollers or is hung from above by chains or rods. Provision is also made in all high-temperature lines for taking up expansion, so that the fittings and valves are not put under strain. This is done by bends or loops in the pipe, by packed expansion joints, by bellows, or packless joints, and sometimes by flexible metal hose.

Prevention of leakage around moving parts   In many kinds of processing equipment it is necessary to have one part move in relation to another part without excessive leakage of a fluid around the moving member. This is true in packed expansion joints and in valves where the stem must enter the valve body and be free to turn without allowing the fluid in the valve to escape. It is also necessary where the shaft of a pump or compressor enters the casing, where an agitator shaft passes through the wall of a pressure vessel, and in other similar places.

Common devices for minimizing leakage while permitting relative motion are stuffing boxes and mechanical seals. Neither completely stops leakage, but if no leakage whatever of the process fluid can be tolerated, it is possible to modify the device to ensure that only innocuous fluids leak into or escape from the equipment. The motion of the moving part may be reciprocating or rotational, or both together; it may be small and occasional, as in a packed expansion joint, or virtually continuous, as in a process pump.

STUFFING BOXES   A stuffing box can provide a seal around a rotating shaft and also around a shaft which moves axially. In this it differs from mechanical seals, which are good only with rotating members. The "box" is a chamber cut into the stationary member surrounding the shaft or pipe, as shown in Fig. 8-1a. Often a boss is provided on the casing or vessel wall to give a deeper chamber. The annular space between the shaft and the wall of the chamber is filled with *packing*, consisting of a rope or rings of inert material, such as asbestos, containing a lubricant such as graphite. The packing, when compressed tightly around the shaft, keeps the fluid from passing out through the stuffing box and yet permits the shaft to turn or move back and forth. The packing is compressed by a follower ring, or gland, pressed into the box by a flanged cap or packing nut. The shaft must have a smooth surface so that it does not

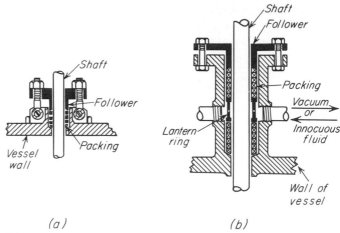

(a) *(b)*

FIGURE 8-1
Stuffing boxes: (*a*) simple form; (*b*) with lantern gland.

wear away the packing; even so, the pressure of the packing considerably increases the force required to move the shaft. A stuffing box, even under ideal conditions, does not completely stop fluid from leaking out; in fact, when the box is operating properly, there should be small leakage. Otherwise the wear on the packing and the power loss in the unlubricated stuffing box are excessive.

When the fluid is toxic or corrosive, means must be provided to prevent it from escaping from the equipment. This can be done by using a *lantern gland* (Fig. 8-1*b*), which may be looked upon as two stuffing boxes on the same shaft, with two sets of packing separated by a lantern ring. The ring is H-shaped in cross section, with holes drilled through the bar of the H in the direction perpendicular to the axis of the shaft. The wall of the chamber of the stuffing box carries a pipe which takes fluid to or away from the lantern ring. By applying vacuum to this pipe, any dangerous fluid which leaks through one set of packing rings is removed to a safe place before it can get to the second set. Or by forcing a harmless fluid, e.g., water, under high pressure into the lantern gland, it is possible to ensure that no dangerous fluid leaks out the exposed end of the stuffing box.

MECHANICAL SEALS In a rotary, or mechanical, seal the sliding contact is between a ring of graphite and a polished metal face, usually of carbon steel. A typical seal is shown in Fig. 8-2. Fluid in the high-pressure zone is kept from leaking out around the shaft by the stationary graphite ring held by springs against the face of the rotating metal collar. A stationary ring of rubber or plastic is set in the space between the body of the seal and the chamber holding it around the shaft; this keeps fluid from leaking past the nonrotating part of the seal and yet leaves the graphite ring free to move axially so that it can be pressed tightly against the collar. Rotary seals require less maintenance than stuffing boxes and have come into wide use in equipment handling highly corrosive fluids.

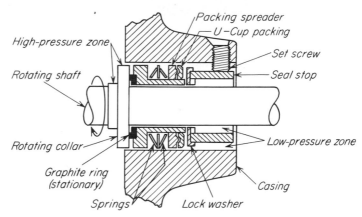

FIGURE 8-2
Mechanical seal.

Valves A typical processing plant contains many thousands of valves, of many different sizes and shapes. Despite the variety in their design, however, all valves have a common primary purpose: to slow down or stop the flow of a fluid. Some valves work best in on-or-off service, i.e., fully open or fully closed. Others are designed to throttle, to reduce the pressure and flow rate of a fluid. Still others permit flow in one direction only or only under certain conditions of temperature and pressure. A steam trap, which is a special form of valve, allows some fluids to pass through while holding others back. Finally, through accessory devices, valves can be made to control the temperature, pressure, liquid level, or other properties of a fluid at points remote from the valve itself.

In all cases, however, the valve initially stops or controls flow. This is done by placing an obstruction in the path of the fluid, an obstruction which can be moved about as desired inside the pipe with little or no leakage of the fluid from the pipe to the outside. Where the resistance to flow introduced by an open valve must be small, the obstruction and the opening which can be closed by it are large. For precise control of flow rate, usually obtained at the price of a large pressure drop, the cross-sectional area of the flow channel is greatly reduced, and a small obstruction is set into the small opening.

GATE VALVES AND GLOBE VALVES The two most common types of valves, gate valves and globe valves, are illustrated in Fig. 8-3. In a gate valve the diameter of the opening through which the fluid passes is nearly the same as that of the pipe, and the direction of flow does not change. As a result, a wide-open gate valve introduces only a small pressure drop. The disk is tapered and fits into a tapered seat; when the valve is opened, the disk rises into the bonnet, completely out of the path of the fluid. Gate valves work best when fully open or fully closed.

Globe valves (so called because in the earliest designs the valve body was spherical) are widely used for controlling the flow rate of a fluid. In a globe valve

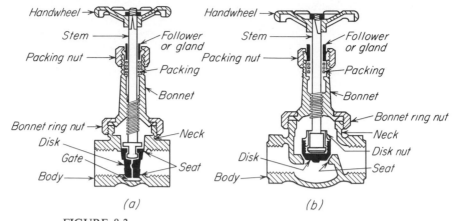

FIGURE 8-3
Common valves: (a) gate valve; (b) globe valve.

the fluid passes through a restricted opening and changes direction several times. This can be seen by visualizing the flow through the valve illustrated in Fig. 8-3b. As a result the pressure drop in this kind of valve is large.

CHECK VALVES   A check valve permits flow in one direction only. It is opened by the pressure of the fluid in the desired direction; when the flow stops or tends to reverse, the valve automatically closes by gravity or by a spring pressing against the disk. Common types of check valves are shown in Fig. 8-4. The movable disk is shown in solid black.

Recommended practice   In designing and installing a piping system many details must be given careful attention, for the successful operation of the entire plant may turn upon a seemingly insignificant feature of the piping arrangement. Some general principles are important enough to warrant mention. In installing pipe, for example, the lines should be parallel and contain, as far as possible, right-angle bends. Provision should be made for opening the lines in order to change the piping or clean it out. This means that unions or flanged connections should be generously included. To facilitate cleaning, tees and crosses with their extra opening closed with plugs should be substituted for elbows in critical places. It then becomes easy to remove a plug and clean out the line with a rod or brush. A process line should always be expected to become clogged, for nearly all process fluids contain some dirt that builds up in the line and eventually stops flow.

In gravity-flow systems the pipe should be oversize and contain as few bends as possible. Fouling of the lines is particularly troublesome where flow is by gravity, since the pressure head on the fluid cannot be increased to keep the flow rate up if the pipe becomes restricted.

Leakage through valves should also be expected. Where complete stoppage of flow is essential, therefore, where leakage past a valve would contaminate a valuable

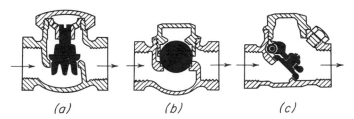

FIGURE 8-4
Check valves: (*a*) lift check; (*b*) ball check; (*c*) swing check.

product or endanger the operators of the equipment, a valve or check valve is inadequate. In this situation a blind flange set between two ordinary flanges will stop all flow; or the line can be broken at a union or pair of flanges and the open ends capped or plugged.

Valves should be mounted vertically with their stems up, if possible. They should be accessible, well within convenient reach of the operator. They should be well supported without strain, with suitable allowance for thermal expansion of the adjacent pipe. Room should be allowed for fully opening the valve and for repacking the stuffing box.

FLUID-MOVING MACHINERY

Fluids are moved through pipe, equipment, or the ambient atmosphere by pumps, fans, blowers, and compressors. Such devices increase the mechanical energy of the fluid. The energy increase may be used to increase the velocity, the pressure, or the elevation of the fluid. In the special case of liquid metals energy may be added by the action of rotating electomagnctic ficlds. Air lifts, jet pumps, and ejectors utilize the energy of a second fluid to move a first. By far the most common method of adding energy is by positive displacement or centrifugal action supplied by outside forces. These methods lead to the two major classes of fluid-moving machinery: (1) those applying direct pressure to the fluid and (2) those using torque to generate rotation. The first group include positive-displacement devices, and the second centrifugal pumps, blowers, and compressors. Also, in positive-displacement devices, the force may be applied to the fluid either by a piston acting in a cylinder or by rotating pressure members. The first group are called reciprocating machines and the second rotary positive-displacement units.

The terms *pump, fan, blower,* and *compressor* do not always have precise meanings. For example, *air pump* and *vacuum pump* designate machines for compressing a gas. Generally, however, a pump is a device for moving a liquid; a fan, a blower, or a compressor adds energy to a gas. Fans discharge large volumes of gas (usually air) into open spaces or large ducts. They are low-speed rotary machines and generate pressures of the order of a few inches of water. Blowers are high-speed rotary devices

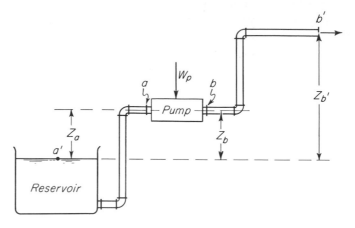

FIGURE 8-5
Pump flow system.

(using either positive displacement or centrifugal force) that develop a maximum pressure of about 2 atm. Compressors discharge at pressures from 2 atm to thousands of atmospheres.

From the standpoint of fluid mechanics the phenomena occurring in these devices can be classified under the usual headings of incompressible and compressible flow. In pumps and fans the density of the fluid does not change appreciably, and in discussing them incompressible-flow theory is adequate. In blowers and compressors the density increase is too great to justify the simplifying assumption of constant density, and compressible-flow theory is required.

In all units certain performance requirements and operating characteristics are important. Flow capacity—measured usually in volumetric flow per unit time at a specified density—power requirements, and mechanical efficiency are obviously important. Reliability and ease of maintenance are also desirable. In small units, simplicity and trouble-free operation are more important than high mechanical efficiency with its savings of a few kilowatts of power.

Pumps

In pumps, the density of the fluid is both constant and large. Pressure differences are usually considerable, and heavy construction is needed.

Developed head   A typical pump application is shown diagrammatically in Fig. 8-5. The pump is installed in a pipeline to provide the energy needed to draw liquid from a reservoir and discharge a constant volumetric flow rate at the exit of the pipeline, $Z_{b'}$ ft above the level of the liquid. At the pump itself, the liquid enters the suction connection at station a and leaves the discharge connection at station b. A Bernoulli equation can be written between stations a and b. Equation (4-34) serves for this.

Since the only friction is that occurring in the pump itself and is accounted for by the mechanical efficiency η, $h_f = 0$. Then Eq. (4-34) can be written

$$\eta W_p = \left(\frac{p_b}{\rho} + \frac{gZ_b}{g_c} + \frac{\alpha_b \overline{V}_b^2}{2g_c}\right) - \left(\frac{p_a}{\rho} + \frac{gZ_a}{g_c} + \frac{\alpha_a \overline{V}_a^2}{2g_c}\right) \tag{8-2}$$

The quantities in the parentheses are called *total heads* and are denoted by H, or

$$H = \frac{p}{\rho} + \frac{gZ}{g_c} + \frac{\alpha \overline{V}^2}{2g_c} \tag{8-3}$$

In pumps usually the difference between the heights of the suction and discharge connections is negligible, and Z_a and Z_b can be dropped from Eq. (8-2). If H_a is the total suction head, H_b the total discharge head, and $\Delta H = H_b - H_a$, Eq. (8-2) can be written

$$W_p = \frac{H_b - H_a}{\eta} = \frac{\Delta H}{\eta} \tag{8-4}$$

POWER REQUIREMENT   The power supplied to the pump drive from an external source is denoted by P_B. It is calculated from Wp by

$$P_B = \dot{m}\, Wp = \frac{\dot{m}\, \Delta H}{\eta} \tag{8-5}$$

where $\dot{m}$ is the mass flow rate.

The power delivered to the fluid is calculated from the mass flow rate and the head developed by the pump. It is denoted by P_f and defined by

$$P_f = \dot{m}\, \Delta H \tag{8-6}$$

From Eqs. (8-5) and (8-6)

$$\frac{P_f}{P_B} = \eta \tag{8-7}$$

Equations (8-2) to (8-7) can also be used for fans by using an average density $\bar{\rho} = (\rho_a + \rho_b)/2$ for ρ.

SUCTION LIFT AND CAVITATION   The power calculated by Eq. (8-4) depends on the difference in pressure between discharge and suction and is independent of the pressure level. From energy considerations it is immaterial whether suction and discharge pressures are below atmospheric pressure or well above it. Practically, the lower limit of the suction pressure is fixed by the vapor pressure of the liquid corresponding to the temperature of the liquid at the suction connection. If the pressure on the liquid reaches the vapor pressure, some of the liquid flashes into vapor, a process called *cavitation*. When cavitation occurs in the suction line, no liquid can be drawn into the pump. Cavitation will not occur if the sum of the velocity and pressure heads at the suction is sufficiently greater than the vapor pressure of the liquid. The excess of the sum of these heads over the vapor pressure is called the *net positive suction head* (NPSH), denoted by H_{sv}. For a pump taking suction from a reservoir like that shown in Fig. 8-5, the NPSH is

$$H_{sv} = \frac{\alpha_a \overline{V}_a^2}{2g_c} + \frac{p_a - p_v}{\rho}$$

(8-8)

where p_v is the vapor pressure. Equation (4-34) written between station a', at the level of the liquid in the reservoir, and station a, at the suction of the pump, gives, assuming $Z_{a'} = 0$ and $\overline{V}_{a'} = 0$,

$$\frac{p_{a'}}{\rho} = \frac{p_a}{\rho} + \frac{\alpha_a \overline{V}_a^2}{2g_c} + h_{fs} + \frac{gZ_a}{g_c}$$

(8-9)

where h_{fs} is the friction in the suction line. Eliminating $p_a/\rho + \alpha_a \overline{V}_a^2/2g_c$ from Eqs. (8-8) and (8-9) gives

$$H_{sv} = \frac{p_{a'} - p_v}{\rho} - h_{sf} - \frac{gZ_a}{g_c}$$

(8-10)

For the special situation where the liquid is nonvolatile ($p_v = 0$), the friction negligible ($h_{fs} = 0$), and the pressure at station a' atmospheric, the NPSH is the barometric head as measured by a column of the liquid and represents the maximum possible suction lift from a tank vented to the atmosphere. For cold water this is about 34 ft or 10 m.

Cavitation may occur within a pump if the pressure at any point reaches that corresponding to $H_{sv} = 0$. This not only prevents normal pump operation but causes severe erosion and mechanical damage. The NPSH must therefore be greater than zero and usually is 2 or 3 m or more.[2] It is the responsibility of the engineer who designs the process to state the value of H_{sv} that is available, and it is the responsibility of the pump supplier to ensure that the pump operates properly at this condition.

EXAMPLE 8-1   Benzene at 100°F (37.8°C) is pumped through the system of Fig. 8-5 at the rate of 40 gal/min (9.09 m³/h). The reservoir is at atmospheric pressure. The gauge pressure at the end of the discharge line is 50 lb$_f$/in.² (345 kN/m²) gauge. The discharge is 10 ft, the pump suction 4 ft, above the level in the reservoir. The discharge line is $1\frac{1}{2}$-in. Schedule 40 pipe. The friction in the suction line is known to be 0.5 lb$_f$/in.² (3.45 kN/m²), and that in the discharge line is 5.5 lb$_f$/in.² (37.9 kN/m²). The mechanical efficiency of the pump is 0.60 (60 percent). The density of benzene is 54 lb/ft³ (865 kg/m³), and its vapor pressure at 100°F (37.8°C) is 3.8 lb$_f$/in.² (26.2 kN/m²). Calculate (a) the developed head of the pump, (b) the total power input, (c) the net positive suction head.

SOLUTION   (a)   The pump work W_p is found by using Eq. (4-34). The downstream station a' is at the level of the liquid in the reservoir, and the upsteam station b' is at the end of the discharge line, as shown in Fig. 8-5. When the level in the tank is chosen as the datum of heights and it is noted that $\overline{V}_{a'} = 0$, Eq. (4-34) gives

$$W_p \eta = \frac{p_{b'}}{\rho} + \frac{gZ_{b'}}{g_c} + \frac{\alpha_{b'} \overline{V}_{b'}^2}{2g_c} + h_f - \frac{p_{a'}}{\rho}$$

The exit velocity $\overline{V}_{b'}$ is found by using data from Appendix 6. For a $1\frac{1}{2}$-in. Schedule 40 pipe, a velocity of 1 ft/s corresponds to a flow rate of 6.34 gal/min, and

$$\overline{V}_{b'} = \frac{40}{6.34} = 6.31 \text{ ft/s}$$

With $\alpha_{b'} = 1.0$, Eq. (4-54) gives

$$W_p\eta = \frac{(14.7 + 50)144}{54} + \frac{g}{g_c} 10 + \frac{6.31^2}{2 \times 32.17} + \frac{(5.5 + 0.5)144}{54} - \frac{14.7 \times 144}{54}$$

$$= 159.9 \text{ ft-lb}_f/\text{lb}$$

By Eq. (8-4) $W_p\eta$ is also the developed head, and

$$\Delta H = H_b - H_a = 159.9 \text{ ft-lb}_f/\text{lb} \ (477.9 \text{ J/kg})$$

(b) The mass flow rate is

$$\dot{m} = \frac{40 \times 54}{7.48 \times 60} = 4.81 \text{ lb/s} \ (2.18 \text{ kg/s})$$

The power input is, from Eq. (8-5),

$$P_B = \frac{4.81 \times 159.9}{550 \times 0.60} - 2.33 \text{ hp} \ (1.74 \text{ kW})$$

(c) The vapor pressure corresponds to a head of

$$\frac{3.8 \times 144}{54} = 10.1 \text{ ft-lb}_f/\text{lb} \ (30.2 \text{ J/kg})$$

The friction in the suction line is

$$h_f = \frac{0.5 \times 144}{54} = 1.33 \text{ ft-lb}_f/\text{lb} \ (3.98 \text{ J/kg})$$

The value of the NPSH is, from Eq. (8-10),

$$H_{sv} = 39.2 - 10.1 - 1.33 - 4\frac{g}{g_c} = 23.77 \text{ ft-lb}_f/\text{lb} \ (71.0 \text{ J/kg}) \qquad ////$$

Positive-displacement pumps   In the first major class of pumps a definite volume of liquid is trapped in a chamber, which is alternately filled from the inlet and emptied at a higher pressure through the discharge. There are two subclasses of positive-displacement pumps. In reciprocating pumps the chamber is a stationary cylinder which contains a piston or plunger; in rotary pumps the chamber moves from inlet to discharge and back to the inlet.

RECIPROCATING PUMPS   Typical reciprocating pumps are shown in Fig. 8-6. In a piston pump (Fig. 8-6a) liquid is drawn through an inlet check valve into the cylinder by the withdrawal of a piston and then forced out through a discharge check valve

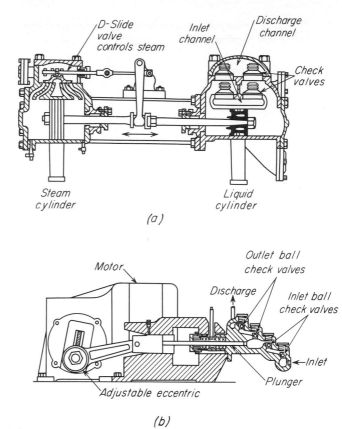

FIGURE 8-6
Positive-displacement reciprocating pumps: (*a*) piston pump; (*b*) plunger pump.

on the return stroke. Most piston pumps are double-acting, i.e., liquid is admitted alternately on each side of the piston, so that one part of the cylinder is being filled while the other is being emptied. Often two or more cylinders are used in parallel with common suction and discharge headers, and the configuration of the pistons is adjusted to minimize fluctuations in the discharge rate. The piston may be motor-driven through reducing gears, or, as shown in Fig. 8-6*a*, a steam cylinder may be used to drive the piston rod directly. The maximum discharge pressure for commercial piston pumps is about 50 atm.

For higher pressures plunger pumps are used. An example is shown in Fig. 8-6*b*. A heavy-walled cylinder of small diameter contains a close-fitting reciprocating plunger, which is merely an extension of the piston rod. At the limit of its stroke the plunger fills nearly all the space in the cylinder. Plunger pumps are single-acting and usually are motor-driven. They can discharge against a pressure of 1,500 atm or more.

The mechanical efficiency of reciprocating pumps varies from 40 to 50 percent for small pumps to 70 to 90 percent for large ones. It is nearly independent of speed

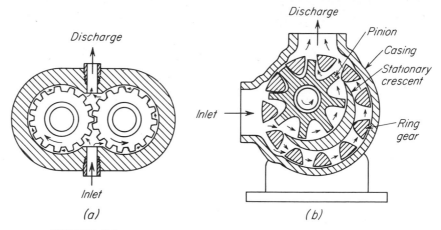

FIGURE 8-7
Gear pumps: (*a*) spur-gear pump; (*b*) internal-gear pump.

within normal operating limits and decreases slightly with increase in discharge pressure, because of added friction and leakage.

VOLUMETRIC EFFICIENCY   The ratio of the volume of fluid discharged to the volume swept by the piston or plunger is called the volumetric efficiency. In positive-displacement pumps the volumetric efficiency is nearly constant with increasing discharge pressure, although it drops a little because of leakage. For a reciprocating pump it varies from about 90 to 100 percent.

ROTARY PUMPS   A wide variety of rotary positive-displacement pumps is available. They bear such names as gear pumps, lobe pumps, screw pumps, cam pumps, and vane pumps. Two examples of gear pumps are shown in Fig. 8-7. Unlike reciprocating pumps, rotary pumps contain no check valves. Close tolerances between the moving and stationary parts minimize leakage from the discharge space back to the suction space; they also limit the operating speed. Rotary pumps operate best on clean, moderately viscous fluids, such as light lubricating oil. Discharge pressures up to 200 atm or more can be attained.

In the spur-gear pump (Fig. 8-7a) intermeshing gears rotate with close clearance inside the casing. Liquid entering the suction line at the bottom of the casing is caught in the spaces between the teeth and the casing and is carried around to the top of the casing and forced out the discharge. Liquid cannot short-circuit back to the suction because of the close mesh of the gears in the center of the pump.

In the internal-gear pump (Fig. 8-7b) a spur gear, or pinion, meshes with a ring gear with internal teeth. Both gears are inside the casing. The ring gear is coaxial with the inside of the casing, but the pinion, which is externally driven, is mounted eccentrically with respect to the center of the casing. A stationary metal crescent fills the space between the two gears. Liquid is carried from inlet to discharge by both gears, in the spaces between the gear teeth and the crescent.

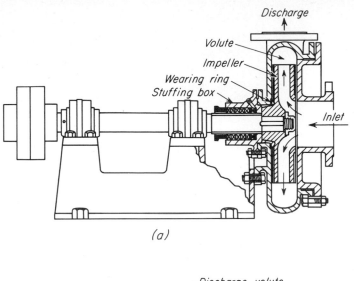

(a)

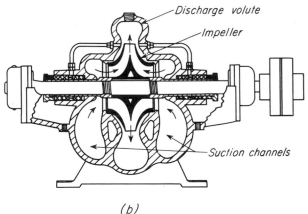

(b)

FIGURE 8-8
Volute centrifugal pumps: (a) single-suction; (b) double-suction.

Centrifugal pumps   In the second major class of pumps the mechanical energy of the liquid is increased by centrifugal action. A simple, but very common, example of a centrifugal pump is shown in Fig. 8-8a. The liquid enters through a suction connection concentric with the axis of a high-speed rotary element called the *impeller*. The impeller carries radial vanes integrally cast in it. Liquid flows outward in the spaces between the vanes and leaves the impeller at a considerably greater velocity with respect to the ground than at the entrance to the impeller. In a properly functioning pump the space between the vanes is completely filled with liquid flowing without cavitation. The liquid leaving the outer periphery of the impeller is collected in a spiral casing called the *volute* and leaves the pump through a tangential discharge connection. In the volute the velocity head of the liquid from the impeller is converted into pressure head. The power is applied to the fluid by the impeller and is transmitted to the impeller by the torque of the drive shaft, which usually is driven by a direct-connected motor at constant speed, commonly at 1,750 r/min.

Under ideal conditions of frictionless flow the mechanical efficiency of a centrifugal pump is, of course, 100 percent, and $\eta = 1$. An ideal pump operating at a given speed delivers a definite discharge rate at each specific developed head. Actual pumps, because of friction and other departures from perfection, fall considerably short of the ideal case.

Centrifugal pumps constitute the most common type of pumping machinery in ordinary plant practice. They come in a number of types other than the simple volute machine shown in Fig. 8-8a. A common type uses a double-suction impeller, which accepts liquid from both sides, as shown in Fig. 8-8b. Also, the impeller itself may be a simple open spider, or it may be enclosed or shrouded. Handbooks, texts on pumps, and especially the catalogs of pump manufacturers show many types, sizes, and designs of centrifugal pumps.

CENTRIFUGAL-PUMP THEORY   The basic equations interrelating the power, developed head, and capacity of a centrifugal pump are derived for the ideal pump from fundamental principles of fluid dynamics. Since the performance of an actual pump differs considerably from that of an ideal one, actual pumps are designed by applying experimentally measured corrections to the ideal situation.

Figure 8-9 shows diagrammatically how the liquid flows through a centrifugal pump. The liquid enters axially at the suction connection, station a. In the rotating eye of the impeller, the liquid spreads out radially and enters the channels between the vanes at station 1. It flows through the impeller, leaves the periphery of the impeller at station 2, is collected in the volute, and leaves the pump discharge at station b.

The performance of the pump is analyzed by considering separately the three parts of the total path: first, the flow from station a to station 1; second, the flow through the impeller from station 1 to station 2; and third, the flow through the volute from station 2 to station b. The heart of the pump is the impeller, and the fluid-mechanical theory of the second section of the fluid path is considered first.

Figure 8-10 shows a single vane, one of the several in an impeller. The vectors represent the various velocities at the stations 1 and 2 at the entrance and exit of the vane, respectively. Consider first the vectors at station 2. By virtue of the design of the pump the tangent to the impeller at its terminus makes an angle β_2 with the tangent to the circle traced out by the impeller tip. Vector v_2 is the velocity of the fluid as seen by an observer fixed at point 2, the end of the vane, and is therefore a *relative* velocity. Two idealizations are now accepted. It is assumed, first, that all liquid flowing across the periphery of the impeller is moving at the same speed, so the numerical value (but not the vector direction) is v_2 at all points; second, it is assumed that the angle between the vector v_2 and the tangent is the actual vane angle β_2. This assumption in turn is equivalent to an assumption that there are an infinite number of vanes, of zero thickness, at an infinitesimal distance apart. This ideal state is referred to as *perfect guidance*. Point 2 at the tip of the blades is moving at peripheral velocity u_2 with respect to the axis. Vector V_2 is the resultant velocity of the fluid stream leaving the impeller as observed from the ground. It is called the *absolute velocity*

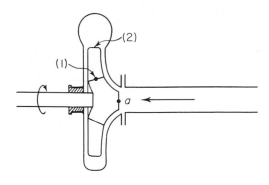

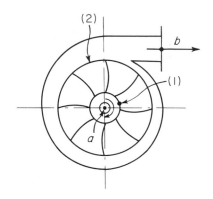

FIGURE 8-9
Centrifugal pump showing Bernoulli
stations.

of the fluid and is, by the parallelogram law, the vector sum of relative velocity v_2 and peripheral velocity u_2. The angle between vectors V_2 and u_2 is denoted by α_2.

A comparable set of vectors applies to the entrance to the vanes at station 1, as shown in Fig. 8-10a. In the usual design, α_1 is nearly 90°, and vector V_1 can be considered radial.

Figure 8-10b is the vector diagram for point 2 that shows the relations between the various vectors in a more useful way. It also shows how the absolute-velocity vector V_2 can be resolved into components, a radial component denoted by V_{r2} and a peripheral component denoted by V_{u2}.

The power input to the impeller, and therefore the power required by the pump, can be calculated from the angular-momentum equation for steady flow, Eq. (4-36). To be consistent with Fig. 8-10, quantity $u_{\theta1}$ in Eq. (4-36) becomes V_{u2}, and $u_{\theta1}$ becomes V_{u1}. Then

$$T g_c = \dot{m}(r_2 V_{u2} - r_1 V_{u1}) \tag{8-11}$$

The momentum correction factors [the β's of Eq. (4-15)] are unity in view of the assumption of perfect guidance. Also, in radial flow, where $\alpha = 90°$, $V_u = 0$. At the entrance, therefore, $r_1 V_{u1} = 0$, the second term in the parentheses of Eq. (8-11) vanishes, and

$$T g_c = \dot{m} r_2 V_{u2} \tag{8-12}$$

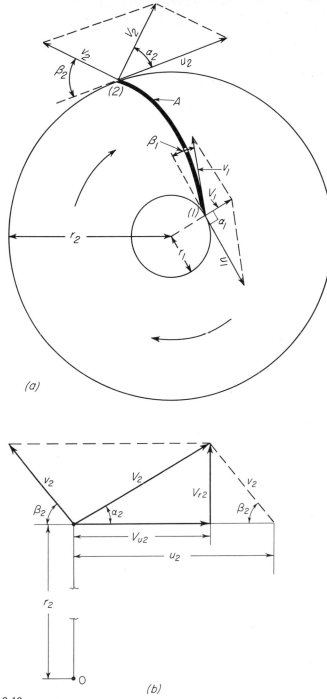

FIGURE 8-10
Velocities at entrance and discharge of vanes in centrifugal pump: (*a*) vectors
and vane; (*b*) vector diagram at tip of vane.

Since $P = T\omega$, the power equation for an ideal pump is

$$P_{fr}g_c = \dot{m}\omega r_2 V_{u2} = \rho q_r \omega r_2 V_{u2} \tag{8-13}$$

where q_r is the volumetric flow rate and subscript r denotes a frictionless pump. The work done per unit mass of fluid is

$$W_{pr} = \frac{P_{fr}}{\dot{m}} = \frac{\omega}{g_c} r_2 V_{u2} \tag{8-14}$$

A Bernoulli equation written between stations 1 and 2, assuming no friction, neglecting $Z_a - Z_b m$ and assuming perfect guidance, gives

$$\frac{\omega}{g_c} r_2 V_{u2} = \frac{p_2}{\rho} - \frac{p_1}{\rho} + \frac{V_2^2}{2g_c} - \frac{V_1^2}{2g_c} \tag{8-15}$$

Also, Bernoulli equations written between stations a and 1 and b and 2, respectively, are

$$\frac{p_a}{\rho} + \frac{\alpha_a \overline{V}_a^2}{2g_c} = \frac{p_1}{\rho} + \frac{V_1^2}{2g_c} \tag{8-16}$$

$$\frac{p_2}{\rho} + \frac{V_2^2}{2g_c} = \frac{p_b}{\rho} + \frac{\alpha_b \overline{V}_b^2}{2g_c} \tag{8-17}$$

Adding Eqs. (8-15), (8-16), and (8-17) gives

$$\frac{p_b}{\rho} + \frac{\alpha_b \overline{V}_b^2}{2g_c} = \frac{p_a}{\rho} + \frac{\alpha_a \overline{V}_a^2}{2g_c} + \frac{\omega}{g_c} r_2 V_{u2} \tag{8-18}$$

Equation (8-18) can be written in the form

$$\Delta H_r = H_b - H_a = \frac{\omega}{g_c} r_2 V_{u2} = W_{pr} \tag{8-19}$$

Equation (8-4) is in agreement with Eq. (8-19) for a frictionless or ideal pump, where $\eta = 1.0$, $\Delta H = \Delta H_r$, and $W_p = W_{pr}$.

THE HEAD-FLOW RATE RELATION OF AN IDEAL PUMP   The radial component V_{r2} is the absolute velocity of the liquid leaving the impeller measured perpendicularly to the periphery of the impeller. The product of V_{r2} and the total cross-sectional area of the channels around the periphery, denoted by A_p, gives the volumetric flow rate through the pump

$$q_r = V_{r2} A_p \tag{8-20}$$

From Fig. 8-10b

$$V_{u2} = u_2 - \frac{V_{r2}}{\tan \beta_2}$$

and, by the definition of angular velocity,

$$\omega = \frac{u_2}{r_2}$$

Substituting for ω and V_{u2} in Eq. (8-19) gives

$$\Delta H_r = \frac{u_2[u_2 - V_{r2}/(\tan \beta_2)]}{g_c} \tag{8-21}$$

and eliminating V_{r2} from Eqs. (8-20) and (8-21) gives

$$\Delta H_r = \frac{u_2(u_2 - q_r/A_p \tan \beta_2)}{g_c} \tag{8-22}$$

Since u_2, A_p, and β_2 are constant, Eq. (8-22) shows that the relation between head and volumetric flow is linear. The slope of the head-flow rate line depends on the sign of $\tan \beta_2$ and therefore varies with angle β_2. If β_2 is less than 90°, as is nearly always the case, the line has a negative slope. Flow in a piping system may become unstable if the line is horizontal or has a positive slope.

ACTUAL PERFORMANCE OF CENTRIFUGAL PUMP   The developed head of an actual pump is considerably less than that calculated from the ideal pump relation of Eq. (8-19). Also, the efficiency is less than unity and the fluid horsepower less than the ideal horsepower. Head losses and power losses will be discussed separately.

LOSS OF HEAD; CIRCULATORY FLOW   A basic assumption in the theory of the ideal pump was that of complete guidance, so the angle between vectors V_2 and u_2 equals the vane angle β_2. The guidance is not perfect in the real pump, and the actual stream of fluid leaves at an angle considerably less than β_2. The physical reason is that the velocity in a given cross section is far from uniform. The effect is the result of an end-to-end circulatory flow of liquid within the impeller channels superimposed on the net flow through the channel. Because of circulation, the resultant velocity V_2 is smaller than the theoretical value. The vector diagram in Fig. 8-11 shows how circulation modifies the theoretical velocities and angles. The full lines and the primed quantities apply to the actual pump and the dotted lines and the unprimed quantities to the theoretical case. The speed of the pump, and hence u_2, and the flow of fluid through the pump, proportional to V_{r2}, are the same in the two cases. It is clear the angle β_2 and the tangential component V_{u2} both decrease. By Eq. (8-19), then, the developed head is also decreased.

Circulatory flow does not cause a mechanical-energy loss, and the power consumption decreases in proportion to the decrease in head, so the efficiency is not affected. Figure 8-12 shows how circulation seriously discounts head.

FLUID FRICTION   Fluid friction lowers head, as shown by the Bernoulli equation. Flow through the passages and channels of the pump is accompanied by friction which is nearly proportional to the square of the velocity and hence to the square of the volumetric flow rate. The head penalty of friction is therefore largest at high flows, as shown in Fig. 8-12.

SHOCK LOSSES   In a volute pump the liquid leaving the impeller in the direction of vector V_2 is suddenly introduced into a stream of liquid traveling circumferentially

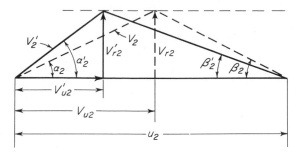

FIGURE 8-11
Vector diagram showing effect of circulation in pump impeller.

around the casing. The sudden change in direction generates turbulence and leads to losses of both head and power. Losses of this kind are called *shock losses*.

MINIMIZING SHOCK LOSSES; DIFFUSER PUMPS   Shock losses are especially large in volute pumps, where no special design elements are used to minimize shock. Large high-speed pumps are designed with a *diffuser*, which is a set of stationary vanes positioned between the periphery of the impeller and the volute. The vanes direct the liquid leaving the impeller with velocity V_2 and change the direction of flow smoothly into the volute with a minimum of shock. Considerable velocity decrease and pressure increase occur in the diffuser.

Figure 8-12 shows the theoretical developed head and head losses in a diffuser pump. The flow rate at point *b* is the design condition, or flow rate at which the pump is rated. The head that remains after the theoretical head is corrected for losses is the actual developed head of the pump, denoted by ΔH. The curve showing the actual head vs. the flow rate, at constant pump speed, is one of the characteristic curves of a pump (Fig. 8-14).

POWER LOSSES   Fluid friction and shock losses lead to losses of power, as they are conversion of mechanical energy into heat.

Three other kinds of loss that do not cause a loss in head but do increase power consumption occur in a centrifugal pump. They are leakage, disk friction, and bearing losses.

The barrier to reverse flow provided by the wearing ring is not perfect, and a certain amount of unavoidable interior *leakage* occurs from the discharge of the impeller back to the suction eye. The effect of leakage is to reduce the volume of actual discharge from the pump per unit of power expended. The extra work used to maintain the leakage is converted into heat and lost.

Disk friction is the friction that occurs between the outer surface of the impeller and the liquid between the impeller and the inside of the casing. It includes the greater friction caused by the pumping action on the liquid that occurs in that same space. Liquid in contact with the rotating impeller is picked up and thrown outward toward the volute. The liquid then flows back along the inside wall of the casing to the shaft,

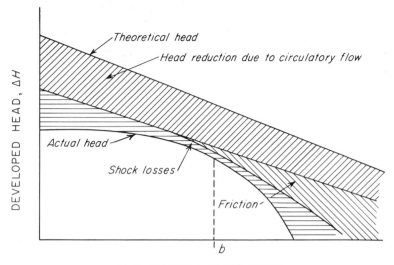

FIGURE 8-12
Theoretical head, actual head, and head losses in centrifugal pump. [*By permission, from A. H. Church, "Centrifugal Pumps and Blowers." Copyright, 1944, John Wiley & Sons, Inc.*]

again to be picked up by the impeller and repumped. Power is required to maintain this useless secondary action.

Bearing losses constitute the power required to overcome mechanical friction in the bearing and stuffing boxes of the pump.

Figure 8-13 shows how the power input to a pump is discounted by the various power losses to give the delivered, or fluid, power. Again, point *b* represents design conditions. The curve to total power input vs. flow rate is another characteristic curve, as shown in Fig. 8-14.

Efficiency   The efficiency, by Eq. (8-7), is the ratio of the delivered power to the total power input, and the relation between efficiency and flow rate is the third characteristic curve of the pump (Fig. 8-14). Efficiency is a maximum at the rated flow rate and decreases at other flow rates, largely because of the variation of shock loss with flow rate shown in Fig. 8-13.

MULTISTAGE CENTRIFUGAL PUMPS   The maximum head that it is practicable to generate in a single impeller is limited by the peripheral speed reasonably attainable. When a head greater than about 70 or 100 ft (20 or 30 m) is needed, two or more impellers can be mounted in series on a single shaft and a multistage pump so obtained. The discharge from the first stage provides suction for the second, the discharge from the second the suction for the third, and so forth. The developed heads of all stages add

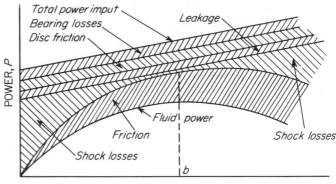

FIGURE 8-13
Fluid power, total power input, and power losses in centrifugal pump. [*By permission, from A. H. Church, "Centrifugal Pumps and Blowers." Copyright, 1944, John Wiley & Sons, Inc.*]

to give a total head several times that of a single stage. The efficiency of a multistage pump is the efficiency (fractional) raised to a power equal to the number of stages and therefore is sensitive to the efficiency of a single stage. Because of their low single-stage efficiencies, volute pumps are seldom multistaged beyond two stages.

PUMP PRIMING Equation (8-19) shows that the theoretical head developed by a centrifugal pump depends on the impeller speed, the radius of the impeller, and the velocity of the fluid leaving the impeller. If these factors are constant, the developed head is the same for fluids of all densities and is the same for liquids and gases. The increase in pressure, however, is the product of the developed head and the fluid density. If a pump develops, say a head of 100 ft, and if the pump is full of water, the increase in pressure is $100 \times 62.3/144 = 43$ lb$_f$/in.^{2}. If the pump is full of air at ordinary density, the pressure increase is about 0.1 lb$_f$/in.^{2}. A centrifugal pump trying to operate on air, then, can neither draw liquid upward from an initally empty suction line nor force liquid along a full discharge line. A pump with air in its casing is *airbound* and can accomplish nothing until the air has been replaced by a liquid. Air can be displaced by priming the pump from an auxiliary priming rank connected to the suction line or by drawing liquid into the suction line by an independent source of vacuum. Also, several types of self-priming pumps are available.

Positive-displacement pumps can compress a gas to a required discharge pressure and are not subject to air binding.

Fans, Blowers, and Compressors

Machinery for compressing and moving gases is conveniently considered from the standpoint of the range of pressure difference produced in the equipment. This order is fans, blowers, compressors.

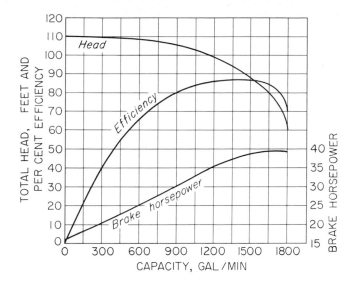

FIGURE 8-14
Characteristic curves for an 8-in. double-section volute pump. [By permission, from F. A. Kristal and F. A. Annett, "Pumps." Copyright, 1953, McGraw-Hill Book Company.]

FANS   Large fans are usually centrifugal fans. Their operating principle is exactly the same as that of centrifugal pumps. Typical fan impellers are shown in Fig. 8-15. The impellers are mounted in light sheet-metal casings. Clearances are large and discharge heads low, from 5 to 60 in. H_2O. Sometimes, as in ventilating fans, nearly all the added energy is converted into velocity energy and almost none into pressure head. In any case, the gain in velocity absorbs an appreciable fraction of the added energy and must be included in estimating efficiency and power. The total efficiency, where the power output is credited with both pressure and velocity heads, is about 70 percent.

Since the change in density in a fan is small, the incompressible-flow equations used in the discussion of centrifugal pumps are adequate. One difference between pumps and gas equipment recognizes the effect of pressure and temperature on the density of the gas entering the machine. Gas equipment is ordinarily rated in terms of standard cubic feet. A volume in standard cubic feet is that measured at a specified temperature and pressure, regardless of the actual temperature and pressure of the gas to the machine. Various standards are used in different industries, but a common one is based on a pressure of 30 in. Hg and a temperature of 60°F (520°R). This corresponds to a molal volume of 378.7 ft^3/lb-mol.

EXAMPLE 8-2   A centrifugal fan is used to take flue gas at rest and at a pressure of 29.0 in. (737 mm) Hg and a temperature of 200°F (93.3°C) and discharge it at a pressure of 30.1 in. (765 mm) Hg and a velocity of 150 ft/s. Calculate the power needed to move 10,000 std ft^3/

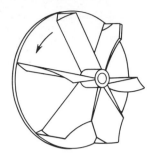 

FIGURE 8-15
Impellers for centrifugal fans.

min (16,990 m³/h) of gas. The efficiency of the fan is 65 percent, and the molecular weight
of the gas is 31.3.

SOLUTION   The actual suction density is

$$\rho_a = \frac{31.3 \times 29.0(460 + 60)}{378.7 \times 30(460 + 200)} = 0.0629 \text{ lb/ft}^3$$

and the discharge density is

$$\rho_b = 0.0629 \frac{30.1}{29.0} = 0.0653 \text{ lb/ft}^3$$

The average density of the flowing gas is

$$\bar{\rho} = \frac{0.0629 + 0.0653}{2} = 0.0641 \text{ lb/ft}^3$$

The mass flow rate is

$$\dot{m} = \frac{10,000 \times 31.3}{378.7 \times 60} = 13.78 \text{ lb/s}$$

The developed pressure is

$$\frac{p_b - p_a}{\bar{\rho}} = \frac{(30.1 - 29)144 \times 14.7}{29.92 \times 0.0641} = 1,214 \text{ ft-lb}_f/\text{lb}$$

The velocity head is

$$\frac{\bar{V}_b^2}{2g_c} = \frac{150^2}{2 \times 32.17} = 349.7 \text{ ft-lb}_f/\text{lb}$$

From Eq. (8-2), calling $\alpha_a = \alpha_b = 1.0$, $\bar{V}_a = 0$, and $Z_a = Z_b$,

$$W_p = \frac{1}{\eta} \left(\frac{p_b - p_a}{\bar{\rho}} + \frac{\bar{V}_b^2}{2g_c} \right) = \frac{1,214 + 349.7}{0.65} = 2,406 \text{ ft-lb}_f/\text{lb}$$

From Eq. (8-5)

$$P_B = \frac{\dot{m}W_p}{550} = \frac{13.78 \times 2,406}{550} = 60.3 \text{ hp (45.0 kW)} \qquad ////$$

Blowers and compressors   When the pressure on a compressible fluid is increased adiabatically, the temperature of the fluid also increases. The temperature rise has a number of disadvantages. Because the specific volume of the fluid increases with temperature, the work required to compress a pound of fluid is larger than if the compression were isothermal. Excessive temperatures lead to problems with lubricants, stuffing boxes, and materials of construction. The fluid may be one that cannot tolerate high temperatures without decomposing.

For the isentropic (adiabatic and frictionless) pressure change of an ideal gas, the temperature relation is, using Eq. (6-24),

$$\frac{T_b}{T_a} = \left(\frac{p_b}{p_a}\right)^{1-1/\gamma} \tag{8-23}$$

where T_a, T_b = inlet and outlet absolute temperatures, respectively
p_a, p_b = corresponding inlet and outlet pressures

For a given gas, the temperature ratio increases with increase in the compression ratio p_b/p_a. This ratio is a basic parameter in the engineering of blowers and compressors. In blowers with a compression ratio below about 3 or 4, the adiabatic-temperature rise is not large, and no special provision is made to reduce it. In compressors, however, where the compression ratio may be as high as 10 or more, the isentropic temperature becomes excessive. Also, since actual compressors are not frictionless, the heat from friction also is absorbed by the gas, and temperatures well above the adiabatic temperature are attained. Compressors, therefore, are cooled by jackets through which cold water or refrigerant is circulated. In small cooled compressors, the exit gas temperature may approach that at the inlet, and isothermal compression achieved. In very small ones, air cooling by external fins cast integrally with the cylinder is sufficient. In larger units, where cooling capacity is limited, a path between those of isothermal and adiabatic compression, called polytropic compression, is followed.

POSITIVE-DISPLACEMENT BLOWERS   A positive-displacement blower is shown in Fig. 8-16. These machines operate like gear pumps except that, because of the special design of the "teeth," the clearance is only a few thousandths of an inch. The relative position of the impellers is maintained precisely by external heavy gears. A single-stage blower can discharge gas at 0.4 to 1 atm, a two-stage blower at 2 atm. The blower shown in Fig. 8-16 is two-lobe. Three-lobe machines also are common.

TURBOBLOWERS   A single-suction turboblower is shown in Fig. 8-17. A turboblower resembles a centrifugal pump in appearance, except that the casing is narrower and the diameters of the casing and discharge scroll are larger than in a pump. The operat-

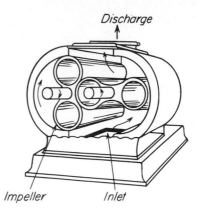

FIGURE 8-16
Two-lobe blower.

ing speed is high, 3,600 r/min or more. Diffusers are necessary both for head and efficiency.

The principle of the operation of a turboblower is the same as that of the centrifugal pump. The ideal power is given by Eq. (8-14). The reason for the high speed and large impeller diameter is that very high heads, measured in feet of low-density fluid, are needed to generate moderate pressure ratios. Thus, the velocities appearing in a vector diagram like Fig. 8-10 are, for a turboblower, approximately tenfold those in a centrifugal pump.

Turbocompressors are multistage units containing a series of impellers on a single shaft, rotating at high speeds in a massive casing.[1,6] Internal channels lead from the discharge of one impeller to the inlet of the next. These machines compress enormous volumes of air or process gas—up to 300,000 ft^3/min at the inlet—to an outlet pressure of 2 atm. Smaller-capacity machines discharge at pressures up to

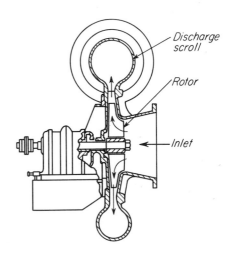

FIGURE 8-17
Single-suction turboblower.

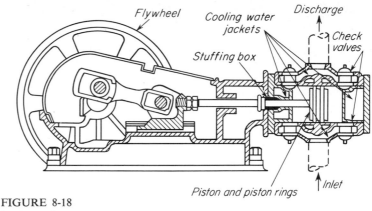

FIGURE 8-18
Reciprocating compressor.

several hundred atmospheres. Interstage cooling is needed on the high-pressure units.

POSITIVE-DISPLACEMENT COMPRESSORS Rotary positive-displacement compressors can be used for discharge pressures to about 6 atm. Most compressors operating at discharge pressures above 3 atm are reciprocating positive-displacement machines. A simple example of single-stage compressor is shown in Fig. 8-18. These machines operate mechanically in the same way as reciprocating pumps, with the important differences that prevention of leakage is more difficult and the temperature rise is important. The cylinder walls and cylinder heads are cored for cooling jackets using water or refrigerant. Reciprocating compressors are usually motor-driven and are nearly always double-acting.

When the required compression ratio is greater than can be achieved in one cylinder, multistage compressors are used. Between each stage are coolers, which are tubular heat exchangers cooled by water or refrigerant. Intercoolers have sufficient heat-transfer capacity to bring the interstage gas streams to the initial suction temperature. Often an aftercooler is used to cool the high-pressure gas from the final stage.

Equation for blowers and compressors Because of the change in density during compressible flow, the integral form of the Bernoulli equation is inadequate. Equation (4-34), however, can be written differentially and used to relate the shaft work to the differential change in pressure head. In blowers and compressors the mechanical kinetic and potential energies do not change appreciably, and the velocity and static-head terms can be dropped. Also, on the assumption that the compressor is frictionless, $\eta = 1.0$ and $h_f = 0$. With these simplifications, Eq. (4-34) becomes

$$dW_{pr} = \frac{dp}{\rho}$$

Integration between the suction pressure p_a and the discharge pressure p_b gives the work of compression of an ideal frictionless gas

$$W_{pr} = \int_{p_a}^{p_b} \frac{dp}{\rho} \tag{8-24}$$

To use Eq. (8-24) the integral must be evaluated, which requires information on the path followed by the fluid in the machine from suction to discharge. The procedure is the same whether the compressor is a reciprocating unit, a rotary positive-displacement unit, or a centrifugal unit, provided only that the flow is frictionless and that in a reciprocating machine the equation is applied over an integral number of cycles, so there is neither accumulation nor depletion of fluid in the cylinders; otherwise the basic assumption of steady flow, which underlies Eq. (4-34), would not hold.

ADIABATIC COMPRESSION   For uncooled units, the fluid follows an isentropic path. For ideal gases, the relation between p and ρ is given by Eq. (6-14), which may be written

$$\frac{p}{\rho^\gamma} = \frac{p_a}{\rho_a^\gamma}$$

or

$$\rho = \frac{p_a}{p_a^{1/\gamma}} p^{1/\gamma} \tag{8-25}$$

Substituting ρ from Eq. (8-25) into Eq. (8-24) and integrating gives

$$W_{pr} = \frac{p_a^{1/\gamma}}{p_a} \int_{p_a}^{p_b} \frac{dp}{p^{1/\gamma}} = \frac{p_a^{1/\gamma}}{(1 - 1/\gamma)\rho_a} (p_b^{1-1/\gamma} - p_a^{1-1/\gamma})$$

By multiplying the coefficient by $p_a^{1-1/\gamma}$ and dividing the terms in the parentheses by the same quantity, this equation becomes

$$W_{pr} = \frac{p_a \gamma}{(\gamma - 1)\rho_a} \left[\left(\frac{p_b}{p_a} \right)^{1-1/\gamma} - 1 \right] \tag{8-26}$$

Equation (8-26) shows the importance of the compression ratio, p_b/p_a.

ISOTHERMAL COMPRESSION   When cooling during compression is complete, the temperature is constant and the process is isothermal. The relation between p and ρ then is, simply,

$$\frac{p}{\rho} = \frac{p_a}{\rho_a} \tag{8-27}$$

Eliminating ρ from Eqs. (8-24) and (8-27) and integrating gives

$$W_{pr} = \frac{p_a}{\rho_a} \int_{p_a}^{p_b} \frac{dp}{p} = \frac{p_a}{\rho_a} \ln \frac{p_b}{p_a} = 2.3026 \frac{p_a}{\rho_a} \log \frac{p_b}{p_a} = \frac{2.3026 R_0 T_a}{M} \log \frac{p_b}{p_a} \tag{8-28}$$

For a given compression ratio and suction condition, the work requirement in isothermal compression is less than that for adiabatic compression. This is one reason why cooling is useful in compressors.

A close relation exists between the adiabatic and isothermal cases. By comparing the integrands in the equation above, it is clear that if $\gamma = 1$, the equations for adiabatic and for isothermal compression are identical.

POLYTROPIC COMPRESSION   In large compressors the path of the fluid is neither isothermal nor adiabatic. The process may still be assumed to be frictionless, however. It is customary to assume that the relation between pressure and density is given by the equation

$$\frac{p}{\rho^n} = \frac{p_a}{\rho_a^n} \tag{8-29}$$

where n is a constant lying between 1.0 and γ. Use of this equation in place of Eq. (8-25) obviously yields Eq. (8-26) with the replacement of γ by n.

The value of n is found empirically by measuring the density and pressure at two points on the path of the process, e.g., at suction and discharge. The value of n is calculated by the equation

$$n = \frac{\log (p_b/p_a)}{\log (\rho_b/\rho_a)}$$

This equation is derived by substituting p_b for p and ρ_b for ρ in Eq. (8-29) and taking logarithms.

COMPRESSOR EFFICIENCY   The ratio of the theoretical work (or fluid power) to the actual work (or total power input) is, as usual, the efficiency and is denoted by η. The maximum efficiency of reciprocating compressor is about 80 to 85 percent.

POWER EQUATION   The power required by an adiabatic compressor is readily calculated from Eq. (8-26). The *dimensional* formula is

$$P_B = \frac{0.0643 T_a \gamma q_0}{520(\gamma - 1)\eta} \left[\left(\frac{p_b}{p_a}\right)^{1-1/\gamma} - 1 \right] \tag{8-30}$$

where P_B = brake horsepower
q_0 = voume of gas compressed, std ft^3/min

For polytropic compression, n is substituted for γ. For isothermal compression

$$P_B = \frac{0.148 T_a q_0}{520\eta} \log \frac{p_b}{p_a} \tag{8-31}$$

EXAMPLE 8-3   A three-stage reciprocating compressor is to compress 180 std ft^3/min (306 m^3/h) of methane from 14 to 900 lb$_f$/in.^{2} (0.95 to 61.3 atm) abs. The inlet temperature is 80°F (26.7°C). For the expected temperature range the average properties of methane are

$$C_p = 9.3 \text{ Btu/lb mol-°F (38.9 J/g mol-°C)} \qquad \gamma = 1.31$$

(a) What is the brake horsepower if the mechanical efficiency is 80 percent? (b) What is the discharge temperature from the first stage? (c) If the temperature of the cooling water is to rise 20°F (6.7°C), how much water is needed in the intercoolers and aftercooler for the compressed gas to leave each cooler at 80°F (26.7°C)? Assume that jacket cooling is sufficient to absorb frictional heat.

SOLUTION   (a) For a multistage compressor it can be shown that the total power is a minimum if each stage does the same amount of work. By Eq. (8-26) this is equivalent to the use of the same compression ratio in each stage. For a three-stage machine, therefore, the compression ratio of one stage should be the cube root of the overall compression ratio, 900/14. For one stage

$$\frac{p_b}{p_a} = \left(\frac{900}{14}\right)^{1/3} = 4$$

The power required for each stage is, by Eq. (8-30),

$$P_B = \frac{(80 + 460)(0.0643) \times 1.31 \times 180}{520(1.31 - 1)(0.80)} (4^{1-1/1.31} - 1) = 24.6 \text{ hp}$$

The total power for all stages is $3 \times 24.6 = 73.8$ hp (55.0 kW).

(b)   From Eq. (8-23), the temperature at the exit of each stage is

$$T_b = (80 + 460)4^{1-1/1.31} = 750°R = 290°F (143.3°C)$$

(c)   Since 1 lb mol = 378.7 std ft^3, the flow rate is

$$\frac{180 \times 60}{378.7} = 28.5 \text{ lb mol/h} (12.9 \text{ kg mol/h})$$

The heat load in each cooler is

$$28.5(290 - 100)(9.3) = 50,400 \text{ Btu/h}$$

The total heat load is $3 \times 50,400 = 151,200$ Btu/h. The cooling-water requirement is

$$\frac{151,200}{20} = 7.560 \text{ lb/h} = 15.1 \text{ gal/min} (3.43 \text{ m}^3/\text{h}) \qquad ////$$

Vacuum pumps   A compressor that takes suction at a pressure below atmospheric and discharges against atmospheric pressure is called a *vacuum pump*. Any type of blower or compressor, reciprocating, rotary, or centrifugal, can be adapted to vacuum practice by modifying the design to accept very low-density gas at the suction and attain the large compression ratios necessary. As the absolute pressure at the suction decreases, the volumetric efficiency drops and approaches zero at the lowest absolute pressure attainable by the pump. The mechanical efficiency also usually is lower than for compressors. The required displacement increases rapidly as the suction pressure falls, so a large machine is needed to move much gas. The compression ratio used in vacuum pumps is higher than in compressors, ranging up to 100 or more, with a correspondingly high adiabatic discharge temperature. Actually, however, the compression is nearly isothermal because of the low mass flow rate and the effective heat transfer from the relatively large area of exposed metal.

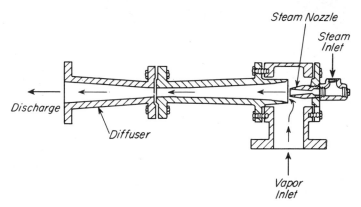

FIGURE 8-19
Steam-jet ejector.

JET EJECTORS   An important kind of vacuum pump that does not use moving parts is the jet ejector, shown in Fig. 8-19, in which the fluid to be moved is entrained in a high-velocity stream of a second fluid.

A common example is a laboratory water jet, in which the water stream entrains air from a suction flask or other vessel. The motive fluid and the fluid to be moved may be the same, as when compressed air is used to move air, but usually they are not. Industrially most use is made of stream-jet ejectors, which are valuable for drawing a fairly high vacuum. As shown in Fig. 8-19, steam at about 7 atm is admitted to a converging-diverging nozzle, from which it issues at supersonic velocity into a diffuser cone. The air or other gas to be moved is mixed with the steam in the first part of the diffuser, lowering the velocity to acoustic velocity or below; in the diverging section of the diffuser the kinetic energy of the mixed gases is converted into pressure energy, so that the mixture can be discharged directly to the atmosphere. Often it is sent to a water-cooled condenser, particularly if more than one stage is used, for otherwise each stage would have to handle all the steam admitted to the preceding stages. As many as five stages are used in industrial processing.

Jet ejectors require very little attention and maintenance and are especially valuable with corrosive gases that would damage mechanical vacuum pumps. For difficult problems the nozzles and diffusers can be made of corrosion-resistant metal, graphite, or other inert material. Ejectors, particularly when multistage, use large quantities of steam and water. They are rarely used to produce absolute pressures below 1 mm Hg.

Comparison of devices for moving fluids   Positive-displacement machines, in general, handle smaller quantities of fluids at higher discharge pressures than centrifugal machines do. Positive-displacement pumps are not subject to air binding and are usually self-priming. In both positive-displacement pumps and blowers the discharge rate is nearly independent of the discharge pressure, so that these machines are extensively used for controlling and metering flow. Reciprocating devices require considerable maintenance but can produce the highest pressures. They deliver a pulsating stream. Rotary pumps work best on fairly viscous lubricating fluids, discharging a steady stream at moderate to high pressures. They cannot be used with slurries. Rotary blowers usually discharge gas at a maximum pressure of 1 atm from a single stage. The discharge line of a positive-displacement pump cannot be closed without stalling or breaking the pump, so that a bypass line with a pressure-relief valve is required.

Centrifugal machines, both pumps and blowers, deliver fluid at a uniform pressure without shocks or pulsations. They run at higher speeds than positive-displacement machines and are connected to the motor drive directly instead of through a gearbox. The discharge line can be completely closed without damage. Centrifugal pumps can handle a wide variety of corrosive liquids and slurries. Centrifugal blowers and compressors are much smaller, for given capacity, than reciprocating compressors and require less maintenance.

For producing vacuum, reciprocating machines are effective for absolute pressures down to 10 mm Hg. Rotary vacuum pumps can lower the absolute pressure to 0.01 mm Hg and over a wide range of low pressures are cheaper to operate than multistage steam-jet ejectors. For very high vacuums, specialized devices such as diffusion pumps are needed.

MEASUREMENT OF FLOWING FLUIDS

To control industrial processes, it is desirable to know the amount of material entering and leaving the process. Because materials are transported in the form of fluids wherever possible, it is important to measure the rate at which a fluid is flowing through a pipe or other channel. Many different types of meters are used industrially, including (1) meters based on direct weighing or measurement of volume, (2) variable-head meters, (3) area meters, (4) current meters, (5) positive-displacement meters, and (6) magnetic meters.

Meters involving weighing or volume measurement are simple and will not be discussed in this text. Current meters, such as cup or vane anemometers, make use of an element which rotates at a speed determined by the velocity of the fluid in which the meter is immersed. Positive-displacement meters include metering pumps of various kinds, which differ in purpose but not in principle from the rotary and reciprocating pumps described earlier in this chapter. Magnetic flowmeters depend on the creation of an electric potential by the motion of a conducting fluid through an externally generated uniform magnetic field. By Faraday's law of electromagnetic induction the voltage created is directly and linearly proportional to the velocity of the flowing fluid. Commercial magnetic flowmeters can measure the velocity of almost all liquids except hydrocarbons, which have too small an electrical conductivity. Since the induced voltage depends on velocity only, changes in the viscosity or density of the liquid have no effect on the meter reading.

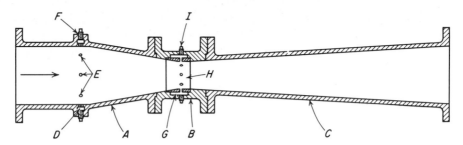

FIGURE 8-20
Venturi meter. *A*, inlet section; *B*, throat section; *C*, outlet section; *D*, *G*, piezometer chambers; *E*, holes to piezometer chambers; *F*, upstream pressure tap; *H*, liner; *I*, downstream pressure tap.

Most widely used for flow measurement are the several types of variable-head meters and area meters. Variable-head meters include venturi meters, orifice meters, and pitot tubes; area meters include rotameters of various designs.

Venturi meter A venturi meter is shown in Fig. 8-20. It is constructed from a flanged inlet section *A*, consisting of a short cylindrical portion and a truncated cone; a flanged throat section *B*; and a flanged outlet section *C*, consisting of a long truncated cone. In the upstream section, at the junction of the cylindrical and conical portions, an annular chamber *D* is provided, and a number of small holes *E* are drilled from the inside of the tube into the annular chamber. The annular ring and the small holes constitute a *piezometer ring*, which has the function of averaging the individual pressures transmitted through the several small holes. The average pressure is transmitted through the upstream pressure connection *F*. A second piezometer ring is formed in the throat section by an integral annular chamber *G* and a liner *H*. The liner is accurately bored and finished to a definite diameter, as the accuracy of the meter is reduced if the throat is not carefully machined to close tolerances. The throat pressure is transmitted through the pressure tap *I*. A manometer or other means for measuring pressure difference is connected between the taps *F* and *I*.

In the venturi meter, the velocity is increased, and the pressure decreased, in the upstream cone. The pressure drop in the upstream cone is utilized as shown below to measure the rate of flow through the instrument. The velocity is then decreased, and the original pressure largely recovered, in the downstream cone. To make the pressure recovery large, the angle of the downstream cone *C* is small, so boundary-layer separation is prevented and friction minimized. Since separation does not occur in a contracting cross section, the upstream cone can be made shorter than the downstream cone with but little friction, and space and material are thereby conserved. Although venturi meters can be applied to the measurement of gas, they are most commonly used for liquids, especially water. The following treatment is limited to incompressible fluids.

The basic equation for the venturi meter is obtained by writing the Bernoulli equation for incompressible fluids between the two pressure stations at D and G. Friction is neglected, the meter is assumed to be horizontal, and there is no pump. If $\overline{V}_a$ and $\overline{V}_b$ are the average upstream and downstream velocities, respectively, and ρ is the density of the fluid, Eq. (4-34) becomes

$$\alpha_b \overline{V}_b^2 - \alpha_a \overline{V}_a^2 = \frac{2g_c(p_a - p_b)}{\rho} \tag{8-32}$$

The continuity relation (4-7) can be written, since the density is constant, as

$$\overline{V}_a = \left(\frac{D_b}{D_a}\right)^2 \overline{V}_b = \beta^2 \overline{V}_b \tag{8-33}$$

where D_a = diameter of pipe
$\quad$ D_b = diameter of throat of meter
$\quad$ β = diameter ratio, D_b/D_a

If $\overline{V}_a$ is eliminated from Eqs. (8-32) and (8-33), the result is

$$\overline{V}_b = \frac{1}{\sqrt{\alpha_b - \beta^4 \alpha_a}} \sqrt{\frac{2g_c(p_a - p_b)}{\rho}} \tag{8-34}$$

VENTURI COEFFICIENT   Equation (8-34) applies strictly to the frictionless flow of noncompressible fluids. To account for the small friction loss between locations a and b, Eq. (8-34) is corrected by introducing an empirical factor C_v and writing

$$\overline{V}_b = \frac{C_v}{\sqrt{1 - \beta^4}} \sqrt{\frac{2g_c(p_a - p_b)}{\rho}} \tag{8-35}$$

The small effects of the kinetic-energy factors α_a and α_b are also taken into account in the definition of C_v. The coefficient C_v is determined experimentally. It is called the *venturi coefficient, velocity of approach not included.* The effect of the approach velocity $\overline{V}_a$ is accounted for by the term $1/\sqrt{1 - \beta^4}$. When D_b is less than $D_a/4$, the approach velocity and the term β can be neglected, since the resulting error is less than 0.2 percent.

For well-designed venturi meters, the constant C_v is about 0.98 for pipe diameters of 2 to 8 in. and about 0.99 for larger sizes.[4]

MASS AND VOLUMETRIC FLOW RATES   The velocity through the venturi throat $\overline{V}_b$ usually is not the quantity desired. The flow rates of practical interest are the mass and volumetric flow rates through the meter. The mass flow rate is calculated by substituting $\overline{V}_b$ from Eq. (8-35) in Eq. (4-6) to give

$$\dot{m} = \overline{V}_b S_b \rho = \frac{C_v S_b}{\sqrt{1 - \beta^4}} \sqrt{2g_c(p_a - p_b)\rho} \tag{8-36}$$

where $\dot{m}$ = mass flow rate
$\quad$ S_b = area of throat

The volumetric flow rate is obtained by dividing the mass flow rate by the density, or

$$q = \frac{\dot{m}}{\rho} = \frac{C_v S_b}{\sqrt{1 - \beta^4}} \sqrt{\frac{2g_c(p_a - p_b)}{\rho}} \tag{8-37}$$

where q is the volumetric flow rate.

PRESSURE RECOVERY   If the flow through the venturi meter were frictionless, the pressure of the fluid leaving the meter would be exactly equal to that of the fluid entering the meter and the presence of the meter in the line would not cause a permanent loss in pressure. The pressure drop in the upstream cone $p_a - p_b$ would be completely recovered in the downstream cone. Friction cannot be completely eliminated, of course, and a permanent loss in pressure and a corresponding loss in power do occur. Because of the small angle of divergence in the recovery cone, the permanent pressure loss from a venturi meter is relatively small. In a properly designed meter, the permanent loss is about 10 percent of the venturi differential $p_a - p_b$, and approximately 90 percent of the differential is recovered.

EXAMPLE 8-4   A venturi meter is to be installed in a Schedule 40 4-in. (100-mm) line to measure the flow of water. The maximum flow rate is expected to be 325 gal/min (73.8 m³/h) at 60°F (15.6°C). The 50-in. (1.27-m) manometer used to measure the differential pressure is to be filled with mercury, and water is to fill the leads about the mercury surfaces. The water temperature is to be 60°F (15.6°C) throughout. What throat diameter should be specified for the venturi, to the nearest $\frac{1}{8}$ in. (3 mm), and what will be the power required to operate the meter at full load?

SOLUTION   Equation (8-37) can be used to calculate the throat size. The quantities to be substituted in the equation are

$$q = \frac{325}{60 \times 7.48} = 0.725 \text{ ft}^3/\text{s} \qquad \rho = 62.37 \text{ lb/ft}^3 \qquad \text{(Appendix 14)}$$

$$C_v = 0.98 \qquad g_c = 32.17 \text{ ft-lb/lb}_f\text{-s}^2$$

$$p_a - p_b = \tfrac{50}{12}(13.6 - 1.0)(62.37) = 3,275 \text{ lb}_f/\text{ft}^2$$

Substituting the above values in Eq. (8-37) gives

$$0.725 = \frac{0.98 S_b}{\sqrt{1 - \beta^4}} \sqrt{\frac{2 \times 32.17 \times 3,275}{62.37}}$$

from which

$$\frac{S_b}{\sqrt{1 - \beta^4}} = 0.01275 = \frac{0.7854 D_b^2}{\sqrt{1 - \beta^4}}$$

As a first approximation, call $\sqrt{1 - \beta^4} = 1.0$. Then

$$D_b = 0.127 \text{ ft} = 1.53 \text{ in.} \qquad \beta = \frac{1.53}{4.026} = 0.380$$

and

$$\sqrt{1 - \beta^4} = \sqrt{1 - 0.38^4} = 0.98$$

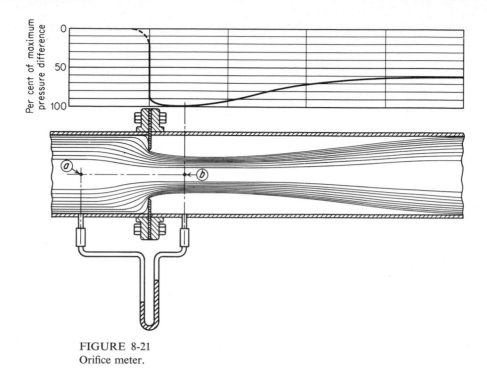

FIGURE 8-21
Orifice meter.

The effect of this term is negligible in view of the desired precision of the final result. To the nearest $\frac{1}{8}$ in., the throat diameter should be $1\frac{1}{2}$ in. (38 mm).

Power loss   The permanent loss in pressure is 10 percent of the differential, or 327.5 lb_f/ft^2. Since the maximum volumetric flow rate is $325/7.48 = 43.4$ ft^3/min, the power required to operate the venturi at full flow is

$$P = \frac{43.4 \times 327.5}{33,000} = 0.43 \text{ hp } (0.32 \text{ kW})$$ ////

Orifice meter   The venturi meter has certain practical disadvantages for ordinary plant practice. It is expensive, it occupies considerable space, and its ratio of throat diameter to pipe diameter cannot be changed. For a given meter and definite manometer system, the maximum measurable flow rate is fixed, so if the flow range is changed, the throat diameter may be too large to give an accurate reading or too small to accommodate the new maximum flow rate. The orifice meter meets these objections to the venturi but at the price of a larger power consumption.

A standard sharp-edged orifice is shown in Fig. 8-21. It consists of an accurately machined and drilled plate mounted between two flanges with the hole concentric with the pipe in which it is mounted. The opening in the plate may be beveled on the downstream side. Pressure taps, one above and one below the orifice plate, are installed and are connected to a manometer or equivalent pressure-measuring device.

The positions of the taps are arbitrary, and the coefficient of the meter will depend upon the position of the taps. Three of the recognized methods of placing the taps are shown in Table 8-2. The taps shown in Fig. 8-21 are vena-contracta taps.

The principle of the orifice meter is identical with that of the venturi. The reduction of the cross section of the flowing stream in passing through the orifice increases the velocity head at the expense of the pressure head, and the reduction in pressure between the taps is measured by the manometer. Bernoulli's equation provides a basis for correlating the increase in velocity head with the decrease in pressure head.

One important complication appears in the orifice meter that is not found in the venturi. Because of the sharpness of the orifice, the fluid stream separates from the downstream side of the orifice plate and forms a free-flowing jet in the downstream fluid. A vena contracta forms, as shown in Fig. 8-21. The jet is not under the control of solid walls, as is the case in the venturi, and the area of the jet varies from that of the opening in the orifice to that of the vena contracta. The area at any given point, e.g., at the downstream tap, is not easily determinable, and the velocity of the jet at the downstream tap is not easily related to the diameter of the orifice. Orifice coefficients are more empirical than those for the venturi, and the quantitative treatment of the orifice meter is modified accordingly.

Extensive and detailed design standards for orifice meters are available in the literature.[3] They must be followed if the performance of a meter is to be predicted accurately without calibration. For approximate or preliminary design, however, it is satisfactory to use an equation similar to Eq. (8-35) as follows:

$$u_o = \frac{C_o}{\sqrt{1 - \beta^4}} \sqrt{\frac{2g_c(p_a - p_b)}{\rho}} \tag{8-38}$$

where u_o = velocity through the orifice
β = ratio of orifice diameter to pipe diameter
p_a, p_b = pressures at stations a and b in Fig. 8-21

In Eq. (8-38) C_o is the *orifice coefficient, velocity of approach not included.* It corrects for the contraction of the fluid jet between the orifice and the vena contracta, for friction, and for α_a and α_b. C_o is always determined experimentally. It varies considerably with changes in β and with Reynolds number at the orifice, $N_{Re,o}$. This

Table 8-2   DATA ON ORIFICE TAPS

Type of tap	Distance of upstream tap from upstream face of orifice	Distance of downstream tap from downstream face
Flange	1 in.	1 in.
Vena contracta	1 pipe diameter (actual inside)	0.3 to 0.8 pipe diameter, depending on β
Pipe	$2\frac{1}{2}$ times nominal pipe diameter	8 times nominal pipe diameter

Reynolds number is defined by

$$N_{Re,o} = \frac{D_o u_o \rho}{\mu} = \frac{4\dot{m}}{\pi D_o \mu} \tag{8-39}$$

where D_o is the orifice diameter.

Equation (8-38) is useful for design because C_o is almost constant and independent of β provided $N_{Re,o}$ is greater than about 20,000. Under these conditions C_o may be taken as 0.61. Furthermore, if β is less than 0.25 the term $\sqrt{1 - \beta^4}$ differs negligibly from unity, and Eq. (8-38) becomes

$$u_o = 0.61 \sqrt{\frac{2g_c(p_a - p_b)}{\rho}} \tag{8-40}$$

The mass flow rate $\dot{m}$ is given by

$$\dot{m} = u_o S_o \rho = 0.61 S_o \sqrt{2g_c(p_a - p_b)\rho} \tag{8-41}$$

Substituting the following equivalent value of S_o, the cross-sectional area of the orifice, in Eq. (8-41) gives

$$S_o = \frac{D_a^2 S_o}{D_a^2} = \frac{D_a^2(\pi/4)D_o^2}{D_a^2} = \frac{\pi}{4}(D_a\beta)^2 \tag{8-42}$$

and solving for β^2 gives the following equation, which can be used to calculate β directly

$$\beta^2 = \frac{4\dot{m}}{0.61\pi D_a^2 \sqrt{2g_c(p_a - p_b)\rho}} \tag{8-43}$$

Unless considerable precision is desired, Eq. (8-43) is adequate for orifice design. A check on the value of the Reynolds number should be made, however, since the coefficient 0.61 is not accurate when $N_{Re,o}$ is less than about 20,000.

It is especially important that enough straight pipe be provided both above and below the orifice to ensure a flow pattern that is normal and undisturbed by fittings, valves, or other equipment. Otherwise the velocity distribution will not be normal, and the orifice coefficient will be affected in an unpredictable manner. Data are available for the minimum length of straight pipe that should be provided upstream and downstream from the orifice to ensure normal velocity distribution.[3b] Straightening vanes in the approach line may be used if the required length of pipe is not available upstream from the orifice.

PRESSURE RECOVERY   Because of the large friction losses from the eddies generated by the reexpanding jet below the vena contracta, the pressure recovery in a orifice meter is poor. The resulting power loss is one disadvantage of the orifice meter. The fraction of the orifice differential that is permanently lost depends on the value of β, and the relationship between the fractional loss and β is shown in Fig. 8-22. For a value of β of 0.5, the lost head is about 73 percent of the orifice differential.

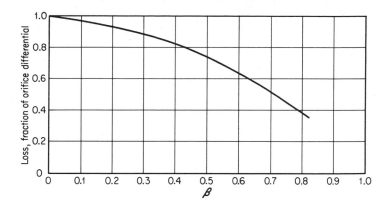

FIGURE 8-22
Overall pressure loss in orifice meters. [*American Society of Mechanical Engineers.*[3a]]

The pressure difference measured by pipe taps, where the downstream tap is eight pipe diameters below the orifice, is really a measurement of permanent loss rather than of the orifice differential.

EXAMPLE 8-5   An orifice meter equipped with flange taps is to be installed to measure the flow rate of topped crude to a cracking unit. The oil is flowing at 100°F (37.8°C) through a Schedule 40 4-in. (100-mm) pipe. An adequate run of straight horizontal pipe is available for the installation. The expected maximum flow rate is 12,000 bbl/d (1 bbl = 42 U.S. gal) (79.5 m^3/h), measured at 60°F (15.6°C). Mercury is to be used as a manometer fluid, and glycol (specific gravity = 1.11) is to be used in the leads as sealing liquid. The maximum reading of the meter is to be 30 in. (762 mm). The viscosity of the oil at 100°F (37.8°C) is 5.45 cP. The specific gravity (160°F/60°F) of the oil is 0.8927. The ratio of the density of the oil at 100°F (37.8°C) to that at 60°F (15.6°C) is 0.984. Calculate (a) the diameter of the orifice and (b) the power loss.

SOLUTION   (a)   Use Eq. (8-43). The values to be substituted are:

Density at 60°F: $\rho_B = 0.8927 \times 62.37 = 55.68$ lb/ft^3

Density at 100°F: $\rho = 0.8927 \times 62.37 \times 0.984 = 54.79$ lb/ft^3

$$\dot{m} = \frac{12,000 \times 42 \times 55.68}{24 \times 3,600 \times 7.48} = 43.42 \text{ lb/s}$$

$$D_a = \frac{4.026}{12} = 0.3355 \text{ ft}$$

$$p_a - p_b = \frac{30}{12}(13.6 - 1.11)62.37 = 1,948 \text{ lb}_f/\text{ft}^2$$

Then, from Eq. (8-43)

$$\beta^2 = \frac{4 \times 43.42}{\pi 0.3355^2 \times 0.61 \sqrt{2 \times 32.17 \times 1,948 \times 54.79}} = 0.3073$$

from which $\beta = 0.554$, and $D_o = 0.554 \times 0.3355 = 0.186$ ft. The orifice diameter is $12 \times 0.186 = 2.23$ in. (56.6 mm). The viscosity is

$$\mu = 5.45 \times 6.72 \times 10^{-4} = 0.00367 \text{ lb/ft-s}$$

The Reynolds number is, by Eq. (8-39),

$$N_{Re,o} = \frac{4 \times 43.42}{\pi 0.186 \times 0.00367} = 81,000$$

The Reynolds number is sufficiently large to justify the value of 0.61 for C_o.

(b) Since $\beta = 0.554$, the fraction of the differential pressure that is permanently lost is, by Fig. 8-22, 68 percent of the orifice differential. The maximum power consumption of the meter is

$$\frac{43.42 \times 1,948 \times 0.68}{0.984 \times 62.37 \times 0.8927 \times 550} = 1.9 \text{ hp (1.4 kW)} \qquad\qquad ////$$

Flow of compressible fluids through venturis and orifices   The preceding discussion of fluid meters was concerned only with the flow of fluids of constant density. When fluids are compressible, similar equations and discharge coefficients for the various meters may be used. Equation (8-36) for venturi meters is modified to the form

$$\dot{m} = \frac{C_v Y S_b}{\sqrt{1 - \beta^4}} \sqrt{2g_c(p_a - p_b)\rho_a} \qquad\qquad (8\text{-}44)$$

For orifice meters Eq. (8-41) becomes

$$\dot{m} = 0.61 Y S_o \sqrt{2g_c(p_a - p_b)\rho_a} \qquad\qquad (8\text{-}45)$$

In Eqs. (8-44) and (8-45) Y is a dimensionless expansion factor, and ρ_a is the density of the fluid under upstream conditions. For the isentropic flow of an ideal gas through a venturi, Y can be calculated theoretically by integrating Eq. (6-25) between stations a and b and combining the result with the definition of Y implied in Eq. (8-44). The result is

$$Y = \left(\frac{p_b}{p_a}\right)^{1/\gamma} \left\{ \frac{\gamma(1 - \beta^4)[1 - (p_b/p_a)^{1-1/\gamma}]}{(\gamma - 1)(1 - p_b/p_a)[1 - \beta^4(p_b/p_a)^{2/\gamma}]} \right\}^{1/2} \qquad (8\text{-}46)$$

Equation (8-46) shows that Y is a function of p_b/p_a, β, and γ. This equation cannot be used for orifices because of the vena contracta. An empirical equation in p_b/p_a, β, and γ for standard sharp-edged orifices is

$$Y = 1 - \frac{0.41 + 0.35\beta^4}{\gamma}\left(1 - \frac{p_b}{p_a}\right) \qquad\qquad (8\text{-}47)$$

Typical values of Y calculated for air by Eqs. (8-46) and (8-47) are shown in Fig. 8-23.

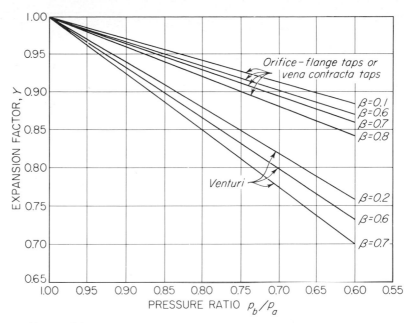

FIGURE 8-23
Expansion factors for air flowmeters. [*By permission, from V. L. Streeter (ed.),* "*Handbook of Fluid Dynamics,*" *p.* **14**–14. *Copyright,* 1961, *McGraw-Hill Book Company.*]

This figure must not be used when p_b/p_a is less than about 0.53, which is the critical pressure ratio at which air flow becomes sonic.

Pitot tube The pitot tube is a device to measure the local velocity along a streamline. The principle of the device is shown in Fig. 8-24. The opening of the impact tube a is perpendicular to the flow direction. The opening of the static tube b is parallel to the direction of flow. The two tubes are connected to the legs of a manometer or equivalent device for measuring small pressure differences. The static tube measures the static pressure p_0 since there is no velocity component perpendicular to its opening. The impact opening includes a stagnation point B at which the streamline AB terminates.

The pressure p_s, measured by the impact tube, is the stagnation pressure of the fluid given for ideal gases by Eq. (7-8). Then Eq. (7-9) applies, where P_o is the static pressure measured by tube b. Solving Eq. (7-9) for u_0 gives

$$u_0 = \left(\frac{2g_c(p_s - p_0)}{\rho_0\{1 + (N_{\text{Ma}}^2/4) + [(2 - \gamma)/24]N_{\text{Ma}}^4 + \cdots\}} \right)^{1/2} \tag{8-48}$$

Since the manometer of the pitot tube measures the pressure difference $p_s - p_0$, Eq. (8-48) gives the local velocity of the point where the impact tube is located. Normally, only the first Mach-number term in the equation is significant.

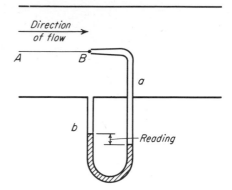

FIGURE 8-24
Principle of pitot tube.

For incompressible fluids, the Mach-number correction factor is unity, and Eq. (8-48) becomes simply

$$u_0 = \sqrt{\frac{2g_c(p_s - p_0)}{\rho}} \tag{8-49}$$

The velocity measured by an ideal pitot tube would conform exactly to Eq. (8-48). Well-designed instruments are in error by not more than 1 percent of theory, but when precise measurements are to be made, the pitot tube should be calibrated and an appropriate correction factor applied. This factor is used as a coefficient before the bracketed terms in Eq. (8-48). It is nearly unity in well-designed pitot tubes.

It should be noted that whereas the orifice and venturi meters measure the *average* velocity of the entire stream of fluid, the pitot tube measures the velocity at *one point only*. As discussed in Chap. 4, this velocity varies over the cross section of the pipe. Consequently, to obtain the true average velocity over the cross section, one of two procedures is used. The tube may be accurately centered at the axis of the pipe and the average velocity calculated from the maximum velocity by means of Fig. 5-7. If this procedure is used, care must be taken to insert the pitot tube at least 100 pipe diameters from any disturbance in the flow, so that the velocity distribution will be normal. The other procedure is to take readings at a number of known locations in the cross section of the pipe and calculate the average velocity $\bar{V}$ for the entire cross section by graphical integration of Eq. (4-4).

The disadvantages of the pitot tube are (1) that it does not give the average velocity directly and (2) that its readings for gases are extremely small. When it is used for measuring low-pressure gases, some form of multiplying gauge, like that shown in Fig. 2-5, must be also used.

EXAMPLE 8-6   Air at 200°F (93.3°C) is forced through a long, circular flue 36 in. (914 mm) in diameter. A pitot-tube reading is taken at the center of the flue at a sufficient distance from flow disturbances to ensure normal velocity distribution. The pitot reading is 0.54 in. (13.7 mm) H_2O, and the static pressure at the point of measurement is 15.25 in. (387 mm) H_2O. The coefficient of the pitot tube is 0.98.

Calculate the flow of air, in cubic feet per minute, measured at 60°F (15.6°C) and a barometric pressure of 29.92 in. (760 mm) Hg.

SOLUTION The velocity at the center of the flue, which is that measured by the instrument, is calculated by Eq. (8-48), using the coefficient 0.98 to correct for imperfections in the flow pattern caused by the presence of the tube. The necessary quantities are as follows. The absolute pressure at the instrument is

$$p = 29.92 + \frac{15.25}{13.6} = 31.04 \text{ in. Hg}$$

The density of the air at flowing conditions is

$$\rho = \frac{29 \times 492 \times 31.04}{359(460 + 200)(29.92)} = 0.0625 \text{ lb/ft}^3$$

From the manometer reading

$$p_s - p_0 = \frac{0.54}{12} 62.37 = 2.81 \text{ lb}_f/\text{ft}^2$$

By Eq. (8-49), the maximum velocity, assuming N_{Ma} is negligible, is

$$u_{max} = 0.98 \sqrt{2 \times 32.174 \frac{2.81}{0.0625}} = 52.7 \text{ ft/s}$$

This is sufficiently low for the Mach-number correction to be negligible. To obtain the average velocity from the maximum velocity, Fig. 5-7 is used. The Reynolds number, $N_{Re,max}$, is calculated as follows. From Appendix 9, the viscosity of air at 200°F is 0.022 cP:

$$N_{Re,} = \frac{(36/12)(52.7) \times 0.0625}{0.022 \times 0.000672} = 670,000$$

From Fig. 5-7

$$\frac{\overline{V}}{u_{max}} = 0.82 \qquad \overline{V} = 0.82 \times 52.7 = 43.2 \text{ ft/s}$$

The volumetric flow rate is

$$q = 43.2 \left(\frac{36}{12}\right)^2 \frac{\pi}{4} \frac{520}{660} \frac{31.04}{29.92} 60 = 14,980 \text{ ft}^3/\text{min} (7.07 \text{ m}^3/\text{s}) \qquad ////$$

Area meters: rotameters In the orifice, nozzle, or venturi, the variation of flow rate through a constant area generates a variable pressure drop, which is related to the flow rate. Another class of meters, called *area meters*, consist of devices in which the pressure drop is constant, or nearly so, and the area through which the fluid flows varies with flow rate. The area is related, through proper calibration, to the flow rate.

The most important area meter is the *rotameter*, which is shown in Fig. 8-25. It consists essentially of a gradually tapered glass tube mounted vertically in a frame with the large end up. The fluid flows upward through the tapered tube and suspends

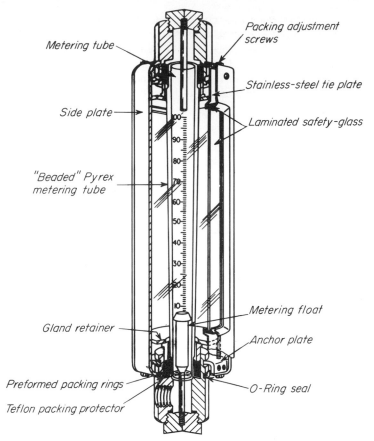

FIGURE 8-25
Rotameter. [*Courtesy, Fischer and Porter Co., Warminster, Penn.*]

freely a float (which actually does not float but is completely submerged in the fluid). The float is the indicating element, and the greater the flow rate, the higher the float rides in the tube. The entire fluid stream must flow through the annular space between the float and the tube wall. The tube is marked in divisions, and the reading of the meter is obtained from the scale reading at the reading edge of the float, which is taken at the largest cross section of the float. A calibration curve must be available to convert the observed scale reading to flow rate. Rotameters can be used for either liquid- or gas-flow measurement.

The bore of a glass rotameter tube is either an accurately formed plain conical taper or a taper with three beads, or flutes, parallel with the axis of the tube. The tube shown in Fig. 8-25 is a fluted tube. For opaque liquids, for high temperatures or pressures, or for other conditions where glass is impracticable, metal tubes are used. Metal tubes are plain tapered. Since in a metal tube the float is invisible, means must be provided for either indicating or transmitting the meter reading.

This is accomplished by attaching a rod, called an *extension*, to the top or bottom of the float and using the extension as an armature. The extension is enclosed in a fluidtight tube mounted on one of the fittings. Since the inside of this tube communicates directly with the interior of the rotameter, no stuffing box for the extension is needed. The tube is surrounded by external induction coils. The length of the extension exposed to the coils varies with the position of the float. This in turn changes the induction of the coil, and the variation of the induction is measured electrically to operate a control valve or to give a reading on a recorder. Also, a magnetic follower, mounted outside the extension tube and adjacent to a vertical scale, can be used as a visual indicator for the top edge of the extension. By such modifications the rotameter has developed from a simple visual indicating instrument using only glass tubes into a versatile recording and controlling device.

Floats may be constructed of metals of various densities from lead to aluminum or from glass or plastic. Stainless-steel floats are common. Float shapes and proportions are also varied for different applications.

THEORY AND CALIBRATION OF ROTAMETERS For a given flow rate, the equilibrium position of the float in a rotameter is established by a balance of three forces: (1) the weight of the float, (2) the buoyant force of the fluid on the float, and (3) the drag force on the float. Force 1 acts downward, and forces 2 and 3 upward. For equilibrium

$$F_D g_c = v_f \rho_f g - v_f \rho g \tag{8-50}$$

where F_D = drag force
  g = acceleration of gravity
  g_c = Newton's-law proportionality factor
  v_f = volume of float
  ρ_f = density of float
  ρ = density of fluid

The quantity v_f can be replaced by m_f/ρ_f, where m_f is the mass of the float, and Eq. (8-50) becomes

$$F_D g_c = m_f g \left(1 - \frac{\rho}{\rho_f} \right) \tag{8-51}$$

For a given meter operating on a definite fluid, the right-hand side of Eq. (8-51) is constant and independent of the flow rate. Accordingly, F_D is also constant, and if the flow rate is changed, the effect of the change in flow rate must be countered by the effect of the change of the position of the float. If, for example, the flow rate is increased, the increase in F_D from the added flow rate is compensated for by the decrease in F_D from the reduction in velocity through the larger space for flow at the new position of equilibrium.

Methods of constructing generalized calibration curves are given in the literature.[5] For a rotameter the relation between meter reading and flow rate is approximately linear, compared with the calibration curve for an orifice meter, for which the flow rate is proportional to the square root of the reading. The calibration of a rotameter, unlike that of an orifice meter, is not sensitive to the velocity distribution in the approaching stream, and neither long, straight approaches nor straightening vanes are necessary.

SYMBOLS

A_p	Cross-sectional area of channels at periphery of pump impeller, ft² or m³
C_o	Orifice coefficient, velocity of approach not included
C_p	Molal specific heat at constant pressure, Btu/lb-°F or J/g mol-°C
C_v	Venturi coefficient, velocity of approach not included
c_p	Specific heat at constant pressure, Btu/lb-°F or J/g-°C
c_v	Specific heat at constant volume, Btu/lb-°F or J/g-°C
D	Diameter, ft or m; D_a, of pipe; D_b, of venturi throat; D_f, of rotameter float; D_o, of orifice
F_D	Drag force, lb$_f$ or N
g	Gravitational acceleration, ft/s² or m/s²
g_c	Newton's-law proportionality factor, 32.174 ft-lb/lb$_f$-s²
H	Total head, ft-lb$_f$/lb or J/kg; H_a, at station a; H_b, at station b; H_{sv}, net positive suction head
h_f	Friction loss, ft-lb$_f$/lb or J/kg; h_{fs}, in pump suction line
M	Molecular weight
m	Mass, lb or kg; m_f, of rotameter float
$\dot{m}$	Mass flow rate, lb/s or kg/s
N_{Ma}	Mach number, dimensionless
$N_{Re,max}$	Maximum local Reynolds number in pipe, $Du_{max}\rho/\mu$
$N_{Re,o}$	Reynolds number at orifice, $D_o u_o \rho/\mu$
n	Constant in Eq. (8-29)
P	Power, ft-lb$_f$/s or W; P_B, power supplied to pump, hp or kW; P_f, fluid power in pump; P_{fr}, in ideal pump
p	Pressure, lb$_f$/ft² or atm; p_a, at station a; p'_a, at station a'; p_b, at station b; p_s, impact pressure; p_v, vapor pressure; p_0, static pressure; p', internal working pressure in pipe, lb$_f$/in.²
q	Volumetric flow rate, ft³/s or m³/s; q_r, through ideal pump; q_0, compressor capacity, std ft³/min
r	Radius, ft or m; r_1, of impeller at suction; r_2, of impeller at discharge
S	Cross-sectional area, ft² or m²; S_b, of venturi throat; S_0, of orifice; also allowable stress, lb$_f$/in.²
T	Absolute temperature, °R or K; T_a, at compressor inlet; T_b, at compressor discharge; also torque, ft-lb$_f$ or J
u	Local fluid velocity, ft/s or m/s; u_{max}, maximum velocity in pipe; u_o, at orifice; u_0, at impact point of pitot tube; u_1, peripheral velocity at inlet of pump impeller; u_2, at impeller discharge
V	Resultant velocity, absolute, in pump impeller, ft/s or m/s; V_{r2}, radial component of velocity V_2; V_u, tangential component; V_{u1}, of velocity V_1; V_{u2}, of velocity V_2; V_1, at suction; V_2, at discharge
$\bar{V}$	Average fluid velocity, ft/s or m/s; $\bar{V}_a$, at station a; $\bar{V}'_a$, at station a'; $\bar{V}_b$, at station b; $\bar{V}'_b$, at station b'
v	Fluid velocity relative to pump impeller, ft/s or m/s; v_1, at suction; v_2, at discharge
v_f	Volume of rotameter float, ft³ or m³
W_p	Pump work, ft-lb$_f$/lb or J/kg; W_{pr}, by ideal pump
Y	Expansion factor, flowmeter
Z	Height above datum plane, ft or m; Z_a, at station a; Z'_a, at station a'; Z_b, at station b; Z'_b, at station b'

Greek letters

α Kinetic-energy correction factor; α_a, at station a; α_b, at station b; $\alpha_{b'}$, at station b'; also angle between absolute and peripheral velocities in pump impeller; α_1, at suction; α_2, at discharge

β Vane angle in pump impeller; β_1, at suction; β_2, at discharge; also ratio, diameter of orifice or venturi throat to diameter of pipe

γ Ratio of specific heats, c_p/c_v

ΔH Head developed by pump; ΔH_r, in frictionless or ideal pump

η Overall mechanical efficiency of pump, fan, or blower

μ Absolute viscosity, lb/ft-s or cP

ρ Density, lb/ft^3 or kg/m^3; ρ_a, at station a; ρ_b, at station b; ρ_f, of rotameter float; ρ_0, of fluid approaching pitot tube; $\bar{\rho}$, average density $(\rho_a + \rho_b)/2$

ω Angular velocity, rad/s

PROBLEMS

8-1 Make a preliminary estimate of the approximate pipe size required for the following services: (*a*) a transcontinental pipeline to handle 6,000 std ft^3/min of natural gas at an average pressure of 50 lb$_f$/in.^{2} abs and an average temperature of 60°F; (*b*) feeding a slurry of *p*-nitrophenol crystals in water to a continuous centrifugal separator at the rate of 2,000 lb/h of solids. The slurry carries 45 percent solids by weight. For *p*-nitrophenol $\rho = 92.2$ lb/ft^3.

8-2 Air entering at 70°F and atmospheric pressure is to be compressed to 4,000 lb$_f$/in.^{2} gauge in a reciprocating compressor at the rate of 125 std ft^3/min. How many stages should be used? What is the theoretical shaft work per standard cubic foot for frictionless adiabatic compression? What is the brake horse-power if the efficiency of each stage is 85 percent? For air $\gamma = 1.40$.

8-3 What is the discharge temperature of the air from the first stage in Prob. 8-2?

8-4 After the installation of the orifice meter of Example 8-5, the manometer reading at a definite constant flow rate is 1.75 in. Calculate the flow through the line in barrels per day measured at 60°F.

8-5 Natural gas having a specific gravity relative to air of 0.60 and a viscosity of 0.011 cP is flowing through a Schedule 40 6-in. pipe in which is installed a standard sharp-edged orifice equipped with flange taps. The gas is at 100°F and 20 lb$_f$/in.^{2} abs at the upstream tap. The manometer reading is 46.3 in. of water at 60°F. The ratio of specific heats for natural gas is 1.30. The diameter of the orifice is 2.00 in. Calculate the rate of flow of gas through the line in cubic feet per minute based on a pressure of 14.4 lb$_f$/in.^{2} and a temperature of 60°F.

8-6 A horizontal venturi meter having a throat diameter of 20 mm is set in a 75-mm-ID pipeline. Water at 15°C is flowing through the line. A manometer containing mercury under water measures the pressure differential over the instrument. When the manometer reading is 500 mm, what is the flow rate in gallons per minute? If 12 percent of the differential is permanently lost, what is the power consumption of the meter?

8-7 It is proposed to pump 10,000 kg/h of toluene at 114°C and 1.1 atm abs pressure from the reboiler of a distillation tower to a second distillation unit/without cooling the toluene before it enters the pump. If the friction loss in the line between the reboiler and pump is 7 kN/m^2 and the density of toluene is 866 kg/m^3, how far above the pump must the liquid level in the reboiler be maintained to give a net positive suction head of 2.5 m?

8-8 Calculate the power required to drive the pump in Prob. 8-7 if the pump is to elevate the toluene 10 m, the pressure in the second unit is atmospheric, and the friction loss in the discharge line is 35 kN/m².

REFERENCES

1 Bauermeister, K. J.: *Chem. Proc. Eng.,* **50**(9):79 (1969).
2 Church, A. H.: "Centrifugal Pumps and Blowers," p. 87, Wiley, New York, 1944.
3 "Fluid Meters: Their Theory and Applications," 5th ed., American Society of Mechanical Engineers, New York, 1959; (*a*) p. 50, (*b*) p. 72.
4 Jorissen, A. L.: *Trans. ASME,* **74**:905 (1952).
5 Martin, J. J.: *Chem. Eng. Prog.,* **45**:338 (1949).
6 Stafford, J. D., Jr.: *Chem. Eng. Prog.,* **67**(4):54 (1971).

9

AGITATION AND MIXING OF LIQUIDS

Many processing operations depend for their success on the effective agitation and mixing of fluids. Though often confused, agitation and mixing are not synonymous. Agitation refers to the induced motion of a material in a specified way, usually in a circulatory pattern inside some sort of container. Mixing is the random distribution, into and through one another, of two or more initially separate phases. A single homogeneous material, such as a tankful of cold water, can be agitated, but it cannot be mixed until some other material (such as a quantity of hot water or some powdered solid) is added to it.

The term *mixing* is applied to a variety of operations, differing widely in the degree of homogeneity of the "mixed" material. Consider, in one case, two gases which are brought together and thoroughly blended, and, in a second case, sand, gravel, cement, and water which are tumbled in a rotating drum for a long time. In both cases the final product is said to be mixed. Yet the products are obviously not equally homogeneous. Samples of the mixed gases—even very small samples—all have the same composition. Small samples of the mixed concrete, on the other hand, obviously differ widely in composition.

This chapter deals with the agitation of liquids of low to moderate viscosity and the mixing of liquids and liquid-solid suspensions. Agitation and mixing of highly viscous liquids, pastes, and dry solid powders are discussed in Chap. 29.

AGITATION OF LIQUIDS

Purposes of agitation   Liquids are agitated for a number of purposes, depending on the objectives of the processing step. These purposes include:

1   Suspending solid particles
2   Blending miscible liquids, e.g., methyl alcohol and water
3   Dispersing a gas through the liquid in the form of small bubbles
4   Dispersing a second liquid, immiscible with the first, to form an emulsion or suspension of fine drops
5   Promoting heat transfer between the liquid and a coil or jacket

Often one agitator serves several purposes at the same time, as in the catalytic hydrogenation of a liquid. In a hydrogenation vessel the hydrogen gas is dispersed through the liquid in which solid particles of catalyst are suspended, with the heat of reaction simultaneously removed by a cooling coil and jacket.

Agitation equipment   Liquids are most often agitated in some kind of tank or vessel, usually cylindrical in form and with a vertical axis. The top of the vessel may be open to the air, or it may be closed. The proportions of the tank vary widely, depending on the nature of the agitation problem. A standardized design such as that shown in Fig. 9-1, however, is applicable in many situations. The tank bottom is rounded, not flat, to eliminate sharp corners or regions into which the fluid currents would not penetrate. The liquid depth is approximately equal to the diameter of the tank. An impeller is mounted on an overhung shaft, i.e., a shaft supported from above. The shaft is driven by a motor, sometimes directly connected to the shaft but more often connected to it through a speed-reducing gearbox. Accessories such as inlet and outlet lines, coils, jackets, and wells for thermometers or other temperature-measuring devices are usually included.

The impeller creates a flow pattern in the system, causing the liquid to circulate through the vessel and return eventually to the impeller. Flow patterns in agitated vessels are discussed in detail later in this chapter.

Impellers   Impeller agitators are divided into two classes: those which generate currents parallel with the axis of the impeller shaft and those which generate currents in a tangential or radial direction. The first are called *axial-flow impellers*, the second *radial-flow impellers*.

The three main types of impellers are propellers, paddles, and turbines. Each type includes many variations and subtypes, which will not be considered here. Other special impellers are also useful in certain situations, but the three main types solve perhaps 95 percent of all liquid-agitation problems.

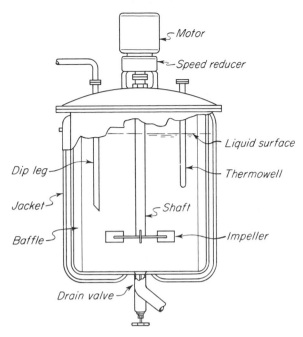

FIGURE 9-1
Typical agitation process vessel.

PROPELLERS A propeller is an axial-flow, high-speed impeller for liquids of low viscosity. Small propellers turn at full motor speed, either 1,150 or 1,750 r/min; larger ones turn at 400 to 800 r/min. The flow currents leaving the impeller continue through the liquid in a given direction until deflected by the floor or wall of the vessel. The highly turbulent swirling column of liquid leaving the impeller entrains stagnant liquid as it moves along, probably considerably more than an equivalent column from a stationary nozzle would. The propeller blades vigorously cut or shear the liquid. Because of the persistence of the flow currents, propeller agitators are effective in very large vessels.

A revolving propeller traces out a helix in the fluid, and if there were no slip between liquid and propeller, one full revolution would move the liquid longitudinally a fixed distance depending on the angle of inclination of the propeller blades. The ratio of this distance to the propeller diameter is known as the *pitch* of the propeller. A propeller with a pitch of 1.0 is said to have *square pitch*.

Various propeller designs are illustrated in Fig. 9-2. Standard three-bladed marine propellers with square pitch are most common; four-bladed, toothed, and other designs are employed for special purposes.

Propellers rarely exceed 18 in. in diameter regardless of the size of the vessel. In a deep tank two or more propellers may be mounted on the same shaft, usually directing the liquid in the same direction. Sometimes two propellers work in opposite

FIGURE 9-2
Mixing propellers: (*a*) standard three-blade; (*b*) weedless; (*c*) guarded.

(a)   *(b)*   *(c)*

directions, or in "push-pull," to create a zone of especially high turbulence between them.

PADDLES   For the simpler problems an effective agitator consists of a flat paddle turning on a vertical shaft. Two-bladed and four-bladed paddles are common. Sometimes the blades are pitched; more often they are vertical. Paddles turn at slow to moderate speeds in the center of a vessel; they push the liquid radially and tangentially with almost no vertical motion at the impeller unless the blades are pitched. The currents they generate travel outward to the vessel wall and then either upward or downward. In deep tanks several paddles are mounted one above the other on the same shaft. In some designs the blades conform to the shape of a dished or hemispherical vessel so that they scrape the surface or pass over it with close clearance. A paddle of this kind is known as an *anchor agitator*. Anchors are useful for preventing deposits on a heat-transfer surface, as in a jacketed process vessel, but they are poor mixers. They nearly always operate in conjunction with a higher-speed paddle or other agitator, usually turning in the opposite direction.

Industrial paddle agitators turn at speeds between 20 and 150 r/min. The total length of a paddle impeller is typically 50 to 80 percent of the inside diameter of the vessel. The width of the blade is one-sixth to one-tenth its length. At very slow speeds a paddle gives mild agitation in an unbaffled vessel; at higher speeds baffles become necessary. Otherwise the liquid is swirled around the vessel at high speed but with little mixing.

TURBINES   Some of the many designs of turbine are shown in Fig. 9-3. Most of them resemble multibladed paddle agitators with short blades, turning at high speeds on a shaft mounted centrally in the vessel. The blades may be straight or curved, pitched or vertical. The impeller may be open, semienclosed, or shrouded. The diameter of the impeller is smaller than with paddles, ranging from 30 to 50 percent of the diameter of the vessel.

Turbines are effective over a very wide range of viscosities. In low-viscosity liquids turbines generate strong currents which persist throughout the vessel, seeking out and destroying stagnant pockets. Near the impeller is a zone of rapid currents, high turbulence, and intense shear. The principal currents are radial and tangential. The tangential components induce vortexing and swirling, which must be stopped by baffles or by a diffuser ring if the impeller is to be most effective.

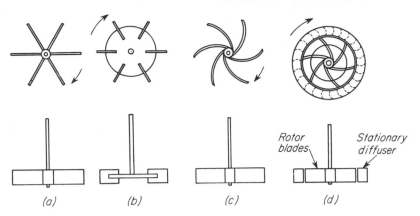

FIGURE 9-3
Turbine impellers: (*a*) open straight blade; (*b*) bladed disk; (*c*) vertical curved blade; (*d*) shrouded curved blade with diffuser ring.

Flow patterns in agitated vessels   The type of flow in an agitated vessel depends on the type of impeller, the characteristics of the fluid, and the size and proportions of the tank, baffles, and agitator. The velocity of the fluid at any point in the tank has three components, and the overall flow pattern in the tank depends on the variations in these three velocity components from point to point. The first velocity component is radial and acts in a direction perpendicular to the shaft of the impeller. The second component is longitudinal and acts in a direction parallel with the shaft. The third component is tangential, or rotational, and acts in a direction tangent to a circular path around the shaft. In the usual case of a vertical shaft, the radial and tangential components are in a horizontal plane, and the longitudinal component is vertical. The radial and longitudinal components are useful and provide the flow necessary for the mixing action. When the shaft is vertical and centrally located in the tank, the tangential component is generally disadvantageous. The tangential flow follows a circular path around the shaft, creates a vortex at the surface of the liquid, as shown in Fig. 9-4, and tends to perpetuate, by a laminar-flow circulation, stratification at the various levels without accomplishing longitudinal flow between levels. If solid particles are present, circulatory currents tend to throw the particles to the outside by centrifugal force, from where they move downward and to the center of the tank at the bottom. Instead of mixing, its reverse, concentration, occurs. Since, in circulatory flow, the liquid flows with the direction of motion of the impeller blades, the relative velocity between the blades and the liquid is reduced, and the power that can be absorbed by the liquid is limited. In an unbaffled vessel circulatory flow is induced by all types of impellers, whether axial flow or radial flow. In fact, if the swirling is strong, the flow pattern in the tank is virtually the same regardless of the design of the impeller. At high impeller speeds the vortex may be so deep that it reaches the impeller, and gas from above the liquid is drawn down into the charge. Generally this is undesirable.

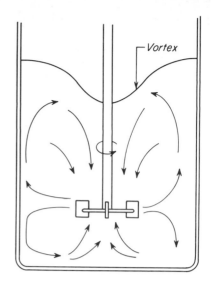

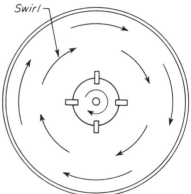

FIGURE 9-4
Vortex formation and circulation pattern in an agitated tank. [*After Lyons.*[17]]

PREVENTION OF SWIRLING   Circulatory flow and swirling can be prevented by any of three methods. In small tanks, the impeller can be mounted off center, as shown in Fig. 9-5. The shaft is moved away from the centerline of the tank, then tilted in a plane perpendicular to the direction of the move. In larger tanks, the agitator may be mounted in the side of the tank, with the shaft in a horizontal plane but at an angle with a radius, as shown in Fig. 9-6.

In large tanks with vertical agitators, the preferable method of reducing swirling is to install baffles, which impede rotational flow without interfering with radial or longitudinal flow. A simple and effective baffling is attained by installing vertical strips perpendicular to the wall of the tank. Baffles of this type and the flow pattern resulting from their use are shown in Fig. 9-7. Except in very large tanks, four baffles are sufficient to prevent swirling and vortex formation. For turbines, the width of the baffle need be no more than one-twelfth the tank diameter; for propellers, no more

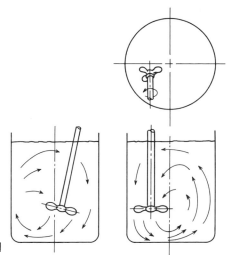

FIGURE 9-5
Off-center impeller. [*After Bissell et al.*[3]]

than one-eighteenth the tank diameter.[3] With side-entering, inclined, or off-center propellers baffles are not needed.

Shrouded impellers and diffuser rings can be used with turbines to stop swirling, in place of baffles, as shown in Fig. 9-3*d*. These devices are adapted from centrifugal-pump practice, but their main effect is to add friction and reduce circulation. Diffuser rings also tend to prevent the currents from reaching the far corners of the tank, and they are difficult to install and maintain. They are useful if intense shear and abnormal turbulence are desired at the impeller discharge.

Once the swirling is stopped, the specific flow pattern in the vessel depends on the type of impeller. Propeller agitators drive the liquid straight down to the bottom of the tank, where the stream spreads radially in all directions toward the wall, flows upward along the wall, and returns to the suction of the propeller from the top. This pattern is shown in Fig. 9-7. Propellers are used when strong vertical currents are desired, e.g., when heavy solid particles are to be kept in suspension. They are not ordinarily used when the viscosity of the liquid is greater than about 50 P.

Paddle agitators give good radial flow in the immediate plane of the impeller blades but are poor in developing vertical currents. This is the main limitation of paddle agitators. As a result paddles are ineffective in suspending solids.

Turbine impellers drive the liquid radially against the wall, where the stream divides, one portion flowing downward to the bottom and back to the center of the impeller from below, and the other flowing upward toward the surface and back to the impeller from above. As shown in Fig. 9-4, two separate circulation currents are generated. Turbines are especially effective in developing radial currents, but they also induce vertical flows, especially when baffled. They are excellent in mixing liquids having about the same specific gravity.

In a vertical cylindrical tank, the depth of the liquid should be equal to, or somewhat greater than, the diameter of the tank. If greater depth is desired, two or

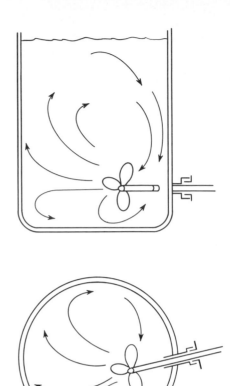

FIGURE 9-6
Side-entering propeller. [*After Bissell et al.*[3]]

more impellers are mounted on the same shaft, and each impeller acts as a separate mixer. Two circulation currents are generated for each impeller, as shown in Fig. 9-8. The bottom impeller, either of the turbine or the propeller type, is mounted about one impeller diameter above the bottom of the tank.

DRAFT TUBES   The return flow to an impeller of any type approaches the impeller from all directions, as it is not under the control of solid surfaces. The flow to and from a propeller, for example, is essentially similar to the flow of air to and from a fan operating in a room. In most applications of impeller mixers this is not a limitation, but when the direction and velocity of flow to the suction of the impeller are to be controlled, draft tubes are used, as shown in Fig. 9-9. These devices may be useful when high shear in the impeller itself is desired, as in the manufacture of certain emulsions, or where solid particles that tend to float on the surface of the liquid in the tank are to be dispersed in the liquid. Draft tubes for propellers are mounted around the impeller, and those for turbines are mounted immediately above the impeller. This is shown in Fig. 9-9. Draft tubes add to the fluid friction in the system, and, for

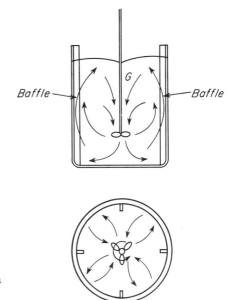

FIGURE 9-7
Flow pattern in a baffled tank with a centrally mounted propeller agitator.

a given power input, they reduce the rate of flow, so they are not used unless they are required.

"Standard" turbine design[24]   The designer of an agitated vessel has an unusually large number of choices to make as to type and location of the impeller, the proportions of the vessel, the number and proportions of the baffles, and so forth. Each of

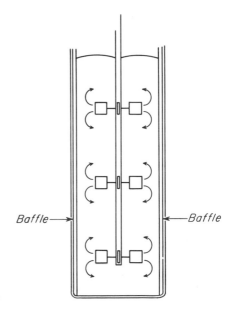

FIGURE 9-8
Multiple turbines in tall tank.

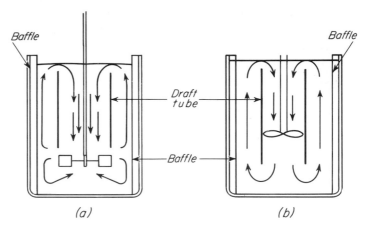

FIGURE 9-9
Draft tubes, baffled tank: (*a*) turbine; (*b*) propeller. [*After Bissell et al.*[3]]

these decisions affects the circulation rate of the liquid, the velocity patterns, and the power consumed. As a starting point for design in ordinary agitation problems, a turbine agitator of the type shown in Fig. 9-10 is commonly used. Typical proportions are

$$\frac{D_a}{D_t} = 1 \qquad \frac{H}{D_t} = 1 \qquad \frac{J}{D_t} = \frac{1}{12}$$

$$\frac{E}{D_a} = 1 \qquad \frac{W}{D_a} = \frac{1}{5} \qquad \frac{L}{D_a} = \frac{1}{4}$$

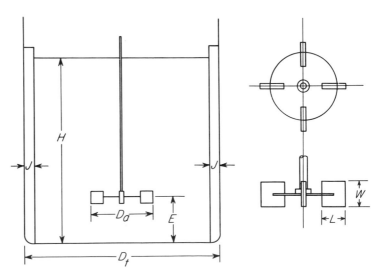

FIGURE 9-10
Measurements of turbine. [*After Rushton.*[26]]

The number of baffles is usually 4; the number of impeller blades ranges from 4 to 16 but is generally 6 or 8. Special situations may, of course, dictate different proportions from those listed above; it may be advantageous, for example, to place the agitator higher or lower in the tank, or a much deeper tank may be needed to achieve the desired process result. The listed "standard" proportions, nonetheless, are widely accepted and are the basis of many published correlations of agitator performance.

CIRCULATION, VELOCITIES, AND POWER CONSUMPTION IN AGITATED VESSELS

For a processing vessel to be effective, regardless of the nature of the agitation problem, the volume of fluid circulated by the impeller must be sufficient to sweep out the entire vessel in a reasonable time. Also, the velocity of the stream leaving the impeller must be sufficient to carry the currents to the remotest parts of the tank. In mixing and dispersion operations the circulation rate is not the only factor, or even the most important one; turbulence in the moving stream often governs the effectiveness of the operation. Turbulence results from properly directed currents and large velocity gradients in the liquid. Circulation and turbulence generation both consume energy; the relations between power input and the design parameters of agitated vessels are discussed later. Some agitation problems, as will be shown, call for large flows and relatively low turbulence, while others require high turbulence and relatively low circulation rates.

Flow number A turbine or propeller agitator is, in essence, a pump impeller operating without a casing and with undirected inlet and output flows. The governing relations are similar to those for centrifugal pumps discussed in chapter 8^{13a}. Consider the flat-bladed turbine impeller shown in Fig. 9-11. The nomenclature is the same as in Fig. 8-10: u_2 is the velocity of the blade tips; V_{u2} and V_{r2} are the tangential velocities of the liquid leaving the blade tips, respectively; and V_2 is the total liquid velocity at the same point. Assume that the tangential liquid velocity is some fraction k of the blade-tip velocity, or

$$V_{u2} = ku_2 = k\pi D_a n \tag{9-1}$$

since $u_2 = \pi D_a n$. The volumetric flow rate through the impeller, from Eq. (8-20), is

$$q = V_{r2} A_p \tag{9-2}$$

Here A_p is taken to be area of the cylinder swept out by the tips of the impeller blades, or

$$A_p = \pi D_a W \tag{9-3}$$

where D_a = impeller diameter
$\qquad W$ = width of blades

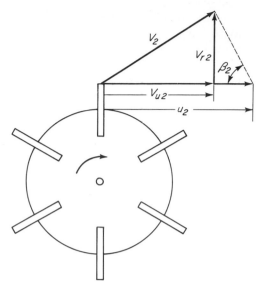

FIGURE 9-11
Velocity vectors at tip of turbine impeller
blade.

From the geometry of Fig. 9-11

$$V_{r2} = (u_2 - V_{u2}) \tan \beta_2 \qquad (9\text{-}4)$$

Substituting for V_{u2} from Eq. (9-1) gives

$$V_{r2} = \pi D_a n (1 - k) \tan \beta_2 \qquad (9\text{-}5)$$

The volumetric flow rate, from Eqs. (9-2) to (9-4), is therefore

$$q = \pi^2 D_a^2 n W (1 - k) \tan \beta_2 \qquad (9\text{-}6)$$

For geometrically similar impellers W is proportional to D_a, and hence, for given values of k and β_2†

$$q \propto n D_a^3$$

The ratio of these two quantities is called the *flow number* N_Q, which is defined by

$$N_Q \equiv \frac{q}{n D_a^3} \qquad (9\text{-}8)$$

Equations (9-6) to (9-8) show that if β_2 is fixed, N_Q is constant. In general, angle β_2, the angle at which liquid leaves the impeller, is not equal to the angle of the tip of the blades. For marine propellers, β_2 may be considered constant; for turbines

† It may be shown that for a given impeller, k and β_2 are related by the equation [13a]

$$\tan \beta_2 = \sqrt{\frac{2k}{1 - k}} \qquad (9\text{-}7)$$

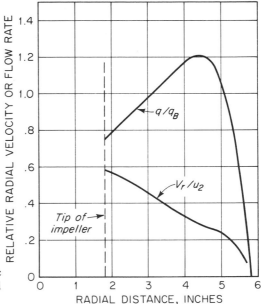

FIGURE 9-12
Radial velocity V_r/u_2 and volumetric flow rate q/q_B in a turbine agitated vessel. [*After Cutter.*[9]]

it is a function of the relative sizes of the impeller and the tank. For design of baffled agitated vessels the following values are recommended:[13b]

For marine propellers: $N_Q = 0.5$

For six-bladed turbines with $W/D_a = 0.2$:

$$N_Q = 0.93 \frac{D_t}{D_a}$$

where D_t is the vessel diameter.

Velocity gradients and velocity patterns The velocity gradient in an agitated vessel varies widely from one point to another in the fluid. For example, if a fast-moving jet issues from the impeller and passes through more or less stagnant liquid, the velocity gradient at the edge of the jet may be very large compared with the gradients in the bulk of the liquid. Furthermore, the gradient at the edge of the jet changes with distance. As the jet travels away from the impeller, it entrains some of the bulk liquid, slowing the jet down, and some of the bulk liquid is accelerated in the direction of motion of the jet. The velocity gradient at the edge of the jet diminishes. Velocity gradients are therefore greatest in the vicinity of the impeller discharge; elsewhere in the vessel, although velocity gradients exist, they are usually small compared with those near the impeller.

The radial component of the fluid velocity V_r, as shown in Fig. 9-12, is about 0.6 times the impeller tip speed u_2 as the liquid leaves the impeller blades; it drops more or less linearly with radial distance to low values near the wall.[9] The volumetric

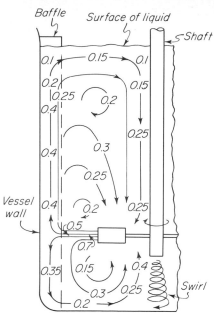

FIGURE 9-13
Velocity patterns in turbine agitator. [*After Morrison et al.*[21]]

flow rate of the impeller stream q at the impeller tips is 0.75 times the flow rate q_B that would exist if the fluid were all moving at the blade-tip velocity across the sides of the cylinder swept out by the blades. Because of entrainment, the total flow in the impeller stream rises to 1.2 times the theoretical value, then drops precipitously near the vessel wall.

Figure 9-13 shows the fluid currents observed with a six-bladed turbine, 6 in. in diameter, turning at 200 r/min in a 12-in. vessel containing cold water.[21] The plane of observation passes through the axis of the impeller shaft and immediately in front of a radial baffle. Fluid leaves the impeller in a radial direction, separates into longitudinal streams flowing upward or downward over the baffle, flows inward toward the impeller shaft, and ultimately returns to the impeller intake. At the bottom of the vessel, immediately under the shaft, the fluid moves in a swirling motion; elsewhere the currents are primarily radial or longitudinal.

The numbers on Fig. 9-13 indicate the scalar magnitude of the fluid velocity at various points, as fractions of the velocity of the tip of the impeller blades. Under the conditions used the tip velocity is 5.2 ft/s. The velocity in the jet quickly drops from the tip velocity to about 0.4 times the tip velocity at the vessel wall. Velocities at other locations in the vessel are of the order of 0.25 times the tip velocity, although there are two toroidal regions of almost stagnant fluid, one above and one below the impeller, in which the velocity is only 0.10 to 0.15 times the tip velocity.

Increasing the impeller speed increases the tip velocity and the circulation rate. It does not, however, increase the fluid velocity at a given location in the same proportion, for a fast-moving jet entrains much more material from the bulk of the liquid than a slower-moving jet does, and the jet velocity drops very quickly with increasing distance from the impeller. At a point $2\frac{1}{2}$ in. from the impeller in the vessel shown in Fig. 9-13, the local fluid velocity was found to increase with the cube root of the impeller speed.

Power consumption An important consideration in the design of an agitated vessel is the power required to drive the impeller. Although this power requirement cannot be estimated theoretically, even in the simplest agitated system, the equations developed in Chap. 8 for a centrifugal pump are a useful starting point. The power input to a frictionless impeller, from Eq. (8-13), is

$$Pg_c = \dot{m}\omega r_2 V_{u2} = \dot{m}\pi D_a n V_{u2} \tag{9-9}$$

Substitution for V_{u2} from Eq. (9-1) and with $\dot{m} = \rho q$ gives

$$Pg_c = \pi^2 \rho q k D_a^2 n^2 \tag{9-10}$$

Substitution of q from Eq. (9-6) gives the result

$$Pg_c = \pi^4 \rho D_a^4 W n^3 (1 - k) \tan \beta_2 \tag{9-11}$$

or, in dimensionless form,

$$\frac{Pg_c}{n^3 D_a^5 \rho} = \pi^4 \frac{W}{D_a} (1 - k) \tan \beta_2 \tag{9-12}$$

The left-hand side of Eq. (9-12) is called the power number, N_P, defined by

$$N_P \equiv \frac{Pg_c}{n^3 D_a^5 \rho} \tag{9-13}$$

As shown by Eq. (9-12), N_P depends on the ratio W/D_a and also on k and β_2. The values of k and β_2 vary with many parameters of the system; they cannot be predicted theoretically but must be found by experiment. Equation (9-12), therefore, cannot be used without further information to predict the power input to an agitated vessel; however, it provides a useful means of correlating experimental data.

The power number, flow number, and k are related as follows. Dividing both sides of Eq. (9-10) by $\rho n^3 D_a^5$ gives

$$\frac{Pg_c}{n^3 D_a^5 \rho} = \frac{\pi^2 kq}{nD_a^3} \tag{9-14}$$

Hence, from Eqs. (9-8) and (9-13)

$$N_P = \pi^2 k N_Q \tag{9-15a}$$

or

$$k = \pi^2 \frac{N_P}{N_Q} \tag{9-15b}$$

Power correlations To estimate the power required to rotate a given impeller at a given speed, empirical correlations of power (or power number) with the other variables of the system are needed. The form of such correlations can be found by dimensional analysis, given the important measurements of the tank and impeller, the distance of the impeller from the tank floor, the liquid depth, and the dimensions of the baffles if they are used. The number and arrangement of the baffles and the number of blades in the impeller must also be fixed. The variables that enter the analysis are the important measurements of tank and impeller, the viscosity μ and the density ρ of the liquid, the speed n, and, because Newton's law applies, the dimensional constant g_c. Also, unless provision is made to eliminate swirling, a vortex will appear at the surface of the liquid. Some of the liquid must be lifted above the average, or unagitated, level of the liquid surface, and this lift must overcome the force of gravity. Accordingly, the acceleration of gravity g must be considered as a factor in the analysis.

The various linear measurements can all be converted to dimensionless ratios, called *shape factors*, by dividing each of them by one of their number which is arbitrarily chosen as a basis. The diameter of the impeller D_a is a suitable choice for this base measurement, and the shape factors are calculated by dividing each of the remaining measurements by the magnitude of D_a. Let the shape factors, so defined, be denoted by $S_1, S_2, S_3, \ldots, S_n$. The impeller diameter D_a is then also taken as the measure of the size of the equipment and used as a variable in the analysis, just as the diameter of the pipe was in the dimensional analysis of friction in pipes. Two mixers of the same geometrical proportions throughout but of different sizes will have identical shape factors but will differ in the magnitude of D_a. Devices meeting this requirement are said to be geometrically similar or to possess geometrical similarity.

When the shape factors are temporarily ignored and the liquid is assumed Newtonian, the power P is a function of the remaining variables, or

$$P = \psi(n, D_a, g_c, \mu, g, \rho) \tag{9-16}$$

Application of the method of dimensional analysis outlined in Appendix 4 gives the result[26]

$$\frac{Pg_c}{n^3 D_a^5 \rho} = \psi\left(\frac{nD_a^2\rho}{\mu}, \frac{n^2 D_a}{g}\right) \tag{9-17}$$

By taking account of the shape factors, Eq. (9-17) can be written

$$\frac{Pg_c}{n^3 D_a^5 \rho} = \psi\left(\frac{nD_a^2\rho}{\mu}, \frac{n^2 D_a}{g}, S_1, S_2, \ldots, S_n\right) \tag{9-18}$$

The first dimensionless group in Eq. (9-17), $Pg_c/n^3 D_a^5 \rho$, is the power number N_P. The second, $nD_a^2\rho/\mu$, is a Reynolds number N_{Re}; the third, $n^2 D_a/g$, is the Froude number N_{Fr}. Equation (9-18) can therefore be written

$$N_P = \psi(N_{Re}, N_{Fr}, S_1, S_2, \ldots, S_n) \tag{9-19}$$

Significance of dimensionless groups[15]   The three dimensionless groups in Eq. (9-17) may be given simple interpretations. Consider the group $nD_a^2\rho/\mu$. Since the peripheral speed of the tip of the impeller u_2 is given by

$$u_2 = \pi D_a n \tag{9-20}$$

then

$$N_{Re} = \frac{nD_a^2 \rho}{\mu} = \frac{(nD_a)D_a\rho}{\mu} \propto \frac{u_2 D_a \rho}{\mu} \tag{9-21}$$

and this group is proportional to a Reynolds number calculated from the diameter and peripheral speed of the impeller. This is the reason for the name of the group.

The power number N_P is analogous to a friction factor or a drag coefficient. It is proportional to the ratio of the drag force acting on a unit area of the impeller and the inertial stress. The inertial stress, in turn, is related to the flow of momentum associated with the bulk motion of the fluid.

The Froude number N_{Fr} is a measure of the ratio of the inertial stress to the gravitational force per unit area acting on the fluid. It appears in fluid-dynamic situations where there is significant wave motion on a liquid surface. It is especially important in ship design.

Since the individual stresses are arbitrarily defined and vary strongly from point to point in the container, their local numerical values are not significant. The magnitudes of the dimensionless groups for the entire system, however, are significant to the extent that they provide correlating magnitudes that yield much simpler empirical equations than those based on Eq. (9-16). The following equations for the power number are examples of such correlations.

Power correlations for specific impellers The various shape factors in Eq. (9-19) depend on the type and arrangement of the equipment. The necessary measurements for a typical turbine-agitated vessel are shown in Fig. 9-10; the corresponding shape factors for this mixer are $S_1 = D_t/D_a$, $S_2 = E/D_a$, $S_3 = L/D_a$, $S_4 = W/D_a$, $S_5 = J/D_t$, and $S_6 = H/D_t$. In addition the number of baffles and the number of impeller blades must be specified. If a propeller is used, its pitch is an important measurement.

The effect of the Froude number appears when there is vortex formation, and then only if the Reynolds number is greater than 300. For baffled tanks, or for side-entering propellers, or for Reynolds numbers less than 300, a vortex does not form, and the Froude number is not a factor. The Froude number is taken into account, when it is a factor, by the exponential equation

$$\frac{N_P}{N_{Fr}^m} = \psi(N_{Re}, S_1, S_2, \ldots, S_n) = \phi \tag{9-22}$$

Quantity ϕ is called the power function.

The exponent m in Eq. (9-22) is, for a given set of shape factors, empirically related to the Reynolds number by the equation[26]

$$m = \frac{a - \log N_{Re}}{b} \tag{9-23}$$

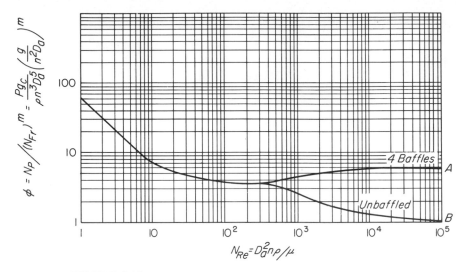

FIGURE 9-14
Power function ϕ versus N_{Re} for six-blade turbine.

where a and b are constants. The magnitudes of a and b for the curves of Figs. 9-14 and 9-15 are given in Table 9-1.

Equation (9-22) is applied by determining the quantity ϕ experimentally as a function of the Reynolds number for constant shape factors and plotting separate curves of ϕ vs. N_{Re} for each set of shape factors. Curves for a number of typical designs and types of impellers have been determined.[1,24,26]

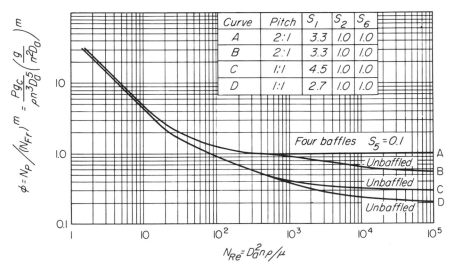

FIGURE 9-15
Power function ϕ versus N_{Re} for three-bladed propellers.[26]

A typical plot of ϕ vs. N_{Re} applying to tanks fitted with centrally located vertical flat-bladed turbines with six blades is shown in Fig. 9-14. The important shape factors are $S_1 = 3$, $S_2 = 1.0$, $S_3 = 0.25$, $S_6 = 1.0$. When baffled with four baffles, each of width one-tenth the tank diameter ($S_5 = 0.1$), curve A applies, and $\phi = N_P$. Without baffles, curve B applies, and the Froude number must be included in ϕ for all Reynolds numbers greater than 300.

Typical curves for three-bladed propeller mixers are shown in Fig. 9-15. For all curves, the propeller is one diameter from the bottom of the tank, and $S_2 = 1.0$. Curves A and B apply for blades having a pitch of 2.0 and a tank-to-propeller ratio $S_1 = 3.3$. Curve A applies to baffled tanks with four baffles, where $S_5 = 0.1$. Curve B applies to unbaffled tanks. Curves C and D show the effect on N_P of changing the scale factor S_1, the tank-to-propeller ratio, when the pitch is about 1.0. At low Reynolds numbers the effect of changing S_1 fades. Curve C applies where $S_1 = 4.5$, and curve D where $S_1 = 2.7$.

Effect of system geometry The effects on N_P of the shape factors $S_1, S_2, \ldots, S_n$ in Eq. (9-19) are sometimes small and sometimes very large. Sometimes two or more of the factors are interrelated; i.e., the effect of changing S_1, say, may depend on the magnitude of S_2 or S_3. With a flat-bladed turbine operating at high Reynolds numbers in a baffled tank, the effects of changing the system geometry may be summarized as follows[1]:

1 Increasing S_1, the ratio of tank diameter to impeller diameter, increases N_P when the baffles are few and narrow and decreases N_P when the baffles are many and wide. Thus shape factors S_1 and S_5 are interrelated. With four baffles and S_5 equal to $\frac{1}{12}$, as is common in industrial practice, changing S_1 has almost no effect on N_P.

2 The effect of changing S_2 depends on the design of the turbine. Increasing S_2 increases N_P for a disk turbine of the type shown in Fig. 9-10. For a pitched-blade turbine increasing S_2 lowers N_P considerably; for an open straight-bladed turbine it lowers N_P slightly.

3 With a straight-bladed open turbine the effect of changing S_4, the ratio of blade width to impeller diameter, depends on the number of blades. For a six-bladed turbine N_P increases directly with S_4; for a four-bladed turbine N_P increases with $S_4^{1.25}$.

Table 9-1 CONSTANTS a AND b of EQ. (9-23)

Fig.	Line	a	b
9–14	B	1.0	40.0
9–15	B	1.7	18.0
9–15	C	0	18.0
9–15	D	2.3	18.0

4   A pitched-blade turbine draws less power than a straight-blade turbine. When the blades make a 45° angle with the impeller shaft, N_P is 0.4 times the value when the blades are parallel with the shaft.

5   Two straight-blade turbines on the same shaft draw about 1.9 times as much power as one turbine alone, provided the spacing between the two impellers is at least equal to the impeller diameter. Two closely spaced turbines may draw as much as 2.4 times as much power as a single turbine.

6   The shape of the tank has relatively little effect on N_P. The power consumed in a horizontal cylindrical vessel, whether baffled or unbaffled, or in a baffled vertical tank with a square cross section is the same as in a vertical cylindrical tank. In an unbaffled square tank the power number is about 0.75 times that in a baffled cylindrical vessel. Circulation patterns, of course, are strongly affected by tank shape, even though power consumption is not.

Calculation of power consumption   The power delivered to the liquid is computed by combining Eq. (9-22) and the definition of N_P to give

$$P = \frac{\phi N_{Fr}^m n^3 D_a^5 \rho}{g_c} \tag{9-24}$$

When the Froude number is not a factor, the power is given by

$$P = \frac{\phi n^3 D_a^5 \rho}{g_c} \tag{9-25}$$

At low Reynolds numbers, the lines of N_P vs. N_{Re} for both baffled and unbaffled tanks coincide, and the slope of the line on logarithmic coordinates is -1. The flow becomes laminar in this range, density is no longer a factor, and Eq. (9-22) becomes

$$N_P N_{Re} = \frac{P g_c}{n^2 D_a^3 \mu} = K_L = \psi_L(S_1, S_2, \ldots, S_n) \tag{9-26}$$

from which
$$P = \frac{K_L n^2 D_a^3 \mu}{g_c} \tag{9-27}$$

Equations (9-26) and (9-27) can be used when N_{Re} is less than 10.

In baffled tanks at Reynolds numbers larger than about 10,000, the power function is independent of the Reynolds number, and viscosity is not a factor. Changes in N_{Fr} have no effect. In this range the flow is fully turbulent and Eq. (9-22) becomes

$$N_P = K_T = \psi_T(S_1, S_2, \ldots, S_n) \tag{9-28}$$

from which
$$P = \frac{K_T n^3 D_a^5 \rho}{g_c} \tag{9-29}$$

Magnitudes of the constants K_T and K_L for various types of impellers and tanks are shown in Table 9-2.

EXAMPLE 9-1 A flat-blade turbine with six blades is installed centrally in a vertical tank. The tank is 6 ft (1.83 m) in diameter; the turbine is 2 ft (0.61 m) in diameter and is positioned 2 ft (0.61 m) from the bottom of the tank. The tank is filled to a depth of 6 ft (1.83 m) with a solution of 50 percent caustic soda, at 150°F (65.6°C), which has a viscosity of 12 cP and a density of 93.5 lb/ft³ (1,498 kg/m³). The turbine is operated at 90 r/min. The tank is unbaffled. What power will be required to operate the mixer?

SOLUTION   Curve B in Fig. 9-14 applies under the conditions of this problem. The Reynolds and Froude numbers are calculated. The quantities for substitution are, in consistent units,

$$D_a = 2 \text{ ft} \qquad n = \tfrac{90}{60} = 1.5 \text{ r/s}$$

$$\mu = 12 \times 6.72 \times 10^{-4} = 8.06 \times 10^{-3} \text{ lb/ft-s}$$

$$\rho = 93.5 \text{ lb/ft}^3 \qquad g = 32.17 \text{ ft/s}^2$$

Then

$$N_{Re} = \frac{D_a^2 n \rho}{\mu} = \frac{2^2 \times 1.5 \times 93.5}{8.06 \times 10^{-3}} = 69,600$$

$$N_{Fr} = \frac{n^2 D_a}{g} = \frac{1.5^2 \times 2}{32.17} = 0.14$$

From Table 9-1, the constants a and b for substitution into Eq. (9-23) are $a = 1.0$ and $b = 40.0$. From Eq. (9-23),

$$m = \frac{1.0 - \log 69,600}{40.0} = -0.096$$

From curve B (Fig. 9-14), for $N_{Re} = 69,600$, $\phi = 1.07$, and from Eq. (9-24),

$$P = \frac{1.07 \times 0.14^{-0.096} \times 93.5 \times 1.5^3 \times 2^5}{32.17} = 406 \text{ ft-lb}_f/s$$

The power requirement is $406/550 = 0.74$ hp (0.55 kW).   ////

Table 9-2   VALUES OF CONSTANTS K_L AND K_T IN EQS. (9-26) AND (9-28) FOR BAFFLED TANKS HAVING FOUR BAFFLES AT TANK WALL, WITH WIDTH EQUAL TO 10 PERCENT OF THE TANK DIAMETER†

Type of impeller	K_L	K_T
Propeller, square pitch, three blades	41.0	0.32
Pitch of 2, three blades	43.5	1.00
Turbine, six flat blades	71.0	6.30
Six curved blades	70.0	4.80
Fan turbine, six blades	70.0	1.65
Flat paddle, two blades	36.5	1.70
Shrouded turbine, six curved blades	97.5	1.08
With stator, no baffles	172.5	1.12

† From J. H. Rushton, *Ind. Eng Chem.*, **44**:2931 (1952).

EXAMPLE 9-2   The tank of Example 9-1 is fitted with four baffles, each having a width of 7.5 in. (190 mm). What power will be required to operate this baffled mixer?

   SOLUTION   Curve A of Fig. 9-14 now applies, and $\phi = N_P$. From curve A, at a Reynolds number of 69,600, $\phi = 6.0$. The Froude number is no longer a factor. Then, from Eq. (9-25),

$$P = \frac{6.0 \times 93.5 \times 1.5^3 \times 2^5}{32.17} = 1,880 \text{ ft-lb}_f/\text{s}$$

The power is $1,880/550 = 3.42$ hp (2.55 kW).

   The increase in power resulting from the installation of the baffles is 360 percent, but the effectiveness of the mixer is also greatly improved. It is important that if baffles are installed in an unbaffled tank operating at high Reynolds numbers, the motor used to drive the mixer be sufficiently powerful. If, for example, a 1-hp motor were installed for operation under the conditions of Example 9-1, and if this motor were not replaced by a larger one when the baffles were installed, the 1-hp motor would probably burn out.   ////

EXAMPLE 9-3   The mixer of Example 9-1 is to be used to mix a rubber-latex compound having a viscosity of 1,200 poises and a density of 70 lb/ft³ (1,120 kg/m³). What power will be required?

   SOLUTION   The Reynolds number is now

$$N_{\text{Re}} = \frac{2^2 \times 1.5 \times 70}{1,200 \times 0.0672} = 5.2$$

This is well within the range of laminar flow, and $\phi = N_P$. For $N_{\text{Re}} = 5.2$, from Fig. 9-14, $N_P = 12.5$, and

$$P = \frac{12.5 \times 70 \times 1.5^3 \times 2^5}{32.17} = 2,940 \text{ ft-lb}_f/\text{s}$$

The power required is $2,940/550 = 5.35$ hp (3.99 kW). This power requirement is independent of whether or not the tank is baffled. There is no reason for baffles in a mixer operated at low Reynolds numbers, as vortex formation does not occur under such conditions.

   Note that a 10,000-fold increase in viscosity increases the power by only about 60 percent over that required by the baffled tank operating on the low-viscosity liquid.   ////

Power consumption in non-Newtonian liquids   In correlating power data for non-Newtonian liquids the power number, $Pg_c/n^3 D_a^5 \rho$, is defined in the same way as for Newtonian fluids. The Reynolds number is not easily defined, since the apparent viscosity of the fluid varies with the velocity gradient and the velocity gradient changes considerably from one point to another in the vessel. Successful correlations have been developed, however, with a Reynolds number defined as in Eq. (9-21) using an average apparent viscosity μ_a calculated from an average velocity gradient $(du/dy)_{\text{av}}$. The Reynolds number is then

$$N_{\text{Re},n} = \frac{nD_a^2 \rho}{\mu_a} \tag{9-30}$$

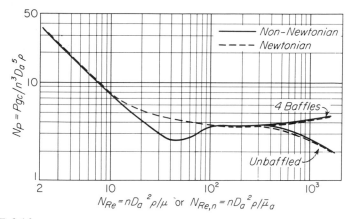

FIGURE 9-16
Power correlation for a six-bladed turbine in non-Newtonian liquids.

For a power-law fluid, as shown by Eq. (3-7), the average apparent viscosity can be related to the average velocity gradient by the equation

$$\mu_a = K' \left(\frac{du}{dy}\right)_{av}^{n'-1} \tag{9-31}$$

Substitution in Eq. (9-30) gives

$$N_{Re,n} = \frac{nD_a^2\rho}{K'(du/dy)_{av}^{n'-1}} \tag{9-32}$$

For a straight-bladed turbine in pseudoplastic liquids it has been shown that the average velocity gradient in the vessel is directly related to the impeller speed. For a number of pseudoplastic liquids a satisfactory, though approximate, relation is[5,11,18]

$$\left(\frac{du}{dy}\right)_{av} = 11n \tag{9-33}$$

Combining Eqs. (9-32) and (9-33) and rearranging gives

$$N_{Re,n} = \frac{n^{2-n'}D_a^2\rho}{11^{n'-1}K'} \tag{9-34}$$

Figure 9-16 shows the power-number–Reynolds-number correlation for a six-bladed turbine impeller in pseudoplastic fluids. The dashed curve is taken from Fig. 9-14 and applies to Newtonian fluids, for which $N_{Re} = nD_a^2\rho/\mu$. The solid curve is for pseudoplastic liquids, for which $N_{Re,n}$ is given by Eqs. (9-30) and (9-34). At Reynolds numbers below 10 and above 100 the results with pseudoplastic liquids are the same as with Newtonian liquids. In the intermediate range of Reynolds numbers between 10 and 100, pseudoplastic liquids consume less power than Newtonian fluids.

The transition from laminar to turbulent flow in pseudoplastic liquids is delayed until the Reynolds number reaches about 40, instead of 10 as in Newtonian liquids.

The flow patterns in an agitated pseudoplastic liquid differ considerably from those in a Newtonian liquid. Near the impeller the velocity gradients are large, and the apparent viscosity of a pseudoplastic is low. As the liquid travels away from the impeller, the velocity gradient decreases, and the apparent viscosity of the liquid rises. The liquid velocity drops rapidly, decreasing the velocity gradients further and increasing the apparent viscosity still more. Even when there is high turbulence near the impeller, therefore, the bulk of the liquid may be moving in slow laminar flow and consuming relatively little power. The toroidal rings of stagnant liquid indicated in Fig. 9-13 are very strongly marked when the agitated liquid is a pseudoplastic.

BLENDING AND MIXING

Mixing is a much more difficult operation to study and describe than agitation. The patterns of flow of fluid velocity in an agitated vessel are complex but reasonably definite and reproducible. The power consumption is readily measured. The results of mixing studies, on the other hand, are seldom highly reproducible and depend in large measure on how mixing is defined by the particular experimenter. Often the criterion for good mixing is visual, as in the use of interference phenomena to follow the blending of gases in a duct[19] or the color change of an acid-base indicator to determine liquid-blending times.[10,22] Other criteria that have been used include the rate of decay of concentration fluctuations following injection of a contaminant in a flowing fluid, the variation in the analyses of small samples taken at random from various parts of the mix, the rate of transfer of a solute from one liquid phase to another, and, in solid-liquid mixtures, the visually observed uniformity of the suspension.

Mixing of miscible liquids   The impeller of an agitator generates a stream of liquid that flows with considerable speed into a relatively slowly moving mass of liquid. Even when there is but one liquid, or when there are two or more miscible liquids, the incorporation of the moving stream into the surrounding liquid mass is far from instantaneous. The moving stream maintains its identity for considerable distance, as seen in Fig. 9-17, which shows the behavior of a circular liquid jet issuing from a nozzle and flowing at high velocity into a stagnant pool of the same liquid. The velocity in the jet issuing from the nozzle is uniform and constant. It remains so in a core, the area of which decreases with distance from the nozzle. The core is surrounded by an expanding turbulent jet, in which the radial velocity decreases with distance from the center line of the jet. The shrinking core disappears at a distance from the nozzle of $4.3D_j$, where D_j is the diameter of the nozzle. The turbulent jet maintains its integrity well beyond the point at which the core has disappeared, but its velocity steadily decreases. The radial decrease in velocity in the jet is accompanied by a pressure increase in accordance with the Bernoulli principle. Fluid flows into the jet and is absorbed, accelerated, and blended into the augmented jet. This

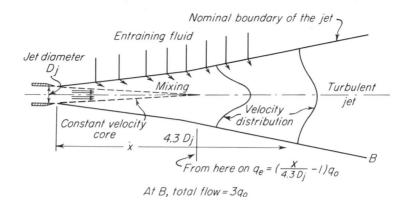

FIGURE 9-17
Flow of submerged circular jet. [*After Rushton et al.*[27]]

process is called *entrainment*. An equation applying over distances larger than $4.3D_j$ is

$$q_e = \left(\frac{X}{4.3D_j} - 1\right)q_0 \tag{9-35}$$

where q_e = volume of liquid entrained per unit time at distance X from nozzle
   q_0 = volume of liquid leaving jet nozzle per unit time

In addition to entrainment, strong shear stresses exist at the boundary between the jet and the surrounding liquid. These stresses tear off eddies at the boundary and generate considerable turbulence, which also contributes to the mixing action.

A large flow of liquid alone does not achieve satisfactory mixing. Enough time and space must be provided for the stream to blend thoroughly into the mass of fluid by the mechanism of entrainment.

Suspension of solid particles   Particles of solids are suspended in liquids for many purposes, perhaps to produce a homogeneous mixture for feeding to a processing unit, to dissolve the solids, or to promote a chemical reaction. Suspension of solids in an agitated vessel is much like fluidization of solids with liquids, as discussed in Chap. 5, except that the energy for suspending the particles comes from the impeller and not from pressure drop in the fluid. Since the particles are kept in motion by the currents of liquid, satisfactory suspension of the particles requires an adequate circulation rate and an appropriate circulation pattern of the liquid in the vessel. Velocity gradients in the liquid are of secondary importance.

The ease with which solids are suspended in the liquid and are subsequently kept in suspension depends chiefly on the settling velocity of the particles. This depends in turn on the size, shape, and density of the particles, the viscosity and density of the liquid, and whether settling is free or hindered. The terminal settling velocity of free-settling particles can be calculated from Eq. (7-46), (7-47), or (7-48), depending on which law of settling governs the motion of the particle. In general, small, light particles with a free-settling velocity below 1 ft/min are readily suspended in an agitated vessel, as they tend to follow the circulation pattern of the liquid. Larger, heavier particles tend to settle out against the fluid motion, but if the free-settling velocity is below about 10 ft/min, it is possible to produce a fairly uniform suspension in an agitated tank. When the free-settling velocity exceeds 10 ft/min, an impeller ordinarily cannot produce a uniform suspension.

Settling rates are also governed by the volume fraction of solids in the suspension. When the solids occupy less than 0.30 times the volume of the suspension, the particles are practically free-settling. When the volume fraction is greater than 0.50, hindered settling occurs, and the slurry behaves more like a viscous non-Newtonian liquid than like separate phases of liquid and solids. With hindered settling the suspended particles normally settle slowly.

POWER REQUIRED TO SUSPEND SOLIDS   The preceding discussion of power consumption indicated how much power a given impeller would deliver to the liquid under specified conditions. No mention was made of how much power is actually required to achieve a given result. Often this power cannot be predicted with any reasonable accuracy, and it must be found by small-scale studies of the specific agitation or mixing problem.

For suspending solid particles in an agitated vessel some approximate correlations for the power requirement are available. In an agitated suspension of free-settling solids, especially at moderate impeller speeds, some of the particles may never reach the liquid surface. Often the solids are lifted only to a maximum height Z_s above the vessel floor. Above this height the liquid is free from solids. As the impeller speed is increased, Z_s increases, sometimes until it equals H, the liquid depth. The power P required to suspend particles to a maximum height Z_s, using a turbine impeller, is given by[28] the empirical equation

$$\frac{Pg_c}{g\rho_m V_m u_t} = (1 - \varepsilon_m)^{2/3} \left(\frac{D_t}{D_a}\right)^{1/2} e^{4.35\beta} \tag{9-36}$$

where

$$\beta = \frac{Z_s - E}{D_t} - 0.1 \tag{9-37}$$

where ρ_m, V_m = density and volume, respectively, of solid-liquid suspension, not including clear liquid in zone above Z_s

u_t = terminal settling velocity of particles calculated from Stokes' law [Eq. (7-46)]

ε_m = volume fraction of liquid in zone occupied by suspension

E = clearance between impeller and vessel floor

The other symbols have their usual meaning.

Tentative correlations are also available for estimating the power required just to lift all the particles from the vessel floor.[14,28]

EXAMPLE 9-4 An agitated vessel 6 ft (1.83 m) in diameter contains 9,000 lb (4,082 kg) of water at 70°F (21.1°C) and 3,000 lb (1,361 kg) of 150-mesh particles of fluorspar having a specific gravity of 3.18. The impeller is a six-bladed straight-blade turbine 2 ft (0.61 m) in diameter, set one impeller diameter above the vessel floor. (a) How much power would be required to suspend the particles to maximum height of 5 ft (1.52 m)? (b) What would be the impeller speed under these conditions?

SOLUTION   (a) Equations (9-36) and (9-37) are used. The quantities needed are

$$D_t = 6 \text{ ft} \qquad D_a = 2 \text{ ft} \qquad Z_s = 5 \text{ ft} \qquad E = 2 \text{ ft}$$

Volume of suspension: $\quad V_m = \dfrac{\pi D_t^2 Z_s}{4} = \pi \times 6^2 \times \dfrac{5}{4} = 141.4 \text{ ft}^3$

Density of liquid: $\qquad\qquad \rho = 62.3 \text{ lb/ft}^3 \qquad$ (Appendix 14)

Density of solid: $\qquad\qquad \rho_p = 3.18 \times 62.3 = 198.1 \text{ lb/ft}^3$

Volume of solids in suspension: $\qquad \dfrac{3,000}{198.1} = 15.1 \text{ ft}^3$

Volume of liquid in suspension: $141.4 - 15.1 = 126.3 \text{ ft}^3$

Density of suspension: $\rho_m = \dfrac{3,000 + (126.3 \times 62.3)}{141.4} = 76.9 \text{ lb/ft}^3$

Volume fraction of liquid in suspension:

$$\varepsilon_m = \frac{126.3}{141.4} = 0.893$$

$$D_p = \frac{0.0041}{12} = 3.42 \times 10^{-4} \text{ ft} \qquad \text{(Appendix 20)}$$

$$\mu = 0.982 \times 6.72 \times 10^{-4} = 6.60 \times 10^{-4} \text{ lb/ft-s} \qquad \text{(Appendix 14)}$$

From Eq. (7-50), the settling velocity is

$$u_t = \frac{32.17(3.42 \times 10^{-4})^2(198.1 - 62.3)}{18 \times 6.60 \times 10^{-4}} = 0.043 \text{ ft/s}$$

From Eq. (9-37)

$$\beta = \frac{5 - 2}{6} - 0.1 = 0.4$$

Substitution in Eq. (9-36), taking g/g_c as unity, gives

$$P = 76.9 \times 141.4 \times 0.043(1 - 0.893)^{2/3}(\tfrac{6}{2})^{1/2} e^{4.35 \times 0.4}$$

$$= 1,039 \text{ ft-lb}_f/\text{s, or } 1.89 \text{ hp } (1.41 \text{ kW})$$

(b) If flow is highly turbulent, the impeller speed required to deliver this much power to the liquid can be found from Eq. (9-29). From Table 9-2, $K_T = 6.30$ for a six-bladed turbine. Then

$$n = \left(\frac{Pg_c}{K_T D_a^5 \rho}\right)^{1/3} = \left(\frac{1,039 \times 32.17}{6.30 \times 2^5 \times 62.3}\right)^{1/3} = 1.386 \text{ r/s} = 83.2 \text{ r/min}$$

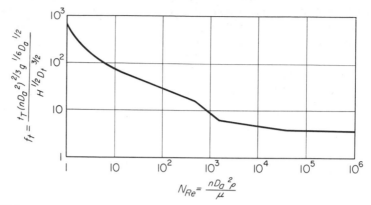

FIGURE 9-18
Correlation of blending times for miscible liquids in a turbine-agitated baffled vessel. [*After Norwood and Metzner.*[22]]

Now check to see whether Eq. (9-29) applies. The Reynolds number is

$$N_{Re} = \frac{nD_a^2 \rho}{\mu} = \frac{1.386 \times 2^2 \times 62.3}{6.60 \times 10^{-4}} = 523,000$$

Since N_{Re} is much greater than 10,000, Eq. (9-29) does apply.   ////

Blending of miscible liquids   The blending of miscible liquids has been extensively studied by adding a small quantity of acid, such as HCl, to an agitated solution containing a stoichiometrically equivalent amount of base such as NaOH and visually observing the time required for the color of an acid-base indicator to change. The results of such a study with six-bladed turbine impellers in baffled tanks are given in Fig. 9-18. The blending time t_T is a complex function of impeller speed n, impeller diameter D_a, vessel diameter D_t, liquid depth H, and the viscosity μ and density ρ of the liquid. In Fig. 9-18 a dimensionless blending-time factor f_t is plotted against the impeller Reynolds number, where

$$f_t = \frac{t_T(nD_a^2)^{2/3}g^{1/6}D_a^{1/2}}{H^{1/2}D_t^{3/2}} = \Phi\left(\frac{nD_a^2\rho}{\mu}\right) = \Phi(N_{Re}) \tag{9-38}$$

Figure 9-18 applies only to the blending of Newtonian liquids.

In baffled vessels fitted with turbine agitators virtually all the blending takes place in the immediate vicinity of the impeller, which indicates that the rate of blending depends chiefly on the steepness of the velocity gradients instead of on the circulation rate. The shape of the curve in Fig. 9-18 shows that there are three regions of blending behavior, depending on the nature of flow. These regions have been named[22] *laminar* ($N_{Re} < 10$), *transition* ($10 < N_{Re} < 10^4$), and *turbulent* ($N_{Re} > 10^4$). There is an approximate correspondence between the curve for blending times and the power-number curve shown in Fig. 9-15.

Curves similar to that in Fig. 9-18 are available for vessels fitted with propellers or using a side-entering jet of liquid.[2,10] In these agitators, unlike turbine impellers, both the rate of circulation and the steepness of the velocity gradients significantly influence the blending rate.

In a pseudoplastic liquid, blending times at Reynolds numbers below about 1,000 are much longer than in Newtonian liquids under the same impeller conditions.[11,20] In the regions of low shear, far from the impeller, the apparent viscosity of the pseudoplastic liquid is greater than it is near the impeller. In these remote regions turbulent eddies decay rapidly, and zones of almost stagnant liquid are often formed. Both effects lead to poor mixing and long blending times. At high Reynolds numbers there is little difference in the mixing characteristics of Newtonian and pseudoplastic liquids.

EXAMPLE 9-5   It is proposed to use the agitated vessel described in Example 9-4 for neutralizing a dilute aqueous solution of NaOH with a stoichiometrically equivalent quantity of concentrated nitric acid (HNO_3). The impeller speed is to be 83.2 r/min. The final depth of liquid in the vessel is to be 6 ft (1.83 m). Assuming that all the acid is added to the vessel at one time, how long will it take for the neutralization to be complete?

SOLUTION   Since the impeller speed and the density and viscosity of the liquid are the same as in Example 9-4, the impeller Reynolds number is also the same. From Fig. 9-18 for $N_{Re} = 523,000$, $f_t = 5$. The other quantities needed are

$$D_a = 2 \text{ ft} \qquad D_t = 6 \text{ ft} \qquad H = 6 \text{ ft} \qquad n = 1.386 \text{ r/s}$$

Solving Eq. (9-38) for t_T gives

$$t_T = \frac{5 \times 6^{1/2} \times 6^{3/2}}{(1.386 \times 2^2)^{2/3} \times 32.17^{1/6} \times 2^{1/2}} = 23 \text{ s}$$

////

Motionless mixers   Gases and nonviscous liquids can often be satisfactorily blended by passing them together through a length of open pipe or a pipe containing orifice plates or segmented baffles. Under appropriate conditions the pipe length may be as short as 5 to 10 pipe diameters, but 50 to 100 pipe diameters is recommended.[16]

More difficult mixing tasks are accomplished by static mixers, commercial devices in which stationary elements successively divide and recombine portions of the fluid stream.[6,7] In the mixer shown in Fig. 9-19 each short helical element divides the stream in two, gives it a 180° twist, and delivers it to the succeeding element, which is set at 90° to the trailing edge of the first element. The second element divides the already divided stream and twists it 180° in the opposite direction. For n elements, then, there are 2^n divisions and recombinations, or over 1 million in a 20-element mixer. The pressure drop is typically 4 times as large as in the same length of empty pipe. In other commercial designs the stream is divided 4 or more times by each element. Mixing, even of highly viscous materials, is excellent after 6 to 20 elements. Static mixers are used for liquid blending, gas and liquid dispersion, chemical reac-

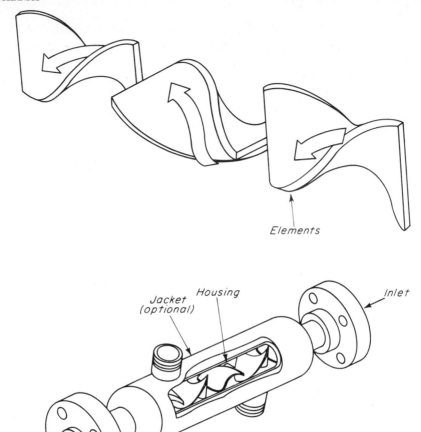

FIGURE 9-19
Static mixer unit. [*STATIC MIXER is a registered Trademark © of Kenics Corporation, N. Andover, Mass.*]

tions, and heat transfer. They are especially effective in mixing low-viscosity fluids with viscous liquids or pastes.

Mixer selection There is not necessarily any direct relation between power consumed and amount or degree of mixing. When a low-viscosity liquid is swirled about in an unbaffled vessel, the particles may follow parallel circular paths almost indefinitely and mix almost not at all. Little of the energy supplied is used for mixing. If baffles are added, mixing becomes rapid; a larger fraction of the energy is used for mixing and relatively less for circulation.

The "best" mixer is the one which mixes in a given time with the smallest

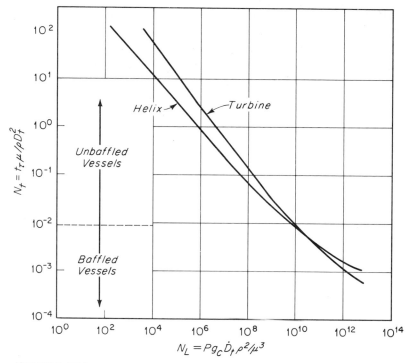

FIGURE 9-20

Mixing-time–power relationships for helical-ribbon and turbine impellers in agitated vessels.[20]

amount of power or which mixes fastest with a given amount of power. Comparative studies of various impellers have shown that there are significant differences between impeller types when blending viscous liquids but comparatively little difference with low-viscosity liquids. Figure 9-20 relates mixing time with power input for two impellers, a turbine and a helical ribbon. In this figure a dimensionless *mixing number* N_t, defined as $t_T \mu / \rho D_t^2$, is plotted against a modified power number N_L, defined as $P g_c \rho^2 D_t / \mu^3$. N_t is formed from the product of the dimensionless groups $n t_T$, N_{Re}, and $(S_1)^{-2}$; N_L is the product of N_P, $(N_{Re})^3$, and S_1. With high viscosities, at values of the N_L of say 10^3 or 10^4, the helix gives much shorter mixing times with the same power input than the turbine; with low-viscosity liquids, where the N_L value is 10^{12} or 10^{14}, the turbine mixes faster than the helix. Baffles are required for good mixing when N_t is less than about 10^{-2}. Data for other impeller types are given in the literature.[12,20,29]

Scale-up of agitator design A major problem in agitator design is to scale up from a laboratory or pilot-plant agitator to a full-scale unit. For some problems generalized correlations such as those shown in Figs. 9-14 to 9-16 and 9-18 are available for scale-up. For many other problems adequate correlations are not available;

for these situations various methods of scale-up have been proposed, all based on geometrical similarity between the laboratory and plant equipment. It is not always possible, however, to have the large and small vessels geometrically similar. Furthermore, even if geometrical similarity is obtainable, dynamic and kinematic similarity are not, so that the results of the scale-up are not always fully predictable. As in most engineering problems the designer must rely on judgment and experience.

Power consumption in large vessels can be accurately predicted from curves of ϕ vs. N_{Re}, as shown in Figs. 9-14 and 9-15. Such curves may be available in the published literature, or they may be developed from pilot-plant studies using small vessels of the proposed design. With low-viscosity liquids the amount of power consumed by the impeller per unit volume of liquid has been used as a measure of mixing effectiveness, based on the reasoning that increased amounts of power mean a higher degree of turbulence and a higher degree of turbulence means better mixing. Studies have shown this to be at least roughly true. In a given mixer the amount of power consumed can be directly related to the rate of solution of a gas or the rate of certain reactions, such as oxidations, that depend on the intimacy of contact of one phase with another. In a rough qualitative way it may be said that $\frac{1}{2}$ to 1 hp per 1,000 gal of thin liquid gives "mild" agitation, 2 to 3 hp per 1,000 gal gives "vigorous" agitation, and 4 to 10 hp per 1,000 gal gives "intense" agitation. These figures refer to the power that is actually delivered to the liquid and do not include power used in driving gear-reduction units or in turning the agitator shaft in bearings and stuffing boxes. The agitator designed in Examples 9-4 and 9-5 would require about $1\frac{1}{2}$ hp per 1,000 gal of liquid and should provide rather mild agitation.

The optimum ratio of impeller diameter to vessel diameter for a given power input is an important factor in scale-up. The nature of the agitation problem strongly influences this ratio: for some purposes the impeller should be small, relative to the size of the vessel; for others it should be large. For dispersing a gas in a liquid, for example, the optimum ratio[25] is about 0.25; for bringing two immiscible liquids into contact, as in liquid-liquid extraction vessels, the optimum ratio is 0.40.[23] For some blending operations the ratio should be 0.6 or even more. In any given operation, since the power input is kept constant, the smaller the impeller, the higher the impeller speed. In general, operations which depend on large velocity gradients rather than on high circulation rates are best accomplished by small, high-speed impellers, as is the case in the dispersion of gases. For operations which depend on high circulation rates rather than steep velocity gradients, a large, slow-moving impeller should be used.

Blending times are usually much shorter in small vessels than in large ones, for it is often impractical to make the blending times the same in vessels of different sizes. This is shown by the following example.

EXAMPLE 9-6   A pilot-plant vessel 1 ft (305 mm) in diameter is agitated by a six-bladed turbine impeller 4 in. (102 mm) in diameter. When the impeller Reynolds number is 10^4, the blending time of two miscible liquids is found to be 15 s. The power required is 2 hp per 1,000 gal (0.4 kW/m^3) of liquid. (a) What power input would be required to give the same blending time in a vessel 6 ft (1,830 mm) in diameter? (b) What would be the blending time in the 6-ft (1,830-mm) vessel if the power input per unit volume was the same as in the pilot-plant vessel?

SOLUTION   (a)   From the definitions of the shape factors given on page 236,

$$D_t = S_1 D_a \qquad H = S_6 D_t = S_1 S_6 D_a$$

The blending-time factor in Eq. (9-38) can therefore be written

$$f_t = \frac{t_T (n D_a^2)^{2/3} g^{1/6} D_a^{1/2}}{(S_1 D_a)^{1/2} (S_1 S_6 D_a)^{3/2}}$$

Since the vessels under consideration are geometrically similar, S_1 and S_6 are constants, and

$$f_t = \frac{c_1 t_T n^{2/3}}{D_a^{1/6}}$$

where c_1 is a constant.

Let n_1, D_{a1}, and t_{T1} be the impeller speed, impeller diameter, and blending time, respectively, in the 1-ft vessel, and n_6, D_{a6}, and t_{T6} be the corresponding quantities in the 6-ft vessel. Then the blending-time factors in the two vessels are

$$f_{t1} = \frac{c_1 t_{T1} n_1^{2/3}}{D_{a1}^{1/6}} \qquad f_{t6} = \frac{c_1 t_{T6} n_6^{2/3}}{D_{a6}^{1/6}}$$

As shown in Fig. 9-18, the blending-time factor at Reynolds numbers of 10^4 and above is almost constant. Since the Reynolds number in the 6-ft vessel will be larger than that in the 1-ft vessel, it is reasonable to assume that $f_{t1} = f_{t6}$. If this is true, then

$$\frac{n_6}{n_1} = \left(\frac{D_{a6}}{D_{a1}}\right)^{1/4} \left(\frac{t_{T1}}{t_{T6}}\right)^{3/2} \tag{9-39}$$

If the blending times in the two vessels are to be the same $(t_{T1} = t_{T6})$, the ratio of impeller speeds in the two vessels is

$$\frac{n_6}{n_1} = \left(\frac{D_{a6}}{D_{a1}}\right)^{1/4} = 6^{1/4} = 1.565$$

The impeller speed in the larger vessel must be 1.585 times that in the smaller vessel.

In geometrically similar vessels the power input per unit volume is proportional to P/D_a^3. At high Reynolds numbers, from Eq. (9-29)

$$\frac{P}{D_a^3} = \frac{K_T n^3 D_a^2 \rho}{g_c}$$

For a liquid of given density this becomes

$$\frac{P}{D_a^3} = c_2 n^3 D_a^2 \tag{9-40}$$

where c_2 is a constant. From this the ratio of power inputs per unit volume in the two vessels is

$$\frac{P_6 / D_{a6}^3}{P_1 / D_{a1}^3} = \left(\frac{n_6}{n_1}\right)^3 \left(\frac{D_{a6}}{D_{a1}}\right)^2$$

Since

$$\frac{n_6}{n_1} = \left(\frac{D_{a6}}{D_{a1}}\right)^{1/4}$$

$$\frac{P_6/D_{a6}^3}{P_1/D_{a1}^3} = \left(\frac{D_{a6}}{D_{a1}}\right)^{11/4} = 6^{11/4} = 138$$

The power per unit volume required in the 6-ft vessel is then $2 \times 138 = 276$ hp per 1,000 gal (55.2 kW/m^3). This is an impractically large amount of power to deliver to a low-viscosity liquid in an agitated vessel.

(*b*) If the power input per unit volume is to be the same in the two vessels, Eq. (9-40) can be solved and rearranged to give

$$\frac{n_6}{n_1} = \left(\frac{D_{a1}}{D_{a6}}\right)^{2/3}$$

Substitution in Eq. (9-39) and rearranging gives

$$\frac{t_{T6}}{t_{T1}} = \left(\frac{D_{a6}}{D_{a1}}\right)^{11/18} = 6^{11/18} = 2.99$$

The blending time in the 6-ft vessel would be $2.99 \times 15 = 45$ s.   ////

Although it is impractical to achieve the same blending time in the full-scale unit as in the pilot-plant vessel, a moderate increase in blending time in the larger vessel reduces power requirement to a reasonable level. Such trade-offs are often necessary in scaling up agitation equipment.

Scale-up for solids suspension In scaling up agitators for suspending solid particles, the use of a constant value of torque per unit volume is recommended.[8] Torque T equals $P/2\pi n$; hence at large Reynolds numbers, where Eq. (9-29) applies, the torque per unit volume, for a liquid of given density, is given by

$$\frac{T}{D_a^3} = c_3 n^2 D_a^2 \tag{9-41}$$

where c_3 is a constant. The recommended method of scale-up thus requires that the quantity $n^2 D_a^2$ be kept constant.

DISPERSION OPERATIONS

In suspending solids, the size and the surface area of the solid particles exposed to the liquid are fixed, as is the total volume of suspended solids. In gas-liquid or liquid-liquid dispersion operations, by contrast, the size of the bubbles or drops and the total interfacial area between the dispersed and continuous phases vary with conditions and degree of agitation. New area must constantly be created against the force of the interfacial tension. Drops and bubbles are continually coalescing and being redispersed. In most gas-dispersion operations, bubbles rise through the liquid pool and escape from the surface, and must be replaced by new ones.

In this dynamic situation the volume of the dispersed phase held up in the liquid pool is also variable, depending on the rate of rise of the bubbles and the volumetric feed rate. Statistical averages are used to characterize the system, since the holdup, interfacial area, and bubble diameter vary with time and with position in the vessel.

Characteristics of dispersed phase; mean diameter Despite these variations, a basic relationship exists between the holdup Ψ (the volume fraction of dispersed phase in the system), the interfacial area a per unit volume, and the bubble or drop diameter D_P. If the total volume of the dispersion is taken as unity, the volume of dispersed phase, by definition, is Ψ. Let the number of drops or bubbles in this volume be N. Then if all the drops or bubbles were spheres of diameter D_p, their total volume would be given by

$$\frac{\pi N D_p^3}{6} = \Psi \tag{9-42}$$

The total surface area of the drops or bubbles in this volume would be

$$\pi N D_p^2 = a \tag{9-43}$$

Dividing Eq. (9-42) by Eq. (9-43) and rearranging gives

$$D_p = \frac{6\Psi}{a} \tag{9-44}$$

Actually, of course, the drops or bubbles differ in size and are not spherical. However, for given values of Ψ and a an equivalent average diameter can be defined by an equation similar to Eq. (9-44), as follows

$$\bar{D}_s \equiv \frac{6\Psi}{a} \tag{9-45}$$

Diameter $\bar{D}_s$ in Eq. (9-45) is known as the *volume-surface mean diameter* or the *Sauter mean diameter*.

Gas dispersion; bubble behavior In a quiescent liquid a single bubble issuing from a submerged circular orifice will be a sphere if the flow rate is small. Under these conditions the bubble diameter D_p can be calculated as follows, by equating the net buoyant force on the bubble to the opposing drag force at the edge of the orifice. The net buoyant force, acting upward, is

$$F_b - F_e = \frac{g}{g_c} \frac{\pi D_p^3}{6} (\rho_L - \rho_V) \tag{9-46}$$

where ρ_L = density of liquid
ρ_V = density of vapor

The drag force is

$$F_D = \pi D_o \sigma \tag{9-47}$$

where D_o = orifice diameter
σ = interfacial tension

When the bubble becomes large enough, the drag force is no longer strong enough to keep the bubble attached to the edge of the orifice; at this point the opposing forces become equal, and the bubble is detached from the orifice. The bubble diameter can be found by combining Eqs. (9-46) and (9-47) and solving for D_p to give

$$D_p = \left[\frac{6D_o\sigma g_c}{g(\rho_L - \rho_V)} \right]^{1/3} \tag{9-48}$$

For Eq. (9-48) to apply, D_o cannot be larger than D_p; hence the maximum orifice diameter for stable operation, found by setting D_o equal to D_p, is

$$D_{o,\text{max}} = \sqrt{\frac{6\sigma g_c}{g(\rho_L - \rho_V)}} \tag{9-49}$$

For air dispersed in water $D_{o,\text{max}}$ is between 6 and 7 mm.

As the gas rate is increased, the bubble diameter increases with flow rate and bubbles may issue in pairs or quartets with considerable coalescence; at still higher rates the gas stream appears to be a continuous jet, which actually consists of large closely spaced irregular bubbles and which disintegrates into a cloud of smaller bubbles 3 or 4 in. above the orifice.

As discussed in Chap. 7, bubbles change in shape from spherical to ellipsoidal to lens-shaped as their diameter increases. Larger bubbles often rise in spiral paths, at terminal velocities which are almost constant and independent of their size (see Fig. 7-7). In clouds or swarms of bubbles there may be considerable coalescence; the rate of rise of clouds of small bubbles is considerably less than that of single bubbles but is not greatly different from that of single bubbles when the bubbles are large.

Gas dispersion in agitated vessels   Gas is normally fed to a processing vessel through the open end of a submerged pipe, through a perforated pipe, or *sparger*, or through a porous ceramic or metal plate. Sometimes the gas by itself provides sufficient agitation of the liquid; more commonly a motor-driven turbine impeller is used to disperse the gas and circulate the liquid and bubbles through the vessel.

For low gas holdups ($\Psi < 0.15$) the following *dimensional* equations are available for gas dispersion in pure liquids by a six-bladed turbine impeller.[4a] The average bubble diameter $\overline{D_s'}$ in millimeters is given by

$$\overline{D_s'} = 4.15 \frac{(\sigma g_c)^{0.6}}{(Pg_c/V)^{0.4}\rho_L^{0.2}} \Psi^{1/2} + 0.9 \tag{9-50}$$

where Pg_c/V is the power input per unit volume of ungassed liquid. The interfacial area a', in mm^{-1}, is given by

$$a' = 1.44 \frac{(Pg_c/V)^{0.4}\rho_L^{0.2}}{(\sigma g_c)^{0.6}} \left(\frac{\overline{V_s}}{u_t} \right)^{1/2} \tag{9-51}$$

where $\overline{V_s}$ = superficial velocity of gas
    = volumetric gas feed rate divided by the cross-sectional area of vessel
u_t = bubble rise velocity

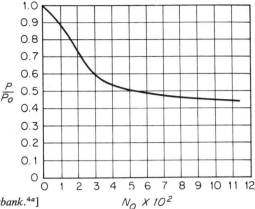

FIGURE 9-21
Power consumption of flat-blade turbine
impellers in aerated vessels. [*After Calderbank.*[4a]]

Combining Eqs. (9-45), (9-50), and (9-51) leads to the following *dimensional* equation
for gas holdup:

$$\Psi = \left(\frac{\overline{V}_s \Psi}{u_t}\right)^{1/2} + 0.0216 \frac{(Pg_c/V)^{0.4}\rho_L^{0.2}}{(\sigma g_c)^{0.6}} \left(\frac{\overline{V}_s}{u_t}\right)^{1/2} \tag{9-52}$$

In Eqs. (9-50) to (9-52) all quantities involving the dimension of length are in
millimeters.

Correlations are also available[4b] for situations in which the gas holdup is high,
between 0.15 and about 0.40. Under these conditions the agitator functions primarily
as a distributor, and the bubble diameter is almost constant and independent of
agitator speed. Then the interfacial area per unit volume is directly proportional to the
holdup.

Gas-handling capacity and loading of turbine impellers   If the gas throughput
to a turbine-agitated vessel is progressively increased, a point is eventually reached
at which the impeller floods; i.e., it is surrounded by so much gas that it can no longer
operate effectively. If the gas flow is then somewhat reduced, the agitator begins to
circulate liquid and disperse the gas once more; this point is known as the *redispersion
point*. The flow number $N_{Q,g}$, based on the gas flow rate q_g at the redispersion point,
has been correlated with the Froude number to give[30]

$$N_{Q,g} \equiv \frac{q_g}{nD_a^3} = 0.194N_{Fr}^{0.75} \tag{9-53}$$

where N_{Fr} equals $n^2 D_a/g$. For Eq. (9-53) to apply, N_{Fr} must be between 0.1 and 2.0.

Power input to turbine dispersers   The power consumed by a turbine impeller
dispersing a gas is less than that indicated by Fig. 9-14 for agitating ungassed liquids.
As shown in Fig. 9-21, this reduction is significant: at high gas rates the required
power is about half that required with an ungassed liquid. It drops still further at the
flooding point, but operation under flooding conditions is of course undesirable.

The power number $N_{P,g}$ at the redispersion point when N_{Fr} is between 0.2 and 1.0 has been found[30] to be

$$N_{P,g} = \frac{P_g g_c}{n^3 D_a^5 \rho_L} = 1.36 N_{Fr}^{-0.56} \qquad (9\text{-}54)$$

Equations (9-50) to (9-54) provide a basis for the design of turbine-agitated vessels for gas dispersion, as illustrated by the following examples.

EXAMPLE 9-7   A baffled cylindrical vessel 2 m in diameter is agitated with a turbine impeller 0.667 m in diameter turning at 180 r/min. The vessel contains water at 20°C, through which 100 m³/h of air is to be passed at atmospheric pressure. Calculate (a) the power input and power input per unit volume of liquid, (b) the gas holdup, (c) the mean bubble diameter, and (d) the interfacial area per unit volume of liquid. For water the interfacial tension is 72.75 dyn/cm. The rise velocity of the air bubbles may be assumed constant at 0.2 m/s.

SOLUTION   (a)   The power input for the ungassed liquid is first calculated; it is then corrected, by Fig. 9-21, for the effect of the gas. From the conditions of the problem

$$D_a = 0.667 \text{ m} \qquad n = 3 \text{ r/s} \qquad \rho = 1{,}000 \text{ kg/m}^3$$

$$\mu = 1 \text{ cP} = 1 \times 10 \text{ kg/m-s} \qquad q_g = 100 \text{ m}^3/\text{h}$$

Hence

$$N_{Re} = \frac{3 \times 0.667^2 \times 1{,}000}{1 \times 10^{-3}} = 1.33 \times 10^6$$

At this large Reynolds number Eq. (9-29) applies. For a flat-bladed turbine, from Table 9-2, $K_T = 6.3$. Hence, from Eq. (9-29), the power required for the ungassed liquid is

$$P_0 = \frac{6.3 \times 3^3 \times 0.667^5 \times 1{,}000}{1{,}000} = 22.45 \text{ kW}$$

The gas flow number is

$$N_{Q,g} = \frac{q_g}{n D_a^3} = \frac{100}{3{,}600 \times 3 \times 0.667^3} = 0.0311$$

From Fig. 9-21, $P/P_0 = 0.585$, and

$$P = 0.585 \times 22.45 = 13.13 \text{ kW}$$

The depth of the liquid, assuming the "standard" design shown in Fig. 9-10, equals D_t or 2 m. The liquid volume is therefore

$$V = \frac{\pi D_t^2 D_t}{4} = 2\pi = 6.28 \text{ m}^3$$

Hence the power input per unit volume is

$$\frac{P}{V} = \frac{13.13}{6.28} = 2.09 \text{ kW/m}^3 \text{ (10.5 hp/1,000 gal)}$$

This is not an unusually high power input for a gas-dispersing agitator. Because of the high tip speeds required for good dispersion, the power consumption is considerably greater than in simple agitation of liquids.

(b)   Since the holdup will probably be low, use Eq. (9-52). The cross-sectional area of the vessel is $\pi D_t^2/4$, or 3.142 m^2; hence the superficial gas velocity is

$$\bar{V}_s = \frac{100}{3{,}600 \times 3.142} = 0.00884 \text{ m/s}$$

For substitution in Eq. (9-52) the following equivalences are helpful:

$$1 \text{ dyn/cm} = 1 \text{ g/s}^2$$
$$1 \text{ kg/m}^3 = 10^{-6} \text{ g/mm}^3$$
$$1 \text{ kW/m}^3 = 10^3 \text{ g/mm-s}^3$$

Hence from the conditions of the problem and the power calculated in part (a),

$$\sigma = 72.75 \text{ g/s}^2 \qquad \rho_L = 10^{-3} \text{ g/mm}^3$$

$$\frac{P}{V} = 2.09 \times 10^3 \text{ g/mm-s}^3$$

Substitution in Eq. (9-52) gives

$$\Psi = \left(\frac{0.00884}{0.2}\Psi\right)^{1/2} + 0.216 \frac{(2.09 \times 10^3)^{0.4}(10^{-3})^{0.2}}{72.75^{0.6}} \left(\frac{0.00884}{0.2}\right)^{1/2}$$

Solving this as a quadratic equation gives $\Psi = 0.0768$.

(c)   The mean bubble diameter is now found from Eq. (9-50). Substitution gives

$$\bar{D}_s' = 4.15 \frac{72.75^{0.6}}{(2.09 \times 10^3)^{0.4}(10^{-3})^{0.2}} 0.0768^{1/2} + 0.9 = 3.7 \text{ mm}$$

(d)   From Eq. (9-45)

$$a' = \frac{6\Psi}{\bar{D}_s'} = \frac{6 \times 0.0768}{3.7} = 0.125 \text{ mm}^{-1} \qquad \qquad ////$$

EXAMPLE 9-8   (a)   Estimate the maximum gas handling capacity of the vessel described in Example 9-7. (b) What is the power consumption under conditions of maximum gas throughput?

SOLUTION   (a)   Use Eq. (9-53). The Froude number is

$$N_{Fr} = \frac{n^2 D_a}{g} = \frac{3^2 \times 0.667}{9.807} = 0.612$$

From Eq. (9-53)

$$q_g = 0.194 \times 0.612^{0.75} \times 3 \times 0.667^3 = 0.1195 \text{ m}^3/\text{s or 430 m}^3/\text{h}$$

(b)   From Eq. (9-54)

$$P_g = 1.36 \times 0.612^{-0.56} \frac{3^3 \times 0.667^5 \times 1{,}000}{1{,}000} = 6.38 \text{ kW}$$

Note that this is less than one-third the power P_0 required by the ungassed liquid.   ////

Dispersion of liquids in liquids One liquid, say benzene, may be dispersed in another liquid, say water, which is immiscible with the first liquid, in various types of equipment. Temporary dispersions of relatively large drops are made in agitated vessels or in pipeline dispersers; stable emulsions of very small droplets require the intense velocity gradients and high shear forces developed in colloid mills. Only temporary dispersions are considered here.

In an agitated vessel the drops of dispersed liquid are normally nearly spherical. Unless the density difference between the liquids is unusually great, the dispersed drops do not rise or settle out of the continuous liquid phase. The fractional holdup of dispersed liquid, therefore, does not depend on the agitator speed but is fixed by the proportions of the two liquids fed to the vessel. The drop size depends primarily on the tip speed of the impeller blades; a dimensionless correlation for drop size in a vessel agitated with a six-bladed turbine, with $D_t/D_a = 3$, is

$$\frac{\bar{D}_s}{D_a} = 0.06(1 + 9\Psi)\left(\frac{\sigma g_c}{n^2 D_a^3 \rho_c}\right)^{0.6} \tag{9-55}$$

where ρ_c is the density of the continuous liquid phase.

In a pipeline disperser the two liquids are passed in turbulent flow through an orifice plate. The mean drop size 300 mm downstream from an orifice plate can be predicted from the relation

$$\frac{\bar{D}_s}{D} = 21.6\left(\frac{D_o}{D}\right)^{3.73}\Psi^{0.121}\left(\frac{D\bar{V}\rho}{\mu_c}\right)^{-0.065}\left(\frac{\sigma g_c}{D\rho\bar{V}^2}\right)^{0.722} \tag{9-56}$$

where D = pipe diameter
  D_o = orifice diameter
  $\bar{V}$ = average liquid velocity through pipe
  ρ_c = density of continuous phase
  μ_c = viscosity of continuous phase

The group $D\rho\bar{V}^2/\sigma g_c$ is called the *Weber number* N_{We}.

EXAMPLE 9-9 It is planned to disperse 10 gal/min (2.271 m³/h) of benzene in 100 gal/min (22.71 m³/h) of water at 20°C with an average drop size of 15 μm. Can this be done in a pipeline disperser? If so, what diameters of the pipe and orifice would be required? The interfacial tension between benzene and water at 20°C is 35 dyn/cm.

SOLUTION From the conditions of the problem

$$\bar{D}_s = 1.5 \times 10^{-2} \text{ mm} \qquad \mu_c = 1 \text{ cP} = 10^{-3} \text{ kg/m-s}$$
$$\rho_c = 1,000 \text{ kg/m}^3 \qquad \sigma = 35 \times 10^{-3} \text{ kg/s}^2$$

The viscosity of the continuous phase is used because the dispersed drops have little effect on the effective viscosity of the mixture unless the volume fraction of the drops is high. Here the volume fraction is

$$\Psi = \frac{10}{100 + 10} = 0.0909$$

The densities of water and benzene at 20°C are 1,000 and 879 kg/m^3, respectively. Hence the density of the mixture is

$$\rho = (0.0909(879) + (1 - 0.0909)(1,000) = 989 \text{ kg/m}^3$$

Next it is necessary to choose a reasonable liquid velocity and the corresponding pipe size. Assume that $\overline{V}$ is about 5 ft/s (1.524 m/s). The total volumetric flow rate is 110 gal/min. Hence, from Appendix 6 a standard 3-in. Schedule 40 pipe will be suitable, for which $D = 3.068$ in. (77.93 mm) and $\overline{V} = 110/23.00 = 4.783$ ft/s (1.458 m/s). Other quantities needed are

$$\frac{D\overline{V}\rho}{\mu} = N_{\text{Re}} = \frac{0.07793 \times 1.458 \times 989}{10^{-3}} = 112,370$$

$$\frac{D\rho\overline{V}^2}{\sigma g_c} = N_{\text{We}} = \frac{0.07793 \times 989 \times 1.458^2}{35 \times 10^{-3}} = 4,681$$

Substituting in Eq. (9-56) and solving for D_0 gives

$$D_o^{3.73} = \frac{(1.5 \times 10^{-2})(77.93^{2.73})(112,370^{0.065})(4,681^{0.722})}{21.6 \times 0.0909^{0.121}} = 128,910$$

$$D_o = 128,910^{1/3.73} = 23.4 \text{ mm } (0.922 \text{ in.})$$

It therefore appears possible to satisfy the conditions of the problem with a pipe diameter of 3.068 in. (77.93 mm) and an orifice diameter of 0.922 in. (23.4 mm).   ////

SYMBOLS

A_p	Area of cylinder swept out by tips of impeller blades, ft^2 or m^2
a	Interfacial area per unit volume, ft^{-1} or m^{-1}; also, constant in Eq. (9-23); a', interfacial area per unit volume calculated from Eq. (9-51), mm^{-1}
b	Constant in Eq. (9-23)
c_1, c_2, c_3	Constants
D	Diameter of pipe, ft or m; D_a, diameter of impeller; D_J, diameter of jet and nozzle; D_o, orifice diameter; D_p, diameter of drops or bubbles; D_t, tank diameter
$\overline{D}_s$	Volume-surface mean diameter of drops or bubbles, ft or mm; $\overline{D}'_s$, mean diameter calculated from Eq. (9-50), mm
E	Height of impeller above vessel floor, ft or m
F	Force, lb$_f$ or N; F_D, drag force; F_b, buoyant force; F_e, gravitational force
f_t	Blending-time factor, dimensionless, defined by Eq. (9-38)
g	Gravitational acceleration, ft/s^2 or m/s^2
g_c	Newton's-law proportionality factor, 32.174 ft-lb/lb$_f$-s^2
H	Depth of liquid in vessel, ft or m
J	Width of baffles, ft or m
K_L, K_T	Constants in Eqs. (9-26) and (9-28), respectively
K'	Flow consistency index of non-Newtonian fluid
k	Ratio of tangential liquid velocity at blade tips to blade-tip velocity
L	Length of impeller blades, ft or m
m	Exponent in Eq. (9-22)
$\dot{m}$	Mass flow rate, lb/s or kg/s
N	Number of drops or bubbles per unit volume
N_{Fr}	Froude number, $n^2 D_a/g$

N_L Modified power number, $Pg_c\rho^2 D_t/\mu^3$, dimensionless

N_P Power number, $Pg_c/n^3 D_a^5\rho$; $N_{p,g}$, at gas redispersion point

N_Q Flow number, q/nD_a^3; $N_{Q,g}$, at gas redispersion point

N_{Re} Agitator Reynolds number, $nD_a^2\rho/\mu$; $N_{Re,n}$, for non-Newtonian fluid, defined by Eq. (9-32)

N_t Mixing number, $t_T\mu/\rho D_t^2$, dimensionless

N_{We} Weber number, $D\rho\bar{V}^2/\sigma g_c$

n Rotational speed, r/s

n' Flow behavior index of non-Newtonian fluid

P Power, ft-lb$_f$/s or kW; P_g, at gas redispersion point; P_0, power consumption in ungassed liquid

q Volumetric flow rate, ft³/s or m³/lb; q_B, leaving impeller; q_e, entrained in jet; q_g, at gas redispersion point; q_0, leaving jet nozzle

r Radial distance from impeller axis; ft or m; r_2, radius of impeller

S Shape factor; $S_1 = D_t/D_a$; $S_2 = E/D_a$; $S_3 = L/D_a$; $S_4 = W/D_a$; $S_5 = J/D_t$; $S_6 = H/D_t$

T Torque, ft-lb$_f$ or N-m

t_T Blending time, s

u Velocity, ft/s or m/s; u_t, terminal velocity of particles, drop, or bubble; u_2, velocity of impeller blade tip; du/dy, velocity gradient or shear rate, s^{-1}

V Resultant velocity, absolute, in impeller, ft/s or m/s; V_r, radial component; V_{r2}, radial component of velocity V_2; V_{u2}, tangential component of velocity V_2; V_2, at impeller blade tips; also, volume, ft³ or m³; V_m, volume of liquid-solid suspension

$\bar{V}$ Average velocity of liquid in pipe, ft/s or m/s; $\bar{V}_s$, superficial velocity of gas in agitated vessel

W Impeller width, ft or m

X Distance from jet nozzle, ft or m

Z_s Maximum height of suspended solids above vessel floor, ft or m

Greek letters

β Exponent defined by Eq. (9-37)

β_2 Angle between the relative velocity vector of fluid and the tangent to the circle traced out by the impeller tip

ε_m Volume fraction of liquid in liquid-solid suspension

μ Absolute viscosity, lb/ft-s or P; μ_a, apparent viscosity of non-Newtonian fluid; μ_c, viscosity of continuous phase in liquid-liquid dispersion

ρ Density, lb/ft³ or kg/m³; ρ_c, of continuous phase in liquid-liquid dispersion; ρ_L, of liquid in gas-liquid dispersion; ρ_V, of gas in gas-liquid dispersion; ρ_m, of liquid-solid suspension

σ Interfacial tension, lb$_f$/ft or dyn/cm

ϕ Power function, N_p/N_{Fr}^m

Ψ Volumetric fractional gas or liquid holdup in dispersion, dimensionless

ψ, Φ Function; ψ_L, for laminar flow; ψ_T, for turbulent flow

ω Angular velocity, rad/s

PROBLEMS

9-1 A tank 4 ft in diameter and 6 ft high is filled to a depth of 4 ft with a latex having a viscosity of 10 P and a density of 47 lb/ft³. The tank is not baffled. A three-bladed 12-in.-diameter propeller is installed in the tank 1 ft from the bottom. The pitch is 1:1 (pitch equals diameter). The motor available develops 10 hp. Is the motor adequate to drive this agitator at a speed of 1,000 r/min?

9-2 What is the maximum speed at which the agitator of the tank described in Prob. 9-1 may be driven if the liquid is replaced by one having a viscosity of 1 P and the same density?

9-3 What power is required for the mixing operation of Prob. 9-1 if a propeller 12 in. in diameter with a 2:1 pitch (pitch twice the diameter) is used and if four baffles, each 5 in. wide, are installed?

9-4 The propeller in Prob. 9-1 is replaced with a six-bladed turbine 16 in. in diameter, and the fluid to be agitated is a pseudoplastic power-law liquid having an apparent viscosity of 15 P when the velocity gradient is 10 s^{-1}. At what speed should the turbine rotate to deliver 5 hp per 1,000 gal of liquid? For this fluid $n' = 0.75$ and $\rho = 60 \text{ lb/ft}^3$.

9-5 A pilot-plant reactor, a scale model of a production unit, is of such size that 1 g charged to the pilot-plant reactor is equivalent to 500 g of the same material charged to the production unit. The production unit is 2 m in diameter and 2 m deep and contains a six-bladed turbine agitator 0.6 m in diameter. The optimum agitator speed in the pilot-plant reactor is found by experiment to be 330 r/min. (*a*) What are the significant dimensions of the pilot-plant reactor? (*b*) If the reaction mass has the properties of water at 70°C and the power input per unit volume is to be constant, at what speed should the impeller turn in the large reactor? (*c*) At what speed should it turn if the mixing time is to be kept constant? (*d*) At what speed should it turn if the Reynolds number is held constant? (*e*) Which basis would you recommend for scale-up? Why?

9-6 The large reactor described in Prob. 9-5 is to be used for suspending 500 kg of 28-mesh galena particles in 4 m^3 of water at 70°C. (*a*) What should the agitator speed be to lift the particles to the liquid surface? (*b*) What power must be supplied to the agitator under these conditions?

9-7 Gaseous ethylene (C_2H_4) is to be dispersed in water in a turbine-agitated vessel at 110°C and an absolute pressure of 3 atm. The vessel is 3 m in diameter with a maximum liquid depth of 3 m. For a flow rate of 1,000 m^3 per pound of ethylene, measured at process conditions, specify (*a*) the diameter and speed of the turbine impeller, (*b*) the power drawn by the agitator, (*c*) the maximum volume of water allowable, and (*d*) the rate at which water is vaporized by the ethylene leaving the liquid surface. Assume that none of the ethylene dissolves in the water.

9-8 For a flow rate of 250 m^3/lb in the vessel described in Prob. 9-7, estimate the gas holdup, mean bubble diameter, and interfacial area per unit volume.

9-9 Design a process vessel to disperse 20 gal of benzene in 200 gal of water at 20°C, with an average drop size of 15 μm. Specify the significant dimensions of the vessel and impeller, the impeller speed, and the power requirement.

9-10 Calculate the power required to accomplish the dispersion described in Example 9-9, and compare it with the power calculated in Prob. 9-9. Which method is better: a tank mixer or a pipeline mixer? Why?

REFERENCES

1 Bates, R. L., P. L. Fondy, and R. R. Corpstein: *Ind. Eng. Chem. Process Des. Dev.,* **2**(4):310 (1963).

2 Biggs, R. D.: *AIChE J.,* **9**:636 (1963).

3 Bissell, E. S., H. C. Hesse, H. J. Everett, and J. H. Rushton: *Chem. Eng. Prog.,* **43**:649 (1947).

4 Calderbank, P. H.: in V. W. Uhl and J. B. Gray (eds.), "Mixing: Theory and Practice," vol. II, Academic, New York, 1967; (*a*) p. 23, (*b*) pp. 28–29, (*c*) p. 73.

5 Calderbank, P. H., and M. B. Moo-Young: *Trans. Inst. Chem. Eng. Lond.,* **37**:26 (1959).

6 Chen, S. J., L. T. Fan, and C. A. Watson: *AIChE J.,* **18**:984 (1972).

7 Chen, S. J., and A. R. MacDonald: *Chem. Eng.,* **80**(7):105 (1973).

8   Connolly, J. R., and R. L. Winter: *Chem. Eng. Prog.,* **65**(8):70 (1969).
9   Cutter, L. A.: *AIChE J.,* **12**:35 (1966).
10 Fox, E. A., and V. E. Gex: *AIChE J.,* **2**:539 (1956).
11 Godleski, E. S., and J. C. Smith: *AIChE J.,* **8**:617 (1962).
12 Gray, J. B.: *Chem. Eng. Prog.,* **59**(3):55 (1963).
13 Gray, J. B.: in V. W. Uhl and J. B. Gray (eds.), "Mixing: Theory and Practice," vol. I, Academic, New York, 1969; (*a*) pp. 181–184, (*b*) p. 208.
14 Hirsekorn, F. S., and S. A. Miller: *Chem. Eng. Prog.,* **49**:459 (1953).
15 Hunsaker, J. C., and B. G. Rightmire: "Engineering Applications of Fluid Mechanics," chap. 7, McGraw-Hill, New York, 1947.
16 Jacobs, L. J.: paper presented at *Eng. Found. Mixing Res. Conf., South Berwick, Maine, Aug. 12-17, 1973.*
17 Lyons, E. J.: *Chem. Eng. Prog.,* **44**:341 (1948).
18 Metzner, A. B., R. H. Feehs, H. L. Ramos, R. E. Otto, and J. D. Tuthill: *AIChE J.,* **7**:3 (1961).
19 Miller, E., S. P. Foster, R. W. Ross, and K. Wohl: *AIChE J.,* **3**:395 (1957).
20 Moo-Young, M., K. Tichar, and F. A. L. Dullien: *AIChE J.,* **18**:178 (1972).
21 Morrison, P. P., H. Olin, and G. Rappe: Chemical Engineering Research Report, Cornell University, June 1962 (unpublished).
22 Norwood, K. W., and A. B. Metzner: *AIChE J.,* **6**:432 (1960).
23 Overcashier, R. H., H. A. Kingsley, Jr., and R. B. Olney: *AIChE J.,* **2**:529 (1956).
24 Perry, J. H. (ed.): "Chemical Engineers' Handbook," 5th ed., pp. **19**-8ff, McGraw-Hill, New York, 1973.
25 Rushton, J. H.: *Chem. Eng. Prog.,* **50**:587 (1954).
26 Rushton, J. H., E. W. Costich, and H. J. Everett: *Chem. Eng. Prog.,* **46**:395, 467 (1950).
27 Rushton, J. H., and J. Y. Oldshue: *Chem. Eng. Prog.,* **49**(4):165 (1953).
28 Weisman, J., and L. E. Efferding: *AIChE J.,* **6**:419 (1960).
29 Zlokarnik, M.: *Chem. Ing. Tech.,* **39**:539 (1967).
30 Zlokarnik, M., and H. Judat: *Chem. Ing. Tech.,* **39**:1163 (1967).

Heat transfer
and its applications

Practically all the operations that are carried out by the chemical engineer involve the production or absorption of energy in the form of heat. The laws governing the transfer of heat and the types of apparatus that have for their main object the control of heat flow are therefore of great importance. This section of the book deals with heat transfer and its applications in process engineering.

Nature of heat flow   When two objects at different temperatures are brought into thermal contact, heat flows from the object at the higher temperature to that at the lower temperature. The net flow is always in the direction of the temperature decrease. The mechanisms by which the heat may flow are three: conduction, convection, and radiation.

CONDUCTION   If a temperature gradient exists in a continuous substance, heat can flow unaccompanied by any observable motion of matter. Heat flow of this kind is called *conduction*. In metallic solids, thermal conduction results from the motion of unbound electrons, and there is close correspondence between thermal conductivity and electrical conductivity. In solids which are poor conductors of electricity, and in most liquids, thermal conduction results from the transport of momentum of individual molecules along the temperature gradient. In gases conduction occurs by the random motion of molecules, so that heat is "diffused" from hotter regions to colder ones. The most common example of conduction is heat flow in opaque solids, as in the brick wall of a furnace or the metal wall of a tube.

CONVECTION When a current or macroscopic particle of fluid crosses a specific surface, such as the boundary of a control volume, it carries with it a definite quantity of enthalpy. Such a flow of enthalpy is called a *convective flow of heat* or simply *convection.* Since convection is a macroscopic phenomenon, it can occur only when forces act on the particle or stream of fluid and maintain its motion against the forces of friction. Convection is closely associated with fluid mechanics. In fact, thermodynamically, convection is not considered as heat flow but as flux of enthalpy. The identification of convection with heat flow is a matter of convenience, because in practice it is difficult to separate convection from true conduction when both are lumped together under the name convection. Examples of convection are the transfer of enthalpy by the eddies of turbulent flow and by the current of warm air flowing across and away from an ordinary radiator.

NATURAL AND FORCED CONVECTION The forces used to create convection currents in fluids are of two types. If the currents are the result of buoyancy forces generated by differences in density and the differences in density are in turn caused by temperature gradients in the fluid mass, the action is called *natural convection.* The flow of air across a heated radiator is an example of natural convection. If the currents are set in motion by the action of a mechanical device such as a pump or agitator, the flow is independent of density gradients, and is called *forced convection.* Heat flow to a fluid pumped through a heated pipe is an example of forced convection. The two kinds of force may be active simultaneously in the same fluid, and natural and forced convection then occur together.

RADIATION Radiation is a term given to the transfer of energy through space by electromagnetic waves. If radiation is passing through empty space, it is not transformed into heat or any other form of energy nor is it diverted from its path. If, however, matter appears in its path, the radiation will be transmitted, reflected, or absorbed. It is only the absorbed energy that appears as heat, and this transformation is quantitative. For example, fused quartz transmits practically all the radiation that strikes it; a polished opaque surface or mirror will reflect most of the radiation impinging on it; a black or matte surface will absorb most of the radiation received by it and will transform such absorbed energy quantitatively into heat.

Monatomic and diatomic gases are transparent to thermal radiation, and it is quite common to find that heat is flowing through masses of such gases both by radiation and by conduction-convection. Examples are the loss of heat from a radiator or unlagged steam pipe to the ambient air of the room and heat transfer in furnaces and other high-temperature gas-heating equipment. The two mechanisms are mutually independent and occur in parallel, so that one type of heat flow can be controlled or varied independently of the other. Conduction-convection and radiation can be studied separately and their separate effects added together in cases where both are important. In very general terms, radiation becomes important at high temperatures and is independent of the circumstances of the flow of the fluid. Conduction-convection is sensitive to flow conditions and is relatively unaffected by temperature level.

Chapter 10 deals with conduction in solids; Chaps. 11 to 13 with heat transfer to fluids by conduction and convection; Chap. 14 with heat transfer by radiation. In Chaps. 15 and 16 the principles developed in the preceding chapters are applied to the design of equipment for heating, cooling, condensing, and evaporating.

HEAT TRANSFER BY CONDUCTION IN SOLIDS

Conduction is most easily understood by considering heat flow in homogeneous isotropic solids because in these there is no convection and the effect of radiation is negligible unless the solid is translucent to electromagnetic waves. First, the general law of conduction is discussed; second, situations of steady-state heat conduction, where the temperature distribution within the solid does not change with time, are treated; third, some simple cases of unsteady conduction, where the temperature distribution does change with time, are considered.

Fourier's law The basic relation of heat flow by conduction is the proportionality between the rate of heat flow across an isothermal surface and the temperature gradient at the surface. This generalization, which applies at any location in a body and at any time, is called *Fourier's law*.[3] It can be written

$$\frac{dq}{dA} = -k\frac{\partial T}{\partial n} \tag{10-1}$$

where A = area of isothermal surface
$\quad n$ = distance measured normally to surface

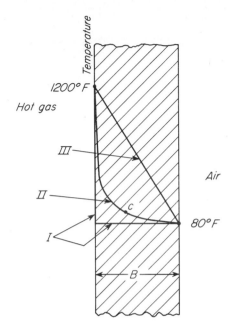

FIGURE 10-1
Temperature distributions, unsteady-
state heating of furnace wall: I, at instant
of exposure of wall to high temperature;
II, during heating at time t; III, at steady
state.

q = rate of heat flow across surface in direction normal to surface
T = temperature
k = proportionality constant

The partial derivative in Eq. (10-1) calls attention to the fact that the temperature may vary with both location and time. The negative sign reflects the physical fact that heat flow occurs from hot to cold and the sign of the gradient is opposite that of the heat flow.

In using Eq. (10-1) it must be clearly understood that the area A is that of a surface perpendicular to the flow of heat and distance n is the length of path measured perpendicularly to area A.

Although Eq. (10-1) applies specifically across an isothermal surface, the same equation can be used for heat flow across any surface, not necessarily isothermal, provided the area A is the area of the surface and the length of the path is measured normally to the surface.[2a] This extension of Fourier's law is vital in the study of two- or three-dimensional flows, where heat flows along curves instead of straight lines. In one-dimensional flow, which is the only situation considered in this text, the normals representing the direction of heat flow are straight. One-dimensional heat flow is analogous to one-dimensional fluid flow, and only one linear coordinate is necessary to measure the length of path.

An example of one-dimensional heat flow is shown in Fig. 10-1, which represents a flat furnace wall. Initially the wall is in temperature equilibrium with the air, at 80°F. The temperature distribution in the wall is represented by line I. At temperature equilibrium, T is independent of both time and position. Assume now that one side of the wall is suddenly exposed to furnace gas at 1200°F. If there is negligible resistance to heat flow between the gas and the wall, the temperature at the gas side of the wall immediately rises to 1200°F and heat flow begins. Momentarily the temperature distribution in the wall is represented by line I in Fig. 10-1. After the elapse of some time, the temperature distribution can be represented by a curve like that of curve II. The temperature at a given distance, e.g., that at point c, is increasing; and T depends upon both time and location. The process is called *unsteady-state conduction*, and Eq. (10-1) applied at each point at each time in the slab. Finally, if the wall is kept in contact with hot gas and cool air for a sufficiently long time, the temperature distribution shown by line III will be obtained; and this distribution will remain unchanged with further elapse of time. Conduction under the condition of constant temperature distribution is called *steady-state conduction*. In the steady state T is a function of position only, and the rate of heat flow at any one point is a constant. For steady one-dimensional flow, Eq. (10-1) may be written

$$\frac{q}{A} = -k \frac{dT}{dn} \tag{10-2}$$

Thermal conductivity   The proportionality constant k is a physical property of the substance called the *thermal conductivity*. It, like the Newtonian viscosity μ, is one of the so-called transport properties of the material. This terminology is based on the analogy between Eqs. (3-4) and (10-2). In Eq. (3-4) the quantity τg_c is a rate of momentum flow per unit area, the quantity du/dy is the velocity gradient, and μ is the required proportionality factor. In Eq. (10-2), q/A is the rate of heat flow per unit area, dT/dn is the temperature gradient, and k is the proportionality factor.

In engineering units, q is measured in Btu/h or watts and dT/dn in °F/ft or °C/m. Then the units of k are Btu/ft^2-h-(°F/ft), which may be written Btu/ft-h-°F, or W/m-°C.

Fourier's law states that k is independent of the temperature gradient but not necessarily of temperature itself. Experiment does confirm the independence of k for a wide range of temperature gradients, except for porous solids, where radiation between particles, which does not follow a linear temperature law, becomes an important part of the total heat flow. On the other hand, k is a function of temperature, but not a strong one. For small ranges of temperature, k may be considered constant. For larger temperature ranges, the thermal conductivity varies linearly with temperature, or

$$k = a + bT \tag{10-3}$$

where a and b are empirical constants. Line III in Fig. 10-1 applies to a solid of constant k, where $b = 0$; the line would show some curvature if k were dependent on temperature.

Thermal conductivities vary over a wide range. They are highest for metals and lowest for finely powdered materials from which air has been evacuated. The thermal conductivity of silver is about 240 Btu/ft-h-°F, and that for evacuated silica aerogel is as low as 0.0012. Solids having low k values are used as heat insulators to minimize the rate of heat flow. Porous insulating materials such as polystyrene foam act by entrapping air and thus eliminating convection. Their k values are about equal to that of air itself. Data showing typical thermal conductivities are given in Appendixes 11 to 14.

STEADY-STATE CONDUCTION

For the simplest case of steady-state conduction, consider a flat slab like that shown in Fig. 10-1. Assume that k is independent of temperature and that the area of the wall is very large in comparison with its thickness, so that heat losses from the edges are negligible. The external surfaces are at right angles to the plane of the illustration, and both are isothermal surfaces. The direction of heat flow is perpendicular to the wall. Also, since in steady state there can be neither accumulation nor depletion of heat within the slab, q is constant along the path of heat flow. If x is the distance from the hot side, Eq. (10-2) can be written

$$\frac{q}{A} = -k \frac{dT}{dx}$$

or

$$dT = -\frac{q}{kA} dx \tag{10-4}$$

Since the only variables in Eq. (10-4) are x and T, direct integration gives

$$\frac{q}{A} = k \frac{T_1 - T_2}{x_2 - x_1} = k \frac{\Delta T}{B} \tag{10-5}$$

where $x_2 - x_1 = B =$ thickness of slab
$T_1 - T_2 = \Delta T =$ temperature drop across slab

When the thermal conductivity k varies linearly with temperature, in accordance with Eq. (10-3), Eq. (10-5) still can be used rigorously by taking an average value $\bar{k}$ for k, which may be found either by using the arithmetic average of the individual values of k for the two surface temperatures, T_1 and T_2, or by calculating the arithmetic average of the temperatures and using the value of k at that temperature.

Equation (10-5) can be written in the form

$$q = \frac{\Delta T}{R} \tag{10-6}$$

where R is the thermal resistance of the solid between points 1 and 2. Equation (10-6) is an instance of the general rate principle, which equates a rate to the ratio of a driving force to a resistance. In heat conduction, q is the rate and ΔT the driving force. The resistance R, as shown by Eq. (10-6) and using $\bar{k}$ for k to account for a linear variation of k with temperature, is $B/\bar{k}A$. The reciprocal of a resistance is called a *conductance*, which for heat conduction is $\bar{k}A/B$. Both resistance and the conductance depend upon the dimensions of the solid as well as on the conductivity k, which is a property of the material.

TEMPERATURE DROPS

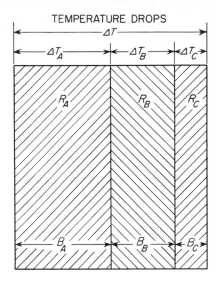

FIGURE 10-2
Thermal resistances in series.

EXAMPLE 10-1   A layer of pulverized cork 6 in. (152 mm) thick is used as a layer of thermal insulation in a flat wall. The temperature of the cold side of the cork is 40°F (4.4°C), and that of the warm side is 180°F (82.2°C). The thermal conductivity of the cork at 32°F (0°C) is 0.021 Btu/ft-h-°F (0.036 W/m-°C), and that at 200°F (93.3°C) is 0.032 (0.055). The area of the wall is 25 ft^2 (2.32 m^2). What is the rate of heat flow through the wall, in Btu per hour (watts)?

SOLUTION   The arithmetic average temperature of the cork layer is $(40 + 180)/2 = 110°F$. By linear interpolation the thermal conductivity at 110°F is

$$\bar{k} = 0.021 + \frac{(110 - 32)(0.032 - 0.021)}{200 - 32}$$

$$= 0.021 + 0.005 = 0.026 \text{ Btu/ft-h-}°F$$

Also, $A = 25 \text{ ft}^2 = \Delta T = 180 - 40 = 140°F$ $B = \tfrac{6}{12} = 0.5 \text{ ft}$

Substituting in Eq. (10-5) gives

$$q = \frac{0.026 \times 25 \times 140}{0.5} = 182 \text{ Btu/h (53.3 W)} \qquad\qquad ////$$

Compound resistances in series   Consider a flat wall constructed of a series of layers, as shown in Fig. 10-2. Let the thicknesses of the layers be B_A, B_B, and B_C and the average conductivities of the materials of which the layers are made be $\bar{k}_A$, $\bar{k}_B$, and $\bar{k}_C$, respectively. Also, let the area of the compound wall, at right angles to the plane of the illustration, be A. Let ΔT_A, ΔT_B, and ΔT_C be the temperature drops across layers A, B, and C, respectively. Assume, further, that the layers are in excellent thermal contact, so that no temperature difference exists across the interfaces between the layers. Then, if ΔT is the total temperature drop across the entire wall,

$$\Delta T = \Delta T_A + \Delta T_B + \Delta T_C \qquad\qquad (10\text{-}7)$$

It is desired, first, to derive an equation for calculating the rate of heat flow through the series of resistances and, second, to show how the rate can be calculated as the ratio of the overall temperature drop ΔT to the overall resistance of the wall.

Equation (10-5) can be written for each layer, using $\bar{k}$ in place of k,

$$\Delta T_A = q_A \frac{B_A}{\bar{k}_A A} \qquad \Delta T_B = q_B \frac{B_B}{\bar{k}_B A} \qquad \Delta T_C = q_C \frac{B_C}{\bar{k}_C A} \qquad (10\text{-}8)$$

Adding Eqs. (10-8) gives

$$\Delta T_A + \Delta T_B + \Delta T_C = \frac{q_A B_A}{A \bar{k}_A} + \frac{q_B B_B}{A \bar{k}_B} + \frac{q_C B_C}{A \bar{k}_C} = \Delta T$$

Since, in steady heat flow, all the heat that passes through the first resistance must pass through the second and in turn pass through the third, q_A, q_B, and q_C are equal and all can be denoted by q. Using this fact and solving for q gives

$$q = \frac{\Delta T}{B_A/\bar{k}_A A + B_B/\bar{k}_B A + B_C/\bar{k}_C A} = \frac{\Delta T}{R_A + R_B + R_C} = \frac{\Delta T}{R} \qquad (10\text{-}9)$$

where R_A, R_B, R_C = resistance of individual layers
R = overall resistance

Equation (10-9) shows that in heat flow through a series of layers the overall thermal resistance equals the sum of the individual resistances.

It is useful to recall the analogies between the flow of heat and the steady flow of electricity through a conductor. The flow of heat is covered by the expression

$$\text{Rate} = \frac{\text{temperature drop}}{\text{resistance}}$$

In the flow of electricity the potential factor is the electromotive force, and the rate of flow is coulombs per second, or amperes. The rate equation for electric flow is

$$\text{Amperes} = \frac{\text{volts}}{\text{ohms}}$$

By comparing this equation with Eq. (10-6) it is evident that rate of flow in Btu per hour is analogous to amperes, temperature drop to voltage, and thermal resistance to electric resistance.

The rate of flow of heat through several resistances in series clearly is analogous to the current flowing through several electric resistances in series. In an electric circuit the potential drop over any one of several resistances is to the total potential drop in the circuit as the individual resistances are to the total resistance. In the same way the potential drops in a thermal circuit, which are the temperature differences, are to the total temperature drop as the individual thermal resistances are to the total thermal resistance. This can be expressed mathematically as

$$\frac{\Delta T}{R} = \frac{\Delta T_A}{R_A} = \frac{\Delta T_B}{R_B} = \frac{\Delta T_C}{R_C} \qquad (10\text{-}10)$$

EXAMPLE 10-2   A flat furnace wall is constructed of a 4.5-in. (114-mm) layer of Sil-o-cel brick, with a thermal conductivity of 0.08 Btu/ft-h-°F (0.138 W/m-°C) backed by a 9-in. (229-mm) layer of common brick, of conductivity 0.8 (1.38). The temperature of the inner face of the wall is 1400°F (760°C), and that of the outer face is 170°F (76.6°C). (a) What is the heat loss through the wall? (b) What is the temperature of the interface between the refractory brick and the common brick? (c) Supposing that the contact between the two brick layers is poor and that a "contact resistance" of 0.50°F-h/Btu (0.948°C/W) is present, what would be the heat loss?

SOLUTION   (a)   Consider 1 ft^2 of wall ($A = 1$ ft^2). The thermal resistance of the Sil-o-cel layer is

$$R_A = \frac{4.5/12}{0.08 \times 1} = 4.687$$

and that of the common brick is

$$R_B = \frac{9/12}{0.8 \times 1} = 0.938$$

The total resistance is

$$R = R_A + R_B = 4.687 + 0.938 = 5.625°\text{F-h/Btu}$$

The overall temperature drop is

$$\Delta T = 1400 - 170 = 1230°\text{F}$$

Substitution in Eq. (10-9) gives, for the heat loss from 1 ft^2 of wall

$$q = \frac{1230}{5.625} = 219 \text{ Btu/h (64.2 W)}$$

(b)   The temperature drop in one of a series of resistances is to the individual resistance as the overall temperature drop is to the overall resistance, or

$$\frac{\Delta T_a}{4.687} = \frac{1230}{5.625}$$

from which

$$\Delta T_a = 1025°\text{F}$$

The temperature at the interface is $1400 - 1025 = 375°$F (190.6°C).
(c)   The total resistance, which now includes a contact resistance, is

$$R = 5.625 + 0.500 = 6.125$$

The heat loss from 1 ft^2 is

$$q = \frac{1230}{6.125} = 201 \text{ Btu/h (58.9 W)} \qquad\qquad ////$$

Heat flow through a cylinder   Consider the hollow cylinder represented by Fig. 10-3. The inside radius of the cylinder is r_i, the outside radius is r_o, and the length of the cylinder is L. The thermal conductivity of the material of which the cylinder is made is k. The temperature of the outside surface is T_o, and that of the inside surface is T_i. It is desired to calculate the rate of heat flow outward for this case.

Consider a very thin cylinder, concentric with the main cylinder, of radius r,

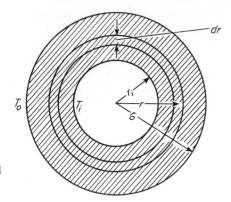

FIGURE 10-3
Flow of heat through thick-walled
cylinder.

where r is between r_i and r_o. The thickness of the wall of this cylinder is dr; and if dr is small enough with respect to r for the lines of heat flow to be considered parallel, Eq. (10-2) can be applied and written in the form

$$q = -k \frac{dT}{dr} 2\pi rL \qquad (10\text{-}11)$$

since the area perpendicular to the heat flow is equal to $2\pi rL$ and the dn of Eq. (10-2) is equal to dr. Rearranging Eq. (10-11) and integrating between limits gives

$$\int_{r_i}^{r_o} \frac{dr}{r} = \frac{2\pi Lk}{q} \int_{T_o}^{T_i} dT$$

$$\ln r_o - \ln r_i = \frac{2\pi Lk}{q}(T_i - T_o)$$

$$q = \frac{k(2\pi L)(T_i - T_o)}{\ln (r_o/r_i)} \qquad (10\text{-}12)$$

Equation (10-12) can be used to calculate the flow of heat through a thick-walled cylinder. It can be put in a more convenient form by expressing the rate of flow of heat as

$$q = \frac{k\bar{A}_L(T_i - T_o)}{r_o - r_i} \qquad (10\text{-}13)$$

This is of the same general form as Eq. (10-5) for heat flow through a flat wall with the exception of $\bar{A}_L$, which must be so chosen that the equation is correct. The term $\bar{A}_L$ can be determined by equating the right-hand sides of Eqs. (10-12) and (10-13) and solving for $\bar{A}_L$.

$$\bar{A}_L = \frac{2\pi L(r_o - r_i)}{\ln (r_o/r_i)} \qquad (10\text{-}14)$$

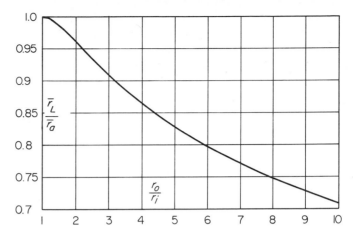

FIGURE 10-4
Relation between logarithmic and arithmetic means.

Note from Eq. (10-14) that $\bar{A}_L$ is the area of a cylinder of length L and radius $\bar{r}_L$, where

$$\bar{r}_L = \frac{r_o - r_i}{\ln(r_o/r_i)} = \frac{r_o - r_i}{2.303 \log(r_o/r_i)} \tag{10-15}$$

The form of the right-hand side of Eq. (10-15) is important enough to repay memorizing. It is known as the *logarithmic mean*, and in the particular case of Eq. (10-15) $\bar{r}_L$ is called the *logarithmic mean radius*. It is the radius which, when applied to the integrated equation for a flat wall, will give the correct rate of heat flow through a thick-walled cylinder.

The logarithmic mean is less convenient than the arithmetic mean, and the latter can be used without appreciable error for thin-walled tubes, where r_o/r_i is nearly 1. The ratio of the logarithmic mean $\bar{r}_L$ to the arithmetic mean $\bar{r}_a$ is a function of r_o/r_i, as shown in Fig. 10-4. Thus, when $r_o/r_i = 2$, the logarithmic mean is $0.96\bar{r}_a$ and the error in the use of the arithmetic mean is 4 percent. The error is 1 percent when $r_o/r_i = 1.4$.

EXAMPLE 10-3 A tube 60 mm (2.36 in.) OD is lagged with a 50-mm (1.97-in.) layer of asbestos, for which the conductivity is 0.21 W/m-°C (0.12 Btu/ft-h-°F), followed with a 40-mm (1.57-in.) layer of cork with a conductivity of 0.05 W/m-°C (0.03 Btu/ft-h-°F). If the temperature of the outer surface of the pipe is 150°C (302°F) and the temperature of the outer surface of the cork is 30°C (86°F), calculate the heat loss in watts per meter of pipe.

SOLUTION These layers are too thick to use the arithmetic mean radius, and the logarithmic mean radius should be used. For the asbestos layer

$$\bar{r}_L = \frac{80 - 30}{\ln(80/30)} = 50.97 \text{ mm}$$

and for the cork layer

$$\bar{r}_L = \frac{120 - 80}{\ln (120/80)} = 98.64 \text{ mm}$$

If asbestos is called substance A and cork substance B, the individual resistances are

$$R_A = \frac{x_A}{k_A \bar{A}_A} = \frac{0.060}{0.21 \times 2\pi(0.05097)L} = \frac{0.892}{L}$$

$$R_B = \frac{x_B}{k_B \bar{A}_B} = \frac{0.040}{0.05 \times 2\pi(0.09864)L} = \frac{1.291}{L}$$

The heat loss is

$$\frac{q}{L} = \frac{150 - 30}{0.892 + 1.291} = 55.0 \text{ W/m (57.2 Btu/ft-h)} \qquad ////$$

UNSTEADY-STATE HEAT CONDUCTION

A full treatment of unsteady-state heat conduction is not in the field of this text.[2-4] A derivation of the partial differential equation for one-dimensional heat flow and the results of the integration of the equations for some simple shapes are the only subjects covered in this section. It is assumed throughout that k is independent of temperature.

Equation for one-dimensional conduction   Consider the solid slab shown in Fig. 10-5. Focus attention on the thin slab of thickness dx located at distance x from the hot side of the slab. The two sides of the element are isothermal surfaces. The temperature gradient at x is, at a definite instant of time, $\partial T/\partial x$, and the heat input in time interval dt at x is $-kA(\partial T/\partial x)\, dt$, where A is the area of the slab perpendicular to the flow of heat and k is the thermal conductivity of the solid. The gradient at distance $x + dx$ is slightly greater than that at x, and may be represented as

$$\frac{\partial T}{\partial x} + \frac{\partial}{\partial x}\frac{\partial T}{\partial x}\, dx$$

The heat flow out of the slab at $x + dx$ is, then,

$$-kA\left(\frac{\partial T}{\partial x} + \frac{\partial}{\partial x}\frac{\partial T}{\partial x}\, dx\right) dt$$

The excess of heat input over heat output, which is the accumulation of heat in layer dx, is

$$-kA\frac{\partial T}{\partial x}\, dt + kA\left(\frac{\partial T}{\partial x} + \frac{\partial^2 T}{\partial x^2}\, dx\right) dt = kA\frac{\partial^2 T}{\partial x^2}\, dx\, dt$$

The accumulation of heat in the layer must increase the temperature of the layer. If c_p and ρ are the specific heat and density, respectively, the accumulation is the product of the mass (volume times density), the specific heat, and the increase in

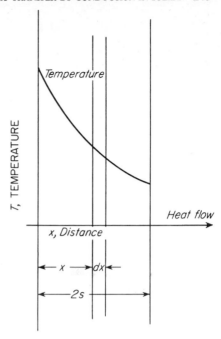

FIGURE 10-5
Unsteady-state conduction in solid slab.

temperature, or $(\rho A\ dx)c_p(\partial T/\partial t)\ dt$. Then, by a heat balance,

$$kA\ \frac{\partial^2 T}{\partial x^2}\ dx\ dt = \rho c_p A\ dx\ \frac{\partial T}{\partial t}\ dt$$

or, after division by $\rho c_p A\ dx\ dt$,

$$\frac{\partial T}{\partial t} = \frac{k}{\rho c_p}\ \frac{\partial^2 T}{\partial x^2} = \alpha\ \frac{\partial^2 T}{\partial x^2} \tag{10-16}$$

The term α in Eq. (10-16) is called the *thermal diffusivity* of the solid and is a property of the material. It has the dimensions of area divided by time.

General solutions of unsteady-state conduction equations are available for certain simple shapes such as the infinite slab, the infinitely long cylinder, and the sphere. For example, the integration of Eq. (10-16) for the heating or cooling of an infinite slab of known thickness from both sides by a medium at constant surface temperature gives

$$\frac{T_s - \bar{T}_b}{T_s - T_a} = \frac{8}{\pi^2}\left(e^{-a_1 N_{Fo}} + \frac{1}{9}e^{-9a_1 N_{Fo}} + \frac{1}{25}e^{-25a_1 N_{Fo}} + \cdots\right) \tag{10-17}$$

where T_s = constant average temperature of surface of slab
T_a = initial temperature of slab
$\bar{T}_b$ = average temperature of slab at time t_T
N_{Fo} = Fourier number, defined as $\alpha t_T/s^2$
α = thermal diffusivity
t_T = time of heating or cooling

s = one-half slab thickness

$a_1 = (\pi/2)^2$

For an infinitely long solid cylinder of radius r_m the average temperature $\overline{T}_b$ is given by the equation[5b]

$$\frac{T_s - \overline{T}_b}{T_s - T_a} = 0.692e^{-5.78N_{Fo}} + 0.131e^{-30.5N_{Fo}} + 0.0534e^{-74.9N_{Fo}} + \cdots \quad (10\text{-}18)$$

where

$$N_{Fo} = \frac{\alpha t_T}{r_m^2}$$

For a sphere of radius r_m the corresponding equation is[2b]

$$\frac{T_s - \overline{T}_b}{T_s - T_a} = 0.608e^{-9.87N_{Fo}} + 0.152e^{-39.5N_{Fo}} + 0.0676e^{-88.8N_{Fo}} + \cdots \quad (10\text{-}19)$$

When N_{Fo} is greater than about 0.1, only the first term of the series in Eqs. (10-17) to (10-19) is significant and the other terms may be ignored. Under these conditions the time required to change the temperature from T_a to $\overline{T}_b$ can be found by rearranging Eq. (10-17), with all except the first term of the series omitted, to give for the slab

$$t_T = \frac{1}{\alpha}\left(\frac{2s}{\pi}\right)^2 \ln \frac{8(T_s - T_a)}{\pi^2(T_s - \overline{T}_b)} \quad (10\text{-}20)$$

For the infinite cylinder the corresponding equation, found from Eq. (10-18), is

$$t_T = \frac{r_m^2}{5.78\alpha} \ln \frac{0.692(T_s - T_a)}{T_s - \overline{T}_b} \quad (10\text{-}21)$$

For a sphere, from Eq. (10-19),

$$t_T = \frac{r_m^2}{9.87\alpha} \ln \frac{0.608(T_s - T_a)}{T_s - \overline{T}_b} \quad (10\text{-}22)$$

Figure 10-6 is a plot of Eqs. (10-17) to (10-19). The ordinate of this figure is known as the *unaccomplished temperature change*, i.e., the fraction of the total possible temperature change that remains to be accomplished at any time.

Equations (10-17) to (10-19) apply only when the surface temperature is constant, so T_s can be set equal to the temperature of the heating or cooling medium only when the temperature difference between the medium and the solid surface is negligible. This implies that there is negligible thermal resistance between the surface and the medium. Equations and graphs[5a] similar to Eqs. (10-17) to (10-19) and Fig. 10-6 are available for local temperatures at points inside slabs and cylinders, for temperatures in spheres and other shapes, and for situations in which the thermal resistance at the surface is large enough to cause variations in the surface temperature. Temperature distributions in heterogeneous solids or bodies of complex shape are found by fluid or electrical analogs or numerical approximation methods.[1]

The total heat Q_T transferred to the solid in time t_T through a unit area of surface is often of interest. From the definition of average temperature the heat

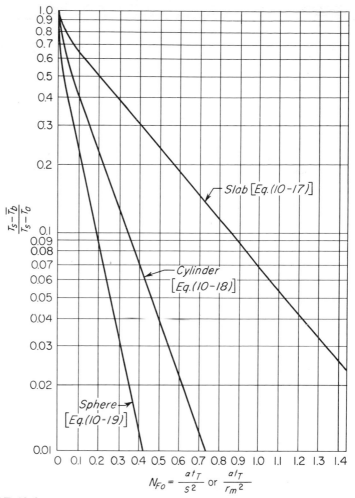

FIGURE 10-6
Average temperatures during unsteady-state heating or cooling of a large slab, an infinitely long cylinder, or a sphere.

required to raise the temperature of a unit mass of solid from T_a to $\overline{T}_b$ is $c_p(\overline{T}_b - T_a)$. For a slab of thickness $2s$ and density ρ the total surface area (one side) of a unit mass is $1/2s\rho$. The total heat transferred per unit area is therefore given by

$$\frac{Q_T}{A} = 2s\rho c_p(\overline{T}_b - T_a) \tag{10-23}$$

The corresponding equation for an infinitely long cylinder is

$$\frac{Q_T}{A} = \frac{r_m\rho c_p(\overline{T}_b - T_a)}{2} \tag{10-24}$$

EXAMPLE 10-4 A flat slab of plastic initially at 70°F (21.1°C) is placed between two platens at 250°F (121.1°C). The slab is 1.0 in. (2.54 cm) thick. (*a*) How long will it take to heat the slab to an average temperature of 210°F (98.9°C)? (*b*) How much heat, in Btu, will be transferred to the plastic during this time per square foot of surface? The density of the solid is 56.2 lb/ft³ (900 kg/m³), the thermal conductivity is 0.075 Btu/ft-h-°F (0.13 W/m-°C); the specific heat is 0.40 Btu/lb-°F (1.67 J/g-°C).

SOLUTION (*a*) The quantities for use with Fig. 10-6 are

$$k = 0.075 \text{ Btu/ft-h-°F} \qquad \rho = 56.2 \text{ lb/ft}^3 \qquad c_p = 0.40 \text{ Btu/lb-°F}$$

$$s = \frac{0.5}{12} = 0.0417 \text{ ft} \qquad T_s = 250°F \qquad T_a = 70°F \qquad \overline{T}_b = 210°F$$

Then

$$\frac{T_s - T_b}{T_s - T_a} = \frac{250 - 210}{250 - 70} = 0.222 \qquad \alpha = \frac{k}{\rho c_p} = \frac{0.075}{56.2 \times 0.40} = 0.00335$$

From Fig. 10-6, for a temperature-difference ratio of 0.222,

$$N_{Fo} = 0.52 = \frac{0.00335 t_T}{0.0417^2}$$

$$t_T = 0.27 \text{ h} = 16 \text{ min}$$

(*b*) Substitution in Eq. (10-23) gives

$$\frac{Q_T}{A} = 2 \times 0.0417 \times 56.2 \times 0.40(210 - 70) = 262 \text{ Btu/ft}^2 \text{ (2975 kJ/m}^2) \qquad ////$$

SEMI-INFINITE SOLID   Sometimes solids are heated or cooled in such a way that the temperature changes in the solid are confined to the region near one surface. Consider, for example, a very thick flat wall of a chimney, initially all at a uniform temperature T_a. Suppose that the inner surface of the wall is suddenly heated to, and held at, a high temperature T_s, perhaps by suddenly admitting hot flue gas to the chimney. Temperatures inside the chimney wall will change with time, rapidly near the hot surface and more slowly farther away. If the wall is thick enough, there will be no measurable change in the temperature of the outer surface for a considerable time. Under these conditions the heat may be considered to be "penetrating" a solid of essentially infinite thickness. Figure 10-7 shows the temperature patterns in such a wall at various times after exposure to the hot gas, indicating the sharp discontinuity in temperature at the hot surface immediately after exposure and the progressive changes at interior points at later times.

For this situation integration of Eq. (10-16) with the appropriate boundary conditions gives, for temperature T at any point a distance x from the hot surface, the equation

$$\frac{T_s - T}{T_s - T_a} = \frac{2}{\sqrt{\pi}} \int_0^Z e^{-Z^2} \, dZ \qquad (10\text{-}25)$$

where $Z = x/2\sqrt{\alpha t}$, dimensionless
α = thermal diffusivity

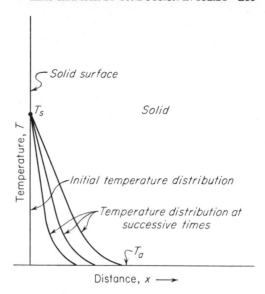

FIGURE 10-7
Temperature distributions in unsteady-state heating of semi-infinite solid.

x = distance from surface
t = time after change in surface temperature, h

The function in Eq. (10-25) is known as the *Gauss error integral* or *probability integral*. Equation (10-25) is plotted in Fig. 10-8.

Equation (10-25) indicates that at any time after the surface temperature is changed there will be some change in temperature at all points in the solid, even points far removed from the hot surface. The actual change at such distant points, however, is negligibly small. Beyond a certain distance from the hot surface not enough heat has penetrated to affect the temperature significantly. This *penetration distance* x_p is arbitrarily defined as that distance from the surface at which the temperature change is 1 percent of the initial change in surface temperature. That is to say, $(T - T_a)/(T_s - T_a) = 0.01$ or $(T_s - T)/(T_s - T_a) = 0.99$. Figure 10-8 shows that the probability integral reaches a value of 0.99 when $Z = 1.82$, from which

$$x_p = 3.64\sqrt{\alpha t} \qquad (10\text{-}26)$$

EXAMPLE 10-5 A sudden cold wave drops the atmospheric temperature to $-20°C$ ($-4°F$) for 12 h. (*a*) If the ground was initially all at $5°C$ ($41°F$), how deep would a water pipeline have to be buried to be in no danger of freezing? (*b*) What is the penetration distance under these conditions? The thermal diffusivity of soil is 0.0011 m²/h (0.0118 ft²/h).

SOLUTION (*a*) Assume that the surface of the ground quickly reaches and remains at $-20°C$. Unless the temperature at the location of the pipe is below $0°C$, there is no danger of freezing. The quantities required for use with Fig. 10-8 are therefore

$$T_s = -20°C \qquad T_a = 5°C \qquad T = 0°C$$

$$t = 12\text{ h} \qquad \alpha = 0.011\text{ m}^2/\text{h}$$

$$\frac{T_s - T}{T_s - T_a} = \frac{-20 - 0}{-20 - 5} = 0.80$$

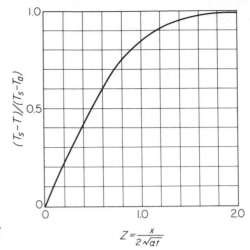

FIGURE 10-8
Unsteady-state heating or cooling of
semi-infinite solid.

$$Z = \frac{x}{2\sqrt{\alpha t}}$$

From Fig. 10-8, $Z = 0.91$. The depth x is therefore

$$x = 2\sqrt{\alpha t} = 2\sqrt{0.0011 \times 12} = 0.23 \text{ m } (0.75 \text{ ft})$$

(*b*)   From Eq. (10-26) the penetration distance is

$$x_p = 3.64\sqrt{0.0011 \times 12} = 0.419 \text{ m } (1.37 \text{ ft}) \qquad ////$$

To find the total heat transferred to a semi-infinite solid in a given time it is necessary to find the temperature gradient and heat flux at the hot surface as a function of time. The temperature gradient at the surface is found by differentiating Eq. (10-25) to give

$$\left(\frac{\partial T}{\partial x}\right)_{x=0} = -\frac{T_s - T_a}{\sqrt{\pi \alpha t}} \tag{10-27}$$

The heat flow rate at the surface is therefore

$$\left(\frac{q}{A}\right)_{x=0} = -k\left(\frac{\partial T}{\partial x}\right)_{=0} = \frac{k(T_s - T_a)}{\sqrt{\pi \alpha t}} \tag{10-28}$$

After substitution of dQ/dt for q, Eq. (10-28) can be integrated to give the total quantity of heat transferred per unit area, Q_T/A, in time t_T, as follows:

$$\frac{Q_T}{A} = \frac{k(T_s - T_a)}{\sqrt{\pi \alpha}} \int_0^{t_T} \frac{dt}{\sqrt{t}} = 2k(T_s - T_a)\sqrt{\frac{t_T}{\pi \alpha}} \tag{10-29}$$

SYMBOLS

A Area, ft² or m²; $\bar{A}_L$, logarithmic mean
a Constant in Eq. (10-3)
a_1 $(\pi/2)^2$

B	Thickness of slab, ft or m; B_A, B_B, B_C, of layers A, B, C, respectively
b	Constant in Eq. (10-3)
c_p	Specific heat at constant pressure, Btu/lb-°F or J/g-°C
e	Base of Naperian logarithms, 2.71828 $\cdots$
g_c	Newton's-law proportionality factor, 32.174 ft-lb$_f$-s²
k	Thermal conductivity, Btu/ft-h-°F or W/m-°C; k_A, k_B, k_C, of layers A, B, C, respectively; $\bar{k}$, average value
L	Length of cylinder, ft or m
N_{Fo}	Fourier number, dimensionless; $\alpha t_T/s^2$ for slab; $\alpha t_T/r_m^2$ for cylinder or sphere
n	Distance measured normally to surface, ft or m
Q	Quantity of heat, Btu or J; Q_T, total quantity transferred
q	Heat flow rate, Btu/h or w; q_A, q_B, q_C, in layers A, B, C, respectively
R	Thermal resistance, °F-h/Btu or °C/W; R_A, R_B, R_C, of layers A, B, C, respectively
r	Radial distance or radius, ft or m; r_i, inside radius; r_m, radius of solid cylinder or sphere; r_o, outside radius; $\bar{r}_L$, logarithmic mean; $\bar{r}_a$, arithmetic mean
s	Half-thickness of slab, ft or m
T	Temperature, °F or °C; T_a, initial temperature; $\bar{T}_b$, average temperature at end of time t_T; T_i, of inside surface; T_o, of outside surface; T_s, surface temperature; T_1, T_2, at locations 1 and 2, respectively
t	Time, s or h; t_T, time required to heat or cool
u	Velocity, ft/s or m/s
x	Distance from surface, ft or m; x_1, x_2, at locations 1 and 2, respectively; x_p, penetration distance in semi-infinite solid
y	Distance, ft or m
Z	$x/2\sqrt{\alpha t}$, dimensionless

Greek letters

α	Thermal diffusivity, $k/\rho c_p$, ft²/h or m²/s
ΔT	Overall temperature drop; ΔT_A, ΔT_B, ΔT_C, in layers A, B, C, respectively
μ	Absolute viscosity, lb/ft-h or P
ρ	Density, lb/ft³ or k/gm³
τ	Shear stress, lb$_f$/ft² or N/m²

PROBLEMS

10-1 A furnace wall consists of 200 mm of refractory fireclay brick, 100 mm of Sil-o-cel brick, and 6 mm of steel plate. The fire side of the refractory is at 1150°C, and the outside of the steel is at 30°C. An accurate heat balance over the furnace shows the heat loss from the wall to be 300 W/m². It is known that there may be thin layers of air between the layers of brick and steel. To how many millimeters of Sil-o-cel are these air layers equivalent? Thermal conductivities are

$$k = \begin{cases} 1.52 \text{ W/m-°C} & \text{for fireclay brick} \\ 0.138 \text{ W/m-°C} & \text{for Sil-o-cel brick} \\ 45 \text{ W/m-°C} & \text{for steel} \end{cases}$$

10-2 A standard 1-in. Schedule 40 steel pipe carries saturated steam at 250°F. The pipe is lagged (insulated) with a 2-in. layer of 85 percent magnesia pipe covering, and outside this magnesia there is a 3-in. layer of cork. The inside temperature of the pipe wall is 249°F, and the outside temperature of the cork is 90°F. Thermal conductivities, in Btu/ft-h-°F, are: for steel, 26; for magnesia, 0.034; for cork, 0.03. Calculate (*a*) the heat loss from 100 ft of pipe, in Btu per hour; (b) the temperatures at the boundaries between metal and magnesia and between magnesia and cork.

10-3   A large sheet of glass 2 in. thick is initially at 300°F throughout. It is plunged into a stream of running water having a temperature of 60°F. How long will it take to cool the glass to an average temperature of 100°F? For glass, $k = 0.40$ Btu/ft-h-°F; $\rho = 155$ lb/ft^3; and $c_p = 0.20$ Btu/lb-°F.

10-4   A very long, wide sheet of plastic 3 mm thick and initially at 40°C is suddenly exposed on both sides to an atmosphere of steam at 105°C. (a) If there is negligible thermal resistance between the steam and the surfaces of the plastic, how long will it take for the temperature at the centerline of the sheet to change significantly? (b) What would be the bulk average temperature of the plastic at this time? For the plastic $k = 0.138$ W/m-°C and $\alpha = 0.00035$ m^2/h.

10-5   A long steel rod, 1 in. in diameter, is initially at a uniform temperature of 1100°F. It is suddenly immersed in a quenching bath of oil at 100°F. In 4 min its average temperature drops to 250°F. How long would it take to lower the temperature from 1100 to 250°F if the rod were $2\frac{1}{2}$ in. in diameter? For steel, $k = 26$ Btu/ft-h-°F; $\rho = 486$ lb/ft^3; $c_p = 0.11$ Btu/lb-°F.

10-6   Steel spheres 3 in. in diameter, heated to 600°F, are to be cooled by immersion in an oil bath at 100°F. If there is negligible thermal resistance between the oil and the steel surfaces, (a) calculate the average temperature of the spheres 10 s and 1 and 6 min after immersion. (b) How long would it take for the unaccomplished temperature change to be reduced to 1 percent of the initial temperature difference? The steel has the same thermal properties as in Prob. 10-5.

10-7   Under the conditions described in Example 10-5, what is the average rate of heat loss per unit area from the ground to the air during the 12-h period? The thermal conductivity of soil is 0.7 W/m-°C.

REFERENCES

1   Carnahan, B., H. A. Luther, and J. O. Wilkes: "Applied Numerical Methods," Wiley, New York, 1969.
2   Carslaw, H. S., and J. C. Jaeger: "Conduction of Heat in Solids," 2d ed., Oxford University Press, Fair Lawn, N.J., 1959; (a) pp. 6–8, (b) p. 234.
3   Fourier, J.: "The Analytical Theory of Heat," trans. by A. Freeman, Dover, New York, 1955.
4   Gebhart, B.: "Heat Transfer," 2d ed., McGraw-Hill, New York, 1971.
5   McAdams, W. H.: "Heat Transmission," 3d ed., McGraw-Hill, New York, 1954; (a) pp. 35–43, (b) p. 232.

11

PRINCIPLES OF HEAT FLOW IN FLUIDS

The flow of heat from a fluid through a solid wall to a cooler fluid is often encountered in chemical engineering practice. The heat transferred may be latent heat accompanying phase changes such as condensation or vaporization, or it may be sensible heat coming from increasing or decreasing the temperature of a fluid without phase change. Typical examples are reducing the temperature of a fluid by transfer of sensible heat to a cooler fluid, the temperature of which is increased thereby; condensing steam by cooling water; and vaporizing water from a solution at a given pressure by condensing steam at a higher pressure. All such cases require that heat be transferred by conduction and convection.

Typical heat-exchange equipment To establish a basis for specific discussion of heat transfer to and from flowing fluids, consider the simple tubular condenser of Fig. 11-1. It consists essentially of a bundle of parallel tubes A, the ends of which are expanded into tube sheets B_1 and B_2. The tube bundle is inside a cylindrical shell C and is provided with two channels D_1 and D_2, one at each end, and two channel covers E_1 and E_2. Steam or other vapor is introduced through nozzle F into the shell-side space surrounding the tubes, condensate is withdrawn through con-

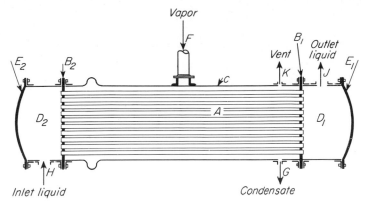

FIGURE 11-1

Single-pass tubular condenser: A, tubes; B_1, B_2, tube sheets; C, shell; D_1, D_2, channels; E_1, E_2, channel covers; F, vapor inlet; G, condensate outlet; H, cold-liquid inlet; J, warm-liquid outlet; K, noncondensed-gas vent.

nection G, and any noncondensable gas that might enter with the inlet vapor is removed through vent K. Connection G leads to a trap, which is a device that allows flow of liquid but holds back vapor. The fluid to be heated is pumped through connection H into channel D_2. It flows through the tubes into the other channel D_1, and is discharged through connection J. The two fluids are physically separated but are in thermal contact with the thin metal tube walls separating them. Heat flows through the tube walls from the condensing vapor to the cooler fluid in the tubes.

If the vapor entering the condenser is not superheated and the condensate is not subcooled below its boiling temperature, the temperature throughout the shell side of the condenser is constant. The reason for this is that the temperature of the condensing vapor is fixed by the pressure of the shell-side space, and the pressure in that space is constant. The temperature of the fluid in the tubes increases continuously as the fluid flows through the tubes.

The temperatures of the condensing vapor and of the liquid are plotted against the tube length in Fig. 11-2. The horizontal line represents the temperature of the condensing vapor, and the curved line below it the rising temperature of the tube-side fluid. In Fig. 11-2, the inlet and outlet fluid temperatures are T_{ca} and T_{cb}, respectively, and the constant temperature of the vapor is T_h. At a length L from the entrance end of the tubes, the fluid temperature is T_c, and the local difference between the temperatures of vapor and fluid is $T_h - T_c$. This temperature difference is called a *point temperature difference* and is denoted by ΔT. The point temperature difference at the inlet of the tubes is $T_h - T_{ca}$, denoted by ΔT_1, and that at the exit end is $T_h - T_{cb}$, denoted by ΔT_2. The terminal point temperature differences ΔT_1 and ΔT_2 are called the *approaches*.

The change in temperature of the fluid, $T_{cb} - T_{ca}$, is called the *temperature range* or, simply, the *range*. In a condenser there is but one range, that of the cold fluid being heated.

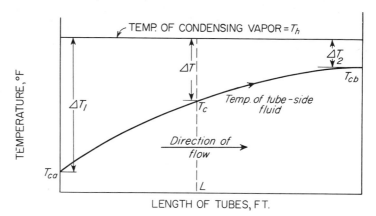

FIGURE 11-2
Temperature-length curves for condenser.

A second example of simple heat-transfer equipment is the double-pipe exchanger shown in Fig. 11-3. It is assembled of standard metal pipe and standardized return bends and return heads, the latter equipped with stuffing boxes. One fluid flows through the inside pipe and the second fluid through the annular space between the outside and the inside pipe. The function of a heat exchanger is to increase the temperature of a cooler fluid and decrease that of a hotter fluid. In a typical exchanger, the inner pipe may be $1\frac{1}{4}$ in. and the outer pipe $2\frac{1}{2}$ in., both IPS. Such an exchanger may consist of several passes arranged in a vertical stack. Double-pipe exchangers are useful when not more than 100 to 150 ft^2 of surface are required. For larger capacities, more elaborate shell-and-tube exchangers, containing up to thousands of square feet of area, and described on pages 399 to 402, are used.

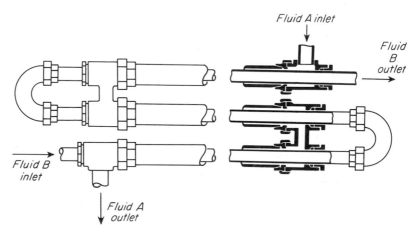

FIGURE 11-3
Double-pipe heat exchanger.

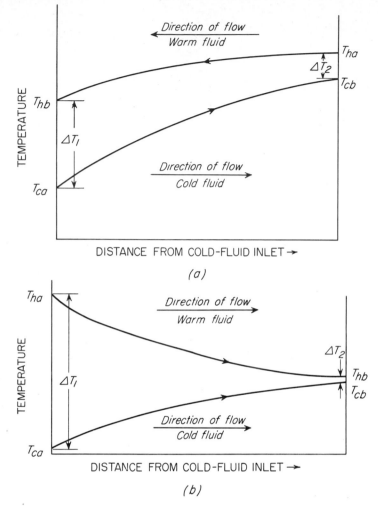

FIGURE 11-4
Temperatures in (*a*) countercurrent and (*b*) parallel flow.

Countercurrent and parallel-current flows The two fluids enter at different ends of the exchanger shown in Fig. 11-3 and pass in opposite directions through the unit. This type of flow is that commonly used and is called *counterflow or countercurrent flow*. The temperature-length curves for this case are shown in Fig. 11-4*a*. The four terminal temperatures are denoted as follows:

Temperature of entering hot fluid, T_{ha}
Temperature of leaving hot fluid, T_{hb}
Temperature of entering cold fluid, T_{ca}
Temperature of leaving cold fluid, T_{cb}

The approaches are

$$T_{ha} - T_{cb} = \Delta T_2 \quad \text{and} \quad T_{hb} - T_{ca} = \Delta T_1 \tag{11-1}$$

The warm-fluid and cold-fluid ranges are $T_{ha} - T_{hb}$ and $T_{cb} - T_{ca}$, respectively.

If the two fluids enter at the same end of the exchanger and flow in the same direction to the other end, the flow is called *parallel*. The temperature-length curves for parallel flow are shown in Fig. 11-4b. Again, the subscript a refers to the entering fluids and subscript b to the leaving fluids. The approaches are $\Delta T_1 = T_{ha} - T_{ca}$ and $\Delta T_2 = T_{hb} - T_{cb}$.

Parallel flow is rarely used in a single-pass exchanger such as that shown in Fig. 11-3 because, as inspection of Figs. 11-4a and b will show, it is not possible with this method of flow to bring the exit temperature of one fluid nearly to the entrance temperature of the other and the heat that can be transferred is less than that possible in countercurrent flow. In the multipass exchangers, described on pages 401 to 402, parallel flow is used in some passes, largely for mechanical reasons, and the capacity and approaches obtainable are thereby affected. Parallel flow is used in special situations where it is necessary to limit the maximum temperature of the cooler fluid or where it is important to change the temperature of at least one fluid rapidly.

ENERGY BALANCES

Quantitative attack on heat-transfer problems is based on energy balances and estimations of rates of heat transfer. Rates of transfer are discussed later in this chapter. Many, perhaps most, heat-transfer devices operate under steady-state conditions, and only this type of operation will be considered here.

Enthalpy balances in heat exchangers In heat exchangers there is no shaft work, and mechanical potential and kinetic energies are small in comparison with the other terms in the energy-balance equation. Equation (1-58) can therefore be applied. It is convenient to convert this equation to an hour basis. Thus, for one stream through the exchanger

$$\dot{m}(H_b - H_a) = q \tag{11-2}$$

where $\quad \dot{m} =$ flow rate of stream
$\qquad q = Q/t =$ rate of heat transfer into stream
$\quad H_a, H_b =$ enthalpies per unit mass of stream at entrance and exit, respectively

Equation (11-2) can be written for each stream flowing through the exchanger.

A further simplification in the use of the heat-transfer rate q is justified. One of the two fluid streams, that outside the tubes, can gain or lose heat by transfer with the ambient air if the fluid is colder or hotter than the ambient. Heat transfer to or from the ambient is not usually desired in practice, and it is usually reduced to a small magnitude by suitable lagging. It is customary to neglect it in comparison with the heat transfer through the walls of the tubes from the warm fluid to the cold fluid, and q is interpreted accordingly.

Accepting the above assumptions, Eq. (11-2) can be written for the warm fluid as

$$\dot{m}_h(H_{hb} - H_{ha}) = q_h \tag{11-3}$$

and for the cold fluid as

$$\dot{m}_c(H_{cb} - H_{ca}) = q_c \tag{11-4}$$

where $\dot{m}_c, \dot{m}_h$ = mass flow rates of cold fluid and warm fluid, respectively

H_{ca}, H_{ha} = enthalpy per unit mass of entering cold fluid and entering warm fluid, respectively

H_{cb}, H_{hb} = enthalpy per unit mass of leaving cold fluid and leaving hot fluid, respectively

q_c, q_h = rates of heat addition to cold fluid and warm fluid, respectively

The sign of q_c is positive, but that of q_h is negative because the warm fluid loses, rather than gains, heat. The heat lost by the warm fluid is gained by the cold fluid, and

$$q_c = -q_h$$

Therefore, from Eqs. (11-3) and (11-4),

$$\dot{m}_h(H_{ha} - H_{hb}) = \dot{m}_c(H_{cb} - H_{ca}) = q \tag{11-5}$$

Equation (11-5) is called the *overall enthalpy balance*.

If constant specific heats are assumed, the overall enthalpy balance for a heat exchanger becomes

$$\dot{m}_h c_{ph}(T_{ha} - T_{hb}) = \dot{m}_c c_{pc}(T_{cb} - T_{ca}) = q \tag{11-6}$$

where c_{pc} = specific heat of cold fluid
c_{ph} = specific heat of warm fluid

Enthalpy balances in total condensers   For a condenser Eq. (1-58) can be written

$$\dot{m}_h \lambda = \dot{m}_c c_{pc}(T_{cb} - T_{ca}) = q \tag{11-7}$$

where $\dot{m}_h$ = rate of condensation of vapor
λ = latent heat of vaporization of vapor

Equation (11-7) is based on the assumption that the vapor enters the condenser as saturated vapor (no superheat) and that the condensate leaves at condensing temperature without being further cooled. If either of these sensible-heat effects is important, it must be accounted for by an added term in the left-hand side of Eq. (11-7). For example, if the condensate leaves at a temperature T_{hb} which is less than T_h, the condensing temperature of the vapor, Eq. (11-7) must be written

$$\dot{m}_h[\lambda + c_{ph}(T_h - T_{hb})] = \dot{m}_c c_{pc}(T_{cb} - T_{ca}) \tag{11-8}$$

where c_{ph} is now the specific heat of the condensate.

RATE OF HEAT TRANSFER

Heat flux Heat-transfer calculations are based on the area of the heating surface and are expressed in Btu per hour per square foot (or watts per square meter) of surface through which the heat flows. The rate of heat transfer per unit area is called the *heat flux*. In many types of heat-transfer equipment the transfer surfaces are constructed from tubes or pipe. Heat fluxes may then be based either on the inside area or the outside area of the tubes. Although the choice is arbitrary, it must be clearly stated, because the numerical magnitude of the heat fluxes will not be the same for both.

Average temperature of fluid stream When a fluid is being heated or cooled, the temperature will vary throughout the cross section of the stream. If the fluid is being heated, the temperature of the fluid is a maximum at the wall of the heating surface and decreases toward the center of the stream. If the fluid is being cooled, the temperature is a minimum at the wall and increases toward the center. Because of these temperature gradients throughout the cross section of the stream, it is necessary, for definiteness, to state what is meant by the temperature of the stream. It is agreed that it is the temperature that would be attained if the entire fluid stream flowing across the section in question were withdrawn and mixed adiabatically to a uniform temperature. The temperature so defined is called the *average* or *mixing-cup stream temperature*. The temperatures plotted in Fig. 11-4 are all average stream temperatures.

Overall Heat-Transfer Coefficient

It is reasonable to expect the heat flux to be proportional to a driving force. In heat flow, the driving force is taken as $T_h - T_c$, where T_h is the average temperature of the hot fluid and T_c is that of the cold fluid. The quantity $T_h - T_c$ is the *overall local temperature difference* ΔT. It is clear from Fig. 11-4 that ΔT can vary considerably from point to point along the tube, and, therefore, since the heat flux is proportional to ΔT, the flux also varies with tube length. It is necessary to start with a differential equation, by focusing attention on a differential area dA through which a differential heat flow dq occurs under the driving force of a local value of ΔT. The local flux is then dq/dA and is related to the local value of ΔT by the equation

$$\frac{dq}{dA} = U\,\Delta T = U(T_h - T_c) \tag{11-9}$$

The quantity U, defined by Eq. (11-9) as a proportionality factor between dq/dA and ΔT, is called the *local overall heat-transfer coefficient*.

　　To complete the definition of U in a given case, it is necessary to specify the area. If A is taken as the outside tube area A_o, U becomes a coefficient based on that area and is written U_o. Likewise, if the inside area A_i is chosen, the coefficient is also based on that area and is denoted by U_i. Since ΔT and dq are independent of the

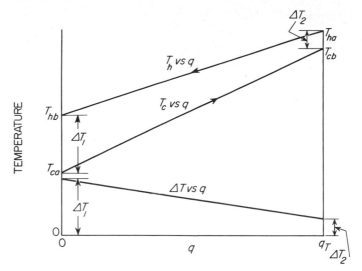

FIGURE 11-5
Temperature vs. heat flow rate in countercurrent flow.

choice of area, it follows that

$$\frac{U_o}{U_i} = \frac{dA_i}{dA_o} = \frac{D_i}{D_o} \tag{11-10}$$

where D_i and D_o are the inside and outside tube diameters, respectively.

Integration over total surface; logarithmic mean temperature difference   To apply Eq. (11-9) to the entire area of a heat exchanger, the equation must be integrated. This can be done formally where certain simplifying assumptions are accepted. The assumptions are (1) the overall coefficient U is constant, (2) the specific heats of the hot and cold fluids are constant, (3) heat exchange with the ambient is negligible, and (4) the flow is steady and either parallel or countercurrent, as shown in Fig. 11-4.

The most questionable of these assumptions is that of a constant overall coefficient. The coefficient does in fact vary with the temperatures of the fluids, but its change with temperature is gradual, so that when the temperature ranges are moderate, the assumption of constant U is not seriously in error.

Assumptions 2 and 4 imply that if T_c and T_h are plotted against q, as shown in Fig. 11-5, straight lines are obtained. Since T_c and T_h vary linearly with q, ΔT does likewise and $d(\Delta T)/dq$, the slope of the graph of ΔT vs. q, is constant. Therefore

$$\frac{d(\Delta T)}{dq} = \frac{\Delta T_2 - \Delta T_1}{q_T} \tag{11-11}$$

where ΔT_1, ΔT_2 = approaches
q_T = rate of heat transfer in entire exchanger

Elimination of dq from Eqs. (11-9) and (11-11) gives

$$\frac{d(\Delta T)}{U \, \Delta T \, dA} = \frac{\Delta T_2 - \Delta T_1}{q_T} \tag{11-12}$$

The variables ΔT and A can be separated, and if U is constant, the equation can be integrated over the limits A_T and 0 for A and ΔT_2 and ΔT_1 for ΔT, where A_T is the total area of the heat-transfer surface. Thus

$$\int_{\Delta T_1}^{\Delta T_2} \frac{d(\Delta T)}{\Delta T} = \frac{U(\Delta T_2 - \Delta T_1)}{q_T} \int_0^{A_T} dA$$

or

$$\ln \frac{\Delta T_2}{\Delta T_1} = \frac{U(\Delta T_2 - \Delta T_1)}{q_T} A_T \tag{11-13}$$

Equation (11-13) can be written

$$q_T = U A_T \frac{\Delta T_2 - \Delta T_1}{\ln (\Delta T_2/\Delta T_1)} = U A_T \, \overline{\Delta T_L} \tag{11-14}$$

where

$$\overline{\Delta T_L} = \frac{\Delta T_2 - \Delta T_1}{\ln (\Delta T_2/\Delta T_1)} = \frac{\Delta T_2 - \Delta T_1}{2.303 \log (\Delta T_2/\Delta T_1)} \tag{11-15}$$

Equation (11-15) defines the *logarithmic mean temperature difference* (LMTD). It is of the same form as that of Eq. (10-15) for the logarithmic mean radius of a thick-walled tube. When ΔT_1 and ΔT_2 are nearly equal, their arithmetic average can be used for $\overline{\Delta T_L}$ within the same limits of accuracy given for Eq. (10-15), as shown in Fig. 10-4.

If one of the fluids is at constant temperature, as in a condenser, no difference exists between countercurrent flow, parallel flow, or multipass flow, and Eq. (11-15) applies to all of them. In countercurrent flow, ΔT_2, the warm-end approach, may be less than ΔT_1, the cold-end approach. In this case, for convenience and to eliminate negative numbers and logarithms, the subscripts in Eq. (11-15) may be interchanged.

Variable overall coefficient The use of Eq. (11-14), which depends on the assumption of the constancy of U, leads to appreciable error if U varies considerably from one end of the exchanger to the other. A more accurate calculation can be made by using Eq. (11-16), which is based on the assumption that U varies linearly with temperature drop over the entire heating surface:[1]

$$q_T = A_T \frac{U_2 \, \Delta T_1 - U_1 \, \Delta T_2}{\ln (U_2 \, \Delta T_1/U_1 \, \Delta T_2)} \tag{11-16}$$

where U_1, U_2 = local overall coefficients at ends of exchanger
$\Delta T_1, \Delta T_2$ = temperature approaches at corresponding ends of exchanger

Equation (11-16) calls for use of a logarithmic mean value of the $U \, \Delta T$ cross product, where the overall coefficient at one end of the exchanger is multiplied by the temperature approach at the other. The derivation of this equation requires that assumptions 2 to 4 above be accepted.

In the completely general case, where none of the assumptions are valid and U varies markedly from point to point, Eq. (11-9) can be integrated by evaluating local values of U, ΔT, and q at several intermediate points in the exchanger. Graphical or numerical evaluation of the area under a plot of $1/U \Delta T$ vs. q, between the limits of zero and q_T, will then give the area A_T of the heat-transfer surface required.

Multipass exchangers In multipass shell-and-tube exchangers the flow pattern is complex, with parallel, countercurrent, and crossflow all present. Under these conditions, even when the overall coefficient U is constant, the LMTD cannot be used. Calculation procedures for multipass exchangers are given in Chap. 15.

Individual Heat-Transfer Coefficients

The overall coefficient depends upon so many variables that it is necessary to break it into its parts. The reason for this becomes apparent if a typical case is examined. Consider the local overall coefficient at a specific point in the double-pipe exchanger shown in Fig. 11-3. For definiteness, assume that the warm fluid is flowing through the inside pipe and that the cold fluid is flowing through the annular space. Assume also that the Reynolds numbers of the two fluids are sufficiently large to ensure turbulent flow and that both surfaces of the inside tube are clear of dirt or scale. If, now, a plot is prepared, as shown in Fig. 11-6, with temperature as the ordinate and distance perpendicular to the wall as the abscissa, several important facts become evident. In the figure, the metal wall of the tube separates the warm fluid on the right from the cold fluid on the left. The change in temperature with distance is shown by the broken line $T_a T_b T_{wh} T_{wc} T_e T_g$. The temperature profile is thus divided into three separate parts, one through each of the two fluids and the other through the metal wall. The overall effect, therefore, should be studied in terms of these individual parts.

It was shown in Chap. 5 that in turbulent flow through conduits three zones exist, even in a single fluid, so that the study of one fluid is, itself, complicated. In each fluid shown in Fig. 11-6 there is a thin sublayer at the wall, a turbulent core occupying most of the cross section of the stream, and a buffer zone between them. The velocity gradients were described in Chap. 5. The velocity gradient is large near the wall, small in the turbulent core, and in rapid change in the buffer zone. It has been found that the temperature gradient in a fluid being heated or cooled when flowing in turbulent flow follows much the same course. The temperature gradient is large at the wall and through the viscous sublayer, small in the turbulent core, and in rapid change in the buffer zone. Basically, the reason for this is that heat must flow through the viscous sublayer by conduction, which calls for a steep temperature gradient in most fluids because of the low thermal conductivity, whereas the rapidly moving eddies in the core are effective in equalizing the temperature in the turbulent zone. In Fig. 11-6 the dotted lines $F_1 F_1$ and $F_2 F_2$ represent the boundaries of the viscous sublayers.

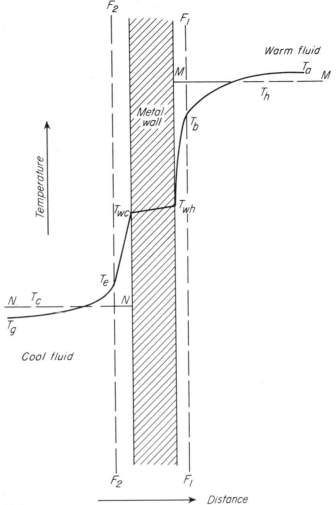

FIGURE 11-6
Temperature gradients in forced convection.

The average temperature of the warm stream is somewhat less than the maximum temperature T_a and is represented by the horizontal line MM, which is drawn at temperature T_h. Likewise, line NN, drawn at temperature T_c, represents the average temperature of the cold fluid.

The overall resistance to the flow of heat from the warm fluid to the cold fluid is a result of three separate resistances operating in series. Two resistances are those offered by the individual fluids, and the third is that of the solid wall. In general, also, as shown in Fig. 11-6, the wall resistance is small in comparison with that of the fluids. The overall coefficient is best studied by analyzing it in terms of the separate resistances and treating each separately. The separate resistances can then be combined to form the overall coefficient. This approach requires the use of individual heat-transfer coefficients for the two fluid streams.

The individual, or surface, heat-transfer coefficient h is defined generally by the equation

$$h = \frac{dq/dA}{T - T_w} \tag{11-17}$$

where dq/dA = local heat flux, based on the area in contact with fluid
T = local average temperature of fluid
T_w = temperature of wall in contact with fluid

A second expression for h is derived from the assumption that there are no velocity fluctuations normal to the wall at the surface of the wall itself. If this assumption is accepted, heat transfer by turbulent convection cannot exist, the mechanism of heat transfer at the wall must be by conduction, and the heat flux is given by Eq. (10-2), noting that the normal distance n may be replaced by y, the normal distance measured into the fluid from the wall. Thus

$$\frac{dq}{dA} = -k \left(\frac{dT}{dy}\right)_w \tag{11-18}$$

The subscript w calls attention to the fact that the gradient must be evaluated at the wall. Eliminating dq/dA from Eqs. (11-17) and (11-18) gives

$$h = -k \frac{(dT/dy)_w}{T - T_w} \tag{11-19}$$

Equation (11-19) can be put into a dimensionless form by multiplying by the ratio of an arbitrary length to the thermal conductivity. The choice of length depends on the situation. For heat transfer at the inner surface of a tube, the tube diameter D is the usual choice. Multiplying Eq. (11-19) by D/k gives

$$\frac{hD}{k} = -D \frac{(dT/dy)_w}{T - T_w} \tag{11-20}$$

A dimensionless group such as hD/k is called a *Nusselt number* N_{Nu}. That shown in Eq. (11-20) is a local Nusselt number based on diameter. The physical meaning of the Nusselt number can be seen by inspection of the right-hand side of Eq. (11-20). The numerator $(dT/dy)_w$ is, of course, the gradient at the wall. The factor $(T - T_w)/D$ can be considered the average temperature gradient across the entire pipe, and the Nusselt number is the ratio of these two gradients.

Equation (11-17), when applied to the two fluids of Fig. 11-6, becomes, for the warm side (inside of tube),

$$h_i = \frac{dq/dA_i}{T_h - T_{wh}} \tag{11-21}$$

and for the cold side (outside of tube),

$$h_o = \frac{dq/dA_o}{T_{wc} - T_c} \tag{11-22}$$

where A_i and A_o are the inside and outside areas of the tube, respectively.

Calculation of overall coefficients from individual coefficients The overall coefficient is constructed from the individual coefficients and the resistance of the tube wall in the following manner.

The rate of heat transfer through the tube wall is given by the differential form of Eq. (10-13),

$$\frac{dq}{d\bar{A}_L} = \frac{k_m(T_{wh} - T_{wc})}{x_w} \tag{11-23}$$

where $T_{wh} - T_{wc}$ = temperature difference through tube wall
$\quad\quad k_m$ = thermal conductivity of wall
$\quad\quad x_w$ = tube-wall thickness
$\quad dq/d\bar{A}_L$ = local heat flux, based on logarithmic mean of inside and outside areas of tube

If Eqs. (11-21) to (11-23) are solved for the temperature differences and the temperature differences added, the result is

$$(T_h - T_{wh}) + (T_{wh} - T_{wc}) + (T_{wc} - T_c) = T_h - T_c = \Delta T$$

$$= dq \left(\frac{1}{dA_i \, h_i} + \frac{x_w}{d\bar{A}_L \, k_m} + \frac{1}{dA_o \, h_o} \right) \tag{11-24}$$

Assume that the heat-transfer rate is arbitrarily based on the outside area. If Eq. (11-24) is solved for dq, and if both sides of the resulting equation are divided by dA_o, the result is

$$\frac{dq}{dA_o} = \frac{T_h - T_c}{(dA_o/dA_i \, h_i) + (x_w/k_m)(dA_o/d\bar{A}_L) + (1/h_o)} \tag{11-25}$$

Now

$$\frac{dA_o}{dA_i} = \frac{D_o}{D_i} \quad \text{and} \quad \frac{dA_o}{d\bar{A}_L} = \frac{D_o}{\bar{D}_L}$$

where D_o, D_i, and $\bar{D}_L$ are the outside, inside, and logarithmic mean diameters of the tube, respectively. Therefore

$$\frac{dq}{dA_o} = \frac{T_h - T_c}{(D_o/D_i h_i) + (x_w/k_m)(D_o/\bar{D}_L) + (1/h_o)} \tag{11-26}$$

Comparing Eq. (11-9) with Eq. (11-26) shows that

$$U_o = \frac{1}{(D_o/D_i h_i) + (x_w/k_m)(D_o/\bar{D}_L) + (1/h_o)} \tag{11-27}$$

If the inside area A_i is chosen as the base area, division of Eq. (11-24) by dA_i gives for the overall coefficient

$$U_i = \frac{1}{(1/h_i) + (D_i/\bar{D}_L)(x_w/k_m) + (D_i/D_o h_o)} \tag{11-28}$$

Resistance form of overall coefficient   A comparison of Eqs. (10-9) and (11-26) suggests that the reciprocal of an overall coefficient can be considered to be an overall resistance, composed of three resistances in series. The total, or overall, resistance is given by the equation

$$\frac{1}{U_o} = \frac{D_o}{D_i h_i} + \frac{x_w}{k_m}\frac{D_o}{\bar{D}_L} + \frac{1}{h_o} \tag{11-29}$$

The individual terms on the right-hand side of Eq. (11-29) represent the individual resistances of the two fluids and of the metal wall. The overall temperature drop is proportional to $1/U$, and the temperature drops in the two fluids and the wall are proportional to the individual resistances, or, for the case of Eq. (11-29),

$$\frac{\Delta T}{1/U_o} = \frac{\Delta T_i}{D_o/D_i h_i} = \frac{\Delta T_w}{(x_w/k_m)(D_o/\bar{D}_L)} = \frac{\Delta T_o}{1/h_o} \tag{11-30}$$

where ΔT = overall temperature drop
$\quad \Delta T_i$ = temperature drop through inside fluid
$\quad \Delta T_w$ = temperature drop through metal wall
$\quad \Delta T_o$ = temperature drop through outside fluid

Fouling factors   In actual service, heat-transfer surfaces do not remain clean. Scale, dirt, and other solid deposits form on one or both sides of the tubes, provide additional resistances to heat flow, and reduce the overall coefficient. The effect of such deposits is taken into account by adding a term $1/dA\ h_d$ to the term in parentheses in Eq. (11-24) for each scale deposit. Thus, assuming that scale is deposited on both the inside and the outside surface of the tubes, Eq. (11-24) becomes, after correction for the effects of scale,

$$\Delta T = dq \left(\frac{1}{dA_i\ h_{di}} + \frac{1}{dA_i\ h_i} + \frac{x_w}{d\bar{A}_L\ k_m} + \frac{1}{dA_o\ h_o} + \frac{1}{dA_o\ h_{do}} \right) \tag{11-31}$$

where h_{di} and h_{do} are the *fouling factors* for the scale deposits on the inside and outside tube surfaces, respectively. The following equations for the overall coefficients based on outside and inside areas, respectively, follow from Eq. (11-31):

$$U_o = \frac{1}{(D_o/D_i h_{di}) + (D_o/D_i h_i) + (x_w/k_m)(D_o/\bar{D}_L) + (1/h_o) + (1/h_{do})} \tag{11-32}$$

and $\quad U_i = \dfrac{1}{(1/h_{di}) + (1/h_i) + (x_w/k_m)(D_i/\bar{D}_L) + (D_i/D_o h_o) + (D_i/D_o h_{do})} \tag{11-33}$

The actual thicknesses of the deposits are neglected in Eqs. (11-32) and (11-33).

Numerical values of fouling factors corresponding to "satisfactory performance in normal operation, with reasonable service time between cleanings"[3] are recommended. They cover a range of approximately 100 to 2000 Btu/ft²-h-°F. Fouling factors for ordinary industrial liquids fall in the range 300 to 1000 Btu/ft²-h-°F. Fouling factors are usually set at values that also provide a safety factor for design.

EXAMPLE 11-1   Methyl alcohol flowing in the inner pipe of a double-pipe exchanger is cooled with water flowing in the jacket. The inner pipe is made from 1-in. (25-mm) Schedule 40 steel pipe. The thermal conductivity of steel is 26 Btu/ft-h-°F (45 W/m-°C). The individual coefficients and fouling factors are given in Table 11-1. What is the overall coefficient, based on the outside area of the inner pipe?

SOLUTION   The diameters and wall thickness of Schedule 40 1-in. pipe, from Appendix 6, are

$$D_i = \frac{1.049}{12} = 0.0874 \text{ ft} \qquad D_o = \frac{1.315}{12} = 0.1096 \text{ ft} \qquad x_w = \frac{0.133}{12} = 0.0111 \text{ ft}$$

The logarithmic mean diameter $\bar{D}_L$ is calculated as in Eq. (10-15), using diameter in place of radius:

$$\bar{D}_L = \frac{D_o - D_i}{2.303 \log (D_o/D_i)} = \frac{0.1096 - 0.0874}{2.303 \log (0.1096/0.0874)} = 0.0983 \text{ ft}$$

The overall coefficient is found from Eq. (11-32).

$$U_o = \cfrac{1}{\cfrac{0.1096}{0.0874 \times 1000} + \cfrac{0.1096}{0.0874 \times 180} + \cfrac{0.0111 \times 0.1096}{26 \times 0.0983} + \cfrac{1}{300} + \cfrac{1}{500}}$$

$$= 71.3 \text{ Btu/ft}^2\text{-h-°F} \ (405 \text{ W/m}^2\text{-°C})$$

////

Special cases of the overall coefficient   Although the choice of area to be used as the basis of an overall coefficient is arbitrary, sometimes one particular area is more convenient than others. Suppose, for example, that one individual coefficient, h_i, is large numerically in comparison with the other, h_o, and that fouling effects are negligible. Also, assuming the term representing the resistance of the metal wall is small in comparison with $1/h_o$, the ratios D_o/D_i and $D_o/\bar{D}_L$ have so little significance

Table 11-1   DATA FOR EXAMPLE 11-1

	Coefficient	
	Btu/ft-h-°F	W/m²-°C
Alcohol coefficient h_i	180	1,020
Water coefficient h_o	300	1,700
Inside fouling factor h_{di}	1000	5,680
Outside fouling factor h_{do}	500	2,840

that they can be disregarded, and Eq. (11-27) can be replaced by the simpler form

$$U_o = \frac{1}{1/h_o + x_w/k_m + 1/h_i} \tag{11-34}$$

In such a case it is advantageous to base the overall coefficient on that area which corresponds to the largest resistance, or the lowest value of h.

For thin-walled tubes of large diameter, flat plates, or any other case where a negligible error is caused by using a common area for A_i, $\bar{A}_L$, and A_o, Eq. (11-34) can be used for the overall coefficient, and U_i and U_o are identical.

Sometimes one coefficient, say, h_o, is so very small in comparison with both x_w/k and the other coefficient h_i that the term $1/h_o$ is very large compared with the other terms in the resistance sum. When this is true, it is sufficiently accurate to equate the overall coefficient to the small individual coefficient, or, in this case, $h_o = U_o$.

Classification of individual heat-transfer coefficients   The problem of predicting the rate of heat flow from one fluid to another through a retaining wall reduces to the problem of predicting the numerical values of the individual coefficients of the fluids concerned in the overall process. A wide variety of individual cases is met in practice, and each type of phenomenon must be considered separately. The following classification will be followed in this text:

1   Heat flow to or from fluids inside tubes, without phase change
2   Heat flow to or from fluids outside tubes, without phase change
3   Heat flow from condensing fluids
4   Heat flow to boiling liquids

Magnitude of heat-transfer coefficients   The ranges of values covered by the coefficient h vary greatly, depending upon the character of the process.[2] Some typical ranges are shown in Table 11-2.

Table 11-2   MAGNITUDES OF HEAT-TRANSFER COEFFICIENTS†

Type of processes	Range of values of h	
	Btu/ft²-h-°F	W/m²-°C
Steam (dropwise condensation)	5,000–20,000	30,000–100,000
Steam (film-type condensation)	1,000–3,000	6,000–20,000
Boiling water	300–9,000	1,700–50,000
Condensing organic vapors	200–400	1,000–2,000
Water (heating or cooling)	50–3,000	300–20,000
Oils (heating or cooling)	10–300	50–1,500
Steam (superheating)	5–20	30–100
Air (heating or cooling)	0.2–10	1–50

† By permission of author and publishers, from W. H. McAdams, "Heat Transmission," 3d ed., p. 5. Copyright by author, 1954, McGraw-Hill Book Company.

Transfer Units in Heat Exchangers

For some purposes, it is convenient to rate heat exchangers in terms of transfer units rather than overall heat-transfer coefficients. The method is based on splitting the length of path of one fluid through the exchanger into two parts, one a dimensionless function of temperatures and temperature driving force only, called the *number of transfer units*, and the other a length which is a function of conditions of flow, called the *length of a transfer unit*. Either fluid may be chosen as a basis, provided its length of path is definitely established by the configuration of the heat exchanger. For example, if the cold fluid is chosen, the number of transfer units and length of one unit are related by

$$L_{Tc} = N_{tc} H_{tc} \tag{11-35}$$

where L_{Tc} = total length of path of cold fluid
N_{tc} = number of transfer units, cold fluid
H_{tc} = length of one transfer unit, cold fluid

The advantage of the use of the transfer unit is that the length of one unit H_{tc} is relatively insensitive to the flow rate through the exchanger, and the effectiveness of the exchanger as a means of changing the fluid temperatures is given directly by the number of transfer units N_{tc} contained in the exchanger.

If the calculation is based on the hot fluid, the equation for length of path of the hot fluid is

$$L_{Th} = N_{th} H_{th} \tag{11-36}$$

If both fluid paths are the same length, as in the simple double-pipe exchanger of Fig. 11-3, then

$$N_{tc} H_{tc} = N_{th} H_{th} \tag{11-37}$$

The definition of transfer unit is completed by adopting the following general equation for N_t:

$$N_t = \int_{T_a}^{T_b} \frac{dT}{T_h - T_c} \tag{11-38}$$

Choice of the fluid stream allows the use of this equation for either.

The relation between H_t and the overall coefficient U is readily found if the simplifying assumptions used in the derivation of the formula for the LMTD [Eq. (11-15)] are accepted. Then, the curve of temperature vs. length for each fluid is linear. For a countercurrent heat exchanger, these lines for the two fluids can be plotted as shown in Fig. 11-7. Assume the cool fluid is chosen as a basis. Then, in Eq. (11-9) the following substitutions may be made:

$$dA_c = \pi D_c \, dL_c \tag{11-39}$$

$$dq = \dot{m}_c c_{pc} \, dT_c \tag{11-40}$$

where $\dot{m}_c$ = rate of flow of cold fluid
c_{pc} = specific heat of cold fluid

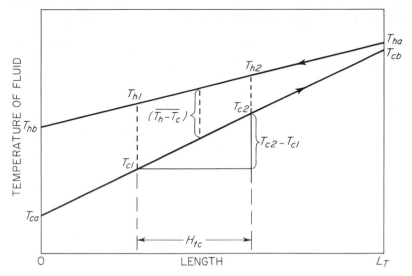

FIGURE 11-7
Temperature patterns in countercurrent double-pipe heat exchanger, showing length of transfer unit H_{tc} when $T_{c2} - T_{c1} = \overline{T_h - T_c}$.

Substituting into Eq. (11-9) gives, after rearrangement,

$$dL_c = \frac{\dot{m}_c c_{pc}}{\pi D_c U_c} \frac{dT_c}{T_h - T_c}$$

Integrating over the entire exchanger gives

$$L_{T_c} = \frac{\dot{m}_c c_{pc}}{\pi D_c U_c} \int_{T_{cb}}^{T_{ca}} \frac{dT_c}{T_h - T_c} \tag{11-41}$$

Here U_c and c_{pc} are both constant and taken outside the integral sign. Comparing Eqs. (11-35), (11-38), and (11-41) shows that

$$H_{tc} = \frac{\dot{m}_c c_{pc}}{\pi D_c U_c} \tag{11-42}†$$

In this equation it is assumed that the overall coefficient is based on the cool side of the heating surface to correspond with the choice of fluid. If the hot fluid is chosen as a basis,

$$H_{th} = \frac{\dot{m}_h c_{ph}}{\pi D_h U_h} \tag{11-43}†$$

Physical significance of transfer unit   When the temperature distributions shown in Fig. 11-7 are used, the concept of a transfer unit can be interpreted physically as

---

† Equations (11-42) and (11-43) indicate that H_t is proportional to the flow rate $\dot{m}$ However, the coefficient U also increases with flow rate, and the ratio $\dot{m}/U$ is relatively insensitive to flow rate.

follows. Consider a section of heat exchanger with the length equal to one cold-fluid transfer unit. As shown in Fig. 11-7, the temperature approaches at the ends of these sections are $T_{h1} - T_{c1}$ and $T_{h2} - T_{c2}$. The temperature range of the cold fluid is $T_{c2} - T_{c1}$. If the average temperature difference, or average driving force, is denoted by $\overline{T_h - T_c}$, this quantity is the reciprocal of the integrated reciprocal of the temperature difference. Since the number of transfer units in the section is unity,

$$N_{tc} = 1 = \frac{T_{c2} - T_{c1}}{\overline{T_h - T_c}} \tag{11-44}$$

Thus, one transfer unit may be viewed as a section of the exchanger in which the change in temperature of one stream is numerically equal to the average driving force in the section. When the two approaches are not greatly different, this quantity is approximately equal to the arithmetic average driving force over the exchanger, or

$$\overline{T_h - T_c} \approx \frac{(T_{h1} - T_{c1}) + (T_{h2} - T_{c2})}{2} \tag{11-45}$$

EXAMPLE 11-2   In the exchanger described in Example 11-1, the flow rate of methyl alcohol is 1,000 lb/h (454 kg/h), and that of water is 1,500 lb/h (680 kg/h). The specific heats are 0.6 and 1.0 Btu/lb-°F (2.51 and 4.19 J/g-°C) for alcohol and water, respectively. The exchanger is 30 ft (9.14 m) long. (a) What are the length of a transfer unit and the number of transfer units based on the cold fluid? (b) What are they based on the hot fluid?

SOLUTION   (a)   Water is the cold fluid in this exchanger. The quantities needed for substitution in Eq. (11-42) are

$$\dot{m}_c = 1{,}500 \text{ lb/h} \qquad c_{pc} = 1.0 \text{ Btu/lb-°F}$$

$$D_c = \frac{1.315}{12} = 0.110 \text{ ft} \qquad \text{(Appendix 6)}$$

$$U_c = U_o = 71.3 \text{ Btu/ft}^2\text{-h-°F}$$

Substitution in Eq. (11-42) gives

$$H_{tc} = \frac{1{,}500 \times 1}{\pi \times 0.110 \times 71.3} = 60.8 \text{ ft (18.53 m)}$$

From Eq. (11-35)

$$N_{tc} = \frac{30}{60.8} = 0.494$$

(b)   For the hot fluid the quantities needed are

$$\dot{m}_h = 1{,}000 \text{ lb/h} \qquad c_{ph} = 0.6 \text{ Btu/lb-°F} \qquad D_h = \frac{1.049}{12} = 0.0874 \text{ ft}$$

$$U_h = U_i = \frac{U_o D_o}{D_i} = 71.3 \times \frac{1.315}{1.049} = 89.5 \text{ Btu/ft}^2\text{-h-°F}$$

By substitution in Eq. (11-43),

$$H_{th} = \frac{1,000 \times 0.6}{\pi \times 0.0874 \times 89.5} = 24.4 \text{ ft (7.44 m)}$$

From Eq. (11-35)

$$N_{th} = \frac{30}{24.4} = 1.23$$

Alternatively, from Eq. (11-37),

$$N_{th} = 0.494 \frac{60.8}{24.4} = 1.23 \qquad \qquad ////$$

SYMBOLS

A Area, ft^2 or m^2; A_T, total area of heat-transfer surface; A_c, on cool-fluid side; A_i, of inside of tube; A_o, of outside of tube; $\bar{A}_L$, logarithmic mean

c_p Specific heat at constant pressure, Btu/lb-°F or J/g-°C; c_{pc}, of cool fluid; c_{ph}, of warm fluid

D Diameter, ft or m; D_c, on cool-fluid side; D_h, on warm-fluid side; D_i, inside diameter of tube; D_o, outside diameter of tube; $\bar{D}_L$, logarithmic mean

H Enthalpy, Btu/lb or J/g; H_a, at entrance; H_b, at exit; H_{ca}, H_{cb}, of cool fluid; H_{ha}, H_{hb}, of warm fluid

H_t Length of transfer unit, ft or m; H_{tc}, based on cool fluid; H_{th}, based on warm fluid

h Individual or surface heat-transfer coefficient, Btu/ft^2-h-°F or W/m^2-°C; h_i, for inside of tube; h_o, for outside of tube

h_d Fouling factor, Btu/ft^2-h-°F or W/m^2-°C; h_{di}, inside tube; h_{do}, outside tube

k Thermal conductivity, Btu/ft-h-°F or W/m-°C; k_m, of tube wall

L Length, ft or m; L_T, total path length; L_{Tc}, of cool fluid; L_{Th}, of warm fluid; L_c, path length of cool fluid

$\dot{m}$ Mass flow rate, lb/h or kg/h; $\dot{m}_c$, of cool fluid; $\dot{m}_h$, of warm fluid

N_{Nu} Nusselt number, hD/k, dimensionless

N_t Number of transfer units; N_{tc}, based on cool fluid; N_{th}, based on warm fluid

q Heat flow rate, Btu/h or w; q_T, total in exchanger; q_c, to cool fluid; q_h, to warm fluid

T Temperature, °F or °C; T_a, at inlet; T_b, at outlet; T_c, of cool fluid; T_{ca}, at cool-fluid inlet; T_{cb}, at cool-fluid outlet; T_{c1}, T_{c2}, at ends of transfer-unit section; T_h, of warm fluid; T_{ha}, at warm-fluid inlet; T_{hb}, at warm-fluid outlet; T_{h1}, T_{h2}, at ends of transfer-unit section; T_w, of tube wall; T_{wc}, on cool-fluid side; T_{wh}, on warm-fluid side

U Overall heat-transfer coefficient, Btu/ft^2-h-°F or W/m^2-°C; U_c, for cool-fluid side; U_h, for warm-fluid side; U_i, based on inside surface area; U_o, based on outside surface area; U_1, U_2, at ends of exchanger

x_w Thickness of tube wall, ft or m

Greek letters

ΔT Overall temperature difference, $T_h - T_c$, °F or °C; ΔT_i, between tube wall and fluid inside tube; ΔT_o, between tube wall and fluid outside tube; ΔT_w, through the tube wall; ΔT_1, ΔT_2, at ends of exchanger; $\Delta \bar{T}_L$, logarithmic mean

λ Latent heat of vaporization, Btu/lb or J/g

PROBLEMS

11-1   Calculate the overall heat-transfer coefficients based on both inside and outside areas for the following cases. All individual coefficients are given in usual units.

Case 1   Water at 50°F flowing in a $\frac{3}{4}$-in. 16 BWG condenser tube at a velocity of 15 ft/s and saturated steam at 220°F condensing on the outside. $h_i = 2,150$. $h_o = 2,500$. $k_m = 69$.

Case 2   Benzene condensing at atmospheric pressure on the outside of a 25-mm steel pipe and air at 15°C flowing within at 6 m/s. The pipe wall is 3.5 mm thick. $h_i = 30$ W/m²-°C. $h_o = 1,200$ W/m²-°C. $k_m = 45$ W/m-°C.

Case 3   Dropwise condensation from steam at a pressure of 50 lb$_f$/in.² gauge on a Schedule 40 1-in. steel pipe carrying oil at a velocity of 3 ft/s. $h_o = 14,000$. $h_i = 130$. $k_m = 26$.

11-2   Calculate the temperatures of the inside and outside surfaces of the metal pipe or tubing in cases 1 and 2 of Prob. 11-1.

11-3   Aniline is to be cooled from 200 to 150°F in a double-pipe heat exchanger having a total outside area of 70 ft². For cooling, a stream of toluene amounting to 8,600 lb/h at a temperature of 100°F is available. The exchanger consists of Schedule 40 1$\frac{1}{4}$-in. pipe in Schedule 40 2-in. pipe. The aniline flow rate is 10,000 lb/h. (*a*) If flow is countercurrent, what are the toluene outlet temperature, the LMTD, and the overall heat-transfer coefficient? (*b*) What are they if flow is parallel?

11-4   In the exchanger described in Prob. 11-2, how much aniline can be cooled if the overall heat-transfer coefficient is 70 Btu/ft²-h-°F?

11-5   (*a*) Under the conditions described in Prob. 11-4, what are the length of a transfer unit and the number of transfer units in the exchanger based on aniline? (*b*) What are they based on toluene?

REFERENCES

1   Colburn, A. P.: *Ind. Eng. Chem.*, **25**:873 (1933).
2   McAdams, W. H.: "Heat Transmission," 3d ed., p. 5, McGraw-Hill, New York, 1954.
3   Perry, J. H. (ed.): "Chemical Engineers' Handbook," 4th ed., p. **10**-20, McGraw-Hill, New York, 1963.

12

HEAT TRANSFER TO FLUIDS WITHOUT PHASE CHANGE

In a large and important class of heat-exchange applications, heat is transferred between fluid streams without phase change in the fluids. Examples are the exchange of heat between a hot petroleum stream and a cooler one, the transfer of heat from a stream of hot gas to cooling water, and the cooling of a hot liquid stream by the ambient atmosphere. In such situations the two streams are separated by a metal wall, which constitutes the heat-transfer surface. The surface may consist of tubes or other channels of constant cross section, of flat plates, or, in such devices as jet engines and advanced power machinery, of special shapes designed to pack a maximum area of transfer surface into a small volume.

Most fluid-to-fluid heat transfer is accomplished in steady-state equipment, but thermal regenerators, in which a bed of solid shapes is alternately heated by a hot fluid and the hot shapes then used to warm a colder fluid, are also used, especially in high-temperature heat transfer. Cyclical unsteady-state processes such as these are not considered in this book.

Regimes of heat transfer in fluids A fluid being heated or cooled may be flowing in laminar flow, in turbulent flow, or in the transition range between laminar and turbulent flow. Also, the fluid may be flowing in forced or in natural convection.

In some instances more than one flow type may occur in the same stream; for instance, in laminar flow at low velocities through large tubes, natural convection may be superimposed on forced laminar flow.

The direction of flow of the fluid may be parallel to that of the heating surface, so that boundary-layer separation does not occur, or the direction of flow may be perpendicular or at an angle to the heating surface, and then boundary-layer separation does take place.

At ordinary velocities the heat generated from fluid friction is negligible in comparison with the heat transferred between the fluids, and frictional heating may be neglected. At high velocities, however, e.g., at Mach numbers above a few tenths, frictional heat becomes appreciable and cannot be ignored. At very high velocities frictional heating may become of controlling importance.

Because the conditions of flow at the entrance to a tube differ from those well downstream from the entrance, the velocity field and the associated temperature field may depend on the distance from the tube entrance. Also, in some situations the fluid flows through a preliminary length of unheated or uncooled pipe so that the fully developed velocity field is established before heat is transferred to the fluid, and the temperature field is created within an existing velocity field.

Finally, the properties of the fluid—viscosity, thermal conductivity, specific heat, and density—are important parameters in heat transfer. Each of these, especially viscosity, is temperature-dependent. Since a temperature field, in which the temperature varies from point to point, exists in a flowing stream undergoing heat transfer, a problem appears in the choice of temperature at which the properties should be evaluated. For small temperature differences between fluid and wall, and for fluids with weak dependence of viscosity on temperature, the problem is not acute, but for highly viscous fluids such as heavy petroleum oils, or where the temperature difference between the tube wall and the fluid is large, the variations in fluid properties within the stream become large, and the difficulty of calculating the heat-transfer rate is increased.

Because of the various effects noted above, the entire subject of heat transfer to fluids without phase change is complex, and in practice is treated as a series of special cases rather than as a general theory. All cases considered in this chapter do, however, have a phenomenon in common: in all of them the formation of a thermal boundary layer, analogous to the hydrodynamic Prandtl boundary layer described in Chap. 3, takes place; it profoundly influences the temperature field and so controls the rate of heat flow.

Thermal boundary layer   Consider a flat plate immersed in a stream of fluid in steady flow and oriented parallel to the plate, as shown in Fig. 12-1a. Assume that the stream approaching the plate does so at velocity u_0 and temperature T_∞ and that the surface of the plate is maintained at a constant temperature T_w. Assume that T_w is greater than T_∞, so that the fluid is heated by the plate. As described in Chap. 3, a boundary layer develops, within which the velocity varies from $u = 0$ at the wall to $u = u_0$ at the outer boundary of the layer. This boundary layer, called the *hydrodynamic boundary layer*, is shown by line OA in Fig. 12-1a. The penetration of

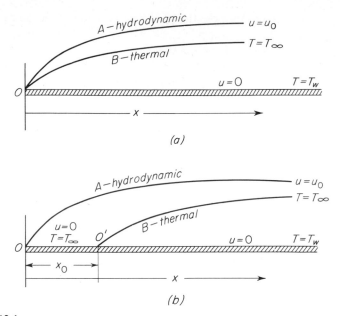

FIGURE 12-1
Thermal and hydrodynamic boundary layers on flat plate: (a) entire plate heated; (b) unheated length = x_0.

heat by transfer from the plate to the fluid changes the temperature of the fluid near the surface of the plate, and a temperature gradient is generated. The temperature gradient also is confined to a layer next to the wall, and within the layer the temperature varies from T_w at the wall to T_∞ at its outside boundary. This layer, called the *thermal boundary layer*, is shown as line OB in Fig. 12-1a. As drawn, lines OA and OB show that the thermal boundary layer is thinner than the hydrodynamic layer at all values of x, where x is the distance from the leading edge of the plate. In most heat-transfer situations this is the case. The actual relationship between the thicknesses of the two kinds of boundary layer at a given point along the plate depends on the dimensionless Prandtl number, defined as $c_p\mu/k$, which appears in Appendix 4 in the dimensional analysis of heat transfer. The two boundary layers are identical when the Prandtl number is unity. When the number is greater than unity, the thermal layer is thinner than the hydrodynamic layer, as shown in Fig. 12-1a, and when the Prandtl number is small, the thermal boundary layer is thicker than the hydrodynamic one. Only in heat transfer to liquid metals is this last situation encountered.

In Fig. 12-1a it is assumed that the entire plate is heated and that both boundary layers start at the leading edge of the plate. If the first section of the plate is not heated, and if the heat-transfer area begins at a definite distance x_0 from the leading edge, as shown by line $O'B$ in Fig. 12-1b, a hydrodynamic boundary layer already exists at x_0, where the thermal boundary layer begins to form.

In flow through a tube, it has been shown (Chap. 3) that the hydrodynamic boundary layer thickens as the distance from the tube entrance increases, and finally the layer reaches the center of the tube. The velocity profile so developed, called *fully developed flow*, establishes a velocity distribution that is unchanged with additional pipe length. The thermal boundary layer in a heated or cooled tube also reaches the center of the tube at a definite length from the entrance of the heated length of the tube, and the temperature profile is fully developed at this point. Unlike the velocity profile, however, the temperature profile flattens as the length of the tube increases, and in very long pipes the entire fluid stream reaches the temperature of the tube wall, the temperature gradients disappear, and heat transfer ceases.

HEAT TRANSFER BY FORCED CONVECTION IN LAMINAR FLOW

In laminar flow, heat transfer occurs only by conduction, as there are no eddies to carry heat by convection across an isothermal surface. The problem is amenable to mathematical analysis based on the partial differential equations for continuity, momentum, and energy. Such treatments are beyond the scope of this book and are given in standard treatises on heat transfer.[6a] Mathematical solutions depend on the boundary conditions established to define the conditions of fluid flow and heat transfer. When the fluid approaches the heating surface, it may have an already completed hydrodynamic boundary layer or a partially developed one. Or the fluid may approach the heating surface at a uniform velocity, and both boundary layers may be initiated at the same time. A simple flow situation, where the velocity is assumed constant in all cross sections and tube lengths, is called *plug* or *rodlike flow*. Independent of the conditions of flow, (1) the heating surface may be isothermal, (2) the heat flux may be equal at all points on the heating surface, in which case the average temperature of the fluid varies linearly with tube length, or (3) the axial temperature of the fluid stream may be required to change linearly. Other combinations of boundary conditions are possible.[6a] The basic differential equation for the several special cases is the same, but the final integrated relationships differ.

Most of the simpler mathematical derivations are based on the assumptions that the fluid properties are constant and temperature-independent and that flow is truly laminar with no cross currents or eddies. These assumptions are valid when temperature changes and gradients are small, but with large temperature changes the simple model is not in accord with physical reality for two reasons. First, variations in viscosity across the tube distort the usual parabolic velocity-distribution profile of laminar flow. Thus, if the fluid is a liquid and is being heated, the layer near the wall has a lower viscosity than the layers near the center and the velocity gradient at the wall increases. A crossflow of liquid toward the wall is generated. If the liquid is being cooled, the reverse effect occurs. Second, since the temperature field generates density gradients, natural convection may set in, which further distorts the flow lines of the fluid. The effect of natural convection may be small or large, depending on a number of factors to be discussed in the section on natural convection.

In this section three types of heat transfer in laminar flow are considered: (1) heat transfer to a fluid flowing along a flat plate, (2) heat transfer in plug flow in tubes, and (3) heat transfer to a fluid stream which is in fully developed flow at the entrance to the tube. In all cases, the temperature of the heated length of the plate or tube is assumed to be constant, and the effect of natural convection is ignored.

Laminar-flow heat transfer to flat plate Consider heat flow to the flat plate shown in Fig. 12-1b. The conditions are assumed to be as follows:

Velocity of fluid approaching plate and at and beyond the edge of the boundary layer OA: u_0.
Temperature of fluid approaching plate and at and beyond the edge of the thermal boundary layer $O'B$: T_∞.
Temperature of plate: from $x = 0$ to $x = x_0$; $T = T_\infty$; for $x > x_0$, $T = T_w$, where $T_w > T_\infty$.
The following properties of the fluid are constant and temperature-independent: density ρ, conductivity k, specific heat c_p, and viscosity μ.

Detailed analysis of the situation yields the equation[2]

$$\left(\frac{dT}{dy}\right)_w = \frac{0.332(T_w - T_\infty)}{\sqrt[3]{1 - (x_0/x)^{3/4}}} \sqrt[3]{\frac{c_p \mu}{k}} \sqrt{\frac{u_0 \rho}{\mu x}} \tag{12-1}$$

where $(dT/dy)_w$ is the temperature gradient at the wall. From Eq. (11-19), the relation between the local heat-transfer coefficient h_x at any distance x from the leading edge and the wall gradient is

$$h_x = \frac{k}{T_w - T_\infty}\left(\frac{dT}{dy}\right)_w \tag{12-2}$$

Eliminating $(dT/dy)_w$ gives

$$h_x = \frac{0.332k}{\sqrt[3]{1 - (x_0/x)^{3/4}}} \sqrt[3]{\frac{c_p \mu}{k}} \sqrt{\frac{u_0 \rho}{\mu x}}$$

This equation can be put into a dimensionless form by multiplying by x/k, giving

$$\frac{h_x x}{k} = \frac{0.332}{\sqrt[3]{1 - (x_0/x)^{3/4}}} \sqrt[3]{\frac{c_p \mu}{k}} \sqrt{\frac{u_0 x \rho}{\mu}} \tag{12-3}$$

The left-hand side of this equation is, from Eq. (11-20), a Nusselt number corresponding to the distance x, or $N_{\mathrm{Nu},x}$. The second group is the Prandtl number N_{Pr}, and the third group is a Reynolds number corresponding to distance x, denoted by $N_{\mathrm{Re},x}$. Equation (12-3) then can be written

$$N_{\mathrm{Nu},x} = \frac{0.332}{\sqrt[3]{1 - (x_0/x)^{3/4}}} \sqrt[3]{N_{\mathrm{Pr}}} \sqrt{N_{\mathrm{Re},x}} \tag{12-4}$$

When the plate is heated over its entire length, as shown in Fig. 12-1a, $x_0 = 0$ and Eq. (12-4) becomes

$$N_{Nu,x} = 0.332 \sqrt[3]{N_{Pr}} \sqrt{N_{Re,x}} \tag{12-5}$$

Equation (12-5) gives the local value of the Nusselt number at distance x from the leading edge. More important in practice is the average value of N_{Nu} over the entire heated length of the plate x_1, defined as

$$N_{Nu} \equiv \frac{hx_1}{k} \tag{12-6}$$

where

$$h = \frac{1}{x_1} \int_0^{x_1} h_x \, dx$$

Equation (12-3) can be written for a plate heated over its entire length, since $x_0 = 0$, as

$$h_x = \frac{C}{\sqrt{x}}$$

where C is a constant containing all factors other than h_x and x. Then

$$h = \frac{C}{x_1} \int_0^{x_1} \frac{dx}{\sqrt{x}} = \frac{2C}{x_1} \sqrt{x_1} = \frac{2C}{\sqrt{x_1}} = 2h_{x_1} \tag{12-7}$$

The average coefficient is clearly twice the local coefficient at the end of the plate, and Eq. (12-5) gives

$$N_{Nu} = 0.664 \sqrt[3]{N_{Pr}} \sqrt{N_{Re,x_1}} \tag{12-8}$$

These equations are valid only for Prandtl numbers larger than about 0.6.

PLUG FLOW   The simplest situation of laminar-flow heat transfer in tubes is defined by the following conditions. The velocity of the fluid throughout the tube and at all points in any cross section of the stream is constant, so that $u = u_0 = \overline{V}$; the wall temperature is constant; and the properties of the fluid are independent of temperature. Mathematically this model is identical to that of heat flow by conduction into a solid rod at constant surface temperature, using as heating time the period of passage of a cross section of the fluid stream at velocity $\overline{V}$ through a tube of length L. This time period is $t_T = L/\overline{V}$. Equation (10-18), then, can be used for plug flow of a fluid by substituting $L/\overline{V}$ for t_T in the Fourier number, which becomes

$$N_{Fo} = \frac{\alpha t_T}{r^2} = \frac{4kt_T}{c_p \rho D^2} = \frac{4kL}{c_p \rho D^2 \overline{V}} \tag{12-9}$$

The Graetz and Peclet numbers   Two other dimensionless groups are commonly used in place of the Fourier number in treating heat transfer to fluids. The *Graetz number* is defined by the equation

$$N_{Gz} \equiv \frac{\dot{m} c_p}{kL} \tag{12-10}$$

where $\dot{m}$ is the mass flow rate. Since $\dot{m} = (\pi/4)\rho\bar{V}D^2$,

$$N_{Gz} = \frac{\pi}{4}\frac{\rho\bar{V}c_pD^2}{kL} \tag{12-11}$$

The *Peclet number* N_{Pe} is defined as the product of the Reynolds number and the Prandtl number, or

$$N_{Pe} \equiv N_{Re}N_{Pr} = \frac{D\bar{V}\rho}{\mu}\frac{c_p\mu}{k} = \frac{\rho\bar{V}c_pD}{k} \tag{12-12}$$

The choice among these groups is arbitrary. They are related by the equations

$$N_{Gz} = \frac{\pi D}{4L}N_{Pe} = \frac{\pi}{N_{Fo}} \tag{12-13}$$

In the following discussion the Graetz number is used.

Equation (10-18) becomes, for plug flow,

$$\frac{T_w - \overline{T}_b}{T_w - T_a} = 0.692e^{-5.78\pi/N_{Gz}} + 0.131e^{-30.5\pi/N_{Gz}} + 0.0534e^{-74.9\pi/N_{Gz}} + \cdots \tag{12-14}$$

Also, T_a and $\overline{T}_b$ now are the inlet and average outlet fluid temperatures, respectively.

Plug flow is not a realistic model for Newtonian fluids, but it does apply to highly pseudoplastic liquids ($n' \approx 0$) or to plastic liquids having a high value of the yield stress τ_0.

FULLY DEVELOPED FLOW   With a Newtonian fluid in fully developed flow, the actual velocity distribution at the entrance to the heated section and the theoretical distribution throughout the tube are both parabolic. For this situation the appropriate boundary conditions lead to the development of another theoretical equation, of the same form as Eq. (12-14). This is[7d]

$$\frac{T_w - \overline{T}_b}{T_w - T_a} = 0.81904e^{-3.657\pi/N_{Gz}} + 0.09760e^{-22.31\pi/N_{Gz}} + 0.01896e^{-53\pi/N_{Gz}} \tag{12-15}$$

Because of the distortions in the flow field from the effects of temperature on viscosity and density, this equation does not give accurate results. In actual fluids a considerably larger heat-transfer rate than that predicted by Eq. (12-15) is obtained.

Both Eqs. (12-14) and (12-15) have the form

$$\frac{T_w - \overline{T}_b}{T_w - T_a} = \chi'(N_{Gz}) \tag{12-16}$$

where the function $\chi'(N_{Gz})$ depends only on the boundary conditions. Equation (12-16) applies to all situations of steady-state heat transfer to fluids in laminar flow in tubes.

Average heat-transfer coefficients in laminar flow   Equations (12-14) and (12-15) permit calculation of temperature changes and heat-transfer rates in fluids in laminar flow. It is convenient to use an average coefficient h_i over the entire tube, defined by the equation

$$h_i = \frac{q}{A_T \overline{\Delta T_i}} \tag{12-17}$$

where q = rate of heat transfer
   A_T = area of heating surface
   $\overline{\Delta T_i}$ = average temperature drop over fluid, from tube wall to average fluid temperature

From an enthalpy balance

$$q = \dot{m}c_p(\overline{T}_b - T_a) \tag{12-18}$$

Also
$$A_T = \pi DL \tag{12-19}$$

Elimination of q and A_T from Eq. (12-17) by Eqs. (12-18) and (12-19) gives

$$h_i = \frac{\dot{m}c_p(\overline{T}_b - T_a)}{\pi DL \overline{\Delta T_i}} \tag{12-20}$$

The left-hand side of Eq. (12-20) can be expressed in the form of a Nusselt number by multiplying through by D/k.

$$N_{\text{Nu}} = \frac{h_i D}{k} = \frac{1}{\pi} \frac{\dot{m}c_p}{kL} \frac{\overline{T}_b - T_a}{\overline{\Delta T_i}} = \frac{N_{\text{Gz}}(\overline{T}_b - T_a)}{\pi \overline{\Delta T_i}} \tag{12-21}$$

In turbulent flow, the coefficient is nearly independent of the tube length, and the LMTD is the logical form for practical use. In laminar flow, however, the average coefficient over the tube depends strongly upon the heated length, and, in fact, $h_i \propto 1/L$. The choice of the LMTD is no longer logical, and the simpler arithmetic mean is more convenient. Thus, let $\overline{\Delta T_i}$ be

$$\overline{\Delta T_i} = \frac{(T_w - T_a) + (T_w - \overline{T}_b)}{2} = \overline{\Delta T_a} \tag{12-22}$$

where $\overline{\Delta T_a}$ is the arithmetic mean temperature drop.

Substitution of $\overline{\Delta T_a}$ for $\overline{\Delta T_i}$ in Eq. (12-21) gives

$$\frac{h_{ia} D}{k} = \frac{2N_{\text{Gz}}}{\pi} \frac{\overline{T}_b - T_a}{(T_w - T_a) + (T_w - \overline{T}_b)} \tag{12-23}$$

The subscript a on h_{ia} calls attention to the fact that the coefficient is based on the arithmetic average temperature drop.

If Eq. (12-16) is solved for $\overline{T}_b$ and the resulting value substituted into the denominator of Eq. (12-23), the result is

$$\frac{h_{ia} D}{k} = \frac{2N_{\text{Gz}}}{\pi} \frac{\overline{T}_b - T_a}{(T_w - T_a)[1 + \chi'(N_{\text{Gz}})]}$$

Then, substituting for $(\overline{T}_b - T_a)/(T_w - T_a)$ from Eq. (12-16),

$$\frac{h_{ia}D}{k} = \frac{2N_{Gz}}{\pi} \frac{1 - \chi'(N_{Gz})}{1 + \chi'(N_{Gz})} = \chi(N_{Gz}) \tag{12-24}$$

where $\chi(N_{Gz})$ is another function of the Graetz number.

For values of N_{Gz} greater than 10 a simpler equation, known as the *Leveque approximation*, is in close agreement with Eq. (12-24). The Leveque approximation is[7e]

$$\frac{h_{ia}D}{k} = N_{Nu,a} = 1.75 N_{Gz}^{1/3} = 1.75 \left(\frac{\dot{m}c_p}{kL}\right)^{1/3} \tag{12-25}$$

Because the simplifying assumptions given on page 312 are not strictly valid, heat-transfer coefficients found experimentally are about 15 percent greater than predicted from theory. To account for this the coefficient in Eq. (12-25) is increased from 1.75 to 2.0. Also, for viscous liquids and large temperature drops, Eq. (12-25) requires modification to account for the difference between heating and cooling. A dimensionless, but empirical, correction factor serves this purpose:

$$\phi_v = \left(\frac{\mu}{\mu_w}\right)^{0.14} \tag{12-26}$$

After these corrections are made,[8] Eq. (12-25) becomes

$$\frac{h_{ia}D}{k} = N_{Nu,a} = 2 \left(\frac{\dot{m}c_p}{kL}\right)^{1/3} \left(\frac{\mu}{\mu_w}\right)^{0.14} = 2N_{Gz}^{1/3} \phi_v \tag{12-27}$$

In Eqs. (12-26) and (12-27) μ is the viscosity at the arithmetic mean temperature of the fluid. This temperature is given by $\overline{T} = (T_a + \overline{T}_b)/2$; μ_w is the viscosity at the wall temperature T_w. For liquids, $\mu_w < \mu$ and $\phi_v > 1.0$ when the liquid is being heated, and $\mu < \mu_w$ and $\phi_v < 1.0$ when the liquid is being cooled. In heating or cooling gases, the inequalities are reversed, because the viscosity of a gas increases with temperature instead of decreasing. The effect of temperature difference on the coefficient is small in heat transfer to and from gases.

Heat transfer to non-Newtonian liquids in laminar flow   For heat transfer to and from liquids that follow the power-law relation [Eq. (3-7)], Eq. (12-27) is modified to[9]

$$\frac{h_{ia}D}{k} = 2\delta^{1/3} \left(\frac{\dot{m}c_p}{kL}\right)^{1/3} \left(\frac{m}{m_w}\right)^{0.14} \tag{12-28}$$

where $\delta = (3n' + 1)/4n'$
   $m = K'8^{n'-1}$, at arithmetic mean temperature
   m_w = value of m at T_w
   K' = flow consistency index
   n' = flow behavior index

Asymptotic limitation   Equation (12-27) is subject to an important limitation, which is a result of the fact that h_{ia} is based on the arithmetic mean temperature drop. If the tube is sufficiently long, the exit fluid temperature $\overline{T}_b$ approaches the wall

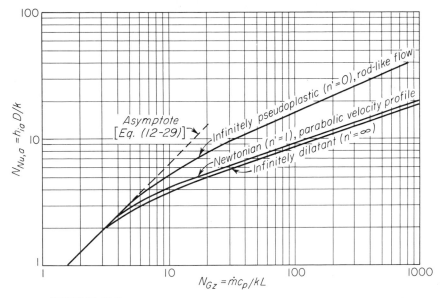

FIGURE 12-2
Heat transfer of fluids in laminar flow inside tubes. [*By permission, from W. L. Wilkinson, "Non-Newtonian Fluids." Copyright, 1960, Pergamon Press.*]

temperature T_w. An asymptotic situation is reached, then, where $\overline{T}_b - T_a = T_w - T_a$ and $T_w - \overline{T}_b = 0$. Equation (12-23) becomes, for this condition,

$$\frac{h_{ia}D}{k} = \frac{2\dot{m}c_p}{\pi kL} = \frac{2}{\pi}N_{Gz} \tag{12-29}$$

The asymptotic condition is closely approached for values of N_{Gz} of about 10, and Eq. (12-29) should be used rather than Eq. (12-27) for values of N_{Gz} less than this. If a heat exchanger is operating under conditions such that the asymptotic range is entered, part of the heating surface is inactive and such surface is wasted.

In Fig. 12-2a are plotted on log-log coordinates curves showing the relation between $N_{Nu,a}$ and N_{Gz} for the following situations: the asymptotic condition given by Eq. (12-29); the rodlike, or plug, flow derived from Eq. (12-14); the fully developed flow of a Newtonian fluid based on Eq. (12-25); and an example of an infinitely dilatant liquid using Eq. (12-28), for which $n' = \infty$ and $\delta = \frac{3}{4}$. In all cases $\phi_v = 1.0$.

HEAT TRANSFER BY FORCED CONVECTION IN TURBULENT FLOW

Perhaps the most important situation in heat transfer is the heat flow in a stream of fluid in turbulent flow in a closed channel, especially in tubes. Turbulence is encountered at Reynolds numbers greater than about 2,100, and since the rate of heat transfer is greater in turbulent flow than in laminar flow, most equipment is operated in the turbulent range.

The earliest approach to this case was based on empirical correlations of test data guided by dimensional analysis. The equations so obtained still are much used in design. Subsequently, theoretical study has been given to the problem. A deeper understanding of the mechanism of turbulent-flow heat transfer has been achieved, and improved equations applicable over wider ranges of conditions have been obtained. In this section the dimensional-empirical approach is discussed first, and then the more theoretical results are briefly considered.

Dimensional-analysis method In Appendix 4, a dimensional analysis of the heat flow to a fluid in turbulent flow through a long straight pipe yields two equivalent dimensionless relations, given as Eqs. (A-19) and (A-20). The equations can be interpreted as giving the local heat flux q/A, sufficiently far from the entrance of the tube for the thermal and hydrodynamic boundary layers to have reached the center of the pipe and the temperature and velocity gradients to be fully established. At and beyond this distance from the entrance q/A is independent of tube length. By inserting the definition of the heat-transfer coefficient h given in Eq. (11-17), Eqs. (A-19) and (A-20) become, respectively,

$$\frac{h_\infty D}{k} = \Phi\left(\frac{D\bar{V}\rho}{\mu}, \frac{c_p\mu}{k}\right) = \Phi\left(\frac{DG}{\mu}, \frac{c_p\mu}{k}\right) \tag{12-30}$$

and

$$\frac{h_\infty}{c_p G} = \Phi_1\left(\frac{DG}{\mu}, \frac{c_p\mu}{k}\right) \tag{12-31}$$

Here the mass velocity G is used in place of its equal, $\bar{V}\rho$, and the subscript ∞ indicates that the equations apply to long pipes at a distance from the entrance.

The three groups in Eq. (12-30) are recognized as the Nusselt, Reynolds, and Prandtl numbers, respectively. The left-hand group in Eq. (12-31) is called the *Stanton number* N_{St}. The four groups are related by the equation

$$N_{St}N_{Re}N_{Pr} = N_{Nu} \tag{12-32}$$

Thus, only three of the four are independent.

Of the two equations (12-30) and (12-31), the second is the more convenient, and will be used in the following treatment.

EFFECT OF TUBE LENGTH Near the tube entrance, where the temperature gradients are still forming, the local coefficient h_x is greater than h_∞. At the entrance itself, where there is no previously established temperature gradient, h_x is infinite. Its value drops rapidly toward h_∞ in a comparatively short length of tube. Dimensionally, the effect of tube length is accounted for by another dimensionless group, x/D, where x is the distance from the tube entrance. The local coefficient approaches h_∞ asymptotically with increase in x, but it is practically equal to h_∞ when x/D is about 50. The average value of h_x over the tube length is denoted by h_i. The value of h_i is found by integrating h_x over the length of the tube. Since $h_x \to h_\infty$ as $x \to \infty$, the relation between h_i and h_∞ is of the form

$$\frac{h_i}{h_\infty} = 1 + \psi\left(\frac{L}{D}\right) \tag{12-33}$$

where L is the length of the pipe. The function $\psi(L/D)$ depends on the extent of development of the hydrodynamic boundary layer and the velocity distribution at the entrance, where $L = 0$. For a number of situations this function is given by[6b]

$$\frac{h_i}{h_\infty} = 1 + \frac{C_1}{L/D} \tag{12-34}$$

where C_1 is a constant. For example, if the entrance is bell-shaped, $C_1 = 1.4$. For tubes with sharp-edged entrances, where the velocity at entrance is uniform over the cross section, Eq. (12-33) becomes

$$\frac{h_i}{h_\infty} = 1 + \left(\frac{D}{L}\right)^{0.7} \tag{12-35}$$

The effect of tube length on h_i fades when L/D becomes greater than about 50. In this respect, heat transfer to fluids in turbulent flow differs from that in laminar flow, where the effect of tube length still is significant at $L/D = 150$.

HEATING AND COOLING   When a fluid is heated or cooled inside a pipe, the viscosity of the fluid immediately adjacent to the pipe wall may differ considerably from the viscosity of the bulk of the fluid. Both the velocity pattern and the rate of heat transfer are affected. The usual method of correcting for the effect of heating or cooling on the rate of heat transfer is to evaluate properties at the mean bulk temperature $T = (T_a + \bar{T}_b)/2$ and use factor ϕ_v from Eq. (12-26).

Empirical equations   To use Eq. (12-30) or (12-31), the function Φ or Φ_1 must be known. A recognized empirical correlation, for sharp-edged entrances and using Eq. (12-30), is

$$\frac{h_i D}{k} = 0.023 \left[1 + \left(\frac{D}{L}\right)^{0.7}\right] \left(\frac{DG}{\mu}\right)^{0.8} \left(\frac{c_p\mu}{k}\right)^{1/3} \left(\frac{\mu}{\mu_w}\right)^{0.14} \tag{12-36}$$

An alternate form of Eq. (12-36) is obtained by dividing both sides by $(DG/\mu)(c_p\mu/k)$ and transposing, to give

$$\frac{h_i}{c_p G} \left(\frac{c_p\mu}{k}\right)^{2/3} \left(\frac{\mu_w}{\mu}\right)^{0.14} = \frac{0.023[1 + (D/L)^{0.7}]}{(DG/\mu)^{0.2}} \tag{12-37}$$

In using these equations the physical properties of the fluid, except for μ_w, are evaluated at the bulk temperature T. Equations (12-36) and (12-37) can be written in more compact form as

$$N_{\mathrm{Nu}} = 0.023 \left[1 + \left(\frac{D}{L}\right)^{0.7}\right] N_{\mathrm{Re}}^{0.8} N_{\mathrm{Pr}}^{1/3} \phi_v \tag{12-38}$$

and

$$N_{\mathrm{St}} N_{\mathrm{Pr}}^{2/3} \phi_v^{-0.14} = 0.023 \left[1 + \left(\frac{D}{L}\right)^{0.7}\right] N_{\mathrm{Re}}^{-0.2} \tag{12-39}$$

For entrances other than sharp-edged, the bracketed term in Eqs. (12-36) to (12-39) is replaced by the relation for h_i/h_∞ given in Eq. (12-34). For long tubes, when L/D is greater than about 50, the bracketed term may be omitted.

Equations (12-36) to (12-39) are not fundamentally different equations; they are merely alternative ways of expressing the same information. They should not be used for Reynolds numbers below 6,000 or for molten metals, which have abnormally low Prandtl numbers.

Average value of h_i in turbulent flow   Since the temperature of the fluid changes from one end of the tube to the other and the fluid properties μ, k, and c_p are all functions of temperature, the local value of h_i also varies from point to point along the tube. This variation is independent of the effect of tube length.

For gases, the temperature effect on h_i is small. The Prandtl number and c_p are nearly independent of temperature, and since G and D are also constant, h_i is proportional to the factor $1/\mu^{0.2}$. For gases, μ increases slowly with temperature, and so h_i changes but little unless the temperature range is large. For example, for air, h_i increases only about 6 percent if the air temperature changes from 100 to 200°F.

For liquids, the temperature effect is much greater than for gases. Equation (12-36) can be condensed to read, for long tubes, and assuming $(\mu/\mu_w) = 1$,

$$h_i = 0.023 \frac{G^{0.8}k^{2/3}c_p^{1/3}}{D^{0.2}\mu^{0.47}} \tag{12-40}$$

Inspection of Eq. (12-40) shows that, for liquids, the effect of a temperature rise is to increase h_i. With increase in temperature, both c_p and k increase slowly and μ decreases rapidly; the increase in h_i with temperature is due principally to the effect of temperature on viscosity. For water, for example, h_i increases about 50 percent over a temperature range from 100 to 200°F.

In practice, unless the variation in h_i over the length of the tube is more than about 2:1, an average value of h_i is calculated and used as a constant in calculating the overall coefficient U. This procedure neglects the variation of U over the tube length and allows the use of the LMTD in calculating the area of the heating surface. The average value of h_i is computed by evaluating the fluid properties c_p, k, and μ at the average fluid temperature, defined as the arithmetic mean between the inlet and outlet temperatures. The value of h_i calculated from Eq. (12-36), using these property values, is called the *average coefficient*. For example, assume that the fluid enters at 100°F and leaves at 200°F. The average fluid temperature is $(100 + 200)/2 = 150°F$, and the values of the properties used to calculate the average value of h_i are those at 150°F.

For larger changes in h_i, two procedures can be used: (1) The values of h_i at the inlet and outlet can be calculated, corresponding values of U_1 and U_2 found, and Eq. (11-16) used. Here the effect of L/D on the entrance value of h_i is ignored. (2) For even larger variations in h_i, and therefore in U, the tube can be divided into sections and an average U used for each section. Then the lengths of the individual sections can be added to account for the total length of the tube.

Estimation of wall temperature T_w   To evaluate μ_w, the viscosity of the fluid at the wall, T_w must be found. The estimation of this temperature requires an iterative calculation based on the resistance equation (11-30). If the individual resistances can be estimated, the total temperature drop ΔT can be split into the individual temperature drops by the use of this equation and an approximate value for the wall temperature found. To determine T_w in this way the wall resistance $(x_w/k_m)(D_o/\bar{D}_L)$ can be neglected, and Eq. (11-30) used as follows.

From the first two members of Eq. (11-30)

$$\Delta T_i = \frac{D_o/D_i h_i}{1/U_o} \Delta T \tag{12-41}$$

Substituting $1/U_o$ from Eq. (11-27) and neglecting the wall-resistance term gives

$$\Delta T_i = \frac{1/h_i}{1/h_i + D_i/D_o h_o} \Delta T \tag{12-42}$$

Use of Eq. (12-42) requires preliminary estimates of the coefficients h_i and h_o. To estimate h_i, Eq. (12-36) can be used. The calculation of h_o will be described later. The wall temperature T_w is then obtained from the following equations:

For heating: $$T_w = T + \Delta T_i \tag{12-43}$$

For cooling: $$T_w = T - \Delta T_i \tag{12-44}$$

where T is the average fluid temperature.

If the first approximation is not sufficiently accurate, a second calculation of T_w based on the results of the first can be made. Unless the factor ϕ_v is quite different from unity, however, the second approximation is unnecessary.

EXAMPLE 12-1   Toluene is being condensed at 230°F (110°C) on the outside of $\frac{3}{4}$-in. (19-mm) 16 BWG copper condenser tubes through which cooling water is flowing at an average temperature of 80°F (26.7°C). Individual heat-transfer coefficients are given in Table 12-1. Neglecting the resistance of the tube wall, what is the tube-wall temperature?

SOLUTION   From Appendix 7, $D_i = 0.620$ in.; $D_o = 0.750$ in. Hence, from Eq. (12-42),

$$T_i = \frac{1/200}{1/200 + 0.620/(0.750 \times 500)} (230 - 80) = 112.7°F$$

Since the water is being heated by the condensing toluene, the wall temperature is found from Eq. (12-43) as follows:

$$T_w = 80 + 112.7 = 192.7°F\ (89.3°C) \qquad ////$$

Table 12-1   DATA FOR EXAMPLE 12-1

	Heat-transfer coefficient	
	Btu/ft²-h-°F	W/m²-°C
For cooling water h_i	200	1,135
For toluene h_o	500	2,840

Cross sections other than circular To use Eq. (12-36) or (12-37) for cross sections other than circular it is only necessary to replace the diameter D in both Reynolds and Nusselt numbers by the equivalent diameter D_e, defined as 4 times the hydraulic radius r_H. The method is the same as that used in calculating friction loss.

EXAMPLE 12-2 Benzene is cooled from 141 to 79°F (60.6 to 26.1°C) in the inner pipe of a double-pipe exchanger. Cooling water flows countercurrently to the benzene, entering the jacket at 60°F (15.6°C) and leaving at 80°F (26.7°C). The exchanger consists of an inner pipe of $\frac{7}{8}$-in. (22.2-mm) 16 BWG copper tubing jacketed with Schedule 40 $1\frac{1}{2}$-in. (38.1-mm) steel pipe. The linear velocity of the benzene is 5 ft/s (1.52 m/s); that of the water is 4 ft/s (1.22 m/s). Neglecting the resistances of the wall and scale films, and assuming $L/D > 150$ for both pipes, compute the film coefficients of the benzene and water and the overall coefficient based on the outside area of the inner pipe.

SOLUTION The average temperature of the benzene is $(141 + 79)/2 = 110°F$; that of the water is $(60 + 80)/2 = 70°F$. The physical properties at these temperatures are given in Table 12-2. The diameters of the inner tube are

$$D_{it} = \frac{0.745}{12} = 0.0621 \text{ ft} \qquad D_{ot} = \frac{0.875}{12} = 0.0729 \text{ ft}$$

The inside diameter of the jacket is, from Appendix 6,

$$D_{ij} = \frac{1.610}{12} = 0.1342 \text{ ft}$$

The equivalent diameter of the annular jacket space is found as follows. The cross-sectional area is $(\pi/4)(0.1342^2 - 0.0729^2)$. The wetted perimeter is $\pi(0.1342 + 0.0729)$. The hydraulic radius is

$$r_H = \frac{(\pi/4)(0.1342^2 - 0.0729^2)}{\pi(0.1342 + 0.0729)} = \tfrac{1}{4}(0.1342 - 0.0729) = \tfrac{1}{4} \times 0.0613 \text{ ft}$$

The equivalent diameter is

$$D_e = 4 \times \tfrac{1}{4} \times 0.0613 = 0.0613 \text{ ft}$$

Table 12-2 DATA FOR EXAMPLE 12-2

Property	Value at average fluid temperature	
	Benzene	Water[†]
Density ρ, lb/ft³	53.1	62.3
Viscosity μ, lb/ft-h	1.16[‡]	2.42 × 0.982 = 2.34
Thermal conductivity k, Btu/ft-h-°F	0.089[§]	0.346
Specific heat c_p, Btu/lb-°F	0.435[¶]	1.000

† Appendix 14.   ‡ Appendix 10.   § Appendix 13.   ¶ Appendix 16.

The Reynolds number and Prandtl number of each stream are next computed:

Benzene: $\quad N_{Re} = \dfrac{D_{it}\bar{V}\rho}{\mu} = \dfrac{0.0621 \times 5 \times 3{,}600 \times 53.1}{1.16} = 5.12 \times 10^4$

$\quad\quad\quad N_{Pr} = \dfrac{c_p\mu}{k} = \dfrac{0.435 \times 1.16}{0.089} = 5.67$

Water: $\quad N_{Re} = \dfrac{D_e\bar{V}\rho}{\mu} = \dfrac{0.0613 \times 4 \times 3{,}600 \times 62.3}{2.34} = 2.35 \times 10^4$

$\quad\quad\quad N_{Pr} = \dfrac{1.00 \times 2.34}{0.346} = 6.76$

Preliminary estimates of the coefficients are obtained from Eq. (12-37), omitting the corrections for tube length and for viscosity ratio:

Benzene: $\quad h_i = \dfrac{0.023 \times 5 \times 3{,}600 \times 53.1 \times 0.435}{(5.12 \times 10^4)^{0.2} \times 5.67^{2/3}} = 346 \text{ Btu/ft}^2\text{-h-}°\text{F}$

Water: $\quad h_o = \dfrac{0.023 \times 4 \times 3{,}600 \times 62.3 \times 1.000}{(2.35 \times 10^4)^{0.2} \times 6.76^{2/3}} = 771 \text{ Btu/ft}^2\text{-h-}°\text{F}$

In these calculations use is made of the fact that $G = \bar{V}\rho$.

The temperature drop over the benzene resistance, from Eq. (12-42), is

$$\Delta T_i = \frac{1/346}{1/346 + 0.0621/(0.0729 \times 771)} (110 - 70) = 29°\text{F}$$

$$T_w = 110 - 29 = 81°\text{F}$$

The viscosities of the liquids at T_w are now found:

$$\mu_w = \begin{cases} 1.45 \text{ lb/ft-h} & \text{for benzene} \\ 0.852 \times 2.42 = 2.06 \text{ lb/ft-h} & \text{for water} \end{cases}$$

The viscosity-correction factors ϕ_v, from Eq. (12-26), are:

$$\phi_v = \begin{cases} \left(\dfrac{1.16}{1.45}\right)^{0.14} = 0.969 & \text{for benzene} \\ \left(\dfrac{2.34}{2.06}\right)^{0.14} = 1.018 & \text{for water} \end{cases}$$

The corrected coefficients are:

Benzene: $\quad h_i = 346 \times 0.969 = 335 \text{ Btu/ft}^2\text{-h-}°\text{F} \ (1{,}900 \text{ W/m}^2\text{-}°\text{C})$

Water: $\quad h_o = 771 \times 1.018 = 785 \text{ Btu/ft}^2\text{-h-}°\text{F} \ (4{,}456 \text{ W/m}^2\text{-}°\text{C})$

The temperature drop over the benzene resistance and the wall temperature become

$$\Delta T_i = \frac{1/335}{1/335 + 0.0621/(0.0729 \times 785)} (110 - 70) = 29.3°\text{F}$$

$$T_w = 110 - 29.3 = 80.7°\text{F}$$

This is so close to the wall temperature calculated previously that a second approximation is unnecessary.

The overall coefficient is found from Eq. (11-26), neglecting the resistance of the tube wall.

$$\frac{1}{U_o} = \frac{0.0729}{0.0621 \times 335} + \frac{1}{785} = 0.00478$$

$$U_o = \frac{1}{0.00478} = 209 \text{ Btu/ft}^2\text{-h-}°\text{F} \ (1{,}186 \text{ W/m}^2\text{-}°\text{C})$$

////

Effect of roughness   For equal Reynolds numbers the heat-transfer coefficient in turbulent flow is somewhat greater for a rough tube than for a smooth one. The effect of roughness on heat transfer is much less than on fluid friction, and economically it is usually more important to use a smooth tube for minimum friction loss than to rely on roughness to yield a larger heat-transfer coefficient. The effect of roughness on h_i is neglected in practical calculations.

Heat transfer at high velocities   When a compressible fluid flows through a tube at high velocities, temperature gradients, caused by friction, appear even when there is no heat transfer through the wall and $q = 0$. The friction, which is a maximum at the wall, raises the temperature of the fluid at the wall above the average fluid temperature.[4c] The temperature difference between wall and fluid causes a flow of heat from wall to fluid, and a steady state is reached when the rate of heat developed by friction at the wall equals the rate of heat transfer back into the fluid stream. The constant wall temperature thus attained is called the *adiabatic wall temperature*. Further treatment of this subject is beyond the scope of this text. The effect becomes appreciable for Mach numbers above approximately 0.4, and appropriate equations must be used in this range of velocities instead of Eqs. (12-36) and (12-37).

The transfer of heat by turbulent eddies and the analogy between the transfer of momentum and heat   On pages 92 to 102 the distribution of velocity and its accompanying momentum flux in a flowing stream in turbulent flow through a pipe was described. Three rather ill-defined zones in the cross section of the pipe were identified. In the first, immediately next to the wall, eddies are rare, and momentum flow occurs almost entirely by viscosity; in the second, a mixed regime of combined viscous and turbulent momentum transfer occurs; in the main part of the stream, which occupies the bulk of the cross section of the stream, only the momentum flow generated by the Reynolds stresses of turbulent flow is important. The three zones are called the *viscous sublayer*, the *buffer zone*, and the *turbulent core*, respectively.

In heat transfer at the wall of the tube to or from the fluid stream, the same hydrodynamic distributions of velocity and of momentum fluxes still persist, and, in addition, a temperature gradient is superimposed on the turbulent laminar velocity field. In the following treatment both gradients are assumed to be completely developed and the effect of tube length negligible.

Throughout the stream of fluid, heat flow by conduction occurs in accordance with the equation

$$\frac{q_c}{A} = -k\,\frac{dT}{dy} \tag{12-45}$$

where
q_c = rate of heat flow by conduction
k = thermal conductivity
A = area of isothermal surface
dT/dy = temperature gradient across isothermal surface

The isothermal surface is a cylinder concentric with the axis of the pipe and located a distance y from the wall, or r from the center of the pipe. Here $r + y = r_w$, where r_w is the radius of the pipe.

In addition to conduction, the eddies of turbulent flow carry heat by convection across each isothermal area. Although both mechanisms of heat flow may occur wherever a temperature gradient exists $(dT/dy \neq 0)$, their relative importance varies greatly with distance from the wall. At the tube wall itself eddies cannot exist, and the heat flux is entirely due to conduction. Equation (12-45) written for the wall is

$$\left(\frac{q}{A}\right)_w = -k\left(\frac{dT}{dy}\right)_w \tag{12-46}$$

where
$(q/A)_w$ = total flux per unit area at wall
$(dT/dy)_w$ = temperature gradient at wall

These quantities are identical with those in Eqs. (11-18) and (11-19).

Within the viscous sublayer heat flows mainly by conduction, but eddies are not completely excluded from this zone, and some convection does occur. The relative importance of turbulent heat flux compared with conductive heat flux increases rapidly with distance from the wall. In ordinary fluids, having Prandtl numbers above about 0.6, conduction is entirely negligible in the turbulent core, but it may be significant in the buffer zone when the Prandtl number is in the neighborhood of unity. Conduction is negligible in this zone when the Prandtl number is large.

The situation is analogous to momentum flux, where the relative importance of turbulent shear to viscous shear follows the same general pattern. Under certain ideal conditions, the correspondence between heat flow and momentum flow is exact, and at any specific value of r/r_w the ratio of heat transfer by turbulence to that by conduction equals the ratio of momentum flux by viscous forces to that by Reynolds stresses. In the general case, however, the correspondence is only approximate and may be greatly in error. The study of the relationship between heat and momentum flux for the entire spectrum of fluids leads to the so-called *analogy theory*, and the equations so derived are called *analogy equations*. A detailed treatment of the theory is beyond the scope of this book, but some of the more elementary relationships are considered.

Since eddies continually cross an isothermal surface from both directions, they carry heat between layers on either side of the surface, which are at different average temperatures. At a given point the temperature fluctuates rapidly about the constant mean temperature at that point, depending on whether a "hot" or a "cold" eddy is crossing through the point. The temperature fluctuations form a pattern with respect to both time and place, just like the fluctuations in velocity and pressure described on pages 53 to 58. The instantaneous temperature T_i at the point can be split into two parts, the constant average temperature T at the point and the fluctuating or deviating temperature T', or

$$T_i = T + T' \tag{12-47}$$

The time average of the deviating temperature T', denoted by $\overline{T'}$, is zero, and the time-average value of the total instantaneous temperature, denoted by $\overline{T_i}$, is T. The average temperature T is that measured by an ordinary thermometer. To measure T_i and so find T' requires special sensing devices that can follow rapid temperature changes.

THE EDDY DIFFUSIVITY OF HEAT   When there is no temperature gradient across the isothermal surface, all eddies have the same temperature independent of the point of origin, $dT/dy = 0$, and no net heat flow occurs. If a temperature gradient exists, an analysis equivalent to that leading to Eq. (3-18) shows that the eddies carry a net heat flux from the higher temperature to the lower, in accordance with the equation

$$\frac{q_t}{A} = -c_p \rho \overline{v'T'} \tag{12-48}$$

where v' is the deviating velocity across the surface and the overline indicates the time average of the product $v'T'$. Although the time averages $\overline{v'}$ and $\overline{T'}$ individually are zero, the average of their product is not, because a correlation exists between these deviating quantities when $dT/dy \neq 0$, in the same way that the deviating velocities u' and v' are correlated when a velocity gradient du/dy exists.

On page 59 an eddy diffusivity for momentum transfer ε_M was defined. A corresponding eddy diffusivity for heat transfer ε_H can be defined by

$$\frac{q_t}{c_p \rho A} \equiv -\varepsilon_H \frac{dT}{dy} = -\overline{v'T'} \tag{12-49}$$

The subscript t refers to the fact that Eq. (12-49) applies to turbulent convection heat transfer. Since conduction also takes place, the total heat flux at a given point, denoted by q, is, from Eqs. (12-45) and (12-49),

$$\frac{q}{A} = \frac{q_c}{A} + \frac{q_t}{A} = -k \frac{dT}{dy} - c_p \rho \varepsilon_H \frac{dT}{dy}$$

or

$$\frac{q}{A} = -c_p \rho (\alpha + \varepsilon_H) \frac{dT}{dy} \tag{12-50}$$

where α is the thermal diffusivity, $k/c_p\rho$. The equation for the total momentum flux corresponding to Eq. (12-50) is Eq. (3-22) written as

$$\frac{\tau g_c}{\rho} = (v + \varepsilon_M) \frac{du}{dy} \tag{12-51}$$

SIGNIFICANCE OF PRANDTL NUMBER; THE EDDY DIFFUSIVITIES   The physical significance of the Prandtl number appears on noting that it is the ratio v/α; it is therefore a measure of the magnitude of the thermal diffusivity relative to that of the momentum diffusivity. Its numerical value depends on the temperature and pressure of the fluid, and therefore it is a true property. The magnitude of the Prandtl numbers encountered in practice covers a wide range. For liquid metals it is of the order 0.01 to 0.04. For diatomic gases it is 0.74, and for water at 160°F it is about unity. For viscous liquids and concentrated solutions it may be as large as 600. Prandtl numbers for various liquids are given in Appendix 18.

The eddy diffusivities for momentum and heat, ε_M and ε_H, respectively, are not properties of the fluid but depend on the conditions of flow, especially in all factors that affect turbulence. For simple analogies, it is usual to assume that ε_M and ε_H are both constant and equal, but when determined by actual velocity and temperature measurements, both are found to be functions of the Reynolds number, the Prandtl number, and position in the tube cross section. Precise measurement of the eddy diffusivities is difficult, and not all reported measurements agree. Results are given in standard treatises.[6c] The ratio $\varepsilon_H/\varepsilon_M$ also varies but is more nearly constant than the individual quantities. The ratio is denoted by ψ. For ordinary liquids, where $N_{Pr} > 0.6$, ψ is close to 1 at the tube wall and in boundary layers generally and approaches 2 in turbulent wakes. For liquid metals ψ approaches zero at the wall, passes through a maximum of about unity at $y/r_w \approx 0.2$, and decreases toward the center of the pipe.[7c].

LINEAR DISTRIBUTION OF HEAT FLUX   It is shown on page 85 that the distribution of shear stress in a stream of fluid in a circular tube is linear with radius for both laminar and turbulent flow. The equation is

$$\frac{\tau}{\tau_w} = \frac{r}{r_w} = \frac{r_w - y}{r_w} = 1 - \frac{y}{r_w} \tag{12-52}$$

An analogous, but approximate, equation holds for the distribution of heat flux across the tube, and

$$\frac{q/A}{(q/A)_w} = \frac{r}{r_w} = \frac{r_w - y}{r_w} = 1 - \frac{y}{r_w} \tag{12-53}$$

Equations (12-52) and (12-53) are useful in the development of analogy equations.

THE REYNOLDS ANALOGY   The simplest and oldest analogy equation is that of Reynolds, which is derived for flow at high Reynolds numbers in straight round tubes. It can be obtained in several ways, one being from the equations for the eddy diffusivities. Dividing Eq. (12-51) by Eq. (12-50) gives

$$\frac{\tau g_c}{q/A} = -\frac{v + \varepsilon_M}{c_p(\alpha + \varepsilon_H)} \frac{du}{dT} \tag{12-54}$$

The basic assumption of the Reynolds analogy is that the ratio of the two molecular diffusivities equals that of the two eddy diffusivities, or

$$\frac{\nu}{\alpha} \equiv N_{Pr} = \frac{\varepsilon_M}{\varepsilon_H}$$

so that

$$\nu = N_{Pr}\alpha \quad \text{and} \quad \varepsilon_M = N_{Pr}\varepsilon_H$$

Eliminating ν and ε_M from Eq. (12-54) gives

$$\frac{\tau g_c}{q/A} = -\frac{N_{Pr}}{c_p}\frac{du}{dT} \tag{12-55}$$

Dividing Eq. (12-52) by Eq. (12-53) and eliminating $\tau g_c/(q/A)$ from Eq. (12-55) gives

$$\frac{\tau g_c}{q/A} = \frac{\tau_w g_c}{(q/A)_w} = -\frac{N_{Pr}}{c_p}\frac{du}{dT} \tag{12-56}$$

Equation (12-56) can be integrated over a range from the tube wall, where $u = 0$ and $T = T_w$, to the point in the stream where $u = \bar{V}$. At this point the temperature may be taken, without serious error, as being equal to the average bulk temperature T, used to define the heat-transfer coefficient in Eq. (11-17). The integration is

$$\int_0^{\bar{V}} du = -\frac{c_p \tau_w g_c}{N_{Pr}(q/A)_w}\int_{T_w}^T dT$$

$$\bar{V} = \frac{c_p \tau_w g_c}{N_{Pr}(q/A)_w}(T_w - T) \tag{12-57}$$

From Eq. (5-6)

$$\tau_w g_c = \frac{f \rho \bar{V}^2}{2} \tag{12-58}$$

and from Eq. (11-17), noting that for fully developed gradients $h_x = h$,

$$\left(\frac{q}{A}\right)_w = h(T_w - T) \tag{12-59}$$

Eliminating $\tau_w g_c$ and $(q/A)_w$ from Eq. (12-57) by Eqs. (12-58) and (12-59), respectively, and rearranging, gives

$$h = \frac{f}{2}\frac{\bar{V}\rho c_p}{N_{Pr}} = \frac{f}{2}\frac{c_p G}{N_{Pr}} \tag{12-60}$$

This may be written in the form

$$\frac{h}{c_p G}N_{Pr} = \frac{f}{2} \tag{12-61}$$

For the special case where $N_{Pr} = 1$ and $\varepsilon_M = \varepsilon_H$,

$$\frac{h}{c_p G} \equiv N_{St} = \frac{f}{2} \tag{12-62}$$

This is the usual form of the Reynolds analogy equation. It agrees well with experimental data for diatomic gases, which have Prandtl numbers of about unity, provided the temperature drop $T_w - T$ is not great.

THE COLBURN ANALOGY; COLBURN j FACTOR   Over a range of Reynolds numbers from 5,000 to 200,000 the friction factor for smooth pipes is adequately given by the empirical equation

$$f = 0.046 \left(\frac{DG}{\mu}\right)^{-0.2} \tag{12-63}$$

Comparison of Eq. (12-63) with Eq. (12-37) for heat transfer in turbulent flow inside long tubes shows that

$$\frac{h}{c_p G} N_{Pr}^{2/3} \left(\frac{\mu_w}{\mu}\right)^{0.14} = j_H = \frac{f}{2} \tag{12-64}$$

Equation (12-64) is a statement of the Colburn analogy between heat transfer and fluid friction. The factor j_H, defined as $(h/c_p G)(c_p \mu/k)^{2/3}(\mu_w/\mu)^{0.14}$, is called the *Colburn j factor*. It is used in a number of other semiempirical equations for heat transfer. While the Reynolds analogy [Eq. (12-62)] applies only to fluids for which the Prandtl number is close to unity, the Colburn analogy [Eq. (12-64)] applies over a range of Prandtl numbers from 0.6 to 120.

Equation (12-37) can be written in j-factor form, for long tubes, as follows:

$$j_H = 0.023 N_{Re}^{-0.2} \tag{12-65}$$

MORE ACCURATE ANALOGY EQUATIONS   A number of more elaborate analogy equations connecting friction and heat transfer in pipes, along flat plates, and in annular spaces have been published. They cover wider ranges of Reynolds and Prandtl numbers than Eq. (12-64) and are of the general form

$$N_{St} = \frac{f/2}{\Phi(N_{Pr})} \tag{12-66}$$

where $\Phi(N_{Pr})$ is a complicated function of the Prandtl number. One example, by Friend and Metzner,[3] applying to fully developed flow in smooth pipe, is

$$N_{St} = \frac{f/2}{1.20 + \sqrt{f/2}\,(N_{Pr} - 1)(N_{Pr})^{-1/3}} \tag{12-67}$$

The friction factor f used in this equation may be that given by Eq. (12-63) or, for a wider range of Reynolds numbers from 3,000 to 3,000,000, by the equation

$$f = 0.00140 + \frac{0.125}{N_{Re}^{0.32}} \tag{12-68}$$

Equation (12-67) applies over a range of Prandtl numbers from 0.46 to 590.

All analogy equations connecting f and h have an important limitation. They apply only to wall, or skin, friction, and must not be used for situations where form drag appears.

Heat transfer in transition region between laminar and turbulent flow
Equation (12-37) applies only for Reynolds numbers greater than 10,000 and Eq. (12-27) only for Reynolds numbers less than 2,100. The range of Reynolds numbers between 2,100 and 10,000 is called the *transition region*, and no simple equation applies here. A graphical method therefore is used. The method is based on graphs of Eqs. (12-27) and (12-37) on a common plot of the Colburn j factor vs. N_{Re}, with lines of constant values of L/D. To obtain an equation for the laminar-flow range it is necessary to transform Eq. (12-27) in the following manner. For the Graetz number is substituted, using Eqs. (12-12) and (12-13), the quantity $(\pi D/4L)N_{Re}N_{Pr}$. The result is

$$\frac{h_{ia}D}{k} = 2\left(\frac{\pi D}{4L}N_{Re}N_{Pr}\right)^{1/3}\left(\frac{\mu}{\mu_w}\right)^{0.14}$$

This relation is multiplied by $(1/N_{Re})(1/N_{Pr})$ to give the j factor. The final equation can be written

$$\frac{h_{ia}}{c_p G}\left(\frac{c_p\mu}{k}\right)^{2/3}\left(\frac{\mu_w}{\mu}\right)^{0.14} \equiv j_H = 1.86\left(\frac{D}{L}\right)^{1/3}\left(\frac{DG}{\mu}\right)_b^{-2/3} \tag{12-69}$$

Equation (12-69) shows that for each value of the length-diameter ratio L/D, a logarithmic plot of the left-hand side vs. N_{Re} gives a straight line with a slope of $-\frac{2}{3}$. The straight lines on the left-hand portion of Fig. 12-3 are plots of this equation for various values of L/D. The lines all terminate at a Reynolds number of 2,100.

Equation (12-37), when plotted for long tubes on the same coordinates, gives a straight line with a slope of -0.20 for Reynolds numbers above 10,000. This line is drawn in the right-hand region of Fig. 12-3.

The curved lines between Reynolds numbers of 2,100 and 10,000 represent the transition region. The effect of L/D is pronounced at the lower Reynolds numbers in this region and fades out as a Reynolds number of 10,000 is approached.

Figure 12.3 is a summary chart which can be used for the entire range of Reynolds numbers from 1,000 to 30,000. Beyond its lower and upper limits, Eqs. (12-27) and (12-37), respectively, can be used.

In using Fig. 12.3 it must be understood that for Reynolds numbers below 2,100 the coefficient h_{ia} is the average value of the coefficient over the entire length of the pipe based on the arithmetic mean temperature drop. For values of N_{Re} above 2,100, the coefficient h_i is either the local value of h_i or the average value of h_i over long tubes, depending on the value of L/D.

EXAMPLE 12-3   A light motor oil with the characteristics given below and in Table 12-3 is to be heated from 150 to 250°F (65.5 to 121.1°C) in a $\frac{1}{4}$-in. (6.35-mm) Schedule 40 pipe 15 ft (4.57 m) long. The pipe wall is at 350°F (176.7°C). How much oil can be heated in this pipe, in pounds per hour? What coefficient can be expected? The properties of the oil are

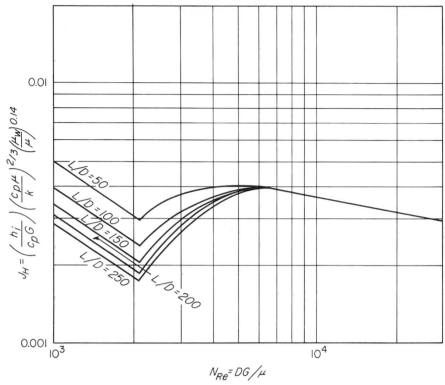

FIGURE 12-3
Heat transfer in transition range. [*By permission of author and publishers, from W. H. McAdams, "Heat Transmission," 3d ed. Copyright by author, 1954, McGraw-Hill Book Company.*]

as follows. The thermal conductivity in 0.082 Btu/ft-h-°F (0.142 W/m-°C). The specific heat is 0.48 Btu/lb-°F (2.01 J/g-°C).

SOLUTION   Assume the flow is laminar. Equation (12-27) can be used to relate the mass flow rate $\dot{m}$ and the coefficient h_{ia}, and Eq. (12-20) is available for a second relationship between these quantities. Simultaneous solution of the two equations gives values of $\dot{m}$ and h_{ia}.

Table 12-3   DATA FOR EXAMPLE 12-3

Temperature		Viscose, cP
°F	°C	
150	65.5	6.0
250	121.1	3.3
350	176.7	1.37

Data for substitution into Eq. (12-27) are

$$\mu = \frac{6.0 + 3.3}{2} = 4.65 \text{ cP} \qquad \mu_w = 1.37 \text{ cP} \qquad D = \frac{0.364}{12} = 0.0303 \text{ ft (Appendix 6)}$$

$$\phi_v = \left(\frac{\mu}{\mu_w}\right)^{0.14} = \left(\frac{4.65}{1.37}\right)^{0.14} = 1.187 \qquad k = 0.082 \qquad c_p = 0.48$$

From Eq. (12-27)

$$\frac{0.0303 h_{ia}}{0.082} = 2 \times 1.187 \left(\frac{0.48 \dot{m}}{0.082 \times 15}\right)^{1/3}$$

From this, $h_{ia} = 4.69 \dot{m}^{1/3}$.

Data for substitution into Eq. (12-20) are

$$\overline{\Delta T_a} = \frac{(350 - 150) + (350 - 250)}{2} = 150°F$$

$$L = 15 \qquad D = 0.0303 \qquad \overline{T_b} - T_a = 250 - 150 = 100°F$$

From Eq. (12-20)

$$h_{ia} = \frac{0.48 \times 100\dot{m}}{\pi 0.0303 \times 15 \times 150} = 0.224\dot{m}$$

Then

$$4.69\dot{m}^{1/3} = 0.224\dot{m}$$

$$\dot{m} = \left(\frac{4.69}{0.224}\right)^{3/2} = 96.0 \text{ lb/h (43.5 kg/h)}$$

and

$$h_{ia} = 0.224 \times 96.0 = 21.5 \text{ Btu/ft}^2\text{-h-°F (12.2 W/m}^2\text{-°C)}$$

To check the assumption of laminar flow, the maximum Reynolds number, which exists at the outlet end of the pipe, should be calculated.

$$N_{Re} = \frac{4 \times 96.0}{\pi 2.42 \times 3.3 \times 0.0303} = 505$$

This is well within the laminar range. ////

Heat transfer to liquid metals   Liquid metals are used for high-temperature heat transfer, especially in nuclear reactors. Liquid mercury, sodium, and a mixture of sodium and potassium called NaK are commonly used as carriers of sensible heat. Mercury vapor is also used as a carrier of latent heat. Temperatures of 1500°F and above are obtainable by using such metals. Molten metals have good specific heats, low viscosities, and high thermal conductivities. Their Prandtl numbers are therefore very low in comparison with those of ordinary fluids.

Equations such as (12-37), (12-39), and (12-67) do not apply at Prandtl numbers below about 0.6, because the mechanism of heat flow in a turbulent stream differs from that in fluids of ordinary Prandtl numbers. In the usual fluid, heat transfer by conduction is limited to the viscous sublayer when N_{Pr} is unity or more and occurs in the buffer zone only when the number is less than unity. In liquid metals, heat

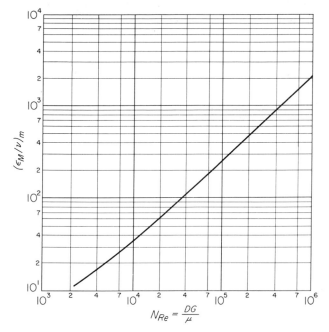

FIGURE 12-4
Values of $(\varepsilon_M/\nu)_m$ for fully developed turbulent flow of liquid metals in circular tubes.

transfer by conduction is important throughout the entire turbulent core and may predominate over convection throughout the tube.

Much study has been given to liquid-metal heat transfer in recent years, primarily in connection with its use in atomic reactors. Design equations, all based on heat-momentum analogies, are available for flow in tubes, in annuli, between plates, and outside bundles of tubes. The equations so obtained are of the form

$$N_{Nu} = \alpha + \beta(\bar{\psi}N_{Pe})^{\gamma} \tag{12-70}$$

where α, β, and γ are constants or functions of geometry and of whether the wall temperature or the flux is constant, and $\bar{\psi}$ is the average value of $\varepsilon_H/\varepsilon_M$ across the stream. For circular pipes, $\alpha = 7.0$, $\beta = 0.025$, and $\gamma = 0.8$. For other shapes, more elaborate functions are needed. A correlation for $\bar{\psi}$ is given by the equation[1]

$$\bar{\psi} = 1 - \frac{1.82}{N_{Pr}(\varepsilon_M/\nu)_m^{1.4}} \tag{12-71}$$

The quantity $(\varepsilon_M/\nu)_m$ is the maximum value of this ratio in the pipe, which is reached at a value of $y/r_w = \frac{5}{9}$. Equation (12-70) becomes, then,

$$N_{Nu} = 7.0 + 0.025 \left[N_{Pe} - \frac{1.82 N_{Re}}{(\varepsilon_M/\nu)_m^{1.4}} \right]^{0.8} \tag{12-72}$$

A correlation for $(\varepsilon_M/\nu)_m$ as a function of the Reynolds number is given in Fig. 12-4.

THE CRITICAL PECLET NUMBER   For a given Prandtl number, the Peclet number is proportional to the Reynolds number, because $N_{Pe} = N_{Pr}N_{Re}$. At a definite value of N_{Pe} the bracketed term in Eq. (12-72) becomes zero. This situation corresponds to the point where conduction controls and the eddy diffusion no longer affects the heat transfer. Below the critical Peclet number, only the first term in Eq. (12-72) is needed, and $N_{Nu} = 7.0$.

For laminar flow at uniform heat flux, by mathematical analysis $N_{Nu} = 48/11 = 4.37$. This has been confirmed by experiment.

HEATING AND COOLING OF FLUIDS IN FORCED CONVECTION OUTSIDE TUBES

The mechanism of heat flow in forced convection outside tubes differs from that of flow inside tubes, because of differences in fluid-flow mechanism. As has been shown on pages 46 and 109, no form drag exists inside tubes except perhaps for a short distance at the entrance end, and all friction is wall friction. Because of the lack of form friction, there is no variation in the local heat transfer at different points in a given circumference, and a close analogy exists between friction and heat transfer. An increase in heat transfer is obtainable at the expense of added friction simply by increasing the fluid velocity. Also, a sharp distinction exists between laminar and turbulent flow, which calls for different treatment of heat-transfer relations for the two flow regimes.

On the other hand, as has been shown on pages 139 to 146, in flow of fluids across a cylindrical shape boundary-layer separation occurs, and a wake develops that causes form friction. No sharp distinction is found between laminar and turbulent flow, and a common correlation can be used for both low and high Reynolds numbers. Also, the local value of the heat flux varies from point to point around a circumference. In Fig. 12-5 the local value of the heat flux is plotted radially for all points around the circumference of the tube. The flux is a maximum at the front and back of the tube and a minimum at the sides. The ratio of the maximum to the minimum is 2.5. In practice, the variations in flux and in local coefficient h_x are often of no importance, and average values based on the entire circumference are used.

Radiation may be important in heat transfer to outside tube surfaces. Inside tubes, the surface cannot see surfaces other than the inside wall of the same tube, and heat flow by radiation does not occur. Outside tube surfaces, however, are necessarily in sight of external surfaces, if not nearby, at least at a distance, and the surrounding surfaces may be appreciably hotter or cooler than the tube wall. Heat flow by radiation, especially when the fluid is a gas, is appreciable in comparison with heat flow by conduction and convection. The total heat flow is then a sum of two independent flows, one by radiation and the other by conduction and convection. The relations given in the remainder of this section have to do with conduction and convection only. Radiation, as such and in combination with conduction and convection, is discussed in Chap. 14.

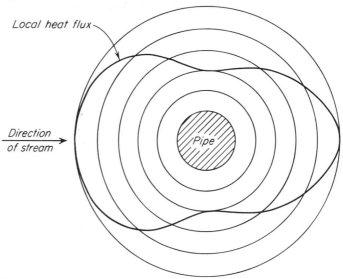

FIGURE 12-5
Heat flow around cylinder, air flow normal to axis. [*By permission of author and publishers, from W. H. McAdams, "Heat Transmission," 3d ed. Copyright by author, 1954, McGraw-Hill Book Company.*]

Fluids flowing normally to a single tube The variables affecting the coefficient of heat transfer to a fluid in forced convection outside a tube, by the same reasoning applied in studying the case of heat flow to fluids inside tubes, are D_o, the outside diameter of the tube; c_p, μ, and k, the specific heat at constant pressure, the viscosity, and the thermal conductivity, respectively, of the fluid; and G, the mass velocity of the fluid approaching the tube. Dimensional analysis gives, then, an equation of the type of Eq. (12-30).

$$\frac{h_o D_o}{k} = \psi_0 \left(\frac{D_o G}{\mu}, \frac{c_p \mu}{k} \right) \tag{12-73}$$

Here, however, ends the similarity between the two types of process—the flow of heat to fluids inside tubes and the flow of heat to fluids outside tubes—and the functional relationships in the two cases differ.

For air, where the Prandtl number is nearly independent of temperature, Eq. (12-73) becomes simply

$$\frac{h_o D_o}{k_f} = \psi_0 \left(\frac{D_o G}{\mu_f} \right) \tag{12-74}$$

Experimental data are correlated by the curve of Fig. 12-6. The effect of radiation is not included in this curve, and radiation must be calculated separately.

The subscript f on the terms k_f and μ_f indicates that in using Fig. 12-6 these terms must be evaluated at the average film temperature T_f midway between the wall

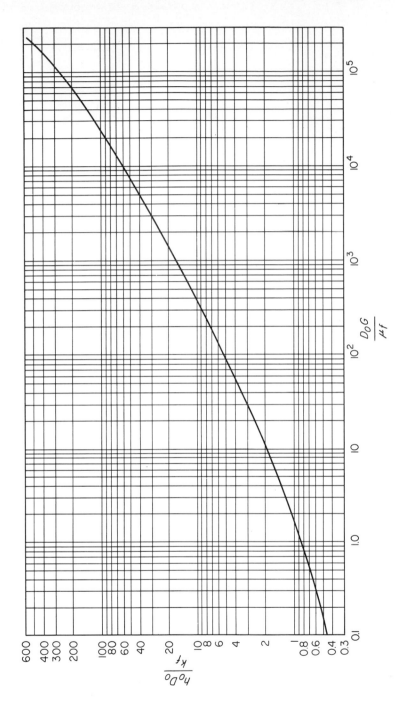

FIGURE 12-6

Heat transfer to air flowing normally to a single tube. *[By permission of author and publishers, from W. H. McAdams, "Heat Transmission," 3d ed. Copyright by author, 1954, McGraw-Hill Book Company.]*

temperature and the mean bulk temperature of the fluid. T_f is therefore given by the equation

$$T_f = \frac{T_w + \overline{T}}{2} \tag{12-75}$$

Figure 12-6 can be used for both heating and cooling.

For heating and cooling liquids flowing normally to single cylinders the following equation is used.[4a]

$$\frac{h_o D_o}{k_f} \left(\frac{c_p \mu_f}{k_f}\right)^{-0.3} = 0.35 + 0.56 \left(\frac{D_o G}{\mu_f}\right)^{0.52} \tag{12-76}$$

Equation (12-76) is plotted in j-factor form in Fig. 22-4, in the section of Chap. 22 dealing with the analogies between heat and mass transfer.

Heat-transfer data for flow normal to cylinders of noncircular cross section are given in the literature.[4b] Banks of tubes across which the fluid flows are common in industrial exchangers. Problems of heat flow in tube banks are discussed in Chap. 15.

Flow past single spheres For heat transfer between a flowing fluid and the surface of a single sphere the following equation is recommended:

$$\frac{h_o D_p}{k_f} = 2.0 + 0.60 \left(\frac{D_p G}{\mu_f}\right)^{0.50} \left(\frac{c_p \mu_f}{k_f}\right)^{1/3} \tag{12-77}$$

where D_p is the diameter of the sphere. In a completely stagnant stream the Nusselt number, $h_o D_p / k_f$, is equal to 2.0. A plot of Eq. (12-77) is given in Fig. 22-5.

Heat transfer in packed beds Data for heat transfer (and mass transfer) between fluids and beds of various kinds of particles are summarized in Fig. 22-6. In this plot a modified j_H factor is plotted against a special Reynolds number based on the square root of the total surface area of an individual particle in the bed.

NATURAL CONVECTION

As an example of natural convection, consider a hot, vertical plate in contact with the air in a room. The temperature of the air in contact with the plate will be that of the surface of the plate, and a temperature gradient will exist from the plate out into the room. At the bottom of the plate, the temperature gradient is steep, as shown by the full line marked "$Z = 10$ mm" in Fig. 12-7. At distances above the bottom of the plate, the gradient becomes less steep, as shown by the full curve marked "$Z = 240$ mm" of Fig. 12-7. At a height of about 600 mm from the bottom of the plate, the temperature-distance curves approach an asymptotic condition and do not change with further increase in height.

The density of the heated air immediately adjacent to the plate is less than that of the unheated air at a distance from the plate, and the buoyancy of the hot air causes an unbalance between the vertical layers of air of differing density. As a result

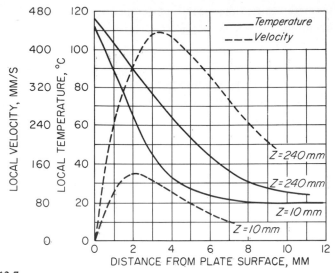

FIGURE 12-7
Velocity and temperature gradients, natural convection from heated vertical plate. [*By permission of author and publishers, from W. H. McAdams, "Heat Transmission," 3d ed. Copyright by author, 1954, McGraw-Hill Book Company.*]

of the unbalanced forces, a circulation is generated, by which hot air near the plate rises and cold air flows toward the plate from the room to replenish the rising airstream. A velocity gradient near the plate is formed. Since the velocities of the air in contact with the plate and that out in the room are both zero, the velocity is a maximum at a definite distance from the wall. The velocity reaches its maximum a few millimeters from the surface of the plate. The dotted curves in Fig. 12-7 show the velocity gradients for heights of 10 and 240 mm above the bottom of the plate. For tall plates, an asymptotic condition is approached.

The temperature difference between the surface of the plate and the air in the room at a distance from the plate causes a transfer of heat by conduction into the current of gas next to the wall, and the stream carries the heat away by convection in a direction parallel to the plate.

The natural convection currents surrounding a hot, horizontal pipe are more complicated than those adjacent to a vertical heated plate, but the mechanism of the process is similar. The layers of air immediately next to the bottom and sides of the pipe are heated, and tend to rise. The rising layers of hot air, one on each side of the pipe, separate from the pipe at points short of the top center of the pipe and form two independent rising currents with a zone of relatively stagnant and unheated air between them.

Natural convection in liquids follows the same pattern, because liquids are also less dense hot than cold. The buoyancy of heated liquid layers near a hot surface generates convection currents just as in gases.

On the assumption that h depends upon pipe diameter, specific heat, thermal conductivity, viscosity, coefficient of thermal expansion, the acceleration of gravity, and temperature difference, dimensional analysis gives

$$\frac{hD_o}{k} = \Phi\left(\frac{c_p\mu}{k}, \frac{D_o^3\rho^2 g}{\mu^2}, \beta\,\Delta T\right) \tag{12-78}$$

Since the effect of β is through buoyancy in a gravitational field, the product $g\beta\,\Delta T$ acts as a single factor, and the last two groups fuse into a dimensionless group called the *Grashof number* N_{Gr}.

For single horizontal cylinders, the heat-transfer coefficient can be correlated by an equation containing three dimensionless groups, the Nusselt number, the Prandtl number, and the Grashof number, or, specifically,

$$\frac{hD_o}{k_f} = \Phi\left(\frac{c_p\mu_f}{k_f}, \frac{D_o^3\rho_f^2\beta g\,\Delta T_o}{u_f^2}\right) \tag{12-79}$$

where h = average heat-transfer coefficient, based on entire pipe surface
   D_o = outside pipe diameter
   k_f = thermal conductivity of fluid
   c_p = specific heat of fluid at constant pressure
   ρ_f = density of fluid
   β = coefficient of thermal expansion of fluid
   g = acceleration of gravity
   ΔT_o = average difference in temperature between outside of pipe and fluid distant from wall

The fluid properties μ_f, ρ_f, and k_f are evaluated at the mean film temperature [Eq. (12-75)]. Radiation is not accounted for in this equation.

The coefficient of thermal expansion β is a property of the fluid, defined as the fractional increase in volume, at constant pressure, of the fluid, per degree of temperature change, or mathematically,

$$\beta = \frac{(\partial v/\partial T)_p}{v} \tag{12-80}$$

where v = specific volume of fluid
   $(\partial v/\partial T)_p$ = rate of change of specific volume with temperature, at constant pressure

For liquids, β can be considered constant over a definite temperature range and Eq. (12-80) written as

$$\beta = \frac{\Delta v/\Delta T}{\bar{v}}$$

where $\bar{v}$ is the average specific volume. In terms of density,

$$\beta = \frac{1/\rho_2 - 1/\rho_1}{(T_2 - T_1)(1/\rho_1 + 1/\rho_2)/2} = \frac{\rho_1 - \rho_2}{\bar{\rho}_a(T_2 - T_1)} \tag{12-81}$$

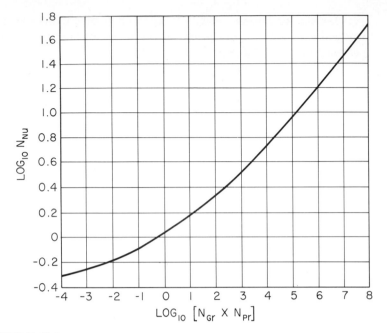

FIGURE 12-8
Heat transfer between single horizontal cylinders and fluids in natural convection.

where $\bar{\rho}_a = (\rho_1 + \rho_2)/2$
ρ_1 = density of fluid at temperature T_1
ρ_2 = density of fluid at temperature T_2

For an ideal gas, since $v = RT/p$,

$$\left(\frac{\partial v}{\partial T}\right)_p = \frac{R}{p}$$

and, using Eq. (12-80),

$$\beta = \frac{R/p}{RT/p} = \frac{1}{T} \tag{12-82}$$

The coefficient of thermal expansion of an ideal gas equals the reciprocal of the absolute temperature.

In Fig. 12-8 is shown a relationship, based on Eq. (12-79), which satisfactorily correlates experimental data for heat transfer from a single horizontal cylinder to liquids or gases. The range of variables covered by the single line of Fig. 12-8 is very great.

For magnitudes of log $N_{Gr}N_{Pr}$ of 4 or more, the line of Fig. 12-8 follows closely the empirical equation[7b]

$$N_{Nu} = 0.53(N_{Gr}N_{Pr})_f^{0.25} \tag{12-83}$$

Natural convection to air from vertical shapes and horizontal planes

Equations for heat transfer in natural convection between fluids and solids of definite geometric shape are of the form

$$\frac{hL}{k_f} = b \left[\frac{L^3 \rho_f^2 g \beta_f \, \Delta T}{\mu_f^2} \left(\frac{c_p \mu}{k} \right)_f \right]^n \tag{12-84}$$

where b, n = const

L = height of vertical surface or length of horizontal square surface, ft[7a]

Properties are taken at the mean film temperature. Equation (12-84) can be written

$$N_{\text{Nu},f} = b(N_{\text{Gr}} N_{\text{Pr}})_f^n \tag{12-85}$$

Values of the constants b and n for various conditions are given in Table 12-4.

Effect of natural convection in laminar-flow heat transfer

In laminar flow at low velocities, in large pipes, and at large temperature drops, natural convection may occur to such an extent that the usual equations for laminar-flow heat transfer must be modified. The effect of natural convection in tubes is found almost entirely in laminar flow, as the higher velocities characteristic of flow in the transition and turbulent regimes overcome the relatively gentle currents of natural convection.

The effect of natural convection on the heat-transfer coefficient of heat transfer to fluids in laminar flow through horizontal tubes can be accounted for by multiplying the coefficient h_{ia}, computed from Eq. (12-69) or Fig. 12-3, by the factor[5]

$$\phi_n = \frac{2.25(1 + 0.010 N_{\text{Gr}}^{1/3})}{\log N_{\text{Re}}} \tag{12-86}$$

Natural convection also occurs in vertical tubes, increasing the rate of heat flow, when the fluid flow is upward, to above that found in laminar flow only. The effect is marked at values of N_{Gr} between 10 and 10,000 and depends on the magnitude of the quantity $N_{\text{Gr}} N_{\text{Pr}} D/L$.[7f]

Table 12-4   VALUES OF CONSTANTS IN EQ. (12-85)†

System	Range of $N_{Gr}N_{Pr}$	b	n
Vertical plates, vertical cylinders	10^4–10^9	0.59	0.25
	10^9–10^{12}	0.13	0.333
Horizontal plates:			
Heated, facing upward or cooled, facing down	10^5–2×10^7	0.54	0.25
	2×10^7–3×10^{10}	0.14	0.333
Cooled, facing upward or heated, facing down	3×10^5–3×10^{10}	0.27	0.25

† By permission of author and publishers, from W. H. McAdams, "Heat Transmission," 3d ed., pp.172, 180. Copyright by author, 1954, McGraw-Hill Book Company.

EXAMPLE 12-4   Air at 1 atm pressure is passed through a horizontal 2-in. (51-mm) Schedule 40 steam-jacketed steel pipe at a velocity of 1.5 ft/s (0.457 m/s) and an inlet temperature of 68°F (20°C). The pipe-wall temperature is 220°F (104.4°C). If the outlet air temperature is to be 188°F (86.7°C), how long must the heated section be?

SOLUTION   To establish the flow regime, the Reynolds number based on the average temperature is calculated. The quantities needed are

$$\bar{T} = \frac{68 + 188}{2} = 128° \qquad D = \frac{2.067}{12} = 0.1723 \text{ ft} \qquad \text{(Appendix 6)}$$

$$\mu \text{ (at } 128°\text{F)} = 0.019 \text{ cP} \qquad \text{(Appendix 9)}$$

$$\rho \text{ (at } 68° \text{)} = \frac{29}{359} \frac{492}{68 + 460} = 0.0753 \text{ lb/ft}^3$$

$$V\rho = G = 1.5 \times 0.0753 \times 3,600 = 406.4 \text{ lb/ft}^2\text{-h}$$

$$N_{Re} = \frac{DG}{\mu} = \frac{0.1723 \times 406.4}{0.019 \times 2.42} = 1,522$$

Hence flow is laminar, and Eq. (12-27) applies. The result may later require correction for the effect of natural convection, using Eq. (12-86). To use Eq. (12-27) the following quantities are needed:

$$c_p \text{ (at } 128°\text{F)} = 0.25 \text{ Btu/lb-}°\text{F} \qquad \text{(Appendix 15)}$$

$$k \text{ (at } 128°\text{F)} = 0.0163 \text{ Btu/ft-h-}°\text{F} \qquad \text{(Appendix 12)}$$

$$\text{(By linear interpolation)}$$

$$\mu_w \text{ (at } 220°\text{F)} = 0.021 \text{ cP} \qquad \text{(Appendix 9)}$$

The internal cross-sectional area of pipe is

$$S = 0.02330 \text{ ft}^2 \qquad \text{(Appendix 6)}$$

The mass flow rate is

$$\dot{m} = GS = 406.4 \times 0.02330 = 9.47 \text{ lb/h}$$

The heat load, from Eq. (12-18), is

$$q = \dot{m}c_p(\bar{T}_b - T_a) = 9.47 \times 0.25(188 - 68) = 284.1 \text{ Btu/h}$$

The average temperature difference is

$$\Delta T_1 = 220 - 188 = 32°\text{F} \qquad \Delta T_2 = 220 - 68 = 152°\text{F}$$

$$\Delta T_i = \Delta T_a = \frac{32 + 152}{2} = 92°\text{F}$$

From the definition of the heat-transfer coefficient, Eq. (12-17), $h = q/A\Delta T_i$. From Eq. (12-19) and Appendix 6, for 2-in. Schedule 40 pipe

$$A = 0.541L$$

Hence

$$h = \frac{284.1}{0.541L \times 92} = \frac{5.708}{L}$$

Also, from Eq. (12-27), the heat-transfer coefficient is

$$
h = \frac{2k}{D} \left(\frac{\dot{m}c_p}{kL}\right)^{1/3} \left(\frac{\mu}{\mu_w}\right)^{0.14}
$$

$$
= \frac{2 \times 0.0163}{0.1723} \left(\frac{9.47 \times 0.25}{0.0163L}\right)^{1/3} \left(\frac{0.019}{0.021}\right)^{0.14} = \frac{0.9813}{L^{1/3}}
$$

Equating the two relationships for h gives

$$
\frac{0.9813}{L^{1/3}} = \frac{5.708}{L}
$$

from which $L = 14.03$ ft (4.276 m).

This result is now corrected for the effect of natural convection, using Eq. (12-86). This requires calculation of the Grashof number, for which the additional quantities needed are

$$
\beta \text{ (at } 128°F) = \frac{1}{460 + 128} = 0.0017°R^{-1}
$$

$$
\Delta T = 220 - 128 = 92°F \qquad \rho \text{ (at } 128°F) = 0.0676 \text{ lb/ft}^3
$$

The Grashof number is therefore

$$
N_{Gr} = \frac{D^3 \rho^2 g \beta \, \Delta T}{\mu^2}
$$

$$
= \frac{0.1723^3 \times 0.0676^2 \times 32.174 \times 0.0017 \times 92}{(0.019 \times 6.72 \times 10^{-4})^2} = 0.7192 \times 10^6
$$

Hence, from Eq. (12-86),

$$
\phi_n = \frac{2.25\,[1 + 0.01(0.7192 \times 10^6)^{1/3}]}{\log 1,522} = 1.34
$$

This factor is used to correct the value of h obtained from Eq. (12-27), namely, $0.9813/L^{1/3}$. Hence

$$
1.34 \times \frac{0.9813}{L^{1/3}} = \frac{5.708}{L}
$$

from which $L = 9.04$ ft (2.76 m).   ////

SYMBOLS

A	Area, ft² or m²; A_T, total area of heat exchanger
b	Constant in Eq. (12-85)
C	Constant; C_1, constant in Eq. (12-34)
c_p	Specific heat at constant pressure, Btu/lb-°F or J/g-°C
D	Diameter, ft or m; D_e, equivalent diameter, $4r_H$; D_i, inside diameter; D_{ij}, inside diameter of jacket; D_{it}, inside diameter of inner tube; $D/$, outside diameter; D_{ot}, outside diameter of inner tube; D_p, of spherical particle; D_L, logarithmic mean

f	Fanning friction factor, dimensionless
G	Mass velocity, lb/ft²-s or kg/m²-s
g	Gravitational acceleration, ft/s² or m/s²
g_c	Newton's-law proportionality factor, 32.174 ft-lb$_f$-s²
h	Individual heat-transfer coefficient, Btu/ft²-h-°F or W/m²-°C; h_i, average over inside of tube; h_{ia}, based on arithmetic mean temperature drop; h_o, for outside of tube or particle; h_x, local value; h_{x1}, at trailing edge of plate; h_∞, for fully developed flow in long pipes
j_H	Colburn j factor, $N_{St}(N\,Pr)^{2/3}$, dimensionless
K'	Flow consistency index of non-Newtonian fluid
k	Thermal conductivity, Btu/ft-h-°F or W/m-°C; k_f, at mean film temperature; k_m, of tube wall
L	Length, ft or m
m	Parameter in Eq. (12-28), $K'8^{n'-1}$; m_w, value at T_w
$\dot{m}$	Mass flow rate, lb/h or kg/h
N_{Fo}	Fourier number, $4kL/c_p\rho D^2\bar{V}$, dimensionless
N_{Gr}	Grashof number, $D^3\rho^2 g\beta\,\Delta T/\mu^2$, dimensionless
N_{Gz}	Graetz number, $\dot{m}c_p/kL$, dimensionless
N_{Nu}	Nusselt number, hD/k, dimensionless; $N_{Nu,a}$, based on arithmetic mean temperature drop; $N_{Nu,f}$, at mean film temperature; $N_{Nu,x}$, local value on flat plate, $h_x x/k$; $N_{Nu,x1}$, at trailing edge of plate
N_{Pe}	Peclet number, $\rho\bar{V}c_p D/k$, dimensionless
N_{Pr}	Prandtl number, $c_p\mu/k$, dimensionless; $N_{Pr,f}$, at mean film temperature
N_{Re}	Reynolds number, DG/μ, dimensionless; $N_{Re,x}$, local value on flat plate, $u_0 x\rho/\mu$; $N_{Re,x1}$, at trailing edge of plate
N_{St}	Stanton number, $h/c_p G$, dimensionless
n	Constant in Eq. (12-85)
n'	Flow behavior index of non-Newtonian fluid, dimensionless
p	Pressure, lb$_f$/ft² or N/m²
q	Heat flow rate, Btu/h or W; q_c, by conduction; q_t, by turbulent convection; q_w, at pipe wall
R	Gas-law constant
r	Radius, ft or m; r_H, hydraulic radius of channel; r_w, radius of pipe
T	Temperature, °F or °C; T_a, at inlet; T_b, at outlet; T_f, mean film temperature; T_i, instantaneous value; T_w, at wall or plate; T_∞, of approaching fluid; $\bar{T}$, average fluid temperature in tube; $\bar{T}_b$, bulk average fluid temperature at outlet; $\bar{T}_i$, time average of instantaneous values; T', fluctuating component; $\bar{T}'$, time average of fluctuating component.
t_T	Total time of heating or cooling, s or h
U	Overall heat-transfer coefficient, Btu/ft²-h-°F or W/m²-°C; U_o, based on outside area; U_1, U_2, at ends of exchanger
u	Fluid velocity, ft/s or m/s; u_0, of approaching fluid; u', fluctuating component
$\bar{V}$	Volumetric average fluid velocity, ft/s or m/s
v	Specific volume, ft³/lb or m³/kg for liquids, ft³/lb mol or m³/kg mol for gases; v, average value
v'	Fluctuating component of velocity in y direction; $\bar{v}'$, time-average value
x	Distance from leading edge of plate or from tube entrance, ft or m; x_w, wall thickness; x_0, at start of heated section; x_1, length of plate
y	Radial distance from wall, ft or m
Z	Height, ft or m

Greek letters

α	Thermal diffusivity, $k/\rho c_p$, ft²/h or m²/h; also constant in Eq. (12-70)
β	Coefficient of volumetric expansion, 1/°R or 1/K; β_f, at mean film temperature T_f; also constant in Eq. (12-70)
γ	Constant in Eq. (12-70)
ΔT	Temperature drop, °F or °C; ΔT_i, from inner wall of pipe to fluid; ΔT_o, from outside surface

to fluid distant from wall; $\overline{\Delta T}_a$, arithmetic mean temperature drop; $\overline{\Delta T}_l$, average temperature drop

δ Parameter in Eq. (12-28), $(3n' + 1)/4n'$

ε Turbulent diffusivity, ft^2/h or m^2/h; ε_H, of heat; ε_M, of momentum

μ Absolute viscosity, lb/ft-s or kg/m-s; μ_f, average value of liquid film; μ_w, value at wall temperature

v Kinematic viscosity, ft^2/h or m^2/h

ρ Density, lb/ft^3 or kg/m^3; ρ_f, of liquid film; $\bar{\rho}_a$, arithmetic average value

τ Shear stress, lb_f/ft^2 or N/m^2; τ_w, shear stress at pipe wall; τ_0, yield stress of plastic fluid

Φ, Φ_1 Function

ϕ_n Natural-convection factor [Eq. (12-86)]

ϕ_v Viscosity-correction factor, $(\mu/\mu_w)^{0.14}$

χ Function in Eq. (12-24); χ', function in Eq. (12-16)

ψ Function in Eq. (12-33); also ratio

        Ratio of turbulent diffusivities, $\varepsilon_H/\varepsilon_M$; $\bar{\psi}$, average value

ψ_0 Function in Eq. (12-73)

PROBLEMS

12-1 Glycerin is flowing at the rate of 500 kg/h through a 25-mm-ID pipe. It enters a heated section 3 m long, the walls of which are at a uniform temperature of 115°C. The temperature of the glycerin at the entrance is 15°C. (*a*) If the velocity profile is parabolic, what would be the temperature of the glycerin at the outlet of the heated section? (*b*) What would the outlet temperature be if flow were rodlike? (*c*) How long would the heated section have to be to heat the glycerin essentially to 115°C?

12-2 Oil at 50°F is heated in a horizontal pipe 50 ft long having a surface temperature of 100°F. The pipe is Schedule 40 2-in. iron pipe. The oil flow rate is 100 gal/h at inlet temperature. What will be the oil temperature as it leaves the pipe and after mixing? What is the average heat-transfer coefficient? Properties of the oil are given in Table 12-5.

12-3 Oil is flowing through a 50-mm-ID iron pipe at 1 m/s. It is being heated by steam outside the pipe, and the steam-film coefficient may be taken as 11 kW/m²-°C. At a particular point along the pipe the oil is at 50°C, its density is 880 kg/m³, its viscosity is 2.1 cP, its thermal conductivity is 0.135 W/m-°C, and its specific heat is 2.17 J/g-°C. What is the overall heat-transfer coefficient at this point, based on the inside area of the pipe? If the steam temperature is 130°C, what is the heat flux at this point, based on the outside area of the pipe?

12-4 Kerosene is heated by hot water in a tube-and-shell heater. The kerosene is inside the tubes, and the water is outside. The flow is countercurrent. The average temper-

Table 12-5 DATA FOR PROB. 12-2

	50°F	100°F
Specific gravity, 60°F/60°F	0.80	0.75
Thermal conductivity, Btu/ft-h-°F	0.072	0.074
Viscosity, cP	20	10
Specific heat, Btu/lb-°F	0.75	0.75

ature of the kerosene is 110°F, and the average linear velocity is 5 ft/s. The properties of the kerosene at 110°F are specific gravity = 0.805, viscosity = 1.5 cP, specific heat = 0.583 Btu/lb-°F, and thermal conductivity = 0.0875 Btu/ft-h-°F. The tubes are low-carbon steel $\frac{3}{4}$ in. OD by 16 BWG. The heat-transfer coefficient on the shell side is 300 Btu/ft²-h-°F. Calculate the overall coefficient, based on the outside area of the tube.

12-5   Assume that the kerosene of Prob. 12-4 is replaced with water at 110°F and flowing at a velocity of 5 ft/s. What percentage increase in overall coefficient may be expected if the tube surfaces remain clean?

12-6   Both surfaces of the tube of Prob. 12-5 become fouled with deposits from the water. The fouling factors are 330 on the inside and 200 on the outside surfaces, both in Btu/ft²-h-°F. What percentage decrease in overall coefficient is caused by the fouling of the tube?

12-7   Water must be heated from 15 to 50°C in a simple double-pipe heat exchanger at a rate of 4 m³/h. The water is flowing inside the inner tube with steam condensing at 110°C on the outside. The tube wall is so thin that the wall resistance may be neglected. Assume that the steam-film coefficient h_0 is 11 kW/m²-°C. What is the length of the shortest heat exchanger that will heat the water to the desired temperature? Average properties of water:

$$\rho = 993 \text{ kg/m}^3 \qquad k = 0.61 \text{ W/m-°C} \qquad \mu = 0.78 \text{ cP} \qquad c_p = 4.19 \text{ J/g-°C}$$

Hint: Find the optimum diameter for the tube.

12-8   A sodium-potassium alloy (78 percent K) is to be circulated through $\frac{1}{2}$-in.-ID tubes in a reactor core for cooling. The liquid-metal inlet temperature and velocity are to be 600°F and 30 ft/s. If the tubes are 2 ft long and have an inside surface temperature of 700°F, find the coolant temperature rise and the energy gain per pound of liquid metal. Properties of NaK (78 percent K):

$$\rho = 45 \text{ lb/ft}^3 \qquad k = 179 \text{ Btu/ft-h-°F} \qquad \mu = 0.16 \text{ cP} \qquad c_p = 0.21 \text{ Btu/lb-°F}$$

12-9   Air is flowing through a steam-heated tubular heater under such conditions that the steam and wall resistances are negligible in comparison with the air-side resistance. Assuming that each of the following factors is changed in turn but that all other original factors remain constant, calculate the percentage variation in $q/\overline{\Delta t_L}$ that accompanies each change. (*a*) Double the pressure on the gas but keep fixed the mass flow rate of the air. (*b*) Double the mass flow rate of the air. (*c*) Double the number of tubes in the heater. (*d*) Halve the diameter of the tubes.

12-10   Water at 60°F is flowing at right angles across a heated 1-in.-OD cylinder, the surface temperature of which is 250°F. The approach velocity of the water is 3 ft/s. (*a*) What is the heat flux, in Btu/per hour per square foot, from the surface of the cylinder to the water? (*b*) What would be the flux if the cylinder were replaced by a 1-in.-OD sphere, also with a surface temperature of 250°F?

12-11   Water is heated from 15 to 65°C in a steam-heated horizontal 50-mm-ID tube. The steam temperature is 120°C. The average Reynolds number of the water is 450. The individual coefficient of the water is controlling. By what percentage would natural convection increase the total rate of heat transfer over that predicted for purely laminar flow? Compare your answer with the increase indicated in Example 12-4.

12-12   A large tank of water is heated by natural convection from submerged horizontal steam pipes. The pipes are Schedule 40 3-in. steel. When the steam pressure is atmospheric and the water temperature 80°F, what is the rate of heat transfer to the water, in Btu per hour per foot of pipe length?

REFERENCES

1 Dwyer, O. E.: *AIChE J.,* **9**:261 (1963).
2 Eckert, E. R. G., and J. F. Gross: "Introduction to Heat and Mass Transfer," pp. 110–114, McGraw-Hill, New York, 1963.
3 Friend, W. L., and A. B. Metzner: *AIChE J.,* **4**:393 (1958).
4 Gebhart, B.: "Heat Transfer," 2d ed., McGraw-Hill, New York, 1971; (*a*) p. 272, (*b*) p. 274, (*c*) p. 283.
5 Kern, D. Q., and D. F. Othmer: *Trans. AIChE,* **39**:517 (1943).
6 Knudsen, J. G., and D. L. Katz: "Fluid Dynamics and Heat Transfer," McGraw-Hill, New York, 1958; (*a*) pp. 361–390, (*b*) pp. 400–403, (*c*) p. 439.
7 McAdams, W. H.: "Heat Transmission," 3d ed., McGraw-Hill, New York, 1954; (*a*) pp. 172, 180, (*b*) p. 177, (*c*) p. 215, (*d*) p. 230, (*e*) p. 232, (*f*) p. 234.
8 Sieder, E. N., and G. E. Tate: *Ind. Eng. Chem.,* **28**:1429 (1936).
9 Wilkinson, W. L.: "Non-Newtonian Fluids," p. 104, Pergamon, London, 1960.

13

HEAT TRANSFER TO FLUIDS WITH PHASE CHANGE

Processes of heat transfer accompanied by phase change are more complex than simple heat exchange between fluids. A phase change involves the addition or subtraction of considerable quantities of thermal energy at constant or nearly constant temperature. The rate of phase change may be governed by the rate of heat transfer, but more often it is governed by the rate of nucleation of bubbles, drops, or crystals and by the behavior of the new phase after it is formed. This chapter covers condensation of vapors and boiling of liquids. Crystallization is discussed in Chap. 28.

HEAT TRANSFER FROM CONDENSING VAPORS

The condensation of vapors on the surfaces of tubes cooler than the condensing temperature of the vapor is important when vapors such as those of water, hydrocarbons, and other volatile substances are processed. Some examples will be met later in this text, in discussing the unit operations of evaporation, distillation, and drying.

The condensing vapor may consist of a single substance, a mixture of condensable and noncondensable substances, or a mixture of two or more condensable vapors. Friction losses in a condenser are normally small, so that condensation is essentially a constant-pressure process. The condensing temperature of a single pure substance depends only on the pressure, and therefore the process of condensation of a pure substance is isothermal. Also, the condensate is a pure liquid. Mixed vapors, condensing at constant pressure, condense over a temperature range and yield a condensate of variable composition until the entire vapor stream is condensed, when the composition of the condensate equals that of the original uncondensed vapor.† A common example of the condensation of one constituent from its mixture with a second noncondensable substance is the condensation of water from a mixture of steam and air.

The condensation of mixed vapors is complicated and beyond the scope of this text.[10a, 11b] The following discussion is directed to the heat transfer from a single volatile substance condensing on a cold tube.

Dropwise and film-type condensation   A vapor may condense on a cold surface in one of two ways, which are well described by the terms *dropwise* and *film-type*. In film condensation, which is more common than dropwise condensation, the liquid condensate forms a film, or continuous layer, of liquid that flows over the surface of the tube under the action of gravity. It is the layer of liquid interposed between the vapor and the wall of the tube which provides the resistance to heat flow and therefore which fixes the magnitude of the heat-transfer coefficient.

In dropwise condensation the condensate begins to form at microscopic nucleation sites. Typical sites are tiny pits, scratches, and dust specks. The drops grow and coalesce with their neighbors to form visible fine drops like those often seen on the outside of a cold-water pitcher in a humid room. The fine drops, in turn, coalesce into rivulets, which flow down the tube under the action of gravity, sweep away condensate, and clear the surface for more droplets. During dropwise condensation, large areas of the cold tube are bare and are directly exposed to the vapor. Because of the absence of the liquid film, the resistance to heat flow at these bare areas is very low and the heat-transfer coefficient correspondingly high. The average coefficient for dropwise condensation may be 5 to 8 times that for film-type condensation. On long tubes, condensation on some of the surface may be film condensation and the remainder dropwise condensation.

The most important and extensive observations of dropwise condensation have been made on steam, but it has also been observed in ethylene glycol, glycerin, nitrobenzene, isoheptane, and some other organic vapors.[14] Liquid metals usually condense in the dropwise manner. The appearance of dropwise condensation depends upon the wetting or nonwetting of the surface by the liquid, and fundamentally, the phenomenon lies in the field of surface chemistry. Much of the experimental work on the dropwise condensation of steam is summarized in the following paragraphs[4]:

1   Film-type condensation occurs on tubes of the common metals if both the steam and the tube are clean, in the presence or absence of air, on rough or on polished surfaces

---

† Exceptions to this statement are found in the condensation of azeotropic mixtures, which are discussed in a later Chapter.

2 Dropwise condensation is obtainable only when the cooling surface is contaminated. It is more easily maintained on a smooth contaminated surface than on a rough contaminated surface.

3 The quantity of contaminant or promoter required to cause dropwise condensation is minute, and apparently only a monomolecular film is necessary.

4 Effective drop promoters are strongly adsorbed by the surface, and substances that merely prevent wetting are ineffective. Some promoters are especially effective on certain metals, e.g., mercaptans on copper alloys; other promoters, such as oleic acid, are quite generally effective. Some metals, e.g., steel and aluminum, are difficult to treat to give dropwise condensation.

5 The average coefficient obtainable in pure dropwise condensation is as high as 20,000 Btu/ft^2-h-°F.

Atlhough attempts are sometimes made to realize practical benefits from these large coefficients by artificially inducing dropwise condensation, this type of condensation is so unstable and the difficulty of maintaining it so great that the method is not common. Also the resistance of the layer of steam condensate even in film-type condensation is ordinarily small in comparison with the resistance inside the condenser tube, and the increase in the overall coefficient is relatively small when dropwise condensation is achieved. For normal design, therefore, film-type condensation is assumed.

Coefficients for film-type condensation The basic equations for the rate of heat transfer in film-type condensation were first derived by Nusselt.[10b, 11a, 13] The Nusselt equations are based on the assumption that the vapor and liquid at the outside boundary of the liquid layer are in thermodynamic equilibrium, so that the only resistance to the flow of heat is that offered by the layer of condensate flowing downward in laminar flow under the action of gravity. It is also assumed that the velocity of the liquid at the wall is zero, that the velocity of the liquid at the outside of the film is not influenced by the velocity of the vapor, and that the temperatures of the wall and the vapor are constant. Superheat in the vapor is neglected, the condensate is assumed to leave the tube at the condensing temperature, and the physical properties of the liquid are taken at mean film temperature.

VERTICAL TUBES In film-type condensation, the Nusselt theory shows that the condensate film starts to form at the top of the tube and that the thickness of the film increases rapidly in the first few inches at the top and then more and more slowly in the remaining tube length. The heat is assumed to flow through the condensate film solely by conduction, and the local coefficient is therefore given by

$$h_x = \frac{k_f}{\delta} \tag{13-1}$$

where δ is the local film thickness.

The local coefficient therefore changes inversely with the film thickness. The variations of both h_x and δ with distance from the top of the tube are shown for a typical case in Fig. 13-1.

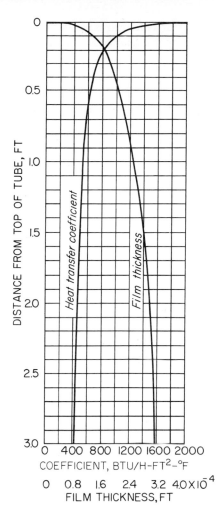

FIGURE 13-1
Film thickness and local coefficients, descending film of condensate. [*By permission, from D. Q. Kern, "Process Heat Transfer." Copyright*, 1950, McGraw-Hill Book Company.]

The film thickness δ can be found from Eq. (5-83). Since there is a temperature gradient in the film, the properties of the liquid are evaluated at the average film temperature T_f, given by Eq. (13-11). For condensation on a vertical surface, for which $\cos \beta = 1$, Eq. (5-83) becomes

$$\delta = \left(\frac{3\Gamma\mu_f}{\rho_f^2 g}\right)^{1/3} \tag{13-2}$$

Substitution for δ in Eq. (13-1) gives for the local heat-transfer coefficient, at a distance L from the top of the vertical surface, the equation

$$h_x = k_f\left(\frac{\rho_f^2 g}{3\Gamma\mu_f}\right)^{1/3} \tag{13-3}$$

For condensation on the outside of a vertical tube the local coefficient is also given by the relations

$$h_x = \frac{dq}{\Delta T_o \, dA_o} = \frac{\lambda \, d\dot{m}}{\Delta T_o \pi D_o \, dL} \tag{13-4}$$

where λ = heat of vaporization
$\dot{m}$ = local flow rate of condensate.

Since $\dot{m}/\pi D_o = \Gamma$, Eq. (13-4) may be written

$$h_x = \frac{\lambda \, d\Gamma}{\Delta T_o \, dL} \tag{13-5}$$

The *average* coefficient h for the entire tube is defined by

$$h \equiv \frac{q_T}{A_o \, \Delta T_o} = \frac{\dot{m}_T \lambda}{\pi D_o L_T \, \Delta T_o} = \frac{\Gamma_b \lambda}{L_T \, \Delta T_o} \tag{13-6}$$

where q_T = total rate of heat transfer
$\dot{m}_T$ = total rate of condensation
L_T = total tube length
Γ_b = condensate loading at the bottom of the tube

Eliminating h_x from Eqs. (13-3) and (13-5) and solving for ΔT_o gives

$$\Delta T_o = \left(\frac{3\Gamma \mu_f}{\rho_f^2 g}\right)^{1/3} \frac{\lambda \, d\Gamma}{k_f \, dL} \tag{13-7}$$

Substituting ΔT_o from Eq. (13-7) into Eq. (13-6) gives

$$h = \frac{\Gamma_b k_f}{L_T} \left(\frac{\rho_f^2 g}{3\mu_f}\right)^{1/3} \frac{dL}{\Gamma^{1/3} \, d\Gamma} \tag{13-8}$$

Rearranging Eq. (13-8) and integrating between limits leads to

$$h \int_0^{\Gamma_b} \Gamma^{1/3} \, d\Gamma = \frac{\Gamma_b k_f}{L_T} \left(\frac{\rho_f^2 g}{3\mu_f}\right)^{1/3} \int_0^{L_T} dL$$

from which

$$h = \frac{4k_f}{3} \left(\frac{\rho_f^2 g}{3\Gamma_b \mu_f}\right)^{1/3} \tag{13-9}$$

Equation (13-9) can be rearranged to give

$$h \left(\frac{\mu_f^2}{k_f^3 \rho_f^2 g}\right)^{1/3} = 1.47 \left(\frac{4\Gamma_b}{\mu_f}\right)^{-1/3} \tag{13-10}$$

The reference temperature for evaluating μ_f, k_f, and ρ_f is defined by the relation[11a]

$$T_f = T_h - \frac{3(T_h - T_w)}{4} = T_h - \frac{3\,\Delta T_o}{4} \tag{13-11}$$

where T_f = reference temperature
T_h = temperature of condensing vapor
T_w = temperature of outside surface of tube wall

Equation (13-10) is often used in an equivalent form, in which the term Γ_b has been eliminated by combining Eqs. (13-6) and (13-10) to give

$$h = 0.943 \left(\frac{k_f^3 \rho_f^2 g \lambda}{\Delta T_o L \mu_f}\right)^{1/4} \tag{13-12}$$

HORIZONTAL TUBES   Corresponding to Eqs. (13-10) and (13-12) for vertical tubes, the following equations apply to single horizontal tubes,

$$h\left(\frac{\mu_f^2}{k_f^3 \rho_f^2 g}\right)^{1/3} = 1.51 \left(\frac{4\Gamma'}{\mu_f}\right)^{-1/3} \tag{13-13}$$

and

$$h = 0.725 \left(\frac{k_f^3 \rho_f^3 g \lambda}{\Delta T_o D_o \mu_f}\right)^{1/4} \tag{13-14}$$

where Γ' is the condensate loading per unit length of tube, in pounds per foot per hour, and all the other symbols have the usual meaning.

The close resemblance between Eqs. (13-10) and (13-13) is obvious. The equations can be made identical by using the coefficient 1.5 for each.

Practical use of Nusselt equations   In the absence of high vapor velocities, experimental data check Eqs. (13-13) and (13-14) well, and these equations can be used as they stand for calculating heat-transfer coefficients for film-type condensation on a single horizontal tube. Also, Eq. (13-14) can be used for film-type condensation on a vertical stack of horizontal tubes, where the condensate falls cumulatively from tube to tube and the total condensate from the entire stack finally drops from the bottom tube. It is necessary only to define Γ_s' as the average loading per tube, based on the total flow from the bottom tube:

$$\Gamma_s' = \frac{\dot{m}_T}{LN} \tag{13-15}$$

where $\dot{m}_T$ = total condensate flow rate for entire stack
L = length of one tube
N = number of tubes in stack

Practically, because of the fact that some condensate splashes away from each individual tube instead of dripping completely to the tube below, it is more accurate to use a value of Γ'_s calculated from the equation[10a]

$$\Gamma'_s = \frac{\dot{m}_T}{LN^{2/3}} \qquad (13\text{-}16)$$

Equation (13-14) becomes

$$h = 0.725 \left(\frac{k_f^3 \rho_f^2 g \lambda}{N^{2/3} \, \Delta T_o D_o \mu_f} \right)^{1/4} \qquad (13\text{-}17)$$

Equations (13-10) and (13-12), for vertical tubes, were derived on the assumption that the condensate flow was laminar. This limits their use to cases where $4\Gamma_b/\mu_f$ is less than 2,100. For long tubes, the condensate film becomes sufficiently thick and its velocity sufficiently large to cause turbulence in the lower portions of the tube. Also, even when the flow remains laminar throughout, coefficients measured experimentally are about 20 percent larger than those calculated from the equations. This is attributed to the effect of ripples on the surface of the falling film. For practical calculations the coefficients in Eqs. (13-10) and (13-12) should be taken as 1.76 and 1.13, respectively. When the velocity of the vapor phase in the condenser is appreciable, the drag of the vapor induces turbulence in the condensate layer. The coefficient of the condensing film is then considerably larger than predicted by Eqs. (13-10) and (13-12).

In Fig. 13-2 is plotted $h(\mu_f^2/k_f^3\rho_f^2 g)^{1/3}$ vs. N_{Re}. Line AA is the theoretical relation for both horizontal and vertical tubes for magnitudes of N_{Re} less than 2,100 [Eqs. (13-10) and (13-13)]. This line can be used directly for horizontal tubes. Line BB, based on Eq. (13-10) with a coefficient of 1.76 instead of 1.47 to allow for the effect of ripples, is recommended for vertical tubes when flow in the condensate film is laminar throughout.

For magnitudes of $4\Gamma_b/\mu_f$ greater than 2,100, the coefficient h increases with increase in Reynolds number. Line CC in Fig. 13-2 can be used for calculation of values of h when the value of $4\Gamma_b/\mu_f$ exceeds 2,100. This value is not reached in condensation on horizontal pipes, and a line for turbulent flow is not needed. The equation for line CC is [6b]

$$h \left(\frac{\mu_f^2}{k_f^3 \rho_f^2 g} \right)^{1/3} = 0.0076 \left(\frac{4\Gamma_b}{\mu_f} \right)^{0.4} \qquad (13\text{-}18)$$

For a given substance over a moderate pressure range the quantity $(k_f^3\rho_f^2 g/\mu_f^2)^{1/3}$ is a function of temperature. Use of Fig. 13-2 is facilitated if this quantity, which can be denoted by ψ_f, is calculated and plotted as a function of temperature for a given substance. Quantity ψ_f has the same dimensions as a heat-transfer coefficient, so that both the ordinate and abscissa scales of Fig. 13-2 are dimensionless. Appendix 14 gives the magnitude of ψ_f for water as a function of temperature. Corresponding tables can be prepared for other substances when desired.

EXAMPLE 13-1   A shell-and-tube condenser with vertical $\frac{3}{4}$-in. (19-mm) 16 BWG copper tubes has chlorobenzene condensing at atmospheric pressure in the shell. The latent heat of condensation of chlorobenzene is 139.7 Btu/lb (324.9 J/g). The tubes are 5 ft (1.5 2m) long.

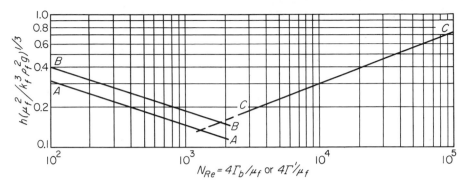

FIGURE 13-2

Heat-transfer coefficients in film condensation: line AA, theoretical relation for laminar flow, horizontal and vertical tubes; also, recommended relation for horizontal tubes; line BB, recommended relation for laminar flow, vertical tubes; line CC, approximate relation for turbulent flow, vertical tubes.

Cooling water at an average temperature of 175°F (79.4°C) is flowing in the tubes. If the water-side coefficient is 400 Btu/ft²-h-°F (2,270 W/m²-°C), (*a*) what is the coefficient of the condensing chlorobenzene; (*b*) what would the coefficient be in a horizontal condenser with the same number of tubes if the average number of tubes in a vertical stack is 6? Neglect fouling factors and tube-wall resistance.

SOLUTION   (*a*)   Equation (13-12) applies, but the properties of the condensate must be evaluated at reference temperature T_f, which is given by Eq. (13-11). In computing T_f the wall temperature T_w must be estimated from h, the condensate-film coefficient. A trial-and-error solution is therefore necessary.

The quantities in Eq. (13-12) that may be specified directly are

$$\lambda = 139.7 \text{ Btu/lb} \qquad g = 4.17 \times 10^8 \text{ ft/h}^2 \qquad L = 5 \text{ ft}$$

The condensing temperature T_h is 267°F.

The wall temperature T_w must be between 175 and 267°F. The resistance of a condensing film of organic liquid is usually greater than the thermal resistance of flowing water; consequently T_w is probably closer to 175°F than it is to 267°F. As a first approximation, T_w will be taken as 205°F.

The temperature difference ΔT_o is 267 − 205 = 62°F.

The reference temperature, from Eq. (13-11), is

$$T_f = 267 - \tfrac{3}{4}(267 - 205) = 220°F$$

The density and thermal conductivity of liquids vary so little with temperature that they may be assumed constant at the following values:

$$\rho_f = 65.4 \text{ lb/ft}^3 \qquad k_f = 0.083 \text{ Btu/ft-h-°F} \qquad \text{(Appendix 13)}$$

The viscosity of the film is

$$\mu_f = 0.30 \times 2.42 = 0.726 \text{ lb/ft-h} \qquad \text{(Appendix 10)}$$

The first estimate of h, using coefficient 1.13, as recommended on page 354, is

$$h = 1.13 \left(\frac{0.083^3 \times 65.4^2 \times 4.17 \times 10^8 \times 139.7}{62 \times 5 \times 0.726} \right)^{1/4} = 179 \ \text{Btu/ft}^2\text{-h-}°\text{F}$$

The corrected wall temperature is found from Eq. (12-42). The outside diameter of the tubes D_o is $0.75/12 = 0.0625$ ft. The inside diameter D_i is $0.0625 - (2 \times 0.065)/12 = 0.0517$ ft.

The temperature drop over the water resistance is, from Eq. (12-42),

$$\Delta T_i = \frac{1/400}{1/400 + 0.0517/(0.0625 \times 179)} (267 - 175) = 32°\text{F}$$

The wall temperature is

$$T_w = 175 + 32 = 207°\text{F}$$

This is sufficiently close to the estimated value, 205°F (96.1°C), to make further calculation unnecessary. The coefficient h is 179 Btu/ft²-h-°F (1,016 W/m²-°C).

(b) For a horizontal condenser, Eq. (13-17) is used. The coefficient of the chlorobenzene is probably greater than in part (a), so the wall temperature T_w is now estimated to be 215°F. The new quantities needed are

$$N = 6 \qquad \Delta T_o = 267 - 215 = 52°\text{F} \qquad D_o = 0.0625 \ \text{ft}$$

$$T_f = 267 - \tfrac{3}{4}(267 - 215) = 228°\text{F}$$

$$\mu_f = 0.280 \times 2.42 = 0.68 \ \text{lb/ft-h} \qquad \text{(Appendix 16)}$$

$$h = 0.725 \left(\frac{0.083^3 \times 65.4^2 \times 4.17 \times 10^8 \times 139.7}{6^{2/3} \times 52 \times 0.0625 \times 0.68} \right)^{1/4} = 271 \ \text{Btu/ft}^2\text{-h-}°\text{F}$$

$$\Delta T_i = \frac{1/400}{1/400 + 0.0517/(0.0625 \times 271)} 92 = 41°\text{F} \qquad T_w = 175 + 41 = 216°\text{F}$$

This checks the assumed value, 215°F (101.7°C), and no further calculation is needed. The coefficient h is 271 Btu/ft²-h-°F (1,538 W/m²-°C).   ////

In general, the coefficient of a film condensing on a horizontal tube is considerably larger than that on a vertical tube under otherwise similar conditions unless the tubes are very short or there are very many horizontal tubes in each stack. Vertical tubes are preferred when the condensate must be appreciably subcooled below its condensation temperature. Mixtures of vapors and noncondensing gases are usually cooled and condensed *inside* vertical tubes, so that the inert gas is continually swept away from the heat-transfer surface by the incoming stream.

Condensation of superheated vapors If the vapor entering a condenser is superheated, both the sensible heat of superheat and the latent heat of condensation must be transferred through the cooling surface. For steam, because of the low specific heat of the superheated vapor and the large latent heat of condensation, the heat of superheat is usually small in comparison with the latent heat. For example, 100°F of superheat represents only 50 Btu/lb, as compared with approximately 1000 Btu/lb for latent heat. In the condensation of organic vapors, such as petroleum fractions, the superheat may be appreciable in comparison with the latent heat. When the heat of superheat is important, either it can be calculated from the degrees of superheat and the specific heat of the vapor and added to the latent heat, or if tables of thermal properties are available, the total heat transferred per pound of vapor can be calculated by subtracting the enthalpy of the condensate from that of the superheated vapor.

The effect of superheat on the rate of heat transfer depends upon whether the temperature of the surface of the tube is higher or lower than the condensation temperature of the vapor. If the temperature of the tube is lower than the temperature of condensation, the tube is wet with condensate, just as in the condensation of saturated vapor, and the temperature of the outside boundary of the condensate layer equals the saturation temperature of the vapor at the pressure existing in the equipment. The situation is complicated by the presence of a thermal resistance between the bulk of the superheated vapor and the outside of the condensate film and by the existence of a temperature drop, which is equal to the degrees of superheat in the vapor, across that resistance. Practically, however, the net effect of these complications is small, and it is satisfactory to assume that the entire heat load, which consists of the heats of superheat and of condensation, is transferred through the condensate film; that the temperature drop is that across the condensate film; and that the coefficient is the average coefficient for condensing vapor as read from Fig. 13-2. The procedure is summarized by the equation

$$q = hA(T_h - T_w) \qquad (13\text{-}19)$$

where q = total heat transferred, including latent heat and superheat
       A = area of heat-transfer surface in contact with vapor
       h = coefficient of heat transfer, from Fig. 13-2
      T_h = saturation temperature of vapor
     T_w = temperature of tube wall

When the vapor is highly superheated and the exit temperature of the cooling fluid is close to that of condensation, the temperature of the tube wall may be greater than the saturation temperature of the vapor, condensation cannot occur, and the tube wall will be dry. The tube wall remains dry until the superheat has been reduced to a point where the tube wall becomes cooler than the condensing temperature of the vapor, and condensation takes place. The equipment can be considered as two sections, one a desuperheater and the other a condenser. In calculations, the two sections must be considered separately. The desuperheater is essentially a gas cooler. The logarithmic mean temperature applies, and the heat-transfer coefficient is that for cooling a fixed gas. The condensing section is treated by the methods described in the previous paragraphs.

Because of the low individual coefficient on the gas side, the overall coefficient in the desuperheater section is small, and the area of the heating surface in that section is large in comparison with the amount of heat removed. This situation should be avoided in practice. Superheat can be eliminated more economically by injection of liquid directly into the superheated vapor, the desuperheating section thereby eliminated, and the high coefficients from condensing vapors so obtained.

Effect of noncondensable gases on rate of condensation   The presence of even small amounts of noncondensing gas in a condensing vapor seriously reduces the rate of condensation. A layer of gas accumulates on the film of condensate and establishes an added thermal resistance which is not present during the condensation of a pure vapor. Also, the vapor of the condensing substance must flow by diffusion through the layer of gas to reach the film of condensate, and the resistance offered by the gas to diffusion further impedes the process of condensation. Rigorous methods of solving the general problem are based on equating the heat flow to the condensate surface at any point to the heat flow away from the surface. This involves trial-and-error solutions for the point temperatures of the condensate surface, and from these an estimation of the point values of the heat flux $U \Delta T$. The values of $1/(U \Delta T)$ for each point are plotted against the heat transferred to that point and the area of the condenser surface found by numerical integration.[2]

HEAT TRANSFER TO BOILING LIQUIDS

Heat transfer to a boiling liquid is a necessary step in the unit operations of evaporation and distillation and also in other kinds of general processing, such as steam generation, petroleum processing, and control of the temperatures of chemical reactions. The boiling liquid may be contained in a vessel equipped with a heating surface fabricated from horizontal or vertical plates or tubes, which supplies the heat necessary to boil the liquid. Or the liquid may flow through heated tubes, under either natural or forced convection, and the heat transferred to the fluid through the walls of the tubes. An important application of boiling in tubes is the evaporation of water from solution. Evaporation is the subject of Chap. 16.

When boiling is accomplished by a hot immersed surface, the temperature of the mass of the liquid is the same as the boiling point of the liquid under the pressure existing in the equipment. Bubbles of vapor are generated at the heating surface, rise through the mass of liquid, and disengage from the surface of the liquid. Vapor accumulates in a vapor space over the liquid; a vapor outlet from the vapor space removes the vapor as fast as it is formed. This type of boiling can be described as the *boiling of saturated liquid* since the vapor leaves the liquid in equilibrium with the liquid at its boiling temperature.

When a liquid is boiled under natural circulation inside a vertical tube, relatively cool liquid enters the bottom of the tube and is heated and partially vaporized as it flows upward. The liquid temperature, initially low, increases to the boiling point under the pressure prevailing in the tube. With forced circulation through horizontal or vertical tubes the liquid may also enter at a fairly low temperature and be heated to its boiling point, changing into vapor near the discharge end of the tube. Sometimes a flow-control valve is placed in the discharge line beyond the tube so that the liquid in the tube may be heated to a temperature considerably above the boiling point corresponding to the downstream pressure. Under these conditions there is no boiling in the tube: the liquid is merely heated, as a liquid, to a high temperature, and

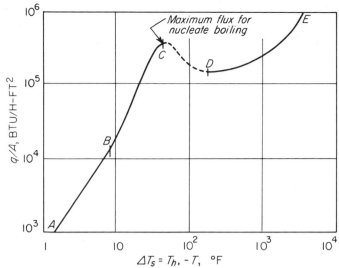

FIGURE 13-3
Heat flux versus temperature drop, boiling water at 212°F: *AB*, natural convec-
tion; *BC*, nucleate boiling; *CD*, transition boiling; *DE*, film boiling.
(*From Reference 12.*)

flashes into vapor as it passes through the valve. Natural- and forced-circulation
boilers are called *calandrias*; they are discussed in detail in Chap. 16.

In some types of forced-circulation equipment the temperature of the mass of
the liquid is below that of its boiling point, but the temperature of the heating surface
is considerably above the boiling point of the liquid. Bubbles form on the heating
surface but on release from the surface are absorbed by the mass of the liquid. This
type of boiling is called *subcooled boiling*. In subcooled boiling, the stream of boiling
liquid is confined by the channel through which the fluid stream is flowing and vapor
separation does not occur, so no vapor space is necessary. The various methods of
boiling will be discussed separately.

Boiling of saturated liquid Consider a horizontal tube immersed in a vessel
containing a boiling liquid. Assume that q/A, the heat flux, and ΔT, the difference
between the temperature of the tube wall and that of the boiling liquid, are measured.
Start with a very low temperature drop ΔT. Increase the temperature drop by steps,
measuring q/F and ΔT at each step, until very large values of ΔT are reached. A plot
of q/A vs. ΔT on logarithmic coordinates will give a curve of the type shown in Fig.
13-3. This curve can be divided into four segments. In the first segment, at low temperature drops, the line AB is straight and has a slope of 1.25. This is consistent with
the equation

$$\frac{q}{A} = a\,\Delta T^{1.25} \tag{13-20}$$

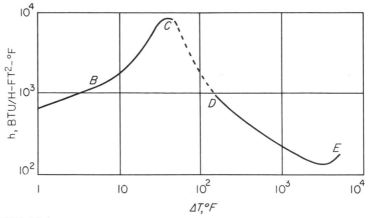

FIGURE 13-4
Heat-transfer coefficients versus ΔT, boiling of water at 1 atm.

where a is a constant. The second segment, line BC, is also approximately straight, but its slope is greater than that of line AB. The slope of the line BC depends upon the specific experiment; it usually lies between 3 and 4. The second segment terminates at a definite point of maximum flux, which is point C in Fig. 13-3. The temperature drop corresponding to point C is called the *critical temperature drop*, and the flux at point C is the *peak flux*. In the third segment, line CD in Fig. 13-3, the flux decreases as the temperature drop rises and reaches a minimum at point D. Point D is called the *Leidenfrost point*. In the last segment, line DE, the flux again increases with ΔT and, at large temperature drops, surpasses the previous maximum reached at point C.

Because, by definition, $h = (q/A)/\Delta T$, the plot of Fig. 13-3 is readily convertible into a plot of h vs. ΔT. This curve is shown in Fig. 13-4. A maximum and a minimum coefficient are evident in Fig. 13-4. They do not, however, occur at the same values of the temperature drop as the maximum and minimum fluxes indicated in Fig. 13-3. The coefficient is normally a maximum at a temperature drop slightly lower than that at the peak flux; the minimum coefficient occurs at a much higher temperature drop than that at the Leidenfrost point. The coefficient is proportional to $\Delta T^{0.25}$ in the first segment of the line in Fig. 13-4 and to between ΔT^2 and ΔT^3 in the second segment.

Each of the four segments of the graph in Fig. 13-4 corresponds to a definite mechanism of boiling. In the first section, at low temperature drops, the mechanism is that of heat transfer to a liquid in natural convection, and the variation of h with ΔT agrees with that given by Eq. (12-83). Bubbles form on the surface of the heater, are released from it, rise to the surface of the liquid, and are disengaged into the vapor space; but they are too few to disturb appreciably the normal currents of free convection.

At larger temperature drops, lying between 9 and 45°F in the case shown in Fig. 13-4, the rate of bubble production is large enough for the stream of bubbles moving up through the liquid to increase the velocity of the circulation currents in the mass of liquid, and the coefficient of heat transfer becomes greater than that in undisturbed natural convection. As ΔT is increased, the rate of bubble formation increases and the coefficient increases rapidly.

The action occurring at temperature drops below the critical temperature drop is called *nucleate boiling*, in reference to the formation of tiny bubbles, or vaporization nuclei, on the heating surface. During nucleate boiling, the bubbles occupy but a small portion of the tube surface at a time, and most of the surface is in direct contact with liquid. The bubbles are generated at localized active sites, usually small pits or scratches on the heating surface. As the temperature drop is raised, more sites become active, improving the agitation of the liquid and increasing the heat flux and the heat-transfer coefficient.

Eventually, however, so many bubbles are present that they tend to coalesce on the heating surface to form a layer of insulating vapor. This layer has a highly unstable surface, from which miniature "explosions" send jets of vapor away from the heating element into the bulk of the liquid. This type of action is called *transition boiling*. In this region increasing the temperature drop increases the thickness of the vapor film and reduces the number of explosions that occur in a given time. The heat flux and heat-transfer coefficient both fall as the temperature drop is raised.

Near the Leidenfrost point another distinct change in mechanism occurs. The hot surface becomes covered with a quiescent film of vapor, through which heat is transferred by conduction and (at very high temperature drops) by radiation. The random explosions characteristic of transition boiling disappear and are replaced by the slow and orderly formation of bubbles at the interface between the liquid and the film of hot vapor. These bubbles detach themselves from the interface and rise through the liquid. Agitation of the liquid is not important, however; all the resistance to heat transfer is offered by the vapor sheath covering the heating element. As the temperature drop increases, the heat flux rises, slowly at first and then more rapidly as radiation heat transfer becomes important. The boiling action in this region is known as *film boiling*.

Film boiling is not usually desired in commercial equipment because the heat-transfer rate is low in view of the large temperature drop and temperature drop is not utilized effectively. Heat-transfer apparatus should be so designed and operated that the temperature drop in the film of boiling liquid is smaller than the critical temperature drop although with cryogenic liquids this is not always feasible.

The effectiveness of nucleate boiling depends primarily on the ease with which bubbles form and free themselves from the heating surface. The layer of liquid next to the hot surface is superheated by contact with the wall of the heater. The superheated liquid tends to form vapor spontaneously and so relieve the superheat. It is the tendency of superheated liquid to flash into vapor that provides the impetus for the boiling process. Physically, the flash can occur only by forming vapor-liquid interfaces in the form of small bubbles. It is not easy to form a small bubble, however, in a superheated liquid, because, at a given temperature, the vapor pressure in a very small bubble is less than that in a large bubble or that from a plane liquid surface. A very small bubble can exist in equilibrium with superheated liquid, and the smaller

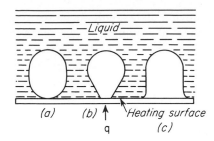

FIGURE 13-5
Effect of interfacial tension on bubble
formation.[9]

the bubble, the greater the equilibrium superheat and the smaller the tendency to flash. By taking elaborate precautions to eliminate all gas and other impurities from the liquid and to prevent shock, it is possible to superheat water by several hundred degrees Fahrenheit without formation of bubbles.

A second difficulty appears if the bubble does not readily leave the surface once it is formed. The important factor in controlling the rate of bubble detachment is the interfacial tension between the liquid and the heating surface. If the interfacial tension is large, the bubble tends to spread along the surface and blanket the heat-transfer area, as shown in Fig. 13-5c, rather than leaving the surface to make room for other bubbles. If the interfacial tension is low, the bubble will pinch off easily, in the manner shown in Fig. 13-5a. An example of intermediate interfacial tension is shown in Fig. 13-5b.

The high rate of heat transfer in nucleate boiling is primarily the result of the turbulence generated in the liquid by the dynamic action of the bubbles.[5]

The coefficient obtained during nucleate boiling is sensitive to a number of variables, including the nature of the liquid, the type and condition of the heating surface, the composition and purity of the liquid, the presence or absence of agitation, and the temperature or pressure. Minor changes in some variables cause major changes in the coefficient. The reproducibility of check experiments is poor. General correlations in this field are available but have low precision.

Qualitatively, the effects of some variables can be predicted from a consideration of the mechanisms of boiling. A roughened surface provides centers for nucleation that are not present on a polished surface. Thus roughened surfaces usually give larger coefficients than smooth surfaces. This effect, however, is often due to the fact that the total surface of a rough tube is larger than that of a smooth surface of the same projected area. A very thin layer of scale may increase the coefficient of the boiling liquid, but even a thin scale will reduce the overall coefficient by adding a resistance that reduces the overall coefficient more than the improved boiling liquid coefficient increases it. Gas or air adsorbed on the surface of the heater or contaminants on the surface often facilitate boiling by either the formation or the disengaging of bubbles. A freshly cleaned surface may give a higher or lower coefficient than the same surface after it has been stabilized by a previous period of operation. This effect is associated with a change in the condition of the heating surface. Agitation increases the coefficient by increasing the velocity of the liquid across the surface, which helps to sweep away bubbles.

The temperature of a boiling liquid can, of course, be changed by changing the pressure. The change in coefficient with change in temperature can be calculated approximately by the equation[3]

$$\log \frac{h_a}{h} = b(T_a - T) \tag{13-21}$$

where T_a = boiling temperature of liquid at 1 atm
$\quad h_a$ = coefficient at temperature T_a
$\quad h$ = coefficient at temperature T
$\quad b$ = constant

The constant b depends on the liquid and on the kind and condition of the heating surface.

Maximum flux and critical temperature drop   The maximum flux $(q/A)_{max}$ depends somewhat on the nature of the boiling liquid and on the type of heating surface but is chiefly sensitive to pressure. It is a maximum at an absolute pressure about one-third of the thermodynamic critical pressure p_e and decreases toward zero at very low pressures and at pressures approaching the critical pressure.[6a] If the maximum flux is divided by the critical pressure of the boiling substance, a curve is obtained, shown in Fig. 13-6, which is about the same for many pure substances and mixtures. The corresponding critical temperature drop (not to be confused with "critical temperature") also varies with pressure, from large values at low pressures to very small ones near the critical pressure. At pressures and temperatures above the critical values, of course, there is no difference between liquid and vapor phases, and "vaporization" has no meaning. At atmospheric pressure the critical temperature drop for water is usually between 70 and 90°F; for common organic liquids it is often between 40 and 50°F.

The critical temperature drop can be exceeded in actual equipment unless precautions are taken to keep the actual temperature drop below the critical. If the source of heat is another fluid, such as condensing steam or hot liquid, the only penalty for exceeding the critical temperature drop is a decrease in flux to a level between that at the peak and that at the Leidenfrost point. If the heat is supplied by an electric heater, exceeding the critical temperature drop usually will burn out the heater, as the boiling liquid cannot absorb heat fast enough at a large temperature drop, and the heater immediately becomes very hot.

The peak flux at the critical temperature drop is large. For water it is in the range 115,000 to 400,000 Btu/ft²-h depending on the purity of water, the pressure, and the type and condition of the heating surface. For organic liquids, the peak flux lies in the range 40,000 to 130,000 Btu/ft²-h. These limits apply to boiling under atmospheric pressure.

It has been proposed that near the critical temperature drop the streams of bubbles characteristic of nucleate boiling are progressively replaced by jets of vapor leaving the heat-transfer surface. These must, of course, be accompanied by streams of liquid flowing toward the surface. At the peak value of the heat flux, the counter-current flows of vapor and liquid reach a limiting condition, the process becomes

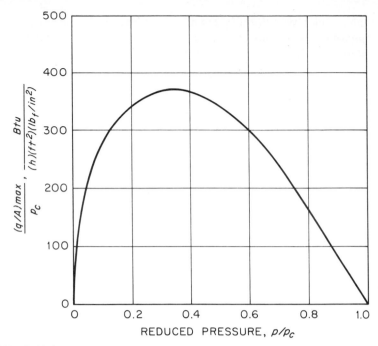

FIGURE 13-6
Maximum boiling heat flux as a function of reduced pressure. [*By permission, from B. Gebhart, "Heat Transfer," 2d ed., p. 424. Copyright, 1971, McGraw-Hill Book Company.*]

unstable, and the jets of vapor collapse to form a continuous vapor sheath. The phenomenon is analogous to the flooding in a packed tower described in Chap. 23.[16] Using this model, Zuber[15] has derived the following dimensionally consistent equation for maximum heat flux, $(q/A)_{max}$:

$$\left(\frac{q}{A}\right)_{max} = \frac{\pi \lambda}{24} [\sigma g_c g(\rho_L - \rho_V)]^{1/4} \rho_v^{1/2} \left(1 + \frac{\rho_V}{\rho_L}\right)^{1/2} \tag{13-22}$$

where σ is the interfacial tension between liquid and vapor, ρ_L and ρ_V are the densities of liquid and vapor, respectively, and the other symbols have their usual meanings. Increasing the pressure on the system increases ρ_V without greatly affecting the other terms in Eq. (13-22) and consequently increases the maximum heat flux. If the pressure is increased sufficiently, the latent heat of vaporization tends toward zero, so that ultimately a reduction in the peak flux occurs, as shown in Fig. 13-6.

Minimum heat flux and film boiling When film boiling is established, undulations of a characteristic wavelength form in the interface between liquid and vapor. These undulations grow into bubbles, which leave the interface at regularly spaced intervals. The diameter of the bubbles is approximately one-half the wavelength of the undulations. Consideration of the dynamics of this process leads to the following equation for the minimum heat flux necessary for stable film boiling on a horizontal plate[6a]

$$\left(\frac{q}{A}\right)_{min} = \frac{\pi \lambda \rho_V}{24} \left[\frac{\sigma g_c g(\rho_L - \rho_V)}{(\rho_L + \rho_V)^2}\right]^{1/4} \tag{13-23}$$

where $(q/A)_{min}$ is the minimum heat flux.

Film boiling is a more orderly process than either nucleate boiling or transition boiling and has been subjected to considerable theoretical analysis. Since the heat-transfer rate is governed solely by the vapor film, the nature of the heating surface has no effect in film boiling. For film boiling on a submerged horizontal tube the following equation applies with considerable accuracy over a wide range of conditions:[1]

$$h_o \left[\frac{\lambda_c \mu_V \, \Delta T}{k_V^3 \rho_V(\rho_L - \rho_V)\lambda' g}\right]^{1/4} = 0.59 + 0.069 \frac{\lambda_c}{D_o} \tag{13-24}$$

where
h_o = heat-transfer coefficient
μ_V = viscosity of vapor
ΔT = temperature drop across vapor film
k_V = thermal conductivity of vapor
ρ_L, ρ_V = densities of liquid and vapor, respectively
D_o = outside diameter of heating tube

In Eq. (13-22) λ' is the change in enthalpy between the liquid and the superheated vapor and is given by

$$\lambda' = \lambda \left(1 + \frac{0.34c_p \, \Delta T}{\lambda}\right)^2 \tag{13-25}$$

where λ = latent heat of vaporization
c_p = specific heat of vapor at constant pressure

The term λ_c in Eq. (13-22) is the wavelength of the smallest wave which can grow in amplitude on a flat horizontal interface. It is related to the properties of the fluid by the equation

$$\lambda_c = 2\pi \left[\frac{\sigma g_c}{g(\rho_L - \rho_V)}\right]^{1/2} \tag{13-26}$$

where σ is the interfacial tension between liquid and vapor. Equation (13-24) does not include the effect of heat transfer by radiation.

Equations have also been developed for film boiling from submerged vertical tubes,[8] but they have less general validity than Eq. (13-24). Vapor disengages from a vertical surface in a more complicated fashion than from a horizontal surface, and the theoretical analysis of the process is correspondingly more difficult.

EXAMPLE 13-2   Freon-11 (CCl_3F) is boiled at atmospheric pressure by a submerged horizontal tube, 1.25 in. (32 mm) OD. The normal boiling point of Freon-11 is 74.8°F

(23.8°C); the tube wall is at 300°F (148.9°C). The properties[7] of Freon-11 are as follows:

$$\rho_L = 91.3 \text{ lb/ft}^3 \ (1{,}462 \text{ kg/m}^3) \qquad \lambda = 78.3 \text{ Btu/lb } (182 \text{ J/g}) \qquad \mu_V = 0.013 \text{ cP}$$

$$\sigma = 19 \text{ dyn/cm} \qquad c_p = 0.145 \text{ Btu/lb-°F } (0.607 \text{ J/g-°C})$$

The thermal conductivity is given by the equation $k_V = 0.00413 + 1.14 \times 10^{-5}T$, where k_V is in Btu/ft-h-°F and T is in °F. Calculate the heat-transfer coefficient h_o and the heat flux q/A.

SOLUTION   Since the temperature difference between the wall and liquid is so large, Eq. (13-24) for film boiling will be used. The quantities needed for substitution are

$$\mu_V = 0.013 \times 2.42 = 0.0314 \text{ lb/ft-h} \qquad \Delta T = 300 - 74.8 = 225.2\text{°F}$$

The average vapor-film temperature is $(300 + 74.8)/2 = 187.4\text{°F}$.

$$k_V = 0.00413 + (1.14 \times 10^{-5} \times 187.4) = 0.00627 \text{ Btu/ft-h-°F}$$

$$\rho_L = 91.3 \text{ lb/ft}^3 \qquad \text{mol. wt.} = 137.5 \qquad D_o = \frac{1.25}{12} = 0.104 \text{ ft}$$

At the average film temperature

$$\rho_V = \frac{137.5 \times 492}{359(187.4 + 460)} = 0.291 \text{ lb/ft}^3$$

At the boiling point

$$\sigma = 19 \frac{2.248 \times 10^{-6}}{0.0328} = 0.0013 \text{ lb}_f/\text{ft}$$

From Eq. (13-25)

$$\lambda' = 78.3 \left(1 + \frac{0.34 \times 0.145 \times 225.2}{78.3}\right)^2 = 102.1 \text{ Btu/lb}$$

From Eq. (13-26), taking g_c/g as unity,

$$\lambda_c = 2\pi \left(\frac{0.0013}{91.3 - 0.291}\right)^{1/2} = 0.0238 \text{ ft}$$

Substitution in Eq. (13-24) gives

$$h_o \left[\frac{0.0238 \times 0.0314 \times 225.2}{0.00627^3 \times 0.291(91.3 - 0.291) \times 102.1 \times 4.17 \times 10^8}\right]^{1/4}$$

$$= 0.59 + \frac{0.069 \times 0.0238}{0.104}$$

from which $h_o = 21.7 \text{ Btu/ft}^2\text{-h-°F } (123 \text{ W/m}^2\text{-°C})$. Then $q/A = 21.7 \times 225.2 = 4887$ Btu/ft²-h (15400 W/m²).   ////

Subcooled boiling   Subcooled boiling can be demonstrated by pumping a gas-free liquid upward through a vertical annular space consisting of a transparent outer tube and an internal heating element and observing the effect on the liquid of a gradual increase in heat flux and temperature of the heating element. It is observed that when the temperature of the element exceeds a definite magnitude, which depends

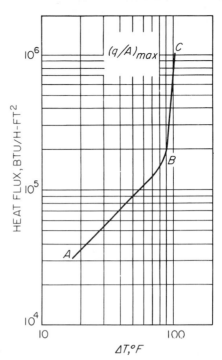

FIGURE 13-7
Example of subcooled boiling.

on the conditions of the experiment, bubbles form, just as in nucleate boiling, and then condense in the adjacent cooler liquid. For a definite velocity and pressure, the thermal flux q/A and the temperature drop from heating element to liquid ΔT can be measured and a logarithmic plot of thermal flux vs. temperature drop prepared. Figure 13-7 shows the results of such an experiment conducted on degassed distilled water flowing at a velocity of 4 ft/s under an absolute pressure of 4 atm² through the annular space between a jacket 0.77 in. ID and an electric heating element 0.55 in. in diameter.

The line in Fig. 13-7 consists of two distinct segments connected by a short transition curve. The lower branch, line AB, covering temperature drops below 80°F is straight and has a slope of 1.0. The flux in this range is proportional to ΔT, and the coefficient h is independent of ΔT. The heat-transfer coefficients in this range agree well with those predicted from the usual equations for heat transfer to non-boiling liquids in turbulent flow, such as Eqs. (12-37) and (12-64). As the temperature drop exceeds 90°F, the curve bends sharply upward when subcooled boiling sets in and section BC is nearly straight. The slope is very great, and a small increase in temperature drop corresponds to a large change in flux. It is this increase in capacity accompanying the appearance of subcooled boiling that makes this type of operation important for heat-transfer equipment that must pack great capacity into small space. When the temperature drop corresponding to point C was reached, the heater became blanketed with vapor and burned out.

The break point in the curve of flux vs. temperature drop for subcooled boiling does not occur at the boiling temperature at the pressure in the equipment. Subcooled boiling does not set in until the liquid boundary layer is appreciably superheated. The bulk liquid may be greatly subcooled.

The most significant characteristic of subcooled boiling is the very high heat flux that can be obtained by this technique, in comparison with the flux obtainable in heat transfer to fluids without change in phase. The maximum flux reached at the burnout point shown in Fig. 13-7 was over 10^6 Btu/ft²-h. A flux as high as 54.8×10^6 Btu/ft²-h has been reported.[6a]

SYMBOLS

A Area, ft² or m²; A_o, outside of tube
a Constant in Eq. (13-20)
b Constant in Eq. (13-21)
c_p Specific heat at constant pressure, Btu/lb-°F or J/g-°C
D Diameter, ft or m; D_o, outside diameter of tube
g Gravitational acceleration, ft/h² or m/h²
g_c Newton's-law proportionality factor, 4.17×10^8 ft-lb/lb$_f$-h²
h Individual heat-transfer coefficient, Btu/ft²-h-°F or W/m²-°C; h_a, of boiling liquid at its normal boiling point; h_o, outside tube; h_x, local value
k Thermal conductivity, Btu/ft-h-°F or W/m-°C; k_v, of vapor; k_f, of condensate film
L Length, ft or m; L_T, total length of tube
$\dot{m}$ Mass flow rate, lb/h or kg/h; $\dot{m}_T$, total condensate flow rate from stack of tubes
N Number of tubes in vertical stack
N_{Re} Reynolds number of condensate film, $4\Gamma/\mu_f$, dimensionless
p Pressure, lb$_f$/in.² or atm; p_c, critical pressure
q Rate of heat transfer, Btu/h or J/s; q_T, total rate of heat transfer in condenser tube
T Temperature, °F or °C; T_a, normal boiling temperature of liquid; T_f, average temperature of condensate film; T_h, saturation temperature of vapor; T_w, wall or surface temperature
U Overall heat-transfer coefficient, Btu/ft²-h-°F or W/m²-°C
Greek letters
β Angle with vertical
Γ Condensate loading, lb/h-ft or kg/h-m; Γ_b, at bottom of vertical tube; Γ', per foot of horizontal tube; Γ'_s, average from bottom tube in vertical stack
ΔT Temperature drop, °F or °C; ΔT_0, drop across condensate film, $T_h - T_w$
δ Thickness of condensate film, ft or m
λ Heat of vaporization, Btu/lb or J/g; λ', change in enthalpy between boiling liquid and super-heated vapor, defined by Eq. (13-25)
λ_c Wavelength of smallest wave which can grow on flat horizontal surface, ft or m [Eq. (13-26)]
μ Absolute viscosity, lb/ft-h or P; μ_v, of vapor; μ_f, of condensate film
ρ Density, lb/ft³ or kg/m³; ρ_L, of liquid; ρ_v, of vapor; ρ_f, of condensate film
σ Interfacial tension between liquid and vapor, lb$_f$/ft or N/m
ψ_f Condensation parameter, $(k_f^3 \rho_f^2 g/\mu_f^2)^{1/3}$, Btu/ft²-h-°F or W/m²-°C

PROBLEMS

13-1 A $1\frac{1}{4}$-in. 14 BWG copper condenser tube, 10 ft long, is to condense *n*-propyl alcohol at atmospheric pressure. Cooling water inside the tube keeps the metal surface at an essentially constant temperature of 25°C. (a) How much vapor, in pounds per hour, will condense if the tube is vertical? (b) How much will condense if the tube is horizontal?

13-2 A horizontal condenser tube is maintained at 30°C on its outside surface. It is surrounded by saturated steam at 0.2 atm abs. Find the outside diameter, in millimeters, necessary to cause a turbulent film of condensate at the bottom of the tube.

13-3 A vertical tubular condenser is to be used to condense 1,500 kg/h of ethyl alcohol, which enters at atmospheric pressure. Cooling water is to flow through the tubes at an average temperature of 30°C. The tubes are 25 mm OD and 21 mm OD. The water-side coefficient is 3000 W/m²-°C. Fouling factors and the resistance of the tube wall may be neglected. If the available tubes are 3 m long, how many tubes will be needed? Data are as follows:

Boiling point of alcohol: $T_h = 78.4°C$

Heat of vaporization: $\lambda = 856 \text{ J/g}$

Density of liquid: $\rho_f = 769 \text{ kg/m}^3$

13-4 A horizontal shell-and-tube condenser is to be used to condense saturated ammonia vapor at 145 $\text{lb}_f/\text{in.}^2$ abs ($T_h = 82°F$). The condenser has seven steel tubes (2 in. OD, 1.8 in. ID), 12 ft long through which cooling water is flowing. The tubes are arranged hexagonally on $2\frac{1}{2}$-in. centers. The latent heat of ammonia at these conditions may be taken as 500 Btu/lb. The cooling water enters at 70°F. Determine the capacity of the condenser for these conditions.

13-5 A study of heat transmission from condensing steam to cooling water in a single-tube condenser gave results for both clean and fouled tubes. For each tube the overall coefficient U was determined at a number of water velocities. The experimental results were represented by the following empirical equations:

$$\frac{1}{U_o} = \begin{cases} 0.00092 + \dfrac{1}{268\overline{V}^{0.8}} & \text{fouled tube} \\[4mm] 0.00040 + \dfrac{1}{268\overline{V}^{0.8}} & \text{clean tube} \end{cases}$$

where U_o = overall heat-transfer coefficient, Btu/ft²-h-°F
   $\overline{V}$ = water velocity, ft/s

The tubes were 0.902 in. ID and 1.000 in. OD, made of admiralty metal, for which $k = 63$ Btu/ft-h-°F. From these data calculate (a) the steam-film coefficient (based on steam-side area), (b) the water-film coefficient when the water velocity is 1 ft/s (based on water-side area), (c) value of h_{do} for the scale in the fouled tube, assuming the clean tube was free of deposit.

13-6 A 25-mm-OD copper tube is to be used to boil water at atmospheric pressure. (a) Estimate the maximum heat flux obtainable as the temperature of the copper surface is increased. (b) If the temperature of the copper surface is 210°C, calculate the boiling film coefficient and the heat flux. The interfacial tension of water at temperatures above 80°C is given by

$$\sigma = 78.38(1 - 0.0025T)$$

where σ = interfacial tension, dyn/cm
   T = temperature, °C

REFERENCES

1   Breen, B. P., and J. W. Westwater: *Chem. Eng. Prog.,* **58**(7): 67 (1962).
2   Colburn, A. P., and O. A. Hougen: *Ind. Eng. Chem.,* **26:**1178 (1934).
3   Cryder, D. S., and A. C. Finalborgo: *Trans. AIChE,* **33:**346 (1937).
4   Drew, T. B., W. M. Nagle, and W. Q. Smith: *Trans. AIChE,* **31:**605 (1935).
5   Forster, H. K., and N. Zuber: *AIChE J.,* **1:**531 (1955).
6   Gebhart, B.: "Heat Transfer," 2d ed., McGraw-Hill, New York, 1971; (*a*) pp. 424–426, (*b*) p. 446.
7   Hosler, E. R., and J. W. Westwater: *ARS J.,* **32:**553 (1962).
8   Hsu, Y. Y., and J. W. Westwater: *Chem. Eng. Prog. Symp. Ser.,* **56**(30): 15 (1960).
9   Jakob, M., and W. Fritz: *VDI-Forschungsh.,* **2:**434 (1931).
10 Kern, D. Q.: "Process Heat Transfer," McGraw-Hill, New York, 1950; (*a*) pp. 256ff., (*b*) pp. 313ff.
11 McAdams, W. H.: "Heat Transmission," 3d ed., McGraw-Hill, New York, 1954; (*a*) pp. 330ff., (*b*) pp. 351ff.
12 McAdams, W. H., J. N. Addoms, P. M. Rinaldo, and R. S. Day: *Chem. Eng. Prog.,* **44:**639 (1948).
13 Nusselt, W.: *VDI Z.,* **60:**541, 569 (1916).
14 Welch, J. F., and J. W. Westwater: *Proc. Int. Heat Transfer Conf. Lond., 1961–1962,* p. 302 (1963).
15 Zuber, N.: *Trans. ASME,* **80:**711 (1958).
16 Zuber, N., J. W. Westwater, and M. Tribus: *Proc. Int. Heat Transfer Conf. Lond. 1961–1962,* p. 230 (1963).

RADIATION HEAT TRANSFER

Radiation, which may be considered to be energy streaming through empty space at the speed of light, may originate in various ways. Some types of material will emit radiation when they are treated by external agencies, such as electron bombardment, electric discharge, or radiation of definite wavelengths. Radiation due to these effects is of minor significance in chemical engineering, and will not be discussed here. All substances at temperatures above absolute zero emit radiation that is independent of external agencies. Radiation that is the result of temperature only is called thermal radiation, and the present discussion is restricted to radiation of this type.

Fundamental facts concerning radiation Radiation moves through space in straight lines, or beams, and only substances in sight of a radiating body can intercept radiation from that body. The fraction of the radiation falling on a body that is reflected is called the *reflectivity* ρ. The fraction that is absorbed is called the *absorptivity* α. The fraction that is transmitted is called the *transmissivity* τ. The sum of these fractions must be unity, or

$$\alpha + \rho + \tau = 1 \tag{14-1}$$

Radiation as such is not heat, and when transformed into heat on absorption, it is no longer radiation. In practice, however, reflected or transmitted radiation usually falls on other absorptive bodies, and is eventually converted into heat, perhaps after many successive reflections.

The maximum possible absorptivity is unity, attained only if the body absorbs all radiation incident upon it and reflects or transmits none. A body meeting this requirement is called a *blackbody*.

The complex subject of thermal radiation transfer has received much study in recent years and is covered in a number of texts.[5,8] The following introductory treatment discusses the following topics: emission of radiation, absorption by opaque solids, radiation between surfaces, radiation to and from semitransparent materials, and combined heat transfer by conduction-convection and radiation.

EMISSION OF RADIATION

The radiation emitted by any given mass of substance is independent of that being emitted by other material in sight of, or in contact with, the mass. The *net* energy gained or lost by a body is the difference between the energy emitted by the body and that absorbed by it from the radiation reaching it from other bodies. Heat flow by conduction and convection may also be taking place independently of the radiation.

When bodies at different temperatures are exposed in sight of one another inside an enclosure, the hotter bodies lose energy by emission of radiation faster than they receive energy by absorption of radiation from the cooler bodies, and the temperatures of the hotter bodies decrease. Simultaneously the cooler bodies absorb energy from the hotter ones faster than they emit energy, and the temperatures of the cooler bodies increase. The process reaches equilibrium when all the bodies reach the same temperature, just as in heat flow by conduction and convection. The conversion of radiation into heat on absorption and the attainment of temperature equilibrium through the net transfer of radiation justify the usual practice of calling radiation "heat."

Wavelength of radiation   Known electromagnetic radiations cover an enormous range of wavelengths, from the short cosmic rays having wavelengths of about 10^{-11} cm to longwave broadcasting waves having lengths of 1,000 m or more.

Radiation of a single wavelength is called *monochromatic*. An actual beam of radiation consists of many monochromatic beams. Although radiation of any wavelength from zero to infinity is, in principle, convertible into heat on absorption by matter, the portion of the electromagnetic spectrum that is of importance in heat flow lies in the wavelength range between 0.5 and 50 μm. Visible light covers a wavelength range of about 0.38 to 0.78 μm, and thermal radiation at ordinary industrial temperatures has wavelengths in the infrared spectrum, which includes waves just longer than the longest visible waves. At temperatures above about 500°C heat radiation in the visible spectrum becomes significant, and the phrases "red heat" and "white heat" refer to this fact. The higher the temperature of the radiating body, the shorter the predominant wavelength of the thermal radiation emitted by it.

For a given temperature, the rate of thermal radiation varies with the state of aggregation of the substance. Monatomic and diatomic gases such as oxygen, argon, and nitrogen radiate weakly, even at high temperatures. Under industrial conditions, these gases neither emit nor absorb appreciable amounts of radiation. Polyatomic gases, including water vapor, carbon dioxide, ammonia, sulfur dioxide, and hydrocarbons, emit and absorb radiation appreciably at furnace temperatures but only in certain bands of wavelength. Solids and liquids emit radiation over the entire spectrum.

EMISSIVE POWER   The monochromatic energy emitted by a radiating surface depends on the temperature of the surface and on the wavelength of the radiation. At constant surface temperature, a curve can be plotted showing the rate of energy emission as a function of the wavelength. Typical curves of this type are shown in Fig. 14-1. Each curve rises steeply to a maximum and decreases asymptotically to zero emission at very large wavelengths. The unit chosen for measuring the monochromatic radiation is based on the fact that, from a small area of a radiating surface, the energy emitted is "broadcast" in all directions through any hemisphere centered on the radiation area. The monochromatic radiation emitted in this manner from unit area in unit time is called the *monochromatic radiating power* W_λ. The ordinates in Fig. 14-1 are values of W_λ.

For the entire spectrum of the radiation from a surface, the total radiating power W is the sum of all the monochromatic radiations from the surface, or, mathematically,

$$W = \int_0^\infty W_\lambda \, d\lambda \qquad (14\text{-}2)$$

Graphically, W is the entire area under the curve of Fig. 14-1 from wavelengths of zero to infinity. Physically, the total radiating power is the total radiation of all wavelengths emitted by unit area in unit time in all directions through a hemisphere centered on the area.

Blackbody radiation; emissivity   As shown later [Eq. (14-10)], a blackbody has the maximum attainable emissive power at any given temperature and is the standard to which all other radiators are referred. The ratio of the total emissive power W of a body to that of a blackbody W_b is by definition the *emissivity* ε of the body. Thus,

$$\varepsilon \equiv \frac{W}{W_b} \qquad (14\text{-}3)$$

The *monochromatic emissivity* ε_λ is the ratio of the monochromatic emissive power to that of blackbody at the same wavelength, or

$$\varepsilon_\lambda \equiv \frac{W_\lambda}{W_{b,\lambda}} \qquad (14\text{-}4)$$

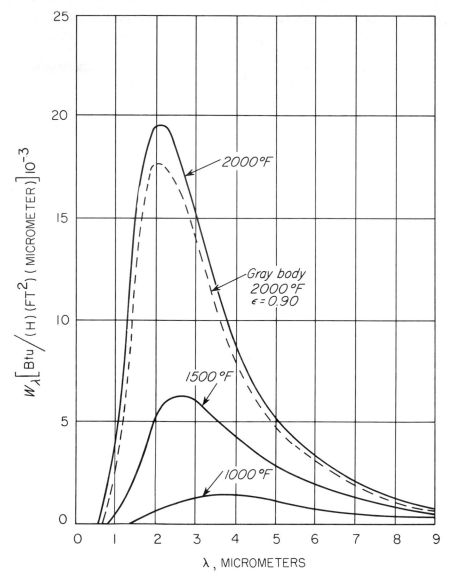

FIGURE 14-1
Energy distribution in spectra of blackbodies and gray bodies.

If the monochromatic emissivity of a body is the same for all wavelengths, the body is called a *gray body*.

Emissivities of solids   Emissivities of solids are tabulated in standard references.[6,7] Emissivity usually increases with temperature. Emissivities of polished metals are low, in the range 0.03 to 0.08. Those of most oxidized metals range from 0.6 to 0.85, those of nonmetals such as refractories, paper, boards, and building materials, from 0.65 to 0.95, and those of paints, other than aluminum paint, from 0.80 to 0.96.

Practical source of blackbody radiation No actual substance is a blackbody, although some materials, such as certain grades of carbon black, do approach blackness. An experimental equivalent of a blackbody is an isothermal enclosure containing a small peephole. If a sight is taken through the peephole on the interior wall of the enclosure, the effect is the same as viewing a blackbody. The radiation emitted by the interior of the walls or admitted from outside through the peephole is completely absorbed after successive reflections, and the overall absorptivity of the interior surface is unity.

LAWS OF BLACKBODY RADIATION A basic relationship for blackbody radiation is the Stefan-Boltzmann law, which states that the total emissive power of a blackbody is proportional to the fourth power of the absolute temperature, or

$$W_b = \sigma T^4 \qquad (14\text{-}5)$$

where σ is a universal constant depending only upon the units used to measure T and W_b. The Stefan-Boltzmann law is an exact consequence of the laws of thermodynamics and electromagnetism.

The distribution of energy in the spectrum of a blackbody is known accurately. It is given by Planck's law,

$$W_{b,\lambda} = \frac{2\pi \mathbf{h} c^2 \lambda^{-5}}{e^{\mathbf{h}c/\mathbf{k}\lambda T} - 1} \qquad (14\text{-}6)$$

where $W_{b,\lambda}$ = monochromatic emissive power of blackbody
$\quad \mathbf{h}$ = Planck's constant
$\quad c$ = speed of light
$\quad \lambda$ = wavelength of radiation
$\quad \mathbf{k}$ = Boltzmann's constant
$\quad T$ = absolute temperature

Equation (14-6) can be written

$$W_{b,\lambda} = \frac{C_1 \lambda^{-5}}{e^{C_2/\lambda T} - 1} \qquad (14\text{-}7)$$

where C_1 and C_2 are constants. The units of the various quantities and magnitudes of the constants in Eqs. (14-5) and (14-6) are given in the list of symbols, page 395.

Plots of $W_{b,\lambda}$ vs. λ from Eq. (14-6) are shown as solid lines in Fig. 14-1 for blackbody radiation at temperatures of 1000, 1500, and 2000°F. The dotted line shows the monochromatic radiating power of a gray body of emissivity 0.9 at 2000°F.

Planck's law can be shown to be consistent with the Stefan-Boltzmann law by substituting $W_{b,\lambda}$ from Eq. (14-6) into Eq. (14-2) and integrating.

At any given temperature, the maximum monochromatic radiating power is attained at a definite wavelength, denoted by λ_{max}. Wien's displacement law states that λ_{max} is inversely proportional to the absolute temperature, or

$$T\lambda_{max} = C \tag{14-8}$$

The constant C is 2,890 when λ_{max} is in micrometers and T is in kelvins or 5,200 when T is in degrees Rankine.

Wien's law also can be derived from Planck's law [Eq. (14-6)] by differentiating with respect to λ, equating the derivative to zero, and solving for λ_{max}.

ABSORPTION OF RADIATION BY OPAQUE SOLIDS

When radiation falls on a solid body, a definite fraction ρ may be reflected and the remaining fraction $1 - \rho$ enters the solid to be either transmitted or absorbed. Most solids (other than glasses, certain plastics, quartz, and some minerals) absorb radiation of all wavelengths so readily that, except in thin sheets, the transmissivity τ is zero, and all nonreflected radiation is completely absorbed in a thin surface layer of the solid. The absorption of radiation by an opaque solid is therefore a surface phenomenon, not a volume phenomenon, and the interior of the solid is not of interest in the absorption of radiation. The heat generated by the absorption can flow into or through the mass of an opaque solid only by conduction.

Reflectivity and absorptivity of opaque solids   Since the transmissivity of an opaque solid is zero, the sum of the reflectivity and the absorptivity is unity, and the factors that influence reflectivity affect absorptivity in the opposite sense. In general, the reflectivity of an opaque solid depends on the temperature and character of the surface, the material of which the surface is made, the wavelength of the incident radiation, and the angle of incidence. Two main types of reflection are encountered, specular and diffuse. The first is characteristic of smooth surfaces such as polished metals; the second is found in reflection from rough surfaces or from dull, or matte, surfaces. In specular reflection, the reflected beam makes a definite angle with the surface, and the angle of incidence equals the angle of reflection. The reflectivity from these surfaces approaches unity, and the absorptivity approaches zero. Matte, or dull, surfaces reflect diffusely in all directions, there is no definite angle of reflection, and the absorptivity can approach unity. Rough surfaces, in which the scale of roughness is large in conparison with the wavelength of the incident radiation, will reflect diffusely even if the radiation from the individual units of roughness is specular. Reflectivities may be either large or small from rough surfaces, depending upon the reflective characteristic of the material itself. Most industrial surfaces of interest to the chemical engineer give diffuse reflection, and in treating practical cases, the important simplifying assumption can be made that reflectivity and absorptivity are independent of angle of incidence. This assumption is equivalent to the *cosine law*, which states that for a perfectly diffusing surface the intensity (or brightness, in the case of visible light) of the radiation leaving the surface is independent of the angle from which the surface is viewed. This is true whether the radiation is emitted by the surface, giving *diffuse radiation*, or is reflected by it, giving *diffuse reflection*.

The reflectivity may vary with the wavelength of the incident radiation, and the absorptivity of the entire beam is then a weighted average of the monochromatic absorptivities and depends upon the entire spectrum of the incident radiation.

The absorptivity of a gray body, like the emissivity, is the same for all wavelengths. If the surface of the gray body gives diffuse radiation or reflection, its monochromatic absorptivity is also independent of the angle of incidence of the radiant beam. The total absorptivity equals the monochromatic absorptivity and is also independent of the angle of incidence.

Kirchhoff's law   An important generalization concerning the radiating power of a substance is Kirchhoff's law, which states that, at temperature equilibrium, the ratio of the total radiating power of any body to the absorptivity of that body depends only upon the temperature of the body. Thus, consider any two bodies in temperature equilibrium with common surroundings. Kirchhoff's law states that

$$\frac{W_1}{\alpha_1} = \frac{W_2}{\alpha_2} \tag{14-9}$$

where W_1, W_2 = total radiating powers of two bodies
α_1, α_2 = the absorptivities of two bodies

This law applies to both monochromatic and total radiation.

If the first body referred to in Eq. (14-9) is a blackbody, $\alpha_1 = 1$, and

$$W_1 = W_b = \frac{W_2}{\alpha_2} \tag{14-10}$$

where W_b denotes the total radiating power of a blackbody. Thus

$$\alpha_2 = \frac{W_2}{W_b} \tag{14-11}$$

But, by definition, the emissivity of the second body ε_2 is

$$\varepsilon_2 = \frac{W_2}{W_b} = \alpha_2 \tag{14-12}$$

Thus, when any body is at temperature equilibrium with its surroundings, its emissivity and absorptivity are equal. This relationship may be taken as another statement of Kirchhoff's law. In general, except for blackbodies or gray bodies, absorptivity and emissivity are not equal if the body is not in thermal equilibrium with its surroundings.

The absorptivity and emissivity, monochromatic or total, of a blackbody are, of course, both unity. The cosine law also applies exactly to a blackbody, as the reflectivity is zero for all wavelengths and all angles of incidence.

Kirchhoff's law applies to volumes as well as to surfaces. Since absorption by an opaque solid is effectively confined to a thin layer at the surface, the radiation

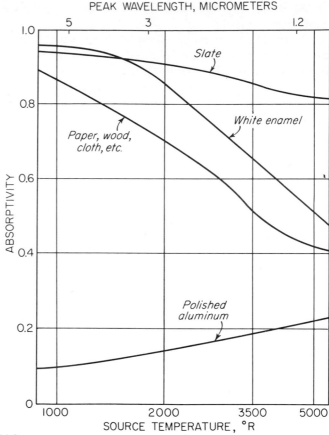

FIGURE 14-2
Absorptivities of various solids versus source temperature and peak wavelength of incident radiation. [*By permission of author and publishers, from H. C. Hottel, p. 62 in W. H. McAdams, "Heat Transmission." Copyright by author, 1954, McGraw-Hill Book Company.*]

emitted from the surface of the body originates in this same surface layer. Radiating substances absorb their own radiation, and radiation emitted by the material in the interior of the solid is also absorbed in the interior and does not reach the surface.

Because the energy distribution in the incident radiation depends upon the temperature and character of the originating surface, the absorptivity of the receiving surface may also depend upon these properties of the originating surface. Kirchhoff's law does not, therefore, always apply to nonequilibrium radiation. If, however, the receiving surface is gray, a constant fraction, independent of wavelength, of the incident radiation is absorbed by the receiving surface, and Kirchhoff's law applies whether or not the two surfaces are at the same temperature.

The majority of industrial surfaces, unfortunately, are not gray, and their absorptivities vary strongly with the nature of the incident radiation. Figure 14-2 shows how the absorptivity of various solids varies with the peak wavelength of the incident radiation and thus with the temperature of the source.[4a] Some solids, such as slate, are almost truly gray, and their absorptivities are almost constant. For polished metallic surfaces the absorptivity α_2 rises with the absolute temperature of the source T_1 and also that of surface T_2 according to the equation

$$\alpha_2 = k_1 \sqrt{T_1 T_2} \tag{14-13}$$

where k_1 is a constant. For most surfaces, however, the absorptivity follows a curve like that indicated for paper, wood, cloth, etc. Such surfaces have high absorptivities for radiation of long wavelengths originating from sources below about 1000°F; as the source temperature rises above this value, the absorptivity falls, sometimes very markedly. With a few materials the absorptivity rises once again when the source temperature is very high.

RADIATION BETWEEN SURFACES

The total radiation from a unit area of an opaque body of area A_1, emissivity ε_1, and absolute temperature T_1 is

$$\frac{q}{A_1} = \sigma \varepsilon_1 T_1^4 \tag{14-14}$$

Practically, however, substances do not emit radiation into empty space that is at the absolute zero of temperature. Even in radiation to a clear night sky, the radiated energy is partially absorbed by the water and carbon dioxide in the atmosphere, and part of the absorbed energy is radiated back to the surface. In the usual situation, the energy emitted by any body is intercepted by other substances in sight of the body, which also are radiators, and their radiation falls on the body, to be absorbed or reflected. For example, a steam line in a room is surrounded by the walls, floor, and ceiling of the room, all of which are radiating to the pipe, and although the pipe loses more energy than it absorbs from its surroundings, the net loss by radiation is less than that calculated from Eq. (14-14).

In furnaces and other high-temperature equipment, where radiation is particularly important, the usual objective is to obtain a controlled rate of net heat exchange between one or more hot surfaces, called *sources*, and one or more cold surfaces, called *sinks*. In many cases the hot surface is a flame, but exchange of energy between surfaces is common, and a flame can be considered to be a special form of translucent surface. The following treatment is limited to the radiant-energy transfer between opaque surfaces in the absence of any absorbing medium between them.

The simplest type of radiation between two surfaces is where each surface can see only the other, e.g., where the surfaces are very large parallel planes, as shown in Fig. 14-3a, and where both surfaces are black. The energy emitted by the first plane is σT_1^4. All the radiation from each of the surfaces falls on the other surface and is

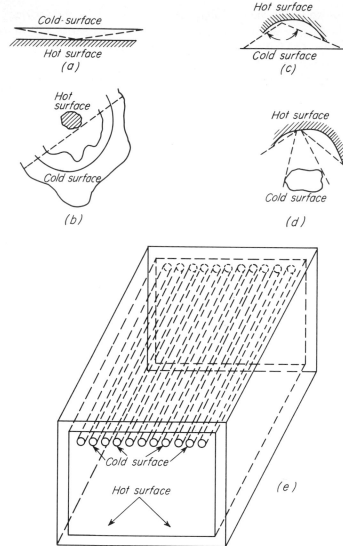

FIGURE 14-3
Angle of vision in radiant heat flow.

completely absorbed, so the net loss of energy by the first plane and the net gain of energy by the second (assuming that $T_1 > T_2$) is $\sigma T_1^4 - \sigma T_2^4$, or $\sigma(T_1^4 - T_2^4)$.

Actual engineering problems differ from this simple situation in the following ways: (1) One or both of the surfaces of interest see other surfaces. In fact, an element of surface in a concave area sees a portion of its own surface. (2) No actual surface is exactly black, and the emissivities of the surfaces must often be considered.

Angle of vision Qualitatively, the interception of radiation from an area element of a surface by another surface of finite size can be visualized in terms of the angle of vision, which is the solid angle subtended by the finite surface at the radiating element. The solid angle subtended by a hemisphere is 2π steradians. This is the maximum angle of vision that can be subtended at any area element by a plane surface in sight of the element. It will be remembered that the total radiating power of an area element is defined to take this fact into account. If the angle of vision is less than 2π steradians, only a fraction of the radiation from the area element will be intercepted by the receiving area and the remainder will pass on to be absorbed by other surfaces in sight of the remaining solid angle. Some of the hemispherical angle of vision of an element of a concave surface is subtended by the originating surface itself.

Figure 14-3 shows several typical radiating surfaces. Figure 14-3a shows how, in two large parallel planes, an area element on either plane is subtended by a solid angle of 2π steradians by the other. The radiation from either plane cannot escape being intercepted by the other. A point on the hot body of Fig. 14-3b sees only the cold surface, and the angle of vision is again 2π steradians. Elements of the cold surface, however, see, for the most part, other portions of the cold surface, and the angle of vision for the hot body is small. This effect of self-absorption is also shown in Fig. 14-3c, where the angle of vision of an element of the hot surface subtended by the cold surface is small. In Fig. 14-3d, the cold surface subtends a small angle at the hot surface, and the bulk of the radiation from the hot surface passes on to some undetermined background. Figure 14-3e shows a simple muffle furnace, in which the radiation from the hot floor, or source, is intercepted partly by the row of tubes across the top of the furnace, which form the sink, and partly by the refractory walls and the refractory ceiling behind the tubes. The refractory in such assemblies is assumed to absorb and emit energy at the same rate, so the net energy effect at the refractory is zero. The refractory ceiling absorbs the energy that passes between the tubes and reradiates it to the backs of the tubes.

If attention is to be focused on the net energy received by the cold surface, the words "hot" and "cold" in Fig. 14-3 may be interchanged, and the same qualitative conclusions hold.

Square-of-the-distance effect The energy from a small surface that is intercepted by a large one depends only upon the angle of vision. It is independent of the distance between the surfaces. The energy received per *unit area* of the receiving surface, however, is inversely proportional to the square of the distance between the surfaces, as shown by the following discussion.

The rate of energy received per unit area of the receiving surface is called the intensity I of the radiation. For diffuse radiation the cosine law can be used to find how the intensity of radiation varies with distance and orientation of the receiving surface from the emitting surface. Consider the element of emitting surface dA_1 shown in Fig. 14-4, at the center of a hemispherical surface A_2 with a radius r. The ring-shaped element of the receiving surface dA_2 has an area $2\pi r^2 \sin\phi \, d\phi$, where ϕ is the angle between the normal to dA_1 and the radius PQ joining dA_1 and dA_2. The intensity at the point directly above dA_1 is denoted by dI_0; that at any other differential element of A_1 is dI. By the cosine law of diffuse radiation

$$dI = dI_0 \cos\phi \tag{14-15}$$

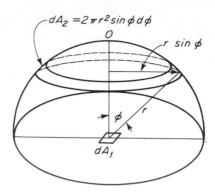

FIGURE 14-4
Diffuse radiation to a hemispherical
surface.

The relation between emissive power W_1 of the emitting surface and the intensity is found as follows. The rate of energy reception by the element of area dA_2, dq_{dA_2}, is, by Eq. (14-15),

$$dq_{dA_2} = dI\, dA_2 = dI_o \cos \phi\, dA_2 \qquad (14\text{-}16)$$

Since $dA_2 = 2\pi r^2 \sin \phi\, d\phi$,

$$dq_{dA_2} = dI_o\, 2\pi r^2 \sin \phi \cos \phi\, d\phi \qquad (14\text{-}17)$$

The rate of emission from area dA_1 must equal the rate at which energy is received by the total area A_2, since all the radiation from dA_1 impinges on some part of A_2. The rate of reception by A_2 is found by integrating dq_{dA_2} over area A_2. Hence,

$$W_1\, dA_1 = \int_{A_2} dq_{dA_2} = \int_0^{\pi/2} 2\pi\, dI_o\, r^2 \sin \phi \cos \phi\, d\phi$$
$$= \pi\, dI_o\, r^2 \qquad (14\text{-}18)$$

Thus
$$dI_0 = \frac{W_1}{\pi r^2}\, dA_1 \qquad (14\text{-}19)$$

By substitution from Eq. (14-15)

$$dI = \frac{W_1}{\pi r^2}\, dA_1 \cos \phi \qquad (14\text{-}20)$$

Quantitative calculation of radiation between black surfaces The above considerations can be treated quantitatively by setting up a differential equation for the net radiation between two elementary areas and integrating the equation for definite types of arrangement of the surfaces. The two plane elements of area dA_1 and dA_2 in Fig. 14-5 are separated by a distance r and are set at any arbitrary orientation to one another that permits a connecting straight line to be drawn between them. In other words, element dA_1 must see element dA_2; at least some of the radiation from dA_1 must impinge upon dA_2. Angles ϕ_1 and ϕ_2 are the angles between the connecting straight line and the normals to dA_1 and dA_2, respectively.

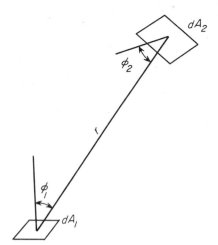

FIGURE 14-5
Differential areas for radiation.

Since PQ is not normal to dA_2 as it is in Fig. 14-4, the rate of energy reception by element dA_2 of radiation originating at dA_1 is

$$dq_{dA_1 \to dA_2} = dI_1 \cos \phi_2 \, dA_2 \qquad (14\text{-}21)$$

where dI_1 is the intensity, at area dA_2, of radiation from area dA_1. From Eqs. (14-20) and (14-21), since element dA_1 is black,

$$dq_{dA_1 \to dA_2} = \frac{W_1}{\pi r^2} \, dA_1 \cos \phi_1 \cos \phi_2 \, dA_2$$

$$= \frac{\sigma T_1^4}{\pi r^2} \cos \phi_1 \cos \phi_2 \, dA_1 \, dA_2 \qquad (14\text{-}22)$$

Similarly, for radiation from dA_2 which impinges on dA_1

$$dq_{dA_2 \to dA_1} = \frac{\sigma T_2^4}{\pi r^2} \cos \phi_1 \cos \phi_2 \, dA_1 \, dA_2 \qquad (14\text{-}23)$$

The net rate of transfer dq_{12} between the two area elements is then found from the difference between the rates indicated in Eqs. (14-22) and (14-23), to give

$$dq_{12} = \sigma \frac{\cos \phi_1 \cos \phi_2 \, dA_1 \, dA_2}{\pi r^2} (T_1^4 - T_2^4) \qquad (14\text{-}24)$$

The integration of Eq. (14-24) for a given combination of finite surfaces is usually a laborious multiple integration based on the geometry of the two planes and their relation to each other. The resulting equation for any of these situations can be written in the form

$$q_{12} = \sigma A F (T_1^4 - T_2^4) \qquad (14\text{-}25)$$

where q_{12} = net radiation between two surfaces
        A = area of either of two surfaces, chosen arbitrarily
        F = dimensionless geometric factor

The factor F is called the view factor or angle factor; it depends upon the geometry of the two surfaces, their spatial relationship with each other, and the surface chosen for A.

If surface A_1 is chosen for A, Eq. (14-25) can be written

$$q_{12} = \sigma A_1 F_{12}(T_1^4 - T_2^4) \tag{14-26}$$

If surface A_2 is chosen,

$$q_{12} = \sigma A_2 F_{21}(T_1^4 - T_2^4) \tag{14-27}$$

Comparing Eqs. (14-26) and (14-27) gives

$$A_1 F_{12} = A_2 F_{21} \tag{14-28}$$

Factor F_{12} may be regarded as the fraction of the radiation leaving area A_1 that is intercepted by area A_2. If surface A_1 can see only surface A_2, the view factor F_{12} is unity. If surface A_1 sees a number of other surfaces, and if its entire hemispherical angle of vision is filled by these surfaces,

$$F_{11} + F_{12} + F_{13} + \cdots = 1.0 \tag{14-29}$$

The factor F_{11} covers the portion of the angle of vision subtended by other portions of body A_1. If the surface of A_1 cannot see any portion of itself, F_{11} is zero. The net radiation associated with an F_{11} factor is, of course, zero.

In some situations the view factor may be calculated simply. For example, consider a small blackbody of area A_2 having no concavities and surrounded by a large black surface of area A_1. The factor F_{21} is unity, as area A_2 can see nothing but area A_1. The factor F_{12} is, by Eq. (14-28),

$$F_{12} = \frac{F_{21} A_2}{A_1} = \frac{A_2}{A_1} \tag{14-30}$$

By Eq. (14-29)

$$F_{11} = 1 - F_{12} = 1 - \frac{A_2}{A_1} \tag{14-31}$$

As a second example consider a long duct, triangular in cross section, with its three walls at different temperatures. The walls need not be flat, but they must have no concavities; i.e., each wall must see no portion of itself. Under these conditions the view factors are given by [4b]

$$F_{12} = \frac{A_1 + A_2 - A_3}{2A_1}$$

$$F_{13} = \frac{A_1 + A_3 - A_2}{2A_1} \tag{14-32}$$

$$F_{23} = \frac{A_2 + A_3 - A_1}{2A_2}$$

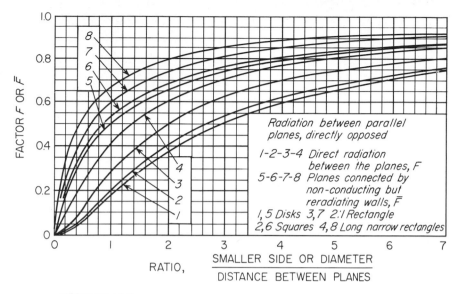

FACTOR F OR $\bar{F}$

RATIO, $\dfrac{\text{SMALLER SIDE OR DIAMETER}}{\text{DISTANCE BETWEEN PLANES}}$

Radiation between parallel planes, directly opposed

1-2-3-4 Direct radiation between the planes, F
5-6-7-8 Planes connected by non-conducting but reradiating walls, $\bar{F}$
1,5 Disks 3,7 2:1 Rectangle
2,6 Squares 4,8 Long narrow rectangles

FIGURE 14-6
View factor and interchange factor, radiation between opposed parallel disks, rectangles, and squares.

The factor F has been determined by Hottel[2] for a number of important special cases. Figure 14-6 shows the F factor for equal parallel planes directly opposed. Line 1 is for disks, line 2 for squares, line 3 for rectangles having a ratio of length to width 2:1, and line 4 is for long, narrow rectangles. In all cases, the factor F is a function of the ratio of the side or diameter of the planes to the distance between them. Figure 14-7 gives factors for radiation to tube banks backed by a layer of refractory which absorbs energy from the rays that pass between the tubes and reradiates the absorbed energy to the backs of the tubes. The factor given in Fig. 14-7 is the radiation absorbed by the tubes calculated as a fraction of that absorbed by a parallel plane of area equal to that of the refractory backing.

Allowance for refractory surfaces When the source and sink are connected by refractory walls in the manner shown in Fig. 14-3e, the factor F can be replaced by an analogous factor, called the *interchange factor* $\bar{F}$, and Eqs. (14-26) and (14-27) written as

$$q_{12} = \sigma A_1 \bar{F}_{12}(T_1^4 - T_2^4) = \sigma A_2 \bar{F}_{21}(T_1^4 - T_2^4) \tag{14-33}$$

The interchange factor $\bar{F}$ has been determined accurately for some simple situations.[3] Lines 5 to 8 of Fig. 14-6 give values of $\bar{F}$ for directly opposed parallel planes connected by refractory walls. Line 5 applies to disks, line 6 to squares, line 7 to 2:1 rectangles, and line 8 to long, narrow rectangles.

An approximate equation for $\bar{F}$ in terms of F is

$$\bar{F}_{12} = \frac{A_2 - A_1 F_{12}^2}{A_1 + A_2 - 2A_1 F_{12}} \tag{14-34}$$

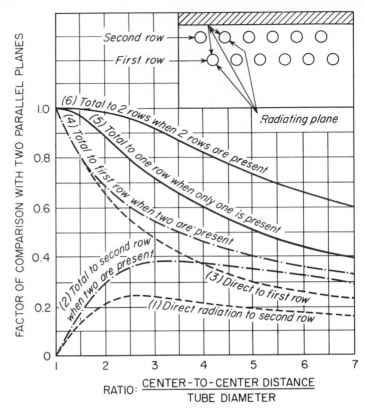

FIGURE 14-7
View factor and interchange factor between parallel plane and rows of tubes.

Equation (14-34) applies where there is but one source and one sink, where neither area A_1 nor A_2 can see itself. It is based on the assumption that the temperature of the refractory surface is constant. This last is a simplifying assumption, as a local temperature of the refractory usually varies between those of the source and of the sink.

Nonblack surfaces The treatment of radiation between nonblack surfaces, in the general case where absorptivity and emissivity are unequal and both depend upon wavelength and angle of incidence, is obviously complicated. Several important special cases can, however, be treated simply.

A simple example is a small body that is not black surrounded by a black surface. Let the areas of the enclosed and surrounding surfaces be A_1 and A_2, respectively, and let their temperatures be T_1 and T_2, respectively. The radiation from surface A_2 falling on surface A_1 is $\sigma A_2 F_{21} T_2^4$. Of this, the fraction α_1, the absorptivity of area A_1 for radiation from surface A_2, is absorbed by surface A_1. The remainder is reflected back to the black surroundings and completely reabsorbed by the area A_2.

Surface A_1 emits radiation in amount $\sigma A_1 \varepsilon_1 T_1^4$, where ε_1 is the emissivity of surface A_1. All this radiation is absorbed by the surface A_2, and none is returned by another reflection. The emissivity ε_1 and the absorptivity α_1 are not in general equal, because the two surfaces are not at the same temperature. The net energy loss by surface A_1 is

$$q_{12} = \sigma \varepsilon_1 A_1 T_1^4 - \sigma A_2 F_{21} \alpha_1 T_2^4 \tag{14-35}$$

But, by Eq. (14-28), $A_2 F_{21} = A_1$, and Eq. (14-35), after elimination of $A_2 F_{21}$, becomes

$$q_{12} = \sigma A_1 (\varepsilon_1 T_1^4 - \alpha_1 T_2^4) \tag{14-36}$$

If surface A_1 is gray, $\varepsilon_1 = \alpha_1$ and

$$q_{12} = \sigma A_1 \varepsilon_1 (T_1^4 - T_2^4) \tag{14-37}$$

In general, for gray surfaces, Eqs. (14-26) and (14-27) can be written

$$q_{12} = \sigma A_1 \mathscr{F}_{12}(T_1^4 - T_2^4) = \sigma A_2 \mathscr{F}_{21}(T_1^4 - T_2^4) \tag{14-38}$$

where $\mathscr{F}_{12}$ and $\mathscr{F}_{21}$ are the *overall interchange factors* and are functions of ε_1 and ε_2.

TWO LARGE PARALLEL PLANES   In simple cases the factor $\mathscr{F}$ can be calculated directly. Consider two large gray parallel planes at absolute temperatures T_1 and T_2, as indicated in Fig. 14-8, with emissivities ε_1 and ε_2, respectively. The energy radiated from a unit area of surface 1 equals $\sigma T_1^4 \varepsilon_1$. Part of this energy is absorbed by surface 2, and part is reflected. The amount absorbed, as shown in Fig. 14-8a, equals $\sigma T_1^4 \varepsilon_1 \varepsilon_2$. Part of the reflected beam is reabsorbed by surface 1 and part is re-reflected to surface 2. Of this re-reflected beam an amount $\sigma T_1^4 \varepsilon_1 \varepsilon_2 (1 - \varepsilon_1)(1 - \varepsilon_2)$ is absorbed. Successive reflections and absorptions lead to the following equation for the total amount of radiation originating at surface 1 that is absorbed by surface 2:

$$q_{1 \to 2} = \sigma T_1^4 \varepsilon_1 \varepsilon_2 [1 + (1 - \varepsilon_1)(1 - \varepsilon_2) + (1 - \varepsilon_1)^2 (1 - \varepsilon_2)^2 + \cdots]$$

Some of the energy originating at surface 2, as shown in Fig. 14-8b, is reflected by surface 1 and returns to surface 2, where part of it is absorbed. The amount of this energy, per unit area, is

$$q_{2 \to 2} = -\sigma T_2^4 [\varepsilon_2 - \varepsilon_2^2 (1 - \varepsilon_1) - \varepsilon_2^2 (1 - \varepsilon_1)^2 (1 - \varepsilon_2) - \cdots]$$

The total amount of energy absorbed by a unit area of surface 2 is therefore

$$\begin{aligned} q_{12} &= q_{1 \to 2} + q_{2 \to 2} \\ &= \sigma T_1^4 \varepsilon_1 \varepsilon_2 [1 + (1 - \varepsilon_1)(1 - \varepsilon_2) + (1 - \varepsilon_1)^2 (1 - \varepsilon_2)^2 + \cdots] \\ &\quad - \sigma T_2^4 \{\varepsilon_2 - \varepsilon_2^2 (1 - \varepsilon_1)[1 + (1 - \varepsilon_1)(1 - \varepsilon_2) + \cdots]\} \end{aligned}$$

Let

$$x = (1 - \varepsilon_1)(1 - \varepsilon_2)$$

Then

$$q_{12} = \sigma T_1^4 \varepsilon_1 \varepsilon_2 (1 + x + x^2 + \cdots) - \sigma T_2^4 [\varepsilon_2 - \varepsilon_2^2 (1 - \varepsilon_1)(1 + x + x^2 + \cdots)]$$

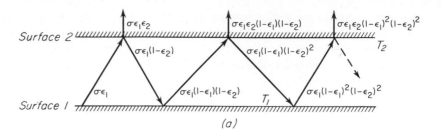

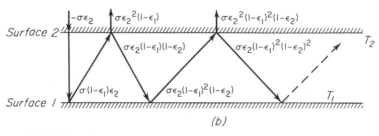

FIGURE 14-8
Evaluation of overall interchange factor for large gray parallel planes: (a) energy originating at surface 1 which is absorbed by unit area of surface 2; (b) energy originating at surface 2 which is reabsorbed by surface 2.

But

$$1 + x + x^2 + \cdots = \frac{1}{1 - x}$$

Thus

$$q_{12} = \sigma T_1^4 \varepsilon_1 \varepsilon_2 \frac{1}{1 - x} - \sigma T_2^4 \left[\varepsilon_3 - \varepsilon_2^2 (1 - \varepsilon_1) \frac{1}{1 - x} \right]$$

Substituting for x and simplifying gives

$$q_{12} = \frac{\sigma(T_1^4 - T_2^4)}{(1/\varepsilon_1) + (1/\varepsilon_2) - 1}$$

Comparison with Eq. (14-38) shows that

$$\mathscr{F}_{12} = \frac{1}{(1/\varepsilon_1) + (1/\varepsilon_2) - 1} \tag{14-39}$$

ONE GRAY SURFACE COMPLETELY SURROUNDED BY ANOTHER   Let the area of the enclosed body be A_1 and that of the enclosure be A_2. The overall interchange factor for this case is given by

$$\mathscr{F}_{12} = \frac{1}{(1/\varepsilon_1) + (A_1/A_2)[(1/\varepsilon_2) - 1]} \tag{14-40}$$

Equation (14-40) applies strictly to concentric spheres or concentric cylinders, but it can be used without serious error for other shapes. The case of a gray body surrounded by a black one can be treated as a special case of Eq. (14-40) by setting $\varepsilon_2 = 1.0$. Under these conditions $\mathscr{F}_{12} = \varepsilon_1$.

For gray surfaces in general the following approximate equation may be used to calculate the overall interchange factor.

$$\mathscr{F}_{12} = \frac{1}{(1/\overline{F}_{12}) + [(1/\varepsilon_1) - 1] + (A_1/A_2)[(1/\varepsilon_2) - 1]} \tag{14-41}$$

where ε_1 and ε_2 are the emissivities of source and sink, respectively. If no refractory is present, F is used in place of $\overline{F}$.

Gebhart[1a] describes a direct method for calculating $\mathscr{F}$ in enclosures of gray surfaces where more than two radiating surfaces are present. Problems involving nongray surfaces are discussed in the literature. [4]

RADIATION TO SEMITRANSPARENT MATERIALS

Many substances of industrial importance are to some extent transparent to the passage of radiant energy. Solids such as glass and some plastics, thin layers of liquid, and many gases and vapors are semitransparent materials. Their transmissivity and absorptivity depend on the length of the path of the radiation and also on the wavelength of the beam.

Attenuation; absorption length As shown below, a material may have very different absorptivities for radiation of different wavelengths. To classify a given material quantitatively as to its ability to absorb or transmit radiation of wavelength λ it is necessary to define an *absorption length* (or *optical path length*) L_λ in the material. This length is the distance of penetration into the material at which the incident radiation has been attenuated a given amount; i.e., the intensity of the radiation has been reduced to a given fraction of the intensity of the incident beam. The fraction usually used is $1/e$, where e is the base of natural logarithms.

The attenuation of an incident radiant beam with a monochromatic intensity $I_{0,\lambda}$ is shown in Fig. 14-9. At distance x from the receiving surface the intensity is reduced to I_λ. By assumption, the attenuation per unit length dI_λ/dx at any given value of x is proportional to the intensity I_λ at that location, or

$$-\frac{dI_\lambda}{dx} = \mu_\lambda I_\lambda \tag{14-42}$$

where μ_λ is the *absorption coefficient* for radiation of wavelength λ. Separating the variables in Eq. (14-42) and integrating between limits, with the boundary condition that $I_\lambda = I_{\lambda,0}$ at $x = 0$, gives

$$\frac{I_\lambda}{I_{o,\lambda}} = e^{-\mu_\lambda x} \tag{14-43}$$

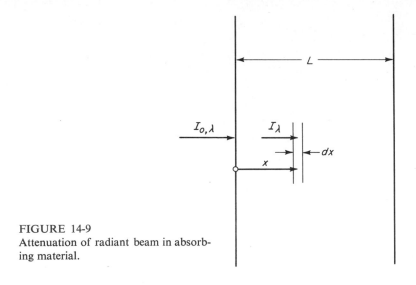

FIGURE 14-9
Attenuation of radiant beam in absorbing material.

The absorption length L_λ is the value of x such that the attenuation is $1/e$. Setting the left-hand side of Eq. (14-43) equal to e^{-1} and setting x equal to L_λ gives

$$L_\lambda = \frac{1}{\mu_\lambda} \tag{14-44}$$

The absorption length is therefore the reciprocal of the absorption coefficient.

If the total thickness L of the material is many times larger than L_λ, the material is said to be opaque to radiation of that wavelength. Most solids are opaque to thermal radiation of all wavelengths. If L is less than a few multiples of L_λ, however, the material is said to be transparent or semitransparent. The absorption length L_λ varies not only with wavelength of the incident radiation but may also vary with the temperature and density of the absorbing material. This is especially true of ab-sorbing gases.

Radiation to layers of liquid or solid In moderately thick layers all solids and liquids are opaque and totally absorb whatever radiation passes into them. In thin layers, however, most liquids and some solids absorb only a fraction of the incident radiation and transmit the rest, depending on the thickness of the layer and the wavelength of the radiation. The variation of absorptivity with thickness and wave-length for thin layers of water is shown in Fig. 14-10. Very thin layers (0.01 mm thick) transmit most of the radiation of wavelengths between 1 and 8 μm, except for ab-sorption peaks at 3 and 6 μm. Layers a few millimeters thick, however, are transparent to visible light (0.38 to 0.78 μm) but absorb virtually all radiant energy with wave-lengths greater than 1.5 μm. For heat-transfer purposes, therefore, such layers of water may be considered to have an absorptivity of 1.0.

Layers of solids such as thin films of plastic behave similarly, but the absorption peaks are usually less well marked. Ordinary glass is also transparent to radiation of

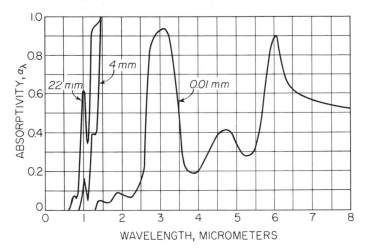

FIGURE 14-10
Spectral distribution of absorptivity of thin layers of water. [*By permission, from H. Gröber, S. Erk, and U. Grigull, "Fundamentals of Heat Transfer," 3d ed., p. 442. Copyright, 1961, McGraw-Hill Book Company.*]

short wavelengths and opaque to that of longer wavelengths. This is the cause of the so-called *greenhouse effect*, in which the contents of a glass-walled enclosure exposed to sunlight become hotter than the surroundings outside the enclosure. Radiation from the sun's surface, at about 10,000°R, is chiefly shortwave radiation and passes readily through the glass; radiation from inside the enclosure, from surfaces at say 80°F (27°C), is of longer wavelength and cannot pass through the glass. The interior temperature rises until convective losses from the enclosure equal the input of radiant energy.

Radiation to absorbing gases   Monatomic and diatomic gases, such as hydrogen, oxygen, helium, argon, and nitrogen, are virtually transparent to infrared radiation. More complex polyatomic molecules, including water vapor, carbon dioxide, and organic vapors, absorb radiation fairly strongly, especially radiation of specific wavelengths. The fraction of the incident radiation absorbed by a given amount of a gas or vapor depends on the length of the radiation path and on the number of molecules encountered by the radiation during its passage, i.e., on the density of the gas or vapor. Thus the absorptivity of a given gas is a strong function of its partial pressure and a weaker function of its temperature.

   If an absorbing gas is heated, it radiates to the cooler surroundings, at the same wavelengths favored for absorption. The emissivity of the gas is also a function of temperature and pressure. Because of the effect of path length, the emissivity and absorptivity of gases are defined arbitrarily in terms of a specific geometry.

   Consider a hemisphere of radiating gas of radius L, with a black element of receiving surface dA_2 located on the base of the hemisphere at its center. The rate of

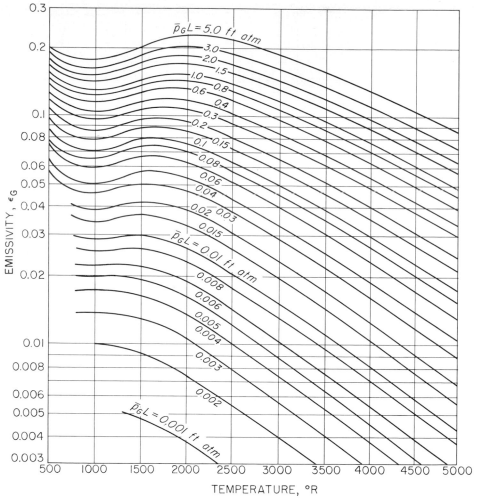

FIGURE 14-11
Emissivity of carbon dioxide at a total pressure of 1 atm. [*By permission of author and publishers, from W. H. McAdams, "Heat Transmission," 3d ed. Copyright by author, 1954, McGraw-Hill Book Company.*]

energy transfer dq_{12} from the gas to the element of area is then

$$\frac{dq_{12}}{dA_2} = \sigma T_G^4 \varepsilon_G \tag{14-45}$$

where T_G = absolute temperature of gas
ε_G = emissivity of gas (by definition)

Emissivity ε_G is therefore the ratio of the rate of energy transfer from the gas to the surface element to the rate of transfer from a black hemispherical surface of radius L and temperature T_G to the same surface element. Figure 14-11 shows how ε_G for

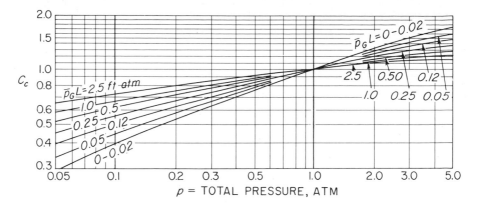

FIGURE 14-12
Correction factor C_c for converting emissivity of CO_2 at 1 atm total pressure to emissivity at p atm. [*By permission of author and publishers, from W. H. McAdams, "Heat Transmission," 3d ed. Copyright by author, 1954, McGraw-Hill Book Company.*]

carbon dioxide varies with radius L and partial pressure $\bar{p}_G$. Figure 14-11 applies at a total pressure p of 1 atm; Fig. 14-12 gives the correction factor C_c for finding ε_G at other total pressures.

When the gas temperature T_G and the surface temperature T_2 are the same, the gas absorptivity α_G, by Kirchhoff's law, equals the emissivity ε_G. When these temperatures differ, α_G and ε_G are not equal; however, Fig. 14-11 can be used to find the gas absorptivity at a total pressure of 1 atm by evaluating α_G as the emissivity at T_2 and at $\bar{p}_G L(T_2/T_G)$, instead of at $\bar{p}_G L$, and multiplying the result by $(T_G/T_2)^{0.65}$. For total pressures other than 1 atm Fig. 14-12 again applies.

Charts similar to Figs. 14-11 and 14-12 are available for water vapor.[4d] When both carbon dioxide and water vapor are present, the total radiation is somewhat less than that calculated from the two gases separately, since each gas is somewhat opaque to radiation from the other. Correction charts to allow for this interaction have been published.[4d]

Effect of geometry on gas radiation The rate of transfer from a radiating gas to the surface of an enclosure depends on the geometry of the system and on the partial pressure of the gas, $\bar{p}_G$. In general the emissivity of a given quantity of gas must be evaluated using a mean beam length L, which is a characteristic of the particular geometry. Length L has been computed for several special cases. For low values of the product $\bar{p}_G L$ it may be shown that L equals $4V/A$, where V is the volume of radiating gas and A is the total surface area of the enclosure. For average values of $\bar{p}_G L$ the mean beam length is 0.85 to 0.90 times that when $\bar{p}_G L = 0$. Values of L for a few geometries are given in Table 14-1. Other cases, including radiation to only part of the enclosing surface, are covered in the literature.[4d]

COMBINED HEAT TRANSFER BY CONDUCTION-CONVECTION AND RADIATION

The total heat loss from a hot body to its surroundings often includes appreciable losses by conduction-convection and radiation. For example, a steam radiator or a hot pipeline in a room loses heat nearly equally by each of the two mechanisms. Since the two types of heat transfer occur in parallel, the total loss is, assuming black surroundings,

$$\frac{q_T}{A} = \frac{q_c}{A} + \frac{q_r}{A} = h_c(T_w - T) + \sigma\varepsilon_w(T_w^4 - T^4) \tag{14-46}$$

where q_T/A = total heat flux
q_c/A = heat flux by conduction-convection
q_r/A = heat flux by radiation
h_c = convective heat-transfer coefficient
ε_w = emissivity of surface
T_w = temperature of surface
T = temperature of surroundings

Equation (14-46) is sometimes written

$$\frac{q_T}{A} = (h_c + h_r)(T_w - T) \tag{14-47}$$

where h_r is a *radiation heat-transfer coefficient*, defined by

$$h_r \equiv \frac{q_r}{A(T_w - T)} \tag{14-48}$$

This coefficient, of course, depends strongly on the temperature difference $T_w - T$ and also on the absolute magnitudes of T_w and T.

Table 14-1   MEAN BEAM LENGTHS FOR GAS RADIATION TO ENTIRE SURFACE AREA OF ENCLOSURE[†]

Shape of enclosure	Mean beam length L	
	When $\bar{p}_G L = 0$	For average values of $\bar{p}_G L$
Sphere, diameter D	0.667D	0.60D
Infinite cylinder, diameter D	D	0.90D
Right circular cylinder, height = diameter = D	0.667D	0.60D
Cube, side D	0.667D	0.57D (CO_2)
Infinite parallel planes, separated by distance D	2D	1.7D (H_2O)
		1.54D (CO_2)

† By permission of author and publishers, from W. H. McAdams, "Heat Transmission," 3d ed., p. 88. Copyright by author, 1954, McGraw-Hill Book Company.

Radiation in film boiling In film boiling at a very hot surface a major fraction of the heat transfer occurs by radiation from the surface to the liquid. Equation (14-46) applies to this situation, since the surrounding liquid, as discussed earlier, has an absorptivity of unity. When radiation is active, the film of vapor sheathing the heating element is thicker than it would be if radiation were absent and the convective heat-transfer coefficient is lower than it would be otherwise. For film boiling at the surface of a submerged horizontal tube, Eq. (13-24) predicts the convective heat-transfer coefficient h_o in the absence of radiation. When radiation is present, the convective coefficient is changed to h_c, which must be found by trial from the equation[1b]

$$h_c = h_o \left(\frac{h_o}{h_c + h_r} \right)^{1/3} \tag{14-49}$$

where h_o is found from Eq. (13-24) and h_r from Eqs. (14-46) and (14-48). Substitution of h_c and h_r into Eq. (14-46) then gives the total rate of heat transfer to the boiling liquid.

EXAMPLE 14-1 In the film boiling of Freon-11, as in Example 13-2, the emissivity of the heating tube may be taken as 0.85. If the vapor film is transparent to radiation and the boiling liquid is opaque, calculate (a) the radiation coefficient h_r; (b) the convective coefficient in the presence of radiation h_c; and (c) the total heat flux, q_T/A.

SOLUTION (a) For this case $T_1 = 300 + 460 = 760°R$, and $T_2 = 74.8 + 460 = 534.8°R$. Assuming that the tube is gray and the surrounding liquid is black, Eq. (14-40) shows that $\mathscr{F}_{12} = \varepsilon_w = 0.85$. Substitution in Eq. (14-37) gives

$$\frac{q_r}{A} = 0.1713 \times 0.85(7.60^4 - 5.348^4) = 367 \text{ Btu/ft}^2\text{-h} \ (1,158 \text{ W/m}^2)$$

From Eq. (14-48)

$$h_r = \frac{367}{760 - 534.8} = 1.63 \text{ Btu/ft}^2\text{-h-}°F$$

(b) From Eq. (14-49), since $h_o = 21.7 \text{ Btu/ft}^2\text{-h-}°F$,

$$h_c = 21.7 \left(\frac{21.7}{h_c + 1.63} \right)^{1/3}$$

By trial, $h_c = 21.2 \text{ Btu/ft}^2\text{-h-}°F \ (120 \text{ W/m}^2\text{-}°C)$.

(c) From Eq. (14-47),

$$\frac{q_T}{A} = (21.2 + 1.63)(760 - 534.8) = 5141 \text{ Btu/ft}^2\text{-h} \ (16,220 \text{ W/m}^2) \qquad ////$$

SYMBOLS

A   Area, ft² or m²; A_1, of surface 1; A_2, of surface 2

C   Constant in Eq. (14-8), 5,200 μm-°R or 2,890 μm-K; C_c, correction factor for radiation to gases at pressures other than 1 atm; C_1, constant in Eq. (14-7), 3.742×10^{-16} W-m²; C_2, constant in Eq. (14-7), 1.439 cm-K

c Speed of light, 9.836×10^8 ft/s or 2.998×10^8 m/s
D Diameter, side of cube, or distance between planes, ft or m
F View factor or angle factor, dimensionless; F_{11}, F_{12}, F_{13}, for radiation from surface 1 to surfaces 1, 2, and 3, respectively; F_{21}, F_{23}, from surface 2 to surfaces 1 and 3, respectively
$\bar{F}$ Interchange factor for systems involving refractory surfaces, dimensionless; $\bar{F}_{12}$, from surface 1 to surface 2; $\bar{F}_{21}$, from surface 2 to surface 1
$\mathscr{F}$ Overall interchange factor, dimensionless; $\mathscr{F}_{12}$, from surface 1 to surface 2; $\mathscr{F}_{21}$, from surface 2 to surface 1
$\mathbf{h}$ Planck's constant, 6.626×10^{-34} J-s
h Individual heat-transfer coefficient, Btu/ft²-h-°F or W/m²-°C; h_c, for convection in the presence of radiation; h_r, for radiation; h_0, for boiling liquid in the absence of radiation
I Radiation intensity, Btu/ft²-h or W/m²; I_0, at point on normal to radiating surface; I_1, at surface 2 of radiation from surface 1
I_λ Monochromatic intensity in absorbing material, Btu/ft²-h or W/m²; $I_{0,\lambda}$, at surface of material
$\mathbf{k}$ Boltzmann constant, 1.380×10^{-23} J/K
k_1 Constant in Eq. (14-13)
L Radius of hemisphere or mean beam length in radiating gas, ft or m
L_λ Absorption length, ft or m
        Pressure, atm; $\bar{p}_G$, partial pressure of radiating gas
q Heat flow rate, Btu/h or W; q_T, total; q_c, by conduction-convection; q_r, by radiation; q_{12}, net exchange between surfaces 1 and 2; $q_{1 \to 2}$, radiation originating at surface 1 which is absorbed by surface 2; $q_{2 \to 2}$, radiation originating at surface 2 which returns to surface 2 and is absorbed
r Radius of hemisphere or length of straight line connecting area elements of radiating surfaces, ft or m
T Temperature, °R or K; T_G, of radiating gas; T_w, of wall or surface; T_1, of surface 1; T_2, of surface 2
V Volume of radiating gas, ft³ or m³
W Total radiating power, Btu/ft²-h or W/m²; W_b, of blackbody; W_1, of surface 1; W_2, of surface 2
W_λ Monochromatic radiating power, Btu/ft²-h-μm or W/m²-μm; $W_{b,\lambda}$, of blackbody
x Distance from surface of absorbing material, ft or m
Greek letters
α Absorptivity, dimensionless; α_G, of gas; α_1, of surface 1; α_2, of surface 2
ε Emissivity, dimensionless; ε_G, of gas; ε_w, of wall; ε_1, of surface 1; ε_2, of surface 2
ε_λ Monochromatic emissivity, dimensionless
λ Wavelength, μm; $\lambda_{\max}$, wavelength at which $W_{b,\lambda}$ is a maximum
μ_λ Absorption coefficient, ft^{-1} or m^{-1}
ρ Reflectivity, dimensionless
σ Stefan-Boltzmann constant, 0.1713×10^{-8} Btu/ft²-h-R⁴ or 1.798×10^{-8} W/m²-K⁴
τ Transmissivity, dimensionless
ϕ Angle with normal to surface; ϕ_1, to surface 1; ϕ_2, to surface 2

PROBLEMS

14-1 Determine the net heat transfer by radiation between two surfaces A and B, expressed as Btu per hour for each square foot of area B, if the temperatures of A and B are 900 and 400°F, respectively, and the emissivities of A and B are 0.90 and 0.25, respectively. Both surfaces are gray. (*a*) Surfaces A and B are infinite parallel planes 10 ft apart. (*b*) Surface A is a spherical shell 10 ft in diameter, and surface B is a similar shell concentric with A and 1 ft in diameter. (*c*) Surfaces A and B are flat parallel squares 5 by 5 ft, one exactly above the other, 5 ft apart. (*d*) Surfaces A and B are concentric cylindrical tubes with diameters of 10 and 9 in., respectively. (*e*) Surface A is an infinite plane, and surface B is an infinite row of 4-in.-OD tubes set on 8-in. centers. (*f*) Same as (*e*) except that 8 in.

above the centerlines of the tubes is another infinite plane having an emissivity of 0.90, which does not transmit any of the energy incident upon it. (*g*) Same as (*f*) except that surface *B* is a double row of 4-in.-OD tubes set on equilateral 8-in. centers.

14-2 A chamber for heat-curing large aluminum sheets, lacquered black on both sides, operates by passing the sheets vertically between two steel plates 150 mm apart. One of the plates is at 300°C and the other, exposed to the atmosphere, is at 25°C. What is the temperature of the lacquered sheet, and what is the heat transferred between the walls when equilibrium has been reached? Neglect convection effects. Emissivity of steel = 0.56; emissivity of lacquered sheets = 1.0.

14-3 The black flat roof of a building has an emissivity of 0.9 and an absorptivity of 0.8 for solar radiation. The sun beats down at midday with an intensity of 300 Btu/ft²-h. (*a*) If the temperature of the air and of the surroundings is 80°F, if the wind velocity is negligible, and if no heat penetrates the roof, what is the equilibrium temperature of the roof? For the rate of heat transfer by conduction-convection use $q/A = 0.38(\Delta T)^{1.25}$, where ΔT is the temperature drop between roof and air in degrees Fahrenheit. (*b*) What fraction of the heat from the roof is lost by radiation?

14-4 The roof of Prob. 14-3 is painted with an aluminum paint, which has an emissivity of 0.9 and an absorptivity for solar radiation of 0.5. What is the equilibrium temperature of the painted roof?

14-5 A 3-in. Schedule 40 iron pipeline carries steam at 90 lb$_f$/in.² gauge. The line is unlagged and is 200 ft long. The surrounding air is at 80°F. The emissivity of the pipe wall is 0.70. How many pounds of steam will condense per hour? What percentage of the heat loss is from conduction-convection?

14-6 A radiant-heating system is installed in the plaster ceiling of a room 15 ft long by 15 ft wide by 8 ft high. The temperature of the concrete floor is maintained at 75°F. Assume that no heat flows through the walls, which are coated with a reradiating material. The temperature of the air passing through the room is held at 75°F. If the required heat supply to the floor is 4000 Btu/h, calculate the necessary temperature at the ceiling surface. How much heat is transferred to the air, in Btu per hour? Emissivity of plaster = 0.93; absorptivity of concrete = 0.63. The convective heat-transfer coefficient between the ceiling and the air is given by the equation $h_c = 0.20(\Delta T)^{1/4}$.

14-7 On a clear night, when the effective blackbody temperature of space is −70°C, the air is at 15°C and contains water vapor at a partial pressure equal to that of ice or liquid water at 0°C. A very thin film of water, initially at 15°C, is placed in a very shallow well-insulated pan, placed in a spot sheltered from the wind with a full view of the sky. If h_c = 2.6 W/m²-°C, state whether ice will form, supporting the conclusion with suitable calculations.

REFERENCES

1 Gebhart, B.: "Heat Transfer," 2d ed., McGraw-Hill, New York, 1971; (*a*) pp. 150ff., (*b*) p. 421.
2 Hottel, H. C.: *Mech. Eng.,* **52**:699 (1930).
3 Hottel, H. C.: "Notes on Radiant Heat Transmission," rev. ed., Department of Chemical Engineering, Massachusetts Institute of Technology, Cambridge, Mass., 1951.
4 Hottel, H. C.: in W. H. McAdams, "Heat Transmission," 3d ed., McGraw-Hill, New York, 1954; (*a*) p. 62, (*b*) pp. 66ff, (*c*) pp. 77ff, (*d*) p. 86.
5 Hottel, H. C., and A. F. Sarofim: "Radiative Transfer," McGraw-Hill, New York, 1967.
6 McAdams, W. H.: "Heat Transmission," 3d ed., pp. 472ff, McGraw-Hill, New York, 1954.
7 Perry, J. H. (ed.): "Chemical Engineers' Handbook," 5th ed., p. 10-46, McGraw-Hill, New York, 1973.
8 Sparrow, E. M., and R. D. Cess: "Radiation Heat Transfer," Brooks/Cole, Belmont, Calif., 1966.

15

HEAT-EXCHANGE EQUIPMENT

In industrial processes heat energy is transferred by a variety of methods, including conduction in electric-resistance heaters; conduction-convection in exchangers, boilers, and condensers; radiation in furnaces and radiant-heat dryers; and by special methods such as dielectric heating. Often the equipment operates under steady-state conditions, but in many processes it operates cyclically, as in regenerative furnaces and agitated process vessels.

This chapter deals with equipment types which are of most interest to a process engineer, namely, tubular exchangers, condensers, scraped-surface exchangers, and agitated vessels. The special devices called evaporators are described in Chap. 16. Information on all types of heat-exchange equipment is given in engineering texts and handbooks.[8, 10, 12, 13]

General design of heat exchange equipment   The design and testing of practical heat-exchange equipment are based on the general principles given in Chaps. 11 to 14. First, material and energy balances are set up. From these results the required heat-transfer area is calculated. The quantities to be calculated are the overall heat-transfer coefficient, the average temperature difference, and, in cyclic equipment, the cycle time. In simple devices these quantities can be evaluated easily and with considerable

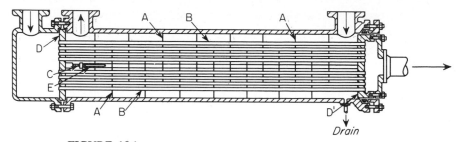

FIGURE 15-1
Single-pass 1-1 counterflow heat exchanger: *A*, baffles; *B*, tubes; *C*, guide rods;
D, *D'*, tube sheets; *E*, spacer tubes.

accuracy, but in complex processing units the evaluation may be difficult and subject
to considerable uncertainty. The final design is nearly always a compromise, based
on engineering judgment, to give the best overall performance in the light of the
service requirements.

Sometimes the design is governed by considerations that have little to do with
heat transfer, such as the space available for the equipment or the pressure drop
which can be tolerated in the fluid streams. Tubular exchangers are, in general,
designed in accordance with various standards and codes, such as the Standards of
the Tubular Exchanger Manufacturers Association (TEMA)[15] and the ASME-API
Unfired Pressure Vessel Code.[1]

The first section of this chapter deals with tubular exchangers in which there is
no phase change; the second section deals with devices in which boiling or condensation occurs; and the third section is a brief treatment of heat transfer in extended-
surface and scraped-surface exchangers and in agitated vessels.

HEAT EXCHANGERS

Heat exchangers are so important and so widely used in the process industries that
their design has been highly developed. Standards devised and accepted by the TEMA
are available covering in detail materials, methods of construction, technique of
design, and dimensions for exchangers.[15] The following sections describe the more
important types of exchanger and cover the fundamentals of their engineering,
design, and operation. Most exchangers are liquid-to-liquid, but gases and noncondensing vapors can also be treated in them.

Single-pass 1–1 exchanger The simple double-pipe exchanger shown in Fig. 11-3
is inadequate for flow rates that cannot readily be handled in a few tubes. If several
double pipes are used in parallel, the weight of metal required for the outer tubes
becomes so large that the shell-and-tube construction, such as that shown in Fig. 15-1,
where one shell serves for many tubes, is more economical. This exchanger, because
it has one shell-side pass and one tube-side pass, is a 1-1 exchanger.

In an exchanger the shell-side and tube-side heat-transfer coefficients are of comparable importance, and both must be large if a satisfactory overall coefficient is to be attained. The velocity and turbulence of the shell-side liquid are as important as those of the tube-side liquid. To prevent weakening of the tube sheets there must be a minimum distance between the tubes, and it is not practicable to space the tubes so closely that the area of the path outside the tubes is as small as that inside the tubes. If the two streams are of comparable magnitude, the velocity on the shell side is low in comparison with that on the tube side. For that reason, baffles are installed in the shell to decrease the cross section of the shell-side liquid and to force the liquid to flow across the tube bank rather than parallel with it. The added turbulence generated in this type of flow further increases the shell-side coefficient.

In the construction shown in Fig. 15-1, the baffles A consist of circular disks of sheet metal with one side cut away. Common practice is to cut away a segment having a height equal to one-fourth the inside diameter of the shell. Such baffles are called *25 percent baffles*. The baffles are perforated to receive the tubes. To minimize leakage, the clearances between baffles and shell and tubes should be small. The baffles are supported by one or more guide rods C, which are fastened between the tube sheets D and D' by setscrews. To fix the baffles in place short sections of tube E are slipped over the rod C between the baffles. In assembling such an exchanger it is necessary to assemble the tube sheets, support rods, spacers, and baffles first and then to install the tubes.

The stuffing box shown at the right-hand end of Fig. 15-1 provides for expansion. This construction is practicable only for small shells.

TUBES AND TUBE SHEETS As described in Chap. 8, tubes are drawn to definite wall thickness in terms of BWG and true outside diameter (OD), and they are available in all common metals. Tables of dimensions of standard tubes are given in Appendix 7. Standard lengths of tubes for heat-exchanger construction are 8, 12, 16, and 20 ft. Tubes are arranged on triangular pitch or square pitch. Unless the shell side tends to foul badly, triangular pitch is used, because more heat-transfer area can be packed into a shell of given diameter than in square pitch. Tubes in triangular pitch cannot be cleaned by running a brush between the rows, as no space exists for cleaning lanes. Square pitch allows cleaning of the outside of the tubes. Also, square pitch gives a lower shell-side pressure drop than triangular pitch.

TEMA standards specify a minimum center-to-center distance 1.25 times the outside diameter of the tubes for triangular pitch and a minimum cleaning lane of $\frac{1}{4}$ in. for square pitch.

SHELL AND BAFFLES Shell diameters are standardized. For shells up to and including 23 in. the diameters are fixed in accordance with ASTM pipe standards. For sizes of 25 in. and above the inside diameter is specified to the nearest inch. These shells are constructed of rolled plate. Minimum shell thicknesses are also specified.

The distance between baffles (center to center) is the *baffle pitch*, or baffle spacing. It should not be less than one-fifth the diameter of the shell or more than the inside diameter of the shell.

Tubes are usually attached to the tube sheets by grooving the holes circumferentially and rolling the tube ends into the holes by means of a rotating tapered mandrel, which stresses the metal of the tube beyond the elastic limit, so the metal flows into the grooves. In high-pressure exchangers, the tubes are welded or brazed to the tube sheet after rolling.

1–2 parallel-counterflow heat exchanger The 1-1 exchanger has limitations. Higher velocities, shorter tubes, and a more satisfactory solution to the expansion problem are realized in multipass construction. Multipass construction decreases the cross section of the fluid path and increases the fluid velocity, with a corresponding increase in the heat-transfer coefficient. The disadvantages are that (1) the exchanger is slightly more complicated and (2) the friction loss through the equipment is increased because of the larger velocities and the multiplication of exit and entrance losses. For example, the average velocity in the tubes of a four-pass exchanger is 4 times that in a single-pass exchanger having the same number and size of tubes and operated at the same liquid flow rate. The tube-side coefficient of the four-pass exchanger is approximately $4^{0.8} = 3.03$ times that for the single-pass exchanger, or even more if the velocity in the single-pass unit is sufficiently low to give laminar flow. The friction loss is approximately $4^{2.8} = 48.5$ times that in the single-pass unit, not including the additional expansion and contraction losses. The most economic design calls for such a velocity in the tubes that the increased cost of power for pumping is offset by the decreased cost of the apparatus. Too low a velocity saves power for pumping but calls for an unduly large (and consequently expensive) exchanger. Too high a velocity saves on the first cost of the exchanger but more than makes up for it in the cost of power.

An even number of tube-side passes is used in multipass exchangers. The shell side may be either single-pass or multipass. A common construction is the 1-2 parallel-counterflow exchanger, in which the shell-side liquid flows in one pass and the tube-side liquid in two or more passes. Two tube-side passes are common. Such an exchanger is shown in Fig. 15-2. In multipass exchangers, floating heads are frequently used, and the bulge in the shell of the condenser in Fig. 11-1 and the stuffing box shown in Fig. 15-1 are unnecessary. The tube-side liquid enters and leaves through the same channel, which is divided by a baffle to separate the entering and leaving tube-side streams.

2–4 exchanger The 1-2 exchanger has an important limitation. Because of the parallel-flow pass, the exchanger is unable to bring the exit temperature of one fluid very near to the entrance temperature of the other. Another way of stating the same limitation is that the heat recovery of a 1-2 exchanger is inherently poor.

A better recovery of heat can be obtained in the 2-4 exchanger, which has two shell-side and four tube-side passes. This type of exchanger also gives higher velocities and a larger overall heat-transfer coefficient than a 1-2 exchanger having two tube-side

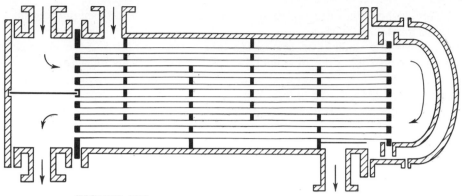

FIGURE 15-2
1-2 parallel-counterflow exchanger.

passes and operating with the same flow rates. An example of a 2-4 exchanger is shown in Fig. 15-3.

Heat-transfer coefficients in shell-and-tube exchanger The heat-transfer co-efficient h_i for the tube-side fluid in a shell-and-tube exchanger can be calculated from Eq. (12-37) or (12-38). The coefficient for the shell-side h_o cannot be so calculated because the direction of flow is partly parallel to the tubes and partly across them and because the cross-sectional area of the stream and the mass velocity of the stream vary as the fluid crosses the tube bundle back and forth across the shell. Also, leakage between baffles and shell and between baffles and tubes short-circuits some of the shell-side liquid and reduces the effectiveness of the exchanger. An approximate but generally useful equation for predicting shell-side coefficients is the *Donohue equation*[4] (15-4), which is based on a weighted average mass velocity G_e of the fluid flowing parallel with the tubes and that flowing across the tubes. The mass velocity

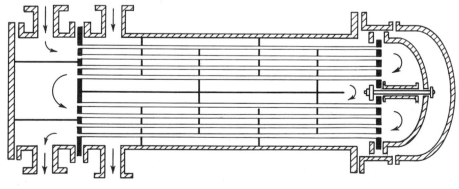

FIGURE 15-3
2-4 exchanger.

G_b parallel with the tubes is the mass flow rate divided by the free area for flow in the baffle window S_b. (The baffle window is the portion of the shell cross section not occupied by the baffle.) This area is the total area of the baffle window less the area occupied by the tubes, or

$$S_b = f_b \frac{\pi D_s^2}{4} - N_b \frac{\pi D_o^2}{4} \qquad (15\text{-}1)$$

where f_b = fraction of the cross-sectional area of shell occupied by baffle window
(commonly 0.1955)
D_s = inside diameter of shell
N_b = number of tubes in baffle window
D_o = outside diameter of tubes

In crossflow the mass velocity passes through a local maximum each time the fluid passes a row of tubes. For correlating purposes the mass velocity G_c for crossflow is based on the area S_c for transverse flow between the tubes in the row at or closest to the centerline of the exchanger. In a large exchanger S_c can be estimated from the equation

$$S_c = PD_s \left(1 - \frac{D_o}{p} \right) \qquad (15\text{-}2)$$

where p — center-to-center distance between tubes
P = baffle pitch

The mass velocities are then

$$G_b = \frac{\dot{m}}{S_b} \qquad \text{and} \qquad G_c = \frac{\dot{m}}{S_c} \qquad (15\text{-}3)$$

The Donohue equation is

$$\frac{h_o D_o}{k} = 0.2 \left(\frac{D_o G_e}{\mu} \right)^{0.6} \left(\frac{c_p \mu}{k} \right)^{0.33} \left(\frac{\mu}{\mu_w} \right)^{0.14} \qquad (15\text{-}4)$$

where $G_e = \sqrt{G_b \, G_c}$. This equation tends to give conservatively low values of h_o, especially at low Reynolds numbers. More elaborate methods of estimating shell-side coefficients are available for the specialist.[13] In j — factor form Eq. (15-4) becomes

$$\frac{h_o}{c_p G_e} \left(\frac{c_p \mu}{k} \right)^{2/3} \left(\frac{\mu_w}{\mu} \right)^{0.14} = j_0 = 0.2 \left(\frac{D_o G_e}{\mu} \right)^{-0.4} \qquad (15\text{-}5)$$

After the individual coefficients are known, the total area required is found in the usual way from the overall coefficient using an equation similar to Eq. (11-14). As discussed in the next section, the LMTD must often be corrected for the effects of cross flow. Shell-and-tube exchangers may also be designed using the concept of transfer units outlined in Chap. 11.

EXAMPLE 15-1   A tubular exchanger, 35 in. (889 mm) ID, contains 828$\frac{3}{4}$-in. (19-mm) OD tubes, 12 ft (3.66 mm) long, on a 1-in. (25-mm) square pitch. Standard 25 percent baffles are spaced 12 in. (305 mm) apart. Liquid benzene at an average bulk temperature of 60°F (15.6°C) is being heated in the shell side of the exchanger at the rate of 100,000 lb/h (45,360 kg/h). If the outside surfaces of the tubes are at 140°F (60°C), estimate the individual heat-transfer coefficient of the benzene.

SOLUTION The shell-side coefficient is found from the Donohue equation (15-4). The sectional areas for flow are first calculated from Eqs. (15-1) and (15-2). The quantities needed are

$$D_o = \frac{0.75}{12} = 0.0625 \text{ ft} \qquad D_s = \frac{35}{12} = 2.9167 \text{ ft}$$

$$p = \frac{1}{12} = 0.0833 \text{ ft} \qquad P = 1 \text{ ft}$$

From Eq. (15-2), the area for cross flow is

$$S_c = 2.9167 \times 1 \left(1 - \frac{0.0625}{0.0833}\right) = 0.7292 \text{ ft}^2$$

The number of tubes in the baffle window is approximately equal to the fractional area of the window f times the total number of tubes. For a 25 percent baffle, $f = 0.1955$. Hence

$$N_b = 0.1955 \times 828 = 161.8, \text{ say 161 tubes}$$

The area for flow in the baffle window, from Eq. (15-1), is

$$S_b = 0.1955 \frac{\pi \times 2.9167^2}{4} - 161 \frac{\pi \times 0.0625^2}{4} = 0.8123 \text{ ft}^2$$

The mass velocities are, from Eq. (15-3),

$$G_c = \frac{100,000}{0.7292} = 137,137 \text{ lb/ft}^2\text{-h} \qquad G_b = \frac{100,000}{0.8123} = 123,107 \text{ lb/ft}^2\text{-h}$$

$$G = \sqrt{G_b G_c} = \sqrt{137,137 \times 123,107} = 129,933 \text{ lb/ft}^2\text{-h}$$

The additional quantities needed for substitution in Eq. (15-4) are

$$\mu \text{ at } 60°F = 0.70 \text{ cP} \qquad \mu \text{ at } 140°F = 0.38 \text{ cP} \qquad \text{(Appendix 10)}$$

$$c_p = 0.41 \text{ Btu/lb-°F} \qquad \text{(Appendix 16)} \qquad k = 0.092 \text{ Btu/ft-h-°F} \qquad \text{(Appendix 13)}$$

From Eq. (15-4)

$$\frac{h_o D_o}{k} = 0.2 \left(\frac{0.0625 \times 129,933}{0.70 \times 2.42}\right)^{0.6} \left(\frac{0.41 \times 0.70 \times 2.42}{0.092}\right)^{0.33} \left(\frac{0.70}{0.38}\right)^{0.14}$$

$$= 68.59$$

Hence

$$h_o = \frac{68.59 \times 0.092}{0.0625} = 101 \text{ Btu/ft}^2\text{-h-°F (573 W/m}^2\text{-°C)}$$

Correction of LMTD for crossflow If a fluid flows perpendicularly to a heated or cooled tube bank, the LMTD, as given by Eq. (11-15), applies only if the temperature of one of the fluids is constant. If the temperatures of both fluids change, the tem-

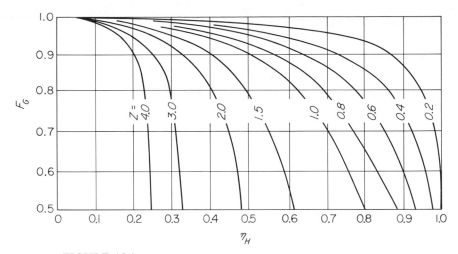

FIGURE 15-4
Correction for LMTD, crossflow. [*From R. A. Bowman, A. C. Mueller, and W. M. Nagle, Trans. ASME,* **62:**283 (1940), *courtesy of American Society of Mechanical Engineers.*]

perature conditions do not correspond to either countercurrent or parallel flow but to a type of flow called *crossflow.*

When flow types other than countercurrent or parallel appear, it is customary to define a correction factor F_G, which is so determined that when it is multiplied by the LMTD for countercurrent flow, the product is the true average temperature drop. Figure 15-4 shows a correlation for F_G for crossflow derived on the assumption that neither stream mixes with itself during flow through the exchanger.[2] The temperatures used in Fig. 15-4 are

Inlet temperature of hot fluid T_{ha}
Outlet temperature of hot fluid T_{hb}
Inlet temperature of cold fluid T_{ca}
Outlet temperature of cold fluid T_{ch}

Each curved line in the figure corresponds to a constant value of the dimensionless ratio Z, defined as

$$Z = \frac{T_{ha} - T_{hb}}{T_{cb} - T_{ca}} \tag{15-6}$$

and the abscissas are values of the dimensionless ratio η_H, defined as

$$\eta_H = \frac{T_{cb} - T_{ca}}{T_{ha} - T_{ca}} \tag{15-7}$$

The factor Z is the ratio of the fall in temperature of the hot fluid to the rise in temperature of the cold fluid. The factor η_H is the *heating effectiveness,* or the ratio of the

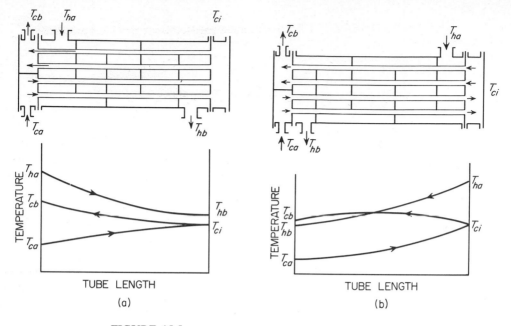

FIGURE 15-5
Temperature-length curves, 1-2 exchanger: (*a*) first nozzle arrangement; (*b*) second nozzle arrangement.

actual temperature rise of the cold fluid to the maximum possible temperature rise obtainable if the warm-end approach were zero (based on countercurrent flow). From the numerical values of η_H and Z the factor F_G is read from Fig. 15-4, interpolating between lines of constant Z where necessary, and multiplied by the LMTD for counterflow to give the true mean temperature drop.

TEMPERATURE-DROP CORRECTION FACTOR FOR 1-2 EXCHANGER Flow in a 1-2 exchanger is partly parallel flow, partly countercurrent flow, and partly crossflow. The temperature-length curves for such an exchanger are shown in Fig. 15-5. Curve $T_{ha} - T_{hb}$ applies to the shell-side fluid, which is assumed to be the hot fluid. Curve $T_{ca} - T_{ci}$ applies to the first pass of the tube-side liquid, and curve $T_{ci} - T_{cb}$ to the second pass of the tube-side liquid. In Fig. 15-5a curves $T_{ha} - T_{hb}$ and $T_{ca} - T_{ci}$ taken together are those of a parallel-flow exchanger, and curves $T_{ha} - T_{hb}$ and $T_{ci} - T_{cb}$ taken together correspond to a countercurrent exchanger. The LMTD, which applies to either parallel or counterflow but not to a mixture of both, cannot be used without correction to calculate the true mean temperature drop. The method used in this situation is analogous to that used to correct the LMTD for crossflow: a factor F_G is so defined that when it is multiplied by the counterflow LMTD, the product is the correct mean temperature drop.

Figure 15-6a shows the factor F_G as a function of the two dimensionless numbers η_H and Z, which have been defined by Eqs. (15-6) and (15-7).

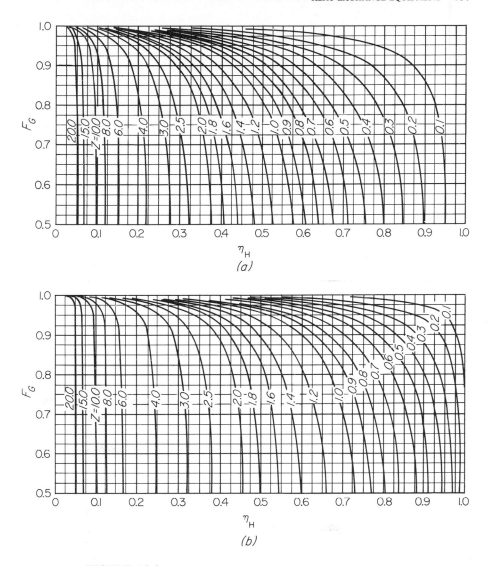

FIGURE 15-6
Correction of LMTD: (*a*) 1-2 exchangers; (*b*) 2-4 exchangers. [*From R. A. Bowman, A. C. Mueller, and W. M. Nagle, Trans. ASME,* **62**:283 (1940), *courtesy of American Society of Mechanical Engineers.*]

Two different arrangements of the inlet and outlet shell nozzles are shown in Figs 15-5*a* and *b*. The factor F_G from Fig. 15-6*a* can be used for either nozzle arrangement.

Factor F_G is always less than unity. The mean temperature drop, and therefore the capacity of the exchanger, is less than that of a countercurrent exchanger having

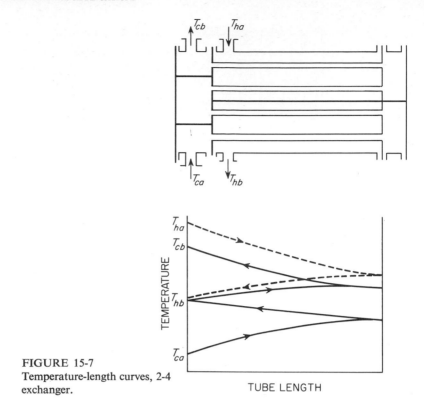

FIGURE 15-7
Temperature-length curves, 2-4
exchanger.

the same LMTD. When F_G is less than approximately 0.75, the 1-2 exchanger should not be used, as the actual exchanger no longer accurately follows the assumptions made in the derivation.

TEMPERATURE-DROP CORRECTION FACTOR FOR 2-4 EXCHANGER   The temperature-length relations in a 2-4 exchanger are shown in Fig. 15-7. The dotted lines refer to the shell-side fluid and the full lines to the tube-side fluid. It is assumed that the hotter fluid is in the shell. The hotter pass of the shell-side fluid is in thermal contact with the two hottest tube-side passes and the cooler shell-side pass with the two coolest tube-side passes. The exchanger as a whole approximates a true counter-current unit more closely than is possible with a 1-2 exchanger.

The correction factor F_G for 2-4 exchangers is given in Fig. 15-6b. This chart can be used for two shell-side and four or more tube-side passes.

Other combinations of shell-side passes and tube-side passes are used, but the 1-2 and the 2-4 types are the most common.

EXAMPLE 15-2   In the 1-2 exchanger sketched in Fig. 15-5a, the values of the temperatures are $T_{ca} = 70°F$ (21.1°C); $T_{cb} = 130°F$ (54.4°C); $T_{ha} = 240°F$ (115.6°C); $T_{hb} = 120°F$ (48.9°C). What is the correct mean temperature drop in this exchanger?

SOLUTION   The correction factor F_G is found from Fig. 15-6a. For this case, from Eqs. (15-6) and (15-7),

$$\eta_H = \frac{130 - 70}{240 - 70} = 0.353 \qquad Z = \frac{240 - 120}{130 - 70} = 2.00$$

From Fig. 15-6a, $F_G = 0.735$. The temperature drops are:

At shell inlet: $\Delta T = 240 - 130 = 110°F$

At shell outlet: $\Delta T = 120 - 70 = 50°F$

$$\overline{\Delta T_L} = \frac{110 - 50}{2.303 \log (110/50)} = 76°F$$

The correct mean is $\overline{\Delta T} = 0.735 \times 76 = 56°F$ (31.1°C). These temperatures are marginal for a 1-2 exchanger.   ////

EXAMPLE 15-3   What is the correct mean temperature difference in a 2-4 exchanger operating with the same inlet and outlet temperatures as in the exchanger in Example 15-2?
    SOLUTION   For a 2-4 exchanger, when $\eta_H = 0.353$ and $Z = 2.00$, the correction factor from Fig. 15-6a is $F_G = 0.945$. The $\overline{\Delta T_L}$ is the same as in Example 15-2. The correct mean $\overline{\Delta T} = 0.945 \times 76 = 72°F$ (40°C).   ////

Plate-type exchangers   For heat transfer between fluids at low or moderate pressures, below about 20 atm, plate-type exchangers are competitive with shell-and-tube exchangers, especially where corrosion-resistant materials are required. Metal plates, usually with corrugated faces, are supported in a frame; hot fluid passes between alternate pairs of plates, exchanging heat with the cold fluid in the adjacent spaces. The plates are typically 5 mm apart. They can be readily separated for cleaning; additional area may be provided simply by adding more plates. Unlike shell-and-tube exchangers, plate exchangers can be used for multiple duty; i.e., several different fluids can flow through different parts of the exchanger and be kept separate from one another. The maximum operating temperature is about 300°F; maximum heat-transfer areas are about 5,000 ft². Plate exchangers are relatively effective with viscous fluids, with viscosities up to about 300 P.
    Other special and *compact* exchangers, which provide large heat-transfer areas in a small volume, are described in the literature.[9]

CONDENSERS

Special heat-transfer devices used to liquefy vapors by removing their latent heats are called *condensers*. The latent heat is removed by absorbing it in a cooler liquid called the *coolant*. Since the temperature of the coolant obviously is increased in a condenser, the unit also acts as a heater, but functionally it is the condensing action that is important, and the name reflects this fact. Condensers fall into two classes.

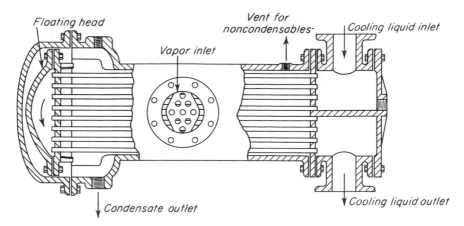

FIGURE 15-8
Two-pass floating-head condenser.

In the first, called shell-and-tube condensers, the condensing vapor and coolant are separated by a tubular heat-transfer surface. In the second, called contact condensers, the coolant and vapor streams, both of which are usually water, are physically mixed and leave the condenser as a single stream.

Shell-and-tube condensers   The condenser shown in Fig. 11-1 is a single-pass unit, since the entire stream of cooling liquid flows through all the tubes in parallel. In large condensers, this type of flow has a serious limitation. The number of tubes is so large that in single-pass flow the velocity through the tubes is too small to yield an adequate heat-transfer coefficient, and the unit is uneconomically large. Also, because of the low coefficient, long tubes are needed if the cooling fluid is to be cooled through a reasonably large temperature range, and such long tubes are not practicable.

To obtain larger velocities, higher heat-transfer coefficients, and shorter tubes the multipass principle used in heat exchangers may also be used for the coolant in a condenser. An example of a two-pass condenser is shown in Fig. 15-8.

PROVISION FOR THERMAL EXPANSION   Because of the differences in temperature existing in condensers, expansion strains may be set up which are sufficiently severe to buckle the tubes or pull them loose from the tube sheets. The most common method of avoiding damage from expansion is the use of the floating-head construction, in which one of the tube sheets (and therefore one end of the tubes) is structurally independent of the shell. This principle is used in the condenser of Fig. 15-8. The figure shows how the tubes may expand or contract, independent of the shell. A perforated plate is set over the vapor inlet to prevent cutting of the tubes by drops of liquid which may be carried by the vapor.

Contact condensers   An example of a contact condenser is shown in Fig. 15-9. Contact condensers are much smaller and cheaper than surface condensers. In the

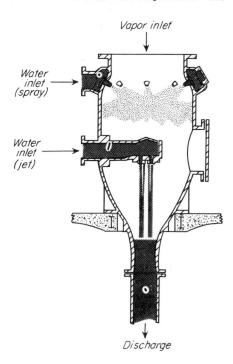

FIGURE 15-9
Contact condenser. [*Schutte and
Koerting Div., Ametek.*]

design shown in Fig. 15-9, part of the cooling water is sprayed into the vapor stream near the vapor inlet, and the remainder is directed into a discharge throat to complete the condensation and to eject the noncondensable gas by the use of a venturi tube. When a shell-and-tube condenser is operated under vacuum, the condensate is usually pumped out, but it may be removed by a barometric leg. This is a vertical tube, about 34 ft (10 m) long, sealed at the bottom by a condensate-receiving tank. In operation the level of liquid in the leg automatically adjusts itself so that the difference in head between levels in leg and tank corresponds to the difference in pressure between the atmosphere and the vapor space in the condenser. Then the liquid flows down the leg as fast as it is condensed without breaking the vacuum. In a direct-contact condenser the pressure regain in the downstream cone of the venturi is often sufficient to eliminate the need for a barometric leg.

Boilers and calandrias   In continous-flow process plants liquids are boiled in kettle-type boilers containing a pool of boiling liquid or in vertical-tube calandrias through which the liquid and vapor flow upward through the tubes. Sometimes, as described in Chap. 16, the liquid may be heated under pressure to a temperature well above its normal boiling point and then allowed to *flash* (partially vaporize) by reducing the pressure at a point external to the heat-transfer equipment.

KETTLE-TYPE BOILERS   A kettle-type boiler, or *reboiler* as it is called when connected to a distillation column, is shown in Fig. 15-10. A horizontal shell contains a relatively small tube bundle, two-pass on the tube side, with a floating head and tube

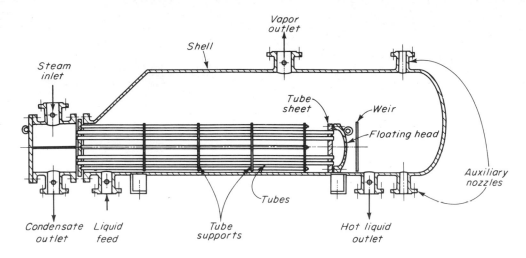

FIGURE 15-10
Kettle-type reboiler.

sheet. The tube bundle is submerged in a pool of boiling liquid, the depth of which
is set by the height of an overflow weir. Feed is admitted to the liquid pool from the
bottom. Vapor escapes from the top of the shell; any unvaporized liquid spills over
the weir and is withdrawn from the bottom of the shell. The heating fluid, usually
steam, enters the tubes as shown; steam condensate is removed through a trap. The
auxiliary nozzles shown in Fig. 15-10 are for inspection, draining, or for insertion of
instrument sensing elements.

CALANDRIAS   Vertical shell-and-tube units, known as calandrias, natural-circulation
or thermosiphon reboilers, are generally the most economical vaporizers for distillation and evaporation operations. A typical arrangement is shown in Fig. 16-3.
Liquid from an evaporator or distillation column enters the bottom of the unit and
is partially vaporized in the heated tubes; the reduction in density causes the vapor-
liquid mixture to rise and draw in additional feed liquid. Liquid and vapor leave the
top of the tubes at high velocity; they are separated and the liquid recycled.

   Configurations and thermal ratings of industrial calandrias fall in a relatively
narrow range. The tubes are typically 1 in. in diameter, sometimes as large as 2 in.
They are 8 to 12 ft long, seldom longer; short tubes 4 to 6 ft long are sometimes
used in vacuum service.

   Heat-transfer rates depend on the properties of the vaporizing liquid, especially
its reduced temperature† and on its tendency to foul the heat-transfer surface.[5]
Typical overall coefficients for steam-heated calandrias are given in Table 15-1. They

---

† Reduced temperature is the ratio of the actual temperature to the critical temperature of the liquid,
both in kelvins or degrees Raukine.

are relatively insensitive to changes in tube length, tube diameter, or static liquid level in the tubes.

For usual applications with saturated steam on the shell side, the heat flux can be estimated from Figure 15-11, which is based on 14 BWG stainless steel 1-in. tubes 8 ft long. The curves are drawn for pure liquids; if mixtures are used, the reduced temperature should be that of the component having the lowest value of T_r. The chart should not be used for absolute pressures below 0.3 atm, nor should the curves be extrapolated.

EXTENDED-SURFACE EQUIPMENT

Difficult heat-exchange problems occur when one of two fluid streams has a much lower heat-transfer coefficient than the other. A typical case is heating a fixed gas, such as air, by means of condensing steam. The individual coefficient for the steam is typically 100 to 200 times that for the airstream; consequently, the overall coefficient is essentially equal to the individual coefficient for the air, the capacity of a unit area of heating surface will be low, and many feet of tube will be required to provide reasonable capacity. Other variations of the same problem are found in heating or cooling viscous liquids or in treating a stream of fluid at low flow rate, because of the low rate of heat transfer in laminar flow.

To conserve space and to reduce the cost of the equipment in these cases, certain types of heat-exchange surfaces, called *extended surfaces*, have been developed in which the outside area of the tube is multiplied, or extended, by fins, pegs, disks, and other appendages and the outside area in contact with the fluid thereby made much larger than the inside area. The fluid stream having the lower coefficient is brought into contact with the extended surface, and flows outside the tubes, while the other fluid, having the high coefficient, flows through the tubes. The quantitative effect of extending the outside surface can be seen from the overall coefficient, written in the following form, in which the resistance of the tube wall is neglected,

$$U_i = \frac{1}{(1/h_i) + (A_i/A_o h_o)} \tag{15-8}$$

Equation (15-8) shows that if h_o is small and h_i large, the value of U_i will be small; but if the area A_o is made much larger than A_i, the resistance $A_i/A_o h_o$ becomes small

Table 15-1   TYPICAL OVERALL COEFFICIENTS IN CALANDRIA BOILERS

Service	Overall coefficient U	
	Btu/ft²-h-°F	W/m²-°C
Heavy organic chemicals	100–160	570–900
Light hydrocarbons	160–220	900–1,250
Water, aqueous solutions	220–350	1,250–2,000

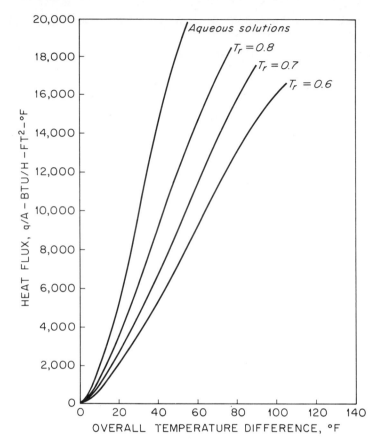

FIGURE 15-11
Heat flux in natural-circulation calandrias.[5]

and U_i increases just as if h_o were increased, with a corresponding increase in capacity per unit length of tube or unit of inside area.

Types of extended surface   Three common types of extended surfaces are available, examples of which are shown in Fig. 15-12. Longitudinal fins are used when the direction of flow of the fluid is parallel to the axis of the tube; transverse fins are used when the direction of flow of the fluid is across the tubes. Spikes, pins, studs, or spines are also used to extend surfaces, and tubes carrying these can be used for either direction of flow. In all types, it is important that the fins be in tight contact with the tube, both for structural reasons and to ensure good thermal contact between the base of the fin and the wall.

Fin efficiency   The outside area of a finned tube consists of two parts, the area of the fins and the area of the bare tube not covered by the bases of the fins. A unit

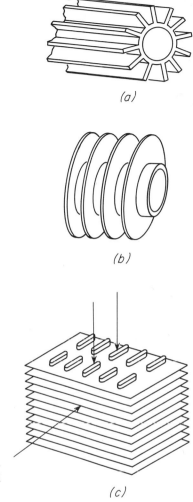

FIGURE 15-12
Types of extended surface: (a) longitudinal; (b) transverse fins; (c) flattened
tubes, continuous fins.

area of fin surface is not so efficient as a unit area of bare tube surface because of the
added resistance to the heat flow by conduction through the fin to the tube. Thus,
consider a single longitudinal fin attached to a tube, as shown in Fig. 15-13, and
assume that the heat is flowing to the tube from the fluid surrounding the fin. Let
the temperature of the fluid be T and that of the bare portion of the tube T_w. The
temperature at the base of the fin will also be T_w. The temperature drop available for
heat transfer to the bare tube is $T - T_w$, or ΔT_o. Consider the heat transferred to the
fin at the tip, the point farthest away from the tube wall. This heat, in order to reach
the wall of the tube, must flow, by conduction, through the entire length of the fin,
from tip to base. Other increments of heat, entering the fin at points intermediate
between tip and base, also must flow through a part of the fin length. A temperature

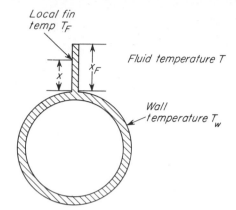

FIGURE 15-13
Tube and single longitudinal fin.

gradient will be necessary, therefore, from the tip of the fin to the base, and the tip will be warmer than the base. If T_F is the temperature of the fin at a distance x from the base, the temperature drop available for heat transfer from fluid to fin at that point will be $T - T_F$. Since $T_F > T_w$, $T - T_F < T - T_w = \Delta T_o$, and the efficiency of any unit area away from the fin base is less than that of a unit area of bare tube. The difference between $T - T_F$ and ΔT_o is zero at the base of the fin and is a maximum at the tip of the fin. Let the average value of $T - T_F$, based on the entire fin area, be denoted by $\overline{\Delta T_F}$. The efficiency of the fin is defined as the ratio of $\overline{\Delta T_F}$ to ΔT_o, and is denoted by η_F. The efficiency can, of course, be expressed on a percentage basis. An efficiency of unity (or of 100 percent) means that a unit area of fin is as effective as a unit area of bare tube, as far as temperature drop is concerned. Any actual fin will have an efficiency less than 100 percent.

Calculations for extended-surface exchangers   Consider, as a basis, a unit area of tube. Let A_F be the area of the fins and A_b the area of the bare tube. Let h_o be the heat-transfer coefficient of the fluid surrounding the fins and tube. Assume that h_o is the same for both fins and tube. An overall coefficient, based on the inside area A_i, can be written

$$U_i = \frac{1}{\{(A_i)/[h_o(\eta_F A_F + A_b)]\} + (x_w D_i/k_m D_L) + (1/h_i)} \qquad (15\text{-}9)$$

To use Eq. (15-9) it is necessary to know the values of the fin efficiency η_F and of the individual coefficients h_i and h_o. The coefficient h_i is calculated by the usual methods. The calculation of the coefficient h_o will be discussed later.

The fin efficiency η_F can be calculated mathematically, on the basis of certain reasonable assumptions, for fins of various types.[6] For example, the efficiency of longitudinal fins is given in Fig. 15-14, in which η_F is plotted as a function of the quantity $a_F x_F$, where x_F is the height of the fin from base to tip and a_F is defined by the equation

$$a_F = \sqrt{\frac{h_o L_p/S}{k_m}} \qquad (15\text{-}10)$$

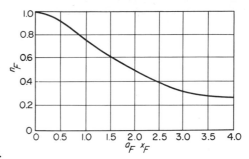

FIGURE 15-14
Fin efficiency, longitudinal fins.

where h_o = coefficient outside tube
k_m = thermal conductivity of metal in fin
L_p = perimeter of fin
S = cross-sectional area of fin

The product $a_F x_F$ is dimensionless.

Fin efficiencies for other types of extended surface are available.[6] Figure 15-14 shows that the fin efficiency is nearly unity when the ratio h_o to k_m is small. Extended surfaces are neither efficient nor necessary if the coefficient h_o is large. Also, fins increase the pressure drop.

The coefficient h_o cannot be accurately found by the use of the equations normally used for calculating the heat-transfer coefficients for bare tubes. The fins change the flow characteristics of the fluid, and the coefficient for an extended surface differs from that for a smooth tube. Individual coefficients for extended surfaces must be determined experimentally and correlated for each type of surface, and such correlations are supplied by the manufacturer of the tubes. A typical correlation for longitudinal finned tubes is shown in Fig. 15-15. The quantity D_e is the equivalent diameter, defined as usual as 4 times the hydraulic radius, which is, in turn, the cross section of the fin-side space divided by the total wetted perimeter of fins and tube calculated as in Example 15-4.

EXAMPLE 15-4 Air is heated in the shell of an extended-surface exchanger. The inner pipe is $1\frac{1}{2}$-in. (38.1-mm) IPS Schedule 40 pipe carrying 28 longitudinal fins $\frac{1}{2}$ in. (12.7 mm) high and 0.035 in. (0.89 mm) thick. The shell is 3-in. (76.2-mm) Schedule 40 steel pipe. The exposed outside area of the inner pipe (not covered by the fins) is 0.416 ft^2 per lineal foot (0.127 m^2/m); the total surface area of the fins and pipe is 2.830 ft^2/ft (0.863 m^2/m). Steam condensing at 250°F (121°C) inside the inner pipe has a film coefficient of 1500 Btu/ft^2-h-°F (8500 W/m^2-°C). The thermal conductivity of steel is 26 Btu/ft-h-°F (45 W/m-°C). The wall thickness of the inner pipe is 0.145 in. (3.68 mm). If the mass velocity of the air is 5,000 lb/h-ft^2 (24,400 kg/h-m^2) and the average air temperature is 130°F (54.5°C), what is the overall heat-transfer coefficient based on the inside area of the inner pipe? Neglect fouling factors.

SOLUTION The film coefficient h_o of the air is found from Fig. 15-15. To use this correlation the Reynolds number of the air must first be calculated, as follows. The viscosity of

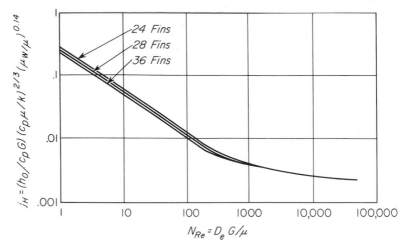

FIGURE 15-15

Heat-transfer coefficients, longitudinal finned tubes; 1½-in. IPS with ½- by 0.035-in. fins in 3-in. IPS shell. [*Brown Fintube Co.*]

air at 130°F is 0.046 lb/ft-h (Appendix 9). The equivalent diameter of the shell space is

$$\text{ID of shell (Appendix 6)} = \frac{3.068}{12} = 0.2557 \text{ ft}$$

$$\text{OD of inner pipe (Appendix 6)} = \frac{1.900}{12} = 0.1583 \text{ ft}$$

The cross-sectional area of the shell space is

$$\frac{\pi(0.2557^2 - 0.1\ 83^2)}{4} - \frac{28 \times 0.5 \times 0.035}{144} = 0.0282 \text{ ft}^2$$

The perimeter of the air space is

$$\pi 0.2557 + 2.830 = 3.633 \text{ ft}$$

The hydraulic radius is

$$r_H = \frac{0.0282}{3.633} = 0.00776 \text{ ft}$$

The equivalent diameter is

$$D_e = 4 \times 0.00776 = 0.0310 \text{ ft}$$

The Reynolds number of air is therefore

$$N_{Re} = \frac{0.0310 \times 5,000}{0.046} = 3.37 \times 10^3$$

From Fig. 15-15, the heat-transfer factor is

$$j_H = \frac{h_o}{c_p G} \left(\frac{c_p \mu}{k}\right)^{2/3} \left(\frac{\mu}{\mu_w}\right)^{-0.14} = 0.0031$$

The quantities needed to solve for h_o are

$$c_p = 0.25 \text{ Btu/lb-°F} \quad \text{(Appendix 15)}$$

$$k = 0.0162 \text{ Btu/ft-h-°F} \quad \text{(Appendix 12)}$$

In computing μ_w the resistance of the wall and the steam film are considered negligible, so $T_w = 250°F$ and $\mu_w = 0.0528 \text{ lb/ft-h}$:

$$\phi_v = \left(\frac{\mu}{\mu_w}\right)^{0.14} = \left(\frac{0.046}{0.0528}\right)^{0.14} = 0.981$$

$$N_{\text{Pr}} = \frac{c_p \mu}{k} = \frac{0.25 \times 0.046}{0.0162} = 0.710$$

$$h_o = \frac{0.0031 \times 0.25 \times 5,000 \times 0.981}{0.710^{2/3}} = 4.78 \text{ Btu/ft}^2\text{-h-°F}$$

For rectangular fins, disregarding the contribution of the ends of the fins to the perimeter, $L_p = 2L$, and $S = Ly_F$, where y_F is the fin thickness and L is the length of the fin. Then, from Eq. (15-10),

$$a_F x_F = x_F \sqrt{\frac{h_o(2L/Ly_F)}{k_m}} = x_F \sqrt{\frac{2h_o}{k_m y_F}} = \frac{0.5}{12} \sqrt{\frac{2 \times 4.78}{26(0.035/12)}} = 0.467$$

From Fig. 15-14, $\eta_F = 0.933$.

The overall coefficient is found from Eq. (15-9). The additional quantities needed are

$$D_i \text{ (Appendix 6)} = \frac{1.610}{12} = 0.1342 \text{ ft}$$

$$\overline{D_L} = \frac{0.1583 - 0.1342}{2.303 \log(0.1583/0.1342)} = 0.1454 \text{ ft}$$

$$A_i = \pi 0.1342 \times 1.0 = 0.422 \text{ ft}^2/\text{lineal ft}$$

$$A_F + A_b = 2.830 \text{ ft}^2/\text{lineal ft}$$

$$A_F = 2.830 - 0.416 = 2.414 \text{ ft}^2/\text{lineal ft}$$

$$L = \frac{0.145}{12} = 0.0121 \text{ ft}$$

$$U_i = \frac{1}{\dfrac{0.422}{4.78(0.933 \times 2.414 + 0.416)} + \dfrac{0.0121 \times 0.1342}{26 \times 0.1454} + \dfrac{1}{1500}}$$

$$= 29.3 \text{ Btu/ft -h-°F} \ (166 \text{ w/m -°C})$$

The overall coefficient, when based on the small inside area of the inner pipe, may be much larger than the air-film coefficient based on the area of the extended surface. ////

Air-cooled exchangers As cooling water has become scarcer and pollution controls more stringent, the use of air-cooled exchangers has increased. These consist of bundles of horizontal finned tubes, typically 1 in. in diameter and 8 to 30 ft long, through which air is circulated by a large fan. Hot process fluids in the tubes, at temperatures from 200 to 800°F or more, can be cooled to about 40°F above the dry-bulb temperature of the air. Heat-transfer areas, figured on the base outside of the tubes, range from 500 to 5,000 ft^2; the fins multiply this by a factor of 7 to 20. Air flows between the tubes at velocities of 10 to 20 ft/s. The pressure drop and power consumption are low, but sometimes to reduce the fan noise to an acceptable level the fan speed must be lower than that for minimum power consumption. In air-cooled condensers the tubes are usually inclined. Detailed design procedures are given in the literature.[11]

SCRAPED-SURFACE EXCHANGERS

Viscous liquids and liquid-solid suspensions are often heated or cooled in scraped-surface exchangers. Typically these are double-pipe exchangers with a fairly large central tube, 4 to 12 in. in diameter, jacketed with steam or cooling liquid. The inside surface of the central tube is wiped by one or more longitudinal blades mounted on a rotating shaft.

The viscous liquid is passed at low velocity through the central tube. Portions of this liquid adjacent to the heat-transfer surface are essentially stagnant, except when disturbed by the passage of the scraper blade. Heat is transferred to the viscous liquid by unsteady-state conduction. If the time between disturbances is short, as it usually is, the heat penetrates only a small distance into the stagnant liquid, and the process is exactly analogous to unsteady-state heat transfer to a semi-infinite solid.

Heat-transfer coefficients in scraped-surface exchangers[7] Assume that the bulk temperature of the liquid at some location along the exchanger is T, and the temperature of the heat-transfer surface is T_w. Assume for the present that $T_w > T$. Consider a small element of area of the heat-transfer surface over which the blade has just passed. Any liquid which was previously on this surface element has been removed by the blade and replaced by other liquid at temperature T. Heat flows from the surface to the liquid during the time interval t_T, which is the time until the next scraper blade passes the surface element, removes the liquid, and redeposits new liquid on the surface.

From Eq. (10-29), the total amount of heat Q_T transferred during time interval t_T is given by

$$\frac{Q_T}{A} = 2k(T_w - T)\sqrt{\frac{t_T}{\pi\alpha}} \qquad (10\text{-}29)$$

where k = thermal conductivity of liquid
  α = thermal diffusivity of liquid
  A = area of heat-transfer surface

The heat-transfer coefficient averaged over each time interval is, by definition,

$$h_i \equiv \frac{Q_T}{t_T A(T_w - T)} \qquad (15\text{-}11)$$

Substitution from Eq. (10-29) into Eq. (15-11), noting that $\alpha = k/\rho c_p$, gives

$$h_i = 2\sqrt{\frac{k\rho c_p}{\pi t_T}} \tag{15-12}$$

The time interval between the passage of successive blades over a given element of area is

$$t_T = \frac{1}{nB} \tag{15-13}$$

where n = agitator speed r/h
B = number of blades carried by shaft

Combining Eqs. (15-12) and (15-13) gives, for the heat-transfer coefficient,

$$h_i = 2\sqrt{\frac{k\rho c_p nB}{\pi}} \tag{15-14}$$

Equation (15-14) shows that the heat-transfer coefficient on a scraped surface depends on the thermal properties of the liquid and the agitator speed and implies that it does not depend on the viscosity of the liquid or its velocity through the exchanger. Actually, although Eq. (15-14) gives a good approximation in many cases, it is somewhat of an oversimplification. The liquid, especially if viscous, is not immediately deposited on the heat-transfer surface behind the scraper blade, as assumed in the development of Eq. (15-14). The heat-transfer coefficient is therefore somewhat affected by changes in liquid viscosity; it also is a function of the liquid velocity in the longitudinal direction and of the diameter and length of the exchanger. An empirical equation for the heat-transfer coefficient incorporating these variables is[14]

$$\frac{h_j D_a}{k} = 4.9 \left(\frac{D_a \bar{V} \rho}{\mu}\right)^{0.57} \left(\frac{c_p \mu}{k}\right)^{0.47} \left(\frac{D_a n}{\bar{V}}\right)^{0.17} \left(\frac{D_a}{L}\right)^{0.37} \tag{15-15}$$

where $\bar{V}$ = bulk average longitudinal velocity
L = length of exchanger
D_a = diameter of scraper (also equal to inside diameter of shell)

Equation (15-15) predicts a smaller effect of agitator speed on the heat-transfer coefficient than Eq. (15-14) does.

HEAT TRANSFER IN AGITATED VESSELS

Heat-transfer surfaces, which may be in the form of heating or cooling jackets or coils of pipe immersed in the liquid, are often used in the agitated vessels described in Chap. 9.

Heat-transfer coefficients In an agitated vessel, as shown in Chap. 9, the dimensionless group $D_a^2 n\rho/\mu$ is a Reynolds number useful in correlating data on power consumption. This same group has been found to be satisfactory as a correlating variable for heat transfer to jackets or coils in an agitated tank. The following equations are typical of those that have been offered for this purpose.

For heating or cooling liquids in a cylindrical tank equipped with a single heating or cooling coil,[3]

$$\frac{h_c D_t}{k} = 1.01 \left(\frac{D_a^2 n \rho}{\mu}\right)^{0.62} \left(\frac{c_p \mu}{k}\right)^{1/3} \left(\frac{\mu}{\mu_w}\right)^{0.14} \tag{15-16}$$

For heating and cooling liquids in a cylindrical vessel equipped with a pitched-blade turbine agitator and a heating or cooling jacket,[16]

$$\frac{h_j D_t}{k} = 0.44 \left(\frac{D_a^2 n \rho}{\mu}\right)^{2/3} \left(\frac{c_p \mu}{k}\right)^{1/3} \left(\frac{\mu}{\mu_w}\right)^{0.24} \tag{15-17}$$

where h_c = individual heat-transfer coefficient between coil surface and liquid
$\quad\ h_j$ = coefficient between liquid and jacketed inner surface of vessel

Equations of this type are not applicable to situations differing significantly from those for which the equations were derived. Equations for various types of agitators and arrangements of heat-transfer surface are given in the literature.[16]

Transient heating or cooling in agitated vessels   Consider a well-agitated vessel containing m lb or kg of liquid of density and specific heat c_p. It contains a heat-transfer surface of area A, heated by a constant-temperature medium such as condensing steam at temperature T_s. If the initial temperature of the liquid is T_a, its temperature T_b at any time t_T can be found as follows. At any instant, the rate of heat transfer is

$$q = \frac{dQ}{dt} = UA(T_s - T) \tag{15-18}$$

Consequently

$$dQ = UA\ (T_s - T)\ dt \tag{15-19}$$

The differential quantity of heat energy dQ raises the liquid temperature according to the equation

$$dQ = m\ c_p\ dT \tag{15-20}$$

Equating the left-hand sides of Eqs. (15-18) and (15-19) and rearranging gives

$$dt = \frac{UA}{mc_p}\frac{dT}{T_s - T} \tag{15-21}$$

If U is constant (usually a reasonable assumption), Eq. (15-21) can be integrated between the limits $t = 0$, $T = T_a$ and $t = t_T$, $T = T_b$, to give

$$\ln \frac{T_s - T_a}{T_s - T_b} = \frac{mc_p t_T}{UA} \tag{15-22}$$

If the heat-transfer medium is not at a constant temperature but is a liquid (such as cooling water) of specific heat c_{pc} entering at temperature T_{ca} and flowing at a constant rate $\dot{m}_c$, the corresponding equation for the liquid temperature is

$$\ln \frac{T_a - T_{ca}}{T_b - T_{ca}} = \frac{\dot{m}_c c_{pc}}{m c_p} \frac{K_1 - 1}{K_1} t_T \qquad (15\text{-}23)$$

where

$$K_1 = \exp \frac{UA}{\dot{m}_c c_{pc}} \qquad (15\text{-}24)$$

Equations for other situations involving transient heat transfer are available in the literature[10a].

SYMBOLS

A Area, ft² or m²; A_F, of fin; A_b, of bare tube; A_i, of inside of tube; A_o, of outside of tube

a_F Fin factor [Eq. (15-10)]

B Number of scraper blades

c_p Specific heat at constant pressure, Btu/lb-°F or J/g-°C; c_{pc}, of cooling liquid

D Diameter, ft or m; D_a, of impeller or scraper; D_e, equivalent diameter of noncircular channel; D_i, diameter of inside of tube; D_0, of outside of tube; D_s, inside diameter of exchanger shell; $\bar{D}_L$, logarithmic mean of inside and outside tube diameters

F_G Correction factor for average temperature difference in crossflow or multipass exchangers, dimensionless

f_b Fraction of cross-sectional area of shell occupied by baffle window

G Mass velocity, lb/ft²-h or kg/m²-s; G_b, in baffle window; G_c, in crossflow; G_e, effective value in exchanger, $\sqrt{G_b G_c}$

h Individual heat-transfer coefficient, Btu/ft²-h-°F or W/m²-°C; h_c, for outside of coil; h_i, for inside of tube; h_j, for inner wall of jacket; h_o, for outside of tube

j j factor, dimensionless; j_0, for shell-side heat transfer

k Thermal conductivity, Btu/ft-h-°F or W/m-°C; k_m, of tube wall

L Length of fin or exchanger, ft or m; L_p, perimeter of fin

m Mass of liquid, lb or kg

$\dot{m}$ Flow rate, lb/h or kg/h; $\dot{m}_c$, of cooling fluid

N_b Number of tubes in baffle window

n Impeller or scraper speed, r/s or r/h

P Baffle pitch, ft or m

p Center-to-center distance between tubes, ft or m

Q Quantity of heat, Btu or J; Q_T, total amount transferred during time interval t_T

q Rate of heat transfer, Btu/h or W

S Cross-sectional area, ft² or m²; S_b, area for flow in baffle window; S_c, area for crossflow in exchanger shell

T Temperature, °F or °C; T_F, at distance x from base of fin; T_a, initial value; T_b, final value; T_{ca}, at cool-fluid inlet; T_{cb}, at cool-fluid outlet; T_{ci}, intermediate cool-fluid temperature; T_{ha}, at warm-fluid inlet; T_{hb}, at warm-fluid outlet; T_r, reduced temperature; T_s, temperature of constant-temperature heating fluid; T_w, of surface or bare portion of finned tube

t Time, s or hr; t_T, length of time interval

U Overall heat-transfer coefficient, Btu/ft²-h-°F or W/m²-°C; U_i, based on inside area

$\bar{V}$ Average fluid velocity in longitudinal direction, ft/s or m/s

x_F Fin height, ft or m; x_w, thickness of tube wall

y_F Fin thickness, ft or m

Z Ratio of temperature ranges in crossflow or multipass exchanger, dimensionless [Eq. (15-6)]

Greek letters

α Thermal diffusivity, ft²/h or m²/h

ΔT Temperature difference, °F or °C; ΔT_0, between fluid and wall of finned tube; $\overline{\Delta T}$, corrected overall average value; $\overline{\Delta T_F}$, average difference between fluid and fins; $\overline{\Delta T_L}$, logarithmic mean value

η_F Fin efficiency, $\overline{\Delta T_F}/\Delta T_0$

η_H Heating effectiveness, dimensionless [Eq. (15-7)]

μ Absolute viscosity, lb/ft-h or cP; μ_w, at wall or surface temperature

ρ Density, lb/ft³ or kg/m³

PROBLEMS

15-1   A vertical-tube two-pass heater is used for heating gas oil. Saturated steam at 50 lb_f/in.² gauge is used as a heating medium. The tubes are 1 in. OD by 16 BWG and are made of mild steel. The oil enters at 60°F and leaves at 150°F. The viscosity-temperature relation is exponential. The viscosity at 60°F is 5.0 cP and at 150°F is 1.8 cP. The oil is 37° API (specific gravity = 0.840) at 60°F. The flow of oil is 120 bbl/h (1 bbl = 42 gal). Assume the steam condenses in film condensation. The thermal conductivity of the oil is 0.078 Btu/ft-h-°F, and the specific heat is 0.480 Btu/lb-°F. The velocity of the oil in the tubes should be approximately 3 ft/s. Calculate the length of tubes needed for this heater.

15-2   By chromium plating the outside of the tubes of Prob. 15-1 and adding mercaptan to the steam, the condensation becomes dropwise, and the condensing-steam coefficient becomes 14,000 Btu/ft²-h-°F. How much oil will the heater heat from 60 to 150°F under these new steam-side conditions?

15-3   Air is blown at a rate of 2.4 m³/s (measured at 0°C and 1 atm) at right angles to a tube bank 10 pipes and 10 spaces wide and 10 rows deep. The length of each pipe is 3 m. The tubes are on triangular centers, and the center-to-center distance is 75 mm. It is desired to heat the air from 20 to 40°C at atmospheric pressure. What steam pressure must be used? The pipes are 25-mm OD steel pipe.

15-4   Crude oil at the rate of 300,000 lb/h is to be heated from 70 to 136°F by heat exchange with the bottom product from a distillation unit. The product at 257,500 lb/h is to be cooled from 295 to 225°F. There is available a tubular exchanger with steel tubes with an inside shell diameter of $23\frac{1}{4}$ in., having one pass on the shell side and two passes on the tube side. It has 324 tubes, $\frac{3}{4}$ in. OD, 14 BWG, 12 ft long arranged on a 1-in.-square pitch and supported by baffles with a 25 percent cut, spaced at 9-in. intervals. Would this exchanger be suitable; i.e., what is the allowable fouling factor? The average properties are shown in Table 15-2.

15-5   A petroleum oil having the properties given in Table 15-3 is to be heated in a horizontal multipass heater with steam at 50 lb_f/in.² gauge. The tubes are to be steel $\frac{3}{4}$ in.

Table 15-2   DATA FOR PROB. 15-4

	Product, outside tubes	Crude, inside tubes
c_p, Btu/lb-°F	0.525	0.475
μ, cP	5.2	2.9
ρ, lb/ft³	54.1	51.5
k, Btu/ft-h-°F	0.069	0.0789

OD by 16 BWG, and their maximum length is 15 ft. The oil enters at 100°F, leaves at 180°F, and enters the tubes at about 3 ft/s. The total flow rate is 150 gal/min. Assuming complete mixing of the oil after each pass, how many passes are required?

15-6 Compare Eqs. (15-14) and (15-15) with respect to the effect of the following variables on the predicted heat-transfer coefficient: (*a*) thermal conductivity, (*b*) specific heat, (*c*) liquid density, (*d*) agitator speed, (*e*) agitator diameter, (*f*) longitudinal fluid velocity.

15-7 A turbine-agitated vessel 2 m in diameter contains 6,200 kg of a dilute aqueous solution. The agitator is $\frac{2}{3}$ m in diameter and turns at 140 r/min. The vessel is jacketed with steam condensing at 110°C; the heat-transfer area is 14 m². The steel walls of the vessel are 10 mm thick. If the solution is at 40°C and the heat-transfer coefficient of the condensing steam is 10 kW/m²-°C, what is the rate of heat-transfer between steam and liquid?

15-8 Under the conditions given in Prob. 15-7, how long would it take to heat the vessel contents (*a*) from 20 to 60°C, (*b*) from 60 to 100°C?

15-9 A steam-heated natural-circulation calandria is to be designed to boil chlorobenzene at atmospheric pressure. (*a*) Approximately how much heat-transfer surface will be required? (*b*) How much area would be required if the average pressure in the calandria were only 0.4 atm? The normal boiling point of chlorobenzene is 132.0°C; its critical temperature is 359.2°C.

REFERENCES

1 American Society of Mechanical Engineers: "Rules for Construction of Unfired Pressure Vessels," New York, 1962.
2 Bowman, R. A., A. C. Mueller, and W. M. Nagle: *Trans. ASME,* **62:**283 (1940).
3 Cummings, G. H., and A. S. West: *Ind. Eng. Chem.,* **42:**2303 (1950).
4 Donohue, D. A.: *Ind. Eng. Chem.,* **41:**2499 (1949).
5 Frank, O., and R. D. Prickett: *Chem. Eng.,* **80**(20):107 (1973).
6 Gardner, K. A.: *Trans. ASME,* **67:**621 (1945).
7 Harriott, P.: *Chem. Eng. Prog. Symp. Ser.,* **29:**137 (1959).
8 Jakob, M.: "Heat Transfer," vol. 2, Wiley, New York, 1957.
9 Kays, W. M., and A. L. London: "Compact Heat Exchangers," 2d ed., McGraw-Hill, New York, 1964.

Table 15-3 DATA FOR PROB. 15-5

Temp., °F	Thermal conductivity Btu/ft-h-°F	Kinematic viscosity 10⁵ ft²/s	Density, lb/ft³	Specific heat Btu/lb-°F
100	0.0739	36.6	55.25	0.455
120	0.0737	21.8	54.81	0.466
140	0.0733	14.4	54.37	0.477
160	0.0728	10.2	53.92	0.487
180	0.0724	7.52	53.48	0.498
200	0.0719	5.70	53.03	0.508
220	0.0711	4.52	52.58	0.519
240	0.0706	3.67	52.13	0.530
260	0.0702	3.07		0.540

10   Kern, D. Q.: "Process Heat Transfer," McGraw-Hill, New York, 1950; (*a*) pp. 626–637.
11   Lerner, J. E.: *Hydrocarbon Proc.,* **51**(2):93 (1972).
12   McAdams, W. H.: "Heat Transmission," 3d ed., McGraw-Hill, New York, 1954.
13   Perry, J. H. (ed.): "Chemical Engineers' Handbook," 5th ed., pp. **11**-3ff, McGraw-Hill, New York, 1973.
14   Skelland, A. H. P.: *Chem. Eng. Sci.,* **7**:166 (1958).
15   Tubular Exchangers Manufacturers Association: "Standards of the TEMA," 5th ed., New York, 1968.
16   Uhl, V. W., and J. B. Gray: "Mixing," vol. 1, p. 284, Academic, New York, 1966.

EVAPORATION

Heat transfer to a boiling liquid has been discussed generally in Chap. 13. A special case occurs so often that it is considered an individual operation. It is called *evaporation* and is the subject of this chapter.

The objective of evaporation is to concentrate a solution consisting of a nonvolatile solute and a volatile solvent. In the overwhelming majority of evaporations the solvent is water. Evaporation is conducted by vaporizing a portion of the solvent to produce a concentrated solution or thick liquor. Evaporation differs from drying in that the residue is a liquid—sometimes a highly viscous one—rather than a solid; it differs from distillation in that the vapor usually is a single component, and even when the vapor is a mixture, no attempt is made in the evaporation step to separate the vapor into fractions; it differs from crystallization in that emphasis is placed on concentrating a solution rather than forming and building crystals. In certain situations, e.g., in the evaporation of brine to produce common salt, the line between evaporation and crystallization is far from sharp. Evaporation sometimes produces a slurry of crystals in a saturated mother liquor. In this book such processes are considered in Chap. 28, which is devoted to crystallization.

Normally, in evaporation the thick liquor is the valuable product and the vapor is condensed and discarded. In one specific situation, however, the reverse is true. Mineral-bearing water often is evaporated to give a solid-free product for boiler feed, for special process requirements, or for human consumption. This technique is often called *water distillation*, but technically it is evaporation. Large-scale evaporation processes are being developed and used for recovering potable water from seawater. Here the condensed water is the desired product. Only a fraction of the total water in the feed is recovered, and the remainder is returned to the sea.

Liquid characteristics   The practical solution of an evaporation problem is profoundly affected by the character of the liquor to be concentrated. It is the wide variation in liquor characteristics (which demands judgment and experience in engineering and operating evaporators) that broadens this operation from simple heat transfer to a separate art. Some of the most important properties of evaporating liquids are as follows.

CONCENTRATION   Although the thin liquor fed to an evaporator may be sufficiently dilute to have many of the physical properties of water, as the concentration increases, the solution becomes more and more individualistic. The density and viscosity increase with solid content until either the solution becomes saturated or the liquor becomes too sluggish for adequate heat transfer. Continued boiling of a saturated solution causes crystals to form; these must be removed or the tubes clog. The boiling point of the solution may also rise considerably as the solid content increases, so that the boiling temperature of a concentrated solution may be much higher than that of water at the same pressure.

FOAMING   Some materials, especially organic substances, foam during vaporization. A stable foam accompanies the vapor out of the evaporator, causing heavy entrainment. In extreme cases the entire mass of liquid may boil over into the vapor outlet and be lost.

TEMPERATURE SENSITIVITY   Many fine chemicals, pharmaceutical products, and foods are damaged when heated to moderate temperatures for relatively short times. In concentrating such materials special techniques are needed to reduce both the temperature of the liquid and the time of heating.

SCALE   Some solutions deposit scale on the heating surfaces. The overall coefficient then steadily diminishes, until the evaporator must be shut down and the tubes cleaned. When the scale is hard and insoluble, the cleaning is difficult and expensive.

MATERIALS OF CONSTRUCTION   Whenever possible, evaporators are made of cast iron and steel. Many solutions, however, attack ferrous metals, or are contaminated by them. Special materials such as copper, nickel, stainless steel, aluminum, impervious graphite, and lead are then used. Since these materials are expensive, high heat-transfer rates become especially desirable to minimize the first cost of the equipment.

Many other liquid characteristics must be considered by the designer of an evaporator. Some of these are specific heat, heat of concentration, freezing point, gas liberation on boiling, toxicity, explosion hazards, radioactivity, and necessity for sterile operation. Because of the variation in liquor properties, many different evaporator designs have been developed. The choice for any specific problem depends primarily on the characteristics of the liquid.

Single- and multiple-effect operation Most evaporators are heated by steam condensing on metal tubes. Nearly always the material to be evaporated flows inside the tubes. Usually the steam is at a low pressure, below 3 atm abs; often the boiling liquid is under a moderate vacuum, up to about 0.05 atm abs. Reducing the boiling temperature of the liquid increases the temperature difference between the steam and the boiling liquid and thereby increases the heat-transfer rate in the evaporator.

When a single evaporator is used, the vapor from the boiling liquid is condensed and discarded. This method is called *single-effect evaporation*, and although it is simple, it utilizes steam ineffectively. To evaporate 1 lb of water from a solution calls for from 1 to 1.3 lb of steam. If the vapor from one evaporator is fed into the steam chest of a second evaporator and the vapor from the second is then sent to a condenser, the operation becomes double-effect. The heat in the original steam is reused in the second effect, and the evaporation achieved by a unit mass of steam fed to the first effect is approximately doubled. Additional effects can be added in the same manner. The general method of increasing the evaporation per pound of steam by using a series of evaporators between the steam supply and the condenser is called *multiple-effect evaporation*.

TYPES OF EVAPORATORS

The chief types of steam-heated tubular evaporators in use today are:

1 Short-tube evaporators
2 Long-tube vertical evaporators
    a Forced-circulation
    b Upward-flow (climbing-film)
    c Downward-flow (falling-film)
3 Agitated-film evaporators

Once-through and circulation evaporators Evaporators may be operated either as once-through or as circulation units. In once-through operation the feed liquor passes through the tubes only once, releases the vapor, and leaves the unit as thick liquor. All the evaporation is accomplished in a single pass. The ratio of evaporation to feed is limited in a single-pass unit; thus these evaporators are well adapted to multiple-effect operation, where the total amount of concentration can be spread over several effects. Agitated-film evaporators are always operated once through; falling-film and climbing-film evaporators can also be operated in this way.

Once-through evaporators are especially useful for heat-sensitive materials. By operating under high vacuum, the temperature of the liquid can be kept low. With a single rapid passage through the tubes the thick liquor is at the evaporation temperature but a short time and can be quickly cooled as soon as it leaves the evaporator.

In circulation evaporators a pool of liquid is held within the equipment. Incoming feed mixes with the liquid from the pool, and the mixture passes through the tubes. Unevaporated liquid discharged from the tubes returns to the pool, so that only part of the total evaporation occurs in one pass. All short-tube and forced-circulation evaporators are operated in this way. Climbing-film evaporators may be either once-through or circulation units.

The thick liquor from a circulation evaporator is withdrawn from the pool. All the liquor in the pool must therefore be at the maximum concentration. Since the liquid entering the tubes may contain several parts of thick liquor for one part of feed, its concentration, density, viscosity, and boiling point are nearly at the maximum. Accordingly, the heat-transfer coefficient tends to be low.

Circulation evaporators are not well suited to concentrating heat-sensitive liquids. With a reasonably good vacuum the temperature of the bulk of the liquid may be nondestructive, but the liquid is repeatedly exposed to contact with hot tubes. Some of the liquid, therefore, may be heated to an excessively high temperature. Although the average residence time of the liquid in the heating zone may be short, part of the liquid is retained in the evaporator for a considerable time. Prolonged heating of even a small part of a heat-sensitive material like a food can ruin the entire product.

Circulation evaporators, however, can operate over a wide range of concentration between feed and thick liquor in a single unit, and are well adapted to single-effect evaporation. They may operate either with natural circulation, with the flow through the tubes induced by density differences, or with forced circulation, with flow provided by a pump.

Short-tube evaporators   In the older types of evaporators the tubes are "short," 4 to 8 ft long, and fairly large, 2 to 4 in. in diameter. In the short-tube vertical evaporator shown in Fig. 16-1 the steam condenses outside the tubes. The tube bundle contains a large central downcomer, the cross-sectional area of which is 25 to 40 percent of the total cross-sectional area of the tubes. Most of the boiling takes place in the smaller tubes, so that the liquid rises through these tubes and returns through the downcomer. Drops of liquid fall through the vapor in the tall space above the tubes, with their removal from the vapor often assisted by baffle plates over the vapor outlet. Thick liquor is withdrawn from the conical bottom of the shell. In this evaporator the driving force for flow of liquid through the tubes is the difference in density between the liquid in the downcomer and the mixture of liquid and vapor in the tubes.

Short-tube vertical evaporators provide moderately good heat transfer at reasonable cost. They are fairly effective with scaling liquids, for the inside of the tubes can be easily cleaned. The heat-transfer coefficients are fairly high with thin

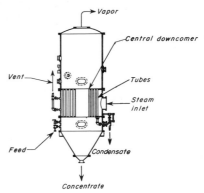

FIGURE 16-1
Short-tube evaporator.

liquids but low when the liquid is viscous. Once considered as "standard" evaporators, short-tube vertical units have been largely displaced by long-tube evaporators and other more specialized designs.

Forced-circulation evaporators In a natural-circulation evaporator[3] the liquid enters the tubes at 1 to 3 ft/s. The linear velocity increases greatly as vapor is formed in the tubes, so that in general the rates of heat transfer are satisfactory. With viscous liquids, however, the overall coefficient in a natural-circulation unit may be uneconomically low. Higher coefficients are obtained in forced-circulation evaporators, an example of which is shown in Fig. 16-2. Here a centrifugal pump forces liquid through the tubes at an entering velocity of 6 to 18 ft/s. The tubes are under sufficient static head to ensure that there is no boiling in the tubes; the liquid becomes superheated as the static head is reduced during flow from the heater to the vapor space, and it flashes into a mixture of vapor and spray in the outlet line from the exchanger just before entering the body of the evaporator. The mixture of liquid and vapor

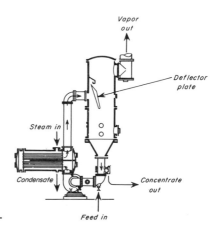

FIGURE 16-2
Forced-circulation evaporator with two-pass horizontal separate heating element.

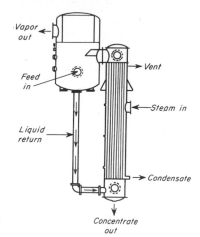

FIGURE 16-3
Climbing-film, long-tube vertical evaporator.

impinges on a deflector plate in the vapor space. Liquid returns to the pump inlet, where it meets incoming feed; vapor leaves the top of the evaporator body to a condenser or to the next effect. Part of the liquid leaving the separator is continuously withdrawn as concentrate.

In the design shown in Fig. 16-2 the exchanger has horizontal tubes and is two-pass on both tube and shell sides. In others vertical single-pass exchangers are used. In both types the heat-transfer coefficients are high, especially with thin liquids, but the greatest improvement over natural-circulation evaporation is with viscous liquids. With thin liquids the improvement with forced circulation does not warrant the added pumping costs over natural circulation, but with viscous material the added costs are justified, especially when expensive metals must be used. An example is caustic soda concentration, which must be done in nickel equipment. In multiple-effect evaporators producing a viscous final concentrate the first effects may be natural-circulation units and the later ones, handling viscous liquid, forced-circulation units. Because of the high velocities in a forced-circulation evaporator, the residence time of the liquid in the tubes is short—about 1 to 3 s—so that moderately heat-sensitive liquids can be concentrated in them. They are also effective in evaporating salting liquors or those that tend to foam.

Long-tube evaporators with upward flow   A typical long-tube vertical evaporator with upward flow of the liquid is shown in Fig. 16-3. The essential parts are (1) a tubular exchanger with steam in the shell and liquid to be concentrated in the tubes, (2) a separator or vapor space for removing entrained liquid from the vapor, and (3) when operated as a circulation unit, a return leg for the liquid from the separator to the bottom of the exchanger. Inlets are provided for feed liquid and steam, and outlets are provided for vapor, thick liquor, steam condensate, and noncondensable gases from the steam.

The tubes are typically 1 to 2 in. in diameter and 12 to 32 ft long. Liquid and vapor flow upward inside the tubes as a result of the boiling action; separated liquid returns to the bottom of the tubes by gravity. Dilute feed, often at about room temperature, enters the system and mixes with liquid returning from the separator. The mixture enters the bottom of the tubes, on the outside of which steam is condensing. For a short distance the feed to the tubes flows upward as liquid, receiving heat from the steam. Bubbles then form in the liquid as boiling begins, increasing the linear velocity and the rate of heat transfer. Near the top of the tubes the bubbles grow rapidly. In this zone bubbles of vapor alternating with slugs of liquid rise very quickly through the tubes and emerge at high velocity from the top.

From the tubes the mixture of liquid and vapor enters the separator. The diameter of the separator is larger than that of the exchanger, so that the linear velocity of the vapor is greatly reduced. As a further aid in eliminating liquid droplets the vapor impinges on, and then passes around, sets of baffle plates before leaving the separator. The evaporator shown in Fig. 16-3 can be operated only as a circulation unit.

Long-tube vertical evaporators are especially effective in concentrating liquids that tend to foam. Foam is broken when the high-velocity mixture of liquid and vapor impinges against the vapor-head baffle.

Falling-film evaporators[8,9]   Concentration of highly heat-sensitive materials such as orange juice requires a minimum time of exposure to a heated surface. This can be done in once-through falling-film evaporators, in which the liquid enters at the top, flows downward inside the heated tubes as a film, and leaves from the bottom. The tubes are large, 2 to 10 in. in diameter. Vapor evolved from the liquid is usually carried downward with the liquid, and leaves from the bottom of the unit. In appearance these evaporators resemble long, vertical, tubular exchangers with a liquid-vapor separator at the bottom and a distributor for the liquid at the top.

The chief problem in a falling-film evaporator is that of distributing the liquid uniformly as a film inside the tubes. This is done by a set of perforated metal plates above a carefully leveled tube sheet, by inserts in the tube ends to cause the liquid to flow evenly into each tube, or by "spider" distributors with radial arms from which the feed is sprayed at a steady rate on the inside surface of each tube. Still another way is to use an individual spray nozzle inside each tube.

When recirculation is allowable without damaging the liquid, distribution of liquid to the tubes is facilitated by a moderate recycling of liquid to the tops of the tubes. This provides a larger volume of flow through the tubes than is possible in once-through operation.

For good heat transfer the Reynolds number $4\Gamma/\mu$ of the falling film should be greater than 2,000 at all points in the tube.[5] During evaporation the amount of liquid is continuously reduced as it flows from top to bottom of the tube, so that the amount of concentration that can be done in a single pass is limited.

Falling-film evaporators, with no recirculation and short residence times, handle sensitive products that can be concentrated in no other way. They are also well adapted to concentrating viscous liquids.

Agitated-film evaporator   The principal resistance to overall heat transfer from the steam to the boiling liquid in an evaporator is on the liquid side. Therefore,

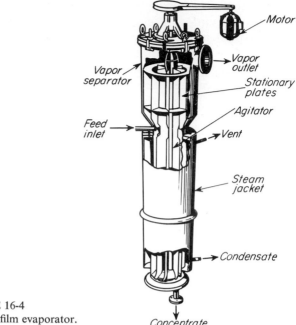

FIGURE 16-4
Agitated-film evaporator.

any method of diminishing this resistance gives a considerable improvement in the overall heat-transfer coefficient. In long-tube evaporators, especially with forced circulation, the velocity of the liquid through the tubes is high. The liquid is highly turbulent, and the rate of heat transfer is large. Another way of increasing turbulence is by mechanical agitation of the liquid film, as in the evaporator shown in Fig. 16-4. This is a modified falling-film evaporator with a single jacketed tube containing an internal agitator. Feed enters at the top of the jacketed section, and is spread out into a thin, highly turbulent, film by the vertical blades of the agitator. Concentrate leaves from the bottom of the jacketed section; vapor rises from the vaporizing zone into an unjacketed separator, which is somewhat larger in diameter than the evaporating tube. In the separator the agitator blades throw entrained liquid outward against stationary vertical plates. The droplets coalesce on these plates and return to the evaporating section. Liquid-free vapor escapes through outlets at the top of the unit.

The chief advantage of an agitated-film evaporator is its ability to give high rates of heat transfer with viscous liquids. The product may have a viscosity as high as 1,000 P at the evaporation temperature. For moderately viscous liquids the heat-transfer coefficient may be estimated from Eq. (15-14). As in other evaporators, the overall coefficient falls as the viscosity rises, but in this design the decrease is slow. With highly viscous materials the coefficient is appreciably greater than in forced-circulation evaporators and much greater than in natural-circulation units. The agitated-film evaporator is particularly effective with such viscous heat-sensitive products as gelatin, rubber latex, antibiotics, and fruit juices. Its disadvantages are high cost; the internal moving parts, which may need considerable maintenance; and the small capacity of single units, which is far below that of multitubular evaporators.

PERFORMANCE OF TUBULAR EVAPORATORS

The principal measures of the performance of a steam-heated tubular evaporator are the capacity and the economy. Capacity is defined as the number of pounds of water vaporized per hour. Economy is the number of pounds vaporized per pound of steam fed to the unit. In a single-effect evaporator the economy is nearly always less than 1, but in multiple-effect equipment it may be considerably greater. The steam consumption, in pounds per hour, is also important. It equals the capacity divided by the economy.

Evaporator Capacity

The rate of heat transfer q through the heating surface of an evaporator, by the definition of the overall heat-transfer coefficient given in Eq. (11-9), is the product of three factors: the area of the heat-transfer surface A, the overall heat-transfer coefficient U, and the overall temperature drop ΔT, or

$$q = UA\,\Delta T \qquad (16\text{-}1)$$

If the feed to the evaporator is at the boiling temperature corresponding to the absolute pressure in the vapor space, all the heat transferred through the heating surface is available for evaporation and the capacity is proportional to q. If the feed is cold, the heat required to heat it to its boiling point may be quite large and the capacity for a given value of q is reduced accordingly, as heat used to heat the feed is not available for evaporation. Conversely, if the feed is at a temperature above the boiling point in the vapor space, a portion of the feed evaporates spontaneously by adiabatic equilibration with the vapor-space pressure and the capacity is greater than that corresponding to q. This process is called *flash evaporation*.

The actual temperature drop across the heating surface depends on the solution being evaporated, the difference in pressure between the steam chest and the vapor space above the boiling liquid, and the depth of liquid over the heating surface. In some evaporators the velocity of the liquid in the tubes also influences the temperature drop because the frictional loss in the tubes increases the effective pressure of the liquid. When the solution has the characteristics of pure water, its boiling point can be read from steam tables if the pressure is known, as can the temperature of the condensing steam. In actual evaporators, however, the boiling point of a solution is affected by two factors, boiling-point elevation and liquid head.

Boiling-point elevation and Dühring's rule The vapor pressure of most aqueous solutions is less than that of water at the same temperature. Consequently, for a

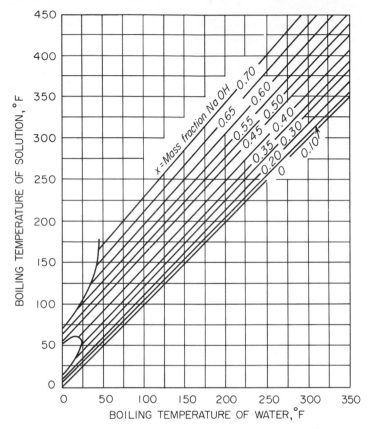

FIGURE 16-5
Dühring lines, system sodium hydroxide—water.[7]

given pressure the boiling point of the solutions is higher than that of pure water. The increase in boiling point over that of water is known as the *boiling-point elevation* of the solution. It is small for dilute solutions and for solution of organic colloids but may be as large as 150°F for concentrated solutions of inorganic salts. The boiling-point elevation must be subtracted from the temperature drop that is predicted from the steam tables.

For strong solutions the boiling-point elevation is best found from an empirical rule known as *Dühring's rule*, which states that the boiling point of a given solution is a linear function of the boiling point of pure water at the same pressure. Thus if the boiling point of the solution is plotted against that of water at the same pressure, a straight line results. Different lines are obtained for different concentrations. Over wide ranges of pressure the rule is not exact, but over a moderate range the lines are very nearly straight, though not necessarily parallel. Figure 16-5 is a set of Dühring lines for solutions of sodium hydroxide in water.[7] The use of this figure may be illustrated by an example. If the pressure over a 25 percent solution of sodium

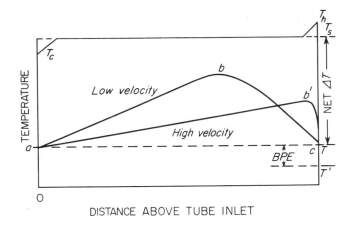

FIGURE 16-6
Temperature history of liquor in tubes and temperature drops in long-tube vertical evaporator.

hydroxide is such that water boils at 180°F, by reading up from the x axis at 180°F to the line for 25 percent solution and then horizontally to the y axis it is found that the boiling point of the solution at this pressure is 200°F. The boiling-point elevation for this solution at this pressure is therefore 20°F.

Effect of liquid head and friction on temperature drop If the depth of liquid in an evaporator is appreciable, the boiling point corresponding to the pressure in the vapor space is the boiling point of the surface layers of liquid only. A drop of liquid at a distance Z ft below the surface is under a pressure of the vapor space plus a head of Z ft of liquid and therefore has a higher boiling point. In addition, when the velocity of the liquid is large, frictional loss in the tubes further increases the average pressure of the liquid. In any actual evaporator, therefore, the average boiling point of the liquid in the tubes is higher than the boiling point corresponding to the pressure in the vapor space. This increase in boiling point lowers the average temperature drop between the steam and the liquid and reduces the capacity. The amount of reduction cannot be estimated quantitatively with precision, but the qualitative effect of liquid head, especially with high liquor levels and high liquid velocities, should not be ignored.

Figure 16-6 relates the temperatures in an evaporator with the distance along the tube, measured from the bottom. The diagram applies to a long-tube vertical evaporator with upflow of liquid. The steam enters the evaporator at the top of the steam jacket surrounding the tubes and flows downward. The entering steam may be slightly superheated at T_h. The superheat is quickly given up, and the steam drops to saturation temperature T_s. Over the greater part of the heating surface this temperature is unchanged. Before the condensate leaves the steam space, it may be cooled slightly to temperature T_c.

The temperature history of the liquor in the tubes is shown by lines abc and $ab'c$ in Fig. 16-6. The former applies at low velocities, about 3 ft/s, and the latter at high velocities, above 10 ft/s, both velocities based on the flow entering the bottom of the tube.[2] It is assumed that the feed enters the evaporator at about the boiling temperature of the liquid at vapor-space pressure, denoted by T. Then the liquid entering the tube is at T, whether the flow is once through or circulatory. At high velocities, the fluid in the tube remains liquid practically to the end of the tube and flashes into a mixture of liquid and vapor in the last few inches of the tube. The maximum liquid temperature occurs at point b', as shown in Fig. 16-6, almost at the exit from the tube.

At lower velocities, the liquid flashes at a point nearer the center of the tube and reaches its maximum temperature, as shown by point b in Fig. 16-6. Point b divides the tube into two sections, a nonboiling section below point b and a boiling section above this point.

At both high and low velocities the vapor and concentrated liquid reach equilibrium at the pressure in the vapor space. If the liquid has an appreciable boiling-point elevation, this temperature T is greater than T', the boiling point of pure water at the vapor-space pressure. The difference between T and T' is the boiling-point elevation (BPE).

The temperature drop, corrected for boiling-point elevation, is $T_s - T$. The true temperature drop, corrected for both boiling-point elevation and static head, is represented by the average distance between T_s and the variable liquid temperature. Although some correlations are available[2] for determining the true temperature drop from the operating conditions, usually this quantity is not available to the designer and the net temperature drop, corrected for boiling-point elevation only, is used.

The pressure history of the fluid in the tube is shown in Fig. 16-7, in which pressure is plotted against distance from the bottom of the tube. The velocity is such that boiling starts inside the tube. The total pressure drop in the tube, neglecting changes in kinetic energy, is the sum of the static head and the friction loss. The mixture of steam and water in the boiling section has a high velocity, and the friction loss is large in this section. As shown by the curves in Fig. 16-7, the pressure changes slowly in the nonboiling section, where the velocity is low, and rapidly in the boiling section, where the velocity is high.

Heat-transfer coefficients   As shown by Eq. (16-1), the heat flux and the evaporator capacity are affected by changes both in the temperature drop and in the overall heat-transfer coefficient. The temperature drop is fixed by the properties of the steam and the boiling liquid, and except for the effect of hydrostatic head is not a function of the evaporator construction. The overall coefficient, on the other hand, is strongly influenced by the design and method of operation of the evaporator.

As shown in Chap. 11 [Eq. (11-32)], the overall resistance to heat transfer between the steam and the boiling liquid is the sum of five individual resistances: the steam-film resistance; the two scale resistances, inside and outside the tubes; the tube-wall resistance; and the resistance from the boiling liquid. The overall coefficient is the reciprocal of the overall resistance. In most evaporators the fouling factor of

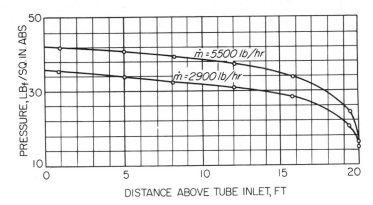

FIGURE 16-7
Pressure history of liquor in tubes of long-tube vertical evaporator. (By permission of author and publishers, from W. H. McAdams, "Heat Transmission," 3d ed., p. 403. Copyright by author, 1954, McGraw-Hill Book Company.)

the condensing steam and the resistance of the tube wall are very small, and they are usually neglected in evaporator calculations. In an agitated-film evaporator the tube wall is fairly thick, so that its resistance may be a significant part of the total.

STEAM-FILM COEFFICIENTS The steam-film coefficient is characteristically high, even when condensation is filmwise. Promoters are sometimes added to the steam to give dropwise condensation and a still higher coefficient. Since the presence of noncondensable gas seriously reduces the steam-film coefficient, provision must be made to vent noncondensables from the steam chest and to prevent leakage of air inward when the steam is at a pressure below atmospheric.

LIQUID-SIDE COEFFICIENTS The liquid-side coefficient depends to a large extent on the velocity of the liquid over the heated surface. In most evaporators, and especially those handling viscous materials, the resistance of the liquid side controls the overall rate of heat transfer to the boiling liquid. In natural-circulation evaporators the liquid-side coefficient for dilute aqueous solutions, as indicated in Chap. 15, is between 200 and 600 Btu/ft^2-h-°F. The heat flux may be conservatively estimated for nonfouling solutions from Fig. 15-11.

Forced circulation gives high liquid-side coefficients even though boiling inside the tubes is suppressed by the high static head. The liquid-side coefficient in a forced-circulation evaporator may be estimated by Eq. (12-37) for heat transfer to a non-boiling liquid if its constant 0.023 is changed to 0.028.[1]

The formation of scale on the tubes of an evaporator adds a thermal-resistance equivalent to a fouling factor.

Overall coefficients Because of the difficulty of measuring the high individual film coefficients in an evaporator, experimental results are usually expressed in terms of overall coefficients. These are based on the net temperature drop corrected for boiling-point elevation. The overall coefficient, of course, is influenced by the same factors influencing individual coefficients; but if one resistance (say that of the liquid film) is controlling, major changes in the other resistances have almost no effect on the overall coefficient.

Typical overall coefficients for various types of evaporators are given in Table 16-1. These coefficients apply to conditions under which the various evaporators are ordinarily used. A horizontal-tube evaporator, for example, gives moderately high coefficients with thin liquids but very low ones with viscous materials. A small accumulation of scale reduces the coefficients to a small fraction of the clean-tube values. An agitated-film evaporator gives a seemingly low coefficient with a liquid having a viscosity of 100 P, but this coefficient is much larger than would be obtained in any other type of evaporator which could handle such a viscous material at all.

In short-tube natural-circulation evaporators the overall coefficient is sensitive to the temperature drop and to the boiling temperature of the solution. With liquids of low viscosity the heat-transfer coefficients are quite high, of the order 1000 to 2000 Btu/ft^2-h-°F for water. Representative data are given in standard handbooks.[6a, 10a]

Evaporator Economy

The chief factor influencing the economy of an evaporator system is the number of effects. By proper design the enthalpy of vaporization in the steam to the first effect can be used one or more times, depending on the number of effects. The economy also is influenced by the temperature of the feed. If the temperature is below the boiling point in the first effect, the heating load uses a part of the enthalpy of

Table 16-1 TYPICAL OVERALL COEFFICIENTS IN EVAPORATORS†

Type	Overall coefficient U	
	Btu/ft^2-h-°F	W/m^2-°C
Long-tube vertical evaporators:		
Natural circulation	200–600	1,000–3,000
Forced circulation	400–2,000	2,000–10,000
Short-tube evaporators:		
Horizontal tube	200–400	1,000–2,000
Calandria type	150–500	800–2,500
Coil evaporators	200–400	1,000–2,000
Agitated-film evaporator, Newtonian liquid, viscosity:		
1 cP	400	2,000
1 P	300	1,500
100 P	120	600

† Data from G. G. Brown et al., "Unit Operations," p. 484, Wiley, New York, 1950; and E. Lindsey, *Chem. Eng.*, **60**(4):227 (1953).

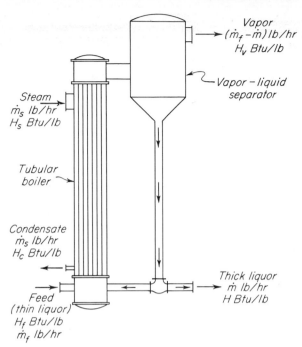

FIGURE 16-8
Material and enthalpy balances in evaporator.

vaporization of the steam and only a fraction is left for evaporation; if the feed is at a temperature above the boiling point, the accompanying flash contributes some evaporation over and above that generated by the enthalpy of vaporization in the steam. Quantitatively, evaporator economy is entirely a matter of enthalpy balances.

Enthalpy balances for single-effect evaporator   In a single-effect evaporator the latent heat of condensation of the steam is transferred through a heating surface to vaporize water from a boiling solution. Two enthalpy balances are needed, one for the steam side and one for the vapor or liquor side.

Figure 16-8 shows diagrammatically a vertical-tube single-effect evaporator. The rate of steam flow and of condensate is $\dot{m}_s$; that of the thin liquor, or feed, is $\dot{m}_f$, and that of the thick liquor is $\dot{m}$. The rate of vapor flow to the condenser, assuming that no solids precipitate from the liquor, is $\dot{m}_f - \dot{m}$. Also, let T_s be the condensing temperature of the steam, T the boiling temperature of the liquid in the evaporator, and T_f the temperature of the feed.

It is assumed that there is no leakage or entrainment, that the flow of non-condensables is negligible, and that heat losses from the evaporator need not be considered. The steam entering the steam chest may be superheated, and the condensate usually leaves the steam chest somewhat subcooled below its boiling point. Both the superheat and the subcooling of the condensate are small, however, and it is acceptable to neglect them in making an enthalpy balance. The small errors made in neglecting them are approximately compensated by neglecting heat losses from the steam chest.

Under these assumptions the difference between the enthalpy of the steam and that of the condensate is simply λ_s, the latent heat of condensation of the steam. The enthalpy balance for the steam side, by Eqs. (1-53) and (1-58), is

$$q_s = \dot{m}_s(H_c - H_s) = -\dot{m}_s\lambda_s \tag{16-2}$$

where q_s = rate of heat transfer through heating surface to steam
  H_s = specific enthalpy of steam
  H_c = specific enthalpy of condensate
  λ_s = latent heat of condensation of steam
  $\dot{m}_s$ = rate of flow of steam

The quantity q_s is inherently negative since, by convention, heat added to the fluid through the control surface is taken as positive and that abstracted from the fluid as negative.

The enthalpy balance for the liquor side is

$$q = (\dot{m}_f - \dot{m})H_v - \dot{m}_f H_f + \dot{m}H \tag{16-3}$$

where q = rate of heat transfer from heating surface to liquid
  H_v = specific enthalpy of vapor
  H_f = specific enthalpy of thin liquor
  H = specific enthalpy of thick liquor

In the absence of heat losses, the heat transferred from the steam to the tubes equals that transferred from the tubes to the liquor, and $-q_s = q$. Thus, by combining Eqs. (16-2) and (16-3),

$$q = \dot{m}_s\lambda_s = (\dot{m}_f - \dot{m})H_v - \dot{m}_f H_f + \dot{m}H \tag{16-4}$$

The liquor-side enthalpies H_v, H_f, and H depend upon the characteristics of the solution being concentrated. Most solutions when mixed or diluted at constant temperature do not give much heat effect. This is true of solutions of organic substances and of moderately concentrated solutions of many inorganic substances. Thus sugar, salt, and papermill liquors do not possess appreciable heats of dilution or mixing. Sulfuric acid, sodium hydroxide, and calcium chloride, on the other hand, especially in concentrated solutions, evolve considerable heat when diluted and so possess appreciable heats of dilution. An equivalent amount of heat is required, in addition to the latent heat of vaporization, when dilute solutions of these substances are concentrated to high densities.

ENTHALPY BALANCE WITH NEGLIGIBLE HEAT OF DILUTION   For solutions having negligible heats of dilution, the enthalpies H_f and H can be calculated from the specific heats of the solution. A datum temperature above which the enthalpies can be calculated is chosen. The thick liquor and the vapor leave the vapor head in equilibrium, and the temperatures of both equal T, the boiling temperature in the evaporator. The most convenient choice of the datum, therefore, is T. If this choice is made, the specific enthalpy H of the thick liquor is zero and the term $\dot{m}H$ vanishes. The specific enthalpy H_f of the thin liquor can then be calculated from the specific heat, which is assumed to be constant over the temperature range from T_f to T. Thus

$$H_f = c_{pf}(T_f - T) \tag{16-5}$$

where c_{pf} = specific heat of thin liquor
  T_f = temperature of thin liquor
  T = temperature of thick liquor

The specific heat of a solution having no heat of mixing is a linear function of concentration. Thus if the specific heat c_{p0} is known for some one aqueous solution containing x_0 weight fraction of solute, the specific heat of any other solution of concentration x weight fraction is

$$c_p = 1 - (1 - c_{p0})\frac{x}{x_0} \tag{16-6}$$

The specific heat of the pure solvent, water, is, of course, 1.

The specific enthalpy H_v to be used in Eq. (16-6) is the enthalpy of the vapor less that of liquid water at the datum temperature T. If the boiling-point elevation of the thick liquor is negligible, this enthalpy difference is simply λ, the latent heat of vaporization of water at the pressure of the vapor space. If the boiling-point elevation of the thick liquor is appreciable, the vapor leaving the solution is superheated by an amount, in degrees, equal to the boiling-point elevation. The specific enthalpy H_v is, accurately, that of the superheated vapor less that of liquid water at the datum temperature T. In practice, however, it is sufficiently accurate, and considerably simpler, to use for H_v the latent heat of vaporization of water λ at the pressure of the vapor space.

Introducing the above assumptions into Eq. (16-4) gives final equations for the enthalpy balances over a single-effect evaporator when the heat of dilution is negligible:

$$q = \dot{m}_s \lambda_s = (\dot{m}_f - \dot{m})\lambda + \dot{m}_f c_{pf}(T - T_f) \tag{16-7}$$

If the temperature T_f of the thin liquor is greater than the datum temperature T, the term $c_{pf}\dot{m}_f(T - T_f)$ is negative and is the enthalpy brought into the evaporator above the datum temperature by the thin liquor. This item is the *flash evaporation*. If the temperature T_f of the thin liquor fed to the evaporator is less than T, the datum temperature, the term $\dot{m}_f c_{pf}(T_f - T)$ is positive and for a given evaporation additional steam will be required to provide this enthalpy. The term $\dot{m}_f c_{pf}(T_f - T)$ is therefore the heating load. In words, Eq. (16-7) states that the heat of vaporization of the steam is utilized (1) to vaporize water from the solution and (2) to heat the feed to the boiling point; if the feed enters above the boiling point in the evaporator, part of the evaporation is from flash.

ENTHALPY BALANCE WITH APPRECIABLE HEAT OF DILUTION; ENTHALPY-CONCENTRATION DIAGRAM. If the heat of dilution of the liquor being concentrated is too large to be neglected, the enthalpy is not linear with concentration at constant temperature.

444

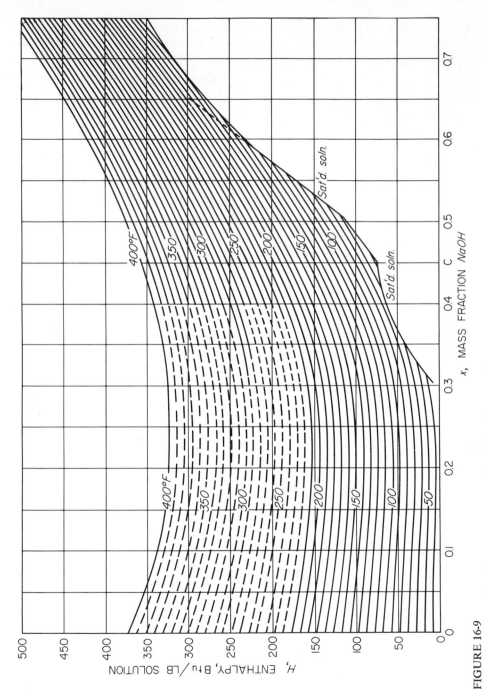

FIGURE 16-9
Enthalpy-concentration diagram, system sodium hydroxide–water.[7]

The most satisfactory source for the values of H_f and H for use in Eq. (16-4) is then an enthalpy-concentration diagram, in which enthalpy, in Btu per pound or joules per gram of solution, is plotted against concentration, in mass fraction or weight percentage of solute.[7] Isotherms drawn on the diagram show the enthalpy as a function of concentration at constant temperature. In using an enthalpy-concentration diagram it is necessary only to read the ordinate of the intersection of the abscissa for the desired concentration and the isotherm for the given temperature.

Figure 16-9 is an enthalpy-concentration diagram for solutions of sodium hydroxide and water. Concentrations are in mass fraction of sodium hydroxide, temperatures in degrees Fahrenheit, and enthalpies in Btu per pound of solution. The enthalpy of water is referred to the same datum as in the steam tables, namely, liquid water at 32°F (0°C), so enthalpies from the figure can be used with those from the steam tables when liquid water or steam is involved in the calculations. In finding data for substitution into Eq. (16-4), values of H_f and H are taken from Fig. 16-9, and the enthalpy H_v of the vapor leaving the evaporator is obtained from the steam tables.

The curved boundary lines on which the isotherms of Fig. 16-9 terminate represent conditions of temperature and concentration under which solid phases form. These are various solid hydrates of sodium hydroxide. The enthalpies of all single-phase solutions lie above this boundary line. The enthalpy-concentration diagram can also be extended to include solid phases.

The isotherms on an enthalpy-concentration diagram for a system with no heat of dilution are straight lines. The curvature of the isotherms in Fig. 16-9 provides a qualitative measure of the effect of the heat of dilution on the enthalpy of solutions of sodium hydroxide and water. Enthalpy-concentration diagrams can be constructed, of course, for solutions having negligible heats of dilution, but they are unnecessary in view of the simplicity of the specific-heat method described in the last section.

There is a close connection between the enthalpy of a solution and its vapor pressure. The vapor pressures of sodium hydroxide solutions, correlated by Dühring lines, are shown in Fig. 16-5, which is consistent with the enthalpy-concentration diagram of Fig. 16-9.

Single-effect calculations   The use of material balances, enthalpy balances, and the capacity equation (16-1) in the design of single-effect evaporators is shown in Examples 16-1 and 16-2.

EXAMPLE 16-1   A solution of organic colloids in water is to be concentrated from 10 to 50 percent solids in a single-effect evaporator. Steam is available at a gauge pressure of 15 $lb_f/in.^2$ (249°F) [1.03 atm (120.5°C)]. A pressure of 4 in. (102 mm) Hg abs is to be maintained in the vapor space; this corresponds to a boiling point for water of 125°F (51.7°C). The feed rate to the evaporator is 55,000 lb/h (24,950 kg/h). The overall heat-transfer coefficient can be taken as 500 Btu/ft²-h-°F (2,800 W/m²-°C). The solution has a negligible elevation in boiling point and a negligible heat of dilution. Calculate the steam consumption, the economy, and the heating surface required if the temperature of the feed is (a) 125°F (51.7°C), (b) 70°F (21.1°C), (c) 200°F (93.3°C). The specific heat of the feed solution is

0.90 Btu/lb-°F (3.77 J/g-°C), and the latent heat of vaporization of the solution may be taken equal to that of water. Radiation losses may be neglected.

SOLUTION   (a)   The flow rate of thin liquor $\dot{m}_f$ is 55,000 lb/h. The amount of water evaporated is found by a material balance from the terminal concentrations and the flow rate of the thin liquor. The feed contains $\frac{90}{10} = 9$ lb of water per pound of solid; the thick liquor contains $\frac{50}{50} = 1$ lb of water per pound of solid. The quantity evaporated, therefore, is $9 - 1 = 8$ lb of water per pound of solid, or

$$8 \times 55,000 \times 0.10 = 44,000 \text{ lb/h}$$

The flow rate of thick liquor $\dot{m}$ is $55,000 - 44,000 = 11,000$ lb/h.
Steam consumption   An enthalpy balance is now written in accordance with Eq. (16-7). The quantities needed are

$$\lambda_s = 946 \text{ Btu/lb} \quad \text{(Appendix 8)} \quad \lambda = 1023 \text{ Btu/lb} \quad \text{(Appendix 8)}$$

$$c_{pf} = 0.90 \text{ Btu/lb-°F} \quad T_f = 125°F \quad T = 125°F$$

The steam requirement, from Eq. (16-7), is

$$\dot{m}_s = \frac{(55,000 - 11,000)1023 - 0}{946} = 47,580 \text{ lb/h (21,580 kg/h)}$$

Economy   The economy is $44,000/47,580 = 0.925$.
Heating surface   This is calculated from Eq. (16-1):

$$A = \frac{q}{U(T_s - T)} = \frac{\dot{m}_s \lambda_s}{U(T_s - T)} = \frac{47,580 \times 946}{500(249 - 125)} = 725 \text{ ft}^2 \ (67.4 \text{ m}^2)$$

(b)   *Steam consumption*   The material balance is the same as in part (a). The only quantity in Eq. (16-7) which changes is T_f, which is now 70°F. Thus

$$\dot{m}_s = \frac{(55,000 - 11,000) \times 1023 - 55,000 \times 0.90(70 - 125)}{946}$$

$$= 50,460 \text{ lb/h (22,890 kg/h)}$$

Economy   The economy is $44,000/50,460 = 0.872$.
Heating surface

$$A = \frac{50,460 \times 946}{500(249 - 125)} = 770 \text{ ft}^2 \ (71.5 \text{ m}^2)$$

(c)   *Steam consumption* T_f is now 200°F.

$$\dot{m}_s = \frac{(55,000 - 11,000) \times 1023 - 55,000 \times 0.90(200 - 125)}{946}$$

$$= 43,660 \text{ lb/h (19,800 kg/h)}$$

Economy   The economy is $44,000/43,660 = 1.008$.
Heating surface

$$A = \frac{43,660 \times 946}{500(249 - 125)} = 665 \text{ ft}^2 \ (61.8 \text{ m}^2)$$

Note that lowering the feed temperature increases the heating area required and reduces the economy. If the feed enters appreciably above its boiling point, the heating area is considerably reduced from that needed with cold feed, and the economy may be greater than unity even in a single-effect evaporator.   ////

EXAMPLE 16-2 A single-effect evaporator is to concentrate 20,000 lb/h (9,070 kg/h) of a 20 percent solution of sodium hydroxide to 50 percent solids. The gauge pressure of the steam is to be 20 $lb_f/in.^2$ (1.37 atm); the absolute pressure in the vapor space is to be 100 mm Hg (1.93 $lb_f/in.^2$). The overall coefficient is estimated to be 250 Btu/ft^2-h-°F (1,400 W/m^2-°C). The feed temperature is 100°F (37.8°C). Calculate the amount of steam consumed, the economy, and the heating surface required.

SOLUTION   The amount of water evaporated is found from a material balance as in Example 16-1. The feed contains $\frac{80}{20}$ = 4 lb of water per pound of solid; the thick liquor contains $\frac{50}{50}$ = 1 lb of water per pound of solid. The quantity evaporated is 4 − 1 = 3 lb of water per pound of solid, or

$$3 \times 20,000 \times 0.20 = 12,000 \text{ lb/h}$$

The flow rate of thick liquor $\dot{m}$ is 20,000 − 12,000 = 8,000 lb/h (3,630 kg/h).
Steam consumption   Since with strong solutions of sodium hydroxide the heat of dilution is not negligible, the rate of heat transfer is found from Eq. (16-4) and Fig. 16-9. The vaporization temperature of the 50 percent solution at a pressure of 100 mm Hg is found as follows.

$$\text{Boiling point of water at 100 mm Hg} = 124°F \qquad \text{(Appendix 8)}$$
$$\text{Boiling point of solution} = 197°F \qquad \text{(Fig. 16-5)}$$
$$\text{Boiling-point elevation} = 197 - 124 = 73°F$$

The enthalpies of the feed and thick liquor are found from Fig. 16-9:

Feed, 20% solids, 100°F: $H_f = 55$ Btu/lb

Thick liquor, 50% solids, 197°F: $H = 221$ Btu/lb

The enthalpy of the vapor leaving the evaporator is found from steam tables. The enthalpy of superheated water vapor at 197°F and 1.93 $lb_f/in.^2$ is 1149 Btu/lb; this is the H_v of Eq. (16-4).

The heat of vaporization of steam λ_s at a gauge pressure of 20 $lb_f/in.^2$ is, from Appendix 8, 939 Btu/lb.

The rate of heat transfer and the steam consumption can now be found from Eq. (16-4):

$$q = (20,000 - 8,000)(1149) + 8,000 \times 221 - 20,000 \times 55 = 14,456,000 \text{ Btu/h}$$

$$\dot{m}_s = \frac{14,456,000}{939} = 15,400 \text{ lb/h (6,990 kg/h)}$$

Economy   The economy is 12,000/15,400 = 0.78.
Heating surface   The condensation temperature of the steam is 259°F. The heating area required is

$$A = \frac{14,456,000}{250(259 - 197)} = 930 \text{ ft}^2 \text{ (86.4 m}^2)$$

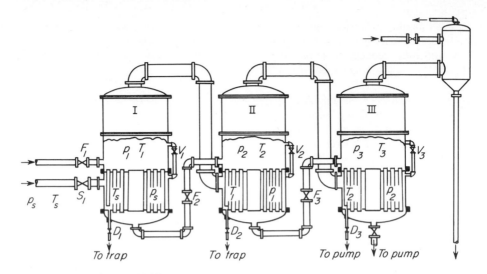

FIGURE 16-10
Triple-effect evaporator: I, II, III, first, second, and third effects; D_1, D_2, D_3, condensate valves; F_1, F_2, F_3, feed- or liquor-control valves; S_1, steam valve; V_1, V_2, V_3, vent valves; p_s, p_1, p_2, p_3, pressures; T_s, T_1, T_2, T_3, temperatures.

If the enthalpy of the vapor H_v were based on saturated vapor at the pressure in the vapor space, instead of on superheated vapor, the rate of heat transfer would be 14,036,000 Btu/h (4,115.7 kW) and the heating area would be 906 ft^2 (84.2 m^2). Thus this approximation would introduce an error of only about 3 percent.   ////

Multiple-effect evaporators   Figure 16-10 shows three short-tube natural-circulation evaporators connected together to form a triple-effect system. Connections are made so that the vapor from one effect serves as the heating medium for the next. A condenser and air ejector establishes a vacuum in the third effect in the series and withdraws noncondensables from the system. The first effect of a multiple-effect evaporator is the effect to which the raw steam is fed and in which the pressure in the vapor space is the highest. The last effect is that in which the vapor-space pressure is a minimum. In this manner the pressure difference between the steam and the condenser is spread across two or more effects in the multiple-effect system. The pressure in each effect is lower than that in the effect from which it receives steam and above that of the effect to which it supplies vapor. Each effect, in itself, acts as a single-effect evaporator, and each has a temperature drop across its heating surface corresponding to the pressure drop in that effect. Every statement that has so far been made about a single-effect evaporator applies to each effect of a multiple-effect system. Arranging a series of evaporator bodies into a multiple-effect system is a matter of interconnecting piping, and not of the structure of the individual units. The numbering of the effects is independent of the order liquor is fed to them. In Fig. 16-10 dilute feed enters the first effect, where it is partly concentrated; it flows to the second effect for additional concentration, and then to the third effect for final concentration. Thick liquor is pumped out of the third effect.

In steady operation the flow rates and evaporation rates are such that neither solvent nor solute accumulates or depletes in any of the effects. The temperature, concentration, and flow rate of the feed are fixed, the pressures in steam inlet and condenser established, and all liquor levels in the separate effects maintained. Then all internal concentrations, flow rates, pressures, and temperatures are automatically kept constant by the operation of the process itself. The concentration of the thick liquor can be changed only by changing the rate of flow of the feed. If the thick liquor is too dilute, the feed rate to the first effect is reduced, and if the thick liquor is too concentrated, the feed rate is increased. Eventually the concentration in the last effect and in the thick-liquor discharge will reach a new steady state at the desired level.

The heating surface in the first effect will transmit per hour an amount of heat given by the equation

$$q_1 = A_1 U_1 \, \Delta T_1 \qquad (16\text{-}8)$$

If the part of this heat that goes to heat the feed to the boiling point is neglected for the moment, it follows that practically all of this heat must appear as latent heat in the vapor that leaves the first effect. The temperature of the condensate leaving through connection D_2 is very near the temperature T_1 of the vapors from the boiling liquid in the first effect. Therefore, in steady operation practically all of the heat that was expended in creating vapor in the first effect must be given up when this same vapor condenses in the second effect. The heat transmitted in the second effect, however, is given by the equation

$$q_2 = A_2 U_2 \, \Delta T_2 \qquad (16\text{-}9)$$

As has just been shown, q_1 and q_2 are nearly equal, and therefore

$$A_1 U_1 \, \Delta T_1 = A_2 U_2 \, \Delta T_2 \qquad (16\text{-}10)$$

This same reasoning may be extended to show that, roughly,

$$A_1 U_1 \, \Delta T_1 = A_2 U_2 \, \Delta T_2 = A_3 U_3 \, \Delta T_3 \qquad (16\text{-}11)$$

It should be understood that Eqs. (16-10) and (16-11) are only approximate equations that must be corrected by the addition of terms which are, however, relatively small compared to the quantities involved in the expressions above.

In ordinary practice the heating areas in all the effects of a multiple-effect evaporator are equal. This is to obtain economy of construction. Therefore, from Eq. (16-11) it follows that, since $q_1 = q_2 = q_3 = q$,

$$U_1 \, \Delta T_1 = U_2 \, \Delta T_2 = U_3 \, \Delta T_3 = \frac{q}{A} \qquad (16\text{-}12)$$

From this it follows that the temperature drops in a multiple-effect evaporator are approximately inversely proportional to the heat-transfer coefficients.

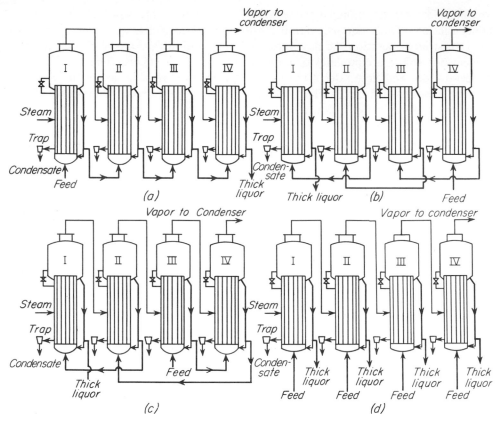

FIGURE 16-11
Patterns of liquor flow in multiple-effect evaporators: (*a*) forward feed; (*b*) backward feed; (*c*) mixed feed; (*d*) parallel feed.

EXAMPLE 16-3 A triple-effect evaporator is concentrating a liquid that has no appreciable elevation in boiling point. The temperature of the steam to the first effect is 227°F (108.3°C); the boiling point of the solution in the last effect is 125°F (51.7°C). The overall heat-transfer coefficients, in Btu/ft²-h-°F, are 500 in the first effect, 400 in the second effect, and 200 in the third effect (2,800, 2,200, and 1,100 W/m²-°C). At what temperatures will the liquid boil in the first and second effects?

SOLUTION The total temperature drop is $227 - 125 = 102$°F. As shown by Eq. (16-12), the temperature drops in the several effects will be approximately inversely proportional to the coefficients. Thus, for example,

$$\Delta T_1 = 102 \, \frac{\frac{1}{500}}{\frac{1}{500} + \frac{1}{400} + \frac{1}{200}} = 21.5°F \ (11.9°C)$$

In the same manner $\Delta T_2 = 26.8$°F (14.9°C) and $\Delta T_3 = 53.7$°F (29.8°C). Consequently the boiling point in the first effect will be 205.5°F (96.4°C), and that in the second effect, 178.7°F (81.5°C). ////

METHODS OF FEEDING   The usual method of feeding a multiple-effect evaporator is to pump the thin liquid into the first effect and send it in turn through the other effects, as shown in Fig. 16-11a. This is called *forward feed*. The concentration of the liquid increases from the first effect to the last. This pattern of liquid flow is the simplest. It requires a pump for feeding dilute solution to the first effect, since this effect is often at about atmospheric pressure, and a pump to remove thick liquor from the last effect. The transfer from effect to effect, however, can be done without pumps, since the flow is in the direction of decreasing pressure, and control valves in the transfer lines are all that is required.

Another common method is *backward feed*, in which dilute liquid is fed to the last effect and then pumped through the successive effects to the first, as shown in Fig. 16-11b. This method requires a pump between each pair of effects in addition to the thick-liquor pump, since the flow is from low pressure to high pressure. Backward feed often gives a higher capacity than forward feed when the thick liquor is viscous, and it gives higher economy than forward feed when the feed liquor is cold.

Other patterns of feed are sometimes used. In *mixed feed* the dilute liquid enters an intermediate effect, flows in forward feed to the end of the series, and is then pumped back to the first effects for final concentration, as shown in Fig. 16-11c. This eliminates some of the pumps needed in backward feed and yet permits the final evaporation to be done at the highest temperature. In crystallizing evaporators, where a slurry of crystals and mother liquor is withdrawn, feed may be admitted directly to each effect to give what is called *parallel feed*, as shown in Fig. 16-11d. In parallel feed there is no transfer of liquid from one effect to another.

Capacity and economy of multiple-effect evaporators   The increase in economy through the use of multiple-effect evaporation is obtained at the cost of reduced capacity. It might be thought that by providing several times as much heating surface the evaporating capacity would be increased, but this is not the case. The total capacity of a multiple-effect evaporator is no greater than that of a single-effect evaporator having a heating surface equal to one of the effects and operating under the same terminal conditions. The amount of water vaporized *per unit area of surface* in an N-effect multiple-effect evaporator is approximately $(1/N)$th that in the single effect. This can be shown by the following analysis.

If the heating load and the heat of dilution are neglected, the capacity of an evaporator is directly proportional to the rate of heat transfer. The heat transferred in the three effects in Fig. 16-10 is given by the equations

$$q_1 = U_1 A_1 \, \Delta T_1 \qquad q_2 = U_2 A_2 \, \Delta T_2 \qquad q_3 = U_3 A_3 \, \Delta T_3 \qquad (16\text{-}13)$$

The total capacity is proportional to the total rate of heat transfer q, found by adding these equations.

$$q = q_1 + q_2 + q_3 = U_1 A_1 \, \Delta T_1 + U_2 A_2 \, \Delta T_2 + U_3 A_3 \, \Delta T_3 \qquad (16\text{-}14)$$

Assume that the surface area is A ft^2 in each effect and that the overall coefficient U is also the same in each effect. Then Eq. (16-14) can be written

$$q = UA(\Delta T_1 + \Delta T_2 + \Delta T_3) = UA \, \Delta T \qquad (16\text{-}15)$$

where ΔT is the total temperature drop between the steam in the first effect and the vapor in the last effect.

Suppose now that a single-effect evaporator with a surface area A is operating with the same total temperature drop. If the overall coefficient is the same as in each effect of the triple-effect evaporator, the rate of heat transfer in the single effect is

$$q = UA \, \Delta T$$

This is exactly the same equation as for the multiple-effect evaporator. No matter how many effects are used, provided the overall coefficients are the same, the capacity will be no greater than that of a single effect having an area equal to that of each effect in the multiple unit. In actuality, because of boiling-point elevation and other factors, the capacity of a multiple-effect evaporator is almost always less than that of the corresponding single effect.

EFFECT OF LIQUID HEAD AND BOILING-POINT ELEVATION   The liquid head and the boiling-point elevation influence the capacity of a multiple-effect evaporator even more than they do that of a single effect. The reduction in capacity caused by the liquid head, as before, cannot be estimated quantitatively. The liquid head, it will be remembered, reduces the temperature drop available in the evaporator. The boiling-point elevation also reduces the available temperature drop in each effect of a multiple-effect evaporator, as follows.

Consider an evaporator that is concentrating a solution with a large boiling-point elevation. The vapor coming from this boiling solution is at the solution temperature and is therefore superheated by the amount of the boiling-point elevation. As discussed in Chap. 13, pages 356 and 357, superheated steam is essentially equivalent to saturated steam at the same pressure when used as a heating medium. The temperature drop in any effect, therefore, is calculated from the temperature of saturated steam at the pressure of the steam chest, and not from the temperature of the boiling liquid in the previous effect. This means that the boiling-point elevation in any effect is lost from the total available temperature drop. This loss occurs in every effect of a multiple-effect evaporator, and the resulting loss of capacity is often important.

The influence of these losses in temperature drop on the capacity of a multiple-effect evaporator is shown in Fig. 16-12. The three diagrams in this figure represent the temperature drops in a single-effect, double-effect, and triple-effect evaporator. The terminal conditions are the same in all three; i.e., the steam pressure in the first effect and the saturation temperature of the vapor evolved from the last effect are identical in all three evaporators. Each effect contains a liquid with a boiling-point elevation. The total height of each column represents the total temperature spread from the steam temperature to the saturation temperature of the vapor from the last effect.

Consider the single-effect evaporator. Of the total temperature drop, the shaded part represents the loss in temperature drop due to boiling-point elevation. The remaining temperature drop, the actual driving force for heat transfer, is represented by the unshaded part. The diagram for the double-effect evaporator shows two

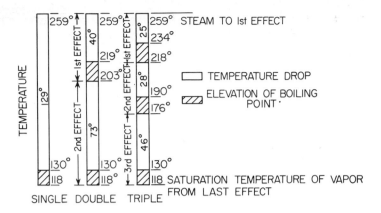

FIGURE 16-12
Effect of boiling-point elevation on capacity of evaporators.

shaded portions because there is a boiling-point elevation in each of the two effects, and the residual unshaded part is smaller than in the diagram for the single effect. In the triple-effect evaporator there are three shaded portions since there is a loss of temperature drop in each of three effects, and the total net available temperature drop is correspondingly smaller.

Figure 16-12 shows that in extreme cases of a large number of effects or very high boiling-point elevations the sum of the boiling-point elevations in a proposed evaporator could be greater than the total temperature drop available. Operation under such conditions is impossible. The design or the operating conditions of the evaporator would have to be revised to reduce the number of effects or increase the total temperature drop.

The economy of a multiple-effect evaporator is not influenced by boiling-point elevations if minor factors, such as the temperature of the feed and changes in the heats of vaporization, are neglected. A pound of steam condensing in the first effect generates about a pound of vapor, which condenses in the second effect generating another pound there, and so on. The *economy* of a multiple-effect evaporator depends on heat-balance considerations and not on the rate of heat transfer. The *capacity*, on the other hand, is reduced by the boiling-point elevation. The capacity of a double-effect evaporator concentrating a solution with a boiling-point elevation is less than half the capacity of two single effects each operating with the same overall temperature drop. The capacity of a triple effect is less than one-third that of three single effects with the same terminal temperatures.

OPTIMUM NUMBER OF EFFECTS   The cost of an evaporator per square foot of surface is a function of the total area of all effects[10b] and decreases with total area, approaching an asymptote for large installations. When the cost of a unit area of heating surface is constant, regardless of the number of effects, the investment required for an N-effect evaporator is N times that for a single-effect evaporator of the same capacity.

The optimum number of effects must be found from an economic balance between the savings in steam obtained by multiple-effect operation and the added investment required.

Multiple-effect calculations   In designing a multiple-effect evaporator the results usually desired are the amount of steam consumed, the area of the heating surface required, the approximate temperatures in the various effects, and the amount of vapor leaving the last effect. As in a single-effect evaporator, these quantities are found from material balances, enthalpy balances, and the capacity equation (16-1). In a multiple-effect evaporator, however, a trial-and-error method is used in place of a direct algebraic solution.

Consider, for example, a triple-effect evaporator. There are seven equations which may be written: an enthalpy balance for each effect, a capacity equation for each effect, and the known total evaporation, or the difference between the thin- and thick-liquor rates. If the amount of heating surface in each effect is assumed to be the same, there are seven unknowns in these equations: (1) the rate of steam flow to the first effect, (2) to (4) the rate of flow from each effect, (5) the boiling temperature in the first effect, (6) the boiling temperature in the second effect, and (7) the heating surface per effect. It is possible to solve these equations for the seven unknowns, but the method is tedious and involved. Another method of calculation is as follows:

1 Assume values for the boiling temperatures in the first and second effects.
2 From enthalpy balances find the rates of steam flow and of liquor from effect to effect.
3 Calculate the heating surface needed in each effect from the capacity equations.
4 If the heating areas so found are not nearly equal, estimate new values for the boiling temperatures and repeat items 2 and 3 until the heating surfaces are equal.

In practice these calculations are best done on a digital computer.

EXAMPLE 16-4   A triple-effect forced-circulation evaporator is to be fed with 60,000 lb/h (27,215 kg/h) of 10 percent caustic soda solution at a temperature 180°F (82.2°C). The concentrated liquor is to be 50 percent NaOH. Saturated steam at 50 $lb_f/in.^2$ (3.43 atm) abs is to be used, and the condensing temperature of vapor from the third effect is to be 100°F (37.8°C). The feed order is II, III, I. Radiation and undercooling of condensate may be neglected. Overall coefficients corrected for boiling-point elevation are given in Table 16-2.

Table 16-2

Effect	Overall coefficient	
	Btu/ft²-h-°F	W/m²-°C
I	700	3,970
II	1,000	5,680
III	800	4,540

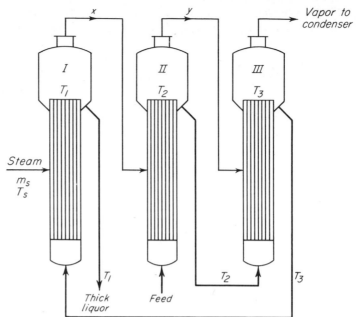

FIGURE 16-13
Flow sheet for Example 16-4.

Calculate (a) the heating surface required in each effect, assuming equal surfaces in each, (b) the steam consumption, (c) the steam economy.

SOLUTION   The total rate of evaporation can be calculated from an overall material balance, assuming that the solids go through the evaporator without loss (Table 16-3).

Figure 16-13 shows a flow diagram for this evaporator. The first, second, and third effects are designated as I, II, and III, respectively. For material and heat balances, let

$\dot{m}_s$ = rate of flow of steam
T_1, T_2, T_3 = boiling temperatures in I, II, and III
q_1, q_2, q_3 = heat-transfer rates in I, II, and III
C_1, C_2, C_3 = concentrations, weight fractions dissolved solids in liquor in I, II, and III
T'_1, T'_2, T'_3 = condensing temperature of vapor from I, II, and III

Table 16-3

Material	Flow rate, lb/h		
	Total	Solid	Water
Feed solution	60,000	6,000	54,000
Thick liquor	12,000	6,000	6,000
Water evaporated	48,000		48,000

To estimate the concentrations and hence the boiling-point elevations, assume, first, equal evaporation rates in each effect, or 48,000/3 = 16,000 lb/h. The intermediate concentrations are then

$$C_1 = \frac{6,000}{60,000 - 16,000} = 0.136 \qquad C_2 = \frac{6,000}{60,000 - 32,000} = 0.214$$

The estimated boiling-point elevations, from Fig. 16-5, are

Effect	I	II	III
Bp elevation, °F	76	7	13

As a guide in the first attempt to distribute the temperature drop among the three effects, the following principles are of value. As discussed on page 449, the temperature drop in any effect is inversely proportional to the overall heat-transfer coefficient. Also, any effect which has an extra load requires a larger proportion of the total temperature drop than the other effects. In this problem the steam temperature, from Appendix 8, is 281°F. The total temperature drop is 281 − 100 = 181°F, but the net temperature drop is only 181 − (76 + 7 + 13) = 85°F. This is the drop available for distribution. The temperature drop in the second effect will be low because of the high coefficient there and because the feed is hot; the drop in the first effect will be slightly larger than in the third effect, because of the difference in coefficients. From these considerations the first assumption is

$$\Delta T_1 = 33°F \qquad \Delta T_2 = 23°F \qquad \Delta T_3 = 29°F$$

The solution temperatures in I (equal to the temperature of the superheated vapor coming from it) is

$$T_1 = 281 - 33 = 248°F$$

The saturation temperature of this vapor is $T_1' = 248 - 76 = 172°F$. In the same way,

$$T_2 = 172 - 23 = 149°F \qquad T_2' = 149 - 7 = 142°F$$

$$T_3 = 142 - 29 = 113°F \qquad T_3' = 113 - 13 = 100°F$$

The enthalpies of the various streams can now be estimated from Fig. 16-9 for the solutions and Appendix 8 for the saturated vapor. The enthalpy of superheated vapor can be estimated from steam tables or by correcting the saturation enthalpy assuming a specific heat of 0.47 Btu/lb-°F for the vapor. The "vapor" from one effect becomes the "steam" for the next effect. Enthalpy balances can now be made to calculate the surface area required. Let x be the true evaporation rate in I; y that in II; and 48,000 − $(x + y)$ that in III. Then temperatures, enthalpies, and flow rates are as given in Table 16-4.

The enthalpy balances are:

Over II:

$$1171x + (60,000)(135) = 1126y + 102(60,000 - y) + 140x$$

$$x - 0.993y = -1,920$$

Over III:

$$1126y + (102)(60,000 - y) = 1111(48,000 - x - y) + (12,000 + x)67 + 110y$$

$$x + 1.940y = 46,000$$

Also
$$x - 0.993y = -1,920$$
$$2.933y = 47,920$$
$$y = 16,340 \text{ lb/h}$$
$$x = 14,300 \text{ lb/h}$$

The evaporation rate in III is

$$48,000 - 14,300 - 16,340 = 17,360 \text{ lb/h}$$

Heat loads:

$$q_1 = (14,300)(1171) + (250)(12,000) - (67)(12,000 + 14,300) = 17,983,000 \text{ Btu/h}$$

$$q_2 = (14,300)(1171 - 140) = 14,743,000 \text{ Btu/h}$$

$$q_3 = (16,340)(1126 - 110) = 16,601,000 \text{ Btu/h}$$

The areas are therefore

$$A_1 = \frac{17,983,000}{700 \times 33} = 778 \text{ ft}^2$$

$$A_2 = \frac{14,743,000}{100 \times 23} = 640 \text{ ft}^2$$

$$A_3 = \frac{16,601,000}{29 \times 800} = 720 \text{ ft}^2$$

$$\text{Average area} = 712 \text{ ft}^2$$

Since the surface areas are not equal, as required, the concentrations are corrected as follows:

$$C_1 = \frac{6,000}{60,000 - 16,340} = 0.137 \qquad C_2 = \frac{6,000}{60,000 - 16,340 - 17,360} = 0.228$$

Table 16-4   TEMPERATURES, ENTHALPIES, AND FLOW RATES FOR EXAMPLE 16-4

Stream	Temp., °F	Saturation temp., °F	Concentration, weight fraction	Enthalpy, Btu/lb	Flow rate, lb/h
Steam	281	281		1174	$\dot{m}_s$
Feed to I	113		0.214	67	$12,000 + x$
Vapor from I	248	172		1171	x
Condensate from I	281			249	$\dot{m}_s$
Liquid from I	248		0.50	250	12,000
Feed to II	180		0.10	135	60,000
Vapor from II	149	142		1126	y
Liquid from II	149		0.136	102	$60,000 - y$
Condensate from II	172			140	x
Vapor from III	113	100		1111	$48,000 - x - y$
Condensate from III	142			110	y

The temperature drops are corrected from the variation in calculated areas as follows:

$$\Delta T_1 = 33 \times \frac{778}{712} = 36°F$$

$$\Delta T_2 = 23 \times \frac{640}{712} = 21°F$$

$$\Delta T_3 = 85 - (36 + 21) = 28°F$$

The corrected temperatures, concentrations, and enthalpies are given in Table 16-5. The revised enthalpy balances are:
Over II:

$$1170x + (60,000)(135) = 1126y + (101)(60,000 - y) + 138x$$

$$x - 0.993y = -1977$$

Over III:

$$1126y + (101)(60,000 - y) = (1111)(48,000 - x - y) + (12,000 + x)(68) + 110y$$

$$x + 1.942y = 46,100$$

$$\underline{x - 0.993y = -1,977}$$

$$2.935y = 48,077$$

$$y = 16,380 \text{ lb/h}$$

$$x = 14,290 \text{ lb/h}$$

The evaporation rate in III is then

$$48,000 - 16,380 - 14,290 = 17,330 \text{ lb/h}$$

The corrected heat loads are

$$q_1 = (14,290)(1170) + (249)(12,000) - (68)(12,000 + 14,290) = 17,920,000 \text{ Btu/h}$$

$$q_2 = (14,290)(1170 - 138) = 14,747,000 \text{ Btu/h}$$

$$q_3 = (16,380)(1126 - 110) = 16,642,000 \text{ Btu/h}$$

Table 16-5   CORRECTED TEMPERATURES, ENTHALPIES, AND FLOW RATES FOR EXAMPLE 16-4

Stream	Temp., °F	Saturation temp., °F	Concentration, weight fraction	Enthalpy, Btu/lb	Flow rate, lb/h
Steam	281	281		1174	$\dot{m}_s$
Feed to I	113		0.228	68	$12,000 + x$
Vapor from I	245	170		1170	x
Condensate from I	281			249	$\dot{m}_s$
Liquid from I	246		0.50	249	12,000
Feed to II	180		0.10	135	60,000
Vapor from II	149	142		1126	y
Liquid from II	149		0.137	101	$60,000 - y$
Condensate from II	170			138	x
Vapor from III	114	100		1111	$48,000 - x - y$
Condensate from III	142			110	y

The areas are

$$A_1 = \frac{17,920,000}{700 \times 36} = 711 \text{ ft}^2$$

$$A_2 = \frac{14,747,000}{1000 \times 21} = 702 \text{ ft}^2$$

$$A_3 = \frac{16,642,000}{800 \times 28} = 743 \text{ ft}^2$$

These areas are close enough to each other to make further calculations unnecessary. The answers to the problem are

(a) *Area per effect*

$$719 \text{ ft}^2 \ (675 \text{ m}^2)$$

(b) *Steam consumption*

$$\dot{m}_s = \frac{17,920,000}{1174 - 249} = 19,370 \text{ lb/h} \ (8,786 \text{ kg/h})$$

(c) *Economy*

$$\frac{48,000}{19,370} = 2.48 \qquad\qquad ////$$

VAPOR RECOMPRESSION

The energy in the vapor evolved from a boiling solution can be used to vaporize more water provided there is a temperature drop for heat transfer in the desired direction. In a multiple-effect evaporator this temperature drop is created by progressively lowering the boiling point of the solution in a series of evaporators through the use of lower absolute pressures. The desired driving force can also be obtained by increasing the pressure (and, therefore, the condensing temperature) of the evolved vapor by mechanical or thermal recompression.[1] The compressed vapor is then condensed in the steam chest of the evaporator from which it came.

Mechanical recompression   The principle of mechanical vapor recompression is illustrated in Fig. 16-14. Cold feed is preheated almost to its boiling point by exchange with hot thick liquor, and is pumped through a heater as in a conventional forced-circulation evaporator. The vapor evolved, however, is not condensed; instead it is compressed to a somewhat higher pressure by a positive-displacement or centrifugal compressor, and becomes the "steam" which is fed to the heater. Since the saturation temperature of the compressed vapor is higher than the boiling point of the feed, heat flows from the vapor to the solution, generating more vapor. A small amount of makeup steam may be necessary. The optimum temperature drop for a typical system is about 10°F. The energy utilization of such a system is very good: based on the steam equivalent of the power required to drive the compressor, the economy corresponds to that of an evaporator containing 10 to 15 effects. The most important applications of mechanical-recompression evaporation are the concentration of very dilute radioactive solutions and the production of distilled water.

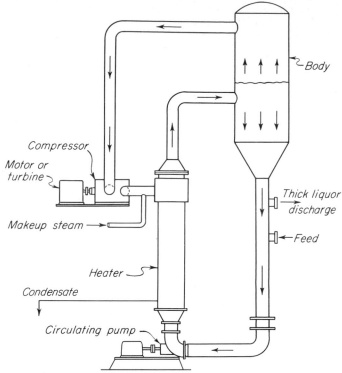

FIGURE 16-14
Mechanical recompression applied to forced-circulation evaporator.

Thermal recompression In a thermal recompression system the vapor is compressed by acting on it with high-pressure steam in a jet ejector. This results in more steam than is needed for boiling the solution, so that excess steam must be vented or condensed. The ratio of motive steam to the vapor from the solution depends on the evaporation pressure; for many low-temperature operations, with steam at 8 to 10 atm pressure, the ratio of steam required to the mass of water evaporated is about 0.5.

Since steam jets can handle large volumes of low-density vapor, thermal recompression is better suited than mechanical recompression to vacuum evaporation. Jets are cheaper and easier to maintain than blowers and compressors. The chief disadvantages of thermal recompression are the low mechanical efficiency of the jets and lack of flexibility in the system toward changed operating conditions.

SYMBOLS

A Area of heat-transfer surface, ft^2 or m^2; A_1, A_2, A_3, in effects I, II, III

c_p Specific heat at constant pressure, Btu/lb-°F or J/g-°C; c_{pf}, of feed; c_{p0}, of solution of concentration x_0, mass fraction of solute

H Enthalpy of thick liquor, Btu/lb or J/g; H_c, of condensate; H_f, of feed; H_s, of saturated steam; H_v, of vapor or superheated steam

$\dot{m}$ Mass flow rate, lb/h or kg/h; of liquor leaving single-effect evaporator; $\dot{m}_f$, of feed; $\dot{m}_s$, of steam and steam condensate

N Number of evaporator effects

p Pressure, lb_f/ft^2 or atm; p_s, of steam; p_1, p_2, p_3, in vapor spaces of effects I, II, III

q Rate of heat transfer, Btu/h or W; q_s, from tubes to steam (inherently negative, by convention); q_1, q_2, q_3, in effects I, II, III

T Temperature, °F or °C; boiling temperature in, and of liquor leaving, single-effect evaporator; T_c, temperature of condensate; T_f, of feed; T_h, of entering steam; T_s, of saturated steam; T', boiling temperature of water at pressure of vapor space; T_1', T_2', T_3', in effects I, II, III; T_1, T_2, T_3, boiling temperature in, and of liquor leaving, effects I, II, III

U Overall heat-transfer coefficient, $Btu/ft^2\text{-}h\text{-}°F$ or $W/m^2\text{-}°C$; U_1, U_2, U_3, in effects I, II, III

x Evaporation rate in effect I, lb/h, Example 16-4; also, mass fraction of dissolved solute; x_0, reference value

y Evaporation rate in effect II, lb/h, Example 16-4

Z Distance below liquid surface, ft or m

Greek letters

Γ Liquid loading on tube, lb/h per foot of perimeter or kg/h per meter of perimeter

ΔT Temperature drop, °F or °C; total overall corrected temperature drop, all effects; ΔT_1, ΔT_2, ΔT_3, temperature drop in effects I, II, III

λ Latent heat, Btu/lb or J/g; λ_s, latent heat of condensation of steam

μ Absolute viscosity, lb/ft-h or cP

PROBLEMS

16-1 A solution of organic colloids is to be concentrated from 20 to 65 percent solids in a vertical-tube evaporator. The solution has a negligible elevation in boiling point, and the specific heat of the feed is 0.93. Saturated steam is available at 0.7 atm abs, and the pressure in the condenser is 100 mm Hg abs. The feed enters at 60°F. The overall coefficient is 1,700 W/m²-°C. The evaporator must evaporate 20,000 kg of water per hour. How many square meters of surface are required, and what is the steam consumption in kilograms per hour?

16-2 A forced-circulation evaporator is to be designed to concentrate 50 percent caustic soda solution to 70 percent. A throughput of 50 tons of NaOH (100 percent basis) per 24 h is needed. Steam at 50 $lb_f/in.^2$ gauge and 95 percent quality is available, and the condenser temperature is 100°F. Feed enters at 100°F. Condensate leaves the heating element at a temperature 20°F below the condensing temperature of the steam. Radiation is estimated to be $1\frac{1}{2}$ percent from the heating element and 2 percent from the vapor head, both based on the enthalpy difference between steam and condensate. Concentrated liquor leaves at the boiling temperature of the liquid in the vapor head. The overall coefficient, based on the outside area, is expected to be 350 Btu/ft²-h-°F. The boiling-point elevation of the 70 percent liquor is 150°F. Calculate (a) the steam consumption in pounds per hour and (b) the number of tubes required if 8-ft by $\frac{7}{8}$-in. 16 BWG tubes are specified.

16-3 A triple-effect evaporator of the long-tube type is to be used to concentrate 35,000 gal/h of a 17 percent solution of dissolved solids to 38 percent dissolved solids. The feed enters at 60°F and passes through three tube-and-shell heaters, *a*, *b*, and *c*, in series, and then through the three effects in order II, III, I. Heater *a* is heated by vapor taken from the vapor line between the third effect and the condenser, heater *b* with vapor from the vapor line between the second and third effects, and heater *c* with vapor from the line between the first effect and the second. In each heater the warm-end temperature approach is 10°F. Other data are given below and in Table 16-6.

Steam to I, 230°F, dry and saturated.

Vacuum on III, 28 in., referred to a 30-in. barometer.

Condensates leave steam chests at condensing temperatures.

Boiling-point elevations, 1°F in II, 5°F in III, 15°F in I.

Coefficients, corrected for boiling-point elevation, 450 in I, 700 in II, 500 in III.

All effects have equal areas of heating surface.

Calculate (a) the steam required in pounds per hour, (b) the heating surface per effect, (c) the economy in pounds per pound of steam, (d) the latent heat to be removed in the condenser.

16-4 It is planned to concentrate 30,000 kg/h of a liquor containing 3 percent solids to a concentration of 48 percent solids. A single evaporator body suitable for this purpose costs $40,000, exclusive of pumps and other accessories common to any number of effects. Fixed charges are 30 percent per year, and steam costs $1.50 per 1,000 kilograms. Assume $0.90N$ lb evaporation per pound of steam, where N is the number of effects. The evaporator is to operate 24 h/d and 300 d/year. Labor costs are independent of the number of effects. How many effects should be used?

16-5 The capacity of a certain quintuple-effect evaporator has fallen considerably below normal. A comparison of the present pressure distribution in the system with that corresponding to normal operation is given in Table 16-7. The steam space of each effect is vented to the vapor space of that effect. The feed is forward and enters the first effect at 240°F. (a) What are the possible causes of the low capacity, and where are these causes located? (b) What is the present capacity, expressed as percent of normal?

16-6 A triple-effect standard vertical-tube evaporator each effect of which has 140 m² of heating surface is to be used to concentrate from 5 percent solids to 40 percent solids a solution possessing a negligible boiling-point elevation. Forward feed is to be used. Steam is available at 120°C, and the vacuum in the last effect corresponds to a boiling temperature of 40°C. The overall coefficients in W/m²-°C are 2,950 in I, 2,670 in II, and 1,360 in III, all specific heats may be taken as 4.2 J/g-°C, and radiation is negligible. Condensates leave at condensing temperature. The feed enters at 90°C. Calculate (a) the kilograms of 5 percent liquor that can be concentrated per hour, (b) the steam consumption in kilograms per hour.

16-7 A solution with a negligible boiling-point elevation is to be concentrated from 10 to 50 percent solids in a triple-effect forward-feed evaporator. Steam is available at 2 atm abs; the absolute pressure in the third effect is to be 0.13 atm. Feed enters at a rate of

Table 16-6

Concentration solids, %	Sp. gr.	Specific heat, Btu/lb-°F
10	1.02	0.98
20	1.05	0.94
30	1.10	0.87
35	1.16	0.82
40	1.25	0.75

20,000 kg/h and a temperature of 15°C. The specific heat of the solution at all concentrations may be taken to be the same as that of water. The overall coefficients are

Effect	I	II	III
Coefficient, W/m²-°C	3,000	2,000	1,100

Each effect is to have the same amount of heating surface. Calculate the area of heating surface required, the steam consumption, the distribution of temperatures, the economy of each effect, and the overall economy.

16-8 A vapor-recompression evaporator is to concentrate a very dilute aqueous solution. The feed rate is to be 30,000 lb/h; the evaporation rate will be 20,000 lb/h. The evaporator will operate at atmospheric pressure, with the vapor mechanically compressed as shown in Fig. 16-14 except that a natural-circulation calandria will be used. If steam costs 70 cents per 1,000 lb, electricity costs 0.8 cents per kilowatthour, and heat-transfer surface in the heater costs $10 per square foot, calculate the optimum pressure to which the vapor should be compressed. The overall compressor efficiency is 70 percent. Assume all other costs are independent of the pressure of the compressed vapor. To how many effects will this evaporator be equivalent?

REFERENCES

1 Beagle, M. J.: *Chem. Eng. Prog.,* **58**(10):79 (1962).
2 Boarts, R. M., W. L. Badger, and S. J. Meisenburg: *Trans. AIChE,* **33**:363 (1937).
3 Foust, A. S., E. M. Baker, and W. L. Badger: *Trans. AIChE,* **35**:45 (1939).
4 Hughmark, G. A.: *Chem. Eng. Prog.,* **60**(7):59 (1964).
5 Lindsey, E.: *Chem. Eng.,* **60**(4):227 (1953).
6 McAdams, W. H.: "Heat Transmission," 3d ed., McGraw-Hill, New York, 1954; (*a*) pp. 398ff, (*b*) p. 403.
7 McCabe, W. L.: *Trans. AIChE,* **31**:129 (1935).
8 Moore, J. G., and W. E. Hesler: *Chem. Eng. Prog.,* **59**(2):87 (1963).
9 Sinek, J. R., and E. H. Young: *Chem. Eng. Prog.,* **58**(12):74 (1962).
10 Standiford, F. C.: in J. H. Perry (ed.), "Chemical Engineers' Handbook," 5th ed., McGraw-Hill, New York, 1973; (*a*) pp. **11**-27ff, (*b*) p. **11**-37.

Table 16-7 **PRESSURE DISTRIBUTION FOR PROB. 16-5**

	Normal		Abnormal	
	Pressure, $lb_f/in.^2$	Temp., °F	Pressure, $lb_f/in.^2$	Temp., °F
Steam	45.4	275	45.4	275
Vapor space				
I	24.5	239	27.3	245
II	15.9	216	19.3	226
III	8.9	188	6.6	174
IV	4.4	157	3.5	148
V	1.4	113	1.4	113

Mass transfer and its applications

INTRODUCTION

A group of operations for separating the components of mixtures is based on the transfer of material from one homogeneous phase to another. Unlike purely mechanical separations, these methods utilize differences in vapor pressure or solubility, not density or particle size. The driving force for transfer is a concentration difference or a concentration gradient, much as a temperature difference or a temperature gradient provides the driving force for heat transfer. These methods, covered by the term *mass-transfer operations*, include such techniques as distillation, gas absorption, dehumidification, liquid extraction, leaching, crystallization, and a number of others not discussed in this book.

The function of *distillation* is to separate, by vaporization, a liquid mixture of miscible and volatile substances into individual components or, in some cases, into groups of components. The separation of a mixture of alchohol and water into its components; of liquid air into nitrogen, oxygen, and argon; and of crude petroleum into gasoline, kerosene, fuel oil, and lubricating stock are examples of distillation.

In *gas absorption* a soluble vapor is absorbed by means of a liquid in which the solute gas is more or less soluble, from its mixture with an inert gas. The washing of ammonia from a mixture of ammonia and air by means of liquid water is a typical example. The solute is subsequently recovered from the liquid by distillation, and the absorbing liquid can be either discarded or reused. When a solute is transferred from the solvent liquid to the gas phase, the operation is known as *desorption* or *stripping*. In *dehumidification* the liquid phase is a pure liquid of the same composition as the component that is being removed from the gas stream; i.e., solvent and solute are the same material. Usually the inert or carrier gas is virtually insoluble in the liquid. Removal of water vapor from air by condensation on a cold surface and the condensation of an organic vapor such as carbon tetrachloride out of a stream of nitrogen are examples of dehumidification. In humidification operations the direction of transfer is from the liquid to the gas phase. In the *drying* of solids, a liquid, usually water, is separated by the use of hot, dry gas (usually air) and so is coupled with the humidification of the gas phase. In *liquid extraction*, sometimes called *solvent extraction*, a mixture of two components is treated by a solvent that preferentially dissolves one or more of the components in the mixture. The mixture so treated is called the *raffinate* and the solvent-rich phase is called the *extract*. The component transferred from raffinate to extract is the *solute*, and the component left behind in the raffinate is the *diluent*. The solvent in the extract leaving the extractor is usually recovered and reused.

In extraction of solids, or *leaching*, soluble material is dissolved from its mixture with an inert solid by means of a liquid solvent. The dissolved material, or solute, can then be recovered by crystallization or evaporation.

Crystallization is used to obtain materials in attractive and uniform crystals of good purity. Since the formation of crystals separates a solute from a melt or a solution and leaves impurities behind, it is a separation operation. Modern theories and models of industrial crystallization from solution, however, are focused on the processes occurring at the crystal-solution interface and the characteristics of particulate solids; this operation is therefore discussed, with other operations on solid particles in Section 5.

The quantitative treatment of mass transfer is based on material and energy balances, equilibria, and rates of heat and mass transfer. Certain concepts applicable generally are discussed here. More specialized topics are discussed in the following chapters.

TERMINOLOGY AND SYMBOLS

It is conventional to refer generally to the two streams in any one operation as the L phase and the V phase. It is also customary to choose the stream having the higher density as the L phase and the one having the lower density as the V phase. An exception may appear in liquid extraction, where the raffinate always is taken as the L phase and the extract as the V phase, even when the raffinate happens to be lighter than the extract. In drying, the L phase is the stream consisting of the solid and the liquid retained in or on the solid. Table A shows how the streams are designated in the various operations.

Note on concentrations Strictly speaking, concentration means mass per unit volume. Mass may be in pounds or kilograms and volume in cubic feet or cubic meters. Pound moles or kilogram moles are often used for mass. In this book the context will make clear what quantity—mole or ordinary mass—is used. It is convenient to extend the use of the word "concentration" to include mole or mass fractions. The relation between concentration and mole or mass fraction of a component i is

$$c_i = \rho x_i$$

where x_i = mole or mass fraction of component i
ρ = molar or mass density of mixture
c_i = corresponding concentration of component i

The mole fraction is the ratio of the moles of the component to the total number of moles in the mixture, with a corresponding definition for mass fraction. By definition, all mole or mass fractions in a mixture sum to unity. If there are r components, $r - 1$ of the mole fractions may be chosen independently; the mole fraction of the remaining component is thereby fixed and equals 1 less the sum of the others.

General symbols are needed for flow rates and concentrations. For all operations, use V and L for the flow rates of V and L phases, respectively. Use A, B, C, etc., to refer to the individual components. If only one component is transferred between phases, choose component A as that component. Use x for the concentration of a component in the L phase, and y for the concentration in the V phase. Thus, y_A is the concentration of component A in a V phase and x_B is that of component B in an L phase. When only two components are present in a phase, the concentration of component A is x or y, and that of component B is $1 - x$ or $1 - y$, and the subscripts A and B are unnecessary.

TERMINAL QUANTITIES Since in steady-flow mass-transfer operations there are two streams and each must enter and leave, there are four terminal quantities. To identify them, use subscript a to refer to that end of the process where the L phase enters and b to refer to that end where the L phase leaves. Then, for *countercurrent* flow, the terminal quantities are as shown in Table B. If there are only two components in a stream, the subscript A can be dropped from the concentration terms.

Table A TERMINOLOGY FOR STREAMS IN MASS-TRANSFER OPERATIONS

Operation	V phase	L phase
Distillation	Vapor	Liquid
Gas absorption, dehumidification	Gas	Liquid
Liquid extraction	Extract	Raffinate
Leaching	Liquid	Solid
Crystallization	Mother liquor	Crystals
Drying	Gas (usually air)	Wet solid

Table B   TERMINAL QUANTITIES FOR
COUNTERCURRENT FLOW

Stream	Flow rate	Concentration component A
L phase, entering	L_a	x_{Aa}
L phase, leaving	L_b	x_{Ab}
V phase, entering	V_b	y_{Ab}
V phase, leaving	V_a	y_{Aa}

DIFFUSIONAL PROCESSES AND EQUILIBRIUM STAGES

Mass-transfer problems can be solved by two distinctly different methods, one utilizing the concept of equilibrium stages, the other based on diffusional rate processes. The choice of the method depends on the kind of equipment in which the operation is carried out. Distillation, leaching, and sometimes liquid extraction are performed in equipment such as mixer-settler trains, diffusion batteries, or plate towers, which contain a series of discrete processing units, and problems in these areas are commonly solved by equilibrium-stage calculations. Gas absorption and other operations which are carried out in packed towers and similar equipment are usually handled using the concept of a diffusional process. All mass-transfer calculations, however, involve a knowledge of the equilibrium relationships between phases. The following chapter (Chap. 17) is devoted to the subject of phase equilibria. The next chapter (Chap. 18) is concerned with equilibrium-stage calculations and their applications to mass-transfer problems. The remaining chapters in this section (Chaps. 19 to 25) apply the principles of diffusion and stage contracts to the various separation operations.

17

PHASE EQUILIBRIA

A limit to mass transfer is reached if the two phases come to equilibrium and the net transfer of material ceases. For a practicable process, which must have a reasonable production rate, equilibrium must be avoided, as the rate of mass transfer at any point is proportional to the driving force, which is the departure from equilibrium at that point. To evaluate driving forces, a knowledge of equilibria between phases is therefore of basic importance. Several kinds of equilibria are important in mass transfer. In all situations, two phases are involved, and all combinations are found except two gas phases or two solid phases. In phases in bulk the effects of surface area or surface curvature are negligible, and the controlling variables are the intensive properties temperature, pressure, and concentrations. Equilibrium data can be shown in tables, equations, or graphs. For most operations considered in this text, the pertinent equilibrium relationships can be shown graphically.

Classification of equilibria To classify equilibria and to establish the number of independent variables or degrees of freedom available in a specific situation, the phase rule is useful. It is

$$\mathscr{F} = \mathscr{C} - \mathscr{P} + 2 \tag{17-1}$$

where $\mathscr{F}$ = number of degrees of freedom
$\mathscr{C}$ = number of components
$\mathscr{P}$ = number of phases

In the following paragraphs the equilibria used in mass transfer are analyzed in terms of the phase rule. Usually there are two phases, and so $\mathscr{F} = \mathscr{C}$.

The number of degrees of freedom, or variance $\mathscr{F}$, is the number of independent intensive variables—temperature, pressure, and concentrations—that must be fixed to define the equilibrium state of the system. If fewer than $\mathscr{F}$ variables are fixed, an infinite number of states fit the assumptions; if too many are arbitrarily chosen, the system will be overspecified and will be unable to reach an equilibrium.

Distillation Assume that there are two components, so $\mathscr{C} = 2$, $\mathscr{P} = 2$, and $\mathscr{F} = 2$. Both components are found in both phases. There are four variables: pressure, temperature, and the concentrations of component A in the liquid and vapor phases (the concentrations of component B are unity less those of component A). If the pressure is fixed, only one variable, e.g., liquid-phase concentration, can be changed independently and temperature and vapor-phase concentration follow. Equilibrium data are often presented in temperature-composition diagrams such as Figs. 17-3 and 17-5, which apply at constant pressure. They can also be given by plotting y_e, the vapor-phase concentration, against x_e, the liquid-phase concentration. Such a plot is called an *equilibrium curve*. Various equilibrium curves are shown in Figs. 17-4, 17-6, and 17-8.

If there are more than two components, the equilibrium relationship cannot be represented by a single curve.

Gas absorption Assume that only one component is transferred between phases. There are three components, and $\mathscr{F} = 3$. Neglect both the solubility of the inert gas in the liquid and the presence of vapor from the liquid in the gas. Then there are four variables: pressure, temperature, and the concentrations of component A in liquid and gas. The temperature and pressure may be fixed: one concentration may be chosen as the remaining independent variable. The other concentration is then determined, and an equilibrium curve y_e vs. x_e plotted. All points on the curve pertain to the same temperature and pressure. An equilibrium curve for sulfur dioxide–water at 1 atm and 30°C is shown in Fig. 17-11. Equilibrium data for various temperatures may also be presented in the form of solubility charts such as those in Fig. 17-10, in which the partial pressure of sulfur dioxide in the gas is plotted as the ordinate.

Dehumidification The solubility of the carrier gas may be neglected. There are two components, the carrier gas and the volatile liquid, so $\mathscr{C} = \mathscr{P} = 2$ and $\mathscr{F} = 2$. One phase, the liquid, is pure, and the variables are temperature, pressure, and the concentration of the vapor in the V phase. If the pressure is held constant, either the temperature or the vapor-phase concentration may be varied and the value of the other variable follows, so that a unique plot of temperature vs. concentration gives an equilibrium relationship. The equilibrium between water and moist air at 1 atm is shown in Fig. 17-12.

Liquid extraction   The minimum number of components is 3, so $\mathscr{F} = 3$. All three components may appear in both phases. The variables are temperature, pressure, and four concentrations. Either temperature may be taken as a constant, and two or more concentrations chosen as independent variables. The pressure usually is assumed constant, and the temperature then varies somewhat. The relations between these variables are given by various graphical methods, examples of which are shown in Figs. 17-16 to 17-21.

Leaching   Two situations are found in leaching. In the first, the solvent available is more than sufficient to dissolve all the solute, and at equilibrium all the solute is in solution. There are, then, two phases, the solid and the solution. The number of components is 3, and $\mathscr{F} = 3$. The variables are temperature, pressure, and concentration of solute in the liquid. All are independently variable.

In the second situation, the solvent available is insufficient to dissolve all the solute, and the excess solute remains as a solid phase in equilibrium with the liquid. Then, $\mathscr{P} = 3$, so $\mathscr{F} = 3$. The variables are temperature, pressure, and concentration of the saturated solution. If the pressure is fixed, the concentration depends on the temperature. This relation is the ordinary solubility curve. Examples are shown in Fig. 17-22.

Drying   In drying a water-wet solid, free liquid water may or may not be present. If it is, there are three phases—vapor, solid, and liquid—and three components, so $\mathscr{F} = 2$. At constant pressure a unique relationship exists between temperature and concentration of water in the vapor, just as in air-water contacts.

The water in hygroscopic solids or in natural materials like wood or leather may be in loose combination with the solid, and no liquid water may be present. Then there are two phases and three components, and $\mathscr{F} = 3$. The variables are temperature, pressure, and the concentrations of water in solid and vapor. If the temperature and pressure are fixed, these two concentrations can be plotted as an equilibrium curve. The curve for air, water, and paper pulp at 1 atm and 25°C is shown in Fig. 17-24.

Basic problem of phase equilibria   The objective of the theory of phase equilibria is to answer this question: Given a two-phase system at equilibrium, if the pressure on the system and the mole fraction of each component in one phase are known, what are (1) the temperature of the system and (2) the mole fractions of all components in the second phase?

Equilibrium between two fluid phases is amenable to treatment based on a firm foundation of thermodynamic and kinetic-molecular principles and extended to practical calculations by semitheoretical correlations and experimental data.[7] Equilibria in solid-fluid systems usually must be evaluated by direct experiment.

The remainder of this chapter is devoted to vapor-liquid, liquid-liquid, gas-solid, and liquid-solid equilibria.

VAPOR-LIQUID EQUILIBRIA†

Consider the situation applying to distillation, a two-phase vapor-liquid equilibrium of two or more components where each component appears in both phases. Three kinds of equilibrium exist, mechanical, thermal, and physicochemical. The criterion for mechanical equilibrium is equality of pressure between phases; that for thermal equilibrium is equality of temperature; but except for two special cases of a pure component or an azeotrope (to be described later), a physicochemical equilibrium is not characterized by equality of concentration. In general, $y_i \neq x_i$ if x_i and y_i are the mass or mole fractions of component i in the liquid and vapor phases, respectively.

Chemical Potential

Thermodynamics provides a criterion for physicochemical equilibrium. Corresponding to temperature and pressure in thermal and mechanical equilibria, there is a quantity, specific to each component, called the *chemical potential* and denoted by μ_i for component i, which is equal for all phases in equilibrium. Therefore the criterion for vapor-liquid equilibrium is

$$\mu_i^L = \mu_i^V = \mu_i \tag{17-2}$$

where superscripts L and V refer to liquid and vapor phases, respectively. Equation (17-2) applies to each component.

One definition of μ_i is

$$\mu_i \equiv \left(\frac{\partial G}{\partial n_i}\right)_{T,p,n_j} \tag{17-3}$$

where G is the Gibbs energy, much used in physical chemistry. The partial differentiation describes the effect on G of an incremental addition of dn_i moles of component i, at constant temperature and pressure, to a mixture initially containing fixed numbers of moles of all components and increasing only component i by the addition. This last constraint is noted by the subscript n_j, with j representing all components other than i.

Direct use of μ is difficult, first because of its abstract nature and little physical meaning, and second because at low concentrations μ_i goes to negative infinity as the mole fraction x_i or y_i goes to zero. Useful auxiliary functions of μ_i, called *fugacity*, *fugacity coefficient*, and *activity coefficient*, have been defined and are used practically in place of μ.

A one-to-one correspondence exists between the chemical potential and the mole fraction of each component in a single phase. The number of independent

† In this section general definitions are given for component i, which may be present in a binary or a multi-component system. Examples are restricted to binary systems. Generalizations to multicomponent systems are discussed in Chap. 21.

intensive variables in a phase is, by Eq. (17-1), 2 more than the number of independent components. For a two-component phase, then, there are three independent intensive variables. The usual choice is temperature, pressure, and one mole fraction. Such a choice fixes all other properties, such as density, specific enthalpy, the second mole fraction, and the two chemical potentials.

Corresponding to the fact that the mole fraction of one component in a phase is dependent on all the remaining mole fractions, there is a relation showing that one chemical potential also is a function of all the others. This is the *Duhem equation*, which for a two-component phase at constant temperature and pressure is

$$x_A \, d\mu_A + x_B \, d\mu_B = 0 \tag{17-4}$$

The corresponding mole-fraction equation is

$$x_A + x_B = 1 \tag{17-5}$$

Subscripts A and B designate the two components. Either x_A (or μ_A), or x_B (or μ_B) may be chosen as the independent mole fraction (or chemical potential), and the other is the dependent one. Equations (17-4) and (17-5) are easily generalized to more than two components.

Fugacity The transformation of chemical potential to mole fraction depends on an intensive property closely related to partial pressure. The fugacity of component i is denoted by f_i and may be defined implicitly by the equation

$$\mu_i = RT \ln f_i + g_i^*(T) \tag{17-6}$$

The quantity $g_i^*(T)$ is a temperature-dependent molar Gibbs energy of the pure gas at the limit of zero pressure. It is the existence of this quantity that circumvents the disadvantage of the zero pressure characteristic of the chemical potential. It is clear from Eq. (17-6) that $g_i^*(T)$ also is the Gibbs energy of pure component i at unit fugacity. It can be calculated from the third law of thermodynamics through the use of statistical mechanics and spectroscopic data and assuming a value for the enthalpy of the gas at $T = 0$. Equation (17-6) may be applied both to pure components, where μ_i may be designated by g_i, and to a component in a mixture.

The fugacity function is extended to the liquid phase by noting that at equilibrium $u_i^V = u_i^L$. Since $g_i^*(T)$ depends only on temperature, it follows that at equilibrium

$$f_i^V = f_i^L \tag{17-7}$$

At equilibrium the fugacity of any one component i is the same in all phases and so serves exactly the same purpose as the chemical potential.

Fugacity coefficient The transformation from chemical potential to mole fraction is completed by the introduction of the fugacity coefficient ϕ_i, defined for a gas by

$$\phi_i \equiv \frac{f_i^V}{p y_i} \tag{17-8}$$

where p is the total pressure on the gas. From Eq. (1-48), Eq. (17-8) can be written in terms of the partial pressure as

$$\phi_i = \frac{f_i^V}{\bar{p}_i}$$

and then

$$f_i^V = \phi_i p y_i = \phi_i \bar{p}_i \qquad (17\text{-}9)$$

Equation (17-6) now becomes

$$\mu_i = RT \ln \phi_i p y_i + g_i^*(T)$$
$$= RT \ln \phi_i \bar{p}_i + g_i^*(T) \qquad (17\text{-}10)$$

PURE GAS   For the special case of a pure gas, $y_i = 1$, $\mu_i = g_i$, and $\phi_i = f_i^V/p°$, where $p°$ is the vapor pressure of pure gas.

IDEAL GAS   When the gas mixture may be considered ideal, all virial coefficients in Eq. (1-44) are zero, $f_i^V = \bar{p}_i$, and ϕ_i is unity. The fugacity coefficient is often called a corrected pressure for a pure gas and a corrected partial pressure for a component of a gas mixture.

Fugacity coefficients from equation of state   The virial equation (1-44) provides a rigorous method for calculating the fugacity coefficients of the components in a mixture of real gases, provided the numerical values of the virial coefficients B, C, etc., can be found. The discussion will be restricted to the second virial coefficient, which is adequate to a pressure about one-half the critical pressure of the mixture; then the virial equation is

$$z = 1 + \rho B \qquad (17\text{-}11)$$

where z is the compressibility factor of the mixture, $p/\rho RT$. For a two-component mixture, the rigorous equation for B is

$$B = y_A^2 B_{AA} + 2 y_A y_B B_{AB} + y_B^2 B_{BB} \qquad (17\text{-}12)$$

where B_{AA}, B_{AB}, and B_{BB} represent the effects of two-molecule collisions, between two A molecules, an A and a B molecule, and two B molecules, respectively. The values of these three coefficients must be known. They are discussed in other texts[7a,10] and tabulated in the literature.[1] Rather complicated estimation methods for calculating these values for various types of molecular and atomic interactions are given in the same sources.

The equations for the fugacity coefficients for the two components are

$$\ln \phi_A z = 2\rho(y_A B_{AA} + y_B B_{AB}) \qquad (17\text{-}13)$$

$$\ln \phi_B z = 2\rho(y_B B_{BB} + y_A B_{AB}) \qquad (17\text{-}14)$$

To use these equations, it is assumed that the temperature, pressure, and mole fractions are fixed. An iteration is needed to match the correct values of ρ and z, easily accomplished for multicomponent systems on a digital computer[9] and by hand calculator for a two-component system.

EXAMPLE 17-1   Calculate the fugacity coefficients of the components in a mixture of 25 mole percent pentane and 75 mole percent n-octane at a pressure of 5 atm and a temperature of 200°C;

$$B_{AA} = -380 \text{ cm}^3/\text{g mol} \qquad B_{BB} = -1{,}940 \text{ cm}^3/\text{g mol} \qquad B_{AB} = -700 \text{ cm}^3/\text{g mol}$$

SOLUTION   From Eq. (17-12)

$$B = 0.25^2(-380) + 2 \times 0.25 \times 0.75(-700) - 0.75^2(1{,}940) = -1{,}377.5 \text{ cm}^3/\text{g mol}$$

For the iteration for the values of z and ρ, $p = 5$ atm, $T = 273 + 200 = 473$ K, and $R = 82.1$ atm-cm^3/g mol-K.

For the first approximation, assume $z = 1$. The first approximation to ρ is

$$\rho = \frac{p}{zRT} = \frac{5}{1 \times 82.1 \times 473} = 1.2876 \times 10^{-4} \text{ g mol/cm}^3$$

From Eq. (17-11), a corrected value for z is

$$z = 1 - 1{,}377.5 \times 1.2876 \times 10^{-4} = 0.8226$$

This calculation is repeated, and the values of z and ρ are found after a few replications:

$$z = 0.7692 \quad \text{and} \quad \rho = 1.678 \times 10^{-4} \text{ g mol/cm}^3$$

Substituting in Eqs. (17-13) and (17-14), using base ten logarithms, gives

$$\log \phi_A z = \frac{(2 \times 1.678 \times 10^{-4})(-0.25 \times 380 - 0.75 \times 700)}{2.3026}$$

$$= -0.09036 = \bar{9}.90964$$

$$\phi_A z = 0.8122 \quad \text{and} \quad \phi_A = \frac{0.8122}{0.7692} = 1.0559$$

$$\log \phi_B z = \frac{(2 \times 1.678 \times 10^{-4})(-0.75 \times 1{,}940 - 0.25 \times 700)}{2.3026}$$

$$= -0.23757 = \bar{9}.76303$$

$$\phi_B z = 0.5795 \quad \text{and} \quad \phi_B = \frac{0.5795}{0.7692} = 0.7534$$

////

Activity Coefficient

The fugacity coefficient ϕ_i can also be used for the liquid phase. This requires a large amount of precise density data for the liquid,[7b] and a better procedure has been found. The fugacity coefficient is replaced by the activity coefficient γ_i, defined by

$$\gamma_i \equiv \frac{f_i^L}{x_i f_i^\circ} \tag{17-15}$$

The parallelism between the fugacity coefficient used in the vapor and the activity coefficient used in the liquid is seen by comparing Eq. (17-8) with Eq. (17-15). For the liquid a special fugacity, denoted by f_i° and called the *standard-state fugacity*, replaces the pressure used in the equation for the fugacity coefficient. To establish the value of γ_i it clearly is necessary to fix that of f_i° ahead of time. At a standard state for component i definite choices must be made for temperature, pressure, and the independent mole fractions of the several components, including i. The choice of the standard state must conform to two requirements: (1) the temperature must be that of the equilibrium, and (2) the standard-state concentrations must not vary with the values of x_i, the mole fractions of the components at equilibrium. Once these requirements are met, only the pressure remains as a free choice. The simplest choice for the standard state is the pure liquid component at the temperature and pressure of the equilibrium. Other choices of pressure and concentration are used, but in the immediate discussion the above choice is assumed.

Basic equation, vapor-liquid equilibria   Since at equilibrium the fugacity of component i is the same in both phases, eliminating f_i^V and f_i^L by Eqs. (17-9) and (17-15) and designating equilibrium mole fractions by subscript e gives

$$\phi_i y_{ie} p = \gamma_i x_{ie} f_i^\circ \tag{17-16}$$

Equation (17-16) is the working equation for vapor-liquid equilibria, and the estimation of numerical values of f_i°, ϕ_i, and γ_i for each component in the mixture is all that is needed to calculate the pressure and the equilibrium composition of one phase from those of the other; or if the factors are known as functions of temperature, the equilibrium temperature can be calculated for a known pressure and the composition of one phase.

Limiting laws of real liquid mixtures   At the extremes of concentration, where $x_i \rightarrow 0$ or $x_i \rightarrow 1.0$, the fugacity of component i follows simple limiting laws in monomeric nonelectrolytic solutions.

Consider a binary phase and assume the temperature and pressure are held constant. Also, assume the pressure is greater than the vapor pressure of the more volatile component, so that vapor does not form. There is one independent variable; choose x_A for this, where component A is the more volatile.

DILUTE SOLUTION   The law applying when x_A is in the neighborhood of zero is

$$f_A^L = H_A x_A \tag{17-17}$$

Here H_A is an empirical constant which depends on temperature and pressure but is independent of x_A over a useful range of values within a domain $0 \leqq x_A \leqq x_A'$, where x_A' is the value of x_A at which H_A begins to vary too much for the desired precision. But, as shown in Fig. 17-1, the graph of f_A^L vs. x_A can be extrapolated linearly to the $x_A = 1$ ordinate, and the intersection of the line with that ordinate measures the constant value of H_A, which can also be interpreted as the slope of the straight line.

HENRY'S LAW   If the dilute solution is in equilibrium with an ideal-gas mixture, the equality of f_i^L and f_i^V requires that

$$H_i x_{ie} = \bar{p}_i = p y_{ie} \tag{17-18}$$

This equation is a statement of Henry's law, and H_i is called the *Henry constant*.

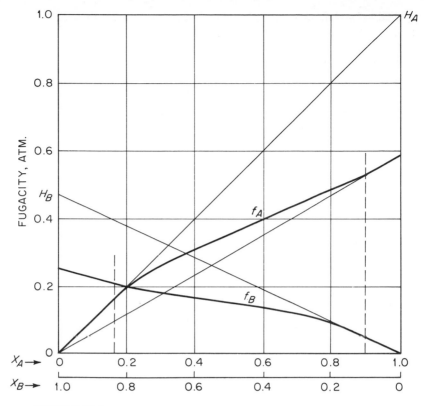

FIGURE 17-1
Fugacity vs. mole fraction, ethyl ether–acetone at 20°C.

CONCENTRATED SOLUTION Consider the behavior of component B in a mixture in which component A is sufficiently dilute to follow Eq. (17-17). It can be shown from Eq. (17-4) that component B then must follow an analogous equation, which is

$$f_B^L = f_B^{\circ} x_B \tag{17-19}$$

The domain of applicability of this relation, which is the law of concentrated solutions, is $x_B' \leq x_B \leq 1$, where $x_B' = 1 - x_A'$. Proportionality factor f_B° is, of course, the standard-state fugacity of component B at the temperature and pressure of the solution. As shown by Fig. 17-1, if either component in a binary solution follows the dilute-solution law, the other follows the concentrated-solution law, and vice versa.

IDEAL SOLUTION Some liquid mixtures obey reasonably well the law of Eq. (17-19) over the entire concentration range from 0 to 1.0 for all components. Solutions following the law exactly are called *ideal solutions*. If one component in a mixture is ideal, so are all the others. In an ideal solution, the laws for dilute and concentrated solutions become one, and $H_i = f_i^{\circ}$.

Comparing Eqs. (17-15) and (17-19) demonstrates the important principle that in an ideal solution all activity coefficients are unity for all components at all concentrations.

Distribution Coefficients

Equation (17-16) provides an exact formula for a useful quantity called the *distribution coefficient* or the *K factor*, defined by

$$K_i \equiv \frac{y_{ie}}{x_{ie}} = \frac{\gamma_i f_i^{\circ}}{p \phi_i} \tag{17-20}$$

Knowledge of K factors provides a method for calculating the composition of one phase from a known composition of the other. For a binary system, Eq. (17-20) gives

$$y_{Ae} = K_A x_{Ae} \qquad y_{Be} = K_B x_{Be} = K_B(1 - x_{Ae}) \tag{17-21}$$

$$x_{Ae} = \frac{K_A}{y_{Ae}} \qquad x_{Be} = \frac{K_B}{y_{Be}} = \frac{K_B}{1 - y_{Ae}} \tag{17-22}$$

Since the mole fractions of all components in any phase sum to unity,

$$y_{Ae} + y_{Be} = K_A x_{Ae} + K_B x_{Be} = K_A + K_B(1 - x_{Ae}) = 1 \tag{17-23}$$

$$x_{Ae} + x_{Be} = \frac{K_A}{y_{Ae}} + \frac{K_B}{1 - y_{Ae}} \tag{17-24}$$

Division of Eq. (17-21) by Eq. (17-23) gives

$$y_{Ae} = \frac{K_A x_{Ae}}{K_A x_{Ae} + K_B(1 - x_{Ae})} = \frac{1}{1 + (K_A/K_B)(1/x_{Ae} - 1)} \tag{17-25}$$

and $$y_{Be} = 1 - y_{Ae}$$

Division of Eq. (17-22) by Eq. (17-24) gives

$$x_{Ae} = \frac{K_A/x_{Ae}}{K_A/y_{Ae} + K_B/(1 - y_{Be})} = \frac{1}{1 + (K_B/K_A)/[y_{Ae}/(1 - y_{Ae})]} \tag{17-26}$$

and $$x_{Be} = 1 - x_{Ae}$$

Relative volatility   The ease of separation of two components by distillation, e.g., component i from component j, is shown quantitatively by the volatility of component i relative to that of component j. This ratio, denoted by a_{ij}, is defined by

$$\alpha_{ij} \equiv \frac{y_{ie}/x_{ie}}{y_{je}/x_{je}} \tag{17-27}$$

By convention, α values are numbered in the direction of decreasing volatility, so $\alpha_{ij} > 1$ when $i < j$. For the special case of binary mixtures,

$$\alpha_{AB} = \frac{y_{Ae}/x_{Ae}}{y_{Be}/x_{Be}} = \frac{K_A}{K_B} = \frac{y_{Ae}/x_{Ae}}{(1 - y_{Ae})/(1 - x_{Ae})} \qquad (17\text{-}28)$$

In nonideal mixtures, α values may vary strongly with concentrations of the components, but in ideal systems they are independent of concentrations, which greatly simplifies calculations of vapor-liquid equilibria. A still easier calculation results from the fact that α values are sometimes practically independent of temperature, especially in close separations where α values are small. Values of unity mean that the mixture is not separable by distillation. Small values, such as $1 < \alpha_{ij} < 1.1$, for example, indicate that the separation is possible but difficult; larger values of α show increasing ease of separation.

Co-ordination of phase equilibria parameters The equalities and one-to-one correspondences of the quantities appearing in the vapor-liquid equilibrium of a general component i are summarized in the following diagram:

Excess G Energy and Activity Coefficients

The estimation of values for activity coefficients has been given much study. Comprehensive books[2,7] should be consulted for details. The following treatment covers the basic concept of excess G energy, and shows one current method of calculating the value of γ from this. The method is restricted to mixtures of nonelectrolytes and monomeric liquids.

The excess Gibbs energy is defined as the difference between the actual Gibbs energy of the mixture and that of a mixture of the same composition calculated on the assumption that the solution is ideal. Naturally, the excess G function of an ideal solution is zero. The excess G energy G^E, which may be either positive or negative, originates in the intermolecular forces in the mixture. It is the variety and complexity of such forces that account for the divergence from ideal behavior.

For a real, nonideal mixture of liquids at a definite temperature and pressure the increase in the G energy of mixing the r pure components at this temperature and pressure is, by the definition of μ given in Eq. (17-3),

$$G_{mix} - G_{comp} = \Delta G = \sum_{i=1}^{r} \mu_i n_i - \sum_{i=1}^{r} g_i n_i = \sum_{i=1}^{r} (\mu_i - g_i)n_i \qquad (17\text{-}29)$$

where g_i is the molar G energy for pure component i.

The definition of γ_i in Eq. (17-15), combined with the definition of fugacity, provides a second equation for ΔG,

$$\Delta G = \sum_{i=1}^{r} n_i RT \ln x_i + \sum_{i=1}^{r} n_i RT \ln \gamma_i \qquad (17\text{-}30)$$

Since the first term in the right-hand side of Eq. (17-30) is the G energy of the ideal solution constructed of the r components, the second term is, by definition, the excess G energy; so

$$G^E = \sum_{i=1}^{r} n_i RT \ln \gamma_i \tag{17-31}$$

Applying this equation to binary mixtures gives

$$G^E = RT(n_A \ln \gamma_A + n_B \ln \gamma_B) \tag{17-32}$$

Differentiating with respect to n_A at constant temperature, pressure, and n_B gives

$$\frac{dG^E}{dn_A} = RT \left(\ln \gamma_A + n_A \frac{d \ln \gamma_A}{dn_A} + n_B \frac{d \ln \gamma_B}{dn_A} \right) \tag{17-33}$$

From the Duhem equation (17-4) it can be shown that the last two terms in this equation sum to zero, and

$$\frac{dG^E}{dn} = \bar{g}_A^E = RT \ln \gamma_A \tag{17-34}$$

where $\bar{g}_A^E$ is a partial molar G, defined generally by Eq. (17-3). The corresponding equation for component B is found by exchanging the subscripts A and B in Eq. (17-34).

Wilson equation   To use Eqs. (17-32) and (17-34) values of $\bar{g}^E$ are needed. Of the many semitheoretical equations that have been proposed for the purpose, one of the best is the Wilson equation,[7c] used here as an example. The equation is, for component A of a binary mixture,

$$\frac{\bar{g}^E}{RT} = -x_{A,e} \ln (x_{A,e} + \Lambda_{AB}) - x_{B,e} \ln (x_{B,e} + \Lambda_{BA}) \tag{17-35}$$

where Λ_{AB} and Λ_{BA} are defined by

$$\Lambda_{AB} \equiv \frac{v_B}{v_A} \exp \left(-\frac{\lambda_{AB} - \lambda_{AA}}{RT} \right) \tag{17-36}$$

and

$$\Lambda_{BA} \equiv \frac{v_A}{v_B} \exp \left(-\frac{\lambda_{AB} - \lambda_{BB}}{RT} \right)$$

Note that $\lambda_{AB} = \lambda_{BA}$ but $\Lambda_{AB} \neq \Lambda_{BA}$. The two parameters $\lambda_{AB} - \lambda_{AA}$ and $\lambda_{AB} - \lambda_{BB}$ are determined experimentally for the mixture and are not calculable from single-component data. Values are given in the literature.[3] An important advantage of the Wilson equation not found in other equations is that since the parameters are not sensitive to temperature, a single value can be used over the temperature range usually found in constant-pressure equipment. The molar volumes of the pure components v_A and v_B are also needed.

The activity coefficients obtained from Eqs. (17-36) are

$$\ln \gamma_A = -\ln (x_{A,e} + \Lambda_{AB}x_{B,e}) + x_{B,e} \left(\frac{\Lambda_{AB}}{x_{A,e} + \Lambda_{AB}x_{B,e}} - \frac{\Lambda_{BA}}{x_{B,e} + \Lambda_{BA}x_{A,e}} \right)$$

$$(17\text{-}37)$$

$$\ln \gamma_B = -\ln (x_{B,e} + \Lambda_{BA}x_{A,e}) - x_{A,e} \left(\frac{\Lambda_{AB}}{x_{A,e} + \Lambda_{AB}x_{B,e}} - \frac{\Lambda_{BA}}{x_{B,e} + \Lambda_{BA}x_{A,e}} \right)$$

$$(17\text{-}38)$$

EXAMPLE 17-2   Calculate the activity coefficients of a mixture of 30 mole percent acetone and 70 mole percent methanol at 55°C. The necessary constants, where acetone is component A, are[3]

$$\lambda_{AB} - \lambda_{AA} = -203.03 \text{ cal/g mol} \qquad \lambda_{AB} - \lambda_{BB} = 666.99 \text{ cal/g mol}$$

$$v_A = 77.36 \text{ cm}^3/\text{g mol} \qquad v_B = 44.28 \text{ cm}^3/\text{g mol}$$

SOLUTION   Use Eq. (17-36):

$$T = 273.1 + 55 = 328.1 \text{ K}$$

From Appendix 2, $R = 1.9872$ cal/K-g mol

$$\Lambda_{AB} = \frac{42.28}{77.36} \exp \left(-\frac{-203.03}{1.9872 \times 328.1} \right) = 0.7462$$

$$\Lambda_{BA} = \frac{77.36}{42.28} \exp \left(-\frac{666.99}{1.9872 \times 328.1} \right) = 0.6578$$

Use Eqs. (17-37) and (17-38):

$$\log \gamma_A = -\log (0.3 + 0.7462 \times 0.7)$$

$$+ \frac{0.7}{2.3026} \left(\frac{0.7462}{0.3 + 0.7462 \times 0.7} - \frac{0.6578}{0.7 + 0.6578 \times 0.3} \right)$$

$$= 0.13754$$

$$\gamma_A = 1.372$$

$$\log \gamma_B = -\log (0.7 + 0.6578 \times 0.3)$$

$$- \frac{0.3}{2.3026} \left(\frac{0.7462}{0.3 + 0.7462 \times 0.7} - \frac{0.6578}{0.7 + 0.6578 \times 0.3} \right)$$

$$= 0.13754$$

$$\gamma_B = 1.372 \qquad\qquad ////$$

Standard-State Fugacity

To use Eq. (17-16) values of f_i°, the fugacity of the pure liquid at the equilibrium temperature T_e and pressure p_e are needed. The equation for this purpose is derived

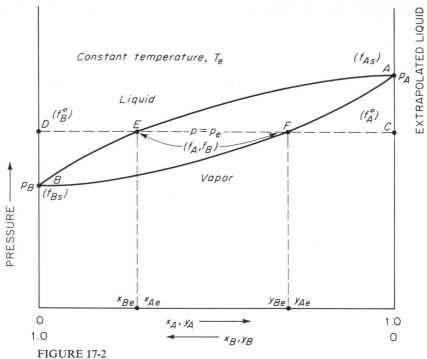

FIGURE 17-2
Isothermal boiling-point diagram.

from the following relation for the effect of pressure on the molar Gibbs energy of a pure liquid at constant temperature:

$$\left(\frac{\partial g}{\partial p}\right)_T = v^L \qquad (17\text{-}39)$$

where v^L is the molar volume. Equation (17-6) written for a pure liquid is

$$\mu_i = g_i = RT \ln f_i^L + g_i^*(T)$$

Equation (17-39) shows that

$$\left(\frac{\partial g}{\partial p}\right)_T = RT\left(\frac{\partial \ln f^L}{\partial p}\right)_T = v^L$$

and, for an isothermal pressure range from p_1 to p_2, integration gives

$$\int_{f_1}^{f_2} d \ln f = \int_{p_1}^{p_2} \frac{v^L}{RT}\, dp = \ln \frac{f_2^L}{f_1^L}$$

and

$$\frac{f_2^L}{f_1^L} = \exp\left(\int_{p_1}^{p_2} \frac{v^L}{RT}\, dp\right) \qquad (17\text{-}40)$$

This relation is the *Poynting equation*.

Vapor-liquid equilibria in binary systems The several fugacities of a two-phase two-component system may be interpreted in terms of the pressures and mole fractions by reference to Fig. 17-2, which is an isothermal diagram for the equilibrium temperature T_e. The coordinates are pressure and mole fraction.† In the figure the equilibrium pressure shown is p_e.

The vapor pressures of the two pure components at temperature T_e are p_A and p_B and are represented by points A and B, respectively. The standard-state fugacities, f_A° and f_B°, are those at points C and D, respectively. These represent pure liquid components at temperature T_e and pressure p_e. Points E and F represent the liquid and vapor phases in equilibrium. The field for the liquid phase lies above the upper curve, which is a plot of the pressure–mole-fraction function of the equilibrium liquid; the field for the vapor lies below the lower curve, which is a plot of the pressure–mole-fraction relation of the equilibrium vapor.

Since points A and B also represent equilibria, i.e., those between pure liquids and their vapors, the fugacities in the two phases are equal and may be denoted by f_{As} and f_{Bs}. These fugacities are determinable in the usual way from the equations of state of the pure components. The standard-state fugacity of component A, for example, can then be calculated by using Eq. (17-40) to link f_{As} with f_A°, from knowledge of v_A^L and setting $f_1 = f_{As}$ and $f_2 = f_A^\circ$. This gives

$$\frac{f_A^\circ}{f_{As}} = \exp\left(\int_{p_A}^{p_e} \frac{v_A^L}{RT}\, dp\right) \equiv \Phi_A \tag{17-41}$$

where Φ_A may be called the Poynting factor of component A.

At temperatures less than about $0.85 T_A^C$, where T_A^C is the critical pressure of component A, v_A^L is nearly independent of pressure and may be considered constant. Then, Eq. (17-41) gives on integration

$$\Phi_A = \exp\left[\frac{v_A^L}{RT}(p_e - p_A)\right] \tag{17-42}$$

At temperatures just below the critical, v_L increases strongly with decrease in pressure, and Eq. (17-41) must be used.

For at least one component in a vapor-liquid equilibrium, the vapor pressure is greater than the equilibrium pressure, as shown by points A and C for component A in Fig. 17-2. The standard state represented by point C actually is in the vapor field, where liquid does not exist at equilibrium. This fact may be ignored in using Eq. (17-42) and the constant value of v_A^L extrapolated over the pressure range from p_e to p_A on the assumption that the component acts like a superheated liquid.

At pressures below about 5 atm $\Phi \approx 1$, and this factor may be disregarded.

Final equation for K From Eq. (17-8), the fugacity of pure component A at pressure p_A is

$$f_{As} = \phi_{As} p_A \tag{17-43}$$

† These coordinates are not those used to interpret actual processes, which are conducted at constant pressure rather than at constant temperature. Figure 17-2 is drawn to clarify the relations between six fugacities, three for each component, e.g., those for component A: f_A, $f_A,^\circ$ and f_{As}.

where ϕ_{As} is the fugacity coefficient of pure component A at temperature T and pressure p_A.

By use of Eqs. (17-20), (17-41), and (17-43), written for general component i,

$$K_i = \frac{y_{ie}}{x_{ie}} = \gamma_i \frac{\phi_{is}}{\phi_i} \frac{p_i}{p} \Phi_i \tag{17-44}$$

Summary of cases Equation (17-44) can be simplified in several special cases by setting one or more of the parameters to unity. A list of such cases, arranged in order of increasing simplicity, is given in Table 17-1.

Raoult's law Cases 5 and 6 in Table 17-1 are described by the isothermal equation

$$K_i = \frac{y_{ie}}{x_{ie}} = \frac{p_i}{p} \tag{17-45}$$

and
$$p_i x_{ie} = p y_{ie} = \bar{p}_i \tag{17-46}$$

Equations (17-45) and (17-46) are equivalent statements of Raoult's law. Where applicable, they give a simple method of calculating vapor-liquid equilibria. The equations apply at any specific temperature and require only values of p_i, the vapor pressures of the components at that temperature. If the vapor-pressure–temperature functions of the components are at hand, Raoult's law is used to calculate the entire boiling-point–mole-fraction relations of both vapor and liquid phases for a constant-pressure system.

Table 17-1 SPECIAL CASES OF VAPOR-LIQUID EQUILIBRIA

Case	Description	Values of parameters in Eq. (17–44)			
1	Real gas and real liquid at high pressure or near critical point	$\gamma_i \neq 1$	$\phi_i \neq 1$	$\phi_{is} \neq 1$	$\Phi_i \neq 1$
2	Real gas, real liquid moderate pressure	$\gamma_i \neq 1$	$\phi_i \neq 1$	$\phi_{is} \neq 1$	$\Phi_i \approx 1$
3	Nearly ideal gas, nonideal liquid, moderate pressure	$\gamma_i \neq 1$	$\phi_i \approx 1$	$\phi_{is} \approx 1$	$\Phi_i \approx 1$
4	Real gas, real liquid, high pressure, or near critical point	$\gamma_i \approx 1$	$\phi_i \neq 1$	$\phi_{is} \neq 1$	$\Phi_i \neq 1$
5	Nearly ideal gas, nearly ideal liquid, moderate pressure	$\gamma_i \approx 1$	$\phi_i \approx 1$	$\phi_{is} \approx 1$	$\Phi_i \approx 1$
6	Same as case 5	Also α_{ij} independent of temperature			

For a two-component mixture, where x and y refer to component A, Eq. (17-46), written for each of the components, gives

$$p_A x_e = p y_e$$
$$p_B(1 - x_e) = p(1 - y_e)$$

which are added to give

$$p_A x_e + p_B(1 - x_e) = p y_e + p(1 - y_e) = p$$

Solving these equations for x_e and y_e yields

$$x_e = \frac{p - p_B}{p_A - p_B} \quad \text{and} \quad y_e = \frac{p_A x}{p} \tag{17-47}$$

Equilibrium curves and boiling-point diagrams Figure 17-3 shows the boiling-point diagram at constant pressure for mixtures of component A, boiling at temperature T_A, and component B, boiling at temperature T_B. Component A is the more volatile. Temperatures are plotted as ordinates and concentrations, preferably in mole fractions, as abscissas. The diagram consists of two curves, the ends of which coincide. Any point, such as point y, on the upper line represents vapor that will just begin to condense at temperature T_1. The concentration of the first drop of liquid is represented by point d. The upper curve is called the *dew-point curve*. Any point, such as point x, on the lower curve represents liquid that will just begin to boil at temperature T_1. The concentration of the first bubble of vapor is represented by point e. The lower curve is called the *bubble-point curve*. Any two points on the same horizontal line, such as points x and y, represent concentrations of liquid and vapor in equilibrium at the temperature given by their ordinate. For all points above the top line, such as point a, the mixture is entirely vapor. For all points below the bottom line, such as point b, the mixture is completely liquid. For points between the two curves, such as point c, the system is partly liquid and partly vapor.

Suppose that a liquid mixture of concentration d is heated slowly. It will begin to boil at temperature T. Although the first bubble of vapor will have the concentration y, represented by e, as soon as an appreciable amount of vapor has been formed, the concentration of the liquid will no longer correspond to d since the vapor is richer in the component A than the liquid from which it is evolved and hence the points x and y both tend to move toward T_B.

When the boiling-point diagram cannot be calculated by the theoretical computations of activity and fugacity coefficients, it must be determined by direct experiments on equilibrium systems. The procedure will not be described here.

For many calculations a simpler diagram, called the *equilibrium diagram*, is sufficient. In this the temperature is eliminated; the equilibrium liquid composition x is plotted as the abscissa and the corresponding vapor composition y as the ordinate, as shown in Fig. 17-4.

EXAMPLE 17-3 The vapor pressures of benzene and toluene are shown in Table 17-2. Assuming that mixtures of these materials follow Raoult's law, calculate and plot the

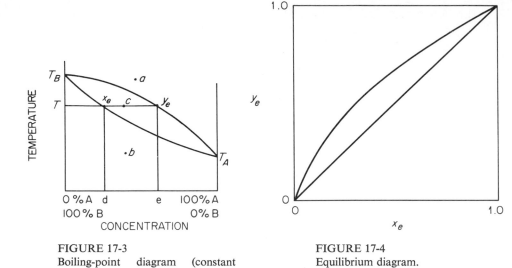

FIGURE 17-3
Boiling-point diagram (constant pressure).

FIGURE 17-4
Equilibrium diagram.

boiling-point and equilibrium curve for the system benzene-toluene at a total pressure of 1 atm.

SOLUTION   Several temperatures between 80.6 and 110.6°C are chosen, and the corresponding data for p_A and p_B are substituted in Eq. (17-47). For example, at 85°C, $p_A = 877$ and $p_B = 345$ mm Hg, and

$$x_e = \frac{760 - 345}{877 - 345} = 0.780 \qquad y_e = \frac{877 \times 0.780}{760} = 0.900$$

Other results are found similarly and are given in Table 17-3. The boiling-point diagram is plotted in Fig. 17-5 and the equilibrium curve in Fig. 17-6.   ////

Azeotropic mixtures   Many systems do not obey Raoult's law and yet have boiling-point and equilibrium curves qualitatively like those of Figs. 17-5 and 17-6.

Table 17-2   VAPOR PRESSURES OF BENZENE AND TOLUENE

| Temperature, °C | Vapor pressure, mm Hg | |
	Benzene	Toluene
80.1	760	
85	877	345
90	1,016	405
95	1,168	475
100	1,344	557
105	1,532	645
110	1,748	743
110.6	1,800	760

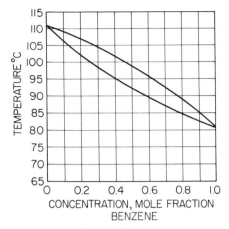

FIGURE 17-5
Boiling-point diagram (system benzene-toluene at 1 atm).

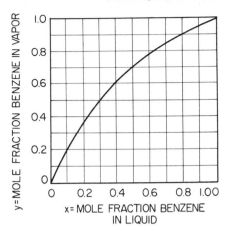

FIGURE 17-6
Equilibrium curve (system benzene-toluene).

That is, the boiling-point curves are bow-shaped, the equilibrium curve is concave downward throughout its length, the bubble and dew points all lie between the boiling temperatures of the pure components, and the vapor in equilibrium with any liquid mixture is richer in component A than the liquid. There are many other important systems that have boiling-point and equilibrium diagrams quite different from those of Figs. 17-5 and 17-6. Figure 17-7a and b gives the boiling-point diagrams for two such mixtures. The first applies to mixtures of chloroform and acetone and the second to mixtures of benzene and ethanol. In Fig. 17-7a the boiling point T_b, corresponding to concentration x_a, is the highest temperature reached by any mixture of these substances and is higher than the boiling temperature of either pure component. In Fig. 17-7b the boiling point T_b, corresponding to concentration x_a, is the lowest boiling temperature reached by any mixture and is lower than that of either pure component. Mixtures of this kind having maximum or minimum boiling points are called azeotropes. Of the two types, minimum-boiling azeotropes are the more

Table 17-3 **BOILING-POINT CURVES FOR EXAMPLE 17-3**

Temperature, °C	Concentration, mole fraction C_6H_6	
	Liquid x_e	Vapor y_e
85	0.780	0.900
90	0.581	0.777
95	0.411	0.632
100	0.258	0.456
105	0.130	0.261
110	0.017	0.039

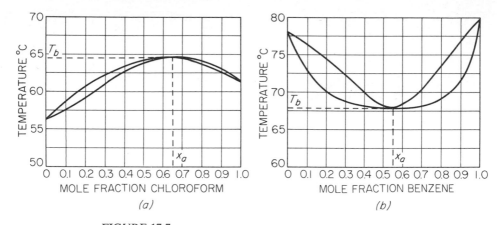

FIGURE 17-7
Boiling-point diagrams for azeotropic systems: (*a*) maximum-boiling azeotrope (system chloroform-acetone); (*b*) minimum-boiling azeotrope (system benzene-ethanol).

common. At an azeotropic concentration, the liquid and vapor curves touch and have a common horizontal tangent. The composition of the vapor produced from an azeotrope is the same as that of the liquid. An azeotrope, then, can be boiled away at constant pressure without change in concentration in either liquid or vapor. Since, under these conditions, the temperature cannot vary, azeotropes are also called constant-boiling mixtures. An azeotrope cannot be separated by constant-pressure distillation into its components. Furthermore, a mixture on one side of the azeotropic composition cannot be transformed by distillation into a mixture on the other side of the azeotrope. If the total pressure is changed, the azeotropic composition is usually shifted and this principle can be utilized to obtain separations under pressure or vacuum that cannot be obtained at atmospheric pressure. The same result is more commonly obtained by adding a third component that destroys the azeotrope.

The equilibrium curves for the systems chloroform-acetone and benzene-ethanol are shown in Figs. 17-8*a* and *b*, respectively. At an azeotropic concentration, the equilibrium curve crosses the $x = y$ diagonal.

Solubility of gases In gas-absorption operations the equilibria of interest are those between a nonvolatile absorbing liquid and a solute gas. The solute is ordinarily removed from its mixtures with relatively large amounts of a carrier gas which does not dissolve in the absorbing liquid. Temperature, pressure, and the concentration of solute in one phase are independently variable. The equilibrium relationship of importance in absorption is again the plot of x_e, the mole fraction of solute in the liquid, against y_e, the mole fraction in the vapor. The data from which this plot is prepared are ordinarily given in tables of physical constants in the form of partial pressures of solute in millimeters of mercury vs. concentrations in the liquid as

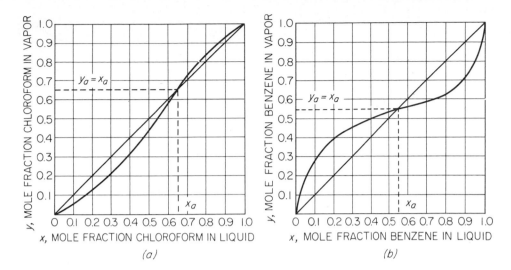

FIGURE 17-8
Equilibrium diagrams for azeotropic systems: (*a*) maximum-boiling azeotrope (system chloroform-acetone); (*b*) minimum-boiling azeotrope (system benzene-ethanol).

grams of solute per 100 g of solvent, equal to pounds of solute per 100 lb of solvent. Thus, for gases that follow Henry's law such as oxygen, carbon dioxide, and air, a Henry's-law constant H'_A is given as defined by the equation $\bar{p}'_A = H'_A c'_A$, where p'_A is the partial pressure of the solute, in millimeters of mercury, and c'_A is the concentration of the solute in the liquid, in grams per 100 g of solvent or pounds per 100 lb of solvent.

To convert H'_A to m, the slope of the equilibrium line in mole-fraction units, the following equation can be used:

$$m = \frac{100 H'_A M_A}{760 p M_B (1 - x_e)} \tag{17-48}$$

where M_A = molecular weight of solute
M_B = molecular weight of liquid solvent
x_e = mole fraction of solute in liquid at equilibrium
p = total pressure, atm

In Eq. (17-48), the quantity $1 - x_e$ may be taken as unity, since Henry's law applies only to dilute solutions. The equilibrium relationship between y_e and x_e for systems which follow Henry's law, therefore, is given by a straight line of slope m, passing through the origin of coordinates.

For more soluble gases, data are required showing partial pressures as functions of concentration and temperature.[6b] Figures 17-9 and 17-10 show such data for

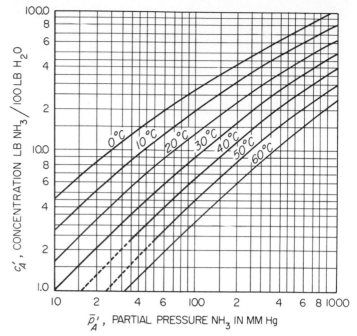

FIGURE 17-9
Solubility of ammonia in water.

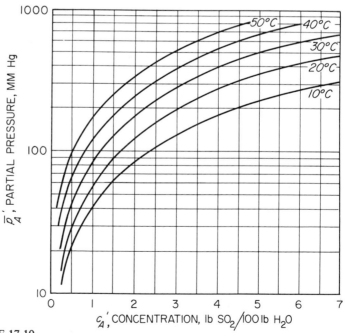

FIGURE 17-10
Partial pressure of sulfur dioxide over aqueous solution.

ammonia in water and sulfur dioxide in water, respectively. In both figures, partial pressures in millimeters of mercury are plotted against concentrations in pounds of solute per 100 lb of water. Points on the curve of x_e vs. y_e can be calculated from these data.

EXAMPLE 17-4   Plot the equilibrium curve of sulfur dioxide–air–water, in mole-fraction units, at 1 atm and 30°C, over the concentration range covered by Fig. 17-10.

SOLUTION   The y_e coordinates are found by dividing the partial pressures taken from Fig. 17-10 by 760. The x_e coordinates are found by noting that, since the molecular weight of SO_2 is 64, c'_A lb of SO_2 per 100 lb H_2O is $c'_A/64$ mol of SO_2 per $100/18 = 5.55$ mol of water. Then, $x_e = (c'_A/64)/(c'_A/64 + 5.55)$. Details of this calculation for several points on the curve of x_e vs. y_e are given in Table 17-4. The equilibrium line is plotted in Fig. 17-11.

Equilibria for dehumidification   In humidification and dehumidification operations the liquid phase is a single pure component. The equilibrium partial pressure of solute in the gas phase is therefore a unique function of temperature when the total pressure on the system is held constant. Also, at moderate pressures the equilibrium partial pressure is almost independent of total pressure and is virtually equal to the vapor pressure of the liquid. By Dalton's law the equilibrium partial pressure may be converted to the equilibrium mole fraction y_e in the gas phase. Since the liquid is pure, x_e is always unity. Equilibrium data are often presented as plots of y_e vs. temperature at a given total pressure, as shown for the system air-water at 1 atm in Fig. 17-12.

For the air-water system (and sometimes for other systems) the concentration of solute in the gas phase is commonly expressed in terms of *humidity* $\mathscr{H}$, which is the pounds of vapor carried by 1 lb of vapor-free gas. Humidity is discussed in Chap. 24.

Enthalpy-concentration diagrams   In many absorption and distillation problems it is necessary to take into account the latent heats, heats of mixing, and sensible heats of the components of the mixture. For binary systems all these quantities, and equilibrium data as well, can be shown on an enthalpy-concentration diagram, as shown in Chap. 16. This diagram may be constructed on either a mass or a mole basis.

Table 17-4   DATA FOR EXAMPLE 17-4

c'_A, lb SO_2/ 100 lb H_2O	$\bar{P}'_A$, mm Hg	$y_e = \dfrac{\bar{P}'_A}{760}$	$x_e = \dfrac{c'_A/64}{c'_A/64 + 5.55}$
1.0	85	0.112	0.0028
2.0	176	0.232	0.0056
3.0	273	0.359	0.0084
4.0	376	0.495	0.0111
5.0	482	0.634	0.0139
6.0	588	0.774	0.0166

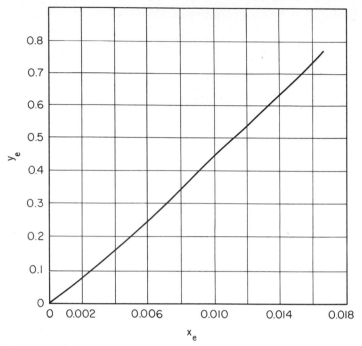

FIGURE 17-11
Equilibrium curve, sulfur dioxide in water at 30°C, 1 atm.

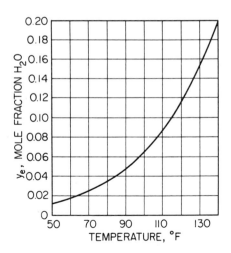

FIGURE 17-12
Equilibria for the system air-water at
1 atm.

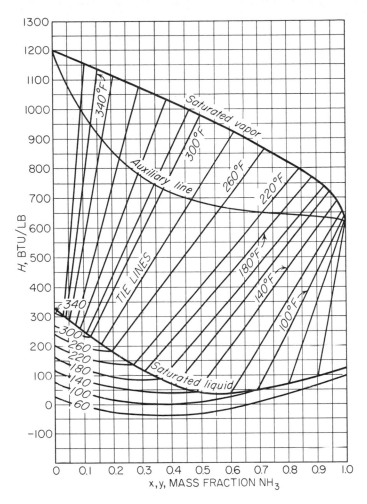

FIGURE 17-13
Enthalpy-concentration diagram (system ammonia-water at 10 atm).

An enthalpy-concentration diagram for ammonia-water mixtures is shown in Fig. 17-13. This chart applies at a constant pressure of 10 atm. The abscissa of any point on Fig. 17-13 is the concentration, in mass fraction of ammonia (the more volatile component) in a mixture of ammonia and water, and the ordinate is the specific enthalpy in Btu per pound of mixture. The enthalpies are referred to arbitrarily chosen standard states. In Fig. 17-13 the standard state of pure liquid water is 32°F and 1 atm. Since at constant pressure the enthalpy of 1 lb of either saturated liquid or saturated vapor depends only on the boiling point and the boiling point only on the concentration, these enthalpies can be plotted as functions of the concentration. Curves showing the specific enthalpies of saturated liquid and saturated vapor are shown in Fig. 17-13. Let the specific enthalpy of saturated vapor be H_y and that of saturated liquid be H_x, both in Btu per pound of mixture. Then the saturated-liquid line is a plot of H_x vs. x, and the saturated-vapor line is a plot of H_y vs. y. The abscissa of the diagram is used for both x and y, the mass fraction of component A in liquid and vapor, respectively. The enthalpy-concentration diagram may be referred to as the Hx diagram. All points above the saturated vapor line represent superheated vapor; all points between the lines are for mixtures of saturated liquid and saturated vapor; and all points below the liquid line pertain to liquids below their boiling temperatures. The isotherms in the liquid region represent enthalpies of liquid mixtures as a function of concentration and temperature.

Equilibrium between liquid and vapor phases is shown by the straight lines connecting the saturated-liquid line and the saturated-vapor line. These straight lines are called *tie lines*. The abscissas of the terminals of any one tie line represent corresponding values of x_e and y_e, the coordinates of a point on the equilibrium curve. Each tie line is an isotherm in the mixed-phase region. The tie lines in Fig. 17-13 correspond to the horizontal lines between bubble-point and dew-point curves of a boiling-point diagram such as that of Fig. 17-2.

There are an infinite number of tie lines. Several are shown in Fig. 17-13. To obtain other tie lines an interpolation may be made, or, more accurately, the construction shown in Fig. 17-14a can be used. For each tie line a point such as point a is obtained by finding the intersection of the vertical line through the lower terminus of the tie line and the horizontal line through the upper terminus. The auxiliary line shown in Fig. 17-13 is located in this manner, and the tie line for any desired x_e or y_e is found by reversing this construction. Another good way of obtaining tie lines is to plot an auxiliary temperature-composition diagram immediately above the enthalpy-composition diagram, as shown in Fig. 17-14b. The tie lines are then found by the construction indicated.

Liquid-Liquid Equilibria

In liquid extraction the phase equilibria of interest are those showing the distribution of a solute a between two immiscible or partially miscible liquids b and s. It is convenient to consider (1) the situation where the mutual solubility of liquids b and s, even in the presence of component a, is negligible and (2) that where this solubility cannot be ignored.

When at equilibrium the layer of b (plus some dissolved a) contains practically no s and the layer of s plus a practically no b, the equilibrium becomes a simple relation between the concentrations of solute in the two phases. This situation is identical with that of gas absorption, and the equilibrium curve is the usual x_e-vs.-y_e plot. In dilute solutions, Henry's law is followed. In some systems the equilibrium line is straight over a wide range of concentrations. As in the solubility of gases in liquids, the equilibrium is sensitive to temperature; but, in contrast to gas-liquid equilibria, liquid-liquid equilibria are nearly independent of pressure.

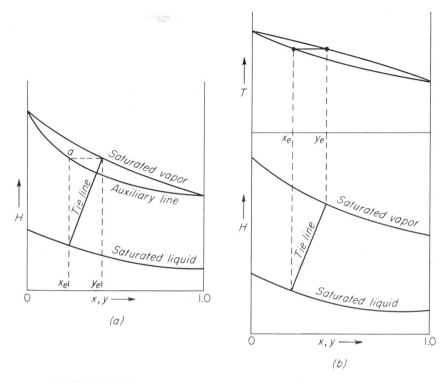

FIGURE 17-14
Methods of tie-line construction: (*a*) using auxiliary line; (*b*) using temperature-composition diagram.

Triangular coordinates   When the mutual solubility of diluent and solvent cannot be neglected, the solubility and equilibrium relationships are often shown on triangular coordinates. In this method of plotting, the composition of any three-component mixture can be shown by a point lying inside an equilateral triangle, as shown in Fig. 17-15. The triangular diagram has certain important characteristics.

Concentrations represented by triangular diagrams are based on the entire mixture, not on one or two components. Then the concentrations of the three components must add to unity, or

$$x_A + x_B + x_S = 1 \qquad \text{and} \qquad y_A + y_B + y_S = 1$$

Either mole fractions or mass fractions may be used. Mass fractions are the more common in these diagrams.

The fact that the sum of the concentrations is unity is correlated with the diagram by the geometric principle that the sum of the perpendicular distances from any point to the three sides of an equilateral triangle equals the altitude of the triangle. Thus, if the altitude is taken as unity, the perpendicular distances from any point

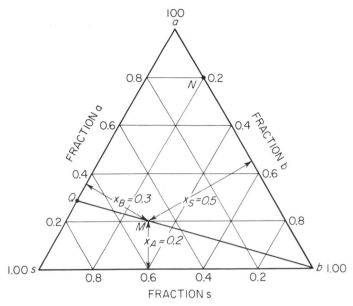

FIGURE 17-15
Triangular coordinates.

to the sides automatically add to unity, and these distances can be used to represent the individual concentrations of the three components. Any apex represents a pure component. Let the apices of the triangle of Fig. 17-15 represent the components *a*, *b*, and *s*, as shown. Then point *M*, for example, is the point for a mixture consisting of 20 percent *a*, 30 percent *b*, and 50 percent *s*.

One side of the triangle represents mixtures of the two components represented by the ends of that side; e.g., point *N* corresponds to a mixture of 80 percent *a* and 20 percent *b*.

Liquid-liquid equilibria with partial miscibility of solvent and diluent
Figure 17-16 is a triangular diagram showing equilibria in the system acetone, methyl isobutyl (MIK), and water at 25°C. The MIK may be considered the solvent for extracting the solute, acetone, from the diluent, water. This figure is an example of the first of two kinds of system important in extraction.

MIK is soluble in water to a concentration of approximately 2 percent MIK, and water is soluble in MIK to about 2 percent water. This is shown by points *B* and *A* in Fig. 17-16. Any mixture of MIK and water of a composition lying between points *A* and *B* forms two liquid layers, represented by these points. The mass ratio of the two layers is found by the center-of-gravity principle.

If acetone is added to a two-layer mixture of MIK and water, the acetone distributes between the layers, and the compositions of the layers follow two solubility curves, one for the water layer, or raffinate phase, and the other for the MIK layer,

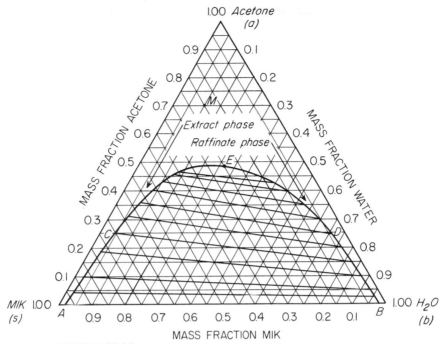

FIGURE 17-16
System acetone-MIK-water at 25°C. (Othmer, White, and Trueger.[5])

or extract phase. Line *BDE* shows compositions of the saturated water layer, and line *ACE* compositions of the saturated MIK layer. As the overall acetone content of the mixture increases, the concentrations of acetone in both layers increase and the solubility curves approach each other. The two phases become identical at point *E* in Fig. 17-16. Such a common point on both curves is called a *plait point*.

The area below the dome-shaped curve *ABEDC* is a composition field within which are located the points for all mixtures that form two liquid layers at equilibrium. All points outside the dome are for single-phase mixtures. Point *M*, for example, is that for a homogeneous mixture of 70 percent acetone, 20 percent MIK, and 10 percent water.

TIE LINES AND DISTRIBUTION CURVES Equilibria in triangular diagrams are shown by tie lines. These are straight lines connecting points on the two solubility curves. The ends of a tie line are the points for two phases in equilibrium. Several tie lines are shown in Fig. 17-16. Tie lines decrease in length as the plait point is approached and vanish at the plait point.

Each point within the dome lies on a tie line, and an infinite number of such lines exists. Graphical methods for interpolating between known tie lines are available.[12] To avoid cluttering the diagram with construction lines, a distribution curve showing y_A, the acetone concentration in one phase, plotted against x_A, the acetone

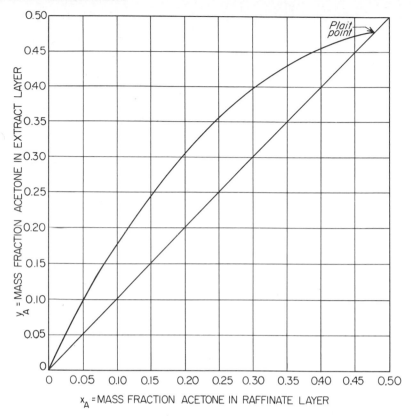

FIGURE 17-17
$x_A y_A$ plot, system acetone-MIK-water at 25°C.

concentration in the other, can be used as source of tie-line data. The distribution curve terminates on the $x = y$ diagonal at the concentration of the plait point. This curve is analogous to the xy curve used in distillation and absorption. Such a curve for the system acetone-MIK-water is shown in Fig. 17-17. To locate a tie line when the acetone concentration of one phase is known, that for the other is found from the distribution curve and used to locate the point on the other solubility curve. To locate the tie line for a given point within the dome trial and error is required.

A second type of equilibrium diagram is shown in Fig. 17-18, which applies to the system aniline, heptane, and methylcyclohexane (MCH) at 25°C. Aniline is the solvent, MCH the solute, and heptane the diluent. Line AB gives the solubility relationships for the extract phase and line CD that for the raffinate phase. Tie lines, as usual, provide phase-equilibrium data. To interpolate between tie lines it is convenient to plot, as a distribution curve, the concentration of MCH in the extract phase against that in the raffinate, as shown in Fig. 17-19.

Systems like that for acetone-MIK-water are called type I, and those like

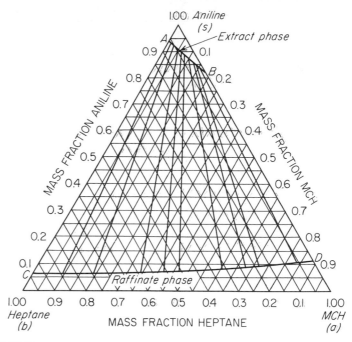

FIGURE 17-18
System aniline–*n*-heptane–MCH at 25°C: *a*, solute, MCH; *b*, diluent, *n*-heptane; *s*, solvent, aniline. (Varteressian and Fenske.[13])

aniline-heptane-MCH are called type II. A type I system may change to a type II system by change in temperature.

Rectangular coordinates Equilateral-triangular diagrams have some practical disadvantages. They require special plotting paper, their scales cannot be changed, and they cannot be enlarged when it is necessary to follow changes in a narrow concentration range. For these reasons, various other kinds of diagrams, which can be plotted on ordinary plotting paper, and the scales of which can be set individually at will, are used. One excellent method is to use a right-angle triangle in place of the equilateral triangle, by making one of the apices of the latter into a right angle.[4] Concentrations of two of the components are then plotted along the rectangular axes and that of the third component calculated by difference. Another method is to use rectangular coordinates and to choose the sum of two of the components as a base mixture for calculations and graphical constructions.

In principle, any pair of components may be chosen as the base mixture. In type I systems, a convenient choice is the combination of diluent and solvent, or $s + b$. Then the abscissa, denoted by X, is defined as the ratio of the diluent to the base mixture, or

$$X = \frac{b}{b + s}$$

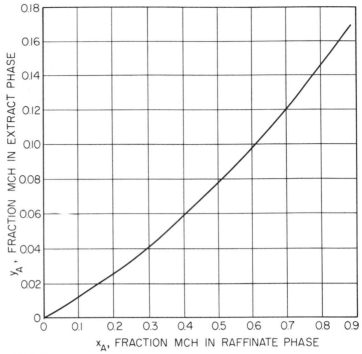

FIGURE 17-19
$x_A y_A$ plot, system aniline–n-heptane–MCH at 25°C.

The ordinate, denoted by Y, is defined as the ratio of the solute to the base mixture, or

$$Y = \frac{a}{b + s}$$

Here a, b, and s refer to either masses or concentrations.

The distribution diagram for finding tie lines is a plot of Y_e for the extract vs. Y_r for the raffinate. The rectangular diagram and distribution curve for the system acetone-water-MIK are given in Fig. 17-20.

In type II systems, the most convenient choice of base mixture is the combination of solute and diluent, or $a + b$. The abscissa X is the ratio of solute to base mixture, or $X = a/(a + b)$. The ordinate Y is the ratio of the solvent to the base mixture, or $Y = s/(a + b)$. The distribution curve is a plot of X_e for the extract vs. X for the raffinate. Plots for the system aniline-heptane-MCH are given in Fig. 17-21. An auxiliary line, shown in Fig. 17-21a, may also be used for locating tie lines.

Gas-Solid and Liquid-Solid Equilibria

Equilibria in systems involving a fluid phase (either liquid or gas) and a solid often are more complicated than vapor-liquid or liquid-liquid equilibria. Not only do the

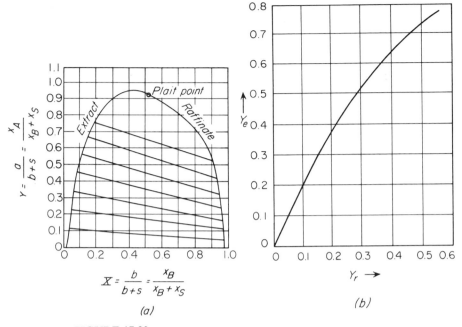

FIGURE 17-20
System acetone-MIK-water at 25°C: (*a*) XY diagram; *a*, solute, acetone; *b*, diluent, water; *s*, solvent, MIK; (*b*) Y_e vs. Y_r.

distribution relationships depend on the temperature and chemical nature of the system components; they also may depend on the physical form, method of preparation, and previous history of the solid. Systems which appear virtually identical may give widely differing results. Theories are of limited value for describing fluid-solid equilibria, and reliance must be placed on empirical correlations and experimental data.

Leaching Two situations must be distinguished in leaching. When the solid is brought into contact with a large amount of liquid solvent, the soluble material in the solid phase completely dissolves in the liquid. At equilibrium the bulk of the solution is unsaturated with respect to solute and has the same composition as the residual liquid held in the solid phase. When the amount of solvent is insufficient to dissolve all the solute, some solid solute remains in the solid phase and the liquid phase is saturated with solute.

The foregoing assumes that the solute is mechanically dispersed through the solid phase and is free to dissolve in the liquid, as in the acid leaching of soluble metal salts from a crushed ore. When solute is held inside liquid-filled cells, as in many natural products, leaching involves diffusion of solute through the cell walls. Under these conditions the equilibrium concentration of solute in the liquid inside the cells may differ from that in the bulk of the solution.

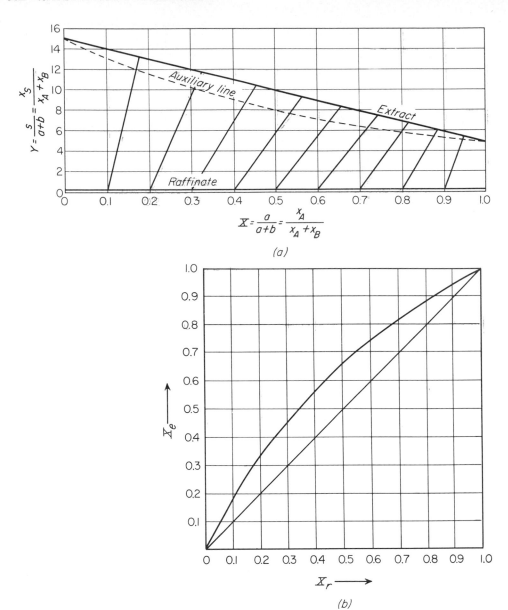

FIGURE 17-21
System aniline–*n*-heptane–MCH at 25°C: (*a*) *XY* diagram; *a*, solute, MCH, *b*, diluent, *n*-heptane; *s*, solvent, aniline; (*b*) equilibrium, raffinate and extract, on solvent-free basis, X_e vs. X_r.

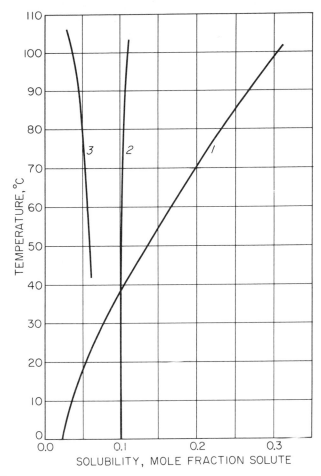

FIGURE 17-22
Solubility curves: curve 1, KNO_3; curve 2, NaCl; curve 3, $MnSO_4 \cdot H_2O$ (all in aqueous solution).

Crystallization   Equilibrium is attained in crystallization when the mother liquor is saturated. Solubility data are given in standard tables.[6a,11] Typical curves showing solubility as a function of temperature are given in Fig. 17-22. Most materials follow curves similar to curve 1; i.e., their solubility increases more or less rapidly with temperature. A few substances follow curves like curve 2, with little change in solubility with temperature; others have what is called an *inverted solubility curve* (curve 3), which means that their solubility decreases as the temperature is raised.

Many important inorganic substances crystallize with water of crystallization. In some systems, several different hydrates are formed, depending on concentration and temperature, and phase equilibria in such systems can be quite complicated. The phase diagram for the system magnesium sulfate–water is shown in Fig. 17-23.

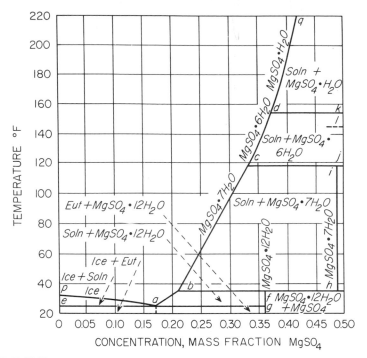

FIGURE 17-23
Phase diagram, system MgSO₄–H₂O. [*By permission, from J. H. Perry (ed.), "Chemical Engineers' Handbook,"* 4th ed. *Copyright*, 1963, *McGraw-Hill Book Company.*]

Concentration in mass fraction of anhydrous magnesium sulfate is plotted against temperature in degrees Fahrenheit. The entire area above and to the left of the broken solid line represents undersaturated solutions of magnesium sulfate in water. The broken line *eagfhij* represents complete solidification of the liquid solution to form various solid phases. The area *pae* represents mixtures of ice and saturated solution. Any solution containing less than 16.5 percent $MgSO_4$ precipitates ice when the temperature reaches line *pa*. Broken line *abcdq* is the solubility curve. Any solution more concentrated than 16.5 percent precipitates, on cooling, a solid when the temperature reaches this line. The solid formed at point *a* is called a *eutectic*. It consists of an intimate mechanical mixture of ice and $MgSO_4 \cdot 12H_2O$. Between points *a* and *b* the crystals are $MgSO_4 \cdot 12H_2O$; between *b* and *c* the solid phase is $MgSO_4 \cdot 7H_2O$ (epsom salt); between *c* and *d* the crystals are $MgSO_4 \cdot 6H_2O$; and above point *d* they are $MgSO_4 \cdot H_2O$. In the area *cihb*, the system at equilibrium consists of mixtures of saturated solution and crystalline $MgSO_4 \cdot 7H_2O$. In area *dljc*, the mixture consists of saturated solution and crystals of $MgSO_4 \cdot 6H_2O$. In area *qdk*, the mixture is saturated solution and $MgSO_4 \cdot H_2O$.

Drying   Equilibrium relationships of importance in drying are those between a solid and moist air (or humid gas). A typical relationship is illustrated in Fig. 17-24,

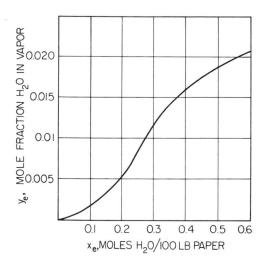

FIGURE 17-24
Equilibrium in system air–water–paper
pulp at 1 atm, 25°C.

which applies to the system air–water–paper *pulp* at 1 atm and 25°C. In this figure moisture contents are expressed in moles per 100 pounds of paper. More commonly, the moisture equilibria for solids are given as relationships between the relative humidity of the air or gas and the moisture content of the solid in pounds of water per pound (or 100 lb) of bone-dry solid. Examples of such relations are given in Fig. 17-25. Curves of this type are nearly independent of temperature. The abscissas of such curves are readily converted to absolute humidities in pounds of water per pound of dry air.

When a wet solid is brought into contact with air of lower humidity than that corresponding to the moisture content of the solid, as shown by the humidity-equilibrium curve, the solid tends to lose moisture and dry to equilibrium with the air. When the air is more humid than the solid in equilibrium with it, the solid absorbs moisture from the air until equilibrium is attained.

EQUILIBRIUM MOISTURE AND FREE MOISTURE   The air entering a dryer is seldom completely dry but contains some moisture and has a definite relative humidity. For air of definite humidity, the moisture content of the solid leaving the dryer cannot be less than the equilibrium-moisture content corresponding to the humidity of the entering air. That portion of the water in the wet solid which cannot be removed by the inlet air, because of the humidity of the latter, is called the *equilibrium moisture*.

The free water is the difference between the total-water content of the solid and the equilibrium-water content. Thus, if X_T is the total moisture content, and if X^* is the equilibrium-moisture content, the free moisture X is

$$X = X_T - X^*$$

It is X, rather than X_T, that is of interest in drying calculations.

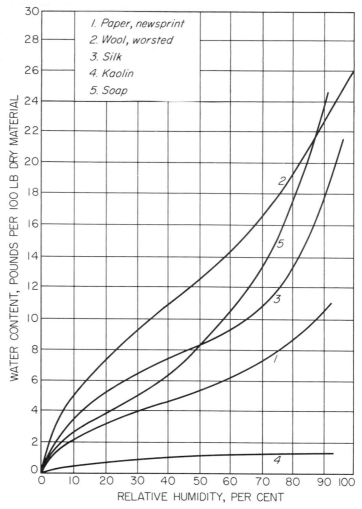

FIGURE 17-25
Equilibrium-moisture curves at 25°C.

BOUND AND UNBOUND WATER   If an equilibrium curve like those in Fig. 17-25 is continued to its intersection with the axis for 100 percent humidity, the moisture content so defined is the minimum moisture this material can carry and still exert a vapor pressure at least as great as that exerted by liquid water at the same temperature. If such a material contains more water than that indicated by this intersection, it can still exert only the vapor pressure of water at the stock temperature. This makes possible a distinction between two types of water held by a given material. The water up to the lowest concentration that is in equilibrium with saturated air (given by the intersection of the curves of Fig. 17-25 with the line for 100 percent humidity) is called *bound water* because it exerts a vapor pressure less than that of liquid water at the same temperature. Substances containing bound water are often called *hygroscopic* substances. The condition at which wood, textiles, and other cellular materials is in equilibrium with saturated air is called the *fiber-saturation point*.

Bound water may exist in several conditions. Liquid water in fine capillaries exerts an abnormally low vapor pressure because of the highly concave curvature of the surface; moisture in cell or fiber walls may suffer a vapor-pressure lowering because of solids dissolved in it; water in natural organic substances is in physical and chemical combination, the nature and strength of which vary with the nature and moisture content of the solid. Unbound water, on the other hand, exerts its full vapor pressure and is largely held in the voids of the solid.

The terms employed in this discussion may be clarified by reference to Fig. 17-25. Consider, for instance, curve 2 for worsted yarns. This intersects the curve for 100 percent humidity at 26 percent moisture; consequently, any sample of wool that contains less than 26 percent water contains only bound water. Any moisture that a sample may contain above 26 percent is unbound water. If the sample contains 30 percent water, for example, 4 percent of this water is unbound and 26 percent bound. Assume, now, that this sample is to be dried with air of 30 percent relative humidity. Curve 2 shows that the lowest moisture content that can be reached under these conditions is 9 percent. This, then, is the equilibrium-moisture content for this particular set of conditions. If a sample containing 30 percent total moisture is to be dried with air at 30 percent relative humidity, it contains 21 percent free water and 9 percent equilibrium moisture. The distinction between bound and unbound water depends on the material itself, while the distinction between free and equilibrium moisture depends on the drying conditions.

SYMBOLS

A', B', C'	Antoine constants [Eq. (17-49)]
B	Second virial coefficient, volume/mole
C	Third virial coefficient (volume/mole)^2
c	Concentration, mass/volume or moles/volume; c'_A, of solute A in liquid, g/100 g solvent
f	Fugacity, atm; f_i^L, of component i in heavy stream or phase; f_i^V, of component i in light stream or phase; f_i°, at standard state of component i
G	Gibbs energy; G_B, excess Gibbs energy; G_{comp}, of pure components before mixing; G_{mix}, of liquid mixture; G°, of ideal mixture
g	Molar Gibbs energy; $\bar{g}_i$, partial molar Gibbs energy of component i; $\bar{g}_i^E$, partial excess Gibbs energy of component i; $g_i^*(T)$, of gas at unit fugacity and temperature T
H	Henry constant, atm/mole fraction; H_i, of component i; H'_A, atm/(g mol A/100 g solvent) [Eq. (17-48)]
K	K factor; for component i, $K_i \equiv y_{ie}/x_{ie}$
L	Flow rate of heavy stream, mass/time or moles/time; L_a, entering stream; L_b, leaving stream
M	Molecular weight, dimensionless
m	Slope of equilibrium line, dy_e/dx_e
n	Number of moles; n_i, moles of component i

p	Pressure, atm; p_e, equilibrium pressure; p', vapor pressure of pure component; p_i, vapor pressure of component i; p_i, partial pressure of component i
R	Gas constant, energy/mole-absolute temperature
T	Temperature, K or °R; T_e, equilibrium temperature; T_i, boiling temperature of component i
V	Flow rate of light stream, mass/time or moles/time; V_a, leaving stream; V_b, entering stream
v	Molar volume, volume/mole; v^L, in liquid phase
X	Mole ratio in L stream; $X_i \equiv$ (moles component i)/(moles base mixture)
x	Mole or mass fraction, in L stream or in liquid phase; x_{ia}, in entering stream; x_{ib}, in leaving stream; x_{ie}, at equilibrium
Y	Mole ratio in V stream; $Y_i \equiv$ (moles component i)/(moles base mixture)
y	Mole or mass fraction, in V stream or in vapor phase; y_{ia}, in leaving stream; y_{ib}, in leaving stream; y_{ie}, at equilibrium
z	Compressibility factor of gas, $p/\rho RT$

Script letters

$\mathscr{C}$	Number of components
$\mathscr{F}$	Phase-rule variance
$\mathscr{H}$	Humidity, mass of vapor per mass of dry gas
$\mathscr{P}$	Number of phases

Greek letters

α	Relative volatility; α_{ij}, of component i relative to component j,

$$\alpha_{ij} \equiv \frac{y_i/x_i}{y_j/x_j} = \frac{K_i}{K_j}$$

γ	Activity coefficient, dimensionless, based on p_e; γ_i, of component i; γ_i^p, based on constant pressure, p_0
Δ	Finite-difference operator
Λ	Parameter, Wilson equation, dimensionless; Λ_{AB}, Λ_{BA}, defined in Eq. (17-36)
λ	Parameter, Wilson equation; λ_{AA}, λ_{AB}, $\lambda_{AB} = \lambda_{BA}$, of components A and B
μ	Chemical potential, energy/mole; μ_i, of component i; μ_i^L, in L stream or liquid; μ_i^V, in V stream or vapor
ρ	Density, mass/volume or moles/volume
Φ	Poynting factor, dimensionless, Eq. (17-41); Φ_A, of component A
ϕ	Fugacity coefficient, dimensionless; ϕ_i, of component i

Superscripts

E	Excess property
L	Heavy stream, or liquid phase
V	Light stream, or vapor phase
$\circ$	Standard state; ideal component or mixture
$*$	Component at unit fugacity

Subscripts

A	Component A
a	Entering L stream, or leaving V stream
B	Component B
b	Entering V stream, or leaving L stream
C	Component C
e	Equilibrium
i	Component i, any one of $\mathscr{C}$ components
j	Component j, any component except i
s	Pure component in equilibrium with vapor
comp	Pure components constituting mixture
mix	Mixture

PROBLEMS

The Antoine equation can be used to estimate the vapor-pressures of pure components. It is

$$\log p' = A' - \frac{B'}{C' + T} \qquad (17\text{-}49)$$

where

p' = vapor pressure, atm
T = boiling point, K
A', B', C' = empirical dimensional constants

17-1 For the system acetone (A), and methanol (B), the Antoine constants are given in Table 17-5. Assuming that Raoult's law applies to this system, plot the isothermal boiling-point diagram for mixtures of acetone and methanol at 60°C.

Table 17-5 ANTOINE CONSTANTS FOR PROB. 17-1[3]

Component	A'	B'	C'
Acetone	4.14366	1,161.0	49
Methanol	4.99782	1,473.11	43

17-2 The Wilson parameters for the acetone-methanol system are[3] $\lambda_{AB} = \lambda_{AA} - 214.95$, $\lambda_{AB} - \lambda_{BB} = 664.08$. The molar volumes of acetone and methanol are[3] $v_A^L = 77.36$ cm³/g mol and $v_B^L = 42.88$ cm³/g mol. Retaining the values $\phi_A = \phi_B = 1$, neglecting the effect of the Poynting correction, and using the values of γ_A and γ_B from the Wilson equation, calculate values of p_e and y_A when $T_e = 60$°C and $x_A = 0.45$.

17-3 The second virial coefficients for acetone and methanol vapors, in cubic centimeters per gram mole, may be taken as $B_{AA} = B_{BB} = -1,000$ and $B_{AB} = -500$. Correct the values of p_e and y_A found in Prob. 17-2 for nonideality of the vapor phase.

17-4 Correct the results of Prob. 17-3 for the Poynting effect.

17-5 Calculate the vapor pressure and vapor composition in equilibrium with a water-n-propanol mixture containing 0.60 mole fraction propanol at 95°C, (a) assuming that the liquid mixture is ideal and (b) assuming that the liquid-phase activity coefficients are given by the van Laar equations

$$\ln \gamma_A = \frac{2.60}{\left[1 + 2.301 \dfrac{x_A}{1 - x_A}\right]^2} \qquad \ln \gamma_B = \frac{1.13}{\left[1 + 0.4346 \dfrac{1 - x_A}{x_A}\right]^2}$$

where subscript A refers to n-propanol and B to water. In both (a) and (b) assume that the vapor-phase fugacity coefficients can be set equal to unity. Vapor pressures of the pure components at 95°C are n-propanol, 0.9230 atm; water, 0.8364 atm.

REFERENCES

1 Dymond, J. H., and E. B. Smith: "The Virial Coefficients of Gases," Clarendon, Oxford, 1969.
2 Hála, Eduard: "Vapor-Liquid Equilibria," Pergamon, New York, 1958.
3 Holmes, M. J., and M. Van Winkle: *Ind. Eng. Chem.*, **62**:21 (1970).

4   Kinney, G. F.: *Ind. Eng. Chem.,* **34:**1102 (1944).
5   Othmer, D. F., R. E. White, and E. Trueger: *Ind. Eng. Chem.,* **33:**1240 (1941).
6   Perry, J. E. (ed.): "Chemical Engineers' Handbook," 5th ed., McGraw-Hill, New York, 1973; (*a*) pp. **3**-92 to **3**-95, (*b*) pp. **3**-96 to **3**-98.
7   Prausnitz, J. M.: "Molecular Thermodynamics of Fluid-Phase Equilibria," Prentice-Hall, Englewood Cliffs, N.J., 1969; (*a*) pp. 89–180, (*b*) p. 181, (*c*) pp. 229–232.
8   Prausnitz, J. M., and P. L. Chueh: "Computer Calculations for High-Pressure Vapor-Liquid Equilibria," Prentice-Hall, Englewood Cliffs, N.J., 1969.
9   Prausnitz, J. M., C. A. Eckert, and J. P. Connell: "Computer Calculations for Multicomponent Vapor-Liquid Equilibria," Prentice-Hall, Englewood Cliffs, N.J., 1967.
10 Reid, R. C., and T. K. Sherwood: "Properties of Gases and Liquids," 2d ed., McGraw-Hill, New York, 1966.
11 Seidell, A.: "Solubilities," 3d ed., Van Nostrand, Princeton, N.J., 1940 (supplement, 1950).
12 Sherwood, T. K., and R. L. Pigford: "Absorption and Extraction," 2d ed., p. 402, McGraw-Hill, New York, 1952.
13 Varteressian, K. A., and M. R. Fenske: *Ind. Eng. Chem.,* **29:**270 (1937).

18

EQUILIBRIUM-STAGE OPERATIONS

One class of mass-transfer devices consists of assemblies of individual units, or stages, interconnected so that the materials being processed pass through each stage in turn. The two streams move countercurrently through the assembly; in each stage they are brought into contact, mixed, and then separated. For mass transfer to take place the streams entering each stage must not be in equilibrium with each other, for it is the departure from equilibrium conditions that provides the driving force for transfer. The leaving streams are usually not in equilibrium either but are much closer to being so than the entering streams are. If, as may actually happen in practice, the mixing in a given stage is so effective that the leaving streams are in fact in equilibrium, the stage is, by definition, an *ideal stage*. Such multistage systems are called *cascades*.

To illustrate the principle of an equilibrium-stage cascade, two typical countercurrent multistage devices, one for distillation and one for leaching, are described here. Other types of mass-transfer equipment are discussed in later chapters.

Typical distillation equipment A plant for continuous distillation is shown in Fig. 18-1. The still A is fed continuously with the liquid mixture to be distilled. The liquid is converted partially into vapor by heat transferred from the heating surface B.

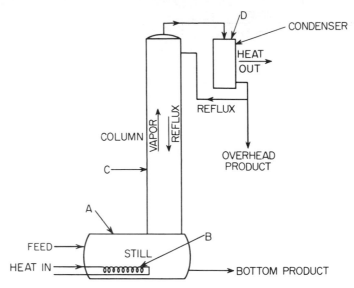

FIGURE 18-1
Still with fractionating column: *A*, still; *B*, heating coil; *C*, column; *D*, condenser.

The vapor formed in the still is richer in low boiler than the unvaporized liquid, but unless the two components differ greatly in volatility, the vapor contains substantial quantities of both components, and if it were condensed, the condensate would be far from pure. To increase the concentration of the low boiler in the vapor and to remove the high boiler from this stream, the vapor stream from the still is brought into intimate countercurrent contact with a descending stream of boiling liquid in the column or tower *C*. A liquid stream, usually concentrated in low boiler, is introduced at the top of the column. This return liquid is called *reflux*.

The reflux entering the top of the column is, if cold, immediately heated to its boiling point by the vapor, and throughout the column liquid and vapor are at their boiling and condensing temperatures, respectively. To obtain an increase in the concentration of the low boiler in the vapor, the reflux must be richer in low boiler than the equilibrium concentration corresponding to the vapor leaving the still. Then, at all levels in the column, some low boiler spontaneously diffuses from the liquid into the vapor, vaporizing as it passes from one phase to the other. The heat of vaporization of this low boiler is supplied by an equal amount of heat of condensation of high boiler. High boiler diffuses spontaneously from vapor to liquid. The net effect, then, is to transfer high boiler from vapor to liquid and to transfer a thermally equivalent amount of low boiler from liquid to vapor. As the vapor rises in the column, it becomes enriched in low boiler, or more volatile component. As the liquid descends the column, its content of high boiler increases. The flow of the bulk of the low boiler is up the column, and that of the high boiler is down the column. Because of the two-way transfer of material between phases, the change in total quantity of vapor is small.

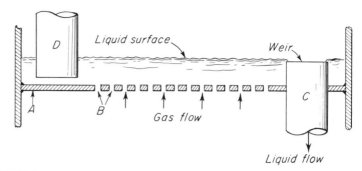

FIGURE 18-2
Sieve plate: *A*, tray or plate; *B*, perforations; *C*, downcomer to plate below; *D*, downcomer from plate above.

The enrichment of the vapor stream as it passes through the column in contact with reflux is called *rectification*. It is immaterial where the reflux originates provided its concentration in low boiler is sufficiently great to give the desired product. The usual source of reflux is the condensate leaving condenser *D*. Part of the condensate is withdrawn as the product, and the remainder returned to the top of the column. Provided an azeotrope is not formed, the vapor reaching the condenser can be brought as close to complete purity as desired by using a tall tower and a large reflux.

From the still, liquid is withdrawn which contains practically all the high boiler, as almost none of this component escapes with the overhead product unless that product is an azeotrope. The liquid from the still, which is called the *bottom product* or *bottoms*, is not nearly pure, however, as there is no provision in the equipment of Fig. 18-1 for rectifying this stream. A method for obtaining nearly pure bottom product by rectification is described in Chap. 19.

The column shown in Fig. 18-1 often contains a number of perforated plates, or trays, stacked one above the other. A cascade of such trays is called a *sieve-plate column*. A single sieve plate is shown in Fig. 18-2. It consists of a horizontal tray *A* carrying a downpipe, or downcomer, *C*, the top of which acts as a weir, and a number of holes *B*. The holes are all of the same size, usually $\frac{1}{4}$ to $\frac{1}{2}$ in. in diameter. The downcomer *D* from the tray above reaches nearly to tray *A*. This construction leads to the following flow of liquid and vapor. Liquid flows from plate to plate down the column, passing through downcomers *D* and *C* and across the plates. The weir maintains a minimum depth of liquid on the tray, nearly independent of the rate of flow of liquid. Vapor flows upward from tray to tray through the perforations. Except at very low vapor rates, well below the normal operating range, the vapor velocity through the perforations is sufficient to prevent leakage or "weeping" of the liquid through the holes. The vapor is subdivided by the holes into many small bubbles and passes in intimate contact through the pool of liquid on the tray. Because of the action of the vapor bubbles, the liquid is actually a boiling, frothy mass. Above the froth and below the next tray is fog from collapsing bubbles. This fog for the most part settles back into the liquid, but some is entrained by the vapor and carried to the plate

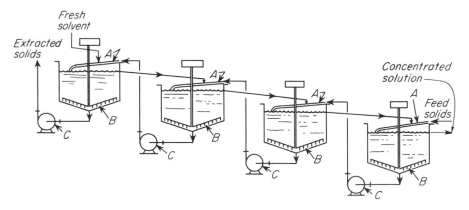

FIGURE 18-3
Countercurrent leaching plant: *A*, launder; *B*, rake; *C*, slurry pump.

above. Sieve-plate columns are representative of an entire class of equipment called *plate columns.*

Typical leaching equipment   In leaching, soluble material is dissolved from its mixture with an inert solid by means of a liquid solvent. A diagrammatic flowsheet of a typical countercurrent leaching plant is shown in Fig. 18-3. It consists of a series of units, in each of which the solid from the previous unit is mixed with the liquid from the succeeding unit and the mixture allowed to settle. The solid is then transferred to the next succeeding unit, and the liquid to the previous unit. As the liquid flows from unit to unit, it becomes enriched in solute, and as the solid flows from unit to unit in the reverse direction, it becomes impoverished in solute. The solid discharged from one end of the system is well extracted, and the solution leaving at the other end is strong in solute. The thoroughness of the extraction depends on the amount of solvent and the number of units. In principle, the unextracted solute can be reduced to any desired amount if enough solvent and a sufficient number of units are used.

Any suitable mixer and settler can be chosen for the individual units in a countercurrent leaching system. In those shown in Fig. 18-3 mixing occurs in launders *A* and in the tops of the tanks, rakes *B* move solids to the discharge, and slurry pumps *C* move slurry from tank to tank.

PRINCIPLES OF STAGE PROCESSES

In the sieve-plate tower and the countercurrent leaching plant shown in Figs. 18-1 and 18-3, the cascade consists of a series of interconnected units, or stages. Study of the assembly as a whole is best made by focusing attention on the streams passing between the individual stages. An individual unit in a stage-contact plant receives

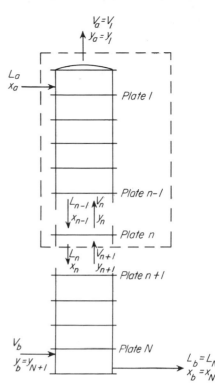

FIGURE 18-4
Material-balance   diagram   for   plate
column.

two streams, one a V phase and one an L phase, from the two units adjacent to it,
brings them into close contact, and delivers L and V phases, respectively, to the same
adjacent units. The fact that the contact units may be arranged either one above the
other, as in the sieve-plate column, or side by side, as in a stage leaching plant, is
important mechanically and may affect some of the details of operation of individual
stages. The same material-balance equations, however, can be used for either arrangement.

Terminology for stage-contact plants   The individual contact units in a cascade
are numbered serially, starting from one end. In this book, the stages are numbered
in the direction of flow of the L phase, and the last stage is that discharging the L
phase. A general stage in the system is the nth stage, which is number n counting from
the entrance of the L phase. The stage immediately ahead of stage n in the sequence
in the $(n - 1)$st stage, and that immediately following it is the $(n + 1)$st stage.
Using a plate column as an example, Fig. 18-4 shows how the units in a stage plant
are numbered. The total number of stages is N, and the last stage in the plant is
therefore the Nth stage.

   To designate the streams and concentrations pertaining to any one stage, all
streams originating in that stage carry the number of the unit as a subscript. Thus, for

a two-component system, y_{n+1} is the mole fraction of component A in the V phase leaving the $(n + 1)$st stage, and L_n is the molal flow rate of the L phase leaving the nth stage. The streams entering and leaving the plant and those entering and leaving stage n in a plate tower are shown in Fig. 18-4. Quantities V_a, L_b, y_a, and x_b in Table B, page 468, are equal to V_1, L_N, y_1, and x_N, respectively. This can be seen by reference to Fig. 18-4.

Material balances   Consider the portion of the cascade that includes stages 1 through n, as shown by the section enclosed by the dotted line in Fig. 18-4. The total input of material to this section is $L_a + V_{n+1}$ mol/h, and the total output is $L_n + V_a$ mol/h. Since, under steady flow, there is neither accumulation nor depletion, the input and the output are equal and

$$L_a + V_{n+1} = L_n + V_a \tag{18-1}$$

Equation (18-1) is a total material balance. Another balance can be written by equating input to output for component A. Since the number of moles of this component in a stream is the product of the flow rate and the mole fraction of A in the stream, the input of component A to the section under study, for a two-component system, is $L_a x_a + V_{n+1} y_{n+1}$ mol/h, the output is $L_n x_n + V_a y_a$ mol/h, and

$$L_a x_a + V_{n+1} y_{n+1} = L_n x_n + V_a y_a \tag{18-2}$$

A material balance can also be written for component B, but such an equation is not independent of Eqs. (18-1) and (18-2), since if Eq. (18-2) is subtracted from Eq. (18-1), the result is the material-balance equation for component B. Equations (18-1) and (18-2) yield all the information that can be obtained from material balances alone written over the chosen section.

Overall balances covering the entire cascade are found in the same manner:

Total material balance: $$L_a + V_b = L_b + V_a \tag{18-3}$$

Component A balance: $$L_a x_a + V_b y_b = L_b x_b + V_a y_a \tag{18-4}$$

Enthalpy balances   In many equilibrium-stage processes the general energy balance [Eq. (1-52)] can be simplified by neglecting the mechanical potential- and kinetic-energy terms. If, in addition, the process is workless and adiabatic, the simple enthalpy balance [Eq. (1-58)] applies. Then, for a two-component system, for the first n stages,

$$L_a H_{L,a} + V_{n+1} H_{V,n+1} = L_n H_{L,n} + V_a H_{V,a} \tag{18-5}$$

where H_L and H_V are the enthalpies per mole of the L phase and V phase, respectively. For the entire cascade,

$$L_a H_{L,a} + V_b H_{V,b} = L_b H_{L,b} + V_a H_{V,a} \tag{18-6}$$

Graphical Methods for Two-Component Systems

For systems containing only two components it is possible to solve many mass-transfer problems graphically. Some methods are based on material balances; some

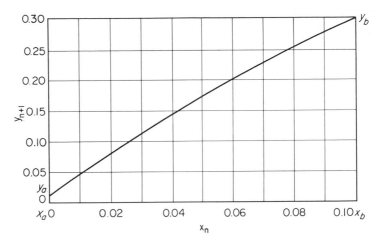

FIGURE 18-5
Solution of Example 18-1.

involve both material and enthalpy balances. The principles underlying these methods are discussed in the following paragraphs. Their detailed applications to specific operations are covered in later chapters.

Material-balance line or operating line   Equation (18-2), the material balance for one component, is a relation between x_n, the concentration of the L phase leaving a stage of the column, and y_{n+1}, the concentration of the V phase entering that stage. Equation (18-2) can be written

$$y_{n+1} = \frac{L_n}{V_{n+1}} \, x_n + \frac{V_a y_a - L_a x_a}{V_{n+1}} \tag{18-7}$$

Considering x_n as the abscissa and y_{n+1} as the ordinate, assume that it is possible to plot this equation on rectangular coordinates, as shown in Fig. 18-5. Then all points in the column that have coordinates like y_{n+1} and x_n must lie on this curve. A line showing such a relation between the concentrations of the V and L streams at each point in a countercurrent cascade is called an *operating line*. Such lines are basic in the graphical treatment of countercurrent contacting devices.

The two ends of the column are represented by two points on the operating line. The coordinates of one point are (x_a, y_a), and those of the other are (x_b, y_b). These points are the terminals of the operating line. This can be shown mathematically as follows. When $x_n = x_a$, $V_{n+1} = V_a$. Substituting into Eq. (18-7) gives

$$y_{n+1} = \frac{L_a x_a}{V_a} + \frac{V_a y_a - L_a x_a}{V_a} = y_a$$

This shows that point (x_a, y_a) lies on the operating line; also, when $x_n = x_b$, $V_{n+1} = V_b$. Substitution into Eq. (18-7) gives

$$y_{n+1} = \frac{L_b x_b}{V_b} + \frac{V_b y_a - L_a x_a}{V_b}$$

Comparing this equation with Eq. (18-4) shows that $y_{n+1} = y_b$, so point (x_b, y_b) also lies on the line.

Construction of operating line   To draw an operating line, information must be available on the stream flows V_{n+1} and L_n. Additional data are required for this. Two simple cases cover many of the situations normally encountered in practice. In the first, which is true usually in rectification, neither L nor V changes appreciably from plate to plate. Then L_n and V_{n+1} in Eq. (18-7) are constant, and the operating line is straight. The line is easily drawn for any two points in it. The two terminal points are usually used.

When the operating line is straight, from inspection of Eq. (18-7), its slope is L/V, or the ratio of the flow of L phase to that of the V phase. To emphasize this fact, the overall balance can be written

$$\frac{y_b - y_a}{x_b - x_a} = \frac{L}{V} \tag{18-8}$$

A second case where the operating line can be drawn is when one component of each stream is not transferred between phases. Then, if L' is the molal flow rate of the inert component in the L phase and V' the moles of the other inert component in the V phase, these quantities are constant throughout the cascade. The relationships between V and V' and L and L' are

$$V_{n+1} = \frac{V'}{1 - y_{n+1}} \qquad L_n = \frac{L'}{1 - x_n} \tag{18-9}$$

Substitution of these relationships in Eq. (18-2) gives

$$L'\left(\frac{x_a}{1 - x_a} - \frac{x_n}{1 - x_n}\right) = V'\left(\frac{y_a}{1 - y_a} - \frac{y_{n+1}}{1 - y_{n+1}}\right) \tag{18-10}$$

No subscripts are needed for the quantities V' and L', as these are constant.

When the phases are dilute, x and y are small in comparison with unity and the operating line becomes practically straight.

EXAMPLE 18-1   By means of a plate column, acetone is absorbed from its mixture with air in a nonvolatile absorption oil. The entering gas contains 30 mole percent acetone, and the entering oil is acetone-free. Of the acetone in the air 97 percent is to be absorbed, and the concentrated liquor at the bottom of the tower is to contain 10 mole percent acetone. Plot the operating line.

SOLUTION   The concentration of acetone in the leaving air is obtained by an acetone balance. A basis of 100 mol of entering air is chosen. Then

$$V' = 0.70 \times 100 = 70 \text{ mol}$$

The acetone in the entering air is 30 mol, of which 3 percent, or $0.03 \times 30 = 0.9$ mol, leaves with the gas, and $30 - 0.9 = 29.1$ mol is absorbed to form a 10 percent solution. Then

$$L' = 29.1 \times \tfrac{90}{10} = 261.9 \text{ mol}$$

The terminal concentrations are

$$y_a = \frac{0.9}{70 + 0.9} = 0.0127 \qquad x_a = 0 \qquad x_b = 0.10 \qquad y_b = 0.30$$

From Eq. (18-10), the equation for the operating line is

$$261.9 \left(0 - \frac{x_n}{1 - x_n} \right) = 70 \left(\frac{0.0127}{1 - 0.0127} - \frac{y_{n+1}}{1 - y_{n+1}} \right)$$

After dropping the subscripts on x and y, this equation can be solved for either $1 - x$ or $1 - y$. Choosing $1 - y$ gives

$$\frac{y}{1 - y} = 0.0129 + 3.74 \frac{x}{1 - x}$$

The operating line passes through points ($x = 0$, $y = 0.0127$) and ($x = 0.10$, $y = 0.30$). Coordinates for other points are found from the above equation. Thus, when $x = 0.03$,

$$\frac{y}{1 - y} = 0.0129 + \frac{3.74 \times 0.03}{0.97} = 0.1286$$

and

$$y = 0.114$$

Likewise, when $x = 0.05$, $y = 0.173$, and when $x = 0.08$, $y = 0.253$. The operating line is plotted in Fig. 18-5. ////

Ideal contact stages   The ideal stage is a standard to which an actual stage may be compared. In an ideal stage, the V phase leaving the stage is in equilibrium with the L phase leaving the same stage. For example, if plate n in Fig. 18-4 is an ideal stage, concentrations x_n and y_n are coordinates of a point on the curve of x_e vs. y_e showing the equilibrium between the phases. In a plate column ideal stages are also called *perfect plates*.

To use ideal stages in design it is necessary to apply a correction factor, called the *stage efficiency* or *plate efficiency*, which relates the ideal stage to an actual one. Plate efficiencies are discussed in Chap. 19, and the present discussion is restricted to ideal stages.

Determining the number of ideal stages   A problem of general importance is that of finding the number of ideal stages required in an actual cascade to cover a desired range of concentration x_a to x_b or its equivalent, y_a to y_b. If this number can be determined, and if information on stage efficiencies is available, the number of actual stages can be calculated. This is the usual method of designing cascades.

The most generally satisfactory methods of determining the number of ideal stages in a cascade are graphical ones. The simplest graphical method, and one that is sufficiently precise for most two-component situations, is based on the use of the

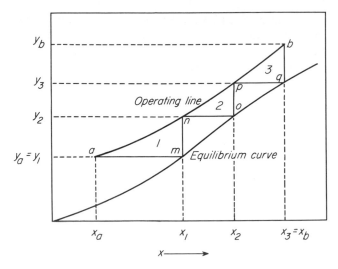

FIGURE 18-6
Graphical determination of number of ideal stages.

operating line in conjunction with the equilibrium line, both plotted on the same coordinates.

Figure 18-6 shows the equilibrium curve and the operating line for a typical stage-contact unit, e.g., a gas absorber. The ends of the operating line are point a, having coordinates (x_a, y_a), and point b, having coordinates (x_b, y_b). The problem of determining the number of ideal stages needed to accomplish the gas-phase concentration change y_b to y_a and the liquid-phase concentration change x_a to x_b is solved as follows.

The concentration of the gas leaving the top stage, which is stage 1, is y_a, or y_1. If the stage is ideal, x_1, the concentration of the liquid leaving this stage, is, by definition of an ideal stage, such that the point (x_1, y_1) must lie on the equilibrium curve. This fact fixes point m, found by moving horizontally from point a to the equilibrium curve. The abscissa of point m is x_1. The operating line is now used. It passes through all points having coordinates of the type (x_n, y_{n+1}), and since x_1 is known, y_2 is found by moving vertically from point m to the operating line at point n, the coordinates of which are (x_1, y_2). The step, or triangle, defined by points a, m, and n represents one ideal stage, the first one in this column. The second stage is located graphically on the diagram by repeating the same construction, passing horizontally to the equilibrium curve at point o, having coordinates (x_2, y_2), and vertically to the operating line again at point p, having coordinates (x_2, y_3). The third stage is found by again repeating the construction, giving triangle pqb. For the situation shown in Fig. 18-6, the third stage is the last, as the concentration of the gas leaving that stage is y_b, and the liquid leaving it is x_b, which are the desired terminal concentrations. Three ideal stages are required for this separation.

Figure 18-6 also shows that x_a is the concentration of the liquid entering an imaginary stage just preceding the first one, or stage 0, and y_b is the concentration of the vapor leaving an imaginary stage just following the last one. Thus, in general, for a cascade of N stages, $x_a = x_0$, $x_b = x_N$, $y_a = y_1$, and $y_b = y_{N+1}$.

The same construction can be used for determining the number of ideal stages needed in any cascade, whether it is used for gas absorption, rectification, leaching, or liquid extraction.†

Absorption-factor method for calculating the number of ideal stages   When the operating and equilibrium lines are both straight over a given concentration range x_a to x_b, the number of ideal stages can be calculated by formula and graphical construction is unnecessary. Formulas for this purpose are derived as follows.

Let the equation of the equilibrium line be

$$y_e = mx_e + B \tag{18-11}$$

where, by definition, m and B are constant. If stage n is ideal,

$$y_n = mx_n + B \tag{18-12}$$

Substitution for x_n into Eq. (18-7) gives, for ideal stages and constant L/V,

$$y_{n+1} = \frac{L(y_n - B)}{mV} + y_a - \frac{Lx_a}{V} \tag{18-13}$$

It is convenient to define an absorption factor A by the equation

$$A \equiv \frac{L}{mV} \tag{18-14}$$

The absorption factor is the ratio of the slope of the operating line to that of the equilibrium line. It is a constant when both of these lines are straight. Equation (18-13) can be written

$$y_{n+1} = A(y_n - B) + y_a - Amx_a$$
$$= Ay_n - A(mx_a + B) + y_a \tag{18-15}$$

The quantity $mx_a + B$ is, by Eq. (18-11), the concentration of the vapor that is in equilibrium with the inlet L phase, the concentration of which is x_a. This can be seen from Fig. 18-7. The symbol y^* is used to indicate the concentration of a V phase in equilibrium with a specified L phase. Then

$$y_a^* = mx_a + B \tag{18-16}$$

and Eq. (18-15) becomes

$$y_{n+1} = Ay_n - Ay_a^* + y_a \tag{18-17}$$

† The graphical step-by-step construction utilizing alternately the operating and equilibrium lines to find the number of ideal stages was first applied to the design of rectifying columns, and is known as the *McCabe-Thiele method*.[3]

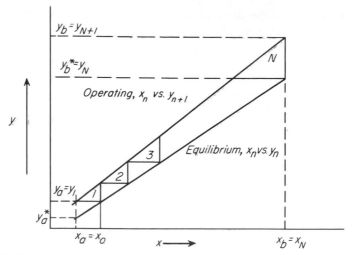

FIGURE 18-7
Derivation of absorption-factor equation.

Equation (18-17) can be used to calculate, step by step, the value of y_{n+1} for each stage starting with stage 1. The method may be followed with the aid of Fig. 18-7. For stage 1, using $n = 1$ in Eq. (18-17) and noting that $y_1 = y_a$ gives

$$y_2 = A y_a - A y_a^* + y_a = y_a(1 + A) - A y_a^*$$

For stage 2, using $n = 2$ in Eq. (18-21) and eliminating y_2 gives

$$y_3 = A y_2 - A y_a^* + y_a = A[y_a(1 + A) - A y_a^*] - A y_a^* + y_a$$
$$= y_a(1 + A + A^2) - y_a^*(A + A^2)$$

These equations can be generalized for the nth stage, giving

$$y_{n+1} = y_a(1 + A + A^2 + \cdots + A^n) - y_a^*(A + A^2 + \cdots + A^n) \quad (18\text{-}18)$$

For the entire cascade, $n = N$, the total number of stages, and

$$y_{n+1} = y_{N+1} = y_b$$

Then

$$y_b = y_a(1 + A + A^2 + \cdots + A^N) - y_a^*(A + A^2 + \cdots + A^N) \quad (18\text{-}19)$$

The sums in the parentheses of Eq. (18-19) are both sums of geometric series. The sum of such a series is

$$S_n = \frac{a_1(1 - r^n)}{1 - r}$$

where s_n = sum of first n terms of series
  a_1 = first term
  r = constant ratio of each term to preceding term

Equation (18-19) can then be written

$$y_b = y_a \frac{1 - A^{N+1}}{1 - A} - y_a^* A \frac{1 - A^N}{1 - A} \tag{18-20}$$

Equation (18-20) is a form of the Kremser equation.[2] It can be used as such or in the form of a chart relating N, A, and the terminal concentrations.[1,4] It can also be put into a simpler form by the following method.

Equation (18-17) is, for stage N,

$$y_b = Ay_N - Ay_a^* + y_a \tag{18-21}$$

It can be seen from Fig. 18-7 that $y_N = y_b^*$ and Eq. (18-21) can be written

$$y_a = y_b - A(y_b^* - y_a^*) \tag{18-22}$$

Collecting terms in Eq. (18-20) containing A^{N+1} gives

$$A^{N+1}(y_a - y_a^*) = A(y_b - y_a^*) + (y_a - y_b) \tag{18-23}$$

Substituting $y_a - y_b$ from Eq. (18-22) into Eq. (18-23) gives

$$A^N(y_a - y_a^*) = y_b - y_a^* - y_b^* + y_a^* = y_b - y_b^* \tag{18-24}$$

Taking logarithms of Eq. (18-24) and solving for N gives

$$N = \frac{\log\left[(y_b - y_b^*)/(y_a - y_a^*)\right]}{\log A} \tag{18-25}$$

and from Eq. (18-22)

$$\frac{y_b - y_a}{y_b^* - y_a^*} = A \tag{18-26}$$

Equation (18-25) can be written

$$N = \frac{\log\left[(y_b - y_b^*)/(y_a - y_a^*)\right]}{\log\left[(y_b - y_a)/(y_b^* - y_a^*)\right]} \tag{18-27}$$

The various concentration differences in Eq. (18-27) are shown in Fig. 18-8.

When the operating line and the equilibrium line are parallel, A is unity and Eqs. (18-25) and (18-27) are indeterminate. When $A = 1$, Eq. (18-19) becomes

$$y_b = y_a(1 + N) - y_a^* N \tag{18-28}$$

and

$$N = \frac{y_b - y_a}{y_a - y_a^*} = \frac{y_b - y_a}{y_b - y_b^*} \tag{18-29}$$

L-phase form of Eq. (18-27) The choice of y as the concentration coordinate rather than x is arbitrary. It is the conventional variable in gas-absorption calculations. It may be used for stripping also, but in practice equations in x are more common. They are

$$N = \frac{\log\left[(x_b - x_b^*)/(x_a - x_a^*)\right]}{\log\left[(x_a^* - x_b^*)/(x_a - x_b)\right]} = \frac{\log\left[(x_b - x_b^*)/(x_a - x_a^*)\right]}{\log S} \tag{18-30}$$

where x^* = equilibrium concentration corresponding to y

S = *stripping factor*

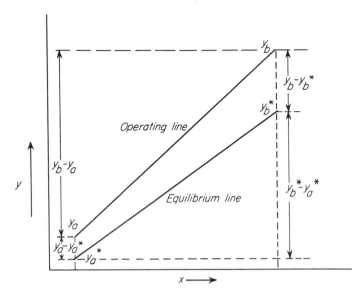

FIGURE 18-8
Concentration differences in Eq. (18-27).

S is defined by

$$S \equiv \frac{1}{A} = \frac{mL}{V} \tag{18-31}$$

The stripping factor is the ratio of the slope of the equilibrium line to that of the operating line. From Eq. (18-11)

$$x^* = \frac{y - B}{m} \tag{18-32}$$

When $S = 1$,

$$N = \frac{x_a - x_b}{x_b - x_b^*} = \frac{x_a - x_b}{x_b - x_a^*} \tag{18-33}$$

The concentration differences in Eq. (18-30) are shown in Fig. 18-9.

In the above derivations concentrations are given in mole fractions, which is the usual choice in distillation. In absorption and stripping, linear equilibrium and operating lines are obtained by using as concentrations the ratio of the mass or moles of the diffusing component to the mass or moles of the inert nondiffusing components in the L and V streams. In liquid or solid streams, where the densities of the streams do not change, mass or moles per unit volume may be used.

As shown in the derivations, it is not assumed that the equilibrium line passes through the origin. It is only necessary that the line be linear in the range where the

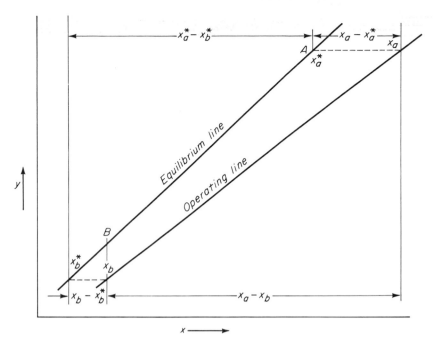

FIGURE 18-9
Concentration differences in Eq. (18-30).

steps representing the stages touch the line, as shown by the line AB in Fig. 18-9. In practice, where the equilibrium is that of an ideal solution following Henry's law, the equilibrium line passes through the origin, so the constant B in Eq. (18-16) is zero and constant m becomes the K factor of the diffusing component.

In the design of a plant, N is calculated from the proposed terminal concentration concentrations by Eq. (18-25), (18-27), (18-29), (18-30), or (18-33). In estimating the effect of a change in operating conditions of an existing plant, Eq. (18-24) or the following equation in x coordinates is used:

$$S^N = \frac{x_b - x_b^*}{x_a - x_a^*} \tag{18-34}$$

Material balances and K factors are available for eliminating unknowns from these equations.

EXAMPLE 18-2   Ammonia is to be scrubbed from a gas containing 4 percent ammonia and 96 percent air by volume in a countercurrent plate absorber. For washing, water containing 0.003 mol NH_3 per mole of H_2O is to be used at a rate of 1.1 mol H_2O per mole of air. Of the ammonia entering in the gas, 90 percent is to be absorbed. At the temperature of the operation, 68°F (20°C), the K factor of the ammonia is

$$K = 0.80 \frac{\text{mol } NH_3/\text{mol air}}{\text{mol } NH_3/\text{mol } H_2O}$$

(a) What is the exit concentration of ammonia in the liquid? (b) How many ideal plates are required?

SOLUTION The exit ammonia concentration is found by an ammonia balance over the absorber, and the number of plates is calculated by use of Eq. (18-25), using mole-ratio units, moles NH_3 per moles of inert carrier (air or water).

(a) Based on 1 mol of air, the ammonia content of the entering air is $Y_b = 4/96 = 0.04167$ mol. Of this, $0.90 \times 0.04167 = 0.03750$ mol is absorbed. This is $0.03750/1.1 = 0.0341$ mol NH_3 per mole of H_2O, and $X_b = 0.003 + 0.0341 = 0.0371$ mol NH_3 per mole of H_2O.

(b) The remaining quantities to be substituted into Eq. (18-25) are

$$Y_a = 0.04167(1 - 0.9) = 0.00417 \text{ mol } NH_3 \text{ per mole air} \qquad A = \frac{1.1}{0.8} = 1.375$$

$$Y_a^* = 0.80 \times 0.00417 = 0.00334 \qquad Y_b^* = 0.80 \times 0.0371 = 0.02968$$

$$N = \frac{\log\left[(0.04167 - 0.02968)/(0.00417 - 0.00334)\right]}{\log 1.375} = 14 +$$

Fifteen plates are needed. ////

Constructions for Mixing and Separation Processes on Enthalpy-Concentration Diagrams

A number of useful graphical constructions can be made directly on diagrams which correlate an additive intensive property, such as specific enthalpy, with concentration. For example, with reference to an enthalpy-concentration diagram, consider the simple mixing process shown in Fig. 18-10a. The process consists of mixing $\dot{m}_R$ lb/h of a stream R with $\dot{m}_S$ lb/h of stream S to form $\dot{m}_R + \dot{m}_S$ lb/h of stream T. The concentrations of streams R, S, and T are x_R, x_S, and x_T, respectively, all in pounds of component A per pound of stream. The specific enthalpies of the same streams are H_R, H_S, and H_T, respectively, all in Btu per pound. Molal units can be used in place of mass units if desired.

Adiabatic process Assume, first, that the process is adiabatic. An enthalpy balance [Eq. (1-56)], neglecting mechanical potential- and kinetic-energy terms, and noting that Q and W_S are zero, is

$$\dot{m}_R H_R + \dot{m}_S H_S = (\dot{m}_R + \dot{m}_S) H_T \tag{18-35}$$

A material balance on component A is

$$\dot{m}_R x_R + \dot{m}_S x_S = (\dot{m}_R + \dot{m}_S) x_T \tag{18-36}$$

Equations (18-35) and (18-36) can each be solved for the ratio $\dot{m}_R/\dot{m}_S$. The results are

$$\frac{\dot{m}_R}{\dot{m}_S} = \frac{H_S - H_T}{H_T - H_R} \tag{18-37}$$

$$\frac{\dot{m}_R}{\dot{m}_S} = \frac{x_S - x_T}{x_T - x_R} \tag{18-38}$$

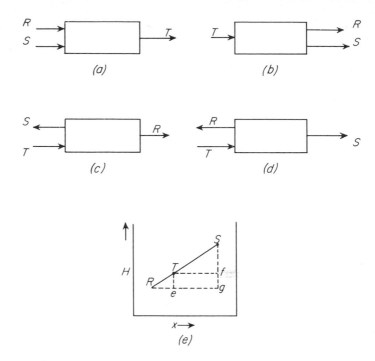

FIGURE 18-10
Adiabatic mixing process: (a) mixing R and S to form T; (b) separating T into R and S; (c) removing S from T to give R; (d) removing R from T to give S; (e) all processes on Hx diagram.

Equating the right-hand sides of Eqs. (18-37) and (18-38) gives

$$\frac{x_S - x_T}{x_T - x_R} = \frac{H_S - H_T}{H_T - H_R} \qquad (18\text{-}39)$$

Each stream can be represented by a point on the chart. The coordinates of the point are the concentration and the enthalpy of the stream. Plot points R, S, and T, each point representing the stream of the same letter, as shown in Fig. 18-10e. Draw lines RT and TS, and plot points $e, f,$ and g, as shown. It is clear from Eq. (18-39) that triangles TSf and RTe are geometrically similar. Then lines RT and TS have the same slope, and line RTS is a single straight line. In an adiabatic mixing process, therefore, the points on the Hx diagram that represent the product stream and the two feed streams lie on the same straight line.

The construction shown in Fig. 18-10e also applies to the three other processes shown in Fig. 18-10b, c, and d. In Fig. 18-10b stream T is separated into streams R and S; in Fig. 18-10c stream S is removed or subtracted from stream T, leaving stream R as a difference; and in Fig. 18-10d stream R is removed from stream T, leaving stream S. Note that for all situations shown in Fig. 18-10 the point representing a

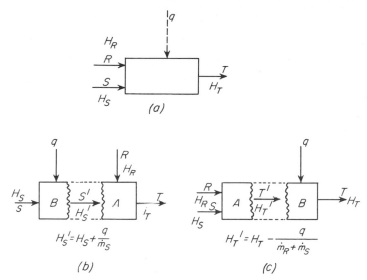

FIGURE 18-11
Nonadiabatic mixing processes: (*a*) actual process; (*b*) equivalent two-step process, heat added to an entering stream; (*c*) equivalent two-step process, heat added to leaving stream; *A*, adiabatic step; *B*, heating step.

single stream entering the process lies between the points representing the two streams leaving the process and that the point for a single stream leaving the process lies between the points for the two streams entering.

Equations (18-37) and (18-38) show that the relative amounts of the three streams are proportional to the lengths of line segments in Fig. 18-10*e*. Let a vinculum over the letters for the terminals of a line symbolize the length of the line. Then $\dot{m}_R/\dot{m}_S$ may be represented by $\overline{Sf}/\overline{Te}$, by $\overline{fT}/\overline{eR}$, or by $\overline{ST}/\overline{TR}$. Also, $\dot{m}_S$ is proportional to $\overline{Te}$, to $\overline{eR}$, or to $\overline{TR}$, and $\dot{m}_R$ is proportional to $\overline{Sf}$, to $\overline{fT}$, or to $\overline{ST}$. Again, the ratio of the mass of stream R to that of stream T is given by $\overline{Sf}/\overline{Sg}$, by $\overline{fT}/\overline{gR}$, or by $\overline{ST}/\overline{SR}$.

If Eq. (18-38) is written as $\dot{m}_R(x_T - x_R) = \dot{m}_S(x_S - x_T)$, it can be seen that point T represents the center of gravity of points R and S, endowed with masses $\dot{m}_R$ and $\dot{m}_S$, respectively. This will be referred to as the *center-of-gravity principle*.

Nonadiabatic process   The construction shown in Fig. 18-10*e* can be extended to nonadiabatic processes by utilizing the following device. Let q be the heat effect accompanying the process, in Btu per hour, as shown in Fig. 18-11*a*. Assume q is positive for an endothermic process, where the process absorbs heat, and negative for an exothermic process, where the process evolves heat. Imagine the actual process to be divided into two steps, an adiabatic step and a heating step, in which either an incoming stream is preheated by adding the heat q *before* the stream enters the adia-

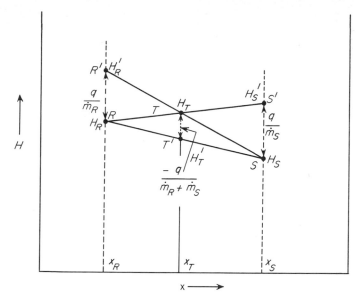

FIGURE 18-12
Nonadiabatic process on Hx diagram.

batic step or in which a leaving stream is heated by adding heat q *after* the stream leaves the adiabatic step. Flow charts for these two situations are shown in Fig. 18-11. In Fig. 18-11b the heat effect is absorbed by stream S before that stream enters the adiabatic step, and in Fig. 18-11c the discharge stream T absorbs heat q after it leaves the adiabatic step. The specific enthalpy of stream S, after being heated, is H'_S, where

$$H'_S = H_S + \frac{q}{\dot{m}_S} \tag{18-40}$$

Then point S', representing the stream after heating, is collinear with points R and T, as shown in Fig. 18-12. The specific enthalpy of stream T, before being heated, is H'_T, where

$$H'_T = H_T - \frac{q}{\dot{m}_R + \dot{m}_S} \tag{18-41}$$

Point T', representing the stream before heating, is collinear with points R and S, representing the feed streams to the process.

If the process is exothermic, q is negative in Eqs. (18-40) and (18-41), and $H'_S < H_S$, $H'_R < H_R$, and $H'_T > H_T$. If the process is endothermic, $H'_S > H_S$, $H'_R > H_R$, and $H'_T < H_T$. The lines in Fig. 18-12 as drawn apply to endothermic processes.

The choice of stream to be corrected for the heat effect is optional. If two heat effects are involved in a single process, it may be convenient to correct one stream for one heat effect and a second stream for the other.

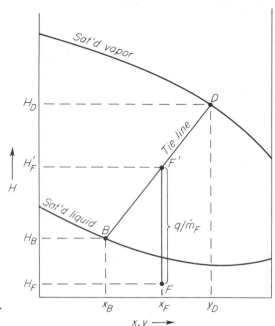

FIGURE 18-13
Partial vaporization and adiabatic separation of liquid feed.

As a specific example, consider an enthalpy-concentration diagram of the type shown in Fig. 17-13 and reproduced schematically in Fig. 18-13. Suppose that a liquid feed of composition x_F and enthalpy H_F is flowing at a mass flow rate of $\dot{m}_F$ into a heater, where it receives heat q and is partially vaporized. After it leaves the heater, the liquid and vapor are separated adiabatically. It is desired to know the compositions and relative flow rates of the liquid and vapor streams, assuming they are in equilibrium with each other.

Unheated feed is represented in Fig. 18-13 by point F. Heated feed is represented by point F', the coordinates of which are H_F' and x_F, where $H_F' = H_F + q/\dot{m}_F$. Since the separation process is adiabatic and vapor and liquid are in equilibrium, the vapor and liquid streams are represented by points D and B, respectively, at the ends of the tie line which passes through point F'. The composition of the vapor is y_D; that of the liquid is x_B. Let $\dot{m}_B$ and $\dot{m}_D$ be the mass flow rates of liquid and vapor, respectively. From the center-of-gravity principle,

$$\frac{\dot{m}_B}{\dot{m}_D} = \frac{\overline{DF'}}{\overline{F'B}} \qquad (18\text{-}42)$$

If $\dot{m}_F$ is known, $\dot{m}_B$ and $\dot{m}_D$ can be found from Eq. (18-42), since $\dot{m}_B + \dot{m}_D = \dot{m}_F$.

Application to multistage systems   The graphical procedures described above can be elaborated to permit calculation of the number of ideal stages in multistage systems.

The detailed application of these methods to distillation problems is discussed in Chap. 19.

Exactly similar constructions on liquid-liquid equilibrium diagrams can be used to solve problems in leaching and liquid-liquid extraction, as shown in Chap. 20.

Constructions on other equilibrium diagrams Graphical mixing rules and the center-of-gravity principle apply on temperature-composition diagrams but only for *isothermal* processes. For example, horizontal tie lines passing through points representing overall composition of mixtures can be drawn on diagrams such as Figs. 17-5 or 17-23. The ends of these tie lines show the compositions of the phases in equilibrium; the lengths of the line segments on each side of the point representing the overall composition indicate the relative quantities of the two phases.

The center-of-gravity principle does *not* apply to constructions on equilibrium *xy* diagrams such as Figs. 17-6, 17-11, or 17-17.

Equilibrium-Stage Calculations for Multicomponent Systems

For systems containing more than two or three components graphical procedures are ordinarily of little value, and the number of ideal stages required in a given problem must be found by algebraic calculations. These involve knowledge and application of the equilibrium relationships [as given by Eq. (17-21), for example]; material balances [Eqs. (18-1) and (18-2)]; and (sometimes) enthalpy balances [Eqs. (18-5) and (18-6)]. The calculations are begun at some point in the cascade where conditions are known. On the basis of certain assumptions the conditions on succeeding stages which satisfy the equilibrium requirements and the material and energy balances are found mathematically, usually by trial and error. The calculations are continued, stage by stage, until the desired terminal conditions are reached or, as often happens, it becomes evident that they cannot be reached. If this occurs, the underlying assumptions are modified and the entire calculation repeated until the problem is solved. The large number of repetitive calculations involved nearly always are done on a digital computer.

For preliminary calculations on multicomponent systems certain approximate methods are available which greatly reduce the labor involved, but the rigorous computation of multicomponent cascades is an important application of computer technique.

SYMBOLS

A Absorption factor, L/mV, dimensionless

a_1 First term of geometric series

B Constant in Eq. (18-11)

H Specific enthalpy, Btu/lb or J/g; H_F, of feed; H_L, of L phase; $H_{L,a}$, at entrance; $H_{L,b}$, at exit; $H_{L,n}$, of L phase leaving stage n; H_R, H_S, H_T, of streams R, S, and T, respectively; H_V, of V phase; $H_{V,a}$, at exit; $H_{V,b}$, at entrance; $H_{V,n+1}$, of V phase leaving stage $n + 1$

K Equilibrium ratio of mole fractions in vapor and liquid

L Flow rate of L phase, lb mol/h or kg mol/h; L_N, from final stage of cascade; L_a, at entrance; L_b, at exit; L_n, from stage n; L', flow rate of inert component in L phase

m Slope of equilibrium curve, dy_e/dx_e

$\dot{m}$ Mass flow rate, lb/h or kg/h; $\dot{m}_B$, of liquid; $\dot{m}_D$, of vapor; $\dot{m}_F$, of feed; $\dot{m}_R$, $\dot{m}_S$, $\dot{m}_T$, of streams R, S, and T, respectively

N Total number of ideal stages

n Serial number of ideal stage, counting from inlet of L phase

Q Quantity of heat, Btu or J

q Rate of heat flow, Btu/h or W

r Ratio of succeeding terms of geometric series

s_n Sum of first n terms of geometric series

V Flow rate of V phase, lb mol/h or kg mol/h; V_a, at exit; V_b, at entrance; V_{n+1}, from stage $n + 1$; V_1, leaving first stage of cascade; V', flow rate of inert component in V phase

W_s Shaft work, ft-lb$_f$ or J

x Mole fraction in L phase; used for component A when only two components are present; x_F, of feed; x_N, in L phase from final stage of cascade; x_R, x_S, x_T, in streams R, S, and T, respectively; x_a, at entrance; x_b, at exit; x_e, at equilibrium; x_n, mole fraction in L phase from stage n; x^*, mole fraction in L phase in equilibrium with specified stream of V phase; x_a^*, in equilibrium with y_a, x_b^*; in equilibrium with y_{bJ}; X Mole ratio in L phase; mol solute/mol inert component: X_a at entrance; X_b at exit

y Mole fraction in V phase; used for component A when only two components are present; y_{N+1}, in V phase entering stage N of cascade; y_a, at exit; y_b, at entrance; y_e, at equilibrium y_n, in V phase from stage n; y^*, mole fraction in V phase in equilibrium with specified stream of L phase; y_a^*, in equilibrium with x_a; y_b^*; in equilibrium with x_{bJ}; Y Mole ratio in V phase; Y_a at exit; Y_b at entrance; Y^* in equilibrium with specified stream of L phase

PROBLEM

18-1 What are the effects on the concentrations of the exit gas and liquid streams of the following changes in the operating conditions of the 15-plate column of Example 18-2? (*a*) A drop in the operating temperature that decreases K from 0.8 to 0.6. Unchanged factors: N, L/V, Y_b and X_a. (*b*) An increase in the L/V ratio from 1.1 to 2.0. Unchanged from original design: temperature, N, Y_b, and X_a. (*c*) An increase in number of stages from 15 to 20. Unchanged from original: temperature, L/V, Y_b, and X_a. (*d*) The use of pure water as the scrubbing liquid. Unchanged from original: temperature, L/V, Y_b.

REFERENCES

1   Brown, G. G., M. Souders, Jr., and H. V. Nyland: *Ind. Eng. Chem.,* **24:**522 (1932).
2   Kremser, A.: *Natl. Petr. News,* **22**(21):42 (May 21, 1930).
3   McCabe, W. L., and E. W. Thiele: *Ind. Eng. Chem.,* **17:**605 (1925).
4   Perry, J. H. (ed.): "Chemical Engineers' Handbook," 5th ed., p. **15**-20, McGraw-Hill, New York, 1973.

19

DISTILLATION

In practice, distillation may be carried out by either of two principal methods. The first method is based on the production of a vapor by boiling the liquid mixture to be separated and condensing the vapors without allowing any liquid to return to the still in contact with the vapors. The second method is based on the return of part of the condensate to the still under such conditions that this returning liquid is brought into intimate contact with the vapors on their way to the condenser. Either of these methods may be conducted as a continuous process or as a batch process. The first sections of this chapter deal with differential distillation and steam distillation, then with continuous steady-state distillation processes, including single-stage partial vaporization without reflux (flash distillation) and continuous distillation with reflux (rectification). The last section is concerned with the design and operation of distillation equipment.

Flash Distillation

Flash distillation consists of vaporizing a definite fraction of the liquid in such a way that the evolved vapor is in equilibrium with the residual liquid, separating the vapor from the liquid, and condensing the vapor. Figure 19-1 shows the elements of a

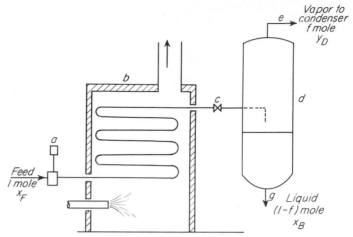

FIGURE 19-1
Plant for flash distillation.

flash-distillation plant. Feed is pumped by pump a through heater b, and the pressure is reduced through valve c. An intimate mixture of vapor and liquid enters the vapor separator d, in which sufficient time is allowed for the vapor and liquid portions to separate. Because of the intimacy of contact of liquid and vapor before separation, the separated streams are in equilibrium. Vapor leaves through line e and liquid through line g.

Flash distillation† of binary mixtures　Consider 1 mol of a two-component mixture fed to the equipment shown in Fig. 19-1. Let the concentration of the feed be x_F, in mole fraction of the more volatile component. Let f be the molal fraction of the feed that is vaporized and withdrawn continuously as vapor. Then $1 - f$ is the molal fraction of the feed that leaves continuously as liquid. Let y_D and x_B be the concentrations of the vapor and liquid, respectively. By a material balance of the more volatile component, based on 1 mol of feed, all of that component in the feed must leave in the two exit streams, or

$$x_F = fy_D + (1 - f)x_B \tag{19-1}$$

There are two unknowns in Eq. (19-1), x_B and y_D. To use the equation a second relationship between the unknowns must be available. Such a relationship is provided by the equilibrium curve, as y_D and x_B are coordinates of a point on this curve. If x_B and y_D are replaced by x and y, respectively, Eq. (19-1) can be written

$$y = -\frac{1 - f}{f}x + \frac{x_F}{f} \tag{19-2}$$

† Flash distillation is used on a large scale in petroleum refining, in which petroleum fractions are heated in pipe stills and the heated fluid flashed into overhead vapor and residual-liquid streams, each containing many components.

Equation (19-2) is the equation of a straight line with a slope of $-(1 - f)/f$ and can be plotted on the equilibrium diagram. The coordinates of the intersection of the line and the equilibrium curve are $x = x_B$ and $y = y_D$. The intersection of this material-balance line and the diagonal $x = y$ can be used conveniently as a point on the line. Letting $x = x_F$ in Eq. (19-2) gives

$$y = -\frac{1 - f}{f} x_F + \frac{x_F}{f}$$

from which $y = x_F = x$. The material-balance line crosses the diagonal at $x = x_F$ for all values of f.

EXAMPLE 19-1 A mixture of 50 mole percent benzene and 50 mole percent toluene is subjected to flash distillation at a separator pressure of 1 atm. Plot the following quantities, all as functions of f, the fractional vaporization: (a) the temperature in the separator, (b) the composition of the liquid leaving the separator, and (c) the composition of the vapor leaving the separator.

SOLUTION For each of several values of f corresponding quantities $- [(1/f) - 1]$ are calculated. Using these quantities as slopes, a series of straight lines each passing through point (x_F, x_F) is drawn on the equilibrium curve of Fig. 17-6. These lines, shown on Fig. 19-2, intersect the equilibrium curve at corresponding values of x_B and y_D. The temperature of each vaporization is then found from Fig. 17-5. The results for the lines shown in Fig. 19-2 are shown in Table 19-1 and plotted in Fig. 19-3. The lines in Fig. 19-3 are nearly straight. The limits for 0 and 100 percent vaporization are the *bubble* and *dew points*, respectively. ////

A flash distillation of a nonideal mixture can be shown graphically on an enthalpy-concentration diagram. An example is shown in Fig. 18-13.

Differential distillation Consider a batch of n_0 mol† of liquid charged to a differential distillation. Assume that at a given instant during the process there are

Table 19-1 DATA FOR EXAMPLE 19-1

Fraction vaporized f	Slope $-\dfrac{1 - f}{f}$	Concentration, mole fraction C_6H_6		Temperature, °C
		Liquid x_B	Vapor y_D	
0	∞	0.50	0.71	92.2
0.2	−4	0.455	0.67	93.7
0.4	−1.5	0.41	0.63	95.0
0.6	−0.67	0.365	0.585	96.5
0.8	−0.25	0.325	0.54	97.7
1.0	0	0.29	0.50	99.0

† Equations (19-3) to (19-5) will hold as well if compositions, including those on the equilibrium curve, are expressed as mass fractions.

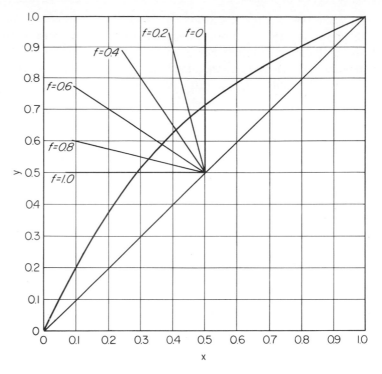

FIGURE 19-2
Graphical construction for Example 19-1.

n mol of liquid left in the still. At this instant let the vapor composition be y and the liquid composition be x. The total number of moles of component A left in the still, n_A, will be

$$n_A = xn \qquad (19\text{-}3)$$

Assume that a small amount of liquid, dn mol, is vaporized. The amount of component A in this vapor, $-dn_A$, will be $-y\, dn$. The minus signs denote the fact that n_A is decreasing during the process. Differentiating Eq. (19-3) gives

$$dn_A = d(xn) = n\, dx + x\, dn \qquad (19\text{-}4)$$

Hence
$$n\, dx + x\, dn = y\, dn$$

By rearrangement,

$$\frac{dn}{n} = \frac{dx}{y - x} \qquad (19\text{-}5)$$

If Eq. (19-5) is integrated between the limits of n_0, the initial charge, and n_1, the final mass of the liquid in moles, on the left-hand side and the limits x_0, the initial con-

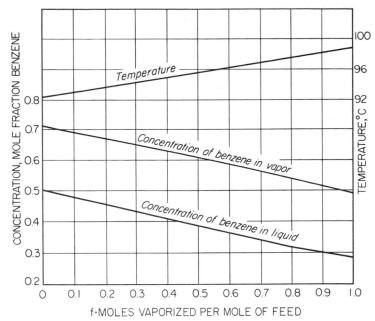

FIGURE 19-3
Results for Example 19-1.

centration of the liquid, and x_1, the final concentration, on the right-hand side, the result is

$$\int_{n_1}^{n_0} \frac{dn}{n} = \int_{x_1}^{x_0} \frac{dx}{y - x} = \ln \frac{n_0}{n_1} \qquad (19\text{-}6)$$

Equation (12-6) is known as the *Rayleigh equation*. The function $dx/(y - x)$ can be integrated numerically from the equilibrium curve, since this curve provides a relationship between x and y.

A second relationship for differential distillation is useful when the relative volatility is constant. Thus, assume that the original charge in the still consists of n_{0A} mol of component A and n_{0B} mol of component B. Assume that the moles of A and B remaining in the still at any instant during the process are n_A and n_B, respectively. Consider now the vaporization of a small amount of vapor from the liquid. As before, the amount of A in this vapor, $-dn_A$, equals $-y\, dn$; since $y_A + y_B = 1$ in the two-component system, the amount of B in the vapor, $-dn_B$, equals $-(1 - y)\, dn$. Thus

$$\frac{-dn_A}{-dn_B} = \frac{y}{1 - y} \qquad (19\text{-}7)$$

From Eq. (17-28) for the relative volatility of binary mixtures, Eq. (19-7) can be written

$$\frac{dn_A}{dn_B} = \alpha_{AB} \frac{x}{1 - x} \qquad (19\text{-}8)$$

Since $x = n_A/(n_A + n_B)$ and $1 - x = n_B/(n_A + n_B)$, Eq. (19-8) becomes, after rearrangement,

$$\frac{dn_A}{n_A} = \alpha_{AB} \frac{dn_B}{n_B} \tag{19-9}$$

If α_{AB} is constant, Eq. (19-9) can be integrated between limits as follows:

$$\frac{1}{\alpha_{AB}} \int_{n_{0A}}^{n_A} \frac{dn_A}{n_A} = \int_{n_{0B}}^{n_B} \frac{dn_B}{n_B}$$

Thus

$$\ln \frac{n_B}{n_{0B}} = \frac{1}{\alpha_{AB}} \ln \frac{n_A}{n_{0A}} \tag{19-10}$$

or

$$\frac{n_B}{n_{0B}} = \left(\frac{n_A}{n_{0A}}\right)^{1/\alpha_{AB}} \tag{19-11}$$

Equation (19-10) shows that if n_B/n_{0B} is plotted as the ordinate vs. n_A/n_{0A} as the abscissa on logarithmic coordinates, a straight line with a slope $1/\alpha_{AB}$ is obtained. Also, since, at the start of the distillation, $n_A = n_{0A}$ and $n_B = n_{0B}$, both n_B/n_{0B} and n_A/n_{0A} initially are unity and the line passes through the point $(1, 1)$. The entire course of any differential distillation at constant α_{BA} is therefore represented by a straight line on logarithmic coordinates with a slope of $1/\alpha_{AB}$ and passing through point $(1, 1)$.

EXAMPLE 19-2 A mixture of 50 mole percent benzene and 50 mole percent toluene is subjected to differential distillation at a pressure of 1 atm. Plot the following quantities, each against f, the mole fraction of charge distilled: (a) the temperature in the still, (b) the composition of liquid in the still, (c) the instantaneous composition of vapor leaving the still, and (d) the cumulative average composition of the condensate.

SOLUTION From Fig. 17-5 the initial boiling temperature of the liquid is 92.2°C. From the vapor-pressure data in Table 17-1, $1/\alpha_{AB}$ is 0.41 at 92.2°C. It will be found that the last liquid to be vaporized is pure toluene, and the temperature at the end of the distillation is therefore 110.6°C, which is the boiling point of toluene. At this temperature $1/\alpha_{AB}$ is 0.42. An average of 0.41 for $1/\alpha_{AB}$ can be used as a constant. Figure 19-4 is a logarithmic plot of n_B/n_{0B} vs. n_A/n_{0A}. A straight line with a slope of 0.41 and passing through point $(1, 1)$ is drawn on Fig. 19-4. After choosing as a basis 1 mol of initial charge and assuming a series of values of n_A, for each choice of n_A the quantity n_A/n_{0A} is calculated, n_B/n_{0B} is read from Fig. 19-4, and n_B is then found. From n_A and n_B the mole fractions of benzene in the liquid are calculated, and the temperatures and instantaneous vapor concentrations are then read from Fig. 17-5. The cumulative average concentrations of the distillate are calculated from the total moles and the moles of benzene vaporized. The results of the calculations are shown in detail in Table 19-2, and the desired curves are plotted in Fig. 19-5. ////

Differential distillation is approached in commercial batch distillations, where the vapor is removed as fast as it is formed without appreciable condensation. Although as a method of separation this process is not effective, many such stills are used, especially where the components to be separated have widely different boiling points and where sharp separations are not required. Also, most laboratory distillations conducted without reflux are of this type.

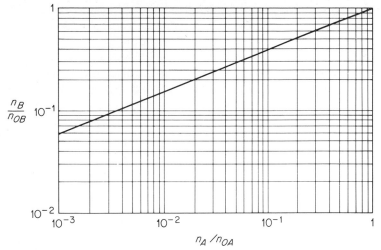

FIGURE 19-4
Differential-distillation plot for Example 19-2.

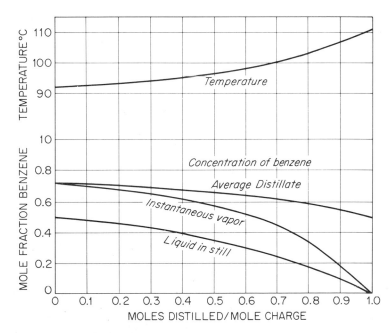

Figure 19-5
Compositions and temperatures for Example 19-2.

Steam distillation A special distillation problem is encountered when the boiling point of the material is so high that the substance decomposes at its atmospheric boiling point or the vaporization temperature cannot easily be reached by steam heat. To distill such a material, either to free it from nonvolatile impurities or to separate it into fractions of differing boiling point, the distillation temperature must be lowered to a point where the liquid can be vaporized without reaching excessive temperatures. One method of accomplishing this is to operate under a vacuum. This method, using high-temperature heating and efficient vacuum-producing equipment, is extensively used in both batch and continuous processes. An important older method which utilizes simpler equipment is steam distillation. In this method, the temperature of the liquid is reduced by vaporizing the material into a stream of carrier vapor, the liquid phase of which is immiscible with the material distilled. A mixed vapor of carrier and vaporized material is taken overhead and condensed, and the liquid layers, each containing one of the components, are separated by gravity. Although any inert carrier can be used, for economy steam is the usual choice. Steam distillation can be combined with rectification, usually in continuous fractionating columns, often under a partial vacuum. Lubricating oils are treated in this manner. The following discussion is limited to batch steam distillation conducted to free a material from a nonvolatile impurity.

THEORY OF STEAM DISTILLATION To conduct a steam distillation the material to be distilled is charged to a batch still, which is equipped with an overhead vapor line, a condenser, a condensate receiver, and gravity separator. Steam is admitted through a perforated pipe in the bottom of the still to give maximum contact between steam and the charge. The process may be operated under either of two conditions. In the first, the only energy supplied to the still is that in the carrier. Some of the carrier is condensed to provide heat to raise the temperature of the still and contents to the operating level, to supply heat of vaporization of the material, and to compensate for heat

Table 19-2 DATA FOR EXAMPLE 19-2

n_A	$\dfrac{n_A}{0.500}$	$\dfrac{n_B}{0.500}$	n_B	$n_A + n_B$	x	T, °C	Instantaneous y	$0.50 - n_A$	Total condensate	y_{cum}
0.50	1.00	1.00	0.50	1.00	0.50	92.2	0.71	0	0	0.71
0.40	0.80	0.91	0.455	0.855	0.47	93.2	0.68	0.10	0.145	0.69
0.30	0.60	0.80	0.40	0.70	0.43	94.5	0.64	0.20	0.30	0.67
0.20	0.40	0.68	0.34	0.54	0.37	96.2	0.59	0.30	0.46	0.65
0.10	0.20	0.51	0.255	0.355	0.28	99.2	0.49	0.40	0.645	0.62
0.05	0.10	0.38	0.19	0.24	0.21	102.0	0.395	0.45	0.76	0.59
0.03	0.06	0.31	0.155	0.185	0.16	104.7	0.325	0.47	0.815	0.58
0.01	0.02	0.20	0.10	0.11	0.091	106.7	0.205	0.49	0.89	0.55
0.005	0.01	0.15	0.075	0.080	0.062	108.2	0.14	0.495	0.92	0.54
0	0	0	0	0	0	110.6	0	0.50	1.00	0.50

losses. Another liquid layer, of condensed carrier, then collects in the still. In the second condition, either the inlet carrier is considerably superheated or additional heat is supplied to the charge by a closed heating coil and formation of liquid thereby prevented. All vapor blown through the liquid then passes out with the product. In either method, when the sum of the partial pressures of carrier and material reaches the total pressure, both substances pass over in the molecular ratio of their partial pressures and nonvolatile impurities remain behind in the still. The mass ratio of carrier to material is

$$\frac{\dot{m}_A}{\dot{m}_B} = \frac{n_A M_A}{n_B M_B} = \frac{\bar{p}_A M_A}{\bar{p}_B M_B} = \frac{(p - \bar{p}_B) M_A}{\bar{p}_B M_B} \tag{19-12}$$

where $\dot{m}_A$ = mass flow rate of carrier in vapor
$\dot{m}_B$ = mass flow rate of material in vapor
$\bar{p}_A$ = partial pressure of carrier
$\bar{p}_B$ = partial pressure of material
M_A = molecular weight of carrier
M_B = molecular weight of material
n_A = moles of carrier in vapor
n_B = moles of material in vapor
p = total pressure

Assume that the effect of the nonvolatile matter on the vaporization of the material may be neglected. Then if two liquid layers are allowed to form, there will be three phases (one vapor and two liquids) and two components (material and carrier). By the phase rule [Eq. (17-1)], the system has one degree of freedom. Either the temperature or the pressure—but not both—may be fixed arbitrarily. If the pressure is atmospheric, the temperature so adjusts itself that the sum of the partial pressures of the two components equals the total pressure of 1 atm. This temperature is lower than the boiling point of either pure component. If steam is the inert carrier and the pressure is atmospheric, the temperature is always less than 100°C and destructive temperatures are not reached.

If the formation of the second liquid layer is prevented, there are two components, two phases, and therefore two degrees of freedom. Both pressure and temperature must be fixed independently before the operation is defined. The temperature of the vapor carrier leaving the still is above that corresponding to its condensing point, and this vapor is therefore superheated. In practice, the pressure is either atmospheric or, if below atmospheric, is controlled by the action of the condenser and vacuum pump, just as in an evaporator operated with a surface condenser. The temperature may be fixed either by a temperature controller or by controlling the pressure of the steam in the closed coil used to supply supplementary heat.

Assuming that Dalton's law is applicable, the equilibrium partial pressure of the material is the vapor pressure of the pure material, and if liquid carrier is also present, its partial pressure is the vapor pressure of the carrier. Both vapor pressures are taken at the temperature of the liquid in the still. In practice, contact between carrier and material is not perfect, and the carrier does not reach equilibrium with the liquid, so the actual partial pressure of the material is less than its vapor pressure. A vaporization efficiency η_s, defined by Eq. (19-13), accounts for the difference between theory and practice:

$$\eta_s = \frac{\bar{p}_B}{p_B} \tag{19-13}$$

where p_B is the vapor pressure of the material. Equation (19-12) can be written

$$\frac{\dot{m}_A}{\dot{m}_B} = \frac{(p - \eta_s p_B)M_A}{\eta_s p_B M_B} = \frac{M_A}{\eta_s M_B}\left(\frac{p}{p_B} - \eta_s\right) \tag{19-14}$$

STEAM CONSUMPTION   The minimum steam consumption (or, more generally, consumption of carrier vapor) is given by Eq. (19-14) when η_s is taken as unity. The actual steam consumption is greater than this because the vaporization efficiency is less than 1, and if supplementary heat is not supplied, steam is also needed to supply the heat requirement of the process. If a second liquid layer forms, the minimum steam consumption per pound of material for the distillation itself is

$$\frac{\dot{m}_A}{\dot{m}_B} = \frac{p_A M_A}{p_B M_B} = \alpha_{AB} \frac{M_A}{M_B} \tag{19-15}$$

where p_A = vapor pressure of carrier
$\quad\quad p_B$ = vapor pressure of material
$\quad\quad \alpha_{AB} = p_A/p_B$

The total pressure is

$$p = p_A + p_B \tag{19-16}$$

Since both p_A and p_B depend only on temperature, a reduction of temperature reduces the total pressure. The change in steam consumption w_A/w_B with pressure depends on the effect of temperature on the relative volatility α_{AB}, which may increase, decrease, or remain practically constant as the temperature is changed. For example, the theoretical steam consumption for steam distilling turpentine is shown in Fig. 19-6. The steam consumption for operation with two liquid phases is given by the top line, which shows that in this system α is practically constant and the theoretical steam consumption is nearly independent of pressure. In other systems α may increase or decrease with an increase in pressure, but the change is small, and the theoretical consumption is not sensitive to change in pressure when two liquid layers are present.

If there is no condensation of carrier, Eq. (19-14) shows that when the total pressure is constant, an increase in temperature always reduces the steam consumption, both theoretical and actual. Likewise, if the temperature is constant and the pressure is lowered, the steam consumption is diminished. For minimum steam consumption, the highest permissible temperature and the lowest practicable pressure should be used. The limit of reduction in pressure is reached when the total pressure is the vapor pressure of the material distilled and the steam consumption vanishes.

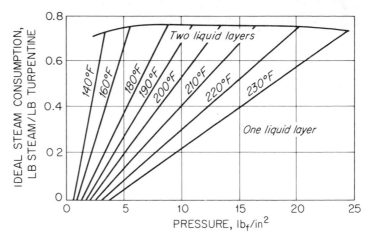

FIGURE 19-6
Steam-consumption diagram for steam distillation of turpentine.

The process then becomes a vacuum distillation. The effects of changes in temperature and pressure on the steam required to distill 1 lb of turpentine under ideal conditions are shown by the slanting straight lines in Fig. 19-6. In practice, a moderate vacuum readily attainable with the available cooling water often is used combined with the use of steam. In some cases, such as the distillation of fatty acids, high vacuum and steam are both required to obtain vaporization at nondestructive temperatures.

EXAMPLE 19-3   A batch of 250 gal (0.946 m³) of turpentine is to be steam-distilled at 1 atm pressure, using saturated steam at a gauge pressure of 5 $lb_f/in.^2$ (3.45 kN/m²). The initial temperature of the charge is 70°F (21.1°C). The vaporization efficiency may be taken as 0.85. The following properties of turpentine may be used:

Molecular weight, 140
Latent heat of vaporization, 128 Btu/lb (298 J/g)
Specific gravity, 0.87
Specific heat, including heat absorbed by still, 0.50 Btu/lb-°F (2.09 J/g-°C)

Table 19-3   VAPOR PRESSURE OF TURPENTINE

Temperature, °F	Vapor pressure, $lb_f/in.^2$	Temperature, °F	Vapor pressure, $lb_f/in.^2$
140	0.51	190	1.59
150	0.65	200	1.97
160	0.81	210	2.42
170	1.01	220	2.96
180	1.28	230	3.64

Neglecting radiation losses, calculate the pounds of steam required (*a*) if no additional heat is supplied and (*b*) if the temperature of the batch is brought to, and held at, the temperature of the inlet steam by means of another heat source. Assume that no condensation occurs in the vapor line.

SOLUTION   (*a*)   First, the temperature in the still is calculated. It can be assumed that the partial pressure of the carrier steam leaving the still is p_A, the vapor pressure of water at the temperature of the liquid in the still. The partial pressure of the turpentine is $0.85p_B$, where p_B is the vapor pressure of turpentine at liquid temperature. Then

$$p = 0.85p_B + p_A = 14.7$$

The vapor pressure of water is given in Appendix 8 and that of turpentine in Table 19-3. Each vapor pressure depends only on temperature. The total pressure p is calculated for several temperatures and plotted against temperature. The temperature at which this line crosses the line $p = 14.7$ is the desired result. A temperature of 205°F is found. The vapor pressure of turpentine at 205°F is 2.18 $lb_f/in.^2$

The steam consumption is calculated in two increments: that required as a carrier and that required for thermal purposes. The turpentine to be vaporized is $250 \times 8.33 \times 0.87 = 1,810$ lb. The steam required for vaporization is, by Eq. (19-14),

$$\frac{m_A}{m_B} = \frac{18}{0.85 \times 140}\left(\frac{14.7}{2.18} - 0.85\right) = 0.89 \text{ lb/lb charge}$$

The total carrier steam is $0.89 \times 1,810 = 1,610$ lb. The heat required in the process is

Heat charge and still, 60 to 140°F: $1,810(205 - 60)(0.50) = 131,000$ Btu

Vaporize turpentine: $1,810 \times 128 = 232,000$

$$\text{Total} = 363,000 \text{ Btu}$$

The carrier steam contributes some heat. From Appendix 8, the enthalpy of 1 lb of saturated steam at a gauge pressure of 5 $lb_f/in.^2$ is 1156.0 Btu, and that of saturated steam at 1 atm is 1147.8 Btu. The heat recovered from the carrier steam is $1,610(1156.0 - 1147.8) = 13,000$ Btu. The heat supplied by heating steam must be $363,000 - 13,000 = 350,000$ Btu.

Each pound of heating steam is condensed to liquid water at 205°F. The enthalpy of water at this temperature is 173.0 Btu/lb. The steam required for heating is

$$\frac{350,000}{1156.0 - 173.0} = 355 \text{ lb}$$

The total steam consumption is $1,610 + 355 = 1,965$ lb. This is $1,965/1,810 = 1.09$ lb per pound of turpentine. The theoretical steam consumption is, from the top line of Fig. 19-6, 0.75 lb per pound of turpentine.

(*b*)   The steam required is only that needed to carry over the vapors, as the thermal requirement is supplied by another heat source. Steam at 5 $lb_f/in.^2$ gauge condenses at 227°F. At this temperature the vapor pressure of turpentine is 3.5 $lb_f/in.^2$ The steam requirement, by Eq. (19-14), is

$$\frac{m_A}{m_B} = \frac{18}{0.85 \times 140}\left(\frac{14.7}{3.5} - 0.85\right) = 0.51 \text{ lb/lb turpentine}$$

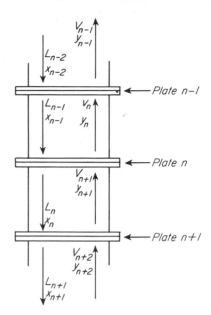

FIGURE 19-7
Material-balance diagram for plate n.

The theoretical steam consumption, computed by setting $\eta_s = 1$, is

$$\frac{m_A}{m_B} = \frac{(14.7 - 3.5)(18)}{3.5 \times 140} = 0.41 \text{ lb/lb turpentine}$$

Of course, any fair comparison of operations (*a*) and (*b*) should take account of the secondary heat source assumed in part (*b*). ////

Continuous Distillation with Rectification

Flash distillation is used most for separating components which boil at widely different temperatures. It is not effective in separating components of comparable volatility, since then both the condensed vapor and residual liquid are far from pure. By many successive redistillations small amounts of some nearly pure components may finally be obtained, but this method is too inefficient for industrial distillations when nearly pure components are wanted. Modern methods used in both laboratory and plant apply the principle of rectification, which is described in this section.

Rectification on an ideal plate Consider a single plate in a column or cascade of ideal plates. Assume that the plates are numbered serially from the top down and that the plate under consideration is the nth plate from the top. It is shown diagrammatically in Fig. 19-7. Then the plate immediately above this plate is plate $n - 1$, and that immediately below it is plate $n + 1$. Subscripts are used on all quantities showing the point of origin of the quantity.

Two fluid streams enter plate n, and two leave it. A stream of liquid, L_{n-1} mol/h, from plate $n - 1$, and a stream of vapor, V_{n+1} mol/h, from plate $n + 1$, are brought

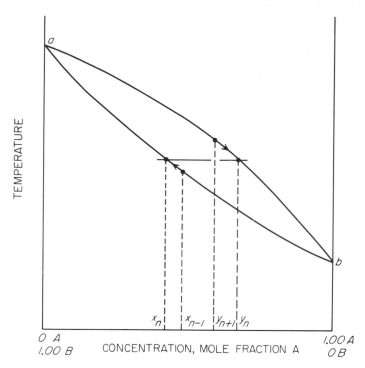

FIGURE 19-8
Boiling-point diagram showing rectification on ideal plate.

into intimate contact. A stream of vapor, V_n mol/h, rises to plate $n - 1$, and a stream of liquid, L_n mol/h, descends to plate $n + 1$. Since the vapor streams are the V phase, their concentrations are denoted by y, and since the liquid streams are the L phase, their concentrations are denoted by x. Then the concentrations of the streams entering and leaving the nth plate are

Vapor leaving plate, y_n
Liquid leaving plate, x_n
Vapor entering plate, y_{n+1}
Liquid entering plate, x_{n-1}

Figure 19-8 shows the boiling-point diagram for the mixture being treated. The four concentrations given above are shown in this figure. By definition of an ideal plate, the vapor and liquid leaving plate n are in equilibrium, so x_n and y_n represent equilibrium concentrations. This is shown in Fig. 19-8. Since concentrations in both phases increase with the height of the column, x_{n-1} is greater than x_n and y_n is greater than y_{n+1}. Although the streams leaving the plate are in equilibrium, those entering it are not. This can be seen from Fig. 19-8. When the vapor from plate $n + 1$ and the liquid from plate $n - 1$ are brought into intimate contact, their concentrations tend to move toward an equilibrium state, as shown by the arrows in Fig. 19-8. Some of the more volatile component A is vaporized from the liquid, decreasing the liquid concentration from x_{n-1} to x_n; and some of the less volatile component B is condensed from the vapor, increasing the vapor concentration from y_{n+1} to y_n. Since the liquid streams are at their bubble points and the vapor streams at their dew points, the heat necessary to vaporize component A must be supplied by the heat released in the condensation of component B. Each plate in the cascade acts as an interchange apparatus in which component A is transferred to the vapor stream and component B to the liquid stream. Also, since the concentration in both liquid and vapor increases with column height, the temperature decreases, and the temperature of plate n is greater than that of plate $n - 1$ and less than that of plate $n + 1$.

Combination rectification and stripping The plant described on pages 511 to 514, in which the feed to the unit is supplied to the still, cannot produce a nearly pure bottom product because the liquid in the still is not subjected to rectification. This limitation is removed by admitting the feed to a plate in the central portion of the column. Then the liquid feed flows down the column to the still, which in this type of plant is called the *reboiler*, and is subjected to rectification by the vapor rising from the reboiler. Since the liquid reaching the reboiler is stripped of component A, the bottom product can be nearly pure B.

A typical continuous fractionating column equipped with the necessary auxiliaries and containing rectifying and stripping sections is shown in Fig. 19-9. The column A is fed near its center with a definite flow of feed of definite concentration. Assume that the feed is a liquid at its boiling point. The action in the column is not dependent on this assumption, and other conditions of the feed will be discussed later. The plate on which the feed enters is called the *feed plate*. All plates above the feed plate constitute the rectifying section, and all plates below the feed, *including the feed plate itself*, constitute the stripping section. The feed flows down the stripping section to the bottom of the column, in which a definite level of liquid is maintained. Liquid flows by gravity to reboiler B. This is a steam-heated vaporizer which generates vapor and returns it to the bottom of the column. The vapor passes up the entire column. At one end of the reboiler is a weir. The bottom product is withdrawn from the pool of liquid on the downstream side of the weir, and flows through the cooler G. This cooler also preheats the feed by heat exchange with the hot bottoms.

The vapors rising through the rectifying section are completely condensed in condenser C, and the condensate is collected in accumulator D, in which a definite liquid level is maintained. Reflux pump F takes liquid from the accumulator and delivers it to the top plate of the tower. This liquid stream is called *reflux*. It provides the downflowing liquid in the rectifying section that is needed to act on the upflowing vapor. Without the reflux no rectification would occur in the rectifying section and the concentration of the overhead product would be no greater than that of the vapor rising from the feed plate. Condensate not picked up by the reflux pump is cooled in heat exchanger E, called the *product cooler*, and withdrawn as the overhead product. If no azeotropes are encountered, both overhead and bottom products may be obtained in any desired purity if enough plates and adequate reflux are provided.

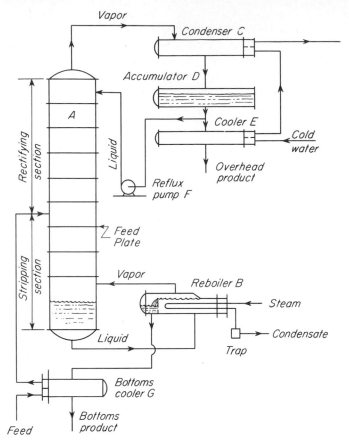

Figure 19-9
Continuous fractionating column with rectifying and stripping sections.

The plant shown in Fig. 19-9 is often simplified, especially in general chemical operations. In place of the reboiler, a heating coil may be placed in the bottom of the column and generate vapor from the pool of liquid there. The condenser is often placed above the top of the column and the reflux pump and accumulator omitted. Reflux then returns to the top plate by gravity. A special valve, called a *reflux splitter*, may be used to control the rate of reflux return. The remainder of the condensate forms the overhead product.

DESIGN AND OPERATING CHARACTERISTICS OF PLATE COLUMNS

Important factors in the design and operation of plate columns are the number of plates required to obtain the desired separation, the diameter of the column, the heat input to the reboiler and the heat output from the condenser, the spacing between the plates, the choice of type of plate, and the details of construction of the plates.

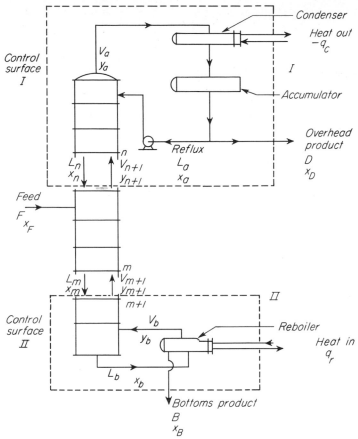

FIGURE 19-10
Material-balance diagram for continuous fractionating column.

In accordance with general principles, the analysis of the performance of plate columns is based on material balances, energy balances, and phase equilibria.

Overall material balances for two-component systems Figure 19-10 is a material-balance diagram for a typical continuous-distillation plant. The column is fed with F mol/h of concentration x_F and delivers D mol/h of overhead product of concentration x_D and B mol/h of bottom product of concentration x_B. Two independent overall material balances can be written:

Total-material balance: $$F = D + B \qquad (19\text{-}17)$$

Component A balance: $$Fx_F = Dx_D + Bx_B \qquad (19\text{-}18)$$

Eliminating B from these equations gives

$$\frac{D}{F} = \frac{x_F - x_B}{x_D - x_B} \qquad (19\text{-}19)$$

Eliminating D gives

$$\frac{B}{F} = \frac{x_D - x_F}{x_D - x_B} \tag{19-20}$$

Equations (19-19) and (19-20) are true for all values of the flows of vapor and liquid within the column.

Net flow rates Quantity D is the difference between the flow rates of the streams entering and leaving the top of the column. A material balance around the condenser in Fig. 19-10 gives

$$D = V_a - L_a \tag{19-21}$$

The difference between the flow rates of vapor and liquid anywhere in the upper section of the column is also equal to D, as shown by considering the part of the plant enclosed by control surface I in Fig. 19-10. This surface includes the condenser and all plates above $n + 1$. A total-material balance around this control surface gives

$$D = V_{n+1} - L_n \tag{19-22}$$

Thus quantity D is the *net flow rate* of material upward in the upper section of the column. Regardless of changes in V and L, their difference is constant and equal to D.

Similar material balances for component A give the equations

$$Dx_D = V_a y_a - L_a x_a = V_{n+1} y_{n+1} - L_n x_n \tag{19-23}$$

Quantity Dx_D is the net flow rate of component A upward in the upper section of the column. It, too, is constant throughout this part of the equipment.

In the lower section of the column the net flow rates are also constant but are in a downward direction. The net flow rate of total material equals B; that of component A is Bx_B. The following equations apply:

$$B = L_b - V_b = L_m - V_{m+1} \tag{19-24}$$

$$Bx_B = L_b x_b - V_b y_b = L_m x_m - V_{m+1} y_{m+1} \tag{19-25}$$

Subscript m is used in place of n to designate a general plate in the stripping section.

Operating lines Because there are two sections in the column, there are also two operating lines, one for the rectifying section and the other for the stripping section. Consider first the rectifying section. As shown in Chap. 18 [Eq. (18-7)], the operating line for this section is

$$y_{n+1} = \frac{L_n}{V_{n+1}} x_n + \frac{V_a y_a - L_a x_a}{V_{n+1}} \tag{19-26}$$

Substitution for $V_a y_a - L_a x_a$ from Eq. (19-23) gives

$$y_{n+1} = \frac{L_n}{V_{n+1}} x_n + \frac{Dx_D}{V_{n+1}} \tag{19-27}$$

The slope of the line defined by Eq. (19-27) is, as usual, the ratio of the flow of liquid stream to that of the vapor stream. For further analysis it is convenient to eliminate V_{n+1} from Eq. (19-27) by Eq. (19-22), giving

$$y_{n+1} = \frac{L_n}{L_n + D} x_n + \frac{Dx_D}{L_n + D} \qquad (19\text{-}28)$$

For the section of the column below the feed plate, Eq. (18-7) becomes

$$y_{m+1} = \frac{L_m}{V_{m+1}} x_m + \frac{V_b y_b - L_b x_b}{V_{m+1}} \qquad (19\text{-}29)$$

As shown by Eq. (19-25), quantity $V_b y_b - L_b x_b$ equals $-Bx_B$, and

$$y_{m+1} = \frac{L_m}{V_{m+1}} x_m - \frac{Bx_B}{V_{m+1}} \qquad (19\text{-}30)$$

This is the equation for the operating line in the stripping section. Again the slope is the ratio of the liquid flow to the vapor flow. Eliminating V_{m+1} from Eq. (19-30) by Eq. (19-24) gives

$$y_{m+1} = \frac{L_m}{L_m - B} x_m - \frac{Bx_B}{L_m - B} \qquad (19\text{-}31)$$

Analysis of Fractioning Columns by McCabe-Thiele Method

When the operating lines represented by Eqs. (19-28) and (19-31) are plotted with the equilibrium curve on the xy diagram, the McCabe-Thiele step-by-step construction can be used to compute the number of *ideal* plates needed to accomplish a definite concentration difference in either the rectifying or the stripping section.[8] It can be seen, however, by inspection of Eqs. (19-28) and (19-31), that unless L_n and L_m are constant, the operating lines are curved and can be plotted only if the change in these internal streams with concentration is known. Enthalpy balances are required in the general case to determine the change in slope of an operating line. With the aid of an enthalpy-concentration chart and its tie lines, an operating line can be located precisely. The method is described later in this chapter.

Constant molal overflow In many distillations the use of mole units for flow quantities and concentrations in enthalpy balances gives operating lines that are nearly straight. On this fact is based the concept of *constant molal overflow*,[6] which means simply that in Eqs. (19-22) to (19-31) subscripts n, $n + 1$, $n - 1$, m, $m + 1$, and $m - 1$ may be dropped. In this simplified model the material-balance equations are linear and the operating lines straight. An operating line can be plotted if the coordinates of two points on it are known. Then the McCabe-Thiele method is used without requiring enthalpy balances.

Reflux ratio The analysis of fractionating columns is facilitated by the use of a quantity called the *reflux ratio*. Two such quantities are used. One is the ratio of the reflux to the overhead product, and the other is the ratio of the reflux to the vapor. Both ratios refer to quantities in the rectifying column. The equations for these ratios are

$$R_D = \frac{L}{D} = \frac{V - D}{D} \quad \text{and} \quad R_V = \frac{L}{V} = \frac{L}{L + D} \tag{19-32}$$

In this text only R_D will be used.

If both numerator and denominator of the terms on the right-hand side of Eq. (19-28) are divided by D, the result is, for constant molal overflow,

$$y_{n+1} = \frac{R_D}{R_D + 1} x_n + \frac{x_D}{R_D + 1} \tag{19-33}$$

Equation (19-33) is an equation for the operating line of the rectifying column. The y intercept of this line is $x_D/(R_D + 1)$. The concentration x_D is set by the conditions of the design; and R_D, the reflux ratio, is an operating variable that can be controlled at will by varying the flow rate of the reflux for a given flow rate of the overhead product.

Condenser and top plate   Equation (19-28), written for constant molal overflow and dropping the subscripts n and $n + 1$, becomes

$$y = \frac{L}{L + D} x + \frac{Dx_D}{L + D} \tag{19-34}$$

The intersection of the operating line represented by Eq. (19-34) and the diagonal represented by the equation $x = y$ is found by simultaneous solution of the two equations. Eliminating y gives

$$x = \frac{L}{L + D} x + \frac{Dx_D}{L + D}$$

On simplification this equation gives

$$x = x_D$$

The operating line for the rectifying column, then, intersects the diagonal at point (x_D, x_D). This is true regardless of the performance of the condenser, only provided that but one overhead product is obtained.

The McCabe-Thiele construction for the top plate does depend on the action of the condenser. Figure 19-11 shows material-balance diagrams for the top plate and the condenser. The concentration of the vapor from the top plate is y_1, and that for the reflux to the top plate is x_c. In accordance with the general properties of operating lines, the upper terminus of the line is at the point (x_c, y_1).

The simplest arrangement for obtaining reflux and liquid product, and one that is frequently used, is the single total condenser shown in Fig. 19-11b, which condenses all the vapor from the column and supplies both reflux and product. When such a single total condenser is used, the concentrations of the vapor from the top plate, of the reflux to the top plate, and of the overhead product are equal and can all be denoted by x_D. The operating terminus of the operating line becomes point (x_D, x_D),

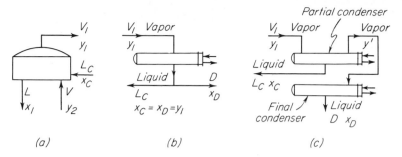

FIGURE 19-11
Material-balance diagrams for top plate and condenser: (*a*) top plate; (*b*) total condenser; (*c*) partial and final condensers.

which is the intersection of the operating line with the diagonal. Triangle *abc* in Fig. 19-12*a* then represents the top plate.

When a partial condenser, or *dephlegmator*, is used, the liquid reflux does not have the same composition as the overhead product; that is, $x_c \neq x_D$. Sometimes two condensers are used in series, first a partial condenser to provide reflux, then a final condenser to provide liquid product. Such an arrangement is shown in Fig. 19-11*c*. Vapor leaving the partial condenser has a composition y', which is the same as x_D. Under these conditions the diagram in Fig. 19-12*b* applies. The operating line passes through the point (x_D, x_D) on the diagonal, but as far as the column is concerned, the operating line ends at point *a′*, which of course has the coordinates (x_c, y_1). Triangle *a′b′c′* in Fig. 19-12*b* represents the top plate in the column. Since the vapor leaving

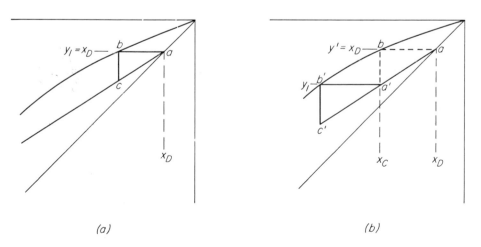

FIGURE 19-12
Graphical construction for top plate: (*a*) using total condenser; (*b*) using partial and final condensers.

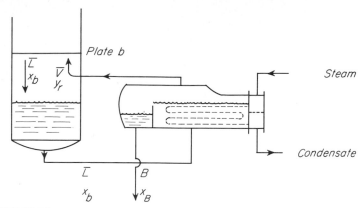

FIGURE 19-13
Material-balance diagram for bottom plate and reboiler.

a partial condenser is normally in equilibrium with the liquid condensate, the vapor composition y' is the ordinate value of the equilibrium curve where the abscissa is x_c, as shown in Fig. 19-12b. The partial condenser, represented by the dotted triangle aba' in Fig. 19-12b, is therefore equivalent to an additional theoretical stage in the distillation apparatus.

It is often assumed that the condenser removes latent heat only and that the condensate is liquid at its bubble point. Then the reflux L is equal to L_c, the reflux from the condenser, and $V = V_1$. If the reflux is cooled below the bubble point, a portion of the vapor coming to plate 1 must condense to heat the reflux; so $V_1 < V$ and $L > L_c$. The assumption of bubble-point reflux is made in the following treatment.

Bottom plate and reboiler   The action at the bottom of the column is analogous to that at the top. Thus, Eq. (19-31), written for constant molal overflow, becomes

$$y = \frac{\bar{L}}{\bar{L} - B} x - \frac{B x_B}{\bar{L} - B}$$

Substitution of $y = x$ in this equation and simplifying gives

$$x = x_B$$

The operating line in the stripping column crosses the diagonal at point (x_B, x_B). This is true regardless of the operation of the reboiler, provided only that but one stream leaves the bottom.

The material-balance diagram for the bottom plate and the reboiler is shown in Fig. 19-13. The terminus of the operating line is point (x_b, y_r), where x_b and y_r are the concentrations of liquid from the bottom plate and of vapor from the reboiler, respectively.

In the common type of reboiler shown in Figs. 19-9 and 19-13, the vapor leaving the reboiler is in equilibrium with the liquid leaving as bottom product. Then x_B and y_r are coordinates of a point on the equilibrium curve, and the reboiler acts as an ideal

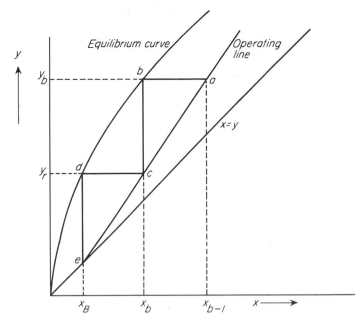

FIGURE 19-14
Graphical construction for bottom plate and reboiler: triangle *cde*, reboiler; triangle *abc*, bottom plate.

plate. In Fig. 19-14 is shown the graphical construction for the reboiler (triangle *cde*) and the bottom plate (triangle *abc*). Such a reboiler is called a *partial reboiler*. Its construction is shown in detail in Fig. 15-10.

Feed plate The addition of the feed increases the reflux in the stripping column, increases the vapor in the rectifying column, or does both. In the rectifying column the vapor flow rate is greater than the liquid flow rate. In the stripping column the liquid flow rate is greater than the vapor flow rate. It follows that the slope of the operating line in the stripping column is greater than 1 and that of the operating line in the rectifying column is less than 1.

The quantitative effects of introducing the feed depend on the condition of the feed. Feed may be introduced as a cold liquid, as a saturated liquid at its bubble point, as a mixture of liquid and vapor, as a saturated vapor at its dew point, or as a superheated vapor.

Figure 19-15 shows diagrammatically the liquid and vapor streams into and out of the feed plate for various feed conditions. In Fig. 19-15a cold feed is assumed. The entire feed stream becomes a part of the reflux in the stripping column. Also, to heat the feed to the bubble point some vapor must condense, and this condensate also becomes a part of $\bar{L}$, which consists of (1) the reflux from the rectifying column, (2) the feed, and (3) the condensate. The vapor to the rectifying column is less, by the amount of condensation, than that in the stripping column.

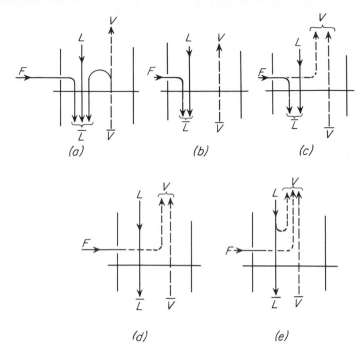

FIGURE 19-15
Flow through feed plate for various feed conditions: (*a*) feed cold liquid; (*b*) feed saturated liquid; (*c*) feed partially vaporized; (*d*) feed saturated vapor; (*e*) feed superheated vapor.

In Fig. 19-15*b* the feed is assumed to be at its bubble point. No condensation is required to heat the feed, so $V = \bar{V}$ and $\bar{L} = F + L$. If the feed is partly vapor, as shown in Fig. 19-15*c*, the liquid portion of the feed becomes part of $\bar{L}$ and the vapor portion becomes part of V. If the feed is saturated vapor, as shown in Fig. 19-15*d*, the entire feed becomes part of V, so $L = \bar{L}$ and $V = F + \bar{V}$. Finally, if the feed is superheated vapor, as shown in Fig. 19-15*e*, part of the liquid from the rectifying column is vaporized to cool the feed to a state of saturated vapor. Then the vapor in the rectifying column consists of (1) the vapor from the stripping column, (2) the feed, and (3) the vaporization. The reflux to the stripping column is less, by the amount of vaporization, than that in the rectifying column.

All five of the feed types can be correlated by the use of a single factor, denoted by q and defined as the moles of liquid flow in the stripping section that result from the introduction of each mole of feed. Then q has the following numerical limits for the various conditions:

Cold feed, $q > 1$
Feed at bubble point (saturated liquid), $q = 1$
Feed partially vapor, $0 < q < 1$
Feed at dew point (saturated vapor), $q = 0$
Feed superheated vapor, $q < 0$

If the feed is a mixture of liquid and vapor, q is the fraction that is liquid. Such a feed may be produced by an equilibrium flash operation, so $q = 1 - f$, where f is the fraction of the original stream vaporized in the flash.

The value of q for cold liquid feed is found from the equation

$$q = 1 + \frac{c_{pL}(T_b - T_F)}{\lambda} \tag{19-35}$$

For superheated vapor the equation is

$$q = -\frac{c_{pV}(T_F - T_d)}{\lambda} \tag{19-36}$$

where c_{pL}, c_{pV} = specific heats of liquid and vapor, respectively
$\quad\quad T_F$ = temperature of feed
$\quad\quad T_b$, T_d = bubble point and dew point of feed, respectively
$\quad\quad \lambda$ = heat of vaporization

Feed line   Equations (19-35) and (19-36) can be used with material balances to find the locus of all points of intersection of the operating lines. The equation for this line of intersections can be found as follows.

The contribution of the feed stream to the internal flow of liquid is qF, so the total flow rate of reflux in the stripping section is

$$\bar{L} = L + qF \quad\quad \text{and} \quad\quad \bar{L} - L = qF \tag{19-37}$$

Likewise, the contribution of the feed stream to the internal flow of vapor is $F(1 - q)$, and so the total flow rate of vapor in the rectifying section is

$$V = \bar{V} + (1 - q)F \quad\quad \text{and} \quad\quad V - \bar{V} = (1 - q)F \tag{19-38}$$

Write Eqs. (19-23) and (19-25) for constant molal overflow as

$$\bar{V}y = Lx + Dx_D \tag{19-39}$$
and
$$\bar{V}y = Lx - Bx_B \tag{19-40}$$

Subtract Eq. (19-40) from Eq. (19-39):

$$y(V - \bar{V}) = (L - \bar{L})x + Dx_D + Bx_B \tag{19-41}$$

From Eq. (19-18), the last two terms in Eq. (19-41) can be replaced by Fx_F. Also, substituting for $L - \bar{L}$ from Eq. (19-37) and for $V - \bar{V}$ from Eq. (19-38) and simplifying leads to the result

$$y = -\frac{q}{1-q}x + \frac{x_F}{1-q} \tag{19-42}$$

Equation (19-42) represents a straight line, called the *feed line*, on which all intersections of the operating lines must fall. The position of the line depends only on x_F and q. The slope of the feed line is $-q/(1 - q)$, and, as can be demonstrated by substituting x for y in Eq. (19-42) and simplifying, the line crosses the diagonal at $x = x_F$.

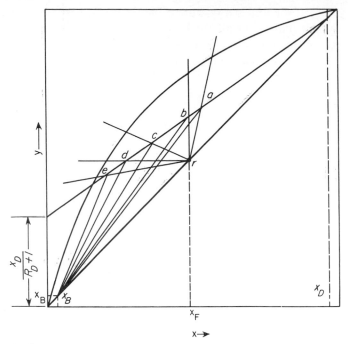

FIGURE 19-16
Effect of feed condition on feed line: *ra*, feed cold liquid: *rb*, feed saturated liquid; *rc*, feed partially vaporized; *rd*, feed saturated vapor; *re*, feed superheated vapor.

Equilibrium in feed plate   In the detailed construction of a feed plate a number of variations in the mixing of the inlet streams to the plate are possible. The simplest model for these, and the one used in this chapter, is that based on the assumption that all three streams entering the feed plate, designated by F, L, and $\overline{V}$, are thoroughly mixed adiabatically and then allowed to separate into the two leaving streams designated by $\overline{L}$ and V. A corollary of the assumption is that the two leaving streams are in equilibrium and the feed plate is a single ideal plate. If the concentrations of the liquid and vapor leaving the plate are x_f and y_f, respectively, the point (x_f, y_f) lies on the equilibrium curve.

Construction of operating lines   The simplest method of plotting the operating lines is (1) locate the feed line; (2) calculate the y-axis intercept $x_D/(R_D + 1)$ of the rectifying line, and plot that line through the intercept and the point (x_D, x_D); (3) draw the stripping line through point (x_B, x_B) and the intersection of the rectifying line with the feed line. The operating lines in Fig. 19-16 show the result of this procedure.

In Fig. 19-16 are plotted operating lines for various types of feed, on the assumption that x_F, x_B, x_D, L, and D are all constant. The corresponding feed lines are shown.

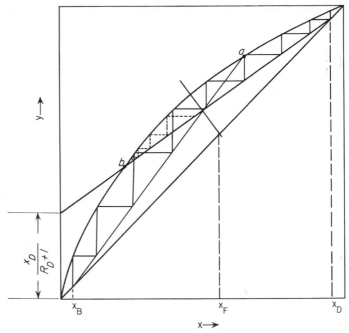

FIGURE 19-17
Feed-plate location.

If the feed is a cold liquid, the feed line slopes upward and to the right; if the feed is a saturated liquid, the line is vertical; if the feed is a mixture of liquid and vapor, the line slopes upward and to the left and the slope is the negative of the ratio of the liquid to the vapor; if the feed is a saturated vapor, the line is horizontal; and, finally, if the feed is a superheated vapor, the line slopes downward and to the left.

Feed-plate location After the operating lines have been plotted, the number of ideal plates is found by the usual step-by-step construction, as shown in Fig. 19-17. The construction can start either at the bottom of the stripping line or at the top of the rectifying line. In the following it is assumed that the construction begins at the top and also that a total condenser is used. As the intersection of the operating lines is approached, it must be decided when the steps should transfer from the rectifying line to the stripping line. The change should be made in such a manner that the maximum enrichment per plate is obtained so that the number of plates is as small as possible. Inspection of Fig. 19-17 shows that this criterion is met if the transfer is made immediately after a value of x is reached that is less than the x coordinate of the intersection of the two operating lines. The feed plate is always represented by the triangle that has one corner on the rectifying line and one on the stripping line. At the optimum position, the triangle representing the feed plate straddles the intersection of the operating lines.

The transfer from one operating line to the other, and hence the feed-plate location, can be made at any location between points a and b in Fig. 19-17, but if the feed plate is placed anywhere but at the optimum point, an unnecessarily large number of plates is called for. It is possible, for example, as shown in Fig. 19-17, to delay transferring to the stripping line until the point b is almost reached. Many small triangles will then be concentrated in the acute angle between the equilibrium and operating lines near point b, and these triangles represent a large number of unnecessary plates in the rectifying section. Likewise, if the transfer from the rectifying line to the stripping line is made nearly at point a, unnecessary plates in the stripping column are called for.

Heating and cooling requirements   Radiation from a column is small, and the column itself is essentially adiabatic. The heat effects of the entire plant are confined to the condenser and the reboiler. If the molal latent heat λ is constant (and the assumption of constant molal latent heat requires this assumption), the heat added in the reboiler q_r is $\overline{V}\lambda$ and that removed in the condenser q_c is $-V\lambda$, both in Btu per hour or watts. Heat effects are considered positive when the heat is added to the process and negative when heat is withdrawn; so q_c is negative.

If saturated steam is used as the heating medium, the steam required at the reboiler is

$$\dot{m}_s = \frac{\overline{V}\lambda}{\lambda_s} \tag{19-43}$$

where $\dot{m}_s$ = steam consumption
$\overline{V}$ = vapor from reboiler
λ_s = latent heat of steam
λ = molal latent heat of mixture

If water is used as the cooling medium in the condenser, the cooling-water requirement is

$$\dot{m}_c = -\frac{-V\lambda}{T_2 - T_1} = \frac{V\lambda}{T_2 - T_1} \tag{19-44}$$

where $\dot{m}_c$ = water consumption
$T_2 - T_1$ = temperature rise of cooling water

EXAMPLE 19-4   A continuous fractionating column is to be designed to separate 30,000 lb/h of a mixture of 40 percent benzene and 60 percent toluene into an overhead product containing 97 percent benzene and a bottom product containing 98 percent toluene. These percentages are by weight. A reflux ratio of 3.5 moles to 1 mol of product is to be used. The molal latent heats of benzene and toluene are 7,360 and 7,960 cal/g mol, respectively. The equilibrium diagram is given in Fig. 17-6. (*a*) Calculate the moles of overhead product and bottom product per hour. (*b*) Determine the number of ideal plates and the positions of the feed plate (*i*) if the feed is liquid and at its boiling point; (*ii*) if the feed is liquid and at 20°C (specific heat = 0.44); (*iii*) if the feed is a mixture of two-thirds vapor and one-third liquid. (*c*) If steam at 20 lb$_f$/in.^{2} gauge is used for heating, how much steam is required per hour for each of the above three cases, neglecting heat losses and assuming the reflux is a saturated liquid? (*d*) If cooling water enters the condenser at 80°F (26.7°C) and leaves at 150°F (65.5°C), how much cooling water is required, in gallons per minute?

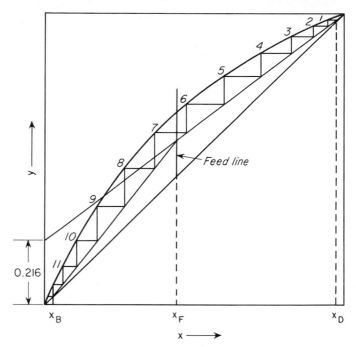

FIGURE 19-18
Example 19-4, part (b) (i).

SOLUTION (a) The molecular weight of benzene is 78 and that of toluene is 92. The concentrations of feed, overhead, and bottoms in mole fraction of benzene are

$$x_F = \frac{40/78}{40/78 + 60/92} = 0.440 \qquad x_D = \frac{97/78}{97/78 + 3/92} = 0.974$$

$$x_B = \frac{2/78}{2/78 + 98/92} = 0.0235$$

The average molecular weight of the feed is

$$\frac{100}{40/78 + 60/92} = 85.8$$

The feed rate F is 30,000/85.8 = 350 lb mol/h. By an overall benzene balance, using Eq. (19-19),

$$D = 350 \frac{0.440 - 0.0235}{0.974 - 0.0235} = 153.4 \text{ lb mol/h } (1.931 \times 10^{-2} \text{ kg mol/s})$$

$$B = 350 - 153.4 = 196.6 \text{ lb mol/h } (2.475 \times 10^{-2} \text{ kg mol/s})$$

(i) The first step is to plot the equilibrium diagram and on it erect verticals at x_D, x_F, and x_B. These should be extended to the diagonal of the diagram. Refer to Fig. 19-18.

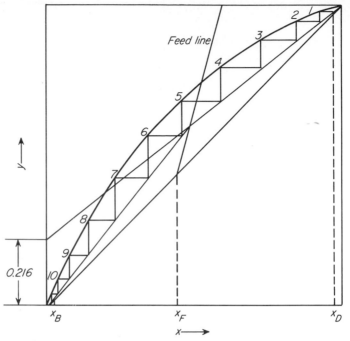

FIGURE 19-19
Example 19-4, part (b) (ii).

The second step is to draw the feed line. Here, $f = 0$, and the feed line is vertical and is a continuation of line $x = x_F$.

The third step is to plot the operating lines. The intercept of the rectifying line on the y axis is, from Eq. (19-33), $0.974/(3.5 + 1) = 0.216$. From the intersection of this operating line and the feed line the stripping line is drawn.

The fourth step is to draw the rectangular steps between the two operating lines and the equilibrium curve. In drawing the steps, the transfer from the rectifying line to the stripping line is at the seventh step. By counting steps it is found that, besides the reboiler, 11 ideal plates are needed and feed should be introduced on the seventh plate from the top.†

(ii) This solution is the same as that of part (b)(i) except for the feed line. From Fig. 17-5, the bubble point of the feed is 95.3°C. The specific heat of the feed, c_{pL}, is 0.44 Btu/lb-°F. The latent heat of vaporization of the feed, λ_L, is $7{,}675 \times 1.8/85.8 = 161.0$ Btu/lb. Substitution in Eq. (19-35) gives

$$q = 1 + \frac{0.44(95.3 - 20)(1.8)}{161.0} = 1.370$$

---

† To fulfill the conditions of the problem literally, the last step, which represents the reboiler, should reach the concentration x_B exactly. This is nearly true in Fig. 19-18. Usually, x_B does not correspond to an integral number of steps. An arbitrary choice of the four quantities x_D, x_F, x_B, and R_D is not necessarily consistent with an integral number of steps. An integral number can be obtained by a slight adjustment of one of the four quantities, but in view of the fact that a plate efficiency must be applied before the actual number of plates is established, there is little reason for making this adjustment.

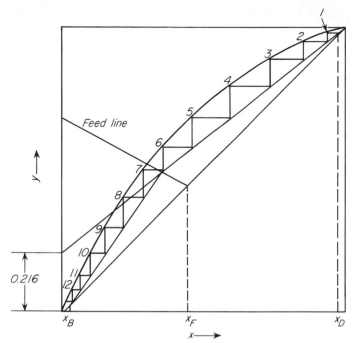

FIGURE 19-20
Example 19-4, part (*b*) (*iii*).

From Eq. (19-42) the slope of the feed line is $-1.370/(1 - 1.370) = 3.70$. When steps are drawn for this case, as shown in Fig. 19-19, it is found that a reboiler and 10 ideal plates are needed and that the feed should be introduced on the fifth plate.

(*iii*)   From the definition of q it follows that for this case $q = \frac{1}{3}$ and the slope of the feed line is -0.5. The solution is shown in Fig. 19-20. It calls for a reboiler and 12 plates, with the feed entering on the seventh plate.

(*c*)   The vapor flow V in the rectifying column, which must be condensed in the condenser, is 4.5 mol per mole of overhead product, or $4.5 \times 153.4 = 690$ mol/h. From Eq. (19-38),

$$\overline{V} = 690 - 350(1 - q)$$

The latent heat of vaporization is 7,960 cal/g mol, or $7,960 \times 1.8 = 14,328$ Btu/lb mol. The heat from 1 lb of steam at 20 $lb_f/in.^2$ gauge is, from Appendix 8, 939 Btu. The steam required is, from Eq. (19-43),

$$\dot{m}_s = \frac{14{,}328}{939} \, \overline{V} = \frac{14{,}328}{939} \, [690 - 350(1 - q)] \qquad \text{lb/h}$$

The results are given in Table 19-4.

(*d*)   The cooling water needed, which is the same in all cases, is, from Eq. (19-44),

$$\dot{m}_c = \frac{14{,}328 \times 690}{150 - 80} = 141{,}230 \text{ lb/h } (17.9 \text{ kg/s})$$

The water requirement is $141{,}230/(60 \times 8.33) = 283$ gal/min   ////

Minimum number of plates   Since the slope of the rectifying line is $R_D/(R_D + 1)$, the slope increases as the reflux ratio increases, until, when R_D is infinite, $V = L$ and the slope is 1. The operating lines then both coincide with the diagonal. This condition is called total reflux. At total reflux the number of plates is a minimum, but the feed and both the overhead and bottom products are zero for any still of finite size. Total reflux represents one limiting case in the operation of fractionating columns. The minimum number of plates required for a given separation may be found by constructing steps on an xy diagram between compositions x_D and x_B, using the 45° line as the operating line for both sections of the column. Since there is no feed in a column operating under total reflux, there is no discontinuity between the upper and lower sections.

For the special situation of ideal mixtures and where the relative volatilities α_{AB} are independent of temperature (case 6 of Table 17-1), a simple equation is available for calculating the value N_{min} from the terminal concentrations x_B and x_D. Since in total reflux $D = 0$, Eq. (19-21) shows that $L_n/V_{n+1} = 1$, and Eq. (19-26) gives $y_{n+1} = x_{n+1}$. Also, from Eq. (17-29),

$$\frac{y_{n+1}}{1 - y_{n+1}} = \alpha_{AB} \frac{x_{n+1}}{1 - x_{n+1}}$$

and therefore

$$\frac{x_n}{1 - x_n} = \alpha_{AB} \frac{x_{n+1}}{1 - x_{n+1}} \tag{19-45}$$

For a total condenser, $x_n = x_D$ and $x_{n+1} = x_1$ so Eq. (19-45) gives

$$\frac{x_D}{1 - x_D} = \alpha_{AB} \frac{x_1}{1 - x_1} \tag{19-46}$$

Writing Eq. (19-45) for a succession of n plates gives

$$\frac{x_1}{1 - x_1} = \alpha_{AB} \frac{x_2}{1 - x_2}$$
$$\cdots \cdots \cdots \cdots \cdots \tag{19-47}$$
$$\frac{x_{n-1}}{1 - x_{n-1}} = \alpha_{AB} \frac{x_n}{1 - x_n}$$

Table 19-4

Case	q	Steam requirement $\dot{m}_s$, lb/h
(c)(i)	0	5,109
(c)(ii)	1.37	12,500
(c)(iii)	0.333	6,560

If Eq. (19-46) and all the equations in the set of Eqs. (19-47) are multiplied together and all the intermediate terms canceled,

$$\frac{x_D}{1 - x_D} = (\alpha_{AB})^n \frac{x_n}{1 - x_n} \qquad (19\text{-}48)$$

To reach the bottom discharge from the column, $N_{\min}$ plates and a reboiler are needed, and Eq. (19-48) gives

$$\frac{x_D}{1 - x_D} = (\alpha_{AB})^{N+1} \frac{x_B}{1 - x_B}$$

Solving the equation for $N_{\min}$ by logarithms gives

$$N_{\min} = \frac{\log\left[x_D(1 - x_B)/x_B(1 - x_D)\right]}{\log \alpha_{AB}} - 1 \qquad (19\text{-}49)$$

Equation (19-49) is the *Fenske-Underwood equation*. If the change in the value of α_{AB} from the bottom of the column to the top is moderate, a geometric mean of the extreme values is recommended for α_{AB}.

Minimum reflux   At any reflux less than total, the number of plates is larger than that at total reflux and increases continuously with increase in the reflux ratio. As the ratio becomes smaller, the number of plates becomes very large, and at a definite minimum, called the *minimum reflux ratio*, the number of plates becomes infinite. All actual columns producing a finite amount of desired top and bottom products must operate at a reflux ratio between the minimum, at which the number of plates is infinity, and infinity, at which the number of plates is a minimum. If L_a/D is the operating reflux ratio and $(L_a/D)_{\min}$ the minimum reflux ratio,

$$\left(\frac{L_a}{D}\right)_{\min} < \frac{L_a}{D} < \infty \qquad (19\text{-}50)$$

The minimum reflux ratio can be found by following the movement of the operating lines as the reflux is reduced. In Fig. 19-21 both operating lines coincide with the diagonal afb at total reflux. For an actual operation lines ae and eb are typical operating lines. As the reflux is further reduced, the intersection of the operating lines moves along the feed line toward the equilibrium curve, the area on the diagram available for steps shrinks, and the number of steps increases. When either one or both of the operating lines touch the equilibrium curve, the number of steps necessary to cross the point of contact becomes infinite. The reflux ratio corresponding to this situation is, by definition, the minimum reflux ratio.

For the normal type of equilibrium curve, which is concave downward throughout its length, the point of contact, at minimum reflux, of the operating and equilibrium lines is at the intersection of the feed line with the equilibrium curve, as shown by lines ad and db in Fig. 19-21. A further decrease in reflux brings the intersection of the operating lines outside of the equilibrium curve, as shown by lines agc and cb. Then even an infinite number of plates cannot pass point g, and the reflux ratio for this condition is less than the minimum.

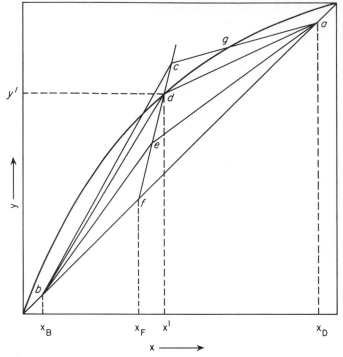

FIGURE 19-21
Minimum reflux ratio.

The slope of operating line *ad* in Fig. 19-21 is such that the line passes through the points (x', y') and (x_D, x_D), where x' and y' are the coordinates of the intersection of the feed line and the equilibrium curve. Let the minimum reflex ratio be R_{Dm}. Then

$$\frac{R_{Dm}}{R_{Dm} + 1} = \frac{x_D - y'}{x_D - x'}$$

or

$$R_{Dm} = \frac{x_D - y'}{y' - x'} \tag{19-51}$$

Equation (19-51) cannot be applied to all systems. Thus, if the equilibrium curve has a concavity upward, e.g., the curve for ethanol and water shown in Fig. 19-22, it is clear that the rectifying line first touches the equilibrium curve between abscissas x_F and x_D and line *ac* corresponds to minimum reflux. Operating line *ab* is drawn for a reflux less than the minimum, even though it does intersect the feed line below point (x', y'). In such a situation the minimum reflux ratio must be computed from the slope of the operating line *ac* that is tangent to the equilibrium curve.

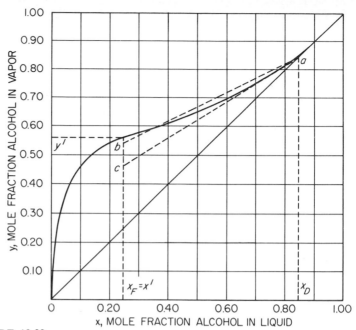

FIGURE 19-22
Equilibrium diagram (system ethanol-water).

Invariant zone At minimum reflux ratio an acute angle is formed at the intersection of an operating line and the equilibrium curve, as shown at point d in Fig. 19-19 or at the point of tangency in Fig. 19-20. In each angle an infinite number of steps is called for, representing an infinite number of ideal plates, in all of which there is no change in either liquid or vapor concentrations from plate to plate, so $x_{n-1} = x_n$, and $y_{n+1} = y_n$. The term *invariant zone* is used to describe these infinite sets of plates. A more descriptive term, *pinch point*, also is used.

With a normal equilibrium curve it is seen from Fig. 19-19 that at minimum reflux ratio the intersection of the q line and the equilibrium curve gives the concentrations of liquid and vapor at the feed plate (and at an infinite number of plates on either side of that plate). So an invariant zone forms at the bottom of the rectifying column and a second one at the top of the stripping column. The two zones differ only in that the liquid-vapor ratio is L/V in one and $\bar{L}/\bar{V}$ in the other.

Optimum reflux ratio As the reflux ratio is increased from the minimum, the number of plates decreases, rapidly at first, and then more and more slowly, until, at total reflux, the number of plates is a minimum. It will be shown later that the cross-sectional area of a column usually is approximately proportional to the flow rate of vapor. As the reflux ratio increases, both V and L increase for a given production, and a point is reached where the increase in column diameter is more rapid than the de-

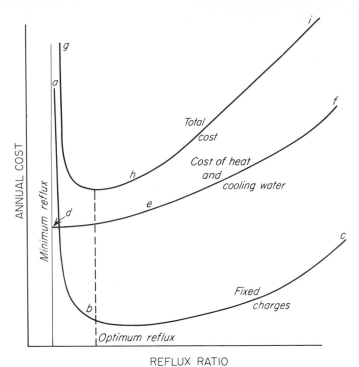

FIGURE 19-23
Optimum reflux ratio.

crease in number of plates. The cost of the unit is roughly proportional to the total plate area, so the fixed charges first decrease and then increase with reflux ratio. This is shown by line *abc* in Fig. 19-23.

The reflux is made by supplying heat at the reboiler and withdrawing it at the condenser. The costs of both heat and cooling water increase with reflux, as shown by curve *def* in Fig. 19-23. The total cost of operation varies with the sum of the fixed charges and the cost of heat and cooling water, as shown by curve *ghi*. Curve *ghi* has a minimum, at a definite reflux ratio, which usually is not much greater than the minimum reflux. This is the point of most economical operation, and this reflux is called the *optimum reflux ratio*. Actually, most plants are operated at refluxes above the optimum, usually in a range of 1.2 to 2.0 times the minimum reflux. The total cost is not very sensitive to reflux ratio in this range, and better operating flexibility is obtained if a reflux greater than the optimum is used.

EXAMPLE 19-5 What is (*a*) the minimum reflux ratio and (*b*) the minimum number of ideal plates for cases (*b*)(*i*), (*b*)(*ii*), and (*b*)(*iii*) of Example 19-4?

SOLUTION (*a*) For minimum reflux ratio use Eq. (19-51). Here $x_D = 0.974$ (see Table 19-5).

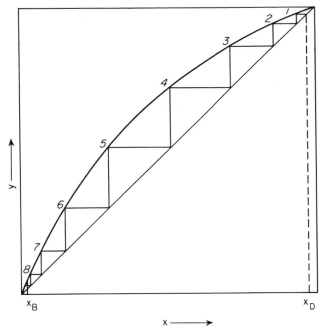

FIGURE 19-24
Example 19-5, part (b).

(b)   For minimum number of plates, the reflux ratio is infinite, the operating lines coincide with the diagonal, and there are no differences between the three cases. The plot is given in Fig. 19-24. A reboiler and eight ideal plates are needed.   ////

Nearly pure products   When either the bottom or overhead product is nearly pure, a single diagram covering the entire range of concentrations is impractical as the steps near $x = 0$ and $x = 1$ become small. This situation may be treated by any one of three methods. In the first, auxiliary diagrams for the ends of the concentration range are prepared, on a large scale, so the individual steps are sufficiently large to be drawn. The operating and equilibrium lines are drawn on the auxiliary plot and the steps transferred to this plot from the main diagram.

Table 19-5

Case	x'	y'	R_{Dm}
(b)(i)	0.440	0.658	1.45
(b)(ii)	0.521	0.730	1.17
(b)(iii)	0.300	0.513	2.16

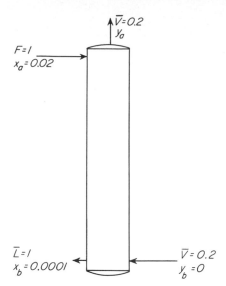

FIGURE 19-25
Material-balance diagram for Example 19-6.

$\bar{V} = 0.2$
y_a

$F = 1$
$x_a = 0.02$

$\bar{L} = 1$
$x_b = 0.0001$

$\bar{V} = 0.2$
$y_b = 0$

The second method for opening up the ends of the diagram is to plot the equilibrium and operating lines on logarithmic coordinates.[12]　The operating line, if straight on rectangular coordinates, is not so on logarithmic coordinates, but calculation of the coordinates of a few points by the material-balance equations allows the curve to be plotted. The steps are distorted on such a diagram, but the number of steps is unaffected by a change from rectangular to logarithmic coordinates. This method is especially effective in the portion of the concentration range near an azeotrope.

The third method of treating nearly pure products is based on the principle that Raoult's law applies near $x = 1$ and Henry's law near $x = 0$. At both ends of the curve of x_e vs. y_e, then, both the equilibrium and the operating lines are straight, so Eq. (18-27) can be used, and no graphical construction is required. The same equation may be used anywhere in the concentration range where both the operating and equilibrium lines are straight or nearly so.

EXAMPLE 19-6　A mixture of 2 mole percent ethanol and 98 mole percent water is to be stripped in a plate column to a bottom product containing not more than 0.01 mole percent ethanol. Steam, admitted through an open coil in the liquid on the bottom plate, is to be used as a source of vapor. The feed is at its boiling point. The steam flow is to be 0.2 mol per mole of feed. For dilute ethanol-water solutions, the equilibrium line is straight, and is given by $y_e = 9.0x_e$. How many ideal plates are needed?

SOLUTION　Since both equilibrium and operating lines are straight, Eq. (18-27) rather than a graphical construction may be used. The material-balance diagram is shown in Fig. 19-25. No reboiler is needed, as the steam enters as a vapor. Also, the liquid flow in the tower equals the feed entering the column. By conditions of the problem

$$F = \bar{L} = 1 \qquad \bar{V} = 0.2 \qquad y_b = 0 \qquad x_a = 0.02$$
$$x_b = 0.0001 \qquad m = 9.0 \qquad y_a^* = 9.0 \times 0.02 = 0.18$$
$$y_b^* = 9.0 \times 0.0001 = 0.0009$$

To use Eq. (18-27) y_a, the concentration of the vapor leaving the column, is needed. This is found by an overall ethanol balance

$$\bar{V}(y_a - y_b) = \bar{L}(x_a - x_b) \qquad 0.2(y_a - 0) = 1(0.02 - 0.0001)$$

from which $y_a = 0.0995$. Substituting into Eq. (18-27) gives

$$N = \frac{\log\,[(0 - 0.0009)/(0.0995 - 0.18)]}{\log\,[(0 - 0.0995)/(0.0009 - 0.18)]} = \frac{\log\,[(0.18 - 0.0995)/0.0009\,]}{\log\,[(0.18 - 0.0009)/0.0995\,]}$$

$$= \frac{\log 89.4}{\log 1.8} = 7.6 \text{ ideal plates} \qquad\qquad ////$$

Analysis of Fractionating Columns by Enthalpy-Concentration Method

The actual variation in the V and L streams in a distillation column is determined by an enthalpy balance, and the limitation imposed by the assumption of constant molal overflow can be removed by a rigorous enthalpy balance used in conjunction with material balances and phase equilibria. A method usually referred to as the *Ponchon-Savarit method* is based on the graphical use of an enthalpy-concentration diagram.[11,12b,15]

In Chap. 17 it was shown that an enthalpy-concentration diagram coordinates latent heats, heats of mixing, and sensible heats and that all these effects are built into the diagram, so that none of them need be computed separately. This diagram can also show equilibrium data in the manner shown by the tie lines in Fig. 17-13. In distillation practice the saturated-vapor line is called the *dew-point line* and the saturated-liquid line the *bubble-point line*.

As shown later, a glance at the enthalpy-concentration diagram tells qualitatively whether an operating line on the equilibrium diagram is straight or bent. The line is rigorously linear if the dew-point and bubble-point lines are straight and parallel. In many distillation systems this is very nearly true, and the McCabe-Thiele method is sufficiently precise to determine the number of ideal plates to an accuracy of a single plate. As shown in Fig. 17-12, the dew- and bubble-point lines may be strongly curved, especially the latter. This is a result of the strong nonideality of these systems. In such cases the Ponchon-Savarit method or its equivalent comes into its own.

Overall enthalpy balance around fractionating column Consider an overall enthalpy balance over the plant shown in Fig. 19-10. In addition to the quantities shown in the figure, let H_F, H_D, and H_B represent the specific enthalpies of the feed, overhead product, and bottom product, respectively, all in energy per mole. Points F, D, and B, representing feed, overhead, and bottoms, are plotted on an enthalpy-concentration diagram in Fig. 19-26. Both products are at their bubble temperatures, so points D and B lie on the saturated-liquid line. The feed may have any thermal condition from cold liquid to superheated vapor. It is shown as a mixture of liquid and vapor in Fig. 19-26. The ratio of vapor to liquid in the feed is, by the center-of-gravity principle, the ratio $\overline{aF}/\overline{Fb}$, where a and b are the intersections of the tie line through point F with the saturated-liquid and saturated-vapor lines, respectively.

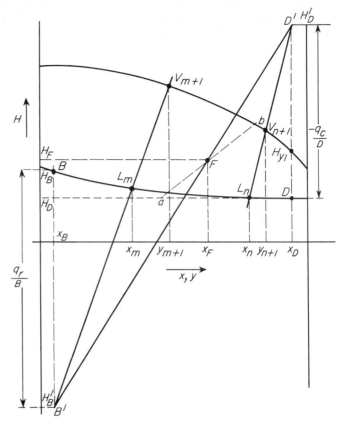

FIGURE 19-26
Ponchon-Savarit method. Construction of overall-enthalpy and enthalpy operating lines.

The overall process is a nonadiabatic splitting of the feed into overhead and bottom products, similar to the processes shown in Figs. 18-11 and 18-12. To use the adiabatic construction on Fig. 19-26, correct stream D for $-q_c$, the heat removed in the condenser, and stream B for q_r, the heat added in the reboiler. Since both streams are product streams, Eq. (18-41) applies and

$$H'_D = H_D - \frac{-q_c}{D} = H_D + \frac{q_c}{D} \tag{19-52}$$

$$H'_B = H_B - \frac{q_r}{B} \tag{19-53}$$

Points D' and B' represent the corrected product streams. Since the process with corrected streams is adiabatic, points D', F, and B' lie on a single straight line, which may be called the *overall enthalpy line*.

For given feed and product streams, only one of the heat effects, $-q_c$ or q_r, is independent and subject to choice by the designer or operator. The usual choice is $-q_c$, which is fixed when the reflux ratio at the top of the column is chosen. The reflux ratio is, by the center-of-gravity principle,

$$R_D = \frac{H'_D - H_{y1}}{H_{y1} - H_D} \tag{19-54}$$

where H_{y1} is the specific enthalpy of the vapor leaving plate 1 and entering the condenser.

Enthalpy balances in rectifying and stripping columns Let the specific enthalpy of the vapor stream rising from plate $n + 1$ to plate n in the rectifying column shown in Fig. 19-10 be $H_{y(n+1)}$, and let that of the liquid stream leaving plate n be H_{xn}. Let point V_{n+1} on Fig. 19-26 represent the vapor from plate $n + 1$, and let point L_n represent the liquid from plate n. Point V_{n+1} lies on the saturated-vapor line and point L_n on the saturated-liquid line. Since the column is adiabatic, points D', V_{n+1}, and L_n are collinear. The straight line through them is called the *enthalpy operating line*. Such a line exists for each plate, and the lines for all plates above the feed pass through point D'. Like the operating lines in the McCabe-Thiele method, enthalpy operating lines relate the concentration of the vapor rising to a plate with that of the liquid leaving it. They also show the ratio of the flow rates of vapor and liquid between plates. By the center-of-gravity principle,

$$\frac{V_{n+1}}{L_n} = \frac{\overline{D'L_n}}{\overline{D'V_{n+1}}} = \frac{x_D - x_n}{x_D - y_{n+1}} \tag{19-55}$$

Let the specific enthalpy of the vapor rising to plate m in the stripping column be $H_{y(m+1)}$, and let that of the liquid from plate m be H_{xm}. Let point L_m in Fig. 19-26 represent the liquid from plate m and V_{m+1} the vapor from plate $m + 1$. Point L_m is on the saturated-liquid line, and point V_{m+1} is on the saturated-vapor line. Points V_{m+1}, L_m, and B' are the terminal points of an adiabatic mixing process, so line $V_{m+1}L_mB'$ is straight and is an enthalpy operating line. There is an enthalpy operating line for each plate below the feed, and all such lines pass through point B'. The interpretation of these lines is identical with that of the operating lines in the rectifying section.

The ratio of the flow rates of vapor and liquid at any point in the stripping section can be found from the equation

$$\frac{V_{m+1}}{L_m} = \frac{x_m - x_B}{y_{m+1} - x_B} \tag{19-56}$$

Net flows of enthalpy On page 550 it was shown that D is the constant net flow rate of material upward in the upper section of the column. In the same way, DH'_D is the net flow rate of enthalpy; i.e., it is the difference between the enthalpy carried upward by the vapor and that carried downward by the liquid. For adiabatic operation this difference is constant, as shown by the following analysis.

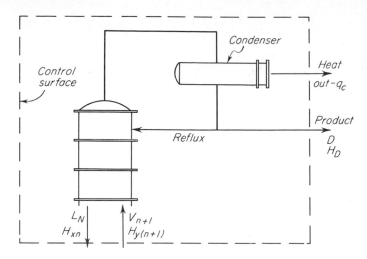

FIGURE 19-27
Enthalpy-balance diagram for upper section of continuous fractionating column.

Consider the section of the column enclosed by the control surface indicated in Fig. 19-27. There is one entering flow stream, V_{n+1}, and two leaving streams, D and L_n. Each of these streams carries enthalpy. In addition, enthalpy is removed from the condenser in an amount q_c. By definition, enthalpy H'_D is given by the equation

$$H'_D = H_D + \frac{q_c}{D}$$

or
$$DH'_D = DH_D + q_c \tag{19-57}$$

An overall enthalpy balance around the control surface gives

$$L_n H_{xn} + DH_D + q_c = V_{n+1} H_{y(n+1)} \tag{19-58}$$

Substituting $(DH_D + q_c)$ from Eq. (19-57) and rearranging gives

$$DH'_D = V_{n+1} H_{y(n+1)} - L_n H_{xn} \tag{19-59}$$

Equation (19-59) shows that DH'_D is the difference between the enthalpies carried by the vapor and liquid streams. Since plate n was not specified, this difference is the same everywhere in the upper section of the column, provided the column operates adiabatically. Since D is constant, H'_D is constant also.

Point D' in Fig. 19-26 thus represents the difference between the streams indicated by points V_{n+1} and L_n, just as point S in Fig. 18-10d and e represents the difference between streams T and R. Point D' is a common operating point for all values of V_{n+1} and L_n in the upper section of the column.

A similar analysis shows the BH'_B is the net enthalpy difference in the lower section of the column. It is given by

$$BH'_B = L_m H_{xm} - V_{m+1} H_{y(m+1)} \tag{19-60}$$

Since B is constant, H'_B is also constant, and point B' is the same for all values of L_m and V_{m+1}.

Number of ideal plates The enthalpy operating lines are used alternately with tie lines to give a step-by-step determination of the number of ideal plates needed to accomplish a given separation. In comparing the Ponchon-Savarit method with the McCabe-Thiele method, the tie lines correspond to the equilibrium curve and the enthalpy operating lines on the Hx diagram to the material-balance operating lines on the xy diagram. The details of the Ponchon-Savarit method are shown in Example 19-7.

EXAMPLE 19-7 A mixture of 35 percent ammonia and 65 percent water is to be separated, at a pressure of 10 atm, into overhead and bottom products of 97.5 and 2.5 percent ammonia, respectively. The feed enters at 60°F (15.6°C). A reflux ratio of 2 is to be used. (*a*) Per pound of overhead product, how much heat must be withdrawn at the condenser and how much added at the reboiler? (*b*) How many ideal plates are needed, and where should the feed be introduced? (*c*) What are the vapor and liquid flows at each point in the plant?

   SOLUTION The Ponchon-Savarit diagram for this problem is shown in Fig. 19-28. The Hx diagram of Fig. 17-13 is used.

   (*a*) Point F is established by the concentration ($x_F = 0.35$) and the temperature (60°F) of the feed. Points D and B are located on the saturated-liquid line at $x_D = 0.975$ and $x_B = 0.025$, respectively. Point D' is found from the specified reflux ratio by use of Eq. (19-54). From Fig. 17-13, $H_{y1} = 690$, and $H_D = 130$. Then

$$2.0 = \frac{H'_D - 690}{690 - 130}$$

From this, $H'_D = 1810$, which is the ordinate of point D'. The abscissa is $x_D = 0.975$. A straight line through points D' and F is the overall-enthalpy line. It intersects the line $x_B = 0.025$ at $H'_B = -1000$.

   For each pound of product, the feed and bottoms are found by Eqs. (19-19) and (19-17).

$$F = \frac{0.975 - 0.025}{0.35 - 0.025} = 2.92 \text{ lb} \qquad B = F - D = 2.92 - 1.0 = 1.92 \text{ lb}$$

The specific enthalpies of the overhead and bottoms are 130 and 300 Btu/lb, respectively. The heat withdrawn at the condenser is $1810 - 130 = 1680$ Btu, and the heat added at the reboiler is $1.92(300 + 1000) = 2500$ Btu (2633 kJ).

   (*b*) Since $x_D = y_1$, the point representing V_1 is on the saturated-vapor line at $x = 0.975$. From point V_1, the appropriate tie line establishes point L_1 on the saturated-liquid line. An operating line through points D' and L_1 intersects the saturated-vapor line at point V_2, and the tie line through this point establishes point L_2.

   A decision must be made where the feed plate should be located. The best location for the point of entry of feed is on the first plate where the liquid concentration is less than the abscissa of the intersection of the saturated-liquid line and the overall-enthalpy line. This means that rectifying operating lines should be used at the right of the overall-enthalpy line and stripping lines to the left.

   Since L_2 is at the left of line $D'FB'$, the next operating line is drawn through point B', thus establishing point V_3 on the saturated-vapor line. From point V_3, point L_3 is located by a tie line; V_4 from an operating line, and L_4 from a tie line. Since x_4 is less than

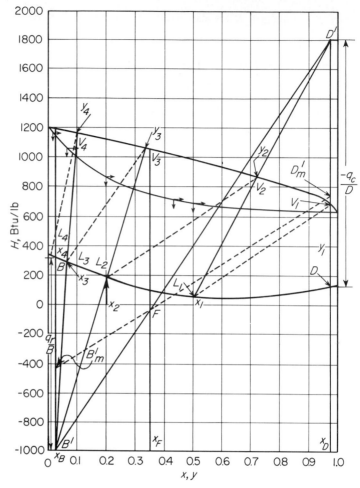

FIGURE 19-28
Solution of Example 19-7.

x_B, four steps are sufficient. A reboiler and three ideal plates with feed admitted to the second plate should be specified.

(c) Equations (19-55) and (19-56) can be used to compute the vapor and liquid streams within the column. A basis of 1 lb of overhead product is taken.

By conditions of the design, the liquid to the top plate is 2.0 lb, and the vapor from the top plate is 3.0 lb. From Fig. 19-28, the concentrations within the column are

$$x_1 = 0.50 \qquad x_2 = 0.20 \qquad x_3 = 0.06 \qquad x_B = x_4 = 0.025†$$
$$x_D = y_1 = 0.975 \qquad y_2 = 0.72 \qquad y_3 = 0.335 \qquad y_4 = 0.080$$

† Arbitrarily equated to x_B to obtain arithmetic consistency in the material and enthalpy balances. A slight adjustment could be made in the reflux ratio that would give integral steps in Fig. 19-28, but this adjustment is unnecessary.

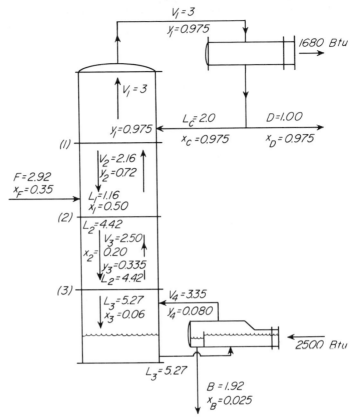

FIGURE 19-29
Results for Example 19-7.

From Eq. (19-55)

$$\frac{V_2}{L_1} = \frac{x_D - x_1}{x_D - y_2} = \frac{0.975 - 0.50}{0.975 - 0.\ 2} = 1.86$$

Since $V_2 = L_1 + D$, $1.86L_1 = L_1 + 1$, so $L_1 = 1.16$ and $V_2 = 2.16$. From Eq. (19-56)

$$\frac{V_3}{L_2} = \frac{x_2 - x_B}{y_3 - x_B} = \frac{0.20 - 0.025}{0.335 - 0.025} = 0.565 \qquad \frac{V_4}{L_3} = \frac{x_3 - x_B}{y_4 - x_B} = \frac{0.06 - 0.025}{0.080 - 0.025} = 0.636$$

Since $L_2 = V_3 + B$, $L_2 = 0.565L_2 + 1.92$, so $L_2 = 4.42$ and $V_3 = 2.50$. Also, $L_3 = V_4 + B$; $L_3 = 0.636L_3 + 1.92$, so $L_3 = 5.27$ and $V_4 = 3.35$. The concentrations, flows, and heat effects throughout the column are shown in Fig. 19-29.   ////

Minimum number of plates   For total reflux, which corresponds to the minimum number of plates, $-q_c$ and q_r are both very large, and the overall-enthalpy-balance and enthalpy operating lines become vertical. Then these vertical lines and the tie

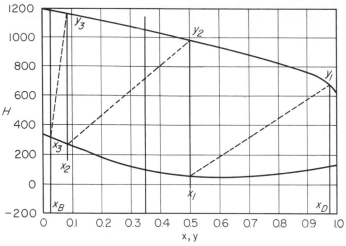

FIGURE 19-30
Minimum number of plates for Example 19-8.

lines are used alternately in the usual manner to determine the minimum number of plates.

Minimum reflux ratio When the reflux ratio is reduced from one that calls for a finite number of plates, a situation is reached in which an enthalpy operating line coincides with a tie line. This may occur either above or below the feed. Then the number of steps becomes infinitely large. The reflux ratio corresponding to this condition is the minimum for the given concentrations of feed and products and the given thermal condition of the feed. Usually, the tie line that is encountered first as the reflux is decreased is the one that passes through point F. This is equivalent to the situation on the McCabe-Thiele diagram for normal equilibrium curves, when the operating lines for minimum reflux pass through the intersection of the feed line and the equilibrium curve. For abnormal equilibrium curves, such as that shown in Fig. 19-22, the condition for minimum reflux ratio may appear elsewhere in the column, either in the rectifying or stripping section. Once the controlling tie line is located, which may require trial and error, it is extended to the line $x = x_D$ and point D'_m located, the ordinate of which is H'_{Dm}. The minimum reflux ratio R_{Dm} can then be calculated from the equation

$$R_{Dm} = \frac{H'_{Dm} - H_{y1}}{H_{y1} - H_D} \tag{19-61}$$

EXAMPLE 19-8 What are the minimum number of plates and the minimum reflux ratio for the column of Example 19-7?

SOLUTION The diagram applying at total reflux is shown in Fig. 19-30. The construction shows that three stages are needed, so a reboiler and two ideal plates are the minimum number that can accomplish the desired separation at total reflux.

The first tie line that coincides with an enthalpy operating line is found to be that passing through point F in Fig. 19-28. This line intercepts line $x = x_D$ at point D'_m, the ordinate of which is $H'_{Dm} = 740$. The line is plotted as line $D'_m FB'_m$. From Eq. (19-61), the minimum reflux ratio is

$$R_{Dm} = \frac{740 - 690}{690 - 130} = 0.089 \text{ lb/lb overhead} \qquad ////$$

Coordination of McCabe-Thiele and Ponchon-Savarit methods The use of the Ponchon-Savarit method for determining the number of plates as shown in Fig. 19-28 is practical when the number of plates is small so that the span of a single plate is large. For a column of many plates the steps become small, and the precision of the graphical construction depends more on accurate draftsmanship than on precise calculation. The two graphical methods can be combined by using the Ponchon-Savarit method for locating operating lines and the McCabe-Thiele method for stepping off the plates. Then the several methods for enlarging the steps or by using absorption factors can be used where precision calls for them.

The coordination of the methods is shown in Fig. 19-31. Any operating ray may be arbitrarily drawn and the values of x_n and y_{n+1} projected as shown to the xy plane. Only a few points are needed even for a strongly bent operating line; of course, these points do not have to represent any actual plate.

Two modifications of this method are useful. First, it is not necessary to plot both diagrams on the same sheet. For any of the values of x_n and y_{n+1}, or of x_m and y_{m+1}, the enthalpy of a point on the operating line may simply be read from the enthalpy chart as shown in Fig. 19-31 and the corresponding point on the operating line plotted on the equilibrium diagram.

The second modification of the standard Ponchon-Savarit method eliminates the need for plotting the net flow points D' and B' and so reduces the height of the enthalpy chart. As shown in Fig. 19-32, the diagram is used only between the base line of enthalpies and the top abscissa, where $H = H'$. The dotted rays above and below these limits are not needed. To establish an operating ray for the rectifying column, point R, whose abscissa, x_n, is the value of a point on the operating line, is placed arbitrarily on the bubble-point curve at the right of the overall-enthalpy line. From knowledge of x_D and H'_D and the ordinate H_{xn} of point R, a linear equation is used to calculate y', the abscissa of point R', where the ray crosses the top of the chart. The straight line RR' is a section of the operating ray, and its intersection with the dew-point line fixes the value of y_{D+1}, which, with the value of x_n, gives the coordinates of a point on the operating line.

In the same manner, to establish a point on the operating line for the stripper, point S, whose abscissa is y_{m+1}, is located arbitrarily on the dew-point curve at the left of the overall-enthalpy line. The ordinate of point S is $H_{y(m+1)}$. Values of x_B and H'_B are used with this number to calculate the abscissa x' of point S', where the ray crosses the base line. Straight line SS' crosses the bubble-point line at $x_1 = x_n$, and the coordinates of a point on the operating line for the stripper are at hand.

The same principles can be used to locate the point on the q line where the two operating lines intersect. The overall-enthalpy-balance line crosses the top and bottom

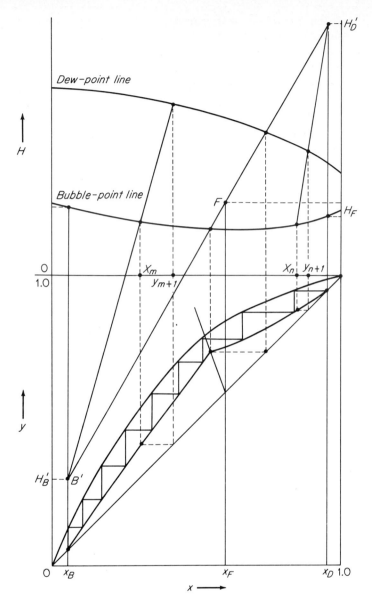

FIGURE 19-31
Coordination of Ponchon-Savarit and McCabe-Thiele diagrams.

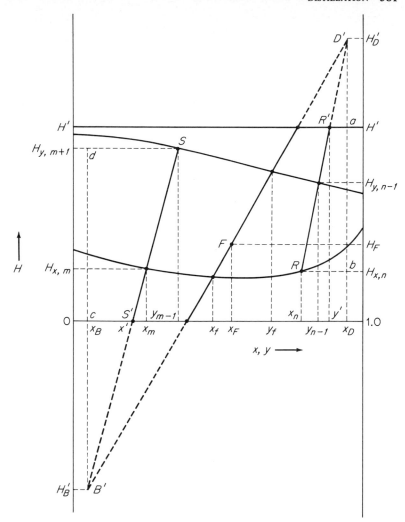

FIGURE 19-32
Use of truncated rays on Ponchon-Savarit diagram.

of the diagram at abscissas y_f and x_f, respectively, which are calculated by use of the enthalpy balance. The straight line through these points intersects the dew-point line at $y = y_f$ and the bubble-point line at $x = x_f$, where x_f and y_f are the coordinates of the desired intersection.

The equation for the rectifying line is, from the similar triangles $\overline{D'aR'}$ and $\overline{D'bR}$,

$$\frac{H'_D - H'}{x_D - y'} = \frac{H'_D - H_{x,n}}{x_D - x_n}$$

Solving for y' gives

$$y' = x_D \frac{H'_D - H_{x,n}}{x_D - x_m}(x_D - x_m) \tag{19-62}$$

In the same manner, from similar triangles $\overline{B'cS'}$ and $\overline{B'dS}$,

$$x' = x_B - H'_B \frac{y_{m+1} - x_B}{H_{y(m+1)} - H'_B} \tag{19-63}$$

Enthalpy-concentration diagram of ideal mixture   In ideal or nearly ideal mixtures the operating line is almost, but not quite, straight, and the McCabe-Thiele method is adequate. In nonideal mixtures with strongly curved isotherms the corresponding operating lines are also curved, though less so than the isotherms.

To demonstrate that straight liquid and vapor lines on the Hx diagram can lead to straight operating lines on the xy diagram, consider, first, the case where the operating and equilibrium lines are not only straight but parallel. Liquid line cd and vapor line ab in Fig. 19-33 are two such lines. Let lines $D'V_nL_{n-1}$ and $D'V_{n+1}L_n$ be any two adjacent enthalpy operating lines in the rectifying section. Since lines ab and cd are parallel, triangles $D'L_{n-1}L_n$ and $D'V_nV_{n+1}$ are geometrically similar. Then

$$\frac{\overline{D'V_n}}{\overline{D'L_{n-1}}} = \frac{\overline{D'V_{n+1}}}{\overline{D'L_n}}$$

By the center-of-gravity principle,

$$\frac{\overline{D'V_n}}{\overline{D'L_{n-1}}} = \frac{L_{n-1}}{V_n} \quad \text{and} \quad \frac{\overline{D'V_{n+1}}}{\overline{D'L_n}} = \frac{L_n}{V_{n+1}}$$

Since $V_n = L_{n-1} + D$ and $V_{n+1} = L_n + D$,

$$\frac{L_{n-1}}{L_{n-1} + D} = \frac{L_n}{L_n + D}$$

and, therefore, $L_n = L_{n-1}$. Since lines $D'V_nL_{n-1}$ and $D'V_{n+1}L_n$ are any two adjacent lines, all enthalpy operating lines in the rectifying section have the same L, and the overflow is constant. An equivalent proof can be given for the stripping section.

Enthalpy and phase equilibrium   The degree of curvature of an isotherm on the enthalpy-concentration chart is quantified by the use of partial molal enthalpies, which are defined in the manner of Eq. (17-3) by

$$\bar{h}_i \equiv \left(\frac{\partial H_T}{\partial n_i}\right)_{T,p,n_j}$$

where H_T is the total enthalpy of a mixture of at least two components and where the numbers of moles of all components except n_i are constant during the addition of an infinitesimal quantity of component i at constant temperature and pressure. For the

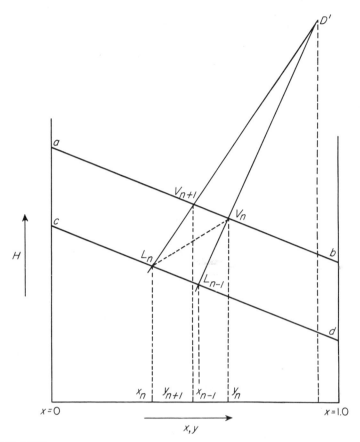

FIGURE 19-33
Straight and parallel liquid and vapor lines on Hx diagram.

special case of two components, if a mixture is prepared by mixing n_A mol of component A and n_B mol of component B, ΔH_T, the increase in enthalpy resulting from the mixing process, is

$$\Delta H_T = (\bar{h}_A n_A + \bar{h}_i n_B) - (H_A^\circ n_A + H_B^\circ n_B) \tag{19-64}$$

where H_A° and H_B° are the enthalpies of 1 mol of pure component A and B, respectively, at the temperature and pressure of the mixture. By the first law of thermodynamics, ΔH_T is the heat that must be added to the mixture to keep its temperature and pressure constant. Thus, if the heat of mixing is zero for any mixture of the components, inspection of Eq. (19-64) shows that

$$\bar{h}_A = H_A^\circ \qquad \text{and} \qquad \bar{h}_B = H_B^\circ \tag{19-65}$$

In general, for all mixtures, the enthalpy of 1 mol of a mixture having a mole fraction of component A of x is

$$H = x\bar{h}_A + (1 - x)\bar{h}_B \tag{19-66}$$

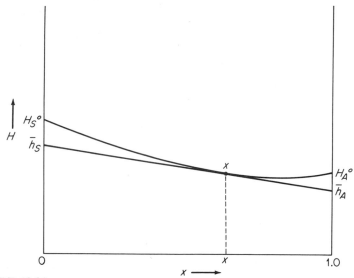

FIGURE 19-34
Enthalpy per mole of binary mixture at constant temperature and pressure.

A typical plot of Eq. (19-66) is shown in Fig. 19-34. Also, a tangent at an arbitrary value of x is drawn. An important property of the partial molal quantities is that the intercepts of such a tangent on the $x = 0$ and the $x = 1$ axes are measures of $\bar{h}_B$ and $\bar{h}_i$, respectively, for the mixture represented by the point of tangency. Therefore, it follows from Eq. (19-65) that if the heat of mixing is zero, the ends of the tangent and of the isotherm coincide and the entire enthalpy line is straight.

Also, the isotherms for two isotherms differing in temperature by ΔT degrees are separated at each end by a distance of $C_p \Delta T$, where C_p is the molal specific heat of the corresponding pure component.

Effect of temperature on activity and fugacity coefficients   The effect of temperature at constant pressure and concentration on the activity coefficient γ_i is given by the exact equation

$$\left(\frac{\partial \ln \gamma_i}{\partial T} \right)_{p,n_i} = - \frac{\bar{h}_i^L - H_i^\circ}{RT^2} \tag{19-67}$$

In an ideal liquid mixture, the activity coefficient is unity at all temperatures, pressures, and concentrations. Therefore it cannot change with temperature, and the left-hand side of Eq. (19-67) is zero, and so is the enthalpy difference on the right-hand side. An ideal mixture has no heat of mixing and has a linear isotherm. The same conclusions apply to the vapor phase when $\phi_i = 1$.

EXAMPLE 19-9   From the data in Table 19-6 construct an enthalpy-concentration diagram for the system benzene-toluene at a pressure of 1 atm. Use the data from Example 17-3 as a source of tie lines, and express enthalpies in SI units.

SOLUTION The conversion factor from calories per gram mole to joules per kilogram mole is 4.184×10^3. The enthalpies of the pure components, based on 0°C, are:

Benzene, liquid phase:

$$H_A^L = 4.184 \times 10^3 \times 33T = 1.381 \times 10^5 T \qquad \text{J/kg mol}$$

Benzene, vapor phase:

$$H_A^V = 4.184 \times 10^3 [33 \times 80.1 + 7,360 + 23(T - 80.1)]$$

$$= 9.623 \times 10^4 (T + 354.8) \qquad \text{J/kg mol}$$

Toluene, liquid phase:

$$H_B^L = 4.184 \times 10^3 \times 40T = 1.674 \times 10^5 T \qquad \text{J/kg mol}$$

Toluene, vapor phase:

$$H_B^V = 4.184 \times 10^3 [40 \times 110.6 + 7,960 + 33(T - 110.6)]$$

$$= 1.382 \times 10^5 (T + 23.5) \qquad \text{J/kg mol}$$

Liquid- and vapor-phase enthalpies are calculated for temperatures of 80, 90, 100, and 110°C and are plotted on the $x = 0$ and $x = 1.0$ ordinates of the enthalpy chart, as shown in Fig. 19-35. Since the system is assumed to be ideal, linear isotherms for liquid and vapor phases are drawn. The bubble-point and dew-point terminals are found by reading the x and y coordinates from Table 17-2, and then the H^L and H^V coordinates are fixed on the isotherms, so the tie lines can be drawn and the bubble-point and dew-point lines established.

Although the isotherms are all linear and the latent heats of the pure components nearly equal, the nonlinearity of the equilibrium curve, as shown by the variation in the slope of the tie lines, gives a slight curvature to the bubble-point line. The dew-point line is quite straight. An operating line on the equilibrium diagram is very slightly bent, but the effect on the number of plates as determined by the McCabe-Thiele construction is negligible.

/////

Design of Sieve-Plate Columns

To translate ideal plates into actual plates, a correction for the efficiency of the plates must be applied. There are other important decisions, some at least as important as fixing the number of plates, that must be made before a design is complete. A mistake in these decisions results in poor fractionation, lower-than-desired capacity, poor

Table 19-6 DATA FOR EXAMPLE 19-9

Component	Enthalpy of vaporization, cal/g mol	Specific heat at constant pressure, cal/g mol-°C		Boiling point, °C
		Liquid	Vapor	
Benzene	7360	33	23	80.1
Toluene	7960	40	33	110.6

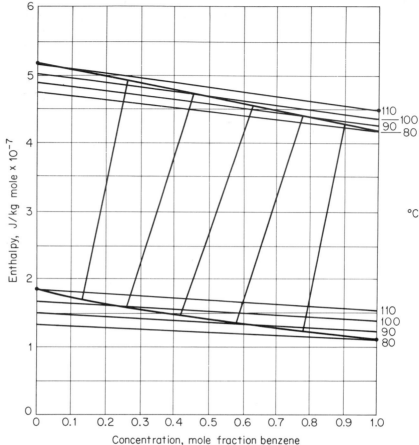

FIGURE 19-35
Enthalpy-concentration diagram for benzene-toluene at 1 atm.

operating flexibility, and, with extreme errors, an inoperative column. Correcting such errors after a plant has been built can be costly. Since many variables that influence plate efficiency depend on the design of the individual plates, the fundamentals of plate design are discussed first.

The extent and variety of rectifying columns and their applications are enormous. The largest units are usually in the petroleum industry, but large and very complicated distillation plants are encountered in fractionating solvents, in treating liquefied air, and in general chemical processing. Tower diameters may range from 1 ft (0.3 m) to more than 30 ft (9 m) and the number of plates from a few plates to scores. Plate spacings may vary from 6 in. or less to several feet. Formerly bubble-cap plates were most common; today most columns contain sieve trays or lift-valve plates. Many types of liquid distribution are specified. Columns may operate at high pressures or low, from temperatures of liquid gases up to 1600°F reached in the rectification of sodium and potassium vapors. The materials distilled can vary greatly in viscosity, diffusivity, corrosive nature, tendency to foam, and complexity of composition. Plate towers are as useful in absorption as in rectification, and the fundamentals of plate design apply to both operations.

Designing fractionating columns, especially large units and those for unusual applications, is best done by experts. Although the number of ideal plates and the heat requirements can be computed quite accurately without much previous experience, other design factors are not precisely calculable, and a number of equally sound designs can be found for the same problem. In common with most engineering activities, sound design of fractionating columns relies on a few principles, on a number of empirical correlations (which are in a constant state of revision), and much experience and judgment.

The following discussion is limited to the usual type of column, equipped with sieve plates, operating at pressures not far from atmospheric, and treating mixtures having ordinary properties.

Normal operation of sieve plate   The task of a sieve plate is to bring streams of liquid and vapor into intimate contact. To accomplish this definite forces are required to move the fluid streams to, through, and away from the mixing zone. Provision must also be made to obtain a reasonably clean separation between the liquid and vapor leaving the mixing zone.

Figure 19-36 shows a plate in a sieve-plate tower in normal operation. The various pressures and other driving forces are indicated on the figure. The driving force that overcomes the resistance to flow of vapor through the holes and the liquid on the plate is pressure. The pressure of the vapor below a plate is greater than that above the plate. This pressure drop across a single plate is usually 50 to 75 mm H_2O. The pressure drop in the entire column is the product of the number of plates and the pressure drop per plate. The pressure drop over a 40-plate column, then, is of the magnitude of 3 m H_2O. Ordinarily this pressure drop is not important, but in vacuum distillation it is the most critical point in the design, because the pressure drop is large with respect to the absolute pressure and the temperature in the base of the column may become too high. A minimum number of plates especially designed for low pressure drop must then be used. The pressure drop in any column is automatically developed by the reboiler, and if the capacity of the reboiler is sufficient to create the required amount of vapor against the back pressure of the column, the pressure drop is not under the control of the operator.

The driving force for the flow of liquid through the downpipe and onto the plate is static head in the downpipe. The driving force for moving the liquid across the plate is gravity, acting through the difference in level between the liquid entering the plate and that leaving it. This head is called the gradient and is denoted by Z_g. Except in very large columns, Z_g is negligible in a sieve-plate column. Finally, the head for carrying the liquid over the exit weir is the weir-crest head, denoted by Z_{cr}, as shown in Fig. 19-36.

It is good practice to provide a calming space between the last row of openings and the exit weir to allow complete release of vapor, so the downpipes receive only clear liquid. Vapor-free liquid then reaches the inlet sides of the plates, and the froth-

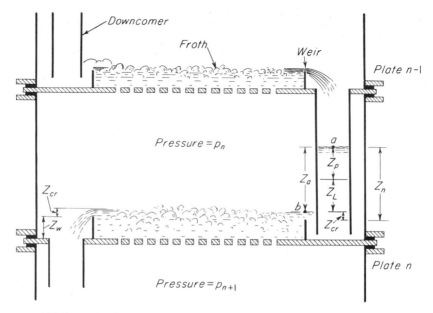

FIGURE 19-36
Normal operation of sieve plate: Z_a, height of station a above datum; Z_{cr}, weir crest; Z_L, liquid-friction head; Z_p, pressure head across plate; Z_n, net head in downpipe; Z_w, weir height.

ing action occurs only above the perforations. Since the density of the froth is much less than that of clear liquid, the depth of the frothy layer may be greater than the depth of clear liquid at either the inlet or outlet of the plate.

Head balance in downpipe   The net static head Z_n, shown in Fig. 19-36, is the difference in level between the surface of the liquid in the downpipe and the top of the exit weir. It is the sum of all driving forces, both liquid and vapor, for the plate. This can be demonstrated as follows.

The pressure drop in the vapor space between plates is negligible, so the entire plate pressure drop can be assumed to occur across the plate perforations and the frothy liquid on the plate. The pressure drop across plate n is $p_{n+1} - p_n$, where p_{n+1} is the pressure in the vapor space above plate $n + 1$ and p_n is that above plate n. To relate the pressure drop across the plate to the liquid level in the downpipe, write a Bernoulli equation for the liquid between station a, at the top of the liquid in the down-pipe, and station b, at the surface of the clear liquid just entering the bubbling section of plate n. Take the datum of heights through station b and neglect the change in velocity head. Then a Bernoulli relation [Eq. (4-30)] gives

$$\frac{p_a}{\rho_L} + \frac{gZ_a}{g_c} = \frac{p_b}{\rho_L} + h_{fL}$$

or

$$p_b - p_a = \frac{\rho_L Z_a g}{g_c} - h_{fL}\rho_L \tag{19-68}$$

where ρ_L = density of liquid

$\quad\quad Z_a$ = height of station a above datum

$\quad\quad h_{fL}$ = friction loss between stations a and b

Pressure drop $p_b - p_a$ is the pressure drop over plate $n - 1$. Since all plates are alike, this is also the pressure drop over plate n, and

$$p_{n+1} - p_n = \frac{\rho_L Z_a g}{g_c} - h_{fL}\rho_L$$

or

$$Z_a = \frac{g_c(p_{n+1} - p_n)}{g\rho_L} + \frac{g_c h_{fL}}{g} \tag{19-69}$$

Define Z_p, the pressure head across the plate, and Z_L, the friction head in the liquid, by the equations

$$Z_p \equiv \frac{g_c(p_{n+1} - p_n)}{g\rho_L} \quad\quad Z_L \equiv \frac{g_c h_{fL}}{g} \tag{19-70}$$

Then $Z_a = Z_p + Z_L$, and, from Fig. 19-36 and Eqs. (19-69) and (19-70), the net static head is

$$Z_n = Z_a + Z_{cr} = Z_p + Z_L + Z_{cr} \tag{19-71}$$

These equations are all in consistent units.

Equation (19-71) is a static-head balance in the downpipe. Head Z_p provides the driving force for overcoming the pressure drop through the plate; head Z_L overcomes the friction in the flow of the liquid through and out of the downpipe; and head Z_{cr} discharges the liquid over the weir. A major task of the designer of a sieve-plate column is to so proportion his design that a high capacity per unit plate area is attained without allowing any of the heads in the right-hand side of Eq. (19-71) to become so large that the operation of the column is impaired.

General approach to plate design The capacity of a column is expressed in terms of the rate of either vapor or liquid flow per unit plate area. Once the reflux ratio is chosen, the liquid flow per unit area is proportional to that of the vapor, and fixing one establishes the other. The general method of plate design is to choose a plate spacing, estimate the permissible vapor load, and calculate a tentative column cross section. Then a preliminary layout of sieve openings, weirs, and downpipes can be prepared. The designer estimates the various liquid heads and pressure drops, checks the level in the downpipe, and if all is in order, the tentative design can be refined and made final. If abnormal heads are found at any point, it may be necessary to modify the original assumptions of reflux ratio, plate spacing, or vapor velocity and to re-design the plate. In some columns, especially ones of large diameter or ones that handle a large ratio of liquid to vapor, the design is controlled by liquid capacity rather than by vapor rates, and to obtain a safe liquid load the final design may be underloaded with respect to vapor flow.

Construction details of plate columns To estimate the heads Z_L, Z_p, and Z_{cr}, several details of construction must be fixed. This discussion is based on the following typical design. Numerous other modifications are used in practice.

The plates are perforated with circular holes, typically $\frac{3}{16}$ in. in diameter, with a center-to-center spacing of $\frac{1}{2}$ to $\frac{3}{4}$ in. Closer spacing leads to excessive coalescence of the vapor bubbles; wider spacing results in inactive areas between openings. Holes are not placed within 1 to 2 in. of the wall of the column or within 2 to 4 in. of down-pipes or weirs.

The flow of liquid is across the plate, from the plate above to the outlet weir and downpipe to the plate below. The liquid flows in alternate directions on adjacent plates.

Two common types of weirs are considered. In small columns, the weir is formed by extending the circular downpipes the distance Z_w, the weir height, above the plate. One or more downpipes may be used on each plate. Seal cups are placed at the bottoms of the downpipes to prevent vapor from entering the pipes at the bottom. Flow in a circular downpipe may occur in various manners, depending on the size of the weir and on the influence of streams from the holes and from other nearby weirs. The type of flow assumed here is that obtained when the diameter of the downpipe is more than 4 times the weir-crest head. Then the liquid flows down the wall of the downpipe as a film until it reaches a definite level of clear liquid in the pipe. For smaller pipes, the pipe may fill with liquid, vapor may be entrained, and a vortex may form. In such situations, the downpipe is reaching a limit of its capacity, and down-pipe area can be the limiting factor in the entire design.

In larger columns, rectangular chord weirs are used. An example is shown in Fig. 19-36. The chord weir is a flat plate that cuts off a sector of the plate. Several circular downpipes, flush with the plate, may be used, or all or nearly all of the sector area may be used for a single large sector-shaped downpipe. An inlet chord weir is often used to provide a liquid seal at the bottom of the downpipes. The area of the plate assigned to the downpipe area behind the weirs may be 15 to 20 percent of the entire plate area.

Limiting vapor velocity At very low vapor velocities the plate efficiency is poor because of poor contact between liquid and vapor. At intermediate velocities in the usual operating range the plate efficiency is at a maximum and is nearly independent of vapor velocity. At high velocities the efficiency drops drastically, and the operating vapor load should be set below this point. Poor efficiencies at high velocities are the result of the following phenomena.

ENTRAINMENT At high velocities the drops of liquid in the vapor space above the liquid cannot settle against the upward flow of the vapor, and are swept out and mixed with the liquid on the next plate above. This dilutes the liquid on that plate and undoes some of the rectification performed by the plate. Unless entrainment is unusually severe, its effect on plate efficiency is small, but entrainment adds to the liquid load, and it may be important on the top plates if the overhead product must be free from color or small concentrations of impurity.

SPLASHING AND FOAMING At low spacings and high velocities liquid may geyser or spout clear across the vapor space into perforations of the plate above. Or if the liquid tends to form a persistent foam, the foam will also accompany the liquid into the next plate. The effect of either of these happenings is equivalent to heavy entrainment. The foam of this kind is not to be confused with the temporary froth that is a normal and important part of action on the plate. Foam is entirely disadvantageous. In the usual column foaming and splashing are unimportant.

CALCULATION OF LIMITING VAPOR VELOCITY Formerly it was thought that entrainment was the prime cause of the decrease of plate efficiency at high vapor flow rates. It was assumed that, with droplets of a definite but unspecified size settling under Newton's law, to prevent entrainment it would be necessary to keep the velocity of the vapor below the terminal velocity of the droplets. On this assumption, from Eq. (7-52) the maximum permissible vapor velocity would be proportional to $\sqrt{(\rho_L - \rho_V)/\rho_V}$,

or

$$u_c = K_V \sqrt{\frac{\rho_L - \rho_V}{\rho_V}} \tag{19-72}$$

where u_c = maximum permissible velocity based on area of bubbling section of plate (total area minus downpipe area)

K_V = empirical constant[14]

In terms of mass velocity, this equation can be written

$$G_c = u_c \rho_V = K_V \sqrt{\rho_V(\rho_L - \rho_V)} \tag{19-73}$$

In spite of the fact that it is now known that entrainment of liquid drops is not the controlling factor in fixing the permissible vapor velocity, Eqs. (19-72) and (19-73) have survived as empirical relationships for calculating limiting vapor flows. Constant K_V has been evaluated from plant data, and various correlations of K_V against operating variables have been made. One such correlation is shown in Fig. 19-37, in which K_V is shown as a function of plate spacing and surface tension of the liquid. The plate spacing is the distance between the top of the clear liquid and the bottom of the next higher plate.

Since ρ_V is usually negligible in comparison with ρ_L, and since ρ_L is roughly the same for most liquids, Eq. (19-73) is often simplified to

$$u_c \sqrt{\rho_V} = K_V \sqrt{\rho_L} = F \tag{19-74}$$

When u_c is in feet per second and ρ_L is in pounds per cubic foot, factor F is between 0.5 and 1.5 in the normal operations of sieve-plate columns [10a]

Pressure head across plate The head Z_p across the plate can be divided into two parts. The first, denoted by Z_V, is the friction head of the vapor flowing through the perforations in the plate. The second, denoted by Z_s, is the static head of liquid over the perforations. Then

$$Z_p = Z_V + Z_s \tag{19-75}$$

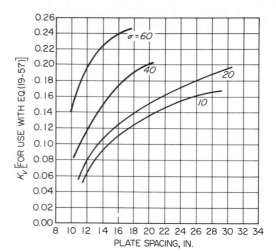

FIGURE 19-37
Sieve-tray column capacities: $\sigma =$ surface tension in dynes per centimeter. [*By permission, from J. H. Perry (ed.), "Chemical Engineers' Handbook," 4th ed. Copyright*, 1963, *McGraw-Hill Book Company*.]

FRICTION LOSS IN PERFORATIONS   This loss is given by the empirical equation

$$Z_V = \frac{8.04}{C_o^2} \frac{u_o^2 \, \rho_V}{g \, \rho_L} \tag{19-76}$$

where u_o is the vapor velocity through the openings. Coefficient C_o depends on the ratio of plate thickness t_p to orifice diameter D_o; C_o ranges from 0.6 for thin plates ($t_p/D_o = 0.1$) to 0.9 for thick plates ($t_p/D_o = 1$). The recommended range[10b] of t_p/D_o is between 0.4 and 0.7.

STATIC HEAD   Head loss Z_s equals the sum of the weir height Z_w, the weir crest Z_{cr}, and one-half the gradient head Z_g (usually negligible). The weir height is 0 to 0.5 in. for vacuum operation; 0.5 to 2 in. for atmospheric-pressure operation; and 1 to 4 in. for distillation under pressure. Crest head Z_{cr}, in feet, over a circular weir at the top of a downpipe is given by the dimensional equations

$$Z_{cr} = 0.46 \left(\frac{q_L}{b} \right)^{0.71} = 0.20 \left(\frac{q_L}{D_d} \right)^{0.71} \tag{19-77}$$

where $q_L =$ liquid flow rate, ft³/s
   $b =$ perimeter of pipe, ft
   $D_d =$ diameter of pipe, ft

For a chord weir, the equation is

$$Z_{cr} = 0.45 \left(\frac{q_L}{b} \right)^{2/3} \tag{19-78}$$

where b is the length of the weir.

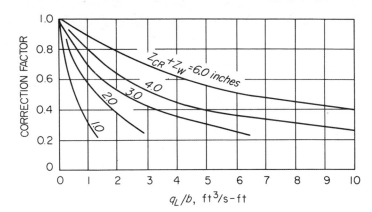

FIGURE 19-38
Correction to chord-weir equation. (*By permission, from C. S. Robinson and E. W. Gilliland, "Elements of Fractional Distillation," 4th ed. Copyright, 1950, McGraw-Hill Book Company.*)

Equation (19-78) applies where the velocity of approach to the weir is negligible. Because of the influence of the wall of the column, this is not true on a sieve plate. Figure 19-38 gives a correction, to be multiplied into the right-hand side of Eq. (19-78), that allows for velocity of approach.[12c] The factor depends on q_L/b and $Z_{cr} + Z_w$.

Friction head in liquid Friction in the liquid is also composed of two parts, the friction in the straight pipe, and the reversal loss at the bottom of the downpipe. For the free flow assumed for downpipes, the friction in the straight pipe is usually negligible. It can be estimated on the basis that 40 diameters of pipe give a friction loss of 1 velocity head.

The reversal loss at the bottom of the downpipe can be taken as three velocity heads, based on the maximum velocity at the base of the downpipe. This maximum velocity may be either that in the downpipe or that at the end of the pipe in the seal cup, depending on which cross section is the smaller. Assuming that the maximum velocity is through the pipe, the friction loss Z_L is

$$Z_L = \left(3 + \frac{L_d}{40D_d}\right)\frac{u_d^2}{2g_c}\frac{g_c}{g} = \left(1.5 + \frac{L_d}{80D_d}\right)\frac{u_d^2}{g} \qquad (19\text{-}79)$$

where u_d = velocity in downpipe
L_d = length of downpipe
D_d = diameter of downpipe
g = acceleration of gravity

If the maximum velocity is elsewhere than in the downpipe, use the maximum velocity instead of u_d and neglect $L_d/80D_d$. The liquid velocity in the downpipe below the liquid level should not be more than about 0.5 ft/s.

When the net head required in the downpipe exceeds the plate spacing, liquid accumulates on the plates and the column floods.

Plate Efficiency

To translate ideal plates into actual plates the plate efficiency must be known. The following discussion applies to both absorption and fractionating columns.

Types of plate efficiency   Three kinds of plate efficiency are used: (1) overall efficiency, which concerns the entire column; (2) Murphree efficiency, which has to do with a single plate; and (3) local efficiency, which pertains to a specific location on a single plate.

The *overall efficiency* η_o is simple to use but is the least fundamental. It is defined as the ratio of the number of ideal plates needed in an entire column to the number of actual plates.[6] For example, if 6 ideal plates are called for and the plate efficiency is 60 percent, the number of actual plates is $6/0.60 = 10$.

The *Murphree efficiency*[9] η_M is defined by

$$\eta_M = \frac{y_n - y_{n+1}}{y_n^* - y_{n+1}} \tag{19-80}$$

where y_n = actual concentration of vapor leaving plate n
      y_{n+1} = actual concentration of vapor entering plate n
      y_n^* = concentration of vapor in equilibrium with liquid leaving downpipe from plate n

The *local efficiency* η' is defined by

$$\eta' = \frac{y_n' - y_{n+1}'}{y_{en}' - y_{n+1}'} \tag{19-81}$$

where y_n' = concentration of vapor leaving specific location in plate n
     y_{n+1}' = concentration of vapor entering plate n at same location
     y_{en}' = concentration of vapor in equilibrium with liquid at same location

Since y_n' cannot be greater than y_{en}', a local efficiency cannot be greater than 1.00, or 100 percent.

RELATION BETWEEN MURPHREE AND LOCAL EFFICIENCIES   In small columns, the liquid on a plate is sufficiently agitated by vapor flow through the perforations for there to be no measurable concentration gradients in the liquid as it flows across the plate. The concentration of the liquid in the downpipe x_n is that of the liquid on the entire plate. Since the concentration of the liquid on the plate is constant, that of the vapor from the plate is also constant and no gradients exist in the vapor streams. A comparison of the quantities in Eqs. (19-80) and (19-81) shows that $y_n' = y_n$, $y_{n+1} = y_{n+1}'$, and $y_n^* = y_{en}'$. Then $\eta_M = \eta'$, and the local and Murphree efficiencies are equal.

In larger columns, liquid mixing in the direction of flow is not complete, and a concentration gradient does exist in the liquid on the plate. The maximum possible variation is from a concentration of x_{n-1} at the liquid inlet to a concentration of x_n at the liquid outlet. To trace the effect of such a concentration gradient on the Murphree efficiency, consider two situations: plate a, where mixing is complete, and plate b, where no liquid mixing occurs. Assume that y_{n+1}, x_n, and x_{n-1} are identical in the two cases. The quantity y_n^* is also the same for both plates, because this is the equilibrium concentration corresponding to liquid concentration x_n. In plate b, however, y_n is greater than that in plate a, because some of the vapor in plate b has the benefit of contact with liquid of concentration x_{n-1}, which is greater than the concentration of any of the liquid on plate a. The Murphree efficiency, by Eq. (19-80), is higher for plate b than for plate a. Also, when mixing does not occur, the Murphree efficiency is greater than the local efficiency. It is possible, on an unmixed plate, for y_n to be greater than y_n^*. The Murphree efficiency is then greater than 100 percent. This has been observed in large columns.[1]

The relation between η_M and η' has been established mathematically for the following three situations:[7]

Case 1 Vapor completely mixed, liquid unmixed

Case 2 Vapor unmixed, liquid unmixed, with the flow of liquid in the same direction on all plates

Case 3 Vapor unmixed, liquid unmixed, with flow of liquid in opposite directions on adjacent plates

The usual sieve-plate column acts in a manner intermediate between cases 1 and 3. Special construction is needed to obtain conditions specified in case 2.

The results of this analysis are shown in Fig. 19-39, in which the ratio η_M/η' is shown as a function of η' and mV/L for each of the three cases. The slope of the equilibrium curve m, the slope of the operating line L/V, and the Murphree efficiency are assumed to be constant.

RELATION BETWEEN MURPHREE AND OVERALL EFFICIENCIES   There is a definite relation between the overall efficiency η_o and the Murphree efficiency η_M when the equilibrium and the operating line are both straight. This is[7]

$$\eta_o = \frac{\log \left[1 + \eta_M(mV/L - 1) \right]}{\log (mV/L)} \tag{19-82}$$

Equation (19-82) is plotted in Fig. 19-40. When the operating and equilibrium lines are parallel, or where both efficiencies are nearly 1.0, $\eta_M = \eta_o$. Otherwise there is a considerable difference between η_o and η_M, the magnitude of which depends on the ratio of the slopes of the two lines. There is no simple relationship between these efficiencies when either the equilibrium line or the operating line is strongly curved.

Use of Murphree efficiency   When the Murphree efficiency is known, it can readily be used in the McCabe-Thiele diagram. The diagram for an actual plate as compared with that for an ideal plate is shown in Fig. 19-41. Triangle acd represents the ideal plate and triangle abe the actual plate. The actual plate, instead of enriching the vapor from y_{n+1} to y_n^*, shown by line segment ac, accomplishes a lesser enrichment,

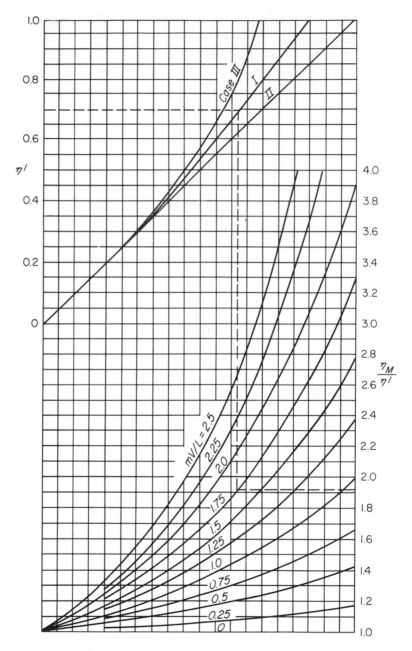

FIGURE 19-39
Relation between local and Murphree efficiencies. (*By permission, from C. S. Robinson and E. W. Gilliland, "Elements of Fractional Distillation," 4th ed. Copyright, 1950, McGraw-Hill Book Company.*)

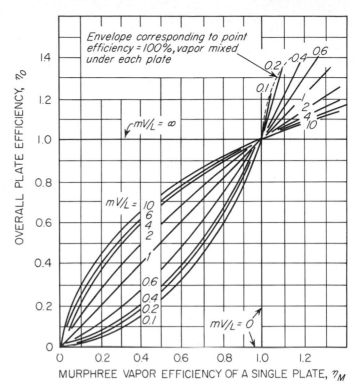

FIGURE 19-40

Relation between Murphree and overall efficiencies. [*By permission, from J. H. Perry (ed.), "Chemical Engineers' Handbook," 3d ed. Copyright, 1950, McGraw-Hill Book Company.*]

$y_n - y_{n+1}$, shown by line segment *ab*. By definition of η_M, the Murphree efficiency is given by the ratio *ab/ac*. To apply a known Murphree efficiency to an entire column, it is necessary only to replace the true equilibrium curve y_e vs. x_e by an effective equilibrium curve y_e' vs. x_e, whose ordinates are calculated from the equation

$$y_e' = y + \eta_M(y_e - y) \tag{19-83}$$

In Fig. 19-41 an effective equilibrium curve for $\eta_M = 0.60$ is shown. Note that the position of the y_e'-vs.-x_e curve depends on both the operating line and the true equilibrium curve. Once the effective equilibrium curve has been plotted, the usual step-by-step construction is made and the number of actual plates determined. The reboiler is not subject to a discount for plate efficiency, and the true equilibrium curve is used for the last step in the stripping column.

An analogous construction using η_M can be used in the enthalpy-concentration method.[2]

When the operating and equilibrium lines are both straight, the number of actual plates can be calculated by either of two equivalent methods. In the first meth-

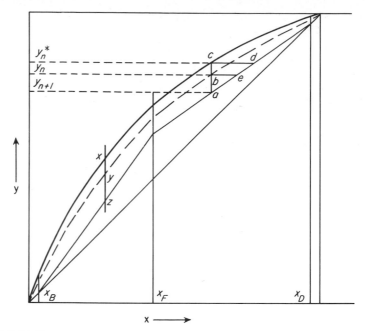

FIGURE 19-41
Use of Murphree efficiency of xy diagram. Dotted line is effective equilibrium curve, y'_e, vs. x_e for $\eta_M = 0.60$; $ba/ca = yz/xz = 0.60$.

od, the number of ideal plates is calculated by Eq. (18-27) and η_o by Eq. (19-82). Dividing the number of ideal plates by η_o gives the number of actual plates. The second method is to correct Eq. (18-27) for η_M and obtain the number of actual plates directly. To obtain positive factors Eq. (18-27) may be written for the fractionating column diagrammed in Fig. 19-42 as

$$N = \frac{\log\left[(y_a^* - y_a)/(y_b^* - y_b)\right]}{\log\left[(y_a^* - y_b^*)/(y_a - y_b)\right]} \tag{19-84}$$

For actual plates, when η_M is constant, the logarithmic numerator in Eq. (19-84) is unchanged since both $y_a^* - y_a$ and $y_b^* - y_b$ are multiplied by η_M and their ratio is unchanged. Also, $y_a - y_b$, the vapor-concentration range covered by the operating line, is unchanged. The correction of Eq. (19-84) for plate efficiency affects only the term $y_a^* - y_b^*$. As shown by Fig. 19-42, this difference is corrected by replacing y_a^* and y_b^* in the denominator by

$$(y_a^*)' = y_a + \eta_M(y_a^* - y_a) \qquad \text{and} \qquad (y_b^*)' = y_b + \eta_M(y_b^* - y_b)$$

When Eq. (19-84) is so corrected, it gives the number of actual plates directly.

Factors influencing plate efficiency   Although thorough studies of plate efficiency have been made,[3-5] the estimation of efficiency is largely empirical. Sufficient data

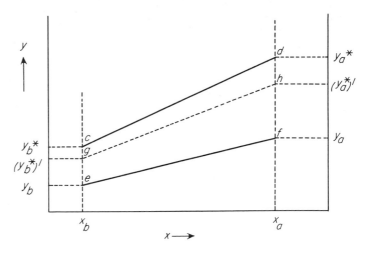

FIGURE 19-42
Linear equilibrium and operating lines for use with η_M: cd, true equilibrium line, y_e vs. x_e; ef, operating line, y_{n+1} vs. x_n; gh, effective equilibrium line, y_e' vs. x_e.

are at hand, however, to show the major factors involved and to provide a basis for estimating the efficiencies for conventional types of columns operating on mixtures of common substances.

The most important requirement for obtaining satisfactory efficiencies is that the plates operate properly. Adequate and intimate contact between vapor and liquid is essential. Any misoperation of the column, such as excessive foaming or entrainment, poor vapor distribution, or short-circuiting, weeping, or dumping of liquid, lowers the plate efficiency.

Plate efficiency is a function of the rate of mass transfer between liquid and vapor. The prediction of mass-transfer coefficients in sieve trays and their relationship to plate efficiency are discussed in Chap. 22. Typical Murphree efficiencies for a sieve-plate column[5] range from 0.60 to 0.75. The vapor velocity has relatively little effect on plate efficiency provided it is high enough to prevent weeping of liquid through the perforations and low enough to avoid excessive entrainment. Detailed information on allowable vapor velocities and efficiencies of sieve plates, bubble-cap plates, and other plate designs is given in the literature.[10c]

Special sieve trays[13]   In some columns equipped with counterflow trays no down-comers are used. There is no crossflow of the liquid; the liquid and vapor pass counter-currently through relatively large holes in the tray. Liquid drains through some holes at one instant, through others somewhat later, with the vapor passing upward through the remainder. Such columns have some advantages in cost over conventional columns and do not foul easily, but their *turndown ratio* (the ratio of the minimum allowable vapor velocity to the lowest velocity at which the column will operate satisfactorily) is low, usually 2 or less.

In a valve-tray column the openings in the plate are quite large, typically $1\frac{1}{2}$ in. (38 mm) for circular perforations and $\frac{1}{2}$ by 6 in. (13 by 150 mm) for rectangular slots. The openings are covered with lids or "valves," which rise and fall as the vapor rate varies, providing a variable area for vapor passage. Downcomers and crossflow of the liquid are used as in the regular sieve trays. Valve trays are more expensive than conventional trays but have the advantage of a large turndown ratio, up to 10 or more, so the operating range of the column is large. Counterflow trays and valve trays are nearly always of proprietary design.

Rectification in packed towers   Packed towers are often used for distillation operations, especially when the separation is relatively easy. Most packed distillation towers are small, 1 m in diameter or less. The principles of their operation are the same as those of absorption columns, as discussed in Chap. 23. As with absorption columns, their performance depends largely on the effectiveness of the liquid distribution over the packing. Details of packed column design are given in Ref. 10d.

SYMBOLS

B Flow rate of bottoms product, mol/h, lb/h, or kg/h

b Perimeter of downpipe or length of weir, ft or m

C_0 Discharge coefficient, flow through perforations of sieve plate

C_p Molal specific heat of pure component at constant pressure

c_p Specific heat at constant pressure; c_{pL}, of liquid; c_{pV}, of vapor

D Flow rate of overhead product, mol/h, lb/h, or kg/h

D_d Inside diameter of downpipe, ft or m

D_0 Diameter of perforation of sieve plate, ft or m

F Feed rate, mol/h, lb/h, or kg/h; also factor for estimating column capacity, defined by Eq. (19-74)

f Fraction of feed that is vaporized

G_c Maximum permissible mass velocity of vapor in column, based on area of bubbling section, lb/ft²-s or kg/m²-s

g Acceleration of gravity, ft/s²

g_c Newton's-law proportionality factor, 32.174 ft-lb/lb$_f$-s²

H Enthalpy, energy per mole or per unit mass; H_B, of bottom product; H_D, of overhead product; H_F, of feed; H_T, total enthalpy of mixture; H_x, of saturated liquid; H_{xm}, of liquid from plate m of stripping column; H_{xn}, of liquid from plate n of rectifying column; H_y, of saturated vapor; $H_{y(m+1)}$, of vapor from plate $m + 1$ of stripping column; $H_{y(n+1)}$, of vapor from plate $n + 1$ of rectifying column; H_{y1}, of vapor from top plate; H'_B, of bottom product, corrected for heat effect q_r; H'_D, of overhead product corrected for heat effect $-q_c$; H'_{Dm}, of overhead product for minimum reflux, corrected for heat effect $-q_c$; H^L, H^V, of liquid and vapor phases, respectively; H°_A, H°_B, H°_i, enthalpies of one mole of pure components A, B, and i, respectively

h_{fL} Friction loss between stations a and b, ft-lb$_f$/lb or J/g

h Partial molal enthalpy, energy per mole; $\bar{h}_A$, $\bar{h}_B$, $\bar{h}_i$, of components A, B, and i, respectively; $\bar{h}^L_i$, of component i in liquid phase

K_V Constant in Eq. (19-72)

L Flow rate of liquid in general or in rectifying column, mol/h, lb/h, or kg/h; L_a, entering top of column; L_b, leaving bottom of column; L_c, of reflux from condenser; L_m, from plate m of stripping column; L_n, from plate n of rectifying column; L, in stripping column

L_d Length of downpipe, ft or m

M Molecular weight; M_A, M_B, of components A and B, respectively

m Serial number of plate in stripping column, counting from feed plate; also, slope of equilibrium curve, dy_e/dx_e

$\dot{m}$ Mass flow rate, lb/h or kg/h; $\dot{m}_A$, $\dot{m}_B$, of components A and B, respectively; $\dot{m}_c$, of cooling water to condenser; $\dot{m}_s$, of steam to reboiler

N Number of ideal plates; $N_{\min}$, minimum number of ideal plates

n Serial number of plate in rectifying column, counting from top; also, number of moles in still or mixture; n_A, n_B, n_i, n_j, of components A, B, i and j, respectively; n_0, moles charged to still; n_{0A}, n_{0B}, of components A and B respectively; n_1, final value

p Pressure, lb_f/ft^2 or N/m^2; p_a, at upstream station a; p_b, at downstream station b; p_{n-1}, p_n, p_{n+1}, in vapor space above plates $n-1$, n, and $n+1$, respectively; also vapor pressure; p_A, p_B, of components A and B, respectively

$\bar{p}$ Partial pressure, lb_f/ft^2 or N/m^2; $\bar{p}_A$, $\bar{p}_B$, of components A and B, respectively

q Rate of heat flow, Btu/h or W; $-q_c$, heat rejected at condenser; q_r, heat added at reboiler; also, moles of liquid to stripping section of column per mole of feed

q_L Volumetric flow rate of liquid in downpipe, ft^3/s or m^3/s

R Reflux ratio; $R_D = L/D$; $R_V = L/V$; R_{Dm}, minimum reflux ratio; also gas-law constant

T Temperature, °F or °C; T_F, feed temperature; T_b, bubble point; T_d, dew point; T_{n-1}, T_n, T_{n+1}, temperature on plates $n-1$, n, $n+1$, respectively; $T_2 - T_1$, temperature rise of cooling water

t_p Thickness of sieve plate, ft or m

u Linear velocity, ft/s or m/s; u_c, maximum permissible vapor velocity, based on area of bubbling section; u_d, velocity of liquid in downpipe; u_0, vapor velocity through perforations in sieve plate

V Flow rate of vapor, in general or in rectifying column, mol/h, lb/h, or kg/h; V_a, from top of column; V_b, entering bottom of column; V_{m+1}, from plate $m+1$ in stripping column; V_n, V_{n+1}, from plates n and $n+1$, respectively, in rectifying column; V_1, from top plate to condenser; $\bar{V}$, in stripping column

x Mole fraction or mass fraction of component A in liquid; x_B, in bottoms product; x_D, in overhead product; x_F, in feed; x_a, in liquid entering single-section column; x_b, in liquid leaving single-section column; x_c, in reflux from condenser; x_e, in equilibrium with vapor of composition y_e; x_f, in liquid leaving feed plate; x_m, in liquid from plate m of stripping column; x_{n-1}, x_n, in liquid from plates $n-1$ and n, respectively, of rectifying column; x_0, x_1, initial and final values, respectively

y Mole fraction or mass fraction of component A in vapor; y_D, in vapor overhead product; y_a, in vapor leaving single-section column; y_b, in vapor entering single-section column; y_e, in equilibrium with liquid of concentration x_e; y_f, in vapor leaving feed plate; y_{m+1}, of vapor from plate $m+1$ in stripping column; y_n, y_{n+1}, in vapor from plates n and $n+1$, respectively, in rectifying column; y_r, in vapor leaving reboiler; y^*, in vapor in equilibrium with specific stream of liquid; y_a^*, in equilibrium with x_a; y_b^*, in equilibrium with x_b; y_n^*, in equilibrium with x_n; y_e', effective equilibrium concentration in vapor corresponding to Murphree efficiency [Eq. (19-83)]; y_{en}', in equilibrium with liquid at specific location on plate n; y_{n+1}', in vapor entering plate n from plate $n+1$, at specific location on plate n

Z Distance above datum, or head, ft or m; Z_L, friction head in liquid; Z_V, friction head, vapor flow through perforations; Z_a, height of station a above datum; Z_{cr}, weir-crest head; Z_g, gradient across plate; Z_n, net static head, from level in downpipe to top of weir, Z_p, pressure head across plate, in vapor stream; Z_s, liquid seal above perforations; Z_w, weir height

Greek letters

α_{AB} Relative volatility, component A relative to component B

γ_i Activity coefficient of component i

η Efficiency; η_M, Murphree plate efficiency; η_o, overall plate efficiency; η_s, vaporization efficiency in steam distillation; η', local plate efficiency

λ Latent heat of vaporization, energy per unit mass; λ_s, of steam

ρ Density, lb/ft^3 or kg/m^3; ρ_L, of liquid; ρ_V, of vapor

σ Surface tension, dyn/cm

ϕ_i Fugacity coefficient of component i

PROBLEMS

19-1 A liquid containing 30 mole percent toluene, 40 mole percent ethylbenzene, and 30 mole percent water is subjected to a continuous flash distillation at a total pressure of 0.5 atm. Vapor-pressure data for these substances are given in Table 19-7. Assuming that mixtures of ethylbenzene and toluene obey Raoult's law and that the hydrocarbons are completely immiscible in water, calculate the temperature and compositions of liquid and vapor phases (*a*) at the bubble point, (*b*) at the dew point, (*c*) at the 50 percent point (one-half of the feed leaves as vapor and the other half as liquid).

19-2 The boiling-point–equilibrium data for the system acetone-methanol at 760 mm Hg are given in Table 19-7. A column is to be designed to separate a feed analyzing 25 mole percent acetone and 75 mole percent methanol into an overhead product containing 78 mole percent acetone and a bottom product containing 1.0 mole percent acetone. The feed enters as an equilibrium mixture of 30 percent liquid and 70 percent vapor. A reflux ratio equal to twice the minimum is to be used. An external reboiler is to be used. Bottom product is removed from the reboiler. The condensate (reflux and overhead product) leaves the condenser at 25°C, and the reflux enters the column at this temperature. The molal latent heats of both components are 7,700 g cal/g mol. The Murphree plate efficiency is 70 percent. Calculate (*a*) the number of plates required above and below the feed; (*b*) the heat required at the reboiler, in Btu per pound mole of overhead product; (*c*) the heat removed in the condenser, in Btu per pound mole of overhead product.

19-3 A plant must distill a mixture containing 80 mole percent methanol and 20 percent water. The overhead product is to contain 99.99 mole percent methanol and the bottom product 0.005 mole percent. The feed is cold, and for each mole of feed 0.4 mol of vapor is condensed at the feed plate. The reflux ratio at the top of the column is 1.35, and the reflux is at its bubble point. Calculate (*a*) the minimum number of plates; (*b*) the minimum reflux ratio; (*c*) the number of plates using a total condenser and a still, assuming an average Murphree plate efficiency of 70 percent; (*d*) the number of plates using a still and a partial condenser operating with the reflux in equilibrium with the vapor going to a final condenser. Equilibrium data are given in Table 19-9.

19-4 An equimolal mixture of benzene and toluene is to be separated in a bubble-plate tower at the rate of 100 lb mol/h at 1 atm pressure. The overhead product must contain at least 98 mole percent benzene. The feed is saturated liquid. A tower is available containing 24 plates. Feed may be introduced either on the eleventh or the seventeenth plate from the top. The maximum vaporization capacity of the reboiler is 120 lb mol/h. The plates are about 50 percent efficient. How many moles per hour of overhead product can be obtained from this tower?

19-5 An aqueous solution of a volatile component *A* containing 7.94 mole percent *A* preheated to its boiling point is to be fed to the top of a continuous stripping column operated at atmospheric pressure. Vapor from the top of the column is to contain 11.25 mole percent *A*. No reflux is to be returned. Two methods are under consideration, both calling for the same expenditure of heat, namely, a vaporization of 0.562 mol per mole feed in each case. Method 1 is to use a still at the bottom of a plate column, generating vapor by use of steam condensing inside a closed coil in the still. In method 2 the still and heating coil are omitted, and live steam is injected directly below the bottom plate. Equilibrium data are given in Table 19-10. The usual simplifying assumptions may be made. What are the advantages of each method?

19-6 A rectifying column containing the equivalent of three ideal plates is to be supplied continuously with a feed consisting of 0.5 mole percent ammonia and 99.5 mole percent

Table 19-7 VAPOR PRESSURES OF ETHYLBENZENE, TOLUENE, AND WATER

Temperature, °C	Vapor pressure, mm Hg		
	Ethylbenzene	Toluene	Water
50	53.8		92.5
60	78.6	139.5	149.4
70	113.0	202.4	233.7
80	160.0	289.4	355.1
90	223.1	404.6	525.8
100	307.0	557.2	760.0
110	414.1		
110.6		760.0	
120	545.9		

Table 19-8 SYSTEM ACETONE-METHANOL

Temperature, °C	Mole fraction acetone		Temperature, °C	Mole fraction acetone	
	Liquid	Vapor		Liquid	Vapor
64.5	0.00	0.000	56.7	0.50	0.586
63.6	0.05	0.102	56.0	0.60	0.656
62.5	0.10	0.186	55.3	0.70	0.725
60.2	0.20	0.322	55.05†	0.80	0.80
58.65	0.30	0.428	56.1	1.00	1.00
57.55	0.40	0.513			

† Azeotrope.

Table 19-9 EQUILIBRIUM DATA FOR METHANOL-WATER

x	0.1	0.2	0.3	0.4	0.5	0.6	0.7	0.8	0.9	1.0
y	0.417	0.579	0.669	0.729	0.780	0.825	0.871	0.915	0.959	1.0

Table 19-10 EQUILIBRIUM DATA IN MOLE FRACTION A

x	0.0035	0.0077	0.0125	0.0177	0.0292	0.0429	0.0590	0.0784
y	0.0100	0.0200	0.0300	0.0400	0.0600	0.0800	0.1000	0.1200

water. Before entering the column the feed is converted wholly into saturated vapor, and it enters between the second and third plates from the top of the column. The vapors from the top plate are totally condensed but not cooled. Per mole of feed, 1.3 mol of condensate is returned to the top plate as reflux, and the remainder of the distillate is removed as overhead product. The liquid from the bottom plate overflows to a reboiler, which is heated by closed steam coils. The vapor generated in the reboiler enters the column below the bottom plate, and bottom product is continuously removed from the reboiler. The vaporization in the reboiler is 0.6 mol per mole of feed. Over the concentration range involved in this problem, the equilibrium relation is given by the equation

$$y = 12.6x$$

Calculate the mole fraction of ammonia in (a) the bottom product from the reboiler, (b) the overhead product, (c) the liquid reflux leaving the feed plate.

19-7 A tower containing six ideal plates, a reboiler, and a total condenser is used to separate, partially, oxygen from air at 65 $lb_f/in.^2$ gauge. It is desired to operate at a reflux ratio (reflux to product) of 2.5 and to produce a bottom product containing 45 weight percent oxygen. The air is fed to the column at 65 $lb_f/in.^2$ gauge and 25 percent vapor by mass. The enthalpy of oxygen-nitrogen mixtures at this pressure is given in Table 19-11. Compute the composition of the overhead if the vapors are just condensed but not cooled.

19-8 It is desired to produce an overhead product containing 80 mole percent benzene from a feed mixture of 70 mole percent benzene and 30 percent toluene. The following methods are considered for this operation. All are to be conducted at atmospheric pressure. For each method calculate the moles of product per 100 mol of feed and the number of moles vaporized per 100 mol feed. (a) Continuous equilibrium distillation. (b) Continuous distillation in a still fitted with a partial condenser, in which 50 mole percent of the entering vapors are condensed and returned to the still. The partial condenser is so constructed that vapor and liquid leaving it are in equilibrium, and holdup in it is negligible.

19-9 The operation of a fractionating column is circumscribed by two limiting reflux ratios: one corresponding to the use of an infinite number of plates and the other a total-reflux, or infinite-reflux, ratio. Consider a rectifying column fed at the bottom with a constant flow of a binary vapor having a constant composition, and assume also that the column has an infinite number of plates. (a) What happens in such a column operating at both extremes

Table 19-11 ENTHALPY OF OXYGEN-NITROGEN AT 65 $lb_f/in.^2$ GAUGE

Temperature, °C	Liquid		Vapor	
	N_2, wt %	H_x, cal/g mol	N_2, wt %	H_y, cal/g mol
−163	0.0	420	0.0	1840
−165	7.5	418	19.3	1755
−167	17.0	415	35.9	1685
−169	27.5	410	50.0	1625
−171	39.0	398	63.0	1570
−173	52.5	378	75.0	1515
−175	68.5	349	86.0	1465
−177	88.0	300	95.5	1425
−178	100.0	263	100.0	1405

at the same time? (*b*) Assume that a finite quantity of product is withdrawn from the top of this column. What happens as more and more product is withdrawn in successive stages if each stage achieves steady state between changes?

19-10 A saturated liquid feed containing 40 mole percent *n*-hexane and 60 mole percent *n*-octane is fed to a small distillation column at a rate of 100 g mol/h. A reflux ratio, $L_a/D = 1.2$, is maintained. The overhead product is 95 mole percent hexane, and the bottom product is 10 mole percent hexane. If each theoretical plate loses 80,000 cal/h, step off the plates on a Ponchon diagram, taking into account the column heat loss. Physicochemical data are given in Tables 19-12 and 19-13. (*a*) Determine the heat added to the reboiler in calories per hour. (*b*) Find the number of theoretical plates excluding the reboiler. Assume that a total condenser is used.

19-11 It is desired to produce an overhead product containing 80 mole percent benzene from a feed mixture of 70 mole percent benzene and 30 mole percent toluene. The following methods are considered for this operation. All are to be conducted at atmospheric pressure. For each method calculate the moles of product per 100 mol of feed and the number of moles vaporized per 100 mol feed.

(*a*) Continuous equilibrium distillation.

(*b*) Differential distillation.

(*c*) Continuous distillation in a still fitted with a partial condenser, in which 50 mole percent of the entering vapors are condensed and returned to the still. The partial condenser is so constructed that vapor and liquid leaving it are in equilibrium, and holdup in it is negligible.

(*d*) Batch distillation, with 50 mole percent of the vapors entering the partial condenser returned to the still. The liquid and vapor leaving the condenser are in equilibrium.

Table 19-12   VAPOR-LIQUID EQUILIBRIUM DATA, 1 atm MOLE FRACTION HEXANE

x, in liquid	0	0.1	0.3	0.5	0.55	0.7	1.0
y, in vapor	0	0.36	0.70	0.85	0.90	0.95	1.0

Table 19-13   ENTHALPY-CONCENTRATION DATA

Concentration, mole fraction hexane	Enthalpy, cal/g mol	
	Saturated liquid	Saturated vapor
0	7000	15,700
0.1	6300	15,400
0.3	5000	14,700
0.5	4100	13,900
0.7	3400	12,900
0.9	3100	11,600
1.0	3000	10,000

19-12 A laboratory still is charged with 10 l of a methanol-water mixture containing 0.70 mole fraction methanol. This is to be distilled batchwise without reflux at 1 atm pressure until 5 l of liquid remains in the still, that is, 5 l has been boiled off. The rate of heat input is constant at 1,000 cal/s. The partial molar volumes are for methanol, 40.5 cm^3/g mol; for water, 18 cm^3/g mol. Neglecting any volume changes on mixing, and using an average heat of vaporization of 9,500 cal/g mol, calculate (*a*) the time t_T required to boil off 5 l; (*b*) the mole fraction of methanol left in the still at times $t_T/2$, $3t_T/4$, and t_T; (*c*) the average composition of the total distillate at time t_T. Equilibrium data for the system methanol-water are given in Table 19-9.

REFERENCES

1   Brown, G. G., M. Souders, Jr., H. V. Nyland, and W. H. Hessler: *Ind. Eng. Chem.,* **27**:383 (1935).
2   Brown, G. G., and associates: "Unit Operations," p. 343, Wiley, New York, 1950.
3   Cheng, S. I., and A. J. Teller: *AIChE J.,* **7**:282 (1961).
4   Gerster, J. A., et al.: Efficiencies in Distillation Columns, *AIChE Res. Comm. Final Rept. Univ. Delaware,* 1958.
5   Jones, J. B., and C. Pyle: *Chem. Eng. Prog.,* **51**:424 (1955).
6   Lewis, W. K.: *Ind. Eng. Chem.,* **14**:492 (1922).
7   Lewis, W. K., Jr.: *Ind. Eng. Chem.,* **28**:399 (1936).
8   McCabe, W. L., and E. W. Thiele: *Ind. Eng. Chem.,* **17**:605 (1925).
9   Murphree, E. V.: *Ind. Eng. Chem.,* **17**:747 (1925).
10   Perry, J. H. (ed.): "Chemical Engineers' Handbook," 5th ed., McGraw-Hill, New York, 1973; (*a*) p. **18**-6, (*b*) p. **18**-8, (*c*) pp. **18**-12 to **18**-19, (*d*) pp. **18**-19 to **18**-49.
11   Randall, M., and B. Longtin: *Ind. Eng. Chem.,* **30**:1063, 1188 (1938).
12   Robinson, C. S., and E. R. Gilliland: "Elements of Fractional Distillation," 4th ed., McGraw-Hill, New York, 1950; (*a*) p. 127, (*b*) p. 146, (*c*) p. 411.
13   Smith, B. D.: "Design of Equilibrium Stage Processes," pp. 565–567, McGraw-Hill, New York, 1963.
14   Souders, M., and G. G. Brown: *Ind. Eng. Chem.,* **26**:98 (1934).
15   Thiele, E. W.: *Ind. Eng. Chem.,* **27**:392 (1932).

20

LEACHING AND EXTRACTION

This chapter discusses the methods of removing one constituent from a solid or liquid by means of a liquid solvent. These techniques fall into two categories. The first, called *leaching* or *solid extraction*, is used to dissolve soluble matter from its mixture with an insoluble solid. The second, called *liquid extraction*, is used to separate two miscible liquids by the use of a solvent which preferentially dissolves one of them. Although the two processes have certain common fundamentals, the differences in equipment and, to some extent, in theory are sufficient to justify separate treatment.

LEACHING

Leaching differs very little from the washing of filtered solids, as discussed in Chap. 30, and leaching equipment strongly resembles the washing section of various filters. In leaching, the amount of soluble material removed is often rather greater than in ordinary filtration washing, and the properties of the solids may change considerably during the leaching operation. Coarse, hard, or granular feed solids may disintegrate into pulp or mush when their content of soluble material is removed.

Leaching Equipment

When the solids form an open, permeable mass throughout the leaching operation, solvent may be percolated through an unagitated bed of solids. With impermeable solids or materials which disintegrate during leaching, the solids are dispersed into the solvent and are later separated from it. Both methods may be either batch or continuous.

Leaching by percolation through stationary solid beds Stationary solid-bed leaching is done in a tank with a perforated false bottom to support the solids and permit drainage of the solvent. Solids are loaded into the tank, sprayed with solvent until their solute content is reduced to the economical minimum, and excavated. In some cases the rate of solution is so rapid that one passage of solvent through the material is sufficient, but countercurrent flow of solvent through a battery of tanks is more common. In this method, fresh solvent is fed to the tank containing the solid that is most nearly extracted; it flows through the several tanks in series and is finally withdrawn from the tank that has been freshly charged. Such a series of tanks is called an *extraction battery*. The solid in any one tank is stationary until it is completely extracted. The piping is arranged so that fresh solvent can be introduced to any tank and strong solution withdrawn from any tank, making it possible to charge and discharge one tank at a time. The other tanks in the battery are kept in counter-current operation by advancing the inlet and drawoff tanks one at a time as the material is charged and removed. Such a process is sometimes called a *Shanks process*.

In some solid-bed leaching the solvent is volatile, necessitating the use of closed vessels operated under pressure. Pressure is also needed to force solvent through beds of some less permeable solids. A series of such pressure tanks operated with countercurrent solvent flow is known as a *diffusion battery*.

Moving-bed leaching[2] In the machines illustrated in Fig. 20-1 the solids are moved through the solvent with little or no agitation. The Bollman extractor (Fig. 20-1a) contains a bucket elevator in a closed casing. There are perforations in the bottom of each bucket. At the top right-hand corner of the machine, as shown in the drawing, the buckets are loaded with flaky solids such as soybeans and are sprayed with appropriate amounts of *half miscella* as they travel downward. Half miscella is the intermediate solvent containing some extracted oil and some small solid particles. As solids and solvent flow concurrently down the righthand side of the machine, the solvent extracts more oil from the beans. Simultaneously the fine solids are filtered out of the solvent, so that clean *full miscella* can be pumped from the right-hand sump at the bottom of the casing. As the partially extracted beans rise through the left side of the machine a stream of pure solvent percolates countercurrently through them. It collects in the left-hand sump and is pumped to the half-miscella storage tank. Fully extracted beans are dumped from the buckets at the top of the elevator into a hopper from which they are removed by paddle conveyors. The capacity of typical units is 50 to 500 tons of beans per 24-h day.

The Hildebrandt extractor shown in Fig. 20-1b consists of a U-shaped screw conveyor with a separate helix in each section. The helices turn at different speeds to

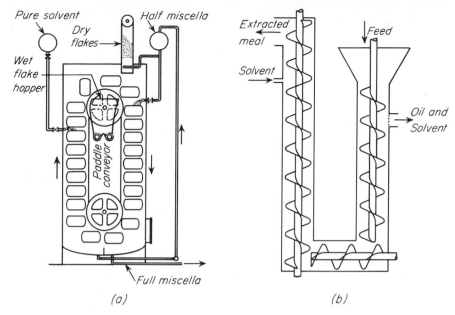

FIGURE 20-1
Moving-bed leaching equipment: (*a*) Bollman extractor; (*b*) Hildebrandt extractor.

give considerable compaction of the solids in the horizontal section. Solids are fed to one leg of the U and fresh solvent to the other to give countercurrent flow.

Dispersed-solid leaching   Solids which form impermeable beds, either before or during leaching, are treated by dispersing them in the solvent by mechanical agitation in a tank or flow mixer. The leached residue is then separated from the strong solution by settling or filtration.

Small quantities can be leached batchwise in this way in an agitated vessel with a bottom drawoff for settled residue. Continuous countercurrent leaching is obtained with several gravity thickeners connected in series as shown in Fig. 18-3 or when the contact in a thickener is inadequate by placing an agitator tank in the equipment train between each pair of thickeners. A still further refinement, used when the solids are too fine to settle out by gravity, is to separate the residue from the miscella in continuous solid-bowl helical-conveyor centrifuges. Many other leaching devices have been developed for special purposes, such as the solvent extraction of various oilseeds, with their specific design details governed by the properties of the solvent and of the solid to be leached.[2] The dissolved material, or solute, is often recovered by crystallization or evaporation.

Principles of Continuous Countercurrent Leaching

The most important method of leaching is the continuous countercurrent method using stages. Even in an extraction battery, where the solid is not moved physically

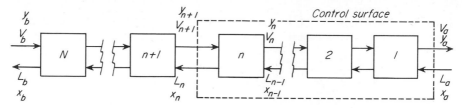

FIGURE 20-2
Countercurrent leaching cascade.

from stage to stage, the charge in any one cell is treated by a succession of liquids of constantly decreasing concentration as if it were being moved from stage to stage in a countercurrent system.

Because of its importance, only the continuous countercurrent method is discussed here. Also, since the stage method is normally used, the differential-contact method is not considered. In common with other stage cascade operations, leaching may be considered, first, from the standpoint of ideal stages and, second, from that of stage efficiencies.

Ideal stages in countercurrent leaching   Figure 20-2 shows a material-balance diagram for a continuous countercurrent cascade. The stages are numbered in the direction of flow of the solid. The V phase is the liquid that overflows from stage to stage in a direction counter to that of the flow of the solid, dissolving solute as it moves from stage N to stage 1. The L phase is the solid flowing from stage 1 to stage N. Exhausted solids leave stage N, and concentrated solution overflows from stage 1.

It is assumed that the solute-free solid is insoluble in the solvent and that the flow rate of this solid is constant throughout the cascade. Let V refer to the overflow solution and L to the liquid retained by the solid, both based on a definite flow of dry solute-free solid. Also, in accordance with standard nomenclature, the terminal concentrations are as follows:

Solution on entering solid x_a
Solution on leaving solid x_b
Fresh solvent entering the system y_b
Concentrated solution leaving the system y_a

As in absorption and distillation, the quantitative performance of a countercurrent system can be analyzed by utilizing an equilibrium line and an operating line, and, as before, the method to be used depends on whether these lines are straight or curved.

EQUILIBRIUM   In leaching, provided sufficient solvent is present to dissolve all the solute in the entering solid and there is no adsorption of solute by the solid, equilibrium is attained when the solute is completely dissolved and the concentration of the solution so formed is uniform. Such a condition may be obtained simply or with difficulty, depending on the structure of the solid. These factors are considered when stage efficiency is discussed. At present, it is assumed that the requirements for equilibrium are met. Then the concentration of the liquid retained by the solid leaving any stage is the same as that of the liquid overflow from the same stage. The equilibrium relationship is, simply, $x_e = y_e$.

OPERATING LINE   The equation for the operating line is obtained by writing material balances for that portion of the cascade consisting of the first n units, as shown by the control surface indicated by the dotted lines in Fig. 20-2. These balances are:

Total solution:
$$V_{n+1} + L_a = V_a + L_n \tag{20-1}$$

Solute:
$$V_{n+1}y_{n+1} + L_a x_a = L_n x_n + V_a y_a \tag{20-2}$$

Eliminating V_{n+1} and solving for y_{n+1} gives

$$y_{n+1} = \frac{1}{1 + (V_a - L_a)/L_n} x_n + \frac{V_a y_a - L_a x_a}{L_n + V_a - L_a} \tag{20-3}$$

This is the equation of the operating line. As usual, the line passes through points (x_a, y_a) and (x_b, y_b).

CONSTANT AND VARIABLE UNDERFLOW   Two cases are to be considered. If the density and viscosity of the solution change considerably with solute concentration, the solids from the lower-numbered stages may retain more liquid than those from the higher-numbered stages. Then, as shown by Eq. (20-3), the slope of the operating line varies from unit to unit. If, however, the mass of solution retained by the solid is independent of concentration, L_n is constant, and the operating line is straight. This condition is called *constant solution underflow*. If the underflow is constant, so is the overflow. Constant and variable overflow are given separate consideration.

NUMBER OF IDEAL STAGES FOR CONSTANT UNDERFLOW   When the operating line is straight, a McCabe-Thiele construction can be used to determine the number of ideal stages, but since in leaching the equilibrium line is always straight, Eq. (18-27) can be used directly for constant underflow. The use of this equation is especially simple here because $y_a^* = x_a$ and $y_b^* = x_b$.

Equation (18-27) cannot be used for the entire cascade if L_a, the solution entering with the unextracted solids, differs from L, the underflows within the system. Equations have been derived for this situation,[1,5] but it is easy to calculate, by material balances, the performance of the first stage separately and then to apply Eq. (18-27) to the remaining stages.

EXAMPLE 20-1   By extraction with kerosene, 2 tons of waxed paper per day is to be dewaxed in a continuous countercurrent extraction system which contains a number of ideal stages. The waxed paper contains, by weight, 25 percent paraffin wax and 75 percent paper pulp. The extracted pulp is put through a dryer to evaporate the kerosene. The pulp, which retains the unextracted wax after evaporation, must not contain over 0.2 lb (0.0907 kg) of wax per 100 lb (46.36 kg) of wax-free pulp. The kerosene used for the extraction contains

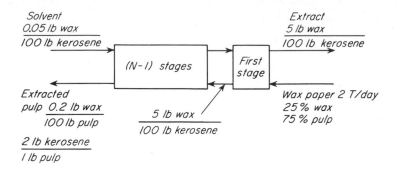

FIGURE 20-3
Material-balance diagram for Example 20-1.

0.05 lb (0.0227 kg) of wax per 100 lb (45.36 kg) of wax-free kerosene. Experiments show that the pulp retains 2.0 lb of kerosene per pound (2.0 kg per kilogram) of kerosene- and wax-free pulp as it is transferred from cell to cell. The extract from the battery is to contain 5 lb (2.27 kg) of wax per 100 lb (45.36 kg) of wax-free kerosene. How many stages are required?

SOLUTION   Any convenient units may be used in Eq. (18-27) as long as the units are consistent and as long as the overflows and underflows are constant. Thus, mole fractions, mass fractions, or mass of solute per mass of solvent are all permissible choices for concentration. The choice should be made that gives constant underflow. In this problem, since it is the ratio of kerosene to pulp that is constant, flow rates should be expressed in pounds of kerosene. Then, all concentrations must be in pounds of wax per pound of wax-free kerosene. The unextracted paper has no kerosene, so the first cell must be treated separately. Equation (18-27) can then be used for calculating the number of remaining units.

The flow quantities and concentrations for this cascade are shown in Fig. 20-3. The kerosene in with the fresh solvent is found by an overall wax balance. Take a basis of 100 lb of wax- and kerosene-free pulp, and let s be the pounds of kerosene fed in with the fresh solvent. The wax balance is, in pounds:

Wax in with pulp, $100 \times \frac{25}{75} = 33.33$
Wax in with solvent, $0.0005s$
    Total wax input, $33.33 + 0.0005s$
Wax out with pulp, $100 \times 0.002 = 0.200$
Wax out with solution, $(s - 200)0.05 = 0.05s - 10$
    Total wax output, $0.05s - 9.80$

Therefore $33.33 + 0.0005s = 0.05s - 9.80$

From this $s = 871$ lb. The kerosene in the exhausted pulp is 200 lb, and that in the strong solution is $871 - 200 = 671$ lb. The wax in this solution is $671 \times 0.05 = 33.55$ lb. The concentration in the underflow to the second unit equals that of the overflow from the first stage, or 0.05 lb of wax per pound of kerosene. The wax in the underflow to unit 2 is $200 \times 0.05 = 10$ lb. The wax in the overflow from the second cell to the first is, by a wax balance over the first unit,

$$10 + 33.55 - 33.33 = 10.22 \text{ lb}$$

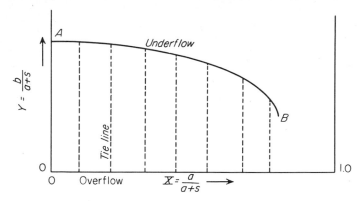

FIGURE 20-4
XY diagram for leaching: line AB, underflow line; dotted lines, vertical tie lines; a, solute; b, solid; s, solvent.

The concentration of this stream is, therefore, $10.22/871 = 0.0117$. The quantities for substitution in Eq. (18-27) are

$$x_a = y_a^* = 0.05 \qquad\qquad y_a = 0.0117$$

$$x_b = y_b^* = \frac{0.2}{200} = 0.001 \qquad y_b = 0.0005$$

Equation (18-27) gives, since stage 1 has already been taken into account,

$$N - 1 = \frac{\log\,[(0.0005 - 0.001)/(0.0117 - 0.05)]}{\log\,[(0.0005 - 0.0117)/(0.001 - 0.050)]}$$

$$= \frac{\log\,[(0.05 - 0.0117)/(0.001 - 0.0005)]}{\log\,[(0.050 - 0.001)/(0.0117 - 0.0005)]} = 3$$

The total number of ideal stages is $N = 1 + 3 = 4$. ////

NUMBER OF IDEAL STAGES FOR VARIABLE UNDERFLOW When the underflow and overflow vary from stage to stage, a modification of the Ponchon-Savarit graphical method may be used for leaching calculations. The modification consists (1) in considering each stream to be a mixture of solid and solution and (2) in using the ratio of solid to solution in place of specific enthalpy. The solution, in turn, is a mixture of solute and solvent. Let a represent the solute, b the solid, and s the solvent. These letters may refer to either concentration or to amounts. Then the abscissa of the Ponchon-Savarit diagram becomes $a/(a + s)$. Denote this by X. The ordinate becomes $b/(a + s)$. Denote this by Y. The XY diagram for a typical system of solid and solution is shown in Fig. 20-4. The curved line corresponds to the saturated-liquid line in the Hx chart. It is a plot of the ratio of the solid to the retained solution as a function of the concentration of the solution. It is the curvature of this line that makes Eq. (18-27) inapplicable. Unlike the corresponding curve on the Hx diagram,

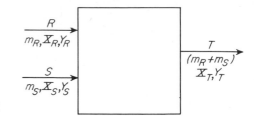

FIGURE 20-5
Mixing process, $R + S = T$.

this line is not a property of the solution itself but depends on the conditions of drainage in the leaching plant. Data for plotting this line must be determined experimentally under the conditions of the extraction.

The XY diagram for leaching is simpler than the Hx diagram for distillation in two ways: (1) since the solid is assumed to be insoluble in the solvent, all points representing overflow streams lie on the axis of abscissas, where $Y = 0$, and there is no line corresponding to the enthalpy of saturated vapor; (2) since at equilibrium the solution concentrations for both underflow and overflow for the same stage are equal, all tie lines are vertical. This is shown in Fig. 20-4.

The basic construction for the adiabatic mixing process on the Hx diagram carries over without change to the XY diagram: the point for a mixture of two streams lies on the straight line connecting the points for the streams that are mixed, and the center-of-gravity principle applies to the line segments between the points. This may be shown as follows.

Figure 20-5 is a flow diagram showing the mixing of streams R and S to form stream T. Let the masses of the solutions in streams R, S, and T be m_R, m_S, and $m_R + m_S$. Let the corresponding solution concentrations be X_R, X_S, and X_T and the corresponding values of the mass of solid per unit mass of solution be Y_R, Y_S, and Y_T. Two independent material balances can be written:

Solute balance:
$$m_R X_R + m_S X_S = (m_R + m_S)X_T \qquad (20\text{-}4)$$

Solid balance:
$$m_R Y_R + m_S Y_S = (m_R + m_S)Y_T \qquad (20\text{-}5)$$

Comparison of these equations with Eqs. (18-35) and (18-36) shows that the only difference is the use of X in Eq. (20-4) in place of x in Eq. (18-36) and Y in Eq. (20-5) instead of H in Eq. (18-35). All conclusions and constructions deduced from Eq. (18-39) for the Hx diagram are therefore applicable to the XY diagram. It is important to note that the line segments on the XY diagram represent not the masses of the total streams but only the masses of the solution in the streams. With this understanding, the center-of-gravity principle applies as before.

OVERALL BALANCES AND POINT J The overall balance for a leaching plant shows four streams, the inlet solvent V_b, the outlet concentrated liquid V_a, the inlet feed L_a, and the outlet exhausted solid L_b. The simple mixing construction shown in Fig. 20-5, which applies only to three streams, must be modified. This may be done, for calculation purposes, by assuming the overall process to be replaced by two fictitious processes

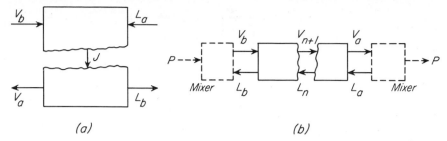

FIGURE 20-6
Definition of points J and P: (a) processes $V_b + L_a = J$ and $J = V_a + L_b$; (b) processes $P = V_a - L_a$ and $P = V_b - L_b$.

in series, in the first of which streams V_b and L_a are mixed to form stream J, and in the second of which stream J is divided into streams V_a and L_b. These processes are shown in Fig. 20-6a and are represented by the equations

$$V_b + L_a = J = V_a + L_b \qquad (20\text{-}6)$$

Five points, labeled V_b, L_a, J, V_a, L_b, each representing a corresponding stream, are plotted in Fig. 20-7. A straight line through points V_a and L_b intersects a straight line through points V_b and L_a at point J. By the center-of-gravity principle, the mass of solution in stream L_a is inversely proportional to the line segment $\overline{JL_a}$; the solution in stream V_a is inversely proportional to segment $\overline{JV_a}$; and so on. If, for example, the mass ratio of the solution in the incoming solvent to that in the feed is known, this ratio equals $\overline{JL_a}/\overline{JV_b}$ and point J can be plotted. Then if the analysis of one other stream, for example, V_a, is known, the line L_bJV_a can be plotted. Since point L_b is on the underflow line, it is also the intersection of the underflow line with line L_bJV_a, and L_b is thereby determined.

THE OPERATING LINE AND POINT P Overall balances can also be treated by assuming a fictitious mixer added to either end of the cascade, as shown by the dotted rectangles in Fig. 20-6b. Imaginary stream P is then assumed to be formed by subtracting L_a from V_a or by subtracting L_b from V_b. These processes are represented by the equations

$$V_a = L_a + P \qquad L_b + P = V_b$$

or

$$P = V_a - L_a = V_b - L_b \qquad (20\text{-}7)$$

These equations are represented on the XY diagram in Fig. 20-7. Since stream P is a mathematical quantity that does not exist physically, either or both of its coordinates may be negative.

A material balance over the first n stages and the fictitious mixer ahead of stage 1 is

$$V_{n+1} = L_n + P \qquad (20\text{-}8)$$

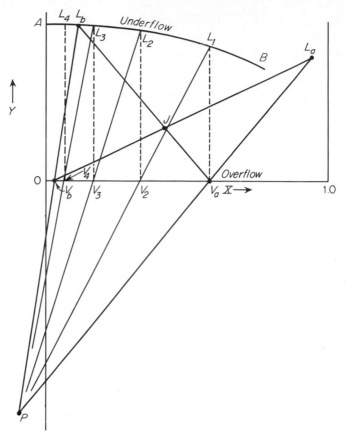

FIGURE 20-7
Leaching on Ponchon-Savarit diagram.

The operating line represented by this equation is a straight line passing through points P, V_{n+1}, and L_n. All other individual operating lines pass through point P, and these operating lines, used alternately with the vertical equilibrium lines, give the Ponchon-Savarit construction shown in Fig. 20-7.

The fictitious quantity P is the *net flow* of all material through the cascade, considered as a single stream, from stage N to stage 1. From Eqs. (20-7) and (20-8),

$$P = V_{n+1} - L_n = V_a - L_a$$

and P is the same for all stages. It is constant throughout the cascade. If $V_a < L_a$, the net flow is negative, and if $V_a > L_a$, P is positive.

EXAMPLE 20-2 Oil is to be extracted from meal by means of benzene, using a continuous countercurrent extractor. The unit is to treat 2,000 lb (907.2 kg) of meal (based on completely exhausted solid) per hour. The untreated meal contains 800 lb (362.9 kg) of oil and 50 lb (22.7 kg) of benzene. The fresh solvent mixture contains 20 lb (9.07 kg) of oil and 1,310 lb

(594.2 kg) of benzene. The exhausted solids are to contain 120 lb (54.4 kg) of unextracted oil. Experiments carried out under conditions identical with those of the projected battery show that the solution retained depends on the concentration of the solution, as shown in Table 20-1. Find (*a*) the concentration of the strong solution, or extract, (*b*) the concentration of the solution adhering to the extracted solids, (*c*) the mass of solution leaving with the extracted meal, (*d*) the mass of extract, (*e*) the number of stages required.

SOLUTION   The data in Table 20-1 provide the coordinates of the *Y*-vs.-*X* line for all underflows. Values of *X* are given in the first column of the table, and corresponding values of *Y* are the reciprocals of the number in the second column. The *YX* line for the underflow is plotted as curve *AB* in Fig. 20-8. From the conditions of the problem, the coordinates of points L_a and V_b are:

Point L_a: $X = \dfrac{800}{800 + 50} = \dfrac{800}{850} = 0.941$ $Y = \dfrac{2,000}{850} = 2.35$

Point V_b: $X = \dfrac{20}{1,310 + 20} = \dfrac{20}{1,330} = 0.015$ $Y = 0$

Points L_a and V_b are plotted on Fig. 20-8, and line L_aV_b drawn. Point *J* lies on this line, a distance $850/(1,330 + 850) = 0.39$ times the distance between points V_b and L_a, measured from V_b. Point *J* is plotted. The ratio of *Y* to *X* for point L_b is the ratio of the solid to the solute in this stream, or $2,000/120 = 16.67$. Point L_b lies on line *AB*. A straight line through the origin with a slope of 16.67 intersects line *AB* at point L_b, and another straight line through points L_b and *J* intersects the $X = 0$ line at point V_a. The *X* coordinate of point L_b is 0.12, and that of point V_a is 0.592.

The total solution input is $1,330 + 850 = 2,180$ lb, and this equals the total solution leaving in streams V_a and L_b. This flow is divided between the two streams in proportion to the line segments on line V_aJL_b. The *X* coordinate of point *J* is 0.372, and, by the center-of-gravity principle,

$$L_b = \frac{0.592 - 0.372}{0.592 - 0.120}\, 2,180 = 1,020 \text{ lb}   V_a = 2,180 - 1,020 = 1,160 \text{ lb}$$

The answers to parts (*a*) to (*d*) are: part (*a*), 0.592; part (*b*), 0.12; part (*c*), 1,020 lb (462.7 kg); part (*d*), 1,160 lb. (526.2 kg).   ////

To determine the number of ideal stages, point *P* is established as the intersection between straight lines through points L_a and V_a and L_b and V_b, as shown in Fig. 20-8. The Ponchon-Savarit construction is shown in the figure, and from this the number of ideal stages is found to be slightly less than four. The answer to part (*e*), then, is four stages.

Table 20-1 DATA FOR EXAMPLE 20-2

Concentration, lb oil/lb solution	Solution retained, lb/lb solid	Concentration, lb oil/ solution	Solution retained, lb/lb solid
0.0	0.500	0.4	0.550
0.1	0.505	0.5	0.571
0.2	0.515	0.6	0.595
0.3	0.530	0.7	0.620

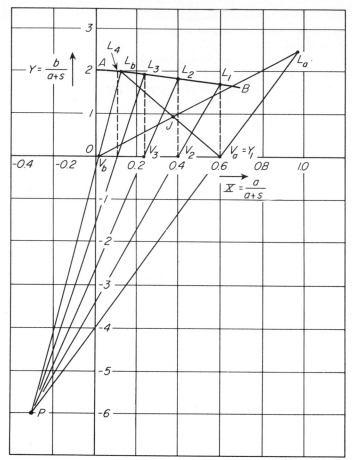

FIGURE 20-8
Solution of Example 23-2.

Saturated concentrated solution A special case of leaching is encountered when the solute is of limited solubility and the concentrated solution reaches saturation. This situation can be treated by the above methods.[4] The solvent input to stage N should be the maximum that is consistent with a saturated overflow from stage 1, and all liquids except that adhering to the underflow from stage 1 should be unsaturated. If saturation is attained in stages other than the first, all but one of the "saturated" stages are unnecessary, and the solute concentration in the underflow from stage N is higher than it needs to be.

If the carrier solid is partially soluble in the solvent, the Ponchon-Savarit method can be extended to take that fact into account. A second line above the X axis is drawn, which is the Y-vs.-X relationship for the overflow, including its content of dissolved solid. Otherwise, the construction is unchanged.

Stage efficiencies In some leaching operations the solid is entirely impervious and inert to the action of the solvent and carries a film of strong solution on its surface. In such a case the process simply involves the equalization of concentrations in the bulk of the extract and in the adhering film. Such a process is rapid, and any reasonable time of contact will bring about equilibrium. The countercurrent leaching process shown in Fig. 18-3 is of this type, and the stage efficiency is taken as unity in calculations for such processes.

In other situations the solute is distributed through a more or less permeable solid. Here the rate of leaching is largely governed by the rate of diffusion through the solid, as discussed in Chap. 22, page 702.

LIQUID EXTRACTION

Liquid extraction is used when distillation and rectification are difficult or ineffective. Close-boiling mixtures or substances that cannot withstand the temperature of vaporization, even under a vacuum, may often be separated by extraction. Extraction utilizes differences in the solubilities of the components rather than differences in their volatilities. Since solubility depends on chemical properties, extraction exploits chemical differences instead of vapor-pressure differences.

When either distillation or extraction may be used, the choice is usually distillation, in spite of the fact that heat and cooling are needed. In extraction the solvent must be recovered for reuse (usually by distillation), and the combined operation is more complicated and often more expensive than ordinary distillation without extraction. In many problems, the choice between methods should be based on a comparative study of both extraction and distillation.

Extraction may be used to separate more than two components; and mixtures of solvents, instead of a single solvent, are needed in some applications. These more complicated methods are not treated in this text.

Extraction Equipment[8]

In liquid-liquid extraction, as in gas absorption and distillation, two phases must be brought into good contact to permit transfer of material and then be separated. In absorption and distillation the mixing and separation are easy and rapid. In extraction, however, the two phases have comparable densities, so that the energy available for mixing and separation—if gravity flow is used—is small, much smaller than when one phase is a liquid and the other is a gas. The two phases are often hard to mix and harder to separate. The viscosities of both phases, also, are relatively high, and linear velocities through most extraction equipment are low. In some types of extractors, therefore, energy for mixing and separation is supplied mechanically.

Extraction equipment may be operated batchwise or continuously. A quantity of feed liquid may be mixed with a quantity of solvent in an agitated vessel, after which the layers are settled and separated into extract and raffinate. This gives about one theoretical contact, which is adequate in simple extractions. The operation may of course be repeated if more than one contact is required, but when the quantities involved are large and several contacts are needed, continuous flow becomes economical. Most extraction equipment is continuous, with either successive stage contacts or differential contacts. Representative types are mixer-settlers, vertical towers of various kinds which operate by gravity flow, agitated tower extractors, and centrifugal extractors. The characteristics of various types of extraction equipment are listed in Table 20-2.

Mixer-settlers For batchwise extraction the mixer and settler may be the same unit. A tank containing a turbine or propeller agitator is most common. At the end of the mixing cycle the agitator is shut off, the layers allowed to separate by gravity, and extract and raffinate drawn off to separate receivers through a bottom drain line carrying a sight glass. The mixing and settling times required for a given extraction can be determined only by experiment; 5 min for mixing and 10 min for settling are typical, but both shorter and much longer times are common.

For continuous flow the mixer and settler must be separate pieces of equipment. The mixer may be a small agitated tank provided with inlets and a drawoff line and baffles to prevent short-circuiting; or it may be a centrifugal pump or other flow mixer. The settler is often a simple continuous gravity decanter. With liquids which emulsify easily and which have nearly the same density it may be necessary to pass the mixer discharge through a screen or pad of glass fiber to coalesce the droplets of the dispersed phase before gravity settling is feasible. For even more difficult separations, tubular or disk-type centrifuges are employed.

If, as is usual, several contact stages are required, a train of mixer-settlers is

Table 20-2 PERFORMANCE OF COMMERCIAL EXTRACTION EQUIPMENT

Type	Liquid capacity of combined streams ft³/ft²-h	HTU,† ft	Plate or stage efficiency, %	Spacing between plates or stages, in.	Typical applications
Mixer-settler			75–100		Duo-Sol lube-oil process
Spray column	50–250	10–20			Ammonia extraction of salt from caustic soda
Packed column	20–150	5–20			Phenol recovery
Perforated-plate column	10–200	1–20	6–24	30–70	Furfural lube-oil process
Baffle column	60–105	4–6	5–10	4–6	Acetic acid recovery
Agitated tower	50–100	1–2	80–100	12–24	Pharmaceuticals and organic chemicals

† HTUs are discussed in Chap. 23, page 729.

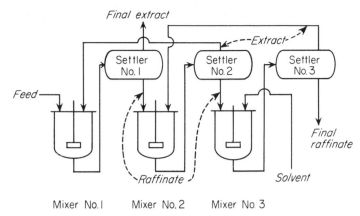

FIGURE 20-9
Mixer-settler extraction system.

operated with countercurrent flow, as shown in Fig. 20-9. The raffinate from each settler becomes the feed to the next mixer, where it meets intermediate extract or fresh solvent. The principle is identical with that of the continuous countercurrent stage leaching system shown in Fig. 18-3.

Spray and packed extraction towers These tower extractors give differential contacts, not stage contacts, and mixing and settling proceed simultaneously and continuously. In the spray tower shown in Fig. 20-10, the lighter liquid is introduced at the bottom and distributed as small drops by the nozzles A. The drops of light liquid rise through the mass of heavier liquid, which flows downward as a continuous stream. The drops are collected at the top and form the stream of light liquid leaving the top of the tower. The heavy liquid leaves the bottom of the tower. In Fig. 20-10, light phase is dispersed and heavy phase is continuous. This may be reversed, and the heavy stream sprayed into the light phase at the top of the column, to fall as dispersed phase through a continuous stream of light liquid. In either case each drop is constantly being "mixed," i.e., brought into fresh contact with the other phase, and is constantly being separated from it. There is continuous transfer of material between phases, and the composition of each phase changes as it flows through the tower. At any given level, of course, equilibrium is not reached; indeed, it is the departure from equilibrium that provides the driving force for material transfer. By using a tall tower, however, the equivalent of a number of perfect stages can theoretically be obtained.

In actual spray towers the contact between the drops and continuous phase is not highly effective except where the drops are initially dispersed. As much as 40 to 45 percent of the total transfer may occur at the point of dispersion.[11] Thus while spray towers are simple to build and easy to operate, they are not highly effective. Adding more height does not always improve results; it is much more effective to

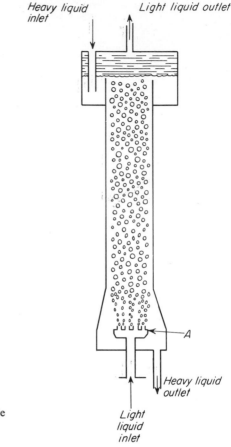

FIGURE 20-10
Spray tower; *A*, nozzle to distribute
light liquid.

redisperse the drops at frequent intervals throughout the tower. This can be done by filling the tower with packing, such as rings or saddles. The packing causes the drops to coalesce and re-form and, as shown in Table 20-2, appreciably increases the effectiveness of the tower. Packed towers approach spray towers in simplicity, and can be made to handle almost any problem of corrosion or pressure at a reasonable cost. Their chief disadvantage is that solids tend to collect in the packing and cause channeling.

FLOODING VELOCITIES IN PACKED TOWERS If the flow rate of either the dispersed phase or the continuous phase is held constant and that of the other phase gradually increased, a point is reached where the dispersed phase coalesces, the holdup of that phase increases, and finally both phases leave together through the continuous-phase outlet. The effect, like the corresponding action in an absorption column, is called flooding. The larger the flow rate of one phase at flooding, the smaller is that of the other. A column obviously should be operated at flow rates below the flooding point.

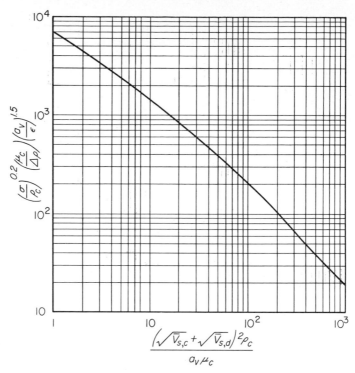

FIGURE 20-11
Flooding velocities in packed extraction towers.

Flooding velocities in packed columns[3] can be estimated from Fig. 20-11. In this figure the abscissa is the group

$$\frac{(\sqrt{\overline{V}_{s,c}} + \sqrt{\overline{V}_{s,d}})^2 \rho_c}{a_v \mu_c}$$

The ordinate is the group

$$\frac{\mu_c}{\Delta\rho}\left(\frac{\sigma}{\rho_c}\right)^{0.2}\left(\frac{a_v}{\varepsilon}\right)^{1.5}$$

where $\overline{V}_{s,c}$, $\overline{V}_{s,d}$ = superficial velocities of continuous and dispersed phases, respectively, ft/h
μ_c = viscosity of the continuous phase, lb/ft-h
σ = interfacial tension between phases, dyn/cm
ρ_c = density of the continuous phase, lb/ft^3
$\Delta\rho$ = density difference between phases, lb/ft^3
a_v = specific surface area of packing, ft^2/ft^3
ε = fraction voids or porosity of packed section

The groups in Fig. 20-11 are not dimensionless, and the proper units must be used.

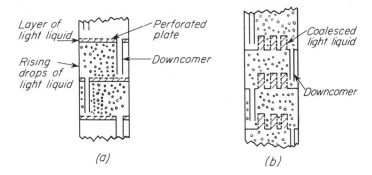

FIGURE 20-12
Perforated-plate extraction towers: (*a*) perforations in horizontal plates; (*b*) perforations in vertical sections.

Perforated-plate towers   Redispersion of liquid drops is also done by transverse perforated plates like those in the sieve-plate distillation tower described in Chap. 18. The perforations in an extraction tower are $1\frac{1}{2}$ to $4\frac{1}{2}$ mm in diameter. Plate spacings are 6 to 24 in. (150 to 600 mm). Usually the light liquid is the dispersed phase, and downcomers carry the heavy continuous phase from one plate to the next. As shown in Fig. 20-12*a*, light liquid collects in a thin layer beneath each plate and jets into the thick layer of heavy liquid above. In many perforated-plate towers the plates are made as shown in Fig. 20-12*b*, with the perforations in vertical sections of the tray.[7] In this design the light liquid jets horizontally into the continuous phase.

Baffle towers   These extraction towers contain sets of horizontal baffle plates. Heavy liquid flows over the top of each baffle and cascades to the one beneath; light liquid flows under each baffle and sprays upward from the edge through the heavy phase. The most common arrangements are disk-and-doughnut baffles and segmental, or side-to-side, baffles. In both types the spacing between baffles is 4 to 6 in. (100 to 150 mm).

Baffle towers contain no small holes to clog or be enlarged by corrosion. They can handle dirty solutions containing suspended solids; one modification of the disk-and-doughnut towers even contains scrapers to remove deposited solids from the baffles. Because the flow of liquid is smooth and even, with no sharp changes in velocity or direction, baffle towers are valuable for liquids that emulsify easily. For the same reason, however, they are not effective mixers, and each baffle is equivalent to only 0.05 to 0.1 ideal stage.[13]

Agitated tower extractors   Mixer-settlers supply mechanical energy for mixing the two liquid phases, but the tower extractors so far described do not. They depend on gravity flow both for mixing and for separation. In some tower extractors, however, mechanical energy is provided by internal turbines or other agitators, mounted on a central rotating shaft. Sets of agitators are often separated by partitions or

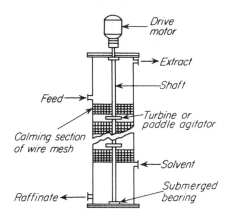

FIGURE 20-13
Agitated extraction tower.

calming sections to give, in effect, a stack of mixer-settlers one above the other. The tower extractor illustrated in Fig. 20-13 is an example. Here the calming sections between the agitators are packed with wire mesh to encourage coalescence and separation of the phases. Most of the extraction takes place in the mixing sections, but some also occurs in the calming sections, so that the efficiency of each mixer-settler unit is sometimes greater than 100 percent. Typically each mixer-settler is 1 to 2 ft high, which means that several theoretical contacts can be provided in a reasonably short column. The problem of maintaining the internal moving parts, however, particularly where the liquids are corrosive, may be a serious disadvantage.

Pulse columns Agitation may also be provided by external means, as in a pulse column. A reciprocating pump "pulses" the entire contents of the column at frequent intervals, so that a rapid reciprocating motion of relatively small amplitude is superimposed on the usual flow of the liquid phases. The tower may contain ordinary packing or special sieve plates. In a packed tower the pulsation disperses the liquids and eliminates channeling, and the contact between the phases is greatly improved. In sieve-plate pulse towers the holes are smaller than in nonpulsing towers, ranging from 1.5 to 3 mm in diameter, with a total open area in each plate of 6 to 23 percent of the cross-sectional area of the tower. Such towers are used almost entirely for processing highly corrosive radioactive liquids. No downcomers are used. Ideally the pulsation causes light liquid to be dispersed into the heavy phase on the upward stroke and the heavy phase to jet into the light phase on the downward stroke. Under these conditions the stage efficiency may reach 70 percent. This is possible, however, only when the volumes of the two phases are nearly the same and where there is almost no volume change during extraction. In the more usual case the successive dispersions are less effective, and there is backmixing of one phase in one direction. The plate efficiency then drops to about 30 percent. Nevertheless, in both packed and sieve-plate pulse columns the height required for a given number of theoretical contacts is often less than one-third that required in an unpulsed column.[10]

Centrifugal extractors The dispersion and separation of the phases may be greatly accelerated by centrifugal force, and several commercial extractors make use of this. In the Podbielniak extractor a perforated spiral ribbon inside a heavy metal casing is wound about a hollow horizontal shaft through which the liquids enter and leave. Light liquid is pumped to the outside of the spiral at a pressure between 3 and 12 atm. to overcome the centrifugal force; heavy liquid is pumped to the center. The liquids flow countercurrently through the passage formed by the ribbon and the casing walls. Heavy liquid moves outward along the outer face of the spiral; light liquid is forced by displacement to flow inward along the inner face. The high shear at the liquid-liquid interface results in rapid mass transfer. In addition, some liquid sprays through the perforations in the ribbon and increases the turbulence. Up to 20 theoretical contacts may be obtained in a single machine, although 3 to 10 contacts are more common. Centrifugal extractors are expensive and find relatively limited use. They have the advantages of providing many theoretical contacts in a small space and of very short holdup times—about 4 s. Thus they are valuable in the extraction of sensitive products such as vitamins and antibiotics.

Auxiliary equipment The dispersed phase in an extraction tower is allowed to coalesce at some point into a continuous layer from which one product stream is withdrawn. The interface between this layer and the predominant continuous phase is set in an open section at the top or bottom of a packed tower; in a sieve-plate tower it is commonly near the middle. The interface level is most often automatically controlled by a vented overflow leg for the heavy phase, as in a continuous gravity decanter. In large columns the interface is sometimes held at the desired point by a level controller actuating a valve in the heavy-liquid discharge line.

   In liquid-liquid extraction the solvent must nearly always be removed from the extract or raffinate, or both. Thus auxiliary stills, evaporators, heaters, and condensers form an essential part of most extraction systems and often cost much more than the extraction device itself. As mentioned at the beginning of this section, if a given separation can be done either by extraction or distillation, economic considerations usually favor distillation. Extraction provides a solution to problems that cannot be solved by distillation alone but does not usually eliminate the need for distillation or evaporation in some part of the separation system.

Principles of Extraction

Since most continuous extraction methods use countercurrent contacts between two phases, one a light liquid and the other a heavier one, many of the fundamentals of countercurrent gas absorption and of rectification carry over into the study of liquid extraction. Thus questions about ideal stages, stage efficiency, minimum ratio between the two streams, and size of equipment have the same importance in extraction as in distillation. The equilibrium relationships in liquid extraction are, as shown in Chap. 17, sometimes more complicated than in the other operation.

   In triangular diagrams, in the same way as in the Hx and the XY diagrams, a point for a mixture of two streams lies on the straight line connecting the points for the two streams, and the center-of-gravity principle applies. This can be demonstrated by the usual method of correlating material-balance equations with similar triangles on the diagram. As a special case of the mixing rule, when one pure component is added to a given mixture, the locus of all points for the resulting solution is the straight line connecting the point for the original mixture with the apex for the added component. Likewise, if one component is stripped from a mixture, the point for the two-component mixture so formed is found by projecting, to the side of the triangle, the straight line connecting the original point and the apex for the component removed. For example, in Fig. 17-15, if the mixture represented by point M is diluted with component b, the points for all mixtures so formed lie on line Mb. If component b is stripped from the original mixture, the b-free solution obtained is represented by point Q.

Ponchon-Savarit method Since the graphical mixing construction and the center-of-gravity principle both apply to triangular diagrams, the complete Ponchon-Savarit method carries over unchanged to these coordinates. Typical constructions for both type I and type II systems are shown in Fig. 20-14. Points P and J have the usual significance, as given by Eqs. (20-6) and (20-7). In using triangular coordinates it must be remembered that stream magnitudes and concentrations are based on total streams. The constructions shown in Fig. 20-14 apply to the flow diagrams of Figs. 20-2 and 20-6. In the situation shown in Fig. 20-14a, three ideal stages are required, and in that of Fig. 20-14b, five stages are called for.

The usual Ponchon-Savarit constructions can also be used on rectangular diagrams for liquid-liquid extraction. It must be remembered, however, that all concentrations and stream magnitudes refer to the base mixture in the stream, not to the total stream. The method is illustrated by the following example.

EXAMPLE 20-3 A countercurrent extraction plant containing three ideal stages is used to extract acetone from its mixture with water by means of MIK at a temperature of 30°C. The feed consists of 40 percent acetone and 60 percent water. Pure solvent equal in mass to the feed is used as the extracting liquid. What is the fraction of the inlet acetone remaining unextracted in the final raffinate? What is the analysis of the final extract? What are the analyses of extract and raffinate if all the solvent is removed from them?

SOLUTION Choose the rectangular coordinates as shown in Fig. 17-20a. Let acetone be component a, water component b, and MIK component s. The combination of water and MIK is the base mixture. Per 100 kg of feed, there are 60 kg of base mixture ($s + b$) and 40 kg of solute a. The coordinates of point L_a, representing the feed, are

$$X = \frac{60}{60} = 1.00 \qquad Y = \frac{40}{60} = 0.667$$

The fresh solvent consists entirely of MIK, and the coordinates of point V_b, representing this stream, are $X = 0$ and $Y = 0$. Points L_a and V_b are plotted on Fig. 20-15, which is an enlargement of the lower part of Fig. 17-20a. Since the ratio of stream L_a to stream V_b is known from the conditions of the problem, point J can be plotted on the straight line connecting these points. The base mixture in the feed is 60 kg, and that in the entering solvent

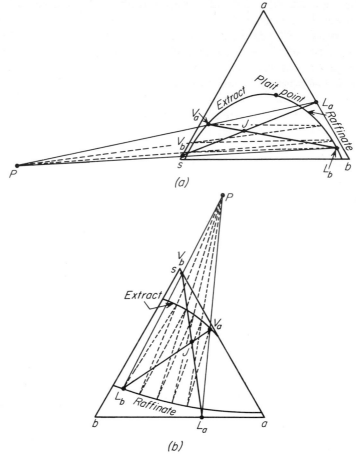

FIGURE 20-14
Ponchon-Savarit construction on triangular diagram: (*a*) type I system; (*b*) type II system.

equals the entire feed, or 100 kg. By the center-of-gravity principle, point J divides the line $V_b L_a$ into segments $\overline{V_b J}$ and $\overline{J L_a}$ in the ratio

$$\frac{\overline{V_b J}}{\overline{J L_a}} = \frac{60}{100} = 0.60$$

The coordinates of point J are $X = 0.375$ and $Y = 0.25$.

The analyses of the final extract and raffinate are unknown. A straight line through points L_b and V_a, which represent these streams, must pass through point J. For any such line there is a point P, which is the intersection of lines $L_a V_a$ and $L_b V_b$. For each P point there is a definite number of ideal stages, found by the Ponchon-Savarit construction. Tie lines are found by the use of Fig. 17-20*b*. The solution to this problem requires a trial-and-

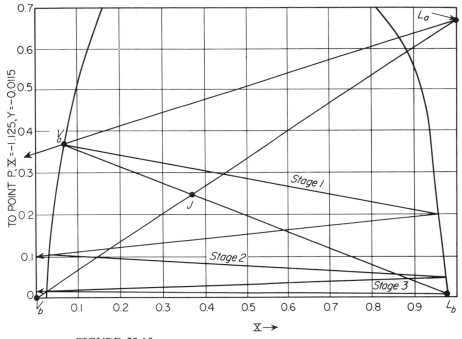

FIGURE 20-15
Solution of Example 20-4.

error determination of the line V_aJL_b, pivoting around the point J that leads to three stages. The final result is shown in Fig. 20-15.

The coordinates of point V_a are $X = 0.070$ and $Y = 0.37$. Those of point L_b are $X = 0.975$ and $Y = 0.010$.

The base mixture in each of the exit streams is found by splitting the entire input of base mixture in proportion to the line segments $\overline{V_aJ}$ and $\overline{JL_b}$. The total input of $s + b$ is $100 + 60 = 160$ kg, and the base mixture in the extract is

$$\frac{0.975 - 0.375}{0.975 - 0.070} \, 160 = 106 \text{ kg}$$

The base mixture in the raffinate is $160 - 106 = 54$ kg, and the unextracted acetone is $54 \times 0.010 = 0.54$ kg, so the fraction of unextracted acetone is $0.54/40 = 0.0135$.

In the final extract, $y_A = 0.37/1.37 = 0.270$, and $y_B = 0.070/1.37 = 0.051$. The analysis of this stream is, then, 27.0 percent acetone, 5.1 percent water, and 67.9 percent MIK.

For the raffinate L_b, $X = 0.975$, and $Y = 0.010$. Then, on a solvent-free basis,

$$\frac{Y}{X} = \frac{x_B}{x_A} = \frac{0.010}{0.975} = \frac{x_B}{1 - x_B}$$

From this, $x_B = 0.010$, and $x_A = 0.990$. The solvent-free raffinate is 1.0 percent acetone and 99.0 percent water.

For the solvent-free extract

$$x_A = \frac{0.270}{0.270 + 0.051} = 0.84 \quad \text{and} \quad x_B = 0.16$$

The solvent-free extract is 84 percent acetone and 16 percent water.   ////

Countercurrent extraction of type II systems using reflux   Just as in rectification, reflux can be used in countercurrent extraction to improve the separation of
the components in the feed. This method is especially effective in treating type II
systems, because with a center-feed cascade and the use of reflux the two feed components can be separated into nearly pure products.

A flow diagram for countercurrent extraction with reflux is shown in Fig. 20-16.
To emphasize the analogy between this method and fractionation it is assumed that
the cascade is a plate column. Any other kind of cascade, however, may be used.
The method requires that sufficient solvent be removed from the extract leaving the
cascade to form a raffinate, part of which is returned to the cascade as reflux, the
remainder being withdrawn from the plant as a product. Raffinate is withdrawn from
the cascade as bottoms product, and fresh solvent is admitted directly to the bottom
of the cascade. None of the bottom raffinate need be returned as reflux, for the
number of stages required is the same whether or not any of the raffinate is recycled
to the bottom of the cascade.[12] The situation is not the same as in continuous distillation, in which part of the bottoms must be vaporized to supply heat to the column.

The solvent separator, which is ordinarily a still, is shown in Fig. 20-16. As
also shown in Fig. 20-16, both products may be stripped of solvent in solvent strippers
to give solvent-free products.

In Fig. 20-16, the feed is denoted by F. It may or may not contain solvent.
The part of the cascade between the feed and the solvent separator is the extract-
enriching section, and the product from this section is denoted by D. The part of the
cascade below the feed is the raffinate-stripping section, and the raffinate product
from this section is denoted by B. Ideal stages are numbered serially from the extract
end. The total number of stages is N. The rectangular diagram of Fig. 20-17 shows
the graphical analysis of the process of Fig. 20-16. A type II system is postulated,
and the coordinates of Fig. 17-21a are chosen. As before, in considering Hx, xy, and
XY diagrams, the same letter is used to designate both the flow rate of a stream and the
point on the diagram representing that stream. Also, since the diagram is on a solvent-
free basis, all flow rates are in pounds or kilograms per hour of $a + b$. Any solvent
that accompanies a stream must be computed separately. For example, assume that
an extract stream V is flowing at a rate of 150 kg/h, total stream, and that the stream
is 20 percent solute, 30 percent diluent, and 50 percent solvent. Then, for this stream,
$V = 150(0.20 + 0.30) = 75$ kg/h. The solvent accompanying this is also 75 kg/h.
The coordinates of point V are

$$X = \frac{0.20}{0.20 + 0.30} = 0.40 \quad \text{and} \quad Y = \frac{0.50}{0.20 + 0.30} = 1.00$$

On the diagram of Fig. 20-17, for pure solvent X is indeterminate and $Y = \infty$.

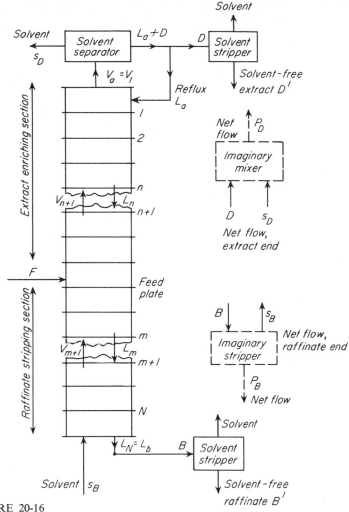

FIGURE 20-16
Countercurrent extraction with reflux.

Any mixing or stripping process involving pure solvent is represented by a vertical straight line connecting the points for the initial and final streams.

The graphical method for countercurrent extraction follows closely that for a center-feed fractionating column.[9,14] Let V_a denote the extract leaving stage 1, and let s_D be the solvent removed from this stream in the solvent separator. Also, assume that the solvent removed in the separator is just sufficient to convert stream V_a into a saturated raffinate. Then product D and reflux L_a have the same composition, and the point representing these streams lies on the saturated-raffinate line. Let L_b denote the raffinate leaving stage N of the cascade. Stream L_b also is the bottom product B. The stream of solvent entering the bottom contains no base mixture, and $V_b = 0$. The flow rate of solvent in this stream is denoted by s_B.

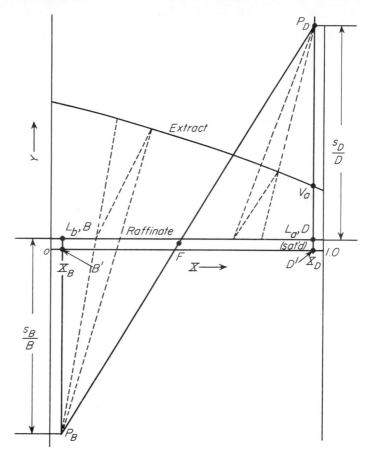

FIGURE 20-17
Extraction with reflux (Ponchon-Savarit construction).

The net flow, denoted by P_D, leaving the extract end of the plant is an imaginary mixture of product D and solvent s_D. If the coordinates of point D are X_D and Y_D, those of point P_D are X_D and $Y_D + s_D/D$. Likewise, the net flow out of the plant at the raffinate end is an imaginary flow of product B less fresh solvent s_B. If the coordinates of point B are X_B and Y_B, those of point P_B are X_B and $Y_B - s_B/B$. Points P_D and P_B are plotted in Fig. 20-17.

Since the feed F is split into two net flows P_D and P_B, points P_D, F, and P_B lie on a common straight line, the overall-balance line, as shown in Fig. 20-17. The number of ideal stages is found by the Ponchon-Savarit method, exactly as in the design of a fractionating column, as shown in Fig. 19-26. Points D' and B', representing solvent-free products leaving the solvent strippers, are also shown in Fig. 20-17.

The following balances, on a solvent-free basis, can be written

$$D = P_D \qquad B = P_B \qquad V_a = L_a + D$$

$$L_b = B \qquad F = P_D + P_B = D + B$$

$$(20\text{-}9)$$

By the center-of-gravity principle, the following equalities between ratios of streams and of line segments can be written

$$\frac{L_a}{P_D} = \frac{L_a}{D} = \frac{\overline{P_D V_a}}{\overline{V_a L_a}} \qquad \frac{D}{F} = \frac{\overline{FP_B}}{\overline{P_D P_B}} \qquad \frac{B}{F} = \frac{\overline{P_D F}}{\overline{P_B P_D}} \qquad (20\text{-}10)$$

The quantity L_a/D is the ratio of the reflux to the extract product. It may be denoted by R_D, and it corresponds to the reflux ratio R_D in a fractionating column.

The close analogy between rectification and extraction, both using reflux, is shown in Table 20-3. Note that the solvent plays the same part in extraction that heat does in rectification.

Limiting reflux ratios   Just as in rectification, two limiting cases exist in operating a countercurrent extractor with reflux. As the reflux ratio R_D becomes very great, the number of stages approaches a minimum, and as R_D is reduced, a minimum value of the reflux ratio is reached where the number of stages becomes infinite. The minimum number of stages and the minimum reflux ratio are found by exactly the same methods used to determine the same quantities in rectification.

Application of the McCabe-Thiele method to extraction   The distribution relation shown in Fig. 17-21b is analogous to the xy diagram used in distillation. By considering this line as an equilibrium curve, the McCabe-Thiele method can be applied to countercurrent extraction with reflux, provided the operating lines can be located.[6] In extraction, the extract and raffinate streams vary from stage to stage, and the McCabe-Thiele operating lines are generally far from straight. The Ponchon-Savarit diagram can be used to establish operating lines on the $X_r X_e$ diagram. It is necessary only to draw, at random, several lines through the net flow points P_D and P_B and transfer the intersections of these lines with the saturated-raffinate and extract lines to the $X_r X_e$ diagram, as shown in Fig. 20-18. The operating lines are then drawn

Table 20-3 COMPARISON OF EXTRACTION WITH
RECTIFICATION, BOTH USING REFLUX

Rectification	Extraction
Vapor flow in cascade V	Extract flow in cascade V
Liquid flow in cascade L	Raffinate flow in cascade L
Overhead product D	Extract product D
Bottom product B	Raffinate product B
Condenser	Solvent separator
Bottom-product cooler	Raffinate solvent stripper
Overhead-product cooler	Extract solvent stripper
Heat to reboiler q_r	Solvent to cascade s_B
Heat removal in condenser $-q_c$	Solvent removal in separator s_D
Reflux ratio $R_D = L_a/D$	Reflux ratio $R_D = L_a/D$
Rectifying section	Extract-enriching section
Stripping section	Raffinate-stripping section

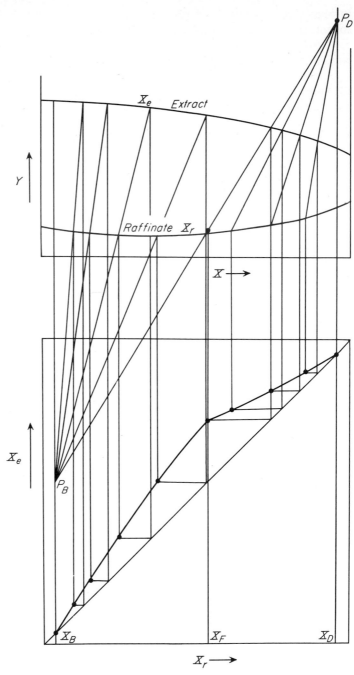

FIGURE 20-18
Construction of operating lines on $X_r X_e$ plane.

through these points. The usual McCabe-Thiele construction for ideal stages can then be applied.†

Whenever both the equilibrium and operating lines on the $X_r X_e$ plane are straight, Eq. (18-27) or (18-30) can, of course, be used.

EXAMPLE 20-4 A mixture of 40 percent MCH and 60 percent heptane is extracted with pure aniline at 25°C in a countercurrent extractor with reflux. The raffinate product is to contain 1 percent MCH and the extract product 98 percent MCH, both on a solvent-free basis. A reflux ratio R_D of 7.7 is to be used. What is the minimum reflux ratio, and how many ideal stages are needed? Per 100 kg of feed, how much solvent is removed in the solvent separator?

SOLUTION The rectangular diagram of Fig. 17-21a is used. Refer to Fig. 20-19. The solvent content of the raffinates is low, and Y_r is considered zero. From the conditions of the problem, $X_F = 0.40$, $X_D = 0.98$, and $X_B = 0.01$. The minimum reflux ratio R_D' is found by locating the overall-balance line, collinear with a tie line, intersecting line $X = X_D$ at the point farthest from the X axis. By trial this is found to be the line collinear with the tie line through the point F. Any other line through a tie line to be used in the Ponchon-Savarit construction intersects the vertical line through X_D at a point nearer the X axis than does line $P_D'FP_B'$, and this line establishes the condition for minimum reflux ratio. The ordinates of points P_D', V_a', and L_a' are, respectively, 34.0, 5.0, and 0. Then the minimum reflux ratio is

$$R_D' = \frac{L_a'}{D} = \frac{34.0 - 5.0}{5.0} = 5.8$$

Since the operating reflux ratio is 7.7, Y, the ordinate of point P_D, is given by the center-of-gravity equation

$$R_D = \frac{Y - 5.0}{5.0} = 7.7 \quad \text{or} \quad Y = 43.5$$

Line $P_D FP_B$, drawn through points P_D and F, establishes the overall-balance line. The Ponchon-Savarit construction is then drawn, as in Fig. 20-19. Tie-line terminals are found from Fig. 17-21b. The total number of ideal stages is 26. The feed should be placed on the eleventh stage.

The solution of this problem by the McCabe-Thiele method is shown in Fig. 20-20. The operating lines in this figure are located by the method of Fig. 20-18.

Per 100 kg of feed, the solvent-free raffinate product is

$$B' = \frac{0.98 - 0.40}{0.98 - 0.01} 100 = 59.8 \text{ kg}$$

The solvent-free extract product is $D' = 100 - 59.8 = 40.2$ kg.

The fresh solvent added at the bottom of the cascade, per kilogram of product B, is the Y coordinate of point P_B, which is -29.2. Then

$$s_B = 29.2 \times 59.8 = 1,750 \text{ kg}$$

---

† Many other graphical constructions have been developed for extraction calculations. For an exhaustive and comprehensive treatment of these, see Ref. 13.

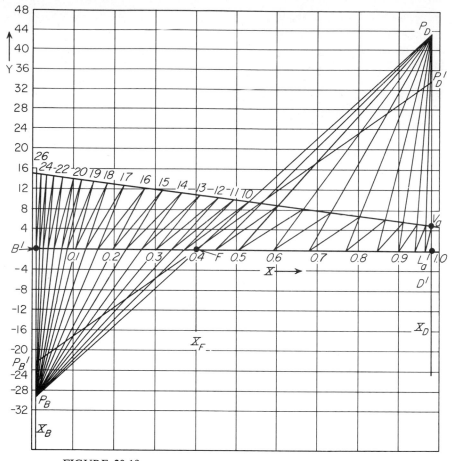

FIGURE 20-19
Example 20-5. Ponchon-Savarit diagram.

The solvent removed in the solvent separator, per kilogram of extract product D, is the ordinate of point P_D, which is 43.5. Then

$$s_D = 43.5 \times 40.2 = 1,750 \text{ kg}$$

Since the raffinates are nearly solvent-free, it is assumed that solvent strippers are not necessary.   ////

Extraction in packed and spray columns   The extraction performance of packed columns and spray columns is governed by the rate of mass transfer between the dispersed and continuous phases. This rate, in turn, depends on the interfacial area of the dispersed phase and the mass-transfer coefficients, as discussed in Chap. 23, page 719.

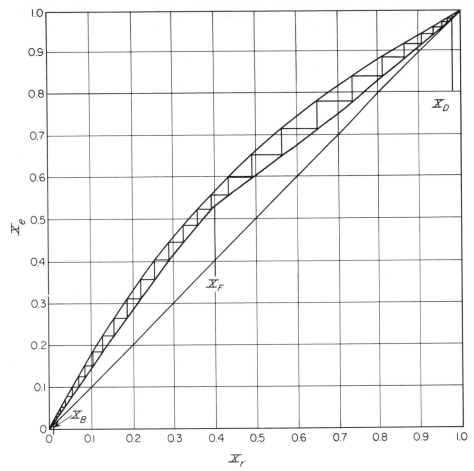

FIGURE 20-20
Example 20-5. McCabe-Thiele construction.

SYMBOLS

a	Solute in stream, concentration or mass
a_v	Specific surface of packing, ft^2/ft^3 or m^2/m^3
B	Bottom product, lb or kg base mixture per hour; B', bottom product leaving stripper
b	Solid in stream, concentration or mass, leaching; diluent in stream, concentration or mass, extraction
D	Overhead product, lb or kg base mixture per hour; D', overhead product leaving solvent stripper
F	Feed to extraction cascade, lb or kg base mixture per hour
H	Enthalpy, Btu/lb or J/g
J	Total input to, and output from, cascade, lb or kg base mixture per hour
L	Underflow, or raffinate phase, lb or kg total or base mixture per hour; L_a, entering cascade; L_b, leaving cascade; L_b', at minimum reflux; L_n, leaving stage n

m Mass flow, lb or kg; m_R, m_S, m_T, for streams R, S, T, respectively

N Number of ideal stages

P Net flow through cascade, in at stage N and out at stage 1, lb or kg base mixture per hour; P_B, net flow leaving extract-enriching section; P_D, net flow leaving raffinate-stripping section; P'_D, at minimum reflux

q Heat added, Btu/lb or J/g; $-q_c$, to condenser; q_r, to reboiler

R_D Reflux ratio; R'_D, minimum

s Solvent in stream, concentration or mass; s_B, solvent added to bottom product, lb/h or kg/h; s_D, solvent removed from overhead product, lb/h or kg/h

V Overflow, or extract phase, mass or moles base mixture per hour; V_a, leaving cascade; V'_a, at minimum reflux; V_b, entering cascade; V_{n+1}, leaving stage $n + 1$

$\bar{V}_s$ Superficial velocity, based on cross section of column, ft/h or m/h; $\bar{V}_{s,c}$ of continuous phase; $\bar{V}_{s,d}$ of dispersed phase

X Abscissa on XY diagram; for leaching, $X = a/(a + s)$; for type I extraction, $X = b/(b + s)$; for type II extraction, $X = a/(a + b)$; X_B, for raffinate product; X_D, for extract product; X_R, X_S, X_T, for underflow streams R, S, T respectively; X_e, for saturated extract in equilibrium with X_r; X_r, for saturated raffinate in equilibrium with X_e

x Mass fraction of solute in underflow or L phase; x_A, x_B, x_S, mass fraction of solute, diluent, solvent, respectively, based on entire L phase; x_a, at entrance; x_b, at exit; x_e, equilibrium value; x_n, in stage n

Y Ordinate on XY diagram; for leaching, $Y = b/(a + s)$; for type I extraction, $Y = a/(b + s)$; for type II extraction, $Y = s/(a + b)$; Y_B, for raffinate product; Y_D, for extract product; Y_R, Y_S, Y_T, for overflow streams R, S, T respectively; Y_e, for saturated extract in equilibrium with Y_r; Y_r, for saturated raffinate in equilibrium with Y_e

y Mass fraction of solute in overflow or V phase; y_A, y_B, y_S, mass fraction of solute, diluent, solvent, respectively, based on entire V phase; y_a, at exit; y_b, at entrance; y_e, equilibrium value; y_{n+1}, in stage $n + 1$

y^* Concentration of overflow solution in equilibrium with specific underflow solution; y^*, in equilibrium with x_a; y^*, in equilibrium with x_b

Greek letters

$\Delta\rho$ Density difference between phases, lb/ft³, or kg/m³

ε Fraction voids in packed section

μ_c Viscosity of continuous phase, lb/ft-h, or kg/m-s

ρ_c Density of continuous phase, lb/ft³, or kg/m³

σ Interfacial tension, dyn/cm, or lb$_f$/m

PROBLEMS

20-1 Roasted copper ore containing the copper as $CuSO_4$ is to be extracted in a countercurrent stage extractor. Each hour a charge consisting of 10 tons of gangue, 1.2 tons of copper sulfate, and 0.5 ton of water is to be treated. The strong solution produced is to consist of 90 percent H_2O and 10 percent $CuSO_4$ by weight. The recovery of $CuSO_4$ is to be 98 percent of that in the ore. Pure water is to be used as the fresh solvent. After each stage, 1 ton of inert gangue retains 2 tons of water plus the copper sulfate dissolved in that water. Equilibrium is attained in each stage. How many stages are required?

20-2 A five-stage countercurrent extraction battery is used to extract the sludge from the reaction

$$Ns_2CO_3 + CaO + H_2O \rightarrow CaCO_3 + 2NaOH$$

The $CaCO_3$ carries with it 1.5 times its weight of solution in flowing from one unit to another. It is desired to recover 98 percent of the NaOH. The products from the reaction enter the first unit with no excess reactants, but with 0.5 kg of H_2O per kilogram of $CaCO_3$. (*a*) How much wash water must be used per kilogram of calcium carbonate? (*b*) What is the concentration of the solution leaving each unit, assuming that $CaCO_3$ is completely insoluble? (*c*) Using the same quantity of wash water, how many units must be added to recover 99.5 percent of the sodium hydroxide?

20-3 In Prob. 20-2 it is found that the sludge retains solution varying with the concentration as shown in Table 20-4. If a 10 percent solution of NaOH, is to be produced, how many stages must be used to recover 95 percent of the NaOH?

20-4 Oil is to be extracted from halibut livers by means of ether in a countercurrent extraction battery. The entrainment of solution by the granulated liver mass was found by experiment to be as shown in Table 20-5. In the extraction battery, the charge per cell is to be 100 lb, based on completely exhausted livers. The unextracted livers contain 0.043 gal of oil per pound of exhausted material. A 95 percent recovery of oil is desired. The final extract is to contain 0.65 gal of oil per gallon of extract. The ether fed to the system is oil-free. (*a*) How many gallons of ether are needed per charge of livers? (*b*) How many extractors are needed?

20-5 In a continuous countercurrent train of mixer-settlers, 100 kg/h of a 40:60 acetone-water solution is to be reduced to 10 percent acetone by extraction with pure 1,1,2-trichloroethane at 25°C. (*a*) Find the minimum solvent rate. (*b*) At 1.6 times the minimum (solvent rate)/(feed rate), find the number of stages required. (*c*) For conditions of part (*b*) find the mass flow rates of all streams. Data are given in Table 20-6.

20-6 Two liquids *A* and *B* which have nearly identical boiling points are to be separated by extraction with a solvent *C*. The data in Table 20-7 represent the equilibrium between the two liquid phases at 95°C. Determine the minimum amount of reflux that must be returned from the extract product and from the raffinate product to produce an extract containing 83 percent *A* and 17 percent *B* (on a solvent-free basis) and a raffinate product containing 10 percent *A* and 90 percent *B* (solvent-free). The feed contains 35 percent *A* and 65 percent *B* on a solvent-free basis and is a saturated raffinate. The raffinate is the heavy

Table 20-4

NaOH, wt %	0	5	10	15	20
Kg solution/kg $CaCO_3$	1.50	1.75	2.20	2.70	3.60

Table 20-5

Solution retained by 1 lb exhausted livers, gal	Solution concentration, gal oil/gal solution	Solution retained by 1 lb exhausted livers, gal	Solution concentration, gal oil/gal solution
0.035	0	0.068	0.4
0.042	0.1	0.081	0.5
0.050	0.2	0.099	0.6
0.058	0.3	0.120	0.68

Table 20-6   EQUILIBRIUM DATA

Limiting solubility curve		
$C_2H_3Cl_3$, wt %	Water, wt %	Acetone, wt %
94.73	0.26	5.01
79.58	0.76	19.66
67.52	1.44	31.04
54.88	2.98	42.14
38.31	6.84	54.85
24.04	15.37	60.59
15.39	26.28	58.33
6.77	41.35	51.88
1.72	61.11	37.17
0.92	74.54	24.54
0.65	87.63	11.72
0.44	99.56	0.00

Tie lines

Weight % in water layer			Weight % in trichloroethane layer		
$C_2H_3Cl_3$	Water	Acetone	$C_2H_3Cl_3$	Water	Acetone
0.52	93.52	5.96	90.93	0.32	8.75
0.73	82.23	17.04	73.76	1.10	25.14
1.02	72.06	26.92	59.21	2.27	38.52
1.17	67.95	30.88	53.92	3.11	42.97
1.60	62.67	35.73	47.53	4.26	48.21
2.10	57.00	40.90	40.00	6.05	53.95
3.75	50.20	46.05	33.70	8.90	57.40
6.52	41.70	51.78	26.26	13.40	60.34

Table 20-7   EQUILIBRIUM DATA

Extract layer			Raffinate layer		
A, %	B, %	C, %	A, %	B, %	C, %
0	7.0	93.0	0	92.0	8.0
1.0	6.1	92.9	9.0	81.7	9.3
1.8	5.5	92.7	14.9	75.0	10.1
3.7	4.4	91.9	25.3	63.0	11.7
6.2	3.3	90.5	35.0	51.5	13.5
9.2	2.4	88.4	42.0	41.0	17.0
13.0	1.8	85.2	48.1	29.3	22.6
18.3	1.8	79.9	52.0	20.0	28.0
24.5	3.0	72.5	47.1	12.9	40.0
31.2	5.6	63.2	Plait point		

liquid. Determine the number of ideal stages on both sides of the feed required to produce the same end products from the same feed when the reflux ratio of the extract, expressed as pounds of extract reflux per pound of extract product (including solvent), is twice the minimum. Calculate the masses of the various streams per 1,000 lb of feed, all on a solvent-free basis.

REFERENCES

1 Baker, E. M.: *Trans. AIChE,* **32:**62 (1936).
2 Cofield, E. P., Jr.: *Chem. Eng.,* **58**(1):127 (1951).
3 Crawford, J. W., and C. R. Wilke: *Chem. Eng. Prog.,* **47:**423 (1951).
4 Elgin, J. C.: *Trans. AIChE,* **32:**451 (1936).
5 Grosberg, J. A.: *Ind. Eng. Chem.,* **42:**154 (1950).
6 Maloney, J. O., and E. A. Schubert: *Trans. AIChE,* **36:**741 (1940).
7 Morello, V. S., and N. Poffenberger: *Ind. Eng. Chem.,* **42:**1021 (1950).
8 Perry, J. H. (ed.): "Chemical Engineers' Handbook," 5th ed., pp. **21**-4 to **21**-29, McGraw-Hill, New York, 1973.
9 Randall, M., and B. Longtin: *Ind. Eng. Chem.,* **30:**1188 (1938).
10 Sage, G., and F. W. Woodfield: *Chem. Eng. Prog.,* **50:**396 (1954).
11 Sherwood, T. K., J. E. Evans, and J. V. A. Longcor: *Ind. Eng. Chem.,* **31:**1144 (1939).
12 Skelland, A. H. P.: *Ind. Eng. Chem.,* **53:**799 (1961).
13 Treybal, R. E.: "Liquid Extraction," 2d ed., McGraw-Hill, New York, 1963.
14 Varteressian, K. A., and M. R. Fenske: *Ind. Eng. Chem.,* **29:**270 (1937).

21

INTRODUCTION TO MULTICOMPONENT-STAGE OPERATIONS

Chapters 18 to 20 are confined to simple processes and binary mixtures in distillation and ternary systems in leaching and extraction. Most chemical engineering separation operations are not so limited. Complicated combinations of several cascades with condensers, reboilers, mixers, dividers, heat-transfer units, and pumps, all operating on multicomponent systems, are common. The combination of these with chemical reactors, which are not discussed in this text, is the basic field of the chemical engineer.

To treat complicated plants in detail is the task of advanced and specialized treatises.[3,5,10] In this chapter some of the fundamentals on which such works are based are introduced to provide a foundation for understanding this large and important technical literature.

In practice the field is dominated by the use of giant digital computers.[2] The mass of numbers needed to quantify the operating and engineering variables and the many iterations required to obtain convergence of the solutions to the equations are usually too numerous for hand calculators. This is not a book on computers, but all computers must be fed with programs based squarely on the principles to which this text is devoted.

In this chapter the procedure for determining the correct number of independent decisions for designing complicated plants is described; multicomponent flash processes are treated; the problems of determining, for multicomponent systems, the minimum number of plates and the minimum reflux ratio are discussed; and an empirical method for determining the number of ideal plates required with an operating reflux ratio is outlined. To keep the discussion to manageable size, simplifying assumptions are accepted. The computer is used in practice to refine the simplified solutions to attain the final desired precision.

COMPLEX SEPARATION PLANTS

The most complicated plants discussed in Chaps. 19 and 20 are those diagrammed in Figs. 19-9 and 20-16. When operated on binary mixtures for distillation or on ternary type II systems for liquid extraction, these plants were looked upon as integrated structures in which all parts are unified by the McCabe-Thiele and Ponchon-Savarit diagrams. In the more general context of this chapter, complex plants are analyzed in terms of their elements. The distillation plant of Fig. 19-10 is so shown in Fig. 21-1. It breaks down into two cascades, a feed plate which is not considered a part of either cascade, a total condenser giving a bubble-point distillate, a reflux divider, and a partial reboiler. This assembly is called a *conventional distillation plant*. It is the classic plant, which when used to separate ethanol and water, brought industrial distillation to maturity. It is the starting point for most advanced discussions of stage operations.

Complex Plants

Three typical complex plants are shown in Figs. 21-2 to 21-4.

Absorber-stripper A method of scrubbing a solute from its mixture in an inert gas and recovering the solvent for further use is shown in Fig. 21-2. In one important application, liquid ethanolamine is used as the solvent to remove carbon dioxide, the solute, from its mixture with air. The plant contains two cascades, a pump, two pressure-reducing valves, and three heat-exchange units—a heater, a cooler, and a liquid-liquid heat exchanger. One cascade is an absorber and the other a stripper. The temperature in the absorber is lower than that in the stripper, and the pressure in the stripper is lower than that in the absorber.

Cold solvent enters the top of the absorber and leaves, with its absorbed carbon dioxide, at the bottom. The stream is heated in the heat exchanger by hot liquid from the bottom of the stripper and further heated in the heater. The pressure on the hot stream is reduced in the reducing valve and the stream fed to the top of the stripper. Steam is admitted through the pressure-reducing valve into the bottom of the stripper. Solvent, nearly free of carbon dioxide, leaves the bottom of the stripper, flows through the pump, is cooled by the heat exchanger and the cooler, and is fed to the top of the

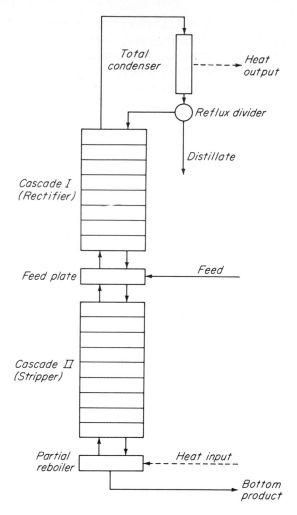

FIGURE 21-1
Conventional distillation plant.

absorber, so completing its circuit. The steam and carbon dioxide are released from the top of the stripper. When the solute is valuable or dangerous, this stream is condensed and the solute recovered in a separate plant.

Azeotropic distillation in twin-column plant The components in an azeotropic mixture cannot be separated in a conventional distillation plant, and more complicated methods must be used. One type of azeotropic plant, shown in Fig. 21-3, includes three cascades, a partial reboiler, a total condenser, a mixer, and a decanter. The entire system operates at the same pressure, usually atmospheric. It may be used when the

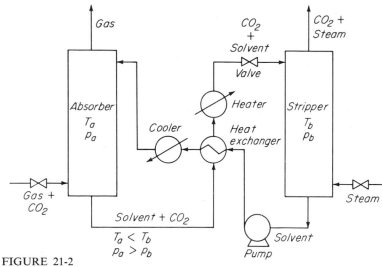

FIGURE 21-2
Absorber-stripper.

azeotropic vapor, on condensation, forms two liquid layers, one of lower and the other of higher concentration than the azeotrope. The system isobutanol-water is of this type. It has a minimum-boiling azeotrope at 90°C consisting of 67 mole percent water and 33 mole percent isobutanol. On condensing it forms two liquid layers, one of 40 mole percent water and the other 95 mole percent.

To separate such a mixture, the feed, which is an isobutanol-rich liquid, is fed to the proper plate in the isobutanol column. Nearly pure isobutanol is drawn from the bottom of this column, which is equipped with a partial reboiler. The vapor from the top of the column, which contains most of the water in the feed, is mixed with the vapor from the top of the water column. The vapor mixture is condensed to two liquid phases in the total condenser and collected in the decanter. The upper, isobutanol-rich layer from the decanter is fed to the top of the isobutanol column, and the heavier, lower, water-rich layer goes to the top of the water column. Live steam is blown into the bottom of the water column to strip the isobutanol from the water layer. Nearly pure water is withdrawn as the bottom product, and the vapor from the top of the column is mixed with that from the isobutanol column, condensed, and returned to the decanter.

Double-column air-distillation plant with side-stream drawoff   One method of fractionating air into nitrogen and oxygen, and also obtaining a stream rich in argon, is to use the doubled column with a side-stream drawoff shown in Fig. 21-4. This plant contains two columns, one operating at high pressure and the other at low pressure; two dividers; two pressure-reducing valves; and a specialized two-pressure heat-transfer unit, functioning as a total condenser in its high-pressure compartment and

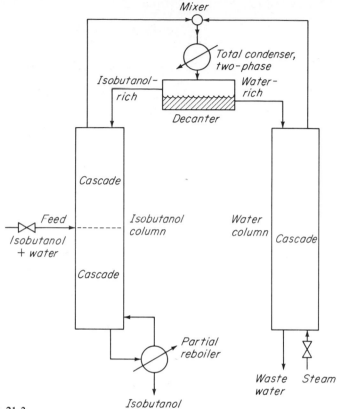

FIGURE 21-3
Twin azeotropic-distillation plant.

a total reboiler in its low-pressure compartment. Nitrogen is the most volatile component and oxygen the least volatile; argon is of intermediate volatility.

Very cold, dry, carbon dioxide–free air enters the bottom of the high-pressure column, which is a rectifier for concentrating nitrogen. The liquid leaving the bottom of the column is partially concentrated oxygen. It flows through a reducing valve to the center of the low-pressure column. The stripping section of this column eliminates nitrogen from oxygen, and the rectifying section concentrates the nitrogen, which leaves the top of the column as a product. A side-stream drawoff plate provides an intermediate liquid stream rich in argon which is available for further processing. The vapor from the high-pressure column is condensed in the condenser-reboiler and is split by a divider to provide reflux for both the high-pressure column and (after pressure reduction) the low-pressure column. The liquid from the bottom of the low-pressure column is vaporized in the condenser-reboiler and split by a divider, giving one stream to provide vapor to the bottom of the low-pressure column and another stream that is withdrawn as the oxygen product.

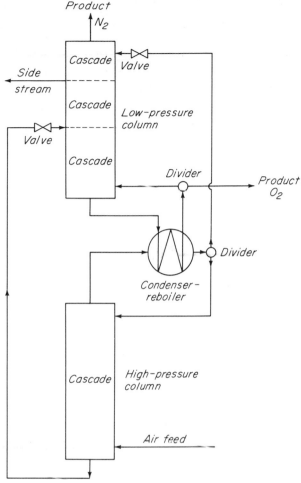

FIGURE 21-4
Double column with side stream for air distillation.

COUNTING VARIABLES

The design of a separation plant calls for the numerical values of many quantities: feed rates, concentrations, pressures, temperatures, heat flow rates, power for pumps, sizes of heat-transfer equipment, and the number of ideal plates in a cascade. These quantities are interrelated, and the logic is highly constrained. Of all the variables, only a few are independent and so available to the decision of the designer. If too few independent variables are fixed, the design is indeterminate and an infinite number of designs will work. If too many variables are chosen, the design is redundant and no design will work. The advent of the computer, a man-made device which cannot think for itself and tries to follow blindly whatever information is fed into it, has sharpened this problem of variable counting. The problem at the first stage of a design is not the choice of magnitudes for variables; it is the number of decisions that must be made by the designer. When this is known exactly, numbers can be assigned to the right number of truly independent variables and the computer directed to make the calculations. The following method of variable counting is a modification of the well-known Kwauk[7] procedure.

Assumptions and principles The Kwauk method of counting variables is applied to continuous steady-flow processes under these assumptions: no chemical reaction, negligible potential and kinetic energy in flow streams, work effect restricted to shaft work, and equilibrium stages in cascades. Under these restrictions the pertinent relations between variables are the following.

MATERIAL BALANCE For each component, from Eq. (1-57), for v streams

$$\sum_{1}^{v} \dot{n}_i = 0 \qquad i = 1, 2, \ldots, \mathscr{C} \tag{21-1}$$

where $\dot{m}_i$ is the molal flow rate of component i and the $\sum$ operator calls for the algebraic sum of the flow rates of all streams entering and leaving the element, taking leaving streams as positive and entering streams as negative. For a total stream balance use Eq. (21-1) and drop subscript i.

ENTHALPY (ENERGY) BALANCE From Eq. (1-56)

$$\sum_{1}^{v} \dot{n}H = q - P \tag{21-2}$$

where H = enthalpy per mole of stream
$\quad q$ = rate of heat transfer from environment into element
$\quad P$ = power flowing from element to environment as shaft work

This equation is applied once over each element.
Equations (21-1) and (21-2) apply to a control volume.

EQUILIBRIA The counting of phase-equilibria variables is subject to the phase rule, Eq. (17-1), written as†

$$\mathscr{F} = \mathscr{C} - \mathscr{P} + 2 \tag{21-3}$$

The phase-rule variance $\mathscr{F}$ is the number of independent temperatures, pressures, and concentrations at equilibrium in a system consisting of $\mathscr{P}$ phases and containing $\mathscr{C}$ components.

General method The general Kwauk method is based on a choice of simple process steps called *elements*. They are the process components like those shown in Figs. 21-1

---

† Script letters are used in the phase rule and in variable counting to reinforce the fact that they designate the integers used for counting and are not symbols for the magnitudes of the variables themselves.

to 21-4. A control surface is thrown around each element. The total number of variables for the element is counted. This number is denoted by $\mathcal{N}_v$. The number of variables that can be eliminated by the use of material and enthalpy balances, the phase rule, and certain other constraints, is counted and denoted by $\mathcal{N}_c$. The difference, $\mathcal{N}_v - \mathcal{N}_c = \mathcal{N}_i$, is the gross number of variables that must be fixed to establish the design. Then, the variables in the set $\mathcal{N}_i$ are sectioned into two subsets. The first subset, $\mathcal{N}_x$, consists of the number of pressure levels in the system (disregarding the small pressure drops from friction) and the number of variables needed to define all feed streams to the plant. The other set, containing the remaining variables in $\mathcal{N}_i$, is denoted by $\mathcal{N}_a$. The following equations show the sorting out of variables:

$$\mathcal{N}_c = \mathcal{N}_v - \mathcal{N}_i \qquad (21\text{-}4)$$

$$\mathcal{N}_a = \mathcal{N}_i - \mathcal{N}_x \qquad (21\text{-}5)$$

The procedure is first applied to the elements constituting a flowsheet. A second procedure is used to calculate $\mathcal{N}_a$ for the whole plant.

Interpretation of $\mathcal{N}_a$ Almost always the variables in $\mathcal{N}_x$ are predetermined by the position of the elements in the general flowsheet of the entire plant and are not open to decision by the designer of the unit under study. If not previously fixed, they are easily spotted and counted. Then $\mathcal{N}_a$ is the number of variables that are truly available for decision. This number has two useful characteristics: for any element it is a very small integer, either -1, 0, or 1; and it is independent of the number of components. The number of components appears only in the feed-stream variables in $\mathcal{N}_x$. The precise calculation of $\mathcal{N}_a$ is the objective of the Kwauk method.

Single-stream variables; state principle To evaluate $\mathcal{N}_v$ and $\mathcal{N}_a$, the number of independent variables needed to fix all the properties of a process stream must be known. This information is given by the fundamental equation of thermodynamics, derived in all texts on the subject. Under the simplifying assumptions used in the analysis the equation is

$$dU = T \, dS - p \, dV + \sum_{i=1}^{\mathscr{C}} \mu_i \, dn_i \qquad (21\text{-}6)$$

where U = internal energy of system
  V = volume of system
  p = pressure on system
  S = entropy of system
  T = temperature of system
  n_i = number of moles of component i
  μ_i = chemical potential, component i

More abstractly, Eq. (21-6) can be written

$$U = f(S, V, n_i) \qquad (21\text{-}7)$$

This states that U is determined if the $\mathscr{C} + 2$ values of S, V, and n_i are known. This theorem is called the *state principle*.[6] All quantities in Eq. (21-7) are extensive properties and are proportional to the mass of the system.

Equation (21-7) applies to a system having any number of phases permitted by the phase rule. This maximum number of phases is found by setting $\mathscr{F} = 0$ in Eq. (21-3), giving

$$\mathscr{P}_{max} = \mathscr{C} + 2 \tag{21-8}$$

The variables in Eq. (21-7) are not convenient for calculation. By transform methods used on thermodynamic equations[11] the following substitutions may be made: nH for U, where $n = \sum n_i$; T for S; and p for V. Then for a steady-flow process n and n_i are replaced by $\dot{n}$ and $\dot{n}_i$, where the dots signify time derivatives. Then

$$\dot{n}H = f(T, p, \dot{n}_i) \qquad i = 1, 2, \ldots, \mathscr{C} \tag{21-9}$$

Although the independent variables in Eq. (21-9) differ from those in Eq. (21-7), there are exactly $\mathscr{C} + 2$ of them.

Other combinations of variables may be used to define a stream. The most common choice is mole flow rate of stream $\dot{n}$; temperature; pressure; and $\mathscr{C} - 1$ mole fractions of components, defined by $x_i = \dot{n}_i/\dot{n}$. The enthalpy of 1 mol of stream H also may be used in place of temperature. In any choice, one of the $\mathscr{C} + 2$ variables must be a flow rate.

Phase equilibria in stream The independence of Eq. (21-6) of the number of phases in the system shows that two or more phases flowing as a single stream have the same number of variables as a single-phase stream. It follows immediately that two or more streams that have been equilibrated, e.g., the two streams leaving a flash distillation or those from an ideal plate, count as only one stream. All such statements as these may be subsumed under the term *conservation of variance*.

Elements

The elements used in separation processes can be separated into two groups: elements in which changes in concentration occur and auxiliary elements in which the concentrations in the streams remain constant. In the second group, a stream has only three independent variables, which may be considered the flow rate, temperature, and pressure. All streams are single-phase, designated 1ϕ.

A selection of auxiliary elements is shown by the simplified flow diagrams in Fig. 21-5, and the systematic Kwauk calculation of $\mathscr{N}_a$ for them is given in Table 21-1. The only variables in $\mathscr{N}_v$ are the three for each stream and possible variables for q and P in Eq. (21-2) when the element is a heat-transfer unit or pump. The conditions relating to $\mathscr{N}_c$ are a single material balance involving the flow rate of the stream, one enthalpy balance, and in some cases one or more equalities of pressure and temperature among the streams or the equality of the temperature of a stream to the

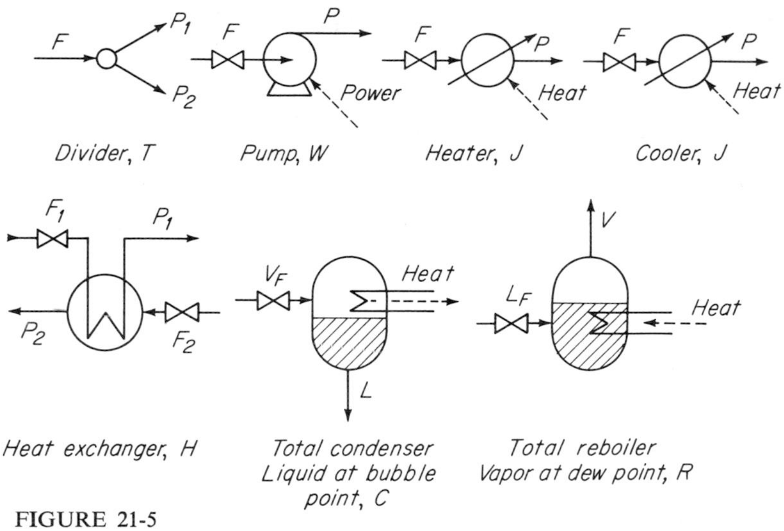

FIGURE 21-5
Auxiliary elements. (Selected and simplified from Ref. 7.)

bubble point or dew point. The system pressures and the variables for all feed streams are the decisions to be made in calculating $\mathcal{N}_x$.

The results of the calculation of $\mathcal{N}_c$ are given in Table 21-1. For the total condenser and total reboiler, $\mathcal{N}_c = 0$, and for the rest, $\mathcal{N}_c = 1$. In the condenser the liquid leaves at the bubble point, and in the reboiler the vapor leaves at the dew point. Each of these facts contributes unity to the equality term in $\mathcal{N}_c$.

Elements in which concentration changes do take place are shown in the flow diagrams in Fig. 21-6 and are analyzed in Table 21-2. In these elements the number of variables in each stream is $\mathscr{C} + 2$, whether the stream is a single phase (designated 1ϕ), a two-phase stream, or two separated streams leaving the control volume after equilibration (both cases designated by 2ϕ). The number of material balances is $\mathscr{C}$, one for each component. Other conditions are as in Table 21-2. The result of the calculations shows that for a partial condenser or partial reboiler $\mathcal{N}_a = 1$; for the other elements in Table 21-2 $\mathcal{N}_a = 0$.

A shortcut for calculating $\mathcal{N}_a$ for single-phase heat-transfer elements (heaters, coolers, heat exchangers, total condensers, and total reboilers) is to count 1 for each element and discount the total by 1 for every stream leaving the element at the bubble or dew point. The total condenser-reboiler shown in Fig. 21-4, for example, has a value of $\mathcal{N}_a = 1 - 1 - 1 = -1$, since both leaving streams are so discounted.

Combining Elements to Form Systems

When elements are connected to form complete plants, a general method is needed for calculating $\mathcal{N}_a$ for the system from the variables of the elements. The difference between an element standing alone and the same element connected to one or more other elements in a system is that in an assembly of elements interstreams connect the elements to make up an integrated plant. If $\mathcal{N}_v$ for the system is calculated by adding the values of $\mathcal{N}_v$ for the elements, the $\mathcal{C} + 2$ variables for each interstream are counted twice. This duplication is removed if these variables also are included in $\mathcal{N}_c$ as

Table 21-1 ELEMENTS WITH NO CONCENTRATION CHANGE*

		Element and symbol				
		Divider T	Pump W	Heater or cooler J	Heat exchanger H	Total condenser or reboiler C or R
Number of streams		3	2	2	4	2
Variables						
Description	Number					
$\mathcal{N}_v$:						
Per stream	3^a	3×3	2×3	2×3	4×3	2×3
Heat	1			1		1
Power	1		1			
Total		9	7	7	12	7
$\mathcal{N}_c$:						
Material balances	1	1	1	1	2^f	1
Enthalpy balance	1	1	1	1	1	1
Equalities	1	2×2^b	1^c			1^d
Total		5	3	2	3	3
$\mathcal{N}_i = \mathcal{N}_v - \mathcal{N}_c$		4	4	5	9	4
$\mathcal{N}_x$:						
Feed streams	3	3	3	3	2×3	3
Pressure	1	0^e	1	1	2^g	1
Total		3	4	4	8	4
$\mathcal{N}_a^h = \mathcal{N}_i - \mathcal{N}_x$		1	0	1	1	0

* Based on Ref. 7.
a Flow rate, temperature, and pressure.
b Temperature and pressure equality in all streams.
c Discharge pressure equal to pressure in receiving element.
d Discharge temperature at bubble or dew point.
e Pressure of inlet equals pressure in element.
f No mass transfer across heating surface. One material balance for each entering stream.
g Pressure in each compartment equal to inlet pressure of feed.
h Examples of choice of variables in $\mathcal{N}_a$:
      Divider: ratio P_1/P_2, or P_2/F.
      Heater, cooler, heat exchanger: one exit temperature, or heating surface of unit.
      Pump: power input.

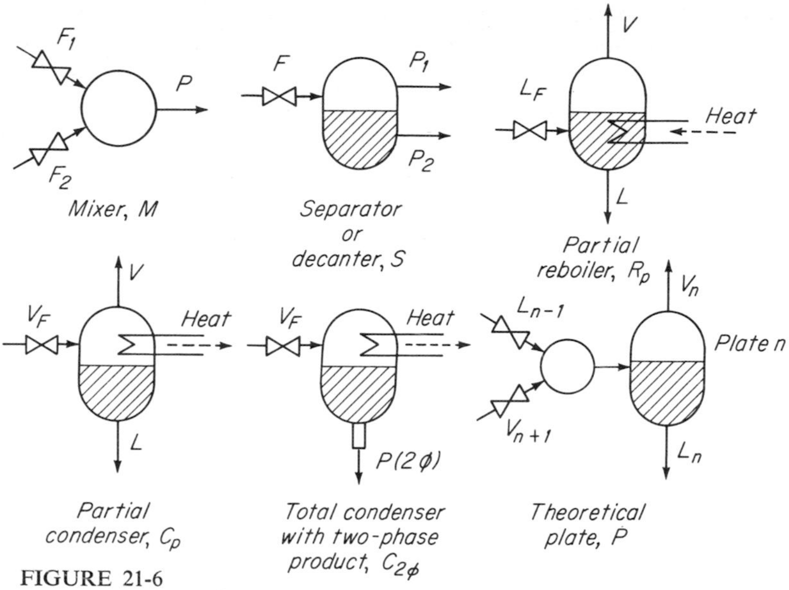

FIGURE 21-6

Elements in which concentration changes occur.

additional constraints, so $\mathscr{N}_i$ will retain only the $\mathscr{C} + 2$ stream variables for the feed streams to the entire plant. This can be done more simply by disregarding all inter-stream variables in the calculation of both $\mathscr{N}_x$ and $\mathscr{N}_c$ for the system. All the $\mathscr{N}_a$ values of the elements survive, and their sum is the $\mathscr{N}_a$ value for the plant. The number of different pressure levels and the stream variables for all streams entering the plant make up $\mathscr{N}_x$.

Complex elements A useful combination of simple elements for analyzing complex plants is called a *complex element*. Flow diagrams for three complex elements are shown in Fig. 21-7.

CASCADE The most important complex element is the cascade, the conventional plate column consisting of a definite number of theoretical plates connected by liquid and vapor interstreams in the usual manner. In calculating the $\mathscr{N}_a$ value for a cascade, the interstreams are neglected, and apparently the value for the cascade is zero. However, there is one question that must be answered: How many plates? This requires a decision, denoted by $\mathscr{N}_\sigma$, and adds unity to the $\mathscr{N}_a$ of the cascade.† Each cascade in a plant, then, contributes unity to the value of $\mathscr{N}_a$.

---

† Another example of this kind of decision arises in the assembly of a freight train from a locomotive, a caboose, and some box cars. The train can be assembled only when someone decides how many box cars to use.

FEED PLATE   A feed plate connects two cascades to provide access for an entering stream other than those admitted to the ends of the cascades. Figure 21-7 shows that a feed plate is a combination of a separator and two mixers, and since $\mathcal{N}_a = 0$ for each of these, it is also zero for the feed plate.

SIDE-STREAM DRAWOFF   A side-stream drawoff plate connects two cascades to permit withdrawing an arbitrary portion of the liquid stream flowing down the cascade. Figure 21-7 shows that such a plate is a combination of theoretical plate and a divider. Since for the divider, $\mathcal{N}_a = 1$ and for a theoretical plate $\mathcal{N}_a = 0$, $\mathcal{N}_a$ for the drawoff plate is unity.

Table 21-2   ELEMENTS WITH CHANGE IN CONCENTRATION*

			Element and symbol			
Description	Number	Mixer M	Separator or decanter S	Partial condenser or reboiler C_p or R_p	Total condenser 2ϕ product $C_{2\phi}$	Theoretical plate P
Number of streams:						
1ϕ		3	1	1	1	2
2ϕ			1	1	1	1
Variables						
$\mathcal{N}_v$:						
Per stream	$\mathscr{C} + 2$	$3(\mathscr{C} + 2)$	$2(\mathscr{C} + 2)$	$2(\mathscr{C} + 2)$	$2(\mathscr{C} + 2)$	$3(\mathscr{C} + 2)$
Heat	1			1	1	
Total		$3\mathscr{C} + 6$	$2\mathscr{C} + 4$	$2\mathscr{C} + 5$	$2\mathscr{C} + 5$	$3\mathscr{C} + 6$
$\mathcal{N}_c$:						
Material balances	$\mathscr{C}$	$\mathscr{C}$	$\mathscr{C}$	$\mathscr{C}$	$\mathscr{C}$	$\mathscr{C}$
Enthalpy balance	1	1	1	1	1	1
Equality	1				1†	
Total		$\mathscr{C} + 1$	$\mathscr{C} + 1$	$\mathscr{C} + 1$	$\mathscr{C} + 2$	$\mathscr{C} + 1$
$\mathcal{N}_i = \mathcal{N}_v - \mathcal{N}_c$		$2\mathscr{C} + 5$	$\mathscr{C} + 3$	$\mathscr{C} + 4$	$\mathscr{C} + 3$	$2\mathscr{C} + 5$
$\mathcal{N}_x$:						
Feed streams	$\mathscr{C} + 2$	$2(\mathscr{C} + 2)$	$\mathscr{C} + 2$	$\mathscr{C} + 2$	$\mathscr{C} + 2$	$2(\mathscr{C} + 2)$
Pressure	1	1	1	1	1	1
Total		$2\mathscr{C} + 5$	$\mathscr{C} + 3$	$\mathscr{C} + 3$	$\mathscr{C} + 3$	$2\mathscr{C} + 5$
$\mathcal{N}_a = \mathcal{N}_i - \mathcal{N}_x$		0	0	1‡	0	0

* Based on Ref. 7.
† Two-phase product is at bubble point.
‡ Typical choice: L in C_p; V in R_p; heat-transfer surface; or heat flow to or from environment.

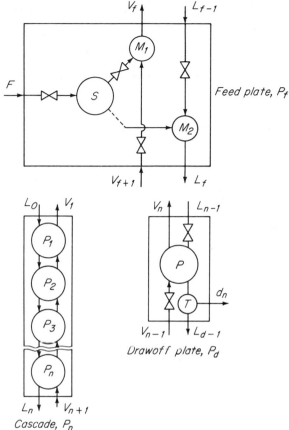

FIGURE 21-7
Complex elements. (Selected and simplified from Ref. 7.)

Summary

The Kwauk method for counting the independent variables for a separation plant of any complexity is reduced to this simple procedure:

1 Count the number of independent pressure levels in the process.
2 Count the $\mathscr{C} + 2$ variables for each feed stream to the plant. Discount this total by one unit if a stream enters at the pressure of the element at which it enters.
3 Count the variables in $\mathscr{N}_a$ by adding (*a*) the number of cascades, (*b*) the number of dividers, (*c*) the number of heat-transfer elements, (*d*) the number of side-stream drawoffs and (*e*) subtracting the number of single-phase streams leaving heat-transfer units at their bubble or dew point.

The variables in categories 1 and 2 are probably fixed, but those in $\mathcal{N}_a$ can and must be chosen by designer decision. The choice of actual variables is open, and individual variables may be traded off at will, but the number $\mathcal{N}_a$ must be conserved.

EXAMPLE 21-1 Analyze the conventional distillation plant of Fig. 21-1 by the Kwauk procedure, and specify a choice of variables in set $\mathcal{N}_a$. Assume two components.

SOLUTION This example is that of the McCabe-Thiele diagram. It has been the subject of much discussion in the literature. ////

Number of variables in $\mathcal{N}_x$:
Pressure level 1
Feed variables; as shown in Fig. 21-1, the feed enters at cascade pressure; the feed variables are $\mathcal{C} + 1 = 2 + 1 = 3$ 3
  Total 4
Number of variables in $\mathcal{N}_a$:
Cascades 2
Reflux splitter 1
Heat-transfer units:
  Condenser 1
  Reboiler 1
Equality, condensate at bubble point −1
  Total 4
Choice of variables:
In feed: flow rate; mole fraction of component A in feed; H of feed (or V/F in feed, or q factor) 3
In $\mathcal{N}_a$:
  Choice 1: number of plates in rectifying column, number of plates in stripping column, reflux ratio in divider, vapor generation in reboiler per mole of feed 4
  Choice 2 (conventional choice): mole fraction component A in distillate, mole fraction component B in bottoms, reflux ratio, transfer of steps in diagram from rectifying line to stripping line 4

Note trade-offs between choices: two concentrations for numbers of plates in two cascades; feedplate location on diagram for the relative amount of vapor generated in reboiler.

EXAMPLE 21-2 Give concise list of $\mathcal{N}_x$ and $\mathcal{N}_a$ variables in plants of (a) Fig. 21-2, (b) Fig. 21-3.

SOLUTION (a) *Absorber-stripper, Fig. 21-2.* Assume $\mathcal{C}$ components in feed to absorber.

$\mathcal{N}_x$: pressures 2
Feed streams:
  Gas $\mathcal{C} + 2$
  Steam 3
  Total $\mathcal{C} + 7$
$\mathcal{N}_a$: cascades 2
Heat-transfer units 3
  Total 5
Choice of variables:
Pressures in the two cascades
Feed streams: gas inlet flow rate; pressure; temperature; $\mathcal{C} - 1$ mole fractions, including mole fraction of CO_2

$\mathcal{N}_a$ variables: number of plates in absorber, number of plates in stripper, one exit temperature for each heat-transfer unit
Other choices: mole fraction CO_2 in gas from absorber; mole fraction CO_2 is gas from stripper; area of three heat-transfer surfaces; also, trade-off flow rate of steam for flow rate of circulating solvent by control of power input to pump

(b) *Twin azeotrope plant, Fig. 20-3*

$\mathcal{N}_x$: pressure	1
Feed streams:	
To plant: flow rate, temperature, pressure, mole fraction isobutanol	4
Steam: flow rate, pressure, temperature	3
Total	8
$\mathcal{N}_a$: cascades	3
Heat-transfer surfaces	2
Equality: condensate at bubble point	−1
Total	4

One choice for the $\mathcal{N}_a$ variables is the number of plates in each of the three cascades and the vapor flow rate from the reboiler. A second choice might be the mole fraction of isobutanol in the bottom of the isobutanol column, the fraction of the isobutanol entering in the feed that is lost in the vapor from the water column, the position of the feed plate in the isobutanol column, and, as in the first choice, the vapor flow rate from the reboiler. Many other sets of four decisions can be selected, and any one decision may appear in more than one set.   ////

MULTICOMPONENT DISTILLATION

Like the distillation of binary mixtures in equilibrium stages, multicomponent distillation calculations use mass and enthalpy balances and liquid-vapor phase equilibria. There is one mass balance for each component, and the single enthalpy balance is identical to that used in the binary case. Phase equilibria, however, are more restricted than the general x_e-vs.-y_e curves used for binary systems. Also, graphical techniques are not generally useful.

Phase equilibria in multicomponent distillation   To save computer time and to allow calculations to be based only on the vapor pressures of the pure components, many distillation programs are written on the assumption that Raoult's law is valid. This places the equilibrium in either case 5 or case 6 of Table 17-1, depending on whether the relative volatilities are temperature-dependent or not.

The K factors are written as $K_i \equiv K_b \alpha_i$, where K_b is that of the arbitrary base component chosen for defining α_{ij}. The usual choice of base component is the least volatile one in the system. Also, the components are numbered serially in the order of their increasing volatility. Then $K_b = K_1$ and $\alpha_{n+1} > \alpha_n$. The second subscript on the α's can be dropped. Then, from Eq. (17-45),

$$\alpha_i = \frac{p_i/p_e}{p_b/p_e} = \frac{p_i}{p_b} \qquad (21\text{-}10)$$

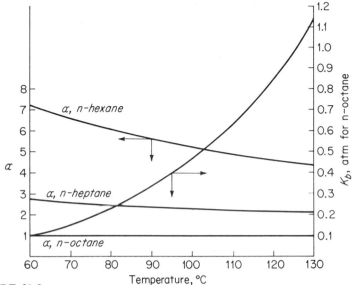

FIGURE 21-8
K and α factors for hydrocarbons.

Values of α_i for individual components are calculated from the vapor-pressure–temperature data of the components and, with values of K_b, may be plotted against temperature on the same diagram. An example of such a plot is given in Fig. 21-8, for three normal paraffinic hydrocarbons from n-hexane to n-octane,† using the n-

Component	A'	B'	C'
n-Hexane	3.99695	1,171.530	− 48.784
n-Heptane	4.02159	1,268.115	− 56.25
n-Octane	4.04296	1,355.126	− 63.633

octane as the base component. Clearly, K_b is a stronger function of temperature than the α values are. If the change of α with temperature is neglected, a single value of α for each component is sufficient and the mixture becomes a case 6 system.

Bubble-point and dew-point calculations   Relative volatilities are used to calculate, at a specific pressure, the bubble-point and dew-point temperatures of a definite multicomponent mixture. The equation on which the calculation is made is

$$y_{ie} = K_i x_{ie} = K_b \alpha_i x_{ie} \tag{21-11}$$

† The curves in Fig. 21-8 are plotted calculations using the Antoine equation (17-49). The parameters are[4] when pressures are in atm, and temperatures in K.

From the definition of mole fraction

$$\sum_{i=1}^{\mathscr{C}} y_{ie} = K_b \sum_{i=1}^{\mathscr{C}} \alpha_i x_{ie} = 1$$

$$\sum_{i=1}^{\mathscr{C}} x_{ie} = \frac{1}{K_b} \sum_{i=1}^{\mathscr{C}} \frac{y_{ie}}{\alpha_i} = 1$$

and

$$K_b = \frac{1}{\sum \alpha_i x_{ie}} = \sum \frac{y_{ie}}{\alpha_i} \tag{21-12}$$

The first summation in Eq. (21-12) is used for a bubble-point calculation and the second for a dew point.

When the relative volatilities are not temperature-dependent, Eqs. (21-12) are used directly to calculate the value of K_b, which is used to find the temperature from the graph or equation for T as a function of K_b. When the α values are functions of temperature, a temperature iteration is necessary.

After the temperature is known, the composition of the vapor at the bubble point and the liquid at the dew point are calculated by the use of the following equations:

Bubble point:

$$y_{ie} = \frac{K_b \alpha_i x_{ie}}{K_b \sum \alpha_i x_{ie}} = \frac{\alpha_i x_{ie}}{\sum \alpha_i x_{ie}} \tag{21-13}$$

Dew point:

$$x_{ie} = \frac{y_{ie}/K_b \alpha_i}{(1/K_b) \sum (y_{ie}/\alpha_i)} = \frac{y_{ie}/\alpha_i}{\sum (y_{ie}/\alpha_i)} \tag{21-14}$$

EXAMPLE 21-3   Find the bubble-point and the dew-point temperatures and the corresponding vapor and liquid compositions for a mixture of 33 mole percent *n*-hexane, 33 mole percent *n*-heptane, and 34 mole percent *n*-octane at 1 atm pressure.

SOLUTION   *Bubble point*   Assume, as the first approximation, a bubble point of $T = 100°C$. From Fig. 21-8, $K_{b,\text{est}} = 0.460$, $\alpha_2 = 2.27$, $\alpha_3 = 5.25$, where the subscripts b, 2, and 3 designate octane, heptane, and butane, respectively. By definition $\alpha_1 = \alpha_b = 1$. Then, using Eq. (21-12) gives

$$K_{b,\text{est}} = \frac{1}{(0.34 \times 1) + (0.33 \times 2.27) + (0.33 \times 5.25)} = 0.354$$

For the second trial, assume $K_{b,\text{est}} = 0.354$. From Fig. 21-8, $T_b = 92.5°C$, $\alpha_2 = 2.33$, $\alpha_3 = 5.55$. Recomputation of K_b gives $K_{b,\text{est}} = 0.336$. Finally, the third trial, using $K_b = 0.336$, reading $T = 90.5°C$, $\alpha_2 = 2.35$, $\alpha_3 = 5.60$, from Fig. 21-8, gives

$$K_b = \frac{1}{0.34 + (0.33 \times 2.35) + (0.33 \times 5.60)}$$

$$= \frac{1}{0.34 + 0.776 + 1.848} = \frac{1}{2.964} = 0.337$$

This is close enough so the bubble point is 90.5°C.

The analysis of the vapor in equilibrium with the bubble-point liquid is calculated from Eq. (21-13). The individual terms in the denominator of this fraction are used.

$$y_1 = \frac{0.34}{2.964} = 0.1147 \text{ mole fraction } n\text{-octane}$$

$$y_2 = \frac{0.776}{2.964} = 0.2618 \text{ mole fraction } n\text{-heptane}$$

$$y_3 = \frac{1.848}{2.964} = 0.6235 \text{ mole fraction } n\text{-hexane}$$

$$\text{Total} \quad 1.0000$$

Dew point Assume $T = 110°C$. Then $K_{b,est} = 0.635$, $\alpha_1 = 2.2$, $\alpha_3 = 4.9$, and, using Eq. (21-13),

$$K_{b,est} = 0.34 + \frac{0.33}{2.20} + \frac{0.33}{4.90} = 0.557$$

After two iterations, $T = 106°C$, $K_b = 0.553$, $\alpha_2 = 2.22$, and $\alpha_3 = 5.04$:

$$K_b = 0.34 + \frac{0.33}{2.22} + \frac{0.33}{5.04} = 0.34 + 0.1486 + 0.0655 = 0.5541$$

The dew point is 106°C.

The analysis of the liquid in equilibrium with the dew-point vapor is, from Eq. (21-14),

$$x_1 = \frac{0.34}{0.5541} = 0.6136 \text{ mole fraction } n\text{-octane}$$

$$x_2 = \frac{0.1486}{0.5541} = 0.2682 \text{ mole fraction } n\text{-heptane}$$

$$x_3 = \frac{0.0655}{0.5541} = 0.1182 \text{ mole fraction } n\text{-hexane}$$

$$\text{Total} \quad 1.0000$$

////

Flash Distillation of Multicomponent Mixtures

Equation (19-2) can be written for each component in a flash distillation in the form

$$y_{Di} = \frac{x_{Fi}}{f} - \frac{1-f}{f} x_{Bi} \tag{21-15}$$

Since the distillate and bottom streams are in equilibrium, this equation may be changed to

$$K_i = \frac{y_{Di}}{x_{Bi}} = \alpha_i K_b = \frac{1}{f}\left(\frac{x_{Fi}}{x_{Bi}} + f - 1\right) \tag{21-16}$$

Solving Eq. (21-16) for x_{Bi} and summing over $\mathscr{C}$ components gives

$$\sum_{i=1}^{\mathscr{C}} x_{Bi} = 1 = \sum_{i=1}^{\mathscr{C}} \frac{x_{Fi}}{f(\alpha_i K_b - 1) + 1} \tag{21-17}$$

This equation is solved by iteration in the same manner as the dew-point calculation using Eq. (21-14), and the final values of T, K_b, and α_i are used to calculate the compositions of the product streams.

EXAMPLE 21-4   The mixture of Example 21-3 is subjected to a flash distillation at 1 atm pressure, and 60 percent of the feed is vaporized. Find the temperature of the flash and the compositions of the liquid and vapor products.

SOLUTION   The flash temperature must lie between the bubble point (90.5°C) and the dew point (106°C). Assume $T = 95°C$. Then, from Fig. 21-8, $K_b = 0.40$, $\alpha_2 = 2.3$, and $\alpha_3 = 5.45$. The value of f is 0.6. The left-hand side of Eq. (21-17) becomes

$$\frac{0.34}{0.60(0.40 - 1) + 1} + \frac{0.33}{0.60[(2.30 \times 0.40) - 1] + 1} + \frac{0.33}{0.60[(5.45 \times 0.40) - 1] + 1}$$

$$= 0.531 + 0.347 + 0.193 = 1.071$$

Additional trials lead to the solution: $T = 100°C$, $K_b = 0.463$, $\alpha_2 = 2.27$, and $\alpha_3 = 5.25$. Then

$$\frac{0.34}{0.60(0.463 - 1) + 1} + \frac{0.33}{0.60[(2.27 \times 0.463) - 1] + 1}$$

$$+ \frac{0.33}{0.60[(5.25 \times 0.463) - 1] + 1} = 0.502 + 0.321 + 0.177 = 1.000$$

The flash temperature is 100°C. The composition of the liquid product is n-hexane 50.2 mole percent, n-heptane 32.1 mole percent, n-octane 17.7 mole percent.

The composition of the vapor product is computed from the values of K_b, α_2, and α_3 using the equation $y_{Di} = \alpha_i K_b x_{Di}$:

$$y_{D1} = 0.502 \times 0.463 \qquad\qquad = 0.232$$

$$y_{D2} = 0.321 \times 0.463 \times 2.27 = 0.338$$

$$y_{D3} = 0.177 \times 0.463 \times 5.25 = 0.430$$

$$\text{Total} \quad\ \ 1.000$$

The vapor product is n-hexane 23.2 mole percent, n-heptane 33.8 mole percent, n-octane 43.0 mole percent.   ////

Fractionation of Multicomponent Mixtures

As in fractionation of binary mixtures, ideal plates are assumed in the design of cascades, and the number of stages is subsequently corrected for plate efficiencies. Also, the two limiting conditions of total reflux and minimum reflux are encountered.

Ordinarily the activity and fugacity coefficients are assumed to be independent of concentrations.

Calculations for distillation plants are made by either of two methods. In the first, a desired separation of components is assumed and the number of plates above and below the feed are calculated from a chosen reflux ratio. In the second, the number of plates above and below the feed is assumed, and the separation of components is calculated using assumed flows of reflux from the condenser and vapor from the reboiler. In binary distillation the first method is the more common; in multicomponent cases the second approach is preferred, especially in computer calculations.

In final computer calculations, neither constant molal overflow nor temperature-independent K factors are assumed, and plate efficiencies also are introduced, but in preliminary estimates the simplifying assumptions are common. When the activity coefficients are assumed to be temperature-independent, group methods are used in which the number of ideal stages in a cascade is the dependent variable. The calculation gives this number without solving for either the plate temperatures or the compositions of the interstreams between plates. If α values are temperature-dependent, these simple methods are not used and plate-by-plate calculations are necessary. The temperature and liquid compositions for plate $n + 1$ are calculated by trial from those already known for plate n, and the calculation proceeds from plate to plate up or down the column.

In the rest of this chapter these methods are sampled in the following way: the estimates of the minimum number of plates at infinite reflux and of minimum reflux with an infinite number of plates are made by group methods on the assumption of constant relative volatilities and are based on a design point of view. An empirical relation for the number of plates at an operating reflux is described.

Key components The objective of distillation is the separation of the feed into streams of pure components. Only at the hypothetical condition of an infinite number of plates can this be done, and actual plants at best can give only nearly pure products. In binary distillation in a conventional plant two nearly pure products can be produced if sufficient plates and adequate reflux are used. The actual products are quantitatively defined by specifying arbitrarily small values of x_A in the bottoms and x_B in the distillate, where component A is more volatile than component B. The assignment of small numbers to these concentrations ensures that the distillate is nearly pure A and the bottoms nearly pure B. Also, when these two values are given, the mass balances can be used to calculate directly the flow rates of the two product streams.

For a conventional plant, it has been shown above that $\mathcal{N}_a = 4$. After the two terminal concentrations have been fixed, two more decisions must be made. The usual choice is the reflux ratio and the location of the feed plate.

In multicomponent distillation $\mathcal{N}_a$ again is only 4, no matter how many components there are. In the design approach, therefore, after the reflux ratio and the feed-plate location have been fixed, there remain only two free choices of concentrations, and these can be assigned to just two components. The choice of these components is a decision made by the designer. When they are so identified, they are called *keys*. Since the keys must differ in volatility, the more volatile, identified by subscript L, is called the *light key*, and the less volatile, identified by subscript H, is called the *heavy key*. Clearly, $\alpha_L > \alpha_H$. Having chosen the keys, the designer arbitrarily assigns small numbers to x_H in the distillate and to x_L in the bottoms just as small numbers are assigned to x_A and x_B in binary distillation. Here, again, the procedure forces most of the light key into the distillate and most of the heavy key into the bottoms. Because of the other components, however, it is impossible to obtain nearly pure products in both distillate and bottoms. The distillate may be nearly pure light key if the keys are the two most volatile components, and the bottoms may be nearly pure heavy key if the keys are the two least volatile components. Usually, if any other component appears in the products under one of these conditions, it is in a very small concentration. Exceptions may be encountered in these cases, especially if adjacent components have very nearly the same α factors. Each case must be considered individually in multicomponent plants.

Unlike the binary case, the choice of two keys does not give determinate mass balances, as not all other mole fractions are calculable by mass balances alone and equilibrium calculations are required to calculate the concentrations of the dew-point vapor from the top plate and the bubble-point liquid leaving the reboiler.

Although any two components can be nominated as keys, usually they are adjacent in the rank order of volatility. Such a choice is called a *sharp separation*. In sharp separations the keys are the only components that appear in both products in appreciable concentrations.

Minimum number of plates The Fenske equation (19-49) applies to any two components, i and j, in a conventional plant at infinite reflux ratio. In this case, the equation has the form

$$N_{\min} = \frac{\log \left(\dfrac{x_{Di}/x_{Bi}}{x_{Dj}/x_{Bj}} \right)}{\log (\alpha_i/\alpha_j)} - 1 \qquad (21\text{-}18)$$

where subscripts D and B refer to the distillate and bottoms, respectively. When the keys have been chosen, components i and j are the light and heavy keys, respectively. Then $x_{Bi} = x_{BL}$ and $x_{Dj} = x_{DH}$ are known. Also, the concentrations of components heavier than the heavy key are nearly zero in the distillate, and those of components lighter than the light key are nearly zero in the bottoms. By assuming these to be zero, overall mass balances can be used to calculate the values of x_{DL} and x_{BH}. Equation (21-18) then gives an excellent first approximation for $N_{\min}$, which is used in Eq. (21-18) for $\mathscr{C} - 2$ other pairs of components to calculate the mole fractions of the remaining $\mathscr{C} - 2$ components. Then, if necessary, more precise values for the keys are used again in Eq. (21-18) to give a final accurate value of $N_{\min}$.

In the derivation of Eq. (21-18), constant α values are assumed. When these values vary with temperature, a constant value corresponding to the feed temperature is a good approximation. If greater precision is wanted, a mean value may be used, defined by

$$\bar{\alpha}_i = \sqrt[3]{\alpha_{Di}\alpha_{Fi}\alpha_{Bi}} \qquad (21\text{-}19)$$

where α_{Di}, α_{Fi}, and α_{Bi} are the volatilities of component i at the bubble points of the distillate, feed, and bottoms, respectively.

EXAMPLE 21-5   The equilibrium vapor-liquid product of the flash process of Example 21-4 is to be used as the feed to a conventional plant. It is desired that the terminal concentrations be

$$x_{D2} = 0.01 \text{ mole fraction } n\text{-heptane in distillate}$$

$$x_{B3} = 0.01 \text{ mole fraction } n\text{-hexane in the bottoms}$$

Calculate the minimum number of ideal plates at infinite reflux ratio and the remaining mole fractions of the components in the products.

SOLUTION   The n-hexane is the light key, the n-pentane is the heavy key, and the n-octane is a heavy nonkey, which goes almost entirely to the bottoms. Using the light key as component i and the heavy key as component j in Eq. (21-18) leads to

$$N_{\min} = \frac{\log \left(\dfrac{x_{D3}/x_{B3}}{x_{D2}/x_{B2}} \right)}{\log (\alpha_3/\alpha_2)} - 1 \tag{21-20}$$

When $x_{B3} = 0.010$, $x_{D2} = 0.010$. From the flash calculation in Example 21-4, $\alpha_2 = 2.27$ and $\alpha_3 = 5.25$, both based on $\alpha_1 = 1$. Concentrations x_{B2} and x_{D3} must be calculated from mass balances. The overall concentrations of the components in the feed are, from Example 21-4,

$$x_{F1} = 0.34 \qquad x_{F2} = 0.33 \qquad x_{F3} = 0.33$$

The total mass balance is

$$F = D + B \tag{21-21}$$

The balance for component 3 is

$$Fx_{F3} = Dx_{D3} + Bx_{B3} \tag{21-22}$$

On the assumption that $x_{D1} \approx 0.0$, $x_{D3} \approx 1.000 - 0.010 = 0.990$. Eliminating B from Eq. (21-22) by Eq. (21-21), on the basis that $F = 100$ mol/h

$$100 \times 0.33 = 0.990D + (100 - D)(0.010)$$

From this $D = 32.65$ mol/h and $B = 100 - 32.65 = 67.35$ mol/h. Equation (21-22) gives

$$100 \times 0.33 = 32.65 \times 0.010 + 67.35x_{B2}$$

which gives $x_{B2} = 0.4851$. The value of x_{B1} can be calculated from the mass balance of the n-octane or by the difference

$$x_{B1} = 1.000 - 0.4851 - 0.010 = 0.5049$$

The mass balance gives, as a check,

$$x_{B1} = 100 \times \frac{0.34}{67.35} = 0.5048$$

which is close enough. Now, N_{min} is calculated by substituting the data into Eq. (21-20) to give

$$N_{min} = \frac{\log\left(\dfrac{0.990/0.010}{0.010/0.4851}\right)}{\log(5.25/2.27)} - 1 = 9.1 \text{ plates}$$

The reboiler acts as another plate.

For components 1 and 2

$$N_{min} + 1 = 10.1 = \frac{\log\left(\dfrac{0.010/0.4851}{x_{D1}/0.5049}\right)}{\log 2.27}$$

From this $x_{D1} = 2.64 \times 10^{-6}$ mole fraction n-octane in the distillate. This clearly is negligible, and the calculation of N_{min} need not be repeated.

A more precise calculation of N_{min} can be obtained by using Eq. (21-19) to calculate revised values of α_2 and α_3. By bubble-point calculations on the distillate and bottoms and the use of data from Example 21-4 the vapor pressures, K factors, and relative volatilities shown in Table 21-3 are obtained, from which

$$\bar{\alpha}_3 = \sqrt[3]{6.70 \times 5.25 \times 4.92} = 5.57$$

$$\bar{\alpha}_2 = \sqrt[3]{2.56 \times 2.27 \times 2.17} = 2.33$$

Using log (5.57/2.33) in the denominator of Eq. (21-20) gives $N_{min} + 1 = 9.7$, and $N_{min} = 8.7$. ////

Minimum Reflux Ratio

In Chap. 19 it was shown that in a binary distillation with normal equilibrium curve in a conventional plant at minimum reflux ratio an invariant zone forms on each side of the feed plate. In multicomponent distillation the condition of minimum reflux ratio also appears, but the presence of nonkey components makes the problem of locating the invariant zones more complicated.

Distributed and undistributed components When there are components lighter than the light key, these, at minimum reflux ratio, may be completely separated and appear entirely in the condenser, with none at all in the bottoms. When there are components heavier than the heavy key, these may appear entirely in the bottoms. Such components are called *separated* or *undistributed components*. Components that

Table 21-3

Stream	Temperature, °C	K_b	α_3	α_2
Distillate	68.9	0.150	6.70	2.56
Feed	100	0.462	5.25	2.27
Bottoms	109.5	0.614	4.92	2.17

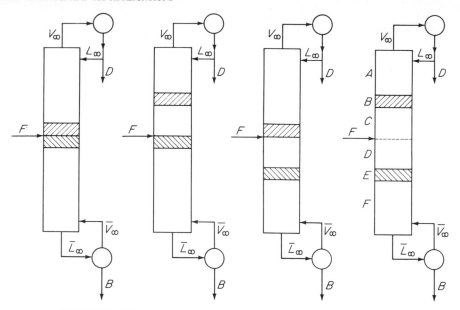

FIGURE 21-9
Invariant zones in multicomponent plant: (*a*) all components distributed; (*b*) no component heavier than heavy key; (*c*) no component lighter than light key; (*d*) undistributed components in both products. Shaded areas represent invariant zones.

appear in both product streams are called *distributed components*. Both keys are, of course, distributed, by definition. If there are components with volatilities between those of the keys, such components distribute; a component just a little more volatile than the light or one just a little heavier than the heavy key also may distribute.

Locations of invariant zones   Figure 21-9 shows the locations of invariant zones for various combinations of components. If all components in the feed are distributed (a rare situation in practice), both invariant zones are adjacent to the feed, as in a binary system (Fig. 21-9*a*). If there are one or more undistributed light components but no components heavier than the heavy key, the upper invariant zone is in the mid-section of the rectifying column (Fig. 21-9*b*). If the bottoms contain undistributed components, the lower invariant zone is in the middle of the stripping column (Fig. 21-9*c*). If there are undistributed components in both product streams, the two invariant zones are in the middle of the two sections of the column (Fig. 21-9*d*).

Zones in columns at minimum reflux   The plant shown in Fig. 21-9*d* has six zones A, B, C, D, E, and F, as identified in the figure. The feed has one heavy nonkey and one light nonkey. Each zone has its own function, which can be followed by reference to Fig. 21-10 giving the distribution profiles of the vapor-phase concentrations in such a plant. The top zone A, the part of the column below the condenser,

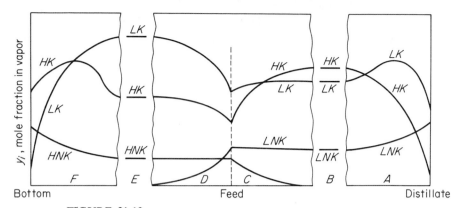

FIGURE 21-10
Concentration profiles of vapor through column at minimum reflux. (Selected and simplified from Ref. 5.)

reduces the concentration of the heavy key to that specified in the design. The concentration of the light key goes through a maximum and then drops to the specified recovery. The light nonkey is concentrated somewhat. Zone B is the upper invariant zone. Zone C, above the feed, eliminates the heavy nonkey from the vapor passing up the rectifying column. Zone D, below the feed, eliminates the light nonkey from the liquid flowing down the stripping column. Zone E is the lower invariant zone, and zone F does the same job with light key as zone A does with the heavy key.

Retrograde fractionation   The task of the entire plant is to increase the ratio of the mole fraction of the light to that of the heavy key from a low value at the reboiler to a large value at the condenser. In zones A and F this occurs, but in zones C and D this ratio decreases as the vapor passes through these zones on either side of the feed, and the work of zones A and F is partially undone. The effect is called *retrograde fractionation*. It can be seen by inspection of the profiles of the key components in zones D and C in Fig. 21-10. The effect persists in actual operating columns at reflux ratios near the minimum but fades out at the usual operating range of actual columns.

Calculation of minimum reflux ratio   Of the several methods of calculating the minimum reflux ratio, the most used is that of Underwood,[9] as explained and amplified by others.[2,5,8] With the simplifying assumptions of constancy of molal overflow and temperature independence of relative volatilities, the method is exact, and further corrections can be made for variable overflow and temperature variations of α factors when the simplifying assumptions are not applicable. Only the method based on the two standard simplifications is discussed here.

Since there may be undistributed components lighter than the light key and heavier than the heavy key, it is convenient to assume that three groups of components may be encountered in the general case. Group I consists of undistributed light components; group II contains the distributed components; and group III consists of the undistributed heavy components. Clearly, group II contains the key components, and the minimum number of components in the group is 2. The most common situation is that where the separation of keys is sharp and only the keys are distributed, so that all other components are in either group I or group II. A binary system is such a case without group I or group III members.

As a simple example of the Underwood method, assume a four-component system, in which components 2 and 3 are the heavy and light components, respectively, component 1 is an undistributed heavy component, and component 4 is an in-distributed light component. Thus, group I contains only component 4, group II consists of components 2 and 3, and group III has only component 1. This assumes that the split of components is predictable in advance. Unless component 1 is mar-ginally heavier than component 2 or component 4 is just a bit more volatile than com-ponent 3, this assumption is justified. In close cases, the assumption can be checked by a more elaborate calculation with either or both of components 1 and 4 placed in group II. The method is outlined later.

Mass balance and equilibrium equations   For constant molal overflow and con-stant relative volatilities, the following mass balances for component i may be written, by reference to Fig. 21-1:

Rectifying column and condenser, above plate n:

$$Vy_{n+1,i} = Lx_{ni} + Dx_{Di} \tag{21-23}$$

Stripping column and reboiler, below plate m:

$$\overline{V}y_{m+1,i} = \overline{L}x_{mi} - B_{xBi} \tag{21-24}$$

By definition of the K factor, in the rectifying column

$$x_{ni} = \frac{y_{ni}}{K_{ni}} \tag{21-25}$$

and in the stripping column

$$x_{mi} = \frac{y_{mi}}{K_{mi}} \tag{21-26}$$

Eliminating x_{ni} from Eq. (21-23) by Eq. (21-25) and x_{mi} from Eq. (21-24) by Eq. (21-26) gives

$$Vy_{n+1,i} = \frac{Ly_{ni}}{K_{ni}} + Dx_{Di} \tag{21-27}$$

$$\overline{V}y_{m+1,i} = \frac{\overline{L}y_{mi}}{K_{mi}} - Bx_{Bi} \tag{21-28}$$

At the condition of minimum reflux ratio, as the number of plates in each column increases without limit at constant V, L, D, B, x_{Di}, and x_{Bi},

$$y_{n+1,i} \to y_{ni} = y_{\infty i} \quad \text{and} \quad y_{m+1,i} \to y_{ni} = y_{\infty i}$$

Then, Eqs. (21-27) and (21-28) can be written

$$V_\infty y_{\infty i} = \frac{L_\infty y_{\infty i}}{K_{\infty i}} + D x_{Di} \tag{21-29}$$

$$\overline{V}_\infty \overline{y}_{\infty i} = \frac{\overline{L}_\infty \overline{y}_{\infty i}}{\overline{K}_{\infty i}} - B x_{Bi} \tag{21-30}$$

Equation (21-29) is further transformed as follows. First, solve for $y_{\infty i}$ to give

$$y_{\infty i} = \frac{D x_{Di}}{V - L/K_{\infty i}} \tag{21-31}$$

Multiply the numerator and denominator of Eq. (21-31) by α_i and then multiply the equation by V_∞; thus

$$V_\infty y_{\infty i} = \frac{\alpha_i D x_{Di}}{\alpha_i - [\alpha_i (L_\infty/V_\infty)/K_{\infty i}]} \tag{21-32}$$

By Eq. (18-14) the quantity $(L_\infty/V_\infty)/K_{\infty i}$ is the absorption factor for component i in the invariant zone of the rectifying column. Denote this by $A_{\infty i}$. Equation (21-32) then can be written

$$V_\infty y_{\infty i} = \frac{\alpha_i D x_{Di}}{\alpha_i - A_{\infty i}} \tag{21-33}$$

By the definitions of K_i and α_i, if the base component for defining the relative volatility (component 1 here) is denoted by subscript b, $\alpha_i = K_i/K_b$, so

$$A_{\infty i} \alpha_i = \frac{L_\infty/V_\infty}{K_{\infty i}} \alpha_i = \frac{L_\infty/V_\infty}{K_{\infty b}} = A_\infty$$

This quantity is the same for all components in the invariant zone, so subscript i can be dropped. Finally, Eq. (21-33) becomes

$$V_{\infty i} y_{\infty i} = \frac{\alpha_i D x_{Di}}{\alpha_i - A_\infty} \tag{21-34}$$

The same manipulation can be carried out on the equation for the stripping column, Eq. (21-30). The result is

$$-\overline{V}_\infty \overline{y}_{\infty i} = \frac{\alpha_i B x_{Bi}}{\alpha_i - \overline{A}_\infty} \tag{21-35}$$

where $\overline{A}_\infty = (\overline{L}_\infty/\overline{V}_\infty)/\overline{K}_{\infty b}$.

Equation (21-34) can be summed over all components that appear in the condenser (components 2, 3, and 4):

$$V_\infty \sum_{i=2}^{4} y_{\infty i} = D \sum_{i=2}^{4} \frac{\alpha_i x_{Di}}{\alpha_i - A_{\infty i}}$$

Since $\sum_{i=2}^{4} y_{\infty i} = 1$,

$$\frac{V_\infty}{D} = \sum_{i=2}^{4} \frac{\alpha_i x_{Di}}{\alpha_i - A_{\infty i}} \tag{21-36}$$

Likewise, Eq. (21-35) is summed over the components that appear in the reboiler (components 1, 2, and 3). Using $\sum \bar{y}_{\infty i} = 1$ gives

$$-\frac{\bar{V}_\infty}{B} = \sum_{i=1}^{3} \frac{\alpha_i x_{Bi}}{\alpha_i - \bar{A}_\infty} \tag{21-37}$$

Mathematical properties of minimum reflux equations   Equations (21-36) and (21-37) are adequate when all components distribute.[8] Except for binary systems, this is rarely true, and a more sophisticated mathematical analysis, by Underwood, is needed when undistributed components are present. Equations (21-36) and (21-37) have the forms

$$\frac{V_\infty}{D} = \sum_{i=2}^{4} \frac{\alpha_i x_{Di}}{\alpha_i - \phi} \tag{21-38}$$

$$-\frac{\bar{V}_\infty}{B} = \sum_{i=1}^{3} \frac{\alpha_i x_{Bi}}{\alpha_i - \Psi} \tag{21-39}$$

Each of these equations contains a variable, ϕ or Ψ, and is a cubic equation in that variable. Each therefore has three roots, one for each component, which may be denoted by ϕ_2, ϕ_3, ϕ_4, and Ψ_1, Ψ_2, and Ψ_3. The properties of these roots are the key to the calculation of minimum reflux for any combination of components in groups I, II, and III.

Characteristics of ϕ and Ψ   Define the function $f(\phi)$ generally by

$$f(\phi) = \frac{\alpha_1 x_{D1}}{\alpha_1 - \phi} + \frac{\alpha_2 x_{D2}}{\alpha_2 - \phi} + \frac{\alpha_3 x_{D3}}{\alpha_3 - \phi} + \frac{\alpha_4 x_{D4}}{\alpha_4 - \phi} \tag{21-40}$$

and function $f(\Psi)$ by

$$f(\Psi) = \frac{\alpha_1 x_{B1}}{\Psi - \alpha_1} + \frac{\alpha_2 x_{B2}}{\Psi - \alpha_2} + \frac{\alpha_3 x_{B3}}{\Psi - \alpha_3} + \frac{\alpha_4 x_{B4}}{\Psi - \alpha_4} \tag{21-41}$$

These equations generalize Eqs. (21-38) and (21-39) to include all components in case all of them distribute. If a heavy component is not found in the distillate, the term in α_1 in Eq. (21-40) is dropped, and if a light component is not found in the bottoms, the term in α_4 in Eq. (21-41) is chopped.

If a value of V_∞/D is plotted as an abscissa in Fig. 21-11a, the intersection on the branches of $f(\phi)$ give the values of the four roots of Eq. (21-40); if a value of $\bar{V}_\infty/B$ is so plotted in part (b), the intersections give the values of the roots of Eq. (21-41).

The ordinate at each value of α in Fig. 21-11 is an asymptote, where $f(\phi) \to \infty$ as $\phi_i \to \alpha_i$ and $f(\phi_{i+1}) \to -\infty$ as $\phi_{i+1} \to \alpha_i$; also, where $f(\Psi_i) \to \infty$ as $\Psi_i \to \alpha_i$ and $f(\Psi_{i-1}) \to -\infty$ as $\Psi_{i-1} \to \alpha_i$.

Figure 21-11c shows the rank order of the roots ϕ and Ψ with the relative volatilities. They are

$$0 < \phi_1 < \alpha_1 < \phi_2 < \alpha_2 < \phi_3 < \alpha_3 < \phi_4 < \alpha_4$$

$$0 < \alpha_1 < \Psi_1 < \alpha_2 < \Psi_2 < \alpha_3 < \Psi_3 < \alpha_4 < \Psi_4$$

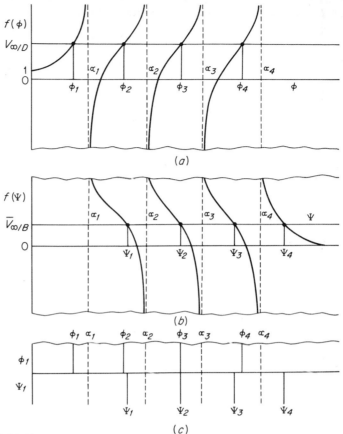

FIGURE 21-11
Roots of Eqs. (21-40) and (21-41): (a) $f(\phi)$ vs. ϕ; (b) $f(\Psi)$ vs. Ψ; (c) comparison of roots and α_1.

Development of Final Equation

For the four-component example with two distributed components, Eqs. (21-38) and (21-39) become

$$V_\infty = D\left(\frac{\alpha_2 x_{D2}}{\alpha_2 - \phi} + \frac{\alpha_3 x_{D3}}{\alpha_3 - \phi} + \frac{\alpha_4 x_{D4}}{\alpha_4 - \phi}\right) \qquad (21\text{-}42)$$

$$-\overline{V}_\infty = B\left(\frac{\alpha_1 x_{B1}}{\alpha_1 - \Psi} + \frac{\alpha_2 x_{B2}}{\alpha_2 - \Psi} + \frac{\alpha_3 x_{B3}}{\alpha_3 - \Psi}\right) \qquad (21\text{-}43)$$

To use these equations, it must be decided which of the six roots that satisfy them should be used. Whatever roots are selected, if they satisfy the individual equations, they also satisfy their sum, which can be written

$$V_\infty - \overline{V}_\infty = \frac{\alpha_1 Bx_{B1}}{\alpha_1 - \Psi} + \frac{\alpha_2 Dx_{D2}}{\alpha_2 - \phi} + \frac{\alpha_2 Bx_{B2}}{\alpha_2 - \Psi} + \frac{\alpha_3 Dx_{D3}}{\alpha_3 - \phi} + \frac{\alpha_3 Bx_{B3}}{\alpha_3 - \Psi} + \frac{\alpha_4 Dx_{D4}}{\alpha_4 - \phi}$$

$$(21\text{-}44)$$

Any choice of roots for use in this equation must be consistent with the four mass-balance equations.

$$Fx_{F1} = Bx_{B1} \qquad (21\text{-}45)$$

$$Fx_{F2} = Dx_{B2} + Bx_{B2} \qquad (21\text{-}46)$$

$$Fx_{F3} = Dx_{D3} + Bx_{B3} \qquad (21\text{-}47)$$

$$Fx_{F4} = Dx_{D4} \qquad (21\text{-}48)$$

It is clear that neither Eq. (21-46) nor Eq. (21-47) is consistent with Eq. (21-44) unless the two terms for component 2 and the two terms for component 3 have the same denominator, and the values of ϕ and Ψ in these terms must be equal. Since there is only one root for ϕ and one root for Ψ which can be used, this means that all roots in Eq. (21-44) must be equal. Figure 21-11c supplies the answer to this problem. When it is noted that, in this example, ϕ_1 and Ψ_4 do not exist, there is only one answer: ϕ_2 and Ψ_3 must be equal. This means that there is one and only one parameter that will validate Eq. (21-44) in the light of Eqs. (21-45) to (21-48). Denote this by Φ, defined by $\phi_2 = \Psi_3 \equiv \Phi$. Using Φ in Eq. (21-44) gives

$$V_\infty - \overline{V}_\infty = \frac{\alpha_1 Bx_{B1}}{\alpha_1 - \Phi} + \frac{\alpha_2}{\alpha_2 - \Phi}(Dx_{D2} + Bx_{B2})$$

$$+ \frac{\alpha_3}{\alpha_3 - \Phi}(Dx_{B3} + Bx_{B3}) + \frac{\alpha_4 Bx_{B4}}{\alpha_4 - \Phi}$$

When the material-balance equations are used, this becomes

$$V_\infty - \overline{V}_\infty = F\left(\frac{\alpha_1 X_{F1}}{\alpha_1 - \Phi} + \frac{\alpha_2 X_{F2}}{\alpha_2 - \Phi} + \frac{\alpha_3 X_{F3}}{\alpha_3 - \Phi} + \frac{\alpha_4 X_{F4}}{\alpha_4 - \Phi}\right)$$

Since $V - \overline{V} = F(1 - q)$, the equation can be written in its final concise form

$$1 - q = \sum_{i=1}^{4} \frac{\alpha_i X_{Fi}}{\alpha_i - \Phi} \qquad (21\text{-}49)$$

The minimum reflux ratio R_∞ is, from Eq. (21-42) using Φ for Ψ,

$$\frac{V_\infty}{D} = \frac{L_\infty + D}{D} = R_\infty + 1 = \sum_{i=2}^{4} \frac{\alpha_i X_{Di}}{\alpha_i - \Phi} \qquad (21\text{-}50)$$

Equations (21-49) and (21-50) are the Underwood equations written for the special case of the four-component mixture with one group I component, one group II component, and a group containing the keys. For the general case Eq. (21-49) is summed over all components, and Eq. (21-50) is summed over the components in groups I and II. This case is covered by using $\mathscr{C}$ in place of 4 in the indices. Usually, Eq. (21-50) is summed over all components, which applies only when all components distribute. When it is written this way, it should be stipulated that the lower index in Eq. (21-50) is the number of the first component that distributes.

EXAMPLE 21-6 Compute the minimum reflux ratio for the plant of Example 21-5.

SOLUTION The values for α should be those at the temperature of the flashed feed. It is clear that the n-octane, component 1, is not distributed and goes entirely into the bottoms. The values for use in Eqs. (21-49) and (21-50) are given in Table 21-4. The light key is n-pentane, the heavy key is n-hexane, and the undistributed heavy is n-octane. The fraction q of liquid in the feed is $1 - 0.6 = 0.4$. Substitution in Eq. (21-49) gives

$$1 - 0.4 = \frac{0.34}{1 - \Phi} + \frac{0.33 \times 2.27}{2.27 - \Phi} + \frac{0.33 \times 5.25}{5.25 - \Phi}$$

Solution of this by iteration gives $\Phi = 3.814$. Substitution in Eq. (21-50) gives

$$R_\infty + 1 = \frac{2.27 \times 0.01}{2.27 - 3.814} + \frac{5.25 \times 0.99}{5.25 - 3.814} = 3.605$$

The minimum reflux ratio is

$$R_\infty = 3.605 - 1 = 2.605 \qquad\qquad \text{////}$$

Solution for more than two distributed components If an additional distributed component appears, with volatility between that of the keys or just heavier or just lighter than one of the keys, the mole fraction of that component in the distillate becomes an unknown. The principle used in determining the value of this unknown is the existence of a second common root between Eqs. (21-38) and (21-39), of the form

$$\phi_{j+1} = \Psi_j \equiv \Phi_j$$

This second parameter can be seen by inspection of Fig. 21-11c. It lies either between the keys or just adjacent to one of them. The second parameter also satisfies Eq. (21-50) and so provides an additional linear equation for simultaneous solution for the unknown concentration. The relation $\sum x_{Di} = 1$ is also needed. The procedure is easily generalized to include any number of distributed components. The number of roots and of equations of the form of Eq. (21-50) is 1 less than the number of components in group II.

Table 21-4 DATA FOR EXAMPLE 21-6

	Component	α_i	X_{Fi}	x_{Di}
1	n-Octane	1	0.34	0
2	n-Hexane	2.27	0.33	0.010
3	n-Pentane	5.25	0.33	0.990

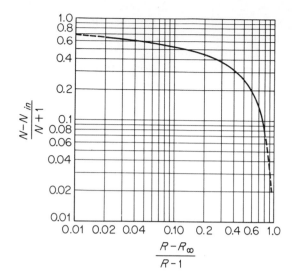

FIGURE 21-12
Gilliland correlation.

An alternate method is to assume that all components distribute and work the problem on this basis. Then if the mole fraction of a component comes out greater than unity, it is an undistributed light component, and if the concentration is negative, it is an undistributed heavy component. These components are placed in group I or group II and the calculation repeated for the remaining components until all qualify as distributed components. Examples of this method are given in Refs. 5 and 8.

Number of ideal plates at operating reflux Although the precise calculation of the number of plates in multicomponent distillation is best accomplished by computer, a simple empirical method due to Gilliland[1] is much used for preliminary estimates. The correlation requires knowledge only of the minimum number of plates at total reflux and the minimum reflux ratio. The correlation is given in Fig. 21-12, and is self-explanatory.

EXAMPLE 21-7 Estimate the number of ideal plates required for the plant of Examples 21-5 and 21-6 if an operating reflux ratio of 3.5 is used.
  SOLUTION From Example 21-5, the minimum number of plates is $N_{min} = 8.7$, and from Example 21-6, the minimum reflux ratio is $R_\infty = 2.605$. The abscissa for Fig. 21-12 is

$$\frac{R - R_\infty}{R + 1} = \frac{3.5 - 2.605}{3.5 + 1} = 0.199$$

The ordinate from Fig. 21-12 is 0.46 and

$$0.46 = \frac{N - 8.7}{N + 1}$$

From this $N = 17$ plates.   ////

SYMBOLS

A	Absorption factor, $(L/V)K$, dimensionless
A', B', C'	Antoine constants [see Eq. (17-49)]
B	Flow rate of heavy or bottoms product, mol/h
C	Symbol for total condenser; C_p, for partial condenser; $C_{2\phi}$, for total condenser with two-phase product
D	Flow rate of light or overhead product, mol/h
F	Feed rate, mol/h
f	Fraction of feed that is vaporized
$f(\phi), f(\Psi)$	Functions defined by Eqs. (21-40) and (21-41)
H	Enthalpy per mole of stream, Btu/lb mol or J/g mol; also symbol for heat exchanger
J	Symbol for heater or cooler
K	Equilibrium ratio, y_e/x_e; K_b, for arbitrary base component chosen for defining relative volatilities; $\bar{K}$, in stripping column
L	Flow rate of liquid, in general or in rectifying column, mol/h; $\bar{L}$, in stripping column
M	Symbol for mixer
N	Number of ideal plates; $N_{\min}$, minimum number at total reflux
n	Number of moles
$\dot{n}$	Molal flow rate, mol/h; $\dot{n}_i$, of component i
P	Power flowing from element to environment as shaft work, ft-lb$_f$/h or W [Eq. (21-2)]; also, symbol for theoretical plate
p	Total pressure on system, lb$_f$/ft^2 or atm; p_b, vapor pressure of base component; p_e, equilibrium pressure; p_i, vapor pressure of component i
q	Rate of heat transfer, Btu/h or W; also, moles of liquid to rectifying section per mole of feed
R	Reflux ratio, L/D; R_∞ minimum reflux ratio; also symbol for reboiler; R_p, for partial reboiler
S	Entropy of system, ft-lb$_f$/°F or J/°C; also, symbol for separator or decanter
T	Temperature, °F, °C, °R, or K; also symbol for divider
U	Internal energy of system, ft-lb$_f$ or J
V	Volume of system, ft^3 or m^3; also, flow rate of vapor, in general or in rectifying section, mol/h; $\bar{V}$, in stripping column
W	Symbol for pump
x	Mole fraction in liquid phase; x_e, of liquid in equilibrium with vapor of composition y_e
y	Mole fraction in vapor phase; $\bar{y}$, in stripping section; y_e, of vapor in equilibrium with liquid of composition x_e

Script letters

$\mathscr{C}$	Number of components, in phase rule
$\mathscr{F}$	Phase-rule variance (number of degrees of freedom)
$\mathscr{N}$	Number of variables in Kwauk method; $\mathscr{N}_a$, number available for decision; $\mathscr{N}_c$, number of constrained variables; $\mathscr{N}_i$, gross number required to establish design; $\mathscr{N}_v$, total number for element; $\mathscr{N}_x$, number needed to define all feed streams and pressure levels; $\mathscr{N}_\sigma$, number of cascades
$\mathscr{P}$	Number of phases, in phase rule; $\mathscr{P}_{\max}$, maximum number

Greek letters

α	Relative volatility, dimensionless; α_H, of heavy key; α_L, of light key; α_{ij}, of component i relative to component j; $\bar{\alpha}$, mean value, defined by Eq. (21-19)
μ	Chemical potential, ft-lb$_f$/lb mol or J/g mol
ν	Number of streams
Φ	Parameter validating Eq. (21-44), equals roots ϕ_2 and Ψ_3; Φ_J, second common root between Eqs. (21-38) and (21-39), equals ϕ_{J+1} and Ψ_J
ϕ	Variable in Eq. (21-38)
Ψ	Variable in Eq. (21-39)

Subscripts

B	In bottoms product
D	In overhead product
e	Equilibrium value
F	In feed
H	Of heavy key
i	Of component i, any one of $\mathscr{C}$ components
j	Of component j, any component except i
L	Of light key
m	On plate m of stripping section
n	On plate n of rectifying section
1ϕ	Single-phase
2ϕ	Two-phase
∞	At the condition of minimum reflux ratio, R_∞

PROBLEMS

21-1 Analyze the double-column air-distillation plant of Fig. 21-4 by the Kwauk procedure and specify a choice of variables in set $\mathscr{N}_a$.

21-2 The feed to a conventional distillation plant and the α factors of the components are shown in Table 21-5. The feed enters as a bubble-point liquid. Of component 2, not more than 2 percent is to leave in the distillate and not more than 2 percent is to leave in the bottoms. Calculate the minimum number of theoretical plates.

21-3 Calculate the minimum reflux ratio of the plant in Prob. 21-2.

21-4 A liquid stream of the composition and volatility pattern of that in Prob. 21-2 is subjected to an equilibrium flash to give 30 percent liquid and 70 percent vapor. Calculate the equilibrium compositions of liquid and vapor. If component 1 has the same vapor-pressure curve as *n*-octane, calculate the temperature of the flash if the total pressure is 2 atm.

Table 21-5

Component	X_{Fi}	α_i
1	0.05	1
2	0.50	1.5
3	0.40	2.5
4	0.05	3.0

REFERENCES

1 Gilliland, E. R.: *Ind. Eng. Chem.*, **32**:110 (1940).
2 Hanson, D. N., J. H. Duffin, and G. F. Somerville: "Computation of Multicomponent Separation Processes," Reinhold, New York, 1962.
3 Holland, D.: "Multicomponent Distillation," Prentice-Hall, Englewood Cliffs, N.J., 1962.

4 Holmes, M. J., and M. Van Winkle: *Ind. Eng. Chem.,* **62:**21 (1970).
5 King, C. J.: "Separation Processes," McGraw-Hill, New York, 1971.
6 Kline, S. J., and F. O. Koenig: *J. Appl. Mech., Trans. ASME,* **79:**29 (1957).
7 Kwauk, M.: *AIChE J.,* **2:**240 (1956).
8 Shiras, R. N., D. N. Hanson, and C. H. Gibson: *Ind. Eng. Chem.,* **42:**871 (1950).
9 Underwood, A. J. V.: *Chem. Eng. Prog.,* **44:**603 (1948).
10 Van Winkle, M.: "Distillation," McGraw-Hill, New York, 1967.
11 Zemansky, M.: "Heat and Thermodynamics," 4th ed., McGraw-Hill, New York, 1967.

22

PRINCIPLES OF DIFFUSION AND
MASS TRANSFER BETWEEN PHASES

Diffusion is the movement, under the influence of a physical stimulus, of an individual component through a mixture. The most common cause of diffusion is a concentration gradient of the diffusing component. A concentration gradient tends to move the component in such a direction as to equalize concentrations and destroy the gradient. When the gradient is maintained by constantly supplying the diffusing component to the high-concentration end of the gradient and removing it at the low-concentration end, the flow of the diffusing component is continuous. This movement is exploited in mass-transfer operations. For example, a salt crystal in contact with a stream of water or a dilute solution sets up a concentration gradient in the neighborhood of the interface, and salt diffuses through the liquid layers in a direction perpendicular to the interface. The flow of salt away from the interface continues until the crystal is dissolved. When the salt is intimately mixed with insoluble solid, the process is an example of leaching.

Although the usual cause of diffusion is a concentration gradient, diffusion can also be caused by a pressure gradient, by a temperature gradient, or by the application of an external force field, as in a centrifuge.[1] Molecular diffusion induced by a pressure gradient (*not* partial pressure) is called *pressure diffusion*, that induced by temperature is *thermal diffusion*, and that from an external field, *forced diffusion*. All three are rare in chemical engineering. Only diffusion under a concentration gradient is considered here.

Diffusion is not restricted to molecular transfer through stagnant layers of solid or liquid. It also takes place in fluid phases by physical mixing and by the eddies of turbulent flow, just as heat flow in a fluid may occur by convection. This is called *eddy diffusion*. Usually the diffusion process is accompanied by bulk flow of the mixture in a direction parallel to the direction of diffusion, and it is often associated with heat flow.

Role of diffusion in mass transfer In all the mass-transfer operations, diffusion occurs in at least one phase and often in both phases. In gas absorption, solute diffuses through the gas phase to the interface between the phases and through the liquid phase from the interface. In rectification, low boiler diffuses through the liquid phase to the interface and away from the interface into the vapor. The high boiler diffuses in the reverse direction and passes through the vapor into the liquid. In leaching, diffusion of solute through the solid phase is followed by diffusion into the liquid. In liquid extraction, the solute diffuses through the raffinate phase to the interface and then into the extract phase. In crystallization, solute diffuses through the mother liquor to the crystals and deposits on the solid surfaces. In humidification there is no diffusion through the liquid phase because the liquid phase is pure and no concentration gradient through it can exist; but the vapor diffuses to or from the liquid-gas interface into or out of the gas phase. In drying, liquid water diffuses through the solid toward the surface of the solid, vaporizes, and diffuses as vapor into the gas. The zone of vaporization may be either at the surface of the solid or within the solid, depending upon factors that are discussed in Chap. 25. When the vaporization zone is in the solid, vapor diffusion takes place in the solid between the vaporization zone and the surface and diffusion of both vapor and liquid occurs within the solid.

THEORY OF DIFFUSION

In this section quantitative relationships for diffusion are discussed. Attention is focused on diffusion in a direction *perpendicular* to the interface between the phases and at a definite location in the equipment. Assuming that a column is being used, the location chosen for analysis is in a definite cross section parallel with the ground. Steady state is assumed, and the concentrations at any point do not change with time. The theory is restricted to binary mixtures.

Comparison of diffusion and heat transfer An analogy exists between flow of heat and diffusion. In each, a gradient is the cause of the flow. In heat transfer a temperature gradient is the driving force; in diffusion the force is a concentration gradient. In each case the flux is directly proportional to the gradient. The analogy cannot be carried further, however, for the reason that heat is not a substance: it is energy in transit. When heat flows from one point to another, it leaves no space behind, nor does it require space in its new location. The velocity of heat flow has no meaning. Diffusion is a physical flow of matter, which occurs at a definite velocity. A diffusion component leaves space behind it, and room must be found for it in its new location.

The material nature of diffusion introduces three complications not found in heat transfer:

1 All components in the phase must be considered. In a two-component mixture, both components *A* and *B* must always be considered.

2 Relative to an interface, only one component may be in motion across the interface, as in gas absorption, where component *A* is transferred from phase to phase; or, as in rectifying cascades, the molal flow rate of component *A* may occur in one direction and an equal molal flow rate of component *B* takes place in the reverse direction, so the net flow is zero. The constancy of molal overflow is thus conserved.

3 The diffusivity, which occupies the same position in molecular diffusion as thermal conductivity does in heat conduction, can only be interpreted in terms of a relative velocity of the diffusing component with respect to some kind of average velocity of the entire stream in a direction normal to the interface. The choice of this velocity is optional, but it must be precisely defined.

Diffusion quantities   Five interrelated concepts are used in diffusion theory:

1 Velocity u, defined as usual by length/time

2 Transfer rate across a plane N, moles/time

3 Flux I, moles/time-area

4 Concentration c and molar density ρ_M, moles/volume (mole fraction may also be used)

5 Concentration gradient dc/db, where b is the length of the path perpendicular to the area across which diffusion is occurring

Appropriate subscripts are used when needed. The equations apply equally well to SI, cgs, and fps units. In some applications, mass, rather than molal units, may be used in flow rates and concentrations.

Velocities in diffusion   Several velocities are needed to describe the movements of individual substances and of the total phase. Since absolute motion has no meaning, any velocity must be based on an arbitrary state of rest. In this discussion "velocity" without qualification refers to the velocity relative to the interface between the phases and is that apparent to an observer at rest with respect to the interface.

The individual molecules of any one component in the mixture are in random motion. If the instantaneous velocities of the component are summed, resolved in the direction perpendicular to the interface, and divided by the number of molecules of the substance, the result is the macroscopic velocity of that component. For component A, for instance, this velocity is denoted by u_A.

Molal flow rate, velocity, and flux If the molal flow rate, in moles per unit time, of the total mixture, in a direction perpendicular to a stationary plane is denoted by N and the volumetric-average velocity by u_0,

$$N = A\rho_M u_0 \qquad (22\text{-}1)$$

where ρ_M is the molar density of the mixture. This relation is equivalent to Eq. (4-2) if $\dot{m}$ is replaced by N, S by A, and ρ by ρ_M.

Likewise for components A and B, crossing a stationary plane, the molal flow rates are

$$N_A = A c_A u_A \qquad (22\text{-}2)$$

$$N_B = A c_B u_B \qquad (22\text{-}3)$$

Diffusivities are defined, not with respect to a stationary plane, but relative to the volume-average velocity u_0. For component A this velocity is designated by $u_A - u_0$. The molal flow rate through unit area of a plane moving with velocity u_0 is called the molal flux of component A and is designated by I_{0A}. From Eqs. (22-2) and (22-3), the fluxes of the components are given by the equations

$$I_{0A} = c_A(u_A - u_0) \qquad (22\text{-}4)$$

$$I_{0B} = c_B(u_B - u_0) \qquad (22\text{-}5)$$

Fick's law of diffusion The diffusivity of component A, in its mixture with component B, is denoted by D_{AB} and defined in terms of the diffusion flux I_{0A} and concentration gradient dc_A/db by

$$I_{0A} = c_A(u_A - u_0) = -D_{AB}\frac{dc_A}{db} \qquad (22\text{-}6)$$

and, for component B,

$$I_{0B} = c_B(u_B - u_0) = -D_{BA}\frac{dc_A}{db} \qquad (22\text{-}7)$$

Equations (22-6) and (22-7) are statements of Fick's first law of diffusion for a binary mixture. Note that this law is based on three decisions:

1 The flux is in moles/area-time.
2 The diffusion velocity is relative to the volume-average velocity.
3 The driving potential is in terms of the molar concentration (moles of component A per unit volume).

Other definitions are found in the literature. The dimensions of D are length^2/time. For component B,

$$I_{0B} = c_B(u_B - u_0) = -D_{BA}\frac{dc_B}{db} \qquad (22\text{-}8)$$

Relations between flow rates and fluxes   Substitution of u_0 from Eq. (22-1) and u_A from Eq. (22-2) into Eq. (22-6) gives

$$I_{0A} = \frac{N_A}{A} - \frac{c_A}{\rho_M} \frac{N_A + N_B}{A} = -D_{AB} \frac{dc_A}{db} \tag{22-9}$$

and, since $c_A/\rho_M = x_A$ and $N_A + N_B = N$, where N is the total molal flow rate of the entire phase,

$$\frac{N_A}{A} = x_A \frac{N}{A} - D_{AB} \frac{dc_A}{db} \tag{22-10}$$

The corresponding equation for component B is

$$\frac{N_B}{A} = x_B \frac{N}{A} - D_{BA} \frac{dc_B}{db} \tag{22-11}$$

When Eqs. (22-10) and (22-11) are added and it is noted that $x_A + x_B = 1$ and that $c_A + c_A = \rho_M$, so $dc_A/db = -dc_B/db$, it follows that

$$I_{0A} = -I_{0B} \quad \text{and} \quad D_{AB} = D_{BA} = D_v$$

Also, the total diffusion flux $I_{0A} + I_{0B} = 0$.

Mole-fraction form of diffusion equation   In engineering usage, the mole fraction is a more convenient unit than molar concentration. If it is assumed that ρ_M is reasonably constant along the path of diffusion, substitution of cx_A for c_A in Eq. (22-10) and use of D_v for diffusivity gives

$$\frac{N_A}{A} = \frac{x_A N}{A} - D_v \frac{dx_A}{db} \tag{22-12}$$

A comparable equation applies to component B.

Interpretation of diffusion equation   Equation (22-12) is the basic relation for mass transfer. It contains all that is essential for following the movement during mass transfer of a typical component designated by subscript A to or from a fixed surface parallel to, and stationary with, the interface between the two phases. It accounts for both the convective bulk flow of the total phase and the molecular diffusion relative to the volume-average velocity of the phase. If the total flux of the component, N_A/A, is denoted by I_A, and if the convective flux, $x_A N/A$, by I_{cA}, Eq. (22-12) may be written

$$I_A = I_{cA} + I_{0A} \tag{22-13}$$

In words, the equation states that the total flux of a component is the convective flux plus the diffusion flux. The vector nature of the fluxes and concentration gradients must be understood, since these quantities are characterized by direction and magnitude. As derived, the positive sense of the vectors is that of the convective flux, which may be either toward or away from the interface. As shown in Eq. (22-7), the sign of the gradient is opposite to the direction of the flux, since physically diffusion is

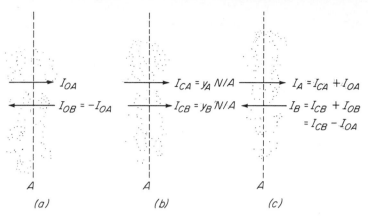

FIGURE 22-1
Diffusion and transport across stationary plane AA: (a) equimolal counter-diffusion (no phase drift); (b) phase drift only, without diffusion; (c) combined diffusion and phase drift. (Vectors pointing left to right are positive.)

"downhill," like the flow of heat "down" a temperature gradient. In the important special case where there is no convective flux, the positive direction is taken in the direction of the diffusion flux of component A.

Equation (22-12) may be used for the vapor phase by substituting y for x.

Convection and diffusion Figure 22-1 shows three cases of diffusion. Figure 22-1a represents the situation in which there is no bulk motion and the two components are diffusing at equal rates in opposite directions. For this case the diffusional flux of A through a stationary plane I_{OA} is given by Eq. (22-1). The flux of B is equal but opposite in direction. Next consider the case in which there is bulk motion of the entire phase but no diffusion at all; this is shown in Fig. 22-1b. Here both A and B are moving in the same direction. The total molal flux through the stationary plane N/A is given by the equation

$$\frac{N}{A} = \frac{N_A + N_B}{A} \tag{22-14}$$

The convective flux of A, I_{cA}, equals the total molal flux times the mole fraction, or

$$I_{cA} = \frac{y_A^N}{A} = \frac{c_A}{\rho_M}\frac{N_A + N_B}{A} \tag{22-15}$$

Finally, Fig. 22-1c shows the situation in which there are both diffusion and bulk motion. Here component A is moving to the right by both mechanisms; its total rate of motion is greater than it would be by diffusion alone. The total flux of A is found by substitution from Eqs. (22-1) and (22-15) into Eq. (22-13), to give

$$I_A = I_{0A} + I_{cA} = -D_v\frac{dc_A}{db}\frac{c_A}{\rho_M}\frac{N_A + N_B}{A} \tag{22-16}$$

Since $I_A = N_A/A$, rearrangement of Eq. (22-16) gives

$$\frac{dc_A}{\rho_M N_A - c_A(N_A + N_B)} = -\frac{db}{D_v \rho_M A} \tag{22-17}$$

Equation (22-17) is generally valid for two-component systems in which there is both bulk motion and also diffusion. Only the two limiting cases of equimolal diffusion and unicomponent diffusion, however, are common.

Equimolal diffusion In equimolal diffusion, often encountered in distillation operations, $N_A = -N_B$ and $N_A + N_B$ is zero. Equation (22-17) can then be integrated over a layer of total thickness B_T as follows:

$$\frac{1}{N_A} \int_{c_{Ai}}^{c_A} dc_A = -\frac{1}{D_v A} \int_0^{B_T} db \tag{22-18}$$

where c_{Ai} = concentration of A at one face of layer
$\quad c_A$ = concentration at other face of layer

Integration of Eq. (22-18) and rearrangement gives

$$I_A = \frac{N_A}{A} = \frac{D_v}{B_T}(c_{Ai} - c_A) = \frac{D_v \rho_M}{B_T}(y_i - y) \tag{22-19}$$

This result is the same as that obtained by direct integration of Eq. (22-1), noting that for equimolal diffusion $I_{0A} = I_A$.

Unicomponent diffusion In Fig. 22-1c, component B tends to move to the left because of diffusion and to the right because of the bulk motion of the entire phase. If the velocity of the bulk phase is just right, the two effects cancel and B does not move at all. B acts as if it were on a treadmill. Here the net flux of B, I_B, is zero and N_B is zero. For this situation Eq. (22-17) can be integrated as follows:

$$\frac{1}{N_A} \int_{c_{Ai}}^{c_A} \frac{dc_A}{\rho_M - c_A} = -\frac{1}{D_v \rho_M A} \int_0^{B_T} db \tag{22-20}$$

This yields, after rearrangement,

$$I_A = \frac{N_A}{A} = \frac{D_v \rho_M}{B_T} \ln \frac{\rho_M - c_A}{\rho_M - c_{Ai}} \tag{22-21}$$

Since $c_A = \rho_M y$ and $c_{Ai} = \rho_M y_i$, Eq. (22-21) can be written

$$I_A = \frac{D_v \rho_M}{B_T} \ln \frac{1 - y}{1 - y_i} \tag{22-22}$$

In unicomponent diffusion the flux of A is greater than in equimolal diffusion because of the bulk motion of the phase. Comparison of Eqs. (22-19) and (22-22)

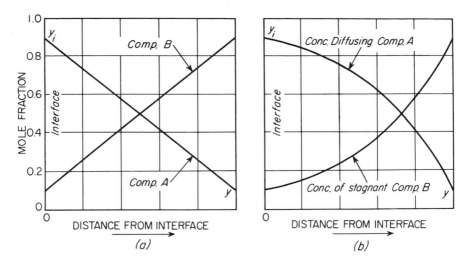

FIGURE 22-2
Concentration gradients for equimolal and unicomponent diffusion: (a) components A and B diffusing at same molal rates in opposite directions; (b) component A diffusing, component B stationary with respect to interface.

shows that the increase is given by the ratio of $\ln\left[(1 - y)/(1 - y_i)\right]$ to $y_i - y$. With dilute mixtures of A in B the increase is small, often negligible; with mixtures containing more than 5 or 10 percent of A, however, the increase is significantly large.

The concentration gradients for components A and B for a typical case of equimolal diffusion are shown in Fig. 22-2a, and those for unicomponent diffusion are shown in Fig. 22-2b. In both cases, concentrations of component A at the interface and at the outer boundary are 0.9 and 0.1, respectively. It is of interest that although in unicomponent diffusion N_B is zero, the concentration gradient of component B is not zero. The gradients are what an observer at rest with respect to the interface would see if the entire phase were moving from left to right just fast enough to prevent component B from moving through the fluid layer to or from the interface.

Prediction of diffusivities Diffusivities are best established by experimental measurements, and where such information is available for the system of interest it should be used directly. Often the desired values are not available, however, and they must be estimated from published correlations. Sometimes a value is available for one set of conditions of temperature and pressure; the correlations are then useful in predicting, from the known value, the desired values for other conditions.

Diffusivities in gases Values of D_v for some common gases diffusing in air at 0°C and 1 atm are given in Appendix 19. At pressures below about 5 atm D_v may be considered to be independent of concentration. Diffusivities in gases can be predicted with considerable accuracy from kinetic theory;[2, 12] the theoretical correlations have been modified in the light of experimental information to give the semiempirical equation[4]

$$D_v = \frac{0.01498 T^{1.81}(1/M_A + 1/M_B)^{0.5}}{p(T_{cA} T_{cB})^{0.1405}(V_{cA}^{0.4} + V_{cB}^{0.4})^2} \tag{22-23}$$

where

$$D_v = \text{diffusivity, cm}^2/\text{s}$$
$$T = \text{temperature, K}$$
$$M_A, M_B = \text{molecular weights of components } A \text{ and } B, \text{ respectively}$$
$$T_{cA}, T_{cB} = \text{critical temperatures of } A \text{ and } B, \text{ K}$$
$$V_{cA}, V_{cB} = \text{critical molar volumes of } A \text{ and } B, \text{ cm}^3/\text{g mol}$$

EXAMPLE 22-1   Estimate the volumetric diffusivity Dv for the system fluorotrichloromethane-nitrogen at 100°C and 10 atm. The critical temperatures and densities are given in Table 22-1.

SOLUTION   Equation (22-23) will be used. The quantities needed for substitution are

$$T = 373.16 \text{ K} \qquad T_{cA} = 471.16 \text{ K} \qquad T_{cB} = 126.16 \text{ K}$$

$$M_A = 137.5 \qquad M_B = 28$$

$$V_{cA} = \frac{137.5}{0.0552} = 2{,}490.9 \text{ cm}^3/\text{g mol}$$

$$V_{cB} = \frac{28}{0.311} = 90.03 \text{ cm}^3/\text{g mol}$$

Substitution in Eq. (22-23) gives

$$D_v = \frac{0.01498 \times 373.16^{1.81}(1/137.5 + 1/28)^{0.5}}{10(471.16 \times 126.16)^{0.1405}(2490.9^{0.4} + 90.02^{0.4})^2}$$

$$= 3.59 \times 10^{-3} \text{ cm}^2/\text{s} \qquad\qquad ////$$

Liquid diffusivities   The theory of diffusion in liquids is not so far advanced nor the experimental data so adequate as in gas diffusion. Diffusivities in liquids are smaller than in gases. Liquid diffusion depends on the energy needed to move the molecules through the liquid, which is a function of the attraction of molecules for each other rather than of molecular velocities. Diffusivities in liquids vary with concentration, approximately linearly, but accurate data showing the extent of the variation are few.

Table 22-1

Component	T_c, °C	ρ_c, g/cm^3
CFCl$_3$	198	0.0552
N$_2$	−147	0.311

Diffusivities for dilute liquid solutions can be calculated approximately from the equation[16]

$$D_v = 7.4 \times 10^{-8} \frac{(\psi_B M_B)^{1/2} T}{\mu V_A^{0.6}}$$

(22-24)

where D_v = diffusivity, cm^2/s

T = absolute temperature, K

μ = viscosity of solution, cP

V_A = molar volume of solute as liquid at its normal boiling point, cm^3/g mol

ψ_B = *association praameter* for solvent

The recommended values of ψ_B are 2.6 for water, 1.9 for methanol, 1.5 for ethanol, and 1.0 for benzene, heptane, ether, and other unassociated solvents. Equation (22-24) does not apply to electrolytes and is valid only at low concentrations.

Turbulent diffusion In a turbulent stream the moving eddies transport matter from one location to another, just as they transport momentum and heat energy. By analogy with Eqs. (3-20) and (12-49) for momentum transfer and heat transfer in turbulent streams, the equation for mass transfer is

$$I_{0A,t} = \frac{N_{0A,t}}{A} = -\varepsilon_N \frac{dc}{db}$$

(22-25)

where $I_{0A,t}$ = molal flux of A, *relative to the phase as a whole*, caused by turbulent action

ε_N = eddy diffusivity

The total molal flux, relative to the entire phase, becomes

$$I_A = \frac{N_{0A}}{A} = -(D_v + \varepsilon_N)\rho_M \frac{dc}{db}$$

(22-26)

Note that matter may be transferred to or away from an interface by turbulent action whether there is a drift of the entire phase relative to the interface or not. The total flux, including that caused by drift, is found by the same method as used in deriving Eq. (22-21) for molecular diffusion alone. The resulting equation is

$$I_A = \frac{N_A}{A} = \frac{(D_v + \varepsilon_N)\rho_M}{B_T} \ln \frac{\rho_M - c_A}{\rho_M - c_{Ai}}$$

(22-27)

Like eddy diffusivity of momentum ε_M and of heat energy ε_H, the eddy diffusivity ε_N depends on the properties of the fluid and also on the fluid velocity and on position in the flowing stream. Thus it cannot be predicted from theory or tabulated for given systems as molecular diffusivities can.

The theory of the eddy diffusivity for mass transfer ε_N follows closely that of the eddy diffusivity for momentum transfer ε_M, discussed in Chap. 3, and that of the eddy diffusivity for heat transfer ε_H, discussed in Chap. 12. The unified treatment of such quantities is the subject of analogy theory.[14] Heat and mass transfer are under the control of the same boundary layer, and ε_H and ε_N are considered equal. As shown in Chap. 12, the thermal boundary layer differs from the momentum boundary layer, and ε_M differs from ε_H. The analogy between heat and mass transfer in turbulent flow holds more accurately than the analogy between either one and momentum transfer. The quantity B_T in Eq. (22-27) is an empirical parameter defined as the thickness of a fictive laminar layer offering the same resistance to mass transfer as the actual boundary layer.

Mass-transfer coefficients When flow past an interface is laminar, Eqs. (22-19) and (22-22), which are based on steady-state molecular diffusion, may be used to predict mass-transfer rates provided D_v and B_T are known. Distance B_T, however, is not always known even when flow is laminar. Furthermore, at Reynolds numbers above the critical the usual buffer zone and turbulent core appear if the stream is passing through well-defined channels, and if the fluid is passing in contact with solid shapes, as in packed towers, separation and wake formation occur. Transfer to and from bubbles or sprays, as in sieve-plate and spray columns, differs even more greatly from the mechanism of simple molecular diffusion. It is doubtful that steady state is ever reached in such apparatus. An additional complication is that the actual interface area A is not often known. In packed towers, the actual interfacial area is considerably less than the geometric area of the packing, and in sprays and bubbles the area is unknown. In principle the mass-transfer rates could often be calculated by Eq. (22-27), but this is not, in fact, true, since neither B_T nor ε_N is known. Mass transfer is therefore treated by the use of mass-transfer coefficients, as defined by the equation

$$k \equiv \frac{N_A}{A(y_i - y)} = \frac{I_A}{y_i - y} \qquad (22\text{-}28)$$

where k is an individual mass-transfer coefficient, analogous to the individual heat-transfer coefficient h. Coefficient k is related to the diffusivities by the equation

$$k = \frac{(D_v + \varepsilon_N)\rho_M}{B_T} \qquad (22\text{-}29)$$

where ε_N may be regarded as the average eddy diffusivity over the (unknown) distance B_T. Thus k is a measure of the resistance to mass transfer by molecular and turbulent diffusion, independent of any drift of the entire phase.

In Eq. (22-28) for the definition of k the concentration y is the average concentration of the entire phase and is that reached if the stream is thoroughly mixed. This is analogous to the usage in heat transfer, where the average temperature of a stream is used in defining h. When concentration is defined in this manner, it is identical with that used in the material balances.

The coefficient k is the *Drew-Colburn coefficient*. Other mass-transfer coefficients are found in the chemical engineering literature, but for simplicity k, as defined by Eq. (22-28), is generally used throughout this book. It is used with a subscript y for a V phase and x for an L phase. The units of k are moles per unit area per hour per unit mole fraction.

Equimolal and unicomponent diffusion With equimolal counterdiffusion of the two components there is no drift of the entire phase, and k as defined by Eq. (22-28) measures the total rate of transfer of A (and of B, in the opposite direction) across a stationary plane or interface. In unicomponent diffusion the rate of mass transfer of A across a stationary plane is increased by the drift of the entire phase. Under these conditions

$$k' = \frac{I_A}{y_i - y} \tag{22-30}$$

where k' is a modified mass-transfer coefficient corrected to allow for drift. In general $k' > k$. Substitution of I_A from Eq. (22-30) into Eq. (22-27) and rearranging gives, since $c = y\rho_M$ and $c_i = y_i\rho_M$,

$$k' = \frac{(D_v + \varepsilon_N)\rho_M}{B_T(y_i - y)} \ln \frac{1 - y}{1 - y_i} \tag{22-31}$$

Substitution from Eq. (22-29) into Eq. (22-31) gives

$$k' = \frac{k}{y_i - y} \ln \frac{1 - y}{1 - y_i} \tag{22-32}$$

Equation (22-32) can be simplified as follows. The mole-fraction difference terms can be written

$$\frac{y_i - y}{\ln \left[(1 - y)/(1 - y_i)\right]} = \frac{(1 - y) - (1 - y_i)}{\ln \left[(1 - y)/(1 - y_i)\right]} = \overline{(1 - y)_L}$$

where $\overline{(1 - y)_L}$ is the logarithmic mean of $1 - y$ and $1 - y_i$, that is, the logarithmic mean concentration of the stagnant component B in the mixture. From Eq. (22 32), then,

$$k' = \frac{k}{\overline{(1 - y)_L}} \tag{22-33}$$

Correlations for predicting mass-transfer coefficients, such as those given later in this chapter, give values of k, not k'. If diffusion is equimolal, k is used directly in Eq. (22-28) to predict the total transfer flux. In unicomponent diffusion with concentrated mixtures the mass-transfer flux is found from Eqs. (22-30) and (22-33) to be

$$I_A = k'(y_i - y) = \frac{k(y_i - y)}{\overline{(1 - y)_L}} \tag{22-34}$$

With dilute mixtures of A, where $\overline{(1 - y)_L}$ approaches unity, the effect of drift is negligible and k' and k are almost equal and may be used interchangeably.

Mass transfer in liquids Equations (22-28) and (22-32) apply in principle to liquids as well as gases. Unlike diffusivities in gases, D_v in liquids varies considerably with concentration; furthermore, most published values for D_v in liquids refer to dilute solutions. For these reasons the effect of drift is usually ignored. If the liquid phase is the L phase, the equation for mass transfer is

$$I_A = \frac{N_A}{A} = k_x(x_i - x) \tag{22-35}$$

where k_x is a liquid-phase coefficient, uncorrected for drift. With concentrated solutions in which drift is significant, usual practice is to measure a corrected coefficient k_x' directly for the specific system, concentrations, and other conditions that are of interest.

Direction of diffusion In the above discussion, it is assumed that component A is diffusing from the interface into the phase. It is equally likely that this component diffuses toward the interface, rather than away from it. The same equations can be used by writing $y - y_i$ or $x - x_i$, for $y_i - y$ or $x_i - x$.

PENETRATION THEORY OF MASS TRANSFER

In heat transfer from a turbulent stream to a stationary surface there are two models on which estimates of heat-transfer rates may be based. One model, the steady-state equivalent-laminar-layer theory, assumes the existence of a fictive film in laminar motion, through which transfer is by conduction only; the other model postulates that elements of fluid move periodically from the turbulent core to the stationary surface, rest there a short time, and are then replaced by other elements from the bulk of the fluid. When the replacement time is governed by the action of a mechanical element, as in a scraped-surface exchanger, time t_T is known and the heat-transfer coefficient is given by Eq. (15-12). In a turbulent stream the residence time is not known, nor is it the same for all elements of the fluid. However, if t_L is defined as the effective average residence time of fluid elements on the surface, Eq. (15-12) may be applied to heat transfer from a turbulent fluid to a solid surface.

Exactly the same reasoning may be applied to processes of mass transfer to interfaces; in fact, it was in connection with mass-transfer problems that the theory was first developed.[10] It is assumed that mass transfer between the fluid elements and the interface takes place exclusively by molecular diffusion. Since conditions are changing with time, the concentration at any particular location within the fluid element varies according to the equation

$$\frac{\partial c}{\partial t} = D_v \frac{\partial^2 c}{\partial b^2} \tag{22-36}$$

where b is the distance from the interface. The particular solution of Eq. (22-36) is the same as that followed with Eq. (10-16) for transient heat transfer to a semi-infinite solid. For mass transfer it is assumed that the concentration at the interface is at all times c_i; that at the instant it comes to rest on the interface the fluid element has a concentration c which is the same as that of the bulk of the fluid; and that the element remains on the interface for so short a time that the composition at the far side of the element, away from the interface, does not change before the element is replaced with a new one.

For simplicity, assume first that diffusion to and from the interface is equimolal and that the phase-drift factor $\overline{(1 - y)_L}$ is unity. Under these conditions the transfer flux is given by the equation

$$I_A = \frac{N_A}{A} = \frac{N_{0A}}{A} = \frac{2(c - c_i)}{\sqrt{\pi}} \sqrt{\frac{D_v}{t_L}} \tag{22-37}$$

where t_L is the average time the fluid elements remain at the interface. From this, the individual mass-transfer coefficient for equimolal diffusion is given by

$$k = \frac{2\rho_M}{\sqrt{\pi}} \sqrt{\frac{D_v}{t_L}} \tag{22-38}$$

The surface-renewal model is not limited to equimolal diffusion. With unicomponent diffusion, the mass-transfer coefficient is given by Eq. (22-38), but the equation for the transfer flux becomes

$$I_A = k \ln \frac{\rho_M - c}{\rho_M - c_i} = \frac{2\rho_M}{\sqrt{\pi}} \sqrt{\frac{D_v}{t_L}} \ln \frac{\rho_M - c}{\rho_M - c_i} \tag{22-39}$$

Time t_L is a function of the fluid velocity, the fluid properties, and the geometry of the system. It must be empirically determined. As with heat transfer, the surface-renewal model in some cases appears to represent the transfer process more realistically than the film theory does. It is especially useful in situations where the transfer surface is periodically disturbed or renewed by some outside agency, as in a scraped-surface unit or by flow of drops or bubbles through a continuous phase. The similarity between Eqs. (22-39) and (15-12) is an example of the close analogy between heat and mass transfer. It is often reasonable to assume that time t_L is the same for both processes and thus to estimate rates of heat transfer from measured mass-transfer rates, or vice versa.

Experimental measurement of mass-transfer coefficients In view of the com. plexity of mass transfer in actual equipment, fundamental equations for mass transfer in actual equipment are not available, and empirical methods, guided by dimensiona-analysis and by semitheoretical analogies, are relied upon to give workable equationsl The approach to the problem has been made in several steps in the following manner.

1 The coefficient k has been studied in experimental devices in which the area of contact between phases is known and where boundary-layer separation does not take place. The wetted-wall tower shown in Fig. 22-3, which is sometimes used practically, is the usual device of this type. It has given valuable information on mass transfer to and from fluids in turbulent flow. A wetted-wall tower is essentially a vertical tube with means for admitting liquid at the top and causing it to flow downward along the inside wall of the tube, under the influence

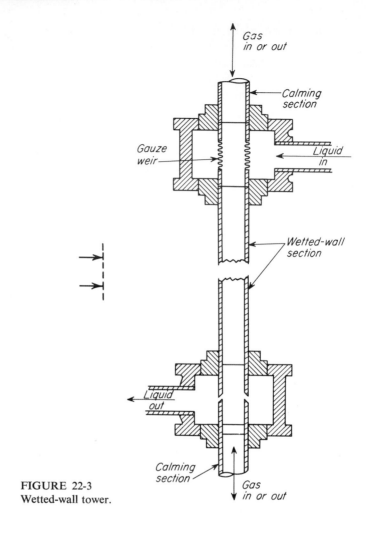

FIGURE 22-3
Wetted-wall tower.

of gravity, and means for admitting gas to the inside of the tube, where it flows through the tower in contact with the liquid. Generally the gas enters the bottom of the tower and flows countercurrent to the liquid, but parallel flow can be used. In the wetted-wall tower, the interfacial area, except for some complications from ripple formation, is known, and form drag is absent.

2   Other experiments have been made where the area of contact is known. The principle used in the work is to measure the rate of evaporation or solution of material from solid shapes in a stream of flowing liquid or gas. Evaporation of liquid from wet solids or evaporation or solution of substances from solid masses of the material itself is used. In this technique, the area is known and is simply the area of the evaporating or dissolving solid. Complications from diffusion in the solid itself are eliminated, as the material being transferred originates at the

interface between the phases. Data obtained in this manner can be used to extend the correlations obtained in wetted-wall experiments to situations where boundary-layer separation is active.

3 Finally, experiments on actual packed towers, sieve plates, and spray devices are made and the data so obtained correlated along lines suggested by the results from the experiments described in paragraphs 1 and 2. In some situations involving bubbles or drops the transfer area *a* per unit volume of the equipment can be measured or predicted with sufficient accuracy for the mass-transfer coefficient *k* to be found from the data. In other situations where the transfer area is not known a combined coefficient *ka*, instead of *k* alone, is used. Correlations are then found not for *k* itself but for the product *ka*.

Experiments conducted for obtaining numerical values for *k* or *ka* consist in measuring experimentally the quantities N_A, A, y_i, and y and calculating *k* by Eq. (22-28) or (22-32) after averaging over the entire length of the apparatus. If *A* or *a* is not known, the total volume of the equipment is used and *ka* calculated. Dimensional analysis is used to plan the experiments and to interpret the results in the form of dimensionless groups and equations. Analogies between friction, heat transfer, and mass transfer are useful guides.

COEFFICIENTS FOR MASS TRANSFER THROUGH KNOWN AREAS

In this section correlations are given for mass transfer between fluids or between fluids and solids where the area *A* is known. Coefficients for equipment in which the area between the phases is not known are discussed in subsequent chapters.

Dimensional analysis From the mechanism of mass transfer, it can be expected that the coefficient *k* would depend on the diffusivity D_v and on the variables that control the character of the fluid flow, namely, the mass velocity *G*, the viscosity μ, the density ρ, and some linear dimension *D*. Since the shape of the interface can be expected to influence the process, a different relation should appear from each shape.

For any given shape of transfer surface

$$k = (D_v, D, G, \mu, \rho)$$

dimensional analysis gives

$$\frac{k\overline{M}}{G} = \psi_1 \left(\frac{DG}{\mu}, \frac{\mu}{\rho D_v} \right) \tag{22-40}$$

Here $\overline{M}$ is the average molecular weight of the entire phase. It is used with *k* for consistency. Equation (22-40) is analogous to the Colburn form of the heat-transfer equations (12-64) and (12-65). A second dimensionless equation, analogous to the

Nusselt form of the heat-transfer equation, is obtained by multiplying Eq. (22-40) by $(DG/\mu)(\mu/\rho D_v)$. This gives

$$\frac{kD\overline{M}}{\rho D_v} = \psi_2 \left(\frac{DG}{\mu}, \frac{\mu}{\rho D_v} \right) \tag{22-41}$$

The dimensionless groups in Eqs. (22-40) and (22-41) have been given names and symbols. The group $kD\overline{M}/\rho D_v$ is called the *Sherwood number* and is denoted by N_{Sh}. This number corresponds to the Nusselt number in heat transfer. The group $\mu/\rho D_v$ is the Schmidt number, denoted by N_{Sc}. It corresponds to the Prandtl number. Typical values of N_{Sc} are given in Appendix 19. The group DG/μ is, of course, a Reynolds number, N_{Re}.

Equation (22-40) is also frequently used in the form of a j factor, analogous to the j_H factor of Eq. (12-65). It is defined by the equation[5]

$$j_M \equiv \frac{k\overline{M}}{G} \left(\frac{\mu}{\rho D_v} \right)^{2/3} \tag{22-42}$$

In general, j_M is a function of N_{Re}.

Wetted-wall towers   Several correlations are available for wetted-wall towers (Fig. 22-3). The Gilliland-Sherwood equation is[8]

$$N_{Sh} = 0.023 N_{Re}^{0.81} N_{Sc}^{0.44} \tag{22-43}$$

where D in N_{Re} and N_{Sh} is the diameter of the tube. This equation applies over a range of Reynolds numbers from 2,000 to 35,000; of Schmidt numbers from 0.60 to 2.5; and over a pressure range of 0.1 to 3 atm.

A second correlation for wetted-wall towers, somewhat less precise than Eq. (22-43), can be written

$$j_M = j_H = \frac{f}{2} = 0.023 N_{Re}^{-0.2} \tag{22-44}$$

where f is the Fanning friction factor for flow in smooth straight pipes. This correlation is satisfactory for both absorption and rectification in wetted-wall towers.[11] The analogy shown in this equation is general for heat and mass transfer in the same equipment. The analogy between heat and mass transfer on the one hand and friction on the other holds only for skin friction. It does not apply to the total friction if there is form drag from separation of flow.[7]

Flow perpendicular to single cylinders   A correlation of j_M vs. N_{Re} for flow perpendicular to single cylinders is shown in Fig. 22-4. The dashed line is a plot of the heat-transfer factor j_H, calculated from Eq. (12-65). It shows the close analogy between heat and mass transfer for this system.

Flow past single spheres   A plot of N_{Sh} vs. $N_{Re}N_{Sc}^{2/3}$ is shown in Fig. 22-5. The line in this figure approaches an asymptote of $N_{Sh} = 2.0$ at low values of $N_{Re}N_{Sc}^{2/3}$.

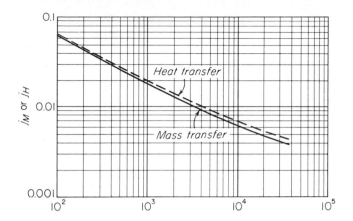

FIGURE 22-4
Heat and mass transfer, flow past single cylinders.

The line is represented by the equation

$$\frac{kD_p\overline{M}}{\rho D_v} = 2 + 0.6\left(\frac{D_pG}{\mu}\right)^{1/2}\left(\frac{\mu}{\rho D_v}\right)^{1/3} \tag{22-45}$$

This equation is the exact analog of Eq. (12-77) for heat transfer. Consequently the curve in Fig. 22-5 applies equally to heat and mass transfer.

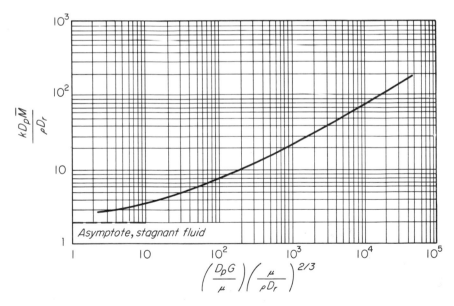

FIGURE 22-5
Heat and mass transfer, flow past single spheres.

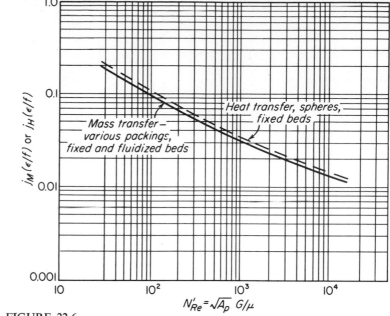

FIGURE 22-6
Heat and mass transfer, flow through beds of solids. (After Gupta and Thodos.[9])

Mass transfer in packed and fluidized beds Data for mass transfer between fluids and beds of various kinds of particles are summarized by the curve in Fig. 22-6. In this plot the j_M factor is multiplied by the porosity ε of the bed and divided by a factor f which measures the area available for transfer. In beds of spheres, either fixed or fluidized, $f = 1.000$; for other particles values of f are given in Table 22-2.

Table 22-2[9]   AREA AVAILABILITY FACTOR f^*

	Factor f	
Particle	Packed beds	Fluidized beds
Spheres	1.000	1.000
Regular cylinders	0.865	1.16
Cubes	0.825	1.24†
Partition rings	1.24	1.40†
Raschig rings	1.34	1.37†
Berl saddles	1.36	1.98†

* Data from Ref. 9
† Calculated from shape of particle. Unmarked values were found from mass- and heat-transfer data.

The abscissa in Fig. 22-6 is a special Reynolds number N'_{Re}, in which the linear dimension is the square root of the total surface area of an individual particle A_p.

The dashed line in Fig. 22-6 is a plot for heat transfer to the surface of spheres in a fixed bed. There is close agreement with the data for mass transfer.

Mass transfer to bubbles and drops When bubbles of gas are dispersed in a liquid in an agitated vessel or a sieve-plate tower, mass transfer to or from the bubbles takes place at a rate governed by the concentration during force, the interfacial area for transfer, and the mass-transfer coefficient in the liquid phase. There may be some resistance to transfer in the gas phase inside the bubble, but this resistance is usually negligible compared with that in the liquid.

LARGE BUBBLES The mass-transfer coefficient in the liquid, k_x, has been found to depend primarily on the physical properties of the gas-liquid system and on the bubble size. "Large" bubbles, with an average diameter greater than about 2.5 mm, are produced when pure liquids are aerated in an agitated vessel or on a sieve plate. For these bubbles k_x is given by[3a]

$$\frac{k_x D_p \overline{M}}{\rho D_v} = 0.42 \left(\frac{\mu}{\rho D_v}\right)^{1/2} \left(\frac{D_p^3 \rho \, \Delta \rho \, g}{\mu^2}\right)^{1/3} \tag{22-46}$$

where D_p = bubble diameter
$\quad D_v$ = diffusivity of dissolved gas in liquid phase
$\quad \overline{M}$ = average molecular weight of liquid
$\quad \rho$ = density of liquid
$\quad \mu$ = viscosity of liquid

Equation (22-46) can be written more compactly,

$$N_{Sh} = 0.42 N_{Sc}^{1/2} N_{Gr}^{1/3} \tag{22-47}$$

where N_{Gr} is a Grashof number, analogous to that discussed in Chap. 12 for heat transfer [see Eq. (12-78)].

It is not necessary to know D_p to calculate k_x from Eq. (22-46); since D_p appears on both sides of the equation, it can be eliminated and the equation rearranged to give

$$k_x \overline{M} \left(\frac{\mu}{\rho D_v}\right)^{1/2} = 0.42(\mu \rho \, \Delta \rho \, g)^{1/3} \tag{22-48}$$

SMALL BUBBLES For small bubbles, with diameters below about 0.5 mm, the following correlation has been recommended:[3a]

$$\frac{k_x D_p \overline{M}}{\rho D_v} = 2.0 + 0.31 \left(\frac{\mu}{\rho D_v}\right)^{1/3} \left(\frac{D_p^3 \rho \, \Delta \rho \, g}{\mu^2}\right)^{1/3} \tag{22-49}$$

or
$$N_{Sh} = 2.0 + 0.31 N_{Sc}^{1/} N_{Gr}^{3 \, 1/3} \tag{22-50}$$

Equation (22-49) applies to the aeration in agitated vessels of aqueous solutions of nonelectrolytes or when liquids generally are aerated with sintered plates or perforated plates with very small openings. Here the bubble diameter D_p appears in the equation for k_x and cannot be eliminated. The quantity 2.0, however, represents the asymptotic condition of molecular diffusion to a stationary bubble in a stagnant medium. In most gas-dispersion operations it can be ignored in comparison with the term $0.31N_{Sc}^{1/3}N_{Gr}^{1/3}$.

Equations (22-46) and (22-49) are remarkable in that they show that the liquid-phase mass-transfer coefficient is independent of impeller speed, impeller diameter, and power input in an agitated vessel and of gas velocity in a sieve-plate tower. In a given system therefore, k_x is approximately constant; it is usually about 5 times greater with "large" bubbles than with "small" bubbles.

Equations (22-46) through (22-50) may also be used for mass transfer to dispersed liquid drops or suspended solids in agitated vessels. Equation (22-49), furthermore, has been shown to apply to heat transfer to small suspended bubbles, drops, and particles; for this purpose it becomes

$$\frac{hD_p}{k} = 2.0 + 0.31 \left(\frac{c_p\mu}{k}\right)^{1/3} \left(\frac{D_p^3\rho\,\Delta\rho\,g}{\mu^2}\right)^{1/3} \tag{22-51}$$

or

$$N_{Nu} = 2.0 + 0.31N_{Pr}^{1/3}N_{Gr}^{1/3} \tag{22-52}$$

TRANSITION RANGE   In the range of bubble diameters between 0.5 and 2.5 mm the transfer coefficient k_x increases approximately linearly with the bubble diameter. To find which regime (large bubbles, small bubbles, or transition) applies in a given situation the bubble or drop diameter may be estimated by (9-50) or (9-55). Since the diameter D_p can usually be eliminated from the equations for k_x, it is not necessary to calculate D_p with great accuracy.

In the aeration of aqueous solutions of electrolytes in agitated vessels, k_x falls as the power input increases and the bubble diameter decreases.[13]

TOTAL TRANSFER RATES   In most operations k_x is not the quantity of interest; it is the total rate of transfer per unit volume of the equipment. This is governed, in part, by the volumetric transfer coefficient $k_x a$, where a is the interfacial area per unit volume. For aeration of pure liquids in a turbine-agitated vessel, a can be estimated from the dimensional equation (9-51)

$$a' = 1.44 \frac{(Pg_c/V)^{0.4}\rho^{0.2}}{(\sigma g_c)^{0.6}} \left(\frac{\overline{V}_s}{u_t}\right)^{1/2}$$

where a' is in mm^{-1}. The result, after conversion to the desired units, may be combined with the value of k_x found from Eq. (22-46) to give the volumetric coefficient $k_x a$. The total transfer rate, therefore, does depend on the power input and the gas velocity, even when k_x is constant.

SIEVE PLATES   For sieve plates k_x can also be found from Eq. (22-46) when the pool of liquid on the plate is shallow and the gas velocity is high. In most distillation operations, however, the plates operate in the transition region, with transfer coefficients between those for "large" and "small" bubbles. For gas absorption and

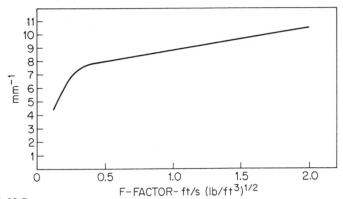

FIGURE 22-7
Gas-liquid interfacial area in sieve-plate columns.

desorption the following dimensional equation has been proposed:[3b]

$$\frac{k_x a \overline{M}}{(1 - \Psi)\rho} = 0.26F + 0.15 \tag{22-53}$$

where Ψ = fractional gas holdup
$F = F$ factor $= \overline{V}_s \sqrt{\rho_v}$ [see Eq. (19-74)]

Area a is given in Fig. 22-7; the holdup Ψ is given by

$$\ln \frac{1}{1 - \Psi} = 0.715F + 0.45 \tag{22-54}$$

EXAMPLE 22-2 (a) Estimate the mass-transfer coefficients in an agitated vessel under the conditions of Example 9-7. (b) If the dispersed gas is pure oxygen, what would the maximum possible rate of oxygen transfer be in gram moles per second per cubic meter of liquid? The solubility of oxygen in water at 20°C and 1 atm is 0.004339 g per 100 g of water. The density of water at 20°C is 0.99823 g/cm³; at 100°C it is 0.95838 g/cm³.

SOLUTION (a) Since a pure liquid is being aerated, use Eq. (22-48). The quantities needed for substitution are

$$\overline{M}_B = 18.02 \qquad \mu = 0.01 \text{ g/cm-s} \qquad \rho = 0.99823 \text{ g/cm}^3 \qquad g = 980.7 \text{ cm/s}^2$$

Since the density of the gas is very small, $\Delta\rho = \rho$. The diffusivity D_v is found from Eq. (22-24), for which

$$T = 20°C = 293 \text{ K} \qquad \mu = 1 \text{ cP}$$

$$V_A = 0.95838 \times 18.02 = 17.27 \text{ cm}^3/\text{g mol} \qquad \psi_B = 2.6$$

Hence, from Eq. (22-24)

$$D_v = \frac{(7.4 \times 10^{-8})(2.6 \times 18.02)^{1/2}(293)}{1 \times 17.27^{0.6}} = 2.69 \times 10^{-5} \text{ cm}^2/\text{s}$$

Substitution in Eq. (22-48) gives

$$k_x = \frac{0.42(0.01 \times 0.99823^2 \times 980.7)^{1/3}}{18.02[0.01/(0.99823 \times 2.69 \times 10^{-5})]^{1/2}}$$

$$= 2.58 \times 10^{-3} \text{ mol/cm}^2\text{-s-unit mole fraction}$$

From part (d), Example 9-7,

$$a = 0.125 \times 10 = 1.25 \text{ cm}^{-1}$$

Hence

$$k_x a = 2.58 \times 10^{-3} \times 1.25$$

$$= 3.23 \times 10^{-3} \text{ mol/cm}^3\text{-s-unit mole fraction}$$

(b)   The maximum oxygen-transfer rate in the liquid will occur when the gas-liquid interface is saturated with oxygen and the bulk of the liquid contains none. Under these conditions

$$x = 0 \qquad x_i = \frac{0.004339/16}{100/18.02 + 0.004339/16} = 4.89 \times 10^{-5}$$

The maximum transfer rate per unit volume V is found from Eq. (22-28), since a, by definition, is the transfer area A per unit volume. Thus

$$\frac{N_A}{V} = k_x a(x_i - x) = 3.23 \times 10^{-3}(4.89 \times 10^{-5} - 0)$$

$$= 1.58 \times 10^{-7} \text{ g mol/cm}^3\text{-s} = 0.158 \text{ g mol/m}^3\text{-s} \qquad ////$$

Liquid-phase and gas-phase coefficients   The foregoing equations are for the liquid-phase mass-transfer coefficient k_x. With sparingly soluble materials or when a pure gas, liquid, or solid is dispersed in a liquid, there is little or no resistance to mass transfer in the dispersed phase and the liquid-phase resistance is said to control. In such cases k_x may be used to predict the rate of mass transfer. When the dispersed bubbles or drops, however, contain a highly soluble gas or liquid, or when the dispersed phase is a permeable solid from which a solute is being leached, the mass-transfer resistance in the dispersed phase must be included in calculations of overall coefficients and mass-transfer rates.

STAGE EFFICIENCIES

The efficiency of a stage or plate in a distillation, absorption, or extraction operation is a function of the mass-transfer rates and transfer coefficients. When material is removed from a permeable solid, as in leaching or drying operations, the transfer rates and sometimes the stage efficiencies can be estimated from diffusion theory.

Distillation and absorption stage efficiency   On a sieve plate, mass transfer occurs in three successive zones: at the point of formation of the bubbles, during the contact between the vapor and liquid on the plate, and during contact between fog and vapor in the space above the liquid. Mass transfer is rapid during bubble formation. The second stage—the transfer of mass between vapor and the pool of liquid on the plate—has been given the most attention, both theoretical and experimental. It has been found that each locality on a plate acts much like a miniature absorber. The overall transfer rate in this zone depends on the relative importance of liquid-phase and vapor-phase resistances. When the liquid resistance is large, the viscosity of the liquid is an important variable. Efficiencies in absorbers are usually less than those in fractionating columns because of the low solubility and the relatively high liquid viscosity.

The concept of a transfer unit in absorption processes is described in Chap. 23. The local efficiency η' is related to the number of transfer units achieved on a given tray by the equations

$$\eta' = 1 - e^{-N_{tOy}} \tag{22-55}$$

$$N_{tOy} = \frac{\overline{K_y a}}{G_{My}} Z_T \tag{22-56}$$

where N_{tOy} = number of overall gas-phase transfer units

$\overline{K_y a}$ = average overall mass-transfer coefficient, based on gas phase

G_{My} = molal mass velocity of gas, based on total tower cross section

Z_T = depth of pool of liquid on plate

The relation of $K_y a$ to the individual coefficients $k_y a$ and $k_x a$ is shown in Chap. 23, page 728.

The prediction of plate efficiency is therefore based on the following pattern:

1 Estimation of the number of transfer units achieved in the vapor and liquid phases
2 Estimation of point efficiency from Eq. (22-55)
3 Estimation of Murphree efficiency η_M from Fig. 19-39
4 Estimation of overall efficiency η_o from Fig. 19-40 or using the procedure illustrated in Fig. 19-41

Stage efficiencies in leaching The stage efficiency in a leaching process depends on the time of contact between the solid and the solution and on the rate of diffusion of the solute through the solid and into the liquid. Consider a particle of solid containing soluble extractable material and in contact with an unsaturated solution. At a given instant t h from the time of initial contact with the liquid, assume that the concentration of the solute in the bulk of the liquid is X_1 and that the average concentration in the solution retained in the solid is $\overline{X}$. The concentration at the interface may lie between X_1 and $\overline{X}$ if both the resistance to diffusion in the solid and that from the interface into the liquid are of comparable magnitude. A general solution, then, requires that both resistances be calculable. This is not possible with present knowledge, but certain special cases may be considered.

First, consider a solid entirely impervious and inert to the action of the solvent with a film of strong solution on its surface. For such a case the process would involve simply the equalization of concentrations in the bulk of the extract and in the adhering film. Such a process is rapid, and any reasonable time of contact will bring about equilibrium. The countercurrent leaching process shown in Fig. 18-3 is of this type, and the stage efficiency is taken as unity in calculations for such processes.

A second case is where the soluble material is uniformly distributed throughout the interior of a permeable solid. Then the rate of diffusion of the solute to the interface may be the rate-controlling factor. Agitation is of little help, but the process is hastened if the solid is finely ground.

A third important situation is that where the solid is impermeable and contains the solute distributed through it in fine particles. Then the surface of the particles of solute may be an extremely small proportion of the total surface of the solid, and the rate of extraction is small. Here the best method is to grind the solid to such a size that the individual particles of solute are released.

Still another case is encountered in the leaching of natural materials such as sugar beets, tanbark, or soybeans. The solute is contained in vegetable cells, and the process requires the diffusion of solute by osmosis through the cell walls. Subdividing the material is ineffective because of the small size of the individual cells, and fairly large particles and long contact times are used. The resistance to diffusion from the interface into the bulk of the solution is small in comparison with that of diffusion in the solid itself, and the concentration at the interface can be assumed equal to that of the bulk of the liquid.

Under certain idealized conditions, the stage efficiency in extracting some (but not all) cellular materials can be estimated from experimental diffusion data obtained under the conditions of temperature and agitation to be used in the plant. The assumptions are as follows:[6]

1 The diffusion rate is represented by the equation

$$\frac{\partial X}{\partial t} = D_v' \frac{\partial^2 X}{\partial b^2} \tag{22-57}$$

where X = concentration of solute in solution within solid
D_v' = diffusivity†
b = distance, measured in direction of diffusion

2 The diffusivity is constant.
3 The solid can be considered to be equivalent to very thin slabs of constant density, size, and shape.
4 The concentration X_1 of the liquid in contact with the solid is constant.
5 The initial concentration in the solid is uniform throughout the solid.

† The diffusivity D_v' differs from the usual volumetric diffusivity D_v: it is based on a gradient expressed in terms of pounds of solute per pound of solute-free solvent rather than mole fraction of solute, and the transfer is in pounds or kilograms rather than moles. Since the diffusivity D_v' can be found only by experiment on the material to be extracted, it is actually used as an empirical constant, and these differences are unimportant in practice.

Under these assumptions, Eq. (22-57) can be integrated in exactly the same manner as Eq. (10-16), for conduction of heat through a slab. The result may be written

$$\frac{\overline{X} - X_1}{X_0 - X_1} = \frac{8}{\pi^2} (e^{-\alpha_1 \beta} + \tfrac{1}{9}e^{-9\alpha_1 \beta} + \tfrac{1}{25}e^{-25\alpha_1 \beta} + \cdots) = \phi(\beta) \qquad (22\text{-}58)$$

where $\overline{X}$ = average concentration of solute in solid at time t
X_0 = uniform concentration of solute in solid at zero time
X_1 = constant concentration of solute in bulk of solution at all times

and

$$\beta = \frac{D'_v t}{r_p^2} \qquad a_1 = \left(\frac{\pi}{2}\right)^2 \qquad (22\text{-}59)$$

where $2r_p$ is the thickness of the particle.

Note that Eq. (22-58) is Eq. (10-17) with T replaced by X, and N_{Fo} by β. Similar equations for long cylindrical particles and for spherical particles can be written by analogy with Eqs. (10-18) and (10-19) for conduction of heat. These equations for slablike, cylindrical, and spherical particles are represented by the curves of Fig. 10-6, and this figure can therefore be used in the solution of diffusion problems.

The Murphree stage efficiency for leaching is given by the equations

$$\eta_M = \frac{X_0 - \overline{X}}{X_0 - X_1} = 1 - \frac{\overline{X} - X_1}{X_0 - X_1} = 1 - \phi(\beta) = 1 - \phi\left(\frac{D'_v t}{r_p^2}\right) \qquad (22\text{-}60)$$

Equation (22-60) follows from the fact that $X_0 - \overline{X}$ is the actual concentration change for the stage and $X_0 - X_1$ is the concentration change that would be obtained if equilibrium were reached.

EXAMPLE 22-3 Assuming, for the meal in Example 20-2, $D'_v = 5 \times 10^{-9}$ cm²/s, that the thickness of the flakes is 0.04 cm, and that the time of contact in each stage is 3 h, estimate the actual number of stages required.

SOLUTION Since $2r_p = 0.04$, $r_p = 0.02$ cm. Also, $t = 3 \times 3,600 = 1.08 \times 10^4$ s. Then

$$\beta = \frac{D'_v t}{r_p^2} = \frac{(5 \times 10^{-9})(1.08 \times 10^4)}{0.02^2} = 0.135$$

From Fig. 10-6, $\phi(\beta) = 0.59$, and the Murphree efficiency is

$$\eta_M = 1 - 0.59 = 0.41$$

Here the average efficiency is nearly equal to the Murphree efficiency. The actual number of stages is $4/0.41 = 9+$. Either 9 or 10 stages should be used.   ////

SYMBOLS

A	Area perpendicular to direction of mass transfer, ft^2 or m^2; A_p, total surface area of particle
a	Area of interface between phases per unit volume of equipment, ft^{-1} or m^{-1}; a', mm^{-1}
a_1	$(\pi/2)^2$ in Eq. (22-59)
B_T	Thickness of layer through which diffusion occurs, ft or m
b	Distance from phase boundary in direction of diffusion, ft or m
c	Concentration, lb mol/ft^3 or g mol/m^3; c_A, of component A; c_{Ai}, of component A at interface; c_B, of component B
c_p	Specific heat at constant pressure, Btu/lb-°F or J/g-°C
D	Linear dimension or diameter, ft or m; D_p, of bubble, drop or particle
D_m	Molal diffusivity, lb mol/ft-h or g mol/m-h; D_{mA}, of component A; D_{mB}, of component B
D_v	Volumetric diffusivity, ft^2/h, m^2/h, or cm^2/s; D_{vA}, of component A; D_{vB}, of component B; D_v', of solute through stationary solution contained in a solid
F	F factor in distillation column, $\bar{V}_s\sqrt{\rho_v}$ [Eq. (22-53)]
f	Fanning friction factor, dimensionless; also, area availability factor (Table 22-2)
G	Mass velocity, lb/ft^2-h or g/m^2-s
G_M	Molal velocity, lb mol/ft^2-h or g mol/m^2-s; G_{My}, of gas phase
g	Gravitational acceleration, ft/s^2 or m/s^2
g_c	Newton's-law proportionality factor, 32.174 ft-lb/lb$_f$-s^2
h	Individual heat-transfer coefficient, Btu/ft^2-h-°F or W/m^2-°C
I	Molal flux, lb mol/ft^2-h or g mol/m^2-s; I_A, of component A; I_B, of component B
I_{cA}	Molal flux of component A caused by bulk motion of the phase, lb mol/ft^2-h or g mol/m^2-s
I_0	Molal flux across plane at rest with respect to the entire phase, lb mol/ft^2-h or g mol/m^2-s; I_{0A}, of component A; I_{0B}, of component B; $I_{0A,t}$, of component A, caused by turbulent action
j_H	Colburn j factor for heat transfer, $(h/c_p G)(c_p\mu/k)^{2/3}$, dimensionless
j_M	Colburn j factor for mass transfer, $(k\bar{M}/G)(\mu/\rho D_v)^{2/3}$, dimensionless
K_y	Overall mass-transfer coefficient in gas phase, lb mol/ft^2-h-unit mole fraction or g mol/m^2-s-unit mole fraction
k	Individual mass-transfer coefficient, lb mol/ft^2-h-unit mole fraction or g mol/m^2-s-unit mole fraction; k_x, in liquid phase; k_y, in gas or vapor phase; k', effective coefficient in unicomponent diffusion; k_x', in liquid phase; also, thermal conductivity, Btu/ft-h-°F or W/m-°C
M	Molecular weight; M_A, M_B, of components A and B, respectively; $\bar{M}$, average molecular weight of phase
N	Rate of mass transfer across a plane or boundary, mol/h; N_A, of component A; N_B, of component B
N_{Gr}	Grashof number, $(D_p^3 \rho\, \Delta\rho\, g)\mu^2$
N_{Nu}	Nusselt number, hD_p/k
N_{Pr}	Prandtl number, $c_p\mu/k$
N_{Re}	Reynolds number, DG/μ; N_{Re}', $\sqrt{A_p}\,G/\mu$
N_{Sc}	Schmidt number, $\mu/\rho D_v$
N_{Sh}	Sherwood number, $kD\bar{M}/\rho D_v$
N_{toy}	Number of overall gas-phase transfer units
N_0	Rate of transfer across plane at rest with respect to entire phase, mol/h; N_{0A}, of component A; N_{0B}, of component B; $N_{0A,t}$, of component A caused by turbulent action
P	Power input to agitated vessel, ft-lb$_f$/s or W
p	Pressure, lb$_f$/ft^2 or atm
q	Heat-transfer rate, Btu/h or W
r_p	One-half particle thickness, ft or m
T	Temperature, °F, °R, °C, or K; T_{cA}, T_{cB}, critical temperatures of components A and B, respectively

t	Time, s or h; t_L, residence time on transfer surface
u	Drift velocity of entire phase, ft/s or m/s; u_t, terminal raise velocity of bubbles
V	Volume of liquid in agitated vessel, ft^3 or m^3
V_A	Molar volume of solute as liquid at its normal boiling point, ft^3/lb mol
V_c	Critical molar volume, cm^3/g mol; V_{cA}, of component A; V_{cB}, of component B
$\bar{V}_s$	Superficial velocity of gas, ft/s or m/s
X	Concentration of solute in solution within solid; X_0, at zero time; X_1, in bulk of solution; X, average concentration in solution within solid at time t
x	Mole fraction in liquid or L phase; used for component A when only two components are present; x_i, at interface between phases
y	Mole fraction in gas or V-phase; used for component A when only two components are present; y_A, of component A; y_B, of component B; y_i, at interface between phases
$\overline{(1 - y)_L}$	Phase-drift factor in unicomponent diffusion; logarithmic mean of $1 - y$ and $1 - y_i$
Z_T	Depth of liquid pool on sieve plate, ft or m

Greek letters

α	Thermal diffusivity, ft^2/h or m^2/h
β	Diffusion group, $D'_v t/r_p^2$, dimensionless
$\Delta \rho$	Density difference between phases, lb/ft^3 or kg/m^3
ε	Eddy diffusivity, ft^2/h or m^2/h; ε_H, of heat; ε_M, of momentum; ε_N, of mass; also, porosity or void fraction, dimensionless
η	Plate or stage efficiency; η_M, Murphree efficiency; η_o, overall efficiency; η', local efficiency
μ	Absolute viscosity, lb/ft-h or cP
ρ	Density, lb/ft^3 or kg/m^3; ρ_c, at critical point; ρ_v, of gas or vapor
ρ_M	Molar density, lb mol/ft^3 or g mol/m^3
σ	Interfacial tension, lb$_f$/ft, N/m, or dyn/cm
ϕ, ψ	Functions; ψ_1, in Eq. (22-40); ψ_2, in Eq. (22-41)
Ψ	Fractional gas holdup in agitated vessel or on sieve plate
ψ_B	Association parameter for solvent [Eq. (22-24)]

PROBLEMS

22-1 Carbon dioxide is diffusing through nitrogen in one direction at atmospheric pressure and 0°C. The mole fraction of CO_2 at point A is 0.2; at point B, 10 ft away, in the direction of diffusion, it is 0.02. The concentration gradient is constant over this distance. Diffusivity D_v is 0.144 cm^2/s. The gas phase as a whole is stationary; i.e., nitrogen is diffusing at the same rate as the carbon dioxide, but in the opposite direction. (*a*) What is the molal flux of CO_2, in pound moles per square foot per hour? (*b*) What is the *net* mass flux, in pounds per square foot per hour? (*c*) At what speed, in meters per second, would an observer have to move from one point to the other so that the net mass flux, *relative to him*, would be zero? (*d*) At what speed would he have to move so that, relative to him, the *nitrogen* is stationary? (*e*) What would be the molal flux of carbon dioxide relative to the observer, under condition (*d*)?

22-2 An open circular tank 20 ft in diameter contains benzene at 22°C which is exposed to the atmosphere in such a manner that the liquid is covered with a stagnant air film estimated to be 5 mm thick. The concentration of benzene beyond the stagnant film is negligible. The vapor pressure of benzene at 22°C is 100 mm Hg. If benzene is worth 40 cents per gallon, what is the value of the loss of benzene from this tank, in dollars per day? The specific gravity of benzene is 0.88.

22-3 Alcohol vapor is being absorbed from a mixture of alcohol vapor and water vapor by means of a nonvolatile solvent in which alcohol is soluble but water is not. The temperature is 97°C, and the total pressure is 760 mm Hg. The alcohol vapor can be considered to be diffusing through a film of alcohol-water-vapor mixture 0.1 mm thick. The mole percent of the alcohol in the vapor at the outside of the film is 80 percent, and that on the inside, next to the solvent, is 10 percent. The volumetric diffusivity of alcohol-water-vapor mixtures at 25°C and 1 atm is 0.15 cm²/s. Calculate the rate of diffusion of alcohol vapor in pounds per hour if the area of the film is 100 ft².

22-4 An alcohol-water-vapor mixture is being rectified by contact with an alcohol-water-liquid solution. Alcohol is being transferred from gas to liquid and water from liquid to gas. The molal flow rates of alcohol and water are equal but in opposite directions. The temperature is 95°C and the pressure 1 atm. Both components are diffusing through a gas film 0.1 mm thick. The mole percentage of the alcohol at the outside of the film is 80 percent, and that on the inside is 10 percent. Calculate the rate of diffusion of alcohol and of water in pounds per hour through a film area of 100 ft².

22-5 A wetted-wall column operating at a total pressure of 518 mm Hg is supplied with water and air, the latter at a rate of 120 g/min. The partial pressure of the water vapor in the airstream is 76 mm, and the vapor pressure of the liquid-water film on the wall of the tower is 138 mm. The observed rate of vaporization of water into the air is 13.1 g/min. The same equipment, now at a total pressure of 820 mm, is supplied with air at the same temperature as before and at a rate of 100 g/min. The liquid vaporized is *n*-butyl alcohol. The partial pressure of the alcohol is 30.5 mm, and the vapor pressure of the liquid alcohol is 54.5 mm. What rate of vaporization, in grams per minute, may be expected in the experiment with *n*-butyl alcohol?

22-6 Air at 100°F and 2.0 atm is passed through a shallow bed of naphthalene spheres $\frac{1}{2}$ in. in diameter at a rate of 5 ft/s, based on the empty cross section of the bed. The vapor pressure of naphthalene is 117 mm. How many pounds per hour of naphthalene will evaporate from 1 ft³ of bed, assuming a bed porosity of 40 percent?

REFERENCES

1 Bird, R. B. : "Advances in Chemical Engineering," vol I, pp. 156–239, Academic, New York, 1956.
2 Bird, R. B., W. E. Stewart, and E. N. Lightfoot: "Transport Phenomena," Wiley, New York, 1960.
3 Calderbank, P. H.: in V. W. Uhl and J. B. Grey (eds.), "Mixing: Theory and Practice," vol. II, Academic, New York, 1967; (a) pp. 64–66, (b) p. 74.
4 Chen, N. H., and D. F. Othmer: *J. Chem. Eng. Data,* **7**:37 (1962).
5 Chilton, T. H., and A. P. Colburn: *Ind. Eng. Chem.,* **26**:1183 (1934).
6 Fan, H. P., J. C. Morris, and H. Wakeman: *Ind. Eng. Chem.,* **40**:195 (1948).
7 Garner, F. H., V. G. Jenson, and R. B. Keey: *Trans. Inst. Chem. Eng. (Lond.),* **37**:191 (1959).
8 Gilliland, E. R., and T. K. Sherwood: *Ind. Eng. Chem.,* **26**:516 (1935).
9 Gupta, A. S., and G. Thodos: *Chem. Eng. Prog.,* **58**(7):58 (1962).
10 Higbie, R.: *Trans. AIChE,* **31**:365 (1935).
11 Johnstone, H. F., and R. L. Pigford: *Trans. AIChE,* **38**:25 (1941).
12 Present, R. D.: "Kinetic Theory of Gases," McGraw-Hill, New York, 1958.
13 Robinson, C. W., and C. R. Wilke: *AIChE J.,* **20**:285 (1974).
14 Sherwood, T. K.: *Chem. Eng. Prog. Symp. Ser.,* **55**(25):71 (1959).
15 Tierney, J. W., L. F. Stutzman, and R. L. Daileader: *Ind. Eng. Chem.,* **46**:1595 (1954).
16 Wilke, C. R., and P. Chang: *AIChE J.,* **1**:264 (1955).

23

GAS ABSORPTION

This chapter deals with the mass-transfer operations known as *gas absorption* and *stripping*, or *desorption*. In gas absorption a soluble vapor is absorbed, by means of a liquid in which the solute gas is more or less soluble, from its mixture with an inert gas. The washing of ammonia from a mixture of ammonia and air by means of liquid water is a typical example. The solute is subsequently recovered from the liquid by distillation, and the absorbing liquid can be either discarded or reused. Sometimes a solute is removed from a liquid by bringing the liquid into contact with an inert gas; such an operation, the reverse of gas absorption, is desorption or gas stripping.

At the end of this chapter is a section on the application of packed towers, used chiefly in gas absorption, to the liquid-liquid extraction operations described in Chap. 20. Also included is a recapitulation of mass-transfer calculation methods for absorption, desorption, rectification, leaching, and liquid extraction.

DESIGN OF PACKED TOWERS

A common apparatus used in gas absorption and certain other operations is the packed tower, an example of which is shown in Fig. 23-1. The device consists of a cylindrical column, or tower, equipped with a gas inlet and distributing space at the bottom; a

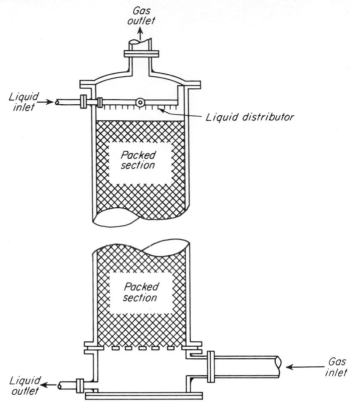

FIGURE 23-1
Packed tower.

liquid inlet and distributor at the top; liquid and gas outlets at the bottom and top, respectively; and a supported mass of inert solid shapes, called *tower packing* or *tower filling*. The inlet liquid, which may be pure solvent or a dilute solution of solute in the solvent and which is called the *weak liquor*, is distributed over the top of the packing by the distributor and, in ideal operation, uniformly wets the surfaces of the packing. The solute-containing gas, or rich gas, enters the distributing space below the packing and flows upward through the interstices in the packing countercurrent to the flow of the liquid. The packing provides a large area of contact between the liquid and gas and encourages intimate contact between the phases. The solute in the rich gas is absorbed by the fresh liquid entering the tower, and dilute, or lean, gas leaves the top. The liquid is enriched in solute as it flows down the tower, and concentrated liquid, called *strong liquor*, leaves the bottom of the tower through the liquid outlet.

Many kinds of tower packing have been invented, and several types are in common use. Packings are divided into those which are dumped at random into the tower and those which must be stacked by hand. Dumped packings consist of units

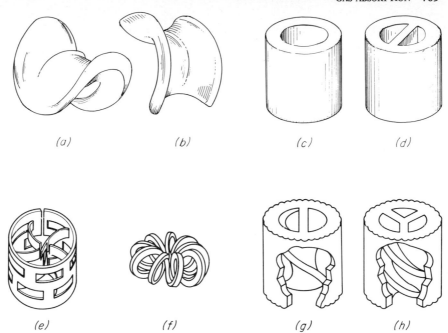

FIGURE 23-2
Typical tower packings: (*a*) Berl saddle; (*b*) Intalox saddle; (*c*) Raschig ring; (*d*) Lessing ring; (*e*) Pall ring; (*f*) Tellerette; (*g*) double-spiral ring; (*h*) triple-spiral ring.

$\frac{1}{4}$ to 2 in. in major dimension, and are used most in the smaller columns. The units in stacked packings are 2 to about 8 in. in size; they are used only in the larger towers. Common packings are illustrated in Fig. 23-2. The rings and saddles shown in Fig. 23-2*a* to *e* and the Tellerettes shown in Fig. 23-2*f* typify dumped packings; the spiral partition rings shown in Fig. 23-2*g* and *h* are manually stacked. Large 2- and 3-in. Raschig rings are also often stacked.

The principal requirements of a tower packing are:

1   It must be chemically inert to the fluids in the tower.
2   It must be strong without excessive weight.
3   It must contain adequate passages for both streams without excessive liquid holdup or pressure drop.
4   It must provide good contact between liquid and gas.
5   It must be reasonable in cost.

Thus most tower packings are made of cheap, inert, fairly light materials such as clay, porcelain, or graphite. Thin-walled metal rings of steel or aluminum are sometimes used. High void spaces and large passages for the fluids are achieved by making the packing units irregular or hollow, so that they interlock into open structures with a porosity of 60 percent or more. The physical characteristics of various packings are given in Table 23-1. Stacked packings, with open channels running uninterruptedly through the packed bed, give lower pressure drops than dumped packings, in which the gas must frequently change in velocity and direction. This advantage is offset, however, by the poorer contact between the fluids in stacked packings.

Contact between liquid and gas The requirement of good contact between liquid and gas is the hardest to meet, especially in large towers. Ideally the liquid, once distributed over the top of the packing, flows in thin films over all the packing surface all the way down the tower. Actually the films tend to grow thicker in some places and thinner in others, so that the liquid collects into small rivulets and flows along localized paths through the packing. Especially at low liquid rates much of the packing surface may be dry or, at best, covered by a stagnant film of liquid. This effect is known as *channeling*; it is the chief reason for the poor performance of large packed towers.

Channeling is most severe in towers packed with stacked packing, less severe in dumped packings of crushed solids, and least severe in dumped packings of regular units such as rings.[11] In towers of moderate size channeling can be minimized by having the diameter of the tower at least 8 times the packing diameter. If the ratio of tower diameter to packing diameter is less than 8:1, the liquid tends to flow out of the packing and down the walls of the column. Even in small towers filled with packings that meet this requirement, however, liquid distribution and channeling

Table 23-1 PHYSICAL CHARACTERISTICS OF TOWER PACKINGS[7,10a]

Type	Material	Dimensions, OD and length, in.	Average bulk density, lb/ft³ of tower volume	Surface area, a_v, ft²/ft³ of tower volume	Porosity ε	Packing factor F_p, ft⁻¹
Dumped packings:						
Berl saddles	Ceramic	$\frac{1}{2}$†	54	142	0.62	240
		1†	45	76	0.68	110
		$1\frac{1}{2}$†	40	46	0.71	65
Intalox saddles	Ceramic	$\frac{1}{2}$†	45	190	0.78	200
		1†	44	78	0.77	98
		$1\frac{1}{2}$†	42	59	0.80	52
Raschig rings	Steel	$\frac{1}{2} \times \frac{1}{2}$	75	122	0.85	410
		1×1	71	56	0.86	137
	Ceramic	$\frac{1}{2} \times \frac{1}{2}$	55	112	0.64	580
		1×1	42	58	0.74	155
		2×2	41	28	0.74	65
Pall rings	Steel	1×1	30	63	0.94	48
Tellerettes	Low-density polyethylene	1†	10	76	0.83	
Stacked packings:						
Single-spiral rings	Stoneware	$3\frac{1}{4} \times 3$	52	34	0.66	
		4×4	55	28	0.67	
		6×6	51	19	0.70	

† Nominal size.

have a major effect on column performance. In tall towers filled with large packing the effect of channeling may be pronounced, and redistributors for the liquid are normally included every 10 or 15 ft in the packed section.

At low liquid rates, regardless of the initial liquid distribution, much of the packing surface is not wetted by the flowing liquid.[4] As the liquid rate rises, the wetted fraction of the packing surface increases, until at a critical liquid rate, which is usually high, all the packing surface becomes wetted and effective. At liquid rates higher than the critical the effect of channeling is not important.

Limiting flow rates; loading and flooding   In a tower containing a given packing and being irrigated with a definite flow of liquid, there is an upper limit to the rate of gas flow. The gas velocity corresponding to this limit is called the *flooding velocity*. It can be found from an inspection of the relation between the pressure drop through the bed of packing and the gas flow rate, from observation of the holdup of liquid, and by the visual appearance of the packing. The flooding velocity, as identified by these three different effects, varies somewhat with the method of identification and appears more as a range of flow rates than as a sharply defined constant.

Figure 23-3 shows the relation between pressure drop and gas flow rate in a packed tower. The pressure drop per unit packing depth comes from fluid friction; it is plotted on logarithmic coordinates against the gas flow rate G_y, expressed in mass of gas per hour per unit of cross-sectional area, based on the empty tower. G_y is therefore related to the superficial gas velocity by the equation $G_y = \overline{V}_{sy}\rho_y$, where ρ_y is the density of the gas. When the packing is dry, the line so obtained is straight and has a slope of about 1.8. The pressure drop therefore increases with the 1.8th power of the velocity, which is consistent with the usual law of friction loss in turbulent flow. If the packing is irrigated with a constant flow of liquid, at a superficial mass velocity G_x, the relation between pressure drop and gas flow rate follows a line such as *bcde* in Fig. 23-3. At low and moderate velocities, the pressure drop is proportional to the 1.8th power of the flow rate but is greater than that in dry packing at the same gas velocity. As the gas velocity increases, however, the line curves upward, starting at point *c* in the figure. Then, at a higher velocity, as shown by the line *cde*, the pressure drop increases sharply at nearly constant gas velocity. The rise may follow a smooth curve, as shown by the solid line *cd*, or may show sharp breaks at points *c* and *d*, as shown by the dotted line. As the pressure drop increases along line *bc*, the amount of liquid held in the packing is constant and independent of the gas velocity. The liquid moves downward through the packing uninfluenced by the motion of the gas. At point *c*, called the *loading point*, the gas flow begins to impede the downward motion of the liquid. Local accumulations of liquid appear here and there in the packing. As the gas rate rises further, the liquid holdup increases and the pressure drop changes along line *cde*, increasing more rapidly with gas velocity than before. At point *e*, the flooding point, the top of the packing is covered with a layer of liquid through which bubbles of gas issue. The liquid can no longer flow down through the packing, and the layer grows until liquid is blown out of the top of the tower with the gas.

The gas velocity in an operating packed tower must obviously be lower than the velocity which will cause flooding. How much lower is a choice to be made by the designer. The lower the velocity, the lower the cost of power and the larger the tower.

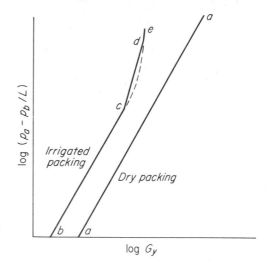

FIGURE 23-3
Pressure drop in packed tower: c, loading point; e, flooding point.

The higher the gas velocity, the larger the power cost and the smaller the tower. Economically, the most favorable gas velocity depends on a balance between the cost of the power and the fixed charges on the equipment. The optimum velocity can be estimated by methods beyond the scope of this book.[13b] It is usually about one-half the flooding velocity.

Packed towers are also commonly designed on the basis of a definite pressure drop per unit height of packing. For absorption towers the design value is usually between 0.25 and 0.5 in. H_2O per foot of packing; for distillation columns it is in the range 0.5 to 0.8 in. H_2O per foot. In most towers packed with rings or saddles loading usually begins at a pressure drop of about 0.5 in. H_2O per foot, and flooding occurs at a pressure drop between 2 and 3 in. H_2O per foot.

Figure 23-4 gives correlations for estimating flooding velocities and pressure drops in packed towers. It consists of a logarithmic plot of

$$\frac{G_y^2 F_p (62.3/\rho_x)\mu_x^{0.2}}{g_c \rho_x \rho_y} \quad \text{vs.} \quad \frac{G_x}{G_y}\sqrt{\frac{\rho_y}{\rho_x}}$$

where G_x = mass velocity of liquid, lb/ft²-s
  G_y = mass velocity of gas, lb/ft²-s
  F_p = packing factor, ft^{-1}
  ρ_x = density of liquid, lb/ft³
  ρ_y = density of gas, lb/ft³
  μ_x = viscosity of liquid, cP
  g_c = Newton's-law proportionality factor, 32.174 ft-lb/lb$_f$-s²

The ordinates of Fig. 23-4 are not dimensionless, and the stated units must be used. The mass velocities are based on the total tower cross section.

EXAMPLE 23-1 A tower packed with 1-in. (25.4-mm) ceramic Raschig rings is to be built to treat 25,000 ft³ (708 m³) of entering gas per hour. The ammonia content of the entering gas is 2 percent by volume. Ammonia-free water is used as absorbent. The temperature is

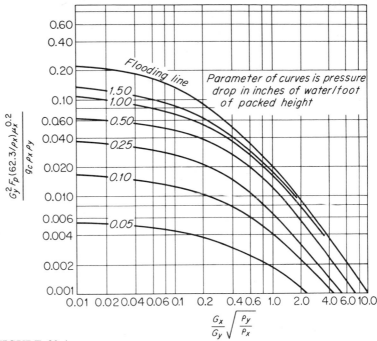

FIGURE 23-4

Generalized correlation for flooding and pressure drop in packed columns after Eckert.[2]

68°F (20°C), and the pressure is 1 atm. The ratio of gas flow to liquid flow is 1 lb of gas per pound of liquid. (*a*) If the gas velocity is to be one-half the flooding velocity, what should be the diameter of the tower? (*b*) What is the pressure drop if the packed section is 20 ft (6.1 m) high?

SOLUTION   The quantities to be used in the groups of Fig. 23-4 are as follows. The average molecular weight of the entering gas is $29 \times 0.98 + 0.02 \times 17 = 28.76$. Then

$$\rho_y = \frac{28.76 \times 492}{359(460 + 68)} = 0.07465 \text{ lb/ft}^3$$

$$\rho_x = 62.3 \text{ lb/ft}^3 \qquad \mu_x = 1 \text{ cP}$$

$$g_c = 32.174 \text{ ft-lb/lb}_f\text{-s}^2 \qquad \frac{G_x}{G_y} = 1.0$$

For 1-in. ceramic rings, $F_p = 155 \text{ ft}^{-1}$ (Table 23-1). Then

$$\frac{G_x}{G_y} \sqrt{\frac{\rho_y}{\rho_x}} = \sqrt{\frac{0.07465}{62.3}} = 0.0346$$

From Fig. 23-4,

$$\frac{G_y^2 F_p (62.3/\rho_x)\mu_x^{0.2}}{g_c \rho_x \rho_y} = 0.19$$

The mass velocity at flooding is

$$G_y = \sqrt{\frac{0.19 \times 32.174 \times 0.07465 \times 62.3}{155 \times 1^{0.2} \times 1}} = 0.428 \text{ lb/ft}^2\text{-s}$$

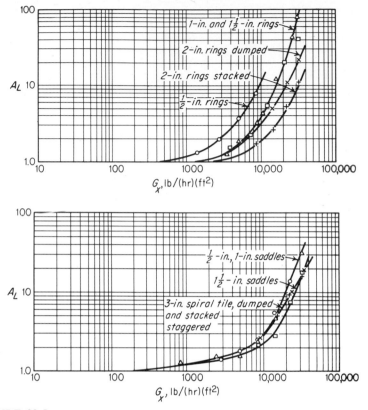

FIGURE 23-5
Pressure drop in irrigated packing. (*By permission, from T. K. Sherwood and R. L. Pigford, "Absorption and Extraction," 2d ed. Copyright, 1952, McGraw-Hill Book Company.*)

(*a*)   The total gas flow is $25,000 \times 0.07465/3600 = 0.518$ lb/s. If the actual velocity is one-half the flooding velocity, the cross-sectional area S of the tower is

$$S = \frac{0.518}{0.428/2} = 2.42 \text{ ft}^2$$

The diameter of the tower is $\sqrt{2.42/0.7854} = 1.76$ ft (536 mm).

(*b*)   At half the flooding velocity, $G_y = 0.428/2 = 0.214$ lb/ft²-s, G_x is unchanged, and the abscissa value in Fig. 23-4 is 0.0692. The ordinate becomes $0.19/4 = 0.0475$. For these conditions the pressure drop is about 0.45 in. H_2O per foot of packed height; the total pressure drop is $20 \times 0.45 = 9$ in. H_2O (16.8 mm H_2O).   ////

Pressure drop in wetted packing   The pressure drop in a bed of solids being irrigated with liquid is considerably greater than that in dry packing at the same gas flow rate. No accurate basic correlation is available for pressure drop through wetted packings, but Fig. 23-5 gives an empirical correction factor for estimating such pressure drops from the corresponding drop through the dry bed.[13] When the water rate G_x, in pounds per square foot per hour, is known, the factor A_L can be read from Fig. 23-5 and used as a multiplier for the pressure drop in dry packing, obtained from Eq. (7-22). Figure 23-5 applies only for flows less than the loading point and for liquids having approximately the viscosity of water. Except for the relatively small pressure drop due to fluid friction, the pressure in a packed tower is constant.

PRINCIPLES OF ABSORPTION

As shown in the previous section, the diameter of a packed absorption tower depends on the quantities of gas and liquid handled, their properties, and the ratio of one stream to the other. The height of the tower, and hence the total volume of packing, depends on the magnitude of the desired concentration changes and on the rate of mass transfer per unit of packed volume. Calculations of the tower height, therefore, rest on material balances, enthalpy balances, and on estimates of driving force and mass-transfer coefficients.

Material balances   In a differential-contact plant such as the packed absorption tower illustrated in Fig. 23-6, there are no sudden discrete changes in composition as in a stage-contact plant. Instead the variations in composition are continuous from one end of the equipment to the other. Material balances for the portion of the column above an arbitrary section, as shown by the dotted line in Fig. 23-6, are as follows:

Total material: $$L_a + V = L + V_a \tag{23-1}$$

Component A: $$L_a x_a + Vy = Lx + V_a y_a \tag{23-2}$$

where V is the molal flow rate of the gas phase and L that of the liquid phase at the same point in the tower. The L-phase and V-phase concentrations x and y apply to this same location.

The overall material-balance equations, based on the terminal streams, are:

Total material: $$L_a + V_b = L_b + V_a \tag{23-3}$$

Component A: $$L_a x_a + V_b y_b = L_b x_b + V_a y_a \tag{23-4}$$

Equations (23-3) and (23-4) are identical with Eqs. (18-3) and (18-4) for a stage-contact column.

The operating-line equations for a differential-contact plant, analogous to Eqs. (18-7) and (18-10) for a stage-contact column, are

$$y = \frac{L}{V} x + \frac{V_a y_a - L_a x_a}{V} \tag{23-5}$$

$$L'\left(\frac{x_a}{1 - x_a} - \frac{x}{1 - x}\right) = V'\left(\frac{y_a}{1 - y_a} - \frac{y}{1 - y}\right) \tag{23-6}$$

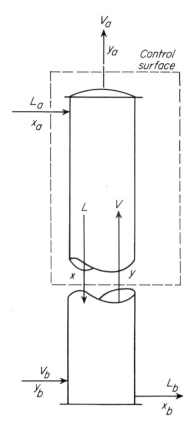

FIGURE 23-6
Material-balance diagram for packed column

In Eqs. (23-5) and (23-6) x and y represent the bulk compositions of the liquid and gas, respectively, in contact with each other at any given section through the column. It is assumed that the compositions at a given elevation are independent of position in the packing. In absorption operations the flow rates V and L usually change appreciably from one point to another in the column, and the operating lines are often strongly curved.

Limiting gas-liquid ratio   Equation (23-5) shows that the average slope of the operating line is L/V, the ratio of the molal flows of liquid and gas. Thus, for a given gas flow, a reduction in liquid flow decreases the slope of the operating line. Consider the operating line ab in Fig. 23-7. Assume that the gas rate and the terminal concentrations x_a, y_a, and y_b are held fast and the liquid flow L decreased. The upper end of the operating line then shifts in the direction of the equilibrium line, and x_b, the concentration of the strong liquor, increases. The maximum possible liquor concentration and the minimum possible liquid rate are obtained when the operating line just touches the equilibrium line, as shown by line ab' in Fig. 23-7. At this condition, an infinitely deep packed section is necessary, as the concentration difference for mass

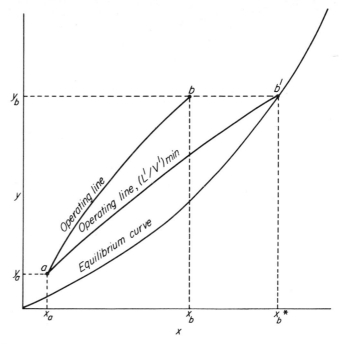

FIGURE 23-7
Limiting gas-liquid ratio.

transfer becomes zero at the bottom of the tower. In any actual tower, the liquid rate must be greater than this minimum if the tower is to operate.

The limiting liquid-gas ratio $(L/V')_{min}$ can be computed from Eq. (23-6) by setting $y = y_b$ and $x = x_b^*$, where x_b^* is the abscissa of the point on the equilibrium line whose ordinate is y_b.

The L/V' ratio is important in the economics of absorption in a countercurrent tower. If the liquid-gas ratio is large, the average distance between the operating and equilibrium line is also large, the concentration difference is favorable throughout the tower, and the tower is short. If the solute gas is to be recovered, however, the cost of recovery is high because of the dilution of the strong liquor. If, on the other hand, concentration costs are reduced by using less liquid, the driving forces in the absorber are reduced, and the tower becomes taller and more expensive. The optimum liquid rate is found by balancing recovery costs against fixed costs of the equipment. In general, the operating line should be approximately parallel to the equilibrium line for economical operation.

Temperature variations in packed towers When rich gas is fed to an absorption tower, the temperature in the tower varies appreciably from bottom to top. This temperature gradient affects the shape of the equilibrium line. The rate of absorption

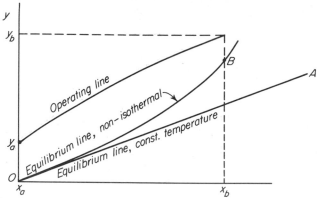

FIGURE 23-8
Effect of temperature gradient on equilibrium line.

is large at the gas inlet, and the heat of condensation and solution of the absorbed constituent may be sufficient to increase the temperature of the liquid considerably. Since the partial pressure of the absorbed component increases with increase in temperature, the concentration of the vapor in equilibrium with a liquid of definite composition also increases with temperature. Even if Henry's law applies, the equilibrium line may be strongly curved in this situation.

The effect on the equilibrium line of a temperature gradient in a tower is shown in Fig. 23-8. Line OA is the equilibrium line for isothermal operation, and line OB is that where the temperature at the bottom of the tower is larger than that at the top. The top temperature is the same in both situations. If the temperature effect is sufficiently great, the equilibrium line may intersect the operating line, and the process is inoperative near the bottom of the tower. Sometimes cooling coils or other cooling means are installed in the tower to reduce this heat effect.

The calculation of the equilibrium line under nonisothermal conditions can be accomplished by enthalpy balances.[13a]

In most towers fed with dilute or moderately strong gas, the temperature gradient in the column is small and the equilibrium line straight or nearly so.

Material balance over a differential section In the flow of the two phases through a differential length of column, V, L, x, and y change slightly. The differential equation showing this is obtained by differentiating Eq. (23-2).

$$d(Lx) = d(Vy) \tag{23-7}$$

This equation is the material balance for component A over a differential section of column.

Each term in Eq. (23-7) is the rate at which component A is transferred from one phase to the other through the area of the interface in the differential section of the column. Thus

$$d(Lx) = d(Vy) = dN_A \tag{23-8}$$

where N_A is the rate of transfer of component A in moles per hour. The relation of the rate of transfer to the properties of the phases and the conditions of operation is the subject of the next section.

RATE OF ABSORPTION

The height of a packed tower depends on the rate of absorption, which in turn is affected by the rate of transfer of mass through the liquid and gas phases. The following treatment has two limitations: no provision is made for possible chemical reactions between the absorbed component and the liquid, and the heat of solution is neglected. Simultaneous mass transfer and heat transfer are considered in Chaps. 24 and 25. Absorption accompanied by chemical reaction in the liquid state is not within the scope of this book.

Double-resistance theory Component A, which is the solute gas transferred from gas to liquid, must pass in series through two diffusional resistances, one in the gas and one in the liquid. It is assumed that the resistance of the interface itself is zero, with the consequence that at the interface (though nowhere else) the gas and liquid phases are in equilibrium. It has been shown experimentally that this assumption is nearly always valid in absorption and desorption operations.

Consider the packed tower shown in Fig. 23-9. Consider the absorption in the short section of height dZ at a level in the tower Z from the top of the packed section. Let the rate of absorption be dN_A mol/h. The interfacial area between phases in this section is dA. The concentrations of the gas and liquid streams, in mole fraction of component A, are y and x, respectively.

The rate of transfer of component A from the bulk of the gas to the interface is, from Eq. (22-34),

$$dN_A = \frac{k_y(y - y_i)}{(1 - y)_L} \, dA \tag{23-9}$$

where k_y = mass-transfer coefficient, gas phase, mol/unit area-h-unit mole-fraction difference

y = mole fraction of A in bulk of gas phase

y_i = mole fraction of A at gas side of interface

The difference $y - y_i$ is the driving force across the gas-phase resistance; $\overline{(1 - y)_L}$ is the logarithmic mean of $1 - y$ and $1 - y_i$.

The rate of transfer of component A in the liquid phase, from the interface to the bulk of the liquid, is, from Eq. (22-35)

$$dN_A = k_x(x_i - x) \, dA \tag{23-10}$$

where k_x = mass-transfer coefficient, liquid phase, mol/unit area-h-unit mole-fraction difference

x_i = mole fraction of component A at liquid side of interface

x = mole fraction of component A in bulk of liquid

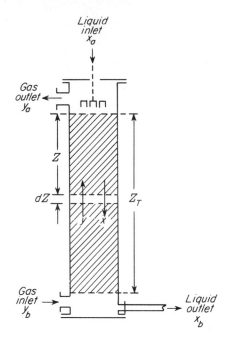

FIGURE 23-9
Diagram of packed absorption tower.

In equations for liquid-phase transfer, phase drift is ignored, for the reasons given in Chap. 22.

If equilibrium at the interface is assumed, x_i and y_i are coordinates of a point on the equilibrium curve and the curve of x_e vs. y_e is also the relationship between x_i and y_i.

Equations (23-9) and (23-10) are analogous to the heat-transfer equations applicable to two fluid phases in contact. Such equations may be written

$$dq = h_1(T_h - T_i) \, dA \qquad \text{and} \qquad dq = h_2(T_i - T_c) \, dA$$

where q = rate of heat transfer
A = area of interface
T_h, T_c = average bulk temperatures of phases 1 and 2
T_i = temperature of interface
h_1, h_2 = individual heat-transfer coefficients for phases 1 and 2

Two complications appear in mass transfer that are not found in heat transfer. (1) Phase drift must be taken into account in mass transfer, and no corresponding effect appears in the flow of heat. In mass transfer, a physical movement of material takes place, while heat transfer is a flow of energy only. (2) The interfacial equilibrium concentrations x_i and y_i are not equal, while in heat transfer the interfacial temperature T_i is the same for both phases.

Because the rate of loss of component A from the gas equals the rate of gain of that component by the liquid, the term dN_A in Eq. (23-9) equals that in Eq. (23-10).

Also, by Eq. (23-8), the transfer rate from the gas is $d(Vy)$, and that to the liquid is $d(Lx)$. Then

$$dN_A = \frac{k_y(y - y_i)}{(1 - y)_L} \, dA = k_x(x_i - x) \, dA = d(Vy) = d(Lx) \qquad (23\text{-}11)$$

In the absence of heat effects and chemical reactions, Eq. (23-11) provides the basis for the theory of mass transfer in a packed tower.

Equation (23-11) contains terms for both gas-phase and liquid-phase resistances. These can be treated separately. The gas-phase equation is

$$\frac{k_y(y - y_i)}{(1 - y)_L} \, dA = d(Vy) \qquad (23\text{-}12)$$

This equation can be transformed for further use. First, since the actual area of transfer in a packed tower cannot readily be measured, the area dA is replaced by the product of the volume of the packed section and the area per unit volume. Also, if S is the cross-sectional area of the tower, the volume of the packing in the section of height dZ is $S \, dZ$, and

$$dA = aS \, dZ \qquad (23\text{-}13)$$

where a is the (unknown) area of interface per unit volume of packed section.

The term $d(Vy)$ also can be reduced to a more convenient form. Let V' be the flow rate of component B in moles per hour. Then

$$V = \frac{V'}{1 - y}$$

Since component B is not absorbed, V' is constant throughout the tower. Therefore,

$$d(Vy) = V'd\left(\frac{y}{1 - y}\right) = V'\frac{dy}{(1 - y)^2} = V\frac{dy}{1 - y} \qquad (23\text{-}14)$$

Substituting $d(Vy)$ from Eq. (23-14) and dA from Eq. (23-13) into Eq. (23-12) and dividing by S gives

$$\frac{k_y a(y - y_i) \, dZ}{(1 - y)_L} = \frac{V}{S}\frac{dy}{1 - y} = G_{My}\frac{dy}{1 - y} \qquad (23\text{-}15)$$

where G_{My}, the molal mass velocity of the gas in moles per unit area per hour, replaces V/S.

A corresponding treatment of the liquid-phase equation

$$k_x(x_i - x) \, dA = d(Lx)$$

gives

$$k_x a(x_i - x) \, dZ = G_{Mx}\frac{dx}{1 - x} \qquad (23\text{-}16)$$

where G_{Mx} is the molal liquid mass velocity. Note that G_{Mx} and G_{My} are based on the total cross section of the tower.

Factors k_y and a, and also factors k_x and a, are considered together as single combined quantities. They are evaluated as such by experimental tests on packed towers.

For practical use, Eqs. (23-15) and (23-16) must be integrated over the total depth of packing Z_T. The technique of this integration depends on three considerations: the shape of the equilibrium line, the change in composition of the stream in the tower, and the relative importance of the two resistances. In the most general situation, the equilibrium line is strongly curved, the inlet gas is concentrated and the outlet gas weak, and both resistances are important. In the simplest case the equilibrium line is straight, the changes in concentrations of both liquid and gas are small, and one or the other of the two resistances may be neglected. Cases of intermediate complexity are also encountered. In the following treatment the most general method is discussed first, followed by various simplified methods.

General case: $\Delta x \, \Delta y$ triangle   In the following construction it is assumed that $k_y a$ and $k_x a$ are known. These coefficients are discussed later. Figure 23-10 shows the operating line CD and the equilibrium curve AB. The operating line is plotted as shown in Example 18-1, from a knowledge of the design conditions. Consider the level in the column where the gas and liquid concentrations are y and x, respectively. These concentrations are the coordinates of point a lying on the operating line CD.

Since the rate of transfer is the same for both phases, $G_{My} \, dy/(1 - y) = G_{Mx} \, dx/(1 - x)$. It therefore follows from Eqs. (23-15) and (23-16) that $[k_y a/(1 - y)_L](y - y_i) \, dZ = k_x a(x_i - x) \, dZ$, or

$$\frac{y - y_i}{x_i - x} = \frac{\overline{(1 - y)_L} k_x a}{k_y a} \tag{23-17}$$

Solving Eq. (23-17) for y gives

$$y = -\frac{k_x a \overline{(1 - y)_L}}{k_y a} x + \left[y_i + \frac{k_x a \overline{(1 - y)_L}}{k_y a} x_i \right] \tag{23-18}$$

Equation (23-18) is that of a straight line, with a slope $-[k_x a \overline{(1 - y)_L}/k_y a]$, passing through points (x, y) and (x_i, y_i). Thus, if $\overline{(1 - y)_L}$, $k_x a$, and $k_y a$ are known, the slope can be calculated and the line ab drawn through point a. The coordinates of point b, which is the intersection of line ab with the equilibrium line, are x_i and y_i. The distance ac is the gas-resistance driving force $y - y_i$, and the distance bc is the liquid-resistance driving force $x_i - x$. The triangle abc may be called the $\Delta x \, \Delta y$ triangle. By constructing several such triangles along the operating line, either Δx or Δy can be determined graphically as a function of y or x.

To use these driving forces, the variables in Eq. (23-15) can be separated and the equation integrated graphically over Z_T, the length of the packed section:

$$\int_{y_a}^{y_b} \frac{\overline{(1 - y)_L} \, dy}{(1 - y)(y - y_i)} = \left(\frac{\overline{k_y a}}{G_{My}} \right) \int_0^{Z_T} dZ = \left(\frac{\overline{k_y a}}{G_{My}} \right) Z_T \tag{23-19}$$

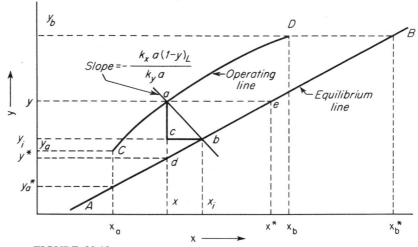

FIGURE 23-10
$\Delta x \, \Delta y$ triangle, packed tower.

It is assumed that Z is measured from the top down, that y_a is the concentration of the exit gas, and that y_b is the concentration of the inlet gas. The terms $\overline{(1 - y)}_L$ and $1 - y$ are retained under the integral sign, as they may vary appreciably with y. The quantity $\overline{k_y a / G_{My}}$ is assumed to be constant. This is an approximation, since G_{My} decreases from bottom to top because of the absorption of component A and $k_y a$ depends on the mass velocity of the gas and also decreases from bottom to top. The effects of changes in these factors tend to compensate, and the ratio $k_y a / G_{My}$ is nearly constant unless the inlet gas is very concentrated. The variation that does occur in $k_y a / G_{My}$ can be taken into account by using an arithmetic mean of the inlet and outlet quantities. This integration also assumes that the effectiveness of the tower is the same at all values of Z.

If the concentration of the gas phase changes appreciably over the length of the tower, $\overline{(1 - y)}_L$ and $k_y a$ vary also. These factors can be evaluated for the two ends of the tower and their arithmetic average used in calculating the slope of the $\Delta x \, \Delta y$ lines. Experimental data may be given either as $k_y a$ or as $k_y a / \overline{(1 - y)}_L$, and must be used accordingly. When $\overline{(1 - y)}_L$ is associated with $k_y a$, it is removed from the integral sign and used on the right-hand side of Eq. (23-19).

The liquid-side equation corresponding to Eq. (23-19) is

$$\int_{x_a}^{x_b} \frac{dx}{(1 - x)(x_i - x)} = \overline{\left(\frac{k_x a}{G_{Mx}}\right)} Z_T \tag{23-20}$$

In using this equation, $x_i - x$ is read from the $\Delta x \, \Delta y$ triangles and used to carry through a graphical or numerical integration of Eq. (23-20).

The choice between Eq. (23-19) and Eq. (23-20) is arbitrary. Both equations give the same result. The precision is better if the larger of the two driving forces is used.

EXAMPLE 23-2   A tower packed with 1-in. (25.4-mm) rings is to be designed to absorb sulfur dioxide from air by scrubbing the gas with water. The entering gas is 20 percent SO_2 by volume, and the leaving gas is to contain not more than 0.5 percent SO_2 by volume. The entering H_2O is SO_2-free. The temperature is 30°C and the total pressure is 2 atm. The water flow is to be twice the minimum. The air flow rate (SO_2-free basis) is to be 200 lb/ft²-h (976 kg/m²-h). What depth of packing is required?

The following equations are available[17] for the mass-transfer coefficients for absorption of SO_2 at 30°C in towers packed with 1-in. rings:

$$k_x a = 0.152 G_x^{0.82} \qquad \frac{k_y a}{(1 - y)_L} = 0.028 G_y^{0.7} G_x^{0.25}$$

where G_x and G_y are the mass velocities of liquid and vapor, respectively, in pounds per square foot per hour, based on the total tower cross section.

SOLUTION   The first step is to plot the equilibrium curve. Since the pressure is 2 atm, the curve of Example 17-4, which applies at 30°C and 1 atm, can be used if the ordinates are halved. This is because the partial pressure of the SO_2 is unchanged by the increase in pressure, and the mole fraction of SO_2 in the gas phase is now $\bar{p}_A/2$, which is one-half that at 1 atm for the same liquid concentration. The equilibrium curve is shown in Fig. 23-11. It is slightly curved at the lower end but is nearly straight when $x > 0.001$.

The minimum water rate is next calculated by using the equation of the operating line passing through points (x_b^*, y_b) and (x_a, y_a). Equation (23-6) can be written over the entire column as

$$G'_{Mx} \left(\frac{x_b}{1 - x_b} - \frac{x_a}{1 - x_a} \right) = G'_{My} \left(\frac{y_b}{1 - y_b} - \frac{y_a}{1 - y_a} \right)$$

where G'_{My} and G'_{Mx} are the molal mass velocities of sulfur dioxide–free air and water, respectively. From the conditions of the design:

$$y_b = 0.20 \qquad x_a = 0 \qquad y_a = 0.005 \qquad G'_{My} = \frac{200}{29} = 6.90 \text{ mol/ft}^2\text{-h}$$

From the equilibrium curve, when $y_b = 0.20$, $x_b^* = 0.0092$, and the minimum water rate G'_{Mx} is given by

$$G'_{Mx} \left(\frac{0.0092}{0.9908} - 0 \right) = 6.90 \left(\frac{0.20}{0.80} - \frac{0.005}{0.995} \right)$$

From this $G'_{Mx} = 182$ mol/ft²-h.

The actual water rate is twice the minimum, or 364 mol/ft²-h, and x_b is calculated from the equation

$$364 \frac{x_b}{1 - x_b} = 6.90 \left(\frac{0.20}{0.80} - \frac{0.005}{0.995} \right)$$

From this, $x_b = 0.00462$.

The equation for the operating line is

$$364 \frac{x}{1 - x} = 6.90 \left(\frac{y}{1 - y} - \frac{0.005}{0.995} \right)$$

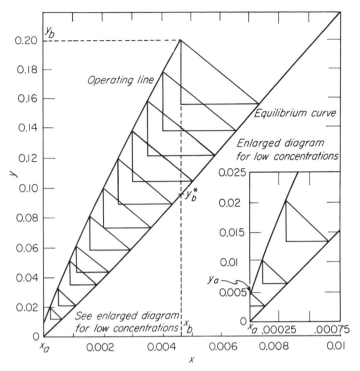

FIGURE 23-11
Equilibrium curve, operating line, $\Delta x \, \Delta y$ triangles for Example 23-2.

Neglecting x in the term $1 - x$, this equation can be solved for x.

$$x = 0.0189 \, \frac{y}{1 - y} - 0.00010$$

By assigning values to y between 0.005 and 0.200 and calculating the corresponding values of x, points on the operating line are obtained. The operating line is plotted in Fig. 23-11. It is slightly concave toward the x axis.

To calculate the mass-transfer coefficients $k_x a$ and $k_y a/(1 - y)_L$, the mass velocities of both gas and liquid streams must be known. They are obtained from the material balances over the tower. The SO_2-free air flow is 6.9 mol/ft²-h. The SO_2 entering with the gas is $(0.20/0.80)(6.9) \times 64.1 = 111$ lb/ft²-h. The total entering gas is, then, $200 + 111 = 311$ lb/ft²-h. The SO_2 leaving with the gas is $(0.005/0.995)(6.9) \times 64.1 = 2$ lb/ft²-h, and the total exit gas is $200 + 2 = 202$ lb/ft²-h. The SO_2 absorbed by the water is $111 - 2 = 109$ lb/ft²-h. The water fed to the top of the tower is $364 \times 18 = 6,550$ lb/ft²-h, and the strong liquor is $6,550 + 109 = 6,659$ lb/ft²-h.

The liquid resistance does not change appreciably from the top to the bottom, and the liquid coefficient can be calculated from the average mass velocity of the liquid, which is $6,550 + 109/2 = 6,605$ lb/ft²-h. Then

$$k_x a = 0.152 \times 6,605^{0.82} = 206$$

The gas resistance at the bottom of the tower is appreciably smaller than that at the top because of the change in G_y. The quantity $k_y a/(1 - y)_L$ is calculated for both ends of the packed section and the arithmetic averaged used as a constant. Then

$$\left[\frac{k_y a}{(1 - y)_L}\right]_b = 0.028 \times 6,659^{0.25} \times 311^{0.7} = 14.1$$

$$\left[\frac{k_y a}{(1 - y)_L}\right]_a = 0.028 \times 6,550^{0.25} \times 202^{0.7} = 10.4$$

The average of these is 12.3. The slope of the hypotenuse of the $\Delta x\, \Delta y$ triangles is $-206/12.3 = -16.7$.

In Fig. 23-11 are plotted the $\Delta x\, \Delta y$ triangles for a number of positions along the operating line. The intersections of the hypotenuses of the triangles with the equilibrium curve give the values of y_i corresponding to the assumed values of y. Table 23-2 shows the calculation of the ordinate $1/(1 - y)(y - y_i)$ for the numerical integration of

$$\int_{0.005}^{0.200} \frac{dy}{(1 - y)(y - y_i)}$$

The integral is equal to 12.72. Also,

$$\left[\frac{k_y a}{(1 - y)_L G_{My}}\right]_a = \frac{10.4}{6.90/0.995} = 1.50 \quad \text{and} \quad \left[\frac{k_y a}{(1 - y)_L G_{My}}\right]_b = \frac{14.1}{6.90/0.80} = 1.63$$

The average of these is 1.56; and, from Eq. (23-19), the height of the packed section is $Z_T = 12.72/1.56 = 8.15$ ft (2.49 m). ////

Table 23-2 NUMERICAL INTEGRATION FOR EXAMPLE 23-2

y	$1 - y$	y_i	$y - y_i$	$\dfrac{1}{(1 - y)(y - y_i)}$
0.005	0.995	0.0027	0.0023	437
0.01	0.99	0.0065	0.0035	288
0.02	0.98	0.0135	0.0065	157
0.03	0.97	0.0205	0.0095	108
0.05	0.95	0.0350	0.015	70.2
0.06	0.94	0.0430	0.017	62.5
0.08	0.92	0.0580	0.022	49.5
0.10	0.90	0.0730	0.027	41.1
0.12	0.88	0.0885	0.0315	36.0
0.14	0.86	0.105	0.035	33.2
0.16	0.84	0.1215	0.0385	30.9
0.18	0.82	0.138	0.042	29.1
0.20	0.80	0.1555	0.0445	28.1

EXAMPLE 23-3   Check the design of the tower of Example 23-2 with respect to flooding.

SOLUTION   For 1-in. rings, $F_p = 155$ ft^{-1} (Table 23-1). Also, at the bottom of the tower, $\overline{M}_y = 0.8 \times 29 + 0.2 \times 64 = 36.0$

$$\rho_y = \frac{36.0 \times 2 \times 273}{359 \times 303} = 0.181 \text{ lb/ft}^3 \qquad \rho_x = 62.3 \text{ lb/ft}^3$$

$$G_x = 1.85 \text{ lb/ft}^2\text{-s} \qquad G_y = 0.0864 \text{ lb/ft}^2\text{-s} \qquad \mu_x = 0.80 \text{ cP}$$

$$\frac{G_x}{G_y} \sqrt{\frac{\rho_y}{\rho_x}} = \frac{1.85}{0.0864} \sqrt{\frac{0.181}{62.3}} = 1.154$$

From Fig. 23-4

$$\frac{G_y^2 F_p (\mu_x)^{0.2}}{g_c \rho_y \rho_x} = 0.018$$

$$G_y = \sqrt{\frac{0.018 \times 32.174 \times 62.3 \times 0.181}{155 \times 0.80^{0.2}}}$$

$$= 0.209 \text{ lb/ft}^2\text{-s} = 756 \text{ lb/ft}^2\text{-h}$$

Since the maximum mass flow rate in the tower is 311 lb/ft^2-h, the loading of the tower is $311/762 = 0.41$, or 41 percent of the flooding velocity. This is safe and is also consistent with sound design.   ////

Simplified methods; overall coefficients   The general method described in the last section is applicable to both curved and straight equilibrium lines. It requires, however, a knowledge of the individual coefficients $k_y a$ and $k_x a$. The experimental determination of coefficients $k_y a$ and $k_x a$ is difficult, and they are not always known for the system and equipment of specific interest. When the equilibrium line is straight, overall coefficients, which are more easily determined by experiment, can be used. Overall coefficients also are simpler to use than individual coefficients because construction of the $\Delta x \, \Delta y$ triangles is unnecessary. The drift factor $\overline{(1 - y)_L}$ is assumed to be unity or is incorporated in the measured overall coefficient.

Overall coefficients are analogous to those used in heat transfer, but because of the change in driving force between phases, they can be defined from the standpoint of either the liquid phase or the gas phase. Each coefficient is based on a calculated overall driving force. Thus, referring to Fig. 23-10, by continuing the vertical line ac to a point d on the equilibrium curve the quantity y^* is found. This is the equilibrium composition of a gas corresponding to a liquid of concentration x. Since, in an actual tower, equilibrium is not achieved at any location, the quantity y^* has no physical significance in the tower itself but is a mathematical fiction. The overall driving force is defined as that measured by line segment ad, or $y - y^*$.

Likewise, if a liquid-side equation is to be used, a horizontal line ae, intersecting the equilibrium line at e, defines the liquid composition x^*, which is the composition a liquid would have if it were equilibrated with gas of composition y. The overall driving force is then represented by line segment ae, or $x^* - x$.

The overall gas-resistance coefficient K_y is defined by the equation

$$K_y \equiv \frac{dN_A/dA}{y - y^*} \tag{23-21}$$

and the overall liquid-resistance coefficient K_x by

$$K_x \equiv \frac{dN_A/dA}{x^* - x} \tag{23-22}$$

Comparison of Eqs. (23-21) and (23-9), assuming $\overline{(1 - y)_L} = 1$, shows that

$$\frac{1}{K_y} = \frac{y - y^*}{k_y(y - y_i)} = \frac{(y - y_i) + (y_i - y^*)}{k_y(y - y_i)} = \frac{1}{k_y} + \frac{y_i - y^*}{k_y(y - y_i)}$$

Eliminating $y - y_i$ by use of Eq. (23-17) gives

$$\frac{1}{K_y} = \frac{1}{k_y} + \frac{y_i - y^*}{k_x(x_i - x)}$$

Reference to Fig. 23-10 shows that $(y_i - y^*)/(x_i - x)$ is the slope of the linear equilibrium line. Call this slope m. Then

$$\frac{1}{K_y} = \frac{1}{k_y} + \frac{m}{k_x} \tag{23-23}$$

Also, dividing by a gives the overall coefficient $K_y a$ based on unit packed volume:

$$\frac{1}{K_y a} = \frac{1}{k_y a} + \frac{m}{k_x a} \tag{23-24}$$

A corresponding derivation gives

$$\frac{1}{K_x a} = \frac{1}{k_x a} + \frac{1}{m k_y a} = \frac{1}{m K_y a} \tag{23-25}$$

where $K_x a$ is the overall liquid-phase coefficient, based on unit packed volume. The units of both $K_y a$ and $K_x a$ are the same as those of $k_y a$ and $k_x a$, namely, moles/unit area-hour-mole fraction.

As shown by Eq. (23-25), the overall coefficients $K_y a$ and $K_x a$ are constant if $k_y a$, $k_x a$, and m are constant. When the equilibrium line is curved, the overall coefficients are not constant and they cannot be used safely in design unless the concentration range covered in the actual tower is nearly the same as that covered by the experiments on which the overall coefficients are based.

Equations (23-19) and (23-20) become, in terms of overall coefficients,

$$\int_{y_a}^{y_b} \frac{dy}{(1 - y)(y - y^*)} = \overline{\left(\frac{K_y a}{G_{My}}\right)} Z_T \tag{23-26}$$

and

$$\int_{x_a}^{x_b} \frac{dx}{(1 - x)(x^* - x)} = \overline{\left(\frac{K_x a}{G_{Mx}}\right)} Z_T \tag{23-27}$$

Either Eq. (23-26) or (23-27) may be used, depending on convenience. The integrals can be evaluated numerically, by reading $y - y^*$ or $x^* - x$ from a plot like Fig. 23-11.

EXAMPLE 23-4 Repeat Example 23-2, assuming a linear equilibrium line and using overall gas and liquid coefficients.

SOLUTION The equilibrium line is nearly straight between $y = 0.01$ and $y = 0.200$. The slope of the line between these points is

$$m = \frac{0.200 - 0.010}{0.0092 - 0.0005} = 21.8$$

The overall gas-side coefficient is, from Eq. (23-24),

$$\frac{1}{K_y a} = \frac{1}{k_y a} + \frac{m}{k_x a} = \frac{1}{12.3} + \frac{21.8}{206} = 0.081 + 0.106 = 0.187$$

$$K_y a = \frac{1}{0.187} = 5.35 \text{ lb mol/ft}^3\text{-h-mole fraction}$$

To use coefficient $K_y a$ the integral

$$\int_{0.005}^{0.200} \frac{dy}{(1 - y)(y - y^*)}$$

is evaluated. The overall driving force $y - y^*$ is read from the vertical distance between the operating and equilibrium lines of Fig. 23-11 for corresponding values of y and the integration performed numerically. The integral is found to be 5.74. At the top of the tower $(G_{My})_a = 6.90/0.995 = 6.93$ lb mol/ft²-h; at the bottom, $(G_{My})_b = 6.90/0.80 = 8.63$ lb mol/ft²-h. The average is 7.78. From Eq. (23-26)

$$Z_T = \frac{5.74 \times 7.78}{5.35} = 8.3 \text{ ft (2.53 m)}$$

The overall liquid-side coefficient is given by Eq. (23-25)

$$\frac{1}{K_x a} = \frac{1}{k_x a} + \frac{1}{mk_y a} = \frac{1}{206} + \frac{1}{21.8 \times 12.3} = \frac{1}{0.00858}$$

and

$$K_x a = 117 \text{ lb mol/ft}^3\text{-h-mole fraction}$$

To use this coefficient, the integral

$$\int_0^{0.00462} \frac{dx}{x^* - x}$$

is evaluated. Since x is small, the term $1 - x$ can be considered unity. The overall driving force $x^* - x$ is read from the horizontal distance between the operating and equilibrium lines and the integral evaluated numerically. The result is an area of 2.35. From Example 23-2, G_{Mx} is 364 lb mol/ft²-h, and from Eq. (23-27),

$$Z_T = \frac{2.35 \times 364}{117} = 7.3 \text{ ft (2.23 m)} \qquad ////$$

The calculation based on the gas phase and the larger driving force gives, in this example, a result close to that found by the general method.

HTU method The concept of a transfer unit, discussed in Chap. 11 for heat transfer, is useful for interpreting and correlating mass-transfer data. It is based on the idea of dividing the packed section into a number of contact units called *transfer units*.

The depth of packing required by a single unit is called the *height of one transfer unit* (HTU). Then the total height of the packed section is

$$Z_T = N_t H \qquad (23\text{-}28)$$

where N_t = number of transfer units
H = HTU

The number of transfer units in a column of total height Z_T ft is defined by any one of the following four equations.

$$N_{ty} = \int_{y_a}^{y_b} \frac{dy}{(1 - y)(y - y_i)} \qquad (23\text{-}29a)$$

$$N_{tOy} = \int_{y_a}^{y_b} \frac{dy}{(1 - y)(y - y^*)} \qquad (23\text{-}29b)$$

$$N_{tx} = \int_{x_a}^{x_b} \frac{dx}{(1 - x)(x_i - x)} \qquad (23\text{-}29c)$$

$$N_{tOx} = \int_{x_a}^{x_b} \frac{dx}{(1 - x)(x^* - x)} \qquad (23\text{-}29d)$$

Each of these numbers differs from the others in any given situation, and the choice among them is a matter of convenience. The differences between them are compensated for by the magnitudes of the corresponding HTUs. The quantities N_{tOy} and N_{tOx} are based on overall driving forces and N_{ty} and N_{tx} on individual driving forces.

These equations correspond to Eq. (11-38) for the number of transfer units in a heat exchanger.

Substitution of N_{ty} from Eq. (23-29a) in Eq. (23-19) gives

$$N_{ty} = \overline{\left(\frac{k_y a}{G_{My}}\right)} Z_T$$

Comparison of this equation with Eq. (23-28) shows that

$$H_y = \overline{\left(\frac{G_{My}}{k_y a}\right)} \qquad (23\text{-}30a)$$

Likewise, using Eqs. (23-20), (23-26), and (23-27) with Eqs. (23-29c), (23-29b), and (23-29d), respectively, gives

$$H_x = \overline{\left(\frac{G_{Mx}}{k_x a}\right)} \qquad (23\text{-}30b)$$

$$H_{Oy} = \overline{\left(\frac{G_{My}}{K_y a}\right)} \qquad (23\text{-}30c)$$

$$H_{Ox} = \overline{\left(\frac{G_{Mx}}{K_x a}\right)} \qquad (23\text{-}30d)$$

Quantities H_{Oy} and H_{Ox} are overall HTUs and H_y and H_x are individual HTUs. Each must be used only with its corresponding N_t.

Equations (23-30c) and (23-30d), based on overall coefficients, are analogous to Eqs. (11-42) and (11-43) for heat transfer. As discussed on page 304, a transfer unit may be viewed as a section in which the change in composition of one stream is numerically equal to the average driving force in the section. In mass transfer any one of four different driving forces may be chosen, giving four different numerical values of N_t and hence of H. The total depth of packing, of course, is the same regardless of what choice is made; thus, by Eq. (23-28), the product of N_t and H is always constant.

The overall HTUs are related to the individual HTUs as follows. Elimination of $K_y a$, $k_y a$, and $k_x a$ from Eq. (23-24) by Eqs. (23-30a), (23-30b), and (23-30c) gives

$$\frac{H_{Oy}}{G_{My}} = \frac{H_y}{G_{My}} + \frac{mH_x}{G_{Mx}}$$

or

$$H_{Oy} = H_y + \frac{mG_{My}}{G_{Mx}} H_x \qquad (23\text{-}31)$$

It is assumed here that $G_{Mx}/k_x a$ and $G_{My}/k_y a$ are constant. Likewise, eliminating $K_x a$, $k_y a$, and $k_x a$ from Eq. (23-25) gives

$$H_{Ox} = H_x + \frac{G_{Mx}}{mG_{My}} H_y = \frac{G_{Mx}}{mG_{My}} H_{Oy} \qquad (23\text{-}32)$$

The overall HTUs, H_{Oy} and H_{Ox}, are constant when the factors m, G_{Mx}, G_{My}, k_y, and k_x are constant throughout the tower. Straight operating and equilibrium lines demand constancy of these same factors, and the straighter the lines, the less the variation in the HTUs. The HTU method is most useful when the equilibrium line is straight and the curvature of the operating line negligible. Equations have been derived for correcting for moderate curvatures in these lines.[1,8]

ADVANTAGES OF HTU METHOD   The HTU is closely related to the mass-transfer coefficient, and the two quantities are essentially equivalent. The HTU is simpler to visualize, as its dimension is simply length and it is measured in feet or meters. The usual order of magnitude of this quantity is 0.5 to 5 ft (0.15 to 1.5 m). The units of the mass-transfer coefficient are more complex, and numerical magnitudes vary over wide limits. Also, since in packed towers $k_y a$ increases with G_{My} and $k_x a$ with G_{Mx}, the ratios $G_{My}/k_y a$ and $G_{Mx}/k_x a$, and therefore H_y and H_x, are nearly independent of the flow rates of liquid and gas. Most experimental data on packed towers are given in HTUs rather than in coefficients.

EXAMPLE 23-5   Repeat Example 23-2, using overall gas and liquid HTUs.

SOLUTION   *Overall gas-resistance HTU* N_{tOy} is defined by the equation

$$N_{tOy} = \int_{0.005}^{0.200} \frac{dy}{(1-y)(y-y^*)}$$

From Example 23-4, the above integral, and therefore N_{tOy}, is 5.74 units. From Eq. (23-30c)

$$H_{Oy} = \frac{G_{My}}{K_y a}$$

From Example 23-4, $G_{My} = 7.78$; $K_y a = 5.35$; and

$$H_{Oy} = \frac{7.78}{5.35} = 1.45 \text{ ft}$$

The total packed height is

$$Z_T = 1.45 \times 5.74 = 8.3 \text{ ft } (2.53 \text{ m})$$

Overall liquid-resistance HTU From Example 23-4, $N_{tOx} = 2.35$ units; $G_{Mx} = 364$; $K_x a = 113$; and, from Eq. (23-30d),

$$H_{Ox} = \frac{364}{113} = 3.11 \text{ ft } (0.95 \text{ m}) \qquad \text{and} \qquad Z_T = 3.11 \times 2.35 = 7.3 \text{ ft } (2.23 \text{ m})$$

Clearly, these results must agree with those of Example 23-4, since the same numbers are used in each example. The HTUs for both gas and liquid are of comparable magnitude, 1.45 and 3.11 ft, respectively, while the corresponding coefficients $K_y a$ and $K_x a$ are of different magnitude, 5.35 and 117, respectively.   ////

Gas resistance or liquid resistance controlling With a slightly soluble gas, m, the slope of the equilibrium line, is large, as the concentration in the gas phase at equilibrium is much greater than that in the liquid phase. From Eq. (23-32) it is apparent that unless the ratio G_{Mx}/G_{My} is also large, the term G_{Mx}/mG_{My} is small and H_{Ox} is nearly equal to H_x. The effect of the gas film can then be ignored, and the liquid resistance is said to control. The situation is analogous to that in heat transfer, when only one thermal resistance is important in comparison with the others and the overall coefficient is nearly equal to an individual coefficient.

For soluble gases, the slope m is small, since at equilibrium a low gas concentration corresponds to a large liquid concentration. Then, unless changes in other factors, such as the ratio G_{My}/G_{Mx}, compensate for the small magnitude of m, the term $mG_{My}H_x/G_{Mx}$ in Eq. (23-31) is small and H_{Oy} is nearly equal to H_y. Then conditions in the liquid are unimportant, and the gas resistance controls.

The conclusion that the gas or liquid resistance controls in a given situation must be drawn cautiously, for the conditions of the operation may be such that a low or high magnitude of m may be compensated by the magnitude of the gas-liquid ratio. For example, when a gas of low solubility is being absorbed, a high liquid rate is used to ensure adequate removal of the solute gas. Then G_{Mx}/G_{My} is large, so the ratio G_{Mx}/mG_{My} may be appreciable in spite of the large value of m, and, as shown by Eq. (23-32), both gas and liquid resistances are important. In most industrial operations both resistances are important, but in experimental work conditions are often deliberately fixed so that one resistance is negligible, making the other resistance easier to study.

Lean gases When y is small throughout the tower, the gas is called *lean*. Several simplifications appear in absorbing such gases. First, the factors $1 - y$ and $\overline{(1 - y)_L}$ are nearly unity, and can be so considered. Equation (23-26), for example, becomes

$$\int_{y_a}^{y_b} \frac{dy}{y - y^*} = \left(\frac{\overline{K_y a}}{G_{My}}\right) Z_T \tag{23-33}$$

Second, the individual quantities $k_y a$, $k_x a$, H_y, and H_x are constant throughout the equipment. Third, the liquid phase is usually also dilute, quantities such as $1 - x$ and $1 - y$ in the equation of the operating line [Eq. (23-6)] become unity, and $L = L'$ and $V = V'$. Thus L and V are constant, and, as shown by Eq. (23-5), the operating line is straight, with a slope of L/V. It is readily plotted by drawing a straight line through the two points (x_a, y_a) and (x_b, y_b).

LINEAR EQUILIBRIUM AND OPERATING LINES; LOGARITHMIC MEAN DRIVING FORCE When both the operating line and the equilibrium line are straight, Eq. (23-33) can be integrated formally. Since y and y^* are both linear with x, their difference is also linear. By using the same method used in Chap. 11 to derive the logarithmic mean temperature difference it is found that

$$G_{My}(y_b - y_a) \equiv K_y a \overline{(y - y^*)}_L Z_T = \frac{N_A}{S} \tag{23-34}$$

where
$$\overline{(y - y^*)}_L \equiv \frac{(y_b - y_b^*) - (y_a - y_a^*)}{\ln\left[(y_b - y_b^*)/(y_a - y_a^*)\right]} = \overline{\Delta y_L} \tag{23-35}$$

The logarithmic mean overall concentration difference is used in the same way the logarithmic mean overall temperature difference is used in heat transfer.

For the liquid phase, the corresponding equation is

$$G_{Mx}(x_b - x_a) \equiv K_x a \overline{(x^* - x)}_L Z_T = \frac{N_A}{S} \tag{23-36}$$

where
$$\overline{(x^* - x)}_L \equiv \frac{(x_b^* - x_b) - (x_a^* - x_a)}{\ln\left[(x_b^* - x_b)/(x_a^* - x_a)\right]} = \overline{\Delta x_L} \tag{23-37}$$

In terms of overall HTUs Eqs. (23-34) and (23-36) can be written

$$N_{tOy} = \frac{y_b - y_a}{\overline{(y - y^*)}_L} = \frac{y_b - y_a}{\overline{\Delta y_L}} \tag{23-38}$$

and
$$N_{tOx} = \frac{x_b - x_a}{\overline{(x^* - x)}_L} = \frac{x_b - x_a}{\overline{\Delta x_L}} \tag{23-39}$$

EXAMPLE 23-6 Repeat Example 23-2 assuming that the operating and equilibrium lines are straight.

SOLUTION From Example 23-4, $K_y a = 5.35$ lb mol/ft^3-h unit mole fraction and $G_{My} = 7.78$ lb mol/ft^2-h. Also, from Fig. 23-11, $y_a = 0.005$, $y_b = 0.200$, $y_a^* = 0$, $y_b^* = 0.0955$. From Eq. (23-35),

$$\overline{\Delta y_L} = \frac{(0.200 - 0.0955) - (0.005 - 0)}{\ln\,[(0.200 - 0.0955)/(0.005 - 0)]} = 0.0327$$

$$\frac{N_A}{S} = 7.78(0.200 - 0.005) = 1.52 \text{ lb mol/ft}^2\text{-h}$$

From Eq. (23-34)

$$Z_T = \frac{N_A/S}{K_y a\,\overline{\Delta y_L}} = \frac{1.52}{5.35 \times 0.0327} = 8.7 \text{ ft (2.65 m)} \qquad ////$$

COEFFICIENTS AND HTUs IN PACKED TOWERS

Although the construction of a packed tower is simple, the action in this equipment is complicated. The process consists of the flow of gas through a bed of solids, the distribution and flow of liquid over the same solids, the interaction of the two fluid streams, and the transfer of the solute component from one stream to the other. Chemical reactions in the liquid also may be important in many situations. It is difficult to measure either resistance independently of the other and so to analyze an overall HTU into the individual ones. The assumption of interfacial equilibrium is questionable, and the two-resistance theory is clearly an oversimplification. The penetration model may be more realistic, but relatively few coefficients have been measured for this model. Under these circumstances, research on tower coefficients has necessarily been highly empirical, although it has been guided by dimensional analysis.

An individual HTU depends upon both the coefficient k and the area per unit volume a. The variables that control the k of a single phase are, by assumption, the mass velocities of both phases, the viscosity, density, and diffusivity of the phase, and the size and shape of the packing. The same variables, with the exception of the diffusivity, affect the quantity a. It is also known that a is affected by the interfacial tension between the fluid phases, by the presence of surface-active agents in the liquid, and by the wetting characteristics of the packing.[11] The magnitude of a does not bear any simple relationship to the area or shape of the units of packing.

When the effects of surface factors are neglected, the HTU of one phase can be written

$$H = \psi(G_y, G_x, \mu, \rho, D_v, D_p, b_1, b_2, \ldots, b_n) \qquad (23\text{-}40)$$

where
$\quad G_y$ = mass velocity of gas
$\quad G_x$ = mass velocity of liquid
$\quad \mu$ = viscosity of phase
$\quad \rho$ = density of phase
$\quad D_v$ = diffusivity of phase
$\quad D_p$ = size of packing
$b_1, b_2, \ldots, b_n$ = lengths sufficient to define geometry of packing and its arrangement in tower

A dimensional analysis, conducted by the usual method, gives for a gas-phase HTU the result

$$\frac{H_y}{D_p} = \psi_y \left(\frac{D_p G_y}{\mu_y}, \frac{G_x}{G_y}, \frac{\mu_y}{\rho_y D_{vy}}, \frac{b_1}{D_p}, \frac{b_2}{D_p}, \cdots, \frac{b_n}{D_p} \right) \tag{23-41}$$

The analogous equation for an HTU based on the liquid is

$$\frac{H_x}{D_p} = \psi_x \left(\frac{D_p G_x}{\mu_x}, \frac{G_y}{G_x}, \frac{\mu_x}{\rho_x D_{vx}}, \frac{b_1}{D_p}, \frac{b_2}{D_p}, \cdots, \frac{b_n}{D_p} \right) \tag{23-42}$$

The Reynolds and Schmidt numbers again appear, as does the ratio of gas rate to liquid rate.

In spite of much careful experimental work on coefficients and HTUs, sufficient data are not available to use Eqs. (23-41) and (23-42) completely, and existing correlations account for only some of the variables. For example, the characteristic particle size D_p for packings of different shapes has not been chosen, and the effects of the geometrical shape factors, such as b_1/D_p, are unknown. Some of the more important correlations of HTU data are given in the following paragraphs.

Liquid-resistance HTU Two kinds of liquid resistances are encountered. For relatively insoluble gases that do not react chemically with the liquid, mass transfer is a physical process. An effective correlation has been found for such gases,[12] which may be written as the *dimensional* equation

$$H_x = \frac{1}{\alpha} \left(\frac{G_x}{\mu_x} \right)^n \left(\frac{\mu_x}{\rho_x D_{vx}} \right)^{0.5} \tag{23-43}$$

where α and n are constants, the magnitudes of which are given in Table 23-3.

Table 23-3 VALUES OF α AND n OF EQ. (23-43)† FOR VARIOUS PACKING MATERIALS AT 77°F[12]

Packing type	Packing size, in.	α	n
Rings	2	80	0.22
	1.5	90	0.22
	1	100	0.22
	0.5	280	0.35
	0.375	550	0.46
Saddles	1.5	160	0.28
	1	170	0.28
	0.5	150	0.28
Tile	3	110	0.28

† All quantities in Eq. (23-43) must be expressed in fps units if these values of α are used.

In the absorption of some common soluble gases in water, ionization and hydrolysis reactions between water and the solute occur which modify the liquid HTU. Chlorine, sulfur dioxide, and probably ammonia show this effect, and Eq. (23-43) does not apply. For example, the following equation for the absorption of sulfur dioxide in water at 77°F is available:[17]

$$k_x a = 0.152 G_x^{0.82} \qquad (23\text{-}44)$$

where G_x is the liquid mass velocity in pounds per square foot per hour.

The liquid-resistance HTU varies with temperature because of the effect of temperature on the Schmidt number. This temperature effect for liquids can be estimated by the equation

$$H_x = H_{x0} e^{-0.013(T-T_0)} \qquad (23\text{-}45)$$

where H_{x0} = HTU at T_0°F
$\qquad H_x$ = HTU at T°F

GAS-RESISTANCE HTU   Correlations of data for H_y are not well established. An equation of the following form has been suggested for correlating experimental data:[10b]

$$H_y = \beta \frac{G_y^q}{G_x^r} \qquad (23\text{-}46)$$

where β, q, and r are empirical constants. The exponents are not really constant, even for a specific system, but vary with the gas and liquid rates. The numerical values of the constants in Eq. (23-46) depend on the choice of units. The gas-phase resistance is not sensitive to temperature because the Schmidt number of gases does not vary greatly with temperature.

For absorption or desorption in towers packed with Raschig rings or Berl saddles, empirical equations are available[14] for k_x, k_y, and a, the area available for mass transfer per unit packed volume, from which the volumetric coefficients ka and the values of H_x and H_y can be estimated.

OVERALL COEFFICIENTS AND HTUS   Data on the overall quantities K_y, K_x, H_{Oy}, and H_{Ox} are given in standard references and handbooks.[10b, 13c] When available, these data are useful, provided they are used under conditions comparable to those under which they were determined. This is especially true when the equilibrium line is strongly curved, because of the importance of m in the relationships of overall and individual factors.

Absorption in plate columns   The calculation of the number of ideal stages in a plate tower is accomplished by the methods of Chap. 18. If either the operating line or the equilibrium line is curved, the McCabe-Thiele method may be used. If both lines are straight, either the McCabe-Thiele or the absorption-factor method [Eq. (18-27)] is applicable. The quotient of the number of ideal stages divided by the plate

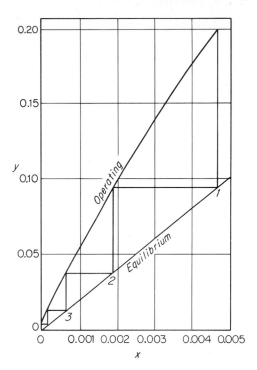

FIGURE 23-12
McCabe-Thiele construction for Example 23-7.

efficiency gives the number of actual plates. Plate efficiencies are discussed in Chaps. 19 and 22. Efficiencies in absorption operations are usually lower than in distillation, chiefly because of the greater viscosity of the liquid and the strong effect of solubility, as measured by the value of m.[13]

EXAMPLE 23-7 Assuming a plate efficiency of 20 percent, how many actual plates are required to accomplish the absorption of sulfur dioxide called for in Example 23-2?

SOLUTION   The McCabe-Thiele construction is shown in Fig. 23-12; 3.8 ideal stages are needed. The actual number of stages is $3.8/0.20 = 19$.

The number of ideal stages also can be approximately estimated from Eq. (18-27), although neither the operating nor the equilibrium line is quite straight. Then

$$y_a = 0.005 \qquad y_b = 0.200 \qquad y_a^* = 0 \qquad y_b^* = 0.0955$$

Substituting these numbers in Eq. (18-27) gives, for the number of ideal stages,

$$N_p = \frac{\log\left[(0.200 - 0.0955)/0.005\right]}{\log\left[(0.200 - 0.005)/0.0955\right]} = 4.26$$

The actual number of plates is $4.26/0.20$, or $21+$. The assumption that the operating and equilibrium lines are straight introduces a positive error of $2+$ plates, at 10 percent.   ////

Desorption This operation, also called *stripping*, is the reverse of absorption and is treated by the same methods. The only differences between absorption and desorption are that, in the latter, (1) the driving potentials Δx and Δy are reversed, and are given by $x - x_i$ and $y_i - y$, respectively, and (2) the operating line lies below the equilibrium line, as in distillation.

MASS TRANSFER IN EXTRACTION COLUMNS

The principles of liquid extraction in countercurrent packed columns are identical with those for absorption in the same kind of equipment. Extraction in spray columns is also treated in the same way. The mass-transfer coefficients $K_x a$, $K_y a$, $k_x a$, and $k_y a$, the HTUs H_{Ox}, H_{Oy}, H_x, and H_y, and the techniques for integrating over the length of the packed section are all the same in extraction as in the other operations. If the operating line is curved, it can be located by the method of Fig. 20-18. When the equilibrium and operating lines are straight, logarithmic mean driving forces, as used in Eqs. (23-34) to (23-39), are applicable.

Coefficients in extraction columns Mass-transfer coefficients and HTUs in extraction equipment depend on the same variables as in liquid-gas contact equipment, and, in addition, they are also sensitive to the details of distribution of the dispersed phase throughout the mass of the continuous one. It is even more difficult to separate the overall coefficients or HTUs into the individual factors than in gas-liquid contacts, and the overall coefficients $K_x a$ and $K_y a$ and the overall HTUs H_{Ox} and H_{Oy} are usually reported. Customarily, these overall quantities are based on the aqueous phase if either the solvent or the diluent is water.

Because of the complexity of the action in extraction equipment and the many factors involved, no basic correlations of extraction coefficients are available. Many data have been reported. They are given in standard references,[9,13,15] which should be consulted for coefficients for specific systems and designs. The following general statements may be made:

1 Coefficients in spray columns are more sensitive to variations in *a* than in *K* or *k*, and this parameter depends on many factors. The size of the drops, the number of drops per unit volume of continuous phase, and the time of residence of the drops in the extractor are especially important. Drops rise (or fall) in general agreement with the laws of terminal velocities discussed in Chap. 7.

2 The relative importance of the two individual resistances depends, as in gas-liquid contacts, on the slope of the equilibrium curve, in accordance with Eqs. (23-24) and (23-25), but this influence of solubility is often obscured by changes in factor *a*. HTUs vary linearly with the factor G_{Mx}/mG_{My} as shown by Eqs. (23-31) and (23-32).[6]

3 End effects are important, especially in short columns. Extraction in long columns is less effective per unit of total height than in short ones. Data on short columns should not, therefore, be extrapolated to tall columns. Figure 23-13 shows how end effects in towers are found and evaluated. In this figure the

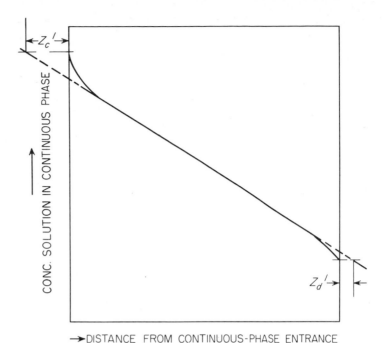

FIGURE 23-13
End effects in extraction column. (After Kreager and Geankopolis.[5])

concentration of solute in the continuous phase is plotted against the distance from the continuous-phase entrance. Near this entrance the rate of extraction is abnormally large, and this is also true, to a lesser extent, near the dispersed-phase inlet. In the middle of the column, the rate, as measured by the slope of the concentration-length line, is constant. If the straight line applying at the center of the column is extrapolated at both ends until it intersects horizontal lines through the terminal concentrations, two lengths Z_c' and Z_d' are determined, as shown in Fig. 23-13. These lengths are fictitious column heights equivalent to the inlet end effects of the continuous and dispersed phases.

4 In packed columns the coefficients are larger for small packings than for large ones, and packed columns give much larger coefficients than unpacked spray columns operated under the same conditions.[6]

5 In packed columns it is important which phase is chosen as the dispersed phase. Preferably, this phase should not wet the packing. If it does so, it tends to flow along the surface of the packing as rivulets rather than as small, discrete droplets and the area factor *a* is small.

6 Except for the condition of paragraph 5, a general rule is to choose the phase flowing in larger volume as the dispersed phase.

7 Especially in spray columns there may be considerable recirculation in the continuous phase between top and bottom.[3] This flattens the operating line and distorts the material balances. If recirculation is neglected in evaluating coefficients, apparent, rather than true, coefficients are found, and these should not be used for columns of sizes or shapes much different from those used in measuring the coefficients.

8 Small quantities of impurities may concentrate at the interface between the two phases and form an additional resistance to mass transfer.[16] There is no way of predicting such effects.

RECAPITULATION OF MASS-TRANSFER-CALCULATION METHODS

In this and previous chapters countercurrent contacting devices have been discussed from the standpoint of the individual operations of absorption, desorption, rectification, leaching, and liquid extraction. General methods, based on a common approach, have been applied in turn to each of these operations. The methods for calculating the number of stages or the height of the unit may now be summarized as follows.

Two types of contacting plant are important, stage contacts and differential contacts. Each is treated by its own methods. The choice of method within each class of unit depends on the curvature, or lack of curvature, of two lines, the equilibrium line and the operating line.

In stage contacts, the absorption-factor method, as given by Eqs. (18-27) and (18-30), is recommended when both lines are straight; the McCabe-Thiele construction is suggested when the equilibrium line is curved and the operating line is straight or when it can easily be plotted by use of material balances; and the Ponchon-Savarit method is available when both lines are curved. In the last situation, an alternative method may be used, namely, the McCabe-Thiele method, utilizing operating lines located by the Ponchon-Savarit construction.

In differential-contact plants, when both lines are straight, overall coefficients or overall HTUs, in combination with logarithmic mean driving forces, as given by Eqs. (23-34), (23-36), (23-38), and (23-39), are used. When the equilibrium line is straight and the operating line curved, graphical integration based on overall mass-transfer coefficients K_x and K_y or the overall HTUs H_{Ox} and H_{Oy} are used. If the operating line is strongly curved, the overall mass-transfer coefficient is, in principle, the more accurate. When the equilibrium line is strongly curved, numerical integration based on individual coefficients k_x and k_y or individual HTUs H_x and H_y should be used whenever data on these individual factors are available or can be estimated. Otherwise K or H_O factors, determined under conditions approaching those for the design, must be used.

If in one part of the diagram the lines are straight and in another part of the same diagram they are curved, the calculation may be made in two parts, using the simpler methods where the lines are straight and more complicated methods where the lines are curved. The results from the two calculations are then added.

SYMBOLS

A	Area of interface between phases, ft² or m²
A_L	Factor in Fig. 23-5
a	Area of interface per unit packed volume, ft²/ft³ or m²/m³; a_v, area of dry packing per unit packed volume
$b_1, b_2, \ldots, b_n$	Defining lengths for packed bed, ft or m
D_p	Linear dimension of unit of packing, ft or m
D_v	Diffusivity, ft²/h, m²/h, or cm²/s; D_{vx}, in liquid; D_{vy}, in gas
F_p	Packing factor, Fig. 23-4, ft⁻¹
G	Mass velocity, based on total tower cross section, lb/ft²-h or kg/m²-h; G_x, of liquid stream; G_y, of gas stream
G_M	Molal mass velocity based on total tower cross section, lb mol/ft²-h or g mol/m²-h; G_{Mx}, of liquid stream; G_{My}, of gas stream; G'_{Mx}, G'_{My}, based on solute-free streams
g_c	Newton's-law proportionality factor, 32.174 ft-lb/lb$_f$-s²
H	Height of transfer unit, ft or m; H_{Ox}, overall, based on liquid phase; H_{Oy}, overall, based on gas phase; H_x, individual, based on liquid phase; H_{x0}, at temperature T_0; H_y, individual, based on gas phase
h	Individual heat-transfer coefficient, Btu/ft²-h-°F or W/m²-°C; h_1, for phase 1; h_2, for phase 2
K	Overall mass-transfer coefficient, lb mol/ft²-h-unit mole fraction or g mol/m²-h-unit mole fraction; K_x, based on liquid phase; K_y, based on gas phase
k	Individual mass-transfer coefficient, lb mol/ft²-h-unit mole fraction or g mol/m²-h-unit mole fraction; k_x, for liquid phase; k_y, for gas phase
L	Molal flow rate of liquid, mol/h; L_a, at liquid inlet; L_b, at liquid outlet; L', molal flow rate of solute-free liquid
$(L'/V')_{min}$	Minimum liquid-gas ratio, based on solute-free streams
M	Molecular weight; $\bar{M}_y$, average molecular weight of gas
m	Slope of equilibrium curve, dy_e/dx_e
N_A	Rate of mass transfer of component A across interface between phases, lb mol/h or g mol/h
N_p	Number of ideal stages
N_t	Number of transfer units; N_{tOx}, overall, based on liquid phase; N_{tOy}, overall, based on gas phase; N_{tx}, individual, liquid phase; N_{ty}, individual, gas phase
n	Exponent in Eq. (23-43)
p	Total pressure, atm; p_a, at inlet; p_b, at outlet; $\bar{p}$, partial pressure
q	Rate of heat transfer, Btu/h or W; also exponent in Eq. (23-46)
r	Exponent in Eq. (23-46)
S	Cross-sectional area of tower, ft² or m²
T	Temperature, °F or °C; T_c, average bulk temperature of cool phase; T_h, of warm phase; T_i, interface temperature; T_0, reference temperature
V	Molal flow rate of gas, mol/h; V_a, at exit; V_b, at entrance; V', molal flow rate of solute-free gas
$\bar{V}_{sy}$	Superficial gas velocity, based on empty tower, ft/s or m/s
x	Mole fraction component A (solute) in liquid; x_a, at liquid inlet; x_b, at liquid outlet; x_e, at equilibrium; x_i, at gas-liquid interface; x_a^*, equilibrium concentration corresponding to gas-phase concentration y_a; x_b^*, equilibrium concentration corresponding to gas-phase concentration y_b
y	Mole fraction component A (solute) in gas; y_a, at gas outlet; y_b, at gas inlet; y_e, at equilibrium; y_i, at gas-liquid interface; y_a^*, equilibrium concentration corresponding to liquid-phase concentration x_a; y_b^*, equilibrium concentration corresponding to liquid-phase concentration x_b
$\overline{(1 - y)_L}$	Phase-drift factor in unicomponent diffusion

Z	Vertical distance below top of packing, ft or m; Z_T, total depth of packed section
Z'	Equivalent column height for end effects in liquid extraction, ft or m; Z'_c, at continuous-phase entrance; Z'_d, at dispersed-phase entrance

Greek letters

α	Constant in Eq. (23-43)
β	Constant in Eq. (23-46)
Δx	Concentration difference over liquid resistance, $x_i - x$, mole fraction of component A; $\overline{\Delta x_L}$, overall logarithmic mean concentration difference over liquid resistance, defined by Eq. (23-37)
Δy	Concentration difference over gas resistance, $y - y_i$, mole fraction of component A; $\overline{\Delta y_L}$, overall logarithmic mean concentration difference over gas resistance, defined by Eq. (23-35)
ε	Porosity, or void fraction, in packed section
μ	Viscosity, lb/ft-h or cP; μ_x, of liquid; μ_y, of gas
ρ	Density, lb/ft^3 or kg/m^3; ρ_x, of liquid; ρ_y, of gas
ψ	Function; ψ_x, in Eq. (23-42); ψ_y, in Eq. (23-41)

PROBLEMS

23-1 A plant design calls for an absorber which is to recover 95 percent of the acetone in an air stream, using water as the absorbing liquid. The entering air contains 14 mole percent acetone. The absorber is to operate at 80°F and 1 atm, and is to produce a product containing 7.0 mole percent acetone. The water fed to the tower contains 0.02 mole percent acetone. The tower is to be designed to operate at 50 percent of the flooding velocity. (*a*) How many pounds per hour of water must be fed to the tower if the gas rate is 500 ft^3/min, measured at 1 atm and 32°F? (*b*) How many transfer units are needed, based on the overall gas-phase driving force? (*c*) If the tower is packed with 1-in. Raschig rings, what should be the packed height?

For equilibrium Assume that $\bar{p}_A = p_A \gamma_A x$, where $\ln \gamma_a = 1.95(1 - x)^2$.
For mass transfer Use Eq. (23-43) for H_x and calculate H_y from $H_y = 1.01 G_y^{0.31}/G_x^{0.33}$. The vapor pressure of acetone at 80°F is 0.33 atm.

23-2 An absorber is to recover 99 percent of the ammonia in the air-ammonia stream fed to it, using water as the absorbing liquid. The ammonia content of the air is 30 mole percent. Absorber temperature is to be kept at 30°C by cooling coils; pressure is 1 atm. (*a*) What is the minimum water rate? (*b*) For a water rate 50 percent greater than the minimum, how many overall gas-phase transfer units are needed?

23-3 A soluble gas is absorbed in water, using a packed tower. The equilibrium relationship may be taken as $Y_e = 0.06 X_e$, where Y_e and X_e are ratios of moles of solute to moles of inert component. Terminal conditions are:

	Top	Bottom
X	0	0.08
Y	0.001	0.009

If $H_x = 0.24$ m and $H_y = 0.36$ m, what is the height of the packed section?

23-4 How many ideal stages are required for the tower of Prob. 23-3?

23-5 A mixture of 5 percent butane and 95 percent air is absorbed in a sieve-plate tower containing eight ideal plates. The absorbing liquid is a heavy, nonvolatile oil having a molecular weight of 250 and a specific gravity of 0.90. The absorption takes place at 1 atm and 15°C. The butane is to be recovered to the extent of 95 percent. The vapor pressure of butane at 15°C is 1.92 atm, and liquid butane has a density of 580 kg/m³ at 15°C. (*a*) Calculate the cubic meters of fresh absorbing oil per cubic meter of butane recovered. (*b*) Repeat, on the assumption that the total pressure is 3 atm and that all other factors remain constant. Assume that Raoult's and Dalton's laws apply.

23-6 An absorption column is fed at the bottom with a gas containing 5 percent benzene and 95 percent air. At the top of the column a nonvolatile absorption oil is introduced, which contains 0.2 percent benzene by weight. Other data are:

Feed, 4,000 lb of absorption oil per hour

Total pressure, 1 atm

Temperature (constant), 80°F

Molecular weight of absorption oil, 230

Vapor pressure of benzene at 80°F, 106 mm Hg

Volume of entering gas, 40,000 ft³/h

Tower packing, Raschig rings, $\frac{1}{2}$ in. nominal size

Fraction of entering benzene absorbed, 0.90

Overall absorption coefficient based on gas phase $K_y a$, 1.8 lb mol/ft³-h-unit mole fraction

Mass velocity of entering gas, 300 lb/ft²-h

Calculate the height and diameter of the packed section of this tower and recommend the total height of the tower.

REFERENCES

1 Colburn, A. P.: *Ind. Eng. Chem.,* **33**:459 (1941).
2 Eckert, J. S.: *Chem. Eng. Prog.,* **66**:(3):39 (1970).
3 Gier, T. E., and J. E. Hougen: *Ind. Eng. Chem.,* **45**:1362 (1953).
4 Gupta, A. S., and G. Thodos: *Chem. Eng. Prog.,* **58**(7):58 (1962); **58**(10):62 (1962).
5 Kreager, R. M., and C. J. Geankopolis: *Ind. Eng. Chem.,* **45**:2156 (1953).
6 Leibson, I., and R. B. Beckmann: *Chem. Eng. Prog.,* **49**:405 (1953).
7 Leva, M.: "Tower Packings and Packed Tower Design," 2d ed., p. 67, U.S. Stoneware Company, Akron, Ohio, 1953.
8 Othmer, D. F., and E G. Scheibel: *Trans. AIChE,* **38**:339 (1942).
9 Perry, J. H.: "Chemical Engineers' Handbook," 4th ed., pp. **14**-66 to **14**-69, McGraw-Hill, New York, 1963.
10 Perry, J. H.: "Chemical Engineers' Handbook," 5th ed., McGraw-Hill, New York, 1973; (*a*) pp. **18**-22 to **18**-24, (*b*), pp. **18**-30 to **18**-54.
11 Sherwood, T. K., and F. A. L. Holloway: *Trans. AIChE,* **36**:21 (1949).
12 Sherwood, T. K., and F. A. L. Holloway: *Trans. AIChE,* **36**:39 (1940).
13 Sherwood, T. K., and R. L. Pigford: "Absorption and Extraction," 2d ed., McGraw-Hill, New York, 1952; (*a*) p. 161, (*b*) p. 245, (*c*) pp. 278–314.
14 Treybal, R. E.: "Mass-Transfer Operations," 2d ed., pp. 165–166, McGraw-Hill, New York, 1968.
15 Treybal, R. E.: "Liquid Extraction," 2d ed., McGraw-Hill, New York, 1963.
16 West, F. B., et al.: *Ind. Eng. Chem.,* **44**:625 (1952).
17 Whitney, R. P., and J. E. Vivian: *Chem. Eng. Prog.,* **45**:323 (1949).

24

HUMIDIFICATION OPERATIONS

Humidification and dehumidification involve the transfer of material between a pure liquid phase and a fixed gas which is insoluble in the liquid. These operations are somewhat simpler than absorption and stripping operations, for when the liquid contains only one component, there are no concentration gradients and no resistance to transfer in the liquid phase. On the other hand, both heat transfer and mass transfer are important and influence one another. In previous chapters they have been treated separately; here and in drying of solids (discussed in Chap. 25) they occur together, and concentration and temperature change simultaneously.

Definitions In humidification operations, especially as applied to the system air-water, a number of rather special definitions are in common use. The usual basis for engineering calculations is a unit mass of vapor-free gas, where "vapor" means the gaseous form of the component which is also present as a liquid and "gas" is the component which is present only in gaseous form. In this discussion a basis of a unit mass of vapor-free gas is used. In the gas phase the vapor will be referred to as component A and the fixed gas as component B. Because the properties of a gas-vapor mixture vary with total pressure, the pressure must be fixed. Unless otherwise specified, a total pressure of 1 atm is assumed. Also, it is assumed that mixtures of gas and vapor follow the ideal-gas laws.

Humidity $\mathcal{H}$ is the mass of vapor carried by a unit mass of vapor-free gas.† So defined, humidity depends only on the partial pressure of the vapor in the mixture when the total pressure is fixed. If the partial pressure of the vapor is $\bar{p}_A$ atm, the molal ratio of vapor to gas is 1 atm at $\bar{p}_A(1 - \bar{p}_A)$. The humidity is therefore

$$\mathcal{H} = \frac{M_A \bar{p}_A}{M_B(1 - \bar{p}_A)} \tag{24-1}$$

where M_A and M_B are the molecular weights of components A and B, respectively.

The humidity is related to the mole fraction in the gas phase by the equation

$$y = \frac{\mathcal{H}/M_A}{1/M_B + \mathcal{H}/M_A} \tag{24-2}$$

Since $\mathcal{H}/M_A$ is usually small compared with $1/M_B$, y may often be considered to be directly proportional to $\mathcal{H}$.

Saturated gas is gas in which the vapor is in equilibrium with the liquid at the gas temperature. By Dalton's law, the partial pressure of vapor in saturated gas equals the vapor pressure of the liquid at the gas temperature. If $\mathcal{H}_s$ is the saturation humidity and p_A the vapor pressure of the liquid,

$$\mathcal{H}_s = \frac{M_A p_A}{M_B(1 - p_A)} \tag{24-3}$$

Relative humidity $\mathcal{H}_R$ is defined as the ratio of the partial pressure of the vapor to the vapor pressure of the liquid at the gas temperature. It is usually expressed on a percentage basis, so 100 percent humidity means saturated gas and 0 percent humidity means vapor-free gas. By definition

$$\mathcal{H}_R = 100 \frac{\bar{p}_A}{p_A} \tag{24-4}$$

Percentage humidity $\mathcal{H}_A$ is the ratio of the actual humidity $\mathcal{H}$ to the saturation humidity $\mathcal{H}_s$ at the gas temperature, also on a percentage basis, or

$$\mathcal{H}_A = 100 \frac{\mathcal{H}}{\mathcal{H}_s} = 100 \frac{\bar{p}_A/(1 - \bar{p}_A)}{p_A/(1 - p_A)} = \mathcal{H}_R \frac{1 - p_A}{1 - \bar{p}_A} \tag{24-5}$$

At all humidities other than 0 or 100 percent, the percentage humidity is less than the relative humidity.

Humid heat c_s is the Btu necessary to increase the temperature of 1 lb or 1 g of gas, plus whatever vapor it may contain, by 1°F or 1°C. Thus

$$c_s = c_{pB} + c_{pA}\mathcal{H} \tag{24-6}$$

where c_{pB} and c_{pA} are the specific heats of gas and vapor, respectively.

---

† A script letter $\mathcal{H}$ is used for humidity, even though it represents the magnitude of the variable, to avoid confusion with enthalpy H.

Humid volume v_H is the total volume of a unit mass of vapor-free gas, plus whatever vapor it may contain, at 1 atm and the gas temperature. From the gas laws, v_H in fps units is related to humidity and temperature by the equation

$$v_H = \frac{359T}{492}\left(\frac{1}{M_B} + \frac{\mathscr{H}}{M_A}\right) \tag{24-7a}$$

where T is the absolute temperature in degrees Rankine. In SI units the equation is

$$v_H = \frac{0.0224T}{273}\left(\frac{1}{M_B} + \frac{\mathscr{H}}{M_A}\right) \tag{24-7b}$$

where v_H is in cubic meters per gram mole and T is in degrees Celsius. For vapor-free gas $\mathscr{H} = 0$, and v_H is the specific volume of the fixed gas. For saturated gas $\mathscr{H} = \mathscr{H}_s$, and v_H becomes the *saturated volume.*

Dew point is the temperature to which a vapor-gas mixture must be cooled (at constant humidity) to become saturated. The dew point of a saturated gas phase equals the gas temperature.

Total enthalpy H_y is the enthalpy of a unit mass of gas, plus whatever vapor it may contain. To calculate H_y two reference states must be chosen, one for gas and one for vapor. Let T_0 be the datum temperature chosen for both components, and base the enthalpy of component B on liquid B at T_0. Let the temperature of the gas be T and the humidity $\mathscr{H}$. The total enthalpy is the sum of three items; the sensible heat of the vapor, the latent heat of the liquid at T_0, and the sensible heat of the vapor-free gas. Then

$$H_y = c_{pB}(T - T_0) + \mathscr{H}\lambda_0 + c_{pA}\mathscr{H}(T - T_0) \tag{24-8}$$

where λ_0 is the latent heat of the liquid at T_0. From Eq. (24-6) this becomes

$$H_y = c_s(T - T_0) + \mathscr{H}\lambda_0 \tag{24-9}$$

Adiabatic-saturation temperature   Consider the process in Fig. 24-1. Gas, with an initial humidity $\mathscr{H}$ and temperature T, flows continuously through the spray chamber A. The chamber is lagged, so the process is adiabatic. Liquid is circulated by pump B from the reservoir in the bottom of the spray chamber through sprays C and back to the reservoir. The gas passing through the chamber is cooled and humidified. The temperature of the liquid reaches a definite steady-state temperature T_s called the *adiabatic-saturation temperature.* Unless the entering gas is saturated, the adiabatic-saturation temperature is lower than the temperature of the entering gas. If contact between liquid and gas is sufficient to bring the liquid and the exit gas into equilibrium, the gas leaving the chamber is saturated at temperature T_s. Since the liquid evaporated into the gas is lost from the chamber, makeup liquid is needed. This is supplied to the reservoir at temperature T_s.

An enthalpy balance can be written over this process. Pump work is neglected, and the enthalpy balance is based on temperature T_s as a datum. Then the enthalpy of the makeup liquid is zero, and the total enthalpy of the entering gas equals that of

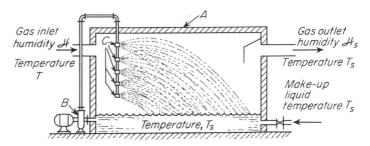

FIGURE 24-1
Adiabatic saturator: A, spray chamber; B, circulating pump; C, sprays.

the leaving gas. Since the latter is at datum temperature, its enthalpy is simply $\mathscr{H}_s\lambda_s$, where $\mathscr{H}_s$ is the saturation humidity and λ_s is the latent heat, both at T_s. From Eq. (24-9) the total enthalpy of the entering gas is $c_s(T - T_s) + \mathscr{H}\lambda_s$, and the enthalpy balance is

$$c_s(T - T_s) + \mathscr{H}\lambda_s = \mathscr{H}_s\lambda_s$$

or

$$\frac{\mathscr{H} - \mathscr{H}_s}{T - T_s} = -\frac{c_s}{\lambda_s} = -\frac{c_{pB} + c_{pA}\mathscr{H}}{\lambda_s} \qquad (24\text{-}10)$$

Humidity chart A convenient diagram showing the properties of mixtures of a permanent gas and a condensable vapor is the humidity chart. A chart for mixtures of air and water at 1 atm is shown in Fig. 24-2. Many forms of such charts have been proposed. Figure 24-2 is based on the Grosvenor[3] chart.

On Fig. 24-2 temperatures are plotted as abscissas and humidities as ordinates. Any point on the chart represents a definite mixture of air and water. The curved line marked "100%" gives the humidity of saturated air as a function of air temperature. By using the vapor pressure of water the coordinates of points on this line are found from Eq. (24-3). Any point above and to the left of the saturation line represents a mixture of saturated air and liquid water. This region is important only in checking fog formation. Any point below the saturation line represents under-saturated air, and a point on the temperature axis represents dry air. The curved lines between the saturation line and the temperature axis marked in even percents represent mixtures of air and water of definite percentage humidities. As shown by Eq. (24-5), linear interpolation between the saturation line and the temperature axis can be used to locate the lines of constant percentage humidity.

The slanting lines running upward and to the left of the saturation line are called *adiabatic-cooling lines*. They are plots of Eq. (24-10), each drawn for a given constant value of the adiabatic-saturation temperature. For a given value of T_s, both $\mathscr{H}_s$ and λ_s are fixed, and the line of $\mathscr{H}$ vs. T can be plotted by assigning values to $\mathscr{H}$ and calculating corresponding values of T. Inspection of Eq. (24-10) shows that the slope of an adiabatic-cooling line, if drawn on truly rectangular coordinates, is $-c_s/\lambda_s$, and, by Eq. (24-6), this slope depends on the humidity. On rectangular coordinates, then, the adiabatic-cooling lines are neither straight nor parallel. In Fig. 24-2 the ordinates

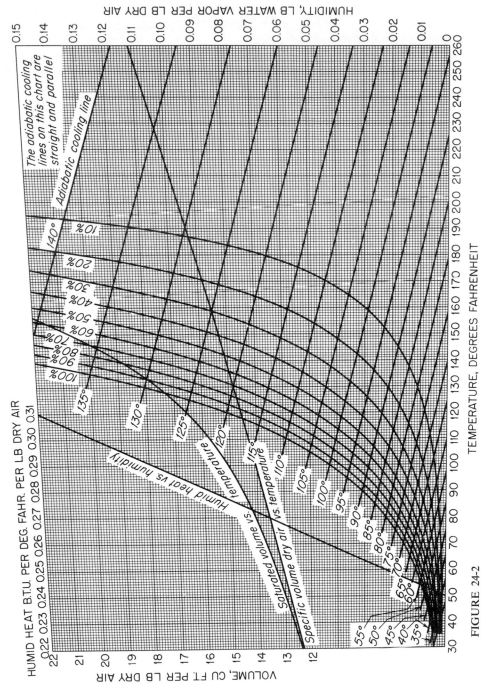

FIGURE 24-2
Humidity chart. Air-water at 1 atm.

are sufficiently distorted to straighten the adiabatics and render them parallel, so interpolation between them is easy. The ends of the adiabatics are identified with the corresponding adiabatic-saturation temperatures.

Lines are shown on Fig. 24-2 for the specific volume of dry air and the saturated volume. Both lines are plots of volume against temperature. Volumes are read on the scale at the right. Coordinates of points on these lines are calculated by use of Eq. (24-7a). Linear interpolation between the two lines, based on percentage humidity, gives the humid volume of unsaturated air. Also, the relation between the humid heat c_s and humidity is shown as a line on Fig. 24-2. This line is a plot of Eq. (24-6). The scale for c_s is at the top of the chart.

Use of humidity chart The usefulness of the humidity chart as a source of data on a definite air-water mixture can be shown by reference to Fig. 24-3, which is a portion of the chart of Fig. 24-2. Assume, for example, that a given stream of undersaturated air is known to have a temperature T_1 and a percentage humidity $\mathscr{H}_{A1}$. Point a represents this air on the chart. This point is the intersection of the constant-temperature line for T_1 and the constant-percentage-humidity line for $\mathscr{H}_{A1}$. The humidity $\mathscr{H}_1$ of the air is given by point b, the humidity coordinate of point a. The dew point is found by following the constant-humidity line through point a to the left to point c on the 100 percent line. The dew point is then read at point d on the temperature axis. The adiabatic-saturation temperature is the temperature applying to the adiabatic-cooling line through point a. The humidity at adiabatic saturation is found by following the adiabatic line through point a to its intersection e on the 100 percent line, and reading humidity $\mathscr{H}_s$ at point f on the humidity scale. Interpolation between the adiabatic lines may be necessary. The adiabatic-saturation temperature T_s is given by point g. If the original air is subsequently saturated at constant temperature, the humidity after saturation is found by following the constant-temperature line through point a to point h on the 100 percent line and reading the humidity at point j.

The humid volume of the original air is found by locating points k and l on the curves for saturated and dry volumes, respectively, corresponding to temperature T_1. Point m is then found by moving along line lk a distance $(\mathscr{H}_A/100)\overline{kl}$ from point l, where $\overline{kl}$ is the line segment between points l and k. The humid volume v_H is given by point n on the volume scale. The humid heat of the air is found by locating point o, the intersection of the constant-humidity line through point a and the humid-heat line, and reading the humid heat c_s at point p on the scale at the top.

EXAMPLE 24-1 The temperature and dew point of the air entering a certain dryer are 150 and 60°F (65.6 and 15.6°C), respectively. What additional data for this air can be read from the humidity chart?

SOLUTION The dew point is the temperature coordinate on the saturation line corresponding to the humidity of the air. The saturation humidity for a temperature of 60°F is 0.011 lb of water per pound (0.011 g per gram) of dry air, and this is the humidity of the air. From the temperature and humidity of the air, the point on the chart for this air

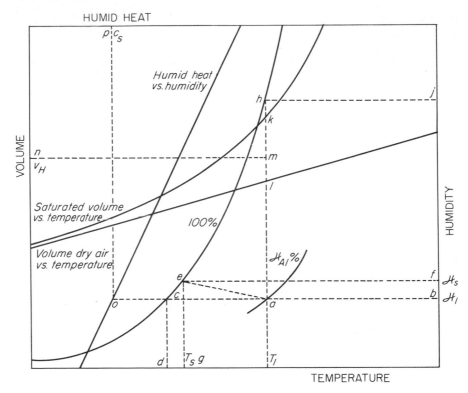

FIGURE 24-3
Use of humidity chart.

is located. At $\mathscr{H} = 0.011$ and $T = 150°F$, the percentage humidity $\mathscr{H}_A$ is found by interpolation to be 5.9 percent. The adiabatic-cooling line through this point intersects the 100
percent line at 85°F (29.4°C), and this is the adiabatic-saturation temperature. The humidity
of saturated air at this temperature is 0.026 lb of water per pound (0.026 g per gram) of dry
air. The humid heat of the air is 0.245 Btu/lb dry air °F (1.03 J/g-°C). The saturated volume
at 150°F is 20.7 ft³ per pound (1.29 m³ per kilogram) of dry air, and the specific volume of
dry air at 150°F is 15.35 ft³/lb (0.958 m³/kg). The humid volume is, then,

$$v_H = 15.35 + 0.059(20.7 - 15.35) = 15.67 \text{ ft}^3/\text{lb dry air } (0.978 \text{ m}^3/\text{kg}) \qquad ////$$

Humidity charts for systems other than air-water A humidity chart may be
constructed for any system at any desired total pressure. The data required are the
vapor pressure and latent heat of vaporization of the condensable component as a
function of temperature, the specific heats of pure gas and vapor, and the molecular
weights of both components. If a chart on a mole basis is desired, all equations can
easily be modified to the use of molal units. If a chart at a pressure other than 1 atm
is wanted, obvious modifications in the above equations may be made. Charts for
several common systems besides air-water have been published[1,6]; Fig. 24-5, to be
discussed later, is a diagram for the system air–benzene vapor.

WET-BULB TEMPERATURE AND MEASUREMENT OF HUMIDITY

The properties discussed above and those shown on the humidity charts are static or equilibrium quantities. Equally important are the rates at which mass and heat are transferred between gas and liquid not in equilibrium. A useful quantity depending on both of these rates is the wet-bulb temperature.

Wet-bulb temperature   The wet-bulb temperature is the steady-state, nonequilibrium temperature reached by a small mass of liquid immersed under adiabatic conditions in a continuous stream of gas. The mass of the liquid is so small in comparison with the gas phase that there is only a negligible change in the properties of the gas, and the effect of the process is confined to the liquid. The method of measuring the wet-bulb temperature is shown in Fig. 24-4. A thermometer, or an equivalent temperature-measuring device such as a thermocouple, is covered by a wick, which is saturated with pure liquid and immersed in a stream of gas having a definite temperature T and humidity $\mathscr{H}$. Assume that initially the temperature of the liquid is about that of the gas. Since the gas is not saturated, liquid evaporates, and because the process is adiabatic, the latent heat is supplied at first by cooling the liquid. As the temperature of the liquid decreases below that of the gas, sensible heat is transferred to the liquid. Ultimately a steady state is reached at such a liquid temperature that the heat needed to evaporate the liquid and heat the vapor to gas temperature is exactly balanced by the sensible heat flowing from the gas to the liquid. It is this steady-state temperature, denoted by T_w, that is called the wet-bulb temperature. It is a function of both T and $\mathscr{H}$.

To measure the wet-bulb temperature with precision, three precautions are necessary: (1) the wick must be completely wet so no dry areas of the wick are in contact with the gas, (2) the velocity of the gas should be large enough to ensure that the rate of heat flow by radiation from warmer surroundings to the bulb is negligible in comparison with the rate of sensible heat flow by conduction and convection from the gas to the bulb; (3) if makeup liquid is supplied to the bulb, it should be at the wet-bulb temperature. When these precautions are taken, the wet-bulb temperature is independent of gas velocity over a wide range of flow rates.

The wet-bulb temperature superficially resembles the adiabatic-saturation temperature T_s. Indeed, for air-water mixtures the two temperatures are nearly equal. This is fortuitous, however, and is not true of mixtures other than air and water. The wet-bulb temperature differs fundamentally from the adiabatic-saturation temperature. In the latter the temperature and humidity of the gas vary during the process, and the end point is a true equilibrium, rather than a dynamic steady state.

Commonly, an uncovered thermometer is used along with the wet bulb to measure T, the actual gas temperature, and the gas temperature is usually called the *dry-bulb temperature*.

Theory of wet-bulb temperature   At the wet-bulb temperature the rate of heat transfer from the gas to the liquid may be equated to the product of the rate of

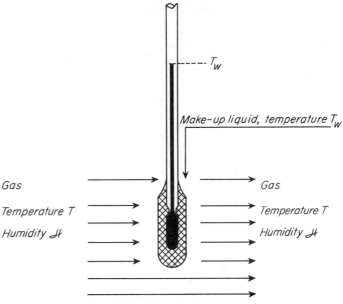

FIGURE 24-4
Principle of wet-bulb thermometer.

vaporization and the sum of the latent heat of evaporation and the sensible heat of the vapor. Since radiation may be neglected, this balance may be written

$$q = M_A N_A [\lambda_w + c_{pA}(T - T_w)] \qquad (24\text{-}11)$$

where q = rate of sensible heat transfer to liquid
   N_A = molal rate of vaporization
   λ_w = latent heat of liquid at wet-bulb temperature T_w

The rate of heat transfer may be expressed in terms of the area, the temperature drop, and the heat-transfer coefficient in the usual way, or

$$q = h_y(T - T_i)A \qquad (24\text{-}12)$$

where h_y = heat-transfer coefficient between gas and surface of liquid
   T_i = temperature at interface
   A = surface area of liquid

The rate of mass transfer may be expressed in terms of the mass-transfer co-efficient, the area, and the driving force in mole fraction of vapor, or

$$N_A = \frac{k_y}{(1 - y)_L}(y_i - y)A \qquad (24\text{-}13)$$

where $\quad N_A$ = molal rate of transfer of vapor

$\qquad y_i$ = mole fraction of vapor at interface

$\qquad y$ = mole fraction of vapor in airstream

$\qquad k_y$ = mass-transfer coefficient, mol/unit area-unit mole fraction

$\overline{(1 - y)}_L$ = phase-drift factor

If the wick is completely wet and no dry spots show, the entire area of the wick is available for both heat and mass transfer and the areas in Eqs. (24-12) and (24-13) are equal. Since the temperature of the liquid is constant, no temperature gradients are necessary in the liquid to act as driving forces for heat transfer within the liquid, the surface of the liquid is at the same temperature as the interior, and the surface temperature of the liquid T_i equals T_w. Since the liquid is pure, no concentration gradients exist, and, granting interfacial equilibrium, y_i is the mole fraction of vapor in saturated gas at temperature T_w. It is convenient to replace the mole-fraction terms in Eq. (24-13) by humidities through the use of Eq. (24-2), noting that y_i corresponds to $\mathscr{H}_w$, the saturation humidity at the wet-bulb temperature. Following this by substitution q from Eq. (24-12) and N_A from Eq. (24-13) into Eq. (24-11) gives

$$h_y(T - T_w) = \frac{k_y}{\overline{(1 - y)}_L} \left(\frac{\mathscr{H}_w}{1/M_B + \mathscr{H}_w/M_A} \right.$$

$$\left. - \frac{\mathscr{H}}{1/M_B - \mathscr{H}/M_A} \right) [\lambda_w + c_{pA}(T - T_w)] \quad (24\text{-}14)$$

Equation (24-14) may be simplified without serious error in the usual range of temperatures and humidities as follows: (1) the phase-drift factor $\overline{(1 - y)}_L$ is nearly unity and can be omitted; (2) the sensible-heat item $c_{pA}(T - T_w)$ is small in comparison with λ_w and can be neglected; (3) the terms $\mathscr{H}_w/M_A$ and $\mathscr{H}/M_A$ are small in comparison with $1/M_B$ and may be dropped from the denominators of the humidity terms. With these simplifications Eq. (24-14) becomes

$$h_y(T - T_w) = M_B k_y \lambda_w (\mathscr{H}_w - \mathscr{H})$$

or
$$\frac{\mathscr{H} - \mathscr{H}_w}{T - T_w} = - \frac{h_y}{M_B k_y \lambda_w} \quad (24\text{-}15)$$

For a given wet-bulb temperature, both λ_w and $\mathscr{H}_w$ are fixed. The relation between $\mathscr{H}$ and T then depends on the ratio h_y/k_y. The close analogy between mass transfer and heat transfer provides considerable information on the magnitude of this ratio and the factors that affect it.

It has been shown in Chap. 12 that heat transfer by conduction and convection between a stream of fluid and a solid or liquid boundary depends on the Reynolds number DG/μ and the Prandtl number $c_p\mu/k$. Also, as shown in Chap. 22, the mass-transfer coefficient depends on the Reynolds number and the Schmidt number $\mu/\rho D v$. As discussed in Chap. 22, the rates of heat and mass transfer, when these processes are under the control of the same boundary layer, are given by equations which are identical in form. For turbulent flow of the gas stream these equations are

$$\frac{h_y}{c_p G} = b N_{Re}^n N_{Pr}^{-m} \tag{24-16}$$

and

$$\frac{\overline{M} k_y}{G} = b N_{Re}^n N_{Sc}^{-m} \tag{24-17}$$

where b, n, m = constants
$\overline{M}$ = average molecular weight of gas stream

Substitution of h_y from Eq. (24-16) and k_y from Eq. (24-17) in Eq. (24-15), assuming $\overline{M} = M_B$, gives

$$\frac{\mathscr{H} - \mathscr{H}_w}{T - T_w} = -\frac{h_y}{M_B k_y \lambda_w} = -\frac{c_p}{\lambda_w}\left(\frac{N_{Sc}}{N_{Pr}}\right)^m \tag{24-18}$$

and

$$\frac{h_y}{M_B k_y} = c_p \left(\frac{N_{Sc}}{N_{Pr}}\right)^m \tag{24-19}$$

Experiment shows that Eq. (24-19) is reasonably accurate if m is taken as $\frac{2}{3}$. This is consistent with the Chilton-Colburn analogy. The experimental value of $h_y/M_B k_y$ for water in air is 0.26.[8] That for organic liquids in air is larger, usually in the range 0.4 to 0.5. The difference, as shown by Eq. (24-19), is the result of the differing ratios of Prandtl and Schmidt numbers for water and for organic vapors. Note that Eq. (24-18) predicts that velocity should not influence the wet-bulb temperature.

Psychrometric line and Lewis relation   For a given wet-bulb temperature, Eq. (24-18) can be plotted on the humidity chart as a straight line having a slope of $-h_y/M_B k_y \lambda_w$ and intersecting the 100 percent line at T_w. This line is called the *psychrometric line*. When both a psychrometric line, from Eq. (24-18), and an adiabatic-saturation line, from Eq. (24-10), are plotted for the same point on the 100 percent curve, the relation between the lines depends on the relative magnitudes of c_s and $h_y/M_B k_y$. The effective specific heat of a unit mass of dry air is the humid heat c_s, which replaces c_p when Eq. (24-18) is used.

For the system air-water at ordinary conditions the following equation is very nearly correct:

$$\frac{h_y}{M_B k_y} = c_s \tag{24-20}$$

Equation (24-20) is known as the *Lewis relation*.[5] When this relation holds, the psychrometric line and the adiabatic-saturation line become essentially the same. In Fig. 24-2 for air-water, therefore, the same line may be used for both. For other systems separate lines must be used for psychrometric lines. Such lines are shown in Fig. 24-5, which is a humidity chart for mixtures of air and benzene vapor at atmospheric pressure. As with nearly all mixtures of air and organic vapors, the psychrometric lines are appreciably steeper than the adiabatic-saturation lines. As a result, the wet-bulb temperature of any mixture other than a saturated one is higher

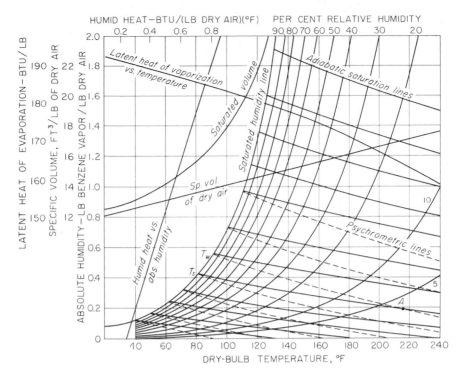

FIGURE 24-5

Humidity chart for air-benzene vapor mixtures. [*By permission, from J. H. Perry (ed.), "Chemical Engineers' Handbook," 3d ed. Copyright, 1950, McGraw-Hill Book Company.*]

than the adiabatic-saturation temperature; for the mixture represented by point A in Fig. 24-5, for example, $T_w = 92°F$ and $T_s = 81°F$.

Measurement of humidity The humidity of a stream or mass of gas may be found by measuring either the dew point or the wet-bulb temperature or by direct absorption methods.

DEW-POINT METHODS If a cooled, polished disk is inserted into gas of unknown humidity and the temperature of the disk gradually lowered, the disk reaches a temperature at which mist condenses on the polished surface. The temperature at which this mist just forms is the temperature of equilibrium between the vapor in the gas and the liquid phase. It is therefore the dew point. A check on the reading is obtained by slowly increasing the disk temperature and noting the temperature at which the mist just disappears. From the observed temperatures of mist formation and disappearance, the humidity can be read from a humidity chart.

PSYCHROMETRIC METHODS   A very common method of measuring the humidity is to determine simultaneously the wet-bulb and dry-bulb temperatures. From these readings the humidity is found by locating the psychrometric line intersecting the saturation line at the observed wet-bulb temperature and following the psychrometric line to its intersection with the ordinate of the observed dry-bulb temperature.

DIRECT METHODS   The vapor content of a gas can be determined by direct analysis, in which a known volume of gas is drawn through an appropriate analytical device.

Equipment for Humidification Operations

When warm liquid is brought into contact with unsaturated gas, part of the liquid is vaporized and the liquid temperature drops. This cooling of the liquid is the purpose behind many gas-liquid contact operations, especially air-water contacts. Water is cooled in large quantities in spray ponds or more commonly in tall towers through which air passes by natural draft or by the action of a fan.

A typical natural-draft cooling tower is shown in Fig. 24-6. The purpose of a cooling tower is to conserve cooling water by allowing the cooled water to be reused many times. Warm water, usually from a condenser or other heat-transfer unit, is admitted to the top of the tower and distributed by troughs and overflows to cascade down over slat gratings, which provide large areas of contact between air and water. Flow of air up through the tower is induced by the buoyancy of the warm air in the tower. A cooling tower is, in principle, a special type of packed tower. The usual packing material is cypress wood, which is the most economical tower filling that withstands the combined action of wind and water. In the tower, part of the water evaporates into the air, and sensible heat is transferred from the warm water to the cooler air. Both processes reduce the temperature of the water. Only makeup water, to replace that lost by evaporation and windage loss, is required to maintain the water balance.

The driving force for evaporation is, very nearly, the difference between the vapor pressure of the water and the vapor pressure it would have at the wet-bulb temperature of the air. Obviously the water cannot be cooled to below the wet-bulb temperature. In practice the discharge temperature of the water must differ from the wet-bulb temperature by at least 4 or 5°F. This difference in temperature is known as the *approach*. The change in temperature in the water from inlet to outlet is known as the *range*. Thus if water were cooled from 95 to 80°F by exposure to air with a wet-bulb temperature of 70°F, the range would be 15°F and the approach 10°F.

The lowest temperature to which water can be cooled throughout the year depends not on the summer dry-bulb temperature but on the maximum wet-bulb temperature. Tables of maximum wet-bulb temperatures have been published for various points in the United States and many other parts of the world.[7]

The loss of water by evaporation during cooling is small. Since roughly 1000 Btu is required to vaporize 1 lb of water, 100 lb must be cooled 10°F to give up enough heat to evaporate 1 lb. Thus for a change of 10°F in the water temperature there is an evaporation loss of 1 percent. In addition there are mechanical spray losses, but in a

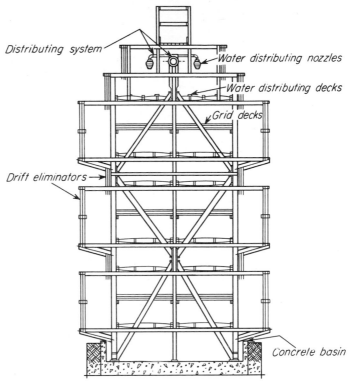

Distributing system

Water distributing nozzles

Water distributing decks

Grid decks

Drift eliminators

Concrete basin

FIGURE 24-6
Natural-draft cooling tower.

well-designed tower these amount to only about 0.2 percent. Under the conditions given above, then, the total loss of water during passage through the cooler would be approximately $15/10 \times 1 + 0.2 = 1.7$ percent. In cooling other liquids by evaporation the vaporization loss, though small, is somewhat greater than with water because of the smaller heat of vaporization.

Humidifiers and dehumidifiers   Gas-liquid contacts are used not only for liquid cooling but also for humidifying or dehumidifying the gas. In a humidifier liquid is sprayed into warm unsaturated gas, and sensible-heat and mass transfer take place in the manner described in the discussion of the adiabatic-saturation temperature. The gas is humidified and cooled adiabatically. It is not necessary that final equilibrium be reached, and the gas may leave the spray chamber at less than full saturation.

Warm saturated gas can be dehumidified by bringing it into contact with cold liquid. The temperature of the gas is lowered below the dew point, liquid condenses, and the humidity of the gas is reduced. After dehumidification the gas can be reheated to its original dry-bulb temperature. Equipment for dehumidification may utilize a direct spray of coarse liquid droplets into the gas, a spray of liquid on refrigerated

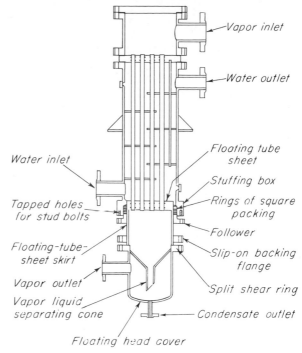

FIGURE 24-7
Outside packed dehumidifying cooler-condenser.

coils or other cold surface, or condensation on a cold surface with no liquid spray. A dehumidifying cooler-condenser is shown in Fig. 24-7. Except for the gas-liquid separator at the bottom it is identical with a shell-and-tube heat exchanger. It is set vertically, not horizontally like most condensers for vapor containing no noncondensable gas; also, vapor is condensed inside the tubes, not outside, and the coolant flows through the shell, not the tubes. This provides a positive sweep of the vapor-gas mixture through the tubes and avoids the formation of any stagnant pockets of inert gas which might blanket the heat-transfer surface.

Theory and Calculation of Humidification Processes

The mechanism of the reaction of unsaturated gas and liquid at the wet-bulb temperature of the gas has been discussed under the description of wet- and dry-bulb thermometry. The process has been shown to be controlled by the flow of heat and the diffusion of vapor through the gas at the interface between the gas and the liquid. Although these factors are sufficient for the discussion of the adiabatic humidifier, where the liquid is at constant temperature, in the case of dehumidifiers and liquid coolers, where the liquid is changing temperature, it is necessary to consider heat flow in the liquid phase also.

In an adiabatic humidifier, where the liquid remains at constant adiabatic-saturation temperature, there is no temperature gradient through the liquid. In dehumidification and in liquid cooling, however, where the temperature of the liquid is changing, sensible heat flows into or from the liquid, and a temperature gradient is thereby set up. This introduces a liquid-phase resistance to the flow of heat. On the other hand, that there can be no liquid-phase resistance to mass transfer in any case, since there can be no concentration gradient in pure liquid.

Mechanism of interaction of gas and liquid   It is important to obtain a correct picture of the relationships of the transfer of heat and of vapor in all situations of gas-liquid contact. In Figs. 24-8, 24-9, 24-11, and 24-12, distances measured perpendicular to the interface are plotted as abscissas, and temperatures and humidities as ordinates. In all figures let

T_x = temperature of bulk of liquid
T_i = temperature at interface
T_y = temperature of bulk of gas
$\mathscr{H}_i$ = humidity at interface
$\mathscr{H}$ = humidity of bulk of gas

Broken arrows represent the diffusion of the vapor through the gas phase, and full arrows represent the flow of heat (both latent and sensible) through gas and liquid phases. In all processes, T_i and $\mathscr{H}_i$ represent equilibrium conditions and are therefore coordinates of points lying on the saturation line on the humidity chart.

The simplest case, that of adiabatic humidification with the liquid at constant temperature, is shown diagrammatically in Fig. 24-8. The latent-heat flow from liquid to gas just balances the sensible-heat flow from gas to liquid, and there is no temperature gradient in the liquid. The gas temperature T_y must be higher than the interface temperature T_i in order that sensible heat may flow to the interface; and $\mathscr{H}_i$ must be greater than $\mathscr{H}$ in order that the gas be humidified.

Conditions at one point in a dehumidifier are shown in Fig. 24-9. Here $\mathscr{H}$ is greater than $\mathscr{H}_i$, and therefore vapor must diffuse to the interface. Since T_i and $\mathscr{H}_i$ represent saturated gas, T_y must be greater than T_i; otherwise the bulk of the gas would be supersaturated with vapor.

This reasoning leads to the conclusion that vapor can be removed from un-saturated gas by direct contact with sufficiently cold liquid without first bringing the bulk of the gas to saturation. This operation is shown in Fig. 24-10. Point A represents the gas that is to be dehumidified. Suppose that liquid is available at such a temperature that saturation conditions at the interface are represented by point B. It has been shown experimentally[4] that in such a process the path of the gas on the humidity chart is nearly a straight line between points A and B. As a result of the humidity and temperature gradients, the interface receives both sensible heat and vapor from the gas. The condensation of the vapor liberates latent heat, and both sensible and latent heat are transferred to the liquid phase. This requires a temperature difference $T_i - T_x$ through the liquid.

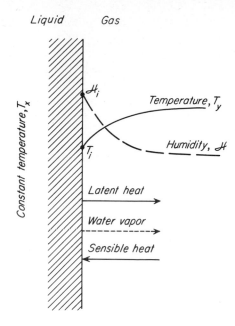

FIGURE 24-8
Conditions in adiabatic humidifier.

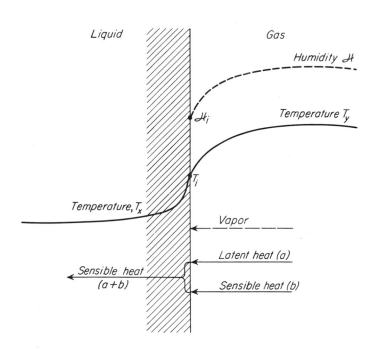

FIGURE 24-9
Conditions in dehumidifier.

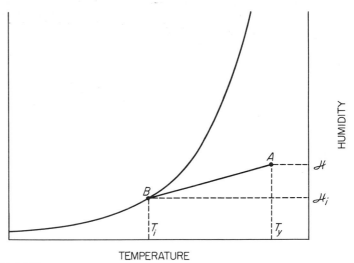

FIGURE 24-10
Dehumidification by cold liquid.

In a cooler-condenser like that shown in Fig. 24-7 the liquid flows over the inner surface of the tubes, which are kept cold by the cooling water outside. Heat is removed from the liquid as it flows downward; this maintains the necessary temperature difference through the liquid and causes progressively more vapor to condense as the gas-vapor mixture passes down through the tubes. The mixture becomes leaner and leaner in condensable material, which it loses to the increasing layer of liquid. Detailed design methods for problems of this kind are given in the literature.[2]

In a countercurrent cooling tower the conditions depend on whether the gas temperature is below or above the temperature at the interface. In the first case, e.g., in the upper part of the cooling tower, the conditions are shown diagrammatically in Fig. 24-11. Here the flow of heat and of vapor (and hence the direction of temperature and humidity gradients) are exactly the reverse of those shown in Fig. 24-9. The liquid is being cooled both by evaporation and by transfer of sensible heat, the humidity and temperature of the gas decrease in the direction of interface to gas, and the temperature drop $T_x - T_i$ through the liquid must be sufficient to give a heat-transfer rate high enough to account for both heat items.

In the lower part of the cooling tower, where the temperature of the gas is above that of the interface temperature, the conditions shown in Fig. 24-12 prevail. Here the liquid is being cooled, hence the interface must be cooler than the bulk of the liquid, and the temperature gradient through the liquid is toward the interface (T_i is less than T_x). On the other hand, there must be a flow of sensible heat from the bulk of the gas to the interface (T_y is greater than T_i). The flow of vapor away from the interface carries, as latent heat, all the sensible heat supplied to the interface from both sides. The resulting temperature profile $T_x T_i T_y$ has a striking V shape, as shown in Fig. 24-12.

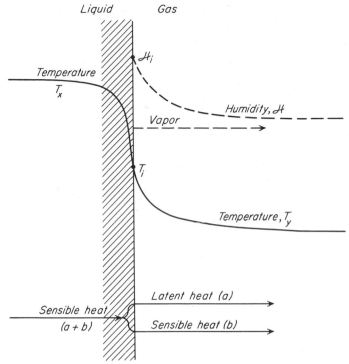

FIGURE 24-11
Conditions in top of cooling tower.

Equations for gas-liquid contacts Consider the countercurrent gas-liquid contactor shown diagrammatically in Fig. 24-13. Gas at humidity $\mathscr{H}_b$ and temperature T_{yb} enters the bottom of the contactor and leaves at the top with a humidity $\mathscr{H}_a$ and temperature T_{ya}. Liquid enters the top of temperature T_{xa} and leaves at the bottom at temperature T_{xb}. The mass velocity of the gas is G'_y, the mass of vapor-free gas per unit area of tower cross section per hour. The mass velocities of the liquid at inlet and outlet are, respectively, G_{xa} and G_{xb}. Let dZ be the height of a small section of the tower at distance Z from the bottom of the contact zone. Let the mass velocity of the liquid at height Z be G_x, the temperatures of gas and liquid be T_y and T_x, respectively, and the humidity be $\mathscr{H}$. At the interface between the gas and the liquid phases, let the temperature be T_i and the humidity be $\mathscr{H}_i$. The cross section of the tower is S, and the height of the contact section is Z_T. Assume that the liquid is warmer than the gas, so the conditions at height Z are those shown in Fig. 24-13. The following equations can be written over the small volume $S\,dZ$.

The enthalpy balance is

$$G'_y \, dH_y = d(G_x H_x) \qquad (24\text{-}21)$$

where H_y and H_x are the total enthalpies of gas and liquid, respectively.

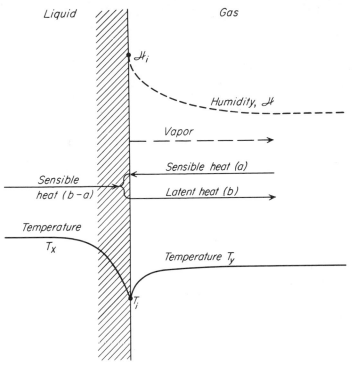

FIGURE 24-12
Conditions in bottom of cooling tower.

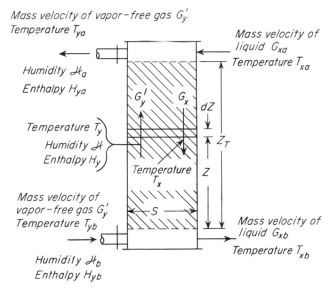

FIGURE 24-13
Countercurrent gas-liquid contactor, flow diagram.

The rate of heat transfer from liquid to interface is

$$d(G_x H_x) = h_x(T_x - T_i)a_H \, dZ \qquad (24\text{-}22)$$

where h_x = heat-transfer coefficient from liquid to interface
a_H = heat-transfer area per unit contact volume

The rate of heat transfer from interface to gas is

$$G_y' c_s \, dT_y = h_y(T_i - T_y)a_H \, dZ \qquad (24\text{-}23)$$

The rate of mass transfer of vapor from interface to gas is

$$G_y' \, d\mathcal{H} = k_y M_B(\mathcal{H}_i - \mathcal{H})a_M \, dZ \qquad (24\text{-}24)$$

where a_M is the mass-transfer arca per unit contact volume. The factors a_M and a_H are not necessarily equal. If the contactor is packed with solid packing, the liquid may not completely wet the packing and the area available for heat transfer, which is the entire area of the packing, is larger than that for mass transfer, which is limited to the surface that is actually wet.

These equations can be simplified and rearranged. First, neglect the change of G_x with height and write for the enthalpy of the liquid

$$H_x = c_L(T_x - T_0) \qquad (24\text{-}25)$$

where c_L = specific heat of liquid
T_0 = base temperature for computing enthalpy

Then

$$d(G_x H_x) = G_x \, dH_x = G_x c_L \, dT_x \qquad (24\text{-}26)$$

Substituting $d(G_x H_x)$ from Eq. (24-26) into Eq. (24-22) gives

$$G_x c_L \, dT_x = h_x(T_x - T_i)a_H \, dZ$$

This may be written

$$\frac{dT_x}{T_x - T_i} = \frac{h_x a_H}{G_x c_L} \, dZ \qquad (24\text{-}27)$$

Second, Eq. (24-23) may be rearranged to read

$$\frac{dT_y}{T_i - T_y} = \frac{h_y a_H}{c_s G_y'} \, dZ \qquad (24\text{-}28)$$

Third, Eq. (24-24) may be written

$$\frac{d\mathcal{H}}{\mathcal{H}_i - \mathcal{H}} = \frac{k_y M_B a_M}{G_y'} \, dZ \qquad (24\text{-}29)$$

Finally, using Eq. (24-21), Eq. (24-26) may be written

$$\frac{dH_y}{dT_x} = \frac{G_x c_L}{G_y'} \qquad (24\text{-}30)$$

Three working equations, which are applied later, can now be derived. First, multiply Eq. (24-24) by λ_0 and add Eq. (24-23) to the product, giving

$$G'_y(c_s\,dT_y + \lambda_0\,d\mathcal{H}) = [\lambda_0 k_y M_B(\mathcal{H}_i - \mathcal{H})a_M + h_y(T_i - T_y)a_H]\,dZ \quad (24\text{-}31)$$

If the packing is completely wet with liquid, so that the area for mass transfer equals that for heat transfer, $a_M = a_H = a$. When the change of c_s with $\mathcal{H}$ is neglected, Eq. (24-9) gives, on differentiation,

$$dH_y = c_s\,dT_y + \lambda_0\,d\mathcal{H} \quad (24\text{-}32)$$

From Eq. (24-32), dH_y can be substituted in the left-hand side of Eq. (24-31). This gives

$$G'_y\,dH_y = [\lambda_0 k_y M_B(\mathcal{H}_i - \mathcal{H})a + h_y(T_i - T_y)a]\,dZ \quad (24\text{-}33)$$

System air-water For the system air-water h_y can be eliminated from Eq. (24-33) using the Lewis relation, Eq. (24-20), to give

$$G'_y\,dH_y = k_y M_B a[(\lambda_0 \mathcal{H}_i + c_s T_i) - (\lambda_0 \mathcal{H} + c_s T_y)]\,dZ \quad (24\text{-}34)$$

If H_i is the enthalpy of the air at the interface,

$$H_i = \lambda_0 \mathcal{H}_i + c_s(T_i - T_0)$$

From this definition of H_i and the expression for H_y given by Eq. (24-9), the bracketed term in Eq. (24-34) is simply $H_i - H_y$. Then Eq. (24-34) becomes

$$G'_y\,dH_y = k_y M_B a(H_i - H_y)\,dZ$$

or

$$\frac{dH_y}{H_i - H_y} = \frac{k_y M_B a}{G'_y}\,dZ \quad (24\text{-}35)$$

Second, from Eqs. (24-21) and (24-22),

$$G'_y\,dH_y = h_x(T_x - T_i)a\,dZ \quad (24\text{-}36)$$

Eliminating dZ from Eqs. (24-35) and (24-36) and rearranging gives, with the aid of Eq. (24-20),

$$\frac{H_i - H_y}{T_i - T_x} = -\frac{h_x}{k_y M_B} = -\frac{h_x c_s}{h_y} \quad (24\text{-}37)$$

Third, Eq. (24-28) is divided by Eq. (24-35), giving

$$\frac{dT_y/(T_i - T_y)}{dH_y/(H_i - H_y)} = \frac{h_y}{k_y M_B c_s}$$

From Eq. (24-20), the right-hand side of this equation equals unity, and

$$\frac{dT_y}{dH_y} = \frac{T_i - T_y}{H_i - H_y} \quad (24\text{-}38)$$

Note that, since the Lewis relation is used in their derivation, Eqs. (24-35), (24-37), and (24-38) apply only to the air-water system and also for situations where $a_M = a_H$. In the following section the discussion is limited to air-water contacts.

Adiabatic humidification Adiabatic humidification corresponds to the process shown in Fig. 24-1 except that the air leaving the humidifier is not necessarily saturated and, for design, rate equations must be used to calculate the size of the contact zone. The inlet and outlet water temperatures are equal. It is assumed in the following that the makeup water enters at adiabatic-saturation temperature and that the volumetric-area factors a_M and a_H are identical. The wet-bulb and adiabatic-saturation temperatures are equal and constant. Then,

$$T_{xa} = T_{xb} = T_i = T_x = T_s = \text{const}$$

where T_s is the adiabatic-saturation temperature of the inlet air. Equation (24-28) then becomes

$$\frac{dT_y}{T_s - T_y} = \frac{h_y a}{c_s G_y'} \, dZ \tag{24-39}$$

With $\bar{c}_s$ used as the average humid heat over the humidifier, Eq. (24-39) can be integrated

$$\int_{T_{yb}}^{T_{ya}} \frac{dT_y}{T_s - T_y} = \frac{h_y a}{\bar{c}_s G_y'} \int_0^{Z_T} dZ$$

$$\ln \frac{T_{yb} - T_s}{T_{ya} - T_s} = \frac{h_y a Z_T}{\bar{c}_s G_y'} = \frac{h_y a S Z_T}{\bar{c}_s G_y' S} = \frac{h_y a V_T}{\bar{c}_s \dot{m}'} \tag{24-40}$$

where $V_T = S Z_T$ = total contact volume
$\dot{m}' = G_y' S$ = total flow of dry air

An equivalent equation, based on mass transfer, can be derived from Eq. (24-29), which is written for adiabatic humidification as

$$\frac{d\mathscr{H}}{\mathscr{H}_s - \mathscr{H}} = \frac{k_y M_B a}{G_y'} \, dZ$$

Since $\mathscr{H}_s$, the saturation humidity at T_s, is constant, this equation can be integrated in the same manner as Eq. (24-39) to give

$$\ln \frac{\mathscr{H}_s - \mathscr{H}_b}{\mathscr{H}_s - \mathscr{H}_a} = \frac{k_y M_B a}{G_y'} Z_T = \frac{k_y M_B a V_T}{\dot{m}'} \tag{24-41}$$

APPLICATION OF HTU METHOD The HTU method is applicable to adiabatic humidification. Thus, by definition,

$$N_t = \ln \frac{\mathscr{H}_s - \mathscr{H}_b}{\mathscr{H}_s - \mathscr{H}_a} \tag{24-42}$$

where N_t is the number of humidity transfer units. From the definition of H_t and Eqs. (24-41) and (24-42),

$$H_t = \frac{Z_T}{N_t} = \frac{G_y'}{k_y M_B a} \tag{24-43}$$

where H_t is the height of one humidity transfer unit.

The number of transfer units can also be defined on the basis of heat transfer, as follows:

$$N_t = \ln \frac{T_{yb} - T_s}{T_{ya} - T_s} \qquad H_t = \frac{G_y' \bar{c}_s}{h_y a} \qquad (24\text{-}44)$$

That H_t given by Eq. (24-43) is the same as that given by Eq. (24-44) can be shown by dividing Eq. (24-43) by the value of H_t from Eq. (24-44). The result is Eq. (24-20). Also, the values of N_t calculated from Eqs. (24-42) and (24-44) are the same, because in both instances $N_t = Z_T/H_t$.

The heat-transfer method, using Eqs. (24-40) and (24-44), is equivalent to the mass-transfer method using Eqs. (24-41) to (24-43). The two methods give the same result.

EXAMPLE 24-2 For a certain process requiring air at controlled temperature and humidity there is needed 15,000 lb (6,804 kg) of dry air per hour at 20 percent humidity and 130°F (54.4°C). This air is to be obtained by conditioning air at 20 percent humidity and 70°F (21.1°C) by first heating, then humidifying adiabatically to the desired humidity, and finally reheating the humidified air to 130°F (54.5°C). The humidifying step is to be conducted in a spray chamber. Assuming the air leaving the spray chamber is to be 4°F (2.22°C) warmer than the adiabatic-saturation temperature, to what temperature should the air be preheated, at what temperature should it leave the spray chamber, how much heat will be required for pre- and reheating, and what should be the volume of the spray chamber? Take $h_y a$ as 85 Btu/ft³-h-°F (1,583 W/m³-°C).

SOLUTION The temperature-humidity path of the air through the heaters and spray chamber is plotted on the section of the humidity chart shown in Fig. 24-14. Air at 20 percent humidity and 130°F has a humidity of 0.022. The air leaving the spray chamber is at this same humidity, and the point representing it on the humidity chart is located by finding the point where the coordinate for $\mathscr{H} = 0.022$ is 4°F from the end of an adiabatic-cooling line. By inspection, it is found that the adiabatic line for the process in the spray chamber corresponds to an adiabatic-saturation temperature of 81°F and that the point on this line at $\mathscr{H} = 0.022$ and $T = 85$°F represents the air leaving the chamber. The original air has a humidity of 0.0030. To reach the adiabatic cooling line for $T_s = 81$°F, the temperature of the air leaving the preheater must be 168°F.

The humid heat of the original air is, from Fig. 24-2, 0.241 Btu/lb-°F. The heat required to preheat the air is, then,

$$0.241 \times 15,000(168 - 70) = 354,000 \text{ Btu/h}$$

The humid heat of the air leaving the spray chamber is 0.250 Btu/lb-°F, and the heat required in the reheater is

$$0.250 \times 15,000(130 - 85) = 169,000 \text{ Btu/h}$$

The total heat required is $354,000 + 169,000 = 523,000$ Btu/h.

To calculate the volume of the spray chamber, Eq. (24-40) may be used. The average humid heat is

$$\bar{c}_s = \frac{0.241 + 0.250}{2} = 0.2455 \text{ Btu/lb dry air-°F}$$

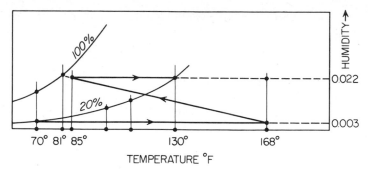

FIGURE 24-14
Temperature-humidity path for Example 24-2.

Substituting in Eq. (24-40) gives

$$2.303 \log \frac{168 - 81}{85 - 81} = \frac{85V_T}{15,000 \times 0.2455}$$

From this, the volume of the spray chamber is $V_T = 134 \text{ ft}^3$ (3.79 m³).   ////

SYMBOLS

A	Surface area of liquid, ft² or m²
a	Transfer area, ft²/ft³ or m²/m³; a_H, for heat transfer; a_M, for mass transfer
b	Constant in Eqs. (24-16), (24-17)
c_p	Specific heat, Btu/lb-°F or J/g-°C; c_L, specific heat of liquid; c_{pA}, c_{pB}, of components A and B, respectively
c_s	Humid heat, Btu/lb-°F or J/g-°C; $\bar{c}_s$, average value
D	Diameter, ft or m
D_v	Diffusivity, ft²/h, m²/h, or cm²/s
G	Mass velocity, lb/ft²-h or kg/m²-h; G_x, of liquid at any point; G_{xa}, of liquid at entrance; G_{xb}, of liquid at exit; G'_y, of gas, mass of vapor-free gas per unit area of tower cross section per hour
H	Enthalpy, Btu/lb or J/g; H_x, of liquid; H_y, of gas
H_t	Height of humidity transfer unit, ft or m
$\mathscr{H}$	Humidity, mass of vapor per unit mass of vapor-free gas; $\mathscr{H}_a$, at top of contactor; $\mathscr{H}_b$, at bottom of contactor; $\mathscr{H}_i$, at gas-liquid interface; $\mathscr{H}_s$, saturation humidity; $\mathscr{H}_w$, saturation humidity at wet-bulb temperature
$\mathscr{H}_A$	Percentage humidity, $100\mathscr{H}/\mathscr{H}_s$
$\mathscr{H}_R$	Relative humidity, $100\bar{p}_A/p_A$
h	Heat-transfer coefficient, Btu/ft²-h-°F or W/m²-°C; h_x, liquid side; h_y, gas side
k	Thermal conductivity, Btu/ft-h-°F or W/m-°C
k_y	Mass-transfer coefficient, lb mol/ft²-h-unit mole fraction or g mol/m²-h-unit mole fraction
M	Molecular weight; M_A, M_B, of components A and B, respectively; $\bar{M}$, average molecular weight of gas stream
m	Exponent in Eqs. (24-16), (24-17)
$\dot{m}'$	Total flow rate of dry air, lb/h or kg/h
N_A	Rate of transfer or vaporization of liquid, mol/h

N_{Pr}	Prandtl number, $c_p\mu/k$, dimensionless
N_{Re}	Reynolds number, DG/μ, dimensionless
N_{Sc}	Schmidt number, $\mu/\rho D_v$, dimensionless
N_t	Number of humidity transfer units
n	Exponent in Eqs. (24-16), (24-17)
p	Pressure, atm; p_A, vapor pressure of liquid; $\bar{p}_A$, partial pressure of vapor
q	Rate of sensible heat transfer to liquid, Btu/h or W
S	Cross-sectional area of tower, ft^2 or m^2
T	Temperature, °F, °R, °C or K; T_i, at gas-liquid interface; T_s, adiabatic-saturation temperature; T_w, wet-bulb temperature; T_x, of bulk of liquid; T_{xa}, of liquid at top of contactor; T_{xb}, of liquid at bottom of contactor; T_y, of bulk of gas; T_{ya}, of gas at top of contactor; T_{yb}, of gas at bottom of contactor; T_0, datum for computing enthalpy
V_T	Total contact volume, ft^3 or m^3
v_H	Humid volume, ft^3/lb or m^3/kg
y	Mole fraction of liquid component in gas stream; y_i, at gas-liquid interface
$(1 - y)_L$	Phase-drift factor
Z	Distance from bottom of contact zone, ft or m; Z_T, total height of contact section
Greek letters	
λ	Latent heat of vaporization, Btu/lb or J/g; λ_s, at T_s; λ_w, at T_w; λ_0, at T_0
μ	Viscosity, lb/ft-h or P
ρ	Density of gas, lb/ft^3 or kg/m^3

PROBLEMS

24-1 One method of removing acetone from cellulose acetate is to blow an airstream over the cellulose acetate fibers. To know the properties of the air-acetone mixtures, the process control department requires a humidity chart for air-acetone. After investigation, it was found that an absolute humidity range of 0 to 6.0 and a temperature range of 5 to 55°C would be satisfactory. Construct the following portions of such a humidity chart for air-acetone at a total pressure of 760 mm Hg: (*a*) percentage humidity lines for 50 and 100 percent, (*b*) saturated volume vs. temperature, (*c*) latent heat of acetone vs. temperature, (*d*) humid heat vs. humidity, (*e*) adiabatic-cooling lines for adiabatic-saturation temperatures of 20 and 40°C, (*f*) wet-bulb temperature (psychrometric) lines for wet-bulb temperatures of 20 and 40°C. The necessary data are given in Table 24-1. For acetone vapor, $c_p = 1.47$ J/g-°C and $h/M_B k_y = 0.95$ J-unit mole fraction/g.

Table 24-1 PROPERTIES OF ACETONE

Temp., °C	Vapor pressure, mm Hg	Latent heat, J/g	Temp., °C	Vapor pressure, mm Hg	Latent heat, J/g
0		564	50	620.9	
10	115.6		56.1	760.0	521
20	179.6	552	60	860.5	517
30	281.0		70	1,189.4	
40	420.1	536	80	1,611.0	495

24-2 The following data were obtained during a test run of a packed cooling tower operating at atmospheric pressure:

Height of packed section, 6 ft
Inside diameter, 12 in.
Average temperature of entering air, 100°F
Average temperature of leaving air, 103°F
Average wet-bulb temperature of entering air, 80°F
Average wet-bulb temperature of leaving air, 96°F
Average temperature of entering water, 115°F
Average temperature of leaving water, 95°F
Rate of entering water, 2,000 lb/h
Rate of entering air, 480 ft³/min

(*a*) Using the entering-air conditions, calculate the humidity of the exit air by means of an enthalpy balance. Compare the result with the humidity calculated from the wet- and dry-bulb readings. (*b*) Assuming that the water-air interface is at the same temperature as the bulk of the water (water-phase resistance to heat transfer negligible), calculate the driving force $\mathscr{H}_i - \mathscr{H}$ at the top and at the bottom of the tower. Using an average $\mathscr{H}_i - \mathscr{H}$, estimate the average value of $k_y a$.

24-3 Air is to be cooled and dehumidified by countercurrent contact with water in a packed tower. The tower is to be designed for the following conditions:

Dry-bulb temperature of inlet air, 28°C
Wet-bulb temperature of inlet air, 25°C
Flow rate of inlet air, 700 kg/h of dry air per hour
Inlet water temperature, 10°C
Outlet water temperature, 18°C

(*a*) For the entering air, find (*i*) the humidity, (*ii*) the percent relative humidity, (*iii*) the dew point, (*iv*) the enthalpy, based on air and liquid water at 0°C. (*b*) What is the maximum water rate which can be used to meet design requirements, assuming a very tall tower? (*c*) Calculate the number of transfer units required for a tower that meets design specifications when 500 kg/h of water is used and if the liquid-phase resistance to heat transfer is negligible.

REFERENCES

1 Brown, G. G., and associates: "Unit Operations," p. 545, Wiley, New York, 1950.
2 Colburn, A. P., and O. A. Hougen: *Ind. Eng. Chem.,* **26:**1178 (1934).
3 Grosvenor, W. M.: *Trans. AIChE,* **1:**184 (1908).
4 Keevil, C. S., and W. K. Lewis: *Ind. Eng. Chem.,* **20:**1058 (1928).
5 Lewis, W. K.: *Trans. AIME,* **44:**325 (1922).
6 Perry, J. H. (ed.): "Chemical Engineers' Handbook," 3d ed., pp. 812–816, McGraw-Hill, New York, 1950.
7 Perry, J. H. (ed.): "Chemical Engineers' Handbook," 5th ed., p. **12**-26, McGraw-Hill, New York, 1973.
8 Sherwood, T. K., and R. L. Pigford: "Absorption and Extraction," 2d ed., pp. 97–101, McGraw-Hill, New York, 1952.

25

DRYING OF SOLIDS

In general, drying a solid means the removal of relatively small amounts of water or other liquid from the solid material, to reduce the content of residual liquid to an acceptably low value. Drying is usually the final step in a series of operations, and the product from a dryer is often ready for final packaging.

Water or other liquids may be removed from solids mechanically by presses or centrifuges or thermally by vaporization. This chapter is restricted to drying by thermal vaporization. It is generally cheaper to remove water mechanically than thermally, and thus it is advisable to reduce the moisture content as much as practicable before feeding the material to a heated dryer.

The moisture content of a dried substance varies from product to product. Occasionally the product contains no water, and is called *bone-dry*. More commonly, the product does contain some water. Dried table salt, for example, contains about 0.5 percent water, dried coal about 4 percent, and dried casein about 8 percent. Drying is a relative term and means merely that there is a reduction in moisture content from an initial value to some acceptable final value.

DRYING EQUIPMENT

Many types of commercial dryers are described in standard references.[10b, 11] Here only a small number of important types are considered. They may be classified according to whether the material is a rigid or granular solid, a semisolid paste, or a liquid solution or slurry; whether or not the material is agitated during drying; and whether the operation is batch or continuous. Still another division is between direct-contact dryers, in which the solids are dried by exposure to hot air or flue gas, and indirect dryers, in which heat is transferred to the material from a heating medium through a metal wall.

Dryers for Solids and Pastes

The nature of the product usually dictates the type of dryer to be used. Coarse, granular inorganic solids may be dried in an agitated dryer by direct contact with hot flue gases; fragile crystals of heat-sensitive organic materials must be dried in unagitated equipment by indirect heating or by contact with warm air at carefully controlled temperature and humidity. Operation under vacuum is often used to reduce drying temperatures. As in other unit operations, batch equipment is favored when production rates are low, holdup times in the unit are long, or many different products are to be dried in the same unit. Continuous dryers are employed when tonnages are large and the drying rate is so rapid that the time of drying is not excessive.

Typical dryers for rigid or granular solids include tray and screen-conveyor dryers for materials that cannot be agitated and rotary, screw-conveyor, tower, and flash dryers for solids that can be agitated. Some of these dryers are batch, some continuous; some are direct-contact, others indirect.

Tray dryers A typical batch tray dryer is illustrated in Fig. 25-1. It consists of a rectangular chamber of sheet metal containing two trucks which support racks H. Each rack carries a number of shallow trays, perhaps 30 in. square and 2 to 6 in. deep, which are loaded with the material to be dried. Heated air is circulated at 7 to 15 ft/s between the trays by fan C and motor D and passes over heaters E. Baffles G distribute the air uniformly over the stack of trays. Some moist air is continuously vented through exhaust duct B; makeup fresh air enters through inlet A. The racks are mounted on truck wheels I, so that at the end of the drying cycle the trucks can be pulled out of the chamber and taken to a tray-dumping station.

Tray dryers are useful when the production rate is less than 50 to 100 lb/h of dry product. They can dry almost anything, but because of the labor required for loading and unloading are expensive to operate. They find most frequent application on valuable products like dyes and pharmaceuticals. Drying by circulation of air across stationary layers of solid is slow, and drying cycles are long: 4 to 48 h per batch. In some tray dryers coarse-grained solids are supported on screen-bottom trays, and the air is passed vertically through them, giving what is known as *through-circulation drying*. Drying in a through-circulation unit is much more rapid than in a cross-circulation unit, but through circulation is usually neither economical nor

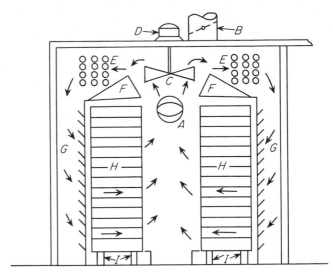

FIGURE 25-1
Tray dryer.

necessary in batch dryers, because shortening the drying cycle does not reduce the labor required for each batch. As mentioned above, when production rates are high and short cycles can be used, a continuous dryer should be selected.

Tray dryers may be operated under vacuum, often with indirect heating. The trays may rest on hollow metal plates supplied with steam or hot water or may themselves have spaces for a heating medium. Water vapor from the solid is removed by an ejector or vacuum pump. *Freeze-drying* is the sublimation of water vapor from ice under high vacuum at temperatures below 0°C. This is done in special vacuum tray dryers for drying vitamins and other heat-sensitive products.

Screen-conveyor dryers   A typical through-circulation screen-conveyor dryer is shown in Fig. 25-2. A layer 1 to 6 in. thick of material to be dried is slowly carried on a traveling metal screen through a long drying chamber or tunnel. The chamber consists of a series of separate sections, each with its own fan and air heater. At the inlet end of the dryer the air usually passes upward through the screen and the solids; near the discharge end, where the material is dry and may be dusty, air is passed downward through the screen. The air temperature and humidity may differ in the various sections, to give optimum conditions for drying at each point.

Screen-conveyor dryers are typically 6 ft wide and 12 to 150 ft long, giving drying times of 5 to 120 min. The minimum screen size is about 30-mesh. Coarse granular, flaky, or fibrous materials can be dried by through circulation without any pretreatment and without loss of material through the screen. Pastes and filter cakes of fine particles, however, must be preformed before they can be handled on a screen-conveyor dryer. Preforming is done by finned drums, extruders, granulators, and

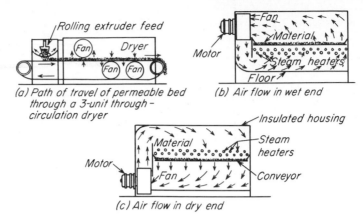

(a) Path of travel of permeable bed through a 3-unit through-circulation dryer

(b) Air flow in wet end

(c) Air flow in dry end

FIGURE 25-2
Through-circulation screen-conveyor dryer.

other devices which aggregate the solid. The aggregates usually retain their shape while being dried and do not dust through the screen except in small amounts. Provision is sometimes made for recovering any fines which do sift through the screen.

Screen-conveyor dryers handle a variety of solids continuously and with a very gentle action; their cost is reasonable; and their steam economy is high. A typical value is 2 lb of steam per pound of water evaporated. Air may be recirculated through, and vented from, each section separately or passed from one section to another countercurrently to the solid. These dryers are particularly applicable when the drying conditions must be appreciably changed as the moisture content of the solid is reduced.

Rotary dryers A rotary dryer consists of a revolving cylindrical shell, horizontal or slightly inclined toward the outlet. Wet feed enters one end of the cylinder; dry material discharges from the other. As the shell rotates, internal flights lift the solids and shower them down through the interior of the shell. Rotary dryers are heated by direct contact of air or gas with the solids, by hot gases passing through an external jacket on the shell, or by steam condensing in a set of longitudinal tubes mounted on the inner surface of the shell. The last of these types is called a steam-tube rotary dryer.

A typical countercurrent direct-contact air-heated dryer is shown in Fig. 25-3. A rotating shell A made of sheet steel is supported on two sets of rollers B and driven by a gear and pinion C. At the upper end is a hood D, which connects through fan E to a stack, and a spout F, which brings in wet material from the feed hopper. Flights G, which lift the material being dried and shower it down through the current of hot air, are welded inside the shell. At the lower end the dried product discharges into a screw conveyor H. Just beyond the screw conveyor is a set of steam-heated extended-surface pipes which preheat the air. The air is moved through the dryer by a fan, which may, if desired, discharge into the air heater so that the whole system is under a positive pressure. Alternatively, the fan may be placed in the stack as shown, so that

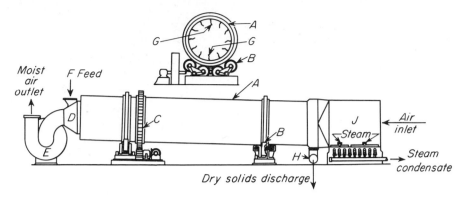

FIGURE 25-3
Countercurrent air-heated rotary dryer: *A*, dryer shell; *B*, shell-supporting rolls;
C, drive gear; *D*, air-discharge hood; *E*, discharge fan; *F*, feed chute; *G*, lifting
flights; *H*, product discharge; *J*, air heater.

it draws the air through the dryer and keeps the system under a slight vacuum. This
is desirable when the material tends to dust. Rotary dryers of this kind are widely
used for salt, sugar, and all kinds of granular and crystalline materials which must
be kept clean and which may not be directly exposed to very hot flue gas.

The allowable mass velocity of the gas in a direct-contact rotary dryer depends
on the dusting characteristics of the solid being dried and ranges from 400 lb/ft²-h
for fine particles to 5,000 lb/ft²-h for coarse, heavy particles. Dryer diameters vary
from 1 to 10 ft.

Screw-conveyor dryers A screw-conveyor dryer is a continuous indirect-heat
dryer, consisting essentially of a horizontal screw conveyor (or paddle conveyor)
enclosed in a cylindrical jacketed shell. Solid fed in one end is conveyed slowly
through the heated zone, and discharges from the other end. The vapor evolved is
withdrawn through pipes set in the roof of the shell. The shell is 3 to 24 in. in diameter
and up to 20 ft long; when more length is required, several conveyors are set one
above another in a bank. Often the bottom unit in such a bank is a cooler, not a
dryer; water or another coolant in the jacket lowers the temperature of the dried solids
before they are discharged.

The rate of rotation of the conveyor is slow, from 2 to 30 r/min. Heat-transfer
coefficients are based on the entire inner surface of the shell, even though the shell
runs only 10 to 60 percent full. The coefficient depends on the loading in the shell
and on the conveyor speed. It ranges, for many solids, between 3 and 10 Btu/ft²-h-°F.

Screw-conveyor dryers handle solids that are too fine and too sticky for rotary
dryers. They are completely enclosed and permit recovery of solvent vapors with little
or no dilution by air. When provided with appropriate feeders, they can be operated
under moderate vacuum. Thus they are adaptable to the continuous removal and
recovery of volatile solvents from solvent-wet solids, such as spent meal, from leaching
operations. For this reason they are sometimes known as *desolventizers*.

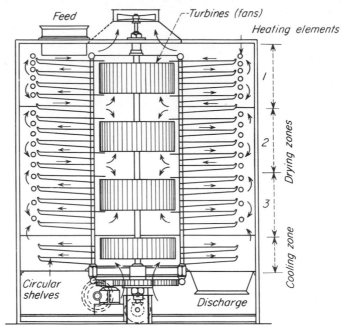

FIGURE 25-4
Turbodryer.

Tower dryers   A tower dryer contains a series of circular trays mounted one above the other on a central rotating shaft. Solid feed dropped on the topmost tray is exposed for a short time to a stream of hot air or gas which passes across the tray. The solid is then scraped off and dropped to the tray below. It travels in this way through the dryer, discharging as dry product from the bottom of the tower. The flow of solids and gas may be either parallel or countercurrent.

The *turbodryer* illustrated in Fig. 25-4 is a tower dryer with internal recirculation of the heating gas. Turbine fans circulate the air or gas outward between some of the trays, over heating elements, and inward between other trays. Gas velocities are commonly 2 to 8 ft/s. The bottom two trays of the dryer shown in Fig. 25-4 constitute a cooling section for dry solids. Preheated air is usually drawn in the bottom of the tower and discharged from the top, giving countercurrent flow. A turbodryer may be considered as a continuous tray dryer, with, however, some agitation of the solids as they tumble from one tray to another.

Flash dryers   In a flash dryer a wet pulverized solid is transported for a few seconds in a hot gas stream. A dryer of this type is shown in Fig. 25-5. Drying takes place during transportation. The rate of heat transfer from the gas to the suspended solid particles is high, and drying is rapid, so that no more than 3 or 4 s is required to evaporate substantially all the moisture from the solid. The temperature of the gas

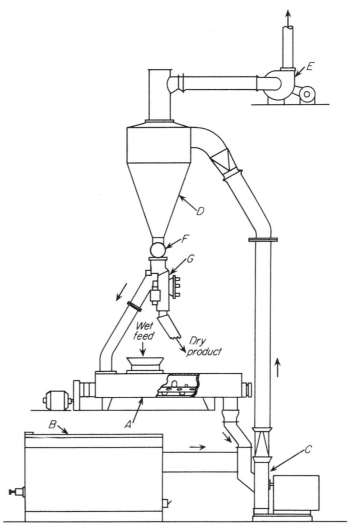

FIGURE 25-5
Flash dryer with disintegrator. *A*, paddle-conveyor mixer; *B*, oil-fired furnace; *C*, hammer mill; *D*, cyclone separator; *E*, vent fan; *F*, star feeder; *G*, solids flow divider and timer.

is high—often about 1200°F at the inlet—but the time of contact is so short that the temperature of the solid rarely rises more than 100°F during drying. Flash drying may therefore be applied to sensitive materials that in other dryers would have to be dried indirectly by a much cooler heating medium.

Sometimes a pulverizer is incorporated in the flash-drying system to give simultaneous drying and size reduction. Such a system is shown in Fig. 25-5. Wet feed enters mixer *A*, where it is blended with enough dry material to make it free-flowing. Mixed material discharges into hammer mill *C*, which is swept with hot flue gas from furnace *B*. Pulverized solid is carried out of the mill by the gas stream through a fairly long duct, in which drying takes place. Gas and dry solid are separated in cyclone *D*, with the cleaned gas passing to vent fan *E*. Solids are removed from the cyclone through star feeder *F*, which drops them into solids divider *G*. This divider is nearly always needed to recycle some dry solid for mixing with the wet feed. It operates by a timer which moves a flapper valve so that for a fixed period of time dry solid returns to the mixer and for another fixed period it is removed as product. Usually more solid is returned to the mixer than is withdrawn. Typical recirculation ratios are 3 to 4 lb of solid returned per pound of product removed from the system.

Fluid-bed dryers Dryers in which the solids are fluidized by the drying gas find applications in a variety of drying problems.[4] The particles are fluidized by air or gas in a boiling-bed unit, as shown in Fig. 25-6. Mixing and heat transfer are very rapid. Wet feed is admitted to the top of the bed; dry product is taken out from the side, near the bottom. Again the drying time is short, but there is a wider distribution of particle residence time than in a flash dryer, and the average time a particle stays in the bed is usually between 30 and 60 s. Some particles stay in the bed considerably longer and are heated to almost the dry-bulb temperature of the suspending gas; consequently, thermally sensitive materials must be dried in a relatively cool suspending medium. If fine particles are present, either from the feed or from particle breakage in the fluidized bed, there may be considerable solids carryover with the exit gas and cyclones and bag filters are needed for fines recovery.

As in flash dryers, part of the dry product may be mixed with wet feed to improve its handling characteristics and reduce stickiness. Sometimes a liquid or slurry feed can be sprayed directly on the particles in the boiling bed without interfering with the fluidizing action.

Dryers of the kind shown in Fig. 25-6 may also be operated batchwise. A charge of wet solids in a perforated container attached to the bottom of the fluidizing chamber is fluidized, heated until dry, then discharged. Such units have replaced tray dryers in many processes.

Dryers for Solutions and Slurries

A few types of dryers evaporate solutions and slurries entirely to dryness by thermal means. Chief among them are drum dryers, spray dryers, and thin-film dryers.

Drum dryers A drum dryer consists of one or more heated metal rolls on the outside of which a thin layer of liquid is evaporated to dryness. Dried solid is scraped off the rolls as they slowly revolve.

A typical drum dryer—a double-drum unit with center feed—is shown in Fig. 25-7. Liquid is fed from a trough or perforated pipe into a pool in the space above and between the two rolls. The pool is confined there by stationary end plates. Heat is transferred by conduction to the liquid, which is partly concentrated in the space

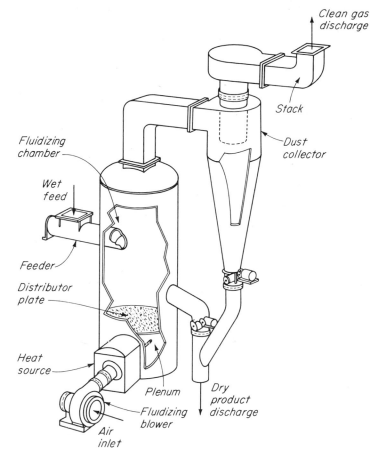

FIGURE 25-6
Continuous fluid-bed dryer.

between the rolls. Concentrated liquid issues from the bottom of the pond as a viscous layer covering the remainder of the drum surfaces. Substantially all the liquid is vaporized from the solid as the drums turn, leaving a thin layer of dry material which is scraped off by doctor blades into conveyors below. Vaporized moisture is collected and removed through a vapor head above the drums.

Double-drum dryers are effective with dilute solutions, concentrated solutions of highly soluble materials, and moderately heavy slurries. They are not suitable for solutions of salts with limited solubility or for slurries of abrasive solids that settle out and create excessive pressure between the drums.

The rolls of a drum dryer are 2 to 10 ft in diameter and 2 to 14 ft long, revolving at 1 to 10 r/min. The time that the solid is in contact with hot metal is 6 to 15 s, which is short enough to result in little decomposition even of heat-sensitive products. The heat-transfer coefficient is high, from 220 to 360 Btu/ft²-h-°F under optimum condi-

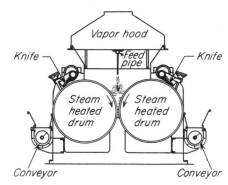

FIGURE 25-7
Double-drum dryer with center feed.

tions, although it may be only one-tenth of these values when conditions are adverse.[11] The drying capacity is proportional to the active drum area; it is usually between 1 and 10 lb of dry product per square foot of drying surface per hour.

Spray dryers   In a spray dryer a slurry or liquid solution is dispersed into a stream of hot gas in the form of a mist of fine droplets. Moisture is rapidly vaporized from the droplets, leaving residual particles of dry solid, which are then separated from the gas stream. The flow of liquid and gas may be cocurrent, countercurrent, or a combination of both in the same unit.

The droplets are formed inside a cylindrical drying chamber, either by spray nozzles or by high-speed spray disks. In either case it is essential to prevent the droplets or wet particles of solid from striking solid surfaces before drying has taken place, so that the drying chambers are necessarily large. Diameters of 8 to 30 ft are common.

In the typical spray dryer shown in Fig. 25-8 the chamber is a cylinder with a short conical bottom. Liquid feed is pumped into a spray-disk atomizer set in the roof of the chamber. In this dryer the spray disk is about 12 in. in diameter and rotates at 5,000 to 10,000 r/min. It atomizes the liquid into tiny drops, which are thrown radially into a stream of hot gas entering near the top of the chamber. Cooled gas is drawn by an exhaust fan through a horizontal discharge line set in the side of the chamber at the bottom of the cylindrical section. The gas passes through a cyclone separator, where any entrained particles of solid are removed. Much of the dry solid settles out of the gas into the bottom of the drying chamber, from which it is removed by a rotary valve and screw conveyor and combined with any solid collected in the cyclone.

The chief advantages of spray dryers are the very short drying time, of the order of 2 to 20 s, which permits drying of highly heat-sensitive materials, and the production of solid or hollow spherical particles. The desired consistency, bulk density, appearance, and flow properties of some products, such as foods or synthetic detergents, may be difficult or impossible to obtain in any other type of dryer. Spray dryers also have the advantage of yielding from a solution, slurry, or thin paste, in a single step, a dry product that is ready for the package. A spray dryer may combine the functions of an evaporator, a crystallizer, a dryer, a size-reduction unit, and a classifier. Where one can be used, the resulting simplification of the overall manufacturing process may be considerable.

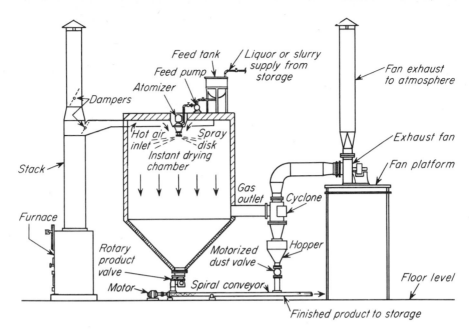

FIGURE 25-8
Spray dryer with parallel flow.

Considered as dryers alone, spray dryers are not highly efficient. Much heat is ordinarily lost in the discharged gases. They are bulky and very large, often 80 ft or more high, and are not always easy to operate. The bulk density of the dry solid—a property of especial importance in packaged products—is often difficult to keep constant, for it may be highly sensitive to changes in the solids content of the feed, to the inlet gas temperature, and to other variables.[10d]

In spray drying solutions the evaporation from the surface of the drops leads to initial deposition of solute at the surface before the interior of the drop reaches saturation. The rate of diffusion of solute back into the drop is slower than the flow of water from the interior to the surface, and the entire solute content accumulates at the surface. The final dry particles are usually hollow, and the product from a spray dryer is very porous.

The performance of a spray dryer depends on the time the drops spend in the drying chamber. This in turn depends on many factors, including the size and shape of the chamber, the size and terminal velocities of the drops, and the flow pattern and velocity of the air. Since the drops formed in commercial atomization equipment cover a wide range of sizes, the time required to reach dryness varies from drop to drop and is, of course, longer for large drops than for small. It is possible to overdry the small drops and at the same time to underdry the large ones, which then leave the dryer in a sticky condition. Although a great deal has been learned about the action in spray dryers,[7, 10d] their design should be left to specialists.

Thin-film dryers Competitive with spray dryers in some situations are thin-film dryers which can accept a liquid or a slurry feed and produce a dry free-flowing solid product. They normally are built in two sections, the first of which is a vertical agitated evaporator-dryer similar to the device illustrated in Fig. 16-4. Here most of the liquid is removed from the feed, and partially wet solid is discharged to the second section; special agitator blades wipe the wall of the first section and remove any sticky solids from the heated surfaces. The second section is a horizontal paddle-conveyor dryer which reduces the moisture content of the material from the first section to the desired value.

The thermal efficiency of thin-film dryers is high, and there is a little loss of solids, since little or no gas need be drawn through the unit. They are useful in removing and recovering solvents from solid products. They are relatively expensive, however, and are limited in heat-transfer area to about 190 ft^2. With dilute feeds they produce about 5 lb of dry solids per square foot of heating surface per hour; with concentrated feeds the production rate is 2 to 3 times greater than with dilute feeds.

PRINCIPLES OF DRYING

Because of the wide variety of materials that are dried in commercial equipment and the many types of equipment that are used, there is no single theory of drying that covers all materials and dryer types. Variations in phase, in shape and size of stock, in moisture equilibria, in the mechanism of flow of moisture through the solid, and the method of providing the heat required for the vaporization—all prevent a unified treatment. General principles, used in a semiquantitative way, are relied upon. Dryers are seldom designed by the user but are bought from companies that specialize in the engineering and fabrication of drying equipment.

Equilibria Equilibria in systems consisting of wet solids and moist air are discussed in Chap. 17. Typical equilibrium relationships are shown in Figs. 17-24 and 17-25. In fluid phases diffusion is governed by concentration differences expressed in mole fractions. In a wet solid, however, the term *mole fraction* may have little meaning, and for ease in drying calculations the moisture content is nearly always expressed in mass of water per unit mass of bone-dry solid. This practice is followed throughout the present chapter.

The driving force for mass transfer in a wet solid is, then, the free moisture content X, which is the difference between the total moisture content X_T and the equilibrium moisture content X^*. It is X, not X_T, that is of interest in drying calculations.

Rate of Drying

The capacity of a thermal dryer depends on both the rate of heat transfer and the rate of mass transfer. Since water vapor must be vaporized, heat of vaporization must be supplied to the vaporization zone, which, depending on the type of material and the conditions of the process, may be either at or near the surface of the solid or within the solid. The moisture must flow either as a liquid or a vapor through the solid and as a vapor from the surface of the solid into the gas stream or airstream.

Heat transfer in dryers   Several methods of heat transfer are used in the dryers described in the last section. When the stock is in the form of lumps or cakes and is not agitated, heat may be supplied either directly by contact with hot gas or indirectly by contact with hot surfaces. In most tray dryers a combination of both methods is used.

When all the heat for vaporizing the water is supplied by direct contact with hot gases and heat transfer by conduction from contact with hot boundaries or from radiation from solid walls is negligible, the process is called *adiabatic drying*. Then the latent heat is obtained entirely from the sensible heat of the gas. From the standpoint of the gas, the process is much like adiabatic humidification. In actual dryers, radiation is active to some extent, and in high-temperature equipment, heated by flue gas or flames, radiation may be of controlling importance.

So long as unbound water is present, the temperature at the zone of vaporization is that corresponding to the temperature of equilibrium between air and liquid water. If only bound water is present, as in the drying of hygroscopic materials, the partial pressure of the water vapor at the zone of vaporization is less than the vapor pressure of water at the temperature of the vaporization zone, so the water vapor is superheated. When the last traces of water are being removed from a hygroscopic solid, the temperature of the solid may be close to that of the air.

In through-circulation drying, as in screen-conveyor dryers, where hot gas is blown through a bed of solid particles or a porous sheet, another kind of heat transfer —that between hot gas and a bed of solids—occurs. Simultaneously, mass transfer between the same phases takes place. The basic mechanism is like that of humidification in a packed tower except that the water diffuses through the solid particles instead of flowing over them.

In flash dryers, spray dryers, fluidized-bed dryers, and rotary dryers the material is in the form of particles or drops. Heat is supplied to the material by allowing it to fall through, or be carried by, a stream of gas. In most of these dryers heat transfer is primarily by conduction and convection, and adiabatic conditions prevail. In indirect rotaries, especially at high temperatures, radiation is important and may be controlling.

Rotary dryers are designed on the basis of heat transfer. Since the particles are falling through the hot gas, the heat-transfer rate in direct rotaries is a function of the surface area of the particles rather than the surface of the shell. This is also true of spray dryers. In indirect rotaries, heat transfer both from the walls and to the surface of the particles must be considered. Empirical equations are available for direct rotary dryers.[10c]

The laws and procedures for heat transfer given in Chaps. 10 to 14 may be used in dryer calculations. Experimental data on the heat-transfer coefficients are sometimes needed for accurate design.

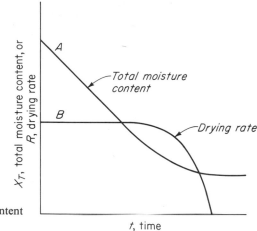

FIGURE 25-9
Typical plots of total moisture content
and drying rate vs. drying time.

Mass transfer from the solid to the gas or air is covered by the relations discussed in Chap. 22. The flow of moisture in the solid itself, however, depends on the structure and character of the solid. To understand this process, knowledge of the internal mechanism of drying is required. This mechanism is the subject of the next section.

Mechanism of drying of solids   Two quite different kinds of solids are found in drying practice. Although many solids are intermediate between the two extremes, it is convenient to assume that the solid is either porous or nonporous. Also, either type may be either hygroscopic or nonhygroscopic.

The details of drying a given solid may be examined from the standpoints of how the drying rate varies with air conditions and what happens in the interior of the solid during drying. For this purpose, consider a slab of stock which is being uniformly dried from both sides but which is protected against moisture loss from the edges. Assume that the temperature, humidity, and velocity and direction of flow of the air across the drying surface are constant. This type of drying is called drying under *constant drying conditions.* Note that only the conditions in the airstream are constant, as the moisture content and other factors in the solid are changing.

Rate-of-drying curves   When a solid is dried under constant drying conditions, the moisture content X_T typically falls as shown by graph A in Fig. 25-9. The graph is linear at first, then curves upward, and eventually levels off. Graph B shows the drying rate: it is horizontal at first, indicating that the drying-rate is constant; then it curves downward and eventually, when the material has reached its equilibrium moisture content, reaches zero.

To study the mechanism of drying under constant conditions, a plot of the drying rate R vs. the free-moisture content $X_T - X^*$ or X, is useful. Such plots for different materials are given in Figs. 25-10 and 25-13 to 25-15. Figure 25-10 is the

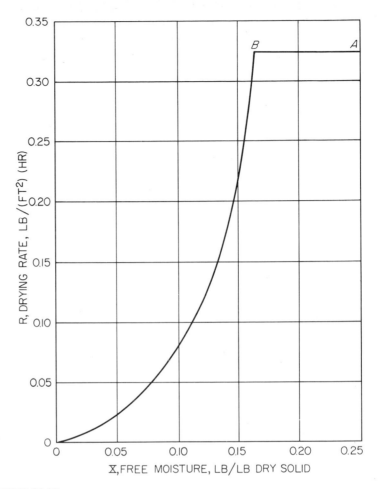

FIGURE 25-10
Drying-rate curves for nonporous clay slab. (After Sherwood.[14])

curve for a nonporous clay slab; Fig. 25-13 is that for a porous, nonhygroscopic ceramic plate having small pores; Fig. 25-14 shows the rate of drying of a bed of sand particles; and Fig. 25-15 one for sheets of paper pulp, which is both hygroscopic and porous. The pronounced differences in the shapes of these curves are the result of differences in the mechanism of moisture flow in the various materials. Experimental study of the distribution of moisture during drying is also illuminating.

CONSTANT-RATE PERIOD   Each drying-rate curve has at least two distinct segments. After a preliminary period (not shown in the various rate-of-drying figures), during which the temperature of the material adjusts itself to the drying conditions, each curve has a horizontal segment *AB*, which pertains to the first major drying period. This period, which may be absent if the initial moisture content of the solid is less than a certain minimum, is called the *constant-rate period*. It is characterized by a rate of drying independent of moisture content. During this period, the solid is so wet that a continuous film of water exists over the entire drying surface, and this water acts as if the solid were not there. If the solid is nonporous, the water removed in this period is mainly superficial water on the surface of the solid. In a porous solid, most of the water removed in the constant-rate period is supplied from the interior of the solid. The evaporation from a porous material is by the same mechanism as that from a wet-bulb thermometer, and the process occurring at a wet-bulb thermometer is essentially one of constant-rate drying. In the absence of radiation or of heat transfer by conduction through direct contact of the solid with hot surfaces, the temperature of the solid during the constant-rate period is the wet-bulb temperature of the air.

CRITICAL MOISTURE CONTENT AND FALLING-RATE PERIOD   As the moisture content decreases, the constant-rate period ends at a definite moisture content, and during further drying the rate decreases. The point terminating the constant-rate period, shown by point *B* in Figs. 25-10 and 25-13 to 25-15, is called the *critical point*.[13] This point marks the instant when the liquid water on the surface is insufficient to maintain a continuous film covering the entire drying area. In nonporous solids, the critical point occurs at about the time when the superficial moisture is evaporated. In porous solids, the critical point is reached when the rate of moisture flow to the surface no longer equals the rate of evaporation called for by the wet-bulb evaporative process.

If the initial moisture content of the solid is below the critical, the constant-rate period does not occur.

The critical moisture content varies with the thickness of the material and with the rate of drying. It is therefore not a property of the material itself. The critical moisture content is best determined experimentally, although some approximate data are available.[10a]

The period subsequent to the critical point is called the *falling-rate period*. It is clear from Figs. 25-10 and 25-13 to 25-15 that the drying-rate curve in the falling-rate period varies from one type of material to another. The shape of the curve also depends on the thickness of the material and on the external variables. In some situations, illustrated in Figs. 25-13 and 25-14, there is a distinct break in the curve during the falling-rate period, signifying a change in the drying mechanism. This break occurs at what is called the *second critical point*, corresponding to the second critical moisture content.

Calculation of drying time under constant drying conditions   In the design of dryers, an important quantity is the time required for drying the material under the conditions existing in the dryer, as this fixes the size of the equipment needed for a given capacity. For drying under constant drying conditions, the time of drying can be determined from the rate of drying curve if it can be constructed. Often the only source of this curve is an experiment on the material to be dried, and this gives the drying time directly. Drying-rate curves for one set of conditions often may be modified to other conditions, and then working back from the drying-rate curve to drying time is useful. For example, consider the drying of a slab of solid of half-thickness s

and face area S, dried from both sides. The total area for drying is $2S$. The ordinate R of the rate-of-drying curve is therefore given by the equation

$$R = -\frac{dm}{2S\,dt} \qquad (25\text{-}1)$$

where m = mass of total moisture in solid
t = drying time

The abscissa of the curve is X, the mass of free moisture per unit mass of bone-dry solid. If ρ_s is the density of the solid, in mass per unit volume of bone-dry material, and if shrinkage is neglected so that ρ_s is constant,

$$dm = 2sS\rho_s\,dX \qquad (25\text{-}2)$$

Combining Eqs. (25-1) and (25-2) gives

$$R = -\frac{s\rho_s\,dX}{dt} \qquad (25\text{-}3)$$

Since the rate-of-drying curve provides a relation between R and X, Eq. (25-3) can be integrated between X_1 and X_2, the initial and final free-moisture contents, respectively,

$$t_T = s\rho_s \int_{X_2}^{X_1} \frac{dX}{R} \qquad (25\text{-}4)$$

where t_T is the time of drying. This equation may be integrated numerically from the rate-of-drying curve or analytically if equations are available giving R as a function of X.

EQUATIONS FOR CONSTANT DRYING CONDITIONS   In two situations equations are available for formally integrating Eq. (25-4). These are during the constant-rate period and during a falling-rate period in which the rate-of-drying curve is linear.

During the constant-rate period, R is constant, and it may be denoted by R_c. Equation (25-4) gives, for t_c, the time of drying in a constant-rate period,

$$t_c = s\rho_s \int_{X_2}^{X_1} \frac{dX}{R_c} = \frac{s\rho_s(X_1 - X_2)}{R_c} \qquad (25\text{-}5)$$

When the rate of drying is linear in X during a falling-rate period,

$$R = aX + b \qquad (25\text{-}6)$$

where a and b are constants. Differentiating Eq. (25-6) gives $dR = a\,dX$. Substituting for dX in Eq. (25-4) gives, for the time required in the falling-rate period,

$$t_f = \frac{s\rho_s}{a} \int_{R_2}^{R_1} \frac{dR}{R} = \frac{s\rho_s}{a} \ln \frac{R_1}{R_2} \qquad (25\text{-}7)$$

where R_1 and R_2 are the ordinates for the initial and final moisture contents, respectively. The constant a is the slope of the rate-of-drying curve and may be written as

$$a = \frac{R_c - R'}{X_c - X'} \qquad (25\text{-}8)$$

where R_c = rate at first critical point
R' = rate at second critical point
X_c = free-moisture content at first critical point
X' = free-moisture content at second critical point

Substitution of a from Eq. (25-8) into Eq. (25-7) gives

$$t_f = \frac{s\rho_s(X_c - X')}{R_c - R'} \ln \frac{R_1}{R_2} \tag{25-9}$$

When the drying process covers both a constant-rate period and a falling-rate period, the X_2 of Eq. (25-5) equals X_c, and the R_1 of Eq. (25-9) equals R_c. The total time of drying t_T is then

$$t_T = t_c + t_f = s\rho_s \left(\frac{X_1 - X_c}{R_c} + \frac{X_c - X'}{R_c - R'} \ln \frac{R_c}{R_2} \right) \tag{25-10}$$

Here X_1 is the moisture content at the start of the entire process, and R_2 is the drying rate at the end of the process.

In some situations, a single straight line passing through the origin adequately represents the entire falling-rate period. The point (X_c, R_c) lies on this line. When this approximation may be made, Eq. (25-10) may be simplified by noting that b, R', X' drop out, that $a = R_c/X_c$, and that $R_c/R_2 = X_c/X_2$. Equation (25-10) then becomes

$$t_T = \frac{s\rho_s}{R_c} \left[(X_1 - X_c) + X_c \ln \frac{X_c}{X_2} \right] \tag{25-11}$$

Here X_2 is the moisture content at the end of the entire process.

ESTIMATION OF DRYING RATES   During the constant-rate period the drying rate R_c can be estimated with fair precision from correlations developed for evaporation from a free liquid surface. In the falling-rate period the behavior of nonporous solids can be correlated, and sometimes predicted, using diffusion theory; that of porous solids can usually be established only by experiment.

CONSTANT-RATE PERIOD   To calculate the rate of drying during the constant-rate period, either the mass-transfer equation (25-12) or the heat transfer equation (25-13) may be used. Using Eq. (24-24) gives

$$\dot{m} = M_B k_y (\mathcal{H}_i - \mathcal{H})A \tag{25-12}$$

$$\dot{m} = \frac{h_y(T - T_i)A}{\lambda_i} \tag{25-13}$$

where $\dot{m}$ = rate of evaporation
   A = drying area
   h_y = heat-transfer coefficient
   k_y = mass-transfer coefficient
   $\mathcal{H}_i$ = humidity of air at interface
   $\mathcal{H}$ = humidity of air
   M_B = molecular weight of air
   T = temperature of air
   T_i = temperature at interface
   λ_i = latent heat at temperature

When the air is flowing parallel with the surface of the solid, the heat-transfer coefficient can be estimated by the *dimensional* equation

$$h_y = 0.0128G^{0.8} \tag{25-14}$$

where h_y = heat-transfer coefficient
G = mass velocity, lb/ft^2-h

When the flow is perpendicular to the surface, the equation is

$$h_y = 0.37G^{0.37} \tag{25-15}$$

In most situations, as previously mentioned, temperature T_i may be assumed to be equal to the wet-bulb temperature of the air. When radiation from hot surroundings and conduction from solid surfaces in contact with the stock are not negligible, however, the temperature of the interface is greater than the wet-bulb temperature and the rate of drying is, by Eq. (25-13), increased accordingly. Methods for estimating these effects are available.[10a, 13]

The rate of drying at the critical point R_c may be estimated from either Eq. (25-12) or (25-13). Choosing the heat-transfer form and assuming drying from both sides of the slab gives

$$R_c = -\frac{dm}{2S\,dt} = \frac{\dot{m}}{A} = \frac{h_y(T - T_i)}{\lambda_i} \tag{25-16}$$

Equation (25-11) may then be written

$$t_T = \frac{s\rho_s\lambda_i}{h_y(T - T_i)}\left[(X_1 - X_c) + X_c \ln \frac{X_c}{X_2}\right] \tag{25-17}$$

Note that in using Eq. (25-17) h_y and $T - T_i$ must be calculated for the constant-rate period; the actual temperature difference may decrease greatly during the falling-rate period, especially toward the end of the process.

Equations (25-5), (25-10), and (25-11) show that the rate of drying of materials following these equations is inversely proportional to the thickness of the solid, except as the thickness affects the critical moisture contents.

EXAMPLE 25-1   A filter cake 24 in. (610 mm) square and 2 in. (51 mm) thick is dried from both sides with air at a wet-bulb temperature of 80°F (26.7°C) and a dry-bulb temperature of 120°F (48.9°C). The air flows parallel with the faces of the cake at a velocity of 2.5 ft/s (0.76 m/s). The dry density of the cake is 120 lb/ft^3 (1,922 kg/m^3). The equilibrium-moisture content is negligible. Under the conditions of drying the critical moisture is 9 percent, dry basis. Assuming the rate of drying in the falling-rate period is proportional to the moisture content, how long will it take to dry this material from an initial moisture content of 20 percent to a final moisture of 2 percent?

SOLUTION   Since the initial moisture content is greater than the critical and the final moisture less, both constant-rate and falling-rate periods are involved. Also, since the rate-of-drying curve passes through the origin, Eq. (25-17) may be used. The interface temperature

T_i is the wet-bulb temperature of the air, 80°F. Then λ_i is, by Appendix 8, 1049 Btu/lb. The mass velocity of the air is

$$G = \frac{2.5 \times 29 \times 492 \times 3,600}{359(460 + 120)} = 617 \ \text{lb/ft}^2\text{-h}$$

The half-thickness is $s = \frac{1}{12}$ ft, and the coefficient h_y is, by Eq. (25-14),

$$h_y = 0.0128 \times 617^{0.8} = 2.18 \ \text{Btu/ft}^2\text{-h-°F}$$

Substituting in Eq. (25-17) gives

$$t_T = \frac{1/12 \times 120 \times 1049}{2.18(120 - 80)} \left((0.20 - 0.09) + 0.09 \times 2.303 \log \frac{0.09}{0.02} \right) = 29.2 \ \text{h}$$

$////$

Falling-rate period The methods used to estimate or predict drying rates and drying times in the falling-rate periods depend on whether the solid is porous or non-porous. In a nonporous material, once there is no more superficial moisture, further drying can occur only at a rate governed by diffusion of internal moisture to the surface. In a porous material other mechanisms appear, and drying may even take place inside the solid instead of at the surface.

Drying of nonporous solids and diffusion theory The moisture distribution in a solid giving a falling-rate curve like that in Fig. 25-10 is shown as the dotted line in Fig. 25-11, where the local-moisture content is plotted against distance from the surface. The curve is concave downward throughout its length. Rate curves like Fig. 25-10 and moisture-distribution curves like Fig. 25-11 are usually found in the drying of nonporous materials like soap, glue, and plastic clay. These substances are essentially colloidal gels of solids and water. They retain considerable amounts of bound water. Dense cellular solids, such as wood and leather, give curves of the same types when they are being dried at moisture contents below the fiber-saturation point.

The shape of the moisture-distribution curves of Fig. 25-11 is qualitatively consistent with that called for by assuming that the moisture is flowing by diffusion through the solid, in accordance with Eq. (22-51). This equation has long been used as a basis for quantitative calculations of the rate of drying of nonporous solids.[9,12] Materials drying in this way are said to be drying by diffusion, although the actual mechanism is probably considerably more complicated than that of simple diffusion.

Diffusion is characteristic of slow-drying materials. The resistance to mass transfer of water vapor from the solid surface to the air is usually negligible, and diffusion in the solid controls the over-all drying rate. The moisture content at the surface, therefore, is at or very near the equilibrium value. The velocity of the air has little or no effect, and the humidity of the air influences the process primarily through its effect on the equilibrium-moisture content. Since diffusivity increases with temperature, the rate of drying increases with the temperature of the solid.

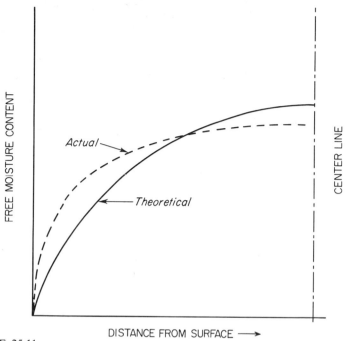

FIGURE 25-11

Moisture distribution in slab dried from both faces. Moisture flow by diffusion.

EQUATIONS FOR DIFFUSION   Assuming that the diffusion law given by Eq. (22-57) does apply, the integrated forms of this equation can be used to relate the time of drying with the initial and final moisture contents. Thus, if the assumptions made in integrating Eq. (22-57) are all accepted, Eq. (22-58) is obtained on integration. For drying, the result can be written

$$\frac{X_T - X^*}{X_{T1} - X^*} = \frac{X}{X_1} = \frac{8}{\pi^2}\left(e^{-a_1\beta} + \frac{1}{9}e^{-9a_1\beta} + \frac{1}{25}e^{-25a_1\beta} + \cdots\right) \quad (25\text{-}18)$$

where $\beta = D'_v t_T / s^2$
   $a_1 = (\pi/2)^2$
   X_T = average total moisture content at time t_T h
   X = average free-moisture content at time t_T h
   X^* = equilibrium-moisture content
   X_{T1} = initial moisture content at start of drying when $t = 0$
   X_1 = initial free-moisture content
   D'_v = diffusivity of moisture through solid
   s = one-half slab thickness

All moisture contents are in mass of water per unit mass of bone-dry solid. Similar equations for drying cylinders or spheres can be written by analogy with Eqs. (10-18) and (10-19) for heat transfer. In Fig. 10-6, with β in place of N_{Fo} and X/X_1 in place of $(T_s - T_b)/(T_s - T_a)$, is a plot of Eq. (25-18) for drying a large flat slab.

The accuracy of the diffusion theory for drying suffers from the fact that the diffusivity usually is not constant but varies with moisture content. It is especially sensitive to shrinkage. The value of D_v' is less at small moisture contents than at large and may be very small near the drying surface. Thus, the moisture distribution called for by the diffusion theory with constant diffusivity is like that shown by the full line in Fig. 25-11. In practice, an average value of D_v', established experimentally on the material to be dried, is used.

When β is greater than about 0.1, only the first term on the right-hand side of Eq. (25-18) is significant, and the remaining terms of the infinite series may be dropped. Solving the resulting equation for the drying time gives

$$t_T = \frac{4s^2}{\pi^2 D_v'} \ln \frac{8X_1}{\pi^2 X} \tag{25-19}$$

Differentiating Eq. (25-19) with respect to time and rearranging gives

$$-\frac{dX}{dt} = \left(\frac{\pi}{2}\right)^2 \frac{D_v'}{s_2} X \tag{25-20}$$

Equation (25-20) shows that when diffusion controls, the rate of drying is inversely proportional to the square of the thickness. Equation (25-19) shows that if time is plotted against the logarithm of the free-moisture content, a straight line should be obtained, from the slope of which D_v' can be calculated.

EXAMPLE 25-2   Planks of wood 1 in. (25.4 mm) thick are dried from an initial moisture content of 25 percent to a final moisture of 5 percent, using air of negligible humidity. If D_v' for the wood is 3.2×10^{-5} ft²/h (8.3×10^{-5} cm²/s), how long should it take to dry the wood?

SOLUTION   Since the air is dry, the equilibrium-moisture content is zero, and $X_1 = 0.25$ and $X = 0.05$. Then $X/X_1 = 0.05/0.25 = 0.20$. From Fig. 10-6, β is read on the abscissa corresponding to the known ordinate of 0.20 and found to be 0.57. Since $s = 0.5/12$ ft,

$$\frac{3.2 \times 10^{-5} t_T}{(0.5/12)^2} = 0.57$$

Solving for t_T gives a drying time of 31.0 h.

Another solution can be found by Eq. (25-19). Substituting the appropriate values in this equation gives

$$t_T = \frac{4(0.5/12)^2 2.303 \log [(8 \times 0.25)/(\pi^2 \times 0.05)]}{\pi^2(3.2 \times 10^{-5})} = 30.8 \text{ h} \qquad \text{////}$$

Shrinkage and casehardening   When bound moisture is removed from a colloidal nonporous solid, the material shrinks. In small pieces this effect may not be important, but in large units improper drying may lead to serious product difficulties which are basically a result of shrinkage. Since the outer layers necessarily lose moisture before the interior portions, the concentration of moisture in these layers is less than that in the interior, and the surface layers shrink against an unyielding, constant-volume core. This surface shrinkage causes checking, cracking, and warping. Also, since the diffusivity is sensitive to moisture concentration and decreases with concentration, the resistance to diffusion in the outer layers is increased by surface dehydration. This accentuates the effect of shrinkage by impeding the flow of moisture to the surface and so increasing the moisture gradient near the surface.

In extreme cases, the shrinkage and drop in diffusivity may combine to give a skin, practically impervious to moisture, which encloses the bulk of the material so the interior moisture cannot be removed. This is called casehardening.

Warping, checking, cracking, and casehardening can be minimized by reducing the rate of drying, thereby flattening the concentration gradients in the solid. Then the shrinkage of the surface is reduced, and the diffusivity throughout the solid is more nearly constant. The moisture gradient at the surface is flattened, and the entire piece is protected against shrinkage.

The rate of drying is controlled most readily by controlling the humidity of the drying air. Since the equilibrium-moisture concentration at the surface is fixed by the humidity of the air, increasing the latter also increases the former. For a given total moisture concentration X_T, the free-moisture concentration X is reduced by increasing the equilibrium moisture X^*. The overall concentration gradient of free water is then reduced, the drying slowed, and the skin effect minimized.

Drying of porous solids and flow of water by capillarity   The flow of water through porous solids does not conform to the laws of diffusion given by Eq. (25-16). This may be seen by comparing the moisture distribution in a solid of this type during drying with that for diffusion. A typical moisture-distribution curve for a porous solid is shown in Fig. 25-12. A point of inflection divides the curve into two parts, one concave upward and one concave downward. This is completely contrary to the distribution called for by the diffusion law as shown in Fig. 25-11.

Moisture flows through porous solids by capillarity.[2,3,5] A porous material contains a complicated network of interconnecting pores and channels, the cross sections of which vary greatly. At the surface are the mouths of pores of various sizes. As water is removed by vaporization, a meniscus across each pore is formed, which sets up capillary forces by the interfacial tension between the water and the solid. The capillary forces possess components in the direction perpendicular to the surface of the solid. It is these forces that provide the driving force for the movement of water through the pores toward the surface.

The strength of capillary forces at a given point in a pore depends on the curvature of the meniscus, which is a function of the pore cross section. Small pores develop greater capillary forces than large ones, and small pores, therefore, can pull water out of the large pores. As the water at the surface is depleted, the large pores tend to empty first. Air must displace the water so removed. This air enters either through the mouths of the larger pores at the drying surface or from the sides and back of the material if drying is from one side only.

The forces developed in fine pores by capillarity are surprisingly large. The flow

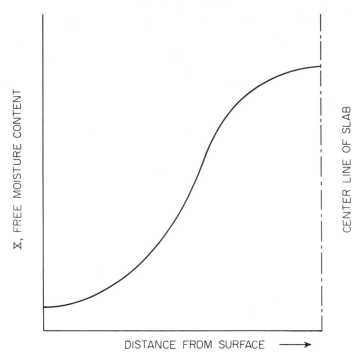

FIGURE 25-12
Moisture distribution in porous slab dried from both faces. Moisture flow by capillarity.

of sap to the tops of tall trees is accomplished by capillarity. In drying unbound water from large pieces of wood, internal forces are developed inside the piece in locations not accessible to air, and the structure of the wood collapses. If the solid is plastic, considerable shrinkage occurs when water is removed by capillarity.

The rate-of-drying curve for a typical porous solid having small pores is shown in Fig. 25-13. As long as the delivery of water from the interior to the surface is sufficient to keep the surface completely wet, the drying rate is constant. The pores are progressively depleted of water, and at the critical point the surface layer of water begins to recede into the solid. This starts with the larger pores. The high points on the surface of the solid emerge like Mt. Ararat after the flood, and the area available for mass transfer from the solid into the air decreases. Then, although the rate of evaporation per unit *wetted* area remains unchanged, the rate based on the *total* surface, including both wet and dry areas, is less than that in the constant-rate period. The rate continues to decrease as the fraction of dry surface increases.

The first section of the falling-rate period is shown by line *BC* in Fig. 25-13. The rate of drying during this period depends on the same factors that are active during the constant-rate period, since the mechanism of evaporation is unchanged and the vaporization zone is at or near the surface. The state of the water during this period is called the *funicular state*. The water in the pores is the continuous phase and the air

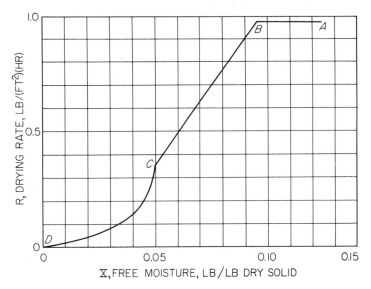

FIGURE 25-13
Drying-rate curve for porous ceramic plate. (After Sherwood and Comings.[15])

the dispersed phase. In the first falling-rate period, the rate-of-drying curve is usually linear.

As the water is progressively removed from the solid, the fraction of the pore volume that is occupied by air increases. When the fraction reaches a certain limit, there is insufficient water left to maintain continuous films across the pores, the interfacial tension in the capillaries breaks, and the pores fill with air, which now becomes the continuous phase. The remaining water is relegated to small isolated pools in the corners and interstices of the pores. This state is called the *pendular state*. When this state appears, the rate of drying again suddenly decreases as shown by line *CD* in Fig. 25-13. The moisture content at which this break appears, shown by point *C* in Fig. 25-13, is called the second critical point, and the period that it initiates is called the second falling-rate period.

In the pendular state the vaporization rate is practically independent of the velocity of the air. The water vapor must diffuse through the solid, and the heat of vaporization must be transmitted to the vaporization zones by conduction through the solid. Temperature gradients are set up in the solid, and the temperature of the solid surface approaches the dry-bulb temperature of the air. For fine pores, the rate-of-drying curve during the second falling-rate period conforms to the diffusion law, and the drying-rate curve is concave upward.

EFFECT OF GRAVITY In quite porous solids, such as beds of sand, the pores are large, the resistance to moisture flow is low, and the capillary forces are small. The force of gravity is then large in comparison with the capillary forces, and there is a directional effect due to gravity.[6]

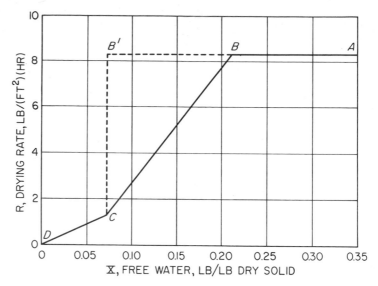

FIGURE 25-14
Drying-rate curves for sand beds. (After Hougen.[6])

Drying-rate curves for a horizontal bed of sand particles are shown in Fig. 25-14. Two broken lines are shown. The full line $ABCD$ is obtained for a sample dried from the top and the dotted line $AB'CD$ for a sample dried from the bottom. In top drying gravity opposes capillarity, and the first critical point is reached early, as shown by point B. Then the usual two falling-rate periods, represented by segments BC and CD, are found. In bottom drying capillary and gravity forces act in the same direction to move water to the drying surface, and the constant-rate period continues until the pendular state appears. The constant-rate period is represented by segment AB'. At the critical point B' the rate drops suddenly to that corresponding to the second critical point found in top drying. Only one critical point and one falling-rate period are obtained.

In very porous solids, the rate-of-drying curve in the second falling-rate period is often straight, and the diffusion equations do not apply.

Drying of hygroscopic porous solids When the solid is both porous and hygroscopic, as in a slab of paper pulp, only the unbound water can form the funicular and pendular states. When the unbound water has been removed, considerable bound water is left. This is then removed by progressive vaporization below the surface of the solid, which is accompanied by diffusion of water vapor through the solid.[8] The second critical point is suppressed, and only one critical point, that at the end of the constant-rate period, is found. The drying-rate curves for several thicknesses of porous paper pulp are shown in Fig. 25-15. It is clear that there is no simple relation between drying rate and slab thickness in drying materials of this character.

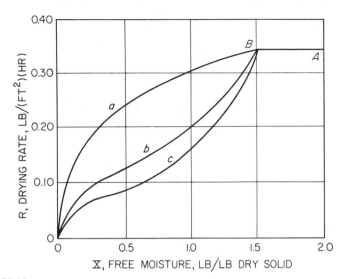

FIGURE 25-15
Drying-rate curves for paper-pulp slabs: (*a*) thickness 0.108 cm; (*b*) thickness
0.648 cm; (*c*) thickness 2.37 cm. (After McCready and McCabe.[8])

Drying time under variable drying conditions Under certain restricted conditions, equations can be derived for calculating the drying time in a continuous or
batch dryer when the humidity and temperature of the air are varying. This involves
an integration over the entire process that takes account of the variations in these
variables. It is generally necessary that the solid be porous and that the drying fall
either in a constant-rate period or in the first falling-rate period.

   When diffusion controls, no satisfactory method of calculation is known for
variable-condition drying. The basic difficulty is that the instantaneous rate is not a
function of the average moisture content of the solid but depends also on the previous
history of the material during drying. Also, the change in solid temperature during
drying cannot be predicted. The most satisfactory method for obtaining drying times
experimentally in such situations is to conduct a small-scale drying experiment on an
actual sample of the material under the same variations in temperature, humidity, and
air velocity that are to be used in the commercial dryer.[1] This gives the drying time
directly.

SYMBOLS

A Area for drying, ft² or m²
a Slope of drying-rate curve [Eq. (25-6)]
a_1 $(\pi/2)^2$
b Constant in Eq. (25-6)
D_v' Volumetric diffusivity of moisture through solid, ft²/h, m²/h, or cm²/s

G Mass velocity of drying air, lb/ft²-h or kg/m²-h

$\mathscr{H}$ Humidity of drying air, mass of vapor per unit mass of dry air; $\mathscr{H}_i$, at solid surface

h_y Heat-transfer coefficient between air and solid surface, Btu/ft²-h-°F or W/m²-°C

k_y Mass-transfer coefficient, from solid surface to air, lb mol/ft²-unit mole-fraction difference or g mol/m²-unit mole-fraction difference

M_B Molecular weight of air

m Mass of total moisture in solid, lb or kg

$\dot{m}$ Rate of evaporation, lb/h or kg/h

R Rate of drying, lb/ft²-h or kg/m²-h; R_c, at first critical moisture content; R_1, at start of drying; R_2, at end of drying; R', at second critical moisture content

S Area of one face of slab, ft² or m²

s One-half slab thickness, ft or m

T Temperature, °F or °C; T_a, initial temperature of slab; T_i, at interface; T_s, at slab surface; $\bar{T}_b$, bulk average temperature of slab

t Drying time, h; t_T, total drying time; t_c, in constant-rate period; t_f, in falling-rate period

X Free moisture content, mass of water per unit mass of dry solid; X_T, total moisture content; X_{T1}, initial total moisture content; X_c, at first critical moisture content; X_1, initial value; X_2, final value; X^*, equilibrium value; X', at second critical moisture content

Greek letters

β Dimensionless group, $D'_v t_T / s^2$, in Eq. (25-16)

λ_i Latent heat of vaporization at interface temperature, Btu/lb or J/g

ρ_s Density of solid, lb dry solid/ft³ or kg dry solid/m³

PROBLEMS

25-1 Wood chips are to be dried in a continuous rotary countercurrent dryer. The chips entering the dryer contain 40 percent moisture (wet basis), and the chips discharged from the dryer contain 15 percent moisture (wet basis). The dryer is to deliver 2,000 lb of dry (15 percent moisture) product per hour. Air enters at 230°F and 0.010 humidity and leaves at 120°F. The chips must remain in the dryer 2 h. Small-scale experiments show that air velocities must not exceed 8 ft/s or chips will be blown from the dryer. The average chip size is 1 by 2 by ¼ in. The ratio of length to diameter of the dryer is to be 5:1. The chips occupy 0.17 ft³ per pound of moisture-free chips. The average pressure in the dryer is 1 atm. What should be the length and diameter of the dryer?

25-2 A porous solid is dried in a batch dryer under constant drying conditions. Six hours are required to reduce the moisture content from 30 to 10 percent. The critical moisture content was found to be 16 percent and the equilibrium moisture 2 percent. All moisture contents are on the dry basis. Assuming that the rate of drying during the falling-rate period is proportional to the free-moisture content, how long should it take to dry a sample of the same solid from 35 to 6 percent under the same drying conditions?

25-3 A slab with a wet weight of 5 kg originally contains 50 percent moisture (wet basis). The slab is 600 by 900 by 75 mm thick. The equilibrium-moisture content is 5 percent of the total weight when in contact with air of 20°C and 20 percent humidity. The drying rate is given in Table 25-1 for contact with air of the above quality at a definite velocity.

Table 25-1   DATA FOR PROB. 25-3

Wet-slab weight, kg	9.1	7.2	5.3	4.2	3.3	2.8	2.5
Drying rate, kg/m²-h	4.9	4.9	4.4	3.9	3.4	2.0	1.0

Drying is from one face. How long will it take to dry the slab to 15 percent moisture content (wet basis)?

25-4 A continuous countercurrent dryer is to be designed to dry 500 lb of wet porous solid per hour from 60 percent moisture to 10 percent, both on the wet basis. Air at 120°F dry bulb and 70°F wet bulb is to be used. The exit humidity is to be 0.012. The average equilibrium-moisture content is 5 percent of the dry weight. The total moisture content (wet basis) at the critical point is 30 percent. The stock may be assumed to remain at a temperature 3°F above that of the wet-bulb temperature of the air throughout the dryer. The mass-transfer coefficient k_y is 50 lb/ft^2-h-unit humidity. The stock has 1.0 ft^2 surface exposed to the air per pound of dry stock. How long must the stock remain in the dryer?

25-5 A film of polymer 2 m wide and 0.76 mm thick leaves the surface of a heated drum containing 35 weight percent (dry basis) of acetone. It is dried by exposing both sides of the film to air at 1 atm and 65°C containing essentially no acetone vapor. The flow of air is across the faces of the film at 3 m/s. The critical acetone content is 25 weight percent acetone (dry basis). The equilibrium acetone content is negligible. The density of the dry solid is 800 kg/m^3. (*a*) What would be the surface temperature of the film during the constant-rate period? (*b*) What would be the constant drying rate, in kilograms per square meter per hour? (*c*) How long would it take to reduce the acetone content from 35 to 1 percent, if the diffusivity of acetone in the solid is 8×10^{-4} cm^2/s? For acetone-air mixtures $h_y/M_B k_y$ is 1.76 when h_y is in W/m^2-°C.

REFERENCES

1 Broughton, D. B., and H. S. Mickley: *Chem. Eng. Prog., 49*:319 (1953).
2 Ceaglske, N. H., and O. A. Hougen: *Trans. AIChE, 33*:283 (1937).
3 Ceaglske, N. H., and F. C. Kiesling: *Trans. AIChE, 36*:211 (1940).
4 Clark, W. E.: *Chem. Eng., 74*(6):177 (1967).
5 Comings, E. W., and T. K. Sherwood: *Ind. Eng. Chem., 26*:1096 (1934).
6 Hougen, O. A.: paper on Dryer Calculations and Design, in "Teaching of Chemical Engineering," *Proc. Chem. Eng. Div., Soc. Prom. Eng. Educ., 2d Summer Sch. Chem. Eng., June,* 1939.
7 Marshall, W. R., Jr.: Atomization and Spray Drying, *Chem. Eng. Prog. Monogr. Ser., 50*(2), 1954.
8 McCready, D. W., and W. L. McCabe: *Trans. AIChE, 29*:131 (1933).
9 Newman, A. B.: *Trans. AIChE, 27*:203, 310 (1931).
10 Perry, J. H. (ed.): "Chemical Engineers' Handbook," 5th ed., McGraw-Hill, New York, 1973; (*a*) pp. **20**-4 to **20**-16, (*b*) pp. **20**-16 to **20**-61, (*c*) pp. **20**-34 to **20**-35, (*d*) pp. **20**-58 to **20**-63.
11 Riegel, E. R.: "Chemical Process Machinery," 2d ed., chap. 17, Reinhold, New York, 1953.
12 Sherwood, T. K.: *Ind. Eng. Chem., 21*:12 (1929).
13 Sherwood, T. K.: *Ind. Eng. Chem., 21*:976 (1929).
14 Sherwood, T. K.: *Ind. Eng. Chem., 24*:307 (1932).
15 Sherwood, T. K., and E. W. Comings: *Trans. AIChE, 28*:118 (1932).

Operations involving particulate solids

Solids, in general, are more difficult to handle than liquids, vapors, or gases. In processing, solids appear in many forms—large angular pieces, wide continuous sheets, finely divided powders. They may be hot, abrasive, fragile, dusty, explosive, plastic, or sticky. Whatever their form, means must be found to manipulate the solids as they occur, and if possible to improve their handling characteristics.

This section is concerned with the properties, methods of formation, modification, and separation of particulate solids. General properties and handling methods are discussed in Chap. 26, and size reduction in Chap. 27. Crystallization is the topic of Chap. 28; mixing, of Chap. 29; and mechanical separations, of Chap. 30.

26

PROPERTIES AND HANDLING OF PARTICULATE SOLIDS

Of all the shapes and sizes that may be found in solids, the most important from a chemical engineering standpoint is the small particle. An understanding of the characteristics of masses of particulate solids is necessary in designing processes and equipment for dealing with streams containing such solids.

CHARACTERIZATION OF SOLID PARTICLES

Individual solid particles are characterized by their size, shape, and density. Particles of homogeneous solids have the same density as the bulk material. Particles obtained by breaking up a composite solid, such as a metal-bearing ore, have various densities, usually different from the density of the bulk material. Size and shape are easily specified for regular particles, such as spheres and cubes, but for irregular particles (such as sand grains or mica flakes) the terms "size" and "shape" are not so clear and must be arbitrarily defined.

Particle shape As discussed in Chap. 7, the shape of an individual particle is conveniently expressed in terms of the sphericity Φ_s, which is independent of particle size.

For a spherical particle of diameter D_p, $\Phi_s = 1$; for a nonspherical particle, the sphericity is defined by the relation

$$\Phi_s \equiv \frac{6v_p}{D_p s_p} \qquad (26\text{-}1)$$

where D_p = equivalent diameter of particle
s_p = surface area of one particle
v_p = volume of one particle

For cubes and cylinders for which the length equals the diameter (provided D_p is taken as the height of the cube or cylinder), Φ_s is also unity. For irregular particles, as shown in Table 26-1, Φ_s is less than unity. For many crushed materials it is between 0.6 and 0.7.

Particle size   In general, "diameters" may be specified for any equidimensional particle. Particles which are not equidimensional, i.e., which are longer in one direction than in others, are often characterized by the *second* longest major dimension.[1] For needlelike particles, for example, D_p would refer to the thickness of the particles, not their length.

By convention, particle sizes are expressed in different units depending on the size range involved. Coarse particles are measured in inches or centimeters; fine particles in terms of screen size; very fine particles in micrometers or nanometers. Ultrafine particles are sometimes described in terms of their surface area per unit mass, usually in square meters per gram.

Standard screen series   Standard screens are used to measure the size (and size distribution) of particles in the size range between about 3 and 0.0015 in. Testing sieves are made of woven wire screens, the mesh and dimensions of which are carefully standardized. The openings are square. Each screen is identified in meshes per inch. The actual openings are smaller than those corresponding to the mesh numbers, however, because of the thickness of the wires. The characteristics of one common series, the Tyler standard screen series, are given in Appendix 20. This set of screens is based on the opening of the 200-mesh screen, which is established at 0.074 mm.

Table 26-1   SPHERICITY OF MISCELLANEOUS MATERIALS†

Material	Sphericity Φ_s	Material	Sphericity Φ_s
Spheres, cubes, short cylinders		Angular sand	0.73
($L = D_p$)	1.0	Crushed glass	0.65
Round sand	0.83	Mica flakes	0.28
Coal dust, pulverized	0.73	Berl saddles	0.3
		Raschig rings	0.3

† By permission from J. H. Perry (ed.), "Chemical Engineers' Handbook," 5th ed., p. 5-53, McGraw-Hill Book Company, New York, 1973.

The area of the openings in any one screen in the series is exactly twice that of the openings in the next smaller screen. The ratio of the actual mesh dimension of any screen to that of the next smaller screen is, then, $\sqrt{2} = 1.41$. For closer sizing, intermediate screens are available, each of which has a mesh dimension $\sqrt[4]{2}$, or 1.189, times that of the next smaller standard screen. Ordinarily these intermediate screens are not used.

Mixed particle sizes and screen analysis   In a sample of uniform particles of diameter D_p the total volume of the particles is m/ρ_p, where m and ρ_p are the total mass of the sample and the density of the particles, respectively. Since the volume of one particle is v_p, the number of particles in the sample N is

$$N = \frac{m}{\rho_p v_p} \tag{26-2}$$

The total surface area of the particles is, from Eqs. (26-1) and (26-2),

$$A = N s_p = \frac{6m}{\Phi_s \rho_p D_p} \tag{26-3}$$

To apply Eqs. (26-2) and (26-3) to mixtures of particles having various sizes and densities, the mixture is sorted into fractions, each of constant density and approximately constant size. Each fraction can then be weighed, or the individual particles in it can be counted or measured by microscopic methods. The above equations can then be applied to each fraction and the results added. Methods for separating mixtures by size and density are discussed in Chap. 30. The simplest and most common method of separating mixtures by size alone is to make a screen analysis using testing sieves. A set of standard screens is arranged serially in a stack, with the smallest mesh at the bottom and the largest at the top. An analysis is conducted by placing the sample on the top screen and shaking the stack mechanically for a definite time. The particles retained on each screen are removed and weighed, and the masses of the individual screen increments are converted to mass fractions or mass percentages of the total sample. Any particles that pass the finest screen are caught in a pan at the bottom of the stack.

The results of a screen analysis are tabulated to show the mass fraction of each screen increment as a function of the mesh size range of the increment. Since the particles on any one screen are passed by the screen immediately ahead of it, two numbers are needed to specify the size range of an increment, one for the screen through which the fraction passes and the other on which it is retained. Thus, the notation 14/20 means "through 14 mesh and on 20 mesh." An analysis tabulated in this manner is called a *differential analysis*. A typical differential analysis is shown in Table 26-2. The symbol $\Delta\phi_n$ is used for the mass fraction of the total sample that is retained by screen n, where the screens are numbered serially, starting at the top of the stack, so screen $n - 1$ is the screen immediately above screen n. The symbol D_{pn} is the particle diameter equal to the mesh opening of screen n.

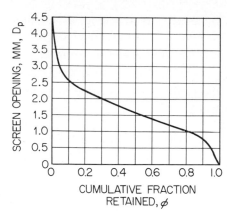

FIGURE 26-1
Cumulative screen analysis.

The second type of screen analysis is the *cumulative analysis*. A cumulative analysis is obtained from a differential analysis by adding, cumulatively, the individual differential increments, starting with that retained on the largest mesh, and tabulating or plotting the cumulative sums against the mesh dimension of the retaining screen of the last to be added. If ϕ is defined by the equation

$$\phi = \Delta\phi_1 + \Delta\phi_2 + \cdots + \Delta\phi_{n_T} = \sum_{n=1}^{n_T} \Delta\phi_n \tag{26-4}$$

the cumulative analysis is a relation between ϕ and D_p, where D_p is the mesh size of screen n. The quantity ϕ is the mass fraction of the sample that consists of particles larger than D_p. The value of ϕ for the entire sample is, of course, unity. The cumulative analysis corresponding to the differential analysis of Table 26-2 is shown in Table 26-3 and is plotted in Fig. 26-1.

CALCULATIONS BASED ON SCREEN ANALYSES   Either the differential or the cumulative analysis can be used to calculate the surface area and particle population of a mixture. If the differential analysis is used, the assumption is made that all particles in a single fraction are equal in size and that the size is the arithmetical mean of the mesh dimensions of the two screens that define the fraction. Thus, the mesh dimensions of

Table 26-2   DIFFERENTIAL SCREEN ANALYSIS

Mesh	$\Delta\phi_n$	D_{pn}, mm	Mesh	$\Delta\phi_n$	D_{pn}, mm
4/6	0.0251	3.327	35/48	0.0102	0.295
6/8	0.1250	2.362	48/65	0.0077	0.208
8/10	0.3207	1.651	65/100	0.0058	0.147
10/14	0.2570	1.168	100/150	0.0041	0.104
14/20	0.1590	0.833	150/200	0.0031	0.074
20/28	0.0538	0.589	Pan	0.0075	
28/35	0.0210	0.417			

standard 10- and 14-mesh screens are 1.651 and 1.168 mm, respectively, and the 10/14 fraction is assumed to consist of uniform particles of diameter $(1.651 + 1.168)/2 = 1.410$ mm. The symbol $\bar{D}$ is used for such arithmetic average diameters. If the cumulative analysis is used, the graph of ϕ vs. D_p is treated as a continuous function, and the methods of the calculus are utilized. In principle, the method based on the cumulative analysis is more precise than that based on the differential analysis, since when the cumulative analysis is used, the assumption that all particles in a single fraction are equal in size is not needed. The accuracy of a screen analysis is not good, however, and the precision of either method of calculation is better than that of the experimental data.

SPECIFIC SURFACE OF MIXTURE It is assumed that the particle density ρ_p and the sphericity ϕ_s are known and that these quantities are independent of particle diameter. If the differential analysis is used, the surface of the particles in each fraction is calculated by Eq. (26-3) and the results for all fractions are added to give A_w, the total surface of one unit mass of sample:

$$A_w = \frac{6\Delta\phi_1}{\Phi_s\rho_p\bar{D}_1} + \frac{6\Delta\phi_2}{\Phi_s\rho_p\bar{D}_2} + \cdots + \frac{6\Delta\phi_{n_T}}{\Phi_s\rho_p\bar{D}_{n_T}} = \frac{6}{\Phi_s\rho_p}\sum_{n=1}^{n_T}\frac{\Delta\phi_n}{\bar{D}_n} \qquad (26\text{-}5)$$

where the subscripts refer to the individual screen increments, n_T is the number of screens, $\bar{D}_n$ is the arithmetic average of D_{pn} and $D_{p(n-1)}$, and the summation means the sum of all the $\Delta\phi_n/\bar{D}_n$ quantities of the individual fractions.

  If the cumulative analysis is used, Eq. (26-3) is written differentially and the total surface found by integrating between the limits $\phi = 0$ and $\phi = 1.0$, or

$$A_w = \frac{6}{\Phi_s\rho_p}\int_0^{1.0}\frac{d\phi}{D_p} \qquad (26\text{-}6)$$

The integration may be conducted by numerical methods. The area per unit mass of sample A_w is called the *specific surface*.

AVERAGE PARTICLE SIZE The average particle size for a mixture of particles may be defined in several different ways. One average, the *volume-surface mean diameter* $\bar{D}_s$

Table 26-3 CUMULATIVE SCREEN ANALYSIS

Mesh	D_p, mm	ϕ	Mesh	D_p, mm	ϕ
4	4.699	0	35	0.417	0.9616
6	3.327	0.0251	48	0.295	0.9718
8	2.362	0.1501	65	0.208	0.9795
10	1.651	0.4708	100	0.147	0.9853
14	1.168	0.7278	150	0.104	0.9894
20	0.833	0.8868	200	0.074	0.9925
28	0.589	0.9406	Pan		1.0000

is related to the specific surface A_w. It is defined by the equation

$$\bar{D}_s = \frac{6}{\Phi_s A_w \rho_p} \tag{26-7}$$

This is the same as the Sauter mean diameter in Eq. (9-45).

Another average, the *arithmetic mean diameter*, is defined by

$$\bar{D}_N = \frac{\int_0^{N_w} D_p \, dN}{N_w} \tag{26-8}$$

where N_w is the number of particles in a unit mass of sample [see Eqs. (26-11) and (26-12)].

The *median diameter* is the diameter that divides the entire number of particles into two equal populations. The *mass mean diameter* is defined by the equation

$$\bar{D}_w = \int_0^{1.0} D_p \, d\phi \tag{26-9}$$

For samples consisting of uniform particles these average diameters are, of course, all the same. For mixtures containing particles of various sizes, however, the several average diameters may differ widely from one another.

NUMBER OF PARTICLES IN MIXTURE   To calculate, from the differential analysis, the number of particles in a mixture, Eq. (26-2) is used to compute the number of particles in each fraction, and N_w, the total population in one mass unit of sample, is obtained by summation over all the fractions. For a given particle shape, the volume of any particle is proportional to its "diameter" cubed, or

$$v_p = a D_p^3 \tag{26-10}$$

where a is the *volume shape factor*. From Eq. (26-2), then, assuming that a is independent of size,

$$N_w = \frac{\Delta\phi_1}{a\rho_p \bar{D}_1^3} + \frac{\Delta\phi_2}{a\rho_p \bar{D}_2^3} + \cdots + \frac{\Delta\phi_{n_T}}{a\rho_p \bar{D}_{n_T}^3} = \frac{1}{a\rho_p} \sum_{n=1}^{n_T} \frac{\Delta\phi_n}{\bar{D}_n^3} \tag{26-11}$$

Using the cumulative analysis, Eq. (26-11) can be interpreted as follows:

$$N_w = \frac{1}{a\rho_p} \int_0^{1.0} \frac{d\phi}{D_p^3} \tag{26-12}$$

The integral may be evaluated graphically or numerically.

Size measurements with fine particles   The sizes of particles too fine for screen analysis are measured by a variety of methods, including differential sedimentation, porosity measurements on settled beds, light absorption of suspensions, adsorption of gases on the particle surfaces, and by visual counting using a microscope.[5] Sometimes it is desirable to extrapolate data from a screen analysis to predict the particle-size distribution in the range of subsieve sizes. Methods of accomplishing this are discussed in Chap. 27.

PROPERTIES OF PARTICULATE MASSES

Masses of solid particles, especially when the particles are dry and not sticky, have many of the properties of a fluid. They exert pressure on the sides and walls of a container; they flow through openings or down a chute. They differ from liquids and gases in several ways, however, because the particles interlock under pressure and cannot slide over one another until the applied force reaches an appreciable magnitude. Unlike most fluids, granular solids and solid masses permanently resist distortion when subjected to at least some distorting force. When the force is large enough, failure occurs and one layer of particles slides over another, but between the layers on each side of the failure there is appreciable friction. There is a close analogy between the flow of particulate solids and that of plastic non-Newtonian liquids.

Solid masses have the following distinctive properties:

1 The pressure is not the same in all directions. In general, a pressure applied in one direction creates some pressure in other directions, but it is always smaller than the applied pressure. It is a minimum in the direction at right angles to the applied pressure.

2 A shear stress applied at the surface of a mass is transmitted throughout a static mass of particles unless failure occurs.

3 The density of the mass may vary, depending on the degree of packing of the grains. The density of a fluid is a unique function of temperature and pressure, as is that of each individual solid particle; but the bulk density of the mass is not. The bulk density is a minimum when the mass is "loose"; it rises to a maximum when the mass is packed by vibrating or tamping.

Depending on their flow properties, particulate solids are divided into two classes, *cohesive* and *noncohesive*. Noncohesive materials like grain, sand, and plastic chips readily flow out of a bin or silo. Cohesive solids, such as wet clay, are characterized by their reluctance to flow through openings.

Pressures in masses of particles The minimum pressure in a solid mass is in the direction normal to that of the applied pressure. In a homogeneous mass the ratio of the normal pressure to the applied pressure is a constant K', which is characteristic of the material. K' depends on the shape and interlocking tendencies of the particles, on the stickiness of the grain surfaces, and on the degree of packing of the material. It is nearly independent of particle size until the grains become very small and the material is no longer free-flowing.

If the applied pressure is p_V and the normal pressure is p_L, the pressure p at any intermediate angle can be found as follows. A right-angled triangular differential section, of thickness b and hypotenuse dL, is shown in Fig. 26-2a. Pressure p_V acts on the base, p_L on the side, and p on the hypotenuse. The angle between base and hypotenuse is θ. At equilibrium the unequal pressures p_V and p_L cannot be balanced by a single pressure p; there must also be a shear stress τ. The forces resulting from these stresses are shown in Fig. 26-2b.

Equating the components of force at right angles to the hypotenuse gives

$$pb \, dL = p_L b \, dL \sin^2 \theta + p_V b \, dL \cos^2 \theta \qquad (26\text{-}13)$$

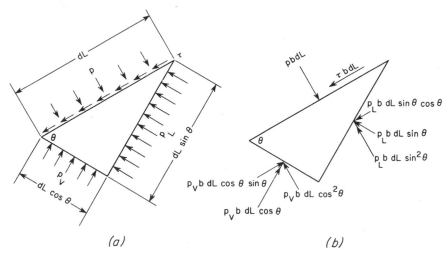

FIGURE 26-2
Stresses and forces in granular solids: (*a*) stresses; (*b*) forces.

Dividing by $b\,dL$, and noting that $\sin^2 \theta = 1 - \cos^2 \theta$, leads to

$$p = (p_V - p_L) \cos^2 \theta + p_L \tag{26-14}$$

Similarly equating forces parallel to the hypotenuse gives

$$\tau = (p_V - p_L) \cos \theta \sin \theta \tag{26-15}$$

When $\theta = 0°$, $p = p_V$; when $\theta = 90°$, $p = p_L$. In both these cases $\tau = 0$. When θ has an intermediate value, there is a shear stress at right angles to p. If corresponding values of p and τ are plotted for all possible values of θ, the resulting graph is a circle with a radius $(p_V - p_L)/2$ and its center on the horizontal axis at $p = (p_V + p_L)/2$. Such a graph is shown in Fig. 26-3; it is known as the *Mohr stress circle*.[7]

The ratio of τ to p for any value of θ is the tangent of an angle α formed by the p axis and the line OX through the origin and the point (p, τ). As θ increases from 0 to 90°, the ratio of τ to p rises to a maximum and then diminishes. It is a maximum when the line through the origin is tangent to the stress circle, as shown by the line OA in Fig. 26-3. Under these conditions α has a maximum value α_m. From Fig. 26-3 it may be seen that

$$\sin \alpha_m = \frac{(p_V - p_L)/2}{(p_V + p_L)/2} = \frac{p_V - p_L}{p_V + p_L} \tag{26-16}$$

The lines OA and OB are tangent to all the stress circles for any valuc of p_V provided the material is noncohesive. They form the *Mohr rupture envelope*. With cohesive solids or solid masses the tangents forming the envelope do not pass through the origin but intercept the vertical axis at points above and below the horizontal axis.[3]

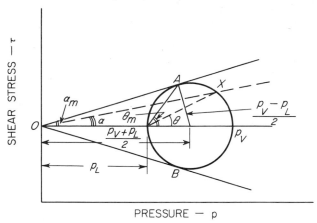

FIGURE 26-3
Mohr stress diagram for noncohesive solids.

The ratio of the normal pressure to the applied pressure p_L/p_V equals K'. Then

$$\sin \alpha_m = \frac{1 - K'}{1 + K'} \tag{26-17}$$

Also

$$K' = \frac{1 - \sin \alpha_m}{1 + \sin \alpha_m} \tag{26-18}$$

ANGLE OF INTERNAL FRICTION AND ANGLE OF REPOSE   The angle α_m is the *angle of internal friction* of the material. The tangent of this angle is the coefficient of friction between two layers of particles. When granular solids are piled up on a flat surface, the sides of the pile are at a definite reproducible angle with the horizontal. This angle, α_r, is the *angle of repose* of the material. Ideally, if the mass were truly homogeneous, α_r would equal α_m. In practice, the angle of repose is smaller than the angle of internal friction because the grains at the exposed surface are more loosely packed than those inside the mass and are often drier and less sticky. The angle of repose is low when the grains are smooth and rounded; it is high with very fine, angular, or sticky particles. K' approaches zero for a cohesive solid. For free-flowing granular materials K' is often between 0.35 and 0.6, which means that α_m is between 15 and 30°.

STORAGE OF SOLIDS

Bulk storage   Pulverized solids like sulfur and coal are usually stored outdoors in large piles, unprotected from the weather. Where hundreds or thousands of tons of material are involved, this is the only economical method. The solids are removed from the pile by dragline or tractor shovel and are delivered to process or to a conveyor. Inventorying is done by estimating the volume of the pile through aerial or ground surveys and multiplying the volume by the bulk density of the material. A pile of finely divided solids is subject to erosion by wind and rain, so that this method of storage is ordinarily restricted to coarse water-insoluble materials. Substances like rock salt, however, are sometimes stored out of doors, usually in a shallow basin, which collects rainwater and its load of dissolved material. Water may be deliberately circulated through and under such a pile of salt to form brine which is pumped to processing operations.

Bin storage   Solids that are too valuable or too soluble to expose in outdoor piles are stored in bins, hoppers, or silos. These are cylindrical or rectangular vessels of concrete or metal. A silo is tall and relatively small in diameter; a bin is not so tall and usually fairly wide. A hopper is a small bin with a sloping bottom, for temporary storage before feeding solids to a process. All these containers are loaded from the top by some kind of elevator; discharging is ordinarily from the bottom. As discussed later, a major problem in bin design is to provide satisfactory discharge.

PRESSURES IN BINS, HOPPERS, AND SILOS   When granular solids are stored in a bin or hopper, the lateral pressure exerted on the walls at any point is less than predicted from the head of material above that point. Furthermore there usually is friction between the wall and the solid grains, and because of the interlocking of the particles, the effect of this friction is felt throughout the mass. The frictional force at the wall tends to offset the weight of the solid and reduces the pressure exerted by the mass on the floor of the container. In the extreme case this force causes the mass to arch, or bridge, so that it does not fall even when the material below is removed.

An expression for the pressure exerted by a granular solid on the floor of a circular bin with vertical walls is derived as follows. Figure 26-4 shows a differential horizontal layer of thickness dZ at a distance Z from the upper surface of the solids. The inside radius of the bin is r; the total height of solids is Z_T. At the level Z, assume that the differential layer is a piston pressing against the solid beneath and that this piston is acted upon by a concentrated vertical force F_V from above. The vertical pressure p_V at level Z therefore is

$$p_V = \frac{F_V}{\pi r^2} \tag{26-19}$$

from which

$$dF_V = \pi r^2 \, dp_V \tag{26-20}$$

The net increase in downward force caused by the differential layer is the force of gravity dF_g minus the frictional force dF_f. Thus

$$dF_V = dF_g - dF_f \tag{26-21}$$

The force of gravity on the layer is $(g/g_c)\pi\rho_b r^2 \, dZ$, where ρ_b is the bulk density of the material. The frictional force is the product of the coefficient of friction μ' at

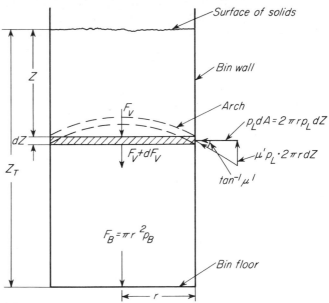

FIGURE 26-4
Forces in circular bin with vertical sides.

the bin wall and the lateral force F_L. The lateral force is, in turn, the product of the lateral pressure p_L and the area on which it acts, $2\pi r\,dZ$. Then

$$dF_V = \pi r^2\,dp_V = \pi r^2 \rho_b \frac{g}{g_c}\,dZ - \mu'(2\pi r p_L\,dZ) \tag{26-22}$$

Dividing through by πr and noting that $p_L/p_V = K'$ gives

$$r\,dp_V = \left(r\rho_b \frac{g}{g_c} - 2\mu' \frac{p_L}{p_V} p_V\right) dZ = \left(r\rho_b \frac{g}{g_c} - 2\mu'K'p_V\right) dZ \tag{26-23}$$

Let p_B equal the vertical pressure on the bin floor. Integrating Eq. (26-23) from top to bottom of the mass leads to

$$\int_0^{Z_T} dZ = \int_0^{p_B} \frac{r\,dp_V}{r\rho_b(g/g_c) - 2\mu'K'p_V}$$

$$Z_T = -\frac{r}{2\mu'K'}\left[\ln\left(r\rho_b \frac{g}{g_c} - 2\mu'K'p_V\right)\right]_0^{p_B} \tag{26-24}$$

Substituting the limits of integration and rearranging gives

$$p_B = \frac{r\rho_b(g/g_c)}{2\mu'K'}\left(1 - e^{-2\mu'K'Z_T/r}\right) \tag{26-25}$$

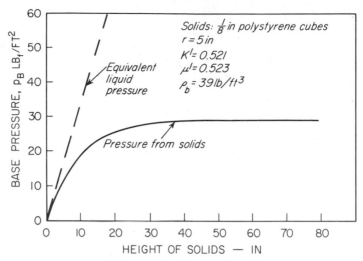

FIGURE 26-5
Base pressure in cylindrical bins. (Rudd.[6])

Equation (26-25) is the *Janssen equation*, which has been well substantiated by experiment. A typical relation of base pressure to height is shown in Fig. 26-5. With many solids, when the height reaches about 3 times the diameter of the bin, additional material has virtually no effect on the pressure at the base.

When the bin is other than circular in cross section, r is replaced by 2 times the hydraulic radius. The coefficient of friction is found experimentally by noting the angle at which the solids just begin to slide on an inclined surface. The coefficient μ' is the tangent of this angle. For granular materials on concrete or smooth metal surface, μ' ranges from 0.35 to 0.55.

EXAMPLE 26-1   A packed absorption tower 6 ft (1.82 m) in diameter and 50 ft (15.24 m) high is to be filled with crushed sized coke. Compute the vertical and lateral pressures at the base caused by the coke. Compare with the pressure that would be exerted by a liquid of the same density. The bulk density and angle of repose are

$$\rho_b = 30 \text{ lb/ft}^3 \text{ (481 kg/m}^3) \qquad \alpha_r = 28°$$

SOLUTION   The quantities needed are

$$\alpha_m \text{ (est.)} = 32° \qquad \sin \alpha_m = 0.5299$$

$$K' \text{ [Eq. (26-18)]} = \frac{1 - 0.5299}{1 + 0.5299} = 0.307$$

$$r = 3 \text{ ft} \qquad Z_T = 50 \text{ ft} \qquad \mu' \text{ (est.)} = 0.5$$

The vertical pressure at the base [from Eq. (26-25)] is

$$p_B = \frac{3 \times 30(g/g_c)}{2 \times 0.5 \times 0.307} (1 - e^{-(2 \times 0.5 \times 0.307 \times 50)/3}) = 291 \text{ lb}_f/\text{ft}^2$$

$$= 2.02 \text{ lb}_f/\text{in.}^2 \text{ (13,930 N/m}^2)$$

The lateral pressure at the base is

$$p_L = K'p_B = 0.307 \times 2.02 = 0.62 \text{ lb}_f/\text{in.}^2 \ (4{,}275 \text{ N/m}^2)$$

The equivalent liquid pressure is

$$p = \frac{\rho Z_T}{144} \frac{g}{g_c} = \frac{30 \times 50}{144} = 10.4 \text{ lb}_f/\text{in.}^2 \ (71{,}700 \text{ N/m}^2) \qquad ////$$

In estimating K' it is best to assume that α_m is fairly large, since this leads to conservative values of p_B.

FLOW OUT OF BINS Solids tend to flow out of any opening near the bottom of a bin but are best discharged through an opening in the floor. As shown in Example 26-1, the pressure at a side outlet is smaller than the vertical pressure at the same level, so that the opening clogs more easily; furthermore, removal of solids from one side of a bin considerably increases the lateral pressure on the other side during the time the solids are flowing. A bottom outlet is less likely to clog and does not induce abnormally high pressures on the wall at any point.

Except in very small bins it is not feasible to open the entire bottom for discharge. Commonly a conical or pyramidal bottom leads to a fairly small circular outlet closed with a valve or to a rotary feeder. The pressure at the bottom of the slope-sided section of the bin is appreciably less than that indicated by Eq. (26-25). In addition, the vertical pressure fluctuates as the material discharges and averages 5 to 10 percent higher than when the mass is stationary.

When the outlet at the bottom of a bin containing free-flowing solids is opened, the material immediately above the opening begins to flow. A central column of solids moves downward without disturbing the material at the sides. Eventually lateral flow begins, first from the topmost layer of solids. A conical depression is formed in the surface of the mass. The solids at the bin floor, at or near the walls, are the last to leave. The material slides laterally into the central column at an angle approximating the angle of internal friction of the solids. If additional material is added at the top of the bin at the same rate as material is flowing out the bottom, the solids near the bin walls remain stagnant and do not discharge no matter how long flow persists.

The rate of flow of granular solids by gravity through a circular opening in the bottom of a bin depends on the diameter of the opening and on the properties of the solid. Within wide limits it does not depend on the height of the bed of solids. An empirical dimensional equation for free-flowing particles of density ρ_p in the size range between 4- and 20-mesh is[2]

$$\dot{m} = \frac{\rho_p D_o^n}{(6.288 \tan \alpha_m + 23.16)(D_p + 1.889) - 44.90} \qquad (26\text{-}26)$$

where $\dot{m}$ = solids flow rate, lb/min
D_o = opening diameter, in.
α_m = angle of internal friction of solids
D_p = particle diameter, in.

Exponent n varies from about 2.8 for angular particles to about 3.1 for spheres. Detailed studies of the flow of particles through openings were made by Laforge and Boruff.[4]

With cohesive solids it is often hard to start flow. Once flow does start, however, it again begins in the material directly above the discharge opening. Frequently the column of solids above the outlet moves out as a plug, leaving a "rathole" with nearly vertical sides. Sticky solids and even some dry powders adhere strongly to vertical surfaces and have enough shear strength to support a plug of considerable diameter above an open discharge. Thus to get flow started and to keep the material moving, hammers and vibrators on the bin walls, internal plows near the bin floor, or jets of air in the discharge opening are often needed. A *poke hole* in the bin wall through which a rod may be inserted is sometimes used to initiate flow.

The discharge opening should be small enough to be readily closed when solids are flowing yet not so small that it will clog. It is best to make the opening large enough to pass the full desired flow when half open. It can then be opened further to clear a partial choke. If the opening is too large, however, the shutoff valve may be hard to close and control of the flow rate will be poor.

ARCHING   The derivation of Eq. (26-25) is based on the assumption that at a bin wall there is a vertical force acting on the material even when the solid mass is stationary. The resultant of the vertical force and the lateral force is at an angle $\tan^{-1} \mu'$ with the horizontal, as shown in Fig. 26-4. This creates an arch, or dome, inside the bin, which in a loose granular solid rests on the material underneath it. In a cohesive solid the arch may be strong enough to support the overlying solid when the discharge is opened. During flow through the discharge the arches collapse and re-form and on occasion may interrupt the flow. Factors influencing the flow of solid materials out of bins have been studied in detail by Jenike et al.[3]

SYMBOLS

A	Area, ft² or m²; total surface area of particles
A_w	Specific surface area of particles, ft²/lb or m²/g
a	Volume shape factor [Eq. (26-10)]
b	Thickness, ft or m
D	Diameter; D_o, diameter of bin opening, inches; D_p, particle size, ft or mm; D_{pn}, mesh opening in screen n
$\bar{D}$	Arithmetic average particle size, ft or mm; $\bar{D}_n$, arithmetic mean diameter [Eq. (26-8)]; $\bar{D}_n$, arithmetic average of $\bar{D}_{pn}$ and $\bar{D}_{p(n-1)}$; $\bar{D}_s$, mean volume-surface diameter [Eq. (26-7)]; $\bar{D}_w$, mass mean diameter; $\bar{D}_1$, for fraction 1; $\bar{D}_2$, for fraction 2
F	Force, lb$_f$ or N; F_B, on bin floor; F_L, lateral component; F_V, vertical component; F_f, frictional force; F_g, gravitational force
g	Gravitational acceleration, ft/s² or m/s²
g_c	Newton's-law proportionality factor, 32.174 ft-lb/lb$_f$-s²
K'	Ratio of pressures, p_V/p_L
L	Length, ft or m
m	Mass of sample, lb or g
$\dot{m}$	Mass flow rate, lb/min

N Number of particles in sample; N_w, number of particles per unit mass

n Screen number, counting from large screen of series; also, exponent in Eq. (26-26); n_T, total number of screens in series

p Pressure, lb_f/ft^2 or N/m^2; p_B, pressure on bin floor; p_L, p_V, directed pressures in stress analysis

r Radius of bin, ft or m

s_p Surface area of particle, ft^2 or mm^2

v_p Volume of particle, ft^3 or mm^3

Z Height, ft or m; Z_T, total height of solids

Greek letters

α Angle; α_m, angle of internal friction; α_r, angle of repose

$\Delta\phi_n$ Mass fraction of total sample retained by screen n and passed by screen $n - 1$

θ Angle; θ_m, maximum value in stress analysis

μ' Coefficient of friction, dimensionless

ρ Density, lb/ft^3 or kg/m^3; ρ_b, bulk density; ρ_p, density of particle

τ Shear stress, lb_f/ft^2 or N/m^2

Φ_s Sphericity of particle, defined by Eq. (26-1)

ϕ Mass fraction of total sample cumulative to size D_{pn} [Eq. (26-4)]

PROBLEMS

26-1   A circular silo 10 ft in diameter contains barley with a bulk density of 39 lb/ft^3. What are the vertical and lateral pressures, in pounds force per square foot, at the base of the silo if the depth of the barley is 40 ft? If it is 80 ft? $K' = 0.40$; $\mu' = 0.45$.

26-2   Draw Mohr stress circles for conditions at the base of the silo for the two depths indicated in Prob. 26-1. What are the values of α_m and θ_m, in degrees, for these two cases?

26-3   Calculate the rate of flow of barley through a circular opening in the floor of a silo, if the opening is 300 mm in diameter. For barley the exponent n in Eq. (26-26) is 2.93, and the particle size is 1 mm.

26-4   Repeat Prob. 26-3 for a 20-mesh sharp sand with a bulk density of 1,600 kg/m^3, for which exponent n is 2.67.

26-5   The mass flow rate of granular solids through a circular opening varies approximately as the cube of the opening diameter. In the steady laminar flow of a Newtonian fluid at constant pressure drop through a cylindrical pipe, how does the mass flow rate vary with pipe diameter? For what kind of non-Newtonian fluid would the mass flow rate, in steady laminar flow, vary as the cube of the pipe diameter? (That is, what would be the value of the flow behavior index n' for a power-law fluid, for this to be the case?)

REFERENCES

1   Brown, G. G., and associates: "Unit Operations," p. 22, Wiley, New York, 1950.
2   Franklin, F. C., and L. N. Johanson: *Chem. Eng. Sci.,* **4:**119 (1955).
3   Jenike, A. W., P. J. Elsey, and R. H. Wooley: *Proc. ASTM,* **60:**1168 (1960).
4   Laforge, R. M., and B. K. Boruff: *Ind. Eng. Chem.,* **56**(2):42 (1964).
5   Orr, C., and J. M. Dalla Valle: "Fine Particle Measurement," Macmillan, New York, 1959.
6   Rudd, J. K.: *Chem. Eng. News,* **32**(4):344 (1954).
7   Taylor, D. W.: "Fundamentals of Soil Mechanics," chap. 13, Wiley, New York, 1948.

27

SIZE REDUCTION

The term *size reduction* is applied to all the ways in which particles of solids are cut or broken into smaller pieces. Throughout the process industries solids are reduced by different methods for different purposes. Chunks of crude ore are crushed to workable size: synthetic chemicals are ground into powder; sheets of plastic are cut into tiny cubes or diamonds. Commercial products must often meet stringent specifications regarding the size and sometimes the shape of the particles they contain. Reducing the particle size also increases the reactivity of solids; it permits separation of unwanted ingredients by mechanical methods; it reduces the bulk of fibrous materials for easier handling.

Solids may be broken in eight or nine different ways, but only four of them are commonly used in size-reduction machines. They are (1) compression, (2) impact, (3) attrition, or rubbing, and (4) cutting. A nutcracker, a sledge hammer, a file, and a pair of shears exemplify these four types of action. In general, compression is used for coarse reduction of hard solids, to give relatively few fines; impact gives coarse, medium, or fine products; attrition yields very fine products from soft, nonabrasive materials. Cutting gives a definite particle size and sometimes a definite shape, with few or no fines.

PRINCIPLES OF COMMINUTION

Criteria for comminution Comminution is a generic term for size reduction; crushers and grinders are types of comminuting equipment. An ideal crusher or grinder would (1) have a large capacity, (2) require a small power input per unit of product, and (3) yield a product of the single size or the size distribution desired. The usual method of studying the performance of process equipment is to set up an ideal operation as a standard, compare the characteristics of the actual equipment with those of the ideal unit, and account for the difference between the two. When this method is applied to comminuting apparatus, the discrepancies between the ideal and the actual are considerable and the gaps have not been completely accounted for, even theoretically. On the other hand, useful quantitative information is obtainable from the incomplete theory now at hand.

The capacities of comminuting machines are best discussed when the individual types of equipment are described. The fundamentals of product size and shape and of energy requirements are, however, common to most machines, and can be discussed more generally.

Characteristics of comminuted products The objective of crushing and grinding is to produce small particles from larger ones. Smaller particles are desired either because of their large surface or because of their shape, size, and number. The energy efficiency of the operation is measured by the new surface created by the reduction in size. For these reasons, the geometric characteristics of particles, both alone and in mixtures, are important in evaluating the product from a crusher or grinder.

Unlike an ideal crusher or grinder, an actual unit does not yield a uniform product, whether the feed is uniformly sized or not. The product always consists of a mixture of particles, ranging in size from a definite maximum to a submicroscopic minimum. Some machines, especially in the grinder class, are designed to control the magnitude of the largest particles in their products, but the fine sizes are not under control. In some types of grinders fines are minimized, but they are not eliminated. If the feed is homogeneous, both in the shapes of the particles and in chemical and physical structure, the shapes of the individual units in the product may be quite uniform; otherwise, the grains in the various sizes of a single product may vary considerably in proportions.

The smallest grain in a comminuted product may be comparable in size to that of a unit crystal,[5a] which is the smallest unit of the material that can exist as an independent crystal. This size is of the order of 1 nm. If, for example, the largest particle in a product just passes a screen having 1-mm openings, the ratio of the diameters of the largest and smallest particles is of the order $10^{-1} : 10^{-7}$, or 10^6. Because of this extreme variation in the sizes of the individual particles, relationships adequate for uniform sizes must be modified when applied to such mixtures. The term "average size," for example, is meaningless until the method of averaging is defined, and, as discussed in Chap. 26, several different average sizes can be calculated.

Unless they are smoothed by abrasion after crushing, comminuted particles resemble polyhedrons with nearly plane faces and sharp edges and corners. The number of major faces may vary, but is usually between 4 and 7. The particles may be compact, with length, breadth, and thickness nearly equal, or they may be platelike or needlelike. A compact grain with several nearly equal faces can be considered to be spherical, and the term "diameter" is generally used for particle size.

Particle-size distribution in comminuted products Particle-size distributions have been discussed in Chap. 26 in terms of cumulative mass fraction ϕ and particle "diameter" D_p. No single distribution applies equally well to all comminuted products, especially in the range of coarser particle sizes. For fine particles, however, it has been found empirically that the slope of the graph of ϕ vs. D_p is an exponential function of the particle diameter D_p,[5b] so

$$-\frac{d\phi}{dD_p} = BD_p^k \tag{27-1}$$

where B and k are constants. The minus sign accounts for the fact that as ϕ increases, D_p decreases. This equation can be used to extrapolate screen-analysis data to sizes too fine to sieve accurately.

Equation (27-1) can be integrated between the limits $\phi = \phi_1$ and $\phi = \phi_2$, and the corresponding limits $D_p = D_{p1}$ and $D_p = D_{p2}$. Integration and substitution of limits gives

$$\phi_2 - \phi_1 = \frac{B}{k+1}(D_{p1}^{k+1} - D_{p2}^{k+1}) \tag{27-2}$$

The constant k depends on the relative importance of the very fine sizes in the sample. It ranges from about -0.5 to about 0.1 for most comminuted products. The larger k is, the less important are the very small particles in the fraction between diameters D_{p1} and D_{p2}. If the product is overground, fine particles become predominant and k is small. The constant B is a measure of the fraction of the entire product that falls between diameters D_{p1} and D_{p2}.

Constants B and k can be obtained from the differential screen analysis by the following method.[5b] It is assumed that the ratio of the screen opening of any screen in the series bears a constant ratio to that of the screen immediately below it. The Tyler screen series conforms to this requirement. If D_{pn} and $D_{p(n-1)}$ are the mesh sizes of the nth and $(n-1)$st screens, respectively, the mass fraction on the nth screen is $\phi_n - \phi_{n-1}$, and Eq. (27-2) can be written over screen n as follows:

$$\phi_n - \phi_{n-1} = \Delta\phi_n = -\frac{B}{k+1}(D_{pn}^{k+1} - D_{p(n-1)}^{k+1}) \tag{27-3}$$

If the constant ratio between $D_{p(n-1)}$ and D_{pn} is r,

$$D_{p(n-1)} = rD_{pn} \tag{27-4}$$

where $r > 1$. Elimination of $D_{p(n-1)}$ from Eq. (27-3) by means of Eq. (27-4) gives

$$\Delta\phi_n = \frac{B(r^{k+1} - 1)}{k+1}D_{pn}^{k+1} = B'D_{pn}^{k+1} \tag{27-5}$$

where

$$B' = \frac{B(r^{k+1} - 1)}{k+1} \tag{27-6}$$

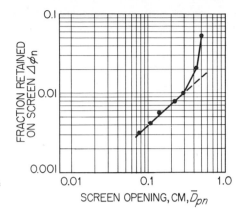

FRACTION RETAINED ON SCREEN $\Delta\phi_n$

SCREEN OPENING, CM, $\bar{D}_{pn}$

FIGURE 27-1
Plot of log $\Delta\phi_n$ vs. log D_{pn} for Example 27-1.

The differential screen analysis provides the necessary relation between $\Delta\phi_n$ and D_{pn}. Equation (27-5) can be written logarithmically as

$$\log \Delta\phi_n = (k + 1) \log D_{pn} + \log B' \qquad (27\text{-}7)$$

The constants B' and k are evaluated by plotting $\Delta\phi_n$ vs. D_{pn} on logarithmic coordinates and drawing the best straight line through the points. The geometric slope of the line is $k + 1$. Once $k + 1$ is known, log B' can be found by use of Eq. (27-7) from the coordinates of any convenient point on the line. The constant B can then be calculated from B' by Eq. (27-6).

Equation (27-1) is used, in combination with Eqs. (26-6) and (26-12), for the calculation of specific surface and particle population. Elimination of $d\phi$ between Eqs. (26-6) and (27-1) and integration gives for A_w, the specific surface,

$$A_w = -\frac{6B}{\Phi_s \rho_p} \int_{D_{p1}}^{D_{p2}} D_p^{k-1} \, dD_p = \frac{6B}{\Phi_s \rho_p^k} (D_{p1}^k - D_{p2}^k) \qquad (27\text{-}8)$$

Equation (27-8) is indeterminate if $k = 0$. In this case,

$$A_w = \frac{6B}{\Phi_s \rho_p} \ln \frac{D_{p1}}{D_{p2}} \qquad (27\text{-}9)$$

Elimination of $d\phi$ between Eqs. (26-12) and (27-1) and integration between limits gives for N_w, the number of particles per unit mass of mixture,

$$N_w = -\frac{B}{a\rho_p} \int_{D_{p1}}^{D_{p2}} \frac{dD_p}{D_p^{3-k}} = \frac{B}{(2-k)a\rho_p} \left(\frac{1}{D_{p2}^{2-k}} - \frac{1}{D_{p1}^{2-k}} \right) \qquad (27\text{-}10)$$

EXAMPLE 27-1   The screen analysis shown in Tables 26-2 and 26-3 applies to a sample of crushed quartz. The density of the particles is 0.00265 g/mm³, and the shape factors are $a = 2$ and $\Phi_s = 0.571$. What is the specific surface, in meters per gram?

SOLUTION   A logarithmic plot of $\Delta\phi_n$ vs. D_{pn} for the finer sizes is shown in Fig. 27-1. It is clear that for diameters smaller than 0.417 mm (35-mesh) the data follow the empirical relationship of Eq. (27-7), so Eqs. (27-2) to (27-10) can be used for calculations in this size range. For all particles larger than 0.417 mm, Eqs. (26-5) to (26-12) can be used.

If the differential analysis is used directly for calculations in the size range between 4.699 and 0.417 mm, Eq. (26-5) can be written

$$A_w = \frac{6}{0.571 \times 0.00265} \sum \frac{\Delta\phi_n}{\bar{D}_n} = 3,965 \sum \frac{\Delta\phi_n}{\bar{D}_n}$$

and Eq. (26-11) as

$$N_w = \frac{1}{2 \times 0.00265} \sum \frac{\Delta\phi_n}{\bar{D}_n^3} = 188.7 \sum \frac{\Delta\phi_n}{\bar{D}_n^3}$$

To obtain $\bar{D}_n$ for a screen increment the arithmetic mean of the mesh dimensions for the defining screens for that increment is calculated from the mesh dimensions given in Appendix 20. Magnitudes of $1/\bar{D}_n$ and $1/\bar{D}_n^3$ for each increment are then calculated and weighted with respect to $\Delta\phi_n$, and the sums of $\Delta\phi_n/\bar{D}_n$ and $\Delta\phi_n/\bar{D}_n^3$ are obtained. These calculations are shown in Table 27-1. Then

$$A_w = 3,965 \times 0.6691 = 2,650 \text{ mm}^2 \qquad N_w = 188.7 \times 0.6110 = 115$$

If the cumulative analysis is used to calculate A_w and N_w, Eqs. (26-6) and (26-12) become

$$A_w = 3,965 \int_0^{0.9616} \frac{d\phi}{D_p} \qquad N_w = 188.7 \int_0^{0.9616} \frac{d\phi}{D_p^3}$$

These equations may be integrated numerically. Integration gives

$$A_w = 2,660 \text{ mm}^2 \qquad N_w = 118$$

These results check closely with those found in calculations based directly on the differential analysis.

To obtain the particle area and population in the fraction passing the 35-mesh screen, the constants k and B of Eq. (27-1) are needed. They are found from the line of Fig. 27-1. The geometric slope of the line is 0.886, which is also the value of $k + 1$; so k is -0.114. The coordinates of any point on the line can be used to evaluate B'. For example, when $\Delta\phi_n$ is 0.004, D_{pn} is 0.1 mm. Then, by Eq. (27-7),

$$\log 0.004 = 0.886 \log 0.1 + \log B'$$

Table 27-1   CALCULATION OF A_W AND N_W FOR EXAMPLE 27-1

Mesh	$\bar{D}_n$, mm	$\Delta\phi_n$	$\dfrac{1}{\bar{D}_n}$	$\dfrac{1}{\bar{D}_n^3}$	$\dfrac{\Delta\phi_n}{\bar{D}_n}$	$\dfrac{\Delta\phi_n}{\bar{D}_n^3}$
4/6	4.013	0.0251	0.249	0.0155	0.0063	0.0004
6/8	2.844	0.1250	0.352	0.0435	0.0440	0.0054
8/10	2.006	0.3207	0.499	0.1239	0.1600	0.0397
10/14	1.409	0.2570	0.710	0.3575	0.1825	0.0918
14/20	1.000	0.1590	1.000	1.0000	0.1590	0.1590
20/28	0.711	0.0538	1.406	2.7822	0.0756	0.1497
28/35	0.503	0.0210	1.988	7.8577	0.0417	0.1650
Total					0.6691	0.6110

from which $B' = 0.0308$. For the Tyler screen series, $r = \sqrt{2} = 1.414$, and, solving Eq. (27-6) for B gives

$$B = \frac{0.0308 \times 0.886}{1.414^{0.886} - 1} = 0.0760$$

The largest particle in the pan fraction passes a mesh size of 0.074 mm. If it is assumed that the relationship of ϕ with D_p in this fraction also conforms to Eq. (27-1), Eq. (27-2) can be used to estimate the diameter of the smallest particle in the pan. This gives

$$0.0075 = \frac{0.0760}{0.886} (0.074^{0.886} - D_{p2}^{0.886})$$

Solution of this equation gives $D_{p2} = 0.0069$ mm.

From Eq. (27-8) the area of the fraction having a size range of 0.417 to 0.0069 mm is

$$A_w = \frac{6 \times 0.0760}{0.00265 \times -0.114} (0.417^{-0.114} - 0.0069^{-0.114}) = 994 \text{ mm}^2$$

The total area of the entire sample is $2,650 + 994 = 3,644$ mm^2/g (17.8 ft^2/lb).

From Eq. (27-10)

$$N_w = \frac{0.0760}{2.144 \times 2 \times 0.00265} (0.0069^{-2.114} - 0.417^{-2.114}) = 251,200$$

The total number of particles is then $118 + 251,200$ or about 251,400. In fine sizes, the number of particles per gram is very large. Calculation shows that approximately 249,500 of the total of 251,400 particles are in the pan fraction. The precision of this calculation of particle population is poor. ////

Energy and power requirements in comminution[4] The cost of power is a major expense in crushing and grinding, so the factors that control this cost are important. During size reduction, the particles of feed material are first distorted and strained. The work necessary to strain them is stored temporarily in the solid as mechanical energy of stress, just as mechanical energy can be stored in a coiled spring. As additional force is applied to the stressed particles, they are distorted beyond their ultimate strength and suddenly rupture into fragments. New surface is generated. Since a unit area of solid has a definite amount of surface energy, the creation of new surface requires work, which is supplied by the release of energy of stress when the particle breaks. By conservation of energy, all energy of stress in excess of the new surface energy created must appear as heat.

CRUSHING EFFICIENCY The ratio of the surface energy created by crushing to the energy absorbed by the solid is the crushing efficiency η_c. If e_s is the surface energy per unit area, in feet times pounds force per square foot, and A_{wb} and A_{wa} are the areas of product and feed, respectively, in square feet per pound, the energy absorbed by the material W_n is, in feet times pounds force per pound,

$$W_n = \frac{e_s(A_{wb} - A_{wa})}{\eta_c} \tag{27-11}$$

The surface energy created by fracture is small in comparison with the toral mechanical energy stored in the material at the time of rupture, and most of the latter is converted into heat. Crushing efficiencies are therefore low. They have been measured experimentally by estimating e_s from theories of the solid state, measuring W_n, A_{wb}, and A_{wa}, and substituting into Eq. (27-11). The precision of the calculation is poor, primarily because of uncertainties in calculating e_s, but the results do show that crushing efficiencies range between 0.1 and 2 percent.[6]

The energy absorbed by the solid W_n is less than that fed to the machine. Part of the total energy input W is used to overcome friction in the bearings and other moving parts, and the rest is available for crushing. The ratio of the energy absorbed to the energy input is η_m, the mechanical efficiency. Then, if W is the energy input,

$$W = \frac{W_n}{\eta_m} = \frac{e_s(A_{wb} - A_{wa})}{\eta_m \eta_c} \tag{27-12}$$

If $\dot{m}$ is the feed rate, the power required by the machine is

$$P = W\dot{m} = \frac{\dot{m}e_s(A_{wb} - A_{wa})}{\eta_c \eta_m} \tag{27-13}$$

Calculation of A_{wb} and A_{wa} from Eq. (26-7) and substitution in Eq. (27-13) gives

$$P = \frac{6\dot{m}e_s}{\eta_c \eta_m \rho_p}\left(\frac{1}{\Phi_b \bar{D}_{sb}} - \frac{1}{\Phi_a \bar{D}_{sa}}\right) \tag{27-14}$$

where $\bar{D}_{sa}$, $\bar{D}_{sb}$ = volume-surface mean diameter of the feed and product, respectively
Φ_a, Φ_b = sphericity of feed and product, respectively

Rittinger's law   A crushing law proposed many years ago by Rittinger states that the work required in crushing is proportional to the new surface created. This law is equivalent to the statement that the crushing efficiency η_c is constant and, for a given machine and feed material, is independent of the sizes of feed and product.[5c] If the sphericities Φ_a and Φ_b are equal and the mechanical efficiency is constant, the various constants in Eq. (27-14) can be combined into a single constant K_r and Rittinger's law written as

$$\frac{P}{\dot{m}} = K_r\left(\frac{1}{\bar{D}_{sb}} - \frac{1}{\bar{D}_{sa}}\right) \tag{27-15}$$

Rittinger's law has been shown to apply reasonably well under conditions where the energy input per unit mass of solid is not too great, and it can be used as a first approximation for actual crushing processes where the constant K_r is determined experimentally in a test on a machine of the type to be used and with the material to be crushed.

EXAMPLE 27-2   A certain crusher accepts a feed of rock having a volume-surface mean diameter of 0.75 in. (19 mm) and discharges a product of volume-surface mean diameter of 0.20 in. (5 mm). The power required to crush 12 tons/h (3 kg/s) is 9.3 hp (6.9 kW). What should be the power consumption if the capacity is reduced to 10 tons/h (2.5 kg/s) and the volume-surface mean diameter to 0.15 in. (3.8 mm)? The mechanical efficiency remains unchanged.

SOLUTION   Since the only variables are the feed rate and the diameters of feed and product, Eq. (27-15) is used, once for the original operation, and once for the new operation. It is unnecessary to convert units to those of Eq. (27-15), as conversion factors can be absorbed into constant K_r. The original operation is

$$\frac{9.3}{12} = K_r \left(\frac{1}{0.20} - \frac{1}{0.75} \right)$$

and the new operation is

$$\frac{P}{10} = K_r \left(\frac{1}{0.15} - \frac{1}{0.75} \right)$$

Dividing the second equation by the first gives

$$\frac{P}{9.3} \frac{12}{10} = \frac{1/0.15 - 1/0.75}{1/0.20 - 1/0.75}$$

from which $P = 11.4$ hp (8.5 kW)   ////

Bond crushing law and work index   Another method for estimating the power required for crushing and grinding is that proposed by Bond.[3] Using semitheoretical reasoning, Bond proposed that the work required to form particles of size D_p from very large feed is proportional to the square root of the surface-to-volume ratio of the product, s_p/v_p. By Eq. (26-1), $s_p/v_p = 6/\Phi_s D_p$, from which it follows that

$$\frac{P}{\dot{m}} = \frac{K_b}{\sqrt{D_p}} \qquad (27\text{-}16)$$

where K_b is a constant which depends on the type of machine and on the material being crushed. This law calls for relatively less energy for the smaller product particles than Rittinger's law. The Bond law is somewhat the more realistic in estimating power requirements of commercial crushers and grinders. Neither law is truly applicable except over a limited range of conditions, however, and more general relationships are being sought.

To use Eq. (27-16), a work index W_i is defined as the gross energy in kilowatt-hours per ton (2,000 lb) of feed needed to reduce a very large feed to such a size that 80 percent of the product passes a 100-μm screen. This definition leads to a relation between K_b and W_i. If D_p is in millimeters, P in kilowatts, and $\dot{m}$ in tons per hour,

$$K_b = \sqrt{100 \times 10^{-3}} \ W_i = 0.3162 W_i \qquad (27\text{-}17)$$

If 80 percent of the feed passes a mesh size of D_{pa} mm and 80 percent of the product a mesh of D_{pb} mm, it follows from Eqs. (27-16) and (27-17) that

$$\frac{P}{\dot{m}} = 0.3162 W_i \left(\frac{1}{\sqrt{D_{pb}}} - \frac{1}{\sqrt{D_{pa}}} \right) \qquad (27\text{-}18)$$

The work index includes the friction in the crusher, and the power given by Eq. (27-18) is gross power.

Table 27-2 gives typical work indexes for some common minerals. These data do not vary greatly among different machines of the same general type and apply to dry crushing or to wet grinding. For dry grinding, the power calculated from Eq. (27-18) is multiplied by $\frac{4}{3}$.

EXAMPLE 27-3 What is the power required to crush 100 tons/h of limestone if 80 percent of the feed passes a 2-in. screen and 80 percent of the product a $\frac{1}{8}$-in. screen?

SOLUTION From Table 27-2, the work index for limestone is 12.74. Other quantities for substitution into Eq. (27-18) are

$$\dot{m} = 100 \text{ tons/h}$$

$$D_{pa} = 2 \times 25.4 = 50.8 \text{ mm} \qquad D_{pb} = 0.125 \times 25.4 = 3.175 \text{ mm}$$

The power required is

$$P = 100 \times 0.3162 \times 12.74 \left(\frac{1}{\sqrt{3.175}} - \frac{1}{\sqrt{50.8}} \right)$$

$$= 169.6 \text{ kW (227 hp)} \qquad\qquad ////$$

Computer simulation of milling operations The size distribution of products from various types of size-reduction equipment can be predicted by a computer simulation of the comminution process.[7a,9] This approach is receiving increased attention, but insufficient information has been developed as yet to make computer simulation a generally useful procedure.

Table 27-2 WORK INDEXES FOR DRY CRUSHING†
OR WET GRINDING‡

Material	Sp. gr.	Work index W_i
Bauxite	2.20	8.78
Cement clinker	3.15	13.45
Cement raw material	2.67	10.51
Clay	2.51	6.30
Coal	1.4	13.00
Coke	1.31	15.13
Granite	2.66	15.13
Gravel	2.66	16.06
Gypsum rock	2.69	6.73
Iron ore (hematite)	3.53	12.84
Limestone	2.66	12.74
Phosphate rock	2.74	9.92
Quartz	2.65	13.57
Shale	2.63	15.87
Slate	2.57	14.30
Trap rock	2.87	19.32

† For dry grinding, multiply by $\frac{4}{3}$.
‡ From Allis-Chalmers Co., Milwaukee, Wis., by permission.

Simulation makes use of two basic concepts, that of a *grinding-rate function* S_u and a *breakage function* $\Delta B_{n,u}$. Consider a stack of n_T standard screens, and let n be the number of a particular screen in the series beginning with the coarsest screen. For any given value of n, let the upper (coarser) screens, larger than screen n, be designated by subscript u. The grinding-rate function S_u (formerly called the selection function) is the fraction of the material of a given size, coarser than that on screen n, which is broken in a given time. If $\Delta\phi_u$ is the mass fraction retained on one of the upper screens, its rate of change by breakage to smaller sizes is

$$\frac{d(\Delta\phi_u)}{dt} = -S_u\,\Delta\phi_u \qquad (27\text{-}19)$$

Suppose, for example, that the coarsest material in the charge to a grinding mill is 4/6 mesh, that the mass fraction of this material, $\Delta\phi_1$, is 0.05, and that one-hundredth of this material is broken every second. Then S_u would be 0.01 s^{-1}, and the $\Delta\phi_1$ would diminish at the rate of $0.01 \times 0.05 = 0.0005$ s^{-1}.

The breakage function $\Delta B_{n,u}$ gives the size distribution resulting from the breakage of the upper material. Some of the 4/6-mesh material, after breaking, would be fairly coarse, some very small, and some in between. Probably very little would be as large as 6/8-mesh, and only a small amount as small as 200-mesh. One would expect sizes in the intermediate range to be favored. Consequently $\Delta B_{n,u}$ varies with both n and u. Furthermore it varies with the composition of the material in the mill, since coarse particles may break differently in the presence of large amounts of fines than they do in the absence of fines. In a batch mill, therefore, $\Delta B_{n,u}$ (and S_u also) would be expected to vary with time as well as with all the other milling variables.

If $\Delta B_{n,u}$ and S_u are known or can be assumed, the rate of change of any given fraction can be found as follows. For any fraction except the coarsest, the initial amount is diminished by breakage to smaller sizes and simultaneously augmented by the creation of new particles from breakage of all coarser fractions. If input and outgo to a given screen are at equal rates, the fraction retained on that screen remains constant. Usually, however, this is not the case, and the mass fraction retained on screen n changes according to the equation

$$\frac{d(\Delta\phi_n)}{dt} = -S_n\,\Delta\phi_n + \sum_{u=1}^{n-1}\Delta\phi_n\,S_u\,\Delta B_{n,u} \qquad (27\text{-}20)$$

Equation (27-20) can be simplified if it is assumed that S_u and $\Delta B_{n,u}$ are constant, but these assumptions are highly unrealistic. In crushing coal, for particles larger than about 28-mesh, S_u has been found to vary with the cube of the particle size[1] and the breakage function to depend on the reduction ratio D_n/D_u according to the equation

$$B_{n,u} = \left(\frac{\bar{D}_n}{\bar{D}_u}\right)^{\beta} \qquad (21\text{-}21)$$

where the exponent β may be constant or may vary with the value of B.

In Eq. (27-21), $B_{n,u}$ is the *total* mass fraction smaller than size $\bar{D}_n$. It is a cumulative mass fraction, in contrast with $\Delta B_{n,u}$, which is the fraction of size $\bar{D}_n$ (retained between screens n and $n+1$) resulting from breakage of particles of size $\bar{D}_u$.

If β in Eq. (27-21) is constant, this equation says that the particle-size distribution of the crushed material is the same for all sizes of the initial material. The value of ΔB_u in crushing 4/6-mesh material to 8/10-mesh will be the same as in crushing 6/8-mesh particles to 10/14-mesh, since the size-reduction ratio is the same.

Usually Eq. (27-20) is solved by the Euler method of numerical approximation, in which the changes in all fractions during successively short time intervals Δt (say 1 s) are calculated by the approximation $d(\Delta\phi_n)/dt = \Delta(\Delta\phi_n)/\Delta t$. Changes in S_u and $\Delta B_{n,u}$ with screen size and (if known) with time can be incorporated. A computer is needed to make the lengthy calculations. The method is illustrated in the following somewhat simplified example.

EXAMPLE 27-4   A batch grinding mill is charged with material of the composition shown in Table 27-1. The grinding-rate function S_u is assumed to be 0.001 s^{-1} for the 4/6-mesh particles. Breakage function B_u is given by Eq. (27-21) with $\beta = 1.3$. Both S_u and B_u are assumed to be independent of time. (a) How long will it take for the fraction of 4/6-mesh material to diminish by 10 percent? (b) Tabulate the individual breakage functions $\Delta B_{n,u}$ for the 14/20-mesh fraction and for all coarser fractions. (c) How will the fraction of 14/20-mesh material vary during the time period defined in part (a)? Use a 10-s interval for Δt in the calculations.

SOLUTION   (a) For the 4/6-mesh material there is no input from coarser material and Eq. (27-19) applies. At the end of time t_T, $\Delta\phi_n$ will be $0.0251 \times 0.9 = 0.02259$. Thus

$$-S_u \int_0^{t_T} dt = \int_{0.0251}^{0.02259} \frac{d(\Delta\phi_n)}{\Delta\phi_n}$$

or
$$t_T = \frac{1}{S_u} \ln \frac{0.0251}{0.02259} = \frac{1}{0.001} \ln 1.111 = 105.3 \text{ s}$$

(b)   Assume S_u varies with D_p^3. Let S_1 and S_2 be the values for 4/6- and 6/8-mesh material respectively. Then $S_1 = 10 \times 10^{-4}$ s^{-1}, and

$$S_2 = S_1 \left(\frac{D_2}{D_1}\right)^3 = 10^{-3} \left(\frac{2.362}{3.327}\right)^3 = 3.58 \times 10^{-4} \text{ s}^{-1}$$

Values of S_3, S_4, and S_5 are calculated similarly; the results are given in Table 27-3.

Table 27-3   CUMULATIVE BREAKAGE FUNCTION $B_{n,u}$ FOR EXAMPLE 27-3

Mesh	$\bar{D}_n$ or $\bar{D}_u$, mm	u	$B_{n,u}$ for $n =$				
			1	2	3	4	5
4/6	3.327	1	1.0	0.641	0.402	0.256	0.165
6/8	2.362	2	0	1.0	0.628	0.400	0.258
8/10	1.651	3	0	0	1.0	0.637	0.411
10/14	1.168	4	0	0	0	1.0	0.644
14/20	0.833	5	0	0	0	0	1.0

The breakage function $B_{n,u}$ is found as follows. When n and u are equal, or whenever $n < u$, $\Delta B_{n,u} = 0$. The total mass fraction smaller than 6/8-mesh resulting from breakage of 4/6-mesh particles, $B_{2,1}$, from Eq. (27-21), is

$$B_{2,1} = \left(\frac{2.362}{3.327}\right)^{1.3} = 0.641$$

Then $\Delta B_{2,1}$, the fraction of broken material retained *on* the 8-mesh screen, is $1 - 0.641$ or 0.359.

The total mass fraction smaller than 8/10-mesh resulting from breakage of 4/6-mesh material, $B_{3,1}$, is

$$B_{3,1} = \left(\frac{1.651}{3.327}\right)^{1.3} = 0.402$$

The mass fraction of the broken 4/6-mesh material retained *on* the 10-mesh screen, $\Delta B_{3,1}$, is $0.641 - 0.402 = 0.239$. Other values of $B_{n,u}$ and $\Delta B_{n,u}$ are found in the same way, to give the results shown in Tables 27-3 and 27-4. Note that when $n = u$, $B_{n,u}$ is unity, by definition. When $u = 1$, as shown in Table 27-3, 0.641 of the broken particles from the 4/6-mesh material is smaller than 8-mesh, 0.402 smaller than 10-mesh, 0.256 smaller than 14-mesh, and 0.165 smaller than 20-mesh.

(c) Let $\Delta\phi_{n,t}$ be the mass fractions retained on the various screens at the end of t time increments Δt. Then $\Delta\phi_{1,0}$, $\Delta\phi_{2,0}$, etc., are the initial mass fractions given in Table 27-1.

The left-hand side of Eq. (27-20) is approximated by $\Delta(\Delta\phi_n)/\Delta t$, where Δt in this example is 10 s and $\Delta(\Delta\phi_n) = \Delta\phi_{n,t+1} - \Delta\phi_{n,t}$. Successive values of $\Delta\phi$ on the various screens can then be calculated from the following form of Eq. (27-20):

$$\Delta\phi_{n,t+1} = \Delta\phi_{n,t} - S_n\,\Delta t\,\Delta\phi_{n,t} + \Delta t \sum_{u=1}^{n-1} \Delta\phi_{u,t}\,S_u\,\Delta B_{n,u}$$

$$= \Delta\phi_{n,t}(1 - S_n\,\Delta t) + \Delta t \sum_{u=1}^{n-1} \Delta\phi_{u,t}\,S_u\,\Delta B_{n,u} \qquad (27\text{-}22)$$

For the top screen $n = 1$, and $\Delta B = 0$. Hence Eq. (27-22) becomes

$$\Delta\phi_{1,t+1} = \Delta\phi_{n,t}(1 - S_n\,\Delta t) = \Delta\phi_{n,t}[1 - (10 \times 10^{-4})(10)] = 0.99\Delta\phi_{1,t}$$

After 10 s, then, the mass fraction on the top screen is

$$\Delta\phi_{1,1} = 0.99 \times 0.0251 = 0.02485$$

Table 27-4 GRINDING-RATE AND BREAKAGE FUNCTIONS FOR EXAMPLE 27-3

Mesh	u	S_n or S_u $\times 10^{-4}$, s^{-1}	$B_{n,u}$ for $n =$					
			1	2	3	4	5	6
4/6	1	10.0	0	0.359	0.239	0.146	0.091	0.165
6/8	2	3.58	0	0	0.372	0.228	0.142	0.258
8/10	3	1.22	0	0	0	0.363	0.226	0.411
10/14	4	0.433	0	0	0	0	0.356	0.644
14/20	5	0.157	0	0	0	0	0	1
-20	6		0	0	0	0	0	0

After 10 s more,

$$\Delta\phi_{1,2} = 0.99 \times 0.02485 = 0.02460$$

and so forth. On the 8-mesh screen ($n = 2$), the mass fractions are, from Eq. (27-22),

$$\Delta\phi_{2,1} = \Delta\phi_{2,0}(1 - S_2\,\Delta t) + \Delta t\,\Delta\phi_{1,0}\,S_1\,\Delta B_{2,1}$$

Substituting the values of S_1, S_2, and $\Delta B_{2,1}$ from Table 27-4 and $\Delta\phi_{1,0}$ from Table 27-1 gives

$$\Delta\phi_{2,1} = \Delta\phi_{2,0}[1 - (3.58 \times 10^{-4}) \times 10] + 10 \times 0.0251 \times 10.0 \times 10^{-4} \times 0.359$$
$$= 0.99642\Delta\phi_{2,0} + 0.00009$$

Thus

$$\Delta\phi_{2,1} = 0.99642 \times 0.1250 + 0.00009 = 0.12464$$

Similarly,

$$\Delta\phi_{2,2} = 0.99642 \times 0.12464 + 10 \times 0.02485 \times 10.0 \times 10^{-4} \times 0.359 = 0.12428$$

The values of $\Delta\phi_{3,t}$, $\Delta\phi_{4,t}$, and $\Delta\phi_{5,t}$ are found in the same way. The results are tabulated in Table 27-5. Note that $\Delta\phi_1$, $\Delta\phi_2$, and $\Delta\phi_3$ all decrease progressively with time and $\Delta\phi_4$ and $\Delta\phi_5$ increase. In many comminution problems the mass fractions of the intermediate sizes often increase and then decrease with time, or vice versa.   ////

SIZE-REDUCTION EQUIPMENT

Size-reduction equipment is divided into crushers, grinders, ultrafine grinders, and cutting machines. *Crushers* do the heavy work of breaking large pieces of solid material into small lumps. A primary crusher operates on run-of-mine material, accepting anything that comes from the mine face and breaking it into 6- to 10-in. lumps. A secondary crusher reduces these lumps to particles perhaps $\frac{1}{4}$ in. in size. *Grinders* reduce crushed feed to powder. The product from an intermediate grinder might

Table 27-5   MASS FRACTIONS, EXAMPLE 27-3

Time, s	4/6 mesh $\Delta\phi_1$	6/8 mesh $\Delta\phi_2$	8/10 mesh $\Delta\phi_3$	10/14 mesh $\Delta\phi_4$	14/20 mesh $\Delta\phi_5$
0	0.02510	0.12500	0.32070	0.25700	0.15900
10	0.02485	0.12464	0.32053	0.25717	0.15919
20	0.02460	0.12428	0.32037	0.25734	0.15938
30	0.02436	0.12393	0.32020	0.25750	0.15957
40	0.02412	0.12357	0.32003	0.25767	0.15975
50	0.02388	0.12322	0.31986	0.25784	0.15994
60	0.02364	0.12286	0.31969	0.25800	0.16013
70	0.02340	0.12251	0.31952	0.25816	0.16031
80	0.02317	0.12215	0.31935	0.25833	0.16050
90	0.02294	0.12180	0.31918	0.25849	0.16068
100	0.02271	0.12144	0.31901	0.25865	0.16087
106†	0.02257	0.12123	0.31890	0.25875	0.16098

† A 6-s interval was used in the final calculation to give a closer approximation to the desired answer.

pass a 40-mesh screen; most of the product from a fine grinder would pass a 200-mesh screen. An *ultrafine grinder* accepts feed particles no larger than $\frac{1}{4}$ in.; the product size is typically 1 to 50 μm. *Cutters* give particles of definite size and shape, 2 to 10 mm in length.

The principal types of size-reduction machines are:

A Crushers (coarse and fine)
    1 Jaw crushers
    2 Gyratory crushers
    3 Crushing rolls
B Grinders (intermediate and fine)
    1 Hammer mills; impactors
    2 Rolling-compression mills
        a Bowl mills
        b Roller mills
    3 Attrition mills
    4 Revolving mills
        a Rod mills
        b Ball mills; pebble mills
        c Tube mills; compartment mills
C Ultrafine grinders
    1 Hammer mills with internal classification
    2 Fluid-energy mills
    3 Agitated mills
D Cutting machines
    1 Knife cutters; dicers; slitters

These machines do their work in distinctly different ways. Slow compression is the characteristic action of crushers. Grinders employ impact and attrition, sometimes combined with compression; ultrafine grinders operate principally by attrition. A cutting action is of course characteristic of cutters, dicers, and slitters.

Crushers Crushers are slow-speed machines for coarse reduction of large quantities of solids. The main types are jaw crushers, gyratory crushers, smooth-roll crushers, and toothed-roll crushers. The first three operate by compression and can break large lumps of very hard materials, as in the primary and secondary reduction of rocks and ores. Toothed-roll crushers tear the feed apart as well as crushing it; they handle softer feeds like coal, bone, and soft shale.

JAW CRUSHERS In a jaw crusher feed is admitted between two jaws, set to form a V open at the top. One jaw, the fixed, or anvil, jaw, is nearly vertical and does not move; the other, the swinging jaw, reciprocates in a horizontal plane. It makes an angle of 20 to 30° with the anvil jaw. It is driven by an eccentric so that it applies great compressive force to lumps caught between the jaws. The jaw faces are flat or slightly bulged; they may carry shallow horizontal grooves. Large lumps caught between the upper parts of the jaws are broken, drop into the narrower space below, and are

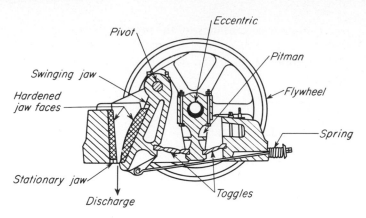

FIGURE 27-2
Blake jaw crusher.

recrushed the next time the jaws close. After sufficient reduction they drop out the bottom of the machine. The jaws open and close 250 to 400 times per minute.

The most common type of jaw crusher is the *Blake crusher*, illustrated in Fig. 27-2. In this machine an eccentric drives a pitman connected to two toggles, one of which is pinned to the frame and the other to the swinging jaw. The pivot point is at the top of the movable jaw or above the top of the jaws on the centerline of the jaw opening. The greatest amount of motion is at the bottom of the V, which means that there is little tendency for a crusher of this kind to choke. Some machines with a 72- by 96-in feed opening can accept rocks 6 ft in diameter and crush 1,000 tons/h to a maximum product size of 10 in. Smaller secondary crushers reduce the particle size of precrushed feed to $\frac{1}{4}$ to 2 in., at much lower rates of throughput.

GYRATORY CRUSHERS   A gyratory crusher may be looked upon as a jaw crusher with circular jaws, between which material is being crushed at some point at all times. A conical crushing head gyrates inside a funnel-shaped casing, open at the top. As shown in Fig. 27-3, the crushing head is carried on a heavy shaft pivoted at the top of the machine. An eccentric drives the bottom end of the shaft. At any point on the periphery of the casing, therefore, the bottom of the crushing head moves toward, and then away from, the stationary wall. Solids caught in the V-shaped space between the head and the casing are broken and rebroken until they pass out the bottom. The crushing head is free to rotate on the shaft and turns slowly because of friction with the material being crushed.

The speed of the crushing head is typically 125 to 425 gyrations per minute. Because some part of the crushing head is working at all times, the discharge from a gyratory is continuous instead of intermittent as in a jaw crusher. The load on the motor is nearly uniform; less maintenance is required than with a jaw crusher; and the power requirement per ton of material crushed is smaller. The biggest gyratories handle up to 3,500 tons/h. The capacity of a gyratory crusher varies with the jaw setting, the impact strength of the feed, and the speed of gyration of the machine.

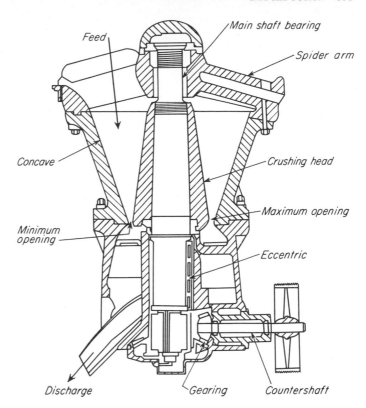

FIGURE 27-3
Gyratory crusher.

The capacity is almost independent of the compressive strength of the material being crushed.

SMOOTH-ROLL CRUSHERS Two heavy smooth-faced metal rolls turning on parallel horizontal axes are the working elements of the smooth-roll crusher illustrated in Fig. 27-4. Particles of feed caught between the rolls are broken in compression and drop out below. The rolls turn toward each other at the same speed. They have relatively narrow faces and are large in diameter so that they can "nip" moderately large lumps. Typical rolls are 24 in. in diameter with a 12-in. face to 78 in. in diameter with a 36-in. face. Roll speeds range from 50 to 300 r/min. Smooth-roll crushers are secondary crushers, with feeds $\frac{1}{2}$ to 3 in., in size and products $\frac{1}{2}$-in. to about 20-mesh.

The limiting size of particles that can be nipped by rolls of a given diameter is discussed below. The particle size of the product depends on the spacing between the rolls, as does the capacity of a given machine. Smooth-roll crushers give few fines and virtually no oversize. They operate most effectively when set to give a reduction ratio of 3 or 4 to 1; that is, the maximum particle diameter of the product is one-third or one-fourth that of the feed. The forces exerted by the roll are very great, from 5,500

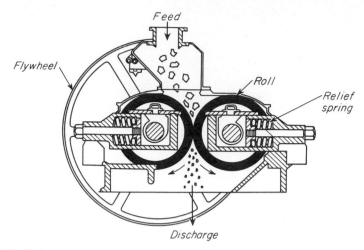

FIGURE 27-4
Smooth-roll crusher.

to 40,000 lb_f per inch of roll width. To allow unbreakable material to pass through without damaging the machine, at least one roll must be spring-mounted.

The angle of nip is the angle between the roll faces at the level where they will just take hold of a particle and draw it into the crushing zone. It is found as follows. Figure 27-5 shows a pair of rolls and a spherical particle just being gripped between them. The radii of rolls and particle are R and r, respectively. The clearance between the rolls is $2d$. Line AB passes through the centers of the left-hand roll and the particle and through point C, the point of contact between roll and particle. Let α be the angle between line AB and the horizontal. Line OE is a tangent to the roll at point C, and this line makes the same angle α with the vertical.

Neglecting gravity, two forces act at point C; the tangential frictional force F_t, having a vertical component $F_t \cos \alpha$, and the radial force F_r, having a vertical component $F_r \sin \alpha$. Force F_t is related to force F_r through the coefficient of friction μ', so $F_t = \mu' F_r$. Force $F_r \sin \alpha$ tends to expel the particle from the rolls, and force $\mu' F_r \cos \alpha$ tends to pull it into the rolls to be crushed. If the particle is to be crushed,

$$F_r \mu' \cos \alpha \geq F_r \sin \alpha$$

or
$$\mu' \geq \tan \alpha \qquad (27\text{-}23)$$

When $\mu' = \tan \alpha$, the angle α is half the angle of nip.

A simple relationship exists between the radius of the rolls, the size of the feed, and the gap between the rolls. Thus, from Fig. 27-5,

$$\cos \alpha = \frac{R + d}{R + r} \qquad (27\text{-}24)$$

The largest particles in the product have a diameter $2d$, and Eq. (27-24) provides a relationship between the roll diameter and the size reduction that can be expected in the mill.

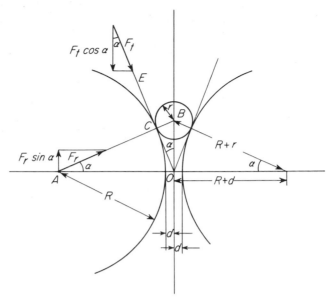

FIGURE 27-5
Angle of nip in crushing rolls.

EXAMPLE 27-5   A pair of rolls is to take a feed equivalent to spheres 1.5 in. (38 mm) in diameter and crush them to spheres having a diameter of 0.5 in. (12.7 mm). If the coefficient of friction is 0.29, what should be the diameter of the rolls?
    SOLUTION   Since $\tan \alpha = 0.29$, $\alpha = 16.2°$ and $\cos \alpha = 0.960$. Then, by Eq. (27-24),

$$0.960 = \frac{R + 0.50/2}{R + 1.50/2}$$

from which $R = 12$ in. The diameter of the rolls should be 24 in., or 2 ft (610 mm).   ////

TOOTHED-ROLL CRUSHERS   In many roll crushers the roll faces carry corrugations, breaker bars, or teeth. Such crushers may contain two rolls, as in smooth-roll crushers, or only one roll working against a stationary curved breaker plate. A single-roll toothed crusher is shown in Fig. 27-6. Machines known as *disintegrators* contain two corrugated rolls turning at different speeds, which tear the feed apart, or a small high-speed roll with transverse breaker bars on its face turning toward a large slow-speed smooth roll. Some crushing rolls for coarse feeds carry heavy pyramidal teeth. Other designs utilize a large number of thin-toothed disks which saw through slabs or sheets of material. Toothed-roll crushers are much more versatile than smooth-roll crushers, within the limitation that they cannot handle very hard solids. They operate by compression, impact, and shear, not by compression alone, as do smooth-roll machines. They are not limited by the problem of nip inherent with smooth rolls and can therefore reduce much larger particles. Some heavy-duty toothed double-roll crushers are used for the primary reduction of coal and similar materials. The particle

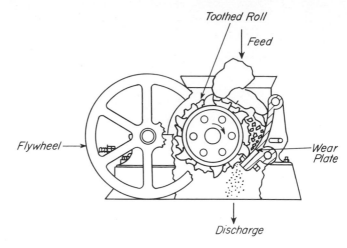

FIGURE 27-6
Single-roll toothed crusher.

size of the feed to these machines may be as great as 20 in.; their capacity ranges up to 500 tons/h.

Grinders The term *grinder* describes a variety of size-reduction machines for intermediate duty. The product from a crusher is often fed to a grinder, in which it is reduced to powder. The chief types of commercial grinders described in this section are hammer mills and impactors, rolling-compression machines, attrition mills, and revolving mills.

HAMMER MILLS AND IMPACTORS These mills all contain a high-speed rotor turning inside a cylindrical casing. The shaft is usually horizontal. Feed dropped into the top of the casing is broken and falls out through a bottom opening. In a hammer mill the particles are broken by sets of swing hammers pinned to a rotor disk. A particle of feed entering the grinding zone cannot escape being struck by the hammers. It shatters into pieces, which fly against a stationary anvil plate inside the casing and break into still smaller fragments. These in turn are rubbed into powder by the hammers and pushed through a grate or screen which covers the discharge opening.

Several rotor disks, 6 to 18 in. in diameter and each carrying four to eight swing hammers, are often mounted on the same shaft. The hammers may be straight bars of metal with plain or enlarged ends or with ends sharpened to a cutting edge. Intermediate hammer mills yield a product 1 in. to 20-mesh in particle size. In hammer mills for fine reduction the peripheral speed of the hammer tips may reach 22,000 ft/min; they reduce 0.1 to 15 tons/h to sizes finer than 200-mesh. Hammer mills grind almost anything—tough fibrous solids like bark or leather, steel turnings, soft wet pastes, sticky clay, hard rock. For fine reduction they are limited to the softer materials.

The capacity and power requirement of a hammer mill vary greatly with the

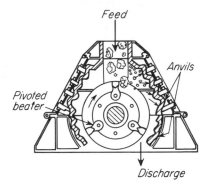

FIGURE 27-7
Impactor.

nature of the feed and cannot be estimated with confidence from theoretical considerations. They are best found from published information[7c] or better from small-scale or full-scale tests of the mill with a sample of the actual material to be ground. Commercial mills typically reduce 100 to 400 lb of solid per hour per horsepower-hour of energy consumed.

An *impactor*, illustrated in Fig. 27-7, resembles a heavy-duty hammer mill except that it contains no grate or screen. Particles are broken by impact alone, without the rubbing action characteristic of a hammer mill. Impactors are often primary-reduction machines for rock and ore, processing up to 600 tons/h. They give particles that are more nearly equidimensional (more "cubical") than the slab-shaped particles from a jaw crusher or gyratory crusher. The rotor in an impactor, as in many hammer mills, may be run in either direction to prolong the life of the hammers.

ROLLING-COMPRESSION MACHINES   In this kind of mill the solid particles are caught and crushed between a rolling member and the face of a ring or casing. The most common types are rolling-ring pulverizers, bowl mills, and roller mills. In the roller mill illustrated in Fig. 27-8, vertical cylindrical rollers press outward with great force against a stationary anvil ring or bull ring. They are driven at moderate speeds in a circular path. Plows lift the solid lumps from the floor of the mill and direct them between the ring and the rolls, where the reduction takes place. Product is swept out of the mill by a stream of air to a classifier separator, from which oversize particles are returned to the mill for further reduction. In a bowl mill and some roller mills the bowl or ring is driven; the rollers rotate on stationary axes, which may be vertical or horizontal. Mills of this kind find most application in the reduction of limestone, cement clinker, and coal. They pulverize up to 50 tons/h. When classification is used, the product may be as fine as 99 percent through a 200-mesh screen.

ATTRITION MILLS   In an attrition mill particles of soft solids are rubbed between the grooved flat faces of rotating circular disks. The axis of the disks is usually horizontal, sometimes vertical. In a single-runner mill one disk is stationary and one rotates; in a double-runner machine both disks are driven at high speed in opposite directions. Feed enters through an opening in the hub of one of the disks; it passes outward

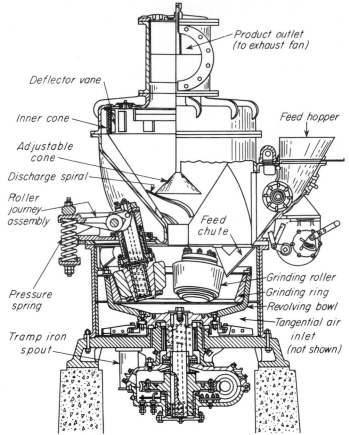

FIGURE 27-8
Roler mill.

through the narrow gap between the disks and discharges from the periphery into a stationary casing. The width of the gap, within limits, is adjustable. At least one grinding plate is spring-mounted so that the disks can separate if unbreakable material gets into the mill. Mills with different patterns of grooves, corrugations, or teeth on the disks perform a variety of operations, including grinding, cracking, granulating, and shredding, and even some operations not related to size reduction at all, such as blending and feather curling.

A single-runner attrition mill is shown in Fig. 27-9. Single-runner mills contain disks of buhrstone or rock emery for reducing solids like clay and talc, or metal disks for solids like wood, starch, insecticide powders, and carnauba wax. Metal disks are usually of white iron, although for corrosive materials disks of stainless steel are sometimes necessary. Double-runner mills, in general, grind to finer products than single-runner mills but process softer feeds. Air is drawn often through the mill to remove the product and prevent choking. The disks may be cooled with water or refrigerated

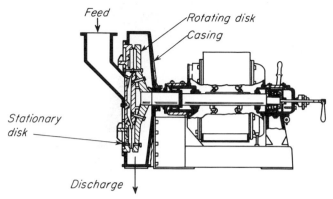

FIGURE 27-9
Attrition mill.

brine to take away the heat generated by the reduction operation. Cooling is essential with heat-sensitive solids like rubber, which would otherwise be destroyed.

The disks of a single-runner mill are 10 to 54 in. in diameter, turning at 350 to 700 r/min. Disks in double-runner mills turn faster, at 1,200 to 7,000 r/min. The feed is precrushed to a maximum particle size of about $\frac{1}{2}$ in. and must enter at a uniform controlled rate. Attrition mills grind from $\frac{1}{2}$ to 8 tons/h to products which will pass a 200-mesh screen. The energy required depends strongly on the nature of the feed and the degree of reduction accomplished and is much higher than in the mills and crushers described so far. Typical values are between 10 and 100 hp-h per ton of product.

REVOLVING MILLS A typical revolving mill is shown in Fig. 27-10. A cylindrical shell slowly turning about a horizontal axis and filled to about half its volume with a solid grinding medium forms a revolving mill. The shell is usually steel, lined with high-carbon steel plate, porcelain, silica rock, or rubber. The grinding medium is metal rods in a rod mill, lengths of chain or balls of metal, rubber, or wood in a ball mill, flint spebble or porcelain or zircon spheres in a pebble mill. For intermediate and fine reduction of abrasive materials revolving mills are unequaled.

Unlike the mills previously discussed, all of which require continuous feed, revolving mills may be continuous or batch. In a batch machine a measured quantity of the solid to be ground is loaded into the mill through an opening in the shell. The opening is then closed and the mill turned on for several hours; it is then stopped and the product discharged. In a continuous mill the solid flows steadily through the revolving shell, entering at one end through a hollow trunnion and leaving at the other end through the trunnion or through peripheral openings in the shell.

In all revolving mills, the grinding elements are carried up the side of the shell nearly to the top, from whence they fall on the particles underneath. The energy expended in lifting the grinding units is utilized in reducing the size of the particles. In some revolving mills, as in a *rod mill*, much of the reduction is done by rolling

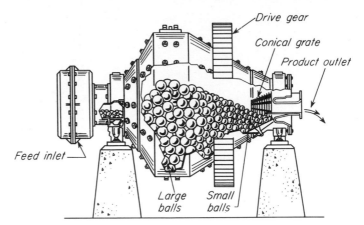

FIGURE 27-10
Conical ball mill.

compression and by attrition as the rods slide downward and roll over one another. The grinding rods are usually steel, 1 to 5 in. in diameter, with several sizes present at all times in any given mill. The rods extend the full length of the mill. They are sometimes kept from twisting out of line by conical ends on the shell. Rod mills are intermediate grinders, reducing a $\frac{3}{4}$-in. feed to perhaps 10-mesh, often preparing the product from a crusher for final reduction in a ball mill. They yield a product with little oversize and a minimum of fines.

In a *ball mill* or *pebble mill* most of the reduction is done by impact as the balls or pebbles drop from near the top of the shell. In a large ball mill the shell might be 10 ft in diameter and 14 ft long. The balls are 1 to 5 in. in diameter; the pebbles in a pebble mill are 2 to 7 in. in size. A *tube mill* is a continuous mill with a long cylindrical shell, in which material is ground for 2 to 5 times as long as in the shorter ball mill. Tube mills are excellent for grinding to very fine powders in a single pass where the amount of energy consumed is not of primary importance. Putting slotted transverse partitions in a tube mill converts it into a *compartment mill*. One compartment may contain large balls, another small balls, and a third pebbles. This segregation of the grinding media into elements of different size and weight aids considerably in avoiding wasted work, for the large, heavy balls break only the large particles, without interference by the fines. The small, light balls fall only on small particles, not on large lumps they cannot break.

Segregation of the grinding units in a single chamber is a characteristic of the *conical ball mill* illustrated in Fig. 27-10. Feed enters from the left through a 60° cone into the primary grinding zone, where the diameter of the shell is a maximum. Product leaves through the 30° cone to the right. A mill of this kind contains balls of different sizes, all of which wear and become smaller as the mill is operated. New large balls are added periodically. As the shell of such a mill rotates, the large balls move toward the point of maximum diameter, and the small balls migrate toward the discharge. The initial breaking of the feed particles, therefore, is done by the largest balls dropping the greatest distance; small particles are ground by small balls dropping a much smaller distance. The amount of energy expended is suited to the difficulty of the breaking operation, increasing the efficiency of the mill.

ACTION IN REVOLVING MILLS The load of balls in a ball or tube mill should be such that when the mill is stopped, the balls occupy somewhat more than one-half the volume of the mill. In operation, the balls are picked up by the mill wall and carried nearly to the top, where they break contact with the wall and fall to the bottom to be picked up again. Centrifugal force keeps the balls in contact with the wall and with each other during the upward movement. While in contact with the wall, the balls do some grinding by slipping and rolling over each other, but most of the grinding occurs at the zone of impact, where the free-falling balls strike the bottom of the mill.

The faster the mill is rotated, the farther the balls are carried up inside the mill and the greater the power consumption. The added power is profitably used because the higher the balls are when they are released, the greater the impact at the bottom and the larger the productive capacity of the mill. If the speed is too high, however, the balls are carried over and the mill is said to be *centrifuging*. The speed at which centrifuging occurs is called the critical speed. Little or no grinding is done when a mill is centrifuging, and operating speeds must be less than the critical.

The speed at which the outermost balls lose contact with the wall of the mill depends on the balance between gravitational and centrifugal forces. This can be shown with the aid of Fig. 27-11. Consider the ball at point A on the periphery of the mill. Let the radii of the mill and of the ball be R and r, respectively. The center of the ball is, then, $R - r$ ft from the axis of the mill. Let the radius AO form the angle α with the vertical. Two forces act on the ball. The first is the force of gravity mg/g_c, where m is the mass of the ball. The second is the centrifugal force $mu^2/(R - r)g_c$, where u is the peripheral speed of the center of the ball. The centripetal component of the force of gravity is $(mg/g_c) \cos \alpha$, and this force opposes the centrifugal force. As long as the centrifugal force exceeds the centripetal force, the particle will not break contact with the wall. As the angle α decreases, however, the centripetal force increases, and unless the speed exceeds the critical, a point is reached where the opposing forces are equal and the particle is ready to fall away. The angle at which this occurs is found by equating the two forces, giving

$$m \frac{g}{g_c} \cos \alpha = \frac{mu^2}{(R - r)g_c}$$

$$\cos \alpha = \frac{u^2}{(R - r)g} \tag{27-25}$$

The speed u is related to the speed of rotation by the equation

$$u = 2\pi n(R - r) \tag{27-26}$$

and Eq. (27-25) can be written

$$\cos \alpha = \frac{4\pi^2 n^2(R - r)}{g} \tag{27-27}$$

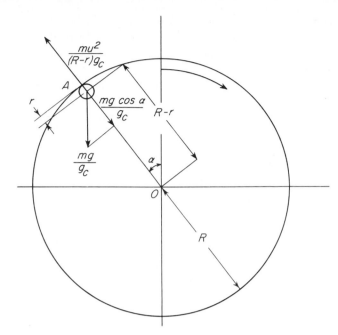

FIGURE 27-11
Forces on ball in ball mill.

At the critical speed, $\alpha = 0$, $\cos \alpha = 1$, and n becomes the critical speed n_c. Then

$$n_c = \frac{1}{2\pi} \sqrt{\frac{g}{R - r}} \qquad (27\text{-}28)$$

CAPACITY AND POWER REQUIREMENT OF REVOLVING MILLS   The maximum amount of energy that can be delivered to the solid being reduced can be computed from the mass of the grinding medium, the speed of rotation, and the maximum distance of fall. In an actual mill the useful energy is much smaller than this, and the total mechanical energy supplied to the mill is much greater. Energy is required to rotate the shell in its bearing supports. Much of the energy delivered to the grinding medium is wasted in overgrinding particles that are already fine enough and in lifting balls or pebbles that drop without doing much if any grinding. Good design, of course, minimizes the amount of wasted energy. Complete theoretical analysis of the many interrelated variables is virtually impossible, so the performance of revolving mills is judged by semiempirical comparisons.[8] Rod mills yield 5 to 200 tons/h of 10-mesh product; ball mills produce 1 to 50 tons/h of powder of which perhaps 70 to 90 percent would pass a 200-mesh screen. The total energy requirement for a typical rod mill grinding hard material is about 5 hp-h/ton; for a ball mill it is about 20 hp-h/ton. Tube mills and compartment mills draw somewhat more power than this. As the product becomes finer, the capacity of a given mill diminishes and the energy requirement increases.

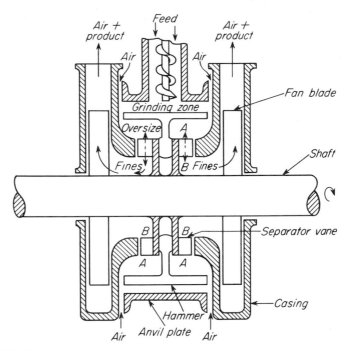

FIGURE 27-12
Principle of Mikro-Atomizer. (*By permission Pulverizing Machinery Div. of Mikro Pul. U.S. Filter Corp.*)

Ultrafine grinders Many commercial powders must contain particles averaging 1 to 20 μm in size, with substantially all particles passing a standard 325-mesh screen which has openings 44 μm wide. Mills which reduce solids to such fine particles are called *ultrafine grinders*. Ultrafine grinding of dry powders is done by grinders, such as high-speed hammer mills, provided with internal or external classification and by fluid-energy or jet mills. Ultrafine wet grinding is done in agitated mills.

A hammer mill with internal classification is the Mikro-Atomizer illustrated in Fig. 27-12. A set of swing hammers is held between two rotor disks, much as in a conventional hammer mill. In addition to the hammers the rotor shaft carries two fans, which draw air through the mill in the direction shown in the figure and discharge into ducts leading to collectors for the product. On the rotor disks are short radial vanes for separating oversize particles from those of acceptable size. In the grinding chamber the particles of solid are given a high rotational velocity. Coarse particles are concentrated along the wall of the chamber because of centrifugal force acting on them. The airstream carries finer particles inward from the grinding zone toward the shaft in the direction *AB*. The separator vanes tend to throw particles outward in the direction *BA*. Whether or not a given particle passes between the separator vanes and out to the discharge depends on which force predominates—the drag exerted by the air or the centrifugal force exerted by the vanes. Acceptably fine particles are carried through; particles that are too large are thrown back for further reduction

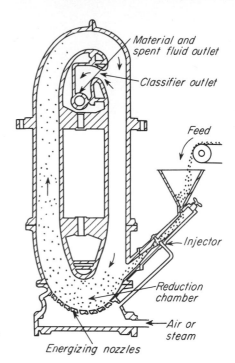

FIGURE 27-13
Fluid-energy mill. (*By permission, Fluid Energy Processing and Equipment Co.*)

in the grinding chamber. The maximum particle size of the product is varied by changing the rotor speed or the size and number of the separator vanes. Mills of this kind reduce 1 or 2 tons/h to an average particle size of 1 to 20 μm, with an energy requirement of about 50 hp-h/ton.

A typical *fluid-energy mill* is shown in Fig. 27-13. In fluid-energy mills the solid particles are suspended in a gas stream and conveyed at high velocity in a circular or elliptical path. Some reduction occurs when the particles strike or rub against the walls of the confining chamber, but most of the reduction is believed to be caused by interparticle attrition. Internal classification keeps the larger particles in the mill until they are reduced to the desired size.

The suspending gas is usually compressed air or superheated steam, admitted at a pressure of 100 lb_f/in.² through energizing nozzles. In the mill shown in the figure the grinding chamber is an oval loop of pipe 1 to 8 in. in diameter and 4 to 8 ft high. Feed enters near the bottom of the loop through a venturi injector. Classification of the ground particles takes place at the upper bend of the loop. As the gas stream flows around this bend at high speed, the coarser particles are thrown outward against the outer wall while the fines congregate at the inner wall. A discharge opening in the inner wall at this point leads to a cyclone separator and a bag collector for the product. The classification is aided by the complex pattern of swirl generated in the gas stream at the bend in the loop of pipe.[2] Fluid-energy mills can accept feed particles as large as $\frac{1}{2}$ in. but are more effective when the feed particles are no larger than

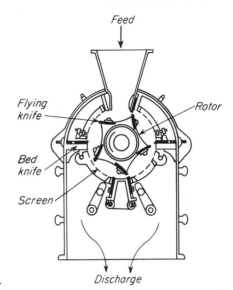

FIGURE 27-14
Rotary knife cutter.

100-mesh. They reduce up to 1 ton/h of nonsticky solid to particles averaging $\frac{1}{2}$ to 10 μm in diameter, using 1 to 4 lb of steam or 6 to 9 lb of air per pound of product.

AGITATED MILLS For some ultrafine grinding operations, small batch nonrotary mills containing a solid grinding medium are available. The medium consists of hard solid elements such as balls, pellets, or sand grains. These mills are vertical vessels 1 to 300 gal in capacity, filled with liquid in which the grinding medium is suspended. In some designs the charge is agitated with a multiarmed impeller; in others, used especially for grinding hard materials (such as silica or titanium dioxide), a reciprocating central column "vibrates" the vessel contents at about 20 Hz. A concentrated feed slurry is admitted at the top and product (with some liquid) is withdrawn through a screen at the bottom. Agitated mills are especially useful in producing particles 1 μm in size or finer.[7b]

Cutting machines[10] In some size-reduction problems the feed stocks are too tenacious or too resilient to be broken by compression, impact, or attrition. In other problems the feed must be reduced to particles of fixed dimensions. These requirements are met by devices which cut, chop, or tear the feed into a product with the desired characteristics. The sawtoothed crushers mentioned above do much of their work in this way. True cutting machines include rotary knife cutters and granulators. These devices find application in a variety of processes but are especially well adapted to size-reduction problems in the manufacture of rubber and plastics.

A *rotary knife cutter*, as shown in Fig. 27-14, contains a horizontal rotor turning at 200 to 900 r/min in a cylindrical chamber. On the rotor are two to twelve flying knives with edges of tempered steel or stellite passing with close clearance over one to seven stationary bed knives. Feed particles entering the chamber from above are cut several hundred times per minute and emerge at the bottom through a screen with 5- to 8-mm openings. Sometimes the flying knives are parallel with the bed knives; sometimes, depending on the properties of the feed, they cut at an angle. Rotary cutters and granulators are similar in design. A granulator yields more or less irregular pieces; a cutter may yield cubes, thin squares, or diamonds.

EQUIPMENT OPERATION

For the proper selection and economical operation of size-reduction machinery, attention must be given to many details of procedure and of auxiliary equipment. A crusher, grinder, or cutter cannot be expected to perform satisfactorily unless (1) the feed is of suitable size and enters at a uniform rate; (2) the product is removed as soon as possible after the particles are of the desired size; (3) unbreakable material is kept out of the machine; and (4), in the reduction of low-melting or heat-sensitive products, the heat generated in the mill is removed. Heaters and coolers, metal separators, pumps and blowers, and constant-rate feeders are therefore important adjuncts to the size-reduction unit.

Open-circuit and closed-circuit operation   In many mills the feed is broken into particles of satisfactory size by passing it once through the mill. When no attempt is made to return oversize particles to the machine for further reduction, the mill is said to be operating in *open circuit*. This may require excessive amounts of power, for much energy is wasted in regrinding particles that are already fine enough. If a 50-mesh product is desired, it is obviously wasteful to continue grinding 100- or 200-mesh material. Thus it is often economical to remove partially ground material from the mill and pass it through a size-separation device. The undersize becomes the product and the oversize is returned to be reground. The separation device is sometimes inside the mill, as in ultrafine grinders; or, as is more common, it is outside the mill. *Closed-circuit operation* is the term applied to the action of a mill and separator connected so that oversize particles are returned to the mill.

For coarse particles the separation device is a screen or grizzly; for fine powders it is some form of classifier.† A typical set of size-reduction machines and separators operating in closed circuit is diagramed in Fig. 27-15. The product from a gyratory crusher is screened into three fractions, fines, intermediate, and oversize. The oversize is sent back to the gyratory; the fines are fed directly to the final reduction unit, a ball mill. Intermediate particles are broken in a rod mill before they enter the ball mill. In the arrangement shown in the diagram the ball mill is grinding wet; i.e., water is pumped through the mill with the solid to carry the broken particles to a centrifugal classifier. The classifier throws down the oversize into a sludge, which is repulped with more water and returned to the mill. The undersize, or product, emerges from

† These devices are discussed in Chap. 30.

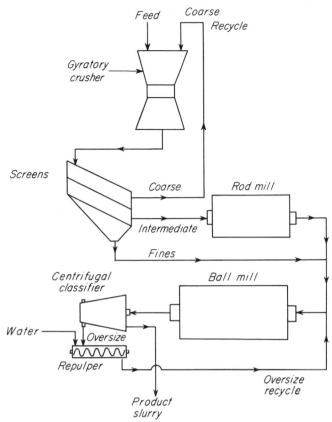

FIGURE 27-15
Flow sheet for closed-circuit grinding.

the classifier as a slurry containing particles of acceptable size. Although screens are simpler to operate than classifiers, they cannot economically make separations when the particles are smaller than about 150- to 200-mesh. It is the overgrinding of precise-ly these fine particles that results in excessive consumption of energy. Closed-circuit operation is therefore of most value in reduction to fine and ultrafine sizes, which demands that the separation be done by wet classifiers or air separators of the types described in Chap. 30. Energy must of course be supplied to drive the conveyors and separators in a closed-circuit system, but despite this, the reduction in total energy requirement over open-circuit grinding often reaches 25 percent.

Feed control Of the operations auxiliary to the size reduction itself, control of the feed to the mill is the most important. The particles in the feed must be of appropriate size. Obviously they must not be so large that they cannot be broken by the mill; if too many of the particles are very fine, the effectiveness of many machines, especially intermediate crushers and grinders, is seriously reduced. With some solids precompression or chilling of the feed before it enters the mill greatly increases the ease with which it can be ground. In continuous mills the feed rate must be controlled within close limits to avoid choking or erratic variations in load and yet make full use of the capacity of the machine. In cutting sheet material into precise squares or thread into uniform lengths for flock, exact control of the feed rate is obviously essential.

Mill discharge   To avoid buildup in a continuous mill the rate of discharge must equal the rate of feed. Furthermore, the discharge rate must be such that the working parts of the mill can operate most effectively on the material to be reduced. In a jaw crusher, for example, particles may collect in the discharge opening and be crushed many times before they drop out. As mentioned before, this is wasteful of energy if many of the particles are crushed more than necessary. Operation of a crusher in this way is sometimes deliberate; it is known as *choke crushing*. Usually, however, the crusher is designed and operated so that the crushed particles readily drop out, perhaps carrying some oversize particles, which are separated and returned. This kind of operation is called *free-discharge crushing* or *free crushing*. Choke crushing is used only in unusual problems, for it requires large amounts of power and may damage the mill.

With fairly coarse comminuted products, as from a crusher, intermediate grinder, or cutter, the force of gravity is sufficient to give free discharge. The product usually drops out the bottom of the mill. In a revolving mill it escapes through openings in the chamber wall at one end of the cylinder (*peripheral discharge*); or it is lifted by scoops and dropped into a cone which directs it out through a hollow trunnion (*trunnion discharge*). A slotted grate or diaphragm keeps the grinding medium from leaving with the product. Peripheral discharge is common in rod mills, trunnion discharge in ball mills and tube mills.

In discharging mills for fine and ultrafine grinding the force of gravity is replaced by the drag of a fluid carrier. The fluid may be a liquid or a gas. Wet grinding with a liquid carrier is common in revolving mills. It gives more wear on the chamber walls and on the grinding medium than dry grinding, but it saves energy, increases capacity, and simplifies handling and classification of the product. Z sweep of air, steam, or inert gas removes the product from attrition mills, fluid-energy mills, and many hammer mills. The powder is taken out of the gas stream by cyclone separators or bag filters.

Removal or supply of heat   Only a very small fraction of the energy supplied to the solid in a grinding mill is used in creating new surface. The bulk of the energy, therefore, is converted to heat, which may raise the temperature of the solid by many degrees. The solid may melt, decompose, or explode unless this heat is removed. For this reason cooling water or refrigerated brine is often circulated through coils or jackets in the mill. Sometimes the air blown through the mill is refrigerated, or solid carbon dioxide (dry ice) is admitted with the feed. Still more drastic temperature reduction is achieved with liquid nitrogen, to give grinding temperatures below $-75°C$. The purpose of such low temperatures is to alter the breaking characteristics of the solid, usually by making it more friable. In this way substances like lard and beeswax become hard enough to shatter in a hammer mill; tough plastics like poly-ethylene, which stall a mill at ordinary temperatures, become brittle enough to be ground without difficulty.

SYMBOLS

A_w Specific surface of particles, ft^2/lb or m^2/g; A_{wa}, for feed; A_{wb}, for product
a Volume shape factor
B Constant in Eq. (27-1); B', constant, defined by Eq. (27-6)
$B_{n,u}$ Total mass fraction of particles smaller than size $\bar{D}_{pn}$ resulting from breakage of particles of size $\bar{D}_{pu}$
b Surface shape factor; breadth of roll face, ft or m
D_p Particle size, ft or mm; D_{pa}, of feed; D_{pb}, of product; D_{p1}, of largest particle in fraction; D_{p2}, of smallest particle in fraction
D_{pn} Mesh opening in screen n, ft or mm; $D_{p(n-1)}$, in screen $n-1$
$\bar{D}_n$ Arithmetic average of D_{pn} and $D_{p(n-1)}$, ft or mm
$\bar{D}_s$ Mean volume-surface diameter, ft or mm; $\bar{D}_{sa}$, of feed; $\bar{D}_{sb}$, of product
$\bar{D}_u$ Average particle diameter on screen u, which is coarser than screen n
d One-half distance between crushing rolls, ft or m
e_s Surface energy per unit area, ft-lb$_f$/ft^2 or J/m^2
F Force, lb$_f$ or N; F_r, radial force; F_t, tangential force
g Acceleration of gravity, ft/s^2 or m/s^2
g_c Newton's-law proportionality factor, 32.174 ft-lb/lb$_f$-s^2
K_b Constant in Bond's law, Eq. (27-16)
K_r Constant in Rittinger's law, Eq. (27-15)
k Constant in Eq. (27-1)
m Mass of ball, lb or g
$\dot{m}$ Feed rate to crusher, tons/h or kg/s
N_w Number of particles per unit mass
n Speed, r/min; screen number, counting from large screen of series; n_T, total number of screens; n_c, critical speed in ball mill
P Power, hp or kW
R Radius of crushing rolls or ball mill, ft or m
r Radius of feed particle, ft or m; radius of balls in ball mill, ft or m; ratio of $D_{p(n-1)}$ to D_{pn} in testing screens
S_u Grinding-rate function for screen u, s^{-1}
s_p Area of particle, ft^2 or m^2
t Time, s; t_T, total grinding time
u Peripheral speed of center of ball in ball mill, ft/s or m/s; also number of screen coarser than screen n
v_p Volume of particle, ft^3 or m^3
W Energy input to crusher, ft-lb$_f$/lb or J/g; W_n, energy absorbed by material during crushing
W_i Bond work index, kWh/ton
Greek letters
α One-half angle of nip, also angle between radius and vertical in ball mill
β Exponent in Eq. (27-21)
$\Delta B_{n,u}$ Breakage function, fraction of particles of size $\bar{D}_u$ which are broken to size $\bar{D}_n$
Δt Time increment, s
$\Delta \phi_n$ Mass fraction of total sample retained by screen n and passed by screen $n-1$
$\Delta \phi_u$ Mass fraction retained on screen u

η_c	Crushing efficiency
η_m	Mechanical efficiency of crusher
μ'	Coefficient of friction
ρ_p	Density of particle, lb/ft^3
Φ_s	Sphericity; Φ_a, of feed; Φ_b, of product
ϕ	Mass fraction of total sample cumulative to size D_{pn}; ϕ_1, of largest particle in fraction; ϕ_2, of smallest particle in fraction

PROBLEMS

27-1 Trap rock is crushed in a gyratory crusher. The feed is nearly uniform 2-in. spheres. The differential screen analysis of the product is given in column 1 of Table 27-6. The power required to crush this material is 430 kW/ton. Of this 10 kW is needed to operate the empty mill. By reducing the clearance between the crushing head and the cone, the differential screen analysis of the product becomes that given in column 2 in Table 27-6. From Rittinger's law, what should the power for the second operation be?

27-2 Using the Bond method, estimate the power necessary in each of the operations in Prob. 27-1 per ton of rock.

27-3 What is the critical speed, in revolutions per minute, of a ball mill 1,200 mm ID charged with 75-mm balls?

Table 27-6 DATA FOR PROB. 27-1

	Product	
Mesh	First grind (1)	Second grind (2)
4/6	3.1	
6/8	10.3	3.3
8/10	20.0	8.2
10/14	18.6	11.2
14/20	15.2	12.3
20/28	12.0	13.0
28/35	9.5	19.5
35/48	6.5	13.5
48/65	4.3	8.5
−65	0.5	
65/100		6.2
100/150		4.0
−150		0.3

REFERENCES

1 Arbiter, N., and C. C. Harris: *Br. Chem. Eng.,* **10**:240 (1965).
2 Berry, C. E.: *Ind. Eng. Chem.,* **38**:672 (1946).
3 Bond, F. C.: *Trans. AIME,* TP-3308B, and *Mining Eng.,* May 1952.
4 Galanty, H. E.: *Ind. Eng. Chem.,* **55**(1):46 (1963).

5   Gaudin, A. M.: "Principles of Mineral Dressing," McGraw-Hill, New York, 1939; (*a*) p. 126, (*b*) p. 129, (*c*) p. 136.
6   Kwong, J. N. S., J. T. Adams, J. F. Johnson, and E. L. Piret: *Chem. Eng. Prog.,* **45**:508, 655, 708 (1949).
7   Perry, J. H. (ed.): "Chemical Engineers' Handbook," 5th ed., McGraw-Hill, New York, 1963; (*a*) pp. **8**-14 to **8**-16, (*b*) p. **8**-29, (*c*) pp. **8**-36 to **8**-40.
8   Piret, E. L.: *Chem. Eng. Prog.,* **49**:56 (1953).
9   Reid, K. J.: *Chem. Eng. Sci.,* **20**(11):953 (1965).
10 Sterrett, K. R., and W. M. Sheldon: *Ind. Eng. Chem.,* **55**(2):46 (1963).

28

CRYSTALLIZATION

Crystallization is the formation of solid particles within a homogeneous phase. It may occur as the formation of solid particles in a vapor, as in snow; as solidification from a liquid melt, as in the freezing of water to ice or the manufacture of large single crystals; or as crystallization from liquid solutions. The present treatment is restricted to the last situation.

Crystallization from solution is important industrially because of the variety of materials that are marketed in the crystalline form. Its wide use has a twofold basis: a crystal formed from an impure solution is itself pure (unless mixed crystals occur), and crystallization affords a practical method of obtaining pure chemical substances in a satisfactory condition for packaging and storing.

MAGMA In industrial crystallization from solution, the two-phase mixture of mother liquor and crystals of all sizes, which occupies the crystallizer and which is withdrawn as a product, is called a *magma*.

Importance of crystal size Clearly, good yield and high purity are important objectives in crystallization, but the appearance and size range of a crystalline product also are significant. If the crystals are to be further processed, reasonable size and size uniformity are desirable for washing, filtering, reacting with other chemicals, transporting, and storing the crystals. If the crystals are to be marketed as a final product, customer acceptance requires individual crystals to be strong, nonaggregated, uniform in size, and noncaking in the package. For these reasons, *crystal size distribution* (*CSD*) must be under control; it is a prime objective in the design and operation of crystallizers.

CRYSTAL GEOMETRY

A crystal is the most highly organized type of nonliving matter. It is characterized by the fact that its constituent particles, which may be atoms, molecules, or ions, are arranged in orderly three-dimensional arrays called *space lattices*. The distances between the particles in a space lattice can be measured by x-ray diffraction, and these lattice constants and the angles between the imaginary planes containing the particles have been determined for common crystalline materials.

Crystal forms: law of Haüy As a result of the space-lattice arrangement of the particles, when crystals are allowed to form without hindrance from other crystals or outside bodies, they appear as polyhedrons having sharp corners and flat sides, or faces. Although the relative sizes of the faces and edges of various crystals of the same material may be widely different, the angles made by corresponding faces of all crystals of the same material are equal and are characteristic of that material.

Crystallographic systems Since all crystals of a definite substance have the same interfacial angles in spite of wide differences in the extent of development of individual faces, crystal forms are classified on the basis of these angles. For example, consider a definite crystal. Choose three important nonparallel faces and call them axial planes. When the crystal is a symmetrical solid, such as a cube, the axial faces are taken parallel with planes of symmetry. The three intersections of the axial planes determine three nonparallel lines, and the three lines parallel with such lines drawn through a single arbitrary point are called the *axes* of the crystal. The axes may all be mutually perpendicular; two of them may be perpendicular to the third but not to each other; they may be equally inclined to each other, with a common angle different from either 60 or 90°; or they may be mutually inclined with three different angles, all differing from 60 or 90°.

In addition to the three axial faces, a fourth fundamental face, intersecting all three axes, is chosen. The lengths of the segments so cut off from the three axes are used as units of length for measurements along the respective axes. The unit lengths so determined may be equal for all axes, equal for two axes but unequal for the third, or unequal for all three.

One class of crystals, showing a hexagonal cross section with 60° angles between the hexagonal sides, is most conveniently referred to four axes instead of the usual three. Three of the axes are at 60° to each other and in the same plane, and the fourth is perpendicular to the plane of the other three.

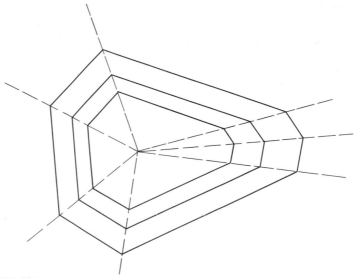

FIGURE 28-1
Growth of invariant crystal.

The combinations of angles and lengths of the axes give rise to seven classes of crystals.

1 Triclinic system   Three mutually inclined and unequal axes, all angles unequal and other than 30, 60, or 90°
2 Monoclinic system   Three unequal axes, two of which are inclined but perpendicular to the third
3 Orthorhombic system   Three unequal rectangular axes
4 Tetragonal system   Three rectangular axes, two of which are equal and different in length from the third
5 Trigonal system   Three equal and equally inclined axes
6 Hexagonal system   Three equal coplanar axes inclined 60° to each other, and a fourth axis different in length from the others and perpendicular to them
7 Cubic system   Three equal rectangular axes

Invariant crystals   Under ideal conditions, a growing crystal maintains geometric similarity during growth. Such a crystal is called *invariant*. Figure 28-1 shows cross sections of an invariant crystal during growth. Each of the polygons in the figure represents the outline of the crystal at a different time. Since the crystal is invariant, these polygons are geometrically similar and the dotted lines connecting the corners of the polygons with the center of the crystal are straight. The center point may be thought of as the location of the original nucleus from which the crystal grew. The rate of growth of any face is measured by the velocity of translation of the face away from the center of the crystal in a direction perpendicular to the face. It may be seen that unless the crystal is a regular polyhedron, the rates of growth of the various faces of an invariant crystal are not equal.

CRYSTAL SIZE AND SHAPE FACTORS   A single dimension can be used as the measure of the size of an invariant crystal of any definite shape. Thus, assume that as a temporary choice, the distance between any two specific corners is used. Denote this by L'. Then the volume and total surface area of that crystal can be written

$$v_p = a(L')^3 \qquad (28\text{-}1)$$

$$A_p = 6b(L')^2 \qquad (28\text{-}2)$$

where v_p = volume of crystal
$\quad A_p$ = area of crystal
$\quad a, b$ = shape factors, depending only on shape of crystal and choice of dimension L'

The ratio of area to volume is

$$\frac{A_p}{v_p} = \frac{6b(L')^2}{a(L')^3} = \frac{6b}{aL'} \qquad (28\text{-}3)$$

Now define L, the final choice of a practical length dimension, as

$$L \equiv \frac{a}{b} L'$$

and then choose the ratio of the shape factors as

$$\frac{a}{b} = 1$$

The primes are now dropped from Eqs. (28-1) and (28-2), and Eq. (28-3) becomes

$$\frac{A_p}{v_p} \equiv \frac{6}{L} \qquad (28\text{-}4)$$

For cubes, L is the length of a side, and $a = b = 1$. For a sphere, L is the diameter, and $a = b = \pi$. For geometric solids in general, L is close to the size determined by screening.[19]

PRINCIPLES OF CRYSTALLIZATION

Crystallization may be analyzed from the standpoints of purity, yield, energy requirements, and rates of nucleation and growth.

Purity of product   A sound, well-formed crystal itself is nearly pure, but it retains mother liquor when removed from the final magma, and if the crop contains crystalline aggregates, considerable amounts of mother liquor may be occluded within the solid mass. When retained mother liquor of low purity is dried on the product, contamination results, the extent of which depends on the amount and degree of impurity of the mother liquor retained by the crystals.

In practice, much of the retained mother liquor is separated from the crystals by filtration or centrifuging, and the balance is removed by washing with fresh solvent. The effectiveness of these purification steps depends on the size and uniformity of the crystals.

Equilibria and yields   Equilibrium in crystallization processes is discussed in Chap. 17. For some systems the phase diagram is complicated, as illustrated by the diagram for the system magnesium sulfate–water shown in Fig. 17-23. Calculations of yields, however, are simplified by the fact that the graphical mixing rule and the center-of-gravity principle apply in such phase diagrams for *isothermal* mixing processes.

In many industrial crystallization processes, the crystals and mother liquor are in contact long enough to reach equilibrium, and the mother liquor is saturated at the final temperature of the process. The yield of the process can then be calculated from the concentration of the original solution and the solubility at the final temperature. If appreciable evaporation occurs during the process, this must be known or estimated.

When the rate of crystal growth is slow, considerable time is required to reach equilibrium. This is especially true when the solution is viscous or where the crystals collect in the bottom of the crystallizer so there is little crystal surface exposed to the supersaturated solution. In such situations, the final mother liquor may retain appreciable supersaturation, and the actual yield will be less than that calculated from the solubility curve.

Crystallizers in which the crystals and mother liquor are not in equilibrium are called class I systems, and those in which the mother liquor is assumed to be saturated are class II systems.[17]

If the crystals are anhydrous, calculation of the yield is simple, as the solid phase contains no solvent. When the crop contains water of crystallization, account must be taken of the water accompanying the crystals, since this water is not available for retaining solute in solution. Solubility data are usually given either in parts by mass of anhydrous material per hundred parts by mass of total solvent or in mass percent anhydrous solute. These data ignore water of crystallization. The key to calculations of yields of hydrated solutes is to express all masses and concentrations in terms of hydrated salt and free water. Since it is this latter quantity that remains in the liquid phase during crystallization, concentrations or amounts based on free water can be subtracted to give a correct result.

EXAMPLE 28-1   A solution consisting of 30 percent $MgSO_4$ and 70 percent H_2O is cooled to 60°F. During cooling, 5 percent of the total water in the system evaporates. How many kilograms of crystals are obtained per kilogram of original mixture?

SOLUTION   From Fig. 17-23 it is noted that the crystals are $MgSO_4 \cdot 7H_2O$ and that the concentration of the mother liquor is 24.5 percent anhydrous $MgSO_4$ and 75.5 percent H_2O. Per 1,000 kg of original solution, the total water is $0.70 \times 1,000 = 700$ kg. The evaporation is $0.05 \times 700 = 35$ kg. The molecular weights of $MgSO_4$ and $MgSO_4 \cdot 7H_2O$ are 120.4 and 246.5, respectively, so the total $MgSO_4 \cdot 7H_2O$ in the batch is $1,000 \times 0.30$ $(246.5/120.4) = 614$ kg, and the free water is $1,000 - 35 - 614 = 351$ kg. In 100 kg of mother liquor, there is $24.5(246.5/120.4) = 50.16$ kg of $MgSO_4 \cdot 7H_2O$ and $100 - 50.16 = 49.84$ kg of free water. The $MgSO_4 \cdot 7H_2O$ in the mother liquor, then, is $(50.16/49.84)351 = 353$ kg. The final crop is $614 - 353 = 261$ kg.

The problem can also be solved by use of the center-of-gravity principle. The overall concentration of the batch after evaporation, in mass fraction of $MgSO_4$, is

$$\frac{0.30 \times 1,000}{1,000 - 35} = 0.311$$

Since the concentration of the mother liquor is 0.245 and that of $MgSO_4 \cdot 7H_2O$ is 0.488 (the abscissa of the ends of line hi in Fig. 28-2), the yield, by the center-of-gravity principle, is

$$965 \frac{0.311 - 0.245}{0.488 - 0.245} = 262 \text{ kg} \qquad \text{////}$$

Enthalpy balances   In heat-balance calculations for crystallizers, the heat of crystallization is important. This is the latent heat evolved when solid forms from a solution. Ordinarily, crystallization is exothermic, and the heat of crystallization varies with both temperature and concentration. Rigorously, the heat of crystallization is related to the heat of solution of the crystals in a large amount of solvent and to the heat of dilution of the solution from saturation to high dilution. Data on heats of solution and of dilution are available,[2] and these, together with data on the specific heats of the solutions and of the crystals, can be used to construct enthalpy-concentration charts like those of Fig. 16-9 and Fig. 17-13 but extended to include solid phases. The diagram is especially useful in calculating enthalpy balances for crystallization processes. An Hx diagram, showing enthalpies of solid phases, for the system $MgSO_4$ and H_2O is given in Fig. 28-2. This diagram is consistent with the phase diagram of Fig. 17-23. As before, enthalpies are given in Btu per pound. They refer to 1 lb of total mixture regardless of the number of phases in the mixture. The area above line $pabcdq$ represents enthalpies of undersaturated solutions of $MgSO_4$ in H_2O, and the isotherms in this area have the same significance as those in Fig. 16-10. The area eap in Fig. 28-2 represents all equilibrium mixtures of ice and freezing $MgSO_4$ solutions. Point n represents pure ice. The isothermal (25°F) triangle age gives the enthalpies of all combinations of ice with partly solidified eutectic or of partly solidified eutectic with $MgSO_4 \cdot 12H_2O$. Area $abfg$ gives the enthalpy-concentration points for all magmas consisting of $MgSO_4 \cdot 12H_2O$ crystals and mother liquor. The isothermal (35.7°F) triangle bhf shows the transformation of $MgSO_4 \cdot 7H_2O$ to $MgSO_4 \cdot 12H_2O$, and this area represents mixtures consisting of a saturated solution containing 21 percent $MgSO_4$, solid $MgSO_4 \cdot 7H_2O$, and solid $MgSO_4 \cdot 12H_2O$. The area $cihb$ represents all magmas of $MgSO_4 \cdot 7H_2O$ and mother liquor. Isothermal (118.8°F) triangle cji represents mixtures consisting of a saturated solution containing

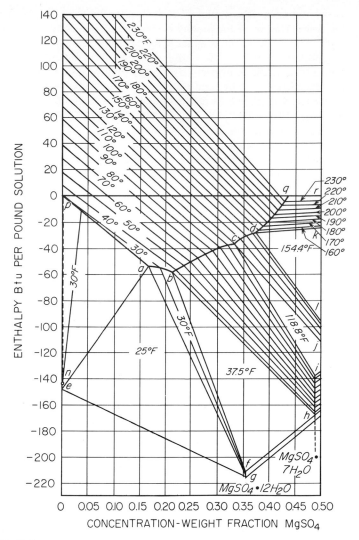

FIGURE 28-2
Enthalpy-concentration diagram, system MgSO$_4$-H$_2$O. [*By permission, from J. H. Perry (ed.), "Chemical Engineers' Handbook," 4th ed. Copyright © 1963, McGraw-Hill Book Company.*]

33 percent MgSO$_4$, solid MgSO$_4 \cdot 6$H$_2$O, and solid MgSO$_4 \cdot 7$H$_2$O. Area *dljc* gives enthalpies of MgSO$_4 \cdot 6$H$_2$O and mother liquor. The isothermal (154.4°F) triangle *dkl* represents mixtures of a saturated solution containing 37 percent MgSO$_4$, solid MgSO$_4 \cdot$ H$_2$O, and solid MgSO$_4 \cdot 6$H$_2$O. Area *qrkd* is part of the field representing saturated solutions in equilibrium with MgSO$_4 \cdot$ H$_2$O. Except for the isotherms in the liquid-phase field and the solubility and freezing curves, all lines on the *Hx* diagram are straight.

The usual constructions for adiabatic and nonadiabatic mixing processes can be made on Fig. 28-2, including use of points in the areas representing more than one phase.

EXAMPLE 28-2   A 32.5 percent solution of $MgSO_4$ at 120°F (48.9°C) is cooled, without appreciable evaporation, to 70°F (21.1°C) in a batch water-cooled crystallizer. How much heat must be removed from the solution per ton of crystals?

SOLUTION   The initial solution is represented by the point on Fig. 28-2 at a concentration of 0.325 in the undersaturated-solution field on a 120°F isotherm. The enthalpy coordinate of this point is -33.0 Btu/lb. The point for the final magma lies on the 70°F isotherm in area *cihb* at concentration 0.325. The enthalpy coordinate of this point is -78.4. Per 100 lb of original solution the heat *absorbed* by the solution is

$$100(-78.4 + 33.0) = -4540 \text{ Btu}$$

This is a heat evolution of 4.540 Btu (1.14×10^6 cal$_{IT}$).

The split of the final magma between crystals and mother liquor can be found by the center-of-gravity principle applied to the 70°F isotherm in either Fig. 17-23 or 28-2. The concentration of the mother liquor is 0.259, and that of the crystals is 0.488. Then, the crystals are

$$100 \frac{0.325 - 0.259}{0.488 - 0.259} = 28.8 \text{ lb/100 lb magma}$$

The heat evolved per ton of crystals is $(4540/28.8)(2,000) = 315,000$ Btu (8.76×10^4 cal$_{IT}$/kg).   ////

Supersaturation   Mass and enthalpy balances shed no light on the CSD of the product from a crystallizer. Conservation laws are unchanged if the product is one huge crystal or a mush of small ones.

In the formation of a crystal two steps are required: (1) the birth of a new particle and (2) its growth to macroscopic size. The first step is called *nucleation*. In a crystallizer, the CSD is determined by the interaction of the rates of nucleation and growth, and the overall process is complicated kinetically. The driving potential for both rates is supersaturation, and neither nucleation nor growth can occur in a saturated or undersaturated solution.

In the theories of nucleation and growth, mole units are used in place of mass units.

Supersaturation may be generated by one or more of three methods. If the solubility of the solute increases strongly with increase in temperature, as is the case with many common inorganic salts and organic substances, a saturated solution becomes supersaturated by simple cooling and temperature reduction. If the solubility is relatively independent of temperature, as is the case with common salt, supersaturation may be generated by evaporating a portion of the solvent. If neither cooling nor evaporation is desirable, as when the solubility is very high, supersaturation may be generated by adding a third component. The third component may act physically by forming, with the original solvent, a mixed solvent in which the solubility of the solute is sharply reduced. This process is called salting. Or, if a nearly complete precipitation is required, a new solute may be created chemically by adding a third component that will react with the original solute and form an insoluble substance. This process is called precipitation. The methods used in wet quantitative analysis are prime examples of precipitation. By the addition of a third component the rapid creation of very large supersaturations is possible.

UNITS FOR SUPERSATURATION   Supersaturation is the concentration difference between that of the supersaturated solution in which the crystal is growing and that of a solution in equilibrium with the crystal. Both phases are assumed to be at the same temperature. Concentrations may be defined either in mole fraction of the solute, denoted by y, or in moles of solute in unit volume of solution, denoted by c. Since only one component is transferred across phase boundaries, component subscripts are omitted. The two supersaturations are defined by the equations

$$\Delta y \equiv y - y_s \tag{28-5}$$

$$\Delta c \equiv c - c_s \tag{28-6}$$

where Δy = supersaturation, mole fraction of solute
$\quad y$ = mole fraction of solute in solution
$\quad y_s$ = mole fraction of solute in saturated solution
$\quad \Delta c$ = molar supersaturation, moles per unit volume
$\quad c$ = molar concentration of solute in solution
$\quad c_s$ = molar concentration of solute in saturated solution

The supersaturations defined by Eqs. (28-5) and (28-6) are related by the equation

$$\Delta c = \rho y - \rho_s y_s \tag{28-7}$$

where ρ and ρ_s are the molar densities of the solution and the saturated solution, respectively. In general, since supersaturations in crystallizers are small, the densities ρ and ρ_s may be considered equal and ρ used to designate both quantities. Then

$$\Delta c = \rho \, \Delta y \tag{28-8}$$

The concentration ratio α and the fractional supersaturation s are defined by

$$\alpha \equiv \frac{c}{c_s} = 1 + \frac{\Delta c}{c_s} = \frac{y}{y_s} = 1 + \frac{\Delta y}{y_s} \equiv 1 + s \tag{28-9}$$

The quantity $100s$ is the percent supersaturation. In practice, it is usually less than about 2 percent.

TEMPERATURE DIFFERENCE AS A POTENTIAL   Supersaturation is often expressed as an equivalent temperature difference. The relation between these driving potentials is shown in Fig. 28-3, which contains a small section of the solubility curve in molar concentrations. The field above the line represents undersaturated solutions and that below the line supersaturated solutions. Point A refers to the saturated solution at temperature T_s, which also is the temperature of the growing crystal, and point B to a saturated solution at temperature T, where $T > T_s$. The supersaturated solution is represented by point C. The temperature of this solution also is T_s. From Eqs. (28-5)

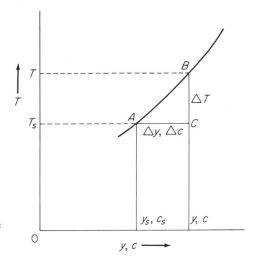

FIGURE 28-3
Supersaturation and temperature
potentials.

and (28-6) the supersaturation potential is represented by line segment $\overline{AC}$. The equivalent temperature driving potential is shown by line segment $\overline{BC}$. Segment $\overline{AB}$ of the solubility curve can be considered linear over the small concentration spanned by line $\overline{AC}$ and the temperature potential defined by

$$\Delta T \equiv T - T_s = \kappa(y - y_s) = \kappa \, \Delta y = \frac{\kappa}{\rho} \Delta c \qquad (28\text{-}10)$$

where κ = slope of T-vs.-y line
ρ = molar density

Nucleation

The rate of nucleation is the number of new particles formed per unit time per unit volume of magma or solids-free mother liquor. This quantity is the first kinetic parameter controlling the CSD.

Origins of crystals in crystallizers If all sources of particles are subsumed under the term nucleation, a number of kinds of nucleation may occur. Many of these are important only as methods to be avoided. They may be classified into three groups: spurious nucleation, primary nucleation, and secondary nucleation.

One origin of crystals is macroscopic attrition, which is more akin to comminution than to real nucleation. Circulating magma crystallizers have internal propeller agitators or external rotary circulating pumps. On impact with these moving parts, soft or weak crystals can break into fragments, form rounded corners and edges, and so give new crystals, both large and small. Such effects also degrade the quality of the product. Attrition is the only source of new crystals that is independent of supersaturation.

Occasionally, especially in experimental work, seed crystals originating in pre-

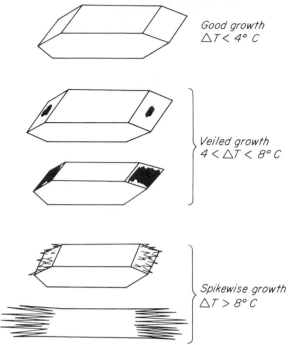

FIGURE 28-4
Illustration of growth regimes (adapted from Ref. 5).

vious crystallizations are added to crystallizing systems. Seeds usually carry on their surfaces many small crystals which were formed during the drying and storage of the seeds after their own manufacture. The small crystals soon wash off and subsequently grow in the supersaturated solution. This is called *initial breeding*.[21] It can be prevented by curing the seed crystals before using, either by contact with undersaturated solution or solvent or by a preliminary growth in a stagnant supersaturated solution.

Growth-related spurious nucleation occurs at large supersaturations or accompanies poor magma circulation. It is characterized by abnormal needlelike and whiskerlike growths from the ends of the crystals, which, under these conditions, may grow much faster than the sides. The spikes are imperfect crystals, which are bound to the parent crystal by weak forces and which break off to give crystals of poor quality. This is called *needle breeding*.[21]

Another growth-related imperfection, unrelated to nucleation, is called *veiled growth* and occurs at moderate supersaturations. It is the result of the occlusion of mother liquor into the crystal face, giving a milky surface and an impure product. The cause of veiled growth is too rapid crystal growth, which traps mother liquor into the crystal faces.

Figures 28-4 and 28-5, which apply to $MgSO_4 \cdot 7H_2O$, show the appearance of good and inferior crystals and the variation in quality of product at various supersaturations.

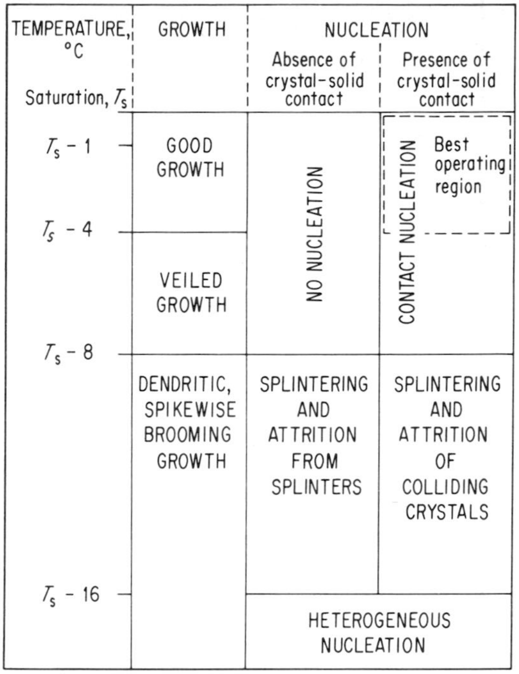

TEMPERATURE, °C Saturation, T_s	GROWTH	NUCLEATION	
		Absence of crystal–solid contact	Presence of crystal–solid contact
$T_s - 1$	GOOD GROWTH	NO NUCLEATION	CONTACT NUCLEATION · Best operating region
$T_s - 4$			
	VEILED GROWTH		
$T_s - 8$			
	DENDRITIC, SPIKEWISE BROOMING GROWTH	SPLINTERING AND ATTRITION FROM SPLINTERS	SPLINTERING AND ATTRITION OF COLLIDING CRYSTALS
$T_s - 16$		HETEROGENEOUS NUCLEATION	

FIGURE 28-5
Effect of supersaturation on crystal growth quality and type of nucleation for $Mg SO_4 \cdot 7H_2O$. (Adapted from Ref. 5.)

All forms of spurious nucleation can be avoided by growing crystals at low supersaturations and by using only well designed and operated pumps and agitators.

Primary nucleation In scientific usage, nucleation refers to the birth of very small bodies of a new phase within a supersaturated homogeneous existing phase. Basically, the phenomenon of nucleation is the same for crystallization from solution, crystallization from a melt, condensation of fog drops in a supercooled vapor, and generation of bubbles in a superheated liquid. In all instances, nucleation is a consequence of rapid local fluctuations on a molecular scale in a homogeneous phase that is in a state of metastable equilibrium.[9] The basic phenomenon is called *homogeneous nucleation*, which is further restricted to the formation of new particles within a phase uninfluenced in any way by solids of any sort, including the walls of the container or even the tiniest submicroscopic particles of foreign substances.

A variant of homogeneous nucleation occurs when solid particles of foreign substances do influence the nucleation process by catalyzing an increase of nucleation rate at a given supersaturation or giving a finite rate at a supersaturation where homogeneous nucleation would occur only after a vast time. This is called *heterogeneous nucleation*.

Homogeneous nucleation In magma crystallization, homogeneous nucleation seldom if ever happens, and if it did, it would not be welcome. The fundamentals of the phenomenon, however, are important in understanding the more useful types of nucleation.

Crystal nuclei may form from molecules, atoms, or ions. These may, in aqueous solutions, be hydrated. Because of their rapid motions such particles are called *kinetic units*. In a very small volume, of the order of 10 nm^3, kinetic theory holds that individual kinetic units vary greatly, in location, time, velocity, energy, and concentration. The apparently stationary values of the intensive properties—density, concentration, and energy—in a macroscopic mass of solution are time-smoothed averages of fluctuations too fast and too small to measure on a macroscopic scale.

Because of fluctuations, an individual kinetic unit often enters another's field of force, and the two particles momentarily join. Usually they break apart immediately, but if they do hang together, they may be joined by a third and subsequent particles. Combinations of this sort are called *clusters*. The adding of particles one by one to a cluster constitutes a chain reaction, which may be considered to be a series of reversible chemical reactions shown in the following scheme:

$$A_1 \ + A_1 \rightleftharpoons A_2$$
$$A_2 \ + A_1 \rightleftharpoons A_3$$
$$\cdots \cdots \cdots \cdots \cdots$$
$$A_{m-1} + A_1 \rightleftharpoons A_m$$

where A_1 is a single kinetic unit and the subscript is the number of units in the cluster.

When m is small, a cluster cannot be thought of as a particle of a new phase with definite identity and boundaries. As m increases, the cluster can be so recognized, and it is then called an *embryo*. Embryos have, for the most part, short lives and break up and revert to clusters or individual units. Depending on the supersaturation, however, some embryos grow to such a size that they attain thermodynamic equilibrium with the solution. Then the embryo is called a *nucleus*. The value of m for a nucleus is in the range of a few to several hundred. The m value for nuclei of liquid water is about 80.

Nuclei are in a state of unstable equilibrium: if a nucleus loses units, it dissolves; if it gains units, it grows and becomes a crystal. The sequence of stages in the evolution of a crystal is, then,

$$\text{Cluster} \rightarrow \text{embryo} \rightarrow \text{nucleus} \rightarrow \text{crystal}$$

Equilibrium Thermodynamically, the difference between a small particle and a large one at the same temperature is that the small particle possesses a significant amount of surface energy per unit mass and the large one does not. One consequence of this difference is that the solubility of a small crystal in the less than micrometer size range is larger than that of the large crystal. Ordinary solubility data apply only to moderately large crystals. A small crystal can be in equilibrium with a supersaturated solution. Such an equilibrium is unstable, because if a large crystal is also present in the solution, the smaller crystal will dissolve and the larger one grow until the small crystal disappears. This phenomenon is called *Ostwald ripening*. The effect of particle size on solubility is the key factor in nucleation.

Surface energy The surface energy of a crystal differs from face to face. Consider face k of a crystal having f faces. Let σ_k be the surface energy per unit area of face k, and assume that this interfacial energy is independent of crystal size. Then U_k, the energy of face k, is

$$U_k = \sigma_k A_k$$

where A_k is the area of face k.

Gibbs-Wulff theorem On the assumption that σ_k is constant and independent of crystal size, thermodynamic analysis shows that if the crystal grows at constant temperature in equilibrium with its mother liquor, the crystal preserves invariancy of shape. This is the *Gibbs-Wulff theorem*. Then, since in such a crystal, the ratio A_k/A_c is constant, where A_c is the total area of the crystal, the surface energy of the crystal can be written

$$U_c = \bar{\sigma} A_c \tag{28-11}$$

where $\bar{\sigma}$ is a weighted average interfacial energy which is independent of crystal size.

If the constant ratio A_k/A_c is denoted by β_k, the total surface energy of the crystal U_c is given by

$$U_c = \sum_{k=1}^{f} \sigma_k A_k = \sum_{k=1}^{f} \sigma_k \beta_k A_c = A_c \sum_{k=1}^{f} \sigma_k \beta_k \tag{28-12}$$

Comparison of Eqs. (28-11) and (28-12) shows that

$$\bar{\sigma} = \sum_{k=1}^{f} \sigma_k \beta_k \tag{28-13}$$

Energy of formation of crystal The formation of a crystal requires two kinds of energy: one to form the surface and one to build the crystal mass. The surface energy is given by Eq. (28-11). From equilibrium thermodynamics, the energy required to build the crystal is, at constant temperature and pressure

$$-(\mu - \mu_\infty)m_c$$

where μ = chemical potential of solute in supersaturated solution
μ_∞ = chemical potential of crystal sufficiently large to be in equilibrium with saturated solution
m_c = molal mass of crystal

The total energy required to build the crystal may be given by the increase in the value of a potential called the work function[27] Ω, so

$$\Delta\Omega = \bar{\sigma} A_c - (\mu - \mu_\infty)m_c = A_c \left[\bar{\sigma} - \frac{m_c}{A_c}(\mu - \mu_\infty) \right] \tag{28-14}$$

If V_M is the molar volume and v_c the volume of the crystal,

$$m_c = \frac{v_c}{V_M} \tag{28-15}$$

Since the crystal is invariant, from Eq. (28-2)

$$A_c = 6bL^2 \tag{28-16}$$

and from Eq. (28-4)

$$v_c = \frac{L}{6} A_c \tag{28-17}$$

Substituting from Eqs. (28-15) to (28-17) into Eq. (28-14) gives

$$\Delta\Omega = 6bL^2 \left[\bar{\sigma} - \frac{L}{6V_M} (\mu - \mu_\infty) \right] \tag{28-18}$$

The thermodynamic condition for equilibrium applicable to this situation is this: for a fixed and constant value of $\mu - \mu_\infty$ (which corresponds to a definite supersaturation),

$$\frac{d(\Delta\Omega)}{dL} = 0$$

Applying this condition to Eq. (28-18) gives for the equilibrium crystal size L_n

$$L_n = \frac{4\bar{\sigma} V_M}{(\mu - \mu_\infty)_n} \tag{28-19}$$

The subscript on the potential difference signifies its value corresponding to the crystal size L_n. This size is called the *critical size* and is, of course, the size of a nucleus.

Kelvin equation   The chemical potential difference is related to the concentrations of the saturated and supersaturated solutions by the equation†

$$(\mu - \mu_\infty)_n = vRT \ln \frac{y}{y_s} = vRT \ln \frac{c}{c_s} = vRT \ln \alpha \tag{28-20}$$

Substituting $\mu - \mu_\infty$ from Eq. (28-20) into Eq. (28-19) gives

$$\frac{4V_M \bar{\sigma}}{vRTL_n} = \ln \alpha \tag{28-21}$$

Equation (28-21) is the Kelvin equation, relating the solubility of a substance to its particle size.

The energy of nucleation $\Delta\Omega_n$ is found by substituting L_n from Eq. (28-19) for L in Eq. (28-18) and then substituting $\mu - \mu_\infty$ from Eq. (28-20) into the resulting equation. After simplification this yields

$$\Delta\Omega_n = \frac{32b\bar{\sigma}^3 V_M^2}{(vRT \ln \alpha)^2} \tag{28-22}$$

---

† This assumes that the activity coefficients of the solute in the two solutions are equal. Since y/y_s is nearly unity, this assumption is reasonable. The factor v is needed to cover the case of an electrolytic solute. It is the number of ions per molecule of solute. For molecular crystals, $v = 1$.

Although crystals are polyhedra, nuclei contain so few particles that geometric shape has little meaning and for simplicity it is assumed that embryos and nuclei are spheres. For a sphere, from Eq. (28-2), if the diameter D is used for length L, $b = \pi/6$, the value of σ is uniform, and Eq. (28-22) may be written

$$\Delta\Omega_n = \frac{16\pi\sigma^3 V_M^2}{3(vRT \ln \alpha)^2} \tag{28-23}$$

Kinetics The energy of nucleation is an energy barrier that controls the kinetics of nucleation. The rate of nucleation, from the theory of chemical kinetics, is

$$B^\circ = C \exp\left(-\frac{\Delta\Omega_n}{kT}\right) = C \exp\left(-\frac{\Delta\Omega_n N_a}{RT}\right) \tag{28-24}$$

Substituting $\Delta\Omega_n$ from Eq. (28-23) into Eq. (28-24) gives

$$B^\circ = C \exp\left[-\frac{16\pi\sigma^3 V_M^2 N_a}{3v^2(RT)^3(\ln \alpha)^2}\right] \tag{28-25}$$

where B° = nucleation rate, number/cm^3-s
  k = Boltzmann constant, 1.3806×10^{-16} erg/g mol-K
  N_a = Avogadro constant, 6.0222×10^{23} molecules/g mol
  R = gas constant, 8.3143×10^7 ergs/g mol-K
  C = frequency factor

The factor C represents the rate at which individual particles strike the surface of the crystal. Its value for nucleation from solutions is not known. From analogy with nucleation of water drops from supersaturated water vapor,[9] it is of the order of 10^{25}. Its accurate value is not important, as the kinetics of nucleation is dominated by the ln α term in the exponent.

Numerical values for σ also are uncertain. The experimental determination of solid-liquid interfacial tensions has not been accomplished. They may be calculated from solid-state theory, using lattice energies. For ordinary salts, σ is of the order[29] of 80 to 100 ergs/cm^2. When the above values of C and σ are used in Eq. (28-25), a value of s can be calculated that will correspond to a nucleation rate of one nucleus per second per cubic centimeter, or $B^\circ = 1$. The calculation gives a very large value of s and one that is impossible for materials of usual solubility. This is one reason for the conclusion that homogeneous nucleation in ordinary crystallization from solution never occurs and that all actual nucleations in these situations are heterogeneous.

In precipitation reactions, where y_s is very small and where large supersaturation ratios can be generated rapidly, homogeneous nucleation probably occurs.[29]

Heterogeneous nucleation The catalytic effect of solid particles on nucleation rate is the reduction of the energy barrier given by Eq. (28-23). One theory of this effect holds that if the nucleus "wets" the surface of the catalyst, the work of nucleus formation is reduced by a factor that is a function of the angle of wetting between the nucleus and the catalyst. Experimental data on the heterogeneous nucleation of potassium chloride solutions[16] show that the nucleation of this substance is consistent with an apparent value of the interfacial tension in the range 2 to 3 ergs/cm^2 for both catalyzed nucleation and nucleation without an added catalyst. If the latter situation was actually a secondary nucleation self-catalyzed by microscopic seeds, the value of σ for seeded solutions of KCl would be 2.8 ergs/cm^2 at a temperature of 300 K. If σ_a is used to denote the apparent interfacial tension, if C is taken as 10^{25}, and if the mathematical approximation $\ln \alpha = \alpha - 1 = s$ is accepted for small values of $\alpha - 1$ Eq. (28-25) may be written

$$B^\circ = 10^{25} \exp\left[-\frac{16\pi V_M^2 N_a \sigma_a^3}{3(RT)^3 vs^2} \right] \tag{28-26}$$

This equation, although it rests on incomplete data, gives results of the correct order and does reflect the very strong effect of supersaturation on nucleation.

EXAMPLE 28-3   Assuming that the rate of heterogeneous nucleation of potassium chloride is consistent with an apparent interfacial tension of 2.5 ergs/cm^2, determine the nucleation rate as a function of s at a temperature of 80°F (300 K).

SOLUTION   Use Eq. (28-26). The molecular weight of KCl is 74.56. The density of the crystal is 1.988 g/cm^3. Since KCl dissociates into two ions, K^+ and Cl^-, $v = 2$. Then

$$V_M = \frac{74.56}{1.988} \text{ cm}^3/\text{g mol} \qquad \sigma_a = 2.5 \text{ ergs/cm}^2$$

The exponent in Eq. (28-26) is

$$-\frac{16\pi(74.56)^2 \times 6.0225 \times 10^{23} \times 2.5^3}{3(300 \times 8.3134 \times 10^7)^3(2^2 s^2)} = -\frac{0.01412}{s^2}$$

For $B^\circ = 1$, the value of s is given by

$$1 = 10^{25} e^{-0.01429/s^2} = e^{57.565} e^{-0.01429/s^2}$$

$$\frac{0.01412}{s^2} = 57.565 \qquad s = \sqrt{\frac{0.01412}{57.565}} = 0.0156$$

From the equation

$$B^\circ = e^{57.565} e^{-0.01412/s^2}$$

the value of B° can be calculated for magnitudes of s around 0.016. The results are shown in Table 28-1. The explosive increase in B° as s increases is apparent.   ////

Table 28-1

s	B°	s	B°
0.014	5.17×10^{-7}	0.016	11.2
0.015	5.57×10^{-3}	0.017	6.04×10^3
0.0158	1	0.018	1.18×10^6

Secondary nucleation The formation of nuclei attributable to the influence of the existing macroscopic crystals in the magma is called *secondary nucleation*.[13,26] Two kinds are known, one attributable to fluid shear and the other to collisions between existing crystals with each other or with the walls of the crystallizer and rotary impellers or agitator blades.

FLUID-SHEAR NUCLEATION This type is known to take place under certain conditions and is suspected in others. When supersaturated solution moves past the surface of a growing crystal at a substantial velocity, the shear stresses in the boundary layer may sweep away embryos or nuclei that would otherwise be incorporated into the growing crystal and so appear as new crystals. This has been reported in work on sucrose crystallization.[15] It also has been demonstrated in the nucleation of $MgSO_4 \cdot 7H_2O$[24] if the solution is subjected to shear at the crystal face at one supersaturation and then quickly cooled to a higher supersaturation and allowed to stand while nuclei grow to macroscopic size.

CONTACT NUCLEATION That secondary nucleation is the most effective and commonest method and that it is influenced by the intensity of agitation has been known for some time; but only recently has the phenomenon of *contact nucleation* been isolated experimentally and shown conclusively to be not only the easiest method of nucleation but also the best. Its advantages are several. (1) It has a low kinetic order and is rate-proportional to the supersaturation s or, in some cases, to $1/(\ln \alpha)$, in contrast to the twentieth or more order of primary nucleation, and so allows easy control without unstable operation. (2) It occurs at low supersaturation, where growth rate of the crystal is at its optimum rate for good quality. (3) The necessary energy at which the crystal is struck is amazingly low, so that the contacts do not give pathological macroscopic attrition. (4) The quantitative fundamentals have been isolated and are rapidly being incorporated into engineering practice.

The position of contact nucleation generally in the fields of crystal growth and nucleation is shown by the dotted area in Fig. 28-5.

The definite existence of the phenomenon was announced by Strickland-Constable and associates in 1963[22] and amplified in 1966.[11] The results were confirmed and extended by McCabe and associates in 1971[5] and 1972,[8] who showed quantitatively how contact nucleation varies with energy and supersaturation. Since then study of the subject has proceeded at an accelerated rate. The basic results are as follows.

VARIABLES IN CONTACT NUCLEATION By the use of contacts of small rods on the selected faces of individual crystals positioned in a flowing supersaturated solution, under such conditions that supersaturation, fluid velocity, and a small amount of impact energy could be measured independently, the variables of contact nucleation were found.[5,8,25] For a given crystal face and fixed magnitudes of the variables, the linear velocity of the liquid was found to have no effect and the number of nuclei per contact depend only on the supersaturation and energy of impact. The following summary of the effects of the variables is based on recent information. Complete quantitative information on the various kinds of crystals is not yet known, but definite results are on hand[25] for two hydrated inorganic crystals, magnesium sulfate heptahydrate, $MgSO_4 \cdot 7H_2O$, and potassium alum, $KAl(SO_4) \cdot H_2O$; one anhydrous inorganic crystal, potassium sulfate, K_2SO_4; and one organic crystal, citric acid, $C_6H_8O_7 \cdot H_2O$. Some data are found in the literature on other inorganic crystals. Important review articles on contact nucleation are available.[3, 10, 23]

EFFECT OF SUPERSATURATION   For the inorganic crystals the nucleation per contact is proportional to the supersaturation. For the organic crystal it is proportional to $1/(\ln \alpha)$. Both are temperature-dependent in accordance with the theory of reaction rates controlled by an activation energy and as given by the Arrhenius equation

$$N = K_n e^{-\Delta E_n / RT} \qquad (28\text{-}27)$$

where N = number of nuclei per contact
ΔE_n = activation energy

EFFECT OF CONTACT ENERGY   The nucleation per contact for the hydrated inorganic crystals and the organic crystal is proportional to the energy of contact E over a range that is significant in practice. For these substances no threshold energy is needed, and the linear plot of N vs. E passes through the origin. The energy required for the anhydrous inorganic crystal is considerably larger than that for the other crystals, and a threshold energy is required before nucleation will take place. In general this crystal is more difficult to nucleate than the others.

In all cases, the energies required are remarkably small, of the magnitude of a few hundred ergs, and no visible effect is observable on the crystal surface. The crystal simply keeps on growing as usual. For higher energies the nucleation per contact is[25] proportional to exp E.

EFFECT OF CRYSTAL ORIENTATION AT IMPACT   The nucleation per contact is extremely sensitive to the accuracy of the impact. For reproducible results the face of the crystal and the end of the rod must be exactly parallel. An edge or corner contact gives many more nuclei than an exact surface contact.

The nucleation varies from face to face on the same crystal, although all faces of the same form give the same result. Nucleation decreases with hardness and smoothness of the crystal, but no conclusive correlation between the kinds of crystals has been found.

Impacts from a rod of rubber, plastic, or other material softer than the crystal face give no nucleation, even when the impact is severe enough to fracture the crystal.

HEALING TIME BETWEEN CONTACTS   An appreciable time between contacts is needed if successive contacts give the same nucleation. For $MgSO_4 \cdot 7H_2O$ the recovery time is about 15 s at a supersaturation of 1°C.[8]

Mechanism of contact nucleation   The precise mechanism of contact nucleation is not known. It is suspected that it is a combination of breakage of microscopic dendritic growths on the surface of the growing crystal and an interference by the contacting object with clusters of solute particles moving to become organized into the crystal. In either picture, contact nucleation is clearly a growth-related process, and it definitely is not macroscopic attrition. The proportionality between N and supersaturation, the effectiveness of low-energy impacts, the increase in N from increase in contact energy with no change in the appearance of the crystal surface, the recovery time between contacts, and the great sensitivity of the nucleation per contact with the kind and direction of the contacts argue against macroattrition as the nucleation source.

SURVIVAL THEORY   A rational concept of contact nucleation is the *survival theory* proposed by Strickland-Constable,[23] which relates the phenomenon to the theory of heterogeneous nucleation. It is hypothesized that the interference with growth by action of the contacting solid deflects or dislodges particles ranging in size from embryos to small crystals much larger than L_n, that is, the size of the nucleus called for by the Kelvin equation (28-21). Particles at least as large as L_n survive and grow, while smaller ones dissolve. This hypothesis is especially attractive in that it is consistent with the prediction that L_n is, by the Kelvin equation, inversely proportional to the supersaturation s. Thus the larger the value of s, the greater the survival of smaller crystals. This is consistent with the experimental fact that N, the number of nuclei per contact, also is proportional to supersaturation.

Crystal Growth

Crystal growth is a diffusional process, modified by the effect of the solid surfaces on which the growth occurs. Solute molecules or ions reach the growing faces of a crystal by diffusion through the liquid phase. The usual mass-transfer coefficient k_y applies to this step. On reaching the surface, the molecules or ions must be accepted by the crystal and organized into the space lattice. The reaction occurs at the surface at a finite rate, and the overall process consists of two steps in series. Neither the diffusional nor the interfacial step will proceed unless the solution is supersaturated.

Overall and individual growth coefficients   Consider a single crystal face, e.g., face i of a growing crystal having a total of n faces. Let the area of the face be A_i. Let y be the concentration of the solution at a distance from the face, y_e the saturation concentration, and y' the concentration of the solution at the interface between crystal and liquid. If there were no interfacial reaction, y' would be equal to y_e, but since such a reaction does exist, $y' > y_e$. The rate of mass transfer from the bulk of the liquid to the interface is given by the equation

$$\frac{dm_i}{A_i\, dt} = \frac{\dot{m}_i}{A_i} = k_y(y - y')$$

(28-28)

where dm_i = moles of solute transferred to face i in time dt h
$\quad\dot{m}_i = dm_i/dt$
$\quad k_y$ = mass-transfer coefficient defined by Eq. (22-28)

Assume that the rate of reaction at the crystal surface is also linear in supersaturation, so that

$$\frac{dm}{A_i\,dt} = \frac{\dot{m}_i}{A_i} = k_{si}(y' - y_s) \tag{28-29}$$

where k_{si} is the coefficient of the surface reaction at face i. Eliminating y' from Eqs. (28-28) and (28-29) gives

$$\frac{\dot{m}_i}{A_i} = \frac{y - y_s}{1/k_y + 1/k_{si}} = K_i(y - y_s) \tag{28-30}$$

where K_i is the overall mass-transfer coefficient for face i.

Rate of growth of entire crystal[28]   Since the ratio of A_i, the area of face i, to A_c, the total area of the crystal, is β_i, Eq. (28-30) for face i can be written

$$\dot{m}_i = A_c\beta_i K_i(y - y_s)$$

The molal growth rate for the entire crystal is

$$\dot{m}_c = A_c(y - y_s)\sum_{i=1}^{f}\beta_i K_i = \bar{K}A_c(y - y_s) \tag{28-31}$$

where $\bar{K}$ is the average overall mass transfer coefficient for the entire crystal. It is assumed that the supersaturation is the same for all faces. For an invariant crystal, $\bar{K}$ is independent of crystal size. From Eq. (28-1), the mass of the crystal is

$$m_c = v_c\rho_c = aL^3\rho_c$$

where ρ_c is the molar density of the crystal. Differentiating this equation yields

$$dm_c = 3aL^2\rho_c\,dL$$

and, from Eq. (28-31),

$$\frac{dm_c}{dt} = \dot{m}_c = 3aL^2\rho_c\frac{dL}{dt} = \bar{K}A_c(y - y_s)$$

When it is noted that $a/b = 1$ and G is used for dL/dt, Eq. (28-2) gives

$$3a\rho_c L^2 G = 6bL^2\bar{K}(y - y_s)$$

$$G = \frac{2\bar{K}(y - y_s)}{\rho_c} \tag{28-32}$$

The average interfacial transfer coefficient for the entire crystal is defined implicitly by

$$\bar{K} = \frac{1}{1/k_y + 1/\bar{k}_s} \tag{28-33}$$

MASS TRANSFER TO CRYSTAL SURFACE   For spherical particles Eq. (22-45) is used to predict the mass-transfer coefficient. The equation shows that the coefficient is in-

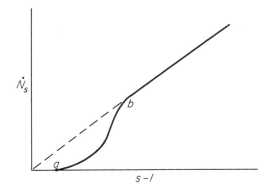

FIGURE 28-6
Rate of two-dimensional nucleation as
functions of supersaturation.

versely proportional to particle size and linear in $\sqrt{N_{Re}}$. But the equation predicts too low a value for the coefficient for crystals, probably because of the different pattern of boundary-layer separations from the flat surfaces and sharp edges. Data on $MgSO_4 \cdot 7H_2O^6$ show a higher level of values of k_y at low velocities and a much stronger increase with velocity. In any case, as the velocity of solution past the crystal increases, the mass-transfer resistance $1/k_y$ goes to zero and the surface resistance k_s controls, so $\overline{K} \approx \overline{k}_s$.

SURFACE-GROWTH COEFFICIENTS   Much research on the growth of crystals by the interfacial reaction has been done and reported in standard monographs on crystallization.[4, 21, 28] Although a coherent theory of crystal growth has evolved, numerical data on $\overline{k}_s$ of a kind that can be used in design are scarce.

One important theory holds that the growth of a face of a crystal occurs layer by layer. Each new layer of particles on a given face begins as a two-dimensional nucleus, one particle thick, attached to the face. Once formed, the nucleus adds particles rapidly at the edge of the nucleus and quickly becomes a complete layer. A new nucleus must of course form to start each new layer. The kinetic equation for two-dimensional nucleation is[28]

$$B_s = Z \exp\left[-\frac{\pi a^2 \sigma^2}{(kT)^2 \ln s}\right] \qquad (28\text{-}34)$$

where B_s = surface nucleation rate, number/cm^2-s
$\quad Z$ = rate of arrival of fresh molecules at the surface, number/cm^2-s
$\quad a^2$ = area covered by one particle, cm^2

The two-dimensional theory, as represented by Eq. (28-34), predicts a rate of growth at low supersaturations of the type shown by the solid curve ab in Fig. 28-6. Under this theory growth does not start until an appreciable supersaturation is reached. At higher supersaturations, beyond that of point b, the rate of growth is nearly linear with supersaturation because the rate at which the particles strike the surface is so great that nuclei are always present. The threshold value for the start of

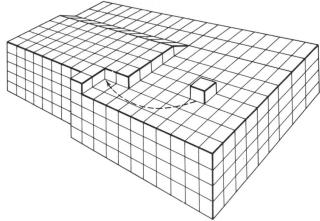

FIGURE 28-7
Screw dislocation in crystal surface and movement of particle into kink.

growth at low supersaturations is the result of the energy barrier implied by Eq. (28-34).

The actual growth rate of most real crystals is shown by the dotted line in Fig. 28-6, which indicates a linear rate at all supersaturations. There are also data showing that the rate of growth sometimes is proportional to s^2 rather than to s.

The difference between the experimental data and the theoretical rate of growth predicted by the two-dimensional nucleation mechanism has been reconciled by the Frank screw-dislocation theory.[7] It is well known that the actual space lattices of real crystals are far from perfect and that crystals have imperfections called *dislocations*. Planes of particles on the surfaces and within the crystals are displaced, and several kinds of dislocations are known. One common dislocation is a screw dislocation (Fig. 28-7), where the individual particles are shown as cubical building blocks. The dislocation is in a shear plane perpendicular to the surface of the crystal, and the slipping of the crystal creates a ramp. The edge of the ramp acts like a portion of a two-dimensional nucleus, and provides a kink into which particles can easily fit. A complete face never can form, and no nucleation is necessary. As growth continues, the edge of the ramp becomes a spiral, and the continued deposit of particles along the edge of the spiral constitutes the mechanism of crystal growth.

Since this mechanism was first suggested, many examples of screw dislocations have actually been observed by electron microscopy and other methods of very high magnification.

GROWTH-RATE EQUATIONS   General methods for estimating values of k_s are nonexistent. For the growth rates of the crystals mentioned in the discussion of nucleation, it was found[25] that, for the hydrated inorganics,

$$G = K_1 s \tag{28-35}$$

For the organic crystal

$$G = Ce^{-K_2/s} \tag{28-36}$$

The growth rates of all crystals were temperature-dependent and followed the Arrhenius equation

$$G = K_3 e^{-\Delta E_s/RT} \qquad (28\text{-}37)$$

In Eqs. (28-35) to (28-37), C, K_1, K_2, and K_3 are constants and ΔE_s is an activation energy. In the experiments on which the equations are based, the mass-transfer resistance was not completely eliminated, but it was not significant.

The ΔL law of crystal growth[12]   If all crystals in magma grow in a uniform supersaturation field and at the same temperature, and if all crystals grow from birth in accordance with either Eq. (28-35) or (28-36), then all crystals are not only invariant but have the same growth rate that is independent of size. This generalization is called the ΔL *law*. When applicable, $G \neq f(L)$, the total growth of each crystal in the magma during the same time interval Δt is the same, and

$$\Delta L = G \, \Delta t \qquad (28\text{-}38)$$

Applicability of ΔL law   The model underlying Eq. (28-38) is highly idealized and is certainly not realistic in all situations. When the law is not applicable, $G = f(L)$ and the growth is called *size-dependent*. Perhaps because of cancellation of errors, the law is sufficiently precise in many situations to be usable, and so greatly simplifies the calculation of industrial processes.

CRYSTALLIZATION EQUIPMENT

Commercial crystallizers may operate either continuously or batchwise. Except for special applications, continuous operation is preferred. The first requirement of any crystallizer is to create a supersaturated solution, as crystallization cannot occur without supersaturation. The means used to produce supersaturation depend primarily on the solubility curve of the solute. Solutes like common salt have solubilities nearly independent of temperature, and others, such as anhydrous sodium sulfate and sodium carbonate monohydrate, have inverted solubility curves and become more soluble as the temperature is lowered. To crystallize these materials, supersaturation must be developed by evaporation. Other materials, such as glauber salt, epsom salt, copperas, and sodium thiosulfate, are much less soluble at low temperatures than at high, and cooling without evaporation produces supersaturation. In intermediate cases, a combination of evaporation and cooling is effective. Sodium nitrate, for example, may be satisfactorily crystallized by cooling without evaporation, evaporation without cooling, or a combination of cooling and evaporation.

Variations in crystallizers   Commercial crystallizers may also be differentiated in several other ways. One important difference is in how the crystals are brought into contact with the supersaturated liquid. In the first technique, called the *circulating-liquid method*, a stream of supersaturated solution is passed through a fluidized bed of growing crystals, within which supersaturation is released by nucleation and growth. The saturated liquid then is pumped through a cooling or evaporating zone, in which supersaturation is generated, and finally the supersaturated solution is recycled through the crystallizing zone. In the second technique, called the *circulating-magma method*, the entire magma is circulated through both crystallization and supersaturation steps without separating the liquid from the solid. Supersaturation as well as crystallization occurs in the presence of crystals. In both methods feed solution is added to the circulating stream between the crystallizing and supersaturating zones.

One type of crystallizer uses size-classification devices designed to retain small crystals in the growth zone for further growth and to allow only crystals of a specified minimum size to leave the unit as product. Ideally, such a crystallizer would produce a classified product of a single uniform size. Other crystallizers are designed to maintain a thoroughly mixed suspension in the crystallizing zone, in which crystals of all sizes from nuclei to large crystals are uniformly distributed throughout the magma. Ideally, the size distribution in the product from a mixed suspension unit is identical to that in the crystallizing magma itself.

Again, some crystallizers are equipped with devices to segregate fine crystals as soon as possible after nucleation and to remove the bulk of them from the crystallizing zone. The objective of this is to reduce sharply the number of nuclei present in order that the remaining ones can grow. The presence of too many nuclei means that the mass of precipitating solute available for each particle is too small to permit adequate growth. The nuclei withdrawn, which actually have a very small mass, are redissolved and returned to the crystallizer. In situations where the size specification for the product is not so severe, removal of nuclei is unnecessary.

Most crystallizers utilize some form of agitation to improve growth rate, to prevent segregation of supersaturated solution that causes excessive nucleation, and to keep crystals in suspension throughout the crystallizing zone. Internal propeller agitators may be used, often equipped with draft tubes and baffles, and external pumps also are common for circulating liquid or magma through the supersaturating or crystallizing zones. The latter method is called *forced circulation*. One advantage of forced-circulation units with external heaters is that several identical units can be connected in multiple effect by using the vapor from one unit to heat the next in line. Systems of this kind are *evaporator-crystallizers*.

Vacuum crystallizers   Most modern crystallizers fall in the category of vacuum units in which adiabatic evaporative cooling is used to create supersaturation. In its original and simplest form, such a crystallizer is a closed vessel in which a vacuum is maintained by a condenser, usually with the help of a steam-jet vacuum pump, or booster, placed between the crystallizer and the condenser. A warm saturated solution at a temperature well above the boiling point at the pressure in the crystallizer is fed to the vessel. A magma volume is maintained by controlling the level of the liquid and crystallizing solid in the vessel and the space above the magma used for release of vapor and elimination of entrainment. The feed solution cools spontaneously to the equilibrium temperature; since both the enthalpy of cooling and the enthalpy of crystallization appear as latent enthalpy of vaporization, a portion of the solvent evaporates. The supersaturation generated by both cooling and evaporation causes nucleation and growth. Product magma is drawn from the bottom of the crystallizer.

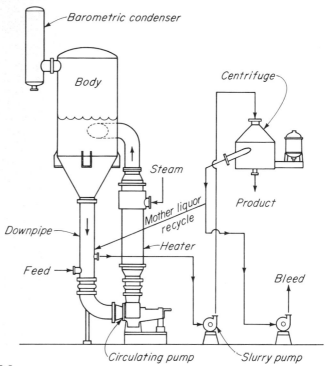

FIGURE 28-8
Continuous crystallizer. [*By permission, from R. C. Bennett and M. Van Buren, Chem. Engr. Symp. Ser. 95,* **65,** 38 (1969).]

The theoretical yield of crystals is proportional to the difference between the concentration of the feed and the solubility of the solute at equilibrium temperature.

Figure 28-8 shows a modern continuous vacuum crystallizer with the conventional auxiliary units for feeding the unit and processing the product magma. The essential action of a single body is much like that of a single-effect evaporator, and in fact these units can be operated in multiple effect. The magma circulates from the cone bottom of the crystallizer body through a downpipe to a low-speed low-head circulating pump, passes upward through a vertical tubular heater with condensing steam in the shell, and thence into the body. The heated stream enters through a tangential inlet just below the level of the magma surface. This imparts a swirling motion to the magma, which facilitates flash evaporation and equilibrates the magma with the vapor through the action of an adiabatic flash. The supersaturation thus generated provides the driving potential for nucleation and growth. The volume of the magma divided by the volumetric flow rate of magma through the system gives the time for growth.

Feed solution enters the downpipe before the suction of the circulating pump. Mother liquor and crystals are drawn off through a discharge pipe positioned above the feed inlet in the downpipe. Mother liquor is separated from the crystals in a continuous centrifuge; the crystals are taken off as a product or for further processing, and the mother liquor is recycled to the downpipe. Some of the mother liquor is bled from the system by a pump to prevent accumulation of impurities.

The simple form of vacuum crystallizer also has serious limitations from the standpoint of crystallization. Under the low pressure existing in the unit, the effect of static head on the boiling point is important; e.g., water at 45°F has a vapor pressure of 0.30 in. Hg, which is a pressure easily obtainable by steam-jet boosters. A static head of 1 ft increases the absolute pressure to 1.18 in. Hg., where the boiling point of water is 84°F. Feed at this temperature would not flash if admitted at any level more than 1 ft below the surface of the magma. Admission of the feed at a point where it does not flash is advantageous in controlling nucleation, but the feed has a tendency to short-circuit to the discharge without releasing its supersaturation.

Because of the effect of static head, evaporation and cooling occur only in the liquid layer near the magma surface, and concentration and temperature gradients near the surface are formed. Also crystals tend to settle to the bottom of the crystallizer, where there may be little or no supersaturation. The crystallizer will not operate satisfactorily unless the magma is well agitated, to equalize concentration and temperature gradients and suspend the crystals. The simple vacuum crystallizer provides no good method for nucleation control, for classification, or for removal of excess nuclei and very small crystals.

Draft-tube–baffle crystallizer   A more versatile and effective equipment is the draft-tube–baffle (DTB) crystallizer shown in Fig. 28-9. The crystallizer body is equipped with a draft tube, which also acts as a baffle to control the circulation of the magma, and a downward-directed propeller agitator to provide a controllable circulation within the crystallizer. An additional circulation system, outside the crystallizer body and driven by a circulating pump, contains the heater and feed inlet. Product slurry is removed through an outlet near the bottom of the conical lower section of the crystallizer body. For a given feed rate both the internal and external circulations are independently variable and provide controllable variables for obtaining the required CSD.

Draft-tube–baffle crystallizers can be equipped with an elutriation leg below the body to classify the crystals by size, and may also be equipped with a baffled settling zone for fines removal. An example of such a unit is shown in Fig. 28-10. Part of the circulating liquor is pumped to the bottom of the leg and used as a hydraulic sorting fluid to carry small crystals back into the crystallizing zone for further growth. The action here is that of hindered-settling classification described in Chap. 30. The discharge slurry is withdrawn from the lower part of the elutriation leg and sent to a rotary filter or centrifuge, and the mother liquor is returned to process.

Unwanted nuclei are removed by providing an annular space, or jacket, by enlarging the cone bottom and using the lower wall of the crystallizer body as a baffle. The annular space provides a settling zone, in which hydraulic classification separates fine crystals from larger ones by floating them in an upward-flowing stream of mother liquor, which is withdrawn from the top of the settling zone. The fine crystals so withdrawn are 60-mesh in size or smaller, and although their number is huge, their

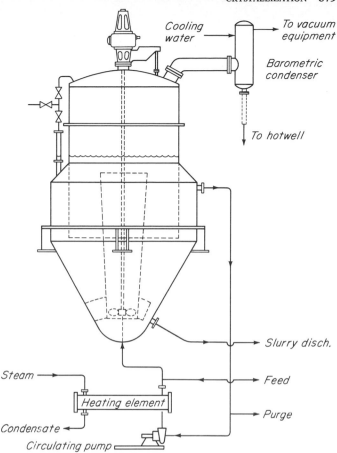

Cooling water

To vacuum equipment

Barometric condenser

To hotwell

Slurry disch.

Steam →

Heating element

Feed

Purge

Condensate ←

Circulating pump

FIGURE 28-9
DTB crystallizer. [*By permission, from R. C. Bennett and M. Van Buren, Chem. Eng. Symp. Ser. 95*, **65**, 45 (1969).]

mass is small, so that the stream from the jacket is nearly solid-free. Larger crystals settle at a larger velocity with respect to the ground than the velocity of the upward flow in the jacket and are retained in the unit. The elutriation leg and the nuclei-removal jacket act analogously: the first allows only large crystals to escape, and the second lets only the smallest particles leave.

By removing a large fraction of the mother liquor from the jacket in this fashion, the magma density is sharply increased. Magma densities of 30 to 50 percent, based on the ratio of the volume of settled crystals to that of the total magma, are achieved.

Additional liquor, not needed for elutriation, is pumped through a steam heater to dissolve entrained small particles, and enters the crystallizer body above the elutriation leg and a little below the bottom of the draft tube. This liquor, now solids-free, is quickly incorporated into the circulating streams within the body of the crystallizer. Reheating the clear-liquor recycle provides yet another way of controlling the CSD of the product.

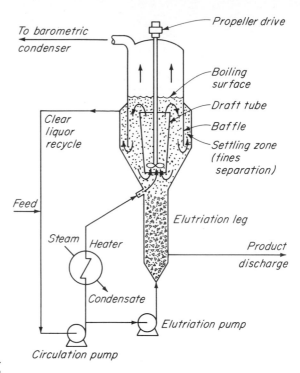

FIGURE 28-10
Draft-Tube–Baffle crystallizer with internal system for fines separation and removal. [*By permission, from A. D. Randolph, Chem. Eng.,* May 1970, p. 86.]

YIELD OF VACUUM CRYSTALLIZER The yield from a vacuum crystallizer can be calculated by enthalpy and material balances. A graphical calculation based on the enthalpy chart of Fig. 28-2 is shown in Fig. 28-11. Since the process is an adiabatic split of the feed into product magma and vapor, points *b* for the feed, *a* for the vapor, and *e* for the magma lie on a straight line. The isotherm connecting point *d* for the crystals with point *f* for the mother liquor also passes through point *e*, and this point is located by the intersection of lines *ab* and *df*. From the line segments and the center-of-gravity principle, the ratios of the various streams are calculated.

EXAMPLE 28-4 A continuous vacuum crystallizer is fed with a 31 percent $MgSO_4$ solution. The equilibrium temperature of the magma in the crystallizer is 86°F (30°C), and the boiling-point elevation of the solution is 2°F (1.11°C). A product magma containing 5 tons (4,536 kg) of $MgSO_4 \cdot 7H_2O$ per hour is obtained. The volume ratio of solid to magma is 0.15. What is the temperature of the feed, the feed rate, and the rate of evaporation?

SOLUTION Figure 28-11 shows the graphical solution of the problem. The vapor leaves the crystallizer at the pressure corresponding to 84°F and carries 2°F superheat, which may be neglected. From steam tables, the enthalpy of the vapor is that of saturated steam at 0.5771 $lb_f/in.^2$, and the coordinates of point *a* are $H = 1098$ Btu/lb and $c = 0$. The enthalpy

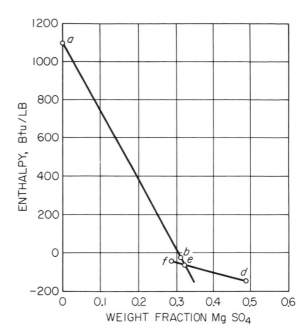

FIGURE 28-11
Solution for Example 28-4.

and average concentration of the product magma are calculated from data given by Fig. 28-2. The straight line *fd* is the 86°F isotherm in the area *bcih* of Fig. 28-2. The coordinates for its terminals are, for point *f*, $H = -43$ Btu/lb and $c = 0.285$, and for point *d*, $H = -149$ Btu/lb, $c - 0.488$. The mass ratio of crystals to mother liquor is

$$\frac{0.15 \times 105}{0.85 \times 82.5} = 0.224$$

where the densities of crystals and mother liquor are taken as 105 and 82.5 lb/ft³, respectively. The rate of production of mother liquor is $10{,}000/0.224 = 44{,}520$ lb/h, and the total magma produced is $10{,}000 + 44{,}520 = 54{,}520$ lb/h. The average concentration of $MgSO_4$ in the magma is

$$\frac{0.224 \times 0.488 + 0.285}{1.224} = 0.342$$

The enthalpy of the magma is

$$\frac{0.224(-149) + (-43)}{1.224} = -68 \text{ Btu/lb}$$

These are the coordinates of point *e*. The point for the feed must lie on the straight line *ae*. Since the feed concentration is 0.31, the enthalpy of the feed is the ordinate of point *b*, or −21 Btu/lb. Point *b* is on the 130°F (94.4°C) isotherm, so this temperature is that of the feed. By the center-of-gravity principle, the evaporation rate is

$$54,520 \, \frac{-21 - (-68)}{1098 - (-21)} = 2290 \text{ lb/h (1040 kg/h)}$$

The total feed rate is $54,520 + 2290 = 56,810$ lb/h (25,770 kg/h).   ////

APPLICATIONS OF PRINCIPLES TO DESIGN

Once the theoretical yield from a crystallizer has been calculated from mass and energy balances, there remains the problem of estimating the CSD of the product from the kinetics of nucleation and growth. An idealized crystallizer model, called the *mixed-suspension–mixed-product-removal model* (MSMPR), has served well as a basis for identifying the kinetic parameters and showing how knowledge of them can be applied to calculate the performance of such a crystallizer.[18-20]

MSMPR Crystallizer

Consider a continuous crystallizer that operates in conformity with the following stringent requirements:[17]

1 The operation is steady state.
2 At all times the crystallizer contains a mixed-suspension magma, with no product classification.
3 At all times uniform supersaturation exists throughout the magma.
4 The ΔL law of crystal growth applies.
5 No size-classified withdrawal system is used.
6 There are no crystals in the feed.
7 The product magma leaves the crystallizer in equilibrium, so the mother liquor in the product magma is saturated.
8 No crystal breakage into finite particle size occurs.

The process is called *mixed-suspension–mixed-product crystallization*. Because of the above restraints, the nucleation rate, in number of nuclei generated in unit time and unit volume of mother liquor, is constant at all points in the magma; the rate of growth, in length per unit time, is constant and independent of crystal size and location; all volume elements of mother liquor contain a mixture of particles ranging in size from nuclei to large crystals; and the particle-size distribution is independent of location in the crystallizer and is identical to the size distribution in the product.

By the use of a generalized population balance the MSMPR model is extended to account for unsteady-state operation, classified product removal, crystals in the feed, crystal fracture, variation in magma volume, and time-dependent growth rate.[19] These variations are not included in the following derivations.

Population-density function   The basic quantity in the theory of the CSD is the population density. To understand the meaning of this variable, assume that a distribution function of the cumulative number of crystals in the magma, in number

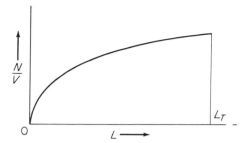

FIGURE 28-12
Cumulative number density vs.
length.

per unit volume of mother liquor, is known as a function of L, the crystal size. Such a function is plotted in Fig. 28-12. The abscissa is L, and the ordinate is N/V, where V is the volume of mother liquor in the magma and N is the number of crystals of size L and smaller in the magma. At $L = 0$, $N = 0$; the total number of crystals is N_t, corresponding to the length L_T of the largest crystal in the magma.

The population density n is defined as the slope of the cumulative distribution curve at size L, or

$$n \equiv \frac{d(N/V)}{dL} = \frac{1}{V}\frac{dN}{dL} \tag{28-39}$$

By assumption 1, V is a constant. Figure 28-13 shows the function $n = f(L)$. It has a maximum value $n°$ where $L = 0$ and is zero where $L = L_T$. In the MSMPR model, the functions of N/V and n with L are invariant in both time and location in the magma. The dimensions of n are number/volume-length.

A growing crystal moves with time along the size axis in the direction of increasing L. A dissolving crystal moves in the other direction. For computations in the MSMPR model a relation between population density n and size L is needed. Such an equation is derived for a crystal of definite size L as shown in Fig. 28-13.

Consider the n crystals of size L per unit volume of magma in the crystallizer. In the MSMPR model, each crystal of length L has the same age, and if t_m is the age of a crystal,

$$L = Gt_m \tag{28-40}$$

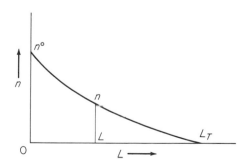

FIGURE 28-13
Population density vs. length.

Assume, now, that of all the n crystals of size L in a unit volume of liquid in the crystallizer Δn are withdrawn as product during time increment Δt. Since the operation is in steady state, withdrawal of product does not affect the size distribution in either magma or product, and since in the MSMPR model the discharge is accurately representative of the magma, it follows that the fraction of particles withdrawn is identical to the ratio of the volume of product liquid taken out in time Δt to the total volume of liquid in the crystallizer. Then if Q is the volumetric flow rate of liquid in the product, V_c is the total volume of liquid in the crystallizer.

$$-\frac{\Delta n}{n} = \frac{Q\,\Delta t}{V_c} \tag{28-41}$$

The time interval Δt also is a period in the life of each crystal, and during this time, by Eq. (28-40), the growth of each crystal is

$$\Delta L = G\,\Delta t \tag{28-42}$$

Eliminating Δt from Eqs. (28-41) and (28-42) gives

$$-\frac{\Delta n}{\Delta L} = \frac{Qn}{GV_c}$$

Letting $\Delta L \to 0$, so that

$$\lim_{\Delta L \to 0} \frac{\Delta n}{\Delta L} = \frac{dn}{dL}$$

leads to

$$-\frac{dn}{dL} = \frac{Qn}{V_c G}$$

The retention time τ of the magma in the crystallizer is defined by

$$\tau \equiv \frac{V_c}{Q}$$

and

$$-\frac{dn}{n} = \frac{1}{G\tau}\,dL \tag{28-43}$$

Integration of Eq. (28-43) gives the function of cumulative population vs. length, where $n°$ is the population density at $L = 0$ and is assumed to represent the nuclei. Thus

$$\int_{n°}^{n} \frac{dn}{n} = -\frac{1}{G\tau} \int_0^L dL$$

$$\ln \frac{n°}{n} = \frac{L}{G\tau} \tag{28-44}$$

The quantity $L/G\tau$ is dimensionless. It may be replaced by x, called the *dimensionless length*, which is defined by

$$x \equiv \frac{L}{G\tau} \tag{28-45}$$

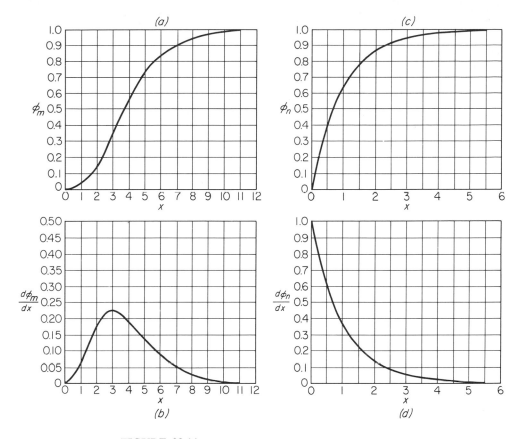

FIGURE 28-14

Size-distribution relations, mixed suspension: (*a*) cumulative mass distribution; (*b*) differential mass distribution; (*c*) cumulative population distribution; (*d*) differential population distribution.

Equation (28-44) may be written

$$n = n°e^{-x} \tag{28-46}$$

When plotted on semilogarithmic coordinates, the linear graph gives $n°$ when $x = 0$ and has a slope of $1/G\tau$. Such a graph is shown in Fig. 28-15 p. 887.

Moment equations   Equation (28-46) is the fundamental relation of the MSMPR crystallizer. From it differential and cumulative equations can be derived for crystal population, crystal length, crystal area, and crystal mass. Also, the kinetic coefficients G and $B°$ are embedded in these equations.

These calculations use moments of the n-vs.-x relation of Eq. (28-46). The normalized jth moment is defined by

$$\mu_j \equiv \frac{\int_0^L nx^j\, dx}{\int_0^\infty nx^j\, dx} \tag{28-47}$$

Using Eq. (28-46) gives

$$\mu_j = \frac{\int_0^x x^j e^{-x}\, dx}{\int_0^\infty x^j e^{-x}\, dx} \tag{28-48}$$

Integrating this equation for values of $j = 0$ through $j = 3$ gives

$$\mu_0 = 1 - e^{-x} \tag{28-49}$$

$$\mu_1 = 1 - (1 - x)e^{-x} \tag{28-50}$$

$$\mu_2 = 1 - (1 + x + \tfrac{1}{2}x^2)e^{-x} \tag{28-51}$$

$$\mu_3 = 1 - (1 + x + \tfrac{1}{2}x^2 + \tfrac{1}{6}x^3)e^{-x} \tag{28-52}$$

The differential distributions are

$$\frac{d\mu_0}{dx} = e^{-x} \tag{28-53}$$

$$\frac{d\mu_1}{dx} = xe^{-x} \tag{28-54}$$

$$\frac{d\mu_2}{dx} = \frac{x^2 e^{-x}}{2} \tag{28-55}$$

$$\frac{d\mu_3}{dx} = \frac{x^3 e^{-x}}{6} \tag{28-56}$$

Equations (28-49) to (28-52) give directly the distributions of the crystals from the idealized MSMPR plant. Thus, μ_0 is the number distribution ϕ_n; μ_1, the size distribution ϕ_L; μ_2, the area distribution ϕ_a; and μ_3 the mass distribution ϕ_m.

Likewise, the differential distributions for number, size, area, and mass are given in Eqs. (28-53) to (28-56). Plots of ϕ_n, ϕ_m, $d\phi_n/dx$, and $d\phi_m/dx$ are given in Fig. 28-14.

The ϕ_m-vs.-x relation is the plot of the cumulative screen analysis. Ordinates on the curve at sizes corresponding to the mesh sizes of the screens are the values of the cumulative analysis.

Since both x and μ are dimensionless, Eqs. (28-49) to (28-56) are universally applicable to MSMPR crystallizers.

Predominant crystal size   Figure 28-14b, the differential mass distribution, shows a node where the value of $d\phi_m/dx$ is a maximum. It is shown by setting the differential coefficient of Eq. (28-56) equal to zero that the node is at the abscissa $x_{pr} = 3$. Thus the most populous size in the product occurs where $L/G\tau = 3$. Nodes also appear in the differential curves for the number and area distributions at $x = 1$ and $x = 2$, respectively.

Kinetic coefficients   Clearly, to use Eq. (28-44) and its consequences in calculations three parameters must be evaluated: the growth rate G, the nucleus density $n°$, and the retention time τ. The last parameter is under the control of the designer, but the other two are dependent variables generated by the crystallizer and its contents and are functions of the engineering details of the plant, the action of the impeller or

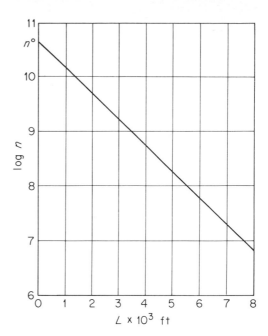

FIGURE 28-15
Population density vs. length
(Example 28-5).

circulating pump, and the nucleating and growth characteristics of the crystallizing system. There is an important equation, readily derivable from the MSMPR model, which connects the nucleation rate $B°$ with the zero-size particle density $n°$. Thus, by formal calculus,

$$\lim_{L \to 0} \frac{dN}{dt} = \lim_{L \to 0} \left(\frac{dL}{dt} \frac{dN}{dL} \right)$$

By definition of the terms in this equation, the limits at $L = 0$ are $dN/dT \to B°$, $dL/dt \to G$, and $dN/dL \to n°$. Therefore,

$$B° = Gn° \tag{28-57}$$

NUMBER OF CRYSTALS PER UNIT MASS   The number of crystals n_c in a unit volume of liquid in either magma or product is

$$n_c = \int_0^\infty n \, dL = n°\tau G \int_0^\infty e^{-x} \, dx = n°\tau G \tag{28-58}$$

The total mass of product crystals in unit volume of liquid is

$$m_c = \int_0^\infty mn \, dL = a\rho_c(G\tau)^3 n° \int_0^\infty x^3 e^{-x} \, dx$$

$$= 6a\rho_c n°(G\tau)^4$$

The number of crystals per unit mass is

$$\frac{n_c}{m_c} - \frac{n \, G\tau}{6a\rho_c n°(G\tau)^4} = \frac{1}{6a\rho_c(G\tau)^3} \tag{28-59}$$

If the predominant size is used as a design parameter, then from the definition $L_{pr} = 3G$ and

$$\frac{n_c}{m_c} = \frac{9}{2a\rho_c L_{pr}^3} \tag{28-60}$$

The nucleation rate must be just sufficient to generate one nucleus for each crystal in the product. If C is the mass production rate of crystals, the required nucleation, in number per unit time and volume of mother liquor, is, from Eqs. (28-59) and (28-60),

$$B° = \frac{Cn_c}{m_c} = \frac{C}{6a\rho_c(G\tau)^3 V_c} = \frac{9C}{2a\rho_c V_c L_{pr}^3} \tag{28-61}$$

EXAMPLE 28-5   In the crystallizer of Example 28-4, a growth rate G of 0.0018 ft/h (0.00055 m/h) is anticipated, and a predominant crystal size of 20-mesh is desired. How large must the magma volume in the crystallizer be; what nucleation rate $B°$ is necessary; and what is the screen analysis of the product, assuming that the operation conforms to all the requirements of the mixed-suspension–mixed-product crystallization?

SOLUTION   The screen opening of a 20-mesh standard screen is, from Appendix 20, 0.0328 in., or 0.00273 ft. This dimension can be used for L, and the shape factor a is assumed to be unity. From Example 28-4, the volume flow rate of mother liquor in the product magma is

$$Q = \frac{44,520}{82.5} = 540 \text{ ft}^3/\text{h}$$

Since, when $x = 3$, $L_{pr} = 0.00273$, Eq. (28-45) gives for the drawdown time and the volume of liquid in the crystallizer

$$\tau = \frac{L_{pr}}{3G} = \frac{0.00273}{3 \times 0.0018} = 0.506 \text{ h} \qquad V_c = 0.506 \times 540 = 273 \text{ ft}^3$$

The total magma volume is $273/0.85 = 321$ ft^3, or 2,400 gal.
   The nucleation rate is, from Eq. (28-61),

$$B° = \frac{9 \times 10,000}{2 \times 105 \times 273 \times 0.00273^3}$$

$$= 7.72 \times 10^7 \text{ nuclei/ft}^3\text{-h} \ (2.74 \times 10^9 \text{ nuclei/m}^3\text{-h})$$

By Eq. (28-57), the zero-size particle density is $n° = 7.72 \times 10^7/0.0018 = 4.289 \times 10^{10}$ (nuclei)/ft^4. The value of $L/G\tau$ is $L/(0.0018)(0.506) = 1.1 \times 10^3 L$. The equation for the number-density distribution is, from Eq. (28-44),

$$\log n = \log n° - \frac{1.1 \times 10^3 L}{2.3026} = 10.6351 - 4.777 \times 10^2 L$$

This equation is plotted in Fig. 28-15.
   The screen analysis is found by reading ordinates from Fig. 24-14c for values of x corresponding to mesh openings. For example, for the 20-mesh point, where $x = 3$, ϕ_m is 0.35. In general,

$$x = \frac{L}{\tau G} = \frac{L}{0.506 \times 0.0018} = 1,098L$$

Details are given in the Table 28-2.

Contact nucleation in crystallizers The contact nucleation discussed on previous pages is based on the single-particle experiments of Clontz and McCabe,[5] which determined the number of crystals generated by a single contact at known supersaturation, energy, and area of contact. The results have been applied in the construction of nucleation models for the design of magma crystallizers. The following treatment follows that proposed by Bennett, Fiedelman, and Randolph.[1] A more elaborate model has been developed by Ottens.[14]

It is apparent that contact nucleation may result from crystal-crystal contacts, from crystal-wall contacts, and from crystal-impeller contacts. Ottens has shown that only the third type of collision is significant, and the following discussion is limited to contacts between the existing crystals in the circulating magma with the blades of the rotating impeller.

The equation of Clontz and McCabe used by Bennett et al. is

$$\frac{N}{A} = \frac{K_4 s E}{\sqrt{A}} \tag{28-62}$$

where N = number of crystals per contact
A = area of contact
s = supersaturation
E = energy of contact
K_4 = dimensional proportionality constant

The correlation is based on the following ideas:

1 The total generation of nuclei is proportional to the sum of the nucleation from all crystals of all sizes each time they pass through the impeller.
2 The driving potential is the supersaturation, which is proportional to growth rate G.

Table 28-2

Mesh	Size			Screen analysis, %	
	ft	mm	x	Cumulative	Differential
8	0.0078	2.37	8.5	97	3
9	0.0065	1.98	7.1	93	4
10	0.0054	1.65	5.9	84	9
12	0.0046	1.40	5.0	74	10
14	0.0038	1.16	4.2	61	13
16	0.0033	1.01	3.6	48	13
20	0.0027	0.82	3.0	35	13
24	0.0023	0.70	2.5	25	10
28	0.0019	0.58	2.1	17	8
32	0.0016	0.49	1.8	11	6
35	0.0014	0.43	1.5	6	5
42	0.0011	0.34	1.2	4	2

3 The energy imparted to a crystal of size L and mass $\rho_c L^3$ is that necessary to accelerate the particle from the speed of the flowing magma to the speed of the tip of the impeller.

4 The area of contact is proportional to L^2.

These assumptions lead directly to the equation for the nucleation rate from contact with the n crystals of length L in a unit volume of liquid in the flowing magma,

$$dB^\circ \sim G\,\frac{1}{t_{To}}\,\frac{m}{2}\,\overline{\Delta u^2}\,nL\;dL$$

where t_{To} = turnover time

$\overline{\Delta u^2}$ = mean square of crystal-impeller velocity

$L = A/\sqrt{A}$

The turnover time is defined as the time between passages of the same crystal between contacts; it is inversely proportional to the rotational speed of the impeller. Term $\overline{\Delta u^2}$ may be replaced by U_T^2, where U_T is the tip speed of the impeller, and m by $\rho_c L^3$. The total nucleation per unit volume of liquid, found by integrating over the range $L = 0$ to $L = \infty$, is

$$B^\circ = \int_0^\infty n\;dB^\circ \sim G\,\frac{U_T^2}{t_{To}}\,\rho_c \int_0^\infty nL^4\;dL$$

The integral calls for the fourth moment. Using Eq. (28-47), integration gives the final equation, after collecting all proportionality factors into the dimensional constant K_N,

$$B^\circ = K_N n^\circ G\,\frac{U_T^2}{t_{To}}\,\rho_c(G\tau)^5 \tag{28-63}$$

Use of this equation in practice requires empirical data from a pilot plant or actual operation of crystallizers of the same design.

SYMBOLS

A	Area, ft² or m²; area of contact between crystals and impeller; A_c, total area of crystal, A_i, area of face i; A_k, of face k; A_p, of particle or crystal
a	Shape factor, defined by Eq. (28-1), dimensionless; a^2, area covered by particle arriving at surface, cm²
B^0	Nucleation rate, number/cm³-s a number/foot³-h
B_s	Nucleation rate for two-dimensional nucleation
b	Shape factor, defined by Eq. (28-2), dimensionless
C	Frequency factor in nucleation, number/cm³-s; constant in Eq. (28-36); also, mass production rate of crystals, lb/h or kg/h
c	Concentration of solution, mole per/unit volume or g mol/m³; c_s, in saturated solution
D	Diameter of particle, ft or m
E	Energy of contact in contact nucleation, ergs
f	Number of crystal faces

G	Growth rate of crystal, ft/h or m/h
K	Overall mass-transfer coefficient, lb mol/ft²-h-unit mole fraction or g mol/m²-s-unit mole fraction; K_i, of face i; $\bar{K}$, average value for entire crystal
K_N	Dimensional constant in Eq. (28-63)
K_n	Constant in Arrhenius equation (28-27)
K_1, K_2, K_3, K_4	Constants in Eqs. (28-35), (28-36), (28-37), and (28-62), respectively
k_s	Coefficient of interfacial reaction, lb mol/ft²-h-unit mole fraction or g mol/m²-s-unit mole fraction; k_{si}, at face i; $\bar{k}_s$, average value for entire crystal
k_y	Mass-transfer coefficient from solution to crystal face, lb mol/ft²-h-unit mole fraction or g mol/m²-s-unit mole fraction
k	Boltzmann constant, 1.3806×10^{-16} erg/g mol-K
L	Linear dimension or size of crystal, ft or m; L_T, maximum size; L_{pr}, predominant size; L', distance between two specific corners
L_n	Critical size; size of nucleus, ft or m
m	Mass, lb mol or g mol; m_c, mass of crystal; also, total mass of product crystals per unit volume of liquid
$\dot{m}$	Molal growth rate, lb mol/h or g mol/s; $\dot{m}_c$, of entire crystal; $\dot{m}_i$, rate of transfer to and growth rate on face i
N	Number of crystals of size L and smaller in crystallizer; N_t, total number of crystals in crystallizer
N_c	Number of nuclei generated in one contact
N_{Re}	Reynolds number, dimensionless
N_a	Avogadro constant, 6.0222×10^{23} molecules/g mol
n	Population density defined by Eq. (28-29), number/ft⁴ or number/m⁴; n^0, maximum value, for nuclei
n_c	Number of crystals per unit mass of crystals
Q	Volumetric flow of liquid in product, ft³/h or m³/h
R	Gas constant, 8.3143×10^7 ergs/g mol-K
s	Fractional supersaturation, defined by Eq. (28-9); $100s$, percent supersaturation
T	Temperature, T_s, of saturated solution and growing crystal
t	Time, t_{To}, turnover time between passages of a given crystal between contacts with rotating impeller
t_m	Age of crystal
U	Surface energy of crystal face, ergs; U_c, total surface energy of crystal; U_k, of face k
U_T	Tip speed of impeller, ft/s or m/s
V	Volume, ft³ or m³; volume of liquid in magma; V_c, total volume of liquid in crystallizer
V_M	Molar volume, $1/\rho$, ft³/lb mol or cm³/g mol
v_c	Volume of crystal, ft³ or m³
x	Dimensionless length, $L/G\tau$; x_{pr}, predominant value
y	Mole fraction of solute in solution, at a distance from crystal face; y_s, in saturated solution; y', at interface between crystal and liquid
Z	Rate of arrival of molecules at crystal surface, number/cm²-s
Abbreviations	
CSD	Crystal size distribution
MSMPR	Mixed-suspension–mixed-product-removal model
Greek letters	
α	Concentration ratio, defined by Eq. (28-9)
β	Ratio of area of face to total surface area of crystal; β_i, of face i; β_k, of face k
Δ_c	Supersaturation, lb mol/ft³ or g mol/m³
ΔE	Activation energy, ergs; ΔE_n, for nucleation; ΔE_s, for crystal growth
ΔL	Increase in crystal size in time increment Δt, ft or m
Δn	Number of crystals of size L withdrawn in time increment Δt
ΔT	Temperature driving potential equivalent to supersaturation potential

Δt	Time increment, h or s
Δu	Crystal-impeller velocity, ft/s or m/s; $\overline{\Delta u^2}$, mean square value
Δy	Supersaturation, mole fraction of solute
$\Delta \Omega$	Increase in work function, ergs; $\Delta \Omega_n$, energy of nucleation
κ	Slope of temperature-concentration line
μ	Distribution of crystals from idealized MSMPR plant; μ_0, number distribution; μ_1, size distribution; μ_2, area distribution; μ_3, mass distribution; also, chemical potential of solute in supersaturated solution; μ_∞, chemical potential of crystal large enough to be in equilibrium with saturated solution
μ_j	Normalized jth moment of crystal distribution, defined by Eq. (28-47)
ν	Number of ions per molecule of solute
ρ	Molar density of solution, lb mol/ft^3 or g mol/cm^3; ρ_c, of crystal; ρ_s, of saturated solution
σ	Interfacial energy, ergs/cm^2; σ_k, of face k; $\bar{\sigma}$, weighted average value for entire crystal
σ_a	Apparent interfacial tension between nucleus and catalyst, ergs/cm^2
τ	Retention time of magma in crystallizer,
ϕ	Distribution relation; ϕ_L, size distribution; ϕ_a, area distribution; ϕ_m, mass distribution; ϕ_n, number distribution
Ω	Work function, ergs

PROBLEMS

28-1 $CuSO_4 \cdot H_2O$ containing 3.5 percent of a soluble impurity is dissolved continuously in sufficient water and recycled mother liquor to make a saturated solution at 80°C. The solution is then cooled to 25°C and crystals of $CuSO_4 \cdot 5H_2O$ thereby obtained. These crystals carry 10 percent of their dry weight as adhering mother liquor. The crystals are then dried to zero free water ($CuSO_4 \cdot 5H_2O$). The allowable impurity in the product is 0.6 percent. Calculate (*a*) the weight of water and of recycled mother liquor required per 100 lb of impure copper sulfate; (*b*) the percentage recovery of copper sulfate, assuming that the mother liquor not recycled is discarded. The solubility of $CuSO_4 \cdot 5H_2O$ at 80°C is 120 g per 100 g of free H_2O and at 25°C is 40 g per 100 g of free H_2O.

28-2 A solution of $MgSO_4$ containing 43 g of solid per 100 g of water is fed to a vacuum crystallizer at 220°F. The vacuum in the crystallizer corresponds to an H_2O boiling temperature of 43°F, and a saturated solution of $MgSO_4$ has a boiling-point elevation of 2°F. How much solution must be fed to the crystallizer to produce 1 ton of epsom salt ($MgSO_4 \cdot 7H_2O$) per hour?

28-3 An ideal product classification in a continuous vacuum crystallizer would achieve the retention of all crystals within the crystallizer until they attained a desired size and then discharge them from the crystallizer.[20] The size distribution of the product would be uniform, and all crystals would have the same value of D_p. Such a process conforms to the other constraints for the mixed-suspension–mixed-product crystallizer except that the magma in the unit is classified by size and each crystal has the same retention or growth time. For such a process show that

$$m_c = \frac{a\rho_c B^\circ V_c L^4}{4G} \quad \text{and} \quad \tau = \frac{L_{pr}}{4G}$$

28-4 Assume that $CuSO_4 \cdot 5H_2O$ is to be crystallized in an ideal product-classifying crystallizer. A 12-mesh product is desired. The growth rate is estimated to be 0.2 $\mu m/s$.

The geometric constant a is 0.20, and the density of the crystal is 143 lb/ft^3. A magma consistency of 0.35 ft^3 of crystals per cubic foot of mother liquor is to be used. What is the production rate, in pounds of crystals per hour per cubic foot of mother liquor? What rate of nucleation, in number per hour per cubic foot of mother liquor, is needed?

28-5 An MSMPR crystallizer produces 1 ton of product per hour having a predominant size of 35-mesh. The volume of crystals per unit volume of magma is 0.15. The temperature in the crystallizer is 120°F, and the retention time is 2.0 h. The densities of crystals and mother liquor are 105 and 82.5 lb/ft^3, respectively. (*a*) Plot the cumulative screen analysis of the theoretical product. (*b*) Determine the required growth rate G and the necessary nucleation rate $B°$.

28-6 Results have been reported on the performance of a crystallizer operating on sodium chloride.[1] Results from one experiment are

$$\text{Tip speed } U_T = 1{,}350 \text{ ft/min} \qquad \text{Retention time} = 1.80 \text{ h}$$

$$\text{Time between passages, or turnover time} = 35 \text{ s}$$

Size-distribution parameters:

$$n° = 1.46 \times 10^6 \text{ number/cm}^4 \qquad B° = 1.84 \text{ number/cm}^3\text{-s}$$

$$\text{Crystal density} = 2.163 \text{ g/cm}^3$$

Calculate the dimensional constant K_N, using the units specified above, and the growth rate in millimeters per hour.

REFERENCES

1 Bennett, R. C., H. Fiedelman, and A. D. Randolph: *Chem. Eng. Prog., 69*(7):86 (1972).
2 Bichowsky, F. R., and F. D. Rossini: "Thermochemistry of Chemical Substances," Reinhold, New York, 1936.
3 Botsaris, G. D., and E. G. Denk: "Annual Reviews of Industrial and Engineering Chemistry," American Chemical Society, Washington, D.C., 1972.
4 Buckley, H. E.: "Crystal Growth," Wiley, New York, 1951.
5 Clontz, N. A., and W. L. McCabe: *AIChE Symp. Ser., No.* 110, **67**:6 (1971).
6 Clontz, N. A. et al.; *Ind. Eng. Chem. Fundam.*, **11**:368 (1972).
7 Frank, F. C.: *Discuss. Faraday Soc.*, **5**:48 (1949).
8 Johnson, R. T., R. W. Rousseau, and W. L. McCabe: *AIChE Symp. Ser., No.* 121, **68**:31 (1972).
9 La Mer, V. K.: *Ind. Eng. Chem.*, **44**:1270 (1952).
10 Margolis, G., and E. B. Gutoff: "Annual Reviews of Industrial and Engineering Chemistry," vol. 2, American Chemical Society, Washington, D.C., 1974.
11 Mason, R. E. A., and R. F. Strickland-Constable: *Trans. Faraday Soc.*, **62**, pt. 2 (1966).
12 McCabe, W. L.: *Ind. Eng. Chem.*, **21**:30, 121 (1929).
13 McCabe, W. L., in J. C. Perry (ed.), "Chemical Engineers' Handbook," 3d ed., p. 1056, McGraw-Hill, New York, 1950.
14 Ottens, E. P. K.: "Nucleation in Continuous Agitated Crystallizers," Technological University, Delft, 1972.
15 Powers, H. E. C.: *Ind. Chem.*, **39**:351 (1963).
16 Preckshot, G. W., and G. G. Brown: *Ind. Eng. Chem.*, **44**:1315 (1950).
17 Randolph, A. D.: *AIChE J.*, **11**:424 (1965).
18 Randolph, A. D., and M. A. Larson: *AIChE J.*, **8**:639 (1962).
19 Randolph, A. D., and M. A. Larson: "Theory of Particulate Processes," Academic, New York, 1971.

20 Saeman, W. C.: *AIChE J.,* **2:**107 (1956).
21 Strickland-Constable, R. F.: "Kinetics and Mechanism of Crystallization," Academic, New York, 1968.
22 Strickland-Constable, R. F., and R. E. A. Mason: *Nature,* **197,** 4870 (1963).
23 Strickland-Constable, R. F.: *AIChE Symp. Ser., No.* 121, **68:**1 (1972).
24 Sung, C. Y., J. Estrin, and G. R. Youngquist: *AIChE J.,* **19:**957 (1973).
25 Tai, C. Y., W. L. McCabe, and R. W. Rousseau, *AIChE J.* **21:** 351 (1975).
26 Ting, H. H., and W. L. McCabe: *Ind. Eng. Chem.,* **26:**1201 (1934).
27 Uhlmann, D. R., and B. Chalmers: *Ind. Eng. Chem.,* **57:**19 (1965).
28 VanHook, A.: "Crystallization, Theory and Practice," Wiley, New York, 1951.
29 Walton, A. G.: *Science,* **148:**601 (1965).

MIXING OF SOLIDS AND PASTES

Mixing is an important, even fundamental, operation in nearly all chemical processes. The blending of liquids is discussed in Chap. 9. Mixing dry solids and heavy, viscous pastes resembles, to some extent, the mixing of liquids of low viscosity. Both processes involve intermingling two or more separate components to form a more or less uniform product. Some of the equipment normally used for blending liquids may, on occasion, mix solids or pastes, and vice versa.

Yet there are significant differences between the two processes. Liquid blending depends on the creation of flow currents, which transport unmixed material to the mixing zone adjacent to the impeller. In heavy pastes or masses of particulate solids no such currents are possible, and mixing is accomplished by other means. In consequence, much more power is normally required in mixing pastes and dry solids than in blending liquids.

Another difference is that in blending liquids a "well-mixed" product usually means a truly homogeneous liquid phase, from which random samples, even of very small size, all have the same composition. In mixing pastes and powders the product often consists of two or more easily identifiable phases, each of which may contain individual particles of considerable size. From a "well-mixed" product of this kind small random samples will differ markedly in composition; in fact, samples from any given such mixture must be larger than a certain critical size (several times the size of the largest individual particle in the mix) if the results are to be significant.

Mixing is harder to define and evaluate with solids and pastes than it is with liquids. Quantitative measures of mixing are discussed later in this chapter, measures which aid in evaluating mixer performance, but in actual practice the proof of a mixer is in the properties of the mixed material it produces. A "well-mixed" paste, for example, is one which does what is required and which has the necessary property— visual uniformity, high strength, uniform burning rate, or other desired characteristic. A good mixer is one which produces this well-mixed product at the lowest overall cost.

Mixing heavy pastes, plastic solids, and rubber is more of an art than a science. The properties of the materials to be mixed vary enormously from one problem to another. Even in a single material they may be widely different at various times during the mixing operation. A batch may start as a dry, free-flowing powder, become pasty on the addition of liquid, stiff and gummy as a reaction proceeds, and then perhaps dry, granular, and free-flowing once more. Indeterminate properties of the material such as "stiffness," "tackiness," and wettability are as significant in these mixing problems as viscosity and density. Mixers for pastes and plastic masses must, above all, be versatile. In a given problem the mixer chosen must handle the material when in its worst condition, and may not be so effective as other designs during other parts of the mixing cycle. As with other equipment, the choice of a mixer for heavy materials is often a compromise.

Types of mixers   Mixing equipment for pastes, rubber, and heavy plastic masses is used when the material is too viscous or plastic to flow readily to the suction side of an impeller and flow currents cannot be created. The material must all be brought to the agitator, or the agitator must visit all parts of the mix. The action in this machinery is well described as a "combination of low-speed shear, smearing, wiping, folding, stretching, and compressing."[2] The mechanical energy is applied by moving parts directly to the mass of material. In the closed types, such as Banbury mixers, the inner wall of the casing acts as part of the mixing means, and all mixing action occurs close to the moving parts. Clearances between mixing arms, rotors, and wall of casing are small. The forces generated in these mixers are large, the machinery must be ruggedly built, and the power consumption is high. The heat evolved per unit mass of material is sufficient to require cooling to prevent the temperature from reaching a level dangerous to the equipment or the material.

Mixers for dry powders include some machines which are also used for heavy pastes and some machines which are restricted to free-flowing powders. Mixing is by slow-speed agitation of the mass with an impeller, by tumbling, or by centrifugal smearing and impact. These mixers are of fairly light construction, at least in comparison with mixers for tough plastic masses, and their power consumption per unit mass of material mixed is moderate.

The boundaries between the fields of application of the various types of mixers are not sharp. Impeller mixers of the kind described in Chap. 9 are not often used when the viscosity is more than about 2 kP, especially if the liquid is not Newtonian. Kneaders and mixer-extruders work on thick pastes and plastic masses; impact wheels

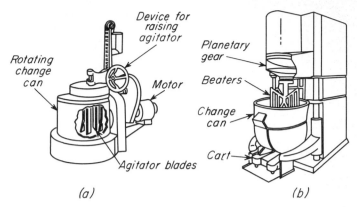

FIGURE 29-1
Double-motion paste mixers: (*a*) pony mixer; (*b*) beater mixer.

are restricted to dry powders. Other mixers, however, can blend liquids, pastes, plastic solids, and powders.

MIXERS FOR PASTES AND PLASTIC MASSES

Mixers described in this section are change-can mixers; kneaders, dispersers, and masticators; continuous kneaders; mixer-extruders; mixing rolls; mullers and pan mixers; and pugmills.

Change-can mixers These devices blend viscous liquids or light pastes, as in food processing or paint manufacture. A small removable can, 5 to 100 gal in size, holds the material to be mixed. In the *pony mixer* shown in Fig. 29-1*a* the agitator consists of several vertical blades or fingers held on a rotating head and positioned near the wall of the can. The blades are slightly twisted. The agitator is mounted eccentrically with respect to the axis of the can. The can rests on a turntable driven in a direction opposite to that of the agitator, so that during operation all the liquid or paste in the can is brought to the blades to be mixed. When the mixing is complete, the agitator head is raised, lifting the blades out of the can; the blades are wiped clean; and the can is replaced with another containing a new batch.

In the beater mixer in Fig. 29-1*b* the can or vessel is stationary. The agitator has a planetary motion; as it rotates, it precesses, so that it repeatedly visits all parts of the vessel. Beaters are shaped to pass with close clearance over the side and bottom of the mixing vessel.

Kneaders, dispersers, and masticators Kneading is a method of mixing used with deformable or plastic solids. It involves squashing the mass flat, folding it over on itself, and squashing it once more. Most kneading machines also tear the mass

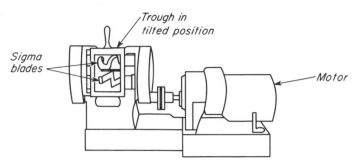

FIGURE 29-2
Two-arm kneader.

apart and shear it between a moving blade and a stationary surface. Considerable energy is required even with fairly thin materials, and as the mass becomes stiff and rubbery, the power requirements become very large.

A *two-arm kneader* handles suspensions, pastes, and light plastic masses. Typical applications are in the compounding of lacquer bases from pigments and carriers and in shredding cotton linters into acetic acid and acetic anhydride to form cellulose acetate. A *disperser* is heavier in construction and draws more power than a kneader; it works additives and coloring agents into stiff materials. A *masticator* is still heavier and draws even more power. It can disintegrate scrap rubber and compound the toughest plastic masses that can be worked at all. Masticators are often called *intensive mixers*.

In all these machines the mixing is done by two heavy blades on parallel horizontal shafts turning in a short trough with a saddle-shaped bottom. The blades turn toward each other at the top, drawing the mass downward over the point of the saddle, then shearing it between the blades and the wall of the trough. The circles of rotation of the blades are usually tangential, so that the blades may turn at different speeds in any desired ratio. The optimum ratio is about $1\frac{1}{2}$:1. In some machines the blades overlap and turn at the same speed or with a speed ratio of 2:1. A small two-arm kneader with tangential blades is sketched in Fig. 29-2.

Designs of mixing blades for various purposes are shown in Fig. 29-3. The common sigma blade shown at the left is used for general-purpose kneading. Its edges may be serrated to give a shredding action. The double-naben, or fishtail, blade in the center is particularly effective with heavy plastic materials. The dispersion blade at the right develops the high shear forces needed to disperse powders or liquids into plastic or rubbery masses. Masticator blades are even heavier than those shown, sometimes being little larger in diameter than the shafts which drive them. Spiral, flattened, and elliptical designs of masticator blades are used.

Material to be kneaded or worked is dropped into the trough and mixed for 5 to 20 min or longer. Sometimes the mass is heated while in the machine, but more commonly it must be cooled to remove the heat generated by the mixing action. The trough is often unloaded by tilting it so that its contents spill out. In kneaders and

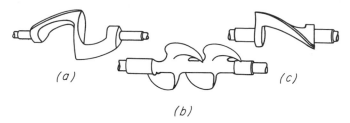

FIGURE 29-3
Kneader and disperser blades: (a) sigma blade; (b) double-naben blade; (c) disperser blade.

some dispersers only one agitator blade is directly driven; the other is turned by timing gears. In masticators both shafts are directly driven, sometimes from both ends, so that the trough cannot be tilted and must be unloaded through an opening in the floor.

In many kneading machines the trough is open, but in some designs, known as *internal mixers*, the mixing chamber is closed during the operating cycle. Thus, a cover, the underside of which conforms to the volume swept out by the blades, can be used on the kneader shown in Fig. 29-2. Such mixers do not tilt. This type is used for dissolving rubber and for making dispersions of rubber in liquids. The most common internal mixer is the *Banbury mixer*, shown in Fig. 29-4. This is a heavy-duty two-arm mixer in which the agitators are in the form of interrupted spirals. The shafts turn at 30 to 40 r/min. Solids are charged in from above and held in the trough during mixing by an air-operated piston under a pressure of 1 to 10 atm. Mixed material is discharged through a sliding door in the bottom of the trough. Banbury mixers compound rubber and plastic solids, masticate crude rubber, devulcanize rubber scrap, and make water dispersions and rubber solutions. They also accomplish the same tasks as kneaders but in a shorter time and with smaller batches. The heat generated in the material is removed by cooling water sprayed on the walls of the mixing chamber and circulated through the hollow agitator shafts.

CONTINUOUS KNEADERS   The machines just described operate batchwise on relatively small amounts of material. The more difficult the material is to mix, the smaller the batch size must be. Many industrial processes are continuous, with steady uniform flow into and out of units of equipment; into such processes batch equipment is not readily incorporated. Continuous kneading machines have been developed which can handle light to fairly heavy materials. In a typical design, the Ko-Kneader, a single horizontal shaft, slowly turning in a mixing chamber, carries rows of teeth arranged in a spiral pattern to move the material through the chamber. The teeth on the rotor pass with close clearance between stationary teeth set in the wall of the casing. The shaft turns and also reciprocates in the axial direction. Material between the meshing teeth is therefore smeared in an axial or longitudinal direction as well as being subjected to radial shear. Solids enter the machine near the driven end of the

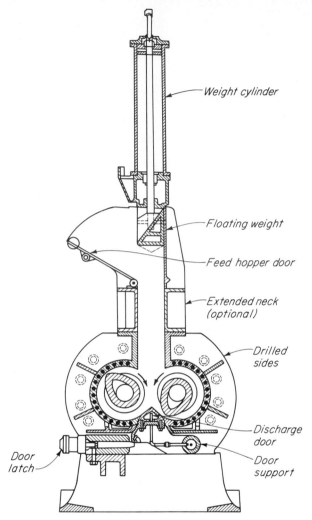

FIGURE 29-4
Banbury internal mixer. (*Farrel Co.*)

rotor, as shown, and discharge through an opening surrounding the shaft bearing in the opposite end of the mixing chamber. The chamber is an open trough with light solids, a closed cylinder with plastic masses. These machines can mix several tons per hour of heavy, stiff, or gummy materials.

Mixer-extruders If the discharge opening of the Ko-Kneader is restricted by covering it with an extrusion die, the pitched blades of the rotor build up considerable pressure in the material. The mix is cut and folded while in the mixing chamber and subjected to additional shear as it flows through the die. Other mixer-extruders

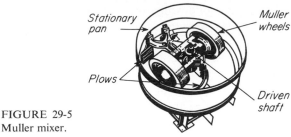

FIGURE 29-5
Muller mixer.

function in the same way. They contain one or two horizontal shafts, rotating but not reciprocating, carrying a helix or blades set in a helical pattern. Pressure is built up by reducing the pitch of the helix near the discharge, reducing the diameter of the mixing chamber, or both. Mixer-extruders continuously mix, compound, and work thermoplastics, doughs, clays, and other hard-to-mix materials. Some also carry a heating jacket and vapor-discharge connections to permit removal of water or solvent from the material as it is being processed.

Mixing rolls Another way of subjecting pastes and deformable solids to intense shear is to pass them between smooth metal rolls turning at different speeds. By repeated passes between such mixing rolls solid additives can be thoroughly dispersed into pasty or stiff plastic materials. Thin fluid pastes in the paint industry are passed through continuous mills containing three to five horizontal rolls, set one above the other in a vertical stack. The paste passes from the slower rolls to successively faster rolls. Rubber products and some plastics are compounded on batch roll mills with two rolls set in the same horizontal plane. Solids are picked up on the faster roll, cut at an angle by the operator, and folded back into the "bite" between the rolls. Additives are sprinkled on the material as it is being worked. Batch roll mills require long mixing times and careful attention by the operator, and in general are less favored for most services than internal mixers.

Muller mixers A muller gives a distinctly different mixing action from that of other machines. Mulling is a smearing or rubbing action similar to that in a mortar and pestle. In large-scale processing this action is given by the wide, heavy wheels of the mixer shown in Fig. 29-5. In this particular design of muller the pan is stationary and the central vertical shaft is driven, causing the muller wheels to roll in a circular path over a layer of solids on the pan floor. The rubbing action results from the slip of the wheels on the solids. Plows guide the solids under the muller wheels, or to an opening in the pan floor at the end of the cycle when the mixer is being discharged. In another design the axis of the wheels is held stationary and the pan is rotated; in still another the wheels are not centered in the pan but are offset, and both the pan and the wheels are driven. Mixing plows may be substituted for the muller wheels to give what is called a *pan mixer*. Mullers are good mixers for batches of heavy solids and pastes; they are especially effective in uniformly coating the particles of a granular solid with a small amount of liquid. Continuous muller mixers with two mixing pans connected in series are also available.

Pugmills In a pugmill the mixing is done by blades or knives set in a helical pattern on a horizontal shaft turning in an open trough or closed cylinder. Solids continuously enter one end of the mixing chamber and discharge from the other. While in the chamber, they are cut, mixed, and moved forward to be acted upon by each succeeding blade. Single-shaft mills utilize an enclosed mixing chamber; open-trough double-shaft mills are used where more rapid or more thorough mixing is required. The chamber of most enclosed mills is cylindrical, but in some it is polygonal in cross section to prevent sticky solids from being carried around with the shaft. Pugmills blend and homogenize clays, break up agglomerates in plastic solids, and mix liquids with solids to form thick, heavy slurries. Sometimes they operate under vacuum to deair clay or other materials. They are built with jackets for heating or cooling.

Power requirements Large amounts of mechanical energy are needed to mix heavy plastic masses. The material must be sheared into elements which are moved relative to one another, folded over, recombined, and redivided. In continuous mixers the material must also be moved through the machine. Only part of the energy supplied to the mixer is directly useful for mixing, and in many machines the useful part is small. Probably mixers which work intensively on small quantities of material, dividing it into very small elements, make more effective use of energy than those which work more slowly on large quantities. Machines which weigh little per pound of material processed waste less energy than heavier machines. Other things being equal, the shorter the mixing time required to bring the material to the desired degree of uniformity, the larger the useful fraction of the energy supplied will be. Regardless of the design of the machine, however, the power needed to drive a mixer for pastes and deformable solids is many times greater than that needed by a mixer for liquids. The energy supplied appears as heat, which must ordinarily be removed to avoid damaging the machine or the material.

Criteria of mixer effectiveness: mixing index The performance of an industrial mixer is judged by the time required, the power load, and the properties of the product. Both the requirements of the mixing device and the properties desired in the mixed material vary widely from one problem to another. Sometimes a very high degree of uniformity is required; sometimes a rapid mixing action; sometimes a minimum amount of power.

The degree of uniformity of a mixed product, as measured by analysis of a number of spot samples, is a valid quantitative measure of mixing effectiveness. Mixers act on two or more separate materials to intermingle them, nearly always in a random fashion. Once a material is randomly distributed through another, mixing may be considered to be complete. Based on these concepts, a statistical procedure for measuring mixing of pastes is as follows.

Consider a paste to which has been added some kind of tracer material for easy analysis. Let the overall average fraction of tracer in the mix be μ. Take a number of small samples at random from various locations in the mixed paste and determine the fraction of tracer x_i in each. Let the number of spot samples be N and the average value of the measured concentrations be $\bar{x}$. When N is very large, $\bar{x}$ will equal μ; when N is small, the two may be appreciably different. If the paste were perfectly mixed (and each analysis were perfectly accurate), every measured value of x_i would equal $\bar{x}$. If mixing is not complete, the measured values of x_i differ from $\bar{x}$ and their standard deviation about the average value of $\bar{x}$ is a measure of the quality of mixing. This standard deviation is estimated from the analytical results by the equation

$$ s = \sqrt{\frac{\sum_{i=1}^{N} (x_i - \bar{x})^2}{N - 1}} \tag{29-1} $$

The value of s is a relative measure of mixing, valid only for a set of tests with a specific material in a specific mixer. A more general measure[4] is the ratio of s to the standard deviation at zero mixing σ_0. Before mixing has begun, the material in the mixer exists as two layers, one of which contains no tracer material and one of which is tracer only. Samples from the first layer would have the analysis $x_i = 0$; in the other layer $x_i = 1$. Under these conditions the standard deviation is given by

$$ \sigma_0 = \sqrt{\mu(1 - \mu)} \tag{29-2} $$

where μ is the overall fraction of tracer in the mix. The mixing index for pastes I_p is then, from Eqs. (29-1) and (29-2),

$$ I_p = \frac{s}{\sigma_0} = \sqrt{\frac{\sum_{i=1}^{N} (x_i - \bar{x})^2}{(N - 1)\mu(1 - \mu)}} \tag{29-3} $$

The calculation of I_p from experimental data is shown by the following example.

EXAMPLE 29-1   A silty soil containing 14 percent moisture was mixed in a large muller mixer with 10.00 weight percent of a tracer consisting of dextrose and picric acid. After 3 min of mixing, 12 random samples were taken from the mix and analyzed colorimetrically for tracer material. The measured concentrations in the sample were, in weight percent tracer, 10.24, 9.30, 7.94, 10.24, 11.08, 10.03, 11.91, 9.72, 9.20, 10.76, 10.97, 10.55. Calculate the mixing index I_p.

SOLUTION   For this test $\mu = 0.10$ and $N = 12$. The calculations to determine $\bar{x}$ and $\sum (x_i - x)^2$ are given in Table 29-1; the values of these quantities are 0.1016 and 0.001189, respectively. Substitution in Eq. (29-3) gives

$$ I_p = \sqrt{\frac{0.001189}{(12 - 1) \times 0.10(1 - 0.10)}} = 0.0347 $$

In any batch mixing process I_p is unity before mixing begins and becomes smaller as mixing proceeds. Typical results found in mixing wet clay in a small two-arm kneader, like that shown in Fig. 29-2, are given in Fig. 29-6. Theoretically I_p would approach zero at long mixing times; in actuality it does not, for two reasons: (1) mixing is never quite complete; (2) unless the analytical methods are extraordinarily precise, the measured values of x_i never agree exactly with each other or

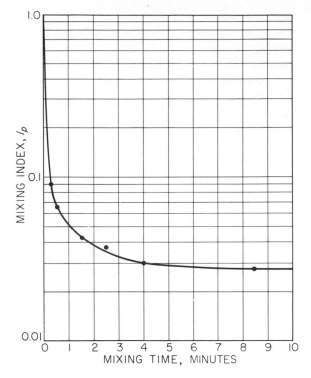

FIGURE 29-6
Mixing plastic clay in a two-arm kneader. (Smith.[6])

Table 29-1   CALCULATIONS FOR EXAMPLE 29-1

Sample No.	x_i	$x_i - \bar{x}$	$(x_i - \bar{x})^2 \times 10^6$
1	0.1024	+0.0008	0.64
2	0.0930	−0.0086	73.96
3	0.0794	−0.0222	492.84
4	0.1024	+0.0008	0.64
5	0.1108	+0.0092	84.64
6	0.1003	−0.0013	1.69
7	0.1191	+0.0175	306.35
8	0.0972	−0.0044	19.36
9	0.0920	−0.0096	92.16
10	0.1076	+0.0060	36.00
11	0.1097	+0.0081	65.61
12	0.1055	+0.0039	15.21
$\sum$	1.2194		1,189.10

$$\bar{x} = \frac{1.2194}{12} = 0.1016 \qquad \sum (x_i - \bar{x})^2 = 0.001189$$

with $\bar{x}$ and I_p is not found to be zero even with perfectly mixed material. The minimum limiting value of I_p for "completely mixed" materials varies with the consistency of the materials being processed, the effectiveness of the mixer, and the precision of the analytical method. Typically it falls in the range between 0.1 and 0.01.

The rate of mixing, as measured by the rate of change of I_p with time, varies greatly with the kind of mixer and the nature of the mixed materials. In a two-arm mixer, for example, dry powdery clay is mixed much more rapidly than very wet, almost liquid clay. Clay with an intermediate moisture content and in a stiff plastic condition is mixed rapidly but with a much greater energy requirement than either very wet or very dry clay.[4]

The effectiveness of a given type of mixer also depends on the nature and consistency of the mixed materials. Muller mixers, for example, give rapid mixing and low limiting values of I_p when operating on sandy granular solids; they give slow, rather poor mixing with plastic or sticky pastes.[6] In contrast to this, some continuous mixers such as the Ko-Kneader mix plastic materials more effectively than they do granular free-flowing solids.

The mixing index is related, in some cases at least, to the physical properties of the mixed material. For example, the compressive strength of clay samples stabilized with portland cement has been related[1] to the value of I_p. Low strengths were indicated when the value of I_p was large; low values of I_p corresponded to extraordinarily high strengths of the stabilized samples.

Axial mixing In the static mixer described in Chap. 9, page 249, two fluids are well mixed radially at any given cross section, but there is little mixing in an axial or longitudinal direction. The fluid behavior approximates that in plug flow, in which there is no axial mixing whatever. In some continuous paste mixers there is also little axial mixing, a desirable characteristic in certain mixing operations or chemical reactions; in others the axial mixing may be significant.

In paste mixers the degree of axial mixing is measured by the injection of a tracer, over a very short time, into the feed, followed by monitoring the concentration of tracer in the outlet stream. Typically the tracer appears at the outlet a little earlier than expected from the mean residence time of the mixer contents. Its outlet concentration rises to a maximum, then decays toward zero as time progresses. The height of the maximum and the length of time required for all (or nearly all) of the tracer to be discharged are measures of the degree of axial mixing.

Results of such tracer tests are normally expressed in terms of a diffusivity E. A low diffusivity means little axial mixing; a high diffusivity means there is a great deal. Obviously a small value of E is desirable when plug flow is best, as in chemical reactors in which mixing of feed and product is to be avoided. A large value of E is desirable when axial mixing is needed to blend successive portions of the mixer feed, e.g., to dampen minor fluctuations in the feed composition or the ratio of the feed components. Equations are available[7] for predicting E from the tracer-time data at the mixer outlet. For two-shaft paddle mixers E typically equals $0.02UL$ to $0.2UL$, where U is the longitudinal velocity of the material in the mixer and L is the mixer length.

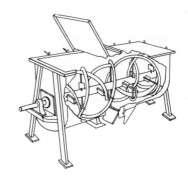

FIGURE 29-7
Ribbon mixer.

The ratio UL/E is known as the *Peclet number* N_{Pe}. Thus for paddle mixers N_{Pe} ranges from 5 to 50. With some agitator designs N_{Pe} is large and falls as the rotor speed increases; with other designs it is small and virtually independent of rotor speed.[7]

MIXERS FOR DRY POWDERS

Many of the machines described in the last section can blend solids when they are dry and free-flowing as well as when they are damp, pasty, rubbery, or plastic. Mullers, pan mixers, and pugmills are examples. Such versatile machines are needed when the properties of the material change markedly during the mixing operation. In general, however, these devices are less effective on dry powders than on other materials and are heavier and more powerful than necessary for free-flowing particulate solids.

The lighter machines discussed here handle dry powders and—sometimes—thin pastes. They mix by mechanical shuffling, as in ribbon blenders; by repeatedly lifting and dropping the material and rolling it over, as in tumbling mixers and vertical screw mixers; or by smearing it out in a thin layer over a rotating disk or impact wheel.

Ribbon blenders A ribbon blender consists of a horizontal trough containing a central shaft and a helical ribbon agitator. A typical ribbon mixer is shown in Fig. 29-7. Two counteracting ribbons are mounted on the same shaft, one moving the solid slowly in one direction, the other moving it quickly in the other. The ribbons may be continuous or interrupted. Mixing results from the "turbulence" induced by the counteracting agitators, not from mere motion of the solids through the trough. Some ribbon blenders operate batchwise, with the solids charged and mixed until satisfactory; others mix continuously, with solids fed in one end of the trough and discharged from the other. The trough is open or lightly covered for light duty, closed and heavy-walled for operation under pressure or vacuum. Ribbon blenders are effective mixers for thin pastes and for powders that do not flow readily. Some batch units are very large, holding up to 9,000 gal of material. The power they require is moderate.

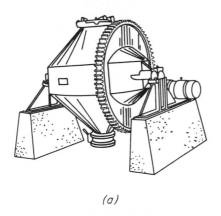

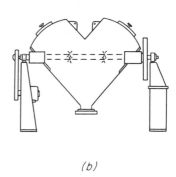

(a) (b)

FIGURE 29-8
Tumbler mixers: (a) double-cone mixer; (b) twin-shell blender.

Tumbling mixers Many materials are mixed by tumbling them in a partly filled container rotating about a horizontal axis. The ball mills described in Chap. 27 are often used as mixers. Most tumbling mills, however, do not contain grinding elements. Tumbling barrels, for example, resemble ball mills without the balls; they effectively mix suspensions of dense solids in liquids and heavy dry powders. Other tumbling blenders, such as those illustrated in Fig. 29-8, handle lighter dry solids only. The double-cone mixer shown at (a) is a popular mixer for free-flowing dry powders. A batch is charged into the body of the machine from above until it is 50 to 60 percent full. The ends of the container are closed and the solids tumbled for 5 to 20 min. The machine is stopped; mixed material is dropped out the bottom of the container into a conveyor or bin. The twin-shell blender shown at (b) is made from two cylinders joined to form a V and rotated about a horizontal axis. Like a double-cone blender, it may contain internal sprays for introducing small amounts of liquid into the mix or mechanically driven devices for breaking up agglomerates of solids. Twin-shell blenders are more effective in some blending operations than double-cone blenders. Tumbling mixers are made in a wide range of sizes and materials of construction. They draw a little less power, ordinarily, than ribbon blenders.

Internal screw mixers Free-flowing grains and other light solids are often mixed in a vertical tank containing a helical conveyor which elevates and circulates the material. Many different designs are commercially available. In the type shown in Fig. 29-9 the double-motion helix orbits about the central axis of a conical vessel, visiting all parts of the mix.

Impact wheels Fine, light powders such as insecticides may be blended continuously by spreading them out in a thin layer under centifugal action. A premix of the several dry ingredients is fed continuously near the center of a high-speed spinning disk 10 to 27 in. in diameter, which throws it outward into a stationary casing. The intense shearing forces acting on the powders during their travel over the disk surface

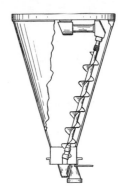

FIGURE 29-9
Internal screw mixer (orbiting type).

thoroughly blend the various materials. The disk in some machines is vertical; in others it is horizontal. The attrition mill shown in Fig. 27-9 is an effective mixer of this type. In some devices, designed for mixing and not size reduction, the premix is dropped onto a horizontal double rotor carrying short vertical pins near its periphery to increase the mixing effectiveness. A 14-in. disk turns at 1,750 r/min for easy problems and 3,500 r/min for materials that are hard to mix. Sometimes several passes through the same machine or through machines in series are necessary. For good results the premix fed to an impact wheel must be fairly uniform, for there is almost no holdup of material in the mixer and no chance for recombining material that has passed through with that which is entering. Impact wheels blend 1 to 25 tons/h of light free-flowing powders.

Mixing index in blending granular solids   The effectiveness of a solids blender is measured by a statistical procedure much like that used with pastes. Spot samples are taken at random from the mix and analyzed. The standard deviation of the analyses s about their average value $\bar{x}$ is estimated, as with pastes, from Eq. (29-1).

   With granular solids the mixing index is based, not on conditions at zero mixing, but on the standard deviation that would be observed with a completely random, fully blended mixture. With pastes, assuming the analyses are perfectly accurate, this value is zero. With granular solids it is not zero.

   Consider, for example, a completely blended mixture of salt and sand grains from which N spot samples, each containing n particles, are taken. Suppose the fraction of sand in each spot sample is determined by counting particles of each kind. Let the overall fraction, by number of particles, of sand in the total mix be μ_p. If n is small (say about 100), the measured fraction x_i of sand in each sample will not always be the same, even when the mix is completely and perfectly blended; there is always some chance that a sample drawn from a random mixture will contain a larger (or smaller) proportion of one kind of particle than the population from which it is taken. Thus for any given size of spot sample there is a theoretical standard deviation for a completely random mixture. This standard deviation σ_e is given by

$$\sigma_e = \sqrt{\frac{\mu_p(1 - \mu_p)}{n}} \qquad (29\text{-}4)$$

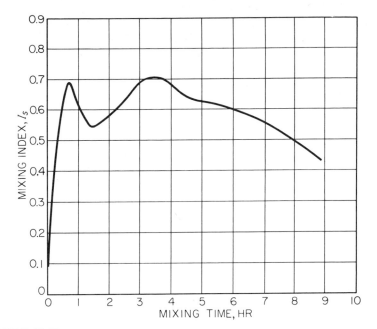

FIGURE 29-10
Blending salt and sand in a tumbling barrel. (*Weidenbaum and Bonilla.*[8])

For granular solids the mixing index I_s is defined as σ_e/s. From Eqs. (29-1) and (29-4) I_s becomes

$$I_s = \frac{\sigma_e}{s} = \sqrt{\frac{\mu_p(1 - \mu_p)(N - 1)}{n \sum_{i=1}^{N} (x_i - \bar{x})^2}} \qquad (29\text{-}5)$$

Typical results for blending salt crystals (NaCl) with Ottawa sand in a small tumbling blender are shown in Fig. 29-10. For the first 40 min the mixing index I_s rose from a very low value to about 0.7; it fluctuated for a time between 0.55 and 0.7 and then, at very long mixing times, began to fall steadily. Mixing is initially rapid, but in this type of mixer it is never perfect. The ingredients of the mix are never blended in a completely random way. Unblending forces, usually electrostatic, are always at work in a dry solids blender,[3] and their effects are especially noticeable here. These forces often prevent the mix from becoming completely blended; when the mixing time is long, they may, as shown by Fig. 29-10, lead to a considerable degree of unmixing and segregation.

Mixing index at zero time The equilibrium standard deviation for complete mixing σ_e is used as a reference with granular solids. With pastes the reference is the standard deviation at zero mixing σ_0. These two references are closely related, as shown by the similarity of Eqs. (29-2) and (29-4). In fact, if n is set equal to 1 in Eq. (29-4), the two equations become identical. This is as it should be, for if samples of one particle each are taken from any mixture of granular solids, the analysis will indicate $x_i = 1$ or $x_i = 0$ and nothing in between. This is the same as with completely unmixed material with any size of sample. Equation (29-2) may therefore be applied to granular solids before mixing begins, and the mixing index at zero mixing becomes

$$I_{s,0} = \frac{\sigma_e}{\sigma_0} = \frac{1}{\sqrt{n}} \tag{29-6}$$

Rate of mixing   In mixing, as in other rate processes, the rate is proportional to the driving force. The mixing index I_s is a measure of how far mixing has proceeded toward equilibrium.

It has been found[8] that for short mixing times the rate of change of I_s is directly proportional to $1 - I_s$ or

$$\frac{dI_s}{dt} = k(1 - I_s) \tag{29-7}$$

where k is a constant. The equilibrium value of I_s is 1; therefore the driving force for mixing at any time can be considered to be $1 - I_s$. With rearranging and integrating between limits, Eq. (29-7) becomes

$$\int_0^t dt = \frac{1}{k} \int_{I_{s,0}}^{I_s} \frac{dI_s}{1 - I_s} \tag{29-8}$$

from which

$$t = \frac{1}{k} \ln \frac{1 - I_{s,0}}{1 - I_s} \tag{29-9}$$

Substitution from Eq. (29-6) gives

$$t = \frac{1}{k} \ln \frac{1 - 1/\sqrt{n}}{1 - I_s} \tag{29-10}$$

Equation (29-10) can be used to calculate the time required for any desired degree of mixing, provided k is known and unblending forces are not active.

SYMBOLS

E Diffusivity in axial mixing, ft²/s or m²/s

I_p Mixing index for pastes, s/σ_0

I_s Mixing index for granular solids σ_e/s; $I_{s,0}$, at zero time or zero mixing

k Constant in Eq. (29-7)

L Mixer length, ft or m

N Number of spot samples

N_{Pe} Peclet number for axial mixing, UL/E, dimensionless

n Number of particles in spot sample

S Strength of mixed sample (Prob. 29-3), $lb_f/in.^2$

s Estimate of standard deviation [Eq. (29-1)]

t Time, s

U Longitudinal velocity of material through mixer, ft/s or m/s
x_i Measured fraction of tracer in spot sample; $\bar{x}$, average measured fraction of tracer
Greek letters
μ True average fraction of tracer in mix; μ_p, true average number fraction of tracer particles
σ Standard deviation; σ_e, equilibrium value for complete mixing of granular solids; σ_0, at zero time or zero mixing

PROBLEMS

29-1   A large Banbury mixer masticates 1,500 lb of scrap rubber with a density of 70 lb/ft^3. The power load is 7,000 hp per 1,000 gal of rubber. How much cooling water, in gallons per minute, is needed to remove the heat generated in the mixer if the temperature of the water is not to rise more than 20°F?

29-2   Data on the rate of mixing of sand and salt particles in an air-fluidized bed are given in Table 29-2. At long mixing times it was shown that the mixing index I_s closely approached 1.0. Assuming Eqs. (29-7) and (29-10) apply to this system, how long will it take for the mixing index to reach 0.95? In each run the initial charge to the reactor was 254.4 g of salt on top of 300.0 g of sand. The average number of particles in each spot sample was 100.

29-3   The compressive strength of a cement-stabilized clay sample from a small two-arm kneader was 100 lb$_f$/in.^{2} when the mixing index I_p was 0.1. The compressive strength S varies with mixing index according to the relation

$$S = -50.07 \ln I_p - 15.29$$

(*a*) From the data in Fig. 29-7, what mixing time would be required to give a product with a compressive strength of 150 lb$_f$/in.^{2}? (*b*) What would be the maximum strength obtainable with prolonged mixing in this mixer?

Table 29-2   DATA ON MIXING OF 35/48-MESH SALT IN SAND IN A 2-in. AIR-FLUIDIZED MIXER[5]

Run no.	Mixing time, s	Number fraction of sand in spot samples									
1	45	0.64	0.68	0.74	0.63	0.73	0.81	0.59	0.65	0.62	0.70
		0.66	0.64	0.77	0.70	0.67	0.58	0.60	0.65	0.87	0.60
		0.49	0.52	0.49	0.54	0.64	0.38	0.32	0.34	0.49	0.52
		0.25	0.32	0.33	0.35	0.48	0.23	0.16	0.32	0.44	0.39
		0.26	0.26	0.21	0.32	0.38	0.22	0.24	0.22	0.15	0.36
2	87	0.53	0.54	0.60	0.60	0.60	0.55	0.56	0.60	0.69	0.63
		0.48	0.67	0.65	0.63	0.62	0.46	0.63	0.58	0.48	0.59
		0.49	0.53	0.46	0.49	0.58	0.34	0.52	0.45	0.50	0.47
		0.42	0.35	0.43	0.49	0.59	0.38	0.39	0.45	0.52	0.39
		0.35	0.36	0.37	0.49	0.48	0.37	0.49	0.32	0.32	0.36

REFERENCES

1   Baker, C. N., Jr.: Strength of Soil Cement as a Function of Degree of Mixing, *33d Annu. Meet. Highw. Res. Board, Washington, D.C., January* 1954.
2   Bullock, H. L.: *Chem. Eng. Prog.,* **47**:397 (1951).
3   Fischer, J. J.: *Chem. Eng.,* **67**(16):107 (1960).
4   Michaels, A. S., and V. Puzinauskas: *Chem. Eng. Prog.,* **50**:604 (1954).
5   Nicholson, W. J.: "The Blending of Dissimilar Particles in a Gas-fluidized Bed," Ph.D. thesis, Cornell University, Ithaca, N.Y., 1965.
6   Smith, J. C.: *Ind. Eng. Chem.,* **47**:2240 (1955).
7   Todd, D. B., and H. F. Irving: *Chem. Eng. Prog.,* **56**(9):84 (1969).
8   Weidenbaum, S. S., and C. F. Bonilla: *Chem. Eng. Prog.,* **51**:27-J (1955).

30

MECHANICAL SEPARATIONS

Frequently it is necessary to separate the components of a mixture into individual fractions. The fractions may differ from each other in particle size, in phase, or in chemical composition. Thus, a crude product may be purified by removing from it contaminating impurities; two or more products in a mixture may be separated into the individual pure products; the stream discharged from a process step may consist of a mixture of product and unconverted raw material, which must be separated and the unchanged raw material recycled to the reaction zone for further processing; or a valuable substance, such as a metallic ore, dispersed in a mass of inert material must be liberated for recovery and the inert material discarded. A number of methods have been invented for accomplishing such separations, and several unit operations are devoted to them. In practice, many separation problems are encountered, and the engineer must choose the method best fitted to the problem.

Procedures for separating the components of mixtures fall into two classes. One class includes methods, called *diffusional operations*, which involve phase changes or transfer of material from one phase to another, as discussed in Chaps 17 to 25. The second class includes methods, called *mechanical separations*, which are useful in separating solid particles or liquid drops. These methods are the subject of the present chapter.

Mechanical separations are applicable to heterogeneous mixtures, not to homogeneous solutions. Colloids, which are an intermediate class of mixtures, are not usually treated by the methods discussed in this chapter, which deals primarily with particles larger than 0.1 μm. The techniques are based on physical differences between the particles such as size, shape, or density. They are applicable to separating solids from gases, liquid drops from gases, solids from solids, and solids from liquids. Two general methods are the use of a sieve, septum, or membrane, such as a screen or a filter, which retains one component and allows the other to pass; and the utilization of differences in the rate of sedimentation of particles or drops as they move through a liquid or gas. For special problems other methods, not discussed here, are used. These special methods exploit differences in the wettability or the electrical or magnetic properties of the substances.

SCREENING

Screening is a method of separating particles according to size alone. In industrial screening the solids are dropped on, or thrown against, a screening surface. The undersize, or *fines*, pass through the screen openings; oversize, or *tails*, do not. A single screen can make but a single separation into two fractions. These are called unsized fractions, because although either the upper or lower limit of the particle sizes they contain is known, the other limit is unknown. Material passed through a series of screens of different sizes is separated into sized fractions, i.e., fractions in which both the maximum and minimum particle sizes are known. Screening is occasionally done wet[2] but much more commonly dry. The discussion in this section is limited to the screening of dry particulate solids.

Industrial screens are made from metal bars, perforated or slotted metal plates, woven wire cloth, or fabric, such as silk bolting cloth. Metals used include steel, stainless steel, bronze, copper, nickel, and monel. The mesh size of woven screens ranges from 4-in. to 400-mesh, but screens finer than 100- or 150-mesh are rarely used.† With very fine particles other methods of separation are usually more economical. Many varieties and types of screens are available for different purposes.[14f] A few representative types are described below.

Types of screens In most screens the particles drop through the screen opening by gravity. In a few designs, which will not be discussed here, they are pushed through by a brush or by centrifugal force. Coarse particles drop quickly and easily through large openings in a stationary surface; with finer particles the screening surface must be agitated in some way. Common ways are by revolving a cylindrical screen about a horizontal axis; or, with flat screens, by shaking, gyrating, or vibrating them mechanically or electrically. Typical screen motions are illustrated in Fig. 30-1.

STATIONARY SCREENS AND GRIZZLIES A grizzly is a grid of parallel metal bars set in an inclined stationary frame. The slope and the path of the material are usually

† Standard screens are discussed in Chap. 26; screen sizes are tabulated in Appendix 20.

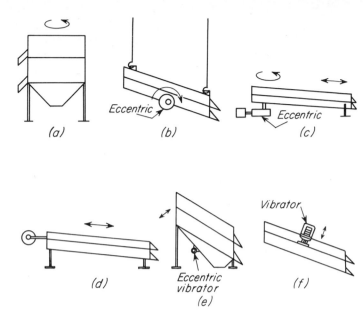

FIGURE 30-1
Motions of screens: (*a*) gyrations in horizontal plane; (*b*) gyrations in vertical plane; (*c*) gyrations at one end, shaking at other; (*d*) shaking; (*e*) mechanically vibrated; (*f*) electrically vibrated.

parallel to the length of the bars. Very coarse feed, as from a primary crusher, falls on the upper end of the grid. Large chunks roll and slide to the tails discharge; small lumps fall through to a separate collector. In cross section the top of each bar is wider than the bottom, so that the bars can be made fairly deep for strength without being choked by lumps passing partway through. The spacing between the bars is 2 to 8 in.

Stationary inclined woven-metal screens operate in the same way, separating particles $\frac{1}{2}$ to 4 in. in size. They are effective only with very coarse free-flowing solids containing few fine particles.

GYRATING SCREENS   In most other screens which produce sized fractions the coarse material is removed first and the fines last. This is illustrated by the gyrating flat screens shown in Fig. 30-2. These machines contain several decks of screens, one above the other, held in a box or casing. The coarsest screen is at the top and the finest at the bottom, with suitable discharge ducts to permit removal of the several fractions. The mixture of particles is dropped on the top screen. Screens and casing are gyrated to sift the particles through the screen openings.

In the design shown in Fig. 30-2a the casing is inclined at an angle between 16 and 30° with the horizontal. The gyrations are in a vertical plane about a horizontal axis. They are caused by an eccentric shaft set in the floor of the casing halfway between the feed point and the discharge. The screens are rectangular and fairly long,

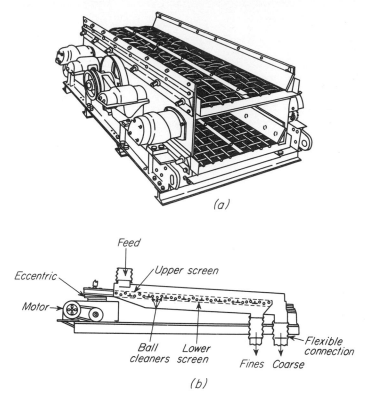

FIGURE 30-2
Gyrating screens: (*a*) heavy-duty vertically gyrated; (*b*) horizontally gyrated.

typically $1\frac{1}{2}$ by 4 ft to 5 by 14 ft. The speed of gyration and the amplitude of throw are adjustable, as is the angle of tilt. One particular combination of speed and throw usually gives the maximum yield of desired product from a given feed. The rate of gyration is 600 to 1,800 r/min; the motor size is 1 to 3 hp. The angle of tilt greatly influences the capacity of the screen. It is best to use the steepest angle possible. Very steep angles, however, can be used only with coarse products; good separation into fine fractions usually requires an angle of not more than 20° with the horizontal.

The screen shown in Fig. 30-2*b* is gyrated in a horizontal plane. It contains rectangular slightly inclined screens which are gyrated at the feed end. The discharge end reciprocates but does not gyrate. This combination of motions stratifies the feed, so that fine particles travel downward to the screen surface, where they are pushed through by the larger particles on top. Often the screening surface is double, as shown in the figure. Between the two screens are rubber balls held in separate compartments. As the screen operates, the balls strike the screen surface and free the openings of any material that tends to plug them. Dry, hard, rounded or cubical grains ordinarily pass without trouble through screens, even fine screens; but elongated, sticky, flaky,

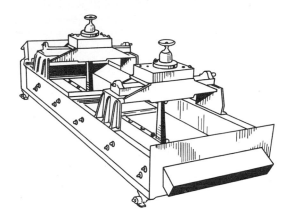

FIGURE 30-3
Electrically vibrated screen.

or soft particles do not. Under the screening action such particles may become wedged into the openings and prevent other particles from passing through. A screen plugged with solid particles is said to be *blinded*.

VIBRATING SCREENS   Screens which are rapidly vibrated with small amplitude are less likely to blind than are gyrating screens. The vibrations may be generated mechanically or electrically. Mechanical vibrations are usually transmitted from high-speed eccentrics to the casing of the unit and from there to steeply inclined screens. Electrical vibrations from heavy-duty solenoids are transmitted to the casing or directly to the screens. Figure 30-3 shows a directly vibrated unit. Ordinarily no more than three decks are used in vibrating screens. Between 1,800 and 3,600 vibrations per minute are usual. A screen 12 in. wide and 24 in. long draws about $\frac{1}{3}$ hp; a 48- by 120-in. screen draws 4 hp.

Comparison of ideal and actual screens   The objective of a screen is to accept a feed containing a mixture of particles of various sizes and separate it into two fractions, an underflow that is passed through the screen and an overflow that is rejected by the screen. Either one, or both, of these streams may be a product, and in the following discussion no distinction is made between the overflow and underflow streams from the standpoint of one's being desirable and the other undesirable.

An ideal screen would sharply separate the feed mixture in such a way that the smallest particle in the overflow would be just larger than the largest particle in the underflow. Such an ideal separation defines a cut diameter D_{pc} which marks the point of separation between the fractions. By diameter is meant a typical particle dimension, chosen as described on page 804. Although the choice of defining length is arbitrary, it is convenient so to choose the diameter that the cut diameter D_{pc} is nearly equal to the mesh opening of the screen.

The performance of an ideal screen in terms of the screen analysis of the feed is shown in Fig. 30-4a. The cut point is point c in the curve. Fraction A consists of all particles larger than D_{pc}, and fraction B consists of all particles smaller than D_{pc}.

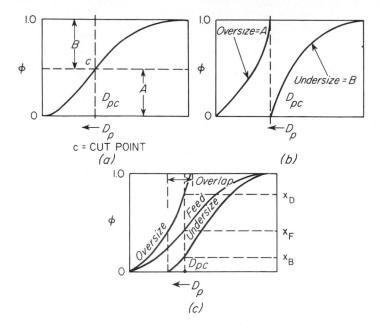

FIGURE 30-4
Ideal vs. actual screening: (*a*) ideal screening; (*b*) screen analyses of products from ideal screening; (*c*) actual screening.

Materials A and B are the overflow and the underflow, respectively. Screen analyses of the ideal fractions A and B are plotted in Fig. 30-4*b*. The first point on the curve for B has the same abscissa as the last point on the curve for A, and there is no overlap of these curves. In these figures diameter D_p increases from right to left.

Actual screens do not give a sharp separation. Instead the screen analyses of the overflow and underflow are like those shown in Fig. 30-4*c*. The overflow has an appreciable content of particles smaller than the desired cut diameter, and the underflow has particles larger than the cut diameter. The two curves overlap. Also, the mass of the two leaving streams will not equal the individual masses of A and B unless it happens that the oversize material in the underflow is equal to the undersize in the overflow.

The closest separations are obtained with spherical particles on standard testing screens, but even here there is an overlap between the smallest particles in the overflow and the largest ones in the underflow. The overlap is especially pronounced when the particles are needlelike or fibrous or where the particles tend to aggregate into clusters that act as large particles. Some long, thin particles may strike the screen surface endwise and pass through easily, while other particles of the same size and shape may strike the screen sidewise and be retained. Commercial screens usually give poorer separations than testing screens of the same mesh opening operating on the same mixture.

MATERIAL BALANCES OVER SCREEN   Simple material balances, of the type described in Chap. 1, can be written over a screen. The equations are useful in calculating the ratios of feed, oversize, and underflow from the screen analyses of the three streams and knowledge of the desired cut diameter. Let

F = mass flow rate of feed
D = mass flow rate of overflow
B = mass flow rate of underflow
x_F = mass fraction of material A in feed
x_D = mass fraction of material A in overflow
x_B = mass fraction of material A in underflow

The mass fractions of material A are shown in Fig. 30-4c. The mass fractions of material B in the feed, overflow, and underflow are $1 - x_F$, $1 - x_D$, and $1 - x_B$.

Since the total material fed to the screen must leave it either as underflow or as overflow,

$$F = D + B \tag{30-1}$$

The material A in the feed must also leave in these two streams, and

$$Fx_F = Dx_D + Bx_B \tag{30-2}$$

Elimination of B from Eqs. (30-1) and (30-2) gives

$$\frac{D}{F} = \frac{x_F - x_B}{x_D - x_B} \tag{30-3}$$

Elimination of D gives†

$$\frac{B}{F} = \frac{x_D - x_F}{x_D - x_B} \tag{30-4}$$

SCREEN EFFECTIVENESS   The effectiveness of a screen (often called the *screen efficiency*) is a measure of the success of a screen in closely separating materials A and B. If the screen functioned perfectly, all of material A would be in the overflow and all of material B would be in the underflow. A common measure of screen effectiveness is the ratio of oversize material A that is actually in the overflow to the amount of A entering with the feed. These quantities are Dx_D and Fx_F, respectively. Thus

$$E_A = \frac{Dx_D}{Fx_F} \tag{30-5}$$

where E_A is the screen effectiveness based on the oversize. Similarly, an effectiveness E_B based on the undersize material is given by

$$E_B = \frac{B(1 - x_B)}{F(1 - x_F)} \tag{30-6}$$

---

† Note the identity of Eqs. (30-3) and (30-4) with Eqs. (19-19) and (19-20) for distillation. Unlike as they are physically, both operations are separation operations, and the same overall material-balance equations apply to them.

A combined overall effectiveness can be defined as the product of the two individual ratios,[8b] and if this product is denoted by E,

$$E = \frac{DBx_D(1 - x_B)}{F^2 x_F(1 - x_F)}$$

Substituting D/F and B/F from Eqs. (30-3) and (30-4) into this equation gives

$$E = \frac{(x_F - x_B)(x_D - x_F)x_D(1 - x_B)}{(x_D - x_B)^2(1 - x_F)x_F} \qquad (30\text{-}7)$$

EXAMPLE 30-1   A quartz mixture having the screen analysis shown in Tables 26-2 and 26-3 is screened through a 10-mesh-per-inch screen constructed of 0.04-in. wire. The cumulative screen analyses of overflow and underflow are given in Table 30-1. Calculate the mass ratios of overflow and underflow to feed and the overall effectiveness of the screen.

SOLUTION   The cumulative analyses of feed, overflow, and product are plotted in Fig. 30-5. The cut-point diameter is the mesh size of the screen, which is $(0.10 - 0.04) \times 25.4 = 1.52$ mm. From Fig. 30-5, for $D_{pc} = 1.52$ mm,

$$x_F = 0.540 \qquad x_D = 0.895 \qquad x_B = 0.275$$

From Eq. (30-3), the ratio of overflow to feed is

$$\frac{D}{F} = \frac{0.540 - 0.275}{0.895 - 0.275} = 0.427$$

The ratio of underflow to feed is

$$\frac{B}{F} = 1 - \frac{D}{F} = 1 - 0.427 = 0.573$$

The overall effectiveness is, from Eq. (30-7),

$$E = \frac{(0.540 - 0.275)(0.895 - 0.540)(1 - 0.275)(0.895)}{(0.895 - 0.275)^2(0.460 \times 0.540)} = 0.64 \qquad ////$$

Table 30-1   SCREEN ANALYSIS FOR EXAMPLE 30-1

Mesh	D_p, mm	ϕ Overflow	ϕ Underflow
4	4.699	0	
6	3.327	0.071	
8	2.362	0.43	0
10	1.651	0.85	0.195
14	1.168	0.97	0.58
20	0.833	0.99	0.83
28	0.589	1.00	0.91
35	0.417		0.94
65	0.208		0.96
Pan			1.00

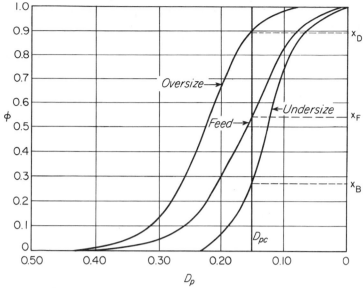

FIGURE 30-5
Analyses for Example 30-1.

Capacity and effectiveness of screens   In addition to effectiveness, capacity is important in industrial screening. The capacity of a screen is measured by the mass of material that can be fed per unit time to a unit area of the screen.

Capacity and effectiveness are opposing factors. To obtain maximum effectiveness the capacity must be small, and large capacity is obtainable only at the expense of a reduction in effectiveness. In practice, a reasonable balance between capacity and effectiveness is desired. Although accurate relationships are not available for estimating these operating characteristics of screens, certain fundamentals apply, which can be used as guides in understanding the basic factors in screen operation.

The capacity of a screen is controlled simply by varying the rate of feed to the unit. The effectiveness obtained for a given capacity depends on the nature of the screening operation. The overall chance of passage of a given undersize particle is a function of the number of times the particle strikes the screen surface and the probability of passage during a single contact. If the screen is overloaded, the number of contacts is small and the chance of passage on contact is reduced by the interference of the other particles. The improvement of effectiveness attained at the expense of reduced capacity is a result of more contacts per particle and better chances for passage on each contact.

Ideally, a particle would have the greatest chance of passing through the screen if it struck the surface perpendicularly, if it were so oriented that its minimum dimensions were parallel with the screen surface, if it were unimpeded by any other particles, and if it did not stick to, or wedge into, the screen surface. None of these conditions apply to actual screening, but this ideal situation can be used as a basis for estimating the effect of mesh size and wire dimensions on the performance of screens.

If the width of the wire in a screen were negligible in comparison with the size of the screen openings, the wires would not interfere with the passing of the particles, and practically the entire screen surface would be active. The probability of passage of a particle that strikes the screen would then be nearly unity. In an actual screen the diameter of the wire, or the fraction of the surface that is not in openings, is appreciable, and the solid meshes strongly affect the performance of the screen, especially in retarding the passage of particles nearly as large as the screen openings.

EFFECT OF MESH SIZE ON CAPACITY OF SCREENS   The probability of passage of a particle through a screen depends on the fraction of the total surface represented by openings, on the ratio of the diameter of the particle to the width of an opening in the screen, and on the number of contacts between the particle and the screen surface. When these factors are all constant, the average number of particles passing through a single screen opening in unit time is a constant, independent of the size of the screen opening. If the size of the largest particle that can just pass through a screen opening is taken equal to the width of a screen opening, both dimensions may be represented by D_{pc}. For a series of screens of different mesh sizes, the number of openings per unit screen area is proportional to $1/D_{pc}^2$. The mass of one particle is proportional to D_{pc}^3. The capacity of the screen, in mass per unit time, is, then, proportional to $(1/D_{pc}^2)D_{pc}^3 = D_{pc}$. Then the capacity of a screen, in mass per unit time, divided by the mesh size should be constant for any specified conditions of operation. This is a well-known practical rule of thumb.[9]

CAPACITIES OF ACTUAL SCREENS   Although the preceding analysis is useful in analyzing the fundamentals of screen operation, in practice a number of complicating factors appear that cannot be treated theoretically. Some of these disturbing factors are the interference of the bed of particles with the motion of any one; blinding; cohesion of particles to each other; the adhesion of particles to the screen surface; and the oblique direction of approach of the particles to the surface. When large and small particles are present, the large particles tend to segregate in a layer next to the screen and so prevent the smaller particles from reaching the screen. All these factors tend to reduce capacity and lower effectiveness. Moisture content of the feed is especially important. Either dry particles or particles moving in a stream of water screen more readily than damp particles, which are prone to stick to the screen surface and to each other and to screen slowly and with difficulty. Capacities of actual screens, in tons/ft²-h-mm mesh size, normally range between 0.05 and 0.2 for grizzlies to 0.2 and 0.8 for vibrating screens. As the particle size is reduced, screening becomes progressively more difficult, and the capacity and effectiveness are, in general, low for particle sizes smaller than about 100-mesh.

FILTRATION

Filtration is the removal of solid particles from a fluid by passing the fluid through a filtering medium, or *septum*, on which the solids are deposited. Industrial filtrations range from simple straining to highly complex separations. The fluid may be a liquid or a gas; the solid particles may be coarse or fine, rigid or plastic, round or elongated, separate individuals or aggregates. The feed suspension may carry a heavy load of solids or almost none. It may be very hot or very cold, or under vacuum or high pressure. Further complexities are introduced by the relative value of the two phases. Sometimes the fluid is the valuable phase; sometimes the solid; sometimes both. In some problems the separation of the phases must be virtually complete; in others only a partial separation is desired. A multitude of filters has therefore been developed to meet the different problems.[3,14b] Some of the commoner types are discussed in the following pages.

Equipment for Liquid-Solid Filtrations

Liquid-solid filters may be divided into four groups, depending on the service they perform: strainers, clarifiers, cake filters, and filter thickeners. A strainer is usually little more than a metal screen set across a flow channel to remove dirt or rust from a flowing liquid. When the screen becomes plugged, it is easily replaced. Clarifiers also remove small quantities of solids, usually to produce sparkling clear liquids, e.g., printing inks or beverages. The removed solids are most often discarded. The filter medium in a clarifier is a septum of cloth or paper or a cartridge of metal disks. Cake filters separate large amounts of solids from a liquid as a cake of crystals or sludge. Often they include provisions for washing the solids and for removing as much residual liquid from the solids as possible before discharge. A filter-thickener gives partial separation of a thin slurry, discharging some clear liquid and a thickened but still flowable suspension of solids.

Liquid flows through a filter medium by virtue of a pressure differential across the medium. Filters are also classified, therefore, into those which operate with a pressure above atmospheric on the upstream side of the filter medium and those which operate with atmospheric pressure on the upstream side and a vacuum on the downstream side. Pressures above atmospheric may be developed by the force of gravity acting on a column of liquid, by a pump, or by centrifugal force. Centrifugal filters are discussed in a later section of this chapter. In a gravity filter the filter medium can be no finer than a coarse screen or a bed of coarse particles like sand. Gravity filters are therefore restricted in their industrial applications to the draining of liquor from very coarse crystals and to the clarification of water.

Most industrial filters are either pressure filters or vacuum filters. They are also either continuous or discontinuous, depending on whether the discharge of filtered solids is steady or intermittent. During much of the operating cycle of a discontinuous filter the flow of liquid through the device is continuous, but it must be interrupted periodically to permit discharging the accumulated solids. In a continuous filter the discharge of both solids and liquid is uninterrupted as long as the equipment is in operation.

Discontinuous pressure filters　Pressure filters can apply a large pressure differential across the septum to give economically rapid filtration of viscous liquids or fine solids. The most common types of pressure filters are filter presses, shell-and-leaf

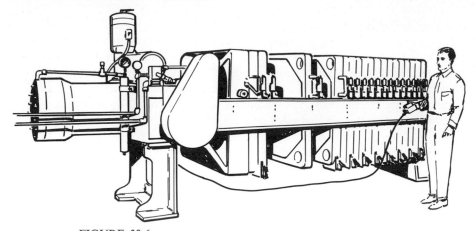

FIGURE 30-6
Filter press equipped for automatic operations. (*T. Shriver and Company, Inc.*)

filters, and cartridge filters. Because of the difficulty of discharging solids against pressures above atmospheric, pressure filters are usually discontinuous. Continuous pressure cake filters find important but limited applications in processing operations.

FILTER PRESS   A filter press contains a set of plates designed to provide a series of chambers or compartments in which solids may collect. The plates are covered with a filter medium such as canvas. Slurry is admitted to each compartment under pressure; liquor passes through the canvas and out a discharge pipe, leaving a wet cake of solids behind.

The plates of a filter press may be square or circular, vertical or horizontal. In some designs the compartments for solids are formed by recesses in the faces of the plates. More commonly, however, they are formed as in the *plate-and-frame press* shown in Fig. 30-6. In this design, square plates 6 to 56 in. on a side alternate with open frames. The plates are $\frac{1}{4}$ to 2 in. thick, the frames $\frac{1}{4}$ to 8 in. thick. Plates and frames sit vertically in a metal rack, with cloth covering the face of each plate, and are squeezed tightly together by a screw or a hydraulic ram. Slurry enters at one end of the assembly of plates and frames. It passes through a channel running lengthwise through one corner of the assembly. Auxiliary channels carry slurry from the main inlet channel into each frame. Here the solids are deposited on the cloth-covered faces of the plates. Liquor passes through the cloth, down grooves or corrugations in the plate faces, and out of the press.

After assembly of the press, slurry is admitted from a pump or blow case under a pressure of 3 to 10 atm. Filtration is continued until liquor no longer flows out the discharge or the filtration pressure suddenly rises. These occur when the frames are full of solid and no more slurry can enter. The press is then said to be *jammed*. Wash liquid may then be admitted to remove soluble impurities from the solids, after which the cake may be blown with steam or air to displace as much residual liquid as

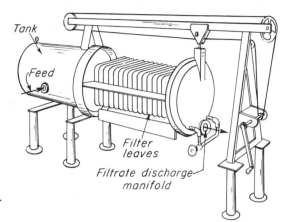

FIGURE 30-7
Horizontal-tank pressure leaf filter.
(*Ametek, Inc.*)

possible. The press is then opened, and the cake of solids scraped off the filter medium and dropped to a conveyor or storage bin. In some filter presses these operations are carried out automatically, as in the press illustrated in Fig. 30-6.

Thorough washing in a filter press may take several hours, for the wash liquid tends to follow the easiest paths and to bypass tightly packed parts of the cake. If the cake is less dense in some parts than in others, as is usually the case, much of the wash liquid will be ineffective. If washing must be exceedingly good, it may be best to reslurry a partly washed cake with a large volume of wash liquid and refilter it or to use a shell-and-leaf filter, which permits more effective washing than a plate-and-frame press.

SHELL-AND-LEAF FILTERS For filtering under higher pressures than are possible in a plate-and-frame press, to economize on labor, or where more effective washing of the cake is necessary, a shell-and-leaf filter may be used. In the horizontal-tank design shown in Fig. 30-7 a set of vertical leaves is held on a retractable rack. The unit shown in the figure is open for discharge; during operation the leaves are inside the closed tank. Feed enters through the side of the tank; filtrate passes through the leaves into a discharge manifold. The design shown in Fig. 30-7 is widely used for filtrations involving filter aids, as discussed later in this chapter.

CARTRIDGE FILTERS A typical cartridge filter, used principally for removing small amounts of solids from process fluids, is the filter illustrated in Fig. 30-8. The filtering cartridge is a series of thin metal disks 3 to 10 in. in diameter set in a vertical stack with very narrow uniform spaces between them. The disks are carried on a vertical hollow shaft, and fit into a closed cylindrical casing. Liquid is admitted to the casing under pressure. It flows inward between the disks to openings in the central shaft and out through the top of the casing. Solids are trapped between the disks and remain in the filter. Since most of the solids are removed at the periphery of the disks, this kind of device is known as an *edge filter*. Periodically the accumulated solids must

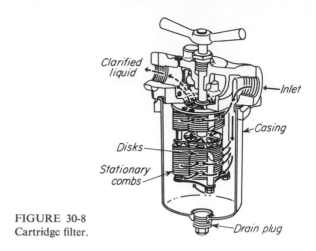

FIGURE 30-8
Cartridge filter.

be dislodged from the cartridge. In the design shown in Fig. 30-8 this is done by giving the cartridge a half turn. The stationary teeth of a comb cleaner pass between the disks, causing the solids to drop to the bottom of the casing, from which they may be removed at fairly long intervals.

Continuous pressure filters Batch filters often require considerable operating labor, which in large-scale processing may be prohibitively expensive. The continuous vacuum filters described below were developed to reduce the labor required for filtration. Sometimes, however, vacuum filtration is not feasible or not economical, as when the solids are very fine and filter very slowly, or when the liquid has a high vapor pressure, has a viscosity greater than 1 P, or is a saturated solution which will crystallize if cooled at all. With slow-filtering slurries the pressure differential across the septum must be greater than can be obtained in a vacuum filter; with liquids that vaporize or crystallize at reduced pressure the pressure on the downstream side of the septum cannot be less than atmospheric. Thus the continuous rotary-drum filters described later are sometimes adapted for operation under positive pressures up to about 3 atm; however, the mechanical problems of discharging the solids from these filters, their expense and complexity, and their small size limit their application to special problems.[13] Where vacuum filtration cannot be used, other means of separation, such as continuous centrifugal filters, should also be considered.

PRESSURE FILTER-THICKENER   The purpose of a filter-thickener is to remove some of the liquid from a thin slurry to produce a thick suspension. The design shown in Fig. 30-9 continuously discharges clear liquor and a stream of thickened slurry. In appearance this device resembles a filter press; it contains no frames, however, and the plates are modified as shown in the detail drawing. Successive plates carry matching channels which form, when the press is assembled, a long winding path for the slurry. The suspension flows continuously all the way through the press, first in the channel

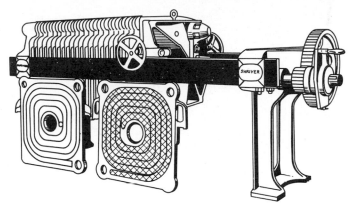

FIGURE 30-9
Pressure filter-thickener. (*T. Shriver and Company, Inc.*)

between the faces of one pair of plates, through an opening, then back between the faces of the next pair. The sides of the channels are covered with filter medium held between the plates. As slurry passes through the channel under pressure, some of the liquid flows through the filter medium into grooves in the plate faces and from there to a clear-liquor discharge manifold. The thickened slurry is kept moving fast enough to ensure that the channel does not clog. The number of plates is chosen so that the pressure drop through the press does not exceed 5 atm. Under these conditions it is usually possible to about double the concentration of the inlet slurry, as for example from 5 percent solids at the inlet to 10 percent solids at the outlet. When a greater degree of concentration than this is needed, the partly thickened slurry from one press is repumped to a second press for additional thickening.

Discontinuous vacuum filters Pressure filters are usually discontinuous; vacuum filters are usually continuous. A discontinuous vacuum filter, however, is sometimes a useful tool. A vacuum *nutsch* is little more than a large Büchner funnel, 3 to 10 ft in diameter and forming a layer of solids 4 to 12 in. thick. Because of its simplicity, a nutsch can readily be made of corrosion-resistant material, and is valuable where experimental batches of a variety of corrosive materials are to be filtered. Nutsches are not recommended for production operations because of the high labor cost involved in digging out the solid cake.

Continuous vacuum filters In all continuous vacuum filters liquor is sucked through a moving septum to deposit a cake of solids. The cake is moved out of the filtering zone, washed, sucked dry, and dislodged from the septum, which then reenters the slurry to pick up another load of solids. Some part of the septum is in the filtering zone at all times, part is in the washing zone, and part is being relieved of its load of solids, so that the discharge of both solids and liquids from the filter is uninterrupted. The pressure differential across the septum in a continuous vacuum filter is not high,

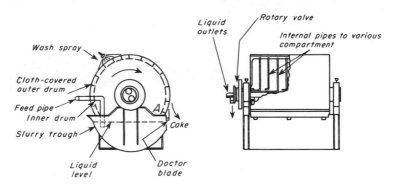

FIGURE 30-10
Continuous rotary vacuum filter.

ordinarily between 10 and 20 in. Hg. Various designs of filter differ in the method of admitting slurry, the shape of the filter surface, and the way in which solids are discharged. Most all, however, apply vacuum from a stationary source to the moving parts of the unit through a rotary valve.

ROTARY-DRUM FILTER   The most common type of continuous vacuum filter is the rotary-drum filter illustrated in Fig. 30-10. A horizontal drum with a slotted face turns at 0.1 to 2 r/min in an agitated slurry trough. A filter medium, such as canvas, covers the face of the drum, which is partly submerged in the liquid. Under the slotted cylindrical face of the main drum is a second, smaller drum with a solid surface. Between the two drums are radial partitions dividing the annular space into separate compartments, each connected by an internal pipe to one hole in the rotating plate of the rotary valve. Vacuum and air are alternately applied to each compartment as the drum rotates. A strip of filter cloth covers the exposed face of each compartment to form a succession of panels.

Consider now the panel shown at *A* in Fig. 30-10. It is just about to enter the slurry in the trough. As it dips under the surface of the liquid, vacuum is applied through the rotary valve. A layer of solids builds up on the face of the panel as liquid is drawn through the cloth into the compartment, through the internal pipe, through the valve, and into a collecting tank. As the panel leaves the slurry and enters the washing and drying zone, vacuum is applied to the panel from a separate system, sucking wash liquid and air through the cake of solids. As shown in the flow sheet of Fig. 30-11, wash liquid is drawn through the filter into a separate collecting tank. After the cake of solids on the face of the panel has been sucked as dry as possible, the panel leaves the drying zone, vacuum is cut off, and the cake is removed by scraping it off with a horizontal knife known as a *doctor blade*. A little air is blown in under the cake to belly out the cloth. This cracks the cake away from the cloth and makes it unnecessary for the knife to scrape the drum face itself. Once the cake is dislodged, the panel reenters the slurry and the cycle is repeated. The operation of any given

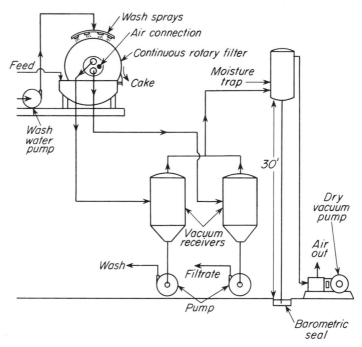

FIGURE 30-11
Flowsheet for continuous vacuum filtration.

panel, therefore, is cyclic, but since some panels are in each part of the cycle at all times, the operation of the filter as a whole is continuous.

Many variations of the rotary-drum filter are commercially available. In some designs there are no compartments in the drum; vacuum is applied to the entire inner surface of the filter medium. Filtrate and wash liquid are removed together through a dip pipe; solids are discharged by air blown through the cloth from a stationary shoe inside the drum, bellying out the filter cloth and cracking off the cake. In other models the cake is lifted from the filter surface by a set of closely spaced parallel strings or by separating the filter cloth from the drum surface and passing it around a small-diameter roller. The sharp change in direction at this roller dislodges the solids. The cloth may be washed as it returns from the roller to the underside of the drum. Wash liquid may be sprayed directly on the cake surface, or, with cakes that crack when air is drawn through them, it may be sprayed on a cloth blanket that travels with the cake through the washing zone and is tightly pressed against its outer surface.

The amount of submergence of the drum is also variable. Most bottom-feed filters operate with about 30 percent of their filter area submerged in the slurry. When high filtering capacity and no washing are desired, a high-submergence filter, with 60 to 70 percent of its filter area submerged, may be used. The capacity of any rotary filter depends strongly on the characteristics of the feed slurry and particularly

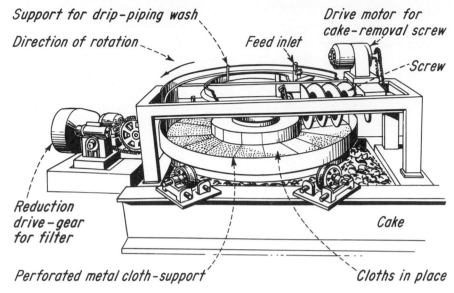

FIGURE 30-12
Continuous horizontal vacuum filter. (*Dorr-Oliver, Inc.*)

on the thickness of the cake which may be deposited in practical operation. The cakes formed on industrial rotary vacuum filters are $\frac{1}{8}$ to about $1\frac{1}{2}$ in. thick. Standard drum sizes range from 1 ft in diameter with a 1-ft face to 10 ft in diameter with a 14-ft face.

Continuous or semicontinuous clarifying devices are exemplified by a *precoat filter*, which is a rotary-drum filter modified for filtering small amounts of fine or gelatinous solids that ordinarily plug a filter cloth. In the operation of this machine a layer of porous filter aid, such as diatomaceous earth, is first deposited on the filter medium. Process liquid is then sucked through the layer of filter aid, depositing a very thin layer of solids. This layer and a little of the filter aid are then scraped off the drum by a slowly advancing knife, which continually exposes a fresh surface of porous material for the subsequent liquor to pass through. A precoat filter may also operate under pressure. In the pressure type the discharged solids and filter aid collect in the housing, to be removed periodically at atmospheric pressure while the drum is being recoated with filter aid. Precoat filters can be used only where the solids are to be discarded or where their admixture with large amounts of filter aid introduces no serious problem. The usual submergence of a precoat-filter drum is 50 percent.

HORIZONTAL FILTER   When free-draining solids are to be thoroughly washed, especially when for any reason the filtrate piping must be easily accessible, a horizontal filter of the type shown in Fig. 30-12 may be used. This is a horizontal ring-shaped trough with a flat cloth-covered false floor. The trough turns slowly about a central axis.

The space beneath the false floor is divided into compartments; a large rotary valve applies vacuum to each compartment and then releases it at appropriate intervals. Filtrate and any wash liquid which may be sprayed on the cake during its travel are sucked through the cloth, into the compartments beneath the false floor, and out through the rotary valve. At the discharge point the vacuum is released and the solids removed radially outward by a helical conveyor-scraper. Each section of a table filter may be considered as a vacuum nutsch, carried successively through filtering, washing, and drying zones to the discharge. On free-draining solids table filters have very high capacity.

Filter media   The septum in any filter must meet the following requirements:

1 It must retain the solids to be filtered, giving a reasonably clear filtrate.
2 It must not plug or blind.
3 It must be resistant chemically and strong enough physically to withstand the process conditions.
4 It must permit the cake formed to discharge cleanly and completely.
5 It must not be prohibitively expensive.

In industrial filtration a common filter medium is canvas cloth, either duck or twill weave. Many different weights and patterns of weave are available for different services. Corrosive liquids require the use of other filter media, such as woolen cloth, metal cloth of monel or stainless steel, glass cloth, or paper. Synthetic fabrics like nylon, polypropylene, Saran, and Dacron are also highly resistant chemically.

In a cloth of a given mesh size, smooth synthetic or metal fibers are less effective than the more ragged natural fibers in removing very fine particles. Ordinarily, however, this is a disadvantage only at the start of filtration, because except with hard, coarse particles containing no fines the actual filtering medium is not the septum but the first layer of deposited solids. Filtrate may first come through cloudy, then grow clear. Cloudy filtrate is returned to the slurry tank for refiltration..

Filter aids   Slimy or very fine solids which form a dense, impermeable cake quickly plug any filter medium that is fine enough to retain them. Practical filtration of such materials requires that the porosity of the cake be increased to permit passage of the liquor at a reasonable rate. This is done by adding a filter aid, such as diatomaceous earth, asbestos, purified wood cellulose, or other inert porous solid, to the slurry before filtration. The filter aid may subsequently be separated from the filter cake by dissolving away the solids or by burning out the filter aid. If the solids have no value, they and the filter aid are discarded together.

Another way of using a filter aid is by precoating, i.e., depositing a layer of it on the filter medium before filtration. In batch filters the precoat layer is usually thin; in a continuous precoat filter, as described previously, the layer of precoat is thick, and the top of the layer is continually scraped off by an advancing knife to expose a fresh filtering surface. Precoats prevent gelatinous solids from plugging the filter medium and give a clearer filtrate. The precoat is really a part of the filter medium rather than of the cake.

Principles of Cake Filtration

From the standpoint of fluid mechanics a filter is a flow device.[17] By means of a pressure difference applied between the slurry inlet and the filtrate outlet, filtrate is forced through the equipment. During filtration the solids in the slurry are retained in the unit, and form a bed of particles through which filtrate must flow.† The filtrate passes through three kinds of resistance in series: (1) resistances of the channels conducting the slurry to the upstream face of the cake and the filtrate away from the filter medium, (2) resistance of the cake, and (3) resistance associated with the filter medium. During the washing of the filter cake the wash water flows through the same resistances that were encountered by the filtrate, although the magnitude of the cake resistance is not the same in both cases.

DISTRIBUTION OF OVERALL PRESSURE DROP   Since the flow is in series, the total pressure drop over the filter can be equated to the sum of the individual pressure drops. In a well-designed filter the resistances of the inlet and outlet connections are small, and can be neglected in comparison with those of the cake and filter medium. In actual operation the resistance associated with the filter medium is greater than that offered by a clean filter medium to flow of clear filtrate. During the first moments of the filtration solid particles become embedded in the meshes of the filter medium and so develop abnormal resistance to the subsequent flow. The entire resistance built up in the filter medium, including that from the embedded particles, is called the *filter-medium resistance*. It is important during the early stages of the filtration. The resistance offered by all solids not associated with the filter medium is called the *cake resistance*. The cake resistance is zero at the beginning of the filtration, and because of the continuous deposition of solids on the medium, this resistance increases steadily with time of filtration. During washing all resistances, including that of the cake, are constant, and that of the filter medium is usually negligible.

Since the resistance of the channels can be neglected, the overall pressure drop at any time is the sum of the pressure drops over medium and cake. If p_a is the inlet pressure, p_b the outlet pressure, and p' the pressure at the boundary between cake and medium,

$$-\Delta p = p_a - p_b = (p_a - p') + (p' - p_b) = -\Delta p_c - \Delta p_m \qquad (30\text{-}8)$$

where $-\Delta p$ = overall pressure drop
$-\Delta p_c$ = pressure drop over cake
$-\Delta p_m$ = pressure drop over medium

Consistent with its usual definition, Δ represents the outlet condition minus the inlet condition. Therefore Δp equals $p_b - p_a$ and is inherently negative.

TYPES OF FILTRATION   For a given slurry in a specific filter the major variable under the control of the operator is the overall pressure drop; e.g., when the outlet pressure

---

† In filtering very dilute suspensions there are insufficient solids to form a cake. Particles are trapped within the filter medium, progressively reducing the paths available for flow of filtrate. This operation is known as *clarifying filtration*. Its effectiveness depends on the properties of the filter medium, and the equations governing flow are not the same as those derived here for cake filtration.[11]

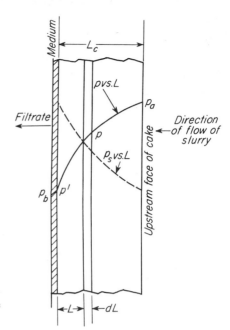

FIGURE 30-13
Section through filter medium and cake, showing pressure gradients: p, fluid pressure; p_s, stress pressure; L, distance from filter medium.

is constant, the pressure drop is controlled by varying the inlet pressure. If the pressure drop is constant, the flow rate is a maximum at the start of the filtration and decreases continuously to the end. This method of operation is called *constant-pressure filtration*. If the pressure drop is varied, usually it is kept small at the start of the filtration and then increased, either continuously or in steps, and is a maximum at the end of the filtration. One method is to maintain the flow rate constant by continuously increasing the inlet pressure. This method is called *constant-rate filtration*. Various combinations of these basic methods are used. A common one is to use constant rate until the inlet pressure reaches a specified maximum and then to continue at constant pressure until the end of the filtration. This minimizes loss of solids through the septum when the resistance is low and avoids packing solids into the septum. In washing both the flow rate and pressure drop are usually constant.

Pressure drop through filter cake   Figure 30-13 shows diagrammatically a section through a filter cake and filter medium at a definite time t from the start of the flow of filtrate. At this time the thickness of the cake, measured from the filter medium, is L_c. The filter area, measured perpendicularly to the direction of flow, is A. Consider the thin layer of cake of thickness dL lying in the cake at a distance L from the medium. Let the pressure at this point be p. This layer consists of a thin bed of solid particles through which the filtrate is flowing. In a filter bed the velocity is sufficiently low to ensure laminar flow. Accordingly, as a starting point for treating the pressure drop through the cake, Eq. (7-18) can be used, noting that $-\Delta p/L = -dp/dL$ and that for laminar flow $k_2 = 0$. If the velocity of the filtrate is designated as u, Eq. (7-18) becomes

$$-\frac{dp}{dL} = \frac{k_1 \mu u(1 - \varepsilon)^2(s_p/v_p)^2}{g_c \varepsilon^3} \qquad (30\text{-}9)$$

where dp/dL = pressure gradient at thickness L
  μ = viscosity of filtrate
  u = linear velocity of filtrate, based on filter area
  s_p = surface of single particle
  v_p = volume of single particle
  ε = porosity of cake
  k_1 = constant
  g_c = Newton's-law proportionality factor

For random packed particles of definite size and shape, $k_1 = 4.167$.
The linear velocity u is given by the equation

$$u = \frac{dV/dt}{A} \qquad (30\text{-}10)$$

where V is the volume of filtrate collected from the start of the filtration to time t. Since the filtrate must pass through the entire cake, V/A is the same for all layers and u is independent of L.

The volume of solids in the layer is $A(1 - \varepsilon) \, dL$, and if ρ_p is the density of the particles, the mass dm of solids in the layer is

$$dm = \rho_p(1 - \varepsilon)A \, dL \qquad (30\text{-}11)$$

Elimination of dL from Eqs. (30-9) and (30-11) gives

$$-dp = \frac{k_1 \mu u(s_p/v_p)^2(1 - \varepsilon)}{g_c \rho_p A \varepsilon^3} \, dm \qquad (30\text{-}12)$$

COMPRESSIBLE AND INCOMPRESSIBLE SLUDGES   In some special situations, e.g., in the filtration of sludges composed of rigid, uniform particles under low pressure drops, all factors on the right-hand side of Eq. (30-12) except m are independent of L, and the equation is integrable directly, over the thickness of the cake. If m_c is the total mass of solids in the cake, the result is

$$-\int_{p_a}^{p'} dp = \frac{k_1 \mu u(s_p/v_p)^2(1 - \varepsilon)}{g_c \rho_p A \varepsilon^3} \int_0^{m_c} dm$$

$$p_a - p' = \frac{k_1 \mu u(s_p/v_p)^2(1 - \varepsilon)m_c}{g_c \rho_p A \varepsilon^3} = -\Delta p_c \qquad (30\text{-}13)$$

Sludges of this type are called *incompressible*.

Most sludges encountered industrially do not have the simple structure of a bed of individual rigid particles. The usual slurry is a mixture of agglomerates, or flocs, consisting of loose assemblies of very small particles, and the resistance characteristics of the cake depend upon the properties of the flocs rather than on the geometry of the individual particles.[10] The flocs are deposited from the slurry on the upstream face of the cake, and form a complicated network of channels to which Eq. (30-12) does not precisely apply. The resistance of such a sludge is sensitive to the method used in preparing the slurry and to the age and temperature of the material. Also, the flocs are distorted and broken down by the forces existing in the cake, and the factors ε, k, and s_p/v_p vary from layer to layer. Sludges the resistance of which vary in this manner are called *compressible sludges*. Although Eq. (30-12) does not apply accurately to compressible sludges, it is a useful guide for developing empirical relationships applicable to such materials. Some sludges are more compressible than others, and a valid treatment should provide for a quantitative measure of compressibility.

MECHANICS OF FILTER CAKES   The variation in resistance of a cake from layer to layer is the result of mechanical effects in the cake. The fluid pressure in a filter cake is a maximum at the upstream face and a minimum at the filter medium. It might seem, therefore, that the porosity of the cake should be a minimum at the upstream face and a maximum at the filter medium. Actually the reverse is true. A compressible cake is relatively open and porous at the upstream face and compressed at the filter medium. The reason for this is as follows. Consider a particle in the bed at distance L from the medium. The stream of filtrate flowing past the particle imparts a drag force on the particle tending to move it in the direction of the medium. This drag is resisted by an equal and opposite force from the particles immediately ahead of the particle under consideration. The particle is also acted upon by forces from those immediately upstream from itself. Since each layer of solids, starting from the upstream face of the cake, transmits its drag force to the particles ahead of itself, the drag forces act cumulatively through the bed and each layer is acted on by a force equal to the sum of the drags of all layers between itself and the upstream face of the cake. Each layer transmits this cumulative drag, augmented by its own drag, to the next layer. These cumulative forces can be converted to pressures by dividing them by A. Such pressures are called *stress pressures*. As shown in Fig. 30-13, the stress pressure is zero at the upstream face of the cake, where the fluid pressure is p_a, and is a maximum at the filter medium, where the fluid pressure is p'. As the fluid pressure decreases, the stress pressure increases. The fluid pressure at a given point acts equally in all directions, but the stress pressure acts in a direction parallel with that of flow and tends to squeeze the particles and flatten them.

Figure 30-14 shows how an undistorted particle, represented by the solid boundary, is distorted by stress pressure to the shape shown by the dotted boundary. The effect is equivalent to squeezing a particle of putty between the fingers while it is immersed under a static head of water.

The cumulative drag at the layer a distance L from the filter medium equals $p_a - p$, the fluid pressure drop in the portion of the cake from the upstream face to the plane at L. It follows that k_1, s_p/v_p, and ε each depend on $p_a - p$ only and that they vary throughout the cake. In general the resistance to flow offered by the cake is smallest at the free cake surface and largest immediately adjacent to the filter

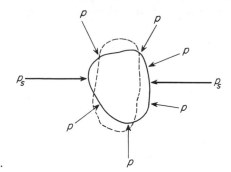

FIGURE 30-14
Distortion of particle by stress pressure.

medium. In practice such local variations are ignored, and an average quantity a, called the *specific cake resistance*, is defined by

$$\alpha = \frac{\overline{k}_1(\overline{s_p/v_p})^2(1 - \bar{\varepsilon})}{\bar{\varepsilon}^3 \rho_p} \tag{30-14}$$

This equation shows how a is related to the *average* values $\overline{k}_1$, $\overline{s_p/v_p}$, and $\bar{\varepsilon}$.

Substitution from Eq. (30-14) into Eq. (30-13) gives

$$p_a - p' = -\Delta p_c = \frac{\mu u m_c \alpha}{g_c A} \tag{30-15}$$

Equation (30-15) is the basic equation for pressure drop through a filter cake. Physically, α is the pressure drop required to give unit velocity of filtrate flow when the viscosity is unity and the cake contains one unit mass of solid per unit filter area. This factor is an average quantity for the entire cake, and it must be measured experimentally for each sludge. For a given sludge, its value depends only on the pressure drop through the cake. The dimensions of α are, from Eq. (30-15), $\overline{L}\overline{M}^{-1}$.

For incompressible sludges α is independent of the pressure drop.

Filter-medium resistance The filter-medium resistance R_m can be defined, by analogy with Eq. (30-15), by the equation

$$\frac{p' - p_b}{R_m} = \frac{-\Delta p_m}{R_m} = \frac{\mu u}{g_c} \tag{30-16}$$

The dimension of R_m is $\overline{L}^{-1}$.

The factors controlling the magnitude of R_m have been studied. It is known that pressure drop, and perhaps flow rate, affect it, and that an old, used filter medium has a much larger resistance than a clean, new one. Since the medium resistance usually is important only during the early stages of the filtration, it is satisfactory to assume that R_m is a constant during any given filtration and to determine its magnitude empirically from experimental data. The filter-medium resistance varies somewhat from experiment to experiment even for the same sludge and filter. When R_m is treated as an empirical constant, it also includes any resistance to flow that may exist in the leads to and from the filter.

From Eqs. (30-15) and (30-16),

$$-\Delta p = -\Delta p_c - \Delta p_m = \frac{\mu u}{g_c}\left(\frac{m_c \alpha}{A} + R_m\right) \qquad (30\text{-}17)$$

Strictly, the cake resistance α is a function of Δp_c rather than of Δp. During the important stage of the filtration, when the cake is of appreciable thickness, $-\Delta p_m$ is small in comparison with $-\Delta p_c$, and the effect on the magnitude of α of carrying the integration of Eqs. (30-13) and (30-14) over a range $-\Delta p$ instead of $-\Delta p_c$ can be safely ignored. In Eq. (30-17), then, α is taken as a function of Δp.

In using Eq. (30-17) it is convenient to replace u, the linear velocity of the filtrate, and m_c, the total mass of solid in the cake, by functions of V, the total volume of filtrate collected to time t. Equation (30-10) relates u and V, and a material balance relates m_c and V. If c is the mass of the particles deposited in the filter per unit volume of filtrate,† the mass of solids in the filter at time t is Vc and

$$m_c = Vc \qquad (30\text{-}19)$$

Substituting u from Eq. (30-10) and m_c from Eq. (30-19) in Eq. (30-17) gives

$$\frac{dt}{dV} = \frac{\mu}{Ag_c(-\Delta p)}\left(\frac{\alpha c V}{A} + R_m\right) \qquad (30\text{-}20)$$

Constant-pressure filtration When Δp is constant, the only variables in Eq. (30-20) are V and t. The equation can be integrated as follows:

$$\int_0^t dt = \frac{\mu}{Ag_c(-\Delta p)}\left(\frac{c\alpha}{A}\int_0^V V\,dV + R_m\int_0^V dV\right)$$

$$t = \frac{\mu}{g_c(-\Delta p)}\left[\frac{c\alpha}{2}\left(\frac{V}{A}\right)^2 + R_m\frac{V}{A}\right] \qquad (30\text{-}21)$$

where V is the total volume of filtrate collected to time t, assuming that time is counted from the instant the first drop of filtrate is obtained, so that when $V = 0$, $t = 0$.

Equation (30-21) is that of a parabola with its vertex displaced from the origin of coordinates ($t = 0$, $V = 0$). A plot of V vs. t for a constant-pressure filtration is therefore a section of this parabola.

---

† The concentration of solid in the slurry fed to the filter is slightly less than c, since the wet cake includes sufficient liquid to fill its pores, and V, the actual volume of filtrate, is slightly less than the total liquid in the original slurry. Correction for this retention of liquid in the cake can be made by material balances if desired. Thus, let m_F be the mass of the wet cake, including the filtrate retained in its voids, and m_c be the mass of the dry cake obtained by washing the cake free of soluble material and drying. Also, let ρ be the density of the filtrate. Then if c_s is the concentration of solids in the slurry in pounds per cubic foot of liquid fed to the filter, material balances give

$$c = \frac{c_s}{1 - [(m_F/m_c) - 1]c_s/\rho} \qquad (30\text{-}18)$$

To evaluate the constants α and R_m for a definite pressure drop, data of V vs. t from an experimental run at that pressure drop are needed. The treatment of such data is facilitated by using Eq. (30-20) in the form

$$\frac{dt}{dV} = K_p V + B \qquad (30\text{-}22)$$

where

$$K_p = \frac{c\alpha\mu}{A^2(-\Delta p)g_c} \qquad (30\text{-}23)$$

and

$$B = \frac{R_m\mu}{A(-\Delta p)g_c} \qquad (30\text{-}24)$$

Assume that a number of observations of V vs. t have been made. Then for any two successive observations, the quantity $\Delta t/\Delta V$ can be calculated, where Δt is the time between observations and ΔV is the increment of filtrate collected over time period Δt. Since, by Eq. (30-22), dt/dV is linear with V, a value of $\Delta t/\Delta V$ is the true slope of the t-vs.-V line at a point $(V_1 + V_2)/2$, or halfway between the observed values of V that define ΔV. Then the arithmetic mean of each of two successive observations of V can be plotted against $\Delta t/\Delta V$ for the same pair of readings. The best straight line is drawn through these points; the slope of the line is K_p, and the ordinate intercept is B. This construction is shown for several constant-pressure runs in Fig. 30-15. Since the various simplifying assumptions made during the derivation of Eq. (30-21) are most questionable during the early stages of the filtration, the points taken during these stages may not fall accurately on the graph, and these points should be given little weight in plotting the line. When K_p and B are known, α and R_m are then calculated from Eqs. (30-23) and (30-24).

Empirical equations for cake resistance   By conducting constant-pressure experiments at various pressure drops, the variation of α with Δp may be found. If α is independent of Δp, the sludge is incompressible. Ordinarily α increases with $-\Delta p$, as most sludges are at least to some extent compressible. For highly compressible sludges, K_p increases rapidly with $-\Delta p$.

Empirical equations may be fitted to observed data for Δp vs. α, the commonest of which is

$$\alpha = \alpha_0(-\Delta p)^s \qquad (30\text{-}25)$$

where α_0 and s are empirical constants. Constant s is the *compressibility coefficient* of the cake. It is zero for incompressible sludges and positive for compressible ones. It usually falls between 0.2 and 0.8. Equation (30-25) should not be used in a range of pressure drops much different from that used in the experiments conducted to evaluate α_0 and s.

EXAMPLE 30-2   Laboratory filtrations conducted at constant pressure drop on a slurry of $CaCO_3$ in H_2O gave the data shown in Table 30-2. The filter area was 440 cm^2, the mass of solid per unit volume of filtrate was 23.5 g/l, and the temperature was 25°C. Evaluate the quantities α and R_m as a function of pressure drop, and fit an empirical equation to the results for α.

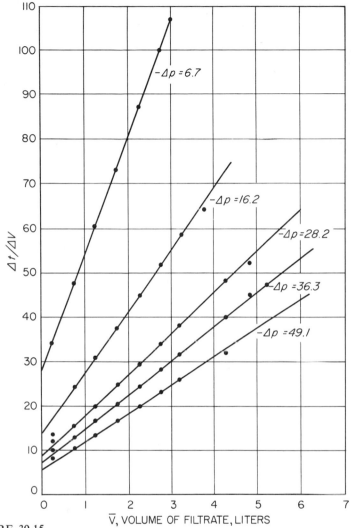

FIGURE 30-15
Plot of $\Delta t/\Delta V$ vs. $\bar{V}$ for Example 30-2.

SOLUTION   The first step is to prepare plots, for each of the five constant-pressure experiments, of $\Delta t/\Delta V$ vs. $\bar{V}$, which is $(V_1 + V_2)/2$ of each increment of filtrate volume. The data and calculations for the first experiment are given in Table 30-3, and the plots for all experiments are shown in Fig. 30-15.

The slope of each line of Fig. 30-15 is K_p, in seconds per liter per liter. To convert to seconds per cubic foot per cubic foot, the conversion factor is $28.31^2 = 801$. The intercept of each line on the axis of ordinates is B, in seconds per liter. The conversion factor to convert this to seconds per cubic foot is 28.31. The slopes and intercepts, in the observed and converted units, are given in Table 30-4.

Table 30-2   VOLUME-TIME DATA[16] FOR EXAMPLE 30-2

Test number	I	II	III	IV	V
Pressure drop $-\Delta p$, $lb_f/in.^2$	6.7	16.2	28.2	36.3	49.1
Filtrate volume V, l	Time, s				
0.5	17.3	6.8	6.3	5.0	4.4
1.0	41.3	19.0	14.0	11.5	9.5
1.5	72.0	34.6	24.2	19.8	16.3
2.0	108.3	53.4	37.0	30.1	24.6
2.5	152.1	76.0	51.7	42.5	34.7
3.0	201.7	102.0	69.0	56.8	46.1
3.5		131.2	88.8	73.0	59.0
4.0		163.0	110.0	91.2	73.6
4.5			134.0	111.0	89.4
5.0			160.0	133.0	107.3
5.5				156.8	
6.0				182.5	

Table 30-3 $\Delta t/\Delta V$ vs. $\bar{V}$ IN TEST I FOR EXAMPLE 30-2

Filtrate volume V, l	Time t, s	Δt	ΔV	$\dfrac{\Delta t}{\Delta V}$	$\bar{V}$
0	0				
0.5	17.3	17.3	0.5	34.6	0.25
1.0	41.3	24.0	0.5	48.0	0.75
1.5	72.0	30.7	0.5	61.4	1.25
2.0	108.3	36.3	0.5	72.6	1.75
2.5	152.1	43.8	0.5	87.6	2.25
3.0	201.7	49.6	0.5	99.2	2.75

Table 30-4   VALUES OF K_p, B, R_m AND α FOR EXAMPLE 30-2

Test	Pressure drop $-\Delta p$		Slope K_p		Intercept B		R_m, ft^{-1} $\times 10^{10}$	α, ft/lb $\times 10^{11}$
	$lb_f/in.^2$	lb_f/ft^2	s/l^2	s/ft^6	s/l	s/ft^3		
I	6.7	965	25.8	20,700	28.5	807	1.99	1.65
II	16.2	2,330	13.9	11,150	13.2	374	2.23	2.15
III	28.2	4,060	9.3	7,450	8.5	241	2.50	2.50
IV	36.3	5,230	7.7	6,170	7.0	198	2.65	2.67
V	49.1	7,070	6.4	5,130	5.5	156	2.82	3.00

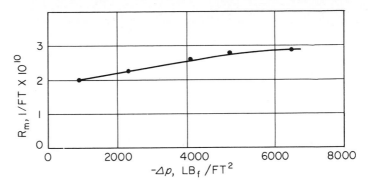

FIGURE 30-16
Plot of R_m vs. $-\Delta p$ for Example 30-2.

The viscosity of water is, from Appendix 14, 0.886 cP, or $0.886 \times 6.72 \times 10^{-4} =$ 5.95×10^{-4} lb/ft-s. The filter area is $440/30.48^2 = 0.474$ ft^2. The concentration c is $(23.5 \times 28.31)/454 = 1.47$ lb/ft^3.

From the values of K_p and B in Table 30-4, corresponding quantities for α and R_m are found from Eqs. (30-23) and (30-24). Thus

$$\alpha = \frac{A^2(-\Delta p)g_c K_p}{c\mu} = \frac{0.474^2 \times 32.17(-\Delta p)K_p}{5.95 \times 10^{-4} \times 1.47} = 8.28 \times 10^3(-\Delta p)K_p$$

$$R_m = \frac{A(-\Delta p)g_c B}{\mu} = \frac{0.474 \times 32.17(-\Delta p)B}{5.95 \times 10^{-4}} = 2.56 \times 10^4(-\Delta p)B$$

Table 30-4 shows the values of α and R_m for each test. Figure 30-16 is a plot of R_m vs. $-\Delta p$. Figure 30-17 is a logarithmic plot of α vs. $-\Delta p$. The points fall on a straight line, so

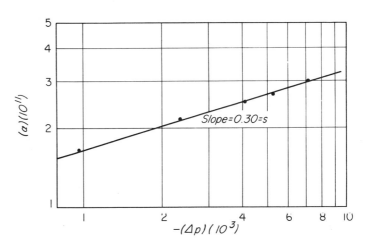

FIGURE 30-17
Logarithmic plot of α vs. $-\Delta p$ for Example 30-2.

Eq. (30-25) is suitable as an equation for α as a function of $-\Delta p$. The slope of the line is 0.30, and this is the value of s for the sludge. The sludge is moderately compressible.

Constant α_0 can be calculated by reading the coordinates of any convenient point on the line of Fig. 30-17 and calculating α_0 by Eq. (30-25). For example, when $-\Delta p = 1,000$, $\alpha = 1.65 \times 10^{11}$, and

$$\alpha_0 = \frac{1.65 \times 10^{11}}{1,000^{0.30}} = 2.08 \times 10^{10} \text{ ft/lb } (1.4 \times 10^{10} \text{ m/kg})$$

Equation (30-25) becomes for this sludge

$$\alpha = 2.08 \times 10^{10}(-\Delta p)^{0.30} \qquad\qquad ////$$

Continuous filtration   In a continuous filter, say of the rotary-drum type, the feed, filtrate, and cake move at steady constant rates. For any particular element of the filter surface, however, conditions are not steady but transient. Follow, for example, an element of the filter cloth from the moment it enters the pond of slurry until it is scraped clean once more. It is evident that the process consists of several steps in series—cake formation, washing, drying, and scraping—and that each step involves progressive and continual change in conditions. The pressure drop across the filter during cake formation is, however, held constant. Thus the foregoing equations for discontinuous constant-pressure filtration may, with some modification, be applied to continuous filters.

In continuous filtration the filter-medium resistance is often negligible in comparison with the resistance of the cake. In Eq. (30-22), therefore, B is negligible and may be omitted. Integrating the modified equation between limits gives

$$\int_0^t dt = K_p \int_0^V V \, dV$$

from which

$$t = \frac{K_p V^2}{2} \qquad\qquad (30\text{-}26)$$

where t is the time required for formation of the cake. In a continuous filter t is always less than the total cycle time t_c; the relationship is

$$t = ft_c \qquad\qquad (30\text{-}27)$$

where f is the fraction of the cycle available for cake formation. In a rotary drum filter f equals the fractional submergence of the drum in the slurry.

Substitution from Eqs. (30-23) and (30-27) in Eq. (30-26) and rearrangement give

$$\frac{V}{A} = \left[\frac{2(-\Delta p)g_c f t_c}{c\alpha\mu}\right]^{1/2} \qquad\qquad (30\text{-}28)$$

Dividing both sides of Eq. (30-28) by t_c gives the average rate of filtrate flow per unit area:

$$\text{Rate} = \frac{V}{A t_c} = \left[\frac{2(-\Delta p)g_c f}{c\alpha\mu t_c}\right]^{1/2} \qquad\qquad (30\text{-}29)$$

If the specific cake resistance varies with pressure drop according to Eq. (30-25), Eq. (30-29) may be modified to

$$
\frac{V}{At_c} = \left[\frac{2(-\Delta p)^{1-s} g_c f}{c\alpha_0 \mu t_c} \right]^{1/2}
\tag{30-30}
$$

When the filter-medium resistance is too large to be ignored, Eq. (30-29) becomes

$$
\frac{V}{At_c} = \frac{[2c\alpha(-\Delta p)g_c f/\mu t_c + (R_m/t_c)^2]^{1/2} - R_m/t_c}{c\alpha}
\tag{30-31}
$$

Equations (30-30) and (30-31) apply both to continuous vacuum filters and to continuous pressure filters. When R_m is negligible, Eq. (30-30) predicts that the filtrate flow rate varies inversely with the square root of the viscosity and of the cycle time. This has been observed experimentally with thick cakes and long cycle times;[13] with short cycle times, however, this is not true and the more complicated relationship shown in Eq. (30-31) must be used.[15] In general, as shown by Eqs. (30-29) and (30-31), the filtration rate increases as the drum speed increases and the cycle time t_c diminishes, because the cake formed on the drum face is thinner than at low drum speeds. At speeds above a certain critical value, however, the filtration rate no longer increases with speed but remains constant, and the cake tends to become wet and difficult to discharge.

The filter area required for a given filtration rate is calculated as shown in Example 30-3.

EXAMPLE 30-3   A rotary drum filter with 30 percent submergence is to be used to filter a concentrated aqueous slurry of $CaCO_3$ containing 14.7 lb of solids per cubic foot of water (236 kg/m³). The pressure drop is to be 20 in Hg. If the filter cake contains 50 percent moisture (wet basis), calculate the filter area required to filter 10 gal/min of slurry when the filter cycle time is 5 min. Assume that the specific cake resistance is the same as in Example 30-2 and that the filter-medium resistance R_m is negligible.

SOLUTION   Equation (30-30) will be used. The quantities needed for substitution are

$$
-\Delta p = 20 \frac{14.69}{29.92} \times 144 = 1{,}414 \; lb_f/ft^2
$$

$$
f = 0.30 \qquad t_c = 5 \times 60 = 300 \; s
$$

From Example 30-2

$$
\alpha_0 = 2.08 \times 10^{10} \; ft/lb \qquad s = 0.30
$$

Assume

$$
\mu = 1 \; cP = 6.72 \times 10^{-4} \; lb/ft\text{-}s \qquad p = 62.3 \; lb/ft^3
$$

Quantity c is found from Eq. (30-18). The slurry concentration c_s is 14.7 lb/ft³. Since the cake contains 50 percent moisture, $m_F/m_c = 2$. Substitution of these quantities in Eq. (30-18) gives

$$
c = \frac{14.7}{1 - (2 - 1)(14.7/62.3)} = 19.24 \; lb/ft^3
$$

Solving Eq. (30-30) for area A gives

$$A = \frac{V}{t_c} \left(\frac{c\alpha_0 \mu t_c}{2(-\Delta p)^{1-s} g_c f} \right)^{1/2} \tag{30-32}$$

The filtrate flow rate is V/t_c. The slurry feed rate is to be 10 gal/min; the filtrate flow rate equals the feed rate times the ratio c_s/c. Thus

$$\frac{V}{t_c} = \frac{10}{60} \frac{1}{7.48} \frac{14.7}{19.24} = 0.0170 \text{ ft}^3/\text{s}$$

Substitution in Eq. (30-32) gives

$$A = 0.0170 \left(\frac{19.3 \times 2.08 \times 10^{10} \times 6.72 \times 10^{-4} \times 300}{2 \times 1,414^{0.70} \times 32.174 \times 0.30} \right)^{1/2}$$

$$= 86.9 \text{ ft}^2 \ (8.07 \text{ m}^2) \qquad\qquad \textit{////}$$

Constant-rate filtration   If filtrate flows at a constant rate, the linear velocity u is constant and

$$u = \frac{dV/dt}{A} = \frac{V}{At} \tag{30-33}$$

Equation (30-15) can be written, after substituting m_c from Eq. (30-19) and u from Eq. (30-33), as

$$\frac{-\Delta p_c}{\alpha} = \frac{\mu c}{tg_c} \left(\frac{V}{A} \right)^2 \tag{30-34}$$

The specific cake resistance α is retained on the left-hand side of Eq. (30-34) because it is a function of Δp for compressible sludges.†

If α is known as a function of $-\Delta p_c$, and if $-\Delta p_m$, the pressure drop through the filter medium, can be estimated, Eq. (30-34) can be used directly to relate the overall pressure drop to time when the rate of flow of filtrate is constant. A more direct use of this equation can be made, however, if Eq. (30-25) is accepted to relate α and $-\Delta p_c$.[12] If α from Eq. (30-25) is substituted in Eq. (30-34), and if $-(\Delta p - \Delta p_m)$ is substituted for $-\Delta p_c$, the result is

$$(-\Delta p_c)^{1-s} = \frac{\alpha_0 \mu c t}{g_c} \left(\frac{V}{At} \right)^2 = [-(\Delta p - \Delta p_m)]^{1-s} \tag{30-35}$$

Again, the simplest method of correcting the overall pressure drop for the pressure drop through the filter medium is to assume the filter medium resistance is constant during a given constant-rate filtration. Then, by Eq. (30-16), $-\Delta p_m$ is also constant in Eq. (30-35).

---

† The concentration c may also vary somewhat with pressure drop. In operation, c_s rather than c is constant and, by Eq. (30-18), since m_F/m_c changes with pressure, c also changes when $[(m_F/m_c) - 1](c_s/\rho)$ is appreciable in comparison with unity. Any such variation in c with pressure drop can be ignored, in view of the other approximations made in the general theory of filtration.

The constants $-\Delta p_m$, α_0, and s can be evaluated by the following method from measurements of t vs. $-\Delta p$. Since the only variables in Eq. (30-35) are $-\Delta p$ and t, the equation can be written

$$[-(\Delta p - \Delta p_m)]^{1-s} = K_r t \tag{30-36}$$

where K_r is defined by

$$K_r = \frac{\mu u^2 c \alpha_0}{g_c} \tag{30-37}$$

Taking logarithms of both sides of Eq. (30-36) gives

$$\log t = (1 - s) \log \left[-(\Delta p - \Delta p_m) \right] - \log K_r \tag{30-38}$$

If $-(\Delta p - \Delta p_m)$ is plotted on logarithmic paper as the abscissa against t as the ordinate, a straight line should be obtained, the slope of which is $1 - s$. Since the actual data are the overall pressure drop $-\Delta p$ and the time t, the pressure drop $-\Delta p_m$ is not known and must be estimated. A tentative magnitude for $-\Delta p_m$ can be found by plotting t vs. $-\Delta p$ on rectangular coordinates, passing a smooth curve through the points, and extrapolating the curve to the pressure axis, where $t = 0$. A tentative result for $-\Delta p_m$ is thus found, which can be used for preparing a plot of $-(\Delta p - \Delta p_m)$ vs. t on logarithmic coordinates. If the line so obtained is straight, the tentative value of $-\Delta p_m$ can be taken as final. If the line is curved, additional approximations for $-\Delta p_m$ can be made until a straight line is achieved.

When a straight line has been obtained on the plot of $-(\Delta p - \Delta p_m)$ vs. t, the constant s is obtained from the slope of the line. The factor K_r is calculated, by means of Eq. (30-36), from the coordinates of any convenient point on the line, and α_0 is then evaluated by means of Eq. (30-37).

EXAMPLE 30-4   Table 30-5 shows data obtained in a constant-rate filtration of a sludge consisting of $MgCO_3$ and H_2O. The rate was 0.100 lb/ft²-s (0.49 kg/m²-s), the viscosity of the filtrate was 0.92 cP, and the concentration of solids in the slurry was 1.08 lb/ft³ (17.3 kg/m³) of filtrate. Evaluate the constants R_m, s, and α_0 for this sludge.

Table 30-5   PRESSURE-DROP TIME DATA IN
CONSTANT-RATE FILTRATION FOR
EXAMPLE 30-4[†]

$-\Delta p$, lb$_f$/in.²	t, s	$-\Delta p$, lb$_f$/in.²	t, s
4.4	10	11.8	70
5.0	20	13.5	80
6.4	30	15.2	90
7.5	40	17.6	100
8.7	50	20.0	110
10.2	60		

[†] From O. D. Hughes, R. W. Ver Hoeve, and C. D. Luke, paper given at meeting of AIChE, Columbus, Ohio, December 1950.

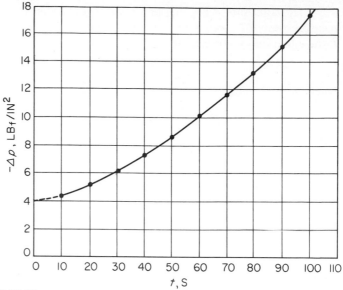

FIGURE 30-18
Plot of $-\Delta p$ vs. t for Example 30-4.

SOLUTION   To obtain a preliminary result for $-\Delta p_m$, $-\Delta p$ is plotted against t on rectangular coordinates, as shown in Fig. 30-18. Extrapolation to the pressure-drop axis gives $-\Delta p_m = 4$ lb$_f$/in.^{2} Figure 30-19 is a logarithmic plot of t vs. $-(\Delta p - 4)$. A satisfactory straight line is established by the points on Fig. 30-20 and the preliminary evaluation of $-\Delta p_m$ can be accepted as final. From Eq. (30-16), since $u = 0.100/62.3 = 0.0016$ ft/s,

$$R_m = \frac{-\Delta p_m g_c}{\mu u} = \frac{4 \times 144 \times 32.17}{0.92 \times 6.72 \times 10^{-4} \times 0.0016} = 1.9 \times 10^{10} \text{ ft}^{-1} \ (6.2 \times 10^{10} \text{ m}^{-1})$$

The slope of the line of Fig. 30-19, equal to $1 - s$, is 0.642, and $s = 0.358$.
From Fig. 30-19, when $-(\Delta p - \Delta p_m) = 10$ lb$_f$/in.^{2} (1,440 lb$_f$/ft^2), $t = 81$ s. From Eq. (30-36)

$$K_r = \frac{1,440^{0.642}}{81} = 1.32$$

From Eq. (30-37)

$$\alpha_0 = \frac{K_r g_c}{\mu u^2 c} = \frac{1.32 \times 32.17}{0.92 \times 6.72 \times 10^{-4} \times 0.100^2 \times 1.08}$$

$$= 6.36 \times 10^6 \text{ ft/lb} \ (4.29 \times 10^6 \text{ m/kg}) \qquad\qquad ////$$

Washing filter cakes   To wash soluble material that may be retained by the filter cake after a filtration, a solvent miscible with the filtrate may be used as a wash. Water is the most common wash liquid. The rate of flow of the wash liquid and the volume of liquid needed to reduce the solute content of the cake to a desired degree are important in the design and operation of a filter. Although the following general

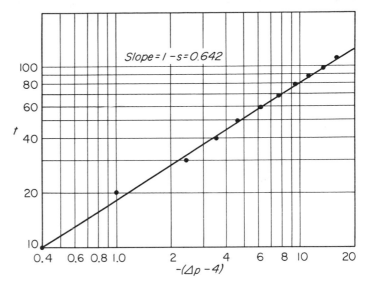

Slope = $1 - s = 0.642$

FIGURE 30-19
Logarithmic plot of t vs. $-(\Delta p - \Delta p_m)$ for Example 30-4.

principles apply to the problem, these questions cannot be completely answered without experiment.[6]

The volume of wash liquid required is related to the concentration-time history of the wash liquid leaving the filter. During washing the concentration-time relation is of the type shown in Fig. 30-20. The first portion of the recovered liquid is represented by segment ab in Fig. 30-20. The effluent consists essentially of the filtrate that was left on the filter, which is swept out by the first wash liquid without appreciable dilution. This stage of washing, called *displacement wash*, is the ideal method of washing a cake. Under favorable conditions where the particle size of the cake is small, as much as 90 percent of the solute in the cake can be recovered during this stage. The volume of wash liquid needed for a displacement wash is equal to the volume of filtrate left in the cake, or $\bar{\varepsilon}AL$, where L is the cake thickness and $\bar{\varepsilon}$ is the average porosity of the cake. The second stage of washing, shown by the segment bc in Fig. 30-20, is characterized by a rapid drop in concentration of the effluent. The volume of wash liquid used in this stage is also of the order of magnitude of that used in the first stage. The third stage is shown by segment cd. The concentration of solute in the effluent is low, and the remaining solute is slowly leached from the cake. If sufficient wash liquid is used, the residual solute in the cake can be reduced to any desired point, but the washing should be stopped when the value of the unrecovered solute is less than the cost of recovering it.

In most filters the wash liquid follows the same path as that of the filtrate.†

---

† In a filter press the wash passes through the entire thickness of the cake. The last filtrate passes through only one-half the final cake.

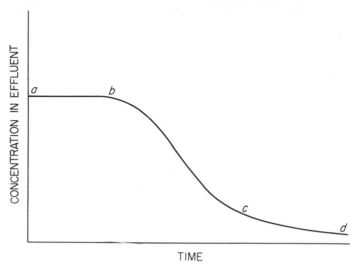

FIGURE 30-20
Washing filter cake.

The rate of flow of the wash liquid is, in principle, equal to that of the last of the filtrate, provided the pressure drop remains unchanged in passing from filtration to washing. If the viscosities of filtrate and wash liquid differ, correction for this difference can be made. This rule is only approximate, however, as during washing the liquid does not actually follow exactly the path of the filtrate because of channeling, formation of cracks in the cake, and short-circuiting.

Centrifugal Filtration

Solids which form a porous cake can be separated from liquids in a filtering centrifugal. Slurry is fed to a rotating basket having a slotted or perforated wall covered with a filter medium such as canvas or metal cloth. Pressure resulting from the centrifugal action forces the liquor through the filter medium, leaving the solids behind. If the feed to the basket is then shut off and the cake of solids spun for a short time, much of the residual liquid in the cake drains off the particles, leaving the solids much "drier" than those from a filter press or vacuum filter. When the filtered material must subsequently be dried by thermal means, considerable savings may result from the use of centrifuge.

The main types of filtering centrifugals are suspended batch machines, which are discontinuous in their operation; automatic short-cycle batch machines; and continuous conveyor centrifuges. In suspended centrifugals the filter media are canvas or other fabric, or woven metal cloth. In automatic machines fine metal screens are used; in conveyor centrifuges the filter medium is usually the slotted wall of the basket itself.

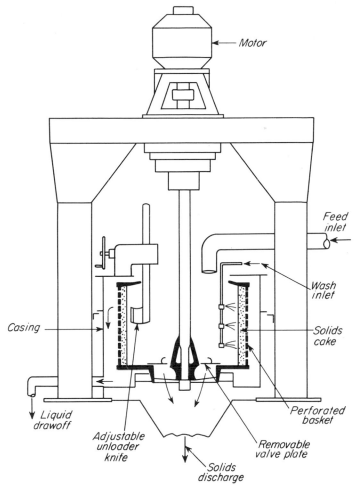

FIGURE 30-21
Top-suspended basket centrifugal.

Suspended batch centrifugals A common type of batch centrifugal in industrial processing is the top-suspended centrifugal shown in Fig. 30-21. The perforated baskets range from 30 to 48 in. in diameter and from 18 to 30 in. deep, and turn at speeds between 600 and 1,800 r/min. The basket is held at the lower end of a free-swinging vertical shaft driven from above. A filter medium lines the perforated wall of the basket. Feed slurry enters the rotating basket through an inlet pipe or chute. Liquor drains through the filter medium into the casing and out a discharge pipe: the solids form a cake 2 to 6 in. thick inside the basket. Wash liquid may be sprayed through the solids to remove soluble material. The cake is then spun as dry as possible, sometimes at a higher speed than during the loading and washing steps. The motor is

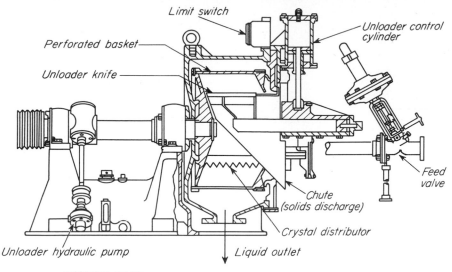

FIGURE 30-22
Automatic batch centrifugal.

shut off and the basket nearly stopped by means of a brake. With the basket slowly turning, at perhaps 30 to 50 r/min, the solids are discharged by cutting them out with an unloader knife, which peels the cake off the filter medium and drops it through an opening in the basket floor. The filter medium is rinsed clean, the motor turned on, and the cycle repeated.

Top-suspended centrifugals are used extensively in sugar refining, where they operate on short cycles of 2 to 3 min per load and produce up to 5 tons/h of crystals per machine. Automatic controls are often provided for some or all of the steps in the cycle. In most processes where large tonnages of crystals are separated, however, other automatic centrifugals or continuous conveyor centrifuges are used.

Another type of batch centrifugal is driven from the bottom, with the drive motor, basket, and casing all suspended from vertical legs mounted on a base plate. Solids are unloaded by hand through the top of the casing or plowed out through openings in the floor of the basket as in top-suspended machines. Except in sugar refining, suspended centrifugals usually operate on cycles of 10 to 30 min per load, discharging solids at a rate of 700 to 4,000 lb/h.

Automatic batch centrifugals A short-cycle automatic batch centrifugal is illustrated in Fig. 30-22. In this machine the basket rotates at constant speed about a horizontal axis. Feed slurry, wash liquid, and screen rinse are successively sprayed into the basket at appropriate intervals for controlled lengths of time. The basket is unloaded while turning at full speed by a heavy knife which rises periodically and cuts the solids out with considerable force through a discharge chute. Cycle timers and solenoid-operated valves control the various parts of the operation: feeding,

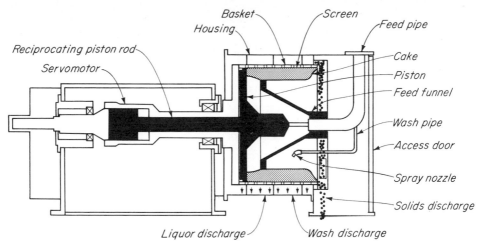

FIGURE 30-23
Reciprocating-conveyor continuous centrifuge. (*Baker-Perkins, Inc.*)

washing, spinning, rinsing, and unloading. Any part of the cycle may be lengthened or shortened as desired.

The basket in these machines is between 20 and 42 in. in diameter. Automatic centrifugals have high productive capacity with free-draining crystals. Usually they are not used when the feed contains many particles finer than 150-mesh. With coarse crystals the total operating cycle ranges from 35 to 90 s, so that the hourly throughput is large. Because of the short cycle and the small amount of holdup required for feed slurry, filtrate, and discharged solids, automatic centrifugals are easily incorporated into continuous manufacturing processes. The small batches of solid can be effectively washed with small amounts of wash liquid, and—as in any batch machine—the amount of washing can be temporarily increased to clean up off-quality material should it become necessary. Automatic centrifugals cannot handle slow-draining solids, which would give uneconomically long cycles, or solids which do not discharge cleanly through the chute. There is also considerable breakage or degradation of the crystals by the unloader knife.

Continuous filtering centrifugals   A continuous centrifugal separator for coarse crystals is the reciprocating-conveyor centrifuge shown in Fig. 30-23. A rotating basket with a slotted wall is fed through a revolving feed funnel. The purpose of the funnel is to accelerate the feed slurry gently and smoothly. Feed enters the small end of the funnel from a stationary pipe at the axis of rotation of the basket. It travels toward the large end of the funnel, gaining speed as it goes, and when it spills off the funnel onto the wall of the basket, it is moving in the same direction as the wall and at very nearly the same speed. Liquor flows through the basket wall, which may be covered with a woven metal cloth. A layer of crystals 1 to 3 in. thick is formed. This layer is moved over the filtering surface by a reciprocating pusher. Each stroke of the

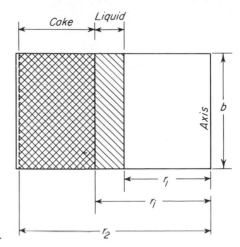

FIGURE 30-24
Centrifugal filter.

pusher moves the crystals a few inches toward the lip of the basket; on the return stroke a space is opened on the filtering surface in which more cake can be deposited. When the crystals reach the lip of the basket, they fly outward into a large casing and drop into a collector chute. Filtrate and any wash liquid that is sprayed on the crystals during their travel leave the casing through separate outlets. The gentle acceleration of the feed slurry and deceleration of the discharged solids minimize breakage of the crystals. Multistage units which minimize the distance of travel of the crystals in each stage are used with solid cakes which do not "convey" properly in a single-stage machine. Reciprocating centrifuges are made with baskets ranging in diameter from 12 to 48 in. They dewater and wash 0.3 to 25 ton/h of solids containing no more than about 10 percent by weight of material finer than 100-mesh.

Principles of centrifugal filtration The basic theory of constant-pressure filtration can be modified to apply to filtration in a centrifuge. The treatment applies after the cake has been deposited and during flow of clear filtrate or fresh water through the cake. Figure 30-24 shows such a cake. In this figure,

r_1 = radius of inner surface of liquid
r = radius of inner face of cake
r_2 = inside radius of basket

The following simplifying assumptions are made. The effects of gravity and of changes in kinetic energy of the liquid are neglected, and the pressure drop from centrifugal action equals the drag of the liquid flowing through the cake; the cake is completely filled with liquid; the flow of the liquid is laminar; the resistance of the filter medium is constant; and the cake is nearly incompressible, so an average specific resistance can be used as a constant.

In the light of these assumptions the flow rate of liquid through the cake is predicted as follows. Assume first that the area A for flow does not change with radius, as would nearly be true with a thin cake in a large-diameter centrifuge. The linear velocity of the liquid is then given by

$$u = \frac{dV/dt}{A} = \frac{q}{A} \tag{30-39}$$

where q is the volumetric flow rate of liquid. Substitution from Eq. (30-39) into Eq. (30-17) gives

$$-\Delta p = \frac{q\mu}{g_c}\left(\frac{m_c\alpha}{A^2} + \frac{R_m}{A}\right) \tag{30-40}$$

The pressure drop from centrifugal action, from Eq. (2-9), is

$$-\Delta p = \frac{\rho\omega^2(r_2^2 - r_1^2)}{2g_c} \tag{30-41}$$

where ω = angular velocity, rad/s
ρ = density of liquid

Combining Eqs. (30-40) and (30-41) and solving for q gives

$$q = \frac{\rho\omega^2(r_2^2 - r_1^2)}{2\mu(\alpha m_c/A^2 + R_m/A)} \tag{30-42}$$

When the change in A with radius is too large to be neglected, it can be shown that Eq. (30-42) should be written[10]

$$q = \frac{\rho\omega^2(r_2^2 - r_1^2)}{2\mu(\alpha m_c/\bar{A}_L\bar{A}_a + R_m/A_2)} \tag{30-43}$$

where A_2 = area of filter medium (inside area of centrifuge basket)
$\bar{A}_a$ = arithmetic mean cake area
$\bar{A}_L$ = logarithmic mean cake area

The average areas $\bar{A}_a$ and $\bar{A}_L$ are defined by the equations

$$\bar{A}_a \equiv (r_i + r_2)\pi b \tag{30-44}$$

$$\bar{A}_L \equiv \frac{2\pi b(r_2 - r_i)}{\ln(r_2/r_i)} \tag{30-45}$$

where b is the height of the basket. Note that Eq. (30-44) applies to a cake of definite mass and is *not* an integrated equation over an entire filtration starting with an empty centrifugal.

SEPARATIONS BASED ON THE MOTION OF PARTICLES THROUGH FLUIDS

Many methods for mechanical separation are based on the movement of solid particles or liquid drops through a fluid. The fluid may be gas or liquid, and it may be flowing or at rest. In some situations the objective of the process is to remove particles

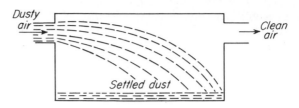

FIGURE 30-25
Dust-settling chamber.

from a stream of fluid in order to eliminate contaminants from the fluid or to recover the particles, as in the elimination of dust and fumes from air or flue gas or the removal of solids from liquid wastes. In other problems, particles are deliberately suspended in fluids to obtain separations of the particles into fractions differing in size or density. The fluid is then recovered, sometimes for reuse, from the fractionated particles.

The principles of particle mechanics that underlie the operations described here are discussed in Chap. 7. It is evident that if a particle starts at rest with respect to the fluid in which it is immersed and is then moved through the fluid by an external force, its motion can be divided into two stages. The first stage is a short period of acceleration, during which the velocity increases from zero to the terminal velocity. The second stage is the period during which the particle is at its terminal velocity.

Since the period of initial acceleration is short, usually of the order of tenths of a second or less, initial-acceleration effects are short-range. Terminal velocities, on the other hand, can be maintained as long as the particle is under treatment in the equipment. Equations such as (7-32), (7-34), and (7-39) apply during the acceleration period, and equations such as (7-49) and (7-57) during the terminal-velocity period. Some separation methods, such as jigging and tabling, depend on differences in particle behavior during the acceleration period. Most common methods, however, including all those described here, make use of the terminal-velocity period only.

Gravity Settling Processes

Removal of solids from gases Dust particles may be removed from gases by a variety of methods.[14c] For coarse solid particles, larger than about 200 mesh, a gravity settling chamber like that shown schematically in Fig. 30-25 is useful. Such a device is a large box at one end of which dust-laden air enters and from the other end of which clarified air leaves. In the absence of air currents, particles settle toward the floor at a speed equal to their terminal velocities. If the air remains in the chamber long enough, the particles reach the floor of the chamber, from which they can subsequently be removed. To prevent the airstream from lifting the particles from the floor and reentraining them, the air velocity should not be greater than about 10 ft/s.

EXAMPLE 30-5   It is desired to remove dust particles 50 μm in diameter from 8,000 ft^3/min (3.8 m^3/s) of air, using a settling chamber for the purpose. The air contains 0.5 grain of dust per cubic foot (1 g/m³) and the temperature and pressure are 70°F (21.1°C) at 1 atm. The particle density is 150 lb/ft³ (2,400 kg/m³). What minimum dimensions of the chamber are consistent with these conditions?

SOLUTION   To establish the range in which the particles are settling, Eq. (7-54) is used. The quantities needed are

$$D_p = \frac{50 \times 10^{-4}}{30.48} = 1.64 \times 10^{-4} \text{ ft}$$

Also, $\mu = 1.21 \times 10^{-5}$ lb/ft-s $\rho = 0.075$ lb/ft³ $\rho_p = 150$ lb/ft³

Then

$$K = \frac{1.64}{10^4} \left(\frac{32.17 \times 0.075 \times 150}{1.21^2 \times 10^{-10}} \right)^{1/3} = 2.22$$

Stokes' law must be used. The terminal velocity is, from Eq. (7-50),

$$u_t = \frac{32.17 \times 1.64^2 \times 10^{-8} \times 150}{18 \times 1.21 \times 10^{-5}} = 0.60 \text{ ft/s}$$

If V_c is the volume of the chamber, the time of residence for the air is $60 V_c/8{,}000 = V_c/133$ s. The maximum distance that a particle can settle during this time is $V_c u_t/133$ ft. If Z_c is the height of the chamber, for complete removal of the particles, those starting from the roof of the chamber must be given sufficient time to fall this same distance, or

$$Z_c = \frac{u_t V_c}{133} = \frac{0.60 V_c}{133} = \frac{V_c}{222} \qquad \text{ft}$$

Since V_c/Z_c is the floor area of the chamber, this area must be at least 222 ft² (20.6 m²).

The minimum cross-sectional area of the chamber is dictated by the maximum permissible velocity of the air. If this is 10 ft/s, the cross-sectional area of the chamber is 8,000/(60 × 10) = 13.3 ft² (1.23 m²).

Since the floor area is 222 ft² and the cross-sectional area is 13.3 ft², the ratio of length to height is 222/13.3 = 16.7. This is the only proportion of the chamber that is fixed by the conditions of the design. Any single dimension can be chosen at will, and the other two will follow. For example, if the height is taken as 3 ft (0.92 m) the length is 50 ft (15.2 m), and the breadth is 13.3/3 = 4.4 ft (1.34 m). The volume of the chamber is 3 × 50 × 4.4 = 660 ft³ (18.7 m³). The smaller the height, the shorter and wider the chamber and the smaller the volume. If the height is too small, however, it is difficult to remove the solids from the floor of the chamber.

The size of the chamber is independent of the dust load of the entering gas, but the heavier the load, the more difficult the cleaning and the practical operation of the chamber become. In this example, the dust load per unit ground area is, since 1 lb is 7,000 grains,

$$\frac{0.5 \times 60 \times 8{,}000}{7{,}000 \times 222} = 0.155 \text{ lb/ft}^2\text{-h} \ (0.756 \text{ kg/m}^2\text{-h})$$

Separation of coarse particles from liquids   A common problem in both chemical and metallurgical practice is that of separating relatively coarse particles, which are called *sands*, from a slurry of fine particles, which are called *slimes*. The most common method is to use continuous settling equipment called *classifiers*. Sufficient

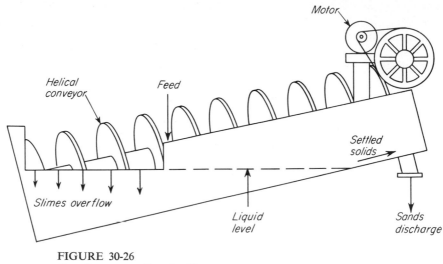

FIGURE 30-26
Crossflow wet-settling classifier.

time is provided to allow the sands to settle to the bottom of the device; the slimes leave in the effluent liquid. Either mechanical or gravitational means can be used to remove the sands continuously from the bottom of the classifier.

A typical classifier for coarse particles is shown in Fig. 30-26. In this device the settling vessel is an inclined trough with a liquid overflow at the lower end. Slurry is continuously fed to the trough at an intermediate point. The flow rate and slurry concentration are adjusted so that the fines do not have time to settle but are carried out with the liquid leaving the classifier. Larger particles sink to the floor of the trough, from which they must be removed. Different types of classifiers differ chiefly in the means by which they do this.

In the crossflow classifier illustrated in Fig. 30-26 the trough is semicylindrical, set at an angle of about 12° with the horizontal. A rotating helical conveyor moves the settled solids upward along the floor of the trough, out of the pool of liquid and up to the sands-discharge chute. Such a classifier works well with coarse particles where exact splits are not required. Typical applications are in connection with ball or rod mills for reduction to particle sizes between 8- and 20-mesh. These classifiers have high capacities. They lift coarse solids for return to the mill, so that auxiliary conveyors and elevators are not needed. For close separations with finer particles, however, other types of classifier must be used.

Removal of fine solids from liquids; sedimentation and thickening To remove relatively coarse sands, which have reasonable settling velocities, gravity classification under free or hindered settling is satisfactory. To remove fine particles of diameters of a few micrometers or less, settling velocities are too low, and for practicable operation the particles must be agglomerated, or flocculated, into large particles that do possess a reasonable settling speed.

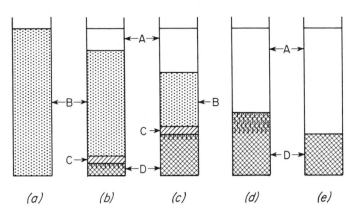

FIGURE 30-27
Batch sedimentation.

FLOCCULATION   Many slimes consist of particles that carry electric charges, which may be either positive or negative, and such particles, because of the mutual repulsion of like charges, tend to remain dispersed. If an electrolyte is added, the ions formed in solution neutralize the charges on the particles. The neutralized particles may then agglomerate to form flocs, each containing many particles. When the original particles are negatively charged, the cation of the electrolyte is effective, and when the charge is positive, the anion is active. In either situation, the greater the valence of the ion, the more effective the ion as a flocculation agent. Other methods of flocculation include the use of surface active agents and the addition of materials, such as glue, lime, alumina, or sodium silicate, that drag down the slime particles with them. In a properly treated slime the flocs are visible to the naked eye.

   Flocculated particles possess two important settling characteristics. The first characteristic is the complicated structure of the flocs. The aggregates are loose, the bond between the particles in them is weak, and they retain a considerable amount of water in their structures, which accompanies the flocs when they settle. Although initially flocs settle in either free or hindered settling and the usual equations apply in principle, it is not practical to use the laws of settling quantitatively, because the diameter and shape of a floc are not readily definable.

   The second characteristic of a flocculated pulp is the complexity of its settling mechanism. The history of the settling of a typical flocculated sludge is as follows.[4] Figure 30-27a shows a sludge uniformly distributed in the liquid and ready to settle. If there are no sands in the mixture, the first appearance of solids is the deposit, on the bottom of the settler, of flocs originating in the lower portion of the mixture. As shown in Fig. 30-27b, these solids, which consist of flocs resting lightly on one another, form a layer, called zone D. Above zone D forms another layer, called zone C, which is a transition layer, the solid content of which varies from that in the original pulp to that in zone D. Above zone C is zone B, which consists of a homogeneous suspension of the same concentration as that of the original pulp. Above zone B is zone A, which, if the particles have been fully flocculated, is a clear liquid. In well-

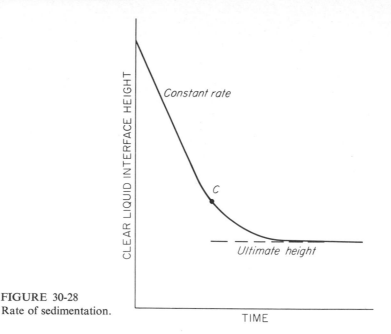

FIGURE 30-28
Rate of sedimentation.

flocculated pulps, the boundary between zones *A* and *B* is sharp. If unagglomerated particles remain, zone *A* is turbid and the boundary between zones *A* and *B* is hazy.

As settling continues, the depths of zones *D* and *A* increase, that of zone *C* remains constant, and that of zone *B* decreases. This is shown in Fig. 30-27*c*. After further settling, zones *B* and *C* disappear, and all the solids are in zone *D*. This is shown in Fig. 30-27*d*. Then a new effect, called *compression*, begins. The moment when compression first is evident is called the *critical point*. In compression, a portion of the liquid which has accompanied the flocs into the compression zone *D* is expelled when the weight of the deposit breaks down the structure of the flocs. During compression, some of the liquid in the flocs spurts out of zone *D* like small geysers, and the thickness of this zone decreases. Finally, when the weight of the solid reaches mechanical equilibrium with the compressive strength of the flocs, the settling process stops, as shown in Fig. 30-27*e*. At this time, the sludge has reached its ultimate height. The entire process shown in Fig. 30-27 is called *sedimentation*.

RATE OF SEDIMENTATION   A typical plot of height of sludge (the boundary between zones *A* and *B*) vs. time is shown in Fig. 30-28. During the early stage of settling the velocity is constant, as shown by the first portion of the curve. As solid accumulates in zone *D*, the rate of settling decreases and steadily drops until the ultimate height is reached. The critical point is reached at point *C* in Fig. 30-28.

Sludges vary greatly in their settling rates and in the relative heights of various zones during settling. Experimental study of each individual sludge is necessary to appraise its settling characteristics accurately.

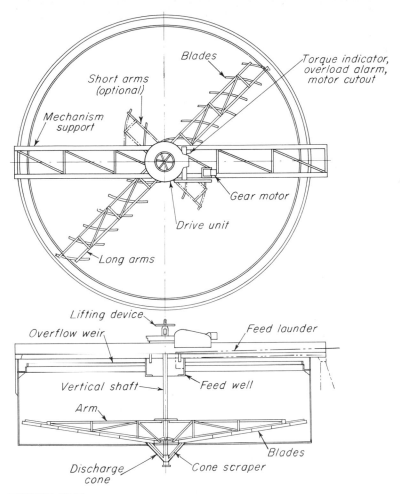

FIGURE 30-29
Gravity thickener. (*Eimco Corp.*)

EQUIPMENT FOR SEDIMENTATION; THICKENERS Industrially, the above process is conducted on a large scale in equipment called *thickeners*. For relatively fast-settling particles a batch settling tank or continuous settling cone may be adequate. For many duties, however, a mechanically agitated thickener like that shown in Fig. 30-29 must be employed. This is a large fairly shallow tank with slow-moving radial rakes driven from a central shaft. Its bottom may be flat or a shallow cone. Dilute feed slurry flows from an inclined trough or launder into the center of the thickener. Liquor moves radially at a constantly decreasing velocity, allowing the solids to settle to the bottom of the tank. Clear liquor spills over the edge of the tank into a launder. The rake arms gently agitate the sludge and move it to the center of the tank, where it flows through a large opening to the inlet of a sludge pump. In some designs of

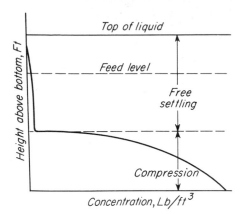

FIGURE 30-30
Solid concentrations in continuous
thickener. (*Comings*. From Ref. 5.)

thickener the rake arms are pivoted so that they can ride over an obstruction, such as
a hard lump of mud, on the tank floor.

Mechanically agitated thickeners are usually large, typically 30 to 300 ft in
diameter and 8 to 12 ft deep. In a large thickener the rakes may revolve once every
30 min. These thickeners are especially valuable when large volumes of dilute slurry
must be thickened, as in cement manufacture or the production of magnesium
from seawater. They are also used extensively in sewage treatment and in water
purification.

The volume of clear liquor produced in a unit time by a continuous thickener
depends primarily on the cross-sectional area available for settling and in industrial
separators is almost independent of the liquid depth. Higher capacities per unit area
of floor area are therefore obtained by using a multiple-tray thickener, with several
shallow settling zones, one above the other, in a cylindrical tank. Rake or scraper
agitators move the settled sludge downward from one tray to the next. Multistage
countercurrent displacement washing is possible in these devices. They are
considerably smaller in diameter, however, than single-stage thickeners.

SEDIMENTATION ZONES IN CONTINUOUS THICKENERS   In a continuous thickener, equipped with rakes to remove the underflow, the feed pulp is admitted at the centerline
of the unit at a depth of 2 or 3 ft below the surface of the liquid. On entrance, the
slurry spreads radially through the cross section of the thickener, the liquid then flows
upward to be withdrawn at the overflow launder, and the solids settle toward the
bottom. Two settling zones are established. The upper zone is free from particles in
the upper levels and increases slightly in solid content below the entrance of the
feed. This is shown by the relationship of solids concentration vs. height in Fig. 30-30.
Particles settle in this zone by free settling. Below the dilute zone is a compression
zone, in which the concentration of solids increases rapidly with distance from the
boundary between the zones. This zone corresponds to the compression zone D
in a batch thickener. Zones B and C of the batch process are not found in a continuous
thickener. The rakes, which operate in the bottom of the compression zone, tend to

break the floc structure and to compact the underflow to a solid content greater than that in the zone D of a batch thickener.

In practice, a clear overflow can be obtained if the upward velocity of the liquid in the dilute zone is less than the minimum terminal velocity of the solid at all points in the zone. The velocity of the liquid is proportional to the overflow rate, and if the overflow rate is too high, a cloudy overflow is obtained. If this happens, the equipment is being overloaded. The solid content and the degree of thickening in the underflow depend on the time of retention of the material in the compression zone, which is, in turn, proportional to the depth of this zone. Since the depth of the zone can be increased only with a deeper and more expensive tank, the economic limit is reached at an underflow concentration somewhat less than the ultimate obtainable with a long process time. Methods are available in the literature for estimating the surface area and depth of a gravity thickener for a given duty.[8a,14a]

Sorting classifiers Devices which separate particles of differing densities are known as *sorting classifiers*. They use one or the other of two principal separation methods—sink-and-float and differential settling.

SINK-AND-FLOAT METHODS A sink-and-float method uses a liquid sorting medium, the density of which is intermediate between that of the light material and that of the heavy. Then the heavy particles settle through the medium, and the lighter ones float, and a separation is thus obtained. This method has the advantage that, in principle, the separation depends only on the difference in the densities of the two substances and is independent of the particle size. This method is also called *heavy-fluid separation*.

Heavy-fluid processes are used to treat relatively coarse particles, usually greater than 10-mesh. The first problem in the use of sink and float is the choice of a liquid medium of the proper gravity to allow the light material to float and the heavy to sink. True liquids can be used, but since the specific gravity of the medium must be in the range 1.3 to 3.5 or greater, there are but few liquids that are sufficiently heavy, cheap, nontoxic, and noncorrosive to be practicable. Halogenated hydrocarbons are used for the purpose. Calcium chloride solutions are used for cleaning coal. A more common choice of medium is a pseudo liquid consisting of a suspension in water of fine particles of a heavy mineral. Magnetite (specific gravity = 5.17), ferro-silicon (specific gravity = 6.3 to 7.0), and galena (specific gravity = 7.5) are used. The ratio of mineral to water can be varied to give a wide range of medium densities. Provision must be made for feeding the mixture to be separated, for removing overflow and underflow, and for recovering the separating fluid, which may be expensive relative to the value of the materials being treated. In the process particles fall or rise at their terminal velocities, and hindered settling is used. Cleaning coal and concentrating ores are the common application of sink and float. Under proper conditions, clean separations between materials differing in specific gravity by only 0.1 have been claimed.[20]

DIFFERENTIAL-SETTLING METHODS Differential settling methods utilize the difference in terminal velocities that can exist between substances of different density. The density of the medium is less than that of either substance. The disadvantage of the method is that since the mixture of materials to be separated covers a range of particle sizes, the larger, light particles settle at the same rate as the smaller, heavy ones and a mixed fraction is obtained.

In differential settling, both light and heavy materials settle through the same medium. This method brings in the concept of equal-settling particles. Consider particles of two materials A and B settling through a medium of density ρ_m. Let material A be the heavier; e.g., component A might be galena (specific gravity $= 7.5$) and component B quartz (specific gravity $= 2.65$). The terminal velocity of a particle of size D_p and of density ρ_p settling under gravity through a medium of density ρ_m is given by Eq. (7-57), using $a_e = g$. This equation can be written, for a galena particle of density ρ_{pA} and diameter D_{pA}, as

$$u_{tA} = \left[\frac{4g D_{pA}^{1+n} \varepsilon^{2-n} (_p\psi)^n (\rho_{pA} - \rho_m)}{3b_1 \rho_m^{1-n} \mu^n} \right]^{1/(2-n)} \tag{30-46}$$

The terminal velocity of a quartz particle of density ρ_{pB} and diameter D_{pB} is

$$u_{tB} = \left[\frac{4g D_{pB}^{1+n} \varepsilon^{2-n} (\psi_p)^n (\rho_{pB} - \rho_m)}{3b_1 \rho_m^{1-n} \mu^n} \right]^{1/(2-n)} \tag{30-47}$$

Although $\rho_{pA} > \rho_{pB}$ and a galena particle of definite size settles faster than a quartz particle of the same size, a larger quartz particle can have the same velocity as a smaller galena one. A relationship between the diameters of such equal-settling particles is found by assuming $u_{tA} = u_{tB}$, equating the right-hand sides of Eqs. (30-46) and (30-47), canceling common factors, and obtaining

$$\frac{D_{pA}}{D_{pB}} = \left(\frac{\rho_{pB} - \rho_m}{\rho_{pA} - \rho_m} \right)^{1/(1+n)} \tag{30-48}$$

Particles of substances A and B whose diameters conform to Eq. (30-48) are called *equal-settling particles*.

If the particles settle in the Stokes'-law range, $n = 1$, and Eq. (30-48) becomes

$$\frac{D_{pA}}{D_{pB}} = \sqrt{\frac{\rho_{pB} - \rho_m}{\rho_{pA} - \rho_m}} \tag{30-49}$$

The significance, in a separation process, of the equal-settling ratio of diameters is shown by Fig. 30-31, in which curves of u_t vs. D_p are plotted for components A and B, for Stokes' law settling. Assume that the diameter range of both substances lies between points D_{p1} and D_{p4} on the size axis. Then, all particles of the light component B having diameters between D_{p1} and D_{p2} will settle more slowly than any particle of the heavy substance A, and can be obtained as a pure fraction. Likewise, all particles of heavy material A having diameters between D_{p3} and D_{p4} settle faster than any particle of substance B, and can also be obtained as a pure fraction. But any light particle having a diameter between D_{p2} and D_{p4} settles at the same speed as a particle of component A in a size range from D_{p1} to D_{p3}, and all particles in these size ranges form a mixed fraction.

Inspection of Eq. (30-48) shows that the sharpness of separation is improved

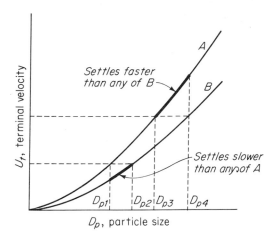

FIGURE 30-31
Equal-settling particles.

if the density of the medium is increased. Since, in hindered settling, the apparent density of the medium ρ_m is that of the entire suspension and is greater than ρ, the density of the liquid itself, hindered settling is more effective than free settling in separation. The rate of hindered settling is less, however, than that of free settling. It is also clear from Fig. 30-31 that the intermediate fraction can be reduced or eliminated by closer sizing of the feed. For example, if the size range of the feed is from D_{p3} to D_{p4} in Fig. 30-31, complete separation is possible.

Centrifugal Settling Processes

A given particle in a given fluid settles under gravitational force at a fixed maximum rate. To increase the settling rate the force of gravity acting on the particle may be replaced by a much stronger centrifugal force. Centrifugal separators have to a considerable extent replaced gravity separators in production operations because of their greater effectiveness with fine drops and particles and their much smaller size for a given capacity.

Separation of solids from gases; cyclones Most centrifugal separators for removing particles from gas streams contain no moving parts. They are typified by the cyclone separator shown in Fig. 30-32. It consists of a vertical cylinder with a conical bottom, a tangential inlet near the top, and an outlet for dust at the bottom of the cone. The inlet is usually rectangular. The outlet pipe is extended into the cylinder to prevent short-circuiting of air from inlet to outlet.

The incoming dust-laden air receives a rotating motion on entrance to the cylinder. The vortex so formed develops centrifugal force, which act to throw the particles radially toward the wall. Basically, a cyclone is a settling device in which a strong centrifugal force, acting radially, is used in place of relatively weak gravitational force acting vertically. The centrifugal force in a cyclone is from 5 times gravity in large, low-velocity units to 2,500 times gravity in small, high-pressure units.[14e]

The path of the air in a cyclone follows a downward vortex, or spiral, adjacent

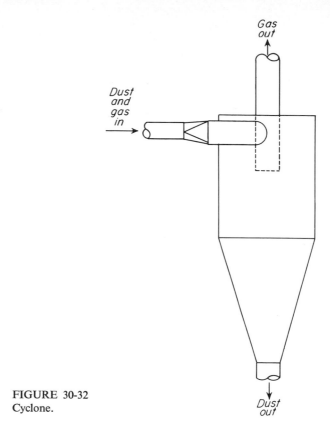

FIGURE 30-32
Cyclone.

to the wall and reaching to the bottom of the cone. The airstream then moves upward in a tighter spiral, concentric with the first, and leaves through the outlet pipe, still whirling. Both spirals rotate in the same direction.

In a cyclone, dust particles quickly reach the terminal velocities corresponding to their sizes and their radial position in the cyclone, in accordance with Eq. (7-50). The radial acceleration in a cyclone depends on the radius of the path being followed by the air, and is given by the empirical equation [19]

$$a_e = \omega^2 r = \frac{b_2}{r^n} \tag{30-50}$$

where b_2 and n are constants and the exponent n lies between 2.0 and 2.4. Combining Eqs. (7-50) and (30-50) gives, for the terminal velocity of a particle of diameter D_p rotating around the axis of the cyclone at a distance r from the center,

$$u_t = \frac{b_2 D_p^2 (\rho_p - \rho)}{18 \mu r^n} \tag{30-51}$$

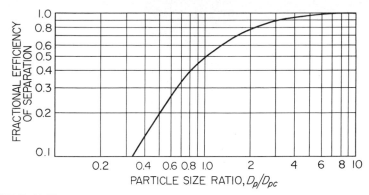

FIGURE 30-33
Efficiency vs. particle-size ratio for cyclones. [*By permission, from J. II. Perry (ed.), "Chemical Engineers' Handbook," 5th ed., p. 20-85. Copyright © 1973, McGraw-Hill Book Company.*]

For a given particle size, the terminal velocity is a maximum in the inner vortex, where r is small, and the finest particles separated from the air are eliminated in the inner vortex. These pass through the outer vortex to the wall of the cyclone and drop through the dust outlet. Smaller particles, which do not have time to reach the wall, are retained by the air and accompany it to the outlet, while larger ones are eliminated early in the travel of the air through the unit. Although the chance of a particle for separation decreases with the square of the particle diameter, the fate of a particle depends also on its position in the cross section of the entering stream and on its history in the cyclone, so the separation according to size is not sharp.

A definite diameter, called the cut diameter, can be defined as that diameter for which one-half the inlet particles, by mass, are separated and the other half retained by the air. Also, the efficiency of separation for a definite particle size can be defined as the mass fraction of the particles of that size that is separated and collected by the unit. A typical relation between efficiency and particle diameter for a definite cyclone is shown in Fig. 30-33. The abscissa is the ratio of the diameter of the particle to the cut diameter, and the ordinate is the efficiency. For example, only 78 percent of the particles twice the size of the cut diameter are removed, and of the particles one-half the cut diameter, 20 percent are separated. Typical cut diameters for high-efficiency cyclones are in the range of 3 to 10 μm.

Cyclones are also extensively used for separating solids from liquids, especially for purposes of classification.[21]

Centrifugal decanters Immiscible liquids are separated industrially in centrifugal decanters. Their operation is similar to that of the gravity decanters described in Chap. 2 except that the separating force is much larger than that of gravity and it acts in the direction away from the axis of rotation instead of downward toward the earth's surface. The main types of centrifugal decanters are tubular centrifuges and disk centrifuges.

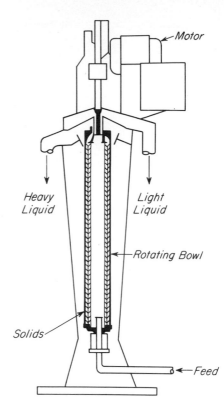

FIGURE 30-34
Tubular centrifuge.

TUBULAR CENTRIFUGE   A tubular liquid-liquid centrifuge is shown in Fig. 30-34. The bowl is tall and narrow, 4 to 6 in. in diameter, and turns in a stationary casing at about 15,000 r/min. Feed enters from a stationary nozzle inserted through an opening in the bottom of the bowl. It separates into two concentric layers of liquid inside the bowl. The inner, or lighter layer spills over a weir at the top of the bowl; it is thrown outward into a stationary discharge cover and from there to a spout. Heavy liquid flows over another weir into a separate cover and discharge spout. The weir over which the heavy liquid flows is removable and may be replaced with another having an opening of a different size. The position of the liquid-liquid interface (the neutral zone) is maintained by a hydraulic balance as in the gravity decanter shown in Fig. 2-6. In some designs the liquids discharge under pressure and the interface position is set by adjusting external valves in the discharge lines.

DISK CENTRIFUGE   For some liquid-liquid separations the disk-type centrifuge illustrated in Fig. 30-35 is highly effective. A short, wide bowl 8 to 20 in. in diameter turns on a vertical axis. The bowl has a flat bottom and a conical top. Feed enters from above through a stationary pipe set into the neck of the bowl. Two liquid layers are formed as in a tubular centrifuge; they flow over adjustable dams into separate discharge spouts. Inside the bowl and rotating with it are closely spaced "disks,"

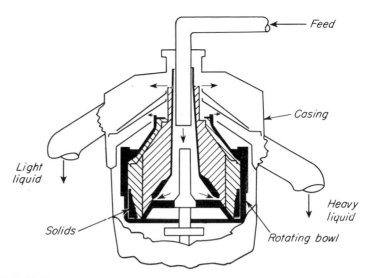

FIGURE 30-35
Disk centrifuge.

which are actually cones of sheet metal set one above the other. Matching holes in the disks about halfway between the axis and the wall of the bowl form channels through which the liquids pass. In operation feed liquid enters the bowl at the bottom, flows into the channels, and upward past the disks. Heavier liquid is thrown outward, displacing lighter liquid toward the center of the bowl. In its travel the heavy liquid very soon strikes the underside of a disk and flows beneath it to the periphery of the bowl without encountering any more light liquid. Light liquid similarly flows inward and upward over the upper surfaces of the disks. Since the disks are closely spaced, the distance a drop of either liquid must travel to escape from the other phase is short, much shorter than in the comparatively thick liquid layers in a tubular centrifuge. In addition, in a disk machine there is considerable shearing at the liquid-liquid interface, as one phase flows in one direction and the other phase in the opposite direction. This shearing helps break certain types of emulsions. Disk centrifuges are particularly valuable where the purpose of the centrifuging is not complete separation but the concentration of one fluid phase, as in the separation of cream from milk and the concentration of rubber latex.

Centrifugal sedimentation   If the liquid fed to a disk or tubular centrifuge contains dirt or other heavy solid particles, the solids accumulate inside the bowl and must periodically be discharged. This is accomplished by stopping the machine, removing and opening the bowl, and scraping out its load of solids. This becomes uneconomical if the solids are more than a few percent of the feed.

Tubular and disk centrifuges are used to advantage for removing traces of solids from lubricating oil, process liquids, ink, and beverages that must be perfectly

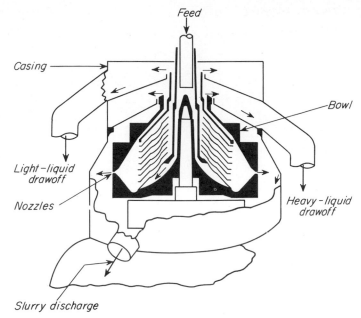

FIGURE 30-36
Nozzle-discharge centrifuge.

clean. They can take out gelatinous or slimy solids that would quickly plug a filter. Usually they clarify a single liquid in a bowl provided with but a single liquid overflow; however, they also may throw down solids while simultaneously separating two liquid phases.

NOZZLE-DISCHARGE CENTRIFUGE   When the feed liquid contains more than a few percent of solids means must be provided for discharging the solids automatically. One way of doing this is shown in Fig. 30-36. This separator is a modified disk-type centrifuge with a double conical bowl. In the periphery of the bowl at its maximum diameter is a set of small holes, or nozzles, perhaps 3 mm in diameter. The central part of the bowl operates in the same way as the usual disk centrifuge, overflowing either one or two streams of clarified liquid. Solids are thrown to the periphery of the bowl and escape continuously through the nozzles, together with considerable liquid. In some designs part of the slurry discharge from the nozzles is recycled through the bowl to increase its concentration of solids; wash liquid may also be introduced into the bowl for displacement washing. In still other designs the nozzles are closed most of the time by plugs or valves which open periodically to discharge a moderately concentrated slurry.

Sludge separators   In a nozzle-discharge centrifuge the solids leave the bowl from below the liquid surface and therefore carry with them considerable quantities of liquid. For separating a feed slurry into a clear liquid fraction and a heavy "dry"

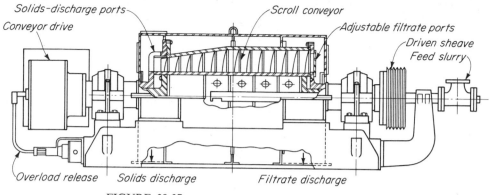

FIGURE 30-37
Cylindrical-conical helical-conveyor centrifuge. (*Bird Machine Co.*)

sludge, the settled solids must be moved mechanically from the liquid and given a chance to drain while still under centrifugal force. This is done in continuous sludge separators, a typical example of which is illustrated in Fig. 30-37. In this helical-conveyor centrifuge a cylindrical bowl with a conical end section rotates about a horizontal axis. Feed enters through a stationary axial pipe, spraying outward into a "pond" or annular layer of liquid inside the cylindrical bowl. Clarified liquid flows through overflow ports in the plate covering the nonconical end of the bowl. The radial position of these ports fixes the thickness of the annular layer of liquid in the bowl. Solids settle through the liquid to the inner surface of the bowl; a helical conveyor turning slightly slower than the bowl moves the solids out of the pond and up the "beach" to discharge openings in the small end of the cone. Wash liquid may be sprayed on the solids as they travel up the beach, to remove soluble impurities. The wash flows into the pond and discharges with the liquor. Drained sludge and clarified liquor are thrown out from the bowl into different parts of the casing, from which they leave through suitable openings.

Helical-conveyor centrifuges are made with maximum bowl diameters from 4 to 54 in. They separate large amounts of material. An 18-in. machine, for example, might handle 1 to 2 tons of solids per hour; a 54-in. machine, 50 tons/h. With thick feed slurries the capacity of a given machine is limited by the allowable torque on the conveyor. With dilute slurries the liquid-handling capacity of the bowl and overflow ports limits the throughput.

Practical operation of a sludge separator, of course, requires that the solids be heavier than the liquid and not be resuspended by the action of the conveyor. The liquid effluent from these machines is usually not completely free from solids, and may require subsequent clarification. Within these restrictions sludge separators solve a wide variety of problems. They separate fine particles from liquids, dewater and wash free-draining crystals, and are often used as classifiers, as discussed later.

Principles of centrifugal sedimentation In a sedimenting centrifuge a particle of given size is removed from the liquid if sufficient time is available for the particle to

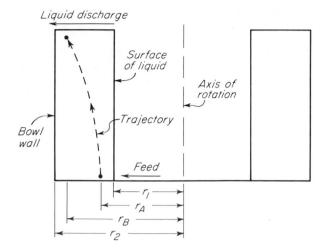

FIGURE 30-38
Particle trajectory in sedimenting centrifuge.

reach the wall of the separator bowl. If it is assumed that the particle is at all times moving radially at its terminal velocity, the diameter of the smallest particle that should just be removed can be calculated.

Consider the volume of liquid in a centrifuge bowl shown in Fig. 30-38. The feed point is at the bottom, and the liquid discharge is at the top. Assume that all the liquid moves upward through the bowl at a constant velocity, carrying solid particles with it. A given particle, as shown in the figure, begins to settle at the bottom of the bowl at some position in the liquid, say at a distance r_A from the axis of rotation. Its settling time is limited by the residence time of the liquid in the bowl; at the end of this time let the particle be at a distance r_B from the axis of rotation. If $r_B < r_2$, the particle leaves the bowl with the liquid; if $r_B = r_2$, it is deposited on the bowl wall and removed from the liquid. If the particle settles in the Stokes-law range, the terminal velocity at radius r is, by Eq. (7-50),

$$u_t = \frac{\omega^2 r(\rho_p - \rho)D_p^2}{18\mu}$$

Since $u_t = dr/dt$,

$$dt = \frac{18\mu}{\omega^2(\rho_p - \rho)D_p^2}\frac{dr}{r} \tag{30-52}$$

Integrating Eq. (30-52) between the limits $r = r_A$ at $t = 0$ and $r = r_B$ at $t = t_T$ gives

$$t_T = \frac{18\mu}{\omega^2(\rho_p - \rho)D_p^2}\ln\frac{r_B}{r_A} \tag{30-53}$$

Residence time t_T is equal to the volume of liquid in the bowl V divided by the volumetric flow rate q. Volume V equals $\pi b(r_2^2 - r_1^2)$. Substitution into Eq. (30-53) and rearrangement gives

$$q = \frac{\pi b \omega^2 (\rho_p - \rho)D_p^2}{18\mu} \frac{r_2^2 - r_1^2}{\ln(r_B/r_A)} \tag{30-54}$$

A cut point can be defined[1] as the diameter of that particle which just reaches one-half the distance between r_1 and r_2. If D_{pc} is the cut diameter, a particle of this size moves a distance $y = (r_2 - r_1)/2$ during the settling time allowed. If a particle of diameter D_{pc} is to be removed, it must reach the bowl wall in the available time. Thus $r_B = r_2$ and $r_A = (r_1 + r_2)/2$. Equation (30-54) then becomes

$$q_c = \frac{\pi b \omega^2 (\rho_p - \rho)D_{pc}^2}{18\mu} \frac{r_2^2 - r_1^2}{\ln[2r_2/(r_1 + r_2)]} \tag{30-55}$$

where q_c is the volumetric flow rate corresponding to the cut diameter. At this flow rate most particles whose diameters are larger than D_{pc} will be eliminated by the centrifuge, and most particles having smaller diameters will remain in the liquid.

If the thickness of the liquid layer is small compared to the radius of the bowl, $r_1 \approx r_2$, and Eq. (30-55) becomes indeterminate. Under these conditions, however, the settling velocity may be considered constant and given by the equation

$$u_t = \frac{D_p^2(\rho_p - \rho)\omega^2 r_2}{18\mu} \tag{30-56}$$

Let the thickness of the liquid layer be s and the settling distance for particles of cut diameter D_{pc} be $s/2$. Then

$$u_t = \frac{s}{2t_T} \tag{30-57}$$

where t_T is the residence time. For a thin layer of liquid t_T is given by

$$t_T = \frac{V}{q_c} = \frac{2\pi r_2 bs}{q_c} \tag{30-58}$$

Combining Eqs. (30-56) to (30-58) and solving for q_c gives

$$q_c = \frac{2\pi b \omega^2 (\rho_p - \rho)D_{pc}^2 r_2^2}{9\mu} \tag{30-59}$$

This analysis is somewhat oversimplified, since the pattern of fluid flow in a centrifuge bowl is more complicated than assumed in Fig. 30-38. In a helical-conveyor bowl, for example, it has been shown[18] that rapid fluid motion is confined to a surface layer of liquid perhaps 1 to 2 mm thick and that the rest of the liquid in the "pond," except for the agitation provided by the conveyor, is essentially stagnant.

Centrifugal classifiers   During the passage of the liquid through a centrifuge bowl the heavier, larger solid particles are thrown out of the liquid. Finer, lighter particles may not settle in the time available and be carried out with the liquid effluent. As in a gravity hydraulic classifier, solid particles can be sorted according to size, shape, or specific gravity.

Continuous helical-conveyor centrifuges are used as classifiers in closed circuit with wet-grinding ball mills. Their operation is like that of the sludge separators except that the bowl speed and slurry feed rate are adjusted and controlled so that acceptably fine particles leave with the liquid in the form of a dilute slurry. Oversize particles are conveyed to the discharge as a sludge, repulped, and returned to the mill for further grinding. Since long settling times and a fairly wet sludge are desirable, the pond of liquid in this type of classifier nearly fills the rotating bowl.

Note the similarity in operating principle of these machines and that of the crossflow classifier shown in Fig. 30-26. In the centrifugal classifier the separating force is of the order of 600 times the force of gravity, permitting sharp separations of particles $1\mu m$ or less in diameter. Much coarser particles than this, however, are also classified in centrifugal machines.

The high settling force in a centrifuge means that practical settling rates can be obtained with much smaller particles than in a gravity classifier. While the higher force does not change the relative settling velocities of small particles, it does overcome the small but disturbing effects of free convection currents and Brownian motion in a gravity classifier, and it permits separation in some cases where none is possible in a gravity unit. With coarse particles the settling regime may be changed, so that a particle that settles by gravity according to Stokes' law may settle according to the intermediate or Newton's law in a centrifuge. Thus mixtures of equal-settling particles from a gravity unit may sometimes be partially separated in a centrifuge. On the other hand, loose flocs or weak agglomerates which settle rapidly in a gravity thickener are often broken up in a centrifugal classifier and settle slowly or not at all despite the increased force for sedimentation.

Impingement Methods

For removing dust or fog from gas streams, a variety of methods is used which depend, entirely or partially, on impingement of the particles against solid surfaces placed in the flowing stream. In these methods, the particles, because of their inertia, are expected to cross the streamlines of the fluid and strike and adhere to the solid, from which they can subsequently be removed. The principle of impingement separation is shown in Fig. 30-39. The solid lines are the streamlines passing around a sphere, and the dotted lines show the paths followed by the particles. Particles initially moving along the streamlines between A and B strike the solid and can be removed if they adhere to the wall and are not reentrained. Particles initially following streamlines outside lines A and B do not strike the solid and cannot be removed from the gas stream by impingement. The *target efficiency* η_t is defined as the fraction of the particles in the gas stream that strike the solid. For particles that would settle through still fluid in the Stokes'-law range, the target efficiencies for ribbons, spheres, and cylinders are shown in Fig. 30-40. The abscissa is the dimensionless group, called the *separation number*, $u_t u_0/g D_b$, where u_t is the terminal velocity of the particle in still

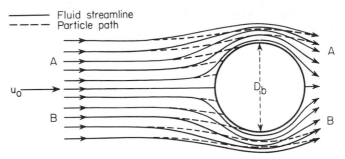

FIGURE 30-39
Principle of impingement. [*By permission, from J. H. Perry (ed.), "Chemical Engineers' Handbook,"* 5th ed., p. **20**-80. *Copyright* © 1973, *McGraw-Hill Book Company.*]

fluid; u_0 is the velocity of the fluid approaching the solid; g is the acceleration of gravity; and D_b is the width of the ribbon or the diameter of the sphere or cylinder.

In settling in the Stokes'-law range, the terminal velocity u_t is proportional to D_p^2. Thus the smaller the particle, the lower the target efficiency. For each solid shape, a minimum separation number exists, below which all particles follow streamlines around the solid, impingement does not occur, and the target efficiency is zero. The target efficiency approaches unity for separation numbers of 100 or more.

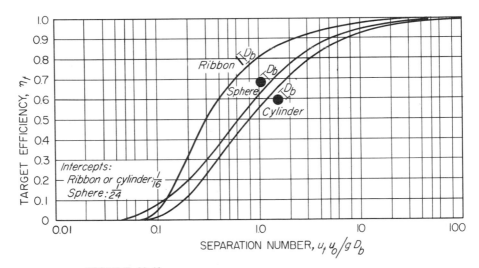

FIGURE 30-40
Target efficiency of spheres, cylinders, and ribbons. [*By permission, from J. H. Perry (ed.), "Chemical Engineers' Handbook,"* 5th ed., p. **20**-81. *Copyright* © 1973, *McGraw-Hill Book Company.*]

Actual equipment using impingement is built in a wide variety of forms. The simplest device, used especially in treating acid mists, consists of a box or tower filled with an inert broken solid, such as coke, through which the gas passes with a low velocity. Tappers are frequently used to clear solids from the impingement surfaces. In bag filters, the initial formation of the cake in the filter cloth or other medium is by impingement. Once the first layer of solid is formed, subsequent deposition of solid is by filtration.

Scrubbers, in which a stream of liquid is passed over the impingement surfaces to wash away the precipitated particles; spray chambers, in which drops of liquid are used as the impingement agent; combination spray chambers and cyclones; and packed towers of the type described in Chap. 23—these constitute a second class of impingement separators.

A third kind of equipment used to remove dust from air includes a variety of impingement filters. Glass fibers, wood shavings, metal screens, or corrugated paper can be used as filter media. In one group, the filter medium is covered with a viscous oil, to retain dust and eliminate reentrainment. Metallic filters can be cleaned and reused, but nonmetallic units are usually discarded without cleaning. The ordinary household air filter is an example of viscous-fluid throwaway filters. Dry filters, which have smaller channels than the viscous type, are used for industrial applications.

SYMBOLS

A Area, ft^2 or m^2; filter area; A_1, area of inner surface of material in centrifuge; A_2, area of outer surface of material in centrifuge; $\bar{A}_L$, logarithmic mean of A_1 and A_2; $\bar{A}_a$, arithmetic mean of A_1 and A_2

a_e Acceleration of particle from external forces, ft/s^2 or m/s^2

B Constant in constant-pressure filtration equation, defined by Eq. (30-24); also, underflow from screen, lb/h or kg/h

b Width of centrifuge basket, ft or m

b_1 Constant in settling law

b_2 Constant in Eq. (30-51)

c Mass of solid deposited in filter per volume of filtrate, lb/ft^3 or kg/m^3

c_s Concentration of solids in slurry to filter, lb/ft^3 or kg/m^3

D Overflow from screen, lb/h or kg/h

D_b Width or diameter of impingement target, ft or m

D_p Particle size, ft or m; D_{pA}, diameter of heavy particle; D_{pB}, diameter of light particle; D_{pc}, cut diameter

E Screen effectiveness, dimensionless; E_A, based on oversize; E_B, based on undersize

F Feed to screen, lb/h or kg/h

f Fraction of filter cycle available for cake formation

g Acceleration of gravity, ft/s^2 or m/s^2

g_c Newton's-law proportionality factor, 32.174 ft-lb/lb$_f$-s^2

K Criterion for settling, dimensionless

K_p Constant in constant-pressure filtration equation, defined by Eq. (30-23)

K_r Constant in constant-rate filtration equation, defined by Eq. (30-37)

k_1 Constant in Eq. (30-9); $\bar{k}_1$, average value

L Distance in cake measured from filter medium, ft or m; L_c, cake thickness

m Mass, lb or kg; m_F, mass of wet filter cake; m_c, mass of solid in filter cake

n Exponent

p — Pressure, lb_f/ft^2 or atm; pressure in cake at distance L from filter medium; p_a, at inlet to filter; p_b, at discharge from filter; p_s, stress pressure; p', at boundary between cake and medium in filter

q — Volumetric flow rate, ft^3/s or m^3/s; q_c, corresponding to removal of particles of cut diameter

R_m — Filter-medium resistance, ft^{-1} or m^{-1}

r — Radius, ft or m; r_i, of interface between cake and liquid layer in centrifuge; r_1, inner radius of material in centrifuge; r_2, outer radius of material in centrifuge

s — Thickness of liquid layer in centrifuge, ft or m; also compressibility coefficient [Eq. (30-25)]

s_p — Surface area of single particle, ft^2 or m^2; $\bar{s}_p$, average value

t — Time, s; t_T, residence time in centrifuge; t_c, total cycle time in continuous filter

u — Linear velocity, ft/s or m/s; of liquid through filter; u_t, terminal velocity of settling particle; u_{tA}, terminal velocity of heavy particle; u_{tB}, terminal velocity of light particle; u_0, of undisturbed fluid approaching solid

V — Volume, ft^3, m^3, or l; of filtrate collected to time t; V_c, of settling chamber; $\bar{V}$, average filtrate (cumulative) in time increment Δt

v_p — Volume of single particle, ft^3 or m^3; $\bar{v}_p$, average value

x — Mass fraction of cut in mixture of particles; x_B, in underflow from screen; x_D, in overflow from screen; x_F, in feed to screen

y — Distance, variable, traveled by particle, ft or m

Z_c — Height of settling chamber, ft or m

Greek letters

α — Specific cake resistance, ft/lb or m/kg; α_0, constant in Eq. (30-25)

$-\Delta p$ — Overall pressure drop through filter, lb_f/ft^2 or atm, $p_a - p_b$; $-\Delta p_c$, pressure drop through cake, $p_a - p'$; $-\Delta p_m$, pressure drop through filter medium, $p' - p_b$

ΔV — Volume increment of filtrate, ft^3 or m^3

Δt — Time increment, s

ε — Porosity or volume fraction voids in bed of solids, or volume fraction of liquid in slurry, dimensionless; $\bar{\varepsilon}$, average porosity of filter cake

η_t — Target efficiency, impingement separator

μ — Viscosity, lb/ft-s or P

ρ — Density, lb/ft^3 or kg/m^3; of fluid or filtrate; ρ_m, of slurry; ρ_p, of particle; ρ_{pA}, of heavy particle; ρ_{pB}, of light particle

ϕ — Mass fraction in screen analysis, cumulative to particle size D_p

ψ_p — Factor in hindered settling [Eq. (30-46)]

ω — Angular velocity, rad/s

PROBLEMS

30-1 It is desired to separate a mixture of crystals into three fractions, a coarse fraction retained on an 8-mesh screen, a middle fraction passing an 8-mesh but retained on a 14-mesh screen, and a fine fraction passing a 14-mesh. Two screens in series are used, an 8-mesh and a 14-mesh, conforming to the Tyler standard. Screen analyses of feed, coarse, medium, and fine fractions are given in Table 30-6. Assuming the analyses are accurate, what do they show about the ratio by weight of each of the three fractions actually obtained? What is the overall effectiveness of each screen?

30-2 The screens used in Prob. 30-1 are shaking screens with a capacity of 4 tons/ft²-h-mm mesh size. How many square feet of screen are needed for each of the screens in Prob. 30-1 if the feed to the first screen is 100 tons/h?

30-3 The data in Table 30-7 were taken in a constant-pressure filtration of a slurry of $CaCO_3$ in H_2O. The filter was a 6-in. filter press with an area of 1.0 ft². The mass fraction

Table 30-6   SCREEN ANALYSES FOR PROB. 30-1

Screen	Feed	Coarse fraction	Middle fraction	Fine fraction
3/4	3.5	14.0		
4/6	15.0	50.0	4.2	
6/8	27.5	24.0	35.8	
8/10	23.5	8.0	30.8	20.0
10/14	16.0	4.0	18.3	26.7
14/20	9.1		10.2	20.2
20/28	3.4		0.7	19.6
28/35	1.3			8.9
35/48	0.7			4.6
Total	100.0	100.0	100.0	100.0

of solids in the feed to the press was 0.139. Calculate the values of α, R_m, and cake thickness for each of the experiments.

30-4   The sludge of Prob. 30-3 is to be filtered in a press having a total area of 10 m^2 and operated at a constant pressure drop of 2 atm. The frames are 40 mm thick. Assume that the filter-medium resistance in the large press is the same as that in the laboratory filter. Calculate the filtration time required and the volume of filtrate obtained in one cycle.

30-5   Assuming that the actual rate of washing is 80 percent of the theoretical rate, how long will it take to wash the cake in the press of Prob. 30-4 with a volume of wash water equal to that of the filtrate?

Table 30-7   DATA FROM CONSTANT-PRESSURE FILTRATION†

5-lb$_f$/in.^{2} pressure drop		15-lb$_f$/in.^{2} pressure drop		30-lb$_f$/in.^{2} pressure drop		50-lb$_f$/in.^{2} pressure drop	
Mass ratio wet cake/ dry cake	1.59		1.47		1.47		1.47
Dry-cake density	63.5		73.0		73.0		73.5
Filtrate, lb	Time, s	Filtrate, lb	Time, s	Filtrate, lb	Time, s	Filtrate, lb	Time, s
0	0	0	0	0	0	0	0
2	24	5	50	5	26	5	19
4	71	10	181	10	98	10	68
6	146	15	385	15	211	15	142
8	244	20	660	20	361	20	241
10	372	25	1,009	25	555	25	368
12	524	30	1,443	30	788	30	524
14	690	35	2,117	35	1,083	35	702
16	888						
18	1,188						

† E. L. McMillen and H. A. Webber, *Trans. AIChE*, **34**:213 (1938).

30-6   A continuous rotary vacuum filter operating with a pressure drop of 0.7 atm is to handle the feed slurry of Prob. 30-3. The drum submergence is to be 25 percent. What total filter area must be provided to match the overall productive capacity of the filter press described in Prob. 30-4?

30-7   The following relation between α and Δp for Superlight $CaCO_3$ has been determined:[10]

$$\alpha = 8.8 \times 10^{10}[1 + 3.36 \times 10^{-4}(-\Delta p)^{0.86}]$$

where $-\Delta p$ is in pounds force per square foot. This relation is followed over a pressure range from 0 to 1,000 $lb_f/in.^2$ A slurry of this material giving 3.0 lb of cake solid per cubic foot of filtrate is to be filtered at a constant pressure drop of 80 $lb_f/in.^2$ and a temperature of 70°F. Experiments on this sludge and the filter cloth to be used gave a value of $R_m = 1.5 \times 10^{10}$ ft^{-1}. A pressure filter of the tank type is to be used. How many square feet of filter surface are needed to give 1,500 gal of filtrate in a 1-h filtration?

30-8   The filter of Prob. 30-7 is washed at 70°F and 80 $lb_f/in.^2$ with a volume of wash water equal to one-third that of the filtrate. The washing rate is 80 percent of the theoretical value. How long should it take to wash the cake?

30-9   The filter of Prob. 30-7 is operated at a constant rate of 0.5 gal/ft²-min from the start of the run until the pressure drop reaches 80 $lb_f/in.^2$ and then at a constant pressure drop of 80 $lb_f/in.^2$ until a total of 1,500 gal of filtrate is obtained. The operating temperature is 70°F. What is the total filtration time required?

30-10   A continuous pressure filter is to yield 1,500 gal/h of filtrate from the slurry described in Prob. 30-7. The pressure drop is limited to a maximum of 50 $lb_f/in.^2$ How much filter area must be provided if the drum submergence is to be 30 percent?

30-11   An aqueous slurry of calcium carbonate is to be settled in a continuous sedimentation unit. The particle size is uniform at 50 μm, and the particles may be considered spheres. The temperature is 25°C. The specific gravity of the particles is 2.93. At what consistency (fraction of solids by volume) will the settling capacity, in pounds per square foot of settling area, be a maximum? What is the capacity under these conditions?

30-12   Air carrying particles of diameter 50 μm and density 60 lb/ft^3 enters a cyclone at a linear velocity of 50 ft/s. The temperature is 70°F. The diameter of the cyclone is 24 in., and the diameter of the outlet is 8 in. The diameter corresponding to the centerline of the inlet connection is 20 in. Assuming that the inlet velocity equals that at the diameter of 20 in. and that the exponent n of Eq. (30-51) is 2.0, what is the terminal velocity of the particles at the radius of the outlet connection?

30-13   Quartz and galena are to be separated by differential hydraulic classification. The mixture has a size range of 65- to 200-mesh. What fractions may be expected if free-settling conditions are used? This mixture is subjected to hindered settling in a sorting bed of specific gravity 2.0 having equal volumes of quartz and galena. What fractions may be expected? Compare quantitatively the velocities of the particles under hindered settling with those under free settling.

30-14   A sink-and-float process is used to separate jasper (quartz) from hematite by means of a ferrosilicon medium. The specific gravities of hematite and ferrosilicon are 5.1 and 6.7, respectively. At what consistency, in volume percent solids, should the medium be maintained?

30-15   The dust-laden air of Prob. 30-12 is passed through an impingement separator at a linear velocity of 20 ft/s. The separator consists essentially of ribbons 1 in. wide. What is the maximum fraction of the particles that can be removed in this manner?

30-16   What is the capacity in cubic meters per hour of a clarifying centrifuge operating under the following conditions?

Diameter of bowl, 600 mm   Specific gravity of liquid, 1.3
Thickness of liquid layer, 75 mm   Specific gravity of solid, 1.6
Depth of bowl, 400 mm   Viscosity of liquid, 3 cP
Speed, 1,000 r/min   Cut size of particles, 30 μm

30-17 A batch centrifugal filter having a bowl diameter of 30 in. and a bowl height of 18 in. is used to filter a suspension having the following properties:

Liquid, water   Final thickness of cake, 6 in.
Temperature, 25°C   Speed of centrifuge, 2,000 r/min
Concentration of solid in feed, 60 g/l   Specific cake resistance, 9.5×10^{10} ft/lb
Porosity of cake, 0.835   Filter medium resistance, 2.6×10^{10} ft^{-1}
Density of dry solid in cake, 125 lb/ft^3

The final cake is washed with water under such conditions that the radius of the inner surface of the liquid is 8 in. Assuming that the rate of flow of wash water equals the final rate of flow of filtrate, what is the rate of washing in gallons per minute?

REFERENCES

1   Ambler, C. M.: *Chem. Eng. Prog.,* **48:**150 (1952).
2   Bullock, H. L.: *Chem. Eng.,* **62**(6):185 (1955).
3   Chalmers, J. M., L. R. Elledge, and H. F. Porter: *Chem. Eng.,* **62**(6):191 (1955).
4   Coe, F. S., and G. H. Clevenger: *Trans. AIME,* **55:**356 (1916).
5   Comings, E. W.: *Ind. Eng. Chem.,* **32:**633 (1940).
6   Crozier, H. E., and L. E. Brownell: *Ind. Eng. Chem.,* **44:**631 (1952).
7   Fitch, B.: *Chem. Eng.,* **78**(19):83 (1971).
8   Foust, A. S., L. A. Wenzel, C. W. Clump, L. Maus, and L. B. Andersen: "Principles of Unit Operations," Wiley, New York, 1960; (*a*) pp. 471–472, (*b*) p. 529.
9   Gaudin, A. M.: "Principles of Mineral Dressing," p. 144, McGraw-Hill, New York, 1939.
10 Grace, H. P.: *Chem. Eng. Prog.,* **49:**303, 367, 427 (1953).
11 Grace, H. P.: *AIChE J.,* **2:**307, 316 (1956).
12 Hughes, O. D., R. W. Ver Hoeve, and C. D. Luke: paper given at meeting of AIChE, Columbus, Ohio, December 1950.
13 Nickolaus, N., and D. A. Dahlstrom: *Chem. Eng. Prog.,* **52**(3):87M (1956).
14 Perry, J. H. (ed.): "Chemical Engineers' Handbook," 5th ed., McGraw-Hill, New York, 1973; (*a*) p. **19**-44, (*b*) p. **19**-63, (*c*) p. **20**-74, (*d*) p. **20**-80, (*e*) pp. **20**-81 to **20**-86, (*f*) p. **21**-39.
15 Rushton, A., and M. S. Hameed: *Filtr. Sep.,* **7:**25 (1970).
16 Ruth, B. F.: personal communication.
17 Scheidegger, A. E.: "The Physics of Flow through Porous Media," Macmillan, New York, 1960.
18 Schnittger, J. R.: *Ind. Eng. Chem. Process Des. Dev.,* **9**(3):407 (1970).
19 Shepherd, C. B., and C. E. Lapple: *Ind. Eng. Chem.,* **31:**972 (1939); **32:**1246 (1940).
20 Taggart, A. F.: "Handbook of Mineral Dressing: Ores and Industrial Minerals," p. **11**-123, Wiley, New York, 1945.
21 Tangel, O. F., and R. J. Brison, *Chem. Eng.,* **62**(6):234 (1955).

Appendixes

CGS AND SI PREFIXES FOR MULTIPLES AND SUBMULTIPLES

Factor	Prefix	Abbreviation	Factor	Prefix	Abbreviation
10^{12}	tera	T	10^{-1}	deci	d
10^9	giga	G	10^{-2}	centi	c
10^6	mega	M	10^{-3}	milli	m
10^3	kilo	k	10^{-6}	micro	μ
10^2	hecto	h	10^{-9}	nano	n
10^1	deka	da	10^{-12}	pico	p
			10^{-15}	femto	f
			10^{-18}	atto	a

APPENDIX 2

VALUES OF GAS CONSTANT

Temperature	Mass	Energy	R
Kelvins	kg mol	J	8,314.3
		cal_{IT}	1.9858×10^3
		cal	1.9872×10^3
		m^3-atm	82.056×10^{-3}
	g mol	cm^3-atm	82.056
Degrees Rankine	lb mol	Btu	1.9858
		ft-lb_f	1,545.3
		Hp-h	7.8045×10^{-4}
		kWh	5.8198×10^{-4}

CONVERSION FACTORS AND CONSTANTS OF NATURE

To convert from	To	Multiply by†
acre	ft^2	43,560*
	m^2	4,046.85
atm	N/m^2	1.01325* $\times$ 10^5
	lb$_f$/in.^{2}	14.696
Avogadro number	particles/g mol	6.022169 $\times$ 10^{23}
barrel (petroleum)	ft^3	5.6146
	gal (U.S.)	42*
	m^3	0.15899
bar	N/m^2	1* $\times$ 10^5
	1b$_f$/in.^{2}	14.504
Boltzmann constant	J/K	1.380622 $\times$ 10^{-23}
Btu	cal$_{IT}$	251.996
	ft-lb$_f$	778.17
	J	1,055.06
	kWh	2.9307 $\times$ 10^{-4}
Btu/lb	cal$_{IT}$/g	0.55556
Btu/lb-°F	cal$_{IT}$/g-°C	1*
Btu/ft^2-h	W/m^2	3.1546
Btu/ft^2-h-°F	W/m^2-°C	5.6783
Btu-ft/ft^2-h-°F	W-m/m^2-°C	1.73073
cal$_{IT}$	Btu	3.9683 $\times$ 10^{-3}
	ft-lb$_f$	3.0873
	J	4.1868*
cal	J	4.184*
cm	in.	0.39370
	ft	0.0328084
cm^3	ft^3	3.531467 $\times$ 10^{-5}
	gal (U.S.)	2.64172 $\times$ 10^{-4}
cP (centipoise)	kg/m-s	1* $\times$ 10^{-3}
	lb/ft-h	2.4191
	lb/ft-s	6.7197 $\times$ 10^{-4}

(Continued overleaf)

To convert from	To	Multiply by†
cSt (centistoke)	m^2/s	$1* \times 10^{-6}$
faraday	C/g mol	9.648670×10^4
ft	m	$0.3048*$
ft-lb$_f$	Btu	1.2851×10^{-3}
	cal$_{IT}$	0.32383
	J	1.35582
ft-lb$_f$/s	Btu/h	4.6262
	hp	1.81818×10^{-3}
ft^2/h	m^2/s	2.581×10^{-5}
	cm^2/s	0.2581
ft^3	cm^3	2.8316839×10^4
	gal (U.S.)	7.48052
	l	28.31684
ft^3-atm	Btu	2.71948
	cal$_{IT}$	685.29
	J	2.8692×10^3
ft^3/s	gal (U.S.)/min	448.83
gal (U.S.)	ft^3	0.13368
	in.^{3}	$231*$
gravitational constant	N-m^2/kg^2	6.673×10^{-11}
gravity acceleration, standard	m/s^2	$9.80665*$
h	min	$60*$
	s	$3,600*$
hp	Btu/h	2,544.43
	kW	0.74570
in.	cm	$2.54*$
in.^{3}	cm^3	16.3871
J	erg	$1* \times 10^7$
	ft-lb$_f$	0.73756
kg	lb	2.20462
kWh	Btu	3,412.1
l	m^3	$1* \times 10^{-3}$
lb	kg	$0.45359237*$
lb/ft^3	kg/m^3	16.018
	g/cm^3	0.016018
lb$_f$/in.^{2}	N/m^2	6.89473×10^3
lb mol/ft^2-h	kg mol/m^2-s	1.3652×10^{-3}
	g mol/cm^2-s	1.3652×10^{-4}
light, speed of	m/s	2.997925×10^8
m	ft	3.280840
	in.	39.3701
m^3	ft^3	35.3147
	gal (U.S.)	264.17
N	dyn	$1* \times 10^5$
	lb$_f$	0.22481
N/m^2	lb$_f$/in.^{2}	1.4498×10^{-4}
Planck constant	J-s	6.626196×10^{-34}
proof (U.S.)	percent alcohol by volume	0.5
ton (long)	kg	1,016
	lb	$2,240*$
ton (short)	lb	$2,000*$
ton (metric)	kg	$1,000*$
	lb	2,204.6
yd	ft	$3*$
	m	$0.9144*$

† Values that end in * are exact, by definition.

DIMENSIONAL ANALYSIS

A physical quantity leads a double life: it has a number giving its magnitude and a unit representing its physical meaning. Dimensional analysis is an algebraic treatment of the symbols for units considered independently of magnitude. Dimensional analysis drastically simplifies the task of fitting experimental data to design equations where a completely mathematical treatment is not possible; it is also useful in checking the consistency of the units in equations, in converting units, and in the scale-up of data obtained in physical models to predict the performance of full-scale equipment. The method is based on the concept of dimension and the use of dimensional formulas.

Primary and secondary quantities   For dimensional analysis, physical quantities are divided into two groups. First, a small number of them are chosen; second, the remainder is expressed in terms of the first group. The quantities of the first kind are called primary quantities, and those of the second kind are called secondary quantities. The choice of primary quantities is quite arbitrary, both in number and in type, and is based largely on convenience. A satisfactory minimum list of primary quantities for all engineering is length, mass, time, temperature, and electric charge. In unit operations electric charge is not needed, and it is convenient, but not essential, to add force and heat to the list. These additions are to facilitate the use of fps units. The SI system does not need them.†

DIMENSIONS

A primary quantity is represented by a letter, called the dimension of that quantity, which symbolizes the quantity generally. It designates the set of all units that have been, are, or may be used to measure that quantity. For example, let $\bar{L}$ be the dimension of length. It is the set

$$\bar{L} = \{\text{foot, inch, meter, mile, rod, yard, cubit, ...}\}$$

where the dots represent all other length units not listed. Similar definitions are made for the dimensions of mass $\bar{M}$, time $\bar{t}$, temperature $\bar{T}$, force $\bar{F}$, and heat $\bar{H}$.

Criterion for primary quantity   The units in the set defining a primary quantity must meet one requirement: a constant conversion factor must exist between any two units contained in the set. This clearly is true of the length units listed in the definition of $\bar{L}$. Another example is the dimension of absolute temperature, which may be written

$$\bar{T} = \{\text{K, °R}\}$$

The units in this set are interconvertible by Eq. (1-28). But

$$\bar{T} \neq \{\text{°C, °F}\}$$

There is no constant k that gives $T°F = kT°C$. The true relation between these quantities is Eq. (1-30), which requires two constants, 32 and 1.8. If, however, temperature differences, rather than absolute temperatures, are needed in a given situation, in view of Eq. (1-31) the dimension of temperature may be defined by

$$\bar{T} = \{\Delta T°C, \Delta T°F\}$$

The mole is included in $\bar{M}$, as the molecular and atomic weights are conversion factors between mass and mole units.

DIMENSIONS OF SECONDARY QUANTITIES

The dimensions of a secondary quantity show how that quantity is constructed from primary quantities. Thus, any velocity, regardless of unit or of magnitude, is found by dividing a length by a time, and the dimensions of velocity are, then, $\bar{L}/\bar{t}$, or $\bar{L}\bar{t}^{-1}$. Likewise, the dimensions of acceleration are $\bar{L}/\bar{t}^2$, or $\bar{L}\bar{t}^{-2}$. The dimensions of pressure p (force divided by area) are $\bar{F}\bar{L}^{-2}$.

DIMENSIONAL FORMULAS   The dimensions of any quantity may be shown by the use of square brackets, thus

$$[L] = \bar{L} \qquad [p] = \bar{F}\bar{L}^{-2}$$

---

† Primary quantities are not to be confused with base units, which are SI units based on standards established by the international standardizing body, The General Conference of Weights and Measures. Primary quantities are the optional choices made for the purpose of a specific dimensional analysis by a given investigator. The dimensions of primary units usually do include the quantities mass, length, time, and sometimes temperature, but this is a matter of convenience.

The second equation, for example, says: The dimensions of pressure are force times length to the minus 2.

Once the primary quantities have been chosen, the dimensional formula for any secondary quantity can be found from its definition, just as those for velocity, acceleration, and pressure were written above. The general result for any quantity G may be written as

$$[G] = \bar{L}^{\alpha}\bar{M}^{\beta}\bar{t}^{\gamma}\bar{F}^{\delta}\bar{T}^{\varepsilon}\bar{H}^{\zeta} \qquad (A\text{-}1)$$

The exponents α, β, γ, δ, ε, and ζ are always positive or negative integers, small positive or negative integral fractions, or zero.

METHOD OF DIMENSIONAL ANALYSIS

A dimensional analysis cannot be made unless enough is known about the physics of the situation to decide what variables are important in the problem and what basic physical laws would be involved in a mathematical solution if one were possible. The basic laws are important because such laws introduce dimensional constants that must be considered along with the list of variables. In the applications of dimensional analysis considered in this book, two such constants may appear if fps units are used. One is g_c, which must be introduced whenever Newton's law [Eq. (1-35)] is involved; the other is J, the mechanical equivalent of heat, which must be used when the heat originating in the conversion of mechanical energy to heat by friction is important to the problem. Deciding what dimensional factors and variables enter the problem is the definitive step in a dimensional analysis. The method is illustrated by the following example.

EXAMPLE A-1   A steady stream of liquid is heated by passing it through a long, straight, heated pipe. The temperature of the wall of the pipe is assumed to be greater by a constant amount than the average temperature of the liquid. The conversion of mechanical energy into heat by friction is negligible in comparison with the heat transferred to the liquid through the wall of the pipe. It is desired to find a relationship that can be used to predict the rate of heat transfer from the wall to the liquid, in Btu per square foot of tube area in contact with the liquid per hour. Assume turbulent flow occurs.

SOLUTION   The mechanism of this process is discussed in Chap. 12. From the known characteristics of the process it may be expected that the rate of heat transfer per unit area q/A depends on a number of quantities, which are listed with their dimensional formulas in Table A-1.

It is known that viscosity forces and forces needed to accelerate and decelerate eddies are involved; so Newton's law enters the situation. The dimensional constant g_c is therefore included in the list of factors. Since conversion of mechanical energy to heat is negligible, J is not needed.

If a theoretical equation for this problem exists, it can be written in the general form

$$\frac{q}{A} = \psi(D,\ \bar{V},\ \rho,\ \mu_F,\ g_c,\ c_p,\ k,\ \Delta T) \qquad (A\text{-}2)$$

where ψ means "function of."

If Eq. (A-2) is a relationship derivable from basic laws, all terms in the function ψ must have the same dimensions as those of the left-hand side of the equation, q/A. Then any term in the function must conform to the dimensional formula

$$\left[\frac{q}{A}\right] = [D]^{a}[\bar{V}]^{b}[\rho]^{c}[\mu_F]^{d}[g_c]^{e}[c_p]^{f}[k]^{g}[\Delta T]^{h} \qquad (A\text{-}3)$$

Substituting the dimensions from Table A-1 gives

$$\bar{H}\bar{L}^{-2}\bar{\imath}^{-1} = \bar{L}^a\bar{L}^b\bar{\imath}^{-b}\bar{M}^c\bar{L}^{-3c}\bar{F}^d\bar{\imath}^d\bar{L}^{-2d}\bar{M}^e\bar{L}^e\bar{F}^{-e}\bar{\imath}^{-2e}\bar{H}^f\bar{M}^{-f}\bar{T}^{-f}\bar{H}^g\bar{L}^{-g}\bar{\imath}^{-g}\bar{T}^{-g}\bar{T}^h \quad \text{(A-4)}$$

Since Eq. (A-2) is assumed to be dimensionally homogeneous, the exponents of the individual primary units on the left-hand side of Eq. (A-4) must equal those on the right-hand side. This gives the following set of equations:

Exponents of $\bar{H}$:	$1 = f + g$	(A-5)
Exponents of $\bar{L}$:	$-2 = a + b - 3c - 2d + e - g$	(A-6)
Exponents of $\bar{\imath}$:	$-1 = -b + d - 2e - g$	(A-7)
Exponents of $\bar{M}$	$0 = c + e - f$	(A-8)
Exponents of $\bar{F}$:	$0 = d - e$	(A-9)
Exponents of $\bar{T}$	$0 = -f - g + h$	(A-10)

Here there are eight unknowns but only six equations. Six of the unknowns may be found in terms of the remaining two. Arbitrarily, two letters must be retained. The final result is equally valid for all choices, but for this problem it is customary to retain the exponents for the velocity $\bar{V}$ and the specific heat c_p. The letters b and f will be retained, and the remaining six eliminated. One method of doing this is as follows. From Eq. (A-5)

$$g = 1 - f \quad \text{(A-11)}$$

From Eqs. (A-10) and (A-11)

$$h = f + g = 1 \quad \text{(A-12)}$$

From Eq. (A-9)

$$d = e \quad \text{(A-13)}$$

From Eqs. (A-7) and (A-13)

$$2e - d = 1 - b - g = e = d \quad \text{(A-14)}$$

From Eqs. (A-11) and (A-14)

$$d = e = 1 - b - 1 + f = f - b \quad \text{(A-15)}$$

Table A-1 QUANTITIES AND DIMENSIONAL FORMULAS FOR EXAMPLE A-1

Quantity	Symbol	Dimensions
Heat flow per unit area	q/A	$\bar{H}\bar{L}^{-2}\bar{\imath}^{-1}$
Diameter of pipe (inside)	D	$\bar{L}$
Average velocity of liquid	$\bar{V}$	$\bar{L}\bar{\imath}^{-1}$
Density of liquid	ρ	$\bar{M}\bar{L}^{-3}$
Viscosity of liquid	μ_F	$\bar{F}\bar{L}^{-2}\bar{\imath}$
Specific heat, at constant pressure, of liquid	c_p	$\bar{H}\bar{M}^{-1}\bar{T}^{-1}$
Thermal conductivity of liquid	k	$\bar{H}\bar{L}^{-1}\bar{\imath}^{-1}\bar{T}^{-1}$
Newton's-law proportionality factor	g_c	$\bar{M}\bar{L}\bar{F}^{-1}\bar{\imath}^{-2}$
Temperature difference between wall and fluid	ΔT	$\bar{T}$

From Eqs. (A-8) and (A-15)

$$c = f - e = f - f + b = b \tag{A-16}$$

From Eqs. (A-6), (A-11), (A-15), and (A-16)

$$
\begin{aligned}
a &= -2 - b + 3c + 2d - e + g \\
&= -2 - b + 3b + 2f - 2b - f + b + 1 - f \\
&= b - 1
\end{aligned}
\tag{A-17}
$$

Equation (A-3) becomes, by substituting values from Eqs. (A-11) to (A-17) for letters a, c, d, e, g, and h,

$$\left[\frac{q}{A}\right] = [D]^{b-1}[\overline{V}]^{b}[\rho]^{b}[\mu_F]^{f-b}[g_c]^{f-b}[c_p]^{f}[k]^{1-f}[\Delta T]$$

Collecting all factors having integral exponents in one group, all factors having exponents b into another group, and those having exponents f into a third gives

$$\left[\frac{qD}{Ak\,\Delta T}\right] = \left[\frac{D\overline{V}\rho}{\mu_F g_c}\right]^{b}\left[\frac{c_p\mu_F g_c}{k}\right]^{f} \tag{A-18}$$

The dimensions of each of the three bracketed groups in Eq. (A-18) are zero, and all groups are dimensionless. Any function whatever of these three groups will be dimensionally homogeneous, and the equation will be a dimensionless one. Let such a function be

$$\frac{qD}{Ak\,\Delta T} = \Phi\left(\frac{D\overline{V}\rho}{\mu_F g_c},\ \frac{c_p\mu_F g_c}{k}\right) \tag{A-19}$$

or

$$\frac{q}{A} = \frac{k\,\Delta T}{D}\,\Phi\left(\frac{D\overline{V}\rho}{\mu_F g_c},\ \frac{c_p\mu_F g_c}{k}\right) \tag{A-20}$$

The relationship given in Eqs. (A-19) and (A-20) is the final result of the dimensional analysis. The form of function Φ must be found experimentally, by determining the effects of the groups in the brackets on the value of the group on the left-hand side of Eq. (A-19). The correlations that have been found for this are given in Chap. 12.

It is usual, in Eqs. (A-19) and (A-20), to use the so-called absolute viscosity μ in place of its equal $\mu_F g_c$. Viscosity is further discussed in Chap. 3.

It is clear that to correlate experimental values of the three groups of variables of Eq. (A-19) is simpler than to attempt to correlate the effects of each of the individual factors of Eq. (A-2).

Formation of other dimensionless groups   If a pair of letters other than b and f is selected for retention, three dimensionless groups are again obtained, but one or more differ from the groups of Eq. (A-19). For example, if b and g are kept, the result is

$$\frac{q}{A\overline{V}\rho c_p\,\Delta T} = \Phi_1\left(\frac{D\overline{V}\rho}{\mu},\ \frac{c_p\mu}{k}\right) \tag{A-21}$$

Other combinations may be found. However, it is unnecessary to repeat the algebra to obtain such additional groups. The three groups in Eq. (A-19) may be combined in any desired manner, by multiplying and dividing them, or reciprocals or multiples of them, together. It is necessary only that each original group be used at least once in finding new groups and that the final assembly contain exactly three groups. For example, Eq. (A-21) is obtained from Eq. (A-19) by multiplying both sides by $(\mu/D\bar{V}\rho)(k/\mu c_p)$

$$\frac{qD}{Ak\,\Delta T}\frac{\mu}{D\bar{V}\rho}\frac{k}{\mu c_p} = \left[\Phi\left(\frac{D\bar{V}\rho}{\mu}, \frac{c_p\mu}{k}\right)\right]\frac{\mu}{D\bar{V}\rho}\frac{k}{\mu c_p} = \Phi_1\left(\frac{D\bar{V}\rho}{\mu}, \frac{c_p\mu}{k}\right)$$

and Eq. (A-21) follows. Note that function Φ_1 is not equal to function Φ. In this way any dimensionless equation may be changed into any number of new ones. This is often useful when it is desired to isolate a single factor in one group. Thus, in Eq. (A-19) c_p appears in only one group, and in Eq. (A-21) k is found in only one. It is shown in Chap. 12 that Eq. (A-21) is more useful for some purposes than Eq. (A-19).

DIMENSIONLESS GROUPS

Symbol	Name	Definition
C_D	Drag coefficient	$\dfrac{2 F_D g_c}{\rho \mu_0^2 A_p}$
f	Fanning friction factor	$\dfrac{-\Delta p_s\, g_c D}{2 L p \bar{V}^2}$
j_H	Heat-transfer factor	$\dfrac{h}{c_p G}\left(\dfrac{c_p \mu}{k}\right)^{2/3}\left(\dfrac{\mu_w}{\mu}\right)^{0.14}$
j_M	Mass-transfer factor	$\dfrac{k\bar{M}}{G}\dfrac{\mu}{D_v \rho}$
N_{Fo}	Fourier number	$\dfrac{\alpha t}{r^2}$
N_{Fr}	Froude number	$\dfrac{u^2}{gL}$
N_{Gr}	Grashof number	$\dfrac{L^3 \rho^2 \beta g\, \Delta T}{\mu^2}$
N_{Gz}	Graetz number	$\dfrac{\dot{m} c_p}{kL}$
N_{Ma}	Mach number	$\dfrac{u}{a}$
N_{Nu}	Nusselt number	$\dfrac{hD}{k}$
N_P	Power number	$\dfrac{P g_c}{\rho n^3 D^5}$
N_{Pe}	Peclet number	$\dfrac{D\bar{V}}{\alpha}$
N_{Pr}	Prandtl number	$\dfrac{c_p \mu}{k}$
N_{Re}	Reynolds number	$\dfrac{DG}{\mu}$
N_{Sc}	Schmidt number	$\dfrac{\mu}{D_v \rho}$
N_{Sh}	Sherwood number	$\dfrac{kD}{D_v \rho}$
N_{We}	Weber number	$\dfrac{D\rho \bar{V}^2}{\sigma g_c}$

APPENDIX 6

DIMENSIONS, CAPACITIES, AND WEIGHTS OF STANDARD STEEL PIPE†

Nominal pipe size, in.	Outside diameter, in.	Schedule no.	Wall thickness, in.	ID, in.	Cross-sectional area of metal, in.²	Inside sectional area, ft²	Circumference, ft, or surface, ft²/ft of length Outside	Inside	Capacity at 1 ft/s velocity U.S. gal/min	Water, lb/h	Pipe weight, lb/ft
⅛	0.405	40	0.068	0.269	0.072	0.00040	0.106	0.0705	0.179	89.5	0.24
		80	0.095	0.215	0.093	0.00025	0.106	0.0563	0.113	56.5	0.31
¼	0.540	40	0.088	0.364	0.125	0.00072	0.141	0.095	0.323	161.5	0.42
		80	0.119	0.302	0.157	0.00050	0.141	0.079	0.224	112.0	0.54
⅜	0.675	40	0.091	0.493	0.167	0.00133	0.177	0.129	0.596	298.0	0.57
		80	0.126	0.423	0.217	0.00098	0.177	0.111	0.440	220.0	0.74
½	0.840	40	0.109	0.622	0.250	0.00211	0.220	0.163	0.945	472.0	0.85
		80	0.147	0.546	0.320	0.00163	0.220	0.143	0.730	365.0	1.09
¾	1.050	40	0.113	0.824	0.333	0.00371	0.275	0.216	1.665	832.5	1.13
		80	0.154	0.742	0.433	0.00300	0.275	0.194	1.345	672.5	1.47
1	1.315	40	0.133	1.049	0.494	0.00600	0.344	0.275	2.690	1,345	1.68
		80	0.179	0.957	0.639	0.00499	0.344	0.250	2.240	1,120	2.17
1¼	1.660	40	0.140	1.380	0.668	0.01040	0.435	0.361	4.57	2,285	2.27
		80	0.191	1.278	0.881	0.00891	0.435	0.335	3.99	1,995	3.00
1½	1,900	40	0.145	1.610	0.800	0.01414	0.497	0.421	6.34	3,170	2.72
		80	0.200	1.500	1.069	0.01225	0.497	0.393	5.49	2,745	3.63
2	2.375	40	0.154	2.067	1.075	0.02330	0.622	0.541	10.45	5,225	3.65
		80	0.218	1.939	1.477	0.02050	0.622	0.508	9.20	4,600	5.02
2½	2.875	40	0.203	2.469	1.704	0.03322	0.753	0.647	14.92	7,460	5.79
		80	0.276	2.323	2.254	0.02942	0.753	0.608	13.20	6,600	7.66
3	3.500	40	0.216	3.068	2.228	0.05130	0.916	0.803	23.00	11,500	7.58
		80	0.300	2.900	3.016	0.04587	0.916	0.759	20.55	10,275	10.25
3½	4.000	40	0.226	3.548	2.680	0.06870	1.047	0.929	30.80	15,400	9.11
		80	0.318	3.364	3.678	0.06170	1.047	0.881	27.70	13,850	12.51
4	4.500	40	0.237	4.026	3.17	0.08840	1.178	1.054	39.6	19,800	10.79
		80	0.337	3.826	4.41	0.07986	1.178	1.002	35.8	17,900	14.98
5	5.563	40	0.258	5.047	4.30	0.1390	1.456	1.321	62.3	31,150	14.62
		80	0.375	4.813	6.11	0.1263	1.456	1.260	57.7	28,850	20.78
6	6.625	40	0.280	6.065	5.58	0.2006	1.734	1.588	90.0	45,000	18.97
		80	0.432	5.761	8.40	0.1810	1.734	1.508	81.1	40,550	28.57
8	8.625	40	0.322	7.981	8.396	0.3474	2.258	2.089	155.7	77,850	28.55
		80	0.500	7.625	12.76	0.3171	2.258	1.996	142.3	71,150	43.39
10	10.75	40	0.365	10.020	11.91	0.5475	2.814	2.620	246.0	123,000	40.48
		80	0.594	9.562	18.95	0.4987	2.814	2.503	223.4	111,700	64.40
12	12.75	40	0.406	11.938	15.74	0.7773	3.338	3.13	349.0	174,500	53.56
		80	0.688	11.374	26.07	0.7056	3.338	2.98	316.7	158,350	88.57

† Based on ANSI B36.10-1959 by permission of ASME.

CONDENSER AND HEAT-EXCHANGER TUBE DATA†

OD, in.	Wall thickness BWG no.	Wall thickness in.	ID, in.	Cross-sectional area metal, in.^{2}	Inside sectional area, ft^2	Circumference, ft, or surface, ft^2/ft or length Outside	Circumference, ft, or surface, ft^2/ft or length Inside	Velocity, ft/s for 1 U.S. gal/min	Capacity at 1 ft/s velocity U.S. gal/min	Capacity at 1 ft/s velocity Water, lb/h	Weight, lb/ft‡
⅝	12	0.109	0.407	0.177	0.000903	0.1636	0.1066	2.468	0.4053	202.7	0.602
	14	0.083	0.459	0.141	0.00115	0.1636	0.1202	1.938	0.5161	258.1	0.479
	16	0.065	0.495	0.114	0.00134	0.1636	0.1296	1.663	0.6014	300.7	0.388
	18	0.049	0.527	0.089	0.00151	0.1636	0.1380	1.476	0.6777	338.9	0.303
¾	12	0.109	0.532	0.220	0.00154	0.1963	0.1393	1.447	0.6912	345.6	0,748
	14	0.083	0.584	0.174	0.00186	0.1963	0.1529	1.198	0.8348	417.4	0.592
	16	0.065	0.620	0.140	0.00210	0.1963	0.1623	1.061	0.9425	471.3	0.476
	18	0.049	0.652	0.108	0.00232	0.1963	0.1707	0.962	1.041	520.5	0.367
⅞	12	0.109	0.657	0.262	0.00235	0.2291	0.1720	0.948	1.055	527.5	0.891
	14	0.083	0.709	0.207	0.00274	0.2291	0.1856	0.813	1.230	615.0	0.704
	16	0.065	0.745	0.165	0.00303	0.2291	0.1950	0.735	1.350	680.0	0.561
	18	0.049	0.777	0.127	0.00329	0.2291	0.2034	0.678	1.477	738.5	0.432
1	10	0.134	0.732	0.364	0.00292	0.2618	0.1916	0.763	1.310	655.0	1.237
	12	0.109	0.782	0.305	0.00334	0.2618	0.2047	0.667	1.499	750.0	1.037
	14	0.083	0.834	0.239	0.00379	0.2618	0.2183	0.588	1.701	850.5	0.813
	16	0.065	0.870	0.191	0.00413	0.2618	0.2278	0.538	1.854	927.0	0.649
1¼	10	0.134	0.982	0.470	0.00526	0.3272	0.2571	0.424	2.361	1,181	1.598
	12	0.109	1.032	0.391	0.00581	0.3272	0.2702	0.384	2.608	1,304	1.329
	14	0.083	1.084	0.304	0.00641	0.3272	0.2838	0.348	2.877	1,439	1.033
	16	0.065	1.120	0.242	0.00684	0.3272	0.2932	0.326	3.070	1,535	0.823
1½	10	0.134	1.232	0.575	0.00828	0.3927	0.3225	0.269	3.716	1,858	1.955
	12	0.109	1.282	0.476	0.00896	0.3927	0.3356	0.249	4.021	2,011	1.618
	14	0.083	1.334	0.370	0.00971	0.3927	0.3492	0.229	4.358	2,176	1.258
2	10	0.134	1.732	0.7855	0.0164	0.5236	0.4534	0.136	7.360	3,680	2.68
	12	0.109	1.782	0.6475	0.0173	0.5236	0.4665	0.129	7.764	3,882	2.22

† Condensed, by permission, from J. H. Perry (ed.), "Chemical Engineers' Handbook," 5th ed., p. 11–12, McGraw-Hill Book Company, Copyright © 1973.

‡ For steel; for copper, multiply by 1.14; for brass, multiply by 1.06.

APPENDIX 8

PROPERTIES OF SATURATED STEAM AND WATER†

Temp. T, °F	Vapor press. p_A lb$_f$/in.^{2}	Specific vol., ft^3/lb		Enthalpy, Btu/lb		
		Liquid v_x	Sat. vapor v_y	Liquid H_x	Vaporization λ	Sat. vapor H_y
32	0.08854	0.01602	3,306	0.00	1075.8	1075.8
35	0.09995	0.01602	2,947	3.02	1074.1	1077.1
40	0.12170	0.01602	2,444	8.05	1071.3	1079.3
45	0.14752	0.01602	2,036.4	13.06	1068.4	1081.5
50	0.17811	0.01603	1,703.2	18.07	1065.6	1083.7
55	0.2141	0.01603	1,430.7	23.07	1062.7	1085.8
60	0.2563	0.01604	1,206.7	28.06	1059.9	1088.0
65	0.3056	0.01605	1,021.4	33.05	1057.1	1090.2
70	0.3631	0.01606	867.9	38.04	1054.3	1092.3
75	0.4298	0.01607	740.0	43.03	1051.5	1094.5
80	0.5069	0.01608	633.1	48.02	1048.6	1096.6
85	0.5959	0.01609	543.5	53.00	1045.8	1098.8
90	0.6982	0.01610	468.0	57.99	1042.9	1100.9
95	0.8153	0.01612	404.3	62.98	1040.1	1103.1
100	0.9492	0.01613	350.4	67.97	1037.2	1105.2
110	1.2748	0.01617	265.4	77.94	1031.6	1109.5
120	1.6924	0.01620	203.27	87.92	1025.8	1113.7
130	2.2225	0.01625	157.34	97.90	1020.0	1117.9
140	2.8886	0.01629	123.01	107.89	1014.1	1122.0
150	3.718	0.01634	97.07	117.89	1008.2	1126.1
160	4.741	0.01639	77.29	127.89	1002.3	1130.2
170	5.992	0.01645	62.06	137.90	996.3	1134.2
180	7.510	0.01651	50.23	147.92	990.2	1138.1
190	9.339	0.01657	40.96	157.95	984.1	1142.0
200	11.526	0.01663	33.64	167.99	977.9	1145.9
210	14.123	0.01670	27.82	178.05	971.6	1149.7
212	14.696	0.01672	26.80	180.07	970.3	1150.4
220	17.186	0.01677	23.15	188.13	965.2	1153.4
230	20.780	0.01684	19.382	198.23	958.8	1157.0
240	24.969	0.01692	16.323	208.34	952.2	1160.5
250	29.825	0.01700	13.821	218.48	945.5	1164.0
260	35.429	0.01709	11.763	228.64	938.7	1167.3
270	41.858	0.01717	10.061	238.84	931.8	1170.6
280	49.203	0.01726	8.645	249.06	924.7	1173.8
290	57.556	0.01735	7.461	259.31	917.5	1176.8
300	67.013	0.01745	6.466	269.59	910.1	1179.7
310	77.68	0.01755	5.626	279.92	902.6	1182.5
320	89.66	0.01765	4.914	290.28	894.9	1185.2
330	103.06	0.01776	4.307	300.68	887.0	1187.7
340	118.01	0.01787	3.788	311.13	879.0	1190.1
350	134.63	0.01799	3.342	321.63	870.7	1192.3
360	153.04	0.01811	2.957	332.18	862.2	1194.4
370	173.37	0.01823	2.625	342.79	853.5	1196.3
380	195.77	0.01836	2.335	353.45	844.6	1198.1
390	220.37	0.01850	2.0836	364.17	835.4	1199.6
400	247.31	0.01864	1.8633	374.97	826.0	1201.0
410	276.75	0.01878	1.6700	385.83	816.3	1202.1
420	308.83	0.01894	1.5000	396.77	806.3	1203.1
430	343.72	0.01910	1.3499	407.79	796.0	1203.8
440	381.59	0.01926	1.2171	418.90	785.4	1204.3
450	422.6	0.0194	1.0993	430.1	774.5	1204.6

† Abstracted from abridged edition of "Thermodynamic Properties of Steam," by Joseph H. Keenan and Fredrick G. Keyes, John Wiley & Sons, Inc., New York, 1937, with the permission of the authors and publisher.

VISCOSITIES OF GASES†

No.	Gas	X	Y	No.	Gas	X	Y
1	Acetic acid	7.7	14.3	29	Freon-113	11.3	14.0
2	Acetone	8.9	13.0	30	Helium	10.9	20.5
3	Acetylene	9.8	14.9	31	Hexane	8.6	11.8
4	Air	11.0	20.0	32	Hydrogen	11.2	12.4
5	Ammonia	8.4	16.0	33	$3H_2 + N_2$	11.2	17.2
6	Argon	10.5	22.4	34	Hydrogen bromide	8.8	20.9
7	Benzene	8.5	13.2	35	Hydrogen chloride	8.8	18.7
8	Bromine	8.9	19.2	36	Hydrogen cyanide	9.8	14.9
9	Butene	9.2	13.7	37	Hydrogen iodide	9.0	21.3
10	Butylene	8.9	13.0	38	Hydrogen sulfide	8.6	18.0
11	Carbon dioxide	9.5	18.7	39	Iodine	9.0	18.4
12	Carbon disulfide	8.0	16.0	40	Mercury	5.3	22.9
13	Carbon monoxide	11.0	20.0	41	Methane	9.9	15.5
14	Chlorine	9.0	18.4	42	Methyl alcohol	8.5	15.6
15	Chloroform	8.9	15.7	43	Nitric oxide	10.9	20.5
16	Cyanogen	9.2	15.2	44	Nitrogen	10.6	20.0
17	Cyclohexane	9.2	12.0	45	Nitrosyl chloride	8.0	17.6
18	Ethane	9.1	14.5	46	Nitrous oxide	8.8	19.0
19	Ethyl acetate	8.5	13.2	47	Oxygen	11.0	21.3
20	Ethyl alcohol	9.2	14.2	48	Pentane	7.0	12.8
21	Ethyl chloride	8.5	15.6	49	Propane	9.7	12.9
22	Ethyl ether	8.9	13.0	50	Propyl alcohol	8.4	13.4
23	Ethylene	9.5	15.1	51	Propylene	9.0	13.8
24	Fluorine	7.3	23.8	52	Sulfur dioxide	9.6	17.0
25	Freon-11	10.6	15.1	53	Toleune	8.6	12.4
26	Freon-12	11.1	16.0	54	2,3,3-Trimethylbutane	9.5	10.5
27	Freon-21	10.8	15.3	55	Water	8.0	16.0
28	Freon-22	10.1	17.0	56	Xenon	9.3	23.0

Coordinates for use with figure overleaf.

† By permission, from J. H. Perry (ed.), "Chemical Engineers' Handbook," 5th ed., pp. 3–210 and 3–211. Copyright, 1973, ©, McGraw-Hill Book Company.

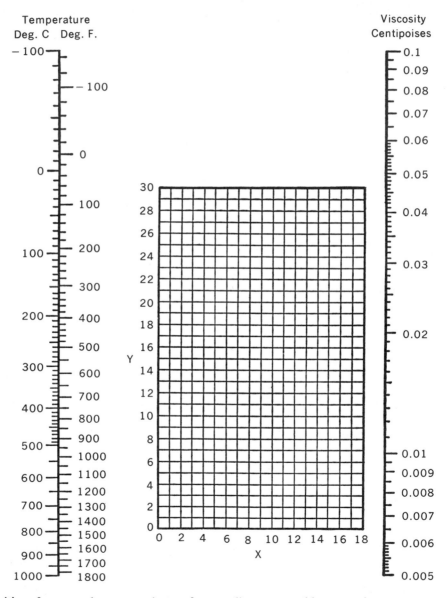

Viscosities of gases and vapors at 1 atm; for coordinates, see table on previous page.

No.	Liquid	X	Y	No.	Liquid	X	Y
1	Acetaldehyde	15.2	4.8	56	Freon-22	17.2	4.7
2	Acetic acid, 100%	12.1	14.2	57	Freon-113	12.5	11.4
3	Acetic acid, 70%	9.5	17.0	58	Glycerol, 100%	2.0	30.0
4	Acetic anhydride	12.7	12.8	59	Glycerol, 50%	6.9	19.6
5	Acetone, 100%	14.5	7.2	60	Heptane	14.1	8.4
6	Acetone, 35%	7.9	15.0	61	Hexane	14.7	7.0
7	Allyl alcohol	10.2	14.3	62	Hydrochloric acid, 31.5%	13.0	16.6
8	Ammonia, 100%	12.6	2.0	63	Isobutyl alcohol	7.1	18.0
9	Ammonia, 26%	10.1	13.9	64	Isobutyric acid	12.2	14.4
10	Amyl acetate	11.8	12.5	65	Isopropyl alcohol	8.2	16.0
11	Amyl alcohol	7.5	18.4	66	Kerosene	10.2	16.9
12	Aniline	8.1	18.7	67	Linseed oil, raw	7.5	27.2
13	Anisole	12.3	13.5	68	Mercury	18.4	16.4
14	Arsenic trichloride	13.9	14.5	69	Methanol, 100%	12.4	10.5
15	Benzene	12.5	10.9	70	Methanol, 90%	12.3	11.8
16	Bimethyl oxalate	12.3	15.8	71	Methanol, 40%	7.8	15.5
17	Biphenyl	12.0	18.3	72	Methyl acetate	14.2	8.2
18	Brine, CaCl₂, 25%	6.6	15.9	73	Methyl chloride	15.0	3.8
19	Brine, NaCl, 25%	10.2	16.6	74	Methyl ethyl ketone	13.9	8.6
20	Bromine	14.2	13.2	75	Naphthalene	7.9	18.1
21	Bromotoluene	20.0	15.9	76	Nitric acid, 95%	12.8	13.8
22	Butyl acetate	12.3	11.0	77	Nitric acid, 60%	10.8	17.0
23	Butyl alcohol	8.6	17.2	78	Nitrobenzene	10.6	16.2
24	Butyric acid	12.1	15.3	79	Nitrotoluene	11.0	17.0
25	Carbon dioxide	11.6	0.3	80	Octane	13.7	10.0
26	Carbon disulfide	16.1	7.5	81	Octyl alcohol	6.6	21.1
27	Carbon tetrachloride	12.7	13.1	82	Pentachloroethane	10.9	17.3
28	Chlorobenzene	12.3	12.4	83	Pentane	14.9	5.2
29	Chloroform	14.4	10.2	84	Phenol	6.9	20.8
30	Chlorosulfonic acid	11.2	18.1	85	Phosphorus tribromide	13.8	16.7
31	o-Chlorotoluene	13.0	13.3	86	Phosphorus trichloride	16.2	10.9
32	m-Chlorotoluene	13.3	12.5	87	Propionic acid	12.8	13.8
33	p-Chlorotoluene	13.3	12.5	88	Propyl alcohol	9.1	16.5
34	m-Cresol	2.5	20.8	89	Propyl bromide	14.5	9.6
35	Cyclohexanol	2.9	24.3	90	Propyl chloride	14.4	7.5
36	Dibromoethane	12.7	15.8	91	Propyl iodide	14.1	11.6
37	Dichloroethane	13.2	12.2	92	Sodium	16.4	13.9
38	Dichloromethane	14.6	8.9	93	Sodium hydroxide, 50%	3.2	25.8
39	Diethyl oxalate	11.0	16.4	94	Stannic chloride	13.5	12.8
40	Dipropyl oxalate	10.3	17.7	95	Sulfur dioxide	15.2	7.1
41	Ethyl acetate	13.7	9.1	96	Sulfuric acid, 110%	7.2	27.4
42	Ethyl alcohol, 100%	10.5	13.8	97	Sulfuric acid, 98%	7.0	24.8
43	Ethyl alcohol, 95%	9.8	14.3	98	Sulfuric acid, 60%	10.2	21.3
44	Ethyl alcohol, 40%	6.5	16.6	99	Sulfuryl chloride	15.2	12.4
45	Ethyl benzene	13.2	11.5	100	Tetrachloroethane	11.9	15.7
46	Ethyl bromide	14.5	8.1	101	Tetrachloroethylene	14.2	12.7
47	Ethyl chloride	14.8	6.0	102	Titanium tetrachloride	14.4	12.3
48	Ethyl ether	14.5	5.3	103	Toluene	13.7	10.4
49	Ethyl formate	14.2	8.4	104	Trichloroethylene	14.8	10.5
50	Ethyl iodide	14.7	10.3	105	Turpentine	11.5	14.9
51	Ethylene glycol	6.0	23.6	106	Vinyl acetate	14.0	8.8
52	Formic acid	10.7	15.8	107	Water	10.2	13.0
53	Freon-11	14.4	9.0	108	o-Xylene	13.5	12.1
54	Freon-12	16.8	5.6	109	m-Xylene	13.9	10.6
55	Freon-21	15.7	7.5	110	p-Xylene	13.9	10.9

Coordinates for use with figure overleaf.

† By permission from J. H. Perry (ed.), "Chemical Engineers' Handbook," 5th ed., pp. 3–212 and 3–213. Copyright, ©, 1973, McGraw-Hill Book Company.

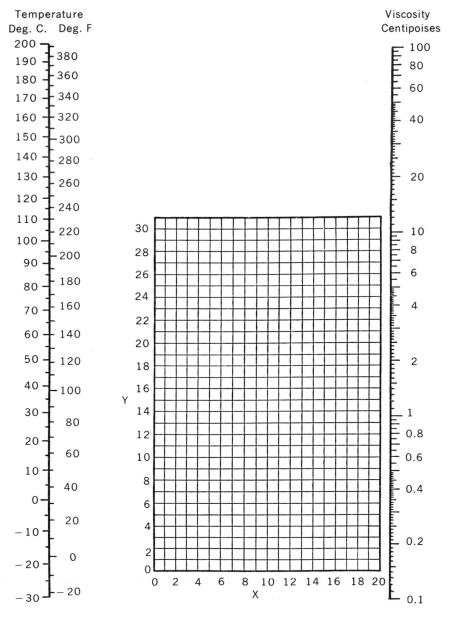

Viscosities of liquids at 1 atm. For coordinates, see table on previous page.

THERMAL CONDUCTIVITIES OF METALS†

Metal	Thermal conductivity k‡		
	32°F	64°F	212°F
Aluminum	117		119
Antimony	10.6		9.7
Brass (70 copper, 30 zinc)	56		60
Cadmium		53.7	52.2
Copper (pure)	224		218
Gold		169.0	170.0
Iron (cast)	32		30
Iron (wrought)		34.9	34.6
Lead	20		19
Magnesium	92	92	92
Mercury (liquid)	4.8		
Nickel	36		34
Platinum		40.2	41.9
Silver	242		238
Sodium (liquid)			49
Steel (mild)			26
Steel (1 % carbon)		26.2	25.9
Steel (stainless, type 304)			9.4
Steel (stainless, type 316)			9.4
Steel (stainless, type 347)			9.3
Tantalum		32	
Tin	36		34
Zinc	65		64

† Based on W. H. McAdams, "Heat Transmission," 3d ed., pp. 445–447, McGraw-Hill Book Company, New York, 1954
‡ k = Btu/ft-h-°F.

THERMAL CONDUCTIVITIES OF GASES AND VAPORS†

Substance	Thermal conductivity k‡	
	32°F	212°F
Acetone	0.0057	0.0099
Acetylene	0.0108	0.0172
Air	0.0140	0.0184
Ammonia	0.0126	0.0192
Benzene	0.0052	0.0103
Carbon dioxide	0.0084	0.0128
Carbon monoxide	0.0134	0.0176
Carbon tetrachloride		0.0052
Chlorine	0.0043	
Ethane	0.0106	0.0175
Ethyl alcohol		0.0124
Ethyl ether	0.0077	0.0131
Ethylene	0.0101	0.0161
Helium	0.0818	0.0988
Hydrogen	0.0966	0.1240
Methane	0.0176	0.0255
Methyl alcohol	0.0083	0.0128
Nitrogen	0.0139	0.0181
Nitrous oxide	0.0088	0.0138
Oxygen	0.0142	0.0188
Propane	0.0087	0.0151
Sulfur dioxide	0.0050	0.0069
Water vapor (at 1 atm abs pressure)		0.0136

† Based on W. H. McAdams, "Heat Transmission," 3d ed., pp. 457–458, McGraw-Hill Book Company, New York, 1954.
‡ k = Btu/ft-h-°F.

THERMAL CONDUCTIVITIES OF LIQUIDS OTHER THAN WATER†

Liquid	Temp., °F	k‡
Acetic acid	68	0.909
Acetone	86	0.102
Ammonia (anhydrous)	5–86	0.29
Aniline	32–68	0.100
Benzene	86	0.092
n-Butyl alcohol	86	0.097
Carbon bisulfide	86	0.093
Carbon tetrachloride	32	0.107
Chlorobenzene	50	0.083
Ethyl acetate	68	0.101
Ethyl alcohol (absolute)	68	0.105
Ethyl ether	86	0.080
Ethylene glycol	32	0.153
Gasoline	86	0.078
Glycerine	68	0.164
n-Heptane	86	0.081
Kerosene	68	0.086
Methyl alcohol	68	0.124
Nitrobenzene	86	0.095
n-Octane	86	0.083
Sulfur dioxide	5	0.128
Sulfuric acid (90%)	86	0.21
Toluene	86	0.086
Trichloroethylene	122	0.080
o-Xylene	68	0.090

† Based on W. H. McAdams, "Heat Transmission," 3d ed., pp. 455–456, McGraw-Hill Book Company, New York, 1954.

‡ k = Btu/ft-h-°F.

APPENDIX 14

PROPERTIES OF LIQUID WATER†

Temperature T, °F	Viscosity† μ', cP	Thermal conductivity‡ k, Btu/ft-h-°F	Density§ ρ, lb/ft^3	$\psi_f = \left(\dfrac{k^3\rho^2 g}{\mu^2}\right)^{1/3}$
32	1.794	0.320	62.42	1,410
40	1.546	0.326	62.43	1,590
50	1.310	0.333	62.42	1,810
60	1.129	0.340	62.37	2,050
70	0.982	0.346	62.30	2,290
80	0.862	0.352	62.22	2,530
90	0.764	0.358	62.11	2,780
100	0.682	0.362	62.00	3,020
120	0.559	0.371	61.71	3,530
140	0.470	0.378	61.38	4,030
160	0.401	0.384	61.00	4,530
180	0.347	0.388	60.58	5,020
200	0.305	0.392	60.13	5,500
220	0.270	0.394	59.63	5,960
240	0.242	0.396	59.10	6,420
260	0.218	0.396	58.51	6,830
280	0.199	0.396	57.94	7,210
300	0.185	0.396	57.31	7,510

† From "International Critical Tables," vol. 5, p. 10, McGraw-Hill Book Company, New York, 1929.
‡ From E. Schmidt and W. Sellschopp, *Forsch. Geb. Ingenieurw.*, **3**:277 (1932).
§ Calculated from J. H. Keenan and F. G. Keyes, "Thermodynamic Properties of Steam," John Wiley & Sons., Inc., New York, 1937.

SPECIFIC HEATS OF GASES†

c_p = Specific heat = Btu/lb-°F = cal/g-°C

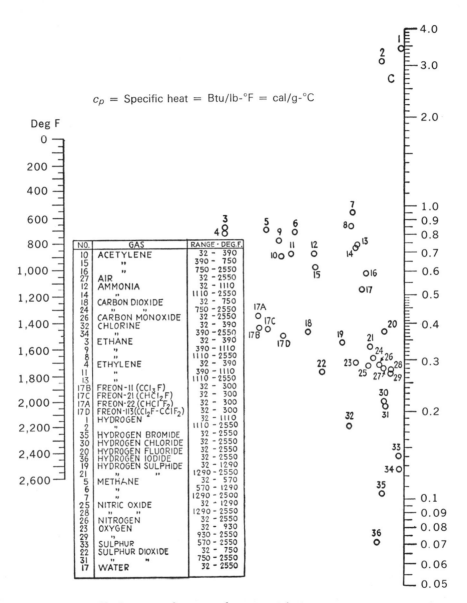

True specific heats c_p of gases and vapors at 1 atm pressure.

† Courtesy of T. H. Chilton.

APPENDIX 16

SPECIFIC HEATS OF LIQUIDS†

Specific heat = Btu/lb-°F = cal/g-°C

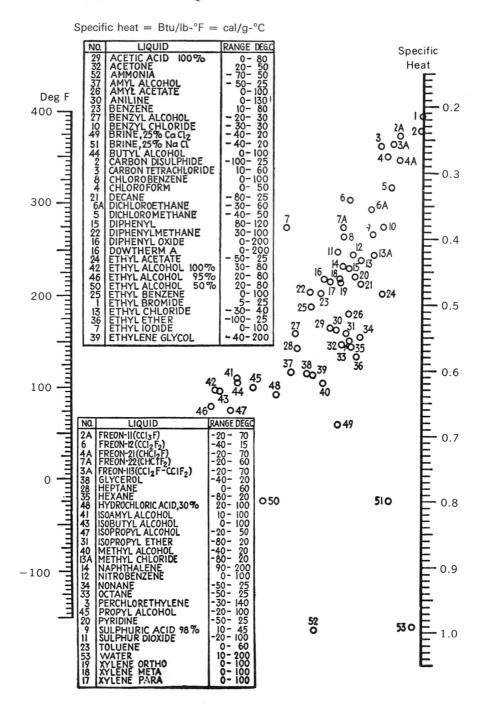

NO.	LIQUID	RANGE DEG.C
29	ACETIC ACID 100%	0- 80
32	ACETONE	20- 50
52	AMMONIA	- 70- 50
37	AMYL ALCOHOL	- 50- 25
26	AMYL ACETATE	0- 100
30	ANILINE	0- 130
23	BENZENE	10- 80
27	BENZYL ALCOHOL	- 20- 30
10	BENZYL CHLORIDE	- 30- 30
49	BRINE, 25% Ca Cl₂	- 40- 20
51	BRINE, 25% Na Cl	- 40- 20
44	BUTYL ALCOHOL	0- 100
2	CARBON DISULPHIDE	-100- 25
3	CARBON TETRACHLORIDE	10- 60
8	CHLOROBENZENE	0- 100
4	CHLOROFORM	0- 50
21	DECANE	- 80- 25
6A	DICHLOROETHANE	- 30- 60
5	DICHLOROMETHANE	- 40- 50
15	DIPHENYL	80- 120
22	DIPHENYLMETHANE	30- 100
16	DIPHENYL OXIDE	0- 200
16	DOWTHERM A	0- 200
24	ETHYL ACETATE	- 50- 25
42	ETHYL ALCOHOL 100%	30- 80
46	ETHYL ALCOHOL 95%	20- 80
50	ETHYL ALCOHOL 50%	20- 80
25	ETHYL BENZENE	0- 100
1	ETHYL BROMIDE	5- 25
13	ETHYL CHLORIDE	- 30- 40
36	ETHYL ETHER	-100- 25
7	ETHYL IODIDE	0- 100
39	ETHYLENE GLYCOL	- 40- 200

NO.	LIQUID	RANGE DEG.C
2A	FREON-11 (CCl₃F)	-20- 70
6	FREON-12 (CCl₂F₂)	-40- 15
4A	FREON-21 (CHCl₂F)	-20- 70
7A	FREON-22 (CHClF₂)	-20- 60
3A	FREON-113 (CCl₂F-CClF₂)	-20- 70
38	GLYCEROL	-40- 20
28	HEPTANE	0- 60
35	HEXANE	-80- 20
48	HYDROCHLORIC ACID, 30%	20- 100
41	ISOAMYL ALCOHOL	10- 100
43	ISOBUTYL ALCOHOL	0- 100
47	ISOPROPYL ALCOHOL	-20- 50
31	ISOPROPYL ETHER	-80- 20
40	METHYL ALCOHOL	-40- 20
13A	METHYL CHLORIDE	-80- 20
14	NAPHTHALENE	90- 200
12	NITROBENZENE	0- 100
34	NONANE	-50- 25
33	OCTANE	-50- 25
3	PERCHLORETHYLENE	-30- 140
45	PROPYL ALCOHOL	-20- 100
20	PYRIDINE	-50- 25
9	SULPHURIC ACID 98%	10- 45
11	SULPHUR DIOXIDE	-20- 100
23	TOLUENE	0- 60
53	WATER	10- 200
19	XYLENE ORTHO	0- 100
18	XYLENE META	0- 100
17	XYLENE PARA	0- 100

Deg F — 400, 300, 200, 100, 0, −100

Specific Heat — 0.2, 0.3, 0.4, 0.5, 0.6, 0.7, 0.8, 0.9, 1.0

PRANDTL NUMBERS FOR GASES AT 1 ATM AND 100°C†

Gas	$N_{Pr} = \dfrac{c_p \mu}{k}$
Air	0.69
Ammonia	0.86
Argon	0.66
Carbon dioxide	0.75
Carbon monoxide	0.72
Helium	0.71
Hydrogen	0.69
Methane	0.75
Nitric oxide, nitrous oxide	0.72
Nitrogen	0.70
Oxygen	0.70
Water vapor	1.06

† Based on W. H. McAdams, "Heat Transmission," 3d ed., p. 471, McGraw-Hill Book Company, New York, 1954.

APPENDIX 18

PRANDTL NUMBERS FOR LIQUIDS†

Liquid	$N_{Pr} = \dfrac{c_p \mu}{k}$	
	60°F	212°F
Acetic acid	14.5	10.5
Acetone	4.5	2.4
Aniline	69	9.3
Benzene	7.3	3.8
n-Butyl alcohol	43	11.5
Carbon tetrachloride	7.5	4.2
Chlorobenzene	9.3	7.0
Ethyl acetate	6.8	5.6
Ethyl alcohol	15.5	10.1
Ethyl ether	4.0	2.3
Ethylene glycol	350	125
n-Heptane	6.0	4.2
Methyl alcohol	7.2	3.4
Nitrobenzene	19.5	6.5
n-Octane	5.0	3.6
Sulfuric acid (98%)	149	15.0
Toluene	6.5	3.8
Water	7.7	1.5

† Based on W. H. McAdams, "Heat Transmission," 3d ed., p. 470, McGraw-Hill Book Company, New York, 1954.

DIFFUSIVITIES AND SCHMIDT NUMBERS FOR GASES IN AIR AT 0°C AND 1 ATM†

Gas	Volumetric diffusivity D_v, ft²/h	$N_{Sc} = \dfrac{\mu}{\rho D_v}$ ‡
Acetic acid	0.413	1.24
Acetone	0.32§	1.60
Ammonia	0.836	0.61
Benzene	0.299	1.71
n-Butyl alcohol	0.273	1.88
Carbon dioxide	0.535	0.96
Carbon tetrachloride	0.24§	2.13
Chlorine	0.36§	1.42
Chlorobenzene	0.24§	2.13
Ethane	0.42§	1.22
Ethyl acetate	0.278	1.84
Ethyl alcohol	0.396	1.30
Ethyl ether	0.302	1.70
Hydrogen	2.37	0.22
Methane	0.61§	0.84
Methyl alcohol	0.515	1.00
Naphthalene	0.199	2.57
Nitrogen	0.52§	0.98
n-Octane	0.196	2.62
Oxygen	0.690	0.74
Phosgene	0.31§	1.65
Propane	0.34§	1.51
Sulfur dioxide	0.40§	1.28
Toluene	0.275	1.86
Water vapor	0.853	0.60

† By permission, from T. K. Sherwood and R. L. Pigford, "Absorption and Extraction," 2d ed., p. 20. Copyright, 1952, McGraw-Hill Book Company.
‡ The value of μ/p is that for pure air, 0.512 ft²/h.
§ Calculated by Eq. (22-23).

TYLER STANDARD SCREEN SCALE

This screen scale has as its base an opening of 0.0029 in., which is the opening in 200-mesh 0.0021-in. wire, the standard sieve, as adopted by the National Bureau of Standards.

Mesh	Clear opening, in.	Clear opening, mm	Approx. opening, in.	Wire diameter, in.
	1.050	26.67	1	0.148
†	0.883	22.43	$\frac{7}{8}$	0.135
	0.742	18.85	$\frac{3}{4}$	0.135
†	0.624	15.85	$\frac{5}{8}$	0.120
	0.525	13.33	$\frac{1}{2}$	0.105
†	0.441	11.20	$\frac{7}{16}$	0.105
	0.371	9.423	$\frac{3}{8}$	0.092
2½†	0.312	7.925	$\frac{5}{16}$	0.088
3	0.263	6.680	$\frac{1}{4}$	0.070
3½†	0.221	5.613	$\frac{7}{32}$	0.065
4	0.185	4.699	$\frac{3}{16}$	0.065
5†	0.156	3.962	$\frac{5}{32}$	0.044
6	0.131	3.327	$\frac{1}{8}$	0.036
7†	0.110	2.794	$\frac{7}{64}$	0.0328
8	0.093	2.362	$\frac{3}{32}$	0.032
9†	0.078	1.981	$\frac{5}{64}$	0.033
10	0.065	1.651	$\frac{1}{16}$	0.035
12†	0.055	1.397		0.028
14	0.046	1.168	$\frac{3}{64}$	0.025
16†	0.0390	0.991		0.0235
20	0.0328	0.833	$\frac{1}{32}$	0.0172
24†	0.0276	0.701		0.0141
28	0.0232	0.589		0.0125
32†	0.0195	0.495		0.0118
35	0.0164	0.417	$\frac{1}{64}$	0.0122
42†	0.0138	0.351		0.0100
48	0.0116	0.295		0.0092
60†	0.0097	0.246		0.0070
65	0.0082	0.208		0.0072
80†	0.0069	0.175		0.0056
100	0.0058	0.147		0.0042
115†	0.0049	0.124		0.0038
150	0.0041	0.104		0.0026
170†	0.0035	0.088		0.0024
200	0.0029	0.074		0.0021

For coarser sizing: 3- to 1½-in. opening

			Approx. opening, in.	Wire diameter, in.
			3	0.207
			2	0.192
			1½	0.148

† These screens, for closer sizing, are inserted between the sizes usually considered as the standard series. With the inclusion of these screens the ratio of diameters of openings in two successive screens is as $1:\sqrt[4]{2}$ instead of $1:\sqrt{2}$.

INDEX

INDEX